A Note from the Authors

You are about to embark on an amazing journey of discovery. The study of life spans from the inner workings of cells to the complex interactions of entire ecosystems, through the information stored in DNA to the ways genetic information evolves over time. At the same time that our understanding of biology is growing in leaps and bounds, so too are great insights into how learners acquire new knowledge and skills. We are thrilled to join Scott Freeman on *Biological Science*, a book dedicated to active, research-based learning and to exploring the experimental evidence that informs what we know about biology. The next few pages highlight the features in this book and in MasteringBiology® that will help you succeed.

From left to right: Michael Black, Emily Taylor, Jon Monroe,
Lizabeth Allison, Greg Podgorski, Kim Quillin

To the Student: How to Use This Book

New chapter-opening Roadmaps visually group and organize information to help you anticipate key ideas as well as recognize meaningful relationships and connections between them.

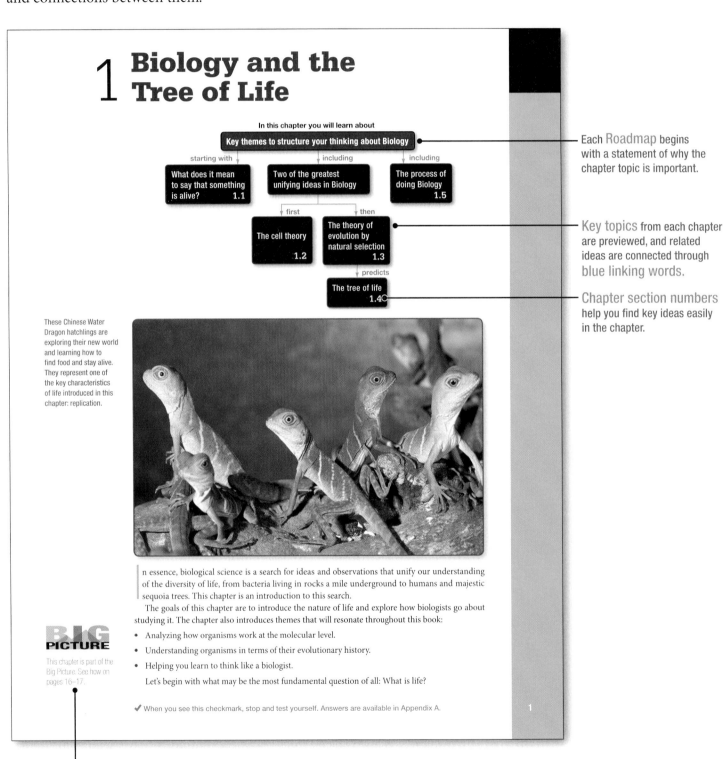

1 Biology and the Tree of Life

In this chapter you will learn about

Key themes to structure your thinking about Biology

starting with

What does it mean to say that something is alive? 1.1

including

Two of the greatest unifying ideas in Biology

including

The process of doing Biology 1.5

first

The cell theory 1.2

then

The theory of evolution by natural selection 1.3

predicts

The tree of life 1.4

Each Roadmap begins with a statement of why the chapter topic is important.

Key topics from each chapter are previewed, and related ideas are connected through blue linking words.

Chapter section numbers help you find key ideas easily in the chapter.

These Chinese Water Dragon hatchlings are exploring their new world and learning how to find food and stay alive. They represent one of the key characteristics of life introduced in this chapter: replication.

In essence, biological science is a search for ideas and observations that unify our understanding of the diversity of life, from bacteria living in rocks a mile underground to humans and majestic sequoia trees. This chapter is an introduction to this search.

The goals of this chapter are to introduce the nature of life and explore how biologists go about studying it. The chapter also introduces themes that will resonate throughout this book:

- Analyzing how organisms work at the molecular level.
- Understanding organisms in terms of their evolutionary history.
- Helping you learn to think like a biologist.

Let's begin with what may be the most fundamental question of all: What is life?

BIG PICTURE

This chapter is part of the Big Picture. See how on pages 16–17.

✔ When you see this checkmark, stop and test yourself. Answers are available in Appendix A.

1

Big Picture Concept Maps are referenced on the opening page of related chapters, pointing you to summary pages that help you synthesize challenging topics.

Big Picture Concept Maps integrate visuals and words to help you synthesize information about challenging topics in biology that span multiple chapters and units.

Three New Big Picture topics
have been added to the Fifth Edition:

- NEW! Doing Biology
- NEW! The Chemistry of Life
- Energy for Life
- Genetic Information
- Evolution
- NEW! Plant and Animal Form and Function
- Ecology

MasteringBiology®

www.masteringbiology.com

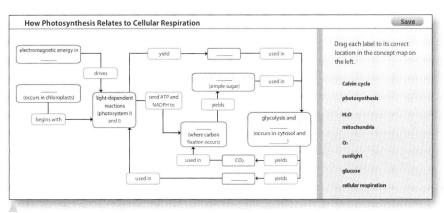

To reinforce the book's Big Picture Concept Maps, your professor may assign Interactive Big Picture Concept Map tutorials.

Practice for success on tests and exams

Intertwined color-coded "active learning threads" are embedded in the text. The gold thread helps you to identify important ideas, and the blue thread helps you to test your understanding.

Blue Thread Caption Questions and Exercises challenge you to critically examine information in figures and tables.

The Gold Thread helps you identify important concepts when reading and reviewing.

Check Your Understanding boxes ask you to work with the important concepts in the chapter.

End-of-Chapter Blue Thread Exercises, integrated in the chapter summary, help you review the major themes of the chapter and synthesize information.

Embedded Blue Thread Questions and Exercises encourage you to stop and test your understanding of challenging topics.

Coupled reaction

ADP–P + ☐ + ■
ATP   A   B

Free energy relative to A + B (kcal/mol)

7.3

Energy released to synthesize BP

ΔG

Exergonic reaction (Energy release)

Activated substrate

☐ + ■–P + ADP
A   BP

Energy released to synthesize AB

ΔG

Exergonic reaction (Energy release)

☐■ + P_i + ADP
AB

ΔG

Uncoupled reaction

☐ + ■
A   B

Endergonic reaction (Energy input)

Energy required to synthesize AB

0

Reactants   Progress of reaction   Products

FIGURE 8.9 Exergonic Phosphorylation Reactions Are Coupled to Endergonic Reactions. In cells, many reactions only occur if one reactant is activated by phosphorylation. The phosphorylated reactant molecule has high enough free energy that the subsequent reaction is exergonic. In this graph, the free energy being tracked on the y-axis represents A, B, and the 7.3 kcal/mol that is released when ATP is hydrolyzed. For simplicity, the free energy in ADP and P_i is not shown. ΔG represents the change in free energy between the reactants and products for each indicated step.

✔ **EXERCISE** Label the ΔG in the uncoupled reaction and the two steps of the coupled reaction to indicate if the change is representing a positive (> 0) or negative (< 0) value.

FIGURE 8.9 graphs how phosphorylation can couple exergonic and endergonic reactions. Note that the reaction between A and B to produce the product AB is endergonic—the ΔG is positive. But after the exergonic transfer of a phosphate group from ATP to B occurs, the free energy of the reactants A and BP is high enough to make the reaction that forms AB exergonic. When reactant molecules in an endergonic reaction are phosphorylated, the free energy released during phosphorylation is coupled to the endergonic reaction to make the combined overall reaction exergonic.

check your understanding

If you understand that . . .

Ⓒ Ⓨ Ⓤ

• When redox reactions occur, electrons change position. Chemical energy is based on the positions of electrons in chemical bonds, so redox reactions usually involve a change in potential energy.
• ATP contains a cluster of three negatively charged phosphate groups.
• When ATP or phosphate groups from ATP bind to substrates, they gain a great deal of potential energy.

✔ **You should be able to . . .**

1. Explain why reduced molecules with many C–H bonds store more potential energy than oxidized molecules with many C–O bonds.
2. Explain why ATP has such high potential energy.

Answers are available in Appendix A.

✔ If you understand the principles of energetic coupling, you should be able to compare and contrast how energy is transferred via redox reactions and ATP hydrolysis.

It is hard to overstate the importance of energetic coupling: Without it, life is impossible. If the cells in your body could no longer drive endergonic reactions by coupling them to exergonic reactions, you would die within minutes.

Now the question is, What role do enzymes play in these reactions?

8.3 How Enzymes Work

Regardless of whether reactions in cells are spontaneous or not, none would occur at the speed required for life without the support of enzymes. How do they do it?

Recall that the initial hypothesis for how enzymes speed up reactions—the "lock-and-key" model—was first proposed in 1894 by Emil Fischer (introduced in Chapter 3). In this model, the substrates would fit into enzymes and react in a manner analogous to a key being inserted into a lock. In other words, enzymes are **catalysts**—they bring substrates together in a precise orientation that makes reactions more likely. Fischer's model also explained why enzymes are specific for a single reaction—specificity is a product of the geometry and chemical properties of the sites where substrates bind.

Enzymes Help Reactions Clear Two Hurdles

Recall that two hurdles must be cleared before reactions can take place: Reactants need to (1) collide in a precise orientation and

---

8.2 Nonspontaneous Reactions May Be Driven Using Chemical Energy

• Redox reactions transfer energy by coupling exergonic oxidation reactions to endergonic reduction reactions.
• High-energy C–H bonds may be formed during the reduction step of a redox reaction when an H⁺ is combined with a transferred electron.
• The hydrolysis of ATP is an exergonic reaction and may be used to drive a variety of cellular processes.
• When a phosphate group from ATP is added to a substrate that participates in an endergonic reaction, the potential energy of the substrate is raised enough to make the reaction exergonic and thus spontaneous.

✔ You should be able to explain what energetic coupling means, and why life would not exist without it.

8.3 How Enzymes Work

• Enzymes are protein catalysts. They speed reaction rates but do not affect the change in free energy of the reaction.
• The structure of an enzyme has an active site that brings sub-

• Protein cleavage and phosphorylation are examples of how enzymes may be regulated by modifying their primary structure.

✔ You should be able to compare and contrast the effect of allosteric regulation versus phosphorylation on enzyme function.

8.5 Enzymes Can Work Together in Metabolic Pathways

• In cells, enzymes often work together in metabolic pathways that sequentially modify a substrate to make a product.
• A pathway may be regulated by controlling the activity of one enzyme, often the first in the series of reactions. Feedback inhibition results from the accumulation of a product that binds to an enzyme in the pathway and inactivates it.
• Metabolic pathways were vital to the evolution of life, and new pathways continue to evolve in cells.

✔ You should be able to predict how the removal of the intermediate in a two-step metabolic pathway would affect the enzymatic rates of the first and last.

(MB) www.masteringbiology.com

1. MasteringBiology Assignments

Identify gaps in your understanding, then fill them

The Fifth Edition provides many opportunities for you to test yourself and offers helpful learning strategies.

Analyze: Can I recognize underlying patterns and structure?

Evaluate: Can I make judgments on the relative value of ideas and information?

Create: Can I put ideas and information together to generate something new?

Apply: Can I use these ideas in the same way or in a new situation?

Understand: Can I explain this concept in my own words?

Remember: Can I recall the key terms and ideas?

◀ **Bloom's Taxonomy** describes six learning levels: Remember, Understand, Apply, Analyze, Evaluate, and Create. Questions in the book span all levels, including self-testing at the higher levels to help you develop higher-order thinking skills that will prepare you for exams.

Steps to Building Understanding
Each chapter ends with three groups of questions that build in difficulty:

✔ TEST YOUR KNOWLEDGE
Begin by testing your basic knowledge of new information.

✔ TEST YOUR UNDERSTANDING
Once you're confident with the basics, demonstrate your deeper understanding of the material.

✔ TEST YOUR PROBLEM-SOLVING SKILLS
Work towards mastery of the content by answering questions that challenge you at the highest level of competency.

BIOSKILL 16 using Bloom's taxonomy

Most students have at one time or another wondered why a particular question on an exam seemed so hard, while others seemed easy. The explanation lies in the type of cognitive skills required to answer the question. Let's take a closer look.

▲ **NEW! BioSkill Covering Bloom's Taxonomy** helps you to recognize question types using the Bloom's cognitive hierarchy, and it provides specific strategies to help you study for questions at all six levels.

Answer Appendix Includes Bloom's Taxonomy Information
Answers to all questions in the text now include the Bloom's level being tested. You can simultaneously practice assessing your understanding of content and recognizing Bloom's levels. Combining this information with the guidance in the BioSkill on Bloom's Taxonomy will help you form a plan to improve your study skills.

▶

✔ Test Your Problem-Solving Skills

13. **analyze** A scientific theory is not a guess—it is an idea whose validity can be tested with data. Both the cell theory and the theory of evolution have been validated by large bodies of observational and experimental data.
14. **apply** If all eukaryotes living today have a nucleus, then it is logical to conclude that the nucleus arose in a common ancestor of all eukaryotes, indicated by the arrow you should have added to the figure. See **FIGURE A1.2**. If it had arisen in a common ancestor of Bacteria or Archaea, then species in those groups would have had to lose the trait—an unlikely event.
15. **evaluate** The data set was so large and diverse that it was no longer reasonable to argue that noncellular life-forms would be discovered. **16.** **apply** b

MasteringBiology®
www.masteringbiology.com

NEW! End-of-chapter questions from the book are now available for your professor to assign as homework in MasteringBiology.

Practice scientific thinking and scientific skills

A unique emphasis on the process of scientific discovery and experimental design teaches you how to think like a scientist as you learn fundamental biology concepts.

RESEARCH

QUESTION: Do horses minimize the cost of locomotion?

HYPOTHESIS: Horses choose gaits that minimize energy use at different speeds.

NULL HYPOTHESIS: Horses do not choose gaits based on cost of locomotion.

EXPERIMENTAL SETUP:

1. Measure oxygen consumption of horses trained to walk, trot, and gallop at a range of speeds on a treadmill. Calculate energy used per distance travelled at different speeds.

5.8 m/s

Oxygen mask

Treadmill

2. Videotape the same horses locomoting freely in the paddock, and measure the gaits and speeds they choose to use naturally.

PREDICTION: For each gait, there is a range of speeds where energy use is minimized. Horses will favor these gaits and speeds.

PREDICTION OF NULL HYPOTHESIS: There will be no correlation between chosen gaits and energy consumption.

RESULTS:

Walking   Trotting   Galloping

Speeds and gaits chosen by free horses

©1981 Nature Publishing Group

Energy/distance (joules per meter) vs Speed (m/s)

CONCLUSION: Horses choose gaits that minimize energy use at different speeds and avoid speeds with high energy consumption.

FIGURE 48.16 Horses Minimize the Cost of Locomotion by Choosing Appropriate Gaits.
SOURCE: Hoyt, D. F., and C. R. Taylor. 1981. Gait and the energetics of locomotion in horses. *Nature* 292: 239–240.

✔**QUANTITATIVE** Use the graph to estimate the relative energy expense of galloping rather than trotting at 3.5 meters/second (m/s).

All of the Research Boxes cite the original research paper and include a question that asks you to analyze the design of the experiment or study.

Research Boxes explain how research studies are designed and give you additional practice interpreting data. Each Research Box consistently models the scientific method, presenting the research question, hypotheses, experimental setup, predictions, results, and conclusion. 15 Research Boxes are new to the Fifth Edition.

MasteringBiology®
www.masteringbiology.com

NEW! Solve It Tutorials are available for homework assignments in MasteringBiology and give you an opportunity to work like a scientist through a simulated investigation that requires you to analyze and interpret data.

Experimental Inquiry Tutorials based on some of biology's most seminal experiments give you a chance to analyze data and the reasoning that led scientists from the data to their conclusions.

Experimental Inquiry tutorial topics include:

- What Can You Learn About the Process of Science from Investigating a Cricket's Chirp?
- Which Wavelengths of Light Drive Photosynthesis?
- What Is the Inheritance Pattern of Sex-Linked Traits?
- Does DNA Replication Follow the Conservative, Semiconservative, or Dispersive Model?
- How Do Calcium Ions Help to Prevent Polyspermy During Egg Fertilization?
- Did Natural Selection of Ground Finches Occur When the Environment Changed?
- What Effect Does Auxin Have on Coleoptile Growth?
- What Role Do Genes Play in Appetite Regulation?
- Can a Species' Niche Be Influenced by Interspecific Competition?
- What Factors Influence the Loss of Nutrients from a Forest Ecosystem?

Build important skills scientists use to perform, evaluate, and communicate scientific research.

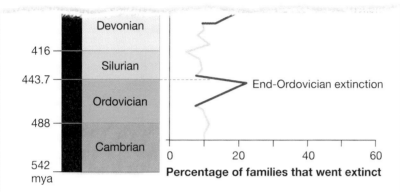

FIGURE 28.14 The Big Five Mass Extinction Events. This graph shows the percentage of lineages called families that went extinct over each interval in the fossil record since the Cambrian explosion. Over 50 percent of families and 90 percent of species went extinct during the end-Permian extinction.

DATA: Benton, M. J., 1995. *Science* 268: 52–58.

✔**QUANTITATIVE** Which extinction event ended the era of the dinosaurs 65 million years ago? About what percentage of families went extinct?

NEW! Graphs and tables now include their data sources, emphasizing the research process that leads to our understanding of biological ideas.

NEW! Quantitative questions are identified throughout the text, helping you practice computational problem solving and data analysis.

Expanded BioSkills Appendix helps you build skills that will be important to your success in biology. At relevant points in the text, you'll find references to the BioSkills appendix that will help you learn and practice foundational skills.

BioSkills Topics include:

- The Metric System and Significant Figures
- Some Common Latin and Greek Roots Used in Biology
- Reading Graphs
- Using Statistical Tests and Interpreting Standard Error Bars
- Combining Probabilities

- Using Logarithms
- Reading a Phylogenetic Tree
- Reading Chemical Structures
- Separating and Visualizing Molecules
- Separating Cell Components by Centrifugation
- Biological Imaging: Microscopy and X-ray Crystallography

- Cell and Tissue Culture Methods
- Model Organisms
- NEW! Primary Literature and Peer Review
- Making Concept Maps
- NEW! Using Bloom's Taxonomy

www.masteringbiology.com

You can access self-paced BioSkills activities in the Study Area, and your instructor can assign additional activities in MasteringBiology.

Visualize biology processes and structures

A carefully crafted visual program helps you gain a better understanding of biology through accurate, appropriately detailed figures.

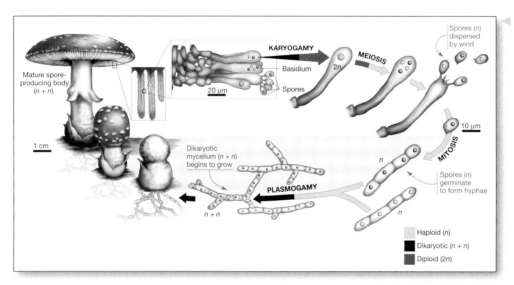

◀ NEW! Redesigned Life Cycle diagrams in Unit 6 and 7 help you compare and contrast processes among different organisms.

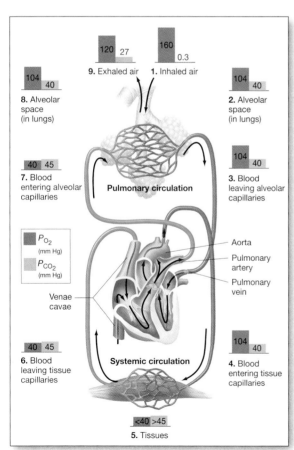

▲ Informative figures help you think through complex biological processes in manageable steps.

▲ Visual Summary Tables pull together important information in a format that allows for easy comparison and review.

Instructor and Student Resources

For Instructors

Instructor Resource DVD-ROM
978-0-321-86112-2 • 0-321-86112-4
Everything you need for lectures in one place, including video segments that demonstrate how to incorporate active-learning techniques into your own classroom. Enhanced menus make locating and assessing the digital resources for each chapter easy. The Instructor Resource CD/DVD-ROM includes PowerPoint® Lecture Outlines that integrate figures and animations for classroom presentations. All textbook figures, art, and photos are in JPEG format, and all PowerPoint slides and JPEGs have editable labels. Over 300 Instructor Animations accurately depict complex topics and dynamic processes described in the book.

Instructor Guide (Download only)
Available in the instructor resource area of MasteringBiology.®

TestGen® (Download only)
All of the exam questions in the Test Bank have been peer reviewed and student tested, providing questions that set the standard for quality and accuracy. To improve the Test Bank, Metadata from MasteringBiology users has been incorporated directly into the software. Test questions that are ranked according to Bloom's taxonomy and improved TestGen® software makes assembling tests that much easier.

For Students

Study Guide
978-0-321-85832-0 • 0-321-85832-8
The Study Guide presents a breakdown of key biological concepts, difficult topics, and quizzes to help students prepare for exams. Unique to this study guide are four introductory, stand-alone chapters that introduce students to foundational ideas and skills necessary for classroom success: Introduction to Experimentation and Research in the Biological Sciences, Presenting Biological Data, Understanding Patterns in Biology and Improving Study Techniques, and Reading and Writing to Understand Biology. "Looking Forward" and "Looking Back" sections help students make connections across the chapters instead of viewing them as discrete entities.

Practicing Biology: A Student Workbook
978-0-321-88647-7 • 0-321-88647-X
This workbook focuses on key ideas, principles, and concepts that are fundamental to understanding biology. A variety of hands-on activities such as mapping and modeling suit different learning styles and help students discover which topics they need more help on. Students learn biology by doing biology. An instructors guide can be downloaded from the Instructor Area of MasteringBiology.

MasteringBiology®

www.masteringbiology.com

MasteringBiology is an online homework, tutorial, and assessment system that delivers self-paced tutorials that provide individualized coaching, focus on your course objectives, and respond to each student's progress. The Mastering system helps instructors maximize class time with customizable, easy-to-assign, and automatically graded assessments that motivate students to learn outside of class and arrive prepared for lecture. MasteringBiology is also available with a complete Pearson eText edition of *Biological Science*.

Highlights of the Fifth Edition Item Library include:

* NEW! assignment options include Solve It activities, end-of-chapter problems, and questions that accompany new BioSkills and new Big Picture Interactive Concept Maps.
* NEW! "best of" homework pre-built assignments help professors assign popular, key content quickly, including a blend of tutorials, end-of-chapter problems, and test bank questions.

* *Get Ready for Biology* and Chemistry Review assignment options help students get up to speed with activities that review chemistry, mathematics, and basic biology.

MasteringBiology® Virtual Labs

978-0-321-88644-6 • 0-321-88644-5
MasteringBiology: Virtual Labs is an online environment that promotes critical-thinking skills using virtual experiments and explorations that might be difficult to perform in a wet-lab environment due to time, cost, or safety concerns. MasteringBiology: Virtual Labs offers unique learning experiences in the areas of microscopy, molecular biology, genetics, ecology, and systematics.

For more information, please visit www.pearsonhighered.com/virtualbiologylabs

BIOLOGICAL SCIENCE

VOLUME 2 Evolution, Diversity, & Ecology

After you explore this book . . .
you should be able to . . .

• Pose an evolutionary hypothesis to explain why meter-long male water dragons are larger and have more colorful throats than the females.

• Propose how DNA sequences could be used to determine the relationships among populations of these lizards in China, India, and Southeast Asia.

• Design an experiment to study the relative importance of swimming, tree climbing, and running to the ability of these semi-aquatic lizards to find food and escape from predators.

• Create questions of your own and suggest methods for finding the answers!

Chinese Water Dragon,
Physignathus cocincinus

BIOLOGICAL SCIENCE

FIFTH EDITION

SCOTT FREEMAN
University of Washington

LIZABETH ALLISON
College of William & Mary

MICHAEL BLACK
California Polytechnic State University in San Luis Obispo

GREG PODGORSKI
Utah State University

KIM QUILLIN
Salisbury University

JON MONROE
James Madison University

EMILY TAYLOR
California Polytechnic State University in San Luis Obispo

PEARSON

Boston   Columbus   Indianapolis   New York   San Francisco   Upper Saddle River
Amsterdam   Cape Town   Dubai   London   Madrid   Milan   Munich   Paris   Montréal   Toronto
Delhi   Mexico City   São Paulo   Sydney   Hong Kong   Seoul   Singapore   Taipei   Tokyo

Editor-in-Chief: Beth Wilbur
Senior Acquisitions Editor: Michael Gillespie
Executive Director of Development: Deborah Gale
Senior Development Editor: Sonia DiVittorio
Project Editor: Anna Amato
Development Editors: Mary Catherine Hager, Moira Lerner Nelson, Bill O'Neal
Art Development Editors: Fernanda Oyarzun, Adam Steinberg
Associate Editor: Brady Golden
Assistant Editor: Leslie Allen
Editorial Assistant: Eddie Lee
Executive Media Producer: Laura Tommasi
Media Producer: Joseph Mochnick
Associate Media Producer: Daniel Ross
Associate Media Project Manager: David Chavez
Text Permissions Project Manager: Joseph Croscup
Text Permissions Specialist: Sheri Gilbert
Director of Production: Erin Gregg
Managing Editor: Michael Early

Production Project Manager: Lori Newman
Production Management: S4Carlisle Publishing Services
Copyeditor: Christianne Thillen
Compositor: S4Carlisle Publishing Services
Design Manager: Mark Ong
Interior Designer: Integra Software Services
Cover Designer: tt eye
Illustrators: Imagineering Media Services
Photo Permissions Management: Phutu
Photo Researchers: Kristin Piljay, Eric Schrader, Maureen Spuhler
Senior Photo Editor: Travis Amos
Manufacturing Buyer: Michael Penne
Director of Marketing: Christy Lesko
Executive Marketing Manager: Lauren Harp
Sales Director for Key Markets: David Theisen

Cover Photo Credit: *Physignathus cocincinus*
Eric Isselée/Fotolia

Credits and acknowledgments for materials borrowed from other sources and reproduced, with permission, in this textbook appear on the appropriate page within the text or beginning on page C:1 of the backmatter.

Library of Congress Cataloging-in-Publication Data
Freeman, Scott, 1955- Biological science / Scott Freeman.—Fifth edition.
    pages cm
  ISBN-13: 978-0-321-74367-1 (student edition)
  ISBN-10: 0-321-74367-9 (student edition)
  ISBN-13: 978-0-321-84159-9 (instructors review copy)
  ISBN-10: 0-321-84159-X (instructors review copy) [etc.]
  1. Biology—Textbooks.   I. Title.
  QH308.2.F73 2014
  570—dc23

4 5 6 7 8 9 10—V011—16 15 14

ISBN 10: 0-321-74367-9; ISBN 13: 978-0-321-74367-1 (Student Edition)
ISBN 10: 0-321-84159-X; ISBN 13: 978-0-321-84159-9 (Instructor's Review Copy)
ISBN 10: 0-321-86216-3; ISBN 13: 978-0-321-86216-7 (Books a la Carte Edition)
ISBN 10: 0-321-84180-8; ISBN 13: 978-0-321-84180-3 (Volume 1)
ISBN 10: 0-321-84181-6; ISBN 13: 978-0-321-84181-0 (Volume 2)
ISBN 10: 0-321-84182-4; ISBN 13: 978-0-321-84182-7 (Volume 3)

www.pearsonhighered.com

Detailed Contents

30 Protists 552

31 Green Algae and Land Plants 577

UNIT 9 ECOLOGY 1059

52 An Introduction to Ecology 1059

53 Behavioral Ecology 1082

54 Population Ecology 1101

About the Authors

A Letter from Scott:

I started working on *Biological Science* in 1997 with a simple goal: To help change the way biology is taught. After just shy of 20,000 hours of work on four editions of this text, that goal still gets me out of bed in the morning. But instead of focusing my energies on textbook writing, I've decided to devote myself full-time to research on student learning and developing new courses for undergraduate and graduate students at the University of Washington.

So with this edition I am passing the torch—to an all-star cast of leading scientists and educators who have enthusiastically taught from, and contributed to, previous editions of *Biological Science*. Working with them, I have seen the new team bring their passion, talent, and creativity to the book, with expertise that spans the breadth of the life sciences. Just as important, they work beautifully together because they think alike. They are driven by a shared concern for student learning, a commitment to the craft of writing, and a background in evidence-based teaching.

These pages provide a brief introduction to Liz Allison, Michael Black, Greg Podgorski, Kim Quillin, Jon Monroe, and Emily Taylor. As a group, they've built on the book's existing strengths and infused this edition with fresh energy, perspective, and ideas. I'm full of admiration for what they have accomplished, and excited about the impact this edition will have on biology students from all over the world.—*Scott Freeman*

Scott Freeman received a Ph.D. in Zoology from the University of Washington and was subsequently awarded an Alfred P. Sloan Postdoctoral Fellowship in Molecular Evolution at Princeton University. He has done research in evolutionary biology on topics ranging from nest parasitism to the molecular systematics of the blackbird family and is coauthor, with Jon Herron, of the standard-setting undergraduate text *Evolutionary Analysis*. Scott is the recipient of a Distinguished Teaching Award from the University of Washington and is currently a Senior Lecturer in the UW Department of Biology, where he teaches introductory biology for majors, a writing-intensive course for majors called The Tree of Life, and a graduate seminar in college science teaching. Scott's current research focuses on how active learning affects student learning and academic performance.

Lizabeth A. Allison is professor and chair of the Biology Department at the College of William & Mary. She received her Ph.D. in Zoology from the University of Washington, specializing in molecular and cellular biology. Before coming to William & Mary, she spent eight years as a faculty member at the University of Canterbury in New Zealand. Liz teaches introductory biology for majors and upper-division molecular biology courses. She has mentored graduate students and more than 80 undergraduate research students, many of them coauthoring papers with her on intracellular trafficking of the thyroid hormone receptor in normal and cancer cells. The recipient of numerous awards, including a State Council for Higher Education in Virginia (SCHEV) Outstanding Faculty Award in 2009, Liz received one of the three inaugural Arts & Sciences Faculty Awards for Teaching Excellence in 2011, and a Plumeri Award for Faculty Excellence in 2012. In addition to her work on this text, she is author of *Fundamental Molecular Biology*, now in its second edition.
Lead Author; Chapter 1 and BioSkills
laalli@wm.edu

Michael Black received his Ph.D. in Microbiology & Immunology from Stanford University School of Medicine as a Howard Hughes Predoctoral Fellow. After graduation, he studied cell biology as a Burroughs Wellcome Postdoctoral Fellow at the MRC Laboratory of Molecular Biology in Cambridge, England. His current research focuses on the use of molecules to identify and track the transmission of microbes in the environment. Michael is a professor of Cell & Molecular Biology at California Polytechnic State University in San Luis Obispo, where he teaches introductory and advanced classes for majors in cell biology and microbiology. In addition to his teaching and research activities, Michael serves as the director of the Undergraduate Biotechnology Lab, where he works alongside undergraduate technicians to integrate research projects and inquiry-based activities into undergraduate classes.
Chapters 2–12, 36, and 51
mblack@calpoly.edu

Greg Podgorski received his Ph.D. in Molecular and Cellular Biology from Penn State University and has been a postdoctoral fellow at the Max Plank Institute for Biochemistry and Columbia University. His research interests are in biology education, developmental genetics, and computational biology. Greg's most recent work has been in mathematical modeling of how patterns of different cell types emerge during development and how tumors recruit new blood vessels in cancer. Greg has been teaching at Utah State University for more than 20 years in courses that include introductory biology for majors and for nonmajors, genetics, cell biology, developmental biology, and microbiology, and he has offered courses in nonmajors biology in Beijing and Hong Kong. He's won teaching awards at Utah State University and has been recognized by the National Academies as a Teaching Fellow and a Teaching Mentor.

Chapters 13–24
greg.podgorski@usu.edu

Jon Monroe is professor of Biology at James Madison University in Harrisonburg, Virginia. Jon completed his undergraduate work in Botany at the University of Michigan and his graduate work in Plant Physiology at Cornell University. He began his current position after a postdoc in biochemistry at Michigan State University. He currently teaches Plant Biology, and Cell and Molecular Biology. Jon's interest in plants is broad, ranging from systematics and taxonomy to physiology and biochemistry. His research, mostly with undergraduates, uses Arabidopsis thaliana to study the functions of a family of β-amylase genes in starch metabolism. Jon has been active in promoting undergraduate research through his work with the American Society of Plant Biologists (ASPB) and the Council on Undergraduate Research. He has received ASPB's Excellence in Teaching award and James Madison University Alumni Association's Distinguished Faculty Award.

Chapters 29–32; 37–41
monroejd@jmu.edu

Kim Quillin received her B.A. in Biology at Oberlin College *summa cum laude* and her Ph.D. in Integrative Biology from the University of California, Berkeley (as a National Science Foundation Graduate Fellow). Kim has worked in the trenches with Scott Freeman on every edition of *Biological Science*, starting with the ground-up development of the illustrations in the first edition in 1999 and expanding her role in each edition, always with the focus of helping students to think like biologists. Kim currently teaches introductory biology at Salisbury University, a member of the University System of Maryland, where she is actively involved in the ongoing student-centered reform of the concepts-and-methods course for biology majors. Her current research focuses on the scholarship of teaching and learning with an emphasis on measuring science process skills and the advantages and pitfalls of active multimedia learning.

Chapters 25–28; 33–35; 48; 52–57
kxquillin@salisbury.edu

Emily Taylor earned a B.A. in English at the University of California, Berkeley followed by a Ph.D. in Biological Sciences from Arizona State University, where she conducted research in the field of environmental physiology as a National Science Foundation Graduate Research Fellow. She is currently an associate professor of Biological Sciences at the California Polytechnic State University in San Luis Obispo, California. Her student-centered research program focuses on the endocrine and reproductive physiology of free-ranging reptiles, especially rattlesnakes. She teaches numerous undergraduate and graduate courses, including introductory biology, anatomy and physiology, and herpetology, and received the California Faculty Association's Distinguished Educator Award in 2010 and Cal Poly's Distinguished Teaching Award in 2012. Her revision of Unit 8 is her first foray into textbook writing.

Chapters 42–50
etaylor@calpoly.edu

Preface to Instructors

The first edition of *Biological Science* was visionary in its unique emphasis on the process of scientific discovery and experimental design—teaching how we know what we know. The goal was for students not only to learn the language of biology and understand fundamental concepts but also to begin to apply those concepts in new situations, analyze experimental design, synthesize results, and evaluate hypotheses and data—to learn how to think like biologists. Each edition since has proudly expanded on this vision. The Fifth Edition is no exception.

A team of six dedicated teacher-scholars has joined Scott to build on and refine the original vision, and by so doing, make the book an even better teaching and learning tool. The pace of biological discovery is rapid, and with each novel breakthrough it becomes even more challenging to decide what is essential to include in an introductory biology text. Pulling together an author team with firsthand expertise from molecules to ecosystems has ensured that the content of the Fifth Edition reflects cutting-edge biology that is pitched at the right level for introductory students and is as accurate and as exciting as ever for instructors and students alike.

New findings from education research continue to inform and inspire the team's thinking about *Biological Science*—we know more today than ever before about how students learn. These findings demand that we constantly look for new ways to increase student engagement in the learning process, and to help instructors align course activities and learning goals with testing strategies.

The New Coauthors

The new coauthor team brings a broad set of talents and interests to the project, motivated by a deep commitment to undergraduate teaching, whether at a small liberal arts college or a large university. Kim Quillin has been a partner in this textbook in every edition. For the Fifth Edition, she revised chapters across three units in addition to spearheading the continued effort to enhance the visual-teaching program. Michael Black, Greg Podgorski, Jon Monroe, and Emily Taylor, who served as unit advisors on the Fourth Edition, were already familiar with the book. And most of the authorial team have been avid users of previous editions for many years.

Core Values

Together, the coauthor team has worked to extend the vision and maintain the core values of *Biological Science*—to provide a book for instructors who embrace the challenge of boosting students to higher levels of learning, and to provide a book for students that helps them each step of the way in learning to think like scientists. Dedicated instructors have high expectations of their students—the Fifth Edition provides scaffolding to help students learn at the level called for by the National Academy of Sciences, the Howard Hughes Medical Institute, the American Association of Medical Academies, and the National Science Foundation.

What's New in This Edition

The Fifth Edition contains many new or expanded features, all of them targeted at ways to help students learn to construct their own knowledge and think like biologists.

- **Road Maps** The new Road Maps at the beginning of each chapter pair with the Big Picture concept maps introduced in the Fourth Edition. Together they help students navigate chapter content and see the forest for the trees. Each Road Map starts with a purpose statement that tells students what they can expect to learn from each chapter. It then goes on to visually group and organize information to help students anticipate key ideas as well as recognize meaningful relationships and connections between the ideas.

- **The Big Picture** Introduced in the Fourth Edition, Big Picture concept maps integrate words and visuals to help students synthesize information about challenging topics that span multiple chapters and units. In response to requests from instructors and students, three new Big Pictures focused on additional tough topics have been added: Doing Biology, The Chemistry of Life, and Plant and Animal Form and Function. In addition, the Ecology Big Picture is completely revised to reflect changes to that unit.

- **New Chapters** Two new chapters are added to better serve instructors and students. Unit 2 now contains a new Chapter 8, Energy and Enzymes: An Introduction to Metabolic Pathways. This chapter consolidates these critical topics in a place where students and instructors need it most—right before the chapters on cellular respiration and photosynthesis. In the Fourth Edition, animal movement was discussed in a chapter largely focused on animal sensory systems. In the Fifth Edition, this important topic is treated in depth in a new Chapter 48, Animal Movement, that explores how muscle and skeletal systems work together to produce locomotion.

- **New BioSkills** Instructors recognize that biology students need to develop foundational science skills in addition to content knowledge. While these skills are emphasized throughout the book, *Biological Science,* beginning with the Third

Edition, has provided a robust set of materials and activities to guide students who need extra help. To promote even fuller use of this resource, the BioSkills are now updated, expanded, and reorganized. New in this edition are a discussion of significant figures within the BioSkills on the Metric System, and two new BioSkills on Primary Literature and Peer Review and Using Bloom's Taxonomy. BioSkills are located in Appendix B, and practice activities can be assigned online in MasteringBiology®.

- **Promotion of Quantitative Skills** Reports like *Biology 2010*, *Scientific Foundations for Future Physicians*, and *Vision and Change* all place a premium on quantitative skills. To infuse a quantitative component throughout the text, new and existing quantitative questions are flagged in each chapter to encourage students to work on developing their ability to read or create a graph, perform or interpret a calculation, or use other forms of quantitative reasoning.

- **Bloom's Taxonomy** In the Fifth Edition, all questions in the text are assigned a Bloom's Taxonomy level to help both students and instructors understand whether a question requires higher-order or lower-order cognitive skills. Questions span all six Bloom's levels. (Bloom's levels are identified in Appendix A: Answers.) The coauthors were trained by experts Mary Pat Wenderoth and Clarissa Dirks[1] to ensure we followed a process that would result in high inter-rater reliability—or agreement among raters—in assigning Bloom's levels to questions. The new BioSkill, Using Bloom's Taxonomy, explains the six Bloom's levels to students and offers a practical guide to the kinds of study activities best suited for answering questions at each level.

- **Expanded Emphasis on "Doing Biology"** A constant hallmark of this text is its emphasis on experimental evidence—on teaching how we know what we know. To reflect the progress of science, in the Fifth Edition, the coauthor team replaced many experiments with fresh examples and added new Research Boxes. And as noted earlier, they added a new Big Picture on Doing Biology, focusing on the process of science and the organizational levels of biology. Data sources are now cited for all graphs and data tables to model the importance of citing data sources to students. Updated Research Box questions continue to encourage students to analyze some aspect of experimental design. Also new to this edition is a BioSkill on Primary Literature and Peer Review.

- **Art Program** The art program is further enhanced in this edition by the addition of more illustrated summary tables. These tables make subject areas more accessible to visual learners and reinforce key concepts of the chapter. Many of the life-cycle figures in Unit 6 are significantly overhauled.

[1] Crowe, A., C. Dirks, and M. P. Wenderoth. 2008. Biology in Bloom: Implementing Bloom's Taxonomy to enhance student learning in biology. *CBE–Life Sciences Education* 7: 368–381.

Updated Blue Thread Scaffolding

In the Third and Fourth editions of *Biological Science*, a metacognitive tool was formulated as the now popular feature known as "Blue Thread"—sets of questions designed to help students identify what they do and don't understand. The fundamental idea is that if students really understand a piece of information or a concept, they should be able to do something with it.

In the Fifth Edition, the Blue Thread is revised to reflect changes in chapter content, and to incorporate user feedback. Blue-Thread questions appear in the following locations:

- **In-text "You should be able to's"** offer exercises on topics that professors and students have identified as the most difficult concepts in each chapter.

- **Caption questions and exercises** challenge students to examine the information in a figure or table critically—not just absorb it.

- **Check Your Understanding boxes** present two to three tasks that students should be able to complete in order to demonstrate a mastery of summarized key ideas.

- **Chapter summaries** include "You should be able to" problems or exercises related to each key concept.

- **End-of-chapter** questions are organized in three levels of increasing difficulty so students can build from lower to higher-order cognitive questions.

Integration of Media

The textbook continues to be supported by MasteringBiology®, the most powerful online homework, tutorial, and assessment system available. Tutorials follow the Socratic method, coaching students to the correct answer by offering feedback specific to a student's misconceptions as well as providing hints students can access if they get stuck. Instructors can associate content with publisher-provided learning outcomes or create their own. Content highlights include the following:

- **NEW! Solve It Tutorials** These activities allow students to act like scientists in simulated investigations. Each tutorial presents an interesting, real-world question that students will answer by analyzing and interpreting data.

- **Experimental Inquiry Tutorials** The call to teach students about the process of science has never been louder. To support such teaching, there are 10 interactive tutorials on classic scientific experiments—ranging from Meselson–Stahl on DNA replication to the Grants' work on Galápagos finches and Connell's work on competition. Students who use these tutorials should be better prepared to think critically about experimental design and evaluate the wider implications of the data—preparing them to do the work of real scientists in the future.

- **BioFlix® Animations and Tutorials** BioFlix are movie-quality, 3-D animations that focus on the most difficult core topics and are accompanied by in-depth, online tutorials that provide hints and feedback to guide student learning. Eighteen BioFlix animations and tutorials tackle topics such as meiosis, mitosis, DNA replication, photosynthesis, homeostasis, and the carbon cycle.

- **NEW! End-of-Chapter Questions** Multiple choice end-of-chapter questions are now available to assign in MasteringBiology.
- **Blue-Thread Questions** Over 500 questions based on the Blue-Thread Questions in the textbook are assignable in MasteringBiology.
- **Big Picture Tutorials** Interactive concept map activities based on the Big Picture figures in the textbook are assignable in MasteringBiology, including tutorials to support the three new Big Pictures: Doing Biology, The Chemistry of Life, and Plant and Animal Form and Function.
- **BioSkills Activities** Activities based on the BioSkills content in the textbook are assignable in MasteringBiology, including activities to support the new BioSkills on Primary Literature and Peer Review and Using Bloom's Taxonomy.

- **Reading Quiz Questions** Every chapter includes reading quiz questions you can assign to ensure students read the textbook and understand the basics. These quizzes are perfect as a pre-lecture assignment to get students into the content before class, allowing you to use class time more effectively.

Serving a Community of Teachers

All of us on the coauthor team are deeply committed to students and to supporting the efforts of dedicated teachers. Doing biology is what we love. At various points along our diverse paths, we have been inspired by our own teachers when we were students, and now are inspired by our colleagues as we strive to become even better teacher-scholars. In the tradition of all previous editions of *Biological Science*, we have tried to infuse this textbook with the spirit and practice of evidence-based teaching. We welcome your comments, suggestions, and questions.

Thank you for your work on behalf of your students.

Content Highlights of the Fifth Edition

As discussed in the preface, a major focus of this revision is to enhance the pedagogical utility of *Biological Science*. Another major goal is to ensure that the content reflects the current state of science and is accurate. The expanded author team has scrutinized every chapter to add new, relevant content, update descriptions when appropriate, and adjust the approach to certain topics to enhance student comprehension. In this section, some of the key content improvements to the textbook are highlighted.

Chapter 1 Biology and the Tree of Life A concept map summarizing the defining characteristics of life is added. The process of doing biology coverage is expanded to include discussion of both experimental and descriptive studies, and more rigorous definitions of the terms hypothesis and theory.

Chapter 2 Water and Carbon: The Chemical Basis of Life A stronger emphasis on chemical evolution is threaded throughout the chapter to bring chemistry to life for the student reader. Two prominent models for chemical evolution are introduced; the historic Miller prebiotic soup experiment was moved here. Advanced discussion of energy and chemical reactions was moved to a new chapter (see Chapter 8).

Chapter 3 Protein Structure and Function The chapter is reorganized to emphasize the link between structure and function, from amino acids to folded proteins. Updated content illustrates that protein shapes are flexible and dynamic, and may remain incompletely folded until the protein interacts with other molecules or ions. Details of how enzymes work were moved to Chapter 8.

Chapter 4 Nucleic Acids and the RNA World New experimental results concerning the synthesis of nucleotides and nucleic acids in a prebiotic environment are discussed. The section on the RNA world is expanded to include the artificial evolution of a novel ribozyme involved in nucleotide synthesis.

Chapter 5 An Introduction to Carbohydrates The molecular basis for resistance of structural polymers, such as cellulose, to degradation is clarified. A new research box illustrates the role of carbohydrates in cellular recognition and attachment using the egg and sperm of mice as a model system.

Chapter 6 Lipids, Membranes, and the First Cells New content on lipid and membrane evolution and the proposed characteristics of the first protocell is introduced. The aquaporin and potassium channel figures are updated; how key amino acids serve as selectivity filters is now highlighted.

Chapter 7 Inside the Cell Several new electron micrographs were selected to more clearly illustrate cell component structure and function. A new figure is added to better depict the pulse–chase assay used to identify the secretory pathway. Coverage of nuclear transport is expanded to differentiate between passive diffusion and active nuclear import. Updated content emphasizes the role of the cytoskeleton in localizing organelles, and how polarity of microtubules and microfilaments influences their growth rate.

Chapter 8 Energy and Enzymes: An Introduction to Pathways This new chapter pulls together concepts in energy, chemical reactions, and enzymes that previously were covered in three different chapters. Oxidation and reduction reactions are emphasized to prepare students for Chapters 9 and 10. The energetics behind ATP hydrolysis and its role in driving endergonic reactions is discussed, and figures are revised to better illustrate the process. Updated content on enzyme regulation and a new process figure show a model for how metabolic pathways may have evolved.

Chapter 9 Cellular Respiration and Fermentation Two new summary tables for glycolysis and the citric acid cycle are added that provide the names of the enzymes and the reaction each catalyzes. New content is introduced to propose a connection between the universal nature of the proton motive force and the story of the chemical evolution of life.

Chapter 10 Photosynthesis More extensive comparison between the chemical reactions in mitochondria and chloroplasts is added. A new figure is introduced to illustrate noncyclic electron flow in the context of the thylakoid membrane. Greater emphasis is placed on the number of ATPs and NADPHs required for each cycle of carbon fixation and reduction.

Chapter 11 Cell–Cell Interactions Coverage of extracellular matrix structure and function is expanded, including its role in intercellular adhesions and cell signaling. The plant apoplast and symplast are now introduced as key terms in the text and illustrated in a new figure. New content and a new figure on unicellular models for intercellular communication via pheromone sensing (yeast) and quorum sensing (slime mold) are added.

Chapter 12 The Cell Cycle A new figure helps explain the pulse–chase assay for identifying phases of the cell cycle. Content is added to the text and to a figure that illustrates the similarities between chromosome segregation in eukaryotes and prokaryotes. A revised description of anaphase emphasizes how microtubule fraying at the kinetochore can drive chromosome movement. The explanation of how phosphorylation and dephosphorylation turns on MPF activity is updated to reflect current research.

Chapter 13 Meiosis To improve the flow of the chapter, the section on advantages of sexual reproduction was moved to before mistakes in meiosis. The discussion of the role and timing of crossing over during meiosis I is updated. A new study that supports the hypothesis that sex evolved in response to the selective pressure of pathogens is introduced.

Chapter 14 Mendel and the Gene   Material on gene linkage is revised to emphasize the importance of genetic mapping. A new matched set of figures on pedigree analysis brings together the various modes of transmission that were previously shown in four individual figures. A new summary table on characteristics of different patterns of inheritance is added.

Chapter 15 DNA and the Gene: Synthesis and Repair   A new research figure is added that focuses on the relationship between telomere length and senescence in cultured somatic cells.

Chapter 16 How Genes Work   Coverage of the evolving concept of the gene and of different types of RNA is expanded. A figure showing the karyotype of a cancer cell is revised to improve clarity.

Chapter 17 Transcription, RNA Processing, and Translation   The sections on transcription in bacteria and eukaryotes are now separated, and content on charging tRNAs was moved to a new section. The discussion of translation is reorganized, first to emphasize the process in bacteria and then to highlight differences in eukaryotes.

Chapter 18 Control of Gene Expression in Bacteria   Coverage of *lac* operon positive regulation is updated to reflect current research. A new section and new process figure on global gene regulation are added, using the *lexA* regulon as an example.

Chapter 19 Control of Gene Expression in Eukaryotes   Extensive updates to the discussion of epigenetics include a new research box and a section on DNA methylation. Coverage of transcription initiation is updated to reflect current science. A new figure illustrates the role of p53 in the cell cycle in normal and cancerous cells.

Chapter 20 Analyzing and Engineering Genes   The material on sequencing the Neanderthal genome is updated, including evidence of limited Neanderthal genetic material in some modern human populations. New information on current generation sequencing technologies and massive parallelism is added. Recent advances in gene therapy are highlighted.

Chapter 21 Genomics and Beyond   Extensive updates throughout reflect recent advances in genomics. Changes include sequence database statistics, genomes that have been sequenced to study evolutionary relationships, and new figures illustrating gene count versus genome size in prokaryotes and eukaryotes and functional classes of human DNA sequences. A new section on systems biology is added. Also included are notes on the discovery of widespread transcription of eukaryotic genomes, deep sequencing, and the spectrum of mutations in human tumors.

Chapter 22 Principles of Development   New information is added on dedifferentiation in induced pluripotent stem cells, maternal genes in *Drosophila* development, how morphogens work, and tool-kit genes. The order of topics in the discussion

of developmental principles is reorganized. The figure on *Hox* genes in *Drosophila* and the mouse is updated.

Chapter 23 An Introduction to Animal Development   The chapter is streamlined by focusing on principles of animal development. The discussion of gametogenesis was moved to Chapter 50 (Animal Reproduction). The presentation of fertilization is simplified, and a new figure summarizing steps of fertilization is added. The figure on gastrulation is modified to better depict the arrangement of the germ layers and their movement.

Chapter 24 An Introduction to Plant Development   The chapter is modified to impart an evolutionary perspective on the similarities and differences in plant and animal development. The chapter also was streamlined by removing material such as details of gametogenesis, which now appears in Chapter 41 (Plant Reproduction).

Chapter 25 Evolution by Natural Selection   Several new key passages are included, among them the use of the Grand Canyon as a context for understanding relative dating of fossils, Darwin's artificial selection experiments with fancy pigeons, and Malthus's concept of struggle for existence. A new example of people living at high altitude in Tibet clarifies the difference between acclimatization and adaptation. An illustrated summary table of common misconceptions is added.

Chapter 26 Evolutionary Processes   Discussion of sexual selection now falls within the section on natural selection, and the terms intersexual and intrasexual selection are added. Several new examples replace those in the Fourth Edition, including inbreeding depression in Florida panthers, gene flow in Oregon steelhead trout, and lateral gene transfer in aphids. An illustrated summary table is added on modes of selection.

Chapter 27 Speciation   Several points are clarified, such as the gradient-like (rather than all-or-nothing) nature of reproductive isolation. The section on polyploidy is reorganized and the figures revised, including a side-by-side comparison of autopolyploidy and allopolyploidy.

Chapter 28 Phylogenies and the History of Life   The phylogenetics section is reorganized and expanded to include three illustrated summary tables and updated life-history timelines. New content is added, including the concept of the Anthropocene, the calendar analogy to the history of the Earth, the Chengjiang fossils, and a Life-in-the-Cambrian illustration. Evidence for the impact hypothesis is combined into an illustrated summary table.

Chapter 29 Bacteria and Archaea   The chapter is updated to include a description of metagenomic experiments with an emphasis on the role of gut bacteria in digestion. A newly recognized phylum of Archaea, the Thaumarchaeota, is included, and the table comparing key characteristics of the Bacteria, Archaea, and Eukarya is streamlined.

Chapter 30 Protists   For simplicity, protist lineages are now referred to throughout the chapter by their more familiar common names. Also, some key lineage boxes were consolidated to trim the number to one box per major lineage. Discussion of the origin of the nuclear envelope and mitochondria is expanded to reflect new thinking on the evolution of eukaryotic cells. Protist life cycle figures are significantly overhauled.

Chapter 31 Green Algae and Land Plants   Coverage of the evolution of land plants is expanded to include the importance of UV light and UV-absorbing molecules on the colonization of land. The "Redwood group" is now referred to as the Cupressophyta. Updates emphasize the role(s) of each stage of a life cycle in dispersal and in increasing genetic variation and individual numbers. A new research box is added, showing the importance of flower color to pollinator preference. Plant life-cycle figures are significantly overhauled.

Chapter 32 Fungi   Coenocytic fungal hyphae are illustrated with a new image showing GFP-labeled nuclei in *Neurospora crassa*. The chapter now points out the similarity between fungal and animal modes of nutrition, in terms of extracellular digestion and absorption of small molecules. The discussion of lignin degradation is updated, and new descriptions of mutualisms of fungi with animals are included. Fungal life cycle figures are significantly overhauled.

Chapter 33 An Introduction to Animals   The chapter is extensively revised to streamline and modernize the presentation, including emphasis on genetic tool kits and symmetry in the phylogeny of animals. The "Themes of Diversification" section is reorganized around five illustrated summary tables. The discussion of life cycles is revised to be more general. Insect metamorphosis has moved to Chapter 34, and a sea urchin life cycle replaces *Obelia*.

Chapter 34 Protostome Animals   Two themes are threaded throughout the chapter: the water-to-land transition and modular body plans. The section on lophotrochozoans emphasizes spiral cleavage, indirect versus direct development, hemocoels, and radulas. The section on ecdysozoans highlights segmentation and *Hox* genes, including discussion of the origin of the wing and metamorphosis. Key lineage boxes include new phylogenies for annelids, crustaceans, and chelicerates.

Chapter 35 Deuterostome Animals   Updates to reflect current research include revised phylogenies, evolution of flight and feathers, *Australopithecus sediba*, human migration out of Africa, and genetic evidence for interbreeding of *Homo neanderthalensis* and *Homo sapiens*.

Chapter 36 Viruses   New content focuses on how viruses contribute to evolution via lateral gene transfer and direct addition of genes to cellular genomes. Content is updated and expanded on viral structure and function, and on lytic and latent infections. Three new figures are added, including a comparison of replication of viruses and cells, how pandemic strains of influenza arise via reassortment, and the devastating impact of the 1918 influenza pandemic.

Chapter 37 Plant Form and Function   The use of terminology is streamlined for consistency and clarity. For example,

"lateral meristem" is replaced with "vascular cambium" to avoid confusion with lateral buds, and the description of bark is clarified to avoid using the term phelloderm. Several complex figures were converted to illustrated summary tables.

Chapter 38 Water and Sugar Transport in Plants   The chapter is revised to improve accuracy, and points out that water loss is also a means for transporting minerals from roots to shoots. Updated content clarifies the role of energy expenditure in moving water across roots, and the Casparian strip as a barrier to the back diffusion of ions and water out of the root. A new research figure shows the importance of the sucrose proton symporter in long-distance transport in *Arabidopsis*.

Chapter 39 Plant Nutrition   In this chapter the coverage of mycorrhizae is modified to emphasize their overall role in nutrient acquisition. In the section on nitrogen fixation, a description of the worldwide practice of crop rotation involving legumes and grains is added.

Chapter 40 Plant Sensory Systems, Signals, and Responses   A new research box is added that reveals the essential role of PHOT1 phosphorylation in phototropism. A section on the effect of day length on flowering was moved here from Chapter 41 and is integrated with the discussion of phytochromes. New content on the role of plasmodesmata in plant action potentials is added. The section on how plants respond to pathogens is simplified and updated with an example of control of stomata during a bacterial infection.

Chapter 41 Plant Reproduction   To provide a clearer example of a gametophyte-dominant life cycle, the liverwort life cycle has been replaced with a moss life cycle. The term "pollination syndrome" is clarified. Also, a research box on how capsaicin prevents seed predation and facilitates dispersal was reinstated from an earlier edition of *Biological Science*.

Chapter 42 Animal Form and Function   The chapter includes a new experiment illustrating physiological trade-offs, along with improved examples of thermoregulatory strategies in animals. Several complex figures were converted to illustrated summary tables.

Chapter 43 Water and Electrolyte Balance in Animals   The chapter is reorganized to better integrate the relationship between excretion and water and electrolyte balance. Osmoregulatory strategies are now organized according to the challenges presented by marine, freshwater, and terrestrial habitats. Coverage of osmoregulation in bony fishes versus cartilaginous fishes, mammalian kidney function, and how nonmammalian vertebrates concentrate their urine is expanded and clarified.

Chapter 44 Animal Nutrition   This chapter contains new information on nutritional imbalances, including diabetes and obesity.

Chapter 45 Gas Exchange and Circulation   Discussion of the insect tracheal system is expanded, including new content on how respiration restricts upper limits of body size of insects.

Details regarding the lymphatic system and heart anatomy in vertebrates are updated, and a new section on cardiovascular disease is added.

Chapter 46 Electrical Signals in Animals This chapter is greatly expanded to reflect recent research and growing interest in neuroscience. New information includes comparative anatomy of vertebrate brains, more case studies of brain injuries or dysfunctions that have led to major discoveries in neuroscience, and the concept of neuroplasticity—especially neurogenesis.

Chapter 47 Animal Sensory Systems Content from the Fourth Edition has been split into two chapters (47, Animal Sensory Systems; 48, Animal Movement). The chapter on sensory systems is now organized by type of sensory reception: mechanoreception (with new coverage on the lateral line system of fishes), photoreception, chemoreception (with new coverage of pheromones), and a new section introducing thermoreception, electroreception, and magnetoreception.

Chapter 48 Animal Movement This new chapter introduces the importance of movement in animals, building from small to large scale. The mechanism of muscle contraction (with revised figures) is covered, followed by discussions of types of muscle tissue (with new content on skeletal-muscle fiber types and parallel- versus pennate-muscle fiber orientation), and skeletal systems (hydrostatic skeletons, exoskeletons, endoskeletons). A completely new final section discusses how biologists study locomotion on land, in the air, and in the water.

Chapter 49 Chemical Signals in Animals Figures and content are updated for clarity. The chapter includes a new section on endocrine disruptors.

Chapter 50 Animal Reproduction New content includes details of sperm and egg structure and function, reproduction in the spotted hyena, and human contraceptive methods.

Chapter 51 The Immune System in Animals Coverage of the innate immune response is expanded to include more detail on Toll-like receptors and how they transmit signals. The section on adaptive immunity is reorganized to improve flow. Updated content on inappropriate immune responses (autoimmunity and allergies) and inadequate responses (immunodeficiency) is grouped together in one section. The hygiene hypothesis is introduced to explain the growing trend of inappropriate immune responses in populations that have reduced exposure to common pathogens and parasites.

Chapter 52 An Introduction to Ecology The first section on levels of ecological study is expanded to include global ecology. The rest of the chapter is reorganized, beginning with the factors that determine the distribution and abundance of organisms (including a new Argentine ant case study) and ending with biomes. The biome section is streamlined with a new emphasis on human impacts, including an introduction to anthropogenic biomes and the Anthropocene.

Chapter 53 Behavioral Ecology The introduction is revised to provide a clearer framework for types of behavior. Sections are now organized around proximate versus ultimate causation. These new examples replace those in the Fourth Edition: Argentine ant territorial behavior (replacing spiny lobsters), optimal foraging in gerbils (replacing white-fronted bee-eaters), sexual selection in *Anolis* lizards (replacing barn swallows), and map orientation in green sea turtles.

Chapter 54 Population Ecology A new introductory section focuses on the distribution of organisms in populations, including dispersion patterns and a consolidated discussion of measurement methods. The human population content is updated and separated into a new section. The quantitative methods boxes and life table are now more student friendly.

Chapter 55 Community Ecology Several changes to content are made, such as a clarification of competitive exclusion versus niche differentiation. New content includes a summary table on constitutive defenses, a discovery story on mimicry (including Bates and Müller), an introduction to food webs, and the process of soil formation in primary succession.

Chapter 56 Ecosystems The title and scope of the chapter are updated to include global ecology. Extensive revisions include many content updates and new figures on the relationship between GPP and NPP, the one-way flow of energy and cycling of nutrients, the food web, open versus closed aquifers, the High Plains Aquifer, the biomagnification of DDT, and the greenhouse effect. The climate change section is expanded and updated, including an illustrated summary table.

Chapter 57 Biodiversity and Conservation Biology Throughout the chapter, there is more emphasis on conserving ecosystem function. Many new examples are added, including Smits's restoration project in Borneo, the Census of Marine Life, the IUCN Red List, the Sinervo lizard extinction experiment, and Florida panther genetic restoration. New summary tables highlight ecosystem services and threats to biodiversity.

Acknowledgments

Reviewers

The peer review system is the key to quality and clarity in science publishing. In addition to providing a filter, the investment that respected individuals make in vetting the material—catching errors or inconsistencies and making suggestions to improve the presentation—gives authors, editors, and readers confidence that the text meets rigorous professional standards.

Peer review plays the same role in textbook publishing. The time and care that this book's reviewers have invested is a tribute to their professional integrity, their scholarship, and their concern for the quality of teaching. Virtually every paragraph in this edition has been revised and improved based on insights from the following individuals.

Tamarah Adair, *Baylor University*
Sandra D. Adams, *Montclair State University*
Marc Albrecht, *University of Nebraska at Kearney*
Larry Alice, *Western Kentucky University*
Leo M. Alves, *Manhattan College*
David R. Angelini, *American University*
Dan Ardia, *Franklin & Marshall College*
Paul Arriola, *Elmhurst College*
Davinderjit K. Bagga, *University of Montevallo*
Susan Barrett, *Wheaton College*
Donald Baud, *University of Memphis*
Vernon W. Bauer, *Francis Marion University*
Robert Bauman, *Amarillo College*
Christopher Beck, *Emory University*
Vagner Benedito, *West Virginia University*
Scott Bingham, *Arizona State University*
Stephanie Bingham, *Barry University*
Wendy Birky, *California State University, Northridge*
Jason Blank, *California Polytechnic State University*
Kristopher A. Blee, *California State University, Chico*
Margaret Bloch-Qazi, *Gustavus Adolphus College*
Lanh Bloodworth, *Florida State College at Jacksonville*
Catherine H. Borer, *Berry College*
James Bottesch, *Brevard Community College*
Jacqueline K. Bowman, *Arkansas Tech University*
John Bowman, *University of California, Davis*
Chris Brochu, *University of Iowa*
Matthew Brown, *Dalhousie University*
Mark Browning, *Purdue University*
Carolyn J. W. Bunde, *Idaho State University*
David Byres, *Florida State College at Jacksonville*
Michael Campbell, *Penn State Erie*
Manel Camps, *University of California, Santa Cruz*
Geralyn M. Caplan, *Owensboro Community and Technical College*
Richard Cardullo, *University of California, Riverside*

David Carlini, *American University*
Dale Casamatta, *University of North Florida*
Deborah Chapman, *University of Pittsburgh*
Joe Coelho, *Quincy University*
Allen Collins, *Smithsonian Museum of Natural History*
Robert A. Colvin, *Ohio University*
Kimberly L. Conner, *Florida State College at Jacksonville*
Karen Curto, *University of Pittsburgh*
Clarissa Dirks, *Evergreen State College*
Peter Ducey, *SUNY Cortland*
Erastus Dudley, *Huntingdon College*
Jeffrey P. Duguay, *Delta State University*
Tod Duncan, *University of Colorado, Denver*
Joseph Esdin, *University of California, Los Angeles*
Brent Ewers, *University of Wyoming*
Amy Farris, *Ivy Tech Community College*
Bruce Fisher, *Roane State Community College*
Ryan Fisher, *Salem State University*
David Fitch, *New York University*
Elizabeth Fitch, *Motlow State Community College*
Michael P. Franklin, *California State University, Northridge*
Susannah French, *Utah State University*
Caitlin Gabor, *Texas State University*
Matthew Gilg, *University of North Florida*
Kendra Greenlee, *North Dakota State University*
Patricia A. Grove, *College of Mount Saint Vincent*
Nancy Guild, *University of Colorado, Boulder*
Cynthia Hemenway, *North Carolina State University*
Christopher R. Herlihy, *Middle Tennessee State University*
Kendra Hill, *South Dakota State University*
Sara Hoot, *University of Wisconsin, Milwaukee*
Kelly Howe, *University of New Mexico*
Robin Hulbert, *California Polytechnic State University*
Rick Jellen, *Brigham Young University*
Russell Johnson, *Colby College*
William Jira Katembe, *Delta State University*
Elena K. Keeling, *California Polytechnic State University*
Jill B. Keeney, *Juniata College*
Greg Kelly, *University of Western Ontario*
Scott L. Kight, *Montclair State University*
Charles Knight, *California Polytechnic State University*
Jenny Knight, *University of Colorado, Boulder*
William Kroll, *Loyola University Chicago*
Dominic Lannutti, *El Paso Community College*
Brenda Leady, *University of Toledo*
David Lindberg, *University of California, Berkeley*
Barbara Lom, *Davidson College*
Robert Maxwell, *Georgia State University*
Marshall D. McCue, *St. Mary's University*
Kurt A. McKean, *SUNY Albany*
Michael Meighan, *University of California, Berkeley*
John Merrill, *Michigan State University*

Richard Merritt, *Houston Community College*
Alan Molumby, *University of Illinois at Chicago*
Jeremy Montague, *Barry University*
Chad E. Montgomery, *Truman State University*
Kimberly D. Moore, *Lone Star College System, North Harris*
Michael Morgan, *Berry College*
James Mulrooney, *Central Connecticut State University*
John D. Nagy, *Scottsdale Community College*
Margaret Olney, *St. Martin's University*
Nathan Okia, *Auburn University at Montgomery*
Robert Osuna, *SUNY Albany*
Daniel Panaccione, *West Virginia University*
Stephanie Pandolfi, *Michigan State University*
Michael Rockwell Parker, *Monell Chemical Senses Center*
Lisa Parks, *North Carolina State University*
Nancy Pelaez, *Purdue University*
Shelley W. Penrod, *Lone Star College System, North Harris*
Andrea Pesce, *James Madison University*
Raymond Pierotti, *University of Kansas*
Melissa Ann Pilgrim, *University of South Carolina Upstate*
Paul Pillitteri, *Southern Utah University*
Debra Pires, *University of California, Los Angeles*
P. David Polly, *Indiana University, Bloomington*
Vanessa Quinn, *Purdue University North Central*
Stacey L. Raimondi, *Elmhurst College*
Stephanie Randell, *McLennan Community College*
Marceau Ratard, *Delgado Community College*
Flona Redway, *Barry University*
Srebrenka Robic, *Agnes Scott College*
Dave Robinson, *Bellarmine University*
George Robinson, *SUNY Albany*
Adam W. Rollins, *Lincoln Memorial University*
Amanda Rosenzweig, *Delgado Community College*
Leonard C. Salvatori, *Indian River State College*
Dee Ann Sato, *Cypress College*
Leena Sawant, *Houston Community College*
Jon Scales, *Midwestern State University*
Oswald Schmitz, *Yale University*
Joan Sharp, *Simon Fraser University*
Julie Schroer, *North Dakota State University*
Timothy E. Shannon, *Francis Marion University*
Lynnette Sievert, *Emporia State University*
Susan Skambis, *Valencia College*
Ann E. Stapleton, *University of North Carolina, Wilmington*
Mary-Pat Stein, *California State University, Northridge*
Christine Strand, *California Polytechnic State University*
Denise Strickland, *Midlands Technical College*
Jackie Swanik, *Wake Technical Community College*
Billie J. Swalla, *University of Washington*
Zuzana Swigonova, *University of Pittsburgh*
Briana Timmerman, *University of South Carolina*
Catherine Ueckert, *Northern Arizona University*
Sara Via, *University of Maryland, College Park*
Thomas J. Volk, *University of Wisconsin–La Crosse*
Jeffrey Walck, *Middle Tennessee State University*
Andrea Weeks, *George Mason University*
Margaret S. White, *Scottsdale Community College*
Steven D. Wilt, *Bellarmine University*
Candace Winstead, *California Polytechnic State University*
James A. Wise, *Hampton University*

Correspondents

One of the most enjoyable interactions we have as textbook authors is correspondence or conversations with researchers and teachers who take the time and trouble to contact us to discuss an issue with the book, or who respond to our queries about a particular data set or study. We are always amazed and heartened by the generosity of these individuals. They care, deeply.

Lawrence Alice, *Western Kentucky University*
David Baum, *University of Wisconsin–Madison*
Meredith Blackwell, *Louisiana State University*
Nancy Burley, *University of California, Irvine*
Thomas Breithaupt, *University of Hull*
Philip Cantino, *Ohio University*
Allen Collins, *Smithsonian Museum of Natural History*
Robert Full, *University of California, Berkeley*
Arundhati Ghosh, *University of Pittsburgh*
Jennifer Gottwald, *University of Wisconsin–Madison*
Jon Harrison, *Arizona State University*
David Hawksworth, *Natural History Museum, London*
Jim Herrick, *James Madison University*
John Hunt, *University of Exeter*
Doug Jensen, *Converse College*
Scott Kight, *Montclair State University*
Scott Kirkton, *Union College*
Mimi Koehl, *University of California, Berkeley*
Rodger Kram, *University of Colorado*
Matthew McHenry, *University of California, Irvine*
Alison Miyamoto, *California State University, Fullerton*
Sean Menke, *Lake Forest College*
Rich Mooi, *California Academy of Sciences*
Michael Oliver, *MalawiCichlids.com*
M. Rockwell Parker, *Monell Chemical Senses Center*
Andrea Pesce, *James Madison University*
Chris Preston, *Monterey Bay Aquarium Research Institute*
Scott Sakaluk, *Illinois State University*
Kyle Seifert, *James Madison University*
Jos Snoeks, *Royal Museum for Central Africa*
Jeffrey Spring, *University of Louisiana*
Christy Strand, *California Polytechnic State University, San Luis Obispo*
Torsten Struck, *University of Osnabrueck, Germany*
Oswald Schmitz, *Yale University*
Ian Tattersall, *American Museum of Natural History*
Robert Turgeon, *Cornell University*
Tom Volk, *University of Wisconsin–La Crosse*
Naomi Wernick, *University of Massachusetts, Lowell*

Supplements Contributors

Instructors depend on an impressive array of support materials—in print and online—to design and deliver their courses. The student experience would be much weaker without the study guide, test bank, activities, animations, quizzes, and tutorials written by the following individuals.

Brian Bagatto, *University of Akron*
Scott Bingham, *Arizona State University*
Jay L. Brewster, *Pepperdine University*

Mirjana Brockett, *Georgia Institute of Technology*
Warren Burggren, *University of North Texas*
Jeff Carmichael, *University of North Dakota*
Tim Christensen, *East Carolina University*
Erica Cline, *University of Washington—Tacoma*
Patricia Colberg, *University of Wyoming*
Elia Crisucci, *University of Pittsburgh*
Elizabeth Cowles, *Eastern Connecticut State University*
Clarissa Dirks, *Evergreen State College*
Lisa Elfring, *University of Arizona, Tucson*
Brent Ewers, *University of Wyoming*
Rebecca Ferrell, *Metropolitan State University of Denver*
Miriam Ferzli, *North Carolina State University*
Cheryl Frederick, *University of Washington*
Cindee Giffen, *University of Wisconsin–Madison*
Kathy M. Gillen, *Kenyon College*
Linda Green, *Georgia Institute of Technology*
Christopher Harendza, *Montgomery County Community College*
Cynthia Hemenway, *North Carolina State University*
Laurel Hester, *University of South Carolina*
Jean Heitz, *University of Wisconsin–Madison*
Tracey Hickox, *University of Illinois, Urbana–Champaign*
Jacob Kerby, *University of South Dakota*
David Kooyman, *Brigham Young University*
Barbara Lom, *Davidson College*
Cindy Malone, *California State University, Northridge*
Jim Manser, retired, *Harvey Mudd College*
Jeanette McGuire, *Michigan State University*
Mark Music, *Indian River State College*
Jennifer Nauen, *University of Delaware*
Chris Pagliarulo, *University of California, Davis*
Stephanie Scher Pandolfi, *Michigan State University*
Lisa Parks, *North Carolina State University*
Debra Pires, *University of California, Los Angeles*
Carol Pollock, *University of British Columbia*
Jessica Poulin, *University at Buffalo, the State University of New York*
Vanessa Quinn, *Purdue University North Central*
Eric Ribbens, *Western Illinois University*
Christina T. Russin, *Northwestern University*
Leonard Salvatori, *Indian River State College*
Joan Sharp, *Simon Fraser University*
Chrissy Spencer, *Georgia Institute of Technology*
Mary-Pat Stein, *California State University, Northridge*
Suzanne Simon-Westendorf, *Ohio University*
Fred Wasserman, *Boston University*
Cindy White, *University of Northern Colorado*
Edward Zalisko, *Blackburn College*

Book Team

Anyone who has been involved in producing a textbook knows that many people work behind the scenes to make it all happen. The coauthor team is indebted to the many talented individuals who have made this book possible.

Development editors Mary Catherine Hager, Moira Lerner-Nelson, and Bill O'Neal provided incisive comments on the revised manuscript. Fernanda Oyarzun and Adam Steinberg used their artistic sense, science skills, and love of teaching to hone the figures for many chapters.

The final version of the text was copyedited by Chris Thillen and expertly proofread by Pete Shanks. The final figure designs were rendered by Imagineering Media Services and carefully proofread by Frank Purcell. Maureen Spuhler, Eric Schrader, and Kristen Piljay researched images for the Fifth Edition.

The book's clean, innovative design was developed by Mark Ong and Emily Friel. Text and art were skillfully set in the design by S4Carlisle Publishing Services. The book's production was supervised by Lori Newman and Mike Early.

The extensive supplements program was managed by Brady Golden and Katie Cook. All of the individuals mentioned—and more—were supported with cheerful, dedicated efficiency by Editorial Assistant Leslie Allen for the first half of the project; Eddie Lee has since stepped in to skillfully fill this role.

Creating MasteringBiology® tutorials and activities also requires a team. Media content development was overseen by Tania Mlawer and Sarah Jensen, who benefited from the program expertise of Caroline Power and Caroline Ross. Joseph Mochnick and Daniel Ross worked together as media producers. Lauren Fogel (VP, Director, Media Development), Stacy Treco (VP, Director, Media Product Strategy), and Laura ensured that the complete media program that accompanies the Fifth Edition, including MasteringBiology, will meet the needs of the students and professors who use our offerings.

Pearson's talented sales reps, who listen to professors, advise the editorial staff, and get the book in students' hands, are supported by tireless Executive Marketing Manager Lauren Harp and Director of Marketing Christy Lesko. The marketing materials that support the outreach effort were produced by Lillian Carr and her colleagues in Pearson's Marketing Comunications group. David Theisen, national director for Key Markets, tirelessly visits countless professors each year, enthusiastically discussing their courses and providing us with meaningful editorial guidance.

The vision and resources required to run this entire enterprise are the responsibility of Vice President and Editor-in-Chief Beth Wilbur, who provided inspirational and focused leadership, and President of Pearson Science Paul Corey, who displays unwavering commitment to high-quality science publishing.

Becky Ruden recruited the coauthor team, drawing us to the project with her energy and belief in this book. The editorial team was skillfully directed by Executive Director of Development Deborah Gale. Finally, we are deeply grateful for three key drivers of the Fifth Edition. Project Editor Anna Amato's superb organizational skills and calm demeanor assured that all the wheels and cogs of the process ran smoothly to keep the mammoth project steadily rolling forward. Supervising Development Editor Sonia DiVittorio's deep expertise, creative vision, keen attention to detail, level, and clarity, and inspiring insistence on excellence kept the bar high for everyone on every aspect of the project. Lastly, Senior Acquisitions Editor Michael Gillespie's boundless energy and enthusiasm, positive attitude, and sharp intellect have fueled and united the team and also guided the book through the hurdles to existence. The coauthor team thanks these exceptional people for making the art and science of book writing a productive and exhilarating process.

1 Biology and the Tree of Life

In this chapter you will learn about

Key themes to structure your thinking about Biology

starting with ↓

What does it mean to say that something is alive? 1.1

including ↓

Two of the greatest unifying ideas in Biology

↓ first

The cell theory 1.2

↓ then

The theory of evolution by natural selection 1.3

↓ predicts

The tree of life 1.4

including ↓

The process of doing Biology 1.5

These Chinese water dragon hatchlings are exploring their new world and learning how to find food and stay alive. They represent one of the key characteristics of life introduced in this chapter—replication.

In essence, biological science is a search for ideas and observations that unify our understanding of the diversity of life, from bacteria living in rocks a mile underground to humans and majestic sequoia trees. This chapter is an introduction to this search.

The goals of this chapter are to introduce the nature of life and explore how biologists go about studying it. The chapter also introduces themes that will resonate throughout this book:

- Analyzing how organisms work at the molecular level.

- Understanding organisms in terms of their evolutionary history.

- Helping you learn to think like a biologist.

Let's begin with what may be the most fundamental question of all: What is life?

This chapter is part of the Big Picture. See how on pages 16–17.

✔ When you see this checkmark, stop and test yourself. Answers are available in Appendix A.

1.1 What Does It Mean to Say That Something Is Alive?

An **organism** is a life-form—a living entity made up of one or more cells. Although there is no simple definition of life that is endorsed by all biologists, most agree that organisms share a suite of five fundamental characteristics.

- *Energy* To stay alive and reproduce, organisms have to acquire and use energy. To give just two examples: plants absorb sunlight; animals ingest food.

- *Cells* Organisms are made up of membrane-bound units called cells. A cell's membrane regulates the passage of materials between exterior and interior spaces.

- *Information* Organisms process hereditary, or genetic, information encoded in units called genes. Organisms also respond to information from the environment and adjust to maintain stable internal conditions. Right now, cells throughout your body are using information to make the molecules that keep you alive; your eyes and brain are decoding information on this page that will help you learn some biology, and if your room is too hot you might be sweating to cool off.

- *Replication* One of the great biologists of the twentieth century, François Jacob, said that the "dream of a bacterium is to become two bacteria." Almost everything an organism does contributes to one goal: replicating itself.

- *Evolution* Organisms are the product of evolution, and their populations continue to evolve.

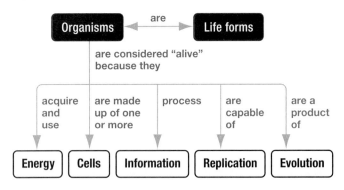

You can think of this text as one long exploration of these five traits. Here's to life!

1.2 The Cell Theory

Two of the greatest unifying ideas in all of science laid the groundwork for modern biology: the cell theory and the theory of evolution by natural selection. Formally, scientists define a **theory** as an explanation for a very general class of phenomena or observations that are supported by a wide body of evidence. The cell theory and theory of evolution address fundamental questions: What are organisms made of? Where do they come from?

When these concepts emerged in the mid-1800s, they revolutionized the way biologists think about the world. They established

two of the five attributes of life: Organisms are cellular, and their populations change over time.

Neither insight came easily, however. The cell theory, for example, emerged after some 200 years of work. In 1665 the Englishman Robert Hooke devised a crude microscope to examine the structure of cork (a bark tissue) from an oak tree. The instrument magnified objects to just 30× (30 times) their normal size, but it allowed Hooke to see something extraordinary. In the cork he observed small, pore-like compartments that were invisible to the naked eye. Hooke coined the term "cells" for these structures because of their resemblance to the cells inhabited by monks in a monastery.

Soon after Hooke published his results, a Dutch scientist named Anton van Leeuwenhoek succeeded in developing much more powerful microscopes, some capable of magnifications up to 300×. With these instruments, van Leeuwenhoek inspected samples of pond water and made the first observations of a dazzling collection of single-celled organisms that he called "animalcules." He also observed and described human blood cells and sperm cells, shown in **FIGURE 1.1**.

In the 1670s an Italian researcher who was studying the leaves and stems of plants with a microscope concluded that plant tissues were composed of many individual cells. By the early 1800s, enough data had accumulated for a German biologist to claim that *all* organisms consist of cells. Did this claim hold up?

All Organisms Are Made of Cells

Advances in microscopy have made it possible to examine the amazing diversity and complexity of cells at higher and higher magnifications. Biologists have developed microscopes that are tens of thousands of times more powerful than van Leeuwenhoek's and have described over a million new species. The basic conclusion made in the 1800s remains intact, however: All organisms are made of cells.

The smallest organisms known today are bacteria that are barely 200 nanometers wide, or 200 *billionths* of a meter. (See **BioSkills 1** in Appendix B to review the metric system and its prefixes.[1]) It would take 5000 of these organisms lined up side by side to span a millimeter. This is the distance between the smallest hash marks on a metric ruler. In contrast, sequoia trees can be over 100 meters tall. This is the equivalent of a 20-story building. Bacteria and sequoias are composed of the same fundamental building block, however—the cell. Bacteria consist of a single cell; sequoias are made up of many cells.

Today a **cell** is defined as a highly organized compartment that is bounded by a thin, flexible structure called a plasma membrane and that contains concentrated chemicals in an aqueous (watery) solution. The chemical reactions that sustain life take place inside cells. Most cells are also capable of reproducing by dividing—in effect, by making a copy of themselves.

The realization that all organisms are made of cells was fundamentally important, but it formed only the first part of the cell

---

[1]BioSkills are located in the second appendix at the back of the book. They focus on general skills that you'll use throughout this course. More than a few students have found them to be a life-saver. Please use them!

(a) van Leeuwenhoek built his own microscopes—which, while small, were powerful. They allowed him to see, for example . . .

(b) . . . human blood cells (this modern photo was shot through one of van Leeuwenhoek's original microscopes) . . .

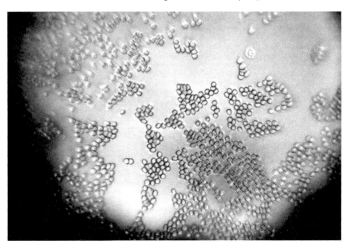

(c) . . . and animal sperm (drawing by van Leeuwenhoek of canine sperm cells on left, human on right).

FIGURE 1.1 Van Leeuwenhoek's Microscope Made Cells Visible.

theory. In addition to understanding what organisms are made of, scientists wanted to understand how cells come to be.

Where Do Cells Come From?

Most scientific theories have two components: The first describes a pattern in the natural world; the second identifies a mechanism or process that is responsible for creating that pattern. Hooke and his fellow scientists articulated the pattern component of the cell theory. In 1858, a German scientist named Rudolph Virchow added the process component by stating that all cells arise from preexisting cells.

The complete **cell theory** can be stated as follows: All organisms are made of cells, and all cells come from preexisting cells.

Two Hypotheses The cell theory was a direct challenge to the prevailing explanation of where cells come from, called spontaneous generation. In the mid-1800s, most biologists believed that organisms could arise spontaneously under certain conditions. For example, the bacteria and fungi that spoil foods such as milk and wine were thought to appear in these nutrient-rich media of their own accord—springing to life from nonliving materials. In contrast, the cell theory maintained that cells do not spring to life spontaneously but are produced only when preexisting cells grow and divide. The all-cells-from-cells explanation was a **hypothesis:** a testable statement to explain a phenomenon or a set of observations.

Biologists usually use the word theory to refer to proposed explanations for broad patterns in nature and prefer hypothesis to refer to explanations for more tightly focused questions. A theory serves as a framework for the development of new hypotheses.

An Experiment to Settle the Question Soon after Virchow's all-cells-from-cells hypothesis appeared in print, a French scientist named Louis Pasteur set out to test its predictions experimentally. An experimental **prediction** describes a measurable or observable result that must be correct if a hypothesis is valid.

Pasteur wanted to determine whether microorganisms could arise spontaneously in a nutrient broth or whether they appear only when a broth is exposed to a source of preexisting cells. To address the question, he created two treatment groups: a broth that was not exposed to a source of preexisting cells and a broth that was.

The spontaneous generation hypothesis predicted that cells would appear in both treatment groups. The all-cells-from-cells hypothesis predicted that cells would appear only in the treatment exposed to a source of preexisting cells.

FIGURE 1.2 (on page 4) shows Pasteur's experimental setup. Note that the two treatments are identical in every respect but one. Both used glass flasks filled with the same amount of the same nutrient broth. Both were boiled for the same amount of time to kill any existing organisms such as bacteria or fungi. But because the flask pictured in Figure 1.2a had a straight neck, it was exposed to preexisting cells after sterilization by the heat treatment. These preexisting cells are the bacteria and fungi that cling to dust particles in the air. They could drop into the nutrient broth because the neck of the flask was straight.

In contrast, the flask drawn in Figure 1.2b had a long swan neck. Pasteur knew that water would condense in the crook of the swan neck after the boiling treatment and that this pool of water

QUESTION: Do cells arise spontaneously or from other cells?

SPONTANEOUS GENERATION HYPOTHESIS: Cells arise spontaneously from nonliving materials.

ALL-CELLS-FROM-CELLS HYPOTHESIS: Cells are produced only when preexisting cells grow and divide.

(a) Pasteur experiment with straight-necked flask:

1. Place nutrient broth in straight-necked flask.

Cells

2. Boil to sterilize the flask (killing any living cells that were in the broth).

No cells   Cells

3. Preexisting cells enter flask from air.

(b) Pasteur experiment with swan-necked flask:

1. Place nutrient broth in swan-necked flask.

Cells

2. Boil to sterilize the flask (killing any living cells that were in the broth).

Condensation settles in neck

No cells   Cells

3. Preexisting cells from air are trapped in swan neck.

PREDICTION OF SPONTANEOUS GENERATION HYPOTHESIS:
Cells will appear in broth.

PREDICTION OF ALL-CELLS-FROM-CELLS HYPOTHESIS:
Cells will appear in broth.

PREDICTION OF SPONTANEOUS GENERATION HYPOTHESIS:
Cells will appear in broth.

PREDICTION OF ALL-CELLS-FROM-CELLS HYPOTHESIS:
Cells will not appear in broth.

RESULTS:

Cells   Both hypotheses supported

No cells   Spontaneous generation hypothesis rejected

CONCLUSION: Cells arise from preexisting cells, not spontaneously from nonliving material.

FIGURE 1.2 The Spontaneous Generation and All-Cells-from-Cells Hypotheses Were Tested Experimentally.

✔**QUESTION** What problem would arise in interpreting the results of this experiment if Pasteur had (1) put different types of broth in the two treatments, (2) heated them for different lengths of time, or (3) used a ceramic flask for one treatment and a glass flask for the other?

would trap any bacteria or fungi that entered on dust particles. Thus, the contents of the swan-necked flask were isolated from any source of preexisting cells even though still open to the air.

Pasteur's experimental setup was effective because there was only one difference between the two treatments and because that difference was the factor being tested—in this case, a broth's exposure to preexisting cells.

One Hypothesis Supported And Pasteur's results? As Figure 1.2 shows, the treatment exposed to preexisting cells quickly filled with bacteria and fungi. This observation was important because it showed that the heat sterilization step had not altered the nutrient broth's capacity to support growth.

The broth in the swan-necked flask remained sterile, however. Even when the flask was left standing for months, no organisms appeared in it. This result was inconsistent with the hypothesis of spontaneous generation.

Because Pasteur's data were so conclusive—meaning that there was no other reasonable explanation for them—the results persuaded most biologists that the all-cells-from-cells hypothesis was correct. However, you will see that biologists now have evidence that life did arise from nonlife early in Earth's history, through a process called chemical evolution (Chapters 2–6).

The success of the cell theory's process component had an important implication: If all cells come from preexisting cells, it follows that all individuals in an isolated population of single-celled organisms are related by common ancestry. Similarly, in you and most other multicellular individuals, all the cells present are descended from preexisting cells, tracing back to a fertilized egg. A fertilized egg is a cell created by the fusion of sperm and egg—cells that formed in individuals of the previous generation. In this way, all the cells in a multicellular organism are connected by common ancestry.

The second great founding idea in biology is similar, in spirit, to the cell theory. It also happened to be published the same year as the all-cells-from-cells hypothesis. This was the realization, made independently by the English scientists Charles Darwin and Alfred Russel Wallace, that all species—all distinct, identifiable types of organisms—are connected by common ancestry.

1.3 The Theory of Evolution by Natural Selection

In 1858 short papers written separately by Darwin and Wallace were read to a small group of scientists attending a meeting of the Linnean Society of London. A year later, Darwin published a book that expanded on the idea summarized in those brief papers. The book was called *The Origin of Species*. The first edition sold out in a day.

What Is Evolution?

Like the cell theory, the theory of evolution by natural selection has a pattern and a process component. Darwin and Wallace's theory made two important claims concerning patterns that exist in the natural world.

1. Species are related by common ancestry. This contrasted with the prevailing view in science at the time, which was that species represent independent entities created separately by a divine being.

2. In contrast to the accepted view that species remain unchanged through time, Darwin and Wallace proposed that the characteristics of species can be modified from generation to generation. Darwin called this process descent with modification.

Evolution is a change in the characteristics of a population over time. It means that species are not independent and unchanging entities, but are related to one another and can change through time.

What Is Natural Selection?

This pattern component of the theory of evolution was actually not original to Darwin and Wallace. Several scientists had already come to the same conclusions about the relationships between species. The great insight by Darwin and Wallace was in proposing a process, called **natural selection,** that explains *how* evolution occurs.

Two Conditions of Natural Selection   Natural selection occurs whenever two conditions are met.

1. Individuals within a population vary in characteristics that are **heritable**—meaning, traits that can be passed on to offspring.

A **population** is defined as a group of individuals of the same species living in the same area at the same time.

2. In a particular environment, certain versions of these heritable traits help individuals survive better or reproduce more than do other versions.

If certain heritable traits lead to increased success in producing offspring, then those traits become more common in the population over time. In this way, the population's characteristics change as a result of natural selection acting on individuals. This is a key insight: Natural selection acts on individuals, but evolutionary change occurs in populations.

Selection on Maize as an Example   To clarify how selection works, consider an example of **artificial selection**—changes in populations that occur when *humans* select certain individuals to produce the most offspring. Beginning in 1896, researchers began a long-term selection experiment on maize (corn).

1. In the original population, the percentage of protein in maize kernels was variable among individuals. Kernel protein content is a heritable trait—parents tend to pass the trait on to their offspring.

2. Each year for many years, researchers chose individuals with the highest kernel protein content to be the parents of the next generation. In this environment, individuals with high kernel protein content produced more offspring than individuals with low kernel protein content.

FIGURE 1.3 shows the results. Note that this graph plots generation number on the *x*-axis, starting from the first generation (0 on the graph) and continuing for 100 generations. The average percentage of protein in a kernel among individuals in this population is plotted on the *y*-axis.

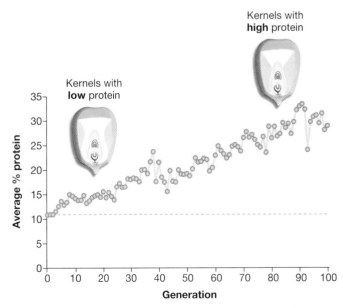

FIGURE 1.3 Response to Selection for High Kernel Protein Content in Maize.

DATA: Moose, S. P., J. W. Dudley, and T. R. Rocheford. 2004. *Trends in Plant Sciences* 9: 358–364; and the Illinois long-term selection experiment for oil and protein in corn (University of Illinois at Urbana–Champaign).

To read this graph, put your finger on the x-axis at generation 0. Then read up the y-axis, and note that kernels averaged about 11 percent protein at the start of the experiment. Now read the graph to the right. Each dot is a data point, representing the average kernel protein concentration in a particular generation. (A generation in maize is one year.) The lines on this graph simply connect the dots, to make the pattern in the data easier to see. During a few years the average protein content goes down, because of poor growing conditions or chance changes in how the many genes responsible for this trait interact. However, at the end of the graph, after 100 generations of selection, average kernel protein content is about 29 percent. (For more help with reading graphs, see **BioSkills 3** in Appendix B.)

This sort of change in the characteristics of a population, over time, is evolution. Humans have been practicing artificial selection for thousands of years, and biologists have now documented evolution by *natural* selection—where humans don't do the selecting—occurring in thousands of different populations, including humans. Evolution occurs when heritable variation leads to differential success in reproduction.

✔ QUANTITATIVE If you understand the concepts of selection and evolution, you should be able to describe how protein content in maize kernels changed over time, using the same x-axis and y-axis as in Figure 1.3, when researchers selected individuals with the *lowest* kernel protein content to be the parents of the next generation. (This experiment was actually done, starting with the same population at the same time as selection for high protein content.)

Fitness and Adaptation   Darwin also introduced some new terminology to identify what is happening during natural selection.

- In everyday English, fitness means health and well-being. But in biology, **fitness** means the ability of an individual to produce viable offspring. Individuals with high fitness produce many surviving offspring.

- In everyday English, adaptation means that an individual is adjusting and changing to function in new circumstances. But in biology, an **adaptation** is a trait that increases the fitness of an individual in a particular environment.

Once again, consider kernel protein content in maize: In the environment of the experiment graphed in Figure 1.3, individuals with high kernel protein content produced more offspring and had higher fitness than individuals with lower kernel protein content. In this population and this environment, high kernel protein content was an adaptation that allowed certain individuals to thrive.

Note that during this process, the amount of protein in the kernels of any individual maize plant did not change within its lifetime—the change occurred in the characteristics of the population over time.

Together, the cell theory and the theory of evolution provided the young science of biology with two central, unifying ideas:

1. The cell is the fundamental structural unit in all organisms.

2. All species are related by common ancestry and have changed over time in response to natural selection.

If you understand that . . .

- Natural selection occurs when heritable variation in certain traits leads to improved success in reproduction. Because individuals with these traits produce many offspring with the same traits, the traits increase in frequency and evolution occurs.
- Evolution is a change in the characteristics of a population over time.

✔ **You should be able to . . .**

Using the graph you just analyzed in Figure 1.3, describe the average kernel protein content over time in a maize population where *no* selection occurred.

Answers are available in Appendix A.

1.4   The Tree of Life

Section 1.3 focuses on how individual populations change through time in response to natural selection. But over the past several decades, biologists have also documented dozens of cases in which natural selection has caused populations of one species to diverge and form new species. This divergence process is called **speciation.**

Research on speciation has two important implications: All species come from preexisting species, and all species, past and present, trace their ancestry back to a single common ancestor.

The theory of evolution by natural selection predicts that biologists should be able to construct a **tree of life**—a family tree of organisms. If life on Earth arose just once, then such a diagram would describe the genealogical relationships between species with a single, ancestral species at its base.

Has this task been accomplished? If the tree of life exists, what does it look like?

Using Molecules to Understand the Tree of Life

One of the great breakthroughs in research on the tree of life occurred when American biologist Carl Woese (pronounced *woze*) and colleagues began analyzing the chemical components of organisms as a way to understand their evolutionary relationships. Their goal was to understand the **phylogeny** of all organisms—their actual genealogical relationships. Translated literally, phylogeny means "tribe-source."

To understand which organisms are closely versus distantly related, Woese and co-workers needed to study a molecule that is found in all organisms. The molecule they selected is called small subunit ribosomal RNA (rRNA). It is an essential part of the machinery that all cells use to grow and reproduce.

Although rRNA is a large and complex molecule, its underlying structure is simple. The rRNA molecule is made up of sequences of four smaller chemical components called ribonucleotides. These ribonucleotides are symbolized by the letters A, U, C, and G. In rRNA, ribonucleotides are connected to one another linearly, like the boxcars of a freight train.

Analyzing rRNA   Why might rRNA be useful for understanding the relationships between organisms? The answer is that the ribonucleotide sequence in rRNA is a trait that can change during the course of evolution. Although rRNA performs the same function in all organisms, the sequence of ribonucleotide building blocks in this molecule is not identical among species.

In land plants, for example, the molecule might start with the sequence A-U-A-U-C-G-A-G (**FIGURE 1.4**). In green algae, which are closely related to land plants, the same section of the molecule might contain A-U-A-U-G-G-A-G. But in brown algae, which are not closely related to green algae or to land plants, the same part of the molecule might consist of A-A-A-U-G-G-A-C.

The research that Woese and co-workers pursued was based on a simple premise: If the theory of evolution is correct, then rRNA sequences should be very similar in closely related organisms but less similar in organisms that are less closely related. Species that are part of the same evolutionary lineage, like the plants, should share certain changes in rRNA that no other species have.

To test this premise, the researchers determined the sequence of ribonucleotides in the rRNA of a wide array of species. Then they considered what the similarities and differences in the sequences implied about relationships between the species. The goal was to produce a diagram that described the phylogeny of the organisms in the study.

A diagram that depicts evolutionary history in this way is called a phylogenetic tree. Just as a family tree shows relationships between individuals, a phylogenetic tree shows relationships between species. On a phylogenetic tree, branches that share a recent common ancestor represent species that are closely related; branches that don't share recent common ancestors represent species that are more distantly related.

The Tree of Life Estimated from Genetic Data   To construct a phylogenetic tree, researchers use a computer to find the arrangement of branches that is most consistent with the similarities and differences observed in the data.

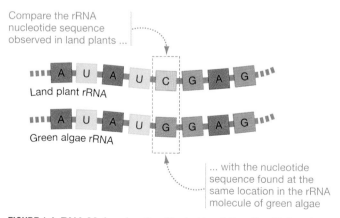

FIGURE 1.4 RNA Molecules Are Made Up of Smaller Molecules.
A complete small subunit rRNA molecule contains about 2000 ribonucleotides; just 8 are shown in this comparison.

✔**QUESTION** Suppose that in the same section of rRNA, molds and other fungi have the sequence A-U-A-U-G-G-A-C. Are fungi more closely related to green algae or to land plants? Explain your logic.

Although the initial work was based only on the sequences of ribonucleotides observed in rRNA, biologists now use data sets that include sequences from a wide array of genetic material. **FIGURE 1.5** shows a recent tree produced by comparing these sequences. Because this tree includes such a diverse array of

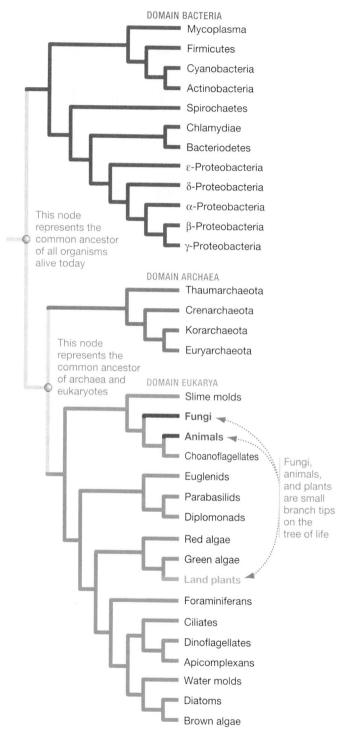

FIGURE 1.5 The Tree of Life. A phylogenetic tree estimated from a large amount of genetic sequence data. The three domains of life revealed by the analysis are labeled. Common names are given for lineages in the domains Bacteria and Eukarya. Phyla names are given for members of the domain Archaea, because most of these organisms have no common names.

species, it is often called the universal tree, or the tree of life. (For help in learning how to read a phylogenetic tree, see **BioSkills 7** in Appendix B.) Notice that the tree's main node is the common ancestor of all living organisms. Researchers who study the origin of life propose that the tree's root extends even further back to the "*last universal common ancestor*" of cells, or **LUCA.**

The tree of life implied by rRNA and other genetic data established that there are three fundamental groups or lineages of organisms: **(1)** the Bacteria, **(2)** the Archaea, and **(3)** the Eukarya. In all **eukaryotes,** cells have a prominent component called the nucleus (**FIGURE 1.6a**). Translated literally, the word eukaryotes means "true kernel." Because the vast majority of bacterial and archaeal cells lack a nucleus, they are referred to as **prokaryotes** (literally, "before kernel"; see **FIGURE 1.6b**). The vast majority of bacteria and archaea are unicellular ("one-celled"); many eukaryotes are multicellular ("many-celled").

When results based on genetic data were first published, biologists were astonished. For example:

- Prior to Woese's work and follow-up studies, biologists thought that the most fundamental division among organisms was between prokaryotes and eukaryotes. The Archaea were virtually unknown—much less recognized as a major and highly distinctive branch on the tree of life.

- Fungi were thought to be closely related to plants. Instead, they are actually much more closely related to animals.

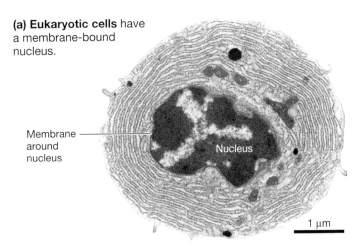

(a) Eukaryotic cells have a membrane-bound nucleus.

Membrane around nucleus

Nucleus

1 μm

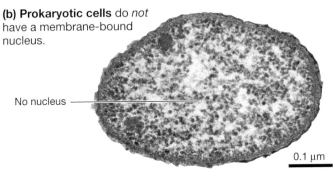

(b) Prokaryotic cells do *not* have a membrane-bound nucleus.

No nucleus

0.1 μm

FIGURE 1.6 Eukaryotes and Prokaryotes.

✔**QUANTITATIVE** How many times larger is the eukaryotic cell in this figure than the prokaryotic cell? (Hint: Study the scale bars.)

- Traditional approaches for classifying organisms—including the system of five kingdoms divided into various classes, orders, and families that you may have learned in high school—are inaccurate in many cases, because they do not reflect the actual evolutionary history of the organisms involved.

The Tree of Life Is a Work in Progress   Just as researching your family tree can help you understand who you are and where you came from, so the tree of life helps biologists understand the relationships between organisms and the history of species. The discovery of the Archaea and the accurate placement of lineages such as the fungi qualify as exciting breakthroughs in our understanding of evolutionary history and life's diversity.

Work on the tree of life continues at a furious pace, however, and the location of certain branches on the tree is hotly debated. As databases expand and as techniques for analyzing data improve, the shape of the tree of life presented in Figure 1.5 will undoubtedly change. Our understanding of the tree of life, like our understanding of every other topic in biological science, is dynamic.

How Should We Name Branches on the Tree of Life?

In science, the effort to name and classify organisms is called **taxonomy.** Any named group is called a **taxon** (plural: **taxa**). Currently, biologists are working to create a taxonomy, or naming system, that accurately reflects the phylogeny of organisms.

Based on the tree of life implied by genetic data, Woese proposed a new taxonomic category called the **domain.** The three domains of life are the Bacteria, Archaea, and Eukarya.

Biologists often use the term **phylum** (plural: **phyla**) to refer to major lineages within each domain. Although the designation is somewhat arbitrary, each phylum is considered a major branch on the tree of life. Within the lineage called animals, biologists currently name 30–35 phyla—each of which is distinguished by distinctive aspects of its body structure as well as by distinctive gene sequences. For example, the mollusks (clams, squid, octopuses) constitute a phylum, as do chordates (the vertebrates and their close relatives).

Because the tree of life is so new, though, naming systems are still being worked out. One thing that hasn't changed for centuries, however, is the naming system for individual species.

Scientific (Latin) Names   In 1735, a Swedish botanist named Carolus Linnaeus established a system for naming species that is still in use today. Linnaeus created a two-part name unique to each type of organism.

- *Genus*   The first part indicates the organism's **genus** (plural: **genera**). A genus is made up of a closely related group of species. For example, Linnaeus put humans in the genus *Homo*. Although humans are the only living species in this genus, at least six extinct organisms, all of which walked upright and made extensive use of tools, were later also assigned to *Homo*.

- *Species*   The second term in the two-part name identifies the organism's species. Linnaeus gave humans the species name *sapiens*.

An organism's genus and species designation is called its **scientific name** or Latin name. Scientific names are always italicized. Genus names are always capitalized, but species names are not—as in *Homo sapiens*.

Scientific names are based on Latin or Greek word roots or on words "Latinized" from other languages. Linnaeus gave a scientific name to every species then known, and also Latinized his own name—from Karl von Linné to Carolus Linnaeus.

Linnaeus maintained that different types of organisms should not be given the same genus and species names. Other species may be assigned to the genus *Homo*, and members of other genera may be named *sapiens*, but only humans are named *Homo sapiens*. Each scientific name is unique.

Scientific Names Are Often Descriptive Scientific names and terms are often based on Latin or Greek word roots that are descriptive. For example, *Homo sapiens* is derived from the Latin *homo* for "man" and *sapiens* for "wise" or "knowing." The yeast that bakers use to produce bread and that brewers use to brew beer is called *Saccharomyces cerevisiae*. The Greek root *saccharo* means "sugar," and *myces* refers to a fungus. *Saccharomyces* is aptly named "sugar fungus" because yeast is a fungus and because the domesticated strains of yeast used in commercial baking and brewing are often fed sugar. The species name of this organism, *cerevisiae*, is Latin for "beer." Loosely translated, then, the scientific name of brewer's yeast means "sugar-fungus for beer."

Scientific names and terms often seem daunting at first glance. So, most biologists find it extremely helpful to memorize some of the common Latin and Greek roots. To aid you in this process, new terms in this text are often accompanied by a translation of their Latin or Greek word roots in parentheses. (A glossary of common root words with translations and examples is also provided in **BioSkills 2** in Appendix B.)

 If you understand that . . .

- A phylogenetic tree shows the evolutionary relationships between species.
- To infer where species belong on a phylogenetic tree, biologists examine genetic and other characteristics of the species involved. Closely related species should have similar characteristics, while less closely related species should be less similar.

✔ **You should be able to . . .**

Examine the following rRNA ribonucleotide sequences and draw a phylogenetic tree showing the relationships between species A, B, and C that these data imply:

Species A: A A C T A G C G C G A T
Species B: A A C T A G C G C C A T
Species C: T T C T A G C G G T A T

Answers are available in Appendix A.

1.5 Doing Biology

This chapter has introduced some of the great ideas in biology. The development of the cell theory and the theory of evolution by natural selection provided cornerstones when the science was young; the tree of life is a relatively recent insight that has revolutionized our understanding of life's diversity.

These theories are considered great because they explain fundamental aspects of nature, and because they have consistently been shown to be correct. They are considered correct because they have withstood extensive testing.

How do biologists go about testing their ideas? Before answering this question, let's step back a bit and consider the types of questions that researchers can and cannot ask.

The Nature of Science

Biologists ask questions about organisms, just as physicists and chemists ask questions about the physical world or geologists ask questions about Earth's history and the ongoing processes that shape landforms.

No matter what their field, all scientists ask questions that can be answered by observing or measuring things—by collecting data. Conversely, scientists cannot address questions that can't be answered by observing or measuring things.

This distinction is important. It is at the root of continuing controversies about teaching evolution in publicly funded schools. In the United States and in Turkey, in particular, some Christian and Islamic leaders have been particularly successful in pushing their claim that evolution and religious faith are in conflict. Even though the theory of evolution is considered one of the most successful and best-substantiated ideas in the history of science, they object to teaching it.

The vast majority of biologists and many religious leaders reject this claim; they see no conflict between evolution and religious faith. Their view is that science and religion are compatible because they address different types of questions.

- Science is about formulating hypotheses and finding evidence that supports or conflicts with those hypotheses.

- Religious faith addresses questions that cannot be answered by data. The questions addressed by the world's great religions focus on why we exist and how we should live.

Both types of questions are seen as legitimate and important.

So how do biologists go about answering questions? After formulating hypotheses, biologists perform experimental studies, or studies that yield descriptive data, such as observing a behavior, characterizing a structure within a cell by microscopy, or sequencing rRNA. Let's consider two recent examples of this process.

Why Do Giraffes Have Long Necks? An Introduction to Hypothesis Testing

If you were asked why giraffes have long necks, you might say based on your observations that long necks enable giraffes to reach food that is unavailable to other mammals. This hypothesis is expressed in African folktales and has traditionally been accepted by many biologists. The food competition hypothesis is so plausible, in fact, that for decades no one thought to test it.

In the mid-1990s, however, Robert Simmons and Lue Scheepers assembled data suggesting that the food competition hypothesis is only part of the story. Their analysis supports an alternative hypothesis—that long necks allow giraffes to use their heads as effective weapons for battering their opponents, and that longer-necked giraffes would have a competitive advantage in fights.

Before exploring these alternative explanations, it's important to recognize that hypothesis testing is a two-step process:

Step 1 State the hypothesis as precisely as possible and list the predictions it makes.

Step 2 Design an observational or experimental study that is capable of testing those predictions.

If the predictions are accurate, the hypothesis is supported. If the predictions are not met, then researchers do further tests, modify the original hypothesis, or search for alternative explanations. But the process does not end here. Biologists also talk to other researchers. Over coffee, at scientific meetings, or through publications, biologists communicate their results to the scientific community and beyond. (You can see the Big Picture of the process of doing biology on pages 16–17.)

Now that you understand more about hypothesis testing, let's return to the giraffes. How did biologists test the food competition hypothesis? What data support their alternative explanation?

The Food Competition Hypothesis: Predictions and Tests The food competition hypothesis claims that giraffes compete for food with other species of mammals. When food is scarce, as it is during the dry season, giraffes with longer necks can reach food that is unavailable to other species and to giraffes with shorter necks. As a result, the longest-necked individuals in a giraffe population survive better and produce more young than do shorter-necked individuals, and average neck length of the population increases with each generation.

To use the terms introduced earlier, long necks are adaptations that increase the fitness of individual giraffes during competition for food. This type of natural selection has gone on so long that the population has become extremely long necked.

The food competition hypothesis makes several explicit predictions. For example, the food competition hypothesis predicts that:

• neck length is variable among giraffes;

• neck length in giraffes is heritable; and

• giraffes feed high in trees, especially during the dry season, when food is scarce and the threat of starvation is high.

The first prediction is correct. Studies in zoos and natural populations confirm that neck length is variable among individuals.

The researchers were unable to test the second prediction, however, because they studied giraffes in a natural population and were unable to do breeding experiments. As a result, they simply had to accept this prediction as an assumption. In

(a) Most feeding is done at about shoulder height.

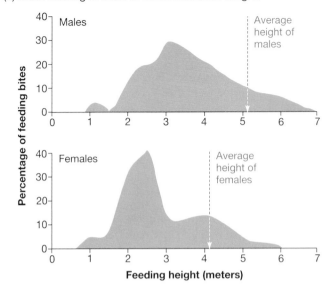

(b) Typical feeding posture in giraffes

FIGURE 1.7 Giraffes Do Not Usually Extend Their Necks Upward to Feed.
DATA: Young, T. P., L. A. Isbell. 1991. *Ethology* 87: 79–89.

general, though, biologists prefer to test every assumption behind a hypothesis.

What about the prediction regarding feeding high in trees? According to Simmons and Scheepers, this is where the food competition hypothesis breaks down.

Consider, for example, data collected by a different research team on the amount of time that giraffes spend feeding in vegetation of different heights. **FIGURE 1.7a** plots the height of vegetation versus the percentage of bites taken by a giraffe, for males and for females from the same population in Kenya. The dashed line on each graph indicates the average height of a male or female in this population.

Note that the average height of a giraffe in this population is much greater than the height where most feeding takes place. In this population, both male and female giraffes spend most of their feeding time eating vegetation that averages just 60 percent of their full height. Studies on other populations of giraffes, during both the wet and dry seasons, are consistent with these data. Giraffes usually feed with their necks bent (**FIGURE 1.7b**).

These data cast doubt on the food competition hypothesis, because one of its predictions does not appear to hold. Biologists have not abandoned this hypothesis completely, though, because feeding high in trees may be particularly valuable during extreme droughts, when a giraffe's ability to reach leaves far above the ground could mean the difference between life and death. Still, Simmons and Scheepers have offered an alternative explanation for why giraffes have long necks. The new hypothesis is based on the mating system of giraffes.

The Sexual Competition Hypothesis: Predictions and Tests Giraffes have an unusual mating system. Breeding occurs year round rather than seasonally. To determine when females are coming into estrus or "heat" and are thus receptive to mating, the males nuzzle the rumps of females. In response, the females urinate into the males' mouths. The males then tip their heads back and pull their lips to and fro, as if tasting the liquid. Biologists who have witnessed this behavior have proposed that the males taste the females' urine to detect whether estrus has begun.

Once a female giraffe enters estrus, males fight among themselves for the opportunity to mate. Combat is spectacular. The bulls stand next to one another, swing their necks, and strike thunderous blows with their heads. Researchers have seen males knocked unconscious for 20 minutes after being hit and have cataloged numerous instances in which the loser died. Giraffes are not the only animals known to fight in this way—male giraffe weevils also use enormously long necks to fight for mating rights.

These observations inspired a new explanation for why giraffes have long necks. The sexual competition hypothesis is based on the idea that longer-necked giraffes are able to strike harder blows during combat than can shorter-necked giraffes. In engineering terms, longer necks provide a longer "moment arm." A long moment arm increases the force of an impact. (Think about the type of sledgehammer you'd use to bash down a concrete wall—one with a short handle or one with a long handle?)

The idea here is that longer-necked males should win more fights and, as a result, father more offspring than shorter-necked males do. If neck length in giraffes is inherited, then the average neck length in the population should increase over time. Under the sexual competition hypothesis, long necks are adaptations that increase the fitness of males during competition for females.

Although several studies have shown that long-necked males are more successful in fighting and that the winners of fights gain access to estrous females, the question of why giraffes have long necks is not closed. With the data collected to date, most biologists would probably concede that the food competition hypothesis needs further testing and refinement and that the sexual competition hypothesis appears promising. It could also be true that both hypotheses are correct. For our purposes, the important take-home message is that all hypotheses must be tested rigorously.

In many cases in biological science, testing hypotheses rigorously involves experimentation. Experimenting on giraffes is difficult. But in the case study considered next, biologists were able to test an interesting hypothesis experimentally.

How Do Ants Navigate? An Introduction to Experimental Design

Experiments are a powerful scientific tool because they allow researchers to test the effect of a single, well-defined factor on a particular phenomenon. Because experiments testing the effect of neck length on food and sexual competition in giraffes haven't been done yet, let's consider a different question: When ants leave their nest to search for food, how do they find their way back?

The Saharan desert ant lives in colonies and makes a living by scavenging the dead carcasses of insects. Individuals leave the burrow and wander about searching for food at midday, when temperatures at the surface can reach 60°C (140°F) and predators are hiding from the heat.

Foraging trips can take the ants hundreds of meters—an impressive distance when you consider that these animals are only about a centimeter long. But when an ant returns, it doesn't follow the same long, wandering route it took on its way away from the nest. Instead, individuals return in a straight line (**FIGURE 1.8**).

Once individuals are close to the nest, they engage in a characteristic set of back-and-forth U-turns until they find their nest hole. How do they do know how far they are from the nest?

The Pedometer Hypothesis Early work on navigation in desert ants showed that they use the Sun's position as a compass—meaning that they always know the approximate direction of the nest relative to the Sun. But how do they know how far to go?

After experiments had shown that the ants do not use landmarks to navigate, Matthias Wittlinger and co-workers set out to test a novel idea. The biologists proposed that Saharan desert ants know how far they are from the nest by integrating information from leg movements.

According to this pedometer hypothesis, the ants always know how far they are from the nest because they track the number

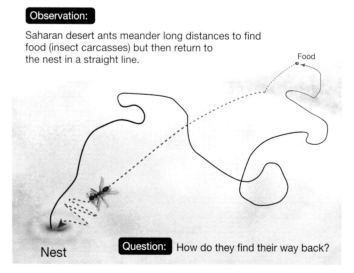

Observation:

Saharan desert ants meander long distances to find food (insect carcasses) but then return to the nest in a straight line.

Food

Nest

Question: How do they find their way back?

FIGURE 1.8 Navigation in Foraging Desert Ants.

of steps they have taken and their stride length. The idea is that they can make a beeline back toward the burrow because they integrate information on the angles they have traveled *and* the distance they have gone—based on step number and stride length.

If the pedometer hypothesis is wrong, however, then stride length and step number should have no effect on the ability of an ant to get back to its nest. This latter possibility is called a **null hypothesis.** A null hypothesis specifies what should be observed when the hypothesis being tested isn't correct.

Testing the Hypothesis To test their idea, Wittlinger's group allowed ants to walk from a nest to a feeder through a channel—a distance of 10 m. Then they caught ants at the feeder and created three test groups, each with 25 individuals (**FIGURES 1.9** and **1.10**):

- *Stumps* By cutting the lower legs of some individuals off, they created ants with shorter-than-normal legs.
- *Normal* Some individuals were left alone, meaning that they had normal leg length.

RESEARCH

QUESTION: How do desert ants find their way back to their nest?

PEDOMETER HYPOTHESIS: Desert ants keep track of stride number and length to calculate how far they are from the nest.

NULL HYPOTHESIS: Stride number and length have nothing to do with navigation (the ants use some other mechanism to navigate).

EXPERIMENTAL SETUP (TEST 1):

1. Ants walk from nest to feeder. 75 ants are collected.

Nest ⟶ Feeder
0 5 10 m

2. Manipulation of legs. Three treatments, 25 ants each.

Cut legs to create "stumps"   Leave legs normal length   Add pig bristles as "stilts"

3. Ants return "home" from feeder and search for nest hole.

Feeder
0 5 10 15 20 m

EXPERIMENTAL SETUP (TEST 2):

1. Recapture manipulated ants from Test 1.

"stumps"   normal   "stilts"

2. The three treatments of ants walk from nest to feeder again.

Nest ⟶ Feeder
0 5 10 m

3. Ants walk back "home" from feeder again.

Feeder
0 5 10 15 20 m

PREDICTION:
Ants with stilts will go too far; ants with stumps will stop short.

PREDICTION OF NULL HYPOTHESIS:
No differences among the three groups.

PREDICTION:
All three groups will start looking for nest after walking 10 m.

PREDICTION OF NULL HYPOTHESIS:
No difference from the observed results in Test 1.

RESULTS:

Stilts
Normal
Stumps
0 5 10 15 20
Homebound run (m)

Stilts
Normal
Stumps
0 5 10 15 20
Homebound run (m)

CONCLUSION: Desert ants use information on stride length and number to calculate how far they are from the nest.

FIGURE 1.9 An Experimental Test: Do Desert Ants Use a "Pedometer"?

SOURCE: Wittlinger, M., R. Wehner, and H. Wolf. 2006. The ant odometer: Stepping on stilts and stumps. *Science* 312: 1965–1967.

✔**QUESTION** What is the advantage of using 25 ants in each group instead of just one?

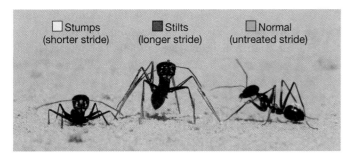

FIGURE 1.10 Manipulation of Desert Ant Stride Length.

Labels in figure:
□ Stumps (shorter stride)   ■ Stilts (longer stride)   □ Normal (untreated stride)

- *Stilts*   By gluing pig bristles onto each leg, the biologists created ants with longer-than-normal legs.

Next they put the ants in a different channel and recorded how far they traveled in a direct line before starting their nest-searching behavior. To see the data they collected, look at the graph on the left side of the "Results" section in Figure 1.9.

- *Stumps*   The ants with stumps stopped short, by about 5 m, before starting to search for the nest opening.
- *Normal*   The normal ants walked the correct distance—about 10 m.
- *Stilts*   The ants with stilts walked about 5 m too far before starting to search for the nest opening.

To check the validity of this result, the researchers put the test ants back in the nest and recaptured them one to several days later, when they had walked to the feeder on their stumps, normal legs, or stilts. Now when the ants were put into the other channel to "walk back," they all traveled the correct distance—10 m—before starting to search for the nest (see the graph on the right side of the "Results" section in Figure 1.9).

The graphs in the "Results" display "box-and-whisker" plots that allow you to easily see where most of the data fall. Each box indicates the range of distances where 50 percent of the ants stopped to search for the nest. The whiskers indicate the lower extreme (stopping short of the nest location) and the upper extreme (going too far) of where the ants stopped to search. The vertical line inside each box indicates the median—meaning that half the ants stopped above this distance and half below. (For more details on how biologists report medians and indicate the variability and uncertainty in data, see **BioSkills 4** in Appendix B.)

Interpreting the Results   The pedometer hypothesis predicts that an ant's ability to walk home depends on the number and length of steps taken on its outbound trip. Recall that a prediction specifies what we should observe if a hypothesis is correct. Good scientific hypotheses make testable predictions—predictions that can be supported or rejected by collecting and analyzing data. In this case, the researchers tested the prediction by altering stride length and recording the distance traveled on the return trip. Under the null hypothesis in this experiment, all the ants—altered and unaltered—should have walked 10 m in the first test before they started looking for their nest.

Important Characteristics of Good Experimental Design   In relation to designing effective experiments, this study illustrates several important points:

- It is critical to include **control** groups. A control checks for factors, other than the one being tested, that might influence the experiment's outcome. In this case, there were two controls. Including a normal, unmanipulated individual controlled for the possibility that switching the individuals to a new channel altered their behavior. In addition, the researchers had to control for the possibility that the manipulation itself—and not the change in leg length—affected the behavior of the stilts and stumps ants. This is why they did the second test, where the outbound and return runs were done with the same legs.

- The experimental conditions must be as constant or equivalent as possible. The investigators used ants of the same species, from the same nest, at the same time of day, under the same humidity and temperature conditions, at the same feeders, in the same channels. Controlling all the variables except one—leg length in this case—is crucial because it eliminates alternative explanations for the results.

- Repeating the test is essential. It is almost universally true that larger sample sizes in experiments are better. By testing many individuals, the amount of distortion or "noise" in the data caused by unusual individuals or circumstances is reduced.

✔ If you understand these points, you should be able to explain: (1) What you would conclude if in the first test, the normal individual had not walked 10 m on the return trip before

check your understanding

  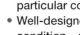

If you understand that . . .

- Hypotheses are proposed explanations that make testable predictions.
- Predictions describe observable outcomes of particular conditions.
- Well-designed experiments alter just one condition—a condition relevant to the hypothesis being tested.

✔ You should be able to . . .

Design an experiment to test the hypothesis that desert ants feed during the hottest part of the day because it allows them to avoid being eaten by lizards. Then answer the following questions about your experimental design:

1. How does the presence of a control group in your experiment allow you to test the hypothesis?

2. How are experimental conditions controlled or standardized in a way that precludes alternative explanations of the data?

Answers are available in Appendix A.

looking for the nest; and (2) What you would conclude if the stilts and stumps ants had not navigated normally during the second test.

From the outcomes of these experiments, the researchers concluded that desert ants use stride length and number to measure how far they are from the nest. They interpreted their results as strong support for the pedometer hypothesis.

The giraffe and ant studies demonstrate a vital point: Biologists practice evidence-based decision making. They ask questions about how organisms work, pose hypotheses to answer those questions, and use experimental or observational evidence to decide which hypotheses are correct.

The data on giraffes and ants are a taste of things to come. In this text you will encounter hypotheses and research on questions ranging from how water gets to the top of 100-meter-tall sequoia trees to how the bacterium that causes tuberculosis has become resistant to antibiotics. As you work through this book, you'll get lots of practice thinking about hypotheses and predictions, analyzing the nature of control treatments, and interpreting graphs.

A commitment to tough-minded hypothesis testing and sound experimental design is a hallmark of biological science. Understanding their value is an important first step in becoming a biologist.

CHAPTER 1　REVIEW

For media, go to MasteringBiology (MB)

If you understand . . .

1.1　What Does It Mean to Say That Something Is Alive?

- There is no single, well-accepted definition of life. Instead, biologists point to five characteristics that organisms share.

✔ You should be able to explain why the cells in a dead organism are different from the cells in a live organism.

1.2　The Cell Theory

- The cell theory identified the fundamental structural unit common to all life.

✔ You should be able to describe the evidence that supported the pattern and the process components of the cell theory.

1.3　The Theory of Evolution by Natural Selection

- The theory of evolution states that all organisms are related by common ancestry.
- Natural selection is a well-tested explanation for why species change through time and why they are so well adapted to their habitats.

✔ You should be able to explain why the average protein content of seeds in a natural population of a grass species would increase over time, if seeds with higher protein content survive better and grow into individuals that produce many seeds with high protein content when they mature.

1.4　The Tree of Life

- The theory of evolution predicts that all organisms are part of a genealogy of species, and that all species trace their ancestry back to a single common ancestor.

- To construct this phylogeny, biologists have analyzed the sequences in rRNA and in an array of genetic material found in all cells.

- A tree of life, based on similarities and differences in these molecules, has three fundamental lineages, or domains: the Bacteria, the Archaea, and the Eukarya.

✔ You should be able to explain how biologists can determine which of the three domains a newly discovered species belongs to by analyzing its rRNA.

1.5　Doing Biology

- Biology is a hypothesis-driven, experimental science.

✔ You should be able to explain (1) the relationship between a hypothesis and a prediction and (2) why experiments are convincing ways to test predictions.

(MB)　MasteringBiology

1. **MasteringBiology Assignments**

   Tutorials and Activities An Introduction to Graphing; Experimental Inquiry: What Can You Learn about the Process of Science from Investigating a Cricket's Chirp?; Introduction to Experimental Design; Levels of Life Card Game; Metric System Review; The Scientific Method

   Questions Reading Quizzes, Blue-Thread Questions, Test Bank

2. **eText** Read your book online, search, take notes, highlight text, and more.

3. **The Study Area** Practice Test, Cumulative Test, BioFlix® 3-D Animations, Videos, Activities, Audio Glossary, Word Study Tools, Art

You should be able to . . .

1. Anton van Leeuwenhoek made an important contribution to the development of the cell theory. How?
   a. He articulated the pattern component of the theory—that all organisms are made of cells.
   b. He articulated the process component of the theory—that all cells come from preexisting cells.
   c. He invented the first microscope and saw the first cell.
   d. He invented more powerful microscopes and was the first to describe the diversity of cells.

2. What does it mean to say that experimental conditions are controlled?
   a. The test groups consist of the same individuals.
   b. The null hypothesis is correct.
   c. There is no difference in outcome between the control and experimental treatment.
   d. All physical conditions except for one are identical for all groups tested.

3. The term *evolution* means that _____ change through time.

4. What does it mean to say that a characteristic of an organism is heritable?
   a. The characteristic evolves.
   b. The characteristic can be passed on to offspring.
   c. The characteristic is advantageous to the organism.
   d. The characteristic does not vary in the population.

5. In biology, to what does the term *fitness* refer?

6. Could *both* the food competition hypothesis and the sexual competition hypothesis explain why giraffes have long necks? Why or why not?
   a. No. In science, only one hypothesis can be correct.
   b. No. Observations have shown that the food competition hypothesis cannot be correct.
   c. Yes. Long necks could be advantageous for more than one reason.
   d. Yes. All giraffes have been shown to feed at the highest possible height and fight for mates.

7. What would researchers have to demonstrate to convince you that they had discovered life on another planet?

8. What did Linnaeus's system of naming organisms ensure?
   a. Two different organisms never end up with the same genus and species name.
   b. Two different organisms have the same genus and species name if they are closely related.
   c. The genus name is different for closely related species.
   d. The species name is the same for each organism in a genus.

9. What does it mean to say that a species is adapted to a particular habitat?

10. Explain how selection occurs during natural selection. What is selected, and why?

11. The following two statements explain the logic behind the use of molecular sequence data to estimate evolutionary relationships:

    "If the theory of evolution is true, then rRNA sequences should be very similar in closely related organisms but less similar in organisms that are less closely related."

    "On a phylogenetic tree, branches that share a recent common ancestor represent species that are closely related; branches that don't share recent common ancestors represent species that are more distantly related."

    Is the logic of these statements sound? Why or why not?

12. Explain why researchers formulate a null hypothesis in addition to a hypothesis when designing an experimental study.

13. A scientific theory is a set of propositions that defines and explains some aspect of the world. This definition contrasts sharply with the everyday usage of the word theory, which often carries meanings such as "speculation" or "guess." Explain the difference between the two definitions, using the cell theory and the theory of evolution by natural selection as examples.

14. Turn back to the tree of life shown in Figure 1.5. Note that Bacteria and Archaea are prokaryotes, while Eukarya are eukaryotes. On the simplified tree below, draw an arrow that points to the branch where the structure called the nucleus originated. Explain your reasoning.

15. The proponents of the cell theory could not "prove" that it was correct in the sense of providing incontrovertible evidence that all organisms are made up of cells. They could state only that all organisms examined to date were made of cells. Why was it reasonable for them to conclude that the theory was valid?

16. Some humans have heritable traits that make them resistant to infection by HIV. In areas of the world where HIV infection rates are high, are human populations evolving? Explain your logic.
    a. No. HIV infection rates would not affect human evolution.
    b. Yes. The heritable traits that confer resistance to HIV should increase over time.
    c. No. The heritable traits that confer resistance to HIV should decrease over time.
    d. Yes. The heritable traits that confer resistance to HIV should decrease over time.

The Big Picture

Biologists study the characteristics of life. The cell theory, the theory of evolution by natural selection, and the tree of life are some of the great ideas in biology that came about by biologists asking questions that can be answered by observing or measuring things—that is, by collecting data.

Notice that the study of life is not a series of linear steps with a beginning and an end. Instead, the process of doing biology is dynamic and ongoing. The answer to one question may lay the foundation for twenty more questions. Working together, biologists from different disciplines integrate data across many levels, from atoms to the biosphere.

Note that the gray numbers in boxes tell you where to go for more information. Also, be sure to do the blue exercises in the Check Your Understanding box.

Characteristics of living things

• Energy
• Cells
• Information
• Replication
• Evolution

1.1

Text section where you can find more information

focuses on

Scientists regularly integrate across many of these levels

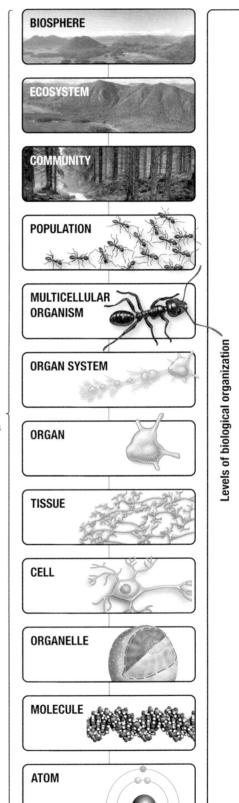

Levels of biological organization

BIOSPHERE

ECOSYSTEM

COMMUNITY

POPULATION

MULTICELLULAR ORGANISM

ORGAN SYSTEM

ORGAN

TISSUE

CELL

ORGANELLE

MOLECULE

ATOM

If you understand the big picture . . .

✔ You should be able to . . .

1. Describe how biologists go about testing their ideas.

2. Provide an example of how an experimental study could span more than one level of biological organization.

3. Compare and contrast a hypothesis with a theory.

4. Propose the next step to take if data support the hypothesis you are testing.

Answers are available in Appendix A.

DOING BIOLOGY

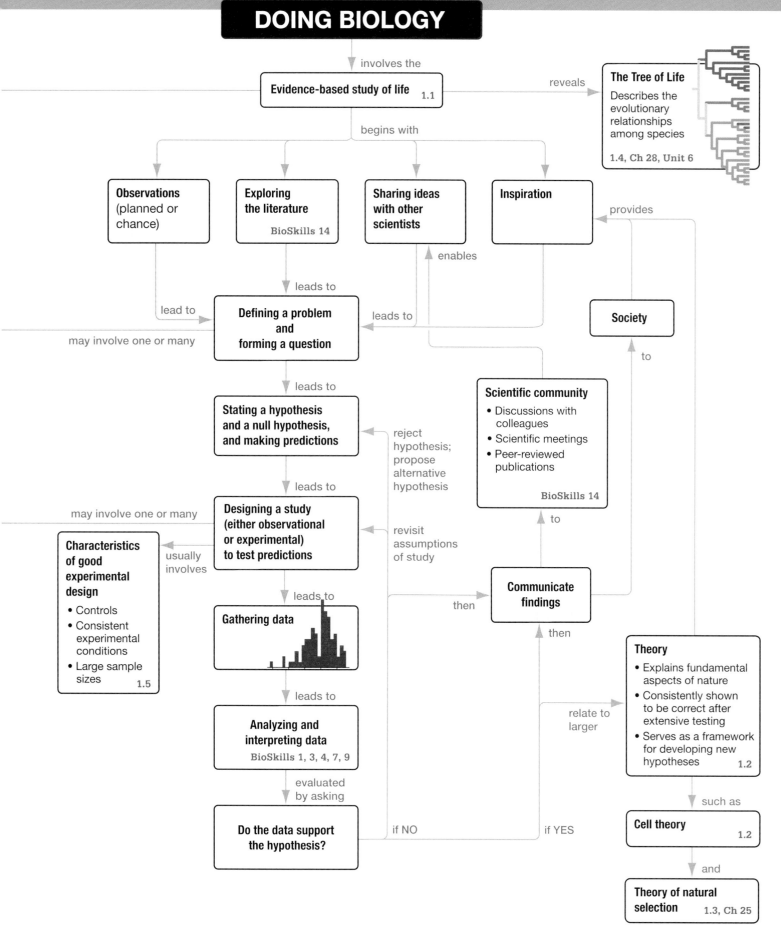

Evidence-based study of life 1.1

involves the

begins with

reveals → **The Tree of Life** Describes the evolutionary relationships among species 1.4, Ch 28, Unit 6

Observations (planned or chance)

Exploring the literature BioSkills 14

Sharing ideas with other scientists

Inspiration

provides

lead to

leads to

may involve one or many

Defining a problem and forming a question

leads to

enables

Society

to

leads to

Stating a hypothesis and a null hypothesis, and making predictions

reject hypothesis; propose alternative hypothesis

Scientific community
- Discussions with colleagues
- Scientific meetings
- Peer-reviewed publications

BioSkills 14

leads to

may involve one or many

Designing a study (either observational or experimental) to test predictions

usually involves

Characteristics of good experimental design
- Controls
- Consistent experimental conditions
- Large sample sizes 1.5

revisit assumptions of study

to

leads to

Gathering data

then

Communicate findings

then

Analyzing and interpreting data BioSkills 1, 3, 4, 7, 9

relate to larger

Theory
- Explains fundamental aspects of nature
- Consistently shown to be correct after extensive testing
- Serves as a framework for developing new hypotheses 1.2

evaluated by asking

such as

Do the data support the hypothesis?

if NO

if YES

Cell theory 1.2

and

Theory of natural selection 1.3, Ch 25

25 Evolution by Natural Selection

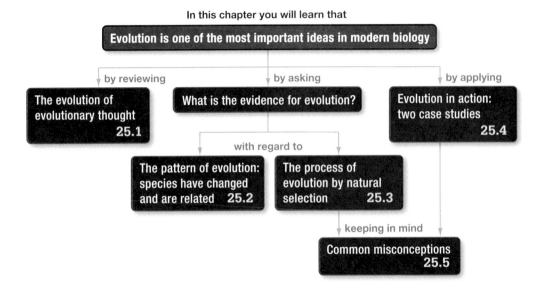

In this chapter you will learn that

Evolution is one of the most important ideas in modern biology

by reviewing

The evolution of evolutionary thought 25.1

by asking

What is the evidence for evolution?

with regard to

The pattern of evolution: species have changed and are related 25.2

The process of evolution by natural selection 25.3

keeping in mind

Common misconceptions 25.5

by applying

Evolution in action: two case studies 25.4

Natural selection acts on individuals in populations such as these sea stars, but only populations evolve. One of Darwin's greatest contributions to science was the introduction of population thinking to the theory of evolution.

This chapter is part of the Big Picture. See how on pages 526–527.

This chapter is about one of the great ideas in science. The theory of evolution by natural selection, formulated independently by Charles Darwin and Alfred Russel Wallace, explains how organisms have come to be adapted to environments ranging from arctic tundra to tropical wet forest. It revealed one of the five key attributes of life: Populations of organisms evolve—meaning that they change through time (Chapter 1).

As an example of a revolutionary breakthrough in our understanding of the world, the theory of evolution by natural selection ranks alongside Copernicus's theory of the Sun as the center of our solar system, Newton's laws of motion and theory of gravitation, the germ theory of disease, the theory of plate tectonics, and Einstein's general theory of relativity. These theories are the foundation stones of modern science; all are accepted on the basis of overwhelming evidence.

✔ When you see this checkmark, stop and test yourself. Answers are available in Appendix A.

Evolution by natural selection is one of the best supported and most important theories in the history of scientific research. But like most scientific breakthroughs, this one did not come easily. When Darwin published his theory in 1859 in a book called *On the Origin of Species by Means of Natural Selection*, it unleashed a firestorm of protest throughout Europe. At that time, the leading explanation for the diversity of organisms was an idea called special creation.

Special creation held that: (1) All species are independent, in the sense of being unrelated to each other; (2) life on Earth is young—perhaps just 6000 years old; and (3) species are immutable, or incapable of change. These beliefs were explained by the instantaneous and independent creation of living organisms by a supernatural being.

Darwin's theory was radically different. How did it differ? Scientific theories usually have two components: a pattern and a process (see Chapter 1):

1. The *pattern component* is a statement that summarizes a series of observations about the natural world. The pattern component is about facts—about how things *are* in nature.

2. The *process component* is a mechanism that produces that pattern or set of observations.

Let's begin with an overview of the evolution of evolutionary thought, and then examine the pattern and process components of the theory of evolution by natural selection.

25.1 The Evolution of Evolutionary Thought

People often use the word revolutionary to describe the theory of evolution by natural selection. Revolutions overturn things—they replace an existing entity with something new and often radically different. A political revolution removes the ruling class or group and replaces it with another. The industrial revolution replaced small shops for manufacturing goods by hand with huge, mechanized assembly lines.

A scientific revolution, in contrast, overturns an existing idea about how nature works and replaces it with another, radically different, idea. The idea that Darwin and Wallace overturned—that species were supernaturally, not naturally, created—had dominated thinking about the nature of organisms in Western civilization for over 2000 years.

Plato and Typological Thinking

The Greek philosopher Plato claimed that every organism was an example of a perfect essence, or type, created by God, and that these types were unchanging. Plato acknowledged that the individual organisms present on Earth might deviate slightly from the perfect type, but he said this deviation was similar to seeing the perfect type in a shadow on a wall. The key to understanding life, in Plato's mind, was to ignore the shadows and focus on understanding each type of unchanging, perfect essence.

Today, philosophers and biologists refer to ideas like this as typological thinking. Typological thinking is based on the idea that species are unchanging types and that variations within species are unimportant or even misleading. Typological thinking also occurs in the Bible's book of Genesis, where God creates each type of organism.

Aristotle and the Great Chain of Being

Not long after Plato developed his ideas, Aristotle ordered the types of organisms known at the time into a linear scheme called the great chain of being, also called the scale of nature (**FIGURE 25.1**). Aristotle proposed that species were organized into a sequence based on increased size and complexity, with humans at the top. He also claimed that the characteristics of species were fixed—they did not change through time.

In the 1700s Aristotle's ideas were still popular in scientific and religious circles. The central claims were that (1) species are fixed types, and (2) some species are higher—in the sense of being more complex or "better"—than others.

Lamarck and the Idea of Evolution as Change through Time

Typological thinking eventually began to break down. In 1809 the biologist Jean-Baptiste de Lamarck proposed a formal theory of **evolution**—that species are not static but change through time. However, the pattern component of Lamarck's theory was initially based on the great chain of being.

When he started his work on evolution, Lamarck claimed that simple organisms originate at the base of the chain by

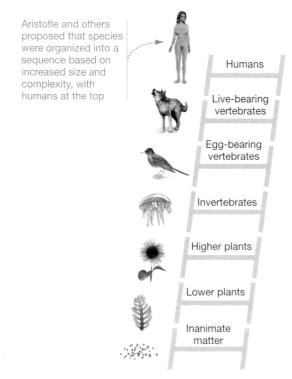

Aristotle and others proposed that species were organized into a sequence based on increased size and complexity, with humans at the top

Humans

Live-bearing vertebrates

Egg-bearing vertebrates

Invertebrates

Higher plants

Lower plants

Inanimate matter

FIGURE 25.1 The Great Chain of Being, or Scale of Nature.

spontaneous generation (see Chapter 1) and then evolve by moving up the chain over time. Thus, Lamarckian evolution is progressive in the sense of always producing larger and more complex, or "better," species. To capture this point, biologists often say that Lamarck turned the ladder of life into an escalator.

Lamarck also contended that species change through time via the inheritance of acquired characters. The idea here is that as an individual develops in response to challenges posed by the environment, its phenotype changes, and it passes on these phenotypic changes to offspring. A classic Lamarckian scenario is that giraffes develop long necks as they stretch to reach leaves high in treetops, and they then produce offspring with elongated necks.

Darwin and Wallace and Evolution by Natural Selection

As his thinking matured, Lamarck eventually abandoned his linear and progressive view of life. Darwin and Wallace concurred. But more important, they emphasized that the process responsible for change through time—evolution—occurs because traits vary among the individuals in a population, and because individuals with certain traits leave more offspring than others do. A **population** consists of individuals of the same species that are living in the same area at the same time.

Darwin and Wallace's proposal was a radical break from the typological thinking that had dominated scientific thought since Plato. Darwin claimed that instead of being unimportant or an illusion, variation among individuals in a population was the key to understanding the nature of species. Biologists refer to this view as **population thinking.**

The theory of evolution by natural selection was revolutionary for several reasons:

1. It overturned the idea that species are static and unchanging.

2. It replaced typological thinking with population thinking.

3. It was scientific. It proposed a mechanism that could account for change through time and made predictions that could be tested through observation and experimentation.

Plato and his followers emphasized the existence of fixed types; evolution by natural selection is all about change and diversity. Now the questions are: What evidence backs the claim that species are not fixed types? What data convince biologists that the theory of evolution by natural selection is correct?

25.2 The Pattern of Evolution: Have Species Changed, and Are They Related?

In *On the Origin of Species*, Darwin repeatedly described evolution as **descent with modification.** He meant that species that lived in the past are the ancestors of the species existing today, and that species change through time.

This view was a radical departure from the independently created and immutable species embodied in Plato's work and in the idea of special creation. In essence, the pattern component of the theory of evolution by natural selection makes two predictions about the nature of species:

1. Species change through time.

2. Species are related by common ancestry.

Let's consider the evidence for each of these predictions in turn.

Evidence for Change through Time

When Darwin began his work, biologists and geologists had just begun to assemble and interpret the fossil record. A **fossil** is any trace of an organism that lived in the past. These traces range from bones and branches to shells, tracks or impressions, and dung. The **fossil record** consists of all the fossils that have been found on Earth and described in the scientific literature.

Why did data in the fossil record support the hypothesis that species have changed through time? And what data from **extant species**—those living today—support the claim that they are modified forms of ancestral species?

The Vastness of Geologic Time Initially, fossils were organized according to their *relative* ages based on a series of principles derived from observations about rock formation. **Sedimentary rocks,** for example, form from sand or mud or other materials deposited at locations such as beaches or river mouths. Sedimentary rocks, along with rocks derived from volcanic ash or lava, are known to form in layers—younger layers are deposited on top of older layers.

Researchers used this information to place fossils in a younger-to-older sequence, based on the fossils' relative position in layers of sedimentary rock (**FIGURE 25.2**). As the scientists observed similarities in rocks and fossils at different sites, they began to create a **geologic time scale:** a sequence of named intervals called eons, eras, and periods that represented the major events in Earth history (see Chapter 28). They also realized that vast amounts of time were required to form the thick layers of sedimentary rock that they were studying, because erosion and deposition of sediments are such slow processes.

This was an important insight. The geologic record indicated that the Earth was much, much older than the 6000 years claimed by proponents of special creation.

After the discovery of radioactivity in the late 1800s, researchers realized that radioactive decay—the steady rate at which unstable or "parent" atoms are converted into more stable "daughter" atoms—furnished a way to assign *absolute* ages, in years, to the relative ages in the geologic time scale.

Radiometric dating is based on three pieces of information:

1. Observed decay rates of parent to daughter atoms

2. The ratio of parent to daughter atoms present in newly formed rocks—such as the amount of uranium atoms versus lead atoms when uranium-containing molten rock first cools (Uranium decays to form lead.)

3. The ratio of parent to daughter atoms present in a particular rock sample

Younger rock layers   Younger fossils

Tracks from a mammal-like reptile

Fern

Trilobite

Older rock layers   Older fossils

FIGURE 25.2 Sedimentary Rocks Reveal the Vastness of Geologic Time. The relative ages of sedimentary rocks are used to determine the relative ages of fossil organisms because younger layers are deposited on top of older ones. The deepest rock layer in the Grand Canyon is over a billion years old, and the top layer is 270 million years old.

Combining information from these two ratios with information on the decay rate allows researchers to estimate how long ago a rock formed. According to data from radiometric dating, Earth is about 4.6 billion years old, and the earliest signs of life appear in rocks that formed 3.4–3.8 billion years ago.

Data from relative and absolute dating techniques agree: Life on Earth is ancient. A great deal of time has gone by for change to occur.

Extinction Changes the Species Present over Time In the early nineteenth century, researchers began discovering fossil bones, leaves, and shells that were unlike structures from any known animal or plant. At first, many scientists insisted that living examples of these species would be found in unexplored regions of the globe. But as research continued and the number and diversity of fossil collections grew, the argument became less and less plausible.

The issue was finally settled in 1812 when Baron Georges Cuvier published a detailed analysis of an **extinct species**—that is, a species that no longer exists—called the Irish "elk." Scientists accepted the fact of extinction because this gigantic deer was judged to be too large to have escaped discovery and too

FIGURE 25.3 Evidence of Extinction. The skeleton of the Irish "elk" dwarfs a human. Scientists agreed that the deer was too large and unique to be overlooked if it were alive; it must have gone extinct.

distinctive to be classified as a large-bodied population of an existing species (**FIGURE 25.3**).

Advocates of special creation argued that fossil species were victims of the flood at the time of Noah. Darwin, in contrast, interpreted extinct forms as evidence that species are not static, immutable entities, unchanged since the moment of special creation. His reasoning was that if species have gone extinct, then the array of species living on Earth has changed through time.

Recent analyses of the fossil record suggest that over 99 percent of all the species that have ever lived are now extinct. The data also indicate that species have gone extinct continuously throughout Earth's history—not just in one or even a few catastrophic events.

Transitional Features Link Older and Younger Species Long before Darwin published his theory, researchers reported striking resemblances between the fossils found in the rocks underlying certain regions and the living species found in the same geographic areas. The pattern was so widespread that it became known as the "law of succession." The general observation was that extinct species in the fossil record were succeeded, in the same region, by similar species.

Early in the nineteenth century, the pattern was simply reported and not interpreted. But later, Darwin pointed out that it provided strong evidence in favor of the hypothesis that species had changed through time. His idea was that the extinct forms and living forms were related—that they represented ancestors and descendants.

As the fossil record expanded, researchers discovered species with characteristics that broadened the scope of the law of succession. A **transitional feature** is a trait in a fossil species that is intermediate between those of ancestral (older) and derived (younger) species. For example, intensive work over the past several decades has yielded fossils that document a gradual change over time from aquatic animals that had fins to terrestrial animals that had limbs (**FIGURE 25.4**). Over a period of about 25 million years, the fins of species similar to today's lungfish changed into limbs similar to those found in today's amphibians, reptiles, and mammals—a group called the tetrapods (literally, "four-footed").

These observations support the hypothesis that an ancestral lungfish-like species first used stout, lobed fins to navigate in shallow aquatic habitats. Then they moved onto land, where their descendants became more and more like today's tetrapods in appearance and lifestyle. Lungfish and tetrapod species have clearly changed through time.

Similar sequences of transitional features document changes that led to the evolution of feathers and flight in birds; stomata

Tulerpeton (~362 mya)

Acanthostega (~365 mya)

Tiktaalik (~375 mya)

Eusthenopteron (~385 mya)

FIGURE 25.4 Transitional Features during the Evolution of the Tetrapod Limb. Fossil species similar to today's lungfish and tetrapods have fin and limb bones that are transitional features. *Eusthenopteron* was aquatic; *Tulerpeton* was probably semiaquatic (mya = million years ago).

✔ **QUESTION** How would observations of transitional features be explained under special creation?

and vascular tissue in plants; upright posture, flattened faces, and large brains in humans; jaws in vertebrates (animals with backbones); the loss of limbs in snakes; and other traits. Data like these are consistent with predictions from the theory of evolution: If the traits observed in more recent species evolved from traits in more ancient species, then transitional forms are expected to occur in the appropriate time sequence.

The fossil record provides compelling evidence that species have evolved. What data from extant forms support the hypothesis that the characteristics of species change through time?

Vestigial Traits Are Evidence of Change through Time Darwin was the first to provide a widely accepted interpretation of vestigial traits. A **vestigial trait** is a reduced or incompletely developed structure that has no function, or reduced function, but is clearly similar to functioning organs or structures in closely related species.

Biologists have documented thousands of examples of vestigial traits.

- Some whales and snakes have tiny hip and leg bones that do not help them swim or slither.

- Ostriches and kiwis have reduced wings and cannot fly.

- Eyeless, blind cave-dwelling fish have eye sockets.

- Even though marsupial mammals give birth to live young, an eggshell forms briefly early in their development; in some species, newborns have a nonfunctioning "egg tooth" similar to those used by birds and reptiles to break open their shells.

- Monkeys and many other primates have long tails; but our coccyx, illustrated in **FIGURE 25.5**, is too small to help us maintain balance or grab tree limbs for support.

- Many mammals, including primates, are able to erect their hair when they are cold or excited. This behavior manifests itself as goose bumps in humans, but goose bumps are largely ineffective in warming us or signalling our emotional state.

FIGURE 25.5 Vestigial Traits Are Reduced Versions of Traits in Other Species. The tailbone is a human trait that has reduced function. It is no longer useful for balance and locomotion.

✔ **QUESTION** How would observations of vestigial traits be explained if evolution occurred via inheritance of acquired characters?

The existence of vestigial traits is inconsistent with the idea of special creation, which maintains that species were perfectly designed by a supernatural being and that the characteristics of species are static. Instead, vestigial traits are evidence that the characteristics of species have changed over time.

Current Examples of Change through Time Biologists have documented hundreds of contemporary populations that are changing in response to changes in their environment. Bacteria have evolved resistance to drugs; insects have evolved resistance to pesticides; weedy plants have evolved resistance to herbicides; the timing of bird migrations, the emergence of insects, and the blooming of flowering plants have evolved in response to climate change. Section 25.4 provides a detailed analysis of research on two examples of evolution in action.

To summarize, change through time continues and can be measured directly. Evidence from the fossil record and living species indicates that life is ancient, that species have changed through the course of Earth's history, and that species continue to change. The take-home message is that species are dynamic—not static, unchanging, and fixed types, as claimed by Plato, Aristotle, and advocates of special creation.

Evidence of Descent from a Common Ancestor

Data from the fossil record and contemporary species refute the hypothesis that species are immutable. What about the claim that species were created independently—meaning that they are unrelated to each other?

Similar Species Are Found in the Same Geographic Area Charles Darwin began to realize that species are related by common ancestry during a five-year voyage he took aboard the English naval ship HMS *Beagle*. While fulfilling its mission to explore and map the coast of South America, the *Beagle* spent a few weeks in the Galápagos Islands off the coast of present-day Ecuador. Darwin had taken over the role of ship's naturalist and, as the first scientist to study the area, gathered extensive collections of the plants and animals found in these islands. Most famous among the birds he collected were the Galápagos finches (featured in Section 25.4) and the Galápagos mockingbirds, pictured in **FIGURE 25.6a**.

Several years after Darwin returned to England, a colleague pointed out that the mockingbirds collected on different islands were distinct species, based on differences in coloration and beak size and shape. This struck Darwin as remarkable. Why would species that inhabit neighboring islands be so similar, yet clearly

(a) Pattern: Although the Galápagos mockingbirds are extremely similar, distinct species are found on different islands.

(b) Recent data support Darwin's hypothesis that the Galápagos mockingbirds share a common ancestor.

FIGURE 25.6 Close Relationships among Island Forms Argue for Shared Ancestry.

Gene:	Amino acid sequence (single-letter abbreviations):
Aniridia (Human)	LQRNRTSFTQEQIEALEKEFERTHYPDVFARERLAAKIDLPEARIQVWFSNRRAKWRREE
eyeless (Fruit fly)	LQRNRTSFTNDQIDSLEKEFERTHYPDVFARERLAGKIGLPEARIQVWFSNRRAKWRREE

FIGURE 25.7 Genetic Homology: Genes from Different Species May Be Similar in DNA Sequence or Other Attributes. Amino acid sequences from a portion of the *Aniridia* gene product found in humans are 90 percent identical to those found in the *Drosophila eyeless* gene product. (For a key to the single-letter abbreviations used for the amino acids, see Figure 3.2.)

distinct? This turns out to be a widespread pattern: In island groups across the globe, it is routine to find similar but distinct species on neighboring islands.

Darwin realized that this pattern—puzzling when examined as a product of special creation—made perfect sense when interpreted in the context of evolution, or descent with modification. The mockingbirds were similar, he proposed, because they had descended from the same common ancestor. That is, instead of being created independently, mockingbird populations that colonized different islands had changed through time and formed new species (**FIGURE 25.6b**).

Recent analyses of DNA sequences in these mockingbirds support Darwin's hypothesis. Researchers have used the DNA sequence comparisons to place the mockingbirds on a **phylogenetic tree**—a branching diagram that describes the ancestor–descendant relationships among species or other taxa (see Chapter 28). As Figure 25.6b shows, the Galápagos mockingbirds are each others' closest living relatives. As Darwin predicted, they share a single common ancestor. (For help with reading phylogenetic trees, see **BioSkills 7** in Appendix B.)

Similar Species Share Homologies Translated literally, homology means "the study of likeness." When biologists first began to study the anatomy of humans and other vertebrates, they were struck by the remarkable similarity of their skeletons, muscles, and organs. But because the biologists who did these early studies were advocates of special creation, they could not explain why striking similarities existed among certain organisms but not others.

Today, biologists recognize that **homology** is a similarity that exists in species because they inherited the trait from a common ancestor. Human hair and dog fur are homologous. Humans have hair and dogs have hair (fur) because they share a common ancestor—an early mammal species—that had hair.

Homology can be recognized and studied at three levels:

1. **Genetic homology** occurs in DNA nucleotide sequences, RNA nucleotide sequences, or amino acid sequences. For example, the *eyeless* gene in fruit flies and the *Aniridia* gene in humans are so similar that their protein products are 90 percent identical in amino acid sequence (**FIGURE 25.7**). Both genes act in determining where eyes will develop—even though fruit flies have a compound eye with many lenses and humans have a camera eye with a single lens.

2. **Developmental homology** is recognized in embryos. For example, early chick, human, and cat embryos have tails and

structures called gill pouches (**FIGURE 25.8**). Later, gill pouches are lost in all three species and tails are lost in humans. But in fish, the embryonic gill pouches stay intact and give rise to functioning gills in adults. To explain this observation, biologists hypothesize that gill pouches and tails exist in chicks, humans, and cats because they existed in the fishlike species that was the common ancestor of today's vertebrates. Embryonic gill pouches are a vestigial trait in chicks, humans, and cats; embryonic tails are a vestigial trait in humans.

3. **Structural homology** is a similarity in adult **morphology,** or form. A classic example is the common structural plan observed in the limbs of vertebrates (**FIGURE 25.9**). In Darwin's own words, "What could be more curious than that the hand of a man, formed for grasping, that of a mole for digging, the leg of the horse, the paddle of the porpoise, and the wing of the bat, should all be constructed on the same pattern, and should include the same bones, in the same relative positions?" An engineer would never use the same underlying structure to design a grasping tool, a digging implement, a walking device, a propeller, and a wing. Instead, the structural homology exists because mammals evolved from the lungfish-like ancestor in Figure 25.4, which had the same general arrangement of bones in its fins.

The three levels of homology interact. Genetic homologies cause the developmental homologies observed in embryos, which then lead to the structural homologies recognized in adults. Perhaps the most fundamental of all homologies is the genetic code.

FIGURE 25.8 Developmental Homology: Structures That Appear Early in Development Are Similar. The early embryonic stages of a chick, a human, and a cat show a strong resemblance.

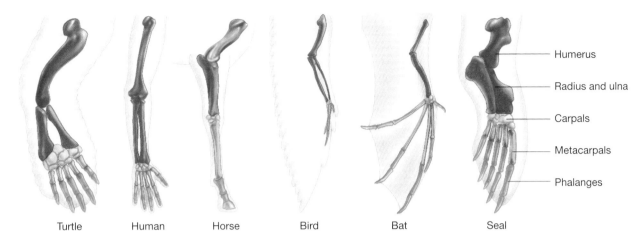

FIGURE 25.9 Structural Homology: Limbs with Different Functions Have the Same Underlying Structure. Even though their function varies, all vertebrate limbs are modifications of the same number and arrangement of bones. (These limbs are not drawn to scale.)

With a few minor exceptions, all organisms use the same rules for transferring the information coded in DNA into proteins (see Chapter 16).

In some cases, hypotheses about homology can be tested experimentally. For example, researchers (**1**) isolated a mouse gene that was thought to be homologous to the fruit fly *eyeless* gene, (**2**) inserted the mouse gene into fruit fly embryos, (**3**) stimulated expression of the foreign gene in locations that normally give rise to appendages, and (**4**) observed formation of eyes on legs and antennae (**FIGURE 25.10**). The function of the inserted gene was identical to that of *eyeless*. This result was strong evidence that the fruit fly and mouse genes are homologous, as predicted from their sequence similarity.

Homology is a key concept in contemporary biology:

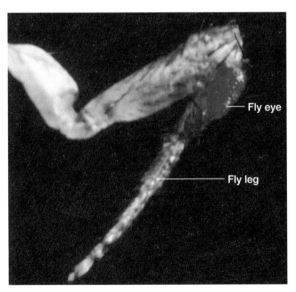

FIGURE 25.10 Evidence for Homology: A Mouse Gene Expressed in Fruit Flies. As an embryo, this fruit fly received a mouse gene that signals where eyes should form. A fruit fly eye formed in the location where the mouse gene was expressed.

- Chemicals that cause cancer in humans can often be identified by testing their effects on mutation rates in bacteria, yeast, zebrafish, mice, and other model organisms because the molecular machinery responsible for copying and repairing DNA is homologous in all organisms (see Chapter 15).

- Drugs intended for human use can be tested on mice or rabbits if the molecules targeted by the drugs are homologous.

- Unknown sequences in the human, rice, or other genomes can be identified if they are homologous to known sequences in yeast, fruit flies, or other well-studied model organisms (see Chapter 21 and **BioSkills 13** in Appendix B).

The theory of evolution by natural selection predicts that homologies will occur. If species were created independently of one another, as special creation claims, these types of similarities would not occur.

Current Examples of Descent from a Common Ancestor Biologists have documented dozens of contemporary populations that are undergoing speciation—a process that results in one species splitting into two or more descendant species. Some populations have served as particularly well-studied examples of speciation in action (see Chapter 27). In most cases, the identity of the ancestral species and the descendant species is known—meaning that biologists have established a direct link between ancestral and descendant species. In addition, the reason for the splitting event is usually known.

The contemporary examples of new species being formed are powerful evidence that species living today are the descendants of species that lived in the past. They support the claim that all organisms are related by descent from a common ancestor.

Evolution's "Internal Consistency"— The Importance of Independent Data Sets

Biologists draw upon data from several sources to challenge the hypothesis that species are immutable and were created independently. The data support the idea that species have descended, with modification, from a common ancestor. **TABLE 25.1** summarizes this evidence.

Perhaps the most powerful evidence for any scientific theory, including evolution by natural selection, is what scientists call internal consistency. This is the observation that data from independent sources agree in supporting predictions made by a theory.

As an example, consider the evolution of whales and dolphins—a group called the cetaceans.

- The fossil record contains a series of species that are clearly identified as cetaceans based on the unusual ear bones found only in this group. Some of the species have the long legs and compact bodies typical of mammals that live primarily on land; some are limbless and have the streamlined bodies typical of aquatic mammals; some have intermediate features.

- A phylogeny of the fossil cetaceans, estimated on the basis of similarities and differences in morphological traits other than limbs and overall body shape, indicates that a gradual transition occurred between terrestrial forms and aquatic, whale-like forms (**FIGURE 25.11**).

SUMMARY TABLE 25.1 **Evidence for Evolution**

Prediction 1: Species Are Not Static, but Change through Time

- Life on Earth is ancient. Most species have gone extinct.
- Fossil (extinct) species frequently resemble living species found in the same area.
- Transitional features document change in traits through time.
- Vestigial traits are common.
- The characteristics of populations vary within species and can be observed changing today.

Prediction 2: Species Are Related, Not Independent

- Closely related species often live in the same geographic area.
- Homologous traits are common and are recognized at three levels:
  1. genetic (gene structure and the genetic code)
  2. developmental (embryonic structures and processes)
  3. structural (morphological traits in adults)
- The formation of new species, from preexisting species, can be observed today.

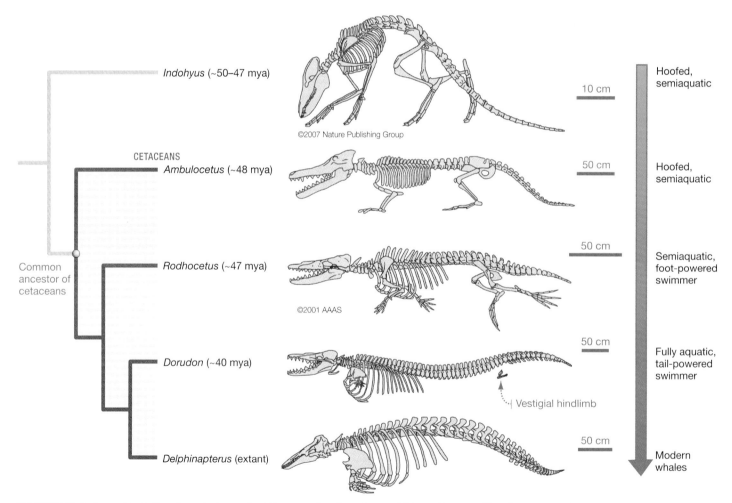

FIGURE 25.11 Data on Evolution from Independent Sources Are Consistent. This phylogeny of fossil cetaceans is consistent with data from relative dating, absolute dating, and phylogenies estimated from molecular traits in living species—all agree that whales evolved from terrestrial ancestors that were related to today's hippos.

- Relative dating, based on the positions of sedimentary rocks where the fossils were found, agrees with the order of species indicated in the phylogeny.

- Absolute dating, based on analyses of radioactive atoms in rocks in or near the layers where the fossils were found, also agrees with the order of species indicated in the phylogeny.

- A phylogeny of living whales and dolphins, estimated from similarities and differences in DNA sequences, indicates that hippos—which spend much of their time in shallow water—are the closest living relative of cetaceans. This observation supports the hypothesis that cetaceans and hippos shared a common ancestor that was semiaquatic.

- Some whales have vestigial hip and limb bones as adults, and some dolphin embryos have vestigial hindlimb buds—outgrowths where legs form in other mammals.

The general message here is that many independent lines of evidence converge on the same conclusion: Whales gradually evolved from a terrestrial ancestor about 50 million years ago.

As you evaluate the evidence supporting the pattern component of the theory of evolution, though, it's important to recognize that no single observation or experiment instantly "proved" the fact of evolution and swept aside belief in special creation. Rather, data from many different sources are much more consistent with evolution than with special creation. Descent with modification is a successful and powerful scientific theory because it explains observations—such as vestigial traits and the close relationships among species on neighboring islands—that special creation does not.

What about the process component of the theory of evolution by natural selection? If the limbs of bats and humans were not created independently and recently, how did they come to be?

25.3 The Process of Evolution: How Does Natural Selection Work?

Darwin's greatest contribution did not lie in recognizing the fact of evolution. Lamarck and other researchers had proposed evolution long before Darwin began his work. Instead, Darwin's crucial insight lay in recognizing a process, called natural selection, that could explain the pattern of descent with modification.

Darwin's Inspiration

How did Darwin arrive at his insight? In part, he turned to pigeon breeding—a model system that would be easier to study and manipulate than populations in the wild. Pigeon breeding was popular in England at the time, offering a wealth of experience for Darwin to tap into. Also, pigeons could be maintained easily, and in Darwin's words, "the diversity of the breeds is something astonishing" (see examples in **FIGURE 25.12**).

Darwin crossbred pigeons and observed how characteristics were passed on to offspring. He could choose certain individuals with desirable traits to reproduce, thus manipulating the composition of the population by a process called **artificial selection.** It was clear to Darwin and other breeders that the diverse varieties were all descended from the wild rock pigeons.

Another influence on Darwin was the fortuitous publication of a book by Thomas Robert Malthus, *An Essay on the Principle of Population*, which inspired a great deal of discussion in England at the time. Malthus's studies of human populations in England and elsewhere led him to a startling conclusion: Since many more individuals are born than can survive, a "struggle for existence" occurs as people compete for food and places to live.

Darwin combined his observations of artificial selection with this notion of "struggle for existence" in natural populations, which he knew—from his extensive studies—contained variation. From this synthesis arose his concept of natural selection. Although both Darwin and Wallace arrived at the same idea, Darwin's name is more closely associated with the concept of natural selection because he thought of it first and provided extensive evidence for it in *On the Origin of Species*.

FIGURE 25.12 Diversity of Pigeon Breeds in Captivity. Darwin used the breeding of pigeons as a model system to study how the characteristics of populations can change over time.

Darwin's Four Postulates

Darwin broke the process of evolution by natural selection into four simple postulates (criteria) that form a logical sequence:

1. The individual organisms that make up a population vary in the traits they possess, such as their size and shape.

2. Some of the trait differences are heritable, meaning that they are passed on to offspring. For example, tall parents tend to have tall offspring.

3. In each generation, many more offspring are produced than can possibly survive. Thus, only some individuals in the population survive long enough to produce offspring, and among the individuals that produce offspring, some will produce more than others.

4. The subset of individuals that survive best and produce the most offspring is not a random sample of the population. Instead, individuals with certain heritable traits are more likely to survive and reproduce. **Natural selection** occurs when individuals with certain characteristics produce more offspring than do individuals without those characteristics. The individuals are selected naturally—meaning, by the environment.

Because the selected traits are passed on to offspring, the frequency of the selected traits increases from one generation to the next. We now know that traits are determined by alleles, particular versions of genes (see Chapter 14). Thus, the outcome of evolution by natural selection is a change in allele frequencies in a population over time.

In studying these criteria, you should realize that variation among individuals in a population is essential if evolution is to occur. Darwin had to introduce population thinking into biology because it is populations that change over time. To come up with these postulates and understand their consequences, Darwin had to think in a revolutionary way.

Today, biologists usually condense Darwin's four postulates into a two-part statement that communicates the essence of evolution by natural selection more forcefully: Evolution by natural selection occurs when (1) heritable variation leads to (2) differential reproductive success.

The Biological Definitions of Fitness, Adaptation, and Selection

To explain the process of natural selection, Darwin referred to successful individuals as "more fit" than other individuals. In doing so, he gave the word fitness a definition different from its everyday English usage. Biological **fitness** is the ability of an individual to produce surviving, fertile offspring relative to that ability in other individuals in the population.

Note that fitness is a measurable quantity. When researchers study a population in the lab or in the field, they can estimate the relative fitness of individuals by counting the number of surviving offspring each individual produces and comparing the data.

The concept of fitness, in turn, provides a compact way of formally defining adaptation. The biological meaning of adaptation,

like the biological meaning of fitness, is different from its normal English usage. In biology, an **adaptation** is a heritable trait that increases the fitness of an individual in a particular environment relative to individuals lacking the trait. Adaptations increase fitness—the ability to produce viable, fertile offspring. You can see the Big Picture of how adaptation and fitness relate to natural selection on pages 526–527.

Lastly, the term selection has a commonsense meaning in the context of artificial selection. Breeders *choose* which characteristics they want to keep or get rid of in their plant and animal breeds. However, the term selection has a very different meaning in the biological context of natural selection. Here, it refers to a passive process—differential reproduction as a result of heritable variation—not a purposeful choice.

25.4 Evolution in Action: Recent Research on Natural Selection

The theory of evolution by natural selection is testable. If the theory is correct, biologists should be able to test the validity of each of Darwin's postulates—documenting heritable variation and differential reproductive success in a wide array of natural populations.

This section summarizes two examples in which evolution by natural selection is being observed in nature. Literally hundreds of other case studies are available, involving a wide variety of traits and organisms. To begin, let's explore the evolution of drug resistance, one of the great challenges facing today's biomedical researchers and physicians.

Case Study 1: How Did *Mycobacterium tuberculosis* Become Resistant to Antibiotics?

Mycobacterium tuberculosis, the bacterium that causes **tuberculosis,** or TB, has long been a scourge of humankind. It usually infects the lungs and causes fever, coughing, sweats, weight loss, and often death. In Europe and the United States, TB was once as great a public health issue as cancer is now. It receded in importance during the early 1900s, though, for two reasons:

1. Advances in nutrition made people better able to fight off most *M. tuberculosis* infections quickly.

2. The development of antibiotics allowed physicians to stop even advanced infections.

In the late 1980s, however, rates of *M. tuberculosis* infection surged in many countries, and in 1993 the World Health Organization declared TB a global health emergency. Physicians were particularly alarmed because the strains of *M. tuberculosis* responsible for the increase were largely or completely resistant to antibiotics that were once extremely effective.

How and why did the evolution of drug resistance occur? The case of a single patient—a young man who lived in Baltimore—illustrates what is happening all over the world.

A Patient History The story begins when the individual was admitted to the hospital with fever and coughing. Chest X-rays, followed by bacterial cultures of fluid coughed up from the lungs, showed that he had an active TB infection. He was given several antibiotics for 6 weeks, followed by twice-weekly doses of the antibiotic rifampin for an additional 33 weeks. Ten months after therapy started, bacterial cultures from the patient's chest fluid indicated no *M. tuberculosis* cells. His chest X-rays were also normal. The antibiotics seemed to have cleared the infection.

Just two months after the TB tests proved normal, however, the young man was readmitted to the hospital with a fever, severe cough, and labored breathing. Despite being treated with a variety of antibiotics, including rifampin, he died of respiratory failure 10 days later. Samples of material from his lungs showed that *M. tuberculosis* was again growing actively there. But this time the bacterial cells were completely resistant to rifampin.

Drug-resistant *M. tuberculosis* cells had killed this patient. Where did they come from? Could a strain that was resistant to antibiotic treatment have evolved *within* him? To answer this question, a research team analyzed DNA from the drug-resistant strain and compared it with stored DNA from *M. tuberculosis* cells that had been isolated a year earlier from the same patient. After examining extensive stretches from each genome, the biologists were able to find only one difference: a point mutation in a gene called *rpoB*.

A Mutation in a Bacterial Gene Confers Resistance   The *rpoB* gene codes for a component of RNA polymerase. This enzyme transcribes DNA to mRNA, so it is essential to the survival and reproduction of bacterial cells (see Chapter 17). In this case, the point mutation in the *rpoB* gene changed a cytosine to a thymine, forming a new allele for the *rpoB* gene (see Chapter 16). This missense mutation caused a change in the amino acid sequence of the RNA polymerase (from a serine to a leucine at the 153rd amino acid)—and a change in its shape.

This shape change proved critical. Rifampin, the antibiotic that was being used to treat the patient, works by binding to the RNA polymerase of *M. tuberculosis* and interfering with transcription. Bacterial cells with the C → T mutation continue to produce offspring efficiently even in the presence of the drug because the drug cannot bind efficiently to the mutant RNA polymerase.

These results suggest the steps that led to this patient's death (**FIGURE 25.13**).

1. By chance, one or a few of the bacterial cells present in the patient, before drug therapy started, happened to have the *rpoB* allele with the C → T mutation. Under normal conditions, mutant forms of RNA polymerase do not work as well as the more common form, so cells with the C → T mutation would not produce many offspring and would stay at low frequency—even while the overall population grew to the point of inducing symptoms that sent the young man to the hospital.

2. Therapy with rifampin began. In response, cells in the population with normal RNA polymerase began to grow much more slowly or to die outright. As a result, the overall bacterial population declined in size so drastically that the patient appeared to be cured—his symptoms began to disappear.

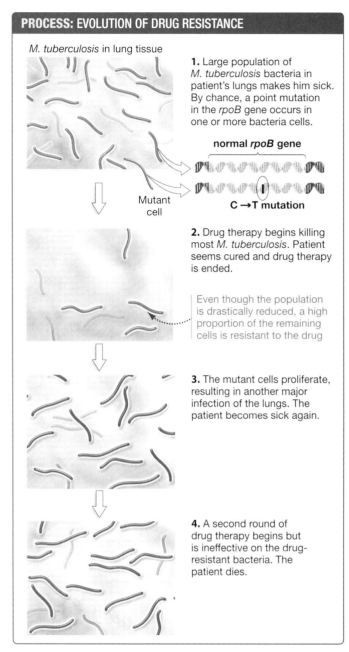

PROCESS: EVOLUTION OF DRUG RESISTANCE

M. tuberculosis in lung tissue

1. Large population of *M. tuberculosis* bacteria in patient's lungs makes him sick. By chance, a point mutation in the *rpoB* gene occurs in one or more bacteria cells.

normal *rpoB* gene

Mutant cell

C → T mutation

2. Drug therapy begins killing most *M. tuberculosis*. Patient seems cured and drug therapy is ended.

Even though the population is drastically reduced, a high proportion of the remaining cells is resistant to the drug

3. The mutant cells proliferate, resulting in another major infection of the lungs. The patient becomes sick again.

4. A second round of drug therapy begins but is ineffective on the drug-resistant bacteria. The patient dies.

FIGURE 25.13 Alleles That Confer Drug Resistance Increase in Frequency When Drugs Are Used.

3. Cells with the C → T mutation continued to increase in number after therapy ended. Eventually the *M. tuberculosis* population regained its former abundance, and the patient's symptoms reappeared.

4. Drug-resistant *M. tuberculosis* cells now dominated the population, so the second round of rifampin therapy was futile.

✔ If you understand these concepts, you should be able to explain: (1) Why the relapse in step 3 occurred, and (2) whether a family member or health-care worker who got TB from this patient at step 3 or step 4 would respond to drug therapy.

Testing Darwin's Postulates   Does the sequence of events illustrated in Figure 25.13 indicate that evolution by natural selection occurred? One way of answering this question is to review Darwin's four postulates and test whether each one was verified:

1. *Did variation exist in the population?*   The answer is yes. Due to mutation, both resistant and nonresistant strains of TB were present before administration of the drug. Most *M. tuberculosis* populations, in fact, exhibit variation for the trait; studies on cultured *M. tuberculosis* show that a mutation conferring resistance to rifampin is present in one out of every 10^7 to 10^8 cells.

2. *Was this variation heritable?*   The answer is yes. The researchers showed that the variation in the phenotypes of the two strains—from drug susceptibility to drug resistance—was due to variation in their genotypes. Because the mutant *rpoB* gene is passed on to daughter cells when a *Mycobacterium* replicates, the allele and the phenotype it produces—drug resistance—are passed on to offspring.

3. *Was there variation in reproductive success?*   The answer is yes. Only a tiny fraction of *M. tuberculosis* cells in the patient survived the first round of antibiotics long enough to reproduce. Most cells died and left no or almost no offspring.

4. *Did selection occur?*   The answer is yes. When rifampin was present, certain cells—those with the drug-resistant allele—had higher reproductive success than cells with the normal allele.

M. tuberculosis individuals with the mutant *rpoB* allele had higher fitness in an environment where rifampin was present. The mutant allele produces a protein that is an adaptation when the cell's environment contains the antibiotic.

This study verified all four postulates and confirmed that evolution by natural selection had occurred. The *M. tuberculosis* population evolved because the mutant *rpoB* allele increased in frequency.

It is critical to note, however, that the individual cells themselves did not evolve. When natural selection occurred, the individual bacterial cells did not change through time; they simply survived or died, or produced more or fewer offspring. This is a fundamentally important point: Natural selection acts on individuals, because individuals experience differential reproductive success. But only populations evolve. Allele frequencies change in populations, not in individuals. Understanding evolution by natural selection requires population thinking.

Drug Resistance: A Widespread Problem   The events reviewed for this single patient have occurred many times in other patients. Recent surveys indicate that drug-resistant strains now account for about 10 percent of the *M. tuberculosis*–causing infections throughout the world.

Unfortunately, the emergence of drug resistance in TB is far from unusual. Resistance to a wide variety of insecticides, fungicides, antibiotics, antiviral drugs, and herbicides has evolved in hundreds of insects, fungi, bacteria, viruses, and plants. In every case, evolution has occurred because individuals with the

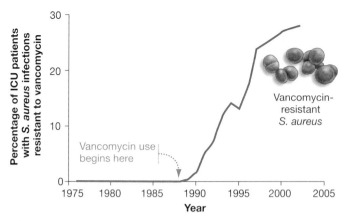

FIGURE 25.14 Trends in Infections Due to Antibiotic-resistant Bacteria. The line indicates changes in the percentage of *S. aureus* infections, acquired in hospitals, that are resistant to the antibiotic vancomycin.

DATA: Centers for Disease Control, 2004.

heritable ability to resist some chemical compound were present in the original population. As the susceptible individuals die from the pesticide, herbicide, or drug, the alleles that confer resistance increase in frequency.

To drive home the prevalence of evolution in response to drugs and other human-induced changes in the environment, consider the data in **FIGURE 25.14**. The graph shows changes through time in the percentage of infections, in intensive care units in the United States, caused by strains of the bacterium *Staphylococcus aureus* that are resistant to the antibiotic vancomycin. Most of these *S. aureus* cells are also resistant to methicillin and other antibiotics—a phenomenon known as multidrug resistance. In some cases, physicians have no effective antibiotics available to treat these infections. ✔ If you understand antibiotic resistance, you should be able to explain why the overprescription of antibiotics by doctors, or the overuse of everyday soaps and cleaners laced with antibiotics, can be a health risk.

Case Study 2: Why Are Beak Size, Beak Shape, and Body Size Changing in Galápagos Finches?

Can biologists study evolution in response to natural environmental change—when humans are not involved? The answer is yes. As an example, consider research led by Peter and Rosemary Grant. For over four decades, these biologists have been investigating changes in beak size, beak shape, and body size that have occurred in finches native to the Galápagos Islands.

Because the island of Daphne Major of the Galápagos is so small—about the size of 80 football fields—the Grants' team has been able to catch, weigh, and measure all the medium ground finches in the island's population (**FIGURE 25.15**) and mark each one with a unique combination of colored leg bands. The medium ground finch makes its living by eating seeds. Finches crack seeds with their beaks.

Early studies of the finch population established that beak size and shape and body size vary among individuals, and that beak

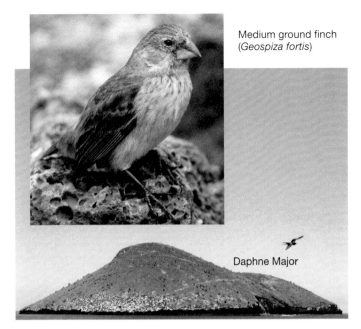

Medium ground finch
(*Geospiza fortis*)

Daphne Major

FIGURE 25.15 Studying Evolution in Action on the Galápagos.

morphology and body size are heritable. Stated another way, parents with particularly deep beaks tend to have offspring with deep beaks. Large parents also tend to have large offspring. Beak size and shape and body size are traits with heritable variation.

Selection during Drought Conditions Not long after the team began to study the finch population, a dramatic selection event occurred. In the annual wet season of 1977, Daphne Major received just 24 millimeters (mm) of rain instead of the 130 mm that normally falls. During the drought, few plants were able to produce seeds, and 84 percent (about 660 individuals) of the medium ground finch population disappeared.

Two observations support the hypothesis that most or all of these individuals died of starvation:

1. The researchers found 38 dead birds, and all were emaciated.

2. None of the missing individuals were spotted on nearby islands, and none reappeared once the drought had ended and food supplies returned to normal.

The research team realized that the die-off was a **natural experiment.** Instead of comparing groups created by direct manipulation under controlled conditions, natural experiments allow researchers to compare treatment groups created by an unplanned change in conditions. In this case, the Grants' team could test whether natural selection occurred by comparing the population before and after the drought.

Were the survivors different from nonsurvivors? The histograms in **FIGURE 25.16** show the distribution of beak sizes in the population before and after the drought. Notice the different scales of the *y*-axes of the two graphs. (For more on how histograms are constructed, see **BioSkills 3** in Appendix B.) On average, survivors tended to have much deeper beaks than did the birds that died.

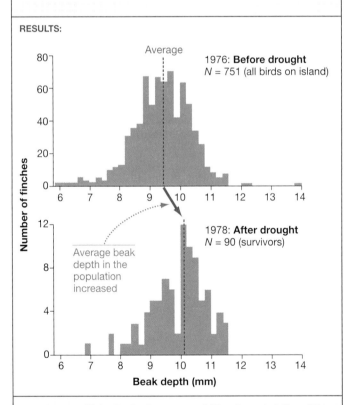

RESEARCH

QUESTION: Did natural selection on ground finches occur when the environment changed?

HYPOTHESIS: Beak characteristics changed in response to a drought.

NULL HYPOTHESIS: No changes in beak characteristics occurred in response to a drought.

EXPERIMENTAL SETUP:

Weigh and measure all birds in the population before and after the drought.

PREDICTION:

PREDICTION OF NULL HYPOTHESIS:

RESULTS:

CONCLUSION: Natural selection occurred. The characteristics of the population have changed.

FIGURE 25.16 A Natural Experiment: Changes in a Medium Ground Finch Population in Response to a Change in the Environment (a Drought). The results show the distribution of beak depth in the population of medium ground finches on Daphne Major before and after the drought of 1977. *N* is the population size.

SOURCE: Boag, P. T., and P. R. Grant. 1981. Intense natural selection in a population of Darwin's finches (Geospizinae) in the Galápagos. *Science* 214: 82–85.

✔ **EXERCISE** Fill in the predictions made by the two hypotheses.

Why were deeper beaks adaptive? At the peak of the drought, most seed sources were absent and the tough fruits of a plant called *Tribulus cistoides* served as the finches' primary food source. These fruits are so difficult to crack that finches ignore them in years when food supplies are normal. The Grants hypothesized that individuals with particularly large and deep beaks were more likely to crack these fruits efficiently enough to survive.

At this point, the Grants had shown that natural selection led to an increase in average beak depth in the population. When breeding resumed in 1978, the offspring produced had beaks that were half a millimeter deeper, on average, than those in the population that existed before the drought. This result confirmed that evolution had occurred.

In only one generation, natural selection led to a measurable change in the characteristics of the population. Alleles that led to the development of deep beaks had increased in frequency in the population. Large, deep beaks were an adaptation for cracking large fruits and seeds.

Continued Changes in the Environment, Continued Selection, Continued Evolution In 1983, the environment on the Galápagos Islands changed again. Over a seven-month period, a total of 1359 mm of rain fell. Plant growth was luxuriant, and finches fed primarily on the small, soft seeds that were being produced in abundance. During this interval, small individuals with small, pointed beaks had exceptionally high reproductive success—meaning that they had higher fitness than those with large, deep beaks—because they were better able to harvest the small seeds. As a result, the characteristics of the population changed again. Alleles associated with small, pointed beaks increased in frequency.

Over subsequent decades, the Grants have documented continued evolution in response to continued changes in the environment. **FIGURE 25.17** documents changes that have occurred in average body size, beak size, and beak shape over 35 years. From 1972 to 2006, average body size got smaller and average beak size initially increased and then declined. In addition, finch beaks got much pointier.

Long-term studies like this are proving to be powerful because they have succeeded in documenting natural selection in response to changes in the environment.

Which Genes Are under Selection? Characteristics like body size, beak size, and beak shape are polygenic, meaning that many genes—each one exerting a relatively small effect—influence the trait (see Chapter 14). Because many genes are involved, it can be difficult for researchers to know exactly which alleles are changing in frequency when polygenic traits evolve.

To explore which of the medium ground finch's genes might be under selection, researchers in Clifford Tabin's lab began studying beak development in an array of Galápagos finch species. More specifically, they looked for variation in the pattern of expression of cell–cell signals that had already been identified as important in the development of chicken beaks. The hope was that homologous genes might affect beak development in finches.

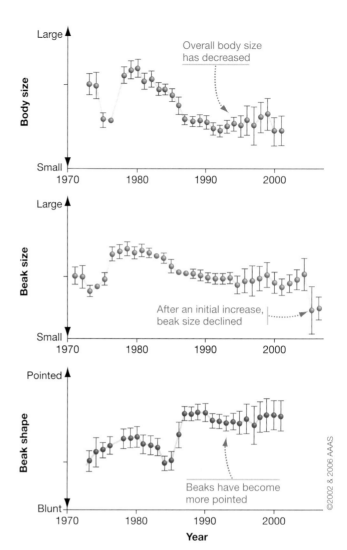

FIGURE 25.17 Body Size, Beak Size, and Beak Shape in Finches Changed over a 35-Year Interval.

DATA: Grant, P. R., and B. R. Grant. 2002. *Science* 296: 707–711; Grant, P. R., and B. R. Grant. 2006. *Science* 313: 224–226.

The researchers struck pay dirt when they carried out in situ hybridization (a technique featured in Chapter 22) showing where a cell–cell signal gene called *Bmp4* is expressed.

- There is a strong correlation between the amount of *Bmp4* expression when beaks are developing in young Galápagos finches and the width and depth of adult beaks (**FIGURE 25.18**).

- When the researchers experimentally increased *Bmp4* expression in young chickens, they found that beaks got wider and deeper than normal.

Similar experiments suggest that variation in alleles for a molecule called calmodulin, which is involved in calcium signaling during development, affects beak length. Based on these data, biologists suspect that alleles associated with *Bmp4* and calmodulin expression may be under selection in the population of medium ground finches that the Grants are studying. If so, the research community will have made a direct connection between natural selection on phenotypes and evolutionary change in genotypes.

Lower *Bmp4* expression
(dark area) in embryo's beak

Higher *Bmp4* expression
(dark area) in embryo's beak

2 mm

2 mm

Shallow
adult beak

Deep
adult
beak

Geospiza fortis

Geospiza magnirostris

FIGURE 25.18 Changes in *Bmp4* Expression Change Beak Depth and Width. These micrographs are in situ hybridizations (see Chapter 22) showing the location and extent of *Bmp4* expression in young *Geospiza fortis* and *G. magnirostris*. In these and four other species that were investigated, the amount of Bmp4 protein produced correlates with the depth and width of the adult beak.

check your understanding

If you understand that . . .

- If individuals with certain alleles produce the most offspring in a population, then those alleles increase in frequency over time. Evolution results from this process of natural selection on heritable variation.

✓ **You should be able to . . .**

1. List Darwin's four postulates in your own words and indicate which are related to heritable variation and which are related to differential reproductive success.

2. Explain how data on beak size and shape of Galápagos finch populations provide examples of heritable variation and differential reproductive success.

Answers are available in Appendix A.

25.5 Common Misconceptions about Natural Selection and Adaptation

Evolution by natural selection is a simple process—just the logical outcome of some straightforward postulates. Ironically, it can be extremely difficult to understand.

Research has verified that evolution by natural selection is often misunderstood. To help clarify how the process works, let's consider four of the most common types of misconceptions about natural selection, summarized in **TABLE 25.2** (on page 460).

Selection Acts on Individuals, but Evolutionary Change Occurs in Populations

Perhaps the most important point to clarify about natural selection is that during the process, individuals do not change—only the population does. During the drought, the beaks of individual finches did not become deeper. Rather, the average beak depth in the population increased over time because deep-beaked individuals produced more offspring than shallow-beaked individuals did. Natural selection acted on individuals, but the evolutionary change occurred in the characteristics of the population.

In the same way, individual *M. tuberculosis* cells did not change when rifampin was introduced to their environment. Each of these bacterial cells had the same RNA polymerase alleles throughout its life. But because the mutant allele increased in frequency in the population over time, the average characteristics of the bacterial population changed.

Natural Selection Is Not "Lamarckian" Inheritance There is a sharp contrast between evolution by natural selection and evolution by the inheritance of acquired characters—the hypothesis promoted by Jean-Baptiste de Lamarck. If you recall, Lamarck proposed that **(1)** individuals change in response to challenges posed by the environment (such as giraffes stretching their necks to reach leaves high in the treetops), and **(2)** the changed traits are then passed on to offspring. The key claim is that the important evolutionary changes occur in individuals.

In contrast, Darwin realized that individuals do not change when they are selected. Instead, they simply produce more or less offspring than other individuals do. When this happens, alleles found in the selected individuals become more or less frequent in the population.

Darwin was correct: There is no mechanism that makes it possible for natural selection to change the nature of an allele inside an individual. An individual's heritable characteristics don't change when natural selection occurs. Natural selection just sorts existing variants—it doesn't change them.

Acclimatization Is *Not* Adaptation The issue of change in individuals is tricky because individuals often *do* change in response to changes in the environment. For example, if you were to travel to the Tibetan Plateau in Asia, your body would experience oxygen deprivation due to the low partial pressure of oxygen at high elevations (see Figure 45.2). As a result, your body would produce more of the oxygen-carrying pigment hemoglobin and more hemoglobin-carrying red blood cells. Your body does not normally produce more red blood cells than it needs, because viscous (thick) blood can cause a disease—chronic mountain sickness—that can lead to heart failure.

The increase in red blood cells is an example of what biologists call **acclimatization**—a change in an individual's phenotype that occurs in response to a change in natural environmental conditions. (When this process occurs in study organisms in a laboratory, it is called **acclimation**.) Phenotypic changes due to acclimatization are not passed on to offspring, because no alleles have changed. As a result, acclimatization does not cause evolution.

Misconception	Example
"Evolutionary change occurs in organisms" CORRECTION: • Natural selection just sorts existing variants in organisms; it doesn't change them • Evolutionary change occurs only in populations • Acclimatization ≠ adaptation Selection does not cause neck length to increase in individual giraffes, only in populations	
"Adaptations occur because organisms want or need them" CORRECTION: • Mutation, the source of new alleles, occurs by chance • Evolution is not goal directed or progressive • There is no such thing as a higher or lower organism Tapeworms are not "lower" than their human hosts, just adapted to a different environment	
"Organisms sacrifice themselves for the good of the species" CORRECTION: • Individuals with alleles that cause self-sacrificing behavior die and do not produce offspring, so these alleles are eliminated from the population Lemmings do not jump off cliffs into the sea to save the species	
"Evolution perfects organisms" CORRECTION: • Some traits are nonadaptive • Some traits cannot be optimized due to fitness trade-offs • Some traits are limited by genetic or historical constraints Finch beaks cannot be both deep and narrow, due to genetic constraints	

In contrast, populations that have lived in Tibet for many generations are adapted to this environment through genetic changes. Among native Tibetans, for example, an allele that increases the ability of hemoglobin to hold oxygen has increased to high frequency. In populations that do not live at high elevations, this allele is rare or nonexistent. ✔ If you understand this concept, you should be able to explain the difference between the biological definition of adaptation and its use in everyday English.

Evolution Is Not Goal Directed

It is tempting to think that evolution by natural selection is goal directed. For example, you might hear a fellow student say that Tibetans "needed" the new hemoglobin allele so that they could survive at high altitudes, or that *M. tuberculosis* cells "wanted" or "needed" the mutant, drug-resistant allele so that they could survive and continue to reproduce in an environment that included rifampin. This purposeful change does not happen. The mutations that created the mutant alleles in both examples occurred randomly, due to errors in DNA synthesis, and they just happened to be advantageous when the environments changed.

Stated another way, mutations do not occur to solve problems. Mutations just happen. Every mutation is equally likely to occur in every environment. There is no mechanism that enables the environment to direct which mistakes DNA polymerase makes when copying genes. Adaptations do not occur because organisms want or need them.

Evolution Is Not Progressive It is often appealing to think that evolution by natural selection is progressive—meaning organisms have gotten "better" over time. (In this context, *better* usually means bigger, stronger, or more complex.) It is true that the groups appearing later in the fossil record are often more morphologically complex than closely related groups that appeared earlier. Flowering plants are considered more complex than mosses, and most biologists would agree that the morphology of mammals is more complex than that of the first vertebrates in the fossil record. But there is nothing predetermined or absolute about this tendency.

In fact, complex traits are routinely lost or simplified over time as a result of evolution by natural selection. You've already analyzed evidence documenting limb loss in snakes (Chapter 22) and whales (this chapter). Populations that become parasitic are particularly prone to loss of complex traits. For example, the tapeworm parasites of humans and other mammals have lost their sophisticated digestive tracts and mouths as a result of natural selection—tapeworms simply absorb nutrients directly from their environment.

There Is No Such Thing as a Higher or Lower Organism The nonprogressive nature of evolution by natural selection contrasts sharply with Lamarck's conception of the evolutionary process, in which organisms progress over time to higher and higher levels on a chain of being (see Figure 25.1).

Under Aristotle's and Lamarck's hypothesis, it is sensible to refer to "higher" and "lower" organisms. But under evolution by natural selection, there is no such thing as a higher or lower organism (**FIGURE 25.19**). Mosses may be a more ancient group than flowering plants, but neither group is higher or lower than the other. Mosses simply have a different suite of adaptations than do flowering plants, so they thrive in different types of environments. A human is no higher than its tapeworm parasite; each is well adapted to its environment.

Organisms Do Not Act for the Good of the Species

Consider the widely circulated story that rodents called lemmings sacrifice themselves for the good of their species. The story claims that when lemming populations are high, overgrazing is so extensive that the entire species is threatened with starvation and extinction. In response, some individuals throw themselves into the sea and drown. This lowers the overall population size and allows the vegetation to recover enough to save the species. Even though individuals suffer, the good-of-the-species hypothesis maintains that the behavior evolved because the group benefits.

The lemming suicide story is false. Although lemmings do disperse from areas of high population density to find habitats with higher food availability, they do not throw themselves into the sea.

To understand why this type of self-sacrificing behavior does not occur, suppose that certain alleles predispose lemmings to sacrifice themselves for others. But consider what happens if alleles exist that prevent this type of behavior—what biologists call

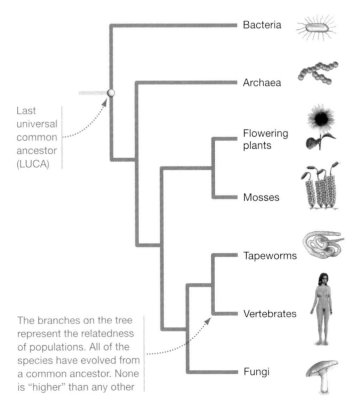

FIGURE 25.19 Evolution Produces a Tree of Life, Not a Progressive Ladder of Life. Under evolution by natural selection, species are related by common ancestry and all have evolved through time. (Not all branches of the tree of life are shown.)

a "selfish" allele. Individuals with self-sacrificing alleles die and do not produce offspring. But individuals with selfish, cheater alleles survive and produce offspring. As a result, selfish alleles increase in frequency while self-sacrificing alleles decrease in frequency. Thus, it is not possible for self-sacrificing alleles to evolve by natural selection.

There Are Constraints on Natural Selection

Although organisms are often exquisitely adapted to their environment, adaptation is far from perfect. A long list of circumstances limits the effectiveness of natural selection; only a few of the most important are discussed here.

Nonadaptive Traits Vestigial traits such as the human coccyx (tailbone) and goose bumps do not increase the fitness of individuals with those traits. The structures are not adaptive. They exist because they were present in the ancestral population.

Vestigial traits are not the only types of structures with no or reduced function. Some adult traits exist as holdovers from structures that appear early in development. For example, human males have nipples despite the absence of mammary glands. Nipples exist in men because they form in the human embryo before sex hormones begin directing the development of male organs instead of female organs.

Perhaps the best example of nonadaptive traits involves evolutionary changes in DNA sequences. A mutation may change a base in the third position of a codon without changing the amino acid sequence of the protein encoded by that gene. Changes such as these are said to be silent. They occur due to the redundancy of the genetic code (see Chapter 16). Silent changes in DNA sequences are extremely common. But because they don't change the phenotype, they can't be acted on by natural selection and are not adaptive.

Genetic Constraints The Grants' team analyzed data on the characteristics of finches that survived the 1977 drought in the Galápagos, and the team made an interesting observation: Although individuals with deep beaks survived better than individuals with shallow beaks, birds with particularly narrow beaks survived better than individuals with wider beaks.

This observation made sense because finches crack *Tribulus* fruits by twisting them. Narrow beaks concentrate the twisting force more efficiently than wider beaks, so they are especially useful for cracking the fruits. But narrower beaks did not evolve in the population.

To explain why, the biologists noted that parents with deep beaks tend to have offspring with beaks that are both deep and wide. This is a common pattern. Many alleles that affect body size have an effect on all aspects of size—not just one structure or dimension. As a result, selection for increased beak depth overrode selection for narrow beaks, even though a deep and narrow beak would have been more advantageous.

The general point here is that selection was not able to optimize all aspects of a trait due to **genetic correlation**. Genetic correlations occur because of pleiotropy, in which a single allele affects multiple traits (see Chapter 14). In this case, selection on alleles for one trait (increased beak depth) caused a correlated, though suboptimal, increase in another trait (beak width).

Genetic correlations are not the only genetic constraint on adaptation. Lack of genetic variation is also important. Consider that salamanders have the ability to regrow severed limbs. Some eels and sharks can sense electric fields. Birds can see ultraviolet light. Even though these traits would possibly confer increased reproductive success in humans, they do not exist—because humans lack the requisite genes.

Fitness Trade-offs   In everyday English, the term trade-off refers to a compromise between competing goals. It is difficult to design a car that is both large and fuel efficient, a bicycle that is both rugged and light, or a plane that is both fast and maneuverable.

In nature, selection occurs in the context of fitness trade-offs. A **fitness trade-off** is a compromise between traits, in terms of how those traits perform in the environment. During the drought in the Galápagos, for example, medium ground finches with large bodies had an advantage because they were able to chase off smaller birds from the few remaining sources of seeds. But individuals with large bodies require large amounts of food to maintain their mass; they also tend to be slower and less nimble than smaller individuals. When food is scarce, large individuals are more prone to starvation. Even if large size is advantageous in an environment, there is always counteracting selection that prevents individuals from getting even bigger.

Biologists have documented trade-offs between the size of eggs or seeds that an individual makes and the number of offspring it can produce, between rapid growth and long life span, and between bright coloration and tendency to attract predators.

The message of this research is simple: Because selection acts on many traits at once, every adaptation is a compromise.

Historical Constraints   In addition to being constrained by genetic correlations, lack of genetic variation, and fitness trade-offs, adaptations are constrained by history. The reason is simple: All traits have evolved from previously existing traits.

Natural selection acts on structures that originally had a very different function. For example, the tiny incus, malleus, and stapes bones found in your middle ear evolved from bones that were part of the jaw and jaw support in the ancestors of mammals. These bones now function in transmitting and amplifying sound from your outer ear to your inner ear. Biologists routinely interpret these bones as adaptations that improve your ability to hear airborne sounds. But are the bones a "perfect" solution to the problem of transmitting sound from the outside of the ear to the inside? The answer is no. They are the best solution possible, given an important historical constraint. Other vertebrates have different structures involved in transmitting sound to the ear. In at least some cases, those structures may be more efficient than our incus, malleus, and stapes.

To summarize, not all traits are adaptive, and even adaptive traits are constrained by genetic and historical factors. In addition, natural selection is not the only process that causes evolutionary change. Three other processes—genetic drift, gene flow, and mutation—change allele frequencies over time (see Chapter 26). Compared with natural selection, these processes have very different consequences. You can see the Big Picture of how natural selection relates to other evolutionary processes on pages 526–527.

CHAPTER 25   REVIEW   *For media, go to MasteringBiology*

If you understand . . .

25.1 The Evolution of Evolutionary Thought

- Plato, Aristotle, and the Bible's book of Genesis consider species as unchanging types. This view is called typological thinking.

- Lamarck proposed a theory of evolution—that species are not static but change through time. He proposed that evolution occurs by the inheritance of acquired characteristics.

- Darwin and Wallace proposed that evolution occurs by natural selection. This was the beginning of population thinking, whereby variation among individuals is the key to understanding evolution.

✔ You should be able to compare and contrast typological thinking and population thinking.

25.2 The Pattern of Evolution: Have Species Changed, and Are They Related?

- Data on (1) the age of the Earth and the fact of extinction; (2) the resemblance of modern to fossil forms; (3) transitional features in fossils; (4) the presence of vestigial traits; and (5) change in contemporary populations show that species change through time.

- Data on (1) the geographic proximity of closely related species; (2) the existence of structural, developmental, and genetic homologies; and (3) the contemporary formation of new species support the consensus that species are related by common ancestry.

- Evidence for evolution is internally consistent, meaning that data from several independent sources are mutually reinforcing.

✔ You should be able to cite examples in support of the statement that species have changed through time and are related by common ancestry.

25.3 The Process of Evolution: How Does Natural Selection Work?

- Darwin developed four postulates that outline the process of evolution by natural selection. These postulates can be summarized by the following statement: Heritable variation leads to differential reproductive success.

- Alleles or traits that increase the reproductive success of an individual are said to increase the individual's fitness. A trait that leads to higher fitness, relative to individuals without the trait, is an adaptation. If a particular allele increases fitness and leads to adaptation, the allele will increase in frequency in the population.

✔ You should be able to explain the difference between the biological and everyday English definitions of fitness.

25.4 Evolution in Action: Recent Research on Natural Selection

- Selection by drugs on the TB bacterium and changes in the size and shape of finch beaks in the Galápagos as a result of seed availability are well-studied examples of evolution by natural selection.

- Both examples demonstrate that evolution can be observed and measured. Evolution by natural selection has been confirmed by a wide variety of studies and has long been considered to be the central organizing principle of biology.

✔ You should be able to predict how changes in *Mycobacterium tuberculosis* populations would be explained under special creation and under evolution by inheritance of acquired characters.

25.5 Common Misconceptions about Natural Selection and Adaptation

- Natural selection acts on individuals, but evolutionary change occurs in populations. Nonheritable changes that occur in individuals are not adaptations and do not result in evolution.

- Evolution is not goal directed and does not lead to perfection. There is no such thing as a higher or lower organism.

- Organisms do not act for the good of the species.

- Not all traits are adaptive, and even adaptive traits are limited by genetic and historical constraints.

✔ You should be able to discuss how adaptations such as the large brains of *Homo sapiens* and the ability of falcons to fly very fast are constrained.

(MB) MasteringBiology

1. MasteringBiology Assignments

Tutorials and Activities Darwin and the Galápagos Islands; Experimental Inquiry: Did Natural Selection of Ground Finches Occur When the Environment Changed?; Natural Selection for Antibiotic Resistance; The Voyage of the *Beagle*: Darwin's Trip Around the World; Reconstructing Forelimbs

Questions Reading Quizzes, Blue-Thread Questions, Test Bank

2. eText Read your book online, search, take notes, highlight text, and more.

3. The Study Area Practice Test, Cumulative Test, BioFlix® 3-D Animations, Videos, Activities, Audio Glossary, Word Study Tools, Art

You should be able to . . .

✔ TEST YOUR KNOWLEDGE

Answers are available in Appendix A

1. How can biological fitness be estimated?
   a. Document how long different individuals in a population survive.
   b. Count the number of healthy, fertile offspring produced by different individuals in a population.
   c. Determine which individuals are strongest.
   d. Determine which phenotype is the most common one in a given population.

2. True or false? Some traits are considered vestigial because they existed long ago.

3. What is an adaptation?
   a. a trait that improves the fitness of its bearer, compared with individuals without the trait
   b. a trait that changes in response to environmental influences within the individual's lifetime
   c. an ancestral trait—one that was modified to form the trait observed today
   d. the ability to produce offspring

4. Why does the presence of extinct forms and transitional features in the fossil record support the pattern component of the theory of evolution by natural selection?
   a. It supports the hypothesis that individuals change over time.
   b. It supports the hypothesis that weaker species are eliminated by natural selection.
   c. It supports the hypothesis that species evolve to become more complex and better adapted over time.
   d. It supports the hypothesis that species change over time.

5. Traits that are derived from a common ancestor, like the bones of human arms and bird wings, are said to be _____.

6. According to data presented in this chapter, which of the following statements is correct?
   a. When individuals change in response to challenges from the environment, their altered traits are passed on to offspring.
   b. Species are created independently of each other and do not change over time.
   c. Populations—not individuals—change when natural selection occurs.
   d. The traits of populations become more perfect over time.

7. Some biologists summarize evolution by natural selection with the phrase "mutation proposes, selection disposes." Mutation is a process that creates heritable variation. Explain what the phrase means.

8. Explain how artificial selection differs from natural selection.

9. Why don't the biggest and strongest individuals in a population always produce the most offspring?
   a. The biggest and strongest individuals always have higher fitness.
   b. In some environments, being big and strong lowers fitness.
   c. Sometimes the biggest and strongest individuals may choose to have fewer offspring.
   d. Sometimes the number of offspring is not related to fitness.

10. **QUANTITATIVE** The graphs in Figure 25.16 show that the average beak depth of medium ground finches increased after the drought.

However, more finches had deep beaks before the drought than after. Explain this seeming contradiction by calculating the percent of finches that survived the drought.

11. Review the section on the evolution of drug resistance in *Mycobacterium tuberculosis*.
    - What evidence do researchers have that a drug-resistant strain evolved in the patient analyzed in their study, and wasn't instead transmitted from another infected individual?
    - If the antibiotic rifampin were banned, would the mutant *rpoB* gene have lower or higher fitness in the new environment? Would strains carrying the mutation continue to increase in frequency in *M. tuberculosis* populations?

12. Describe how Darwin's four postulates would apply to a population of rabbits sharing a meadow with foxes.

13. Scientists have observed white deer mice living on coastal beaches in Florida and brown deer mice living in nearby forests. Compare and contrast how the theory of evolution by natural selection, special creation, and evolution by inheritance of acquired characters might explain this observation.

14. The average height of humans in industrialized nations has increased steadily for the past 100 years. This trait has clearly changed over time. Most physicians and human geneticists hypothesize that the change is due to better nutrition and a reduced incidence of disease. Has human height evolved?
    a. Yes, because average height has changed over time.
    b. No, because changes in height due to nutrition and reduced incidence of disease are not heritable.
    c. Yes, because height is a heritable trait.
    d. No, because height is not a heritable trait.

15. Scientists hypothesize that humans and chimpanzees diverged from a common ancestor that lived in Africa about 6–7 million years ago. What evidence would support this hypothesis?

16. The geneticist James Crow wrote that successful scientific theories have the following characteristics: (1) They explain otherwise puzzling observations; (2) they provide connections between otherwise disparate observations; (3) they make predictions that can be tested; and (4) they are heuristic, meaning that they open up new avenues of theory and experimentation. Crow added two other elements of scientific theories that he considered important on a personal, emotional level: (5) They should be elegant, in the sense of being simple and powerful; and (6) they should have an element of surprise. How well does the theory of evolution by natural selection fulfill these six criteria?

26 Evolutionary Processes

In this chapter you will learn that

Four evolutionary processes change allele frequencies in populations

↓ by starting with a

Null hypothesis: The Hardy–Weinberg Principle   26.1

which makes five assumptions with respect to a particular gene . . .

Nonrandom mating   26.2   no

no → **Natural selection   26.3**

no → **Genetic drift   26.4**

no → **Gene flow   26.5**

no → **Mutation   26.6**

Four evolutionary processes

This albino sea turtle carries rare alleles that prevent pigment formation. The frequency of these alleles changes in the sea turtle population over time due to the evolutionary processes discussed in this chapter.

S ince evolution is one of the five key attributes of life, understanding evolutionary processes is essential to understanding living things. How did the diversity of organisms around us (including us) come to be?

It turns out that natural selection (see Chapter 25) is only one of four processes that can shift allele frequencies in populations over time, causing evolution:

1. *Natural selection* increases the frequency of certain alleles—the ones that contribute to reproductive success in a particular environment. Natural selection is the only one of the four processes that leads to adaptation.

2. *Genetic drift* causes allele frequencies to change randomly. In some cases, drift may cause alleles that decrease fitness to increase in frequency.

This chapter is part of the Big Picture. See how on pages 526–527.

✔ When you see this checkmark, stop and test yourself. Answers are available in Appendix A.

3. *Gene flow* occurs when individuals leave one population, join another, and breed. Allele frequencies may change when gene flow occurs, because arriving individuals introduce alleles to their new population and departing individuals remove alleles from their old population.

4. *Mutation* modifies allele frequencies by continually introducing new alleles. The alleles created by mutation may be beneficial or deleterious (detrimental) or have no effect on fitness.

This chapter has two fundamental messages: Natural selection is not the only agent responsible for evolution, and each of the four evolutionary processes has different consequences for genetic variation and fitness.

The first few decades of the 1900s were pivotal for the study of evolutionary processes. Biologists began to apply Mendelian genetics (Chapter 14) to Darwinian evolution (Chapter 25), resulting in an era known as the Modern Synthesis. Evolutionary biologists, mathematicians, and geneticists collaborated to make huge leaps in *quantifying* evolution. One product of this era was a model called the **Hardy–Weinberg principle,** which serves as a mathematical null hypothesis for the study of evolutionary processes. Let's examine the logic of this model and then consider each of the four evolutionary processes in turn.

26.1 Analyzing Change in Allele Frequencies: The Hardy–Weinberg Principle

In 1908, a British mathematician, G. H. Hardy, and a German doctor, Wilhelm Weinberg, each published a major result independently. At the time, it was commonly believed that changes in allele frequency occur simply as a result of sexual reproduction—meiosis followed by the random fusion of gametes (eggs and sperm) to form offspring. Some biologists claimed that dominant alleles inevitably increase in frequency when gametes combine at random. Others predicted that two alleles of the same gene inevitably reach a frequency of 0.5.

To test these hypotheses, Hardy and Weinberg analyzed the frequencies of alleles when many individuals in a population mate and produce offspring. Instead of thinking about the consequences of a mating between two parents with a specific pair of genotypes (as you did with Punnett squares in Chapter 14), Hardy and Weinberg wanted to know what happened in an entire population when *all* of the individuals—and thus all possible genotypes—bred.

Like Darwin, Hardy and Weinberg were engaged in population thinking. A **population** is a group of individuals from the same species that live in the same area at the same time and can interbreed—and that vary in the traits they possess (Chapter 25).

The Gene Pool Concept

To analyze the consequences of matings among all of the individuals in a population, Hardy and Weinberg invented a novel approach: They imagined that all of the alleles from all the gametes produced in each generation go into a single group called the **gene pool** and then combine at random to form offspring. Something similar to this happens in species like clams and sea stars, which release their gametes into the water where the gametes mix randomly before combining to form zygotes.

To determine which genotypes would be present in the next generation and in what frequencies, Hardy and Weinberg calculated what happens when two gametes are plucked at random out of the gene pool, many times, and each of these gamete pairs then combines to form offspring. These calculations predict the genotypes of the offspring that would be produced, as well as the frequency of each genotype.

Quantitative Methods 26.1 walks through Hardy's and Weinberg's calculations by focusing on just one gene with two alleles, A_1 and A_2. The letter p is used to symbolize the frequency of A_1 alleles in the gene pool, and q is used to symbolize the frequency of A_2 alleles in the same gene pool. **FIGURE 26.1** illustrates the same calculations a little differently. The figure uses a Punnett square to predict the outcome of random mating—meaning, random combinations of all gametes in a population. (Recall that in Chapter 14 you used Punnett squares to predict the outcome of a mating between two individuals.) The outcome of this analysis is the same as in Quantitative Methods 26.1.

The resulting Hardy–Weinberg principle makes two fundamental claims:

1. If the frequencies of alleles A_1 and A_2 in a population are given by p and q, then the frequencies of genotypes A_1A_1,

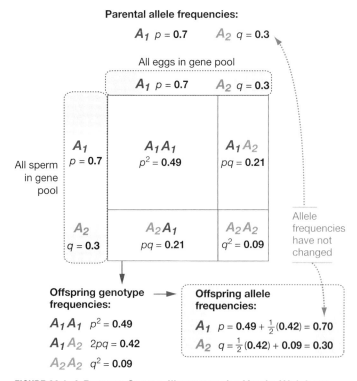

Parental allele frequencies:

A_1 $p = 0.7$ A_2 $q = 0.3$

All eggs in gene pool

	A_1 $p = 0.7$	A_2 $q = 0.3$
A_1 $p = 0.7$	A_1A_1 $p^2 = 0.49$	A_1A_2 $pq = 0.21$
A_2 $q = 0.3$	A_2A_1 $pq = 0.21$	A_2A_2 $q^2 = 0.09$

All sperm in gene pool

Allele frequencies have not changed

Offspring genotype frequencies:

A_1A_1 $p^2 = 0.49$
A_1A_2 $2pq = 0.42$
A_2A_2 $q^2 = 0.09$

Offspring allele frequencies:

A_1 $p = 0.49 + \frac{1}{2}(0.42) = 0.70$
A_2 $q = \frac{1}{2}(0.42) + 0.09 = 0.30$

FIGURE 26.1 A Punnett Square Illustrates the Hardy–Weinberg Principle.

Hardy and Weinberg began by analyzing the simplest situation possible—where just two alleles of a particular gene exist in a population. Let's call these alleles A_1 and A_2. The letter p is used to symbolize the frequency of A_1 alleles in the gene pool and q is used to symbolize the frequency of A_2 alleles in the same gene pool. Because there are only two alleles, the two frequencies must add up to 1:

$$p + q = 1$$

Now follow the steps in **FIGURE 26.2**:

Parental Allele Frequencies

Although p and q can have any value between 0 and 1, let's suppose that the initial frequency of A_1 is 0.7 and that of A_2 is 0.3. In this gene pool, 70 percent of the gametes carry A_1 and 30 percent carry A_2.

Offspring Genotype Frequencies

Each time a gamete is involved in forming an offspring, there is a 70 percent chance that it carries A_1 and a 30 percent chance that it carries A_2. In general, there is a p chance that it carries A_1 and a q chance that it carries A_2.

Because only two alleles are present, three genotypes are possible in the offspring generation: A_1A_1, A_1A_2, and A_2A_2. What will the frequency of these three genotypes be? According to the logic of Hardy's and Weinberg's result:

- Frequency of the A_1A_1 genotype: p^2
- Frequency of the A_1A_2 genotype: $2pq$
- Frequency of the A_2A_2 genotype: q^2

The genotype frequencies in the offspring generation must add up to 1, which means that:

$$p^2 + 2pq + q^2 = 1$$

In this numerical example:

$$0.49 + 0.42 + 0.09 = 1$$

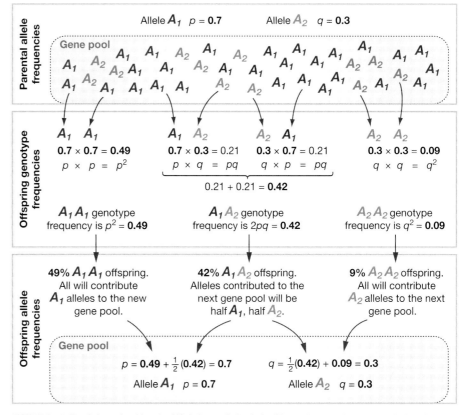

FIGURE 26.2 Deriving the Hardy–Weinberg Principle. To understand the logic behind calculating the frequency of A_1A_2 genotypes, see **BioSkills 5** in Appendix B.

Offspring Allele Frequencies

The easiest way to calculate the allele frequencies in the offspring is to imagine that they form gametes that go into a gene pool. All of the gametes from A_1A_1 individuals carry A_1, so 49 percent (p^2) of the gametes in the gene pool will carry A_1. But half of the gametes from A_1A_2 will also carry A_1, so an additional ½(0.42) = 0.21 (this is ½ × $2pq$ = pq) gametes in the gene pool will carry A_1, for a total of 0.49 + 0.21 = 0.70 or $p^2 + pq = p(p + q) = p$. Use the same logic to figure out the frequency of A_2.

In this example, the frequency of allele A_1 in the offspring generation is still 0.7 and the frequency of allele A_2 is still 0.3. Thus, the

frequency of allele A_1 is still p and the frequency of allele A_2 is still q.

No change in allele frequency occurred. Even if A_1 is dominant to A_2, it does not increase in frequency. And there is no trend toward both alleles reaching a frequency of 0.5. This result is called the Hardy–Weinberg principle.

✓**QUANTITATIVE** If you understand these calculations, you should be able to: (1) calculate the frequencies of the three offspring genotypes if $p = 0.6$ and $q = 0.4$ in the parental gene pool; (2) use your answer to calculate the allele frequencies in the offspring gene pool; (3) determine whether evolution occurred.

A_1A_2, and A_2A_2 will be given by p^2, $2pq$, and q^2, respectively, for generation after generation. That is:

$$\text{Allele frequencies: } p + q = 1$$
$$\text{Genotype frequencies: } p^2 + 2pq + q^2 = 1$$

2. When alleles are transmitted via meiosis and random combinations of gametes, their frequencies do not change over time. For evolution to occur, some other factor or factors must come into play.

What are these other factors?

The Hardy–Weinberg Model Makes Important Assumptions

The Hardy–Weinberg model is based on important assumptions about how populations and alleles behave. These assumptions helped to define the four processes of evolution that can be acting on the population. In addition, the model assumes that mating is random with respect to the gene in question. Here are the five assumptions that must be met:

1. *Random mating* This condition was enforced by picking gametes from the gene pool at random. Individuals were not allowed to *choose* a mate.

2. *No natural selection* The model assumed that *all* members of the parental generation survived and contributed equal numbers of gametes to the gene pool, no matter what their genotype.

3. *No genetic drift (random allele frequency changes)* The model assumed that alleles were picked in their exact frequencies p and q, and not at some different values caused by chance—that is, the model behaved as though the population was infinitely large. For example, allele A_1 did not "get lucky" and get drawn more than 70 percent of the time.

4. *No gene flow* No new alleles were added by immigration or lost through emigration. As a result, all of the alleles in the offspring population came from the original population's gene pool.

5. *No mutation* The model didn't consider that new A_1s or A_2s or other, new alleles might be introduced into the gene pool.

The Hardy–Weinberg principle tells us what to expect if selection, genetic drift, gene flow, and mutation are not affecting a gene, *and* if mating is random with respect to that gene. Under these conditions, the genotypes A_1A_1, A_1A_2, and A_2A_2 should be in the Hardy–Weinberg proportions p^2, $2pq$, and q^2, respectively, and no evolution will occur.

How Does the Hardy–Weinberg Principle Serve as a Null Hypothesis?

Recall that a null hypothesis predicts there are no differences among the treatment groups in an experiment (see Chapter 1). Biologists often want to test whether nonrandom mating is occurring, natural selection is acting on a particular gene, or one of the other evolutionary processes is at work. In addressing questions like these, the Hardy–Weinberg principle functions as a null hypothesis.

Given a set of allele frequencies, the Hardy–Weinberg principle predicts what genotype frequencies will occur when mating is random with respect to that gene, and when natural selection, genetic drift, gene flow, and mutation are not affecting the gene. If biologists observe genotype frequencies that do not conform to the Hardy–Weinberg prediction, it means that something interesting is going on: Either nonrandom mating is occurring (which changes genotype frequencies but not allele frequencies), or allele frequencies are changing for some reason. Further research is then needed to determine which of the five Hardy–Weinberg conditions is being violated.

Let's consider two examples to illustrate how the Hardy–Weinberg principle is used as a null hypothesis: MN blood types and *HLA* genes, both in humans.

Case Study 1: Are MN Blood Types in Humans in Hardy–Weinberg Proportions? One of the first genes that geneticists could analyze in natural populations was the MN blood group of humans. Most human populations have two alleles, designated *M* and *N*, at this gene.

Because the *MN* gene codes for a protein found on the surface of red blood cells—the *M* allele codes for the M version, and the *N* allele codes for the N version—researchers could determine whether individuals are *MM*, *MN*, or *NN* by treating blood samples with antibodies to each protein (this technique was first introduced in Chapter 11). The *M* and *N* alleles are codominant—meaning that heterozygotes have both M and N versions of the protein on their red blood cells.

To estimate the frequency of each genotype in a population, geneticists obtain data from a large number of individuals and then divide the number of individuals with each genotype by the total number of individuals in the sample.

TABLE 26.1 shows *MN* genotype frequencies for populations from throughout the world and illustrates how observed genotype frequencies are compared with the genotype frequencies expected if the Hardy–Weinberg principle holds. The analysis is based on the following steps:

Step 1 Estimate genotype frequencies by observation—in this case, by testing many blood samples for the *M* and *N* alleles. These frequencies are given in the rows labeled "observed" in Table 26.1.

Step 2 Calculate observed allele frequencies from the observed genotype frequencies. In this case, the frequency of the *M* allele is the frequency of *MM* homozygotes plus half the frequency of *MN* heterozygotes; the frequency of the *N* allele is the frequency of *NN* homozygotes plus half the frequency of *MN* heterozygotes. (You can review the logic behind this calculation in Quantitative Methods 26.1.)

Step 3 Use the observed allele frequencies to calculate the genotypes expected according to the Hardy–Weinberg principle. Under the null hypothesis of random mating and no evolution, the expected genotype frequencies are $p^2 : 2pq : q^2$.

Step 4 Compare the observed and expected values. Researchers use statistical tests to determine whether the differences between the observed and expected genotype frequencies are small enough to be due to chance or large enough to reject the null hypothesis of no evolution and random mating.

Although using statistical testing is beyond the scope of this text (see **BioSkills 4** in Appendix B for a brief introduction to the topic), you should be able to inspect the numbers and comment on them. In these populations, for example, the observed and expected *MN* genotype frequencies are almost identical. (A statistical test shows that the small differences observed are probably due to chance.) For every population surveyed, genotypes

TABLE 26.1 The MN Blood Group of Humans: Observed and Expected Genotype Frequencies

The expected genotype frequencies are calculated from the observed allele frequencies, using the Hardy–Weinberg principle.

Population and Location	Data Type	Genotype Frequencies			Allele Frequencies	
		MM	*MN*	*NN*	*M*	*N*
Inuit (Greenland)	Observed	0.835	0.156	0.009	0.913	0.087
	Expected	0.834	0.159	0.008		
Native Americans (U.S.)	Observed	0.600	0.351	0.049	0.776	0.224
	Expected	0.602	0.348	0.050		
Caucasians (U.S.)	Observed	0.292	0.494	0.213	0.540	0.460
	Expected	0.290	0.497	0.212		
Aborigines (Australia)	Observed	0.025	0.304	0.672	0.176	0.824
	Expected	0.031	0.290	0.679		
Ainu (Japan)	Observed — Step ❶→	0.179	0.502	0.319 —Step ❷→	▓▓▓	▓▓▓
	Expected — Step ❹↕	▓▓▓	▓▓▓	▓▓▓ ←Step ❸		

DATA: W. C. Boyd. 1950. Boston: Little, Brown and Company.

✔**QUANTITATIVE** Fill in the values for observed allele frequencies and expected genotype frequencies for the Ainu people of Japan.

at the *MN* gene are in Hardy–Weinberg proportions. As a result, biologists conclude that the assumptions of the Hardy–Weinberg model are valid for this locus.

The results imply that when these data were collected, the *M* and *N* alleles in these populations were not being affected by the four evolutionary processes and that mating was random with respect to this gene—meaning that humans were not choosing mates based on their *MN* genotype.

Before moving on, however, it is important to note that a study such as this does not mean that the *MN* gene has never been subject to nonrandom mating, or has never been under selection or genetic drift. Even if selection has been very strong for many generations, one generation of random mating and no evolution will result in genotype frequencies that conform to Hardy–Weinberg expectations.

Case Study 2: Are *HLA* Genes in Humans in Hardy–Weinberg Equilibrium?
Geneticist Therese Markow and colleagues collected data on the genotypes of 122 individuals from the Havasupai tribe native to Arizona. These biologists were studying two genes that are important in the functioning of the human immune system. More specifically, the genes that they analyzed code for proteins that help immune system cells recognize and destroy invading bacteria and viruses.

Previous work had shown that different alleles exist at both the *HLA-A* and *HLA-B* genes, and that the alleles at each gene code for proteins that recognize proteins from slightly different disease-causing organisms. Like the *M* and *N* alleles, *HLA* alleles are codominant.

As a result, the research group hypothesized that individuals who are heterozygous at one or both of these genes may have a strong fitness advantage. The logic is that heterozygous people possess a wider variety of HLA proteins, so their immune systems can recognize and destroy more types of bacteria and viruses. They should be healthier and have more offspring than homozygous people do.

To test this hypothesis, Markow and her colleagues used their data on observed genotype frequencies to determine the frequency of each allele present. When they used these allele frequencies to calculate the expected number of each genotype according to the Hardy–Weinberg principle, they found the observed and expected values reported in **TABLE 26.2**.

When you inspect these data, notice that there are many more heterozygotes and many fewer homozygotes than expected under Hardy–Weinberg conditions. Statistical tests show it is extremely unlikely that the difference between the observed and expected numbers could occur purely by chance.

These results supported the team's prediction and indicated that one of the assumptions behind the Hardy–Weinberg

TABLE 26.2 *HLA* Genes of Humans: Observed and Expected Genotypes

The expected numbers of homozygous and heterozygous genotypes are calculated from observed allele frequencies, according to the Hardy–Weinberg principle.

Gene	Data Type	Genotype Counts (*n* = 122)	
		Homozygotes	Heterozygotes
HLA-A	Observed	38	84
	Expected	48	74
HLA-B	Observed	21	101
	Expected	30	92

DATA: Markow, T., P. H. Hedrick, K. Zuerlein, et al. 1993. *Journal of Human Genetics* 53: 943–952, Table 3.

principle was being violated. But which assumption? The researchers argued that mutation, gene flow, and drift are negligible in this case and offered two explanations for their data:

1. *Mating may not be random with respect to the HLA genotype.* Specifically, people may subconsciously prefer mates with *HLA* genotypes unlike their own and thus produce an excess of heterozygous offspring. This hypothesis is plausible. For example, experiments have shown that college students can distinguish each others' genotypes at genes related to *HLA* on the basis of body odor. Individuals in this study were more attracted to the smell of genotypes unlike their own. If this is true among the Havasupai, then nonrandom mating would lead to an excess of heterozygotes compared with the proportion expected under the Hardy–Weinberg principle.

2. *Heterozygous individuals may have higher fitness.* This hypothesis is supported by data collected by a different research team, who studied the Hutterite people living in South Dakota. In the Hutterite population, married women who have the same *HLA*-related alleles as their husbands have more trouble getting pregnant and experience higher rates of spontaneous abortion than do women with *HLA*-related alleles different from those of their husbands. The data suggest that homozygous fetuses have lower fitness than do fetuses heterozygous at these genes. If this were true among the Havasupai, selection would lead to an excess of heterozygotes relative to Hardy–Weinberg expectations.

Which explanation is correct? It is possible that both are. Using the Hardy–Weinberg principle as a null hypothesis allowed biologists to detect an interesting pattern in a natural population. Research continues on the question of why the pattern exists.

Now let's consider each of the processes that can violate the Hardy–Weinberg assumptions—and thus influence evolution.

check your understanding

If you understand that . . .

- The Hardy–Weinberg principle functions as a null hypothesis when researchers test whether nonrandom mating or evolution is occurring at a particular gene.

✔ QUANTITATIVE You should be able to . . .

Analyze whether a gene suspected of causing hypertension in humans is in Hardy–Weinberg proportions. In one study, the observed genotype frequencies were A_1A_1 0.574; A_1A_2 0.339; A_2A_2 0.087. (Note: The sample size in this study was so large that a difference of 3 percent or more in any of the observed versus expected frequencies indicated a statistically significant difference—meaning, a difference that was not due to chance.)

Answers are available in Appendix A.

26.2 Nonrandom Mating

In the Hardy–Weinberg model, gametes were picked from the gene pool at random and paired to create offspring genotypes. In nature, however, matings between individuals may not be random with respect to the gene in question. Even in species like clams that simply broadcast their gametes into the surrounding water, gametes from individuals that live close to each other are more likely to combine than gametes from individuals that live farther apart.

The most intensively studied form of nonrandom mating is called **inbreeding,** the mating between relatives. Since relatives share a recent common ancestor, individuals that inbreed are likely to share alleles they inherited from their common ancestor.

How Does Inbreeding Affect Allele Frequencies and Genotype Frequencies?

To understand how inbreeding affects populations, let's follow the fate of alleles and genotypes when inbreeding occurs. As before, focus on a single locus with two alleles, A_1 and A_2, and suppose that these alleles initially have equal frequencies of 0.5.

Now imagine that the gametes produced by individuals in the population don't go into a gene pool. Instead, individuals self-fertilize. Many flowering plants, for example, contain both male and female organs and routinely self-pollinate. Self-fertilization, or selfing, is the most extreme form of inbreeding.

As **FIGURE 26.3a** shows, homozygous parents that self-fertilize produce all homozygous offspring. Heterozygous parents, in contrast, produce homozygous and heterozygous offspring in a $1:2:1$ ratio.

FIGURE 26.3b shows the outcome for the population as a whole. In this figure, the width of the boxes represents the frequency of the three genotypes, which start out at the Hardy–Weinberg ratio of $p^2:2pq:q^2$. Notice that the homozygous proportion of the population increases with each generation, while the heterozygous proportion is halved. At the end of the four generations illustrated, heterozygotes are rare. The same outcomes occur more slowly with less extreme forms of inbreeding.

This simple exercise demonstrates two fundamental points about inbreeding:

1. Inbreeding increases homozygosity. In essence, inbreeding takes alleles from heterozygotes and puts them into homozygotes.

2. Inbreeding itself does not cause evolution, because allele frequencies do not change in the population as a whole.

Nonrandom mating changes only genotype frequencies, not allele frequencies, so is not an evolutionary process itself. ✔ If you understand this concept, you should be able to predict how observed genotype frequencies and allele frequencies should differ from those expected under the Hardy–Weinberg principle when inbreeding is occurring.

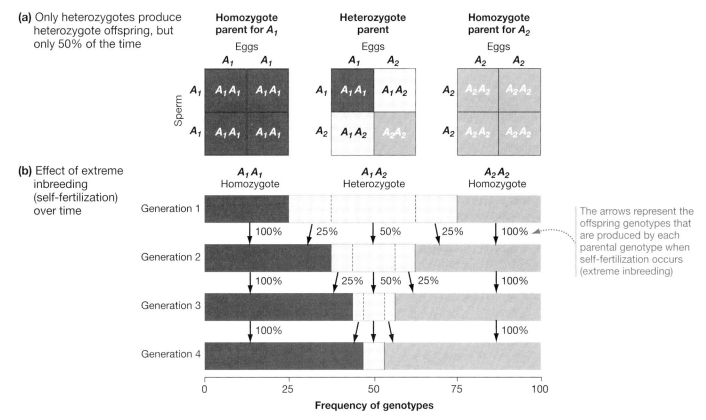

(a) Only heterozygotes produce heterozygote offspring, but only 50% of the time

(b) Effect of extreme inbreeding (self-fertilization) over time

The arrows represent the offspring genotypes that are produced by each parental genotype when self-fertilization occurs (extreme inbreeding)

Frequency of genotypes

FIGURE 26.3 Inbreeding Increases Homozygosity and Decreases Heterozygosity. (a) Heterozygous parents produce homozygous and heterozygous offspring in a 1 : 2 : 1 ratio. **(b)** The width of the boxes corresponds to the frequency of each genotype.

How Does Inbreeding Influence Evolution?

The trickiest point to grasp about inbreeding is that even though it does not cause evolution directly—because it does not change allele frequencies—it can speed the rate of evolutionary change. More specifically, it increases the rate at which natural selection eliminates recessive **deleterious** alleles—alleles that lower fitness—from a population.

Consider **inbreeding depression**—the decline in average fitness that takes place when homozygosity increases and heterozygosity decreases in a population. Inbreeding depression results from two causes:

1. *Many recessive alleles represent loss-of-function mutations.*
Because these alleles are usually rare, there are normally very few homozygous recessive individuals in a population. Instead, most loss-of-function alleles exist in heterozygous individuals. The alleles have little or no effect when they occur in heterozygotes, because one normal allele usually produces enough functional protein to support a normal phenotype. But inbreeding increases the frequency of homozygous recessive individuals. Loss-of-function mutations are usually deleterious or even lethal when they are homozygous—they are quickly eliminated by selection.

2. *Many genes—especially those involved in fighting disease—are under intense selection for heterozygote advantage, a selection process that favors genetic diversity.* If an individual is homozygous at these genes, then fitness declines.

The upshot here is that the offspring of inbred matings are expected to have lower fitness than the offspring of outcrossed matings. This prediction has been verified in a wide variety of species. **FIGURE 26.4** shows results from recent efforts to reduce illness in a small population of endangered Florida panthers by introducing females from Texas. Note that the two sets of data points compare the fitnesses of offspring from non-inbred

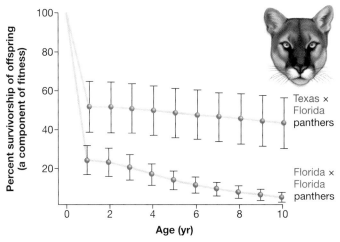

FIGURE 26.4 Inbreeding Depression Occurs in Florida Panthers. Inbreeding depression is the fitness difference between non-inbred and inbred individuals.

DATA: Johnson, W. E., E. P. Onorato, M. E. Roelke, et al. 2010. *Science* 329: 1641–1645.

Texas × Florida matings and inbred Florida × Florida matings. Inbreeding depression is represented by the vertical distance between data points at each age.

Several studies show similar results on inbreeding depression in human populations around the world. Children of first-cousin marriages consistently have a higher mortality rate than children of marriages between nonrelatives. Because inbreeding has such deleterious consequences in humans, it is not surprising that many human societies have laws forbidding marriages between individuals who are related as first cousins or closer.

In insects, vertebrates, and many other animals where inbreeding is uncommon, females often don't mate at random but actively choose certain males, and/or males compete among themselves to secure mates. This form of nonrandom mating is fundamentally different from inbreeding because it *does* lead to changes in allele frequencies in the population, and thus is a form of natural selection—called **sexual selection.** According to the Hardy–Weinberg assumptions, sexual selection violates the "no natural selection" hypothesis, as discussed in the next section.

26.3 Natural Selection

Evolution by natural selection occurs when heritable variation leads to differential success in survival and reproduction (see Chapter 25). Stated another way, natural selection occurs when individuals with certain phenotypes produce more surviving offspring than individuals with other phenotypes do. If certain alleles are associated with the favored phenotypes, they increase in frequency while other alleles decrease in frequency. The result is evolution—a violation of the assumptions of the Hardy–Weinberg model.

How Does Selection Affect Genetic Variation?

When biologists analyze the consequences of selection, they often focus on **genetic variation**—the number and relative frequency of alleles that are present in a particular population. The reason is simple: Lack of genetic variation in a population is usually a bad thing.

To understand why this is so, recall that selection can occur only if heritable variation exists in a population (Chapter 25). If genetic variation is low and the environment changes—perhaps due to the emergence of a new disease-causing virus, a rapid change in climate, or a reduction in the availability of a particular food source—no alleles that have high fitness under the new conditions are likely to be present. As a result, the average fitness of the population will decline. If the environmental change is severe enough, the population may even be faced with extinction.

Natural selection occurs in a wide variety of patterns, or modes, each with different consequences to genetic variation:

- *Directional selection* changes the average value of a trait.

- *Stabilizing selection* reduces variation in a trait.

- *Disruptive selection* increases variation in a trait.

- *Balancing selection* maintains variation in a trait.

Let's take a closer look at each of these four modes in turn.

Mode 1: Directional Selection When **directional selection** occurs, the average phenotype of a population changes in one direction.

To get a sense of how directional selection affects genetic variation, look at the top graph in **FIGURE 26.5a**, which plots the value of a trait on the *x*-axis and the number of individuals with a particular value of that trait on the *y*-axis. (In a graph like this, the *y*-axis could also plot the frequency of individuals with a

(a) Directional selection changes the average value of a trait.

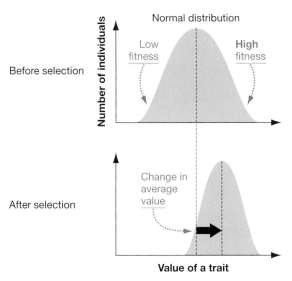

(b) For example, directional selection caused average body size to increase in a cliff swallow population.

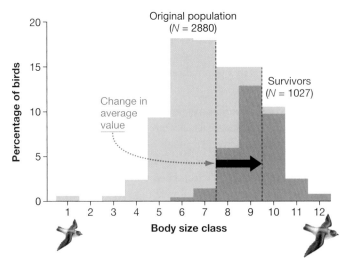

FIGURE 26.5 Directional Selection. When directional selection acts on traits that have a normal distribution, individuals at one end of the distribution experience poor reproductive success.

DATA: Brown, C. R., and M. B. Brown. 1998. *Evolution* 52: 1461–1475.

particular trait value—see **BioSkills 3** in Appendix B.) Note that the trait in question is determined by multiple genes, so it has a bell-shaped, normal distribution in this hypothetical population (Chapter 14).

The second graph in the figure shows what happens when directional selection acts on this trait. Directional selection tends to reduce the genetic diversity of populations. If directional selection continues over time, the favored alleles will eventually approach a frequency of 1.0 while disadvantageous alleles will approach a frequency of 0.0, a clear violation of the assumptions of the Hardy–Weinberg model. Alleles that reach a frequency of 1.0 are said to be fixed; those that reach a frequency of 0.0 are said to be lost. When disadvantageous alleles decline in frequency, **purifying selection** is said to occur.

Now consider data from real organisms in the wild. In 1996 a population of cliff swallows native to the Great Plains of North America endured a six-day period of exceptionally cold, rainy weather. Cliff swallows feed by catching mosquitoes and other insects in flight. Insects disappeared during this cold snap, however, and the biologists recovered the bodies of 1853 swallows that died of starvation.

As soon as the weather improved, the researchers caught and measured the body size of 1027 survivors from the same population. As the histograms in **FIGURE 26.5b** show, survivors were much larger on average than the birds that died.

Directional selection, favoring large body size, had occurred. To explain this observation, the investigators suggested that larger birds survived because they had larger fat stores and did not get as cold as the smaller birds. As a result, the larger birds were less likely to die of exposure to cold and more likely to avoid starvation until the weather warmed up and insects were again available.

If variation in swallow body size was heritable, and if this type of directional selection continued, then alleles that contribute to small body size would quickly decline in frequency in the cliff swallow population.

As it turned out in this case, directional selection was short-lived because the severe weather event was temporary. In other cases, directional selection can be more persistent, such as the directional selection that occurs during global climate change.

Mode 2: Stabilizing Selection   When cliff swallows were exposed to cold weather, selection greatly reduced one extreme in the range of phenotypes and resulted in a directional change in the average characteristics of the population. But selection can also reduce both extremes in a population, as illustrated in **FIGURE 26.6a**. This mode of selection is called **stabilizing selection.** It has two important consequences: There is no change in the average value of a trait over time, and genetic variation in the population is reduced.

FIGURE 26.6b shows a classical data set in humans that illustrates stabilizing selection. Biologists who analyzed birth weights and mortality in 13,730 babies born in British hospitals in the 1940s found that babies of average size (slightly over 7 pounds)

survived best. Mortality was high for very small babies and very large babies. This is persuasive evidence that birth weight was under strong stabilizing selection in this population. Alleles associated with high birth weight or low birth weight were subject to purifying selection, and alleles associated with intermediate birth weight increased in frequency.

Mode 3: Disruptive Selection   **Disruptive selection** has the opposite effect of stabilizing selection. Instead of favoring phenotypes near the average value and eliminating extreme phenotypes, it eliminates phenotypes near the average value and favors

(a) Stabilizing selection reduces the amount of variation in a trait.

(b) For example, very small and very large babies are the most likely to die, leaving a narrower distribution of birthweights.

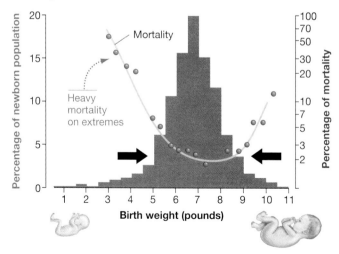

FIGURE 26.6 Stabilizing Selection. When stabilizing selection acts on normally distributed traits, individuals with extreme phenotypes experience poor reproductive success.

DATA: Karn, M. N., H. Lang-Brown, J. J. MacKenzie, et al. 1951. *Annals of Eugenics* 15: 306–322.

extreme phenotypes (**FIGURE 26.7a**). When disruptive selection occurs, the overall amount of genetic variation in the population is increased.

Recent research has shown that disruptive selection is responsible for the striking distribution of bills of black-bellied seedcrackers (**FIGURE 26.7b**). The data plotted in the graph show that individuals with either short or very long beaks survive best and that birds with intermediate phenotypes are at a disadvantage.

In this case, the agent that causes natural selection is food. At a study site in south-central Cameroon, West Africa, a researcher found that only two sizes of seed are available to the seedcrackers: large and small. Birds with small beaks crack and eat small

(a) Disruptive selection increases the amount of variation in a trait.

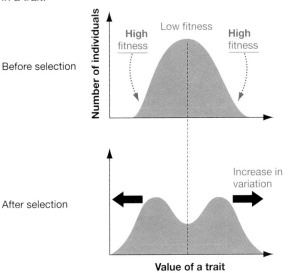

(b) For example, only juvenile black-bellied seedcrackers that had short or extremely long beaks survived long enough to breed.

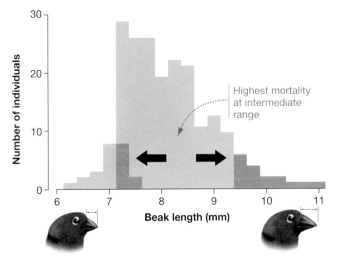

FIGURE 26.7 Disruptive Selection. When disruptive selection occurs on traits with a normal distribution, individuals with extreme phenotypes experience high reproductive success.
DATA: Smith, T. B. 1987. *Nature* 329: 717–719.

seeds efficiently. Birds with large beaks handle large seeds efficiently. But birds with intermediate beaks have trouble with both, so alleles associated with medium-sized beaks are subject to purifying selection. Disruptive selection maintains high overall variation in this population.

Disruptive selection is important because it sometimes plays a part in speciation, or the formation of new species. If small-beaked seedcrackers began mating with other small-beaked individuals, their offspring would tend to be small-beaked and would feed on small seeds. Similarly, if large-beaked individuals chose only other large-beaked individuals as mates, they would tend to produce large-beaked offspring that would feed on large seeds.

In this way, selection would result in two distinct populations. Under some conditions, the populations may eventually form two new species. (See Chapter 27 for a detailed discussion of the process of species formation, which is based on disruptive selection and other mechanisms.)

Mode 4: Balancing Selection   Directional selection, stabilizing selection, and disruptive selection describe how natural selection can act on one polygenic trait in a single generation or episode. Another mode of selection, called **balancing selection,** occurs when no single allele has a distinct advantage. Instead, there is a balance among several alleles in terms of their fitness and frequency. Balancing selection occurs when:

1. Heterozygous individuals have higher fitness than homozygous individuals do, a pattern called **heterozygote advantage**. This pattern of selection is one explanation for the data in Table 26.2, where heterozygotes for the *HLA* genes have a fitness advantage compared to homozygotes. The consequence of this pattern is that genetic variation is maintained in populations.

2. The environment varies over time or in different geographic areas occupied by a population—meaning that certain alleles are favored by selection at different times or in different places. As a result, overall genetic variation in the population is maintained or increased.

3. Certain alleles are favored when they are rare, but not when they are common—a pattern known as **frequency-dependent selection**. For example, rare alleles responsible for coloration in guppies are favored because predators learn to recognize common color patterns. Alleles for common colors get eliminated; alleles for rare colors increase in frequency. As a result, overall genetic variation in the population is maintained or increased.

TABLE 26.3 summarizes the four modes of natural selection. No matter how natural selection occurs, though, its most fundamental attribute is the same: It increases fitness and leads to adaptation. Natural selection results in allele frequencies that deviate from those predicted by the Hardy–Weinberg principle because selection favors certain alleles over others. ✔ If you understand this concept, you should be able to predict how genotype frequencies differ from Hardy–Weinberg proportions under directional selection.

Mode of Selection	Effect on Phenotype	Effect on Genetic Variation
Directional selection	Favors one extreme phenotype, causing the average phenotype in the population to change in one direction.	Genetic variation is reduced.
Stabilizing selection	Favors phenotypes near the middle of the range of phenotypic variation, maintaining average phenotype.	Genetic variation is reduced.
Disruptive selection	Favors extreme phenotypes at both ends of the range of phenotypic variation.	Genetic variation is increased.
Balancing selection	No single phenotype is favored in all populations of a species at all times.	Genetic variation is maintained.

Sexual Selection

Darwin was the first to recognize that selection based on success in courtship is a mechanism of evolutionary change. Although he introduced the idea of sexual selection in *On the Origin of Species* in 1859, it was not until 12 years later that he detailed his views on sexual selection in *The Descent of Man, and Selection in Relation to Sex.*

Darwin was initially perplexed by seemingly nonadaptive traits such as the extravagantly long and iridescent trains of peacocks. Why was it adaptive for the peacock to spend energy producing a structure that would increase the peacock's vulnerability to predation? Darwin concluded that the peacock's train is adaptive because it helps the peacock to attract mates, thereby increasing its reproductive success.

The peacock example demonstrates mate choice, sometimes referred to as *inter*sexual selection—the selection of an individual of one gender for mating by an individual of the other gender. Darwin also recognized that individuals sometimes compete with one another to obtain mates. This form of selection *within* a gender is referred to as *intra*sexual selection.

Theory: The Fundamental Asymmetry of Sex Why would the extravagant trains of peacocks occur in the males but not the peahens? A. J. Bateman addressed this question in the 1940s, followed by Robert Trivers three decades later. The Bateman–Trivers theory contains an observed pattern and a hypothesized process:

- *Pattern* Sexual selection usually acts on males much more strongly than on females. Traits that attract members of the opposite sex are much more highly elaborated in males.

- *Process* The energetic cost of creating a large egg is enormous, whereas a sperm contains few energetic resources. That is, "eggs are expensive, but sperm are cheap."

In most species, females invest much more in their offspring than do males. This phenomenon is called the fundamental asymmetry of sex. It is characteristic of almost all sexual species and has two important consequences:

1. Because eggs are large and energetically expensive, females produce relatively few young over the course of a lifetime. A female's fitness is limited not by her ability to find a mate but primarily by her ability to gain the resources needed to produce more eggs and healthier young.

2. Sperm are so energetically inexpensive to produce that a male can father an almost limitless number of offspring. Thus, a male's fitness is limited not by the ability to acquire the resources needed to produce sperm but by the number of females he can mate with.

The Bateman–Trivers theory of sexual selection makes strong predictions:

- If females invest a great deal in each offspring, then they should protect that investment by being choosy about their mates. Conversely, if males invest little in each offspring, then they should be willing to mate with almost any female.

- If there are an equal number of males and females in the population, and if males are trying to mate with any female possible, then males will compete with each other for mates.

- If male fitness is limited by access to mates, then any allele that increases a male's attractiveness to females or success in male–male competition should increase rapidly in the population, violating the assumptions of Hardy–Weinberg. Thus, sexual selection should act more strongly on males than on females.

Do data from experimental or observational studies agree with these predictions? Let's consider each of them in turn.

Female Choice for "Good Alleles" If females are choosy about which males they mate with, what criteria do females use to make their choice? Recent experiments have shown that in several bird species, females prefer to mate with males that are well fed and in good health. These experiments were motivated by three key observations:

1. In many bird species, the existence of colorful feathers or a colorful beak is due to the presence of the red and yellow pigments called carotenoids.

2. Carotenoids protect tissues and stimulate the immune system to fight disease more effectively.

3. Animals usually cannot synthesize their own carotenoids, but plants can. To obtain carotenoids, animals have to eat carotenoid-rich plant tissues.

These observations suggest that the healthiest and best-nourished birds in a population have the most colorful beaks and feathers. Sick birds have dull coloration because they are using all of their carotenoids to stimulate their immune system. Poorly fed birds have dull coloration because they have few carotenoids available. By choosing a colorful male as the father of her offspring, a female is likely to have offspring with alleles that will help the offspring fight disease effectively and feed efficiently.

To test the hypothesis that females prefer colorful males, a team of researchers experimented with zebra finches (**FIGURE 26.8a**). They identified pairs of brothers and randomly assigned one brother to the treatment group and one brother to the control group. They fed the treatment group a diet that was heavily supplemented with carotenoids, and they fed the control group a diet that was similar in every way except for the additional carotenoids.

As predicted, the males eating the carotenoid-supplemented diet developed more colorful beaks than did the males fed the carotenoid-poor diet. When given a choice of perching with either of the two brothers, 9 out of the 10 females tested preferred the more-colorful male (**FIGURE 26.8b**). These results support the hypothesis that females of this species are choosy about their mates and that they prefer to mate with healthy, well-fed males.

Enough experiments have been done on other bird species to support a general conclusion: Colorful beaks and feathers, along with songs and dances and other types of courtship displays, carry the message "I'm healthy and well fed because I have good alleles. Mate with me."

Female Choice for Paternal Care   Choosing "good alleles" is not the entire story in sexual selection via female choice. In many species, females prefer to mate with males that care for young or that provide the resources required to produce eggs.

Brown kiwi females make an enormous initial investment in their offspring—their eggs routinely represent over 15 percent of the mother's total body weight (**FIGURE 26.9**)—but choose to mate with males that take over all of the incubation and other care of the offspring. It is common to find that female fish prefer to mate with males that protect a nest site and care for the eggs until they hatch. In humans and many species of birds, males provide food, protection, and other resources for rearing young.

To summarize, females may choose mates on the basis of (**1**) physical characteristics that signal male genetic quality, (**2**) resources or parental care provided by males, or (**3**) both. In some species, however, females do not have the luxury of choosing a male. Instead, competition among males is the primary cause of sexual selection.

Male–Male Competition   As an example of research on how males compete for mates, consider data from a long-term study of a northern elephant seal population breeding on Año Nuevo Island, off the coast of California.

Elephant seals feed mostly on marine fish, squid, and octopus, and spend most of the year in the water. But when females are ready to mate and give birth, they haul themselves out of the

(a) Male zebra finch

(b) Female choice for colorful beaks

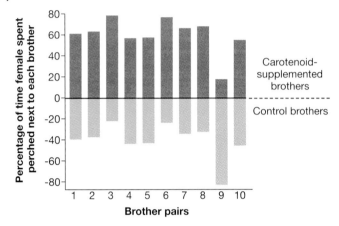

FIGURE 26.8 Female Zebra Finches Prefer Males with Colorful Beaks.

DATA: Blount, J. D., N. B. Metcalf, T. R. Birkhead, et al. 2003. *Science* 300: 125–127.

water onto land. Females prefer to give birth on islands, where newborn pups are protected from terrestrial and marine predators. Because elephant seals have flippers that are ill-suited for walking, females can haul themselves out of the water only on the few beaches that have gentle slopes. As a result, large numbers of females congregate in tiny areas to breed.

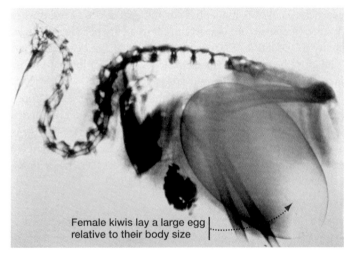

FIGURE 26.9 In Many Species, Females Make a Large Investment in Each Offspring. X-ray of a female kiwi, ready to lay an egg.

Male elephant seals establish territories on breeding beaches by fighting (**FIGURE 26.10a**). A **territory** is an area that is actively defended and that provides exclusive or semi-exclusive use by the owner. Males that win battles with other males monopolize matings with the females residing in their territories. Males that lose battles are relegated to territories with fewer females or are excluded from the beach. Fights are essentially slugging contests and are usually won by the larger male. The males stand face to face, bite each other, and land blows with their heads.

Based on these observations, it is not surprising that male northern elephant seals frequently weigh 2700 kg (5940 lbs) and are over four times more massive, on average, than females. The logic here runs as follows:

- Males that dominate beaches with large congregations of females father large numbers of offspring. Males that lose fights father few or no offspring.

- The alleles of dominant males rapidly increase in frequency in the population.

- If the ability to win fights and produce offspring is determined primarily by body size, then alleles for large body size have a significant fitness advantage, leading to the evolution of large male size. The fitness advantage is due to sexual selection, and the consequence is directional selection on large body size.

FIGURE 26.10b provides evidence for intense sexual selection in males. Biologists have marked many of the individuals in the seal population on Año Nuevo to track the lifetime reproductive success of a large number of individuals. The *x*-axis indicates fitness, plotted as the number of offspring produced over a lifetime. The *y*-axis indicates the percentage of males in the population that achieved each category of offspring production (0, 1–10, and so on). As the data show, in this population a few males father a large number of offspring, while most males father few or none.

Among females, variation in reproductive success is also high, but it is much lower than in males (**FIGURE 26.10c**). In this species, most sexual selection appears to be driven more by male–male competition than by female choice.

Sexual Dimorphism Results from Sexual Selection In elephant seals and most other animals studied, most females that survive to adulthood get a mate. In contrast, many males do not. Because sexual selection tends to be much more intense in males than females, males tend to have many more traits that function only in courtship or male–male competition. Stated another way, sexually selected traits often differ sharply between the sexes.

Sexual dimorphism (literally, "two-forms") refers to any trait that differs between males and females. Such traits range from weapons that males use to fight over females, such as antlers and

(a) Males compete for the opportunity to mate with females.

(b) Variation in reproductive success is high in males.

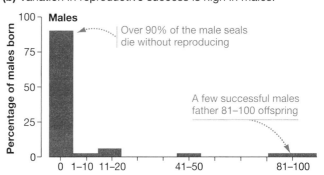

(c) Variation in reproductive success is relatively low in females.

FIGURE 26.10 Intense Sexual Selection in Male Elephant Seals. The histograms show that variation in lifetime reproductive success is higher in **(b)** male northern elephant seals than it is in **(c)** females.

DATA: Le Boeuf, B. J., and R. S. Peterson. 1969. *Science* 163: 91–93.

horns (**FIGURE 26.11a**), to the elaborate ornamentation and behavior used in courtship displays (**FIGURE 26.11b**). Humans are sexually dimorphic in size, distribution of body hair, and many other traits. ✔ If you understand sexual selection, you should be able to (1) define the fundamental asymmetry of sex, and (2) explain why males are usually the sex with exaggerated traits used in courtship.

Take-Home Messages Female choice and male–male competition illustrate how selection can favor certain phenotypes in a population. The adaptive alleles responsible for these phenotypes increase in frequency over time. Thus, evolution occurs and the assumptions of the Hardy–Weinberg model are violated.

Sexual selection is just one type of natural selection. All the other types, sometimes referred to collectively as **ecological (or**

(a) Red deer males use weaponry to compete for mates.

(b) Male raggiana birds of paradise display for females.

FIGURE 26.11 Sexually Selected Traits Are Used to Compete for Mates. Males often have exaggerated traits that they use in fighting or courtship. In many species, females lack these traits.

environmental) selection, favor traits that enable organisms to do things other than obtain mates—such as survive in their physical and biological environments. Different agents of natural selection can act on organisms simultaneously, sometimes favoring the same traits, other times resulting in fitness trade-offs (Chapter 25). For example, sexual selection may favor long trains in peacocks, while ecological selection may favor shorter trains that make the peacocks less vulnerable to predators. The relative importance of different agents of selection can change over time and in space.

Natural selection is the only evolutionary process that results in adaptation, but it is not the only evolutionary process that violates the Hardy–Weinberg assumptions. Let's consider what happens when random changes in allele frequency occur.

26.4 Genetic Drift

Genetic drift is defined as any change in allele frequencies in a population that is due to chance. The process is aptly named, because it causes allele frequencies to drift up and down randomly over time. When drift occurs, allele frequencies change due to blind luck—what is known in statistics as **sampling error.** Sampling error occurs when the allele frequencies of a chosen subset of a population (the sample) are different from those in the total population, by chance. Drift occurs in every population, in every generation, but especially in small populations.

Simulation Studies of Genetic Drift

To understand why genetic drift occurs, imagine a couple marooned on a deserted island. Suppose that at gene A, the wife's genotype is $A_T A_H$ and the husband is also $A_T A_H$. In this population, the two alleles are each at a frequency of 0.5.

Now suppose that the couple produce five children over their lifetime. Half of the eggs produced by the wife carry allele A_T and half carry allele A_H. Likewise, half of the sperm produced by the husband carry allele A_T and half carry allele A_H. To simulate which sperm and which egg happen to combine to produce each of the five offspring, you can flip a coin for each sperm and each egg, where tails represents allele A_T and heads represents allele A_H.

The following coin flips were done by a pair of students in a recent biology class:

	Sperm	Egg	Genotype
First offspring	A_H	A_H	$A_H A_H$
Second offspring	A_T	A_T	$A_T A_T$
Third offspring	A_T	A_H	$A_H A_T$
Fourth offspring	A_H	A_H	$A_H A_H$
Fifth offspring	A_T	A_H	$A_H A_T$

When the parents die, there are a total of 10 alleles in the population. But note that the allele frequencies have changed. In this generation, six of the 10 alleles (60 percent) are A_H; four of the 10 alleles (40 percent) are A_T. Evolution—a change in allele frequencies in a population—occurred due to genetic drift.

Instead of each allele being sampled in exactly its original frequency when offspring formed, as the Hardy–Weinberg principle assumes, a chance sampling error occurred. Allele A_H "got lucky," and allele A_T was "unlucky."

Computer Simulations   **FIGURE 26.12** shows what happens when a computer simulates the same process of random combinations in gametes over time. The program that generated the graphs combines the alleles in a gene pool at random to create an offspring generation, calculates the allele frequencies in the offspring generation, and uses those allele frequencies to create a new gene pool.

In this example, the process was continued for 100 generations. The *x*-axis on each graph plots time in generations; the *y*-axis plots the frequency of one of the two alleles present at the *A* gene in a hypothetical population.

The top graph shows eight replicates of this process with a population size of 4; the bottom graph shows eight replicates with a population of 400. Notice (**1**) the striking differences

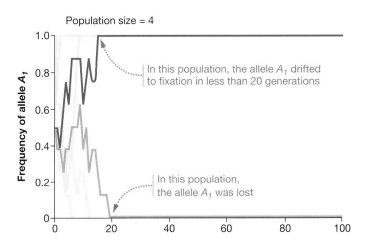

Population size = 4

In this population, the allele A_1 drifted to fixation in less than 20 generations

In this population, the allele A_1 was lost

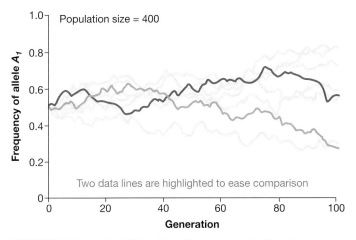

Population size = 400

Two data lines are highlighted to ease comparison

Generation

FIGURE 26.12 Genetic Drift Is More Pronounced in Small Populations than in Large Populations.

✔**EXERCISE** Draw a graph predicting what these simulation results would look like for a population size of 4000.

between the effects of drift in the small versus large population and (**2**) the consequences for genetic variation when alleles drift to fixation or loss.

Given enough time, drift can be an important factor even in large populations. For example, alleles containing silent mutations, usually in the third position of a codon, do not change the gene product (Chapter 16). As a result, most of these alleles have little or no effect on the phenotype. Yet these alleles routinely drift to high frequency or even fixation over time.

✔ If you understand genetic drift, you should be able to examine the MN blood group genotype frequencies in Table 26.1 and describe how drift could explain differences in genotype frequencies among populations. Note that there are no data indicating a selective advantage for different MN genotypes in different environments.

Key Points About Genetic Drift   These simulated matings illustrate three important points about genetic drift:

1. Genetic drift is random with respect to fitness. The changes in allele frequency that it produces are not adaptive.

2. Genetic drift is most pronounced in small populations. In the computer simulation, allele frequencies changed much less in the large population than in the small population. And if the couple on the deserted island had produced 50 children instead of five, allele frequencies in the next generation almost certainly would have been much closer to 0.5.

3. Over time, genetic drift can lead to the random loss or fixation of alleles. In the computer simulation with a population of 4, it took at most 20 generations for one allele to be fixed or lost. When random loss or fixation occurs, genetic variation in the population declines.

The importance of drift in small populations is a particular concern for conservation biologists because many populations are being drastically reduced in size by habitat destruction and other human activities. Small populations that occupy nature reserves or zoos are particularly susceptible to genetic drift. If drift leads to a loss of genetic diversity, it could darken the already bleak outlook for some endangered species.

Experimental Studies of Genetic Drift

Research on genetic drift began in the 1930s and 1940s with theoretical work that used mathematical models to predict the effect of genetic drift on allele frequencies and genetic variation. In the mid-1950s, Warwick Kerr and Sewall Wright did an experiment to determine how drift works in practice.

Kerr and Wright started with a large laboratory population of fruit flies that contained a **genetic marker**—a specific allele that causes a distinctive phenotype. In this case, the marker was the morphology of bristles. Fruit flies have bristles on their bodies that can be either straight or bent. This difference in bristle phenotype depends on a single gene. Kerr and Wright's lab population contained just two alleles—normal (straight) and "forked" (bent), designated as A_N and A_F respectively. Since the trait is

QUESTION: Does genetic drift in lab populations work as predicted by mathematical models?

HYPOTHESIS: Genetic drift causes alleles to be fixed or lost over time.

NULL HYPOTHESIS: Allele frequencies do not change; they stay in Hardy–Weinberg proportions.

EXPERIMENTAL SETUP:

Bristle shape

Normal bristles

Forked bristles

1 mm

4 males
4 females

4 males
4 females

x16

96 x Generation 1

Generation 2

1. Set up 96 cages of fruit flies with starting frequencies of both alleles = 0.5. Allow flies to breed.

2. From the F₁ offspring, randomly choose four males and four females. Allow them to breed.

3. After 16 generations, count the number of populations with either or both of the alleles.

PREDICTION: The number of fruit fly populations with fixed or lost alleles will increase over time. The number of populations with both alleles will decrease.

PREDICTION OF NULL HYPOTHESIS: Both alleles will remain present in all fruit fly populations for all generations.

RESULTS:

Genetic drift reduced allelic diversity in most populations:

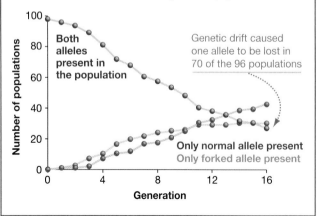

Both alleles present in the population

Genetic drift caused one allele to be lost in 70 of the 96 populations

Only normal allele present
Only forked allele present

CONCLUSION: Genetic drift occurs in laboratory populations as predicted by mathematical models.

---

sex-linked, males have only one allele (see Chapter 14). In females, the forked allele is recessive to the normal allele, but Kerr and Wright were able to distinguish homozygous females ($A_N A_N$) from heterozygotes ($A_N A_F$).

Kerr and Wright studied drift in these alleles as shown in the experimental setup in **FIGURE 26.13**:

Step 1 The researchers set up 96 small populations in their lab, each consisting of four adult females and four adult males of the fruit fly *Drosophila melanogaster*. They chose individual flies to begin these experimental populations so that the frequency of the normal and forked alleles in each of the 96 starting populations was 0.5. The two alleles do not affect the fitness of flies in the lab environment, so Kerr and Wright could be confident that if changes in the frequency of normal and forked phenotypes occurred, they would not be due to natural selection.

Step 2 After these first-generation adults bred, Kerr and Wright reared their offspring. In the offspring (F₁) generation, they randomly chose four males and four females—meaning that they simply grabbed individuals without caring whether their bristles were normal or forked—from each of the 96 offspring populations and allowed them to breed and produce the next generation.

Step 3 They repeated this procedure until all 96 populations had undergone a total of 16 generations. They then counted the number of populations that had both alleles still present, only the normal allele present, or only the forked allele present.

During the entire course of the experiment, no migration from one population to another occurred. Previous studies had shown that mutations from normal to forked bristles (and forked to normal) are rare. Thus, the only evolutionary process operating during the experiment was genetic drift. It was as if random accidents claimed the lives of all but eight individuals in each generation, so that only eight bred.

Their result? After 16 generations, both alleles were still present in only 26 of the 96 populations—significantly less than the full 96 expected by Hardy–Weinberg proportions. Forked bristles were found on all of the individuals in 29 of the experimental populations. Due to drift, the forked allele had been fixed in these 29 populations and the normal allele had been lost. In 41 other populations, however, the opposite was true: All individuals had normal bristles. In these populations, the forked allele had been lost due to chance.

The message of the study is startling: In 73 percent of the experimental populations (70 out of the 96), genetic drift had reduced allelic diversity at this gene to zero.

FIGURE 26.13 An Experiment on the Effects of Genetic Drift in Small Populations.

SOURCE: Kerr, W. E., and S. Wright. 1954. Experimental studies of the distribution of gene frequencies in very small populations of *Drosophila melanogaster*: I. Forked. *Evolution* 8: 172–177.

✔ **QUESTION** Why do you think the researchers decided to start each generation with only eight individuals?

As predicted, genetic drift decreased genetic variation within populations and increased genetic differences between populations. Is drift important in natural populations as well?

What Causes Genetic Drift in Natural Populations?

The random sampling process that takes place during fertilization occurs in every population in every generation in every species that reproduces sexually. Similarly, accidents that remove individuals at random occur in every population in every generation.

It is important to realize, though, that because drift is caused by sampling error, it can occur by *any* process or event that involves sampling—not just the sampling of gametes that occurs during fertilization or the loss of unlucky individuals due to accidents. Let's consider two special cases of genetic drift, called founder effects and bottlenecks.

Founder Effects on the Green Iguanas of Anguilla When a group of individuals immigrates to a new geographic area and establishes a new population, a founder event is said to occur. If the new population is small enough, the allele frequencies in the new population are almost guaranteed to be different from those in the source population—due to sampling error. A change in allele frequencies that occurs when a new population is established is called a **founder effect** (**FIGURE 26.14a**).

In 1995, fishermen on the island of Anguilla in the Caribbean witnessed a founder event involving green iguanas. A few weeks after two major hurricanes swept through the region, a large raft composed of downed logs tangled with other debris floated onto a beach on Anguilla. The fishermen noticed green iguanas on the raft and several onshore. Because green iguanas had not previously been found on Anguilla, the fishermen notified biologists. The researchers were able to document that at least 15 individuals had arrived; two years later, they were able to confirm that at least some of the individuals were breeding. A new population had formed.

During this founder event, it is extremely unlikely that allele frequencies in the new Anguilla population of green iguanas exactly matched those of the source population, which was thought to be on the islands of Guadeloupe.

Colonization events like these have been the major source of populations that occupy islands all over the world, as well as island-like habitats such as mountain meadows, caves, and ponds. Each time a founder event occurs, a founder effect is likely to accompany it, changing allele frequencies through genetic drift.

Genetic Bottleneck on Pingelap Atoll If a large population experiences a sudden reduction in size, a population bottleneck is said to occur. The term comes from the metaphor of a few individuals of a population fitting through the narrow neck of a bottle, by chance. Disease outbreaks, natural catastrophes such as floods or fires or storms, or other events can cause population bottlenecks.

Genetic bottlenecks follow population bottlenecks, just as founder effects follow founder events. A **genetic bottleneck**

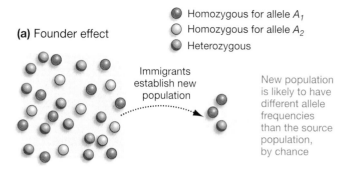

(a) Founder effect

- Homozygous for allele A_1
- Homozygous for allele A_2
- Heterozygous

Immigrants establish new population

New population is likely to have different allele frequencies than the source population, by chance

(b) Genetic bottleneck

High mortality strikes individuals at random

Bottlenecked population is likely to have different allele frequencies than original population, by chance

Time 1   Time 2

FIGURE 26.14 Two Causes of Genetic Drift in Natural Populations. The smaller the new population, the higher the likelihood that genetic drift will result in differences in allele frequencies as well as in loss of alleles.

✔ QUANTITATIVE The original population in **(a)** consists of 9 A_1A_1, 11 A_1A_2, and 7 A_2A_2 individuals (this hypothetical population is very small for simplicity). Compare the frequencies of the A_1 allele in the original and new populations.

is a sudden reduction in the number of alleles in a population (**FIGURE 26.14b**). Genetic drift occurs during genetic bottlenecks and causes a change in allele frequencies.

For an example from a natural population, consider the humans who occupy Pingelap Atoll in the South Pacific. On this island, only about 20 people out of a population of several thousand managed to survive the effects of a typhoon and a subsequent famine that occurred around 1775. The survivors apparently included at least one individual who carried a loss-of-function allele at a gene called *CNGB3*, which codes for a protein involved in color vision.

The *CNGB3* allele is recessive, and when it is homozygous it causes a serious vision deficit called achromatopsia. People with this condition have poor vision and are either totally or almost totally color-blind. The condition is extremely rare in most populations, and the frequency of the *CNGB3* allele is estimated to be under 1.0 percent.

In the population that survived the Pingelap Atoll disaster, however, the loss-of-function allele was at a frequency of about 1/40, or 2.5 percent. If the allele was at the typical frequency of 1.0 percent or less before the population bottleneck, then a large frequency change occurred during the bottleneck, due to drift.

In today's population on Pingelap Atoll, over 1 in 20 people are afflicted with achromatopsia, and the allele that causes the affliction is at a frequency of well over 20 percent. Because the loss-of-function allele is extremely unlikely to be favored by directional selection or heterozygote advantage, researchers hypothesize that the frequency of the allele in this small population has continued to increase over the past 230 years due to drift.

26.5 Gene Flow

When an individual leaves one population, joins another, and breeds, **gene flow** is said to occur—the movement of alleles between populations. Note that while allele flow might be a more apt description, gene flow is the term that evolutionary biologists traditionally use. Also note that while organisms can *emigrate* from a source population or *immigrate* to a new population, the movement of their alleles is called gene flow.

As an evolutionary process, gene flow usually has one outcome: It equalizes allele frequencies between the source population and the recipient population. When alleles move from one population to another, the populations tend to become more alike. To capture this point, biologists say that gene flow homogenizes allele frequencies among populations (**FIGURE 26.15**).

Measuring Gene Flow between Populations

The presence or absence of gene flow has particularly important implications for the conservation of threatened and endangered species. Numerous studies have documented the decline of gene flow between wild populations that have been isolated from one another—for example, by habitat fragmentation. Many studies have also documented the effects of gene flow between wild populations and captive populations.

For example, the captive breeding of fish is increasing around the world as wild populations are being depleted. In some cases, wild fish are used to start captive populations, which are then kept isolated in "farms" until they go to market—except when they escape. In other cases, captive-bred fish are purposefully released into the wild in an effort to supplement the size of wild

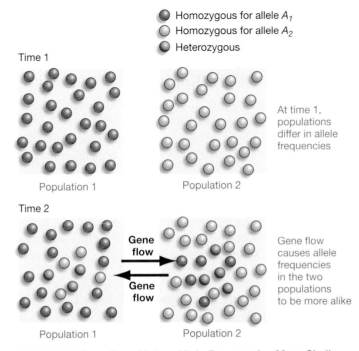

FIGURE 26.15 **Gene Flow Makes Allele Frequencies More Similar between Populations.** Gene flow can occur in one or both directions. If gene flow continues and no selection occurs, allele frequencies will eventually be identical.

populations. What are the effects of gene flow between captive-bred populations and wild populations?

A team of biologists recently studied steelhead trout (*Oncorhynchus mykiss*) in the Hood River of Oregon to answer this question (**FIGURE 26.16a**). Some of the trout in the Hood River are wild, while others were raised in a hatchery and released to supplement the diminishing wild population. The researchers were able to catch, sample, and release all of the steelhead trout in the Hood River each year for three years by trapping them in a dam near the mouth of the river as the fish traveled upstream to their wintering streams.

The researchers extracted DNA from tissue samples and calculated genotypes using eight microsatellite alleles (see Chapter 21) to determine whether the parents of each fish were wild or captive-bred. The team then compared the reproductive fitness—the number of surviving adult offspring—of the trout in three groups:

1. Individuals with two wild parents (Wild × Wild)
2. Individuals with one wild parent and one captive-bred parent (Wild × Captive)
3. Individuals with two captive-bred parents (Captive × Captive)

The graph in **FIGURE 26.16b** shows the fitness results for female steelhead trout. Individuals with one or two captive-bred parents are assigned a value from 0 to 1 relative to the fitness of trout with two wild parents (fitness = 1). Individuals with one captive-bred parent have 16 percent lower fitness than wild trout on average, while individuals with two captive-bred parents have significantly lower fitness, 38 percent lower than wild-bred trout,

(a) Wild steelhead trout populations are declining

(b) Captive-bred trout reduce the fitness of wild populations

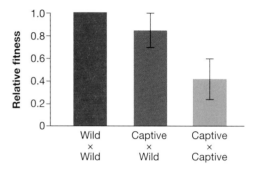

FIGURE 26.16 Gene Flow Reduces Fitness in a Population of Steelhead Trout. In this study, the fitness of female steelhead trout with one or two captive-bred parents is shown relative to females with two wild parents. Data for males are comparable (not shown).

DATA: Araki, H., B. Cooper, and M. S. Blouin. 2009. *Biology Letters* 5(5): 621–624.

on average. The researchers observed similar results for males (not shown).

These data demonstrate that gene flow is occurring from the captive-bred population to the wild population and is causing a reduction in fitness in that population. Related research on farmed Atlantic salmon in Ireland demonstrated that farmed salmon produce smaller eggs, spawn at different times, and have different predator-avoidance behaviors than wild salmon. When farmed salmon were introduced into wild populations, the reproductive success of the wild population decreased.

The conclusion of these studies on steelhead trout and Atlantic salmon is that gene flow between captive-bred and wild populations reduces fitness in the wild populations. As a result, efforts to augment wild populations with captive-bred fish may ultimately result in a *decline* in population size—contrary to the intentions of fisheries biologists.

Gene Flow Is Random with Respect to Fitness

It is not true, though, that gene flow always reduces fitness in the receiving population. If a population has lost alleles due to genetic drift, then the arrival of new alleles via gene flow should increase genetic diversity. If increased genetic diversity results in better

resistance to infections by bacteria or viruses or other parasites, for example, gene flow would increase the average fitness of individuals. This was the result when gene flow was used to aid an endangered population of the Florida panther (see Figures 26.4 and 57.14).

Gene flow is random with respect to fitness—the arrival or departure of alleles can increase or decrease average fitness, depending on the situation. But in every case, movement of alleles between populations tends to reduce their genetic differences.

This latter generalization is particularly important in our own species right now. Large numbers of people from Africa, the Middle East, Mexico, Central America, and Asia are immigrating to the countries of the European Union and the United States. Because individuals from different cultural and ethnic groups are intermarrying and having offspring, allele frequencies are becoming more similar across human populations.

26.6 Mutation

To appreciate the role of mutation as an evolutionary force, let's return to one of the central questions that biologists ask about an evolutionary process: How does it affect genetic variation in a population? Recall that:

- Natural selection often favors certain alleles and leads to a decrease in overall genetic variation.

- Genetic drift tends to decrease genetic diversity over time, as alleles are randomly lost or fixed.

- Gene flow increases genetic diversity in a recipient population if new alleles arrive with immigrating individuals. But gene flow may decrease genetic variation in the source population if alleles leave with emigrating individuals.

If most of the evolutionary processes lead to a loss of genetic diversity over time, what restores it? In sexually reproducing organisms, independent assortment and recombination (crossing over) are major sources of genetic diversity (see Chapter 13). But these processes create new combinations of existing alleles. Where do entirely new alleles come from? The answer to this question is **mutation.**

Mutations can occur in a number of ways (Chapters 16 and 21):

- *Point mutations*   If a change in nucleotide sequence occurs in a stretch of DNA that codes for a protein, the new allele may result in a polypeptide with a novel amino acid sequence. If the mutation occurs in a stretch of DNA that codes for regulatory RNA, the new allele may result in a change in regulation of the expression of other alleles.

- *Chromosome-level mutations*   One consequence of chromosome mutation is gene duplication, which increases the number of copies of a gene. If duplicated genes diversify via point mutations, they can lose their function or create new alleles.

- *Lateral gene transfer (also known as horizontal gene transfer)*   New studies suggest that the transfer of genes from one species to another, rather than from parent to offspring, might be a more important source of genetic variation than previously realized (Chapter 21).

Because errors and chromosome damage are inevitable, mutation constantly introduces new alleles into populations in every generation. Mutation is an evolutionary process that increases genetic diversity in populations.

Even though it consistently leads to an increase in genetic diversity in a population, mutation is random with respect to the affected allele's impact on the fitness of the individual. Changes in the makeup of chromosomes or in specific DNA sequences do not occur in ways that tend to increase fitness or decrease fitness. Mutation just happens.

But because most organisms are well adapted to their current habitat, random changes in genes usually result in products that do not work as well as the alleles that currently exist. Stated another way, most mutations in sequences that code for a functional protein or RNA result in deleterious alleles, which lower fitness. Deleterious alleles tend to be eliminated by purifying selection.

On rare occasions, however, mutation in these types of sequences produces a **beneficial** allele—an allele that allows individuals to produce more offspring. Beneficial alleles should increase in frequency in the population due to natural selection.

Because mutation produces new alleles, it can in principle change the frequencies of alleles through time. But does mutation *alone* occur often enough to make it an important factor in changing allele frequencies? The short answer is no.

Mutation as an Evolutionary Process

To understand why mutation is not a significant mechanism of evolutionary change by itself, consider that the highest mutation rates that have been recorded at individual genes in humans are on the order of 1 mutation in every 10,000 gametes produced by an individual. This rate means that for every 10,000 alleles produced, on average one will have a mutation at the gene in question.

When two gametes combine to form an offspring, then, at most about 1 in every 5000 offspring will carry a mutation at a particular gene. Now suppose that 195,000 humans live in a population, that 5000 offspring are born one year, and that at the end of that year, the population numbers 200,000. Humans are diploid, so in a population this size there are, in total, 400,000 copies of each gene. Only one of them is a new allele created by mutation, however. Over the course of a year, the allele frequency change introduced by mutation is 1/400,000, or 0.0000025 (2.5×10^{-6}). At this rate, it would take 4000 years for mutation to produce a change in allele frequency of 1 percent.

These calculations support the conclusion that mutation does little on its own to change allele frequencies. Although mutation can be a significant evolutionary process in bacteria and archaea, which have extremely short generation times, mutation in eukaryotes rarely causes a change from the genotype frequencies expected under the Hardy–Weinberg principle. As an evolutionary process, mutation is slow compared with selection, genetic drift, and gene flow.

However, mutation can have a very large effect on evolution when *combined* with genetic drift, gene flow, and selection. Let's consider two examples.

Experimental Studies of Mutation

Consider a lab experiment designed by Richard Lenski and colleagues to evaluate the role that mutation plays over many generations.

Experimental Evolution Lenski's group focused on *Escherichia coli*, a bacterium that is a common resident of the human intestine. To begin, they set up a large series of populations, each founded with a single cell, and allowed them to replicate for over four years—about 10,000 generations. The strain of *E. coli* used in the experiment is completely asexual and reproduces by cell division. Thus, mutation was the only source of genetic variation in these populations.

The biologists saved a sample of cells from each population at regular intervals during the experiment and stored them in a freezer. Because frozen *E. coli* cells resume growth when they are thawed, the frozen cells serve as an archive of individuals from different generations.

Were cells from the older and newer generations of each population different? Lenski's group used competition experiments to address this question. They grew cells from two different generations on the same plate and compared their growth rates. The populations of cells that were more numerous had grown the fastest, meaning that they were better adapted to the experimental environment.

In this way the researchers could measure the fitness of descendant populations relative to ancestral populations. Relative fitness values greater than 1 meant that recent-generation cells outnumbered older-generation cells following the competition.

Fitness Increased in Fits and Starts The data from the competition experiments are graphed in **FIGURE 26.17**. Notice that relative fitness increased dramatically—almost 30 percent—over time. But notice also that fitness increased in fits and starts. This pattern is emphasized by the solid line on the graph, which represents a mathematical function fitted to the data points.

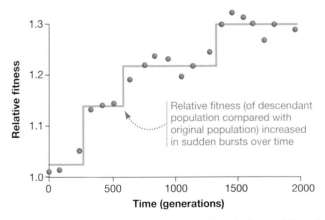

FIGURE 26.17 Evidence of Mutation in Lab Populations of *E. coli*.
DATA: Elena, S. F., V. S. Cooper, and F. E. Lenski. 1996. *Science* 272: 1802–1804.

What caused this stair-step pattern? Lenski's group hypothesized that genetic drift was relatively unimportant in this experiment because population sizes were so large. Instead, they proposed that each jump was caused by a novel mutation that conferred a fitness benefit under selection. Their interpretation was that cells that happened to have the beneficial mutation grew rapidly and came to dominate the population.

After a beneficial mutation occurred, the fitness of the population stabilized—sometimes for hundreds of generations—until another random but beneficial mutation occurred and produced another jump in fitness. These results demonstrate the combined effects of mutation and natural selection.

Studies of Mutation in Natural Populations

Lenski's experiment demonstrates the effect of cumulative mutations in laboratory populations. Now consider a recent study showing how several forms of mutation have combined with natural selection to create color variations, or polymorphisms, in pea aphids.

Lateral Gene Transfer Pea aphids (*Acyrthosiphon pisum*) are small insects that feed on plant sap. They occur in two colors in the wild, red and green, and both colors coexist within populations (**FIGURE 26.18a**). The two phenotypes are maintained in the population due to frequency-dependent selection—ladybird beetles are more likely to prey on red aphids, whereas parasitoid wasps are more likely to lay their eggs in green aphids. The balancing selection on both colors preserves genetic variation in the populations.

The color of aphids is determined by carotenoid pigments, the same group of pigments that give zebra finches their bright orange beaks. Finches and other animals get their carotenoid pigments from the food they eat, mostly from plants.

But there is a twist. Pea aphids do *not* acquire their carotenoids from the plant sap they eat. So where do the pigments come from? Researchers recently examined several bacteria associated with pea aphids and concluded that the carotenoids did not originate from these symbionts either. Instead, other researchers found that the aphids are generating their *own* carotenoids—the first animals ever discovered to have this ability.

If the ancestors of aphids did not have the genes that code for the necessary enzymes for carotenoid biosynthesis, how did the pea aphids obtain this pathway? The answer is by mutation, more specifically by lateral gene transfer—from the genome of a fungal symbiont to the genome of a recent ancestor of the pea aphids. This result was surprising—clear evidence of lateral gene transfer from one eukaryote to another.

Gene Duplication Lateral gene transfer is only part of the mutation story in pea aphids. Carotenoids are a family of molecules with similar pigment and a number of intermediate forms. Two types of enzymes are primarily responsible for the biosynthesis of these molecules in both pea aphids and their fungal ancestor.

A comparison of aphid and fungal genome sequences suggests that after the fungal genes for these two types of enzymes

(a) Red–green color polymorphism in pea aphids

(b) Origin of genes for carotenoid synthesis enzymes in aphids

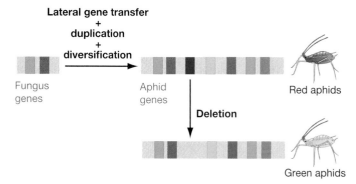

FIGURE 26.18 Evidence of Mutation in Pea Aphids.

were transferred to aphids, they underwent further mutations—duplications, sequence diversification by point mutations, and deletion. As a result, red pea aphids have the enzymes necessary to synthesize yellow, green, and red carotenoid pigments, while green pea aphids have the enzymes necessary to synthesize only the yellow and green ones (**FIGURE 26.18b**). The deletion of a gene in the green aphid genome illustrates an important point: Sometimes a loss-of-function mutation can be adaptive. In this case, being green helps aphids to avoid predation by ladybird beetles.

Take-Home Messages

The research on mutations in *E. coli* populations in the lab and the origin of color polymorphism in pea aphids reinforces three fundamental messages about the role of mutation in evolution.

1. Mutation is the ultimate source of genetic variation. Crossing over and independent assortment shuffle existing alleles into new combinations, but only mutation creates new alleles. Mutations just happen—organisms cannot create mutations because they "want" or "need" them.

2. If mutation did not occur, evolution would eventually stop. Recall that natural selection and genetic drift tend to eliminate alleles. Without mutation, eventually there would be no variation for selection and drift to act on.

	Definition/Description	Effect on Genetic Variation	Effect on Average Fitness
Natural selection	Certain alleles are favored	Can maintain, increase, or reduce genetic variation	Can produce adaptation, increasing fitness
Genetic drift	Random changes in allele frequencies; most important in small populations	Tends to reduce genetic variation via loss or fixation of alleles	Random with respect to fitness; usually reduces average fitness
Gene flow	Movement of alleles between populations; reduces differences between populations	May increase genetic variation by introducing new alleles; may decrease it by removing alleles	Random with respect to fitness; may increase or decrease average fitness by introducing high- or low-fitness alleles
Mutation	Production of new alleles	Increases genetic variation by introducing new alleles	Random with respect to fitness; most mutations in coding sequences lower fitness

3. Mutation *alone* is usually inconsequential in changing allele frequencies at a particular gene. However, when considered across the genome and when combined with natural selection, genetic drift, and gene flow, mutation becomes an important evolutionary process.

TABLE 26.4 summarizes these points and similar conclusions about the four evolutionary processes. Each of the four evolutionary processes has different consequences for allele frequencies, all of them violating Hardy–Weinberg predictions. These processes interact to create new species—that is, new branches on the tree of life (Chapter 27). The ultimate result is biological diversity (Chapter 28 and Unit 6). You can see how nonrandom mating and the four evolutionary processes fit into the Big Picture of Evolution on pages 526–527.

CHAPTER 26 REVIEW

For media, go to MasteringBiology

If you understand . . .

26.1 Analyzing Change in Allele Frequencies: The Hardy–Weinberg Principle

- The Hardy–Weinberg principle played an important role in the synthesis of Mendelian genetics and Darwinian evolution.

- The Hardy–Weinberg principle can serve as a null hypothesis in evolutionary studies because it predicts what genotype and allele frequencies are expected to be if mating is random with respect to the gene in question and if none of the four evolutionary processes is operating on that gene.

✔ **QUANTITATIVE** You should be able to write out the Hardy–Weinberg equations for allele and genotype frequencies and calculate the genotype frequencies expected if the frequency of A_1 is 0.2.

26.2 Nonrandom Mating

- Nonrandom mating changes only genotype frequencies, not allele frequencies, so is not an evolutionary process itself.

- Inbreeding, or mating among relatives, is a form of nonrandom mating. It leads to an increase in homozygosity and a decrease in heterozygosity.

- Inbreeding can accelerate natural selection and can cause inbreeding depression.

✔ You should be able to predict how extensive inbreeding during the 1700s and 1800s affected the royal families of Europe.

26.3 Natural Selection

- Natural selection is the only evolutionary process that produces adaptation.

- Directional selection favors phenotypes at one end of a distribution. It decreases allelic diversity in populations.

- Stabilizing selection eliminates phenotypes with extreme characteristics. It decreases allelic diversity in populations.

- Disruptive selection favors extreme phenotypes and thus maintains genetic variation in populations. Disruptive selection sometimes leads to the formation of new species.

- Balancing selection occurs when no single phenotype is favored; there is a balance among alleles in terms of fitness and frequency. Balancing selection preserves genetic variation.

- Sexual selection is a type of natural selection that leads to the evolution of traits that help individuals attract mates. It usually has a stronger effect on males than on females.

✔ You should be able to explain why natural selection violates the Hardy–Weinberg principle.

26.4 Genetic Drift

- Genetic drift causes random changes in allele frequencies.
- Genetic drift is particularly important in small populations, and it tends to reduce overall genetic diversity.
- Genetic drift can result from random fusion of gametes at fertilization, founder events, and population bottlenecks.

✔ You should be able to suggest how genetic drift could be important to the management of endangered species.

26.5 Gene Flow

- Gene flow equalizes allele frequencies between populations.
- Gene flow can introduce alleles from one population to another when individuals migrate among populations.
- The introduced alleles may have a beneficial, neutral, or deleterious effect.

✔ You should be able to suggest how gene flow could be important to the management of endangered species.

26.6 Mutation

- Mutation is the only evolutionary process that creates new alleles. They may be beneficial, neutral, or deleterious.
- Mutation occurs too infrequently to be a major cause of change in allele frequency alone, but it is important when combined with natural selection, genetic drift, and gene flow.

✔ You should be able to suggest how mutation could be important to the management of endangered species.

(MB) MasteringBiology

1. MasteringBiology Assignments

Tutorials and Activities Causes of Evolutionary Change, Hardy–Weinberg Principle, Mechanisms of Evolution, Three Modes of Natural Selection

Questions Reading Quizzes, Blue-Thread Questions, Test Bank

2. eText Read your book online, search, take notes, highlight text, and more.

3. The Study Area Practice Test, Cumulative Test, BioFlix® 3-D Animations, Videos, Activities, Audio Glossary, Word Study Tools, Art

You should be able to . . .

✔ TEST YOUR KNOWLEDGE
Answers are available in Appendix A

1. In what sense is the Hardy–Weinberg principle a null hypothesis?

2. Why isn't inbreeding considered an evolutionary process?
   a. It does not change genotype frequencies.
   b. It does not change allele frequencies.
   c. It does not occur often enough to be important in evolution.
   d. It does not violate the assumptions of the Hardy–Weinberg principle.

3. Why is genetic drift aptly named?
   a. It causes allele frequencies to drift up or down randomly.
   b. It occurs when alleles from one population drift into another.
   c. It occurs when mutations drift into a genome.
   d. It occurs when populations drift into new habitats.

4. What does it mean when an allele reaches "fixation"?
   a. It is eliminated from the population.
   b. It has a frequency of 1.0.

   c. It is dominant to all other alleles.
   d. It is adaptive.

5. True or false? Gene flow can either increase or decrease the average fitness of a population.

6. Mutation is the ultimate source of genetic variability. Why is this statement correct?
   a. DNA polymerase (the enzyme that copies DNA) is remarkably accurate.
   b. "Mutation proposes and selection disposes."
   c. Mutation is the only source of new alleles.
   d. Mutation occurs in response to natural selection. It generates the alleles that are required for a population to adapt to a particular habitat.

✔ TEST YOUR UNDERSTANDING
Answers are available in Appendix A

7. **QUANTITATIVE** In a population of 2500, how many babies would you expect to have cystic fibrosis, a homozygous recessive condition, if the frequency of the dominant allele is 0.9 and the population is at Hardy–Weinberg equilibrium?
   a. $0.9 \times 2500 = 2025$
   b. $2 \times 0.9 \times 0.1 \times 2500 = 800$
   c. $0.9 \times 0.1 \times 2500 = 400$
   d. $0.1 \times 0.1 \times 2500 = 25$

8. Suggest why inbreeding could cause recessive deleterious alleles for cystic fibrosis to be "purged" from a population.

9. Determine what is incorrect in the following statement: Red aphids mutated their genes so that they could be green and avoid predation by ladybird beetles.

10. Why does sexual selection often lead to sexual dimorphism?

11. Consider an allele that increases reproductive success in elephant seal males versus an allele that increases reproductive success in females. Which allele will increase in frequency faster, and why?

12. Draw a small concept map (**BioSkills 15**) showing how selection, genetic drift, gene flow, and mutation relate to genetic variation.

13. QUANTITATIVE In humans, albinism is caused by loss-of-function mutations in genes involved in the synthesis of melanin, the dark pigment in skin. Only people homozygous for a loss-of-function allele (genotype *aa*) have the albino phenotype. In Americans of northern European ancestry, albino individuals are present at a frequency of about 1 in 10,000 (or 0.0001). Assuming that genotypes are in Hardy–Weinberg proportions, what is the frequency of Caucasians in the United States who carry an allele for albinism?

14. A group of researchers presented artificial calls of the male cricket *Teleogryllus commodus* to female crickets in the lab to measure selection for male calls. They used artificial calls so that they could vary properties such as length of the pause between calls and number of trills in each call. When positive selection was plotted as a function of call characteristics, the shapes of the selection curves started low, peaked in the middle, and ended low. What kind of selection is occurring?
a. sexual selection
b. balancing selection
c. sexual selection and stabilizing selection
d. sexual selection and balancing selection

15. Suppose you were studying several species of human. In one, males never lifted a finger to help females raise children. In another, males provided just as much parental care as females except for actually carrying the baby during pregnancy. How does the fundamental asymmetry of sex compare in the two species? How would you expect sexual dimorphism to compare between the two species?

16. You are a conservation biologist charged with creating a recovery plan for an endangered species of turtle. The turtle's habitat has been fragmented by suburbanization and highway construction into small, isolated, but protected areas. Some evidence indicates that certain turtle populations are adapted to typical freshwater marshes, whereas others are adapted to acidic wetlands or salty habitats. Further, some turtle populations number less than 25 breeding adults, making genetic drift and inbreeding a major concern. In creating a recovery plan, the tools at your disposal are captive breeding, the capture and transfer of adults to create gene flow, or the creation of habitat corridors between wetlands to make migration possible. How would you use gene flow to help this species?

27 Speciation

In this chapter you will learn how

Speciation creates new branches on the tree of life

by asking

How is a species defined and identified? 27.1

by asking

How do the isolation and divergence of populations produce new species?

then

What happens when isolated populations make contact? 27.4

in

Allopatry 27.2

in

Sympatry 27.3

New sunflower species have formed recently in southwestern North America. This chapter explains how.

This chapter is part of the Big Picture. See how on pages 526–527.

lthough Darwin called his masterwork *On the Origin of Species by Means of Natural Selection*, he actually had little to say about how new species arise. Instead, his data and analyses focused on the process of natural selection and the changes that occur *within* populations over time.

Recall that there are four evolutionary processes—natural selection, genetic drift, mutation, and gene flow—and that gene flow makes allele frequencies more similar among populations (see Chapter 26). If gene flow ends, allele frequencies in isolated populations are free to diverge—meaning that the populations begin to evolve independently of each other. For example, when a new mutation creates an allele that changes the phenotype of individuals in one population, there is no longer any way for that allele to appear in the other population. If mutation, selection, and genetic drift cause isolated populations to diverge sufficiently, distinct types, or species, form—that is, the process of speciation takes place.

✔ When you see this checkmark, stop and test yourself. Answers are available in Appendix A.

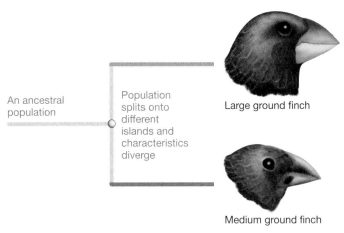

An ancestral population

Population splits onto different islands and characteristics diverge

Large ground finch

Medium ground finch

FIGURE 27.1 Speciation Creates Evolutionarily Independent Populations. The large ground finch and medium ground finch are derived from the same ancestral population. This ancestral population split into two populations isolated by lack of gene flow. Because the populations began evolving independently, over time they acquired the distinctive characteristics observed today.

In essence, speciation results from *genetic isolation* and *genetic divergence*. Genetic isolation results from lack of gene flow, and genetic divergence occurs because selection, genetic drift, and mutation proceed independently in the isolated populations.

Speciation is a splitting event that creates two or more distinct species from a single ancestral species (**FIGURE 27.1**). When speciation is complete, a new branch has been added to the tree of life. Thus, the study of speciation is critical to understanding how life evolved and has many practical implications for the preservation of biodiversity on Earth (see Chapter 57). You can see how speciation fits into the Big Picture of Evolution on pages 526–527.

Let's begin by considering the *pattern* component of studying speciation—how can we define and identify species? Then let's explore the *process* of speciation—how do new species form?

27.1 How Are Species Defined and Identified?

If your friend were to tell you that she was planning to study polar bears and grizzly bears during her summer research project, you would immediately recognize these animals as distinct species. Polar bears and grizzly bears are distinguishable from one another in appearance, behavior, habitat use, and other traits. But what if your friend was going to compare elephants that live in the forests and savannas of Africa? Are these elephants the same species or two different species?

Evolutionary biologists have been wrangling with the definition of species for decades—how can you reliably distinguish two or more species of bears, elephants, or bacteria in the field or fossil record? Although there is no single, universal answer, scientists do agree there is a distinction between the *definition* of species in general and the criteria that can be used to *identify* species in particular cases.

In general, a **species** is defined as an evolutionarily independent population or group of populations. Of the many criteria that can be used to identify species, the three most common ones are (**1**) the biological species concept, (**2**) the morphospecies concept, and (**3**) the phylogenetic species concept.

The Biological Species Concept

According to the **biological species concept,** the main criterion for identifying species is reproductive isolation. This is a logical yardstick because no gene flow occurs between populations that are reproductively isolated from each other. Specifically, if two different populations do not interbreed in nature, or if they fail to produce viable and fertile offspring when matings take place, then they are considered distinct species. The influential evolutionary biologist Ernst Mayr strongly promoted the biological species concept because the criterion of reproductive isolation enables clear evidence of evolutionary independence.

Reproductive isolation can result from a wide variety of events and processes. To organize the various mechanisms that stop gene flow between populations, biologists distinguish between:

- **prezygotic** (literally, "before-zygote") **isolation,** which prevents individuals of different species from mating; and

- **postzygotic** ("after-zygote") **isolation,** in which the offspring of matings between members of different species do not survive or reproduce.

TABLE 27.1 outlines some of the more important mechanisms of prezygotic and postzygotic isolation.

Although the biological species concept has a strong theoretical foundation, it has disadvantages. The criterion of reproductive isolation cannot be evaluated in fossils or in species that reproduce asexually. What's more, the concept is difficult to apply when closely related populations do not happen to overlap with each other geographically. In this case, biologists are left to guess whether interbreeding and gene flow would occur if the populations happened to come into contact.

Further, reproductive isolation can sometimes be a complex gradient rather than an all-or-nothing scenario. For example, if a male lion and female tiger produce a rare "liger" that survives and reproduces in captivity (the ranges of these cats do not overlap in nature), are lions and tigers necessarily the same species?

The Morphospecies Concept

How do biologists identify species when the criterion of reproductive isolation cannot be applied? Under the **morphospecies** ("form-species") **concept,** researchers identify evolutionarily independent lineages by differences in size, shape, or other morphological features. The logic behind the morphospecies concept is that distinguishing features are most likely to arise if populations are independent and isolated from gene flow.

The morphospecies concept is compelling simply because it is so widely applicable. It is a useful criterion when biologists have no data on the extent of gene flow, and it is equally applicable to sexual, asexual, or fossil species. Its disadvantages

TABLE 27.1 Mechanisms of Reproductive Isolation

	Process	Example
Prezygotic Isolation		
Temporal	Populations are isolated because they breed at different times.	Bishop pines and Monterey pines release their pollen at different times of the year.
Habitat	Populations are isolated because they breed in different habitats.	Parasites that begin to exploit new host species are isolated from their original population.
Behavioral	Populations do not interbreed because their courtship displays differ.	To attract male fireflies, female fireflies give a species-specific sequence of flashes.
Gametic barrier	Matings fail because eggs and sperm are incompatible.	In sea urchins, a protein called bindin allows sperm to penetrate eggs. Differences in the amino acid sequence of bindin cause matings to fail between closely related populations.
Mechanical	Matings fail because male and female reproductive structures are incompatible.	In alpine skypilots (a flowering plant), the length of the floral tube varies. Bees can pollinate in populations with short tubes, but only hummingbirds can pollinate in populations with long tubes.
Postzygotic Isolation		
Hybrid viability	Hybrid offspring do not develop normally and die as embryos.	When ring-necked doves mate with rock doves, less than 6 percent of eggs hatch.
Hybrid sterility	Hybrid offspring mature but are sterile as adults.	Eastern meadowlarks and western meadowlarks are almost identical morphologically, but their hybrid offspring are usually infertile.

are that **(1)** it may lead to the naming of two or more species when there is only one **polymorphic species** with differing phenotypes, such as the spotted and black morphs of jaguars; **(2)** it cannot identify **cryptic species,** which differ in traits other than morphology, such as the meadowlarks in Table 27.1; and **(3)** the morphological features used to distinguish species are subjective. In the worst case, different researchers working on the same populations disagree on the characters that distinguish species, such as when paleontologists disagree on the identity of an extinct human species based on measurements of fossil bone fragments.

The Phylogenetic Species Concept

The **phylogenetic species concept** identifies species based on the evolutionary history of populations. The reasoning behind the phylogenetic species concept begins with Darwin's theory that all species are related by common ancestry. In modern terms, all species form a monophyletic ("one-tribe") group—the tree of life.

A **monophyletic group,** also called a **clade** or **lineage,** consists of an ancestral population, all of its descendants, and *only* those descendants. On any given evolutionary tree, there are many monophyletic groups (**FIGURE 27.2a**).

(a) Monophyletic groups

━ **Monophyletic group**: an ancestral population and all descendants

╫ **Synapomorphy**: trait unique to a monophyletic group

FIGURE 27.2 The Phylogenetic Species Concept Is Based on Monophyletic Groups. (a) Monophyletic groups can be mapped on phylogenetic trees. **(b)** Although there are several isolated populations of elephants on two continents, the phylogenetic species concept identifies no more or less than three species of elephants. This tree was created by comparing DNA sequences. Thus, the synapomorphies in this case are DNA sequences unique to each species.

(b) Phylogenetic species: smallest monophyletic groups

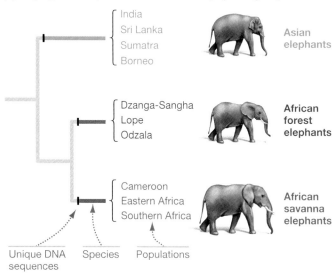

India
Sri Lanka
Sumatra
Borneo
— Asian elephants

Dzanga-Sangha
Lope
Odzala
— African forest elephants

Cameroon
Eastern Africa
Southern Africa
— African savanna elephants

Unique DNA sequences   Species   Populations

Monophyletic groups, in turn, are identified by traits called synapomorphies ("unique-forms"). A **synapomorphy** is a trait that is found in certain groups of organisms and their common ancestor, but is missing in more distant ancestors. Fur and lactation, for example, are synapomorphies that identify mammals as a monophyletic group.

Synapomorphies are homologous traits that can be identified at the genetic, developmental, or structural level. In many cases, researchers use DNA sequence data to identify synapomorphies and estimate phylogenetic trees. (Chapter 28 explores the data and logic that biologists use to reconstruct phylogenies, and **BioSkills 7** in Appendix B reviews how to read phylogenetic trees.)

Under the phylogenetic species concept, species are defined as the smallest monophyletic groups on the tree of life. Phylogenetic species are made up of populations that share one or more synapomorphies.

The tree in **FIGURE 27.2b,** for example, is based on DNA sequence data from an array of elephant populations. Each of the populations is not distinctive enough to represent separate branches. The phylogenetic species on the tree are labeled Asian elephants, African forest elephants, and African savanna elephants. These are the smallest monophyletic groups on the tree.

The names at some of the tips on this tree (Cameroon, Eastern Africa, Southern Africa, etc.) represent populations within species. These populations may be separated geographically, but their characteristics are so similar that they do not form independent branches on the tree. They are simply part of the same monophyletic group containing other populations.

The phylogenetic species concept has two distinct advantages: (1) It can be applied to any population (fossil, asexual, or sexual), and (2) it is logical because different species have different synapomorphies only if they are isolated from gene flow and have evolved independently. The approach has a distinct disadvantage, however: Carefully estimated phylogenies are available only for a tiny (though growing) subset of populations on the tree of life.

Critics of the phylogenetic species concept also point out that it would probably lead to recognition of many more species than either the morphospecies or biological species concepts.

Proponents counter that, far from being a disadvantage, the recognition of increased numbers of species might better reflect the extent of life's diversity.

The three major species concepts are summarized in **TABLE 27.2**. In practice, researchers use a combination of species concepts to identify evolutionarily independent populations in nature. Conflicts have occurred, however, as in the case of the dusky seaside sparrow.

Species Definitions in Action: The Case of the Dusky Seaside Sparrow

Seaside sparrows live in salt marshes along the Atlantic and Gulf Coasts of the United States. The scientific name of this species is *Ammodramus maritimus*. Recall that scientific names consist of a genus name followed by a species name (Chapter 1).

Using the morphospecies and biological species concepts, researchers had traditionally named six seaside sparrow "subspecies" (**FIGURE 27.3a**). **Subspecies** are populations that live in discrete geographic areas and have distinguishing features, such as coloration or calls, but are not considered distinct enough to be called separate species. For example, *A. m. nigrescens* (*nigrescens* indicates the subspecies name) lives on the Atlantic Coast of Florida, while *A. m. peninsulae* lives on the Gulf Coast of Florida. These subspecies can interbreed if the geographical barriers to their isolation are removed.

By the late 1960s, biologists began to be concerned about the future of some seaside sparrow subspecies because their habitats were increasingly threatened by agriculture and ocean-front housing. Further, since the seaside sparrow subspecies are physically isolated from one another, and because young seaside sparrows tend to breed near where they hatched, little or no gene flow occurs among populations.

A subspecies called the dusky seaside sparrow (*A. m. nigrescens*) was in particular trouble; by 1980 only six individuals from this population remained. All were males. At this point government and private conservation agencies sprang into action under the Endangered Species Act, a law whose goal is to prevent the extinction of species and, as in this case, subspecies.

SUMMARY TABLE 27.2 **Three Most Common Species Concepts**

	Criterion for Identifying Populations as Species	Advantages	Disadvantages
Biological species	Reproductive isolation between populations (they don't breed and don't produce viable, fertile offspring)	Reproductive isolation = evolutionary independence	Not applicable to asexual or fossil species; difficult to assess if populations do not overlap geographically.
Morphospecies	Morphologically distinct populations	Widely applicable	Subjective (researchers often disagree about how much or what kinds of morphological distinction indicate speciation); misidentifies polymorphic species; misses cryptic species.
Phylogenetic species	Smallest monophyletic group on phylogenetic tree	Widely applicable; based on testable criteria	Relatively few well-estimated phylogenies are currently available.

(a) Each subspecies of seaside sparrow has a restricted range.

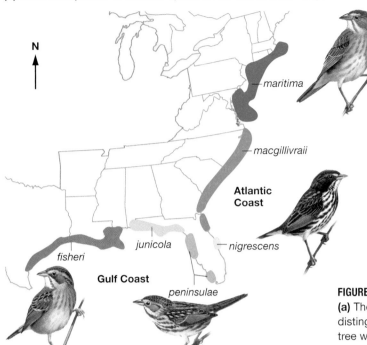

(b) The six subspecies form two monophyletic groups when DNA sequences are compared.

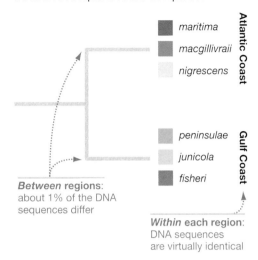

Between regions: about 1% of the DNA sequences differ

Within each region: DNA sequences are virtually identical

FIGURE 27.3 Seaside Sparrows Form Two Monophyletic Groups.
(a) The subspecies of seaside sparrows named on this map are distinguished by their distinctive coloration and songs. **(b)** This phylogenetic tree was constructed by comparing DNA sequences. The tree shows that seaside sparrows represent two distinct monophyletic groups, one native to the Atlantic Coast and the other native to the Gulf Coast.

To launch the rescue program, the remaining male dusky seaside sparrows were taken into captivity and bred with females from the nearby subspecies, *A. m. peninsulae*. Officials planned to use these hybrid offspring as breeding stock for a reintroduction program. The goal was to preserve as much genetic diversity as possible by reestablishing a healthy population of dusky-like birds.

The plan was thrown into turmoil, however, when a different group of biologists estimated the phylogeny of the seaside sparrows by comparing gene sequences. The tree in **FIGURE 27.3b** shows that seaside sparrows represent just two distinct monophyletic groups: one native to the Atlantic Coast and the other native to the Gulf Coast. Under the phylogenetic species concept, the dusky sparrow is part of the same monophyletic group that includes the other Atlantic Coast sparrows.

Officials had unwittingly crossed the dusky males, *A. m. nigrescens* from the Atlantic Coast, with females from the Gulf Coast lineage—a population that may have been geographically isolated from dusky sparrows for more than 250,000 years. Because the goal of the conservation effort was to preserve existing genetic diversity, this was the wrong population to use.

The researchers who did the phylogenetic analysis maintained that the biological and morphospecies concepts had misled a well-intentioned conservation program. Under the phylogenetic species concept, they claimed that officials should have allowed the dusky sparrow to go extinct and then concentrated their efforts on preserving one or more populations from each coast. In this way, the two monophyletic groups of sparrows—and the most genetic diversity—would be preserved.

Under the morphospecies concept, however, officials did the right thing by attempting to preserve all subspecies. They argue that dusky seaside sparrows had distinctive heritable traits, like coloration and songs, that are better "watered down" by crossbreeding than lost completely.

When it comes to conservation, life-and-death decisions like these are crucial. Now our task is to consider a fundamental question about process: How do isolation and divergence produce the event called speciation?

check your understanding

C Y U

If you understand that . . .
* Species are evolutionarily independent because no gene flow occurs between them and other species.
* Biologists use an array of criteria to identify evolutionarily independent groups.

✔ **You should be able to . . .**

1. Explain why the criteria invoked by the biological species, morphospecies, and phylogenetic species concepts allow biologists to identify evolutionarily independent groups.

2. Describe the disadvantages of the biological species, morphospecies, and phylogenetic species concepts.

Answers are available in Appendix A.

27.2 Isolation and Divergence in Allopatry

Speciation begins when gene flow between populations is reduced or eliminated, causing genetic isolation. Genetic isolation happens routinely when populations become geographically separated. Populations that are geographically separated are said to be in **allopatry** ("different-homeland"). Thus, speciation that begins with geographic isolation is known as **allopatric speciation.**

Geographic isolation, in turn, occurs in one of two ways: dispersal or **vicariance,** the physical splitting of a habitat. As **FIGURE 27.4a** illustrates, a population can disperse to a new habitat, colonize it, and found a new population. Alternatively, a new geographic barrier can split a population into two or more subgroups that are physically isolated from each other (**FIGURE 27.4b**).

The case studies that follow address two questions: First, how do dispersal and vicariance events occur? Answering this question takes us into the field of **biogeography**—the study of how species and populations are distributed geographically (see Chapter 52). And second, once populations are physically isolated, how do mutation, genetic drift, and selection produce divergence?

Allopatric Speciation by Dispersal

Peter Grant and Rosemary Grant had the good fortune to witness a colonization event while working in the Galápagos Islands off the coast of South America. The Grants have been studying finches on the island of Daphne Major since 1971 (Chapter 25).

In 1973 the Grants began to observe a yearly migration of large ground finches from their home island to Daphne Major for a few months between breeding seasons. Finally, in 1982, thirteen colonists stayed on Daphne Major throughout the breeding season, and five of them produced offspring. These colonists represented a new population, allopatric with their migrant source population, because the two populations bred and nested on different islands.

Could this dispersal event lead to speciation? To evaluate this question, the Grants caught, weighed, and measured most of the parents and offspring produced on Daphne Major over the succeeding 12 years. When they compared these data with measurements of large ground finches in the migrant population, they discovered that the average beak size in the colonist population was much larger.

Two evolutionary processes could be responsible for the change in beak size:

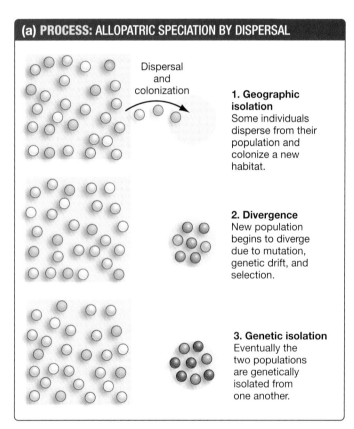

(a) PROCESS: ALLOPATRIC SPECIATION BY DISPERSAL

1. Geographic isolation
Some individuals disperse from their population and colonize a new habitat.

2. Divergence
New population begins to diverge due to mutation, genetic drift, and selection.

3. Genetic isolation
Eventually the two populations are genetically isolated from one another.

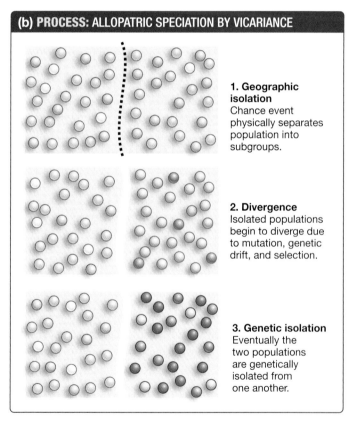

(b) PROCESS: ALLOPATRIC SPECIATION BY VICARIANCE

1. Geographic isolation
Chance event physically separates population into subgroups.

2. Divergence
Isolated populations begin to diverge due to mutation, genetic drift, and selection.

3. Genetic isolation
Eventually the two populations are genetically isolated from one another.

FIGURE 27.4 Allopatric Speciation Begins via Dispersal or Vicariance. (a) When dispersal occurs, colonists establish a new population in a novel location. **(b)** In vicariance, a widespread population becomes fragmented into isolated subgroups. Different colors represent genetic variation.

✔**QUESTION** A continuous salt marsh once existed across what is now central Florida, but it was disrupted by geologic events. How does this discovery support the phylogenetic species concept in seaside sparrows (see Figure 27.3)?

1. The founder event produced a colonizing population that happened to have particularly large beaks relative to the migrant population. This is an example of genetic drift (Chapter 26).

2. Natural selection in the new environment could favor alleles associated with large beaks.

The Grants concluded from their detailed observations that both genetic drift and natural selection were at play.

The new population of large ground finches is not yet considered a separate species from the migrant population, because some gene flow continues to occur—such as when a new migrant joins the colonist population. Given enough time, however, the populations may continue to diverge. Dispersal and colonization, followed by genetic drift and natural selection, are thought to be responsible for speciation in Galápagos finches and many other island groups.

Allopatric Speciation by Vicariance

If a new physical barrier such as a mountain range or river splits the geographic range of a species, vicariance has taken place. Such changes are commonplace on Earth due to continental drift (Chapter 28), climate fluctuations, and many other factors (see Chapter 52). For example, geologists estimate that the Isthmus of Panama closed about 3 million years ago, forming a land bridge between North and South America. This vicariance event split the Caribbean and Pacific populations of many marine species, including snapping shrimp (**FIGURE 27.5a**).

A phylogenetic tree of the snapping shrimp shows that many of the species found on either side of the isthmus are **sister species**—meaning that they are each other's closest relative (**FIGURE 27.5b**). This is exactly the pattern predicted if the populations of many species were split in two: the populations on either side of the isthmus subsequently diverged to form distinct species.

To summarize, geographic isolation of populations via dispersal or vicariance produces genetic isolation due to the interruption of gene flow—the first requirement of speciation. When genetic isolation is accompanied by genetic divergence due to mutation, selection, and genetic drift, speciation results.

27.3 Isolation and Divergence in Sympatry

When populations or species live in the same geographic area, or at least close enough to one another to make interbreeding possible, biologists say that they live in **sympatry** ("together-homeland"). Traditionally, evolutionary biologists such as Ernst Mayr predicted that speciation could not occur among sympatric populations because gene flow would easily overwhelm any differences among populations created by genetic drift and natural selection.

But recently, a number of studies have countered this prediction. **Sympatric speciation**—speciation that occurs even

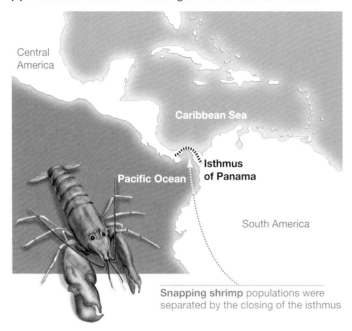

(a) **Vicariance event:** The closing of the Isthmus of Panama

Central America

Caribbean Sea

Pacific Ocean

Isthmus of Panama

South America

Snapping shrimp populations were separated by the closing of the isthmus

(b) **Result:** Pairs of sister species straddling the isthmus

Pacific #1
Caribbean #1

Pacific #2
Caribbean #2

Pacific #3
Caribbean #3

Pacific #4
Caribbean #4

Pacific #5
Caribbean #5

Pacific #6
Caribbean #6

Snapping shrimp species are more closely related to sister species on the other side of the Isthmus than to other species on the same side

FIGURE 27.5 Evidence for Speciation by Vicariance in Snapping Shrimp. (a) The Isthmus of Panama separated the Caribbean Sea from the Pacific Ocean about 3 million years ago. **(b)** The most logical interpretation of this phylogenetic tree is that 6 species of snapping shrimp split into 12 species when the Isthmus of Panama formed.

though populations live within the same geographical area (see **FIGURE 27.6** on page 496)—can indeed occur, and serves as a source of new branches on the tree of life. How does it work?

Two types of events can initiate the process of sympatric speciation: **(1)** external events, such as disruptive selection for extreme phenotypes based on different ecological niches (see Chapter 26 for a review of disruptive selection), and **(2)** internal events, such as chromosomal mutations. Let's consider each one in turn.

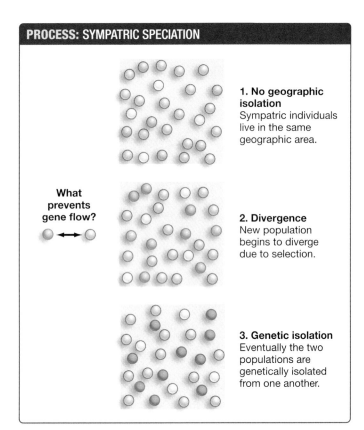

PROCESS: SYMPATRIC SPECIATION

1. No geographic isolation Sympatric individuals live in the same geographic area.

What prevents gene flow?

2. Divergence New population begins to diverge due to selection.

3. Genetic isolation Eventually the two populations are genetically isolated from one another.

FIGURE 27.6 Sympatric Speciation Has Long Perplexed Researchers. Recent studies have documented several mechanisms that reduce gene flow and result in sympatric speciation. Different colors represent genetic variation.

Sympatric Speciation by Disruptive Selection

Biologists use the term **niche** (pronounced *nitch*) to describe the range of ecological resources that a species can use and the range of conditions that it can tolerate (see Chapter 55). A key realization in the study of sympatric speciation was that even though sympatric populations are not geographically isolated, they may become reproductively isolated by adapting to different ecological niches via disruptive selection.

For example, apple maggot flies rely on apples to complete their life cycle—apples are an important part of their niche. Male and female apple maggot flies usually court and mate on apple fruits. The female then lays a fertilized egg inside the fruit, which will be the food source for the growing larva. After the fruit drops off the tree, the larva burrows into the ground and pupates—meaning that it secretes a protective case and undergoes metamorphosis (see Chapter 33). The new adult then emerges the following spring, starting the cycle anew.

Apple trees were introduced to North America from Europe less than 300 years ago, however. Where did apple maggot flies come from? Did they arrive with the apple trees?

Evolutionary biologist Jeffrey Feder and colleagues have been researching this mystery in a long-term study of apple maggot flies and their closest relatives. Phylogenetic trees, estimated from synapomorphies in DNA sequence data, indicate that apple maggot flies are very closely related to hawthorn maggot flies,

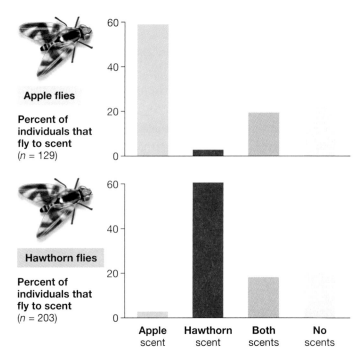

Apple flies

Percent of individuals that fly to scent (*n* = 129)

Hawthorn flies

Percent of individuals that fly to scent (*n* = 203)

Apple scent | Hawthorn scent | Both scents | No scents

FIGURE 27.7 Disruptive Selection on Fruit Preference in Flies. Each fly was tested with four types of scent, one at a time, in a lab setting.

DATA: Dambroski, H. R., C. Linn Jr., S. H. Berlocher, et al. 2005. *Evolution* 59: 1953–1964.

✔**QUESTION** Apple maggot flies mate on their food plants. How is this behavior relevant to the population's divergence from hawthorn flies?

which are native to North America. These data suggest that apple maggot flies originated from hawthorn maggot flies following the introduction of apples. Hawthorn flies lay their eggs in hawthorn fruits and apple flies lay their eggs in apples, even though hawthorn trees and apple trees often grow almost side by side. Do the apple flies and hawthorn flies interbreed?

By following marked individuals in the field, Feder's team determined that only about 6 percent of the matings observed are between apple flies and hawthorn flies. The data in **FIGURE 27.7** show why. The bars on the graphs indicate the percentage of apple flies (top) or hawthorn flies (bottom) that land on a surface containing scents from apple, hawthorn, both apple and hawthorn, or neither apple nor hawthorn, in laboratory tests. Notice that

- Apple flies respond most strongly to apple scents; hawthorn flies respond most strongly to hawthorn scents.

- In both types of flies, there is no difference in the response to a mix of both scents and no scent at all.

- Apple flies *avoid* hawthorn scent, and hawthorn flies *avoid* apple scent.

Other experiments have established that (**1**) a fly's ability to discriminate scents has a genetic basis—meaning that apple flies and hawthorn flies have different alleles associated with attraction to fruit; (**2**) specific odor receptor cells are responsible for the difference in scent response; and (**3**) hybrid individuals do not orient to fruit scents as well as their parents.

The upshot is that although apple flies and hawthorn flies live in the same geographic area, prezygotic reproductive isolation is occurring as a result of natural selection for adaptations to two different niches. Apple flies mate on apples (avoiding hawthorn fruits) and hawthorn flies mate on hawthorn fruits (avoiding apples). Hybrid flies have lower fitness due to their reduced success in finding fruits and thus mates, an indication that disruptive selection is occurring (see Chapter 26).

Although they are not yet separate species on the basis of the biological, morphological, or phylogenetic species concepts, apple flies and hawthorn flies are diverging. They are currently in the process of becoming distinct species.

Follow-up studies have shown that the genetic divergence of apple flies and hawthorn flies is not limited to a few alleles associated with attraction to fruit. The divergence of their genomes is widespread—remarkable, given the short time since they began to diverge.

✔ If you understand the sympatric speciation process occurring in these flies, you should be able to hypothesize why natural selection would favor divergence based on the observation that apple fruits drop about 3–4 weeks earlier in the fall than hawthorn fruits.

Although the apple maggot fly's story might seem localized and specific, the events may be common. Biologists currently estimate that over 3 million insect species exist. Most of these species are associated with specific host plants, and thus they occupy specific niches. Based on the data from apple maggot flies, it is reasonable to hypothesize that switching host plants has been a major trigger for speciation throughout the course of insect evolution. Further, speciation itself can trigger more speciation. Researchers have described a species of parasitic wasp that specializes on apple maggot flies.

Sympatric Speciation by Polyploidization

Based on the theory and data reviewed thus far, it is clear that gene flow, genetic drift, and natural selection play important roles in speciation. Can the fourth evolutionary process—mutation—influence speciation as well?

Even though mutation is the ultimate source of genetic variation in populations, it is an inefficient mechanism of evolutionary change on its own—the rate of mutation tends to be low (Chapter 26). Natural selection and genetic drift are usually responsible for amplifying the effects of mutations.

One particular type of mutation, though, turns out to be extremely important in speciation—especially in plants. **Polyploidy** occurs when an error in meiosis or mitosis results in a doubling of the chromosome number—a massive mutation. There are two types of polyploids:

1. **Autopolyploid** ("same-many-form") individuals are produced when a mutation results in a doubling of chromosome number and the chromosomes all come from the same species (**FIGURE 27.8a**).

(a) Formation of **autopolyploid**

(b) Formation of **allopolyploid**

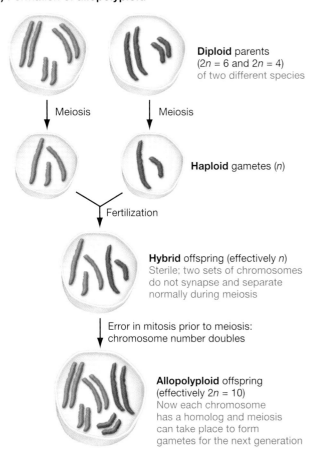

FIGURE 27.8 Autopolyploids and Allopolyploids. (a) The key event leading to autopolyploidy is nondisjunction during meiosis, resulting in diploid gametes rather than haploid gametes. **(b)** Allopolyploidy can occur by different methods; one example is shown. The key in this case is an error during mitosis that leads to the duplication of chromosomes, enabling the pairing of homologs. In both cases, the polyploids can produce gametes that can self-fertilize.

2. **Allopolyploid** ("different-many-form") individuals are created when parents that belong to different species mate and produce an offspring with two different sets of chromosomes (**FIGURE 27.8b**).

How can these two types of polyploidy lead to speciation? Let's consider a specific example of each.

Autopolyploidy Autopolyploidy is thought to be much less common than allopolyploidy. One example is the maidenhair fern, native to woodlands across North America.

Biologists initially set out to do a routine survey of allelic diversity in a population of these ferns. By chance, they stumbled upon individuals that could produce offspring that were tetraploid (4*n*) rather than diploid—that is, polyploid mutants within a normal population.

To follow up on the observation, researchers studied the parents of the mutant individuals. The parents turned out to have a defect in meiosis that caused nondisjunction of chromosomes (introduced in Chapter 13). Instead of producing normal, haploid cells as a result of meiosis, the mutant individuals produced diploid cells. These diploid cells eventually led to the production of diploid gametes (see Figure 27.8a). Because maidenhair ferns can self-fertilize, the diploid gametes could combine to form tetraploid offspring.

How could this event lead to speciation? The tetraploid offspring can self-fertilize or mate with their tetraploid parent or each other, but not with other diploids in the population. Consider what happens when that tetraploid individual mates with a diploid individual (**FIGURE 27.9**).

- During meiosis, tetraploid individuals produce diploid gametes, and diploid individuals produce haploid gametes. These gametes unite to form a triploid (3*n*) zygote.

- Even if this offspring develops normally and reaches sexual maturity, its three homologous chromosomes cannot synapse and separate correctly during meiosis. Thus, the chromosomes are distributed to daughter cells unevenly.

- Because almost all of its gametes contain an uneven number of chromosomes, the triploid individual is virtually sterile.

Polyploid individuals are reproductively isolated from the original diploid population and thus evolutionarily independent, because breeding between diploids and tetraploids generally results in sterile offspring. Thus, according to the biological species concept, speciation occurs in a single generation—instantaneous in evolutionary terms.

✔ If you understand how autopolyploidy works, you should be able to create a scenario explaining how the process gave rise to a tetraploid grape with extra-large fruit, from a diploid population with smaller fruit. (You've probably seen both types of fruit in the supermarket.)

Allopolyploidy As Figure 27.8b illustrates, new polyploid species may be created when two species hybridize. The drawing shows a hybrid offspring forming from a mating between two diploid species. Because the offspring has chromosomes that do

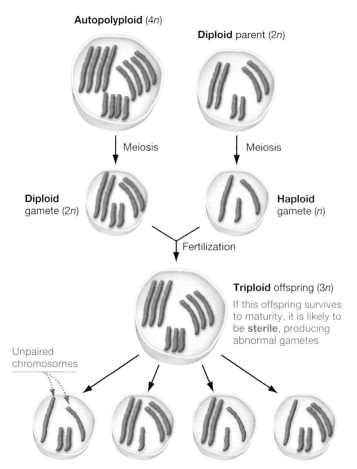

FIGURE 27.9 Polyploidy Can Lead to Reproductive Isolation. The mating diagrammed here illustrates why tetraploid individuals are reproductively isolated from diploid individuals.

not pair normally during meiosis, it is sterile. But if a mutation occurs during mitosis that doubles the chromosome number in this individual before meiosis, then each chromosome gains a homolog and meiosis can proceed normally.

Notice that the allopolyploid offspring in Figure 27.8b appears diploid in the sense that it has two copies of each chromosome (2*n* = 10). However, it is functionally polyploid because it has two copies each of *two sets* of chromosomes (indicated by blue and red). In terms of genetic information, this offspring has four copies of genes that are common to both parent species, instead of just two.

Exactly this chain of events occurred after three European species of weedy plants in the genus *Tragopogon* were introduced to western North America in the early 1900s. In 1950 a botanist described the first of two new allopolyploid species (**FIGURE 27.10**). Based on an analysis of chromosomes, both were clearly the descendants of matings between the introduced diploids.

Follow-up work by plant geneticists Pam and Doug Soltis has shown that at least one of the new allopolyploid species is expanding its geographic range and outcompeting its parental species. Further, this polyploidization event has occurred multiple times between the same parent species. The Soltises raise an

Tragopogon dubius
(2n = 12)

Tragopogon porrifolius
(2n = 12)

Diploid species
introduced to
North America

Allopolyploid species
This polyploidization event
occurred multiple times

Tragopogon mirus
(2n = 24)

FIGURE 27.10 Allopolyploids Can Form New Species.

interesting question: Should the resulting polyploid populations be considered the same or different daughter species?

✔ If you understand how allopolyploidy works, you should be able to create a scenario explaining how a cross between a tetraploid population called Emmer wheat and a wild, diploid wheat gave rise to the hexaploid bread wheat grown throughout the world today.

Why Is Speciation by Polyploidy so Common in Plants? The claim that speciation by polyploidization has been particularly important in plants is backed by the observation that many diploid species have close relatives that are polyploid. Why have polyploids been so successful? The Soltises have addressed this question by examining the genetic implications of polyploidy:

1. Polyploids have higher levels of heterozygosity than their diploid relatives.

2. Polyploids can tolerate higher levels of self-fertilization because they are not as affected by inbreeding depression as their diploid relatives. (See Chapter 26 to review inbreeding.)

3. Genes on duplicated chromosomes can diverge independently, increasing genetic variation in the population.

Genetic variation is a prerequisite for evolution (Chapter 26). Thus, the high genetic diversity of polyploids has enabled a rapid diversification of plant species.

To summarize, speciation by polyploidization is driven by chromosome-level mutations and occurs in sympatry. Compared to the gradual process of speciation by geographic isolation or by disruptive selection in sympatry, speciation by polyploidization is virtually instantaneous. It is fast, sympatric, and common.

If you understand that . . .

• Speciation occurs when populations become isolated genetically and then diverge due to selection, genetic drift, or mutation.

✔ **You should be able to . . .**

Give an example of an event that can lead to the genetic isolation of populations, and explain why selection, drift, and mutation would cause the populations to diverge.

Answers are available in Appendix A.

27.4 What Happens When Isolated Populations Come into Contact?

Suppose two populations that have been isolated come into contact again. If divergence has taken place and if divergence has affected when, where, or how individuals in the populations mate, then interbreeding is unlikely to take place. In cases such as this, prezygotic isolation exists. When it does, mating between the populations is rare, gene flow is minimal, and the populations continue to diverge.

But what if prezygotic isolation does not exist, and the populations begin interbreeding? The simplest outcome is that the populations fuse over time, as gene flow erases any distinctions between them. Several other possibilities exist, however. Let's explore three of them: reinforcement, hybrid zones, and speciation by hybridization.

Reinforcement

If two populations have diverged extensively and are distinct genetically, it is reasonable to expect that their hybrid offspring will have lower fitness than their parents. This phenomenon has been observed in many cases, such as in the apple maggot fly example. If two populations are well adapted to different habitats, then hybrid offspring will not be well adapted to either habitat. If the two populations have diverged enough genetically, hybrid offspring also may fail to develop normally or may be infertile (see postzygotic isolation in Table 27.1).

When postzygotic isolation occurs, there should be strong selection against interbreeding because hybrid offspring represent a wasted effort on the part of parents—especially for females, due to their typically higher investment in offspring. Individuals that do not interbreed, because of different courtship ritual or pollination system or other form of prezygotic isolation, should be favored because they produce more viable offspring.

Natural selection for traits that isolate populations in this way is called **reinforcement.** The name is descriptive because the selected traits reinforce differences that evolved while the populations were isolated from one another.

Some of the best data on reinforcement come from laboratory studies of closely related fruit fly species in the genus *Drosophila*. Evolutionary biologists Jerry Coyne and Allen Orr analyzed a large series of experiments that tested whether members of closely related fly species are willing to mate with one another. They found an interesting pattern:

- If closely related species are sympatric—meaning that they live in the same area—individuals from the two species are seldom willing to mate with one another.

- If the species are allopatric—meaning that they live in different areas—then individuals from the two species are often willing to mate with one another.

The pattern is logical because natural selection can act to reduce mating between species only if their ranges overlap. Thus, it is reasonable to find that sympatric species exhibit prezygotic isolation but that allopatric species do not. There is a long-standing debate, however, over just how important reinforcement is in other species.

Hybrid Zones

Hybrid offspring are not always dysfunctional. In some cases they are capable of mating and producing viable offspring that have features that are intermediate between those of the two parental populations. When this is the case, hybrid zones can form. A **hybrid zone** is a geographic area where interbreeding occurs and hybrid offspring are common.

Depending on the fitness of hybrid offspring and the extent of breeding between parental species, hybrid zones can be narrow or wide, and long or short lived. As an example of how researchers analyze the dynamics of hybrid zones, let's consider recent work on two bird species.

Townsend's warblers and hermit warblers live in the coniferous forests of North America's Pacific Northwest. In western Washington State, where their ranges overlap, the two species hybridize extensively. As the drawings in **FIGURE 27.11** show, hybrid offspring have characteristics that are intermediate relative to the two parental species.

To explore the dynamics of this hybrid zone, ornithologist Sievert Rohwer teamed up with population geneticist Eldredge Bermingham and others to examine gene sequences in the mitochondrial DNA (mtDNA) of a large number of Townsend's, hermit, and hybrid warblers collected from forests throughout the region. The team found that each of the parental species has certain species-specific mtDNA sequences. This result allowed the researchers to infer how hybridization was occurring.

To grasp the reasoning here, it is critical to realize that mtDNA is maternally inherited in most animals and plants. If a hybrid individual has Townsend's mtDNA, its mother had to be a Townsend's warbler while its father had to be a hermit warbler. In this way, identifying mtDNA types allowed the research team to infer whether Townsend's females were mating with hermit males, or vice versa, or both.

Their data presented a clear pattern: Most hybrids form when Townsend's males mate with hermit warbler females.

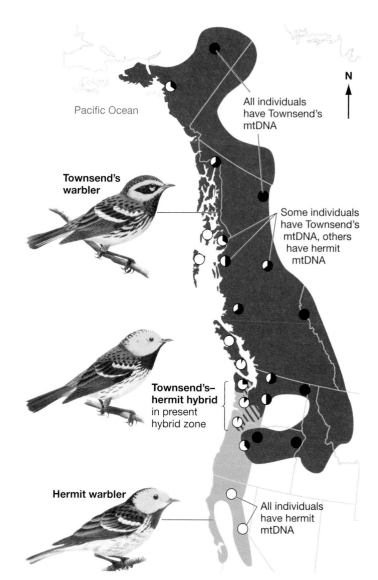

FIGURE 27.11 Analyzing a Hybrid Zone. When Townsend's warblers and hermit warblers hybridize, the offspring have intermediate characteristics. The current ranges of Townsend's warblers (in red) and hermit warblers (in orange) are shown in the map. The small pie charts show the percentage of individuals with Townsend's warbler mtDNA (in black) and hermit warbler mtDNA (in white).

DATA: Rohwer, S., E. Bermingham, and C. Wood. 2001. *Evolution* 55: 405–422.

Rohwer followed up on this result with experiments showing that Townsend's males are extremely aggressive in establishing territories and that they readily attack hermit warbler males. The data suggest that Townsend's males invade hermit territories, drive off the hermit males, and mate with hermit females.

The team also found something completely unexpected. When they analyzed the distribution of mtDNA types along the Pacific Coast and in the northern Rocky Mountains, they discovered that many Townsend's warblers actually had hermit mtDNA. Figure 27.11 shows that in some regions—such as the larger islands off the coast of British Columbia—*all* of the warblers had hermit mtDNA (indicated on the map by white pie charts), even though they looked like full-blooded Townsend's warblers (indicated on the map by the red distribution).

How could this be? To explain the result, the team hypothesized that hermit warblers were once found as far north as Alaska and that Townsend's warblers have gradually taken over their range. Their logic is that repeated mating with Townsend's warblers over time made the hybrid offspring look more and more like Townsend's, even while maternally inherited mtDNA kept the genetic record of the original hybridization event intact.

If this hypothesis is correct, then the hybrid zone should continue moving south. If it does so, hermit warblers may eventually become extinct. In some cases, however, hybridization does not lead to extinction but rather leads to the opposite—the creation of new species.

New Species through Hybridization

If two species interbreed and produce hybrid offspring that can not only survive and reproduce but also possess a unique combination of traits that happen to be adaptive in their particular environment, a new species may result. How would you identify such a species in nature?

Evolutionary biologist Loren Rieseberg and colleagues recently examined the relationships of three sunflower species native to the Southwest: *Helianthus annuus* and *H. petiolaris* are known to hybridize in regions where their ranges overlap, and *H. anomalus* resembles these hybrids (see photo at the start of this chapter). How are these species related?

DNA sequencing studies have shown that some gene regions in *H. anomalus* are remarkably similar to those found in *H. annuus*, while other *H. anomalus* gene sequences are almost identical to those found in *H. petiolaris*. Based on these data, Rieseberg hypothesized that *H. anomalus* originated in hybridization between *H. annuus* and *H. petiolaris*. All three species have the same number of chromosomes, so neither allopolyploidy nor autopolyploidy was involved. Instead, the chromosomes of *H. annuus* and *H. petiolaris* must be similar enough that they can synapse and undergo normal meiosis in hybrid offspring.

The specific hypothesis here is that *H. annuus* and *H. petiolaris* were isolated and diverged as separate species, and later began interbreeding. The hybrid offspring created a third, new species that had unique combinations of alleles from each parental species and therefore different characteristics. This hypothesis is supported by the observation that *H. anomalus* grows in much drier habitats than either of the parental species—suggesting that a unique combination of alleles allowed *H. anomalus* to thrive in dry habitats.

Rieseberg and his colleagues set out to test the hybridization hypothesis by trying to re-create the speciation event experimentally (**FIGURE 27.12**).

Step 1 They mated individuals from the two parental species and raised the offspring in a greenhouse.

Step 2 When these hybrid individuals were mature, the researchers either mated the plants to other hybrid individuals or "backcrossed" them to individuals from one of the parental species.

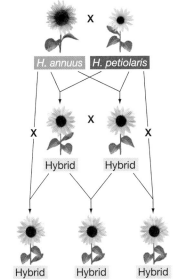

RESEARCH

QUESTION: Can new species arise by hybridization between existing species?

HYPOTHESIS: *Helianthus anomalus* originated by hybridization between *H. annuus* and *H. petiolaris*.

NULL HYPOTHESIS: *Helianthus anomalus* did not originate by hybridization between *H. annuus* and *H. petiolaris*.

EXPERIMENTAL SETUP:

H. annuus X *H. petiolaris*

1. Mate *H. annuus* and *H. petiolaris* and raise offspring.

X   Hybrid   X   Hybrid   X

2. Mate F$_1$ hybrids or backcross F$_1$s to parental species; raise offspring.

Hybrid   Hybrid   Hybrid

3. Repeat for four more generations.

PREDICTION: Experimental hybrids will have the same mix of *H. annuus* and *H. petiolaris* genes as natural *H. anomalus*.

PREDICTION OF NULL HYPOTHESIS: Experimental hybrids will not have the same mix of *H. annuus* and *H. petiolaris* genes as natural *H. anomalus*.

RESULTS:
DNA comparison of a chromosomal region called S:

(Only colored portions of S region were analyzed)

Experimental hybrid (cross of *H. annuus* and *H. petiolaris*)

Alignment

H. anomalus (naturally occurring species)

Gene region similar to *H. annuus*   Gene region similar to *H. petiolaris*

CONCLUSION: New species may arise via hybridization between existing species.

FIGURE 27.12 Experimental Evidence That New Species Can Originate in Hybridization Events.

SOURCE: Rieseberg, L. H., B. Sinervo, C. R. Linder, et al. 1996. Role of gene interactions in hybrid speciation: Evidence from ancient and experimental hybrids. *Science* 272: 741–745.

✔QUESTION Why it is valid to use experiments with living organisms—like this one—to infer what happened during historical speciation events?

	Process	Example
Fusion of the populations	The two populations freely interbreed.	Occurs whenever populations of the same species come into contact.
Reinforcement of divergence	If hybrid offspring have low fitness, natural selection favors the evolution of traits that prevent interbreeding between the populations.	Appears to be common in fruit fly species that occupy the same geographic areas.
Hybrid zone formation	Hybridization occurs in a well-defined geographic area. This area may move over time or be stable.	Many stable hybrid zones have been described; the hybrid zone between hermit and Townsend's warblers appears to have moved over time.
Extinction of one population	If one population or species is a better competitor for shared resources, then the poorer competitor may be driven to extinction.	Townsend's warblers may be driving hermit warblers to extinction.
Creation of new species	If the combination of genes in hybrid offspring allows them to occupy distinct habitats or use novel resources, they may form a new species.	Hybridization between sunflowers gave rise to a new species with unique characteristics.

Step 3 This breeding program continued for four more generations before the experiment ended. Ultimately, the experimental lines were backcrossed twice, and they were mated to other hybrid offspring three times.

The goal of these crosses was to simulate matings that might have occurred naturally.

The experimental hybrids looked like the natural hybrid species, but did they resemble them genetically? To answer this question, Rieseberg's team constructed genetic maps of each population, using a large series of genetic markers (see Chapter 20). Because each parental population had many unique markers in its genomes, the research team hoped to identify which genes found in the experimental hybrids came from which parental species.

Some of their data are diagrammed in the "Results" section of Figure 27.12. The bottom bar in the illustration represents a region called S in the genome of the naturally occurring species *Helianthus anomalus*. As the legend indicates, this region contains three sections of sequences that are also found in *H. petiolaris* (indicated with the color red) and two that are also found in *H. annuus* (indicated with the color orange). The top bar shows the composition of this same region in the genome of the experimental hybrid lines.

The key observation here is that the genetic composition of the synthesized hybrids matches that of the naturally occurring hybrid species. In effect, the researchers succeeded in re-creating a speciation event. Their results provide strong support for the hybridization hypothesis for the origin of *H. anomalus*. ✔ If you understand this experiment, you should be able to suggest one result that would have caused the researchers to reject the hybridization hypothesis.

Secondary contact of two populations can produce a dynamic range of possible outcomes: fusion of the populations, reinforcement of divergence, founding of stable hybrid zones, extinction of one population, or the creation of new species. **TABLE 27.3** summarizes the outcomes of secondary contact.

The study of speciation provides an essential link between the processes of evolution (Chapters 25 and 26) and the tree of life (Chapter 28 and Unit 6). Research on speciation is accelerating, in part due to availability of new genetic and phylogenetic tools, and in part due to the study of human impacts on biodiversity (see Chapter 57). Although human activities usually result in the extinction of species, they can also result in the creation of new species. Humans are changing the shape of the tree of life.

If you understand . . .

27.1 How Are Species Defined and Identified?

- A species is defined as an evolutionarily independent population or group of populations. Researchers use several criteria to identify whether populations represent distinct species.

- The biological species concept uses reproductive isolation as a criterion to identify species.

- The morphospecies concept identifies species using distinctive morphological traits.

- The phylogenetic species concept identifies species as the smallest monophyletic groups on the tree of life.

✓ You should be able to explain whether human populations would be considered separate species under the biological species, morphospecies, and phylogenetic species concepts.

27.2 Isolation and Divergence in Allopatry

- Speciation is a splitting event in which one lineage gives rise to two or more independent descendant lineages.

- Speciation occurs when populations of the same species become genetically isolated by lack of gene flow and then diverge from each other due to selection, genetic drift, and mutation.

- Allopatric speciation occurs when populations diverge in geographic isolation.

- Geographic isolation occurs through dispersal, when small groups of individuals colonize a new habitat, or through vicariance, when a large, continuous population becomes fragmented into isolated habitats.

✓ You should be able to design an experiment that would, given enough time, result in the production of two species from a single ancestral population.

27.3 Isolation and Divergence in Sympatry

- Sympatric speciation occurs when populations diverge genetically despite living in the same geographic area.

- Sympatric speciation can occur when disruptive selection favors individuals that breed in different ecological niches.

- Mutations that produce polyploidy can trigger rapid speciation in sympatry because they lead to reproductive isolation between diploid and tetraploid populations.

✓ You should be able to evaluate whether your experiment on speciation (see previous exercise under 27.2) represents a case of dispersal, vicariance, different habitat use, or polyploidy.

27.4 What Happens When Isolated Populations Come into Contact?

- If gene flow occurs, populations that have diverged may fuse into a single species.

- If prezygotic isolation exists, populations that come back into contact will probably continue to diverge.

- Secondary contact can lead to reinforcement—the evolution of mechanisms that prevent hybridization.

- Gene flow between different species can lead to the formation of hybrid zones that move over time or are stable.

- In some cases, hybridization between species can create new species with unique combinations of traits.

✓ You should be able to predict how a hybrid zone will change over time when hybrid offspring have higher fitness than the parental populations.

(MB) MasteringBiology

1. **MasteringBiology Assignments**

   Tutorials and Activities Allopatric Speciation, Defining Species, Polyploid Plants, Speciation by Changes in Ploidy

   Questions Reading Quizzes, Blue-Thread Questions, Test Bank

2. **eText** Read your book online, search, take notes, highlight text, and more.

3. **The Study Area** Practice Test, Cumulative Test, BioFlix® 3-D Animations, Videos, Activities, Audio Glossary, Word Study Tools, Art

You should be able to . . .

1. What distinguishes a morphospecies?
   a. It has distinctive characteristics, such as size, shape, or coloration.
   b. It represents a distinct twig in a phylogeny of populations.
   c. It is reproductively isolated from other species.
   d. It is a fossil from a distinct time in Earth history.

2. Which of the following describes vicariance?
   a. Small populations coalesce into one large population.
   b. A population is fragmented into isolated subpopulations.
   c. Individuals colonize a novel habitat.
   d. Individuals disperse and found a new population.

3. Why are genetic isolation and genetic divergence occurring in apple maggot flies, even though populations occupy the same geographic area?
   a. Different populations feed and mate on different types of fruit.
   b. One population is tetraploid; others are diploid.
   c. The introduction of a nonnative host plant caused vicariance.
   d. Responses to scents have changed due to disruptive selection.

4. The biological species concept can be applied only to which of the following groups?
   a. bird species living today
   b. dinosaurs and bird species living today
   c. dinosaurs
   d. bacteria

5. Why is "reinforcement" an appropriate name for the concept that natural selection should favor divergence and genetic isolation if populations experience postzygotic isolation?
   a. Selection should reinforce high fitness for hybrid offspring.
   b. Selection should reinforce the fact that the diverging populations are "good species" under the morphological species concept.
   c. Hybrid offspring do not develop at all or are sterile when mature because selection reinforces only the success of purebred species.
   d. The selected traits reinforce differences that evolved while the populations were isolated from one another.

6. True or False? Gene flow increases the divergence of populations.

7. Which studies in this chapter represent direct observation of speciation? Which are indirect studies of historical speciation events?

8. When the ranges of two different species meet, a stable "hybrid zone" occupied by hybrid individuals may form. How is this possible?
   a. Hybrid individuals may have intermediate characteristics that are advantageous in a given region.
   b. Hybrid individuals are always allopolyploid and are thus unable to mate with either of the original species.
   c. Hybrid individuals may have reduced fitness and thus be strongly selected against.
   d. One species has a selective advantage, so as hybridization continues, the other species will go extinct.

9. In the case of the seaside sparrow, how did the species identified by the biological species concept, the morphospecies concept, and the phylogenetic species concept conflict?

10. Explain how isolation and divergence are occurring in apple maggot flies. Of the four evolutionary processes (mutation, gene flow, genetic drift, and selection), which two are most important in causing this event?

11. A large amount of gene flow is now occurring among human populations due to intermarriage among people from different ethnic groups and regions of the world. Is this phenomenon increasing or decreasing racial differences in our species? Explain.

12. **QUANTITATIVE** If one species ($2n = 10$) crosses with another species ($2n = 18$), producing an allopolyploid offspring, what is the ploidy of the offspring?
    a. $2n = 10$
    b. $2n = 18$
    c. $2n = 18 + 10 = 28$
    d. $4n = 36 + 20 = 56$

13. Unlike animal gametes, plant reproductive cells do not differentiate until late in life. Why are plants much more likely than animals to produce diploid gametes and polyploid offspring?
    a. Cells that do not divide for long periods accumulate mutations.
    b. Cells that differentiate later in life undergo many rounds of mitosis, so are more likely to accumulate mutations.
    c. Cells that differentiate later in life undergo many rounds of meiosis, so are more likely to accumulate mutations.
    d. Undifferentiated cells are more prone to mutations than differentiated cells.

14. Sexual selection is a type of natural selection that favors individuals with traits that increase their ability to obtain mates,

such as female choice for bright orange beaks in zebra finches. Propose a scenario where sexual selection could contribute to divergence in sympatric speciation.

15. A friend says that apple maggot flies prefer apple fruit scents because they need to, in order to survive. Another agrees and adds that the flies acquire the ability to distinguish the apple scents by spending time on the fruit, and that's why their offspring prefer apples. What's wrong with these statements?

16. All over the world, natural habitats are being fragmented into tiny islands as suburbs, ranches, and farms expand, and roads are built to connect them. Explain why this fragmentation process could lead to extinction. Then explain how it could lead to speciation.

28 Phylogenies and the History of Life

In this chapter you will learn that

Vast amounts of change have occurred in the 3.6-billion-year history of life

by asking

by asking

How can we study the history of life?

How does the diversity of life undergo big changes?

via | via | via | via

Phylogenetic trees 28.1

The fossil record 28.2

Adaptive radiation 28.3

Mass extinction 28.4

A fossilized trilobite. The last trilobites disappeared during a mass extinction event analyzed in Section 28.4.

This chapter is part of the Big Picture. See how on pages 526–527.

This chapter is about time and change. More specifically, it's about vast amounts of time and profound change in organisms. Both of these topics can be difficult for humans to grasp. Our lifetimes are measured in decades, and our knowledge of history is usually measured in centuries or millennia. But this chapter analyzes events that occurred over millions and even billions of years.

It takes practice to get comfortable analyzing the profound changes that occur in organisms over deep time. To help you get started, the chapter begins by introducing the two major analytical tools that biologists use to reconstruct the history of life: phylogenetic trees and the fossil record. The remaining two sections explore evolutionary episodes called adaptive radiations and mass extinctions, which can cause major changes to the tree of life. The Big Picture of Evolution on pages 526–527 shows how the tree of life relates to speciation (Chapter 27) and the four evolutionary processes (Chapters 25 and 26).

✔ When you see this checkmark, stop and test yourself. Answers are available in Appendix A.

The evolutionary history of a group of organisms is called its **phylogeny.** A **phylogenetic tree** is a graphical summary of this history, showing the ancestor–descendant relationships among populations, species, or higher taxa, and clarifying who is related to whom. The **tree of life** is the most universal of all phylogenetic trees, depicting the evolutionary relationships among all living organisms on Earth (see Chapter 1). Phylogenetic trees have revolutionized the study of evolution and many related fields. For example:

- Phylogenetic trees can be used in taxonomy to define species (Chapter 27).

- Phylogenetic trees provide evidence for the endosymbiosis theory—that mitochondria and chloroplasts originated as free-living bacteria (Chapter 30).

- Phylogenetic trees can be used to examine how the HIV virus jumped from other primates to humans (Chapter 36).

- Phylogenetic trees can aid in identifying species that are a priority for conservation (Chapter 57).

Phylogenetic trees are extremely effective at summarizing data and testing hypotheses. However, they are unusual diagrams—it can take practice to interpret them correctly.

Take a moment to review the parts of a phylogenetic tree in **TABLE 28.1**. Notice that the tree in this table, like all the trees used in this text, is rooted, which means that the **branches** on the left are more ancient, or ancestral, while the ones on the right are most recent, or derived. Notice also that the **nodes** (forks) represent *hypothetical* common ancestors—the taxa themselves are always located on the branch **tips,** never within the tree. That is, none of the taxa are presumed to be ancestors of others, even if some of the taxa are extinct.

Let's practice reading phylogenetic trees by focusing on how biologists go about building them.

How Do Biologists Estimate Phylogenies?

Like any other pattern or measurement in nature, from the average height of a person in a particular human population to the speed of a passing flock of geese, the historical relationships among species cannot be known with absolute certainty. In addition, humans have not been around to observe most evolutionary events directly. Thus, the relationships depicted in phylogenetic trees must be *estimated* from the best available data.

Creating the Data Matrix The first step in inferring evolutionary relationships is to decide which populations, species, or higher taxa to compare and which characteristics to use—usually morphological or genetic characteristics, or both. For example, to reconstruct relationships among fossil species of humans (there were many; see Chapter 35), scientists analyze aspects of tooth, jaw, and skull structure, since this is often the only

---

Graphical Symbols and Definitions

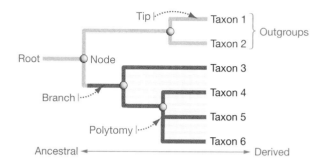

Branch Represents a population through time

Node (fork) A point within the tree where a branch splits into two or more branches

Outgroup A taxon that diverged prior to the taxa that are the focus of the study; helps to root the tree

Root The most ancestral branch in the tree

Polytomy A node that depicts an ancestral branch dividing into three or more (rather than two) descendant branches; usually indicates that insufficient data were available to resolve which taxa are more closely related

Tip (terminal node) Endpoint of a branch; represents a living or extinct group of genes, species, families, phyla, or other taxa

---

sort of evidence available. To reconstruct relationships among contemporary human populations, investigators usually compare the sequences of bases in a particular gene.

Consider the simple case of estimating a phylogenetic tree for four closely related species using a short DNA sequence as the character set, where a **character** is any genetic, morphological, physiological, or behavioral characteristic to be studied. The data matrix for this group is shown in **FIGURE 28.1a**. The sequence consists of six characters (nucleotides 1, 2, 3, 4, 5, and 6), each with four possible character states: A, T, C, and G.

Notice that a fifth species is included in the matrix as an **outgroup**—a species that is closely related to the group being studied, but not part of it. Outgroups are used to establish the polarity of each trait—that is, whether a character state is ancestral or derived. An **ancestral trait** is a characteristic that existed in an ancestor; a **derived trait** is one that is a modified form of the ancestral trait, found in a descendant. Derived traits originate via mutation, selection, and genetic drift (see Chapter 26).

It's important to recognize that ancestral and derived traits are relative. If you are comparing mammals with the fossil forms called mammal-like reptiles, then fur and lactation are derived traits. But if you are comparing whales and humans, then fur and lactation are ancestral traits. In the example in Figure 28.1a, the ancestral DNA sequence is estimated to be AAA GCT based on the outgroup; any nucleotides that differ from those in this sequence are considered derived. Since outgroup lineages can

(a) Data matrix

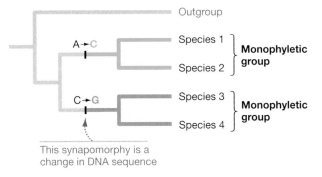

DNA nucleotide sequence

1	2	3	4	5	6	
A	A	A	G	C	T	Outgroup
A	A	C	G	C	T	Species 1
A	A	C	G	C	T	Species 2
A	A	A	G	G	T	Species 3
A	A	A	G	G	T	Species 4

(b) Phylogenetic tree inferred from the data

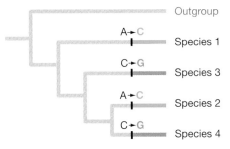

Outgroup

A→C

Species 1 ⎤ **Monophyletic**
Species 2 ⎦ **group**

C→G

Species 3 ⎤ **Monophyletic**
Species 4 ⎦ **group**

This synapomorphy is a change in DNA sequence

(c) One of many alternate trees

Outgroup

A→C
Species 1

C→G
Species 3

A→C
Species 2

C→G
Species 4

FIGURE 28.1 Constructing a Phylogenetic Tree. Cladistic analysis **(a)** begins with a data matrix and **(b)** builds trees based on the principle that closely related species are likely to share derived traits, called synapomorphies. **(c)** The tree that best fits the data must be selected from many alternate trees (only one shown here).

✔ **EXERCISE** Draw another alternate tree for the same data set.

evolve too, multiple outgroups are often used to estimate ancestral character states (for simplicity, only one is used here).

Using the Data Matrix to Estimate a Tree The most commonly used method of inferring phylogenetic trees, called cladistics, was introduced by German biologist Willi Hennig in the 1960s. The **cladistic approach** is based on the principle that relationships among species can be reconstructed by identifying shared derived characters—synapomorphies (introduced in Chapter 27).

A **synapomorphy** is a trait found in two or more taxa that is present in their most recent common ancestor but is missing in more distant ancestors. Synapomorphies allow biologists to recognize monophyletic groups—also called **clades** or **lineages.** A **monophyletic group** is an evolutionary unit that includes an ancestral population and all of its descendants, but no others.

Graphical Symbol	Definition
	Homology Similarity in organisms due to common ancestry (trait **A** is homologous among the red branches)
	Monophyletic group An evolutionary unit that includes an ancestral population and all of its descendants but no others (also called a lineage, or clade)
Synapomorphy	**Synapomorphy** A shared, *derived* trait (trait **A** occurs *only* in the red branches)
	Homoplasy Similarity in organisms due to reasons other than common ancestry (trait **B** arose twice independently)
	Polyphyletic group An unnatural group (shown in purple) that does not include the most recent common ancestor (see segmented animals in Chapter 33)
Loss of trait	**Paraphyletic group** A group (shown in blue) that includes an ancestral population and some of its descendents, but not all (see dicots in Chapter 31, and fish and reptiles in Chapter 35)

See **TABLE 28.2** for a comparison of monophyletic groups and two types of non-monophyletic groups.

In the case of the character set in Figure 28.1a, species 1 and 2 can be recognized as a monophyletic group based on their shared, derived trait, a C in the third position of the DNA sequence. Similarly, species 3 and 4 can be recognized as a different monophyletic group based on their shared, derived trait, a G in the fifth position of the sequence. This logic is summarized by the tree in **FIGURE 28.1b**.

In this example, it is relatively easy to infer a tree that represents the data set in a logical way. In common practice, there are several possible complications to resolve. For example:

- Traits can be similar in two species not because those traits were present in a common ancestor, but because similar traits evolved independently in two different lineages. For example, the A → C changes in DNA sequence at nucleotide 3 may have arisen separately in species 1 and 2 rather than once in their common ancestor (**FIGURE 28.1c**).

- Sometimes a reversal in a change occurs, such as an A → C transition in a nucleotide in one branch followed by a C → A change in the same nucleotide in a subsequent branch, thus creating the appearance that no change occurred.

- Sometimes the species forms a monophyletic group one way according to one trait in the matrix and a different way according to a different trait in the matrix (as would occur if species 1 and 3 had a unique base change at position 1).

- The example in Figure 28.1 shows a small number of species with a small number of traits. The larger the number of species and traits added to the matrix, the greater the number of possible trees and character combinations. There are 15 possible trees for a phylogeny of 4 taxa but over 100,000 trees for a phylogeny of 8 taxa.

How can researchers identify the tree with the best fit to the data? And how can they avoid an erroneous conclusion? Biologists often invoke the logical principle of **parsimony** to address these questions. Under parsimony, the most likely explanation or pattern is the one that implies the least amount of change.

To implement a parsimony analysis, biologists use computer programs that compare the branching patterns that are theoretically possible in a phylogenetic tree and count the number of changes in DNA sequences required to produce each pattern. For example, the tree in Figure 28.1b requires two changes in base sequence; the tree in Figure 28.1c requires four.

Since independent evolution of traits should be rare compared with similarity due to shared descent, the tree that implies the fewest overall evolutionary changes is hypothesized to be the one that most accurately reflects what really happened during evolution. If the branching pattern in Figure 28.1b is the most parsimonious of all possible trees, then biologists conclude that it is the most likely representation of actual phylogeny based on the data in hand.

There are methods other than parsimony to find the tree with the best fit to the data, each guided by different assumptions. However, all the methods are similar in serving as "filters" to identify optimal trees out of a pool of hundreds, thousands, or millions of possible trees. Some researchers use multiple methods to gain more confidence in their results.

It's important to recognize that the trees created using cladistic analysis (called cladograms) focus on branching *patterns*—the branch *lengths* themselves are arbitrary. Some analyses produce trees whose branch lengths represent genetic distance or time since divergence. You can identify these trees by the presence of scale bars (see **TABLE 28.3**). (If you feel that you need more practice interpreting phylogenetic trees, you can turn to **BioSkills 7** in Appendix B at any time.)

How Can Biologists Distinguish Homology from Homoplasy?

The cladistic approach is a powerful method of estimating phylogenies because it is based on one of the central concepts in evolutionary biology—homology (see Table 28.2). **Homology** (literally, "same-source") occurs when traits are similar due to shared ancestry; in contrast, **homoplasy** ("same-form") occurs when traits are similar for reasons other than common ancestry.

SUMMARY TABLE 28.3   Branch Lengths in Phylogenetic Trees

Graphical Symbol	Meaning of Branch Lengths
	Branch lengths are *arbitrary*. Emphasis is on the branching *pattern*, which estimates evolutionary relationships among populations. This is the type of tree used in this book.
0.1 — Number of base substitutions per site	Horizontal branch lengths show the extent of *genetic difference* among populations. A scale bar is included.
4 3 2 1 0 — Millions of years ago	Horizontal branch lengths show the extent of *evolutionary time* between nodes. A scale bar is included.

A common cause of homoplasy is convergent evolution. **Convergent evolution** occurs when natural selection favors similar solutions to the problems posed by a similar way of making a living in different species. Since species have many characteristics, some may be homologous with characteristics in other species, while others may be convergent. Let's consider how to distinguish homology and homoplasy.

Are *Hox* Genes in Fruit Flies and Humans Homologous or Convergent? Consider the *Hox* genes of insects and vertebrates (introduced in Chapter 22). Even though insects and vertebrates last shared a common ancestor some 600–700 million years ago, biologists argue that their *Hox* genes are derived from the same ancestral sequences—that is, they are homologous. Several lines of evidence support this hypothesis:

- Groups of *Hox* genes are organized on chromosomes in a similar way. **FIGURE 28.2** shows that *Hox* genes in both insects and vertebrates are found in gene clusters, and similar genes are often found in the same order on the chromosome. Genes with these characteristics are called gene families (Chapter 21). The organization of the gene families is nearly identical among insect and vertebrate species.

- All *Hox* genes share a 180-base-pair sequence called the homeobox. The portion of a protein encoded by the homeobox (introduced in Chapter 22) is almost identical in insects and vertebrates and has a similar function. It binds to DNA and regulates the expression of other genes.

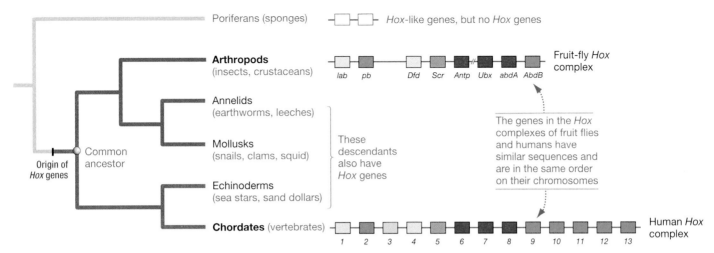

FIGURE 28.2 Homology: Similarities Are Inherited from a Common Ancestor. All animals except sponges have *Hox* complexes that are similar to those illustrated for fruit flies and humans. (Human *Hox* genes occur in four clusters but are summarized here in one for simplicity.)

• The products of the *Hox* genes have similar functions (see Chapter 22): identifying the locations of cells in embryos. They are also expressed in similar patterns in time and space.

Further, if traits found in distantly related lineages are similar due to common ancestry, then similar traits should be found in many intervening lineages on the tree of life—because all of the species in question inherited the trait from the same common ancestor. Indeed, all other lineages of animals that branched off between insects and mammals have similar *Hox* genes. Sponges serve as an outgroup to the clade since they appear to have split off before *Hox* genes arose.

The internal consistency of the genetic, developmental, and phylogenetic data provide strong evidence that *Hox* genes in fruit flies and humans are homologous.

Are Streamlined Bodies in Dolphins and Ichthyosaurs Homologous or Convergent? The extinct aquatic reptiles called ichthyosaurs were strikingly similar to modern dolphins (**FIGURE 28.3**). Both are large marine predators with streamlined bodies and large dorsal fins. Are these characteristics homologous in dolphins and ichthyosaurs?

Figure 28.3 shows a phylogenetic tree based on skeletal characters, such as the number and placement of openings in the skull (since ichthyosaurs are known only from their fossilized skeletons). This analysis shows that dolphins occur within the mammal clade, whereas ichthyosaurs are most closely related to lizards. Notice that the sister groups to dolphins and ichthyosaurs—that is, the most closely related branches on the tree—do not possess streamlined bodies, flippers, and fins.

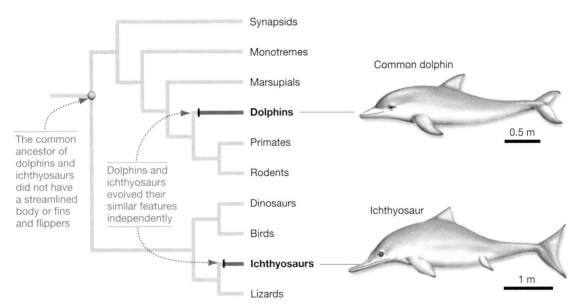

FIGURE 28.3 Homoplasy: Traits Are Similar but Not Inherited from a Common Ancestor. Dolphins and ichthyosaurs look similar but are not closely related.

Based on these data, it is logical to argue that the similarities between ichthyosaurs and dolphins result from convergent evolution. The convergent traits do not occur in the common ancestor of the similar species. Streamlined bodies, flippers, and dorsal fins are adaptations that help *any* aquatic species—whether it is a reptile or a mammal—chase down fish in open water.

Whale Evolution: A Case Study

Let's apply what you've learned about phylogenetic trees and the concept of homology to an interesting evolutionary case study: the origin of a unique clade of mammals, the whales. Scientists since Darwin have puzzled over the rightful place of whales (including dolphins) on the tree of life. Data from the fossil record support the hypothesis that whales evolved from terrestrial mammals. More specifically, hippos, which are semiaquatic, appear to be the closest living relative of today's whales. How can phylogenetic trees be used to test this claim?

Data Set 1: A Phylogeny Based on Morphological Traits Hippos, along with cows, deer, pigs, and camels, are artiodactyls. Members of this mammal group have hooves and an even number of toes. They also share another feature: the unusual pulley shape of an ankle bone called the astragalus. Along with having a unique foot morphology, the shape of the astragalus is a synapomorphy that identifies the artiodactyls as a monophyletic group. Members of the outgroup, perissodactyls (including horses and rhinos), do not possess a pulley-shaped astragalus.

These data support the tree shown in **FIGURE 28.4a**. Notice that whales do not have an astragalus (indeed, no ankles at all) and are shown as an outgroup to the artiodactyls on this tree based on other morphological characteristics. It is logical to map the gain of the pulley-shaped astragalus in the ancestral population, marked by a black bar in Figure 28.4a, because all descendants of that ancestor have the trait, but members of the outgroup do not.

Data Set 2: A Phylogeny Based on DNA Sequence Data When researchers began comparing DNA sequences of artiodactyls and other species of mammals, however, the data showed that whales share many similarities with hippos. These results supported the tree shown in **FIGURE 28.4b**.

Here, it is still logical to map the evolution of the pulley-shaped astragalus at the same ancestor as in Figure 28.4a. But to recognize that the trait had to have been lost in an ancestor of today's whales—which do not have a pulley-shaped astragalus—the tree contains another black bar mapping the trait loss.

The tree supported by the DNA data conflicts with the tree supported by morphological data because it implies that the pulley-shaped astragalus evolved in artiodactyls and then was lost during whale evolution. These two changes (a gain *and* a loss of the astragalus) are less parsimonious than just one change (a gain only).

Data Set 3: Transposable Elements The conflict between the two phylogenies was resolved when researchers analyzed the distribution of the parasitic DNA sequences called **SINEs (short interspersed nuclear elements),** which occasionally insert themselves into the genomes of mammals. (SINEs are transposable elements, similar to the LINEs introduced in Chapter 21.) SINEs provide convincing evidence of relationships because the chance of the same SINE inserting itself in exactly the same place in the genomes of two species is astronomically small.

As the data in **FIGURE 28.4c** show, whales and hippos share several types of SINEs (4, 5, 6, and 7) that are not found in other groups. Other SINEs are present in some artiodactyls but not in others. Camels have no SINEs at all.

To explain these data, biologists hypothesize that no SINEs were present in the population that is ancestral to all of the species in the study. Then, after the branching event that led to the split between the camels and all the other artiodactyls, different SINEs inserted themselves into the genomes of descendant populations.

If this hypothesis is correct, then the presence of a particular SINE represents a derived character. Because whales and hippos share four of these unique derived characters, it is logical to conclude that these animals are closely related.

✔ If you understand this concept, you should be able to explain why SINEs numbered 4–7 are synapomorphies that identify whales and hippos as a monophyletic group, and why the similarity in these SINEs is unlikely to represent homoplasy.

Conclusion: Whales Are Closely Related to Hippos If a phylogenetic tree were estimated based on the SINE data, it would have the same branching pattern as the one shown in Figure 28.4b. (To convince yourself, try the exercise in the figure caption). Thus, most biologists accept the phylogeny shown in Figure 28.4b as the most accurate estimate of whale evolutionary history. According to this phylogeny, whales are artiodactyls and share a relatively recent common ancestor with hippos. This observation inspired the hypothesis that both whales and dolphins are descended from a population of artiodactyls that spent most of their time feeding in shallow water, much as hippos do today.

In 2001, the discovery of fossil artiodactyls supported this hypothesis in spectacular fashion. These fossil species were clearly related to whales—they have an unusual ear bone found only in whales—and yet had a pulley-shaped astragalus.

The combination of DNA sequence data and data from the fossil record has clarified how a particularly interesting group of mammals evolved. What else do fossils have to say?

check your understanding

If you understand that . . .

- Phylogenies can be estimated by finding synapomorphies that identify monophyletic groups.

✔ **You should be able to . . .**

Determine whether the following traits represent homology or homoplasy: hair in humans and whales; extensive hair loss in humans and whales; limbs in humans and whales; social behavior in humans and certain whales (e.g., dolphins).

Answers are available in Appendix A.

(a) Data set 1 (morphological traits):
Whales diverged before the origin of artiodactyls.

(b) Data set 2 (DNA sequences):
Whales and hippos share a common ancestor.

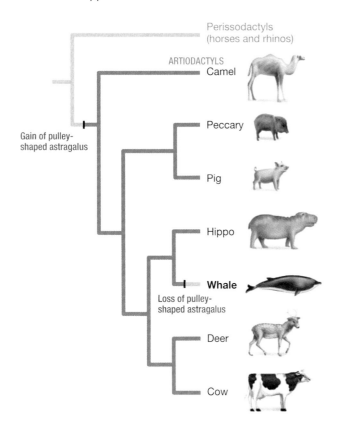

Perissodactyls (horses and rhinos)

Whale

ARTIODACTYLS
Camel

Peccary

Pig

Gain of pulley-shaped astragalus

Hippo

Astragalus (ankle bone)

Deer

Cow

Perissodactyls (horses and rhinos)

ARTIODACTYLS
Camel

Peccary

Pig

Gain of pulley-shaped astragalus

Hippo

Whale
Loss of pulley-shaped astragalus

Deer

Cow

(c) Data set 3 (presence and absence of SINEs): Supports the close relationship between whales and hippos.

Locus	1	2	3	4	5	6	7	8	9	10	11	12	13	14	15	16	17	18	19	20
Cow	0	0	0	0	0	0	0	1	1	1	1	1	1	1	1	1	1	1	0	0
Deer	0	0	0	0	0	0	0	1	?	1	1	1	1	1	1	?	1	1	0	0
Whale	1	1	1	1	1	1	1	0	?	1	0	1	1	0	0	0	?	1	0	0
Hippo	0	?	0	1	1	1	1	0	1	1	0	1	1	0	0	0	?	1	0	0
Pig	0	0	0	?	0	0	0	0	?	0	0	0	?	?	0	0	0	1	1	1
Peccary	?	?	?	?	?	?	?	?	?	?	?	?	?	?	?	?	?	?	1	1
Camel	0	0	0	0	0	0	0	0	0	0	0	0	0	0	0	0	0	0	0	0

1 = SINE present
0 = SINE absent
? = still undetermined

Whales and hippos share four *unique* SINEs (4, 5, 6, and 7)

FIGURE 28.4 Where Do Whales Belong on the Tree of Life?

DATA for (c): Nikaido, M., A. P. Rooney, and N. Okada. 1999. *Proceedings of the National Academy of Sciences, USA* 96: 10261–10266.

✔**EXERCISE** Map the origin of the following groups of SINEs on the tree in part (b): Group (4, 5, 6, 7) identifies hippos and whales as part of a monophyletic group; (8, 11, 14, 15, 17) identifies deer and cows as part of a monophyletic group; (10, 12, 13) identifies hippos, whales, deer, and cows as part of a monophyletic group. What two species does SINEs group (19, 20) identify as part of a monophyletic group?

28.2 Tools for Studying History: The Fossil Record

Phylogenetic analyses are powerful ways to infer the order in which events occurred during evolution and to understand how particular groups of species are related. But only the fossil record provides *direct* evidence about what organisms that lived in the past looked like, where they lived, and when they existed.

A **fossil** is a piece of physical evidence from an organism that lived in the past. The **fossil record** is the total collection of fossils that have been found throughout the world. The fossil record is housed in thousands of private and public collections.

Let's review how fossils form, analyze the strengths and weaknesses of the fossil record, and then summarize major events that have taken place in life's approximately 3.5-billion-year history.

How Do Fossils Form?

Most of the processes that form fossils begin when part or all of an organism is buried in ash, sand, mud, or some other type of sediment. For example, imagine that leaves, pollen, seeds, branches, or entire trees fall onto a muddy swamp, where they are buried by soil and debris before they decay. Once burial occurs, several things can happen.

- If decomposition does not occur, the organic remains can be preserved intact—like the fossil pollen in **FIGURE 28.5a**.

- If sediments accumulate on top of the material and become cemented into rocks such as mudstone or shale, the sediments' weight can compress the organic material below into a thin, carbonaceous film. This happened to the leaf in **FIGURE 28.5b**.

- If the remains decompose *after* they are buried—as did the branch in **FIGURE 28.5c**—the hole that remains can fill with dissolved minerals and create an accurate **cast** of the remains.

- If the remains rot extremely slowly, dissolved minerals can gradually infiltrate the interior of the cells and harden into stone, forming a permineralized fossil, such as petrified wood (**FIGURE 28.5d**).

After many centuries have passed, fossils can be exposed at the surface by erosion, a road cut, quarrying, or other processes. If researchers find a fossil, they can prepare it for study by painstakingly clearing away the surrounding rock.

If the species represented is new, researchers describe its morphology in a scientific publication, name the species, estimate the fossil's age based on dates assigned to nearby rock layers, and add the specimen to a collection so that it is available for study by other researchers. It is now part of the fossil record.

This scenario is based on conditions that are ideal for fossilization: The trees fell into an environment where burial was rapid and decomposition was slow. In most habitats the opposite situation occurs—burial is slow and decomposition is rapid. In reality, then, fossilization is an extremely rare event.

To appreciate this point, consider that there are 10 specimens of the first bird-like dinosaur to be discovered, *Archaeopteryx*. If you accept an estimate that crow-sized birds native to wetland habitats would have a population size of around 10,000 and a life span of 10 years, and if you accept the current estimate that the species existed for about 2 million years, then you can calculate that about 2 billion *Archaeopteryx* lived. But as far as researchers currently know, only 1 out of every 200,000,000 individuals were fossilized and discovered. For this species, the odds of becoming part of the fossil record were almost 40 times worse than your odds are of winning the grand prize in a state lottery.

Limitations of the Fossil Record

Before looking at how the fossil record is used to answer questions about the history of life, it is essential to review the nature of this archive and recognize several features.

Habitat Bias Because burial in sediments is so crucial to fossilization, there is a strong habitat bias in the database. Organisms that live in areas where sediments are actively being deposited—including beaches, mudflats, and swamps—are much more likely to form fossils than are organisms that live in other habitats.

Within these habitats, burrowing organisms such as clams are already underground—pre-buried—at death and are therefore much more likely to fossilize. Organisms that live aboveground in dry forests, grasslands, and deserts are much less likely to fossilize.

Taxonomic and Tissue Bias Slow decay is almost always essential to fossilization, so organisms with hard parts such as bones or shells are most likely to leave fossil evidence. This requirement introduces a strong taxonomic bias into the record. Clams, snails, and other organisms with hard parts have a much higher tendency to be preserved than do worms.

A similar bias exists for tissues within organisms. For instance, pollen grains are encased in a tough outer coat that resists decay, so they fossilize much more readily than do flowers. Shark teeth are abundant in the fossil record, but shark skeletal elements, which are made of cartilage, are almost nonexistent.

Temporal Bias Recent fossils are much more common than ancient fossils. This causes a temporal bias in the fossil record. The reason is straightforward—the older a fossil is, the longer it has been exposed to potentially destructive forces.

Older fossils usually occur in sedimentary rock layers deep below newer layers (see Chapter 25). These fossils are more vulnerable to crushing, heating, melting, and distortion by various chemical and physical processes. Older fossils are also more likely to be subducted further into the Earth's interior by the sliding and collision of Earth's tectonic plates. The fossils that survive these destructive forces will be discovered by scientists only if the

(a) Intact fossil (pollen)   **(b)** Compression fossil (leaf)   **(c)** Cast fossil (bark)   **(d)** Permineralized fossil (tree trunk)

10 μm

FIGURE 28.5 Fossils Are Formed in Several Ways. Different preservation processes give rise to different types of fossils.

deep layers of sedimentary rock they occupy are exposed to the surface by erosion (such as in the Grand Canyon) or by uplift.

Abundance Bias The fossil record has an abundance bias; it is weighted toward common species. Organisms that are abundant, widespread, and/or present on Earth for long periods of time leave evidence much more often than do species that are rare, local, or ephemeral.

To summarize, the fossil database represents a highly nonrandom sample of the past. **Paleontologists**—scientists who study fossils—recognize that they are limited to asking questions about tiny and scattered segments on the tree of life. ✔ If you understand the biases of the fossil record, you should be able to predict at least two ways that paleontologists could be misled by their observations of fossils.

And yet, as this chapter shows, the fossil record is a scientific treasure trove. Analyzing fossils is the only way scientists have of examining the physical appearance of extinct forms and inferring how they lived. The fossil record is like an ancient library, filled with volumes that give us glimpses of what life was like millions of years before humans appeared.

Life's Time Line

It is very difficult to date the origin of life precisely. Our best estimate is that the Earth started to form about 4.6 billion years ago, and that life began before 3.5 billion years ago—probably in the warm, hydrothermal vents at the bottom of the ocean (see Chapter 2). To organize the tremendous sweep of time between then and now, researchers divide Earth history into segments called eons, eras, and periods.

Originally, geologists used distinctive rock formations or fossilized organisms to identify the boundaries between named time intervals. Later, researchers were able to use radiometric dating to assign absolute dates—expressed as years before the present—to events in the fossil record. Radiometric dating is based on the well-studied decay rates of certain radioactive isotopes (see Chapter 25). By dating rocks near fossils, researchers can also assign an absolute age to many of the species in the fossil record.

To summarize the history of life, researchers create time lines that record key "evolutionary firsts"—the appearance of important new lineages or innovations. It's important to recognize, though, that the times assigned to these first appearances underestimate when lineages appeared and events occurred. The reason is simple: A particular species or lineage can exist for millions of years before leaving fossil evidence. The fossil record and efforts to date fossils are constantly improving, so time lines are always a work in progress.

Precambrian The interval between the formation of Earth about 4.6 billion years ago and the appearance of most animal groups about 542 million years ago (abbreviated mya) is called the **Precambrian** (**FIGURE 28.6**); it is divided into the Hadean, Archaean, and Proterozoic eons.

FIGURE 28.6 Major Events of the Precambrian. Life, photosynthesis, and the oxygen atmosphere all originated in the Precambrian.

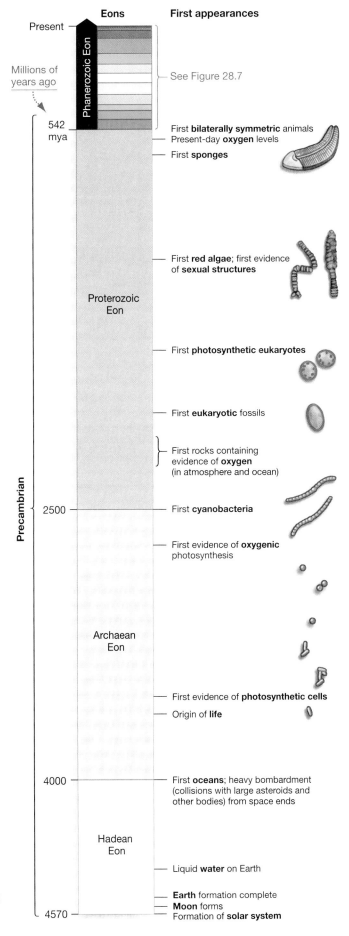

Eons | First appearances

Present

Millions of years ago

Phanerozoic Eon

See Figure 28.7

542 mya — First **bilaterally symmetric** animals
Present-day **oxygen** levels
First **sponges**

Proterozoic Eon

First **red algae**; first evidence of **sexual structures**

First **photosynthetic eukaryotes**

First **eukaryotic** fossils

First rocks containing evidence of **oxygen** (in atmosphere and ocean)

2500 — First **cyanobacteria**

First evidence of **oxygenic** photosynthesis

Archaean Eon

First evidence of **photosynthetic cells**

Origin of **life**

4000 — First **oceans**; heavy bombardment (collisions with large asteroids and other bodies) from space ends

Hadean Eon

Liquid **water** on Earth

Earth formation complete
Moon forms
4570 — Formation of **solar system**

Precambrian

Periods	Epoch	First appearances	Representative organisms

Present
2.6

Quaternary — Holocene / Pleistocene — *Homo sapiens;* first chimpanzees

Neogene — Pliocene / Miocene — Earliest hominins

23.0 mya — Oldest pollen from daisy-family plants

Oligocene — First apes

Paleogene — Eocene — First fully aquatic whales

First horses; first primates; first rabbits/hares

65.5 — Paleocene

→ Cretaceous extinction

Cenozoic Era

Cretaceous
- First bee; first ant
- First magnolia-family plants
- First water lilies
- First centric diatoms
- First angiosperm (flowering plant)

145.5
- First bird-like reptile
- First placental mammals
- First tyrannosaurid dinosaur

Jurassic

199.6 → Triassic extinction

Triassic
- First mammals
- First dinosaurs
- First nectar-drinking insects

Mesozoic Era

251 → Permian extinction

Permian
- First vessels in plants
- First basidiomycete fungi

299

Carboniferous — Pennsylvanian — Origin of amniotes

Mississippian

359.2 → Devonian extinction

Devonian
- First seed plants; first plants with leaves
- First tetrapods (amphibians)
- First winged insects
- First tree-sized plants
- First ferns, vascular plants, ascomycete fungi, lichens
- First fish with jaws
- First insects

416

Silurian
- First bony fish

443.7 → Ordovician extinction

Ordovician
- First mycorrhizal fungi (Glomales); first cartilaginous fish
- First land plants
- First bryozoans (most recent origin of an animal phylum)

488

Cambrian
- Arthropods diversify; first echinoderm
- First arthropods, mollusks, vertebrates, other phyla

542

Paleozoic Era

Phanerozoic Eon

FIGURE 28.7 Major Events of the Phanerozoic Eon (at left). The Phanerozoic began with the initial diversification of animals, continued with the evolution and early diversification of land plants and fungi, and includes the subsequent movement of animals to land. A total of five mass extinctions occurred during the eon.

Here are the important things to know about the Precambrian:

• Life was exclusively unicellular for most of Earth's history.

• Oxygen was virtually absent from the oceans and atmosphere for almost 2 billion years after the origin of life. Photosynthetic bacteria were responsible for the creation of the oxygen atmosphere (see Chapter 29).

Phanerozoic Eon   The interval between 542 mya and the present is called the Phanerozoic eon and is divided into three eras (**FIGURE 28.7**). Each of these eras is further divided into intervals called periods, which are themselves divided into epochs.

1. The **Paleozoic** ("ancient life") **era** begins with the appearance of most major animal lineages and ends with the obliteration of almost all multicellular life-forms at the end of the Permian period. The Paleozoic includes the origin and initial diversification of the animals, land plants, and fungi, as well as the appearance of land animals.

2. The **Mesozoic** ("middle life") **era** begins with the end-Permian extinction events and ends with the extinction of the dinosaurs (except birds) and other groups at the boundary between the Cretaceous period and Paleogene period. In terrestrial environments of the Mesozoic, gymnosperms were the most dominant plants and dinosaurs were the most dominant vertebrates.

3. The **Cenozoic** ("recent life") **era** is divided into the Paleogene, Neogene, and Quaternary periods. On land, angiosperms were the most dominant plants and mammals were the largest vertebrates.

At the top of Figure 28.7 you can see that the periods of the Cenozoic are divided into epochs, and that we are presently in the Holocene epoch. However, many scientists have called for the delineation of a new epoch—the Anthropocene ("new human epoch")—to reflect the dramatic physical, chemical, and biological changes that humans are causing on Earth, especially since the industrial revolution. Much discussion will take place before any official change is made.

Changes in the Oceans and Continents   The changes in environments and life-forms that are recorded in Figures 28.6 and 28.7 took place in the context of radical changes in climate and in the extent and location of Earth's oceans and continents. Earth's crust is broken into enormous plates that are in constant motion, driven by heat rising from the planet's core.

FIGURE 28.8 summarizes the most recent data on how the positions of the continents and oceans have changed through time,

FIGURE 28.8 Continental Positions during the Phanerozoic (at right). Dramatic changes in the extent and position of the continents took place during the Phanerozoic. These changes affected the total amount of land area, the relative amounts of land in the tropics versus northern latitudes, and the nature of ocean currents.

North and South America joined by land bridge. Worldwide glaciation.

Pangaea breaks apart. Climate mild, temperate.

Supercontinent Pangaea assembles. Climate warm.

Supercontinent of Laurentia to the north and Gondwana to the south. Climate mild.

Extensive ice on Gondwana. Oceans cover much of today's North America.

starting at the beginning of the Phanerozoic Eon. The figure also includes notes on the prevailing climate. Over the past 542 million years alone, terrestrial and marine landforms and climates have changed radically.

Taken together, the fundamental message in time-line data is a story of constant change. The changes are well documented, but the sweep of time involved is still difficult for the human mind to comprehend. If the history of the Earth were scaled to fit in one calendar year, our species, *Homo sapiens*, would not appear until one hour before the stroke of midnight on December 31st.

28.3 Adaptive Radiation

When biologists consider the history of life, two of the most compelling types of evolutionary events to study are (1) periods when species originate and diversify rapidly, and (2) periods when species go extinct rapidly. Let's focus on dramatic events that create biological diversity first; the chapter's concluding section analyzes how that diversity gets wiped out.

When a single lineage rapidly produces *many* descendant species with a wide range of adaptive forms, biologists say that an **adaptive radiation** has occurred. Adaptive radiations are noteworthy because they cause dramatic changes to the tree of life. They can be observed as a sudden appearance of related, diverse species in the fossil record and also by phylogenetic analysis.

FIGURE 28.9 provides an example: the Hawaiian silverswords. The 30 species in this lineage evolved from a species of tarweed, native to California, that colonized the islands about 5 million years ago. This is an example of allopatric speciation by dispersal

(see Chapter 27). Compared to speciation rates in other groups, this rate is extremely rapid. Further, today's silverswords vary from mosslike mats to vines and rosettes. They live in habitats ranging from lush rain forests to austere lava flows.

The Hawaiian silverswords fulfill the three hallmarks of an adaptive radiation: (1) they are a monophyletic group, (2) they speciated rapidly, and (3) they diversified ecologically—meaning, in terms of the resources they use and the habitats they occupy.

Biologists use the term niche to describe the range of resources that a species can use and the range of conditions that it can tolerate (see Chapter 27). Silverswords occupy a wide array of niches.

Why Do Adaptive Radiations Occur?

Adaptive radiations are a major pattern in the history of life. But why do some lineages diversify rapidly, while others do not?

Two general mechanisms can trigger adaptive radiations: new resources, and new ways to exploit resources. Let's consider each in turn.

Ecological Opportunity Ecological opportunity—meaning the availability of new or novel types of resources—has driven a wide array of adaptive radiations. For example, biologists explain the diversification of silverswords by hypothesizing that few other flowering plant species were present on the Hawaiian Islands 5 million years ago. With few competitors, the descendants of the colonizing tarweed were able to grow in a wide range of habitats. Over time—by the processes of mutation, genetic drift, and natural selection (discussed in Chapter 26)—some became specialized for growth in different niches.

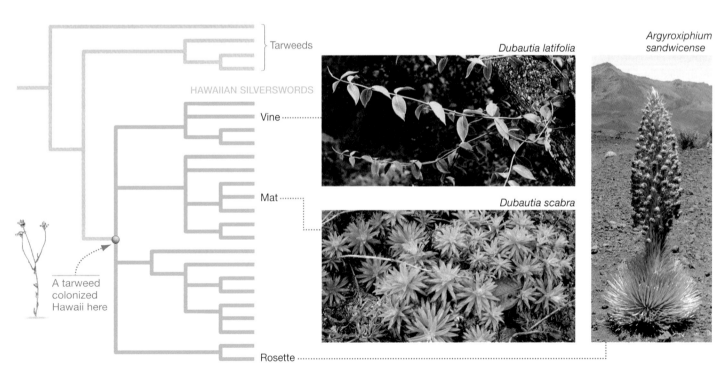

FIGURE 28.9 An Adaptive Radiation. This tree shows a subset of the Hawaiian silverswords, illustrating the extent of morphological divergence.

(a) Short-legged lizard species spend most of their time on the twigs of trees and bushes.

Anolis insolitus
(twig anole)

(b) Long-legged lizard species live on tree trunks and the ground.

Anolis cybotes
(trunk/ground anole)

FIGURE 28.10 Adaptive Radiations of *Anolis* Lizards. (a, b) Species of *Anolis* lizards vary in leg length and tail length. **(c)** Evolutionary relationships among lizard species on the islands of Hispaniola and Jamaica. The initial colonist species was different on these islands, but in terms of how they look and where they live, a similar suite of four species evolved.

(c) The same adaptive radiation of *Anolis* has occurred on different islands, starting from different types of colonists.

The same type of ecological opportunity was hypothesized to explain the adaptive radiation of *Anolis* lizards on islands in the Caribbean. The lineage includes 150 species. They thrive in a wide array of habitats and have diverse body sizes and shapes. And in most cases, a lizard's size and shape are correlated with the habitat it occupies. For example:

- Species that live on tree twigs have short legs and tails that allow them to move efficiently on narrow surfaces (**FIGURE 28.10a**).

- Species that spend most of their time clinging to broad tree trunks or running along the ground have long legs and tails, making them fast and agile on broad surfaces (**FIGURE 28.10b**).

Most islands in the Caribbean have a distinct suite of lizard species. And in most cases, each island has a species that lives only in the twigs, on the ground, or in other distinctive habitats. The classical explanation for this pattern was that a mini-radiation occurred on each island: An original colonizing group encountered no competitors, and diversified in a way that led to efficient use of the available resources by a group of descendant species.

To test this hypothesis rigorously, Jonathan Losos and his colleagues estimated the phylogeny of *Anolis* from DNA sequence data. The results supported a key claim of the ecological opportunity hypothesis: The lizards on each island were monophyletic.

The critical data, though, are shown in **FIGURE 28.10c**. These phylogenetic trees, for species found on two different islands, are typical. The key observation is that the original colonist on each island was specialized for a different niche. The initial species on Hispaniola lived on the trunks and crowns of trees, while the original colonist on Jamaica occupied twigs.

From different evolutionary starting points, then, an adaptive radiation filled the same niches on both islands. This is exactly what the ecological opportunity hypothesis predicts. In *Anolis* lizards in the Caribbean, a series of mini-radiations were triggered when colonists arrived on an island that had food, space, and other resources, but lacked competitors. ✔ If you understand this concept, you should be able to generate a hypothesis to explain why tarweeds in California and *Anolis* lizards on the mainland are not particularly species-rich or ecologically diverse.

Morphological Innovation   The evolution of a key morphological trait—one that allowed descendants to live in new areas, exploit new sources of food, or move in new ways—triggered many of the important diversification events in the history of life. For example:

- *Flowers* are a unique reproductive structure that helped trigger the diversification of angiosperms (flowering plants). Because flowers are particularly efficient at attracting pollinators, the evolution of the flower made angiosperms more efficient in reproduction. Today angiosperms are far and away the most species-rich lineage of land plants. Over 250,000 species are known.

- *Feathers and wings* gave some dinosaurs the ability to fly. Feathers originally evolved for display or insulation; later they were used in gliding and in powered flight. Today the lineage called birds contains about 10,000 species, whose representatives live in virtually every habitat on the planet.

In each case, the evolution of new morphological features is hypothesized to have supported rapid speciation and ecological divergence.

The Cambrian Explosion

Almost all life-forms were unicellular for almost 3 *billion* years after the origin of life. The exceptions were several lineages of small multicellular algae, which show up in the fossil record about 1 billion years ago. Then, the first animals—early sponges—appear, perhaps as early as 635 million years ago.

In a relatively short time, creatures with shells, exoskeletons, internal skeletons, legs, heads, tails, eyes, antennae, jaw-like mandibles, segmented bodies, muscles, and brains had evolved. It was arguably the most spectacular period of evolutionary change in the history of life. Because much of it occurred during the Cambrian period, this adaptive radiation of animals is called the **Cambrian explosion.**

All organisms were small for 3 billion years; then life got big, in just 50 million years—1/60th of the total time life had existed. Before asking how this happened, let's explore the fossilized evidence for what happened.

The Doushantuo, Ediacaran, and Early Cambrian Fossils The Cambrian explosion is documented by three major fossil assemblages that record the state of animal life near the beginning of the Cambrian. As **FIGURE 28.11a** indicates, the species collected from each of these intervals are referred to respectively as the Doushantuo fossils (named after the Doushantuo formation in China), Ediacaran fossils (named after the Ediacara Hills, Australia), and Cambrian fossils (from the Burgess Shale in Canada, the Chengjiang region of China, and several other sites around the world).

Each fossil assemblage records a distinctive **fauna**—or collection of animal species:

- *Doushantuo microfossils* include tiny sponges and corals, less than 1 mm across. These creatures—the first animals on Earth—probably made their living by filtering organic debris from the water.

- *Ediacaran faunas* include sponges, jellyfish, and comb jellies as well as fossilized burrows, tracks, and other traces from unidentified animals. None of these organisms have shells, limbs, heads, or feeding appendages (**FIGURE 28.11b, bottom**). Ediacaran animals likely filtered or absorbed organic material as they burrowed in sediments, sat immobile on the seafloor, or floated in the water.

- *Cambrian faunas* are among the most sensational additions ever made to the fossil record. Sponges, jellyfish, and comb jellies are abundant in these rocks, but so are several worm-like creatures and arthropods (**FIGURE 28.11b, top**), in addition to mollusks, echinoderms, and virtually all the other major animal lineages (see Chapter 33). This list includes our own lineage—the chordates—represented by early vertebrate fossils with streamlined, fishlike bodies and a skull made of cartilage (**FIGURE 28.12**). A tremendous increase in the size and morphological complexity of animals occurred, accompanied by diversification in how they made a living. Species in these faunas swam, burrowed, walked, ran, slithered, clung, or floated; there were predators, scavengers, filter feeders, and grazers. The diversification filled many of the ecological niches found in marine habitats today.

(a) A time line of early animal evolution

(b) Early animals from two major macroscopic fossil assemblages

Cambrian fossils are diverse, large, and have hard parts.

Ediacaran fossils are soft-bodied.

FIGURE 28.11 Fossils Document the Cambrian Explosion. The origin of animals and their diversification during the Cambrian explosion is documented by three major fossil assemblages, two of them represented by macroscopic fossils.

Early vertebrate

FIGURE 28.12 Life in the Cambrian. This artist's interpretation of animal diversity during the Cambrian is based on fossils found in the Chengjiang formation of China. Chengjiang is famous for its unique fossils of fishlike early vertebrates.

What Triggered the Cambrian Explosion? The Doushantuo, Ediacaran, and Cambrian faunas document what happened during the Cambrian explosion. Now the question is: *How* did all this speciation, morphological change, and ecological diversification come about?

To answer this question, biologists point to an array of data sets and hypotheses:

- *Higher oxygen levels* By analyzing the composition of rocks formed during the Proterozoic, geologists have established that oxygen levels gradually rose in the atmosphere and ocean. Increased oxygen levels make aerobic respiration more efficient (see Chapter 9); increased aerobic respiration is required to support larger bodies and more active movements. Some biologists suggest that oxygen levels reached a critical threshold at the start of the Cambrian explosion, making the evolution of big, mobile animals possible.

- *The evolution of predation* Before the Cambrian explosion, animals made their living by eating organic material that settled on the seafloor or filtering cells and debris from the water. But Cambrian fossils include shelled animals with holes in the shells—evidence that a predator drilled through and ate the animal inside. When predators evolved, they exerted selection pressure for shells, hard exoskeletons, rapid movement, and other adaptations for prey to defend themselves. This selection would drive morphological divergence.

- *New niches beget more new niches* The vast majority of Doushantuo and Ediacaran animals lived on the ocean floor—in what biologists call benthic habitats. Once animals could move off this substrate, they could exploit algae and other resources that were available above the ocean floor. The presence of animals at an array of depths created selection pressure for the evolution of species that could eat them. In this way, the ability to exploit new niches created new niches, driving speciation and ecological diversification.

- *New genes, new bodies* *Hox* genes play a key role in organizing the development of the body in diverse animals by signaling where cells are in the embryo (Chapter 22). **FIGURE 28.13** summarizes the *Hox* genes found in some major animal

FIGURE 28.13 *Hox* Genes Played an Important Role in Animal Diversification.

groups. The key observation is that the earliest animals in the fossil record had few or no *Hox* genes; most groups that appeared later have more. The idea here is that gene duplication and diversification increased the number of *Hox* genes in animals and made it possible for larger, more complex bodies to evolve.

It's important to recognize that most or all of these hypotheses could be correct. They are not mutually exclusive. If increased oxygen levels made larger bodies and more rapid movement possible, then animals could move into new habitats off the seafloor and become large enough to eat smaller animals. Selection would favor individuals with mutations in *Hox* genes and other DNA sequences that make the development of a large, complex body possible.

About 100 million years after the Cambrian explosion, a similar adaptive radiation occurred after the first plants—the descendants of green algae—adapted to life on land. In the span of about 56 million years, an array of growth forms and most major lineages of plants appear in the fossil record. This "Devonian explosion" resulted from the morphological innovations that allowed green plants to thrive on land, coupled with the ecological opportunity provided by an environment awash in sunlight (Chapter 31).

Adaptive radiations occur when species originate and diversify rapidly. Now let's consider how dramatic events can *wipe out* diversity.

 check your understanding

 If you understand that . . .
- Adaptive radiations are triggered by ecological opportunity and morphological innovation.

 ✔ **You should be able to . . .**
Explain the role of ecological opportunity and morphological innovation in the Cambrian explosion.

Answers are available in Appendix A.

28.4 Mass Extinction

A **mass extinction** refers to the rapid extinction of a large number of lineages scattered throughout the tree of life. More specifically, a mass extinction occurs when at least 60 percent of the species present are wiped out within 1 million years.

Mass extinction events are evolutionary hurricanes. They buffet the tree of life, snapping twigs and breaking branches. They are catastrophic episodes that wipe out huge numbers of species and lineages in a short time, giving the tree of life a drastic pruning. They are the polar opposite of adaptive radiation.

Before analyzing two of the best-studied mass extinctions in the fossil record, let's step back and ask how they differ from normal extinction events.

How Do Mass Extinctions Differ from Background Extinctions?

Background extinction refers to the lower, average rate of extinction observed when a mass extinction is not occurring. Although there is no hard-and-fast rule for distinguishing between background and mass extinction rates, paleontologists traditionally recognize and study five historic mass extinction events. **FIGURE 28.14**, for example, plots the percentage of plant and animal lineages called families that died out during each stage in the geologic time scale since the Cambrian explosion. Five spikes in the graph—denoting a large number of extinctions within a short time—are drawn in red. These are referred to as "The Big Five."

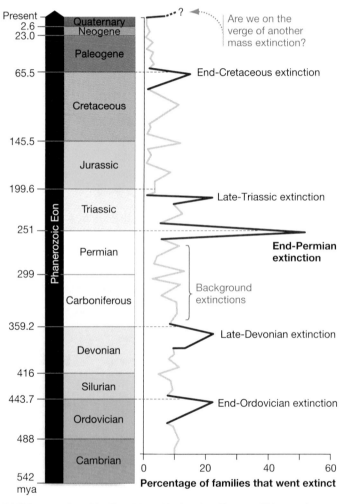

FIGURE 28.14 The Big Five Mass Extinction Events. This graph shows the percentage of lineages called families that went extinct over each interval in the fossil record since the Cambrian explosion. Over 50 percent of families and 90 percent of species went extinct during the end-Permian extinction.

DATA: Benton, M. J., 1995. *Science* 268: 52–58.

✔**QUANTITATIVE** Which extinction event ended the era of the dinosaurs 65 million years ago? About what percentage of families went extinct?

Biologists are interested in distinguishing between background and mass extinctions because these events have contrasting causes and effects.

- *Background extinctions* are thought to occur when normal environmental change, emerging disease, predation pressure, or competition with other species reduces certain populations to zero.

- *Mass extinctions* result from extraordinary, sudden, and temporary changes in the environment. During a mass extinction, species do not die out because individuals are poorly adapted to normal or gradually changing environmental conditions. Rather, species die out from exposure to exceptionally harsh, short-term conditions—such as huge volcanic eruptions or catastrophic sea-level changes.

To drive these points home, consider a mass extinction event that nearly uprooted the tree of life entirely. The end-Permian extinction, which occurred about 251 million years ago, came close to ending multicellular life on Earth.

The End-Permian Extinction

The end-Permian has been called the Mother of Mass Extinctions. To appreciate the scale of the event, imagine that while walking along a seashore, you identified 100 different species of algae and animals living on the beach and in tide pools and the shallow water offshore. Now imagine that you snapped your fingers, and 90 of those species disappeared forever. Only 10 species are left. An area that was teeming with diverse forms of life would look barren.

This is what happened, all over the world, during the end-Permian extinction. The event was a catastrophe of almost unimaginable proportions. On a personal level, it would be like nine of your ten best friends dying.

Although biologists have long appreciated the scale of the end-Permian extinction, research on its causes is ongoing. Consider the following:

- Flood basalts are outpourings of molten rock that flow across Earth's surface. The largest flood basalts on Earth, called the Siberian traps, occurred during the end-Permian. They added enormous quantities of heat, CO_2, and sulfur dioxide to the atmosphere.

- High levels of atmospheric CO_2 trapped heat, causing intense global warming, the stratification of the oceans, and reaction of deep-ocean sulfur dioxide with water to form sulfuric acid, which is toxic to most organisms.

- Rocks that formed during the interval indicate that the oceans became completely or largely anoxic—meaning that they lacked oxygen. These conditions are fatal to organisms that rely on aerobic respiration.

- There is convincing evidence that sea level dropped dramatically during the extinction event, reducing the amount of habitat available for marine organisms.

- Terrestrial animals may have been restricted to small patches of low-elevation habitats, due to low oxygen concentrations and high CO_2 levels in the atmosphere.

In short, both marine and terrestrial environments deteriorated dramatically for organisms that depend on oxygen to live. A prominent researcher has captured this point by naming the suite of killing mechanisms the "world went to hell hypothesis."

What biologists don't understand is *why* the environment changed so radically, and so quickly. For example, no one has yet been able to establish a convincing connection between the eruption of the Siberian traps and the other, more global, environmental changes that occurred.

The cause of the end-Permian extinction may be the most important unsolved question in research on the history of life, aside from the origin of life itself (see Chapter 2). The cause of the dinosaur's demise, in contrast, is settled.

What Killed the Dinosaurs?

The end-Cretaceous extinction of 65 million years ago is as satisfying a murder mystery as you could hope for, but the butler didn't do it. The **impact hypothesis** for the extinction of the dinosaurs, first put forth in the early 1970s by father-and-son team Luis and Walter Alvarez, proposed that an asteroid struck Earth and snuffed out an estimated 60–80 percent of the multicellular species alive.

Evidence for the Impact Hypothesis   The impact hypothesis was intensely controversial at first. As researchers set out to test its predictions, however, support began to grow (**TABLE 28.4**):

- Worldwide, sedimentary rocks that formed at the Cretaceous–Paleogene (K–P) boundary were found to contain extremely high quantities of the element iridium. Iridium is vanishingly rare in Earth rocks, but it is an abundant component of asteroids and meteorites.

- Shocked quartz and microtektites are minerals that are found only at documented meteorite impact sites. In Haiti and an array of other locations, both shocked quartz and microtektites have been discovered in abundance in rock layers dated to 65 million years ago.

- A crater larger than New Jersey was found just off the northwest coast of Mexico's Yucatán peninsula. Microtektites are abundant in sediments from the crater's walls, and the crater dates to the K–P boundary.

Taken together, these data provided conclusive evidence in favor of the impact hypothesis. Researchers agree that the mystery is solved.

Nature of the Impact   The shape of the Yucatán crater and the distribution of shocked quartz and microtektites dated to 65 mya indicates that the asteroid hit Earth at an angle and splashed material over much of southeastern North America. Based on

Type of Evidence	Explanation
	Spike of iridium in 65-million-year-old rocks worldwide Iridium is very rare on Earth but abundant in asteroids and meteors
100 µm	**Spike of shocked quartz in 65-million-year-old rocks worldwide** Shocked quartz forms only under extremely high pressures created from events like impacts (not from volcanoes)
1 mm	**Spike of micro-tektites in 65-million-year-old rocks in Gulf of Mexico** These glass droplets form only under extremely high temperatures and pressures as rock is splashed, molten, from an impact site
Gulf of Mexico / Crater / Yucatán Peninsula of Mexico	**Huge, 65-million-year-old crater discovered off coast of Mexico** The crater is 180 kilometers in diameter (absolutely massive), so an impact of this size would have far-reaching effects

currently available data, astronomers and paleontologists estimate that the asteroid was about 10 km across.

To get a sense of the event's scale, consider that Mt. Everest is about 10 km above sea level and that planes cruise at an altitude of about 10 km. Imagine Earth being hit by a rock the size of Mt. Everest, or a rock that would fill the space between you and a jet in the sky.

According to both computer models and geologic data, the consequences of the K–P asteroid strike were nothing short of devastating.

- A tremendous fireball of hot gas would have spread from the impact site; large soot and ash deposits in sediments dated to 65 million years ago testify to catastrophic wildfires, worldwide.

- The largest tsunami in the last 3.5 billion years would have disrupted ocean sediments and circulation patterns.

- The impact site itself is underlain by a sulfate-containing rock called anhydrite. The SO_4^{2-} released by the impact would have reacted with water in the atmosphere to form sulfuric acid, triggering extensive acid rain.

- Massive quantities of dust, ash, and soot would have blocked the Sun for long periods, leading to rapid global cooling and a crash in plant productivity.

Selectivity of the Extinctions The asteroid impact did not kill indiscriminately. Perhaps by chance, certain lineages escaped virtually unscathed while others vanished. Among vertebrates, for example, most dinosaurs (except birds), pterosaurs (flying reptiles), and all of the large-bodied marine reptiles (mosasaurs and plesiosaurs) expired; birds, mammals, crocodilians, amphibians, and turtles survived.

Why? Answering this question has sparked intense debate. For years the leading hypothesis was that the K–P extinction event was size selective. The logic here was that the extended darkness and cold would affect large organisms disproportionately, because they require more food than do small organisms. But extensive data on the survival and extinction of marine clams and snails have shown no hint of size selectivity, and small-bodied dinosaurs (except for birds) perished along with large-bodied forms.

One hypothesis currently being tested is that organisms that were capable of inactivity for long periods—by hibernating or resting as long-lived seeds or spores—were able to survive the catastrophe. Also, recent data show selectivity according to habitat; organisms in the open oceans suffered higher extinction rates than those in coastal waters. The reason for selectivity of the extinctions is still unsolved.

Recovery from the Extinction After the K–P extinction, fern fronds and fern spores dominate the plant fossil record from North America and Australia. These data suggest that extensive stands of ferns replaced diverse assemblages of cone-bearing and flowering plants after the impact. The fundamental message here is that terrestrial ecosystems around the world were radically simplified by the extinction event. In marine environments, some invertebrate groups do not exhibit typical levels of species diversity in the fossil record until 4–8 million years past the K–P boundary. Recovery was slow.

The organisms present in the Paleogene were markedly different from those of the preceding period. The lineage called

Mammalia had begun to diversify before the K–P extinction, during a period of rapid diversification of flowering plants. But early mammals were mostly rat-sized predators and scavengers in the heyday of the dinosaurs. It wasn't until after the dinosaurs became extinct that each of the major lineages of mammals diversified into the orders we recognize today—from bats to bears, and pigs to primates—and average body size increased exponentially. Why? A major branch on the tree of life had disappeared. With competitors removed from many ecological niches, mammals flourished.

The change in the terrestrial vertebrate fauna was not due to a competitive superiority conferred by adaptations such as fur and lactation. Rather, it was due to a chance event: a once-in-a-billion-years collision with a massive rock from outer space.

The Sixth Mass Extinction?

Are mass extinctions events of the distant past? Not likely. Many scientists now propose that life on Earth is on the verge of the sixth mass extinction, precipitated by human impacts such as habitat loss, pollution, overfishing, invasive species, and climate change. Could human activities really cause such dire consequences as massive as those caused by an asteroid impact?

There is mounting evidence that the answer to this question is yes. The exponential increase of human populations, combined with our hunger for resources, has put the diversity of life as we know it at risk.

Unit 6 explores the diversity of life itself. Then Unit 9 examines how human actions are rapidly pruning branches from the tree of life during your lifetime.

CHAPTER 28 REVIEW

For media, go to MasteringBiology (MB)

If you understand . . .

28.1 Tools for Studying History: Phylogenetic Trees

- Phylogenetic trees show the evolutionary history of a group of organisms—who is most closely related to whom. They have revolutionized the study of evolution and many related fields.

- Phylogenetic trees are usually estimated by analyzing shared, derived characters (synapomorphies) that identify monophyletic groups.

- Researchers often use the principle of parsimony to decide which of the many possible trees is most likely to reflect actual evolutionary history. Parsimony assumes that the most likely explanation or pattern is the one that implies the least amount of change.

- Homology occurs when traits are shared due to common ancestry. Homoplasy occurs when traits are similar due to reasons other than common ancestry, such as convergent evolution.

✔ You should be able to explain why upright posture and bipedalism (walking on two legs) is a synapomorphy that defines a monophyletic group called hominins, which includes humans.

28.2 Tools for Studying History: The Fossil Record

- The fossil record is the only direct source of data about what extinct organisms looked like and where they lived.

- The fossil record is biased: Common, recent, and abundant species that burrow and that have hard parts are most likely to be present in the record.

- During the Precambrian, which started with the formation of the Earth 4.6 billion years ago and ended with the origin of most animals about 542 million years ago, life was almost exclusively unicellular.

- Oxygen was virtually absent from the oceans and atmosphere for almost 2 billion years after the origin of life. The oxygen atmosphere was generated by photosynthetic bacteria.

- During the Phanerozoic, which began about 542 million years ago and continues to the present, animals originated and diversified first in the ocean and then on land. Land plants and fungi also originated and diversified.

- If the history of the Earth were scaled to fit in one calendar year, our species, *Homo sapiens*, would not appear until one hour before the stroke of midnight on December 31st.

✔ You should be able to explain why some paleontologists use snail fossils rather than other types of fossils to estimate how biodiversity has changed over time.

28.3 Adaptive Radiation

- Adaptive radiations occur when a single lineage rapidly produces many descendant species with a wide range of adaptive forms.

- Adaptive radiations can be triggered by ecological opportunity, such as the availability of unoccupied niches on an island, or morphological innovations, such as flowers and wings.

- The Cambrian explosion was one of the most spectacular adaptive radiations of all time, producing virtually all of the major animal lineages existing today. During this event, the size and morphological complexity of animals increased tremendously, and animals diversified in how they made a living.

✔ You should be able to explain why the evolution of animals that could swim free of benthic habitats—such as fishlike early vertebrates—created an ecological opportunity.

28.4 Mass Extinction

- Mass extinctions are short-term environmental catastrophes that eliminate most of the species alive. Mass extinctions have altered the course of evolutionary history at least five times.

- Mass extinctions such as the end-Permian and end-Cretaceous prune the tree of life more or less randomly and have marked the end of several branches within prominent lineages and the subsequent rise of new branches.

- After the devastation of a mass extinction, the fossil record indicates that it can take 10–15 million years for ecosystems to recover their former levels of diversity.

- A mass extinction may presently be under way due to the drastic effects of humans on the Earth in recent history.

✔ You should be able to evaluate this statement: Environmental changes caused by humans are eliminating species that are poorly adapted to their natural environment.

MasteringBiology

1. MasteringBiology Assignments

Tutorials and Activities Adaptive Radiation, Constructing Phylogenetic Trees, Fossil Record, History of Life, Overview of Macroevolution, Scrolling Geologic Record

Questions Reading Quizzes, Blue-Thread Questions, Test Bank

2. eText Read your book online, search, take notes, highlight text, and more.

3. The Study Area Practice Test, Cumulative Test, BioFlix® 3-D Animations, Videos, Activities, Audio Glossary, Word Study Tools, Art

You should be able to . . .

✔ TEST YOUR KNOWLEDGE
Answers are available in Appendix A

1. Choose the best definition of a fossil.
   a. any trace of an organism that has been converted into rock
   b. a bone, tooth, shell, or other hard part of an organism that has been preserved
   c. any trace of an organism that lived in the past
   d. the process that leads to preservation of any body part from an organism that lived in the past

2. What is the difference between a branch and a node on a phylogenetic tree?
   a. A branch is a population through time; a node is where a population splits into two independent populations.
   b. A branch represents a common ancestor; a node is a species.
   c. A branch is a lineage; a node is any named taxonomic group.
   d. A branch is a population through time; a node is the common ancestor of all species or lineages present on a particular tree.

3. Which of the following best characterizes an adaptive radiation?
   a. Speciation occurs extremely rapidly, and descendant populations occupy a large geographic area.

   b. A single lineage diversifies rapidly, and descendant populations occupy many habitats and ecological roles.
   c. Natural selection is particularly intense, because disruptive selection occurs.
   d. Species recover after a mass extinction.

4. True or false? The dinosaurs (other than birds) went extinct during the asteroid impact at the K–P boundary because they were poorly adapted to their ecological niches.

5. Which of the following is an example of homoplasy?
   a. hair in humans and fur in mice
   b. astragalus ankle bones in hippos and deer
   c. *Hox* genes in humans and flies
   d. streamlined bodies in dolphins and ichthyosaurs

6. True or false? Monophyletic groups are identified by shared, derived traits.

✔ TEST YOUR UNDERSTANDING
Answers are available in Appendix A

7. The text claims that the fossil record is biased in several ways. What are these biases? If the database is biased, is it still an effective tool to use in studying the diversification of life? Why or why not?

8. The initial diversification of animals took place over some 50 million years, mostly during the Cambrian period. Propose an analogy from everyday life to illustrate the significance of this event.

9. Why is parsimony a useful tool for assessing which phylogenetic tree is most accurate? Why was parsimony misleading in the case of the astragalus during the evolution of artiodactyls?

10. Give an example of an adaptive radiation that occurred after a colonization event and another that occurred after a key innovation. In each case, provide a hypothesis to explain why the adaptive radiation occurred.

11. QUANTITATIVE The end-Permian extinction was devastating. If each student in your graduating class represented a different species during the Permian and there were 1880 students in your graduating class, about how many students would make it to graduation?

12. **BioSkills 7** in Appendix B recommends a "one-snip test" to identify monophyletic groups—meaning that if you cut any branch on a tree, everything that "falls off" is a monophyletic group. Why is this valid?
    a. Monophyletic groups are nested on a tree—meaning that they are hierarchical.
    b. Monophyletic groups can also be called clades or lineages.
    c. Species are the smallest monophyletic groups on the tree of life.
    d. One snip gets an ancestor and all of its descendants.

13. Suppose that the dying wish of a famous eccentric was that his remains be fossilized. His family has come to you for expert advice. What method would you recommend as most likely to maximize the chances that his wish will be fulfilled?

a. Place the corpse in an environment where decomposition is rapid (such as a forest).

b. Place the corpse in an environment where decomposition is slow (such as a swamp or bog).

c. Bury only the bones, since soft parts don't fossilize well.

d. Place the corpse in an environment where plenty of oxygen is available.

14. Use the data matrix below to draft at least three alternate phylogenetic trees. Label the outgroup. Indicate which tree is most parsimonious, and explain why. How could you change your analysis to increase your confidence that the tree represents true evolutionary relationships?

Presence or Absence of *Hox* Genes*

	Hox1	Hox2	Hox3	Hox4	Hox12
Mouse	1	1	1	1	1
Snail	1	1	1	1	0
Fly	1	1	1	1	0
Sea anemone	1	1	0	0	0
Sponge	0	0	0	0	0

*1 = gene present; 0 = gene absent.

15. Some researchers contend that the end-Permian extinction event was caused by an impact with a large extraterrestrial object. Describe three forms of evidence that you would like to see before you accept this hypothesis.

16. One of the "triggers" proposed for the Cambrian explosion is a dramatic rise in oxygen concentrations that occurred in the oceans about 800 mya. (Review material in Chapter 9 on how oxygen compares with other atoms or compounds as an electron acceptor during cellular respiration; look near the end of Section 9.6, where aerobic and anaerobic respiration are compared.) Make a small concept map outlining the logic behind the "oxygen-trigger" hypothesis for the Cambrian explosion. (See **BioSkills 15** in Appendix B if you need help making concept maps.)

Geneticist and evolutionary biologist Theodosius Dobzhansky said that "Nothing in biology makes sense except in the light of evolution." Use this concept map to study how ideas introduced in Unit 5 fit together.

The key is to connect the four evolutionary processes that work at the level of populations—natural selection, genetic drift, mutation, and gene flow—to processes, events, and outcomes at higher levels of organization: speciation, adaptive radiation, mass extinction, and the tree of life.

It's all about changes in allele frequencies. Over time, small changes that occur between populations lead to large changes that distinguish major lineages on the tree of life.

Note that each box in the concept map indicates the chapter and section where you can go for review. Also, be sure to do the blue exercises in the Check Your Understanding box below.

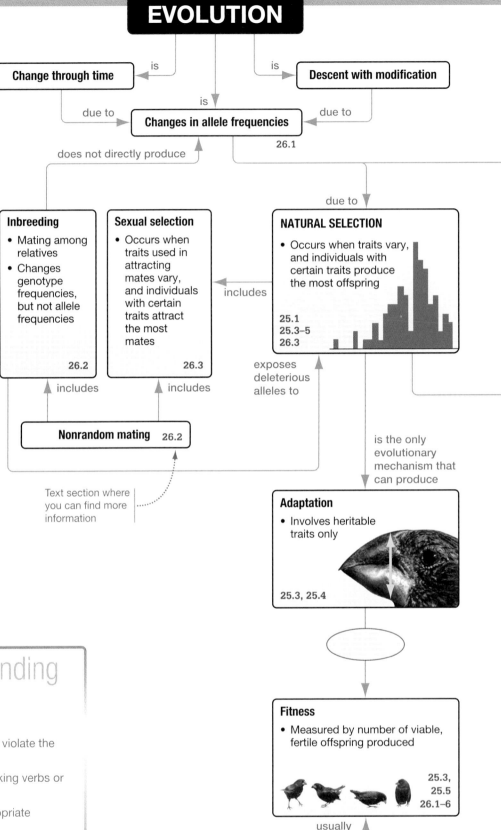

EVOLUTION

Change through time ← is

is → **Descent with modification**

is

due to

due to

Changes in allele frequencies
26.1

does not directly produce

due to

Inbreeding
- Mating among relatives
- Changes genotype frequencies, but not allele frequencies

26.2

Sexual selection
- Occurs when traits used in attracting mates vary, and individuals with certain traits attract the most mates

26.3

includes

NATURAL SELECTION
- Occurs when traits vary, and individuals with certain traits produce the most offspring

25.1
25.3–5
26.3

exposes deleterious alleles to

includes

Nonrandom mating 26.2

includes

Text section where you can find more information

is the only evolutionary mechanism that can produce

Adaptation
- Involves heritable traits only

25.3, 25.4

usually reduces

Fitness
- Measured by number of viable, fertile offspring produced

25.3, 25.5
26.1–6

check your understanding

If you understand the big picture . . .

✔ **You should be able to . . .**

1. Draw a circle around the processes that violate the Hardy–Weinberg principle.
2. Fill in the blue ovals with appropriate linking verbs or phrases.
3. Add a box for "Fossil record" with appropriate connections.
4. Draw arrows linking genetic drift, mutation, and gene flow to the appropriate box using the linking phrase "is random with respect to."

Answers are available in Appendix A.

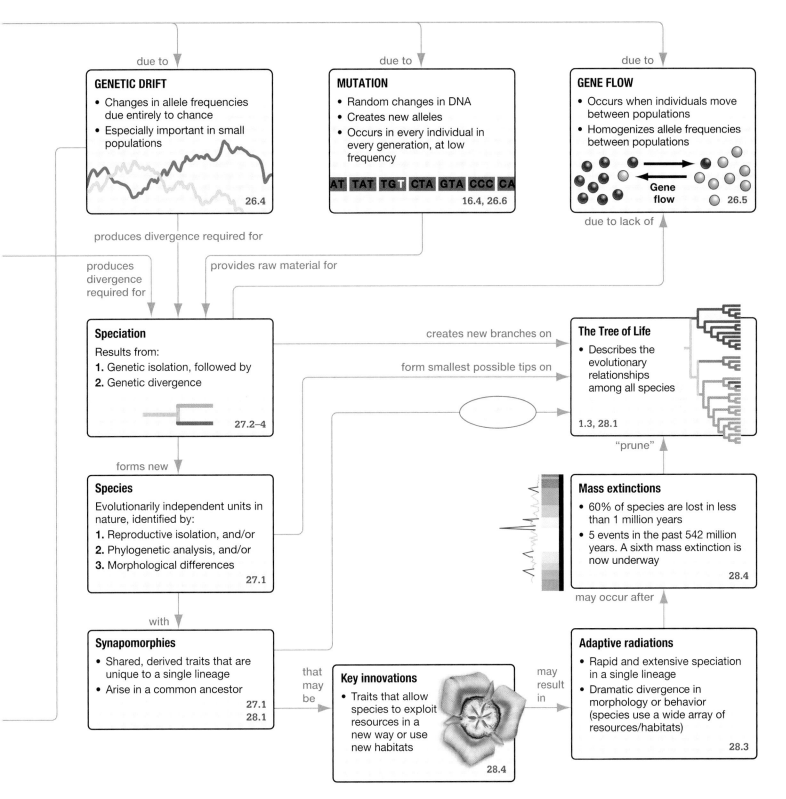

GENETIC DRIFT

• Changes in allele frequencies due entirely to chance
• Especially important in small populations

26.4

MUTATION

• Random changes in DNA
• Creates new alleles
• Occurs in every individual in every generation, at low frequency

AT TAT TGT CTA GTA CCC CA

16.4, 26.6

GENE FLOW

• Occurs when individuals move between populations
• Homogenizes allele frequencies between populations

Gene flow

26.5

due to ▼   *due to* ▼   *due to* ▼

due to lack of ▲

produces divergence required for

produces divergence required for   *provides raw material for*

Speciation

Results from:
1. Genetic isolation, followed by
2. Genetic divergence

27.2–4

creates new branches on

form smallest possible tips on

The Tree of Life

• Describes the evolutionary relationships among all species

1.3, 28.1

"prune" ▲

forms new ▼

Species

Evolutionarily independent units in nature, identified by:
1. Reproductive isolation, and/or
2. Phylogenetic analysis, and/or
3. Morphological differences

27.1

Mass extinctions

• 60% of species are lost in less than 1 million years
• 5 events in the past 542 million years. A sixth mass extinction is now underway

28.4

may occur after ▲

with ▼

Synapomorphies

• Shared, derived traits that are unique to a single lineage
• Arise in a common ancestor

27.1
28.1

that may be

Key innovations

• Traits that allow species to exploit resources in a new way or use new habitats

28.4

may result in

Adaptive radiations

• Rapid and extensive speciation in a single lineage
• Dramatic divergence in morphology or behavior (species use a wide array of resources/habitats)

28.3

29 Bacteria and Archaea

In this chapter you will learn that

Bacteria and archaea are the most abundant and diverse forms of life on Earth

Prokaryotes — which are

in contrast to

Eukaryotes Ch. 30–36

by asking → **Why and how do biologists study these organisms?** 29.1, 29.2

by examining → **Three themes in their diversification** 29.3 → Metabolism / Morphology / Ecological diversity

then, looking closer at → **Key lineages of bacteria and archaea** 29.4

Although this hot spring looks devoid of life, it is actually teeming with billions of bacterial and archaeal cells.

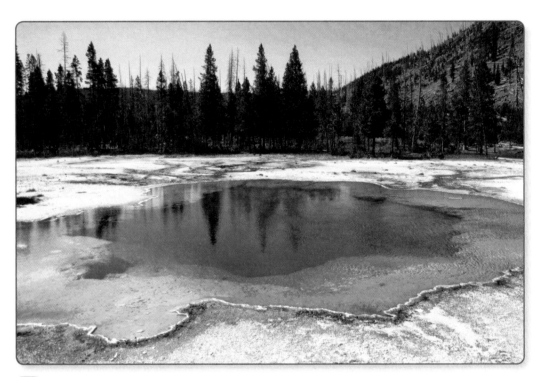

Bacteria and Archaea (usually pronounced *ar-KEE-ah*) form two of the three largest branches on the tree of life (**FIGURE 29.1**). The third major branch, or domain, consists of eukaryotes and is called the Eukarya. Virtually all members of the Bacteria and Archaea domains are unicellular, and all are prokaryotic—meaning that they lack a membrane-bound nucleus.

Although their relatively simple morphology makes bacteria and archaea appear similar to the untrained eye, in some ways they are strikingly different at the molecular level (**TABLE 29.1**). Organisms in the Bacteria and Archaea domains are distinguished by the types of molecules that make up their plasma membranes and cell walls:

- **Bacteria** have a unique compound called peptidoglycan in their cell walls (see Chapter 5).

✔ When you see this checkmark, stop and test yourself. Answers are available in Appendix A.

FIGURE 29.1 Bacteria, Archaea, and Eukarya Are the Three Domains of Life. Archaea are more closely related to eukaryotes than they are to bacteria.

✔**QUESTION** Was the common ancestor of all species living today prokaryotic or eukaryotic? Explain your reasoning.

- **Archaea** have unique phospholipids in their plasma membranes—the hydrocarbon tails of the phospholipids are made from isoprene (see Chapter 6).

If you were unicellular, bacteria and archaea would look as different to you as mammals and fish.

In addition, the machinery that bacteria and archaea use to process genetic information is strikingly different. More specifically, the DNA polymerases, RNA polymerases, transcription-initiation proteins, and ribosomes found in Archaea and Eukarya are distinct from those found in Bacteria and similar to each other. These differences have practical consequences: Antibiotics that poison bacterial ribosomes do not affect the ribosomes of archaea or eukaryotes. If all ribosomes were identical, these antibiotics would kill you along with the bacterial species that was supposed to be targeted.

As an introductory biology student, you may know less about bacteria and archaea than about any other group on the tree of life. Before taking this course, it's likely that you'd never even heard of the Archaea. You might be surprised to find out that biologists suspect that the first eukaryotic cells may have formed as a result of a symbiotic event between these two groups of prokaryotes. Notice in Table 29.1 that cells in the domain Eukarya share some features with both Bacteria and Archaea. Studying these shared features has helped biologists understand the early evolution of life on Earth.

This chapter's goal is to convince you that even though bacteria and archaea are tiny, they have an enormous impact on you and the planet in general. By the time you finish reading the chapter, you should understand why a researcher summed up the bacteria and archaea by claiming, "They run this joint."

29.1 Why Do Biologists Study Bacteria and Archaea?

Biologists study bacteria and archaea for the same reasons they study any organisms. First, they are intrinsically fascinating. Discoveries such as finding bacterial cells living a kilometer underground or in 95°C hot springs keep biologists awake at night, staring at the ceiling. They can't wait to get into the lab in the morning and figure out how those cells stay alive.

SUMMARY TABLE 29.1 **Characteristics of Bacteria, Archaea, and Eukarya**

	Bacteria	Archaea	Eukarya
DNA enclosed by a nuclear envelope? (see Chapter 7)	No	No	Yes
Circular chromosome present?	Yes (but linear in some species)	Yes	No (linear)
Organelles enclosed by membranes present? (see Chapter 7)	No	No	Yes
Rotating flagella present? (see Chapter 7)	Yes	Yes	No (flagella and cilia undulate)
Multicellular species?	No (with some exceptions)	No	Yes
Plasma membrane lipids composed of glycerol bonded to unbranched fatty acids by ester linkages? (see Chapter 6)	Yes	No (branched lipids bonded by ether linkages)	Yes
Cell walls, when present, contain peptidoglycan? (see Chapter 5)	Yes	No	No
RNA polymerase composed of >10 subunits?	No (only 5 subunits)	Yes	Yes
Translation initiated with methionine? (see Chapter 17)	No (initiated with N-formylmethionine; f-met)	Yes	Yes

✔**EXERCISE** Using the data in this table, add labeled marks to Figure 29.1 indicating where the following traits evolved: peptidoglycan in cell wall, archaeal-type plasma membrane, archaeal- and eukaryotic-type translation initiation, and nuclear envelope.

Second, there are practical benefits to understanding the species that share the planet with us. Understanding bacteria and archaea is particularly important—in terms of both understanding life on Earth and improving human health and welfare.

Biological Impact

The lineages in the domains Bacteria and Archaea are ancient, diverse, abundant, and ubiquitous. The oldest fossils of any type found to date are 3.5-billion-year-old carbon-rich deposits derived from bacteria. Because eukaryotes do not appear in the fossil record until 1.75 billion years ago, biologists infer that prokaryotes were the only form of life on Earth for at least 1.7 billion years.

Just how many bacteria and archaea are alive today? Although a mere 5000 species have been formally named and described to date, it is virtually certain that millions exist. Consider that over 1000 species of prokaryotes are living in your large intestine right now, and another 700 species are living in your mouth. Well-known microbiologist Norman Pace points out that there may be tens of millions of different insect species but notes, "If we squeeze out any one of these insects and examine its contents under the microscope, we find hundreds or thousands of distinct microbial species." Most of these **microbes** (microscopic organisms) are bacteria or archaea. Virtually all are unnamed and undescribed. If you want to discover and name new species, then study bacteria or archaea.

Abundance Besides recognizing how diverse bacteria and archaea are in terms of numbers of species, it's critical to appreciate their abundance.

- The approximately 10^{13} (10 trillion) cells in your body are outnumbered ten to one by the bacterial and archaeal cells living on and in you. You are a walking, talking habitat—one that is teeming with bacteria and archaea.

- A mere teaspoon of good-quality soil contains *billions* of microbial cells, most of which are bacteria and archaea.

- In sheer numbers, species in a lineage of marine archaea may be the most successful organisms on Earth. Biologists routinely find these cells at concentrations of over 10,000 individuals per milliliter in most of the world's oceans. At these concentrations, one liter of seawater contains a population equivalent to that of the largest human cities. Yet this lineage was not described until the early 1990s.

- Recent research has found enormous numbers of bacterial and especially archaeal cells in rocks and sediments as much as 1600 meters underneath the world's oceans. Although recently discovered, the bacteria and archaea living under the ocean may make up 10 percent of the world's total mass of living material.

- Biologists estimate the total number of individual bacteria and archaea alive today at over 5×10^{30}. If they were lined up end to end, these cells would make a chain longer than the Milky Way galaxy. They contain 50 percent of all the carbon and 90 percent of all the nitrogen and phosphorus found in organisms.

In terms of the total volume of living material on our planet, bacteria and archaea are dominant life-forms.

Habitat Diversity Bacteria and archaea are found almost everywhere. They live in environments as unusual as oxygen-free mud, hot springs, and salt flats. In seawater they are found from the surface to depths of 10,000 meters (m), and at temperatures over 120°C (well above water's boiling point) near submarine volcanoes.

Although there are far more prokaryotes than eukaryotes, much more is known about eukaryotic diversity than about prokaryotic diversity. Researchers who study prokaryotic diversity are exploring one of the most wide-open frontiers in all of science. So little is known about the extent of these domains that recent collecting expeditions have turned up entirely new **phyla** (singular: **phylum**). These are names given to major lineages within each domain. To a biologist, this achievement is equivalent to the sudden discovery of a new group of eukaryotes as distinctive as flowering plants or animals with backbones.

The physical world has been explored and mapped, and many of the larger plants and animals are named. But in **microbiology**—the study of organisms that can be seen only with the aid of a microscope—this is an age of exploration and discovery.

Some Microbes Thrive in Extreme Environments

Bacteria or archaea that live in high-salt, high-temperature, low-temperature, or high-pressure habitats are **extremophiles** ("extreme-lovers"). Studying them has been extraordinarily fruitful for understanding the tree of life, developing industrial applications, and exploring the structure and function of enzymes.

As an example of these habitats, consider hydrothermal vents at the bottom of the ocean, where water as hot as 300°C emerges and mixes with 4°C seawater. At locations like these, archaea are abundant forms of life.

Researchers recently discovered an archaeon that grows so close to these hydrothermal vents that its surroundings are at 121°C—a record for life at high temperature. This organism can live and grow in water that is heated past its boiling point (100°C) and at pressures that would instantly destroy a human. Since high temperature breaks non-covalent bonds holding macromolecules together, extreme heat usually denatures proteins, makes membranes leaky, and separates the strands of the DNA double helix. Biologists are intrigued by how these unusual archaeal cells can thrive under such extreme conditions.

Other discovered bacteria and archaea can grow

- at a pH less than 1.0;

- at temperatures of 0°C under Antarctic ice;

- in water that is virtually saturated with salt (**FIGURE 29.2**).

Extremophiles have become a hot area of research. The genomes of a wide array of extremophiles have been sequenced, and expeditions regularly seek to characterize new species. Why?

- *Origin of life* Based on models of conditions that prevailed early in Earth's history, it appears likely that the first forms of life lived at high temperature and pressure in environments

Archaeal cells tint these salt crystals different colors

Salt ponds where seawater is evaporating

FIGURE 29.2 Some Archaeal Cells Live in High-Salt Habitats. The evaporating water is colored red by these photosynthetic cells.

that lacked oxygen—conditions that humans would call extreme. Thus, understanding extremophiles may help explain how life on Earth began.

- *Extraterrestrial life?* In a similar vein, many astrobiologists ("space-biologists") use extremophiles as model organisms in the search for extraterrestrial life. The idea is that if bacteria and archaea can thrive in extreme habitats on Earth, cells might possibly be found in similar environments on other planets or moons of planets.

- *Commercial applications* Because enzymes that function at extreme temperatures and pressures are useful in many industrial processes, extremophiles are of commercial interest as well. Recall that *Taq* polymerase—a DNA polymerase that is stable up to 95°C—is used to run the polymerase chain reaction (PCR) in research and commercial settings (see Chapter 20). This enzyme was isolated from a bacterium called *Thermus aquaticus* ("hot water"), which was discovered in a hot spring in Yellowstone National Park.

Bacteria and archaea may be small, but they thrive in an amazing range of conditions.

Medical Importance

The first paper documenting that an archaeon was associated with a human disease—a dental condition called periodontitis—was published in 2004. But biologists have been studying disease-causing bacteria for over a century.

Of the thousands of bacterial species living in and on your body, only a tiny fraction can disrupt normal body functions enough to cause illness. Bacteria that cause disease are said to be **pathogenic** (literally, "disease-producing"). Pathogenic bacteria have been responsible for some of the most devastating epidemics in human history.

TABLE 29.2 lists some of the bacteria that cause illness in humans. Here are the important things to note:

- Pathogenic forms come from several different lineages in the domain Bacteria.

TABLE 29.2 Some Bacteria That Cause Illness in Humans

Lineage	Species	Tissues Affected	Disease
Firmicutes	*Clostridium tetani*	Wounds, nervous system	Tetanus
	Staphylococcus aureus	Skin, urogenital canal	Acne, boils, impetigo, toxic shock syndrome
	Streptococcus pneumoniae	Respiratory tract	Pneumonia
	Streptococcus pyogenes	Respiratory tract	Strep throat, scarlet fever
Spirochaetes	*Borrelia burgdorferi*	Skin and nerves	Lyme disease
	Treponema pallidum	Urogenital canal	Syphilis
Actinobacteria	*Mycobacterium tuberculosis*	Respiratory tract	Tuberculosis
	Mycobacterium leprae	Skin and nerves	Leprosy
	Propionibacterium acnes	Skin	Acne
Chlamydiales	*Chlamydia trachomatis*	Urogenital canal	Genital tract infection
ε-Proteobacteria	*Helicobacter pylori*	Stomach	Ulcer
β-Proteobacteria	*Neisseria gonorrhoeae*	Urogenital canal	Gonorrhea
γ-Proteobacteria	*Haemophilus influenzae*	Ear canal, nervous system	Ear infections, meningitis
	Pseudomonas aeruginosa	Urogenital canal, eyes, ear canal, lungs	Infections of eye, ear, urinary tract, lungs
	Salmonella enterica	Gastrointestinal tract	Food poisoning
	Yersinia pestis	Lymph and blood	Plague

- Pathogenic bacteria tend to affect tissues at the entry points to the body, such as wounds or pores in the skin, the respiratory and gastrointestinal tracts, and the urogenital canal.

Koch's Postulates   Robert Koch was the first person to establish a link between a particular species of bacterium and a specific disease. When Koch began his work on the nature of disease in the late 1800s, microscopists had confirmed the existence of the particle-like organisms people now call bacteria, and Louis Pasteur had shown that bacteria and other microorganisms are responsible for spoiling milk, wine, broth, and other foods. Koch hypothesized that bacteria might also be responsible for causing infectious diseases, which spread by being passed from an infected individual to an uninfected individual.

Koch set out to test this hypothesis by identifying the organism that causes anthrax. Anthrax is a disease of cattle and other grazing mammals that can result in fatal blood poisoning. The disease also occurs infrequently in humans and mice.

To establish a causative link between a specific microbe and a specific disease, Koch proposed that four criteria had to be met:

1. The microbe must be present in individuals suffering from the disease and absent from healthy individuals. By careful microscopy, Koch was able to show that the bacterium *Bacillus anthracis* was always present in the blood of cattle suffering from anthrax, but absent from healthy individuals.

2. The organism must be isolated and grown in a pure culture away from the host organism. Koch was able to grow pure colonies of *B. anthracis* in glass dishes on a nutrient medium, using gelatin as a substrate.

3. If organisms from the pure culture are injected into a healthy experimental animal, the disease symptoms should appear. Koch demonstrated this crucial causative link in mice injected with *B. anthracis*. The symptoms of anthrax infection appeared and then the infected mice died.

4. The organism should be isolated from the diseased experimental animal, again grown in pure culture, and demonstrated by its size, shape, and color to be the same as the original organism. Koch did this by purifying *B. anthracis* from the blood of diseased experimental mice.

These criteria, now called **Koch's postulates,** are still used to confirm a causative link between new diseases and a suspected infectious agent. Microbiologists now recognize that many bacteria cannot be grown in culture, so they use other means of detection for those organisms.

The Germ Theory   Koch's experimental results were the first test of the **germ theory of disease.**

- The pattern component of this theory is that certain diseases are infectious—meaning that they can be passed from person to person.

- The process responsible for this pattern is the transmission and growth of certain bacteria and viruses.

Viruses are acellular particles that parasitize cells (see Chapter 36).

The germ theory of disease laid the foundation for modern medicine. Initially its greatest impact was on sanitation—efforts to prevent transmission of pathogenic bacteria. During the American Civil War, for example, records indicate that more soldiers died of bacterial infections contracted from drinking water contaminated with human feces than from wounds in battle. Also during that conflict it was common for surgeons to sharpen their scalpels on their shoe leather, after walking in horse manure.

Fortunately, improvements in sanitation and nutrition have caused dramatic reductions in mortality rates due to infectious diseases in the industrialized countries.

What Makes Some Bacterial Cells Pathogenic?   Virulence, or the ability to cause disease, is a heritable trait that varies among individuals in a population. Most *Escherichia coli*, for example, are harmless inhabitants of the gastrointestinal tract of humans and other mammals. But some *E. coli* cells cause potentially fatal food poisoning.

What makes some cells of the same species pathogenic, while others are harmless? Biologists have answered this question for *E. coli* by sequencing the entire genome of a harmless lab strain and the pathogenic strain called O157:H7, which is harmful to humans. The genome of the pathogenic strain is slightly larger because it has acquired virulence genes, including one coding for a protein **toxin.** After entering a host cell, this toxin binds to ribosomes and inhibits protein synthesis, killing the host cells. Because of key differences between the ribosomes of bacteria and eukaryotic cells, only host-cell protein synthesis is blocked by the toxin. Cells lining the blood vessels near the host's intestinal epithelium are most affected by the toxin, and the resulting damage leads to bloody diarrhea and possible death. If sanitation is poor, the pathogenic bacteria are likely to infect many new hosts.

Similar types of studies are identifying the genes responsible for virulence in a wide array of pathogenic bacteria.

The Past, Present, and Future of Antibiotics   **Antibiotics** are molecules that kill bacteria or stop them from growing. They are produced naturally by a wide array of soil-dwelling bacteria and fungi. In these environments, antibiotics are hypothesized to help cells reduce competition for nutrients and other resources.

The discovery of antibiotics in 1928, their development over subsequent decades, and widespread use starting in the late 1940s gave physicians effective tools to combat many bacterial infections.

Unfortunately, extensive use of antibiotics in the late twentieth century in clinics and animal feed led to the evolution of drug-resistant strains of pathogenic bacteria (see Chapter 25). One study found that there are now soil-dwelling bacteria in natural environments that—far from being killed by antibiotics—actually use them as food.

Coping with antibiotic resistance in pathogenic bacteria has become a great challenge of modern medicine. Some researchers even claim that humans may be entering the "post-antibiotic era" in medicine.

New research indicates that bacteria have another advantage: They usually grow as **biofilms,** dense bacterial colonies enmeshed

FIGURE 29.3 Biofilm Growing Inside a Catheter. This micrograph shows *Staphylococcus aureus* cells growing inside a catheter—a tube inserted into a body so that fluids can be withdrawn or injected. Bacterial communities that secrete polysaccharides and adhere to surfaces are sometimes more resistant to antibiotics.

FIGURE 29.4 Bacteria and Archaea Can Play a Role in Cleaning Up Pollution. These clean-up workers are spraying nitrogenous nutrients to encourage bacterial and archaeal growth following the 1989 *Exxon Valdez* oil spill in Alaska.

in a polysaccharide-rich matrix that helps shield the bacteria from antibiotics. Antibiotic-resistant biofilms on medical devices such as catheters are a growing problem in hospitals (**FIGURE 29.3**).

Role in Bioremediation

Bacteria are often in the news because of their dire medical effects. However, only a tiny proportion of bacteria and archaea actually cause disease in humans or other organisms. In the vast majority of cases, bacteria and archaea either have no direct impact on humans or are beneficial. For example, microbes play an important role in wastewater treatment efforts, and researchers are using bacteria and archaea to clean up sites polluted with organic solvents—an effort called **bioremediation.**

Throughout the industrialized world, some of the most serious pollutants in soils, rivers, and ponds consist of organic compounds that were originally used as solvents or fuels but leaked or were spilled into the environment. Most of these compounds are highly hydrophobic. Because they do not dissolve in water, they tend to accumulate in sediments. If the compounds are subsequently ingested by burrowing worms or clams or other organisms, they can be passed along to fish, insects, humans, birds, and other species.

At moderate to high concentrations, these pollutants are toxic to eukaryotes. Petroleum from oil spills and compounds that contain ring structures and chlorine atoms, such as the family of compounds called dioxins, are particularly notorious because of their toxicity to humans.

Fortunately, naturally existing populations of bacteria and archaea can grow in spills and degrade the toxins. This growth can be enhanced using two complementary bioremediation strategies:

1. *Fertilizing contaminated sites to encourage the growth of existing bacteria and archaea that degrade toxic compounds.* After several oil spills, researchers added nitrogen to affected sites as a fertilizer (**FIGURE 29.4**). Dramatic increases

occurred in the growth of bacteria and archaea that use hydrocarbons in cellular respiration, probably because the cells used the added nitrogen to synthesize enzymes and other key compounds. In at least some cases, the fertilized shorelines cleaned up much faster than unfertilized sites.

2. *"Seeding," or adding, specific species of bacteria and archaea to contaminated sites.* Seeding shows promise of alleviating pollution in some situations. For example, researchers have recently discovered bacteria that are able to render certain chlorinated, ring-containing compounds harmless. Instead of being poisoned by the pollutants, these bacteria use the chlorinated compounds as electron acceptors during cellular respiration. In at least some cases, the by-product is dechlorinated and nontoxic to humans and other eukaryotes.

To follow up on these discoveries, researchers are now growing the bacteria in quantity, to test the hypothesis that seeding can speed the rate of decomposition in contaminated sediments. Initial reports suggest that seeding may help clean up at least some polluted sites.

29.2 How Do Biologists Study Bacteria and Archaea?

Biologists' understanding of the domains Bacteria and Archaea is advancing more rapidly right now than at any time during the past 100 years—and perhaps faster than our understanding of any other lineages on the tree of life.

As an introduction to the domains Bacteria and Archaea, let's examine a few of the techniques that biologists use to answer questions about them. Some of these research strategies have been used since bacteria were first discovered; some were invented less than 10 years ago.

Using Enrichment Cultures

Which species of bacteria and archaea are present at a particular location, and what do they use as food? To answer questions like these, biologists rely heavily on their ability to culture organisms in the lab. Of the 5000 species of bacteria and archaea described to date, almost all were discovered when they were isolated from natural habitats and grown under controlled conditions in the laboratory.

One classical technique for isolating new types of bacteria and archaea is called **enrichment culture**. Enrichment cultures are based on establishing a specified set of growing conditions—temperature, lighting, substrate, types of available food, and so on. Cells that thrive under the specified conditions increase in numbers enough to be isolated and studied in detail.

To appreciate how this strategy works in practice, consider research on bacteria that live deep below Earth's surface. One study began with samples of rock and fluid from drilling operations in Virginia and Colorado. The samples came from sedimentary rocks at depths ranging from 860 to 2800 meters below the surface, where temperatures are between 42°C and 85°C. The questions posed in the study were simple: Is anything alive down there? If so, what do the organisms use to fuel cellular respiration?

The research team hypothesized that if organisms were living deep below the surface of the Earth, the cells might use hydrogen molecules (H_2) as an electron donor and the ferric ion (Fe^{3+}) as an electron acceptor (**FIGURE 29.5**). Recall that most eukaryotes use sugars as electron donors and use oxygen as an electron acceptor during cellular respiration (see Chapter 9). Fe^{3+} is the oxidized form of iron, and it is abundant in the rocks the biologists collected from great depths. It exists at great depths below the surface in the form of ferric oxyhydroxide, $Fe(OH)_3$. The researchers predicted that if an organism in the samples reduced the ferric ions during cellular respiration, then a black, oxidized, and magnetic mineral called magnetite (Fe_3O_4) would start appearing in the cultures as a by-product of cellular respiration.

What did their enrichment cultures produce? In some cultures, a black compound began to appear within a week. A variety of tests confirmed that the black substance was indeed magnetite. As the "Results" section of Figure 29.5 shows, microscopy revealed the organisms themselves—previously undiscovered bacteria. Because they grow only when incubated at 45°C–75°C, these organisms are considered **thermophiles** ("heat-lovers"). The discovery was spectacular—it was one of the first studies demonstrating that Earth's crust is teeming with organisms to depths of over a mile below the surface. Enrichment culture continues to be a productive way to isolate and characterize new species of bacteria and archaea.

Using Metagenomics

Researchers estimate that of all the bacteria and archaea living today, less than 1 percent have been grown in culture. To augment research based on enrichment cultures, researchers are

RESEARCH

QUESTION: Can bacteria live a mile below Earth's surface?

HYPOTHESIS: Bacteria are capable of cellular respiration deep below Earth's surface by using H_2 as an electron donor and Fe^{3+} as an electron acceptor.

NULL HYPOTHESIS: Bacteria from this environment are not capable of using H_2 as an electron donor and Fe^{3+} as an electron acceptor.

EXPERIMENTAL SETUP:

1. Prepare enrichment culture abundant in H_2 and Fe^{3+}; raise temperatures above 45°C.

2. Add rock and fluid samples extracted from drilling operations at depths of about 1000 m below Earth's surface.

PREDICTION: Black, magnetic grains of magnetite (Fe_3O_4) will accumulate because Fe^{3+} is reduced by growing cells and shed as a waste product. Cells will be visible.

PREDICTION OF NULL HYPOTHESIS: No magnetite will appear. No cells will grow.

RESULTS: Cells are visible, and magnetite is detectable.

CONCLUSION: At least one bacterial species that can live deep below Earth's surface grew in this enrichment culture. Different culture conditions might result in the enrichment of different species present in the same sample.

FIGURE 29.5 Enrichment Cultures Isolate Large Populations of Cells That Grow under Specific Conditions.

✔ **QUESTION** Suppose no organisms had grown in this culture. Explain why the lack of growth would be either strong or weak evidence on the question of whether organisms live a mile below the Earth's surface.

SOURCE: Liu, S. V., J. Zhou, C. Zhang, et al. 1997. Thermophilic Fe(III)-reducing bacteria from the deep subsurface: the evolutionary implications. *Science* 277: 1106–1109.

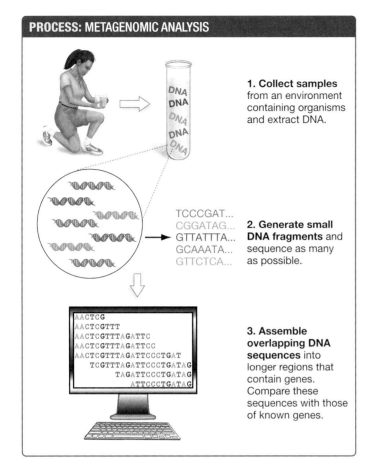

PROCESS: METAGENOMIC ANALYSIS

1. Collect samples from an environment containing organisms and extract DNA.

TCCCGAT...
CGGATAG...
GTTATTTA...
GCAAATA...
GTTCTCA...

2. Generate small DNA fragments and sequence as many as possible.

```
AACTCG
AACTCGTTT
AACTCGTTTAGATTC
AACTCGTTTAGATTCC
AACTCGTTTAGATTCCCTGAT
    TCGTTTAGATTCCCTGATAG
         TAGATTCCCTGATAG
             ATTCCCTGATAG
```

3. Assemble overlapping DNA sequences into longer regions that contain genes. Compare these sequences with those of known genes.

FIGURE 29.6 Metagenomics Allows Researchers to Identify Species That Have Never Been Seen. Metagenomic analysis is used to generate DNA sequences from an environmental sample. That information can then be used to identify novel species and investigate biological processes.

employing a technique called **metagenomics** or **environmental sequencing** (see Chapter 21). Metagenomics is employed in part to document the presence of bacteria and archaea in an environmental sample that cannot be grown in culture. It is based on extracting and sequencing much of the DNA from a sample and then identifying species and biochemical pathways by comparing the DNA sequences with those of known genes. **FIGURE 29.6** outlines the steps performed in a metagenomics study.

Metagenomic analysis allows biologists to rapidly identify and characterize organisms that have never been seen. The technique has revealed huge new branches on the tree of life and produced revolutionary data on the habitats where bacteria and archaea are found.

In one recent study biologists extracted DNA from 125 human fecal samples and generated over 500 billion base pairs of sequence, over 150 times more than the entire human genome. The results they obtained are fascinating:

- In total, the samples contained about 1000 different species of bacteria. Some species were found in most of the samples while others were found in only a few of the humans sampled;

- the vast majority of bacterial species identified were from three phyla: Bacteroidetes, Firmicutes, and Actinobacteria;

- the identified bacterial genes that were shared by all of the human subjects suggest that bacteria play important roles in human physiology, including digestion of complex carbohydrates and synthesis of essential amino acids and vitamins.

Results like these make it clear that humans harbor a diverse ecosystem of symbiotic bacteria. Some bacteria may make us sick, but we depend on many others to stay healthy.

In combination with **direct sequencing**—a technique based on isolating and sequencing a specific gene from organisms found in a particular habitat—metagenomics is revolutionizing biologists' understanding of bacterial and archaeal diversity.

Evaluating Molecular Phylogenies

To put data from enrichment culture and metagenomic studies into context, biologists depend on the accurate placement of species on phylogenetic trees. Recall that phylogenetic trees illustrate the evolutionary relationships among species and lineages (see Chapter 1, Chapter 28, and **BioSkills 7** in Appendix B). They are a pictorial summary of which species are more closely or distantly related to others.

Some of the most useful phylogenetic trees for the Bacteria and the Archaea have been based on studies of the RNA molecules found in the small subunit of ribosomes, or what biologists call 16S and 18S RNA. (See Chapter 17 for more information on the structure and function of ribosomes.) In the late 1960s Carl Woese and colleagues began a massive effort to determine and compare the base sequences of 16S and 18S RNA molecules from a wide array of species. The result of their analysis was the **tree of life,** illustrated in Figure 29.1.

check your understanding

If you understand that . . .

- Enrichment cultures isolate cells that grow in response to specific conditions. They create an abundant sample of bacteria that thrive under particular conditions, allowing further study.
- Metagenomics is based on isolating DNA from samples taken directly from the environment, generating random DNA fragments for sequencing, and then analyzing the DNA sequences to identify the organisms and genes present.

✔ **You should be able to . . .**

1. Design an enrichment culture that would isolate species that could be used to clean up oil spills.
2. Outline a study designed to identify the bacterial and archaeal species present in a soil sample near the biology building on your campus.

Answers are available in Appendix A.

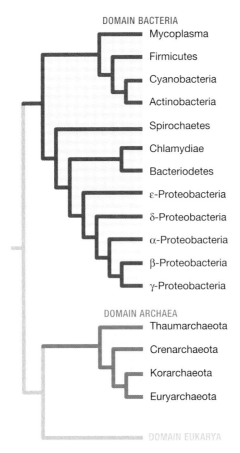

DOMAIN BACTERIA
- Mycoplasma
- Firmicutes
- Cyanobacteria
- Actinobacteria
- Spirochaetes
- Chlamydiae
- Bacteriodetes
- ε-Proteobacteria
- δ-Proteobacteria
- α-Proteobacteria
- β-Proteobacteria
- γ-Proteobacteria

DOMAIN ARCHAEA
- Thaumarchaeota
- Crenarchaeota
- Korarchaeota
- Euryarchaeota

DOMAIN EUKARYA

FIGURE 29.7 Phylogeny of Some Major Lineages in Bacteria and Archaea.

Woese's tree is now considered a classic result. Before its publication, biologists thought that the major division among organisms was between prokaryotes and eukaryotes. But based on data from ribosomal RNA molecules, the major divisions of life-forms are actually the Bacteria, Archaea, and Eukarya. Tracing the early evolutionary history of these domains is extremely difficult since the events distinguishing the lineages took place so long ago. In addition, lateral gene transfer (described in Chapter 21) has blurred the boundaries of the domains. However, recent studies suggest that Eukarya may have formed from a symbiosis between an archaeal cell and a bacterial cell. Traits of both lineages are found within Eukarya (see Chapter 30).

More recent analyses of morphological and molecular characteristics have succeeded in identifying a large series of monophyletic groups within the domains. Recall that a **monophyletic group** consists of an ancestral population and all of its descendants (see Chapter 28). Monophyletic groups can also be called clades or lineages.

The phylogenetic tree in **FIGURE 29.7** summarizes recent results but is still considered highly provisional. Work on molecular phylogenies continues at a brisk pace. Section 29.4 explores some of these lineages in detail, but for now let's turn to the question of how all this diversification took place.

29.3 What Themes Occur in the Diversification of Bacteria and Archaea?

At first, the diversity of bacteria and archaea can seem almost overwhelming. To make sense of the variation among lineages and species, biologists focus on two themes in diversification: morphology and metabolism. Regarding metabolism, the key question is which molecules are used as food. Bacteria and archaea are capable of living in a wide array of environments because they vary in cell structure and in how they make a living.

Morphological Diversity

Because we humans are so large, it is hard for us to appreciate the morphological diversity that exists among bacteria and archaea. To us, they all look small and similar. But at the scale of a bacterium or archaean, different species are wildly diverse in morphology.

Size, Shape, and Motility   To appreciate how diverse these organisms are in terms of morphology, consider bacteria alone:

- *Size*   Bacterial cells range in size from the smallest of all free-living cells—bacteria called mycoplasmas with volumes as small as 0.15 μm³—to the largest bacterium known, *Thiomargarita namibiensis*, which has volumes as large as 200×10^6 μm³. Over a billion *Mycoplasma* cells could fit inside an individual *T. namibiensis* (**FIGURE 29.8a**).

- *Shape*   Bacterial cells range in shape from filaments, spheres, rods, and chains to spirals (**FIGURE 29.8b**).

- *Motility*   Many bacterial cells are motile; their swimming movements are powered by rotating flagella. Depending on the direction of rotation, cells either swim ahead or tumble, which allows them to change direction. Gliding movement, which enables cells to creep along a surface, also occurs in several groups, though the molecular mechanism responsible for this form of motility is still unknown (**FIGURE 29.8c**).

Cell-Wall Composition   For single-celled organisms, the composition of the plasma membrane and cell wall are particularly important. The introduction to this chapter highlights the dramatic differences between the plasma membranes and cell walls of bacteria versus archaea.

Within bacteria having cell walls, biologists distinguish two general types of wall using a dyeing system called the **Gram stain.** As **FIGURE 29.9a** shows, Gram-positive cells look purple but Gram-negative cells look pink.

At the molecular level, most cells that are **Gram-positive** have a plasma membrane surrounded by a cell wall with extensive peptidoglycan (**FIGURE 29.9b**). You might recall that peptidoglycan is a complex substance composed of carbohydrate strands that are cross-linked by short chains of amino acids (see Chapter 5). Most cells that are **Gram-negative,** in contrast, have a plasma

(a) Size varies.

Most bacteria are about 1 μm in diameter, but some are much larger.

Smallest (*Mycoplasma mycoides*)

0.3 μm

Compare sizes

100 μm

Largest (*Thiomargarita namibiensis*)

(b) Shape varies...

... from rods to spheres to spirals. In some species, cells adhere to form chains.

Rods, chains of spheres (compost bacteria)

Spirals (*Campylobacter jejuni*)

(c) Motility varies.

Some bacteria are nonmotile, but swimming and gliding are common.

Swimming (*Pseudomonas aeruginosa*)

Gliding (*Oscillatoria limosa*)

FIGURE 29.8 Morphological Diversity among Bacteria Is Extensive. Some of the cells in these micrographs have been colorized to make them more visible.

membrane surrounded by a cell wall that has two components—a thin gelatinous layer containing peptidoglycan and an outer phospholipid bilayer (**FIGURE 29.9c**).

Analyzing cell cultures with the Gram stain can be an important preliminary step in treating bacterial infections. Because they contain so much peptidoglycan, Gram-positive cells may respond to treatment by penicillin-like drugs that disrupt peptidoglycan synthesis. Gram-negative cells, in contrast, are more

likely to be affected by erythromycin or other types of drugs that poison bacterial ribosomes.

To summarize, members of the Bacteria and the Archaea are remarkably diverse in their overall size, shape, and motility as well as in the composition of their cell walls and plasma membranes. But when asked to name the innovations that were most responsible for the diversification of these two domains, biologists do not point to their morphological diversity. Instead, they

(a) Gram-positive cells stain more than Gram-negative cells.

Gram-positive   Gram-negative

(b) Gram-positive cell wall

(c) Gram-negative cell wall

Cell wall

Polysaccharides

Outer membrane

Peptidoglycan

Plasma membrane

Protein   Protein

Cell wall

FIGURE 29.9 Gram Staining Distinguishes Two Types of Cell Walls in Bacteria. Cells with extensive peptidoglycan retain a large amount of stain and look purple; others retain little stain and look pink, as can be seen in part (a).

point to metabolic diversity—variation in the chemical reactions that go on inside these cells.

Metabolic Diversity

The most important thing to remember about bacteria and archaea is how diverse they are in the types of compounds they can use as food. Bacteria and archaea are the masters of metabolism. Taken together, they can subsist on almost anything—from hydrogen molecules to crude oil. Bacteria and archaea look small and relatively simple to us in their morphology, but their biochemical capabilities are dazzling.

Just how varied are bacteria and archaea when it comes to making a living? To appreciate the answer, recall that all organisms have two fundamental nutritional needs—acquiring chemical energy in the form of adenosine triphosphate (ATP) and obtaining molecules with carbon–carbon bonds that can be used as building blocks for the synthesis of fatty acids, proteins, DNA, RNA, and other large, complex compounds required by the cell.

Bacteria and archaea produce ATP in three ways:

1. **Phototrophs** ("light-feeders") use light energy to excite electrons. ATP is produced by photophosphorylation (see Chapter 10).

2. **Chemoorganotrophs** oxidize organic molecules with high potential energy, such as sugars. ATP may be produced by cellular respiration—with sugars serving as electron donors—or via fermentation pathways (see Chapter 9).

3. **Chemolithotrophs** ("rock-feeders") oxidize inorganic molecules with high potential energy, such as ammonia (NH_3) or methane (CH_4). ATP is produced by cellular respiration, and inorganic compounds serve as the electron donor.

Bacteria and archaea fulfill their second nutritional need—obtaining building-block compounds with carbon–carbon bonds—in two ways:

1. By synthesizing their own compounds from simple starting materials such as CO_2 and CH_4. Organisms that manufacture their own building-block compounds are termed **autotrophs** ("self-feeders").

2. By absorbing ready-to-use organic compounds from their environment. Organisms that acquire building-block compounds from other organisms are called **heterotrophs** ("other-feeders").

Because there are three distinct ways of producing ATP and two general mechanisms for obtaining carbon, there are a total of six methods for producing ATP and obtaining carbon. The names that biologists use for organisms that employ these six "feeding strategies" are given in **TABLE 29.3**.

Of the six possible ways of producing ATP and obtaining carbon, just two are observed among eukaryotes. But bacteria and archaea do them all. In addition, certain species can switch among modes of living, depending on environmental conditions. In their metabolism, eukaryotes are simple compared with bacteria and archaea. ✔ If you understand the essence of metabolic diversity in bacteria and archaea, you should be able to match the six example species described in **TABLE 29.4** to the appropriate category in Table 29.3.

What makes this remarkable diversity possible? Bacteria and archaea have evolved dozens of variations on the basic processes of respiration and photosynthesis (see Chapters 9 and 10). They use compounds with high potential energy to produce ATP via cellular respiration (electron transport chains) or fermentation, they use light to produce high-energy electrons, and they reduce carbon from CO_2 or other sources to produce sugars or other building-block molecules with carbon–carbon bonds.

The story of bacteria and archaea can be boiled down to two sentences: The basic chemistry required for photosynthesis, cellular respiration, and fermentation originated in these lineages. Then the evolution of variations on each of these processes allowed prokaryotes to diversify into millions of species that occupy diverse habitats. Let's take a closer look.

Producing ATP Through Cellular Respiration: Variation in Electron Donors and Acceptors   Millions of bacterial, archaeal, and eukaryotic species—including animals and some plants—are chemoorganotrophs. These organisms obtain the energy required to make ATP by breaking down organic compounds such as sugars, starch, or fatty acids.

SUMMARY TABLE 29.3 **Six General Methods for Obtaining Energy and Carbon–Carbon Bonds**

		Source of C–C Bonds (for synthesis of complex organic compounds)	
		Autotrophs: self-synthesized from CO_2, CH_4, or other simple molecules	Heterotrophs: from molecules produced by other organisms
Source of Energy (for synthesis of ATP)	**Phototrophs:** from sunlight	photoautotrophs	photoheterotrophs
	Chemoorganotrophs: from organic molecules	chemoorganoautotrophs	chemoorganoheterotrophs
	Chemolithotrophs: from inorganic molecules	chemolitho[auto]trophs	chemolithotrophic heterotrophs

TABLE 29.4 Six Examples of Metabolic Diversity

Example	How ATP Is Produced	How Building-Block Molecules Are Synthesized
Cyanobacteria	via oxygenic photosynthesis	from CO_2 via the Calvin cycle
Clostridium aceticum	via fermentation of glucose	from CO_2 via reactions called the acetyl-CoA pathway
Ammonia-oxidizing archaea (e.g., Nitrosopumilus sp.)	via cellular respiration, using ammonia (NH_3) as an electron donor	from CO_2 via the Calvin cycle
Helicobacteria	via anoxygenic photosynthesis	absorb carbon-containing building-block molecules from the environment
Escherichia coli	via fermentation of organic compounds or cellular respiration, using organic compounds as electron donors	absorb carbon-containing building-block molecules from the environment
Beggiatoa	via cellular respiration, using hydrogen sulfide (H_2S) as an electron donor	absorb carbon-containing building-block molecules from the environment

Cellular enzymes can strip electrons from organic molecules that have high potential energy and then transfer these high-energy electrons to the electron carriers NADH and $FADH_2$ (see Chapter 9). These compounds feed electrons to an electron transport chain (ETC), where electrons are stepped down from a high-energy state to a low-energy state (**FIGURE 29.10a**). In eukaryotic cells the ETC is located in the highly folded inner mitochondrial membrane. In bacteria and archaea, this membrane is the plasma membrane.

The energy that is released allows components of the ETC to generate a proton gradient across the plasma membrane (**FIGURE 29.10b**). The resulting flow of protons back through the enzyme ATP synthase results in the production of ATP, via the process called chemiosmosis.

The essence of this process, called **cellular respiration,** is that a molecule with high potential energy serves as an original electron donor and is oxidized, while a molecule with low potential energy serves as a final electron acceptor and becomes reduced. The potential energy difference between the electron donor and electron acceptor is eventually transformed into chemical energy in the form of ATP or is used for other processes (see Chapter 9).

(a) Model of Electron Transport Chain (ETC)

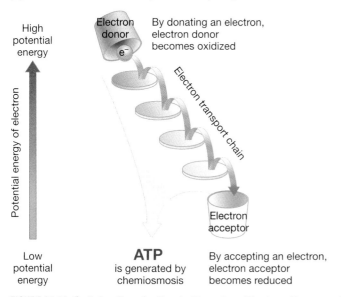

(b) ETC generates proton gradient across plasma membrane.

FIGURE 29.10 Cellular Respiration Is Based on Electron Transport Chains. Protons may diffuse away from the cell but a gradient will still form.

✔**EXERCISE** In part (a), add the chemical formula for a specific electron donor, electron acceptor, and reduced by-product for a species of bacteria or archaea. Then write in the electron donor, electron acceptor, and reduced by-product observed in humans.

Most eukaryotes carry out aerobic respiration:

- Organic compounds with high potential energy—often glucose—serve as the original electron donor. When cellular respiration is complete, glucose is completely oxidized to CO_2, which is given off as a by-product.

- Oxygen is the final electron acceptor, and water is also produced as a by-product.

Many bacteria and archaea also rely on these molecules.

It is common, however, to find bacteria and archaea that employ an electron donor other than sugars and an electron acceptor other than oxygen during cellular respiration. These species produce by-products other than carbon dioxide and water (TABLE 29.5):

- Molecules with high potential energy serve as electron donors. The substances used as electron donors range from hydrogen molecules (H_2) and hydrogen sulfide (H_2S) to ammonia (NH_3) and methane (CH_4).

- Compounds with relatively low potential energy—including sulfate (SO_4^{2-}), nitrate (NO_3^-), carbon dioxide (CO_2), or ferric ions (Fe^{3+})—act as electron acceptors.

It is only a slight exaggeration to claim that researchers have found bacterial and archaeal species that can use almost any compound with relatively high potential energy as an electron donor and almost any compound with relatively low potential energy as an electron acceptor.

Because the electron donors and electron acceptors used by bacteria and archaea are so diverse, one of the first questions biologists ask about a species is whether it undergoes cellular respiration—and if so, how. The best way to answer this question is through the enrichment culture technique introduced in Section 29.2. Recall that in an enrichment culture, researchers supply specific electron donors and electron acceptors in the medium and try to isolate cells that can use those compounds to support growth.

The remarkable metabolic diversity of bacteria and archaea explains why they play such a key role in cleaning up some types of pollution. Species that use organic solvents or petroleum-based fuels as electron donors or electron acceptors may excrete waste products that are less toxic than the original compounds.

Producing ATP via Fermentation: Variation in Substrates One strategy for making ATP that does not involve electron transport chains is called **fermentation** (see Chapter 9). In fermentation, no outside electron acceptor is used.

Because fermentation is a much less efficient way to make ATP compared with cellular respiration, in many species it occurs as an alternative metabolic strategy when no electron acceptors are available to make cellular respiration possible. In other species, fermentation does not occur at all. But in many bacteria and archaea, fermentation is the only way that cells make ATP.

Although some eukaryotic organisms can ferment glucose to ethanol or lactic acid (see Chapter 9), some bacteria and archaea are capable of using other organic compounds as the starting point for fermentation. Bacteria and archaea that produce ATP via fermentation are still classified as organotrophs, but they are much more diverse in the substrates used. For example:

- The bacterium *Clostridium aceticum* can ferment ethanol, acetate, and fatty acids as well as glucose.

- Other species of *Clostridium* ferment complex carbohydrates (including cellulose or starch), proteins, purines, or amino acids. Species that ferment amino acids produce by-products with names such as cadaverine and putrescine. These molecules are responsible for the odor of rotting flesh.

- Other bacteria can ferment lactose, a prominent component of milk. In some species this fermentation has two end products: propionic acid and CO_2. Propionic acid is responsible for the taste of Swiss cheese; the CO_2 produced during fermentation creates the holes in cheese.

- Many bacterial species in the human digestive tract ferment complex carbohydrates in our diet. The human cells then absorb the by-products and extract even more energy from them using O_2 as the final electron acceptor.

The diversity of enzymatic pathways observed in bacterial and archaeal fermentations extends the metabolic repertoire of these organisms. The diversity of substrates that are fermented also

TABLE 29.5 **Some Electron Donors and Acceptors Used by Bacteria and Archaea**

Electron Donor	By-Product from Electron Donor	Electron Acceptor	By-Product from Electron Acceptor	Category*
Sugars	CO_2	O_2	H_2O	Organotrophs
H_2 or organic compounds	H_2O or CO	SO_4^{2-}	H_2S or S^{2-}	Sulfate reducers
H_2	H_2O	CO_2	CH_4	Methanogens
CH_4	CO_2	O_2	H_2O	Methanotrophs
H_2S or S^{2-}	SO_4^{2-}	O_2	H_2O	Sulfur bacteria
Organic compounds	CO_2	Fe^{3+}	Fe^{2+}	Iron reducers
NH_3	NO_2^-	O_2	H_2O	Ammonia oxidizers
Organic compounds	CO_2	NO_3^-	N_2O, NO, or N_2	Nitrate reducers

*The name biologists use to identify species that use a particular metabolic strategy.

supports the claim that as a group, bacteria and archaea can use virtually any molecule with relatively high potential energy as a source of high-energy electrons for producing ATP.

Producing ATP via Photosynthesis: Variation in Electron Sources and Pigments Instead of using molecules as a source of high-energy electrons, phototrophs pursue a radically different strategy: **photosynthesis.** Among bacteria and archaea, photosynthesis can happen in three different ways:

1. Light activates a pigment called bacteriorhodopsin, which uses the absorbed energy to transport protons across the plasma membrane and out of the cell. The resulting flow of protons back into the cell drives the synthesis of ATP via chemiosmosis (see Chapter 9).

2. A recently discovered bacterium that lives near hydrothermal vents on the ocean floor performs photosynthesis not by absorbing light, but by absorbing geothermal radiation.

3. Pigments that absorb light raise electrons to high-energy states. As these electrons are stepped down to lower energy states by electron transport chains, the energy released is used to generate ATP.

An important feature of this last mode of photosynthesis is that the process requires a source of electrons (see Chapter 10). Recall that in cyanobacteria and plants, the required electrons come from water. When these organisms "split" water molecules apart to obtain electrons, they generate oxygen as a by-product. Species that use water as a source of electrons for photosynthesis are said to complete **oxygenic** ("oxygen-producing") photosynthesis.

In contrast, many phototrophic bacteria use a molecule other than water as the source of electrons. In some cases, the electron donor is hydrogen sulfide (H_2S); a few species can use the ion known as ferrous iron (Fe^{2+}). Instead of producing oxygen as a by-product of photosynthesis, these cells produce elemental sulfur (S) or the ferric ion (Fe^{3+}). This type of photosynthesis is said to be **anoxygenic** ("no oxygen-producing").

The photosynthetic pigments found in plants are chlorophylls *a* and *b* (see Chapter 10). Cyanobacteria have these two pigments. But researchers have isolated seven additional chlorophylls from different lineages of bacterial phototrophs. Each lineage has one or more of these distinctive chlorophylls, and each type of chlorophyll absorbs light best at a different wavelength. ✔ If you understand that different photosynthetic bacteria contain different kinds of light-absorbing pigments, you should be able to explain how several different photosynthetic species can live in the same habitat without competing for light.

Obtaining Building-Block Compounds: Variation in Pathways for Fixing Carbon In addition to acquiring energy, organisms must obtain building-block molecules that contain carbon–carbon bonds. Organisms use two mechanisms to procure usable carbon—either making their own or getting it from other organisms (see Chapters 9 and 10). Autotrophs make their own building-block compounds; heterotrophs consume them.

In many autotrophs, including cyanobacteria and plants, the enzymes of the Calvin cycle transform carbon dioxide (CO_2) into organic molecules that can be used in synthesizing cell material. The carbon atom in CO_2 is reduced during the process and is said to be "fixed." Animals and fungi, in contrast, obtain carbon from living plants or animals, or by absorbing the organic compounds released as dead tissues decay.

Bacteria and archaea pursue these same two strategies. Some interesting twists occur among bacterial and archaeal autotrophs, however. Not all of them use the Calvin cycle to make building-block molecules, and not all start with CO_2 as a source of carbon atoms. For example, consider these biochemical pathways:

- Some proteobacteria are called **methanotrophs** ("methane-eaters") because they use methane (CH_4) as their carbon source. (They also use CH_4 as an electron donor in cellular respiration.) Methanotrophs process CH_4 into more complex organic compounds via one of two enzymatic pathways, depending on the species.

- Some bacteria can use carbon monoxide (CO) or methanol (CH_3OH) as a starting material.

These observations drive home an important message from this chapter: Compared with eukaryotes, the metabolic capabilities of bacteria and archaea are remarkably complex and diverse.

Ecological Diversity and Global Impacts

The metabolic diversity observed among bacteria and archaea explains why these organisms can thrive in such a wide array of habitats.

- The array of electron donors, electron acceptors, and fermentation substrates exploited by bacteria and archaea allows the heterotrophic species to live just about anywhere.

- The evolution of three distinct types of photosynthesis—based on bacteriorhodopsin, geothermal energy, or pigments that donate high-energy electrons to ETCs—extends the types of habitats that can support phototrophs.

The complex chemistry that these cells carry out, combined with their numerical abundance, has made them potent forces for global change throughout Earth's history. Bacteria and archaea have altered the chemical composition of the oceans, atmosphere, and terrestrial environments for billions of years. They continue to do so today.

The Oxygen Revolution Today, oxygen represents almost 21 percent of the molecules in Earth's atmosphere. But researchers who study the composition of the atmosphere are virtually certain that no free molecular oxygen (O_2) existed for the first 2.3 billion years of Earth's existence. This conclusion is based on two observations:

1. There was no plausible source of oxygen at the time the planet formed.

2. Chemical analysis of the oldest Earth rocks suggests that they formed in the absence of atmospheric oxygen.

Cyanobacteria are releasing bubbles of O_2

FIGURE 29.11 Cyanobacteria Were the First Organisms to Perform Oxygenic Photosynthesis.

FIGURE 29.12 Cellular Respiration Can Produce More Energy When Oxygen Is the Final Electron Acceptor. More potential energy in glucose can be released when oxygen is the final acceptor compared to other molecules or ions.

DATA: Wilson, D. F., M. Erecińska, and P. L. Dutton. 1974. *Annual Review of Biophysics and Bioengineering 3*: 203–230; Tables 1 and 3.

✔QUESTION Which organisms grow faster—those using aerobic respiration or those using anaerobic respiration? Explain your reasoning.

Early in Earth's history, the atmosphere was dominated by nitrogen and carbon dioxide. Where did the oxygen we breathe come from? The answer is cyanobacteria.

Cyanobacteria is a lineage of photosynthetic bacteria. According to the fossil record, species of cyanobacteria first became numerous in the oceans about 2.7–2.5 billion years ago. Their appearance was momentous because cyanobacteria were the first organisms to perform oxygenic ("oxygen-producing") photosynthesis (**FIGURE 29.11**).

The fossil record and geologic record indicate that oxygen concentrations in the oceans and atmosphere began to increase 2.3–2.1 billion years ago. Once oxygen was common in the oceans, cells could begin to use it as the final electron acceptor during cellular respiration. **Aerobic** respiration was now a possibility. Before that, organisms had to use compounds other than oxygen as a final electron acceptor—only **anaerobic** respiration was possible. (Aerobic and anaerobic respiration are introduced in Chapter 9.)

The evolution of aerobic respiration was a crucial event in the history of life. Because oxygen is extremely electronegative, it is an efficient electron acceptor. Much more energy is released as electrons move through electron transport chains with oxygen as the ultimate acceptor than is released with other substances as the electron acceptor.

To drive this point home, study the graph in **FIGURE 29.12**. Notice that the vertical axis plots free energy changes; the graph shows the energy released when glucose is oxidized with iron, nitrate, or oxygen as the final electron acceptor. Once oxygen was available, then, cells could produce much more ATP for each electron donated by NADH or FADH$_2$. As a result, the rate of energy production rose dramatically.

To summarize, data indicate that cyanobacteria were responsible for a fundamental change in Earth's atmosphere—a high concentration of oxygen. Never before, or since, have organisms done so much to alter the nature of our planet.

Nitrogen Fixation and the Nitrogen Cycle In many environments, fertilizing forests or grasslands with nitrogen results in increased growth. Researchers infer from these results that plant growth is often limited by the availability of nitrogen.

Organisms must have nitrogen to synthesize proteins and nucleic acids. Although molecular nitrogen (N_2) is extremely abundant in the atmosphere, most organisms cannot use it because of the strong triple bond linking the nitrogen atoms. To incorporate nitrogen atoms into amino acids and nucleotides, all eukaryotes and many bacteria and archaea have to obtain N in a form such as ammonia (NH_3) or nitrate (NO_3^-).

Certain bacteria and archaea are the only species that are capable of converting molecular nitrogen to ammonia. The steps in the process, called **nitrogen fixation,** are highly endergonic reduction-oxidation (redox) reactions (see Chapter 9). The key enzyme that catalyzes the reaction—nitrogenase—is found only in selected bacterial and archaeal lineages. Many of these organisms are free living, but some form important relationships with plants:

• Some species of cyanobacteria live in association with a water fern that grows in rice paddies and helps fertilize the plants.

• In terrestrial environments, nitrogen-fixing bacteria live in close association with plants—often taking up residence in special root structures called nodules (see Chapter 39).

Why is nitrogenase not found in all organisms? The answer lies in an interesting property of the enzyme. When exposed to O_2, nitrogenase is irreversibly poisoned and is degraded. The only organisms with the nitrogenase gene are those that live in anaerobic habitats or are able to protect the enzyme from O_2.

Nitrogen fixation is only the beginning of the story, however. A quick glance back at Table 29.5 should convince you that bacteria and archaea use a wide array of nitrogen-containing

FIGURE 29.13 Bacteria and Archaea Drive the Movement of Nitrogen Atoms through Ecosystems. Nitrogen atoms cycle in different molecular forms.

✔**EXERCISE** Add arrows and labels to indicate that animals ingest amino groups from plants or other animals and release amino groups or ammonia.

compounds as electron donors and electron acceptors during cellular respiration.

To understand why this is important, consider that the nitrite (NO_2^-) produced by some bacteria as a by-product of respiration does not build up in the environment. Instead, other species of bacteria and archaea use it as an electron donor, and it is oxidized to molecular nitrate (NO_3^-). Nitrate, in turn, is reduced to molecular nitrogen (N_2) by yet another suite of bacterial and archaeal species. In this way, bacteria and archaea are responsible for driving the movement of nitrogen atoms through ecosystems around the globe in a process called the **nitrogen cycle** (**FIGURE 29.13**).

Similar types of interactions occur with molecules that contain phosphorus, sulfur, and carbon. In this way, bacteria and archaea play a key role in the cycling of nitrogen and other nutrients. ✔ If you understand the role of bacteria and archaea in the nitrogen cycle, you should be able to provide a plausible explanation of what the composition of the atmosphere and what the nitrogen cycle might be like if bacteria and archaea did not exist.

Nitrate Pollution Most crop plants—including corn, rice, and wheat—do not live in association with nitrogen-fixing bacteria. To increase yields of these crops, farmers use fertilizers that are high in nitrogen. In parts of the world, massive additions of nitrogen in the form of ammonia are causing serious pollution problems.

FIGURE 29.14 shows why. When ammonia is added to a cornfield—in midwestern North America, for example—much of

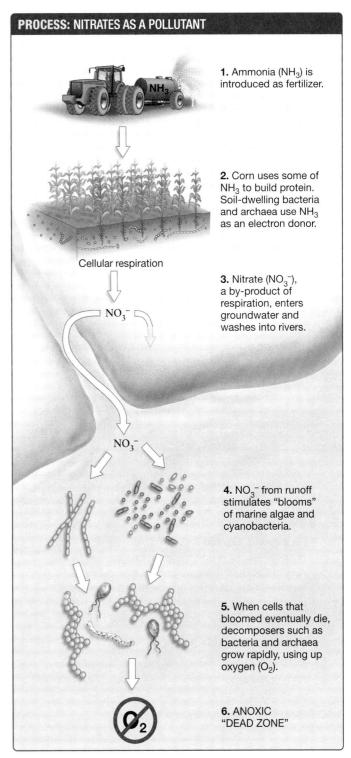

PROCESS: NITRATES AS A POLLUTANT

1. Ammonia (NH_3) is introduced as fertilizer.

2. Corn uses some of NH_3 to build protein. Soil-dwelling bacteria and archaea use NH_3 as an electron donor.

Cellular respiration

NO_3^-

3. Nitrate (NO_3^-), a by-product of respiration, enters groundwater and washes into rivers.

NO_3^-

4. NO_3^- from runoff stimulates "blooms" of marine algae and cyanobacteria.

5. When cells that bloomed eventually die, decomposers such as bacteria and archaea grow rapidly, using up oxygen (O_2).

6. ANOXIC "DEAD ZONE"

FIGURE 29.14 Nitrates Act as a Pollutant in Aquatic Ecosystems.

it never reaches the growing corn plants. Instead, bacteria and archaea in the soil use a significant fraction of the ammonia as food. Microbes that use ammonia as an electron donor to fuel cellular respiration release nitrite (NO_2^-) as a waste product. Other microbes use nitrite as an electron donor and release nitrate (NO_3^-). Nitrate molecules are extremely soluble in water and tend to be washed out of soils into groundwater or streams.

From there they eventually reach the ocean, where they can cause pollution.

To understand why nitrates can pollute the oceans, consider the Gulf of Mexico:

1. Nitrates carried by the Mississippi River are used as a nutrient by cyanobacteria and algae that live in the Gulf.

2. These cells explode in numbers in response.

3. When they die and sink to the bottom of the Gulf, bacteria and archaea and other decomposers use them as food.

4. The decomposers use so much oxygen as an electron acceptor in cellular respiration that oxygen levels in the sediments and even in Gulf waters decline.

Nitrate pollution has been so severe that large areas in the Gulf of Mexico are anoxic (lacking in oxygen). The oxygen-free "dead zone" in the Gulf of Mexico is devoid of fish, shrimp, and other organisms that require oxygen.

Lately, the dead zone has encompassed about 22,000 square kilometers (km^2)—roughly the size of New Jersey. Similar problem spots are cropping up in other parts of the world. Virtually every link in the chain of events leading to nitrate pollution involves bacteria and archaea.

The general message of this section is simple: Bacteria and Archaea may be small in size, but because of their abundance, ubiquity, and ability to do sophisticated chemistry, they have an enormous influence on the global environment.

check your understanding

If you understand that . . .
- As a group, Bacteria and Archaea can use a wide array of electron donors and acceptors in cellular respiration and a diverse set of compounds in fermentation, perform anoxygenic as well as oxygenic photosynthesis, and fix carbon from several different sources via a variety of pathways.

✔ **You should be able to . . .**

Defend the claim that the metabolism of bacteria and archaea is much more sophisticated than that of eukaryotes.

Answers are available in Appendix A.

29.4 Key Lineages of Bacteria and Archaea

In the decades since the phylogenetic tree identifying the three domains of life was first published, dozens of studies have confirmed the result. It is now well established that all organisms alive today belong to one of the three domains, and that archaea and eukaryotes are more closely related to each other than either group is to bacteria.

Although the relationships among the major lineages within Bacteria and Archaea are still uncertain in some cases, many of the lineages themselves are well studied. Let's survey the attributes of species from selected major lineages within the Bacteria and Archaea, with an emphasis on themes explored earlier in the chapter: their morphological and metabolic diversity, their impacts on humans, and their importance to other species and to the environment.

Bacteria

The name *bacteria* comes from the Greek root *bacter*, meaning "rod" or "staff." The name was inspired by the first bacteria to be seen under a microscope, which were rod shaped. But as the following descriptions indicate, bacterial cells come in a wide variety of shapes.

Biologists who study bacterial diversity currently recognize at least 21 lineages, or phyla, within the domain. Some of these lineages were recognized by distinctive morphological characteristics and others by phylogenetic analyses of gene sequence data. The lineages reviewed here are just a sampling of bacterial diversity.

- Bacteria > Firmicutes
- Bacteria > Cyanobacteria
- Bacteria > Actinobacteria
- Bacteria > Spirochaetes (Spirochetes)
- Bacteria > Chlamydiae
- Bacteria > Proteobacteria

Archaea

The name *archaea* comes from the Greek root *archae*, for "ancient." The name was inspired by the hypothesis that this is a particularly ancient group, which turned out to be incorrect. Also incorrect was the initial hypothesis that archaeans are restricted to hot springs, salt ponds, and other extreme habitats. Archaea live in virtually every habitat known.

Recent phylogenies based on DNA sequence data indicate that the domain is composed of at least four major phyla, called the Thaumarchaeota, Crenarchaeota, Euryarchaeota, and Korarchaeota. The Korarchaeota are known only from direct sequencing studies. They have never been grown in culture, and almost nothing is known about them.

- Archaea > Thaumarchaeota
- Archaea > Crenarchaeota
- Archaea > Euryarchaeota

The Firmicutes have also been called "low-GC Gram positives" because their cell walls react positively to the Gram stain and because their DNA contains a relatively low percentage of guanine and cytosine (G and C). In some species, G and C represent less than 25 percent of the bases present. There are over 1100 species. ✔ You should be able to mark the origin of the Gram-positive cell wall on Figure 29.7.

Morphological diversity Most are rod shaped or spherical. Some of the spherical species form chains or tetrads (groups of four cells). A few form a durable resting stage called a spore. One subgroup synthesizes a cell wall made of cellulose.

Metabolic diversity Some species can fix nitrogen; some perform anoxygenic photosynthesis. Others make all of their ATP via various fermentation pathways; still others perform cellular respiration, using hydrogen gas (H_2) as an electron donor.

Human and ecological impacts Recent metagenomic studies have shown that members of this lineage are extremely common in the human gut. Species in this group also cause a variety of diseases, including anthrax, botulism, tetanus, walking pneumonia, boils, gangrene, and strep throat. *Bacillus thuringiensis* produces a toxin that is one of the most important insecticides currently used in farming. Species in the lactic acid bacteria group are used to ferment milk products into yogurt or cheese (**FIGURE 29.15**).

Lactobacillus bulgaricus (rods) and *Streptococcus thermophilus*

FIGURE 29.15 Firmicutes in Yogurt. (The cells in this scanning electron micrograph have been colorized.)

The cyanobacteria were formerly known as the "blue-green algae"—even though algae are eukaryotes. Only about 80 species of cyanobacteria have been described to date, but they are among the most abundant organisms on Earth. In terms of total mass, cyanobacteria dominate the surface waters in many marine and freshwater environments.

Morphological diversity Cyanobacteria may be found as independent cells, in chains that form filaments (**FIGURE 29.16**), or in the loose aggregations of individual cells called colonies. The shape of *Nostoc* colonies varies from flat sheets to ball-like clusters of cells.

Metabolic diversity All perform oxygenic photosynthesis; many can also fix nitrogen. Because cyanobacteria can synthesize virtually every molecule they need, they can be grown in culture media that contain only CO_2, N_2, H_2O, and a few mineral nutrients. Some species associate with fungi, forming lichens, while others form associations with protists, sponges, or plants. In each case the cyanobacterium provides some form of nutritional benefit to the host. ✔ You should be able to mark the origin of oxygenic photosynthesis on Figure 29.7.

Human and ecological impacts If cyanobacteria are present in high numbers, their waste products can make drinking water smell bad. Some species release molecules called microcystins that are toxic to plants and animals. Cyanobacteria were responsible for the origin of the oxygen atmosphere on Earth. Today they still produce much of the oxygen and nitrogen and many of the organic compounds that feed other organisms in freshwater and marine environments.

Nostoc species

FIGURE 29.16 Cyanobacteria Contain Chlorophyll and Are Green.

Actinobacteria are sometimes called the "high-GC Gram positives" because (1) their cell-wall material appears purple when treated with the Gram stain—meaning that they have a peptidoglycan-rich cell wall and lack an outer membrane—and (2) their DNA contains a relatively high percentage of guanine and cytosine. In some species, G and C represent over 75 percent of the bases present. Over 1100 species have been described to date (**FIGURE 29.17**). ✔ You should be able to mark the origin of this high-GC genome on Figure 29.7.

Morphological diversity Cell shape varies from rods to filaments. Many of the soil-dwelling species are found as chains of cells that form extensive branching filaments called **mycelia.** Because of their morphology they were initially misclassified as fungi, and the incorrect name Actinomyces persists.

Metabolic diversity Many are heterotrophs that use an array of organic compounds as electron donors and oxygen as an electron acceptor. There are several parasitic species that get most of their nutrition from host organisms.

Human and ecological impacts Two serious human diseases, tuberculosis and leprosy, are caused by parasitic *Mycobacterium* species. Over 500 distinct antibiotics have been isolated from species in the genus *Streptomyces*; 60 of these—including streptomycin, neomycin, tetracycline, and erythromycin—are now actively prescribed to treat diseases in humans or domestic livestock. One

actinobacterium species is critical to the manufacture of Swiss cheese. Species in the genus *Streptomyces* and *Arthrobacter* are abundant in soil and are vital as decomposers of dead plant and animal material. Some species live in association with plant roots and fix nitrogen; others can break down toxins such as herbicides, nicotine, and caffeine.

Streptomyces griseus

5 µm

FIGURE 29.17 A *Streptomyces* Species That Produces the Antibiotic Streptomycin.

Based on numbers of species, the spirochetes are one of the smaller bacterial phyla: Only 13 genera and a total of 62 species have been described to date.

Morphological diversity Spirochetes are distinguished by their unique corkscrew shape and flagella (**FIGURE 29.18**). Instead of extending into the water surrounding the cell, spirochete flagella are contained within a structure called the outer sheath, which surrounds the cell. When these flagella beat, the cell lashes back and forth and swims forward. ✔ You should be able to mark the origin of the spirochete flagellum on Figure 29.7.

Metabolic diversity Most spirochetes manufacture ATP via fermentation. The substrate used in fermentation varies among species and may consist of sugars, amino acids, starch, or the pectin found in plant cell walls. A spirochete that lives only in the hindgut of termites can fix nitrogen.

Human and ecological impacts The sexually transmitted disease syphilis is caused by a spirochete. Syphilis is thought to have been

brought by European explorers to the Western hemisphere, where it was responsible for killing tens of millions of native people. Lyme disease, also caused by a spirochete, is transmitted to humans by deer ticks. Spirochetes are extremely common in freshwater and marine habitats; many live only under anaerobic conditions.

Treponema pallidum

1 µm

FIGURE 29.18 Spirochetes Are Corkscrew-Shaped Cells Inside an Outer Sheath.

Bacteria > Chlamydiae

In terms of numbers of species living today, Chlamydiae may be the smallest of all major bacterial lineages. Although the group is highly distinct phylogenetically, only 13 species are known. All are Gram-negative.

Morphological diversity Chlamydiae are spherical. They are tiny, even by bacterial standards.

Metabolic diversity All known species in this phylum live as parasites *inside* host cells and are termed **endosymbionts** ("inside-together-living"). Chlamydiae acquire almost all of their nutrition from their hosts. In **FIGURE 29.19**, the chlamydiae have been colored red; the animal cells that they live in are colored brown. ✔ You should be able to mark the origin of the endosymbiotic lifestyle in this lineage on Figure 29.7. (The endosymbiotic lifestyle has also arisen in other bacterial lineages, independently of Chlamydiae.)

Human and ecological impacts *Chlamydia trachomatis* infections are the most common cause of blindness in humans. When the same organism is transmitted from person to person via sexual intercourse, it can cause serious urogenital tract infections. If untreated in women, this disease can lead to ectopic pregnancy, premature births, and infertility. One species causes epidemics of a pneumonia-like disease in birds.

Chlamydia trachomatis

FIGURE 29.19 Chlamydiae Live Only Inside Animal Cells.

Bacteria > Proteobacteria

The approximately 1200 species of proteobacteria form five major subgroups, designated by the Greek letters α (alpha), β (beta), γ (gamma), δ (delta), and ε (epsilon). Because they are so diverse in their morphology and metabolism, the lineage is named after the Greek god Proteus, who could assume many shapes.

Morphological diversity Proteobacterial cells can be rods, spheres, or spirals. Some form stalks (**FIGURE 29.20a**). Some are motile. In one group, cells may move together to form colonies, which then transform into the specialized cell aggregate shown in **FIGURE 29.20b**. This structure is known as a **fruiting body.** At their tips, the fruiting bodies produce cells that are surrounded by a durable coating. These spores sit until conditions improve, and then they resume growth.

Metabolic diversity Proteobacteria make a living in virtually every way known to bacteria—except that none perform oxygenic photosynthesis. Various species may perform cellular respiration by using organic compounds, nitrite, methane, hydrogen gas, sulfur, or ammonia as electron donors and oxygen, sulfate, or sulfur as electron acceptors. Some perform anoxygenic photosynthesis.

Human and ecological impacts *Escherichia coli* may be the best studied of all organisms and is a key species in biotechnology (see Chapter 20, and **BioSkills 13** in Appendix B). Pathogenic proteobacteria cause Legionnaire's disease, cholera, food poisoning, plague, dysentery, gonorrhea, Rocky Mountain spotted fever, typhus, ulcers, and diarrhea. *Wolbachia* infections are common in insects and are often transmitted from mothers to offspring

(a) Stalked bacterium

Caulobacter crescentus

(b) Fruiting bodies

Chondromyces crocatus

Spores

Stalk

FIGURE 29.20 Some Proteobacteria Grow on Stalks or Form Fruiting Bodies. The stalked bacterium (a) has been colorized.

via eggs. Biologists use *Agrobacterium* cells to transfer new genes into crop plants. Certain acid-loving species of proteobacteria are used in the production of vinegars. Species in the genus *Rhizobium* (α-proteobacteria) live in association with plant roots and fix nitrogen. A group in the δ-proteobacteria, the bdellovibrios, are predators—they drill into other proteobacterial cells and digest them. Because some species use nitrogen-containing compounds as electron acceptors, proteobacteria are critical players in the cycling of nitrogen atoms through terrestrial and aquatic ecosystems.

The Thaumarchaeota were recently recognized as a monophyletic, ancient lineage of archaea. Members of this phylum are extremely abundant in oceans, estuaries, and terrestrial soils. Unlike the extremophiles, species in this lineage are considered mesophilic because they grow best at moderate temperatures.

Morphological diversity Only a few members of this group have been observed, and all consist of rod-shaped cells that are less than 1 micrometer (μm) in length, smaller than typical prokaryotes. One species, *Nitrosopumilus maritimus* (**FIGURE 29.21**), is so abundant it is estimated to constitute possibly 25 percent of the total prokaryotic cell biomass in open oceans.

Metabolic diversity Members of this phylum are called ammonia oxidizers because they use ammonia as a source of electrons and generate nitrite as a by-product. They use the energy from ammonia oxidation to fix CO_2.

Human and ecological impacts Because of their abundance, these organisms are thought to play a major role in Earth's nitrogen and carbon cycles (see Chapter 56). Their presence in deep ocean waters that lack reduced carbon and sunlight may help explain the productivity of these habitats. The species *Cenarchaeum symbiosum* lives as an endosymbiont inside a marine sponge.

Nitrosopumilus maritimus

0.1 μm

FIGURE 29.21 This Species of Thaumarchaeota Has Rod-Shaped Cells and Is Extremely Abundant in the Open Ocean.

✔**QUANTITATIVE** Use the scale bar in the figure to measure the length of a *N. maritimus* cell. If these cells were placed end to end, how many of them would fit along a meter stick?

The Crenarchaeota got their name because they are considered similar to the oldest archaeans; the word root *cren–* refers to a source or fount. Biologists have named only 37 species so far, but they are virtually certain that thousands are yet to be discovered.

Morphological diversity Crenarchaeota cells can be shaped like filaments, rods, discs, or spheres. One species that lives in extremely hot habitats has a tough cell wall consisting solely of glycoprotein.

Metabolic diversity Depending on the species, cellular respiration can involve organic compounds, sulfur, hydrogen gas, or Fe^{2+} ions as electron donors and oxygen, nitrate, sulfate, sulfur, carbon dioxide, or Fe^{3+} ions as electron acceptors. Some species make ATP exclusively through fermentation pathways.

Human and ecological impacts Crenarchaeota have yet to be used in the manufacture of commercial products. In certain extremely hot, high-pressure, cold, or acidic environments, crenarchaeota may be the only life-form present (**FIGURE 29.22**). Acid-loving species thrive in habitats with pH 1–5; some species are found in ocean sediments at depths ranging from 2500 to 4000 m below the surface.

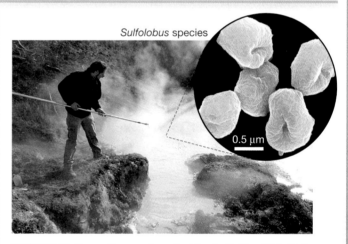

Sulfolobus species

0.5 μm

FIGURE 29.22 Some Crenarchaeota Live in Sulfur-Rich Hot Springs. The cells in the micrograph have been colorized to make them more visible.

The Euryarchaeota are aptly named, because the word root *eury*– means "broad." Members of this phylum live in every conceivable habitat. Some species are adapted to high-salt habitats with pH 11.5—almost as basic as household ammonia. Other species are adapted to acidic conditions with a pH as low as 0. Species in the genus *Methanopyrus* live near hot springs called black smokers that are 2000 m (over 1 mile) below sea level (**FIGURE 29.23**).

Morphological diversity Euryarchaeota cells can be spherical, filamentous, rod shaped, disc shaped, or spiral. Rod-shaped cells may be short or long or arranged in chains. Spherical cells can be found in ball-like aggregations. Some species have several flagella. Some species lack a cell wall; others have a cell wall composed entirely of glycoproteins.

Metabolic diversity The group includes a variety of methane-producing species. These **methanogens** can use up to 11 different organic compounds as electron acceptors during cellular respiration; all produce CH_4 as a by-product of respiration. In other species of Euryarchaeota, cellular respiration is based on hydrogen gas or Fe^{2+} ions as electron donors and nitrate or sulfate as electron acceptors. Species that live in high-salt environments can use the molecule retinal—which is responsible for light reception in your eyes—to capture light energy and perform photosynthesis.

Human and ecological impacts Species in the genus *Ferroplasma* live in piles of waste rock near abandoned mines. As a by-product

of metabolism, they produce acids that drain into streams and pollute them. Methanogens live in the soils of swamps and the guts of mammals (including yours). They are responsible for adding about 2 billion tons of methane to the atmosphere each year. A methanogen in this phylum was also recently implicated in gum disease.

Methanopyrus kandleri

1 μm

FIGURE 29.23 Some Euryarchaeota Cells Live in the Chimneys of "Black Smokers" on the Seafloor.

CHAPTER 29   REVIEW   *For media, go to MasteringBiology* (MB)

If you understand . . .

29.1 Why Do Biologists Study Bacteria and Archaea?

- Bacteria and archaea are the most abundant organisms on Earth and are found in every habitat that has been sampled.

- Bacteria and archaea are very small, prokaryotic cells, and most are unicellular.

- Bacteria and archaea can be distinguished by their different kinds of membrane lipids and cell walls as well as by their different transcription machinery.

- Bacteria play many beneficial roles in animal digestion, bioremediation, and the production of antibiotics.

- Bacteria cause some of the most dangerous human diseases, including plague, syphilis, botulism, cholera, and tuberculosis.

✔ You should be able to explain the difference between a bacterium, an archaeon, and a eukaryote.

29.2 How Do Biologists Study Bacteria and Archaea?

- Enrichment cultures are used to grow large numbers of bacterial or archaeal cells that thrive under specified conditions.

- Using metagenomic analysis, biologists can study bacteria and archaea that cannot be cultured by extracting DNA directly from an environment and then sequencing and characterizing DNA fragments. Information obtained is used to identify biochemical processes and novel organisms that are then placed on the tree of life.

✔ You should be able to explain how metagenomic analysis might be used to reveal whether bacteria carry out nitrogen fixation in the gut of an insect.

29.3 What Themes Occur in the Diversification of Bacteria and Archaea?

- Metabolic diversity and complexity are the hallmarks of the bacteria and archaea, just as morphological diversity and complexity are the hallmarks of the eukaryotes.

- Among bacteria and archaea, a wide array of inorganic or organic compounds with high potential energy may serve as electron donors in cellular respiration, and a wide variety of inorganic or organic molecules with low potential energy may serve as electron acceptors. Dozens of distinct organic compounds are fermented.

- Photosynthesis is widespread in bacteria. In cyanobacteria, water is used as a source of electrons and oxygen gas is generated as a by-product. But in other species, the electron excited by photon capture comes from a source other than water, and no oxygen is produced.

- To acquire building-block molecules containing carbon–carbon bonds, some bacteria and archaea species use the enzymes of the Calvin cycle to reduce CO_2. But several other biochemical pathways found in bacteria and archaea can also reduce simple organic compounds to sugars or carbohydrates.

- Because of their metabolic diversity, bacteria and archaea play a large role in carbon and nitrogen cycling and alter the global atmosphere, oceans, and terrestrial environments.

- Nitrogen-fixing species provide nitrogen in forms that can be used by many other species, including plants and animals.

✔ You should be able to explain why species that release H_2S as a by-product and that use H_2S as an electron donor often live side by side.

29.4 Key Lineages of Bacteria and Archaea

- Prokaryotes can be divided into two lineages, the Bacteria and the Archaea, based on a wide variety of morphological, biochemical, and molecular characters.

- Bacteria are divided into 21 major lineages including organisms that play major roles in ecosystems as primary producers, decomposers, and parasites.

- Archaea are divided into four major lineages and were thought to exist only in extreme environments; they are now recognized to be widespread.

(MB) MasteringBiology

1. **MasteringBiology Assignments**

   Tutorials and Activities Classification of Prokaryotes, Diversity in Bacteria, Tree of Life, Water Pollution from Nitrates

   Questions Reading Quizzes, Blue-Thread Questions, Test Bank

2. **eText** Read your book online, search, take notes, highlight text, and more.

3. **The Study Area** Practice Test, Cumulative Test, BioFlix® 3-D Animations, Videos, Activities, Audio Glossary, Word Study Tools, Art

You should be able to . . .

1. How do the molecules that function as electron donors and those that function as electron acceptors differ?
   a. Electron donors are almost always organic molecules; electron acceptors are always inorganic.
   b. Electron donors are almost always inorganic molecules; electron acceptors are always organic.
   c. Electron donors have relatively high potential energy; electron acceptors have relatively low potential energy.
   d. Electron donors have relatively low potential energy; electron acceptors have relatively high potential energy.

2. What do some photosynthetic bacteria use as a source of electrons instead of water?
   a. oxygen (O_2)
   b. hydrogen sulfide (H_2S)
   c. organic compounds (e.g., CH_3COO^-)
   d. nitrate (NO_3^-)

3. What is distinctive about the chlorophylls found in different photosynthetic bacteria?
   a. their membranes
   b. their role in acquiring energy
   c. their role in carbon fixation
   d. their absorption spectra

4. What are organisms called that use inorganic compounds as electron donors in cellular respiration?
   a. phototrophs
   b. heterotrophs
   c. organotrophs
   d. lithotrophs

5. True or False. Certain aerobic bacteria in the presence of oxygen can convert nitrogen gas to ammonia.

6. Unlike plant cell walls that contain cellulose, bacterial cell walls are composed of _____.

7. What has metagenomic analysis allowed researchers to do for the first time?
   a. sample organisms from an environment and grow them under defined conditions in the lab
   b. isolate organisms from an environment and sequence their entire genome
   c. study organisms that cannot be cultured (grown in the lab)
   d. identify important morphological differences among species

8. Biologists often use the term energy source as a synonym for "electron donor." Why?

9. The text claims that the tremendous ecological diversity of bacteria and archaea is possible because of their impressive metabolic diversity. Do you agree with this statement? Why or why not?

10. Would you predict that disease-causing bacteria, such as those listed in Table 29.2, obtain energy from light, organic molecules, or inorganic molecules? Explain your answer.

11. The text claims that the evolution of an oxygen atmosphere paved the way for increasingly efficient cellular respiration and higher growth rates in organisms. Explain.

12. From what we know about the evolutionary relationship between the three largest domains, as depicted in Figure 29.1, explain the statement, "Prokaryotes are a paraphyletic group."

13. When using Koch's postulates, which of the following is an essential requirement for the suspected pathogen?
    a. It is present in all organisms with the disease.
    b. It can be cultured on an agar plate.
    c. It is pathogenic on a wide variety of organisms.
    d. It can reproduce sexually within the host.

14. The researchers who observed that magnetite was produced by bacterial cultures from the deep subsurface carried out a follow-up experiment. These biologists treated some of the cultures with a drug that poisons the enzymes involved in electron transport chains. In cultures where the drug was present, no more magnetite was produced. Does this result support or undermine their hypothesis that the bacteria in the cultures perform cellular respiration? Explain your reasoning.

15. *Streptococcus mutans* obtains energy by oxidizing sucrose. This bacterium is abundant in the mouths of Western European and North American children and is a prominent cause of cavities. The organism is virtually absent in children from East Africa, where tooth decay is rare. Propose a hypothesis to explain this observation. Outline the design of a study that would test your hypothesis.

16. Suppose that you've been hired by a firm interested in using bacteria to clean up organic solvents found in toxic waste dumps. Your new employer is particularly interested in finding cells that are capable of breaking a molecule called benzene into less toxic compounds. Where would you go to look for bacteria that can metabolize benzene as an energy or carbon source? How would you design an enrichment culture capable of isolating benzene-metabolizing species?

30 Protists

In this chapter you will learn that

Protists are a large and diverse group that tell us about the evolution of life on Earth

by asking

How and why do biologists study these organisms?
30.1, 30.2

by examining

Major themes in their diversification
30.3

Innovations in morphology: Endosymbiosis theory and evolution of multicellularity

Innovations in feeding, movement, and reproduction

then, looking closer at

Key lineages of protists
30.4

The brown alga (giant kelp) shown here lives attached to rocks in ocean waters. This algae and other species featured in this chapter are particularly abundant in the world's oceans.

T his chapter introduces the third domain on the tree of life: the **Eukarya.** Eukaryotes range from single-celled organisms that are the size of bacteria to sequoia trees and blue whales. The largest and most morphologically complex organisms on the tree of life—algae, plants, fungi, and animals—are eukaryotes.

Although eukaryotes are astonishingly diverse, they share fundamental features that distinguish them from bacteria and archaea:

- Compared to bacteria and archaea, most eukaryotic cells are large, have many more organelles, and have a much more extensive system of structural proteins called the cytoskeleton.

✔ When you see this checkmark, stop and test yourself. Answers are available in Appendix A.

- The nuclear envelope is a **synapomorphy** that defines the Eukarya. Recall that a synapomorphy is a shared, derived trait that distinguishes major monophyletic groups (see Chapter 28).

- Multicellularity is rare in bacteria and unknown in archaea, but has evolved multiple times in eukaryotes.

- Bacteria and archaea reproduce asexually by fission; many eukaryotes reproduce asexually via mitosis and cell division.

- Many eukaryotes undergo meiosis and reproduce sexually.

One of this chapter's fundamental goals is to explore how these morphological innovations—features like organelles and the nuclear envelope—evolved. Another goal is to analyze how morphological innovations allowed eukaryotes to pursue novel ways of performing basic life tasks such as feeding, moving, and reproducing.

In introducing the Eukarya, this chapter focuses on a diverse collection of lineages known as the protists. The term **protist** refers to all eukaryotes that are not land plants, fungi, or animals. Protist lineages are colored orange in **FIGURE 30.1**.

Protists do not make up a monophyletic group. Instead, they refer to a **paraphyletic group**—they represent some, but not all, of the descendants of a single common ancestor. No synapomorphies define the protists. Protists have no trait that is found only in protists and in no other organisms.

By definition, then, the protists are a diverse lot. The common feature among protists is that they tend to live in environments where they are surrounded by water most of the time (**FIGURE 30.2**). Most plants, fungi, and animals are terrestrial, but protists are found in wet soils and aquatic habitats or inside the bodies of other organisms—including, perhaps, you.

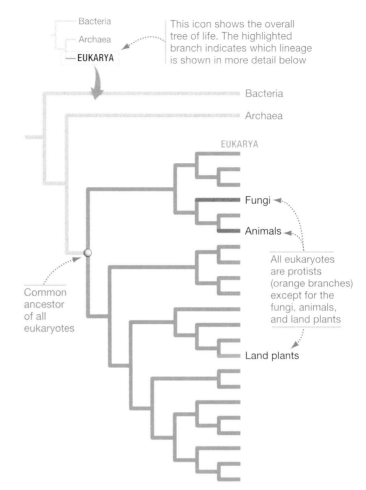

FIGURE 30.1 Protists Are Paraphyletic. The protists include some, but not all, descendants of a single common ancestor.

Open ocean: Surface waters teem with microscopic protists, such as these diatoms.

Shallow coastal waters: Gigantic protists, such as these kelp, form underwater forests.

Intertidal habitats: Protists such as these sea palms are particularly abundant in tidal habitats.

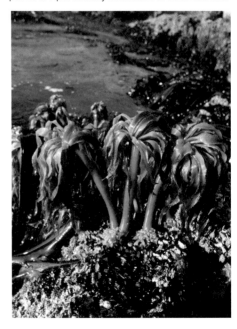

FIGURE 30.2 Protists Are Particularly Abundant in Aquatic Environments.

30.1 Why Do Biologists Study Protists?

Biologists study protists for three reasons, in addition to their intrinsic interest: **(1)** they are important medically, **(2)** they are important ecologically, and **(3)** they are critical to understanding the evolution of plants, fungi, and animals. The rest of the chapter focuses on why protists are interesting in their own right and how they evolved. Let's first consider their impact on human health and the environment.

Impacts on Human Health and Welfare

The most spectacular crop failure in history, the Irish potato famine, was caused by a protist. In 1845 most of the 3 million acres that had been planted to grow potatoes in Ireland became infested with *Phytophthora infestans*—a parasite that belongs to a group of protists called water molds. Potato tubers that were infected with *P. infestans* rotted in the fields or in storage.

As a result of crop failures in Ireland for two consecutive years, an estimated 1 million people out of a population of less than 9 million died of starvation or starvation-related illnesses. Several million others emigrated. Many people of Irish heritage living in North America, New Zealand, and Australia trace their ancestry to relatives who left Ireland to evade the famine. As devastating as the potato famine was, however, it does not begin to approach the misery caused by the protist *Plasmodium*.

Malaria Physicians and public health officials point to three major infectious diseases that are currently afflicting large numbers of people worldwide: tuberculosis, HIV, and malaria. Tuberculosis is caused by a bacterium (see Chapter 25); HIV is caused by a virus (see Chapter 36). Malaria is caused by a protist—specifically, by apicomplexan species in the eukaryotic lineage Alveolata.

Malaria ranks as one of the world's worst infectious diseases. In India alone, over 30 million people each year suffer from debilitating fevers caused by malaria. At least 300 million people worldwide are sickened by it each year, and nearly 1 million die from the disease annually. The toll is equivalent to eight 747s, loaded with passengers, crashing every day. Most of the dead are children of preschool age.

Five species of the protist *Plasmodium* are capable of parasitizing humans. Infections start when *Plasmodium* cells enter a person's bloodstream during a mosquito bite. As **FIGURE 30.3** shows, *Plasmodium* initially infects liver cells; later, some of the *Plasmodium* cells change into a distinctive cell type that infects the host's red blood cells. The *Plasmodium* cells multiply asexually inside the host cells, which are killed as parasite cells exit to infect additional liver cells or red blood cells. Large numbers of *Plasmodium* cells in a human host's blood increase the chance that a mosquito's blood meal will contain some of them.

After blood from an infected human is transferred to a mosquito during a bite, *Plasmodium* cells differentiate to form gametes. Inside the mosquito, gametes fuse to form a diploid zygote, which undergoes meiosis. Sexual reproduction in the mosquito is beneficial to the parasite because it generates genetic diversity to help evade the host's immune system. The haploid cells that result from meiosis can infect a human when the mosquito bites again.

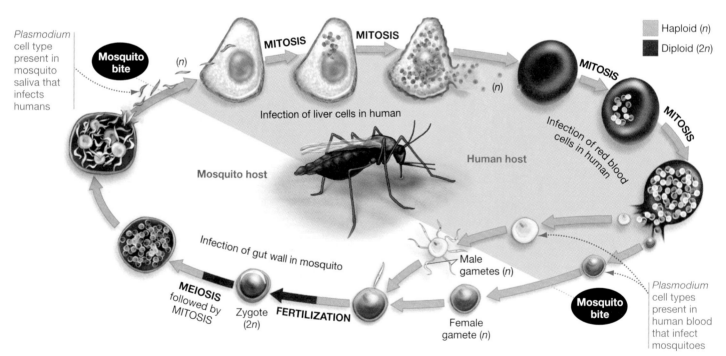

FIGURE 30.3 *Plasmodium* Lives in Mosquitoes and in Humans, Where It Causes Malaria. Over the course of its life cycle, *Plasmodium falciparum* alternates between a mosquito host where sexual reproduction takes place, and a human host where asexual reproduction takes place. In humans, it infects and kills liver cells and red blood cells, contributing to anemia and high fever. In mosquitoes, the protist lives in the gut and salivary glands.

TABLE 30.1 Some Human Health Problems Caused by Protists

Species	Disease
Five species of *Plasmodium,* primarily *P. falciparum* and *P. vivax*	Malaria has the potential to affect 40 percent of the world's total population.
Toxoplasma gondii	Toxoplasmosis may cause eye and brain damage in infants and in AIDS patients.
Many species of dinoflagellates	Toxins released during harmful algal blooms accumulate in clams and mussels and poison people if eaten.
Many species of *Giardia*	Diarrhea due to giardiasis (beaver fever) can last for several weeks.
Trichomonas vaginalis	Trichomoniasis is a reproductive tract infection and one of the most common sexually transmitted diseases. About 2 million young women are infected in the United States each year; some of them become infertile.
Several species of *Leishmania*	Leishmaniasis can cause skin sores or affect internal organs—particularly the spleen and liver.
Trypanosoma gambiense and *T. rhodesiense*	Trypanosomiasis ("sleeping sickness") is a potentially fatal disease transmitted through bites from tsetse flies. Occurs in Africa.
Trypanosoma cruzi	Chagas disease affects 16–18 million people and causes 50,000 deaths annually, primarily in South and Central America.
Entamoeba histolytica	Amoebic dysentery results from severe infections.
Phytophthora infestans	An outbreak of this protist wiped out potato crops in Ireland in 1845–1847, causing famine.

Although *Plasmodium* is arguably the best studied of all protists, researchers have still not been able to devise effective and sustainable measures to control it.

- *Plasmodium* has evolved resistance to some of the drugs used to control its growth in infected people.

- Efforts to develop a vaccine against *Plasmodium* have been fruitless to date, in part because the parasite evolves so quickly. Flu virus and HIV pose similar problems (see Chapter 51).

- Natural selection has favored mosquito strains that are resistant to the insecticides that have been sprayed in their breeding habitats in attempts to control malaria's spread.

Unfortunately, malaria is not the only important human disease caused by protists. **TABLE 30.1** lists some protists that cause human suffering and economic losses.

Harmful Algal Blooms When a unicellular species experiences rapid population growth and reaches high densities in an aquatic environment, it is said to "bloom." Unfortunately, a handful of the many protist species involved in blooms can be harmful.

Harmful algal blooms are usually due to photosynthetic protists called dinoflagellates. Certain dinoflagellates synthesize toxins to protect themselves from predation by small animals called copepods. Because these organisms sometimes have high concentrations of accessory pigments called xanthophylls, their blooms can sometimes discolor seawater (**FIGURE 30.4**).

Algal blooms can be harmful to people because clams and other shellfish filter photosynthetic protists out of the water as food. During a bloom, high levels of toxins can build up in the flesh of these shellfish. Typically, the shellfish themselves are not harmed. But if a person eats contaminated shellfish, several types of poisoning can result.

Paralytic shellfish poisoning, for example, occurs when people eat shellfish that have fed heavily on protists that synthesize poisons called saxitoxins. Saxitoxins block ion channels that have to open for electrical signals to travel through nerve cells

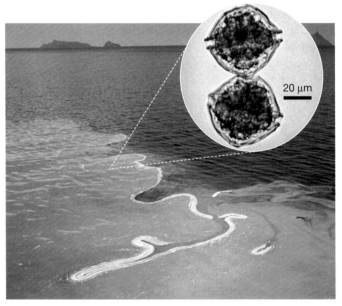

20 µm

FIGURE 30.4 Harmful Algal Blooms Are Caused by Dinoflagellates.

(see Chapter 46). In humans, high dosages of saxitoxins cause unpleasant symptoms such as prickling sensations in the mouth or even life-threatening symptoms such as muscle weakness and paralysis.

But like the bacteria and archaea, most species of protists are not harmful. Let's have a look at some protists that are essential to healthy ecosystems.

Ecological Importance of Protists

As a whole, the protists represent just 10 percent of the total number of named eukaryote species. Although the number of named species of protists is relatively low, their abundance is extraordinarily high. The number of individual protists found in some habitats is astonishing.

- One milliliter of pond water can contain well over 500 single-celled protists that swim with the aid of flagella.

- Under certain conditions, dinoflagellates can reach concentrations of 60 thousand cells per milliliter of seawater.

Why is this important?

Protists Play a Key Role in Aquatic Food Chains Photosynthetic protists take in carbon dioxide from the atmosphere and reduce, or "fix," it to form sugars or other organic compounds with high potential energy (see Chapter 10). Photosynthesis transforms some of the energy in sunlight into chemical energy that organisms can use to grow and produce offspring.

Species that produce chemical energy in this way are called **primary producers.** Diatoms, in the Stramenopila lineage, are photosynthetic protists that rank among the leading primary producers in the oceans simply because they are so abundant. Primary production in the ocean represents almost half of the total carbon dioxide that is fixed on Earth.

Diatoms and other organisms that drift in the open oceans or lakes are called **plankton.** The sugars and other organic compounds produced by photosynthetic plankton are the basis of food chains in freshwater and marine environments.

A **food chain** describes nutritional relationships among organisms, and thus how chemical energy flows within ecosystems. In this case, photosynthetic protists and other primary producers are eaten by primary consumers, many of which are protists. Primary consumers are eaten by fish, shellfish, and other secondary consumers, which in turn are eaten by tertiary consumers—whales, squid, and large fish (such as tuna).

Many of the species at the base of food chains in aquatic environments are protists.

Could Protists Help Limit Global Climate Change? Carbon dioxide levels in the atmosphere are increasing rapidly because humans are burning fossil fuels and forests (see Chapter 56). Carbon dioxide traps heat that is radiating from Earth, so high CO_2 levels in the atmosphere contribute to the rise in temperatures associated with global climate change—an issue that many observers consider today's most pressing environmental problem.

The carbon atoms in carbon dioxide molecules move to organisms on land or in the oceans and then back to the atmosphere, in what researchers call the **global carbon cycle.** To restrict global climate change, researchers are trying to figure out ways to decrease carbon dioxide concentrations in the atmosphere and increase the amount of carbon stored in terrestrial and marine environments.

To understand how this might be done, consider the marine carbon cycle diagrammed in **FIGURE 30.5**. The cycle starts when CO_2 from the atmosphere dissolves in water and is taken up by primary producers—photosynthetic plankton called phytoplankton—and converted to organic matter. The plankton are eaten by primary consumers, die and are consumed by decomposers or scavengers, or die and sink to the bottom of the ocean. There they may enter one of two long-lived repositories:

1. *Sedimentary rocks* Several lineages of protists have shells made of calcium carbonate ($CaCO_3$). When these shells rain down from the ocean surface and settle in layers at the bottom, the deposits that result are compacted by the weight of the water and by sediments accumulating above them. Eventually the deposits turn into rock. The limestone used to build the pyramids of Egypt consists largely of protist shells.

2. *Petroleum* Although the process of petroleum (oil) formation is not well understood, it begins with accumulations of dead bacteria, archaea, and protists at the bottom of the ocean.

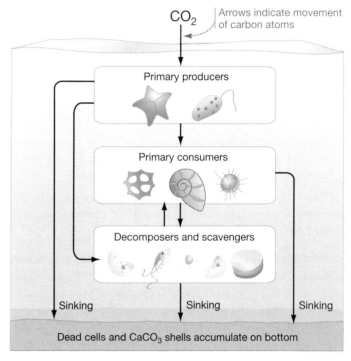

FIGURE 30.5 Protists Play a Key Role in the Marine Carbon Cycle. At the surface, carbon atoms tend to shuttle quickly among organisms. But if carbon atoms sink to the bottom of the ocean in the form of shells or dead cells, they may be locked up for long periods in carbon sinks. (The bottom of the ocean may be miles below the surface.)

Recent experiments have shown that the carbon cycle speeds up when habitats in the middle of the ocean are fertilized with iron. Iron is a critical component of the electron transport chains responsible for photosynthesis and respiration, but it is in particularly short supply in the open ocean. After iron is added to ocean waters, it is not uncommon to see populations of protists and other primary producers increase by a factor of 10.

Some researchers hypothesize that when these blooms occur, the amount of carbon that rains down into carbon sinks in the form of shells and dead cells may increase. If so, then fertilizing the ocean to promote blooms might be an effective way to reduce CO_2 concentrations in the atmosphere.

The effectiveness of iron fertilization is hotly debated, however. Fertilizing the ocean with iron might lead to large accumulations of dead organic matter and the formation of anaerobic dead zones (see Chapter 29). But if further research shows that iron fertilization is safe and effective, it could be added to the list of approaches to limit global climate change.

check your understanding

If you understand that . . .

- Malaria is caused by a protist that lives in mosquitoes and in humans in different parts of its life cycle.
- Harmful algal blooms are caused by protists that produce a toxin as a defense against predation.
- Protists are key primary producers in aquatic environments. As a result, they play a key role in the global carbon cycle.

✔ **You should be able to . . .**

1. Explain why public health workers are promoting the use of insecticide-treated sleeping nets as a way of reducing malaria.

2. Make a flowchart showing the chain of events that would start with massive iron fertilization and end with large deposits of carbon-containing compounds on the ocean floor.

Answers are available in Appendix A.

30.2 How Do Biologists Study Protists?

Although biologists have made great strides in understanding pathogenic protists and the role that protists play in the global carbon cycle, they have found it extremely difficult to gain any sort of solid insight into how the group as a whole diversified over time. The problem is that the eukaryotic lineages split over a billion years ago and diverged so much that it is not easy to find any overall patterns in the evolution of the group.

Recently, researchers have made progress in understanding protist diversity by combining data on the morphology of key groups with phylogenetic analyses of DNA sequences. However, despite the progress, significant questions remain about how the Eukarya diversified. Let's analyze how this work is being done, beginning with classical results on the morphological traits that distinguish major eukaryote groups.

Microscopy: Studying Cell Structure

Using light microscopy (see **BioSkills 11** in Appendix B), biologists were able to identify and name many of the protist species known today. For example, the early microscopist Anton van Leeuwenhoek identified the parasite *Giardia intestinalis* by examining samples of his own feces.

When transmission electron microscopes became available, a major breakthrough in understanding protist diversity occurred: Detailed studies revealed that protists could be grouped according to overall cell structure, according to organelles with distinctive features, or both.

For example, both light and electron microscopy confirmed that the species that caused the Irish potato famine has reproductive cells with an unusual type of **flagellum.** Flagella are organelles that project from the cell and whip back and forth to produce swimming movements (see Chapter 7). In reproductive cells of this species, one of the two flagella present has tiny, hollow, hairlike projections. Biologists noted that kelp and other species of brown algae also have cells with this type of flagellum.

To make sense of these results, researchers interpreted these types of distinctive morphological features as synapomorphies. Species that have a flagellum with hollow, hairlike projections became known as stramenopiles (literally, "straw-hairs"); the hairs typically have branches at the tip (**FIGURE 30.6**).

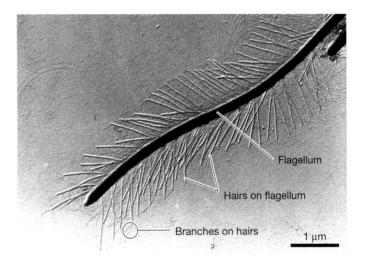

FIGURE 30.6 Species in the Lineage Called Stramenopila Have a Distinctive Flagellum. The unusual, hollow "hairs" that decorate the flagella of stramenopiles often have three branches at the tip.

Lineage	Distinguishing Morphological Features (synapomorphies)
Amoebozoa	Cells lack cell walls. When portions of the cell extend outward to move the cell, they form large lobes.
Opisthokonta	Reproductive cells have a single flagellum at their base. The cristae inside mitochondria are flat, not tube-shaped as in other eukaryotes. (This lineage includes protists as well as the fungi and the animals. Fungi and animals are discussed separately (see Chapters 32 through 35).
Excavata	Most cells have a pronounced "feeding groove" where prey or organic debris is ingested. Most species lack typical mitochondria, although genes derived from mitochondria are found in the nucleus.
Plantae	Cells have chloroplasts with a double membrane.
Rhizaria	Cells lack cell walls, although some produce an elaborate shell-like covering. When portions of the cell extend outward to move the cell, they are slender in shape.
Alveolata	Cells have sac-like structures called alveoli that form a continuous layer just under the plasma membrane. Alveoli are thought to provide support.
Stramenopila	If flagella are present, cells usually have two—one of which is covered with hairlike projections.

In recognizing the Stramenopila lineage, investigators hypothesized that because an ancestor had evolved a distinctive flagellum, all or most of its descendants also had this trait. The qualifier *most* is important, because it is not unusual for certain subgroups within a lineage to lose particular traits over the course of evolution, much as humans are gradually losing fur and tailbones.

Eventually, seven major groups of eukaryotes came to be identified on the basis of diagnostic morphological characteristics. These groups and the synapomorphies that identify them are listed in **TABLE 30.2**. Notice that in almost every case, the synapomorphies that define eukaryotic lineages are changes in structures that protect or support the cell or that influence the organism's ability to move or feed. The plants, fungi, and animals represent subgroups within two of the seven major eukaryotic lineages (see Chapters 31 through 35).

Evaluating Molecular Phylogenies

When researchers began using DNA sequence data to estimate the evolutionary relationships among eukaryotes, the analyses suggested that the seven groups identified on the basis of distinctive morphological characteristics were indeed monophyletic. This was important support for the hypothesis that the distinctive morphological features were shared, derived characters that existed in a common ancestor of each lineage.

The phylogenetic tree in **FIGURE 30.7** is the current best estimate of the eukaryotes' evolutionary history. As you read this tree, note that:

- The Amoebozoa and the Opisthokonta—which include fungi and animals—form a monophyletic group called the Unikonta.

- The other five major eukaryotic lineages form a monophyletic group called the Bikonta.

Understanding where the root or base of the tree of Eukarya lies has been more problematic. Finding it will help researchers understand what the common ancestor looked like and how eukaryotic lineages evolved over time. One hypothesis suggested that the first eukaryotic split was between unikonts and bikonts—meaning, between eukaryotes that have one flagellum and those with two flagella. This hypothesis is tentative and controversial, however, given the data analyzed to date. Researchers continue to work on the issue of placing the root of the Eukarya.

Discovering New Lineages via Direct Sequencing

The effort to refine the phylogeny of the Eukarya is ongoing. But of all the research frontiers in eukaryotic diversity, the most exciting may be the one based on the technique called direct sequencing.

Direct sequencing is based on sampling soil or water, analyzing the DNA sequence of specific genes in the sample, and using the data to place the organisms in the sample on a phylogenetic tree. Direct sequencing and metagenomics have led to the discovery of previously unknown but major lineages of Archaea (see Chapter 29). To the amazement of biologists all over the world, the same thing happened when researchers used direct sequencing to survey eukaryotes.

The first direct sequencing studies that focused on eukaryotes were published in 2001. One study sampled organisms at depths from 250 to 3000 meters (m) below the surface in waters off Antarctica; another focused on cells at depths of 75 m in the Pacific Ocean, near the equator. Both studies detected a wide array of species that were new to science.

Investigators who examined the samples microscopically were astonished to find that many of the newly discovered eukaryotes were tiny—from 0.2 micrometers (μm) to 5 μm in diameter. Subsequent work has confirmed the existence of many,

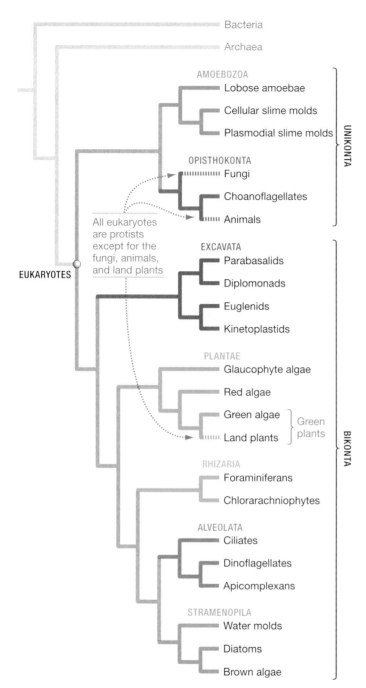

FIGURE 30.7 **Phylogenetic Analyses Have Identified Seven Major Lineages of Eukaryotes.** This tree shows selected subgroups from the seven major lineages discussed in this chapter. Kinetoplastids are not discussed in detail elsewhere in the chapter, but they include the pathogens *Trypanosoma* and *Leishmania* described in Table 30.1.

diverse species of protists that are less than 0.2 μm in diameter. These cells overlap in size with bacteria, which typically range from 0.5 μm to 2 μm in diameter.

Eukaryotic cells are much more variable in size than previously imagined. A whole new world of tiny protists has only recently been discovered.

If you understand that . . .

- According to the most recent analyses, the domain Eukarya comprises seven major lineages. Members of each lineage have distinctive aspects of cell structure.
- Direct sequencing has allowed investigators to recognize large numbers of previously undescribed eukaryotes, some of which are extremely small.

✔ **You should be able to . . .**

1. Explain why opisthokonts, alveolates, and stramenopiles got their names. (The root *opistho* refers to the back of a cell; the root *kont* refers to a flagellum.)
2. Explain why direct sequencing studies allow researchers to characterize species that have never been seen before.

Answers are available in Appendix A.

30.3 What Themes Occur in the Diversification of Protists?

The protists range in size from bacteria-sized single cells to giant kelp. They live in habitats from the open oceans to the guts of termites. They are almost bewildering in their morphological and ecological diversity. Because they are a paraphyletic group, they do not share derived characteristics that set them apart from all other lineages on the tree of life.

One general theme that can help tie protists together is their amazing diversity. Once an important new innovation arose in protists, it triggered the evolution of species that live in a wide array of habitats and make a living in diverse ways.

What Morphological Innovations Evolved in Protists?

Virtually all bacteria and all archaea are unicellular. Given the distribution of multicellularity in eukaryotes, it is logical to conclude that the first eukaryote was also a single-celled organism.

Further, all eukaryotes alive today have (**1**) either mitochondria or genes that are normally found in mitochondria, (**2**) a nucleus and endomembrane system, and (**3**) a cytoskeleton. Based on the distribution of cell walls in living eukaryotes, it is likely that the first eukaryotes lacked this feature.

Using these observations, biologists hypothesize that the earliest eukaryotes were probably single-celled organisms with mitochondria, a nucleus and endomembrane system, and a cytoskeleton, but no cell wall.

It is also likely that the first eukaryotic cells swam using a novel type of flagellum. Eukaryotic flagella are completely different structures from bacterial flagella and evolved independently.

The eukaryotic flagellum is made up of microtubules, and dynein is the major motor protein. An undulating motion occurs as dynein molecules walk down microtubules (see Chapter 7). The flagella of bacteria and archaea, in contrast, are composed primarily of a protein called flagellin. Instead of undulating, prokaryotic flagella rotate to produce movement.

✔ If you understand the synapomorphies that identify the eukaryotes as a monophyletic group, you should be able to map the origin of the nuclear envelope and the eukaryotic flagellum on Figure 30.7. Once you've done that, let's consider how several of these key new morphological features arose and influenced the subsequent diversification of protists, beginning with one of the traits that define the Eukarya.

Endosymbiosis and the Origin of the Mitochondrion   Mitochondria are organelles that generate adenosine triphosphate (ATP) using pyruvate as an electron donor and oxygen as the ultimate electron acceptor (see Chapter 9).

In 1981 evolutionary biologist Lynn Margulis expanded on a radical hypothesis—first proposed in the nineteenth century—to explain the origin of mitochondria. The **endosymbiosis theory** proposes that mitochondria originated when a bacterial cell took up residence inside another cell about 2 billion years ago.

The theory's name was inspired by the Greek word roots *endo*, *sym*, and *bio* (literally, "inside-together-living"). **Symbiosis** is said to occur when individuals of two different species live in physical contact; **endosymbiosis** occurs when an organism of one species lives inside the cells of an organism of another species.

How and when this momentous event occurred is hotly debated. Some scientists think a primitive amoeba-like eukaryote engulfed a bacterium and simply failed to digest it in its lysosome, allowing it to take up residence. Evidence that was used to support this theory includes the similar feeding habits of today's amoeba, and the existence of some protist species that seem to lack mitochondria.

Recent evidence, however, indicates that all protists originally had mitochondria, and some lost them. A new idea gaining support is that the first eukaryote may have been formed as a result of an endosymbiosis between two prokaryotes—an archaeal host and a bacterium (**FIGURE 30.8**). After this chance union occurred, cells developed nuclei and became much larger. Both of these changes appear to have been triggered by the bacterial invader.

The relationship between the archaeal host and the engulfed bacterial cell was presumed to be stable because a mutual advantage existed between them: The host supplied the bacterium with protection and carbon compounds from its prey, while the bacterium produced much more ATP than the host cell could synthesize on its own.

When Margulis first began promoting the endosymbiosis theory, it met with a storm of criticism—largely because it seemed slightly preposterous. But gradually biologists began to examine it rigorously. For example, endosymbiotic relationships between protists and bacteria exist today. Among the α-proteobacteria alone (see Chapter 29), three major groups are found *only* inside eukaryotic cells. In many cases, the bacterial cells are transmitted to offspring in eggs or sperm and are required for survival.

PROCESS: THE ENDOSYMBIOSIS THEORY

Bacterium

Archaeon

1. Archaeal cell surrounds and engulfs bacterium.

2. Bacterium lives within host archaeal cell.

Pyruvate and O₂

ATP

3. Endosymbiosis: Host cell supplies bacterium with protection and carbon compounds. Bacterium supplies host cell with ATP.

FIGURE 30.8 Proposed Initial Steps in the Evolution of the Mitochondrion.

✔ **QUESTION** According to this hypothesis, how many membranes should surround a mitochondrion? Explain your logic.

Several observations about the structure of mitochondria are also consistent with the endosymbiosis theory:

- Mitochondria are about the size of an average α-proteobacterium.

- Mitochondria replicate by fission, as do bacterial cells. The duplication of mitochondria takes place independently of division by the host cell. When eukaryotic cells divide, each daughter cell receives some of the many mitochondria present.

- Mitochondria have their own ribosomes and manufacture some of their own proteins. Mitochondrial ribosomes closely resemble bacterial ribosomes in size and composition and are poisoned by antibiotics such as streptomycin that inhibit bacterial, but not eukaryotic, ribosomes.

- Mitochondria have double membranes, consistent with the engulfing mechanism of origin illustrated in Figure 30.8.

- Mitochondria have their own genomes, which are organized as circular molecules—much like a bacterial chromosome. Mitochondrial genes code for a few of the proteins needed to conduct electron transport and RNAs needed to translate the mitochondrial genome.

Although these data are impressive, they are only consistent with the endosymbiosis theory. Stated another way, they do not exclude other explanations. This is a general principle in science: Evidence is considered strong when it cannot be explained by reasonable alternative hypotheses. In this case, the key was to find data that tested predictions made by Margulis's idea against predictions made by an alternative theory: that mitochondria evolved within eukaryotic cells, separately from bacteria.

A breakthrough occurred when researchers realized that according to the "within-eukaryotes" theory, the genes found in mitochondria are derived from nuclear genes in ancestral eukaryotes. Margulis's theory, in contrast, proposed that the genes found in mitochondria were bacterial in origin.

These predictions were tested by studies on the phylogenetic relationships of mitochondrial genes. Specifically, researchers compared gene sequences isolated from eukaryotic mitochondrial DNA with sequences of similar genes isolated from eukaryotic nuclear DNA and with DNA from several species of bacteria. Exactly as the endosymbiosis theory predicted, mitochondrial gene sequences are much more closely related to the sequences from the α-proteobacteria than to sequences from the nuclear DNA of eukaryotes.

The result diagrammed in **FIGURE 30.9** was considered overwhelming evidence that the mitochondrial genome came from an α-proteobacterium rather than from a eukaryote. The endosymbiosis theory was the only reasonable explanation for the data.

The results were a stunning vindication of a theory that had once been intensely controversial. Mitochondria evolved via endosymbiosis between an α-proteobacterium and an archaeal host.

One intriguing feature of mitochondrial genomes is that they typically encode less than 50 genes, whereas the genomes of their bacterial cousins encode about 1500 genes. Most of the genes from the endosymbiotic bacterium moved into the nuclear genome in what was one of the most spectacular **lateral gene transfer** events in the history of life (see Chapter 21). Some unnecessary genes were lost, but many still encode proteins that are synthesized by cytosolic ribosomes and imported into mitochondria.

✔ If you understand the endosymbiosis theory and the evidence for it, you should be able to describe how the chloroplast—the organelle that is the site of photosynthesis in eukaryotes—could arise via endosymbiosis.

Now let's turn to another distinguishing feature of eukaryotic cells—the nucleus, with its unique double membrane called the nuclear envelope. The origin of the nuclear envelope is also a subject of debate.

The Nuclear Envelope   The leading hypothesis to explain the origin of the nuclear envelope is based on infoldings of the plasma membrane. As the drawings in **FIGURE 30.10** show, a stepwise process could give rise to small infoldings that were elaborated

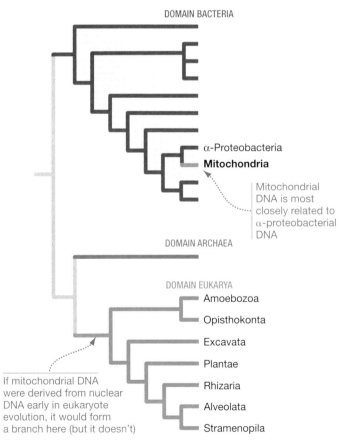

FIGURE 30.9 Phylogenetic Data Support the Endosymbiosis Theory.

FIGURE 30.10 A Hypothesis for the Origin of the Nuclear Envelope. Infoldings of the plasma membrane, analogous to those shown here, have been observed in bacteria living today.

by mutation and natural selection over time, and the infolding could eventually become detached from the plasma membrane. Note that the infoldings would have given rise to the nuclear envelope and the endoplasmic reticulum (ER) together.

Two lines of evidence support this hypothesis: Infoldings of the plasma membrane occur in some bacteria living today, and the nuclear envelope and ER of today's eukaryotes are continuous (see Chapter 7). ✔ If you understand the infolding hypothesis, you should be able to explain why these observations support it.

According to current thinking, the evolution of the nuclear envelope was advantageous because it separated transcription and translation. Recall that RNA transcripts are processed inside the nucleus but translated outside the nucleus (see Chapter 17). In bacteria and archaea, transcription and translation occur together.

With a simple nuclear envelope in place, alternative splicing and other forms of RNA processing could occur—giving the early eukaryotes a novel way to control gene expression (see Chapter 19). The take-home message here is that an important morphological innovation gave the early eukaryotes a new way to manage and process genetic information.

Once a nucleus had evolved, it underwent diversification. In some cases, unique types of nuclei are associated with the founding of important lineages of protists.

- Ciliates (Alveolata) have a diploid micronucleus that is involved only in reproduction, and a polyploid macronucleus where transcription occurs.

- Diplomonads (Excavata) have two nuclei that look identical; it is not known how they interact.

- In foraminifera (Rhizaria), red algae (Plantae), and plasmodial slime molds (Amoebozoa), certain cells contain many nuclei.

- Dinoflagellates (Alveolata) have chromosomes that lack histones and attach to the nuclear envelope.

In each case, the distinctive structure of the nucleus is a synapomorphy that allows biologists to recognize these lineages as distinct monophyletic groups.

Structures for Support and Protection   Many protists have cell walls outside their plasma membrane; others have hard external structures called a **shell;** others have rigid structures inside the plasma membrane. In many cases, these novel structures represent synapomorphies that identify monophyletic groups among protists. For example:

- Diatoms (Stramenopila) are surrounded by a glass-like, silicon-dioxide cell wall (**FIGURE 30.11a**). The cell wall is made up of two pieces that fit together in a box-and-lid arrangement, like the petri plates you may have seen in lab.

- Dinoflagellates (Alveolata) have a cell wall made up of cellulose plates (**FIGURE 30.11b**).

- Within the foraminiferans (Rhizaria), some subgroups secrete an intricate, chambered calcium carbonate shell (**FIGURE 30.11c**).

- Members of other foraminiferan subgroups, and some amoebae, cover themselves with tiny pebbles.

- The parabasalids (Excavata) have a unique internal support rod, consisting of cross-linked microtubules that run the length of the cell.

- The euglenids (Excavata) have a collection of protein strips located just under the plasma membrane. The strips are supported by microtubules and stiffen the cell.

- All alveolates (Alveolata) have distinctive sac-like structures called alveoli, located just under the plasma membrane, that help stiffen the cell.

In many cases, the diversification of protists has been associated with the evolution of innovative structures for support and protection.

Multicellularity   One of the most significant changes in the history of life on Earth occurred when organisms containing more than one cell evolved. The mutations leading to **multicellularity** probably first caused cells to simply stick together after cell division. Selection pressures could then act on these larger, colonial organisms, allowing them to evolve and diversify.

(a) Diatom

50 µm

Glassy cell wall made of silicon dioxide

(b) Dinoflagellate

10 µm

Tough plates in cell wall made of cellulose

(c) Foraminiferan

50 µm

Chalky, chambered shell made of calcium carbonate

FIGURE 30.11 Hard Outer Coverings in Protists Vary in Composition.

Eventually cells became specialized for different functions. In the simplest multicellular species, certain cells are specialized for producing or obtaining food while other cells are specialized for reproduction. The key point about multicellularity is that not all cells express the same genes.

A few species of bacteria are capable of aggregating and forming structures called fruiting bodies (see Chapter 29). Because cells in the fruiting bodies of these bacteria differentiate into specialized stalk cells and spore-forming cells, they are considered multicellular. But the vast majority of multicellular species are members of the Eukarya.

Multicellularity arose independently in a wide array of eukaryotic lineages, including the green plants, fungi, animals, brown algae, slime molds, and red algae.

To summarize, an array of novel morphological traits played a key role as protists diversified: the mitochondrion, the nucleus and endomembrane system, structures for protection and support, and multicellularity. Evolutionary innovations allowed protists to build and manage the eukaryotic cell in new ways.

Once a new type of eukaryotic cell or multicellular individual existed, subsequent diversification was often triggered by novel ways of finding food, moving, or reproducing. Let's consider each of these life processes in turn.

How Do Protists Obtain Food?

Bacteria and archaea can use a wide array of molecules as electron donors and electron acceptors during cellular respiration (see Chapter 29). Some get these molecules by absorbing them directly from the environment. Other bacteria don't absorb their nutrition—instead, they make their own food via photosynthesis. Many groups of protists are similar to bacteria in the way they find food: They perform photosynthesis or absorb their food directly from the environment.

But one of the most important stories in the diversification of protists was the evolution of a novel method for finding food.

Many protists ingest their food—they eat bacteria, archaea, or even other protists whole. This process is called **phagocytosis** (see Chapter 7). When phagocytosis occurs, an individual takes in packets of food much larger than individual molecules. Thus, protists feed by either (**1**) ingesting packets of food, (**2**) absorbing organic molecules directly from the environment, or (**3**) performing photosynthesis.

Some protists ingest food as well as performing photosynthesis—meaning that they use a combination of feeding strategies. It's also important to recognize that all three lifestyles—ingestive, absorptive, and photosynthetic—can occur within a single lineage.

To drive this last point home, consider the monophyletic group called the alveolates. This lineage has three major subgroups—dinoflagellates, apicomplexans, and ciliates. About half of the dinoflagellates are photosynthetic, but many others are parasitic and absorb nutrients from their hosts. Apicomplexans are parasitic. Ciliates include many species that ingest prey, but some ciliates live in the guts of cattle or the gills of fish. Other ciliate species make a living by holding algae or other types of photosynthetic endosymbionts inside their cells.

The punch line? Within each of the seven major lineages of eukaryotes, different methods for feeding helped trigger diversification.

Ingestive Feeding Ingestive lifestyles are based on eating live or dead organisms or on scavenging loose bits of organic debris. Protists such as the cellular slime mold *Dictyostelium discoideum* are large enough to engulf bacteria and archaea; many protists are large enough to surround and ingest other protists or even microscopic animals.

Feeding by phagocytosis is possible in protists that lack a cell wall. A flexible membrane and dynamic cytoskeleton give these species the ability to surround and "swallow" prey using long, fingerlike projections called **pseudopodia** ("false-feet") (**FIGURE 30.12a**). Phagocytosis was a prerequisite for the endosymbiosis event that led to chloroplasts.

(a) Pseudopodia engulf food.

(b) Ciliary currents sweep food into gullet.

FIGURE 30.12 Ingestive Feeding. Methods of prey capture vary among ingestive protists. **(a)** Some predators engulf prey with pseudopodia; **(b)** other predators sweep them into their gullets with water currents set up by the beating of cilia. Note that the cells in part (a) have been colorized.

Although many ingestive feeders actively hunt down prey and engulf them, others do not. Instead of taking themselves to food, these species attach themselves to a surface. Protists that feed in this way have cilia that surround the mouth and beat in a coordinated way. The motion creates water currents that sweep food particles into the cell (**FIGURE 30.12b**).

Absorptive Feeding Absorptive feeding occurs when nutrients are taken up directly from the environment, across the plasma membrane, usually through transport proteins. Absorptive feeding is common among protists.

Some protists that live by absorptive feeding are **decomposers,** meaning that they feed on dead organic matter, or **detritus.** But many of the protists that absorb their nutrition directly from the environment live inside other organisms. If an absorptive species damages its host, that species is called a **parasite.**

Photosynthesis—Endosymbiosis and the Origin of Chloroplasts

You might recall that photosystems I and II evolved in bacteria, and that both photosystems occur in cyanobacteria (see Chapter 10). None of the basic machinery required for photosynthesis evolved in eukaryotes. Instead, they "stole" it—via endosymbiosis.

The endosymbiosis theory contends that the eukaryotic chloroplast originated when a protist engulfed a cyanobacterium. Once inside the protist, the photosynthetic bacterium provided its eukaryotic host with oxygen and glucose in exchange for protection and access to light. If the endosymbiosis theory is correct, today's chloroplasts trace their ancestry to cyanobacteria.

All of the photosynthetic eukaryotes have both chloroplasts and mitochondria. The evidence for an endosymbiotic origin for the chloroplast is persuasive.

- Chloroplasts have the same list of bacteria-like characteristics presented earlier for mitochondria.

- Many examples of endosymbiotic cyanobacteria are living inside the cells of protists or animals today.

- Chloroplasts contain a circular DNA molecule containing genes that are extremely similar to cyanobacterial genes.

- The photosynthetic organelle of one group of protists, the glaucophyte algae, has an outer layer containing the same constituent (peptidoglycan) found in the cell walls of cyanobacteria.

Like mitochondria, the chloroplast genome is very small compared with genomes of living cyanobacteria; most of the original genes were either lost or transferred to the nucleus.

✔If you understand the evidence for the endosymbiotic origin of the chloroplast, you should be able to add a label indicating the location of chloroplast genes on the phylogenetic tree in Figure 30.7.

Photosynthesis—Primary versus Secondary Endosymbiosis

Which eukaryote originally obtained a photosynthetic organelle? Because all species in the Plantae have chloroplasts with two membranes, biologists infer that the original, or primary,

endosymbiosis occurred in these species' common ancestor. That species eventually gave rise to all subgroups in the Plantae lineage—the glaucophyte algae, red algae, and green plants (green algae and land plants).

But chloroplasts also occur in four of the other major lineages of protists—the Excavata, Rhizaria, Alveolata, and Stramenopila. In these species, the chloroplast is surrounded by more than two membranes—usually four.

To explain this observation, researchers hypothesize that the ancestors of these groups acquired their chloroplasts by ingesting photosynthetic protists that already had chloroplasts. This process, called secondary endosymbiosis, occurs when an organism engulfs a photosynthetic eukaryotic cell and retains its chloroplasts as intracellular symbionts. Secondary endosymbiosis is illustrated in **FIGURE 30.13**.

FIGURE 30.14 shows where primary and secondary endosymbiosis occurred on the phylogenetic tree of eukaryotes. Once

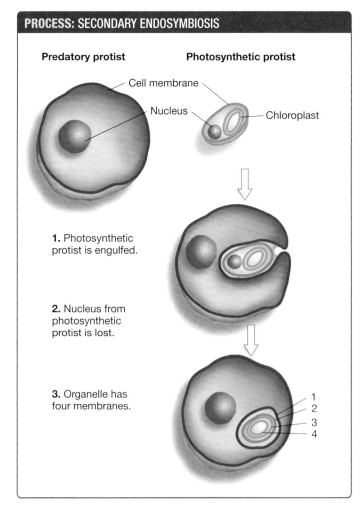

PROCESS: SECONDARY ENDOSYMBIOSIS

Predatory protist   Photosynthetic protist

Cell membrane

Nucleus   Chloroplast

1. Photosynthetic protist is engulfed.

2. Nucleus from photosynthetic protist is lost.

3. Organelle has four membranes.

1
2
3
4

FIGURE 30.13 Secondary Endosymbiosis Leads to Organelles with Four Membranes. The chloroplasts found in some protists have four membranes and are hypothesized to be derived by secondary endosymbiosis. In species where chloroplasts have three membranes, biologists hypothesize that secondary endosymbiosis was followed by the loss of one membrane.

protists obtained the chloroplast, it was "swapped around" to new lineages via secondary endosymbiosis.

✔ If you understand endosymbiosis, you should be able to assess and explain why the primary and secondary endosymbiosis events introduced in this chapter represent the most massive lateral gene transfers in the history of life, in terms of the number of genes moved at once. (To review lateral gene transfer, see Chapter 21.)

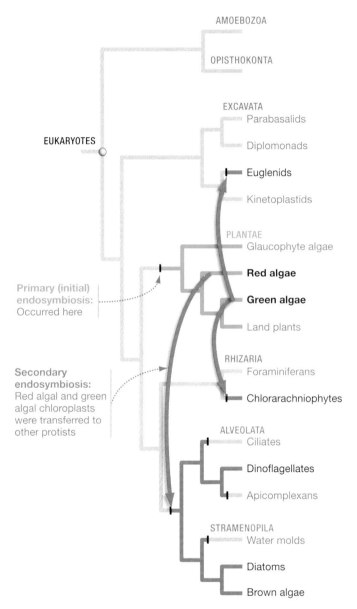

FIGURE 30.14 **Photosynthesis Arose in Protists by Primary Endosymbiosis and Then Spread among Lineages via Secondary Endosymbiosis.** Biochemical similarities link the chloroplasts found in alveolates and stramenopiles with red algae and the chloroplasts found in euglenids and chlorarachniophytes with green algae. Notice that among the alveolates and stramenopiles, only the dinoflagellates, diatoms, and brown algae have photosynthetic species. In ciliates, apicomplexans, and water molds, the chloroplast has been lost or has changed function.

How Do Protists Move?

Many protists actively move to find food. Predators such as the slime mold *Dictyostelium discoideum* crawl over a substrate in search of prey. Most of the unicellular, photosynthetic species are capable of swimming to sunny locations, though others drift passively in water currents. How are these crawling and swimming movements possible?

Amoeboid motion is a sliding movement observed in some protists. In the classic mode illustrated in **FIGURE 30.15**, pseudopodia stream forward over a substrate, and the rest of the cytoplasm, organelles, and plasma membrane follow. The motion requires ATP and involves interactions between proteins called actin and myosin inside the cytoplasm. The mechanism is related to muscle movement in animals (see Chapter 48). But at the level of the whole cell, the precise sequence of events during amoeboid movement is still uncertain. The issue is attracting researchers' attention, because key immune system cells in humans use amoeboid motion as they hunt down and destroy disease-causing agents.

The other major mode of locomotion in protists involves flagella or cilia (**FIGURE 30.16**, see page 566). Recall that flagella and cilia both consist of nine sets of doublet (paired) microtubules arranged around two central, single microtubules (see Chapter 7). Flagella are long and are usually found alone or in pairs; cilia are short and usually occur in large numbers on any one cell.

FIGURE 30.15 **Amoeboid Motion Is Possible in Species That Lack Cell Walls.** In amoeboid motion, long pseudopodia stream out from the cell. The rest of the cytoplasm, organelles, and external membrane follow.

(a) Swimming via **flagella**

(b) Swimming via **cilia**

FIGURE 30.16 Many Protists Swim Using Flagella or Cilia.
(a) Flagella are long and few in number. **(b)** Cilia are short and numerous. In many cases they are used in swimming.

Even closely related protists can use radically different forms of locomotion. For example, within the Alveolata, the ciliates swim by beating their cilia, the dinoflagellates swim by whipping their flagella, and mature apicomplexan cells move by amoeboid motion (though their gametes swim via flagella). It is also common to find protists that do not exhibit active movement, but instead float passively in water currents.

Movement is yet another example of the extensive diversification that occurred within each of the seven major monophyletic groups of eukaryotes.

How Do Protists Reproduce?

Sexual reproduction originated in protists. Sexual reproduction can best be understood in contrast to asexual reproduction (see Chapter 13). The key issues are the type of nuclear division involved and the consequences for genetic diversity in the offspring produced.

- Asexual reproduction is based on mitosis and cell division in eukaryotic organisms and on fission in bacteria and archaea. It results in daughter cells that are genetically identical to the parent.

- Sexual reproduction is based on meiosis and fusion of gametes. It results in daughter cells that are genetically different from their parents and from each other.

Most protists undergo asexual reproduction routinely. But sexual reproduction occurs only intermittently in many protists—often at one particular time of year, or when individuals are stressed by overcrowding or scarce food supplies.

The evolution of sexual reproduction ranks among the most significant evolutionary innovations observed in eukaryotes.

Sexual versus Asexual Reproduction   The leading hypothesis to explain why meiosis evolved states that genetically variable offspring may be able to thrive if the environment changes. For example, offspring with genotypes different from those of their parents may be better able to withstand attacks by parasites that successfully attacked their parents (see Chapter 13). This is a key point because many types of parasites, including bacteria and viruses, have short generation times and evolve quickly.

Because the genotypes and phenotypes of parasites are constantly changing, natural selection favors host individuals with new genotypes. The idea is that novel genotypes in offspring generated by meiosis may contain combinations of alleles that allow hosts to withstand attack by new strains of parasites. In short, many biologists view sexual reproduction as an adaptation to fight disease.

✔ If you understand the changing-environment hypothesis for the evolution of sex, you should be able to evaluate and explain whether it is consistent with the observation that many protists undergo meiosis when food is scarce or the population density is high.

Life Cycles—Haploid versus Diploid Dominated

A life cycle describes the sequence of events that occur as individuals grow, mature, and reproduce. The evolution of meiosis introduced a new event in the life cycle of protist species; what's more, it created a distinction between haploid and diploid phases in the life of an individual.

Recall that diploid individuals have two of each type of chromosome inside each cell (see Chapter 13). Haploid individuals have just one of each type of chromosome inside each cell. When meiosis occurs in diploid cells, it results in the production of haploid cells.

The life cycle of most bacteria and archaea is extremely simple: A cell divides, feeds, grows, and divides again. Bacteria and archaea are always haploid.

In contrast, virtually every aspect of a life cycle is variable among protists—whether meiosis occurs, whether asexual reproduction occurs, and whether the haploid or the diploid phase of the life cycle is the longer and more prominent phase.

FIGURE 30.17 illustrates some of this variation. Figure 30.17a depicts the haploid-dominated life cycle observed in many unicellular protists. The specific example given here is the dinoflagellate *Gyrodinium uncatenum*.

To analyze a life cycle, start with **fertilization**—the fusion of two gametes to form a diploid zygote. Then trace what happens to the zygote. In this case, the diploid zygote undergoes meiosis. The haploid products of meiosis then grow into mature cells that eventually undergo asexual reproduction or produce gametes by mitosis and cell division.

Now contrast the dinoflagellate life cycle in Figure 30.17a with the diploid-dominated life cycle in Figure 30.17b. The specific organism shown here is the diatom *Thalassiosira punctigera*. Notice that after fertilization, the diploid zygote develops into a sexually mature, diploid adult cell that can reproduce asexually by dividing mitotically. Meiosis occurs in the adult and results in the formation of haploid gametes, which then fuse to form a diploid zygote. The important contrasts with the

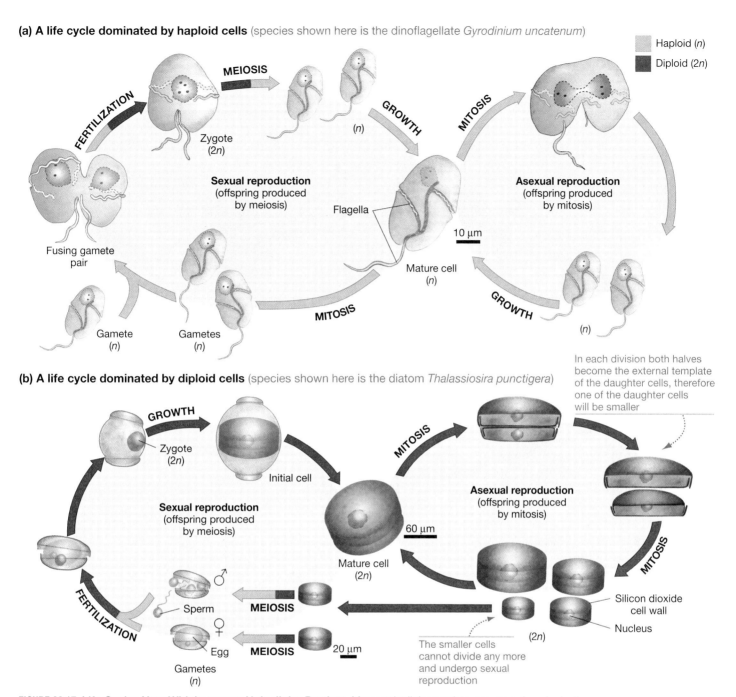

(a) A life cycle dominated by haploid cells (species shown here is the dinoflagellate *Gyrodinium uncatenum*)

Haploid (*n*)
Diploid (2*n*)

MEIOSIS

FERTILIZATION

Zygote
(2*n*)

GROWTH

(*n*)

MITOSIS

MITOSIS

Sexual reproduction
(offspring produced
by meiosis)

Asexual reproduction
(offspring produced
by mitosis)

Flagella

10 μm

Fusing gamete
pair

Mature cell
(*n*)

Gamete
(*n*)

Gametes
(*n*)

MITOSIS

GROWTH

(*n*)

(b) A life cycle dominated by diploid cells (species shown here is the diatom *Thalassiosira punctigera*)

GROWTH

Zygote
(2*n*)

MITOSIS

In each division both halves
become the external template
of the daughter cells, therefore
one of the daughter cells
will be smaller

Initial cell

Sexual reproduction
(offspring produced
by meiosis)

Asexual reproduction
(offspring produced
by mitosis)

60 μm

MITOSIS

Mature cell
(2*n*)

Silicon dioxide
cell wall

Nucleus

Sperm ♂

MEIOSIS

Egg ♀

MEIOSIS

20 μm

(2*n*)

FERTILIZATION

Gametes
(*n*)

The smaller cells
cannot divide any more
and undergo sexual
reproduction

FIGURE 30.17 Life Cycles Vary Widely among Unicellular Protists. Many unicellular protists can reproduce by both sexual reproduction and asexual reproduction. The cell may be **(a)** haploid for most of its life or **(b)** diploid for most of its life.

haploid-dominated life cycle are that **(1)** meiosis occurs in the adult cell rather than in the zygote, and **(2)** gametes are the only haploid cells in the life cycle.

Life Cycles—Alternation of Generations In contrast to the relatively simple life cycles of single-celled protists shown in Figure 30.17, many multicellular protists have a multicellular haploid form in one phase of their life cycle alternating with a multicellular diploid form in another phase. This alternation of multicellular haploid and diploid forms is known as **alternation of generations.**

- The multicellular haploid form is called a **gametophyte,** because specialized cells in this individual produce gametes by mitosis and cell division.

- The multicellular diploid form is called a **sporophyte** because it has specialized cells that undergo meiosis to produce haploid cells called spores.

- A **spore** is a single haploid cell that divides mitotically to form a multicellular, haploid gametophyte. A gamete is also a single haploid cell, but its role is to fuse with another gamete to form a zygote.

When alternation of generations occurs, a diploid sporophyte undergoes meiosis and cell division to produce haploid spores. Each spore then divides by mitosis to form a haploid, multicellular gametophyte. The haploid gametes produced mitotically by gametophytes then fuse to form a diploid zygote, which grows into the diploid, multicellular sporophyte.

Gametophytes and sporophytes may be identical in appearance, as in the brown alga called *Ectocarpus siliculosus* (**FIGURE 30.18a**). In many cases, however, the gametophyte and sporophyte look different, as in the brown algae in the genus *Laminaria* (**FIGURE 30.18b**). Among the protists, alternation of generations evolved independently in brown algae, red algae, and other groups.

(a) Alternation of generations in which multicellular haploid and diploid forms look identical (*Ectocarpus siliculosus*)

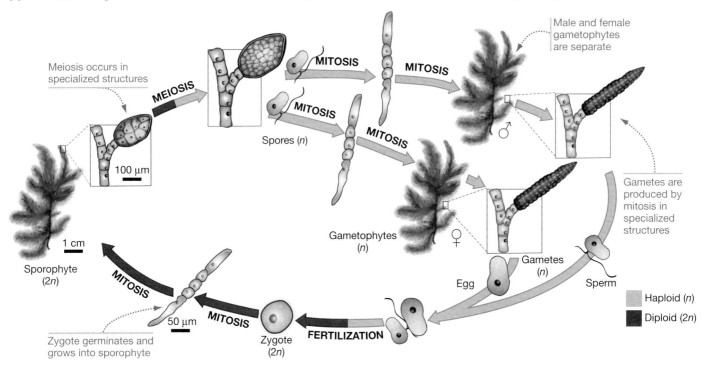

(b) Alternation of generations in which multicellular haploid and diploid forms look different (*Laminaria* sp.)

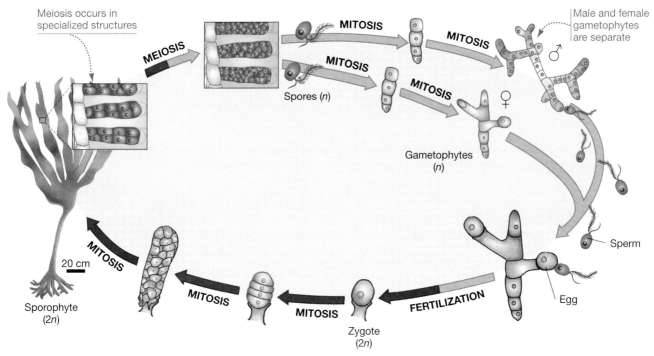

FIGURE 30.18 Alternation of Generations Occurs in Many Multicellular Protists. Compared here, two species of brown algae.

To understand alternation of generations, some students find it helpful to memorize this sentence: Sporophytes produce spores by meiosis; gametophytes produce gametes by mitosis.

✔ If you understand how life cycles vary among multicellular protists, you should be able to (1) define the terms alternation of generations, gametophyte, sporophyte, spore, zygote, and gamete; and (2) diagram a life cycle where alternation of generations occurs, without looking at Figure 30.18.

Why does so much variation occur in the types of life cycles observed among protists? The answer is not known. Variation in life cycles is a major theme in the diversification of protists and is discussed further in the chapter on green algae and land plants (see Chapter 31). Explaining why that variation exists remains a topic for future research.

30.4 Key Lineages of Protists

Each of the seven major Eukarya lineages has at least one distinctive morphological characteristic. But once an ancestor evolved a distinctive cell structure, its descendants diversified into a wide array of lifestyles. For example, parasitic species evolved independently in all seven major lineages. Photosynthetic species exist in most of the seven, and multicellularity evolved independently in at least four. Similar statements could be made about the evolution of life cycles and modes of locomotion.

In effect, each of the seven lineages represents a similar radiation of species into a wide array of lifestyles. Let's take a more detailed look at some representative taxa from six of the seven major lineages, starting with an overview of each one. The seventh major lineage—the opisthokonts—is featured in the chapters on fungi and the animals (see Chapters 32–35). The lineage called green plants is analyzed separately (see Chapter 31).

Amoebozoa

Species in the Amoebozoa lack cell walls and take in food by engulfing it. They move via amoeboid motion and produce large, lobe-like pseudopodia. They are abundant in freshwater habitats and in wet soils; some are parasites of humans and other animals.

Major subgroups in the lineage are lobose amoebae, cellular slime molds, and plasmodial slime molds. The cellular slime mold *Dictyostelium discoideum* is described in detail in **BioSkills 13** in Appendix B. ✔ You should be able to mark the origin of this lineage's amoeboid form on Figure 30.7 and explain whether it evolved independently of the amoeboid form in Rhizaria.

● Amoebozoa > Plasmodial Slime Molds

Excavata

The unicellular species that form the Excavata are named for the morphological feature that distinguishes them—an "excavated" feeding groove found on one side of the cell. Because some lack recognizable mitochondria, excavates were once thought to trace their ancestry to eukaryotes that existed before the origin of mitochondria. But researchers have found that excavates either have

(1) genes in their nuclear genomes that are of mitochondrial origin, or (2) unusual organelles that appear to be vestigial mitochondria. These observations support the hypothesis that the ancestors of excavates had mitochondria, but that these organelles were lost or reduced over time in this lineage. ✔ You should be able to mark the loss of the mitochondrion on Figure 30.7.

● Excavata > Parabasalids, Diplomonads, and Euglenids

Plantae

Biologists are beginning to use the name **Plantae** to refer to the monophyletic group that includes glaucophyte algae, red algae, green algae, and land plants.

The glaucophyte algae are unicellular or colonial. They live as plankton or attached to substrates in freshwater environments—particularly in bogs or swamps. Some glaucophyte species have flagella or produce flagellated spores, but sexual reproduction has never been observed in the group. The chloroplasts of glaucophytes have a distinct bright blue-green color.

All subgroups within Plantae are descended from a common ancestor that engulfed a cyanobacterium, beginning the endosymbiosis that led to the evolution of the chloroplast—their distinguishing morphological feature. This initial endosymbiosis probably occurred in an ancestor of today's glaucophyte algae. To support this hypothesis, biologists point to several important similarities between cyanobacterial cells and the chloroplasts that are found in the glaucophytes.

- Both have cell walls that contain peptidoglycan and that can be disrupted by the antibacterial compounds lysozyme and penicillin.

- The glaucophyte chloroplast also has a membrane outside its wall that is similar to the membrane found in Gram-negative bacteria (see Chapter 29).

Consistent with these observations, phylogenetic analyses of DNA sequence data place the glaucophytes as the sister group to all other members of the Plantae. ✔ You should be able to mark the origin of the plant chloroplast on Figure 30.7.

Green algae and land plants are described later (see Chapter 31); here we consider just the red algae in detail.

● Plantae > Red Algae

Rhizaria

The rhizarians are single-celled amoebae that lack cell walls, though some species produce elaborate shell-like coverings. They move by amoeboid motion and produce long, slender pseudopodia. Over 11 major subgroups in this lineage have been identified and named, including the planktonic organisms called actinopods—which synthesize glassy, silicon-rich skeletons—and the chlorarachniophytes, which obtained a chloroplast via secondary endosymbiosis and are photosynthetic. The best-studied and most abundant group is the foraminiferans. ✔ You should be able to mark the origin of the rhizarian amoeboid form on Figure 30.7.

● Rhizaria > Foraminiferans

Alveolata

Alveolates are distinguished by small sacs, called alveoli, that are located just under their plasma membranes. Although all members of this lineage are unicellular, the ciliates, the dinoflagellates, and the apicomplexans are remarkably diverse in morphology and lifestyle. ✔ You should be able to mark the origin of alveoli on Figure 30.7.

● Alveolata > Ciliates, Dinoflagellates, and Apicomplexans

Stramenopila (Heterokonta)

Stramenopiles are sometimes called heterokonts, which translates as "different hairs." At some stage of their life cycle, all stramenopiles have flagella that are covered with distinctive hollow "hairs." The structure of these flagella is unique to the stramenopiles (Figure 30.6). ✔ You should be able to mark the origin of the "hairy" flagellum on Figure 30.7. The lineage includes a large number of unicellular forms, although the brown algae are multicellular and include the world's tallest marine organisms, the kelp.

● Stramenopila > Water Molds, Diatoms, and Brown Algae

Amoebozoa > Plasmodial Slime Molds

The plasmodial slime molds got their name because individuals form a large, weblike structure that consists of a single cell containing thousands of diploid nuclei (**FIGURE 30.19**). Like water molds, the plasmodial slime molds were once considered fungi on the basis of their general morphological similarity (see Chapter 32).

Morphology The huge "supercell" form, with many nuclei in a single cell, occurs in few protists other than plasmodial slime molds. ✔ You should be able to mark the origin of the supercell on Figure 30.7.

Feeding and locomotion Plasmodial slime molds feed on microorganisms associated with decaying vegetation and move by amoeboid motion or cytoplasmic streaming.

Reproduction Some evidence suggests that when food becomes scarce, part of the amoeba forms a stalk topped by a ball-like structure in which nuclei undergo meiosis and form spores. The spores are then dispersed to new habitats by the wind or small animals. After spores germinate to form amoebae, two amoebae fuse to form a diploid cell that begins to feed and eventually grows into a supercell.

Human and ecological impacts Like cellular slime molds, plasmodial slime molds influence nutrient cycling by feeding on microorganisms. Some species are important model organisms for the study of cell biology and the origin of multicellularity.

Physarum polycephalum

Single cell with many nuclei

5 mm

FIGURE 30.19 Plasmodial Slime Molds Are Important Decomposers in Forests.

The lineage Excavata includes free-living and symbiotic species, and some that are parasites. All are single-celled, and many lack functional mitochondria.

All of the 300 known species of parabasalids live inside animals; some live only in the guts of termites, where they aid in the digestion of cellulose. The relationship between termites and parabasalids is considered mutualistic, because both parties benefit from the symbiosis.

About 100 species of diplomonads have been named. Many live in the guts of animal species without causing harm to their host; other species live in stagnant-water habitats.

There are about 1000 known species of euglenids. Although most live in freshwater, a few are found in marine habitats.

Morphology Parabasalid cells have a distinctive rod of cross-linked microtubules that runs the length of the cell. The rod is attached to the basal bodies where a cluster of flagella arise (**FIGURE 30.20**). Diplomonad cells have two nuclei, which resemble eyes when viewed under the microscope (**FIGURE 30.21**). Each nucleus is associated with four flagella, for a total of eight flagella per cell. Euglenid cells have a unique system of interlocking proteins just inside the plasma membrane that stiffen and support the cell. Some euglenid cells have a light-sensitive "eyespot" and use flagella to swim toward light (**FIGURE 30.22**).

Feeding and locomotion Some euglenids have chloroplasts and perform photosynthesis, and some diplomonads are parasitic, but most other members of the Excavata feed by engulfing bacteria, archaea, and organic matter. Most swim using their flagella. ✔ You should be able to explain how the observation of ingestive feeding in euglenids relates to the hypothesis that this lineage gained chloroplasts via secondary endosymbiosis.

Giardia intestinalis

FIGURE 30.21 *Giardia* Causes Intestinal Infections in Humans.

Reproduction Nearly all members of the Excavata are known to reproduce only asexually. Sexual reproduction has been observed in just a few species of parabasalids.

Human and ecological impacts Infections of *Trichomonas*, a parabasalid, can sometimes cause reproductive tract problems in humans. However, members of this genus may also live in the gut or mouth of humans without causing harm. The diplomonad *Giardia intestinalis* is a common intestinal parasite in humans and causes giardiasis, or beaver fever. Euglenids are important components of freshwater plankton and food chains.

Trichomonas vaginalis

FIGURE 30.20 *Trichomonas* Causes the Sexually Transmitted Disease Trichomoniasis. The cell in this micrograph has been colorized.

Euglena velata

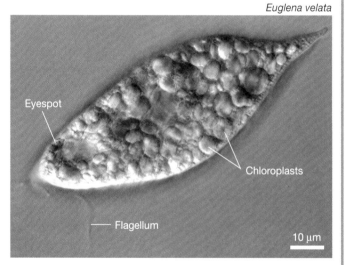

FIGURE 30.22 Euglenids Are Common in Ponds and Lakes.

The 6000 species of red algae live primarily in marine habitats. One species lives over 200 m below the surface; another is the only eukaryote capable of living in acidic hot springs. Although their color varies, many species are red because their chloroplasts contain large amounts of the accessory pigment phycoerythrin, which absorbs strongly in the blue and green portions of the visible spectrum. Because blue light penetrates water better than other wavelengths, red algae are able to live at considerable depth in the oceans. ✔ You should be able to mark the origin of high phycoerythrin concentrations on Figure 30.7.

Morphology Red algae cells have walls that are composed of cellulose and other polymers. A few species are unicellular, but most are multicellular. Many of the multicellular species are filamentous; others grow as thin, hard crusts on rocks or coral. Some species have erect, leaf-like structures called thalli (**FIGURE 30.23**).

Feeding and locomotion The vast majority of red algae are photosynthetic, though a few parasitic species have been identified. Red algae are the only type of algae that lack flagella.

Reproduction Asexual reproduction occurs through production of spores by mitosis. Alternation of generations is common, but the types of life cycles observed in red algae are extremely variable.

Human and ecological impacts On coral reefs, some red algae become encrusted with calcium carbonate. These species contribute to reef building and help stabilize the entire reef structure. Cultivation of *Porphyra* (or nori) for sushi and other foods is a billion-dollar-per-year industry in East Asia. Microbiological agar is also derived from the cell walls of various species of red algae.

Mesophyllum lichenoides

1 cm

FIGURE 30.23 Red Algae Adopt an Array of Growth Forms.

Foraminiferans, or forams, got their name from the Latin *foramen*, meaning "hole." Forams produce shells that have holes (see **FIGURE 30.24**) through which pseudopodia emerge. ✔ You should be able to mark the origin of the foram shell on Figure 30.7. Fossil shells of foraminifera are abundant in marine sediments—a continuous record of fossilized forams dates back 530 million years. A foram species was recently found living in sediments at a depth of 11,000 m below sea level. They are abundant in marine plankton as well as bottom habitats.

Morphology Foraminiferan cells generally have multiple nuclei. The shells of forams are usually made of organic material stiffened with calcium carbonate ($CaCO_3$), and most species have several chambers. One species known from fossils was 12 cm long, but most species are much smaller. The size and shape of the shell are traits that distinguish foram species from each other.

Feeding and locomotion Like other rhizarians, forams feed by extending their pseudopodia and using them to capture and engulf bacterial and archaeal cells or bits of organic debris, which are digested in food vacuoles. Some species have symbiotic algae that perform photosynthesis and contribute sugars to their host. Forams do not move actively but simply float in the water.

Reproduction Members of the foraminifera reproduce asexually by mitosis. When meiosis occurs, the resulting gametes are released into the open water, where pairs fuse to form new diploid individuals.

Human and ecological impacts The shells of dead forams commonly form extensive sediment deposits when they settle out of the water, producing layers that eventually solidify into chalk, limestone, or marble. Geologists use the presence of certain foram species to date rocks—particularly during petroleum exploration.

Globigerinoides ruber

0.1 mm

FIGURE 30.24 Forams Are Shelled Amoebae.

Members of the lineage Alveolata comprise three subgroups of single-celled species that are diverse in morphology and nutrition. All alveolates contain flattened, membrane-bound vesicles called alveoli that support the plasma membrane.

Ciliates are named for the cilia that cover them. Some 12,000 species are known from freshwater, marine, and wet soil environments. Most of the 4000 known species of dinoflagellates are marine or freshwater plankton. All of the 5000 known species of apicomplexans are parasitic.

Morphology Ciliate cells have two distinctive nuclei: a polyploid macronucleus that is actively transcribed and a diploid micronucleus that is involved only in gene exchange between individuals (both are stained dark red in **FIGURE 30.25**). ✔ You should be able to mark the origin of the macronucleus/micronucleus structure on Figure 30.7.

Each dinoflagellate species has a distinct shape, in some species maintained by cellulose plates. Apicomplexan cells lack cilia or flagella.

Feeding and locomotion Ciliates may be filter feeders, predators, or parasites. They use cilia to swim and have a mouth area where food is ingested. About half of the dinoflagellates are photosynthetic. They are distinguished by the arrangement of their two flagella, which allows them to swim in a spinning motion: One flagellum projects out from the cell while the other runs around the cell. Some dinoflagellate species are capable of **bioluminescence,** meaning they emit light via an enzyme-catalyzed reaction (**FIGURE 30.26**). Being parasitic, all apicomplexans absorb nutrition directly from their host. Some species can move by amoeboid motion.

Reproduction Ciliates undergo an unusual type of gene exchange called conjugation, in which micronuclei are passed between individuals. Both asexual and sexual reproduction occur in dinoflagellates. In some apicomplexan species, the life cycle involves two distinct hosts, and cells must be transmitted from one host to the next.

Human and ecological impacts Ciliates are abundant marine plankton and are also common in the digestive tracts of mammalian grazers, where they feed on plant matter and help the host animal digest it. Photosynthetic dinoflagellates are important primary producers in marine ecosystems. A few species are responsible for harmful algal blooms (see Figure 30.4). Apicomplexan species in the genus *Plasmodium* cause malaria. *Toxoplasma* is an important pathogen in people infected with HIV (**FIGURE 30.27**).

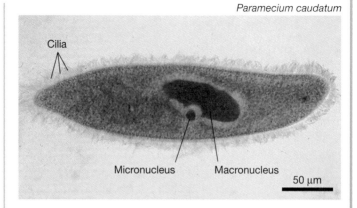
Paramecium caudatum

Cilia
Micronucleus Macronucleus
50 μm

FIGURE 30.25 Ciliates Are Abundant in Freshwater Plankton.

Noctiluca scintillans

Light produced by cells
0.5 mm

FIGURE 30.26 Some Dinoflagellate Species Are Bioluminescent.

Toxoplasma gondii

Remnants of dead host cell
Toxoplasma cells
5 μm

FIGURE 30.27 *Toxoplasma* **Causes Infections in AIDS Patients.**

The stramenopiles are a large and morphologically diverse group of eukaryotes. They include parasitic, saprophytic, and photosynthetic species, and they range in size from tiny, single-celled members to some of the largest multicellular organisms in the oceans.

Several morphological features are shared among the subgroups of stramenopiles, including flagella that are covered with tiny, hollow projections (see Figure 30.6). Photosynthetic diatoms and brown algae contain chloroplasts that originated from a secondary endosymbiosis involving a red alga (see Figure 30.13).

Morphology Some water molds are unicellular and some form long, branching filaments called **hyphae** (**FIGURE 30.28**). Most diatoms are unicellular (**FIGURE 30.29**), often in a box-and-lid arrangement (see Figure 30.11a). All species of brown algae are multicellular; the body typically consists of leaflike blades, a stalk known as a stipe, and a rootlike holdfast (**FIGURE 30.30**). ✔ You should be able to mark the origin of brown algal multicellularity on Figure 30.7.

Feeding and locomotion Most species of water molds feed on decaying organic material; a few are parasitic. Mature individuals are **sessile**—that is, permanently fixed to a substrate. Both diatoms and brown algae are photosynthetic, but diatoms are planktonic whereas brown algae are sessile.

Reproduction Many species of stramenopiles exhibit diploid-dominant life cycles. Several produce spores via asexual or sexual reproduction. Water mold spores form in special structures, like those shown in Figure 30.28. Most species of brown algae exhibit alternation of generations.

Human and ecological impacts Water molds are extremely important decomposers in freshwater ecosystems. Parasitic forms that affect humans indirectly include the organism that caused the Irish potato famine. Other water mold species are responsible for epidemic diseases of trees, including the diebacks currently occurring in oaks.

Diatoms are considered the most important primary producer in fresh and salt water. Their glassy cell walls settle into massive accumulations that are mined and sold as diatomaceous earth, which is used in filtering applications and as an ingredient in polishes, paint, cosmetics, and other products.

In many coastal areas with cool water, brown algae form kelp forests that are important habitats for a wide variety of animals.

Phytophthora infestans

FIGURE 30.28 *Phytophthora Infestans* **Infects Potatoes.** This is a colorized scanning electron micrograph.

Isthmia nervosa

FIGURE 30.29 Diatoms Have Glass-Like Cell Walls.

Durvillaea potatorum

FIGURE 30.30 Many Brown Algae Have a Holdfast, Stalk (Stipe), and Leaflike Blades.

If you understand . . .

30.1 Why Do Biologists Study Protists?

- Protists include all eukaryotes except the land plants, fungi, and animals.

- Protists are often tremendously abundant in marine and freshwater plankton and other habitats. Protists provide food for many organisms in aquatic ecosystems and fix so much carbon that they have a large impact on the global carbon budget.

- Parasitic protists cause several important diseases in humans, including malaria.

- Toxin-producing protists that grow to high densities can result in a harmful algal bloom.

✔ You should be able to explain what is "primary" about primary production by photosynthetic protists.

30.2 How Do Biologists Study Protists?

- Microscopic investigation of protists led to the understanding that many are single celled. Unlike bacteria, they contain mitochondria, a nucleus and endomembrane system, and a cytoskeleton.

- Direct sequencing has revealed important new groups of protists.

- Phylogenies based on molecular data confirmed that there are seven major lineages of protists and that the group is paraphyletic.

✔ You should be able to describe the major differences between eukaryotic cells and prokaryotic cells.

30.3 What Themes Occur in the Diversification of Protists?

- The first eukaryote probably resulted from an ancient endosymbiosis that occurred when a prokaryotic host cell (likely an archaeal cell) engulfed an α-proteobacterium. This bacterium evolved into today's mitochondria.

- The nuclear envelope probably evolved to spatially separate the processes of transcription and translation.

- Multicellularity evolved in several different protist groups independently.

- Protists vary widely in the way they obtain food. They exhibit predatory, parasitic, or photosynthetic lifestyles, which evolved in many groups independently.

- The evolution of ingestive feeding was important for two reasons: (1) It allowed eukaryotes to obtain resources in a new way—by eating bacteria, archaea, and other eukaryotes; and (2) it made endosymbiosis and the evolution of mitochondria and chloroplasts possible.

- The chloroplast's size and its circular chromosome, ribosomes, and double membrane are consistent with the hypothesis that this organelle originated as an endosymbiotic cyanobacterium.

- After primary endosymbiosis occurred, chloroplasts were "passed around" to new lineages of protists via secondary endosymbiosis.

- Protists vary widely in the way they reproduce. They undergo cell division based on mitosis and reproduce asexually. Many protists also undergo meiosis and sexual reproduction at some phase in their life cycle.

- Alternation of generations is common in multicellular protist species—meaning the same species has separate haploid and diploid forms. When alternation of generations occurs, haploid gametophytes produce gametes by mitosis; diploid sporophytes produce spores by meiosis.

✔ You should be able to explain the original source of (1) the two membranes in a plant chloroplast and (2) the four membranes in the chloroplast-derived organelle of *Plasmodium*.

30.4 Key Lineages of Protists

- Protists are a highly diverse group of eukaryotic species organized into seven lineages.

- Protist lineages are defined by DNA sequence evidence and morphological traits such as the presence of chloroplasts or unique flagella.

- A wide range of traits including parasitism, autotrophy, and multicellularity evolved independently in several different lineages of protists.

MasteringBiology

1. **MasteringBiology Assignments**

   Tutorials and Activities Alternation of Generations in a Protist; Life Cycles of Protists; Tentative Phylogeny of Eukaryotes

   Questions Reading Quizzes, Blue-Thread Questions, Test Bank

2. **eText** Read your book online, search, take notes, highlight text, and more.

3. **The Study Area** Practice Test, Cumulative Test, BioFlix® 3-D Animations, Videos, Activities, Audio Glossary, Word Study Tools, Art

You should be able to . . .

1. Why are protists considered paraphyletic?
   a. They include many extinct forms, including lineages that no longer have any living representatives.
   b. They include some but not all descendants of their most recent common ancestor.
   c. They represent all of the descendants of a single common ancestor.
   d. Not all protists have all of the synapomorphies that define the Eukarya, such as a nucleus.

2. What material is *not* used by protists to manufacture hard outer coverings?
   a. cellulose
   b. lignin
   c. glass-like compounds that contain silicon
   d. mineral-like compounds such as calcium carbonate ($CaCO_3$)

3. What does amoeboid motion result from?
   a. interactions among actin, myosin, and ATP
   b. coordinated beats of cilia
   c. the whiplike action of flagella
   d. action by the mitotic spindle, similar to what happens during mitosis and meiosis

4. According to the endosymbiosis theory, what type of organism is the original ancestor of the chloroplast?
   a. a photosynthetic archaean
   b. a cyanobacterium
   c. an algal-like, primitive photosynthetic eukaryote
   d. a modified mitochondrion

5. True or False: All protists are unicellular.

6. The most important primary producers in marine ecosystems are _____.

7. Which of the following is true of the parasite that causes malaria in humans?
   a. It undergoes meiosis immediately following zygote formation.
   b. It is a complex prokaryote.
   c. It makes gametes by meiosis.
   d. It has a diploid-dominant life cycle.

8. Explain the logic behind the claim that the nuclear envelope is a synapomorphy that defines eukaryotes as a monophyletic group.

9. Consider the endosymbiosis theory for the origin of the mitochondrion. What did each partner provide the other, and what did each receive in return?

10. Why was finding a close relationship between mitochondrial DNA and bacterial DNA considered particularly strong evidence in favor of the endosymbiosis theory?

11. The text claims that the evolutionary history of protists can be understood as a series of morphological innovations that established seven distinct lineages, each of which subsequently diversified based on innovative ways of feeding, moving, and reproducing. Explain how the Alveolata support this claim.

12. Multicellularity is defined in part by the presence of distinctive cell types. At the cellular level, what does this criterion imply?
    a. Individual cells must be extremely large.
    b. The organism must be able to reproduce sexually.
    c. Cells must be able to move.
    d. Different cell types express different genes.

13. Consider the following:
    • *Plasmodium* has an unusual organelle called an apicoplast. Recent research has shown that apicoplasts are derived from chloroplasts via secondary endosymbiosis and have a large number of genes related to chloroplast DNA.
    • Glyphosate is one of the most widely used herbicides. It works by poisoning an enzyme located in chloroplasts.
    • Biologists are testing the hypothesis that glyphosate could be used as an antimalarial drug in humans.
    How are these observations connected?

14. Suppose a friend says that we don't need to worry about the rising temperatures associated with global climate change. Her claim is that increased temperatures will make planktonic algae grow faster and that carbon dioxide (CO_2) will be removed from the atmosphere faster. According to her, this carbon will be buried at the bottom of the ocean in calcium carbonate shells. As a result, the amount of carbon dioxide in the atmosphere will decrease and global warming will decline. Comment.

15. Biologists are beginning to draw a distinction between "species trees" and "gene trees." A species tree is a phylogeny that describes the actual evolutionary history of a lineage. A gene tree, in contrast, describes the evolutionary history of one particular gene, such as a gene required for the synthesis of chlorophyll *a*. In some cases, species trees and gene trees don't agree with each other. For example, the species tree for green algae indicates that their closest relatives are protists and plants. But the gene tree based on chlorophyll *a* from green algae suggests that this gene's closest relative is a bacterium, not a protist. What's going on? Why do these types of conflicts exist?

31 Green Algae and Land Plants

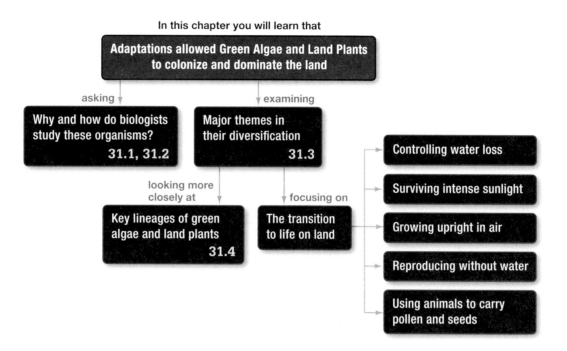

In this chapter you will learn that

Adaptations allowed Green Algae and Land Plants to colonize and dominate the land

asking ↓

Why and how do biologists study these organisms? 31.1, 31.2

examining ↓

Major themes in their diversification 31.3

looking more closely at ↓

Key lineages of green algae and land plants 31.4

focusing on ↓

The transition to life on land

- Controlling water loss
- Surviving intense sunlight
- Growing upright in air
- Reproducing without water
- Using animals to carry pollen and seeds

Earth has been called the Blue Planet, but it would be just as accurate to call it the Green Planet.

n terms of their total mass and their importance to other organisms, the green plants dominate terrestrial and freshwater habitats. When you walk through a forest or meadow, you are surrounded by green plants. If you look at pond or lake water under a microscope, green plants are everywhere.

The green plants comprise two major types of organisms: the green algae and the land plants. Green algae are important photosynthetic organisms in aquatic habitats—particularly lakes, ponds, and other freshwater settings. Land plants are the key photosynthesizers in terrestrial environments.

✓ When you see this checkmark, stop and test yourself. Answers are available in Appendix A.

Although green algae have traditionally been considered protists, it is logical to study them along with land plants for two reasons: (1) They are the closest living relatives to land plants and form a monophyletic group with them, and (2) the transition from aquatic to terrestrial life occurred when land plants evolved from green algae.

Land plants were the first organisms that could thrive with their tissues completely exposed to the air instead of being partially or completely submerged. Before they evolved, likely the only life on the continents consisted of bacteria, archaea, and single-celled protists that thrive in wet soils. By colonizing the continents, plants transformed the nature of life on Earth. It was, in the words of plant evolutionist Karl Niklas, "One of the greatest adaptive events in the history of life." Land plants made the Earth green.

31.1 Why Do Biologists Study the Green Algae and Land Plants?

Biologists study the green plants because they are fascinating, and because we couldn't live without them. Along with most other animals and fungi, humans are almost completely dependent on land plants for food. People rely on plants for other necessities of life as well—oxygen, fuel, the fibers used to make clothing and other products, building materials, and pharmaceuticals. But we also rely on land plants for important intangible value such as the aesthetic appeal of landscaping and bouquets. To drive this point home, consider that the sale of cut flowers generates over $1 billion each year in the United States alone.

Based on these observations, it is not surprising that agriculture, forestry, and horticulture are among the most important endeavors supported by biological science. Tens of thousands of biologists are employed in research designed to better understand plants, to increase their productivity, and to create new ways of using them for the benefit of people. Research programs also focus on two types of land plants that cause problems for people: weeds that decrease the productivity of crop plants, and newly introduced species that invade and then degrade natural areas.

Plants Provide Ecosystem Services

An **ecosystem** consists of all the organisms in a particular area, along with physical components of the environment such as the atmosphere, precipitation, surface water, sunlight, soil, and nutrients. Green algae and land plants provide **ecosystem services** because they enhance the life-supporting attributes of the atmosphere, surface water, soil, and other physical components of an ecosystem. Green plants alter the environment in ways that benefit many other organisms.

Plants Produce Oxygen Recall that plants perform oxygenic (literally, "oxygen-producing") photosynthesis (see Chapter 10). In this process, electrons that are removed from water molecules are used to reduce carbon dioxide (CO_2). In the process of

stripping electrons from water, plants release oxygen molecules (O_2) as a by-product.

Oxygenic photosynthesis evolved in cyanobacteria and was responsible for the origin of an oxygen-rich atmosphere (see Chapter 29). The evolutionary success of plants continued this trend because plants add huge amounts of oxygen to the atmosphere.

Without the green plants, we and other aerobic organisms would be in danger of suffocating for lack of oxygen.

Plants Build and Hold Soil Fallen leaves, roots, and stems that are not eaten when they are alive provide food for worms, fungi, bacteria, archaea, protists, and other decomposers in the soil. These organisms add organic matter to the soil, which changes the soil structure and the ability of soils to hold nutrients and water. These changes are beneficial to other organisms.

The extensive network of fine roots produced by trees, grasses, and other land plants helps hold soil particles in place. And by taking up nutrients in the soil, plants prevent the nutrients from being blown or washed away.

When areas are devegetated by grazing, farming, logging, or suburbanization, large quantities of soil and nutrients are lost to erosion by wind and water (**FIGURE 31.1**).

Plants Hold Water and Moderate Climate Plant tissues take up and retain water. Intact forests, prairies, and wetlands also prevent rain from quickly running off a landscape: Plant leaves soften the physical impact of rainfall on soil, and plant organic matter builds the soil's water-holding capacity.

By providing shade, plants reduce temperatures beneath them and increase relative humidity. They also reduce the impact of winds that dry out landscapes.

When plants are removed from landscapes to make way for farms or suburbs, habitats become much drier and are subject to more extreme temperature swings.

Plants are Primary Producers Land plants are the dominant primary producers in terrestrial ecosystems. Primary producers convert energy in sunlight into chemical energy (see Chapter 30). The

FIGURE 31.1 Devegetation Leads to Erosion.

sugars that land plants produce by photosynthesis support virtually all of the other organisms present in terrestrial habitats.

Plants are eaten by herbivores ("plant-eaters"), which range in size from insects to elephants. These consumers are eaten by carnivores ("meat-eaters"), ranging in size from the tiniest spiders to polar bears. Humans are an example of omnivores ("all-eaters"), organisms that eat both plants and animals. Omnivores feed at several different levels in the food chain. For example, people consume plants, herbivores such as chicken and cattle, and carnivores such as salmon and tuna.

Because of their role in primary production, land plants are the key to the carbon cycle on the continents. Plants take CO_2 from the atmosphere and reduce it to make sugars. Although green algae and land plants also produce a great deal of CO_2 as a result of cellular respiration, they fix much more CO_2 than they release.

The loss of plant-rich prairies and forests due to suburbanization or conversion to agriculture has contributed to increased concentrations of CO_2 in the atmosphere. Higher carbon dioxide levels, in turn, are partly responsible for the rising temperatures associated with global climate change (see Chapter 56).

Plants Provide Humans with Food, Fuel, Fiber, Building Materials, and Medicines

It is difficult to overstate the importance of plant research to the well-being of human societies. Plants provide our food supply as well as a significant percentage of the fuel, fibers, building materials, and medicines that we use.

Food Agricultural research began with the initial domestication of crop plants, which occurred independently at several locations around the world starting about 12,000 years ago (**FIGURE 31.2a**). By actively selecting individuals with the largest and most nutritious seeds, leaves, or other plant parts year after year, our ancestors gradually changed the characteristics of certain wild species. This process is called **artificial selection** (see Chapter 25).

FIGURE 31.2b compares kernel size in modern maize and its wild ancestor—a grass called teosinte that is native to Central America. Artificial selection has been responsible for dramatic increases in the protein content of maize kernels over the past 100 years (see data in Chapter 1). A current focus in agricultural research is the improvement of crop varieties through genetic engineering (see Chapter 20).

Fuel For perhaps 100,000 years, wood burning was the primary source of energy used by all humans. Wood has now been replaced in the industrialized countries by fossil fuels, including coal and petroleum or natural gas. Each of these forms of energy, however, is derived from plants that either grew during our lifetime or when dinosaurs roamed the earth.

Starting in the mid-1800s, people in England, Germany, and the United States began to mine coal deposits that originally formed during the Carboniferous period some 359–299 million years ago. The coal fueled blast furnaces that smelted vast quantities of iron ore into steel and powered the steam engines that sent trains

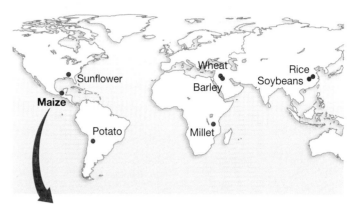

(a) Plants were domesticated at an array of locations.

(b) Artificial selection changes the traits of domesticated species.

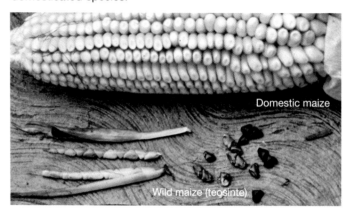

FIGURE 31.2 Crop Plants Are Derived from Wild Species via Artificial Selection. (a) Crop plants have originated on virtually every continent. **(b)** In some cases, artificial selection has changed domesticated species so radically that they bear little resemblance to their wild ancestors.

streaking across Europe and North America. It is no exaggeration to claim that the organic compounds synthesized by plants more than 300 million years ago fueled the industrial revolution.

Coal still supplies about 20 percent of the energy used in Japan and Western Europe and 80 percent of the energy used in China. Current research is focused on finding cleaner and more efficient ways of mining and burning coal.

Fiber and Building Materials Although nylon and polyester derived from petroleum are increasingly important in manufacturing, cotton and other types of plant fibers are still important sources of raw material for clothing, rope, and household articles like towels and bedding.

Woody plants also provide lumber for houses and furniture. Relative to its density, wood is a stiffer and stronger building material than concrete, cast iron, aluminum alloys, or steel.

Woody plants also provide most of the fibers used in papermaking. The cellulose fibers obtained from trees or bamboo and then used in paper manufacturing are stronger under tension (pulling) than nylon, silk, chitin, collagen, tendon, or bone—even though cellulose is 25 percent less dense. One line

TABLE 31.1 **Some Drugs Derived from Land Plants**

Source	Compound	Use
Belladonna plant	Atropine	Dilating pupils during eye exams
Opium poppy	Codeine, morphine	Pain relief, cough suppressant (codeine)
Foxglove	Digitalin	Heart medication
Ipecac	Ipecac	Treating amoebic dysentery, poison control
Peppermint	Menthol	Cough suppressant, relief of stuffy nose
Papaya	Papain	Reducing inflammation, treating wounds
Quinine tree	Quinine	Malaria prevention
	Quinidine	Heart medication
Aspen, willow trees	Salicin	Pain relief (aspirin)
Wild yams	Steroids	Precursor compounds for manufacture of birth control pills and cortisone (to treat inflammation)
Pacific yew	Taxol	Treating ovarian cancer
Curare vine	Tubocurarine	Muscle relaxant used in surgery
Rosy periwinkle	Vinblastine, vincristine	Treating leukemia (cancer of blood)

of research is focused on bioengineering to reduce lignin content of certain woody species, so fewer pollutants are generated in extracting lignin during the papermaking process.

Pharmaceuticals In both traditional and modern medicine, plants are a key source of drugs. **TABLE 31.1** lists some of the more familiar medicines derived from land plants; overall, an estimated 25 percent of the prescriptions written in the United States each year include at least one molecule derived from plants.

In most cases, plants synthesize these compounds in order to repel insects, deer, or other types of herbivores. For example, experiments have confirmed that morphine, cocaine, nicotine, caffeine, and other toxic compounds found in plants are effective deterrents to herbivores. Researchers continue to isolate and test new plant compounds for medicinal use in humans and domesticated animals.

31.2 How Do Biologists Study Green Algae and Land Plants?

Given the importance of plants to the planet in general and humans in particular, it is not surprising that knowing as much as possible about plants, including how they evolved, is a key component of biological science. To understand how green plants

originated and diversified, biologists analyze (**1**) morphological traits, (**2**) the fossil record, and (**3**) phylogenetic trees estimated from similarities and differences in DNA sequences from homologous genes and whole genomes.

These approaches are complementary and have produced a remarkably clear picture of how land plants evolved from green algae and then diversified. Let's consider each of these research strategies.

Analyzing Morphological Traits

The green algae include species that are unicellular, colonial, or multicellular and that live in marine, freshwater, or moist terrestrial habitats. The vast majority are aquatic. Although some land plants also live in ponds or lakes or rivers, the vast majority live on land.

Similarities Between Green Algae and Land Plants The green algae have long been hypothesized to be closely related to land plants, because key traits are similar in the two groups.

- Their chloroplasts contain the photosynthetic pigments chlorophyll *a* and *b* and the accessory pigment β-carotene.

- They have similar arrangements of the internal, membrane-bound sacs called thylakoids (see Chapter 10).

- Their cell walls, sperm, and peroxisomes are similar in structure and composition. (Recall that peroxisomes are organelles in which specialized oxidation reactions take place; see Chapter 7.)

- Their chloroplasts synthesize starch as a storage product.

Of all the green algal groups, the two most similar to land plants are the Coleochaetophyceae (coleochaetes) and Charophyceae (stoneworts; **FIGURE 31.3**). Because the species that make up these groups are multicellular and live in ponds and other types of freshwater environments, biologists hypothesize that land plants evolved from multicellular green algae that lived in freshwater habitats.

Major Morphological Differences among Land Plants Based on morphology, the land plants were traditionally clustered into three broad categories:

1. **Nonvascular plants** (also called bryophytes) lack **vascular tissue**—specialized groups of cells that conduct water and

Coleochaetophyceae (coleochaetes) Charophyceae (stoneworts)

FIGURE 31.3 Most Green Algae Are Aquatic. Examples of species from the green algal lineages closely related to the land plants.

(a) Nonvascular plants

5 mm

Do not have vascular tissue to conduct water and provide support (for example, mosses)

(b) Seedless vascular plants

Have vascular tissue but do not make seeds (for example, ferns)

(c) Seed plants

Have vascular tissue and make seeds (for example, flowering plants, or angiosperms)

FIGURE 31.4 Morphological Diversity in Land Plants.

nutrients from one part of the plant body to another. Mosses are one group of nonvascular plants (**FIGURE 31.4a**).

2. Seedless vascular plants have vascular tissue but do not make seeds. Instead, they make microscopic spores that are carried by wind to new habitats. Ferns are an example of seedless vascular plants (**FIGURE 31.4b**).

3. Seed plants have vascular tissue. A **seed** consists of an embryo and a store of nutritive tissue, surrounded by a tough protective layer. The flowering plants, or **angiosperms** (literally, "encased seeds"), are a group of seed plants (**FIGURE 31.4c**).

How are nonvascular plants, seedless vascular plants, and seed plants related to each other, and to green algae?

Using the Fossil Record

The first green plants in the fossil record are green algae, found in rocks from 700 to 725 million years old. The first land plant fossils are found in rocks that are about 475 million years old. Because green algae appear long before land plants, the data support the hypothesis that land plants are derived from green algae.

At roughly the same time that green algae appeared and began to diversify, the oceans and atmosphere were starting to become

oxygen rich—as never before in Earth's history. Based on this time correlation, it is reasonable to hypothesize that the evolution of green algae contributed to the rise of oxygen levels on Earth. The origin of the oxygen atmosphere occurred not long before the appearance of animals in the fossil record and may have played a role in their origin and early diversification.

The fossil record of the land plants themselves is massive. In an attempt to organize and synthesize the database, **FIGURE 31.5** breaks it into five time intervals—each encompassing a major event in the diversification of land plants.

Origin of Land Plants The oldest interval begins about 475 million years ago (mya), spans some 60 million years, and documents the origin of the group.

Most of the fossils dating from this period are microscopic. They consist of the reproductive cells called spores and sheets of a waxy coating called cuticle. Several observations support the hypothesis that these fossils came from green plants that were growing on land.

1. Cuticle is a watertight barrier that coats the aboveground parts of today's land plants and helps them resist drying.

2. The fossilized spores are surrounded by a sheetlike coating. Under the electron microscope, the material appears almost

FIGURE 31.5 The Fossil Record of Land Plants Can Be Broken into Five Major Intervals. Note that the first insects are found in the fossil record at about 400 mya, the first terrestrial vertebrate animals at 365 mya, and the first mammals at 190 mya.

identical in structure to a waxy material called **sporopollenin**. Sporopollenin also encases spores and pollen from modern land plants and helps them resist drying.

3. Fossilized spores that are 475 million years old have recently been found in association with spore-producing structures, called **sporangia** (singular: **sporangium**), that are similar to sporangia observed in some nonvascular plants.

Silurian–Devonian Explosion The second major interval in the fossil record of land plants is called the "Silurian–Devonian explosion." In rocks dated 416–359 mya, biologists find fossils from most of the major plant lineages. Virtually all of the adaptations that allow plants to occupy dry, terrestrial habitats are present, including water-conducting tissue and roots.

According to the fossil record, plants colonized the land in conjunction with fungi that grew in a mutually beneficial symbiosis. The fungi grew belowground and helped provide land plants with nutrients from the soil; in return, the plants provided the fungi with sugars and other products of photosynthesis (see Chapters 32 and 39).

The Carboniferous Period The third interval in the fossil history of plants spans the aptly named Carboniferous period. In sediments dated from about 359 to 299 mya, biologists find extensive deposits of coal. Coal is a carbon-rich rock packed with fossil spores, branches, leaves, and tree trunks.

Most of these fossils are derived from seedless vascular plants in the Lycophyta (the lycophytes or club mosses), Equisetophyta (horsetails), and Pteridophyta (ferns). Although the only living lycophytes and horsetails are small, during the Carboniferous these groups were species rich and included a wide array of tree-sized forms.

Because coal formation is thought to start only in the presence of water, the Carboniferous fossils indicate the presence of extensive forested swamps.

Diversification of Gymnosperms The fourth interval in land plant history is characterized by seed plants called **gymnosperms** ("naked-seeds"). Five major groups of gymnosperms are living today: Ginkgophyta (ginkgoes), Cycadophyta (cycads), Cupressophyta (redwoods, junipers, and yews), Pinophyta (pines, spruces, and firs), and Gnetophyta (gnetophytes).

Because gymnosperms grow readily in dry habitats, biologists infer that both wet and dry environments on the continents became blanketed with green plants for the first time during this interval. Gymnosperms are prominent in the fossil record from 299 mya to 145 mya.

Diversification of Angiosperms The fifth interval in the history of land plants is still under way. This is the age of flowering plants—the angiosperms. The first flowering plants in the fossil record appear about 150 mya. The plants that produced the first flowers are the ancestors of today's grasses, orchids, daisies, oaks, maples, and roses.

According to the fossil record, then, the green algae appear first, followed by the nonvascular plants, seedless vascular plants, and seed plants. Organisms that appear late in the fossil record are often less dependent on moist habitats than are groups that appear earlier. For example, the sperm cells of mosses and ferns swim to accomplish fertilization, while gymnosperms and angiosperms produce pollen grains that are transported via wind or insects and that then produce sperm.

These observations support the hypotheses that green plants evolved from green algae and that land plants evolved to colonize dry habitats.

To test the validity of these observations, biologists analyze data sets that are independent of the fossil record. Foremost among these are DNA sequences used to infer phylogenetic trees. Does the phylogeny of land plants confirm or contradict the patterns in the fossil record?

Evaluating Molecular Phylogenies

The phylogenetic tree in **FIGURE 31.6** is a recent version of results emerging from laboratories around the world. The black bars across some branches show when key innovations occurred.

Each of the following bullets states a key observation about this tree, followed by an interpretation—often, a hypothesis that is supported by the observation.

- The green plants are monophyletic.
  Interpretation: A single common ancestor gave rise to all of the green algae and land plants.

- Green algae are paraphyletic.
  Interpretation: The green algae include some but not all of the descendants of a single common ancestor.

- Charophyceae are the closest living relative to land plants.
  Interpretation: Land plants evolved from a multicellular green alga that lived in freshwater habitats.

check your understanding

If you understand that . . .

- Biologists use the fossil record and phylogenetic analyses to study how green plants diversified.
- The data analyzed to date support the hypotheses that green plants are monophyletic and that land plants evolved from multicellular green algae that inhabited freshwater.
- The fossil record and molecular analyses agree that the nonvascular plants evolved first, followed by the seedless vascular plants and the seed plants.

✔ **You should be able to . . .**

Explain why (1) morphological data, (2) the fossil record, and (3) molecular phylogenies all support the hypothesis that land plants evolved from green algae.

Answers are available in Appendix A.

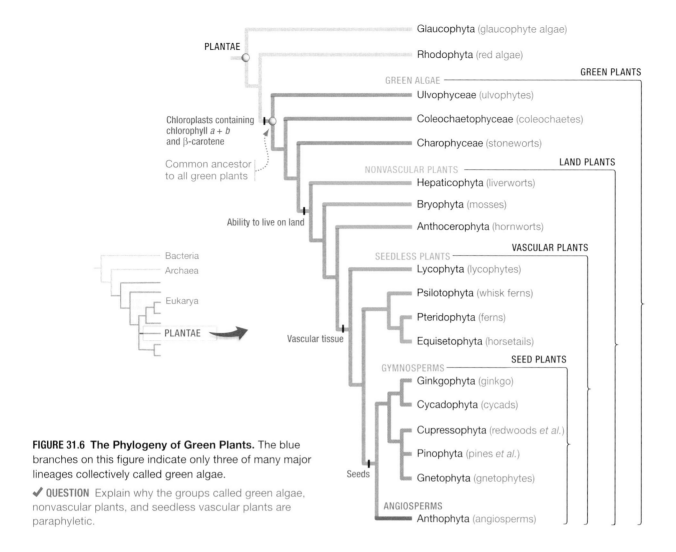

FIGURE 31.6 The Phylogeny of Green Plants. The blue branches on this figure indicate only three of many major lineages collectively called green algae.

✔ **QUESTION** Explain why the groups called green algae, nonvascular plants, and seedless vascular plants are paraphyletic.

- Land plants are monophyletic.
  Interpretation: There was only one successful transition from freshwater environments to land.

- The nonvascular plants are the earliest-branching groups among land plants.
  Interpretation: The nonvascular plants are the most ancient living group of land plants.

- The nonvascular plants are paraphyletic.
  Interpretation: The nonvascular plants include some but not all descendants of a single common ancestor.

- The seedless vascular plants are paraphyletic, but the vascular plants as a whole are monophyletic.
  Interpretation: Vascular tissue evolved once.

- The seed plants—the gymnosperms plus angiosperms—are monophyletic.
  Interpretation: The seed evolved once.

- The gymnosperms are a monophyletic group, as are the angiosperms.

Interpretation: Among seed plants, there was a major divergence in how seeds develop—either "naked" (in gymnosperms) or protected inside a capsule (in angiosperms).

Although the tree in Figure 31.6 will undoubtedly change and improve as additional data accumulate, biologists are confident about its most fundamental message: The fossil record and the phylogenetic tree based on DNA sequence data agree on the order in which groups appeared. Land plant evolution began with nonvascular plants, proceeded to seedless vascular plants, and continued with the evolution of seed plants.

31.3 What Themes Occur in the Diversification of Land Plants?

Land plants have evolved from algae that grew on the muddy shores of ponds 475 million years ago to organisms that enrich the soil, produce much of the oxygen you breathe and most of the food you eat, and serve as symbols of health, love, and beauty. How did this happen?

Answering this question begins with recognizing the most striking trend in the phylogeny and fossil record of green plants: The most ancient groups in the lineage are dependent on wet habitats, while the more recently evolved groups can live and reproduce in dry—or even desert—conditions. The story of land plants is the story of adaptations that allowed photosynthetic organisms to move from aquatic to terrestrial environments.

Let's first consider adaptations that allowed plants to grow in dry conditions without drying out and dying, and then analyze the evolution of traits that allowed plants to reproduce efficiently on land. This section closes with a brief look at the radiation of flowering plants, which are the most important plants in many of today's terrestrial environments.

The Transition to Land, I: How Did Plants Adapt to Dry Conditions with Intense Sunlight?

For aquatic green algae, terrestrial environments are deadly. Compared with a habitat in which the entire organism is bathed in water, in terrestrial environments only a portion, if any, of the plant's tissues are in direct contact with water. Tissues that are exposed to air tend to dry out and die.

Once green plants made the transition to survive out of water, though, growth on land offered a bonanza of resources.

- *Light* The water in ponds, lakes, and oceans absorbs and reflects light. As a result, the amount of light available to drive photosynthesis is drastically reduced even a meter or two below the water's surface.

- *Carbon dioxide* CO_2—the most important molecule required by photosynthetic organisms—is more abundant in the atmosphere and diffuses more readily there than it does in water.

Natural selection favored early land plants with adaptations that solved the drying problem. These adaptations arose in three steps: (1) preventing water loss, which kept cells from drying out and dying; (2) providing protection from harmful ultraviolet (UV) radiation; and (3) moving water from tissues with direct access to water to tissues without direct access. Let's examine each step in turn.

Preventing Water Loss: Cuticle and Stomata Section 31.2 points out that sheets of the waxy substance called cuticle are present early in the fossil record of land plants, along with encased spores. This observation is significant because the presence of cuticle in fossils is a diagnostic indicator of land plants.

Cuticle is a watertight sealant that covers the aboveground parts of plants and gives them the ability to survive in dry environments (**FIGURE 31.7a**). If biologists had to point to one innovation that made the transition to land possible, it would be the random mutations that led to the production of cuticle.

Covering surfaces with wax creates a problem, however, regarding the exchange of gases across those surfaces. Plants need to take in carbon dioxide (CO_2) from the atmosphere in order to perform photosynthesis. But cuticle is almost as impervious to CO_2 as it is to water.

Most plants solve this problem with a structure called a **stoma** ("mouth"; plural: **stomata**). A stoma consists of an opening surrounded by specialized **guard cells** (**FIGURE 31.7b**). The opening, called a **pore,** opens or closes as the guard cells change shape. When guard cells lose water, they become flaccid or "limp" and the stomata close. Pores are normally closed at night to limit water loss from the plant when CO_2 uptake is not needed. But when

(a) Cuticle is a waxy layer that prevents water loss from stems and leaves.

Leaf cross section

Cuticle

Moist photosynthetic cells

25 μm

(b) Stomata have pores that allow gas exchange in photosynthetic tissues.

Pore

Guard cell

Cuticle

Stoma

FIGURE 31.7 Cuticle and Stomata Are the Most Fundamental Plant Adaptations to Life on Land. In these micrographs, leaf cells have been stained blue to make their structure more visible. **(a)** The interior of plant leaves and stems is extremely moist; cuticle (stained red) prevents water from evaporating away. **(b)** Stomata, which are opened and closed by guard cells, create pores to allow CO_2 to diffuse into the interior of leaves and stems where cells are actively photosynthesizing. Stomata are often on the underside of leaves.

guard cells absorb water and then become turgid or "taut," they open the pore. Open stomata allow CO_2 to diffuse into the interior of leaves and stems where cells are actively photosynthesizing. (The mechanism behind guard-cell movement is explored in Chapter 40.)

Stomata are present in all land plants except the liverworts, which have pores but no guard cells. These data suggest that the earliest land plants evolved pores that allowed gas exchange to occur at breaks in the cuticle-covered surface. Later, the evolution of guard cells gave land plants the ability to regulate gas exchange—and control water loss—by opening and closing their pores.

Providing Protection From UV Irradiation

Life out of water gave land plants a distinct advantage. Because they were exposed to higher light intensities, the plants could carry out photosynthesis faster. However, out of water they were also exposed to the UV rays of the sun. UV light is known to damage DNA by causing thymine dimers (see Chapter 15). Water absorbs UV light, so algae did not face this problem to the same extent.

Researchers think that many species of algae probably colonized the wet soil near their home ponds, but only some were able to survive the harsh sunlight. The plants that survived were those that by chance made compounds that absorb UV light. Most plants today accumulate UV-absorbing compounds, called flavonoids, to protect their DNA from damage. These flavonoids are plant pigments that function as a sunscreen for leaves and stems. Humans use similar compounds to manufacture sunscreen.

The first plants had a cuticle to keep from drying out, and UV-absorbing compounds to act as sunscreen, so they could grow on the water-saturated soils of lake or pond edges. The next challenge? Defying gravity.

The Importance of Upright Growth

Multicellular green algae can grow erect because they float. They float because the density of their cells is similar to water's density. But outside of water, the body of a multicellular green alga collapses. The water that fills its cells is 1000 times denser than air. Although the cell walls of green algae are strengthened by the presence of cellulose, their bodies lack the structural support to withstand the force of gravity and to keep an individual erect in air.

Based on these observations, biologists hypothesize that the first land plants were small or had a low, sprawling growth habit. Besides lacking rigidity, the early land plants would have had to obtain water through pores or through cells that lacked cuticle—meaning they would have had to grow in a way that kept many or most of their tissues in direct contact with moist soil. If this hypothesis is correct, then competition for space and light would have become intense soon after the first plants began growing on land.

In a terrestrial environment, individuals that can grow erect have much better access to sunlight than individuals that are incapable of growing erect. But two problems have to be overcome for a plant to grow erect: (1) transporting water from tissues that are in contact with wet soil to tissues that are in contact with dry air, against the force of gravity; and (2) becoming rigid enough to avoid falling over in response to gravity and wind. As it turns out, vascular tissue solved both problems.

The Origin of Vascular Tissue

Paul Kenrick and Peter Crane explored the origin of water-conducting cells and erect growth in plants by examining the extraordinary fossils found in a rock formation in Scotland called the Rhynie Chert. These rocks formed about 400 million years ago and contain some of the first large plant specimens in the fossil record—as opposed to the microscopic spores and cuticle found in older rocks.

The Rhynie Chert contains numerous plants that fossilized in an upright position. This indicates that many or most of the Rhynie plants grew erect. How did they stay vertical?

By examining fossils with the electron microscope, Kenrick and Crane established that species from the Rhynie Chert contained elongated cells that were organized into tissues along the length of the plant. Based on these data, the biologists hypothesized that the elongated cells were part of water-conducting tissue and that water could move from the base of the plants upward to erect portions through these specialized water-conducting cells.

- Some of the fossilized water-conducting cells had simple, cellulose-containing cell walls like the water-conducting cells found in today's mosses (**FIGURE 31.8a**, see page 586).

- Some of the water-conducting cells present in the early fossils had cell walls with thickened rings containing a molecule called lignin (**FIGURE 31.8b**).

Lignin is a complex polymer built from six-carbon rings. It is extraordinarily strong for its weight and is particularly effective in resisting compressing forces such as gravity.

These observations inspired the following hypothesis: The evolution of lignified cell walls gave stem tissues the strength to remain erect in the face of wind and gravity. Today, the presence of lignin in the cell walls of water-conducting cells is considered the defining feature of vascular tissue. The evolution of vascular tissue allowed early plants to support erect stems and transport water from roots to aboveground tissues. (See Chapter 38 for additional details on the structure and function of plant vascular tissue.)

Elaboration of Vascular Tissue: Tracheids and Vessels

Once simple water-conducting tissues evolved, evolution by natural selection favored more complex tissues that were more efficient in providing support and transport.

In fossils that are about 380 million years old, biologists find the advanced water-conducting cells called tracheids. **Tracheids** are long, thin, tapering cells that have

- a thickened, lignin-containing **secondary cell wall** in addition to a cellulose-based **primary cell wall**; and

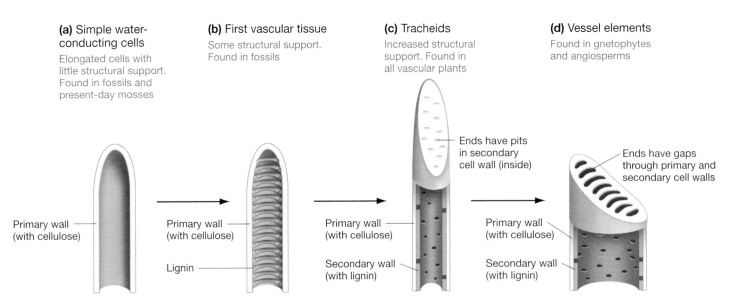

(a) Simple water-conducting cells
Elongated cells with little structural support. Found in fossils and present-day mosses

Primary wall (with cellulose)

(b) First vascular tissue
Some structural support. Found in fossils

Primary wall (with cellulose)

Lignin

(c) Tracheids
Increased structural support. Found in all vascular plants

Ends have pits in secondary cell wall (inside)

Primary wall (with cellulose)

Secondary wall (with lignin)

(d) Vessel elements
Found in gnetophytes and angiosperms

Ends have gaps through primary and secondary cell walls

Primary wall (with cellulose)

Secondary wall (with lignin)

FIGURE 31.8 Evolutionary Sequence Observed in Water-Conducting Cells. According to the fossil record and the phylogeny of green plants, water-conducting cells became stronger over time due to the evolution of lignin and secondary cell walls. Efficient water transport was maintained through pits where the secondary cell wall is missing, or gaps where both the secondary and primary cell wall are absent.

✔ **QUESTION** Biologists claim that vessels are more efficient than tracheids at transporting water, in part because vessels are shorter and wider than tracheids. Why does this claim make sense?

- pits in the sides and ends of the cell where the secondary cell wall is absent, where water can flow efficiently from one tracheid to the next (**FIGURE 31.8c**).

The secondary cell wall gave tracheids the ability to provide better structural support, but water could still move easily through the cells because of the pits. Today, all vascular plants contain tracheids.

In fossils dated to 250–270 million years ago, biologists have documented the most advanced type of water-conducting cells observed in plants. **Vessel elements** are shorter and wider than tracheids, and their upper and lower ends have gaps where both the primary and secondary cell wall are missing. The width of vessels and the presence of open gaps reduce resistance and make water movement more efficient (**FIGURE 31.8d**). In vascular tissue, vessel elements are lined up end to end to form a continuous pipelike structure.

In the stems and branches of some vascular plant species, tracheids or a combination of tracheids and vessels form the extremely strong support material called **wood.** (The anatomy of wood is explained in detail in Chapter 37.) The ability of plants to make lignified vascular tissues allowed them to grow tall and transport water from soil all the way to the top—a remarkable engineering feat.

Mapping Evolutionary Changes on the Phylogenetic Tree

Cuticle, stomata, and vascular tissue were key adaptations that allowed early plants to colonize and thrive on land. **FIGURE 31.9** summarizes how land plants adapted to dry conditions by mapping on the phylogenetic tree where major innovations occurred as the group diversified.

As you study the tree in Figure 31.9, note that fundamentally important adaptations to dry conditions—such as cuticle, pores, stomata, vascular tissue, and tracheids—evolved just once. But convergent evolution also occurred (see Chapter 28). When convergence occurs, similar traits evolve independently in two distinct lineages. Vessels evolved independently in gnetophytes and angiosperms.

The evolution of cuticle, stomata, and vascular tissue made it possible for plants to avoid drying out and to grow upright, while moving water from the base of the plant to its apex. Plants gained adaptations that allowed them not only to survive on land, but also to flourish. Now the question is, how did they reproduce?

The Transition to Land, II: How Do Plants Reproduce in Dry Conditions?

Life cycles of sexually reproducing eukaryotes, including plants, serve several different functions:

- increase genetic variability as a result of meiosis and fertilization;

- increase the number of individuals; and

- disperse individuals to new habitats.

Each of these functions is necessary if a species is to survive. To understand the role of each stage of a life cycle and how life cycles evolve, researchers ask a series of key questions. Is a species mobile or sessile? Does it live in water or on land?

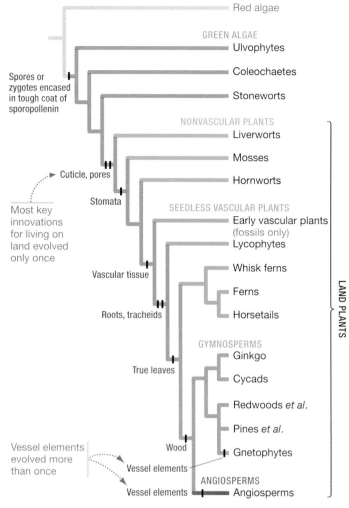

FIGURE 31.9 A Series of Evolutionary Innovations Allowed Plants to Adapt to Life on Land.

✔ **QUESTION** Explain the logic that biologists use to map the location of the innovations on phylogenetic trees.

In some organisms, all three functions are accomplished at one stage of the life cycle. For example, after meiosis coral polyps release many gametes into the water, where they fuse and disperse. Even though adult corals are sessile animals, they use water currents to move their genetically variable gametes and larvae. Terrestrial animals are mobile, so adults can easily disperse and join other adults for mating.

How do green plants reproduce? Keep in mind two important points—the impact of the transition from water to land on the dispersal stage of the life cycle, and the fact that land plants are sessile. Each of these factors has profoundly affected how life cycles of land plants evolved.

Section 31.2 introduces one of the key adaptations for reproducing on land: spores that resist drying because they are encased in a tough coat of sporopollenin. Sporopollenin-like compounds are found in the walls of some green algal zygotes, and thick-walled, sporopollenin-rich spores appear early in

the fossil record of land plants. Based on these observations, biologists infer that sporopollenin-encased spores were one of the innovations that made the initial colonization of land possible.

Two other innovations that occurred early in land plant evolution were instrumental for efficient reproduction in a dry environment: (1) Gametes were produced in complex, multicellular structures; and (2) the embryo was retained on the parent (mother) plant and was nourished by it.

Producing Gametes in Protected Structures The fossilized gametophytes of early land plants contain specialized reproductive organs called **gametangia** (singular: **gametangium**). Although members of the green algae group Charophyceae (stoneworts) also develop gametangia, the gametangia found in land plants are larger and more complex.

The evolution of an elaborate gametangium was important because it protected gametes from drying and from mechanical damage. Gametangia are present in all land plants living today except angiosperms, where structures inside the flower perform the same functions.

In both the Charophyceae and the land plants, individuals produce distinctive male and female gametangia (**FIGURE 31.10**, see page 588).

- The sperm-producing gametangium is called an **antheridium** (plural: **antheridia**).

- The egg-producing gametangium is called an **archegonium** (plural: **archegonia**).

In terms of their function, antheridia and archegonia are analogous to the testes and ovaries of animals.

Retaining Offspring: Land Plant Embryos Are Nourished by Their Parent The second innovation that occurred early in land plant evolution involved the eggs that formed inside archegonia. Instead of shedding their eggs into the water or soil, land plants retain them.

Eggs are also retained in the green algal lineages that are most closely related to land plants: In Charophyceae and other closely related groups, sperm swim to the egg, fertilization occurs, and the resulting zygote stays attached to the parent. Either before or after fertilization, the egg or zygote receives nutrients from the mother plant. But the parent plant dies each autumn as the temperature drops. The zygote remains on the dead parental tissue, settles to the bottom of the lake or pond, and overwinters. In spring, meiosis occurs, and the resulting spores develop into haploid adult plants.

In land plants, the zygote is also retained on the parent plant after fertilization. But in contrast to the zygotes of most green algae, the zygotes of all land plants begin to develop on the living parent plant, forming a multicellular embryo that remains attached to the parent and can be nourished by it. This is important because land plant embryos do not have to manufacture

FIGURE 31.10 In All Land Plant Groups but Angiosperms, Gametes Are Produced in Gametangia. Gametangia such as in this moss (*Polytrichium*) are complex, multicellular structures that protect developing gametes from drying and mechanical damage.

their own food early in life. Instead, they receive most or all of their nutrients from the parent plant.

The retention of the embryo was such a key event in land plant evolution that the formal name of the group is **Embryophyta**—literally, the "embryo-plants." The retention of the fertilized egg in **embryophytes** is analogous to pregnancy in mammals, where offspring are retained by the mother and nourished through the initial stages of growth. Thick-walled spores, elaborate gametangia, and the embryophyte condition weren't the only key innovations associated with reproducing on land, though. In addition, all land plants undergo the phenomenon known as **alternation of generations** (see Chapter 30).

Alternation of Generations When alternation of generations occurs, individuals represent a multicellular haploid phase or a multicellular diploid phase. The multicellular haploid stage is called the **gametophyte;** the multicellular diploid stage is called the **sporophyte**. The two phases of the life cycle are connected by distinct types of reproductive cells—gametes and spores.

Although alternation of generations is observed in a wide array of eukaryotic lineages and in some groups of green algae, it does not occur in the algal groups most closely related to land plants. In the coleochaetes and stoneworts, the multicellular form is haploid. Only the zygote is diploid. As **FIGURE 31.11** shows, the zygote undergoes meiosis to form haploid spores. After dispersing with the aid of flagella, the spores begin dividing by mitosis and eventually grow into adult, haploid individuals. You might recall that this haploid-dominant life cycle is common in protists (see Chapter 30).

These data suggest that alternation of generations originated in land plants independently of its evolution in other groups of eukaryotes, and that it originated early in their history—soon after they evolved from green algae. Why was this strategy successful?

Multicellular sporophytes can make many spores and easily disperse them through wind currents. Male gametes produced by multicellular gametophytes swim through moist soil or water droplets on the surface of the gametophyte. As you'll see below, when male gametes no longer needed to swim, there was no need for a free-living gametophyte stage in the life cycle.

Alternation of generations always involves the same basic sequence of events, illustrated in **FIGURE 31.12**. To review how this type of life cycle works, put your finger on the sporophyte in the figure and trace the cycle clockwise to find the following five key events in a land plant:

1. The sporophyte produces spores by meiosis. Spores are haploid.

2. Spores germinate and divide by mitosis to develop into multicellular, haploid gametophytes.

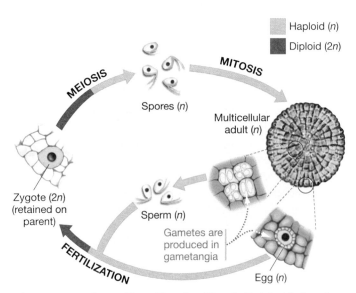

FIGURE 31.11 In Green Algae That Are Closely Related to Land Plants, Only the Zygote Is Diploid. The coleochaetes and stoneworts do not have alternation of generations. The multicellular stage is haploid.

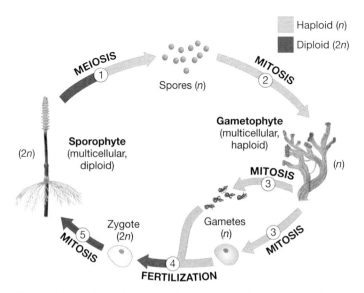

FIGURE 31.12 All Land Plants Undergo Alternation of Generations. Alternation of generations always involves the same sequence of five events.

Haploid (n)
Diploid (2n)

Spores (n)

Sporophyte
(multicellular,
diploid)

Gametophyte
(multicellular,
haploid)

Gametes
(n)

Zygote
(2n)

MEIOSIS
1

MITOSIS
2

MITOSIS
3

MITOSIS
3

FERTILIZATION
4

MITOSIS
5

- Zygotes and spores are both single cells that divide by mitosis to form a multicellular individual. Zygotes develop into sporophytes; spores develop into gametophytes.
- Zygotes are diploid and spores are haploid.
- Zygotes result from the fusion of two haploid cells, such as a sperm and an egg, but spores are not formed by the fusion of gametes.
- Spores are produced by meiosis inside structures called sporangia; gametes are produced by mitosis inside gametangia.

Alternation of generations can be a difficult topic to master, for two reasons: (**1**) It is unfamiliar because it does not occur in humans or other animals, and (**2**) gamete formation results from mitosis—not meiosis, as it does in animals (see Chapter 13).

✔If you understand the basic principles of alternation of generations, you should be able to describe this type of life cycle and explain one hypothesis for why this life cycle might have evolved in land plants.

The Gametophyte-Dominant to Sporophyte-Dominant Trend in Life Cycles

The five steps illustrated in Figure 31.12 occur in all species with alternation of generations. But in land plants, the relationship between the gametophyte and sporophyte is highly variable.

In nonvascular plants such as mosses, the sporophyte is small and short lived and is largely dependent on the gametophyte for nutrition (**FIGURE 31.13**). When you see leafy-looking mosses growing on a tree trunk or on rocks, you are looking at gametophytes. Because the gametophyte is long lived and produces most of the food required by the individual, it is considered the dominant part of the life cycle.

3. Gametophytes produce gametes by mitosis. Both the gametophyte and the gametes are haploid, but gametophytes are multicellular while gametes are unicellular.

4. Two gametes unite during fertilization to form a diploid zygote.

5. The zygote divides by mitosis and develops into a multicellular, diploid sporophyte.

Once you've traced the cycle successfully, take a moment to compare and contrast zygotes and spores.

FIGURE 31.13 Gametophyte-Dominated Life Cycles Evolved Early. Like today's mosses, the earliest land plants in the fossil record have gametophytes that are much larger and longer lived than the sporophyte. The sporophyte depends on the gametophyte for nutrition.

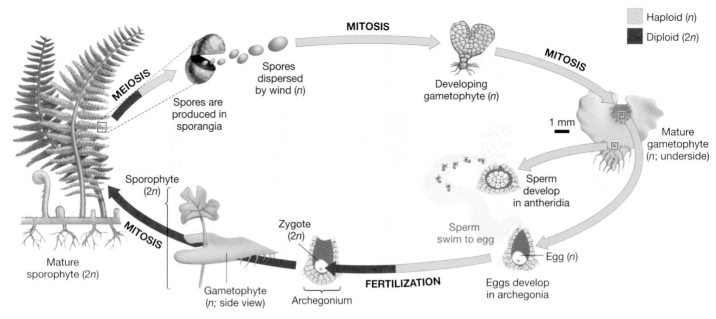

Haploid (n)

Diploid (2n)

MITOSIS

Spores dispersed by wind (n)

MITOSIS

Developing gametophyte (n)

MITOSIS

MEIOSIS

Spores are produced in sporangia

1 mm

Mature gametophyte (n; underside)

Sporophyte (2n)

MITOSIS

Sperm develop in antheridia

Zygote (2n)

Sperm swim to egg

Egg (n)

Mature sporophyte (2n)

FERTILIZATION

Gametophyte (n; side view)

Archegonium

Eggs develop in archegonia

FIGURE 31.14 Sporophyte-Dominated Life Cycles Evolved Later. In lineages that evolved later, such as ferns, the sporophyte is much larger and longer lived than the gametophyte. However, when young, the sporophyte also depends on the gametophyte for nutrition.

✔ **QUESTION** How can you tell that alternation of generations occurs in mosses (Figure 31.13) and ferns (Figure 31.14)?

In contrast, in ferns and other vascular plants the sporophyte is much larger and longer lived than the gametophyte (**FIGURE 31.14**). The ferns you see growing in gardens or forests are sporophytes. You'd have to hunt on your hands and knees to find their free-living gametophytes, which are typically just a few millimeters in diameter. As you'll learn later in the chapter, the gametophytes of gymnosperms and angiosperms are even smaller—they are microscopic—and are retained within the sporophyte. Ferns and other vascular plants are said to have a sporophyte-dominant life cycle.

✔ If you understand the difference between gametophyte-dominant and sporophyte-dominant life cycles, you should be able to examine the photos of hornworts (a nonvascular plant) and horsetails (a seedless vascular plant) in **FIGURE 31.15** and identify which is the gametophyte and which is the sporophyte.

The transition from gametophyte-dominated life cycles to sporophyte-dominated life cycles is one of the most striking of all trends in land plant evolution. To explain why it occurred, biologists hypothesize that sporophyte-dominated life cycles were advantageous because diploid cells can respond to varying environmental conditions more efficiently than haploid cells can—particularly if the individual is heterozygous at many genes. This idea has yet to be tested rigorously, however.

Heterospory In addition to sporophyte-dominated life cycles, another important innovation found in seed plants is called **heterospory**—the production of two distinct types of spores by different structures.

All of the nonvascular plants and most of the seedless vascular plants are homosporous. **Homospory** is the production of a single type of spore. (Among the seedless vascular plants, some lycophytes and a few ferns are heterosporous.) Homosporous species produce spores that develop into bisexual gametophytes that produce both eggs and sperm (**FIGURE 31.16a**). If these gametophytes are isolated in nature, they can self-fertilize and produce offspring. If two bisexual gametophytes are close enough for the exchange of sperm, outcrossing is favored because it increases genetic variation (see Chapter 13). The mechanisms that prevent self-fertilization in this case are poorly understood.

The two types of spore-producing structures in heterosporous species are often found on the same individual (**FIGURE 31.16b**).

1. **Microsporangia** are spore-producing structures that produce microspores. **Microspores** develop into male gametophytes, which produce the small gametes called sperm.

2. **Megasporangia** are spore-producing structures that produce megaspores. **Megaspores** develop into female gametophytes, which produce the large gametes called eggs.

Thus, the gametophytes of seed plants are either male or female, but never both.

The evolution of heterospory was a key event in land plant evolution because it made possible one of the most important adaptations for life in dry environments—pollen.

Pollen The nonvascular plants and the seedless vascular plants have sperm that swim to the egg to perform fertilization. For a sperm cell to swim, there has to be a continuous sheet of water

(a) Hornwort gametophytes and sporophytes

(b) Horsetail gametophyte and sporophyte

FIGURE 31.15 The Reduction of the Gametophyte Is One of the Strongest Trends in Land Plant Evolution. (a) A hornwort with spike-like sporophytes emerging from the gametophyte. **(b)** A horsetail species, both as a tiny, microscopic gametophyte and as a large, macroscopic sporophyte.

between the male and female gametophytes, or a raindrop has to splash sperm onto a female gametophyte.

In species that live in dry environments, these conditions are rare. The land plants made their final break with their aquatic origins and were able to reproduce efficiently in dry habitats when a structure evolved that could move their gametes without the aid of water.

In heterosporous seed plants, the microspore germinates to form a tiny male gametophyte that is surrounded by a tough coat of sporopollenin, resulting in a **pollen grain**.

(a) Nonvascular plants and most seedless vascular plants are **homosporous.**

(b) Seed plants are **heterosporous.**

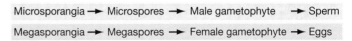

FIGURE 31.16 Unlike Homosporous Plants, Heterosporous Plants Produce Male and Female Spores That Are Morphologically Distinct.

Pollen grains can be exposed to the air for long periods of time without dying from dehydration. They are also tiny enough to be carried to female gametophytes by wind or animals. Upon landing near the egg, the male gametophyte releases the sperm cells that accomplish fertilization.

When pollen evolved, then, heterosporous plants lost their dependence on water to accomplish fertilization. Instead of swimming to the egg as a naked sperm cell, their tiny gametophytes took to the skies.

Seeds The evolution of large gametangia protected the eggs and sperm of land plants from drying. Embryo retention allowed offspring to be nourished directly by their parent, and pollen enabled fertilization to occur in the absence of water.

Retaining embryos has a downside, however: In ferns and horsetails, sporophytes have to live in the same place as their parent gametophyte. Seed plants overcome this limitation. Their embryos are portable and can disperse to new locations. The dispersal stage of the life cycle thus shifted from the haploid spore to the young diploid sporophyte.

A seed is a structure that, like a bird's egg, includes an embryo and a food supply surrounded by a tough coat that allows for effective dispersal of embryos (**FIGURE 31.17**). Spores are an effective dispersal stage in nonvascular plants and seedless vascular plants, but they lack the stored nutrients found in seeds.

FIGURE 31.17 Seeds Contain an Embryo and a Food Supply and Can Be Dispersed. (This *Pinus* (pine) specimen has been stained.)

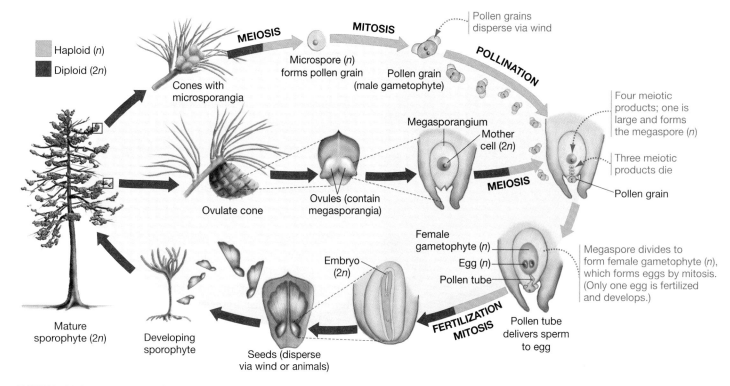

FIGURE 31.18 Heterospory in Gymnosperms: Microspores Produce Pollen Grains; Megaspores Produce Female Gametophytes.

✔ **QUESTION** Compare the life cycle of the pine tree in Figure 31.18 with that of the fern pictured in Figure 31.14. Is the gymnosperm gametophyte larger than, smaller than, or about the same size as a fern gametophyte? Compared with ferns, is the gymnosperm gametophyte more or less dependent on the sporophyte for nutrition?

The evolution of heterospory, pollen, and seeds triggered a dramatic radiation of seed plants starting about 299 million years ago. To make sure that you understand these key processes and structures, study the life cycle of the pine tree pictured in **FIGURE 31.18**.

1. Starting with the sporophyte on the left, note that this and many other gymnosperm species have separate structures, called cones, where microsporangia and megasporangia develop. In this case, the two types of spores associated with heterospory develop in separate cones.

2. The microsporangia contain a cell that divides by meiosis to form microspores, which then divide by mitosis to form pollen grains—tiny male gametophytes.

3. Megasporangia are found inside protective structures called ovules. Megasporangia contain a mother cell that divides by meiosis to form a megaspore.

4. The megaspore undergoes mitosis to form the female gametophyte, which contains egg cells.

5. The female gametophyte stays attached to the sporophyte as pollen grains arrive and produce sperm that fertilize the eggs.

6. Seeds mature as the embryo develops. Inside the seed, cells derived from the female gametophyte become packed with nutrients provided by the sporophyte.

When the seed disperses and germinates, the cycle of life begins anew.

Flowers Flowering plants, or angiosperms, are the most diverse land plants living today. About 250,000 species have been described, and more are discovered each year. Their success in terms of geographical distribution, number of individuals, and number of species revolves around a reproductive organ—the **flower.**

Most flowers contain two key reproductive structures: the stamens and carpels illustrated on the left-hand side of **FIGURE 31.19**. Stamens and carpels are responsible for heterospory.

- A **stamen** includes a structure called an anther, where microsporangia develop. Meiosis occurs inside the microsporangia, forming microspores. Microspores then divide by mitosis to form pollen grains.

- A **carpel** contains a protective structure called an **ovary,** where the ovules are found.

The presence of enclosed ovules inspired the name angiosperm ("encased-seed") as opposed to gymnosperm ("naked-seed").

As in gymnosperms, ovules contain the megasporangia. A cell inside the megasporangium divides by meiosis to form the megaspore, which then divides by mitosis to form the female

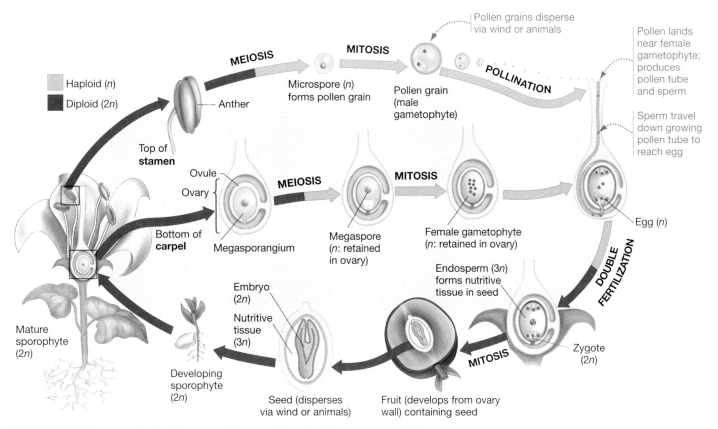

FIGURE 31.19 Heterospory in Angiosperms: Flowers Contain Microspores and Megaspores.

✔ **QUESTION** Gymnosperm pollen grains typically contain from 4 to 40 cells; mature angiosperm pollen grains contain three cells. Gymnosperm female gametophytes typically contain hundreds of cells; angiosperm female gametophytes typically contain seven. Do these observations conflict with the trend of reduced gametophytes during land plant evolution, or are they consistent with it? Explain your logic.

gametophyte. After a pollen grain lands on a carpel and produces sperm, fertilization takes place, as shown on the right-hand side of Figure 31.19.

Angiosperm fertilization is unique, however, because it involves *two* sperm cells. One sperm fuses with the egg to form the diploid (2*n*) zygote, while a second sperm fuses with two nuclei in the female gametophyte to form a triploid (3*n*) nutritive tissue called **endosperm.** The involvement of two sperm nuclei is called **double fertilization.**

The evolution of the flower, then, is an elaboration of heterospory. The key innovation was the evolution of the ovary, which helps protect female gametophytes from insects and other predators.

Double fertilization is another striking innovation associated with the flower, but its adaptive significance is still not well understood. Explaining the significance of double fertilization is a major challenge for biologists interested in understanding how land plants diversified.

Pollination by Insects and Other Animals The story of the flower doesn't end with the ovary. Once stamens and carpels evolved, they became enclosed by modified leaves called **sepals** and **petals.** The

four structures then diversified to produce a fantastic array of sizes, shapes, and colors—from red roses to blue violets. Specialized cells inside flowers also began producing a wide range of scents.

To explain these observations, biologists hypothesize that flowers are adaptations to increase the probability that an animal will perform **pollination**—the transfer of pollen from one individual's stamen to another individual's carpel. Instead of leaving pollination to an undirected agent such as wind, the hypothesis is that natural selection favored structures that reward an animal—usually an insect—for carrying pollen directly from one flower to another.

Under the directed-pollination hypothesis, natural selection has favored flower colors and shapes and scents that are successful in attracting particular types of pollinators. A pollinator is an animal that disperses pollen. Pollinators are attracted to flowers because flowers provide the animals with food in the form of protein-rich pollen or a sugar-rich fluid known as **nectar.** In this way, the relationship between flowering plants and their pollinators is mutually beneficial. The pollinator gets food; the plant gets sex (fertilization).

What evidence supports the hypothesis that flowers vary in size, structure, scent, and color in order to attract specific pollinators?

The first type of evidence on this question is correlational in nature. In general, the characteristics of a flower correlate closely with the characteristics of its pollinator. A few examples of these so-called **pollination syndromes** will help drive this point home:

- *Scent* The carrion flower in **FIGURE 31.20a** produces molecules that smell like rotting flesh. The scent attracts carrion flies, which normally lay their eggs in animal carcasses. In effect, the plant tricks the flies. While looking for a place to lay their eggs on a flower, the flies become dusted with pollen. If the flies are already carrying pollen from a visit to a different carrion flower, they are likely to deposit pollen grains near the plant's female gametophyte. In this way, the carrion flies pollinate the plant.

- *Flower shape* Flowers that are pollinated by hummingbirds typically have petals that form a long, tubelike structure corresponding to the size and shape of a hummingbird's beak (**FIGURE 31.20b**). Nectar-producing cells are located at the base of the tube. When hummingbirds visit the flower, they insert their beaks and harvest the nectar. In the process, they transfer pollen grains attached to their throats or faces.

- *Flower color* Hummingbird-pollinated flowers tend to have red petals (**FIGURE 31.20b**), while bee-pollinated flowers tend to be purple or yellow (**FIGURE 31.20c**). Hummingbirds are attracted to red; bees have excellent vision in the purple and ultraviolet end of the spectrum.

The directed-pollination hypothesis also has strong experimental support. Consider, for example, recent work on South American petunias. Field observations led researchers to conclude that a white-flowered species (*Petunia axillaris*) was primarily pollinated at night by hawk moths, while a purple-flowered species (*P. integrifolia*) was pollinated during the day by bumblebees. (Note that white is highly visible at night, and purple is the color most easily seen by bees.)

Was flower color the key to flower choice by these insects? This is difficult to determine because of the many other differences in flower characteristics between two plant species, such as flower shape, odor, and nectar composition. In this experiment, researchers identified a gene named *AN2* responsible for the pigmentation in the purple petunia and introduced it into the genome of the white petunia, producing new plants that now had purple flowers. The researchers then documented that other than color, the wild-type and genetically altered flowers were identical.

In a controlled, greenhouse experiment, hawk moths and bumblebees were allowed to choose between the two *P. axillaris* plants, and as predicted by the directed-pollination hypothesis, flower color was key to pollinator preference (**FIGURE 31.21**). These data strongly support the hypothesis that flower color is an adaptation that increases the frequency of pollination by particular insects.

Based on results like this, biologists contend that the spectacular diversity of angiosperms resulted, at least in part, from natural selection exerted by the equally spectacular diversity of insect, mammal, and bird pollinators. Mutations that change flower color can change gene flow and start speciation (see Chapter 27).

Fruits The evolution of the ovary was an important event in land plant diversification, but not only because it protected the female gametophytes of angiosperms. It also made the evolution of fruit possible.

A **fruit** is a structure that is derived from the ovary and encloses one or more seeds (**FIGURE 31.22a**). Tissues derived from the ovary are often nutritious and brightly colored (**FIGURE 31.22b**). Animals eat these types of fruits, digest the nutritious tissue

(a) Carrion flowers: Smell like rotting flesh and attract carrion flies

(b) Hummingbird-pollinated flowers: Red, long tubes with nectar at the base

(c) Bee-pollinated flowers: Often bright purple

FIGURE 31.20 Flowers with Different Fragrances, Shapes, and Colors Attract Different Pollinators.

HYPOTHESIS: Hawk moths prefer white petunias while bumblebees prefer purple, even when the flowers are otherwise identical.

NULL HYPOTHESIS: Neither insect has a clear preference for white- or purple-flowered petunias.

EXPERIMENTAL SETUP:

1. **Isolate** *AN2* **gene** for purple flower color from a related species (*P. integrifolia*) and insert it into *P. axillaris*.

Wild-type *P. axillaris*

Genetically altered *P. axillaris*

2. **Grow** wild-type (white) and genetically altered (purple) *P. axillaris* in controlled greenhouse conditions.

3. **Count number of visits** by hawk moths and bumblebees to white and purple flowers.

PREDICTION: Hawk moths will visit white-flowered *P. axillaris* more frequently, while bumblebees will visit purple-flowered *P. axillaris* more frequently.

PREDICTION OF NULL HYPOTHESIS: Both insects will visit white and purple flowers equally.

RESULTS:

Hawk moths visited white flowers 4 times more frequently than purple flowers

Bumblebees visited purple flowers 7 times more frequently than white flowers

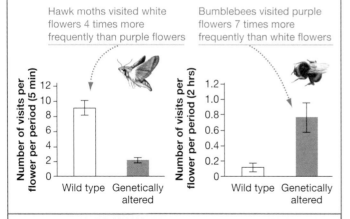

CONCLUSION: Petunia flower preference among bumblebees and hawk moths is significantly influenced by flower color.

FIGURE 31.21 The Adaptive Significance of Flower Color.
SOURCE: Hoballah, M. E., T. Gübitz, J. Stuurman, et al. 2007. Single gene-mediated shift in pollinator attraction in *Petunia*. *The Plant Cell* 19: 779–790.

✔ **QUESTION** Why did the researchers measure pollination in a greenhouse? Why did they plot flower visits on different scales?

(a) Fruits are derived from ovaries and contain seeds.

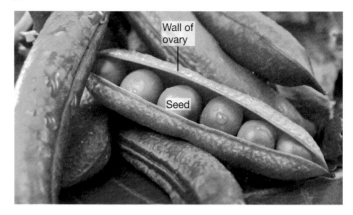

Wall of ovary

Seed

(b) Many fruits are dispersed by animals.

FIGURE 31.22 Fruits Are Derived from Ovaries Found in Angiosperms. (a) A pea pod is one of the simplest types of fruit. **(b)** The ovary wall often becomes thick, fleshy, and nutritious enough to attract animals that disperse the seeds inside.

around the seeds, and disperse seeds in their feces. In other cases, the tissues derived from the ovary help fruits disperse via wind or water. The evolution of flowers made efficient pollination possible; the evolution of fruits made efficient seed dispersal possible.

The list of adaptations that allow land plants to reproduce in dry environments is impressive; **FIGURE 31.23** (see page 596) summarizes them. Once land plants had vascular tissue and could grow efficiently in dry habitats, the story of their diversification revolved around traits that allowed sperm cells to reach eggs efficiently and helped seeds disperse to new locations.

The Angiosperm Radiation

For the past 125 million years, land plant diversification has really been about angiosperms. The Anthophyta or angiosperms represent one of the great adaptive radiations in the history of

FIGURE 31.23 Evolutionary Innovations Allowed Plants to Reproduce Efficiently on Land.

✔ **EXERCISE** Redwoods, pines, gnetophytes, and angiosperms are the only land plants that do not have flagellated sperm that swim to the egg (at least a short distance). Mark the loss of flagellated sperm on Figure 31.6.

life. An **adaptive radiation** occurs when a single lineage produces a large number of descendant species that are adapted to a wide variety of habitats (see Chapter 28).

The diversification of angiosperms is associated with three key adaptations: (**1**) water-conducting vessels, (**2**) flowers, and (**3**) fruits. In combination, these traits allow angiosperms to transport water, pollen, and seeds efficiently. Based on these observations, it is not surprising that most land plants living today are angiosperms.

On the basis of morphological traits, the 250,000 species of angiosperms identified to date have traditionally been classified into two major groups: the monocotyledons, or **monocots,** and the dicotyledons, or **dicots.** Some familiar monocots include the grasses (such as corn and wheat), orchids, palms, and lilies;

familiar dicots include beans, roses, buttercups, daisies, oaks, and maples.

The names of the two groups were inspired by differences in a structure called the cotyledon. A **cotyledon** ("seed-leaf") stores nutrients and supplies them to the developing embryonic plant. As **FIGURE 31.24** shows, monocots have a single cotyledon (hence the "mono") while dicots have two cotyledons (hence the "di"). The figure also highlights other major morphological differences observed in monocots and dicots, including the arrangement of vascular tissue and leaf veins and the characteristics of flowers.

It would be misleading, however, to think that all species of flowering plants fall into one of these two groups—either monocots or dicots. Recent work has shown that dicots do not form a monophyletic group consisting of a common ancestor and all of its descendants.

To drive this point home, consider the phylogeny illustrated in **FIGURE 31.25**. These relationships were estimated by comparing the sequences of several genes that are shared by all angiosperms. Notice that species with dicot-like characters are found in multiple lineages of the angiosperm phylogenetic tree. Based on this analysis, biologists have concluded that although monocots are monophyletic, dicots are not. Dicots are paraphyletic.

Biologists have adjusted the names assigned to angiosperm lineages to reflect this new knowledge of phylogeny. The most important of these changes was identifying the **eudicots** ("true dicots") as a monophyletic lineage that includes most of the plants once considered dicots. One flowering plant that is neither a monocot nor a eudicot is the magnolia tree. Plant systematists continue to work toward understanding relationships throughout the angiosperm phylogenetic tree; there will undoubtedly be more name changes as knowledge grows.

check your understanding

C Y U

If you understand that . . .

- Land plants were able to make the transition to growing in terrestrial environments, where sunlight and carbon dioxide are abundant, based on a series of evolutionary innovations.
- Adaptations for growing on land included cuticle, stomata, UV-absorbing compounds, and vascular tissue.
- Adaptations for effective reproduction on land included gametangia, the retention of embryos on the parent, pollen, seeds, flowers, and fruits.

✔ **You should be able to . . .**

1. Explain why the evolution of cuticle, UV-absorbing compounds, and vascular tissue was important in survival or reproduction.

2. On Figure 31.6, map where the origin of pollen, flowers, and fruits occurred.

Answers are available in Appendix A.

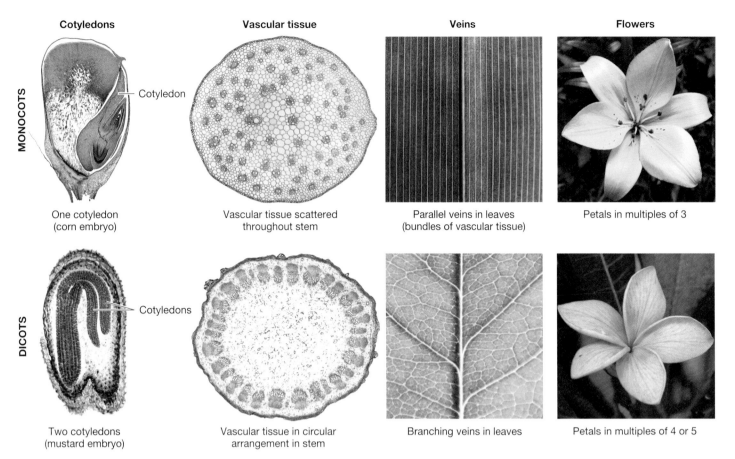

Cotyledons | **Vascular tissue** | **Veins** | **Flowers**

MONOCOTS

One cotyledon (corn embryo) | Vascular tissue scattered throughout stem | Parallel veins in leaves (bundles of vascular tissue) | Petals in multiples of 3

DICOTS

Two cotyledons (mustard embryo) | Vascular tissue in circular arrangement in stem | Branching veins in leaves | Petals in multiples of 4 or 5

FIGURE 31.24 Four Morphological Differences between Monocots and Dicots. (The stem cross sections, showing vascular tissue and cotyledons, have been stained.)

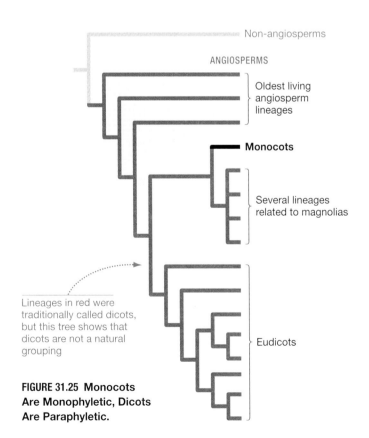

Non-angiosperms

ANGIOSPERMS

Oldest living angiosperm lineages

Monocots

Several lineages related to magnolias

Eudicots

Lineages in red were traditionally called dicots, but this tree shows that dicots are not a natural grouping

FIGURE 31.25 Monocots Are Monophyletic, Dicots Are Paraphyletic.

31.4 Key Lineages of Green Algae and Land Plants

The evolution of cuticle, pores, stomata, UV-absorbing compounds, and water-conducting tissues allowed green plants to grow on land, where resources for photosynthesis are abundant. Once the green plants were on land, the evolution of gametangia, retained embryos, pollen, seeds, and flowers enabled them to reproduce efficiently even in dry environments. The adaptations reviewed in Section 31.3 allowed the land plants to make the most important water-to-land transition in the history of life.

To explore green plant diversity in more detail, let's take a closer look at some major groups of green algae and land plants. The first part of this section considers broad groupings of lineages; the "index cards" that follow focus on particular monophyletic groups.

Green Algae

The **green algae** are a paraphyletic group that totals about 7000 species. Their bright green chloroplasts are similar to those found in land plants. Specifically, green algal chloroplasts have a double membrane and contain chlorophylls *a* and *b*. And like land plants, green algae synthesize starch in the chloroplast as a

(a) Green algae with red carotenoid pigments are responsible for pink snow.

(b) Many unicellular protists harbor green algae.

(c) Most lichens are an association between fungi and microscopic green algae.

Algal cells

10 μm

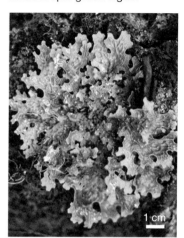

1 cm

FIGURE 31.26 Some Green Algae Live in Unusual Environments.

storage product of photosynthesis. They also have a cell wall that is composed primarily of cellulose.

Green algae are important primary producers in nearshore ocean environments and in all types of freshwater habitats. They are also found in several types of more exotic environments, including snowfields at high elevations, pack ice, and ice floes. These habitats are often splashed with bright colors due to large concentrations of unicellular green algae (**FIGURE 31.26a**). Although these cells live at near-freezing temperatures, they make all their own food via photosynthesis.

In addition, green algae live in close association with an array of other organisms.

- Unicellular green algae are common endosymbionts in planktonic protists that live in lakes and ponds (**FIGURE 31.26b**). The association is considered mutually beneficial: The algae supply the protists with food; the protists provide protection to the algae.

- **Lichens** are stable associations between green algae and fungi, or between cyanobacteria and fungi, and are often found in terrestrial environments that lack soil, such as tree bark or bare rock (**FIGURE 31.26c**). The algae or cyanobacteria in a lichen are protected from drying by the fungus; in return, they provide sugars produced by photosynthesis.

Of the 17,000 species of lichens described to date, about 85 percent involve green algae. The green algae that are involved are unicellular or grow in long filaments. (Lichens are explored in more detail in Chapter 32.)

Green algae are a large and fascinating group of organisms. In this final section of the chapter, let's take a closer look at just three of the many lineages.

- Green Algae > Ulvophyceae (Ulvophytes)
- Green Algae > Coleochaetophyceae (Coleochaetes)
- Green Algae > Charophyceae (Stoneworts)

Nonvascular Plants

The initial lineages to branch off the phylogeny of living land plants are sometimes known as nonvascular plants.

All of the nonvascular plant species present today have a low, sprawling growth habit. In fact, it is unusual to find species that are more than 5 to 10 centimeters tall.

Individuals are anchored to soil, rocks, or tree bark by structures called **rhizoids.** Although simple water-conducting cells and tissues are found in some mosses, nonvascular plants lack vascular tissue with lignin-reinforced cell walls.

All nonvascular plants have flagellated sperm that swim to eggs through raindrops or small puddles on the plant surface. Spores are dispersed by wind.

- Nonvascular Plants > Hepaticophyta (Liverworts)
- Nonvascular Plants > Bryophyta (Mosses)
- Nonvascular Plants > Anthocerophyta (Hornworts)

Seedless Vascular Plants

The seedless vascular plants are a paraphyletic group between the nonvascular plants and the seed plants. All species of seedless vascular plants have conducting tissues with cells that are reinforced with lignin. Tree-sized lycophytes and horsetails are abundant in the fossil record, and tree ferns are still common inhabitants of certain habitats, such as in New Zealand and on mountain slopes in the tropics.

The sporophyte is the larger and longer-lived phase of the life cycle in all of the seedless vascular plants. The gametophyte is physically independent of the sporophyte, however. Eggs are retained on the gametophyte, and sperm swim to the egg with the aid of flagella. Thus, seedless vascular plants depend on the presence of water for reproduction—they need enough water to form a continuous layer that "connects" gametophytes and allows sperm to swim to eggs. Sporophytes develop on the gametophyte and are nourished by the gametophyte when they are small.

- Seedless Vascular Plants > Lycophyta (Lycophytes, or Club Mosses)
- Seedless Vascular Plants > Psilotophyta (Whisk Ferns)
- Seedless Vascular Plants > Pteridophyta (Ferns)
- Seedless Vascular Plants > Equisetophyta (or Sphenotophyta) (Horsetails)

Seed Plants

The seed plants are a monophyletic group consisting of the gymnosperms—ginkgo, cycads, redwoods, pines, and gnetophytes—and the angiosperms. The group is defined by two key synapomorphies, traits common to a lineage and found in no other lineages: the production of seeds and the production of pollen grains.

1. Seeds are a specialized structure for dispersing embryonic sporophytes to new locations. Seeds are the mature form of a fertilized ovule, the female reproductive structure that encloses the female gametophyte and egg cell.

2. Pollen grains are tiny, sperm-producing gametophytes that are easily dispersed through air as opposed to water.

Seed plants are found in virtually every type of habitat, and they adopt every growth habit known in land plants. Their forms range from mosslike mats to shrubs and vines to 100-meter-tall trees. Seed plants can be **annual** (have a single growing season) or **perennial** (live for many years), with life spans ranging from a few weeks to almost five thousand years. (See Chapters 37 through 41 to further explore the structure and function of seed plants.)

- Seed Plants > Gymnosperms > Ginkgophyta (Ginkgoes)
- Seed Plants > Gymnosperms > Cycadophyta (Cycads)
- Seed Plants > Gymnosperms > Cupressophyta (Redwoods, Junipers, Yews)
- Seed Plants > Gymnosperms > Pinophyta (Pines, Spruces, Firs)
- Seed Plants > Gymnosperms > Gnetophyta (Gnetophytes)
- Seed Plants > Anthophyta (Angiosperms)

Green Algae > Ulvophyceae (Ulvophytes)

The Ulvophyceae are a monophyletic group composed of several diverse and important subgroups, with a total of about 4000 species. Members of this lineage range from unicellular to multicellular.

Many of the large green algae in habitats along ocean coastlines are members of the Ulvophyceae. *Ulva*, the sea lettuce (**FIGURE 31.27**), is a representative marine species. But freshwater lakes and streams also contain large numbers of unicellular or small multicellular species that are free-floating plankton.

Reproduction Most ulvophytes reproduce both asexually and sexually. Asexual reproduction often involves production of spores that swim with the aid of flagella. Sexual reproduction usually results in production of a resting stage—a cell that is dormant in winter. In many species the gametes are not called eggs and sperm, because they are the same size and shape. In most species gametes are shed into the water, so fertilization takes place away from the parent plants.

Life cycle Many unicellular forms are diploid only as zygotes. Alternation of generations occurs in multicellular species. When alternation of generations occurs, gametophytes and sporophytes may look identical or different.

Human and ecological impacts Ulvophyceae are important primary producers in freshwater environments and in coastal areas of the oceans.

Ulva lactuca

5 cm

FIGURE 31.27 Green Algae Are Important Primary Producers in Aquatic Environments.

This group includes 19 species. Most coleochaetes are barely visible to the unaided eye and grow as flat sheets of cells (**FIGURE 31.28**). They are considered multicellular because they have specialized photosynthetic and reproductive cells and because they contain **plasmodesmata** that connect adjacent cells in plants (see Chapter 11).

The coleochaetes are strictly freshwater algae. They grow attached to aquatic plants such as water lilies and cattails or over submerged rocks in lakes and ponds. When they grow near beaches, they are often exposed to air when water levels drop in late summer.

Reproduction Asexual reproduction is common in coleochaetes and involves production of flagellated spores. During sexual reproduction, eggs are retained on the parent and are nourished after fertilization with the aid of transfer cells—a situation similar to that observed in land plants. In some species certain individuals are male and produce only sperm, while other individuals are female and produce only eggs.

Life cycle Alternation of generations does not occur. Multicellular individuals are haploid; the only diploid stage in the life cycle is the zygote.

Human and ecological impacts Because they are closely related to land plants, coleochaetes are studied intensively by researchers interested in how land plants made the water-to-land transition.

Coleochaete orbicularis

20 µm

FIGURE 31.28 Coleochaetes Are Thin Sheets of Cells.

This group includes several hundred species. They are collectively known as stoneworts, because they commonly accumulate crusts of calcium carbonate ($CaCO_3$) over their surfaces. Like the coleochaetes, they have plasmodesmata and are multicellular. ✔ You should be able to mark the origin of plasmodesmata on Figure 31.6. (They do not occur in ulvophytes.) Some species of stonewort can be a meter or more in length.

The stoneworts are freshwater algae. Certain species are specialized for growing in relatively deep waters, though most live in shallow water near lake beaches or pond edges.

Reproduction Sexual reproduction is common and involves production of prominent, multicellular gametangia similar to those observed in early land plants. In stoneworts, as in coleochaetes, the eggs are retained on the parent plant, which supplies eggs with nutrients before fertilization. ✔ You should be able to mark the origin of egg retention on Figure 31.6. (Egg retention does not occur in ulvophytes.)

Life cycle Alternation of generations does not occur. Multicellular individuals are haploid; the only diploid stage in the life cycle is the zygote.

Human and ecological impacts Some species form extensive beds in lake bottoms or ponds and provide food for ducks and geese as well as food and shelter for fish (**FIGURE 31.29**). They are a good indicator that water is not polluted.

Chara species

5 cm

FIGURE 31.29 Stoneworts Can Form Beds on Lake Bottoms.

Liverworts got their name because some species native to Europe have liver-shaped leaves. According to the medieval *Doctrine of Signatures*, God indicated how certain plants should be used by giving them a distinctive appearance. Thus, liverwort teas were hypothesized to be beneficial for liver ailments. (They are not.) About 6500 species are known. They are commonly found growing on damp forest floors or riverbanks, often in dense mats (**FIGURE 31.30**), or on the trunks or branches of tropical trees.

Adaptations to land Liverworts are covered with cuticle. Some species have pores that allow gas exchange; in species that lack pores, the cuticle is thin.

Reproduction Asexual reproduction occurs when fragments of a plant are broken off and begin growing independently. Some species also produce small structures called **gemmae** asexually, during the gametophyte phase. Mature gemmae are knocked off the parent plant by rain and grow into independent gametophytes. During sexual reproduction, sperm and eggs are produced in gametangia.

Life cycle The gametophyte is the largest and longest-lived phase in the life cycle. Sporophytes are small, grow directly from the gametophyte, and depend on the gametophyte for nutrition. Spores are shed from the sporophyte and are carried away by wind or rain.

Human and ecological impacts When liverworts grow on bare rock or tree bark, their dead and decaying body parts contribute to the initial stages of soil formation. The liverwort *Marchantia polymorpha*, shown in Figure 31.30, is an important model organism in plant biology.

Marchantia polymorpha

1 cm

FIGURE 31.30 Liverworts Thrive in Moist Habitats.

Over 12,000 species of mosses have been named and described to date, and more are being discovered every year—particularly in the tropics.

Although mosses are common in moist forests, they can also be found in more extreme environments, such as deserts and windy, treeless habitats in the Arctic, Antarctic, or on mountaintops. In these severe conditions, mosses are able to thrive because their bodies can become extremely dry without dying. When the weather makes photosynthesis difficult, individuals dry out and become dormant, or inactive. Then when rains arrive or temperatures warm, the plants rehydrate and begin photosynthesis and reproduction.

Adaptations to land One subgroup of mosses contains simple conducting tissues consisting of cells that are specialized for the transport of water or food. But because these cells do not have walls that are reinforced by lignin, they are not considered true vascular tissue. ✔ You should be able to mark the origin of the simple water-conducting cells and tissues in this moss subgroup on Figure 31.6. Because they lack true vascular tissue, most mosses are not able to grow much taller than a few centimeters.

Reproduction Asexual reproduction often occurs by fragmentation, meaning that pieces of gametophytes that are broken off by wind or a passing animal can begin growing independently. In many species, sexual reproduction cannot involve self-fertilization because the sexes are separate—meaning that an individual plant produces only eggs, in archegonia, or only sperm, in antheridia. A typical sporophyte produces up to 50 million tiny spores. Spores are usually distributed by wind.

Life cycle The moss life cycle is similar to that of liverworts and hornworts: The sporophyte is retained on the much larger and longer-lived gametophyte and gets most of its nutrition from the gametophyte.

Human and ecological impacts Species in the genus *Sphagnum* are often the most abundant plant in wet habitats of northern environments (**FIGURE 31.31a**, page 602). Because *Sphagnum*-rich environments account for 1 percent of Earth's total land area, equivalent to half the area of the United States, *Sphagnum* species are among the most abundant plants in the world. *Sphagnum*-rich

(Continued on next page)

(a) *Sphagnum* moss is abundant in northern wet habitats.

(b) Semi-decayed *Sphagnum* moss forms peat.

5 mm

FIGURE 31.31 *Sphagnum* **Mosses Are among the Most Abundant Plants in the World.**

habitats are waterlogged, nitrogen poor, and often anaerobic, however, so the decomposition of dead mosses and other plants is slow. As a result, large deposits of semi-decayed organic matter, known as **peat,** accumulate. Researchers estimate that the world's peatlands store about 400 billion metric tons of carbon. If peatlands begin to burn or decay rapidly due to rising temperatures associated with global climate change, the CO_2 released will exacerbate the warming trend (see Chapter 56).

Peat is harvested as a traditional heating and cooking fuel in some countries (**FIGURE 31.31b**). It is also widely used as a soil additive in gardening, because *Sphagnum* can absorb up to 20 times its dry weight in water. This high water-holding capacity is due to the presence of large numbers of dead cells in the leaves of these mosses, which readily fill with water via pores in their walls.

Nonvascular Plants > Anthocerophyta (Hornworts)

Hornworts got their name because their sporophytes have a horn-like appearance (**FIGURE 31.32**) and because "wort" is the Anglo-Saxon word for plant. About 100 species have been described to

Anthoceros sp.

2 cm

FIGURE 31.32 Hornworts Have Horn-Shaped Sporophytes.

date. Hornworts can be found worldwide but are more common in tropical regions in damp soil or on the bark of trees.

Adaptations to land In most species of hornworts, sporophytes have stomata that function much like those of vascular plants. Sporophytes have cell layers that change shape when exposed to low humidity, facilitating the release of spores.

Reproduction Depending on the species, gametophytes may contain only egg-producing archegonia or only sperm-producing antheridia, or both. Stated another way, individuals of some species are either female or male, while in other species each individual has both types of reproductive organs.

Life cycle The gametophyte is the longest-lived phase in the life cycle. Although sporophytes grow directly from the gametophyte, they are green because their cells contain chloroplasts. Sporophytes manufacture some of their own food but also get nutrition from the gametophyte. Spores disperse from the parent plant via wind or rain.

Human and ecological impacts Some species harbor symbiotic cyanobacteria that fix nitrogen.

Although the fossil record documents lycophytes that were 2 m wide and 40 m tall, the 1000 species of lycophytes living today are all small in stature (**FIGURE 31.33**). Most lycophytes live on the forest floor or on the branches or trunks of tropical trees. Because of their appearance, they are often called ground pines or club mosses—even though they are neither pines nor mosses.

Adaptations to land Lycophytes are the most ancient land plant lineage with **roots**—a belowground system of tissues that anchors the plant and is responsible for absorbing water and mineral nutrients. Roots differ from the rhizoids observed in nonvascular plants, because roots contain vascular tissue and thus are capable of conducting water and nutrients to the upper reaches of the plant. Unusual leaves called microphylls, which extend from the stems, are a synapomorphy found in lycophytes. ✔ You should be able to mark the origin of microphylls on Figure 31.6.

Reproduction Asexual reproduction can occur by fragmentation or gemmae. During sexual reproduction, spores of some species give rise to bisexual gametophytes—meaning that each gametophyte produces both eggs and sperm. Self-fertilization is extremely rare, however. In the genera *Selaginella* and *Isoetes*, in contrast, heterospory occurs and gametophytes are either male or female.

Life cycle The gametophytes of some species live entirely underground and get their nutrition from symbiotic fungi. In certain species, gametophytes live 6 to 15 years and give rise to a large number of sporophytes over time.

Human and ecological impacts Tree-sized lycophytes were abundant in the coal-forming forests of the Carboniferous period. In coal-fired power plants today, electricity is being generated by burning fossilized lycophyte and fern tissues. Spores of some members are highly flammable. Before the invention of electric flashbulbs, these spores were used as flash powder in early photography.

Lycopodium species

5 cm

FIGURE 31.33 Lycophytes Living Today Are Small in Stature.

Only two genera of whisk ferns are living today, and there are perhaps six distinct species. Whisk ferns are restricted to tropical regions and have no fossil record. They are extremely simple morphologically; their aboveground parts consist of branching stems that have tiny, scale-like outgrowths instead of leaves (**FIGURE 31.34**).

Because they lack roots and leaves, the whisk ferns were considered an evolutionary "throwback"—a group that retained traits found in the earliest vascular plants. However, molecular phylogenies support an alternative hypothesis: The morphological simplicity of whisk ferns is a derived trait—meaning that complex structures have been lost in this lineage.

Adaptations to land Whisk ferns lack roots. Some species gain most of their nutrition from fungi that grow in association with the whisk ferns' underground stems, called **rhizomes.** Other species grow in rock crevasses or as **epiphytes** ("upon-plants"), meaning that they grow on the branches of other plants—in this case, on the branches of tree ferns. ✔ After reviewing where leaves and roots originated during land plant evolution (see Figure 31.9), you should be able to mark the loss of leaves and roots in whisk ferns on Figure 31.6.

Psilotum nudum

1 cm

FIGURE 31.34 Psilotophytes Are Extremely Simple Morphologically.

Reproduction Asexual reproduction occurs via the extension of rhizomes and the production of new aboveground stems. When spores mature, they are dispersed by wind and germinate into gametophytes that contain both archegonia and antheridia.

Life cycle Sporophytes may be up to 30 cm tall, but gametophytes are less than 2 mm long and live under the soil surface. Gametophytes absorb nutrients directly from the soil or from symbiotic fungi. Fertilization takes place inside the archegonium, and the sporophyte develops directly on the gametophyte.

Human and ecological impacts Some whisk fern species are popular landscaping plants, particularly in Japan. The same species can be a serious pest in greenhouses.

With 12,000 species, ferns are by far the most species-rich group of seedless vascular plants. They are particularly abundant in the tropics. About a third of the tropical species are epiphytes, usually growing on the trunks or branches of trees. Species that can grow epiphytically live high above the forest floor, where competition for light is reduced, without making wood and growing tall themselves. The growth habits of ferns are highly variable among species, however, and ferns range in size from rosettes the size of your smallest fingernail to 20-meter-tall trees.

Adaptations to land Ferns are the only seedless vascular plants that have large, well-developed leaves—commonly called **fronds.** Young fronds are coiled and because of their shape are called "fiddleheads"; they uncoil as they develop (**FIGURE 31.35a**). Leaves give the plant a large surface area, allowing it to capture sunlight for photosynthesis efficiently.

Reproduction In a few species, gametophytes reproduce asexually via production of gemmae. Typically, species that can reproduce via gemmae never produce gametes or sporophytes. In most species, however, sexual reproduction is the norm. Ferns are homosporous, but gametophytes develop as males or are bisexual, depending on light intensity, proximity to other gametophytes, and other environmental conditions. In bisexual gametophytes, maturation of male and female gametes is staggered to minimize self-fertilization.

Life cycle Although fern gametophytes contain chloroplasts and are photosynthetic, the sporophyte is typically the larger and longer-lived phase of the life cycle. In mature sporophytes, sporangia are usually found in clusters called **sori** on the undersides of leaves (**FIGURE 31.35b**). The structure of the sporangia is a distinctive feature of ferns. When mature, the side wall of a sporangium ruptures and a thickened row of cells along the other side of the sporangium shrinks, pulling the open sporangium back. When the humidity is low the sporangium springs back, catapulting the spores into the dry air currents. ✔ You should be able to mark the evolution of the distinctive fern sporangium on Figure 31.6.

Human and ecological impacts In many parts of the world, people gather the fiddleheads of ferns in spring as food. Ferns are also widely used as ornamental plants in landscaping.

(a) Fern fronds develop as coiled "fiddleheads."

Sadleria cyatheoides

(b) Fern sporangia are often in clusters (sori) on the underside of fronds.

Polypodium vulgare

Sori

5 mm

FIGURE 31.35 Fern Fronds Uncoil as They Develop and Form Clusters of Sporangia (Sori) When They Mature.

Although horsetails are prominent in the fossil record of land plants, just 15 species are known today. All 15 are in the genus *Equisetum*. Translated literally, *Equisetum* means "horse-bristle." Both the scientific name and the common name, horsetail, come from the brushy appearance of the stems and branches in some species (**FIGURE 31.36a**). Horsetails may be locally abundant in wet habitats such as stream banks or marsh edges.

Adaptations to land Horsetails have an adaptation that allows them to flourish in waterlogged, oxygen-poor soils: Their stems are hollow, so oxygen readily diffuses down the stem to reach roots that cannot obtain oxygen from the surrounding soil. Horsetails are also distinguished by having whorled leaves and branches. Horsetail spores have appendages, called elaters, that change position with changes in humidity and cause spores to fall over wet habitats where gametophytes are more likely to survive (**FIGURE 31.36b**). ✔ You should be able to mark the origin of the distinctive spore elaters on Figure 31.6.

Reproduction Asexual reproduction is common in sporophytes and occurs via fragmentation or the extension of rhizomes. From these rhizomes, two types of erect, specialized stems may grow—stems that contain tiny leaves and chloroplast-rich branches and that are specialized for photosynthesis, or stems that bear clusters of sporangia and produce huge numbers of spores by meiosis.

Life cycle Gametophytes perform photosynthesis but are small and short lived. They normally produce both antheridia and archegonia, but in most cases the sperm-producing structure matures first. This pattern is thought to be an adaptation that minimizes self-fertilization and maximizes cross-fertilization.

Human and ecological impacts Horsetail stems are rich in silica granules. The glass-like deposits not only strengthen the stem but also made these plants useful for scouring pots and pans before the invention of other scrubbing tools—hence these plants are often called "scouring rushes."

(a) Fertile shoots bear conelike clusters of sporangia.

(b) As spores inside sporangia dry out, their elaters unfurl, tossing out the spores.

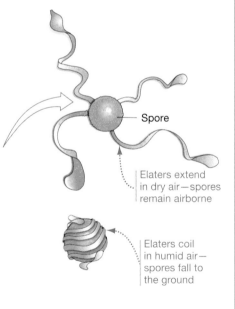

FIGURE 31.36 Horsetails Have a Distinctive Brushy Appearance and Unusual Spores.

Although ginkgoes have an extensive fossil record, just one species is alive today. Leaves from the ginkgo, or maidenhair, tree are virtually identical in size and shape to those observed in fossil ginkgoes that are 150 million years old (**FIGURE 31.37**).

(a) Fossil ginkgo
Ginkgo huttoni

(b) Living ginkgo
Ginkgo biloba

2 cm

5 cm

FIGURE 31.37 The Ginkgo Tree Is a "Living Fossil."

Adaptations to land Unlike most gymnosperms, the ginkgo is **deciduous**—meaning that it loses its leaves each autumn. This adaptation allows plants to be dormant during the winter, when photosynthesis and growth are difficult.

Reproduction and life cycle Sexes are separate—individuals are either male or female. Pollen is transported by wind. Sperm have flagella, however. Once pollen grains land near the female gametophyte and mature, the sperm cells leave the pollen grain and swim to the egg cells.

Human and ecological impacts Although today's ginkgo trees are native to southeast China, they are planted widely as an ornamental. They are especially popular in urban areas because they are tolerant of air pollution. Male trees are usually planted because the seeds produced by female trees contain butyric acid, making them smell like rancid butter. However, in some countries, the inside of the seed is eaten as a delicacy.

The cycads are so similar in overall appearance to palm trees, which are angiosperms, that cycads are sometimes called "sago palms." Although cycads were extremely abundant when dinosaurs were present on Earth 150–65 million years ago, only about 140 species are living today. Most are found in the tropics (**FIGURE 31.38**).

Adaptations to land Cycads do not make wood but are supported by stiff stems. They are unique among gymnosperms in having compound leaves—meaning that each leaf is divided into many smaller leaflets. ✔ You should be able to mark the origin of the distinctive cycad leaf on Figure 31.6.

Reproduction and life cycle In cycads each sporophyte individual bears either microsporangia or megasporangia, but not both. Like other seed plants, cycads are heterosporous. Pollen is carried by insects (usually beetles or weevils) or, in some species, wind. Cycad seeds are large and often brightly colored. The colors attract birds and mammals, both of which eat and disperse the seeds.

Human and ecological impacts Cycads harbor large numbers of symbiotic cyanobacteria in specialized, aboveground root structures. The cyanobacteria are photosynthetic and fix nitrogen. The nitrogen acts as an important nutrient for nearby plants as well as for the cycads themselves. Cycads are popular landscaping plants in some parts of the world.

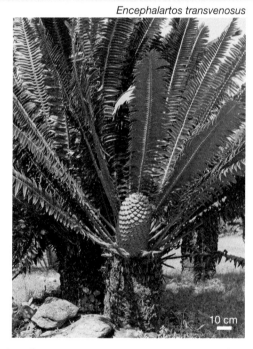

Encephalartos transvenosus

10 cm

FIGURE 31.38 Cycads Resemble Palms but Are Not Closely Related to Them.

Seed Plants > Gymnosperms > Cupressophyta (Redwoods, Junipers, Yews)

The species in this lineage vary in growth form from sprawling juniper shrubs to the world's largest plants. Redwood trees growing along the Pacific Coast of North America can reach heights of over 115 m (375 ft) and have trunk diameters of over 9.5 m (30 ft). These species were recently confirmed as a monophyletic group, independent of the Pinophyta, which also reproduce via cone-bearing structures.

Adaptations to land All of the species in this lineage are trees or large shrubs. Most have narrow leaves, which in many cases are arranged in overlapping scales (**FIGURE 31.39**). Narrow leaves have a small amount of surface area, which is not optimal for capturing sunlight and performing photosynthesis. But because the small surface area reduces water loss from leaves, many species in this lineage thrive in dry habitats or in cold environments where water is often frozen.

Thuja plicata

5 cm

FIGURE 31.39 Some Species in the Cupressophyta Have Scale-like Leaves.

Reproduction and life cycle The species in this group are wind pollinated. As in all seed plants, the female gametophyte is retained on the parent. Thus, fertilization and seed development take place in the female cone. As in gymnosperms, seeds do not form inside an encapsulated structure. Depending on the species, the seeds are dispersed by wind or by seed-eating birds or mammals.

Human and ecological impacts Redwoods, red cedar, white cedar, and yellow cedar have wood that is highly resistant to decay and thus prized for making furniture, decks, and house siding, or for other applications where wood is exposed to the weather. Yew wood is often preferred for making traditional archery bows, and the berry-like cones of juniper are used to flavor gin. The anticancer drug taxol was originally found in Pacific yew trees.

Seed Plants > Gymnosperms > Pinophyta (Pines, Spruces, Firs)

The gymnosperms include two major lineages of cone-bearing species: the pines and allies discussed here, and the group already described that includes redwoods, junipers, and yews. Species in both lineages have a reproductive structure called the cone, which produces microsporangia or megasporangia (**FIGURE 31.40**).

Pinophyta include the familiar pines, spruces, firs, Douglas fir, tamaracks, and true cedars. These are among the largest and most abundant trees on the planet, as well as some of the most long lived. One of the bristlecone pines native to southwestern North America is at least 4750 years old.

Adaptations to land Pinophyta have needle-like leaves whose small surface area allows them to thrive in habitats where water is scarce. Pines are common on sandy soils that have poor water-holding capacity, and spruces and firs are common in cold environments where water is often frozen. All of the living species make wood as a support structure.

Reproduction and life cycle Sexes are separate, and pollen is transferred to female cones by the wind. Female cones take two years to mature and are usually found high in trees. Male cones are usually found lower in the tree—possibly to reduce self-pollination from falling pollen.

Human and ecological impacts In terms of biomass, pines, spruces, firs, and other species in this group dominate forests that grow at high latitudes and high elevations as well as at sandy sites in warmer regions. Their seeds are key food sources for a variety of birds, squirrels, and mice, and their wood is the basis of the building products and paper industries in many parts of the world. The paper in this book was made from species in this group.

(a) Cones that produce microsporangia and pollen

Picea abies

Pollen

1 cm

(b) Cones that produce megasporangia and eggs

Picea abies

5 cm

FIGURE 31.40 Pollen-Bearing Cones Produce Microsporangia; Ovulate Cones Produce Megasporangia.

Seed Plants > Gymnosperms > Gnetophyta (Gnetophytes)

The gnetophytes comprise about 70 species in three genera: *Gnetum* comprises vines and trees from the tropics; *Ephedra* is made up of desert-dwelling shrubs, including what may be the most familiar gnetophyte—the shrub called Mormon tea, which is common in the deserts of southwestern North America; *Welwitschia* contains a single species that probably qualifies as the world's most bizarre plant (**FIGURE 31.41**), *W. mirabilis*, which is native to the deserts of southwest Africa. Although it has large belowground structures, the aboveground part consists of just two strap-like leaves, which grow continuously from the base and die at the tips. The leaves also split lengthwise as they grow and age.

Adaptations to land Gnetophytes have vessel elements in addition to tracheids. All of the living species make wood as a support structure.

Reproduction and life cycle The microsporangia and megasporangia are arranged in clusters at the end of stalks, similar to the way flowers are clustered in some angiosperms. Pollen is transferred by the wind or by insects. Double fertilization occurs in two of the three genera. As in other gymnosperms, seeds do not form inside an encapsulated structure.

Human and ecological impacts The drug ephedrine was originally isolated from species of *Ephedra* that are native to northern China and Mongolia. Ephedrine is used in the treatment of hay fever, colds, and asthma.

Welwitschia mirabilis

0.5 m

FIGURE 31.41 *Welwitschia* **Is an Unusual Plant.**

Seed Plants > Anthophyta (Angiosperms)

The flowering plants, or angiosperms, are far and away the most species-rich lineage of land plants. Over 250,000 species have already been described. They range in size from *Lemna gibba*—a floating, aquatic species that is less than half a millimeter wide—to massive oak trees. Angiosperms thrive in desert to freshwater to rain forest environments and are found in virtually every habitat except the deep oceans. They are the most common and abundant plants in most terrestrial environments.

The defining adaptation of angiosperms is the flower. Flowers are reproductive structures that hold either pollen-producing microsporangia or the megasporangia that produce megaspores and eggs, or both. Nectar-producing cells are often present at the base of the flower, and the color of petals helps attract insects, birds, or bats that carry pollen from one flower to another (**FIGURE 31.42a**). Some angiosperms are pollinated by wind, however. Wind-pollinated flowers lack both colorful petals and nectar-producing cells (**FIGURE 31.42b**).

Adaptations to land In addition to flowers, angiosperms evolved water-conducting vessels, the conducting cells that make water transport particularly efficient. Most angiosperms contain both tracheids and vessels.

Reproduction and life cycle Unlike gymnosperms, angiosperms have a carpel, a structure within the flower that contains an ovary. The ovary encloses one or more ovules, each of which encloses a female gametophyte. In most cases, male gametophytes are carried to female gametophytes by animal pollinators that are inadvertently dusted with pollen as they visit flowers to find food. Depending on the angiosperm species, self-fertilization may be common or absent. When the egg produced by the female gametophyte is fertilized, the ovule develops into a seed. When the ovary matures, it forms a fruit that contains the seed or seeds.

Human and ecological impacts It is almost impossible to overstate the importance of angiosperms to humans and other organisms. In most terrestrial habitats today, angiosperms supply the food that supports virtually every other species. For example, many insects eat flowering plants. Historically, the diversification of angiosperms correlated closely with the diversification of insects, which are by far the most species-rich lineage on the tree of life. It is not unusual for a single tropical tree to support dozens or even hundreds of insect species. Angiosperm seeds and fruits have also supplied the staple foods of virtually every human culture that has ever existed.

(a) Animal-pollinated flower (this species produces both pollen and eggs in the same flower)

Ornithogalum dubium

0.5 cm

(b) Wind-pollinated flower (this species has separate male and female flowers)

Acer negundo

Male flower

1 cm

Acer negundo

Female flower

1 cm

FIGURE 31.42 Wind-Pollinated Flowers Lack the Colorful Petals and Nectar Found in Most Animal-Pollinated Species.

CHAPTER 31 REVIEW

For media, go to MasteringBiology

If you understand . . .

31.1 Why Do Biologists Study the Green Algae and Land Plants?

- Plants are important primary producers and a source of oxygen, and they hold soil and water in place.
- Plants improve the quality of the environment for other organisms.
- Humans depend on plants for food, fiber, pharmaceuticals, building materials, and fuel.

✔ You should be able to predict how the ongoing and massive loss of plant species and plant communities will affect soils around the world.

31.2 How Do Biologists Study Green Algae and Land Plants?

- Scientists analyze morphological traits, examine the fossil record, and evaluate molecular phylogenies to study and compare green algae and land plants.
- Morphological differences suggest that land plants evolved from multicellular, freshwater green algae from the Charophyceae.
- The fossil record reveals that the first, simple land plants evolved about 475 million years ago and then diversified into plants with vascular systems, seeds, and flowers.

✔ You should be able to describe the synapomorphies that define all green algae and land plants.

31.3 What Themes Occur in the Diversification of Land Plants?

- The evolution of cuticle allowed plant tissues to be exposed to air without dying.
- UV-absorbing compounds allowed the first land plants to tolerate the harsh UV rays from the sun.
- The evolution of stomata, breaks in the cuticle consisting of pores controlled by guard cells, allowed plants to maximize gas exchange and minimize water loss.
- Vascular tissue conducts water and has secondary cell walls reinforced with lignin that provides structural support allowing plants to grow upright. This adaptation reduces competition for light.
- Tracheids are water-conducting cells found in all vascular plants; in addition, angiosperms and gnetophytes have water-conducting cells called vessels.
- All land plants are embryophytes, meaning that eggs and embryos are retained on the parent plant. Consequently, the developing embryo can be nourished by its mother.
- All land plants have alternation of generations. Over the course of land plant evolution, the gametophyte phase became reduced in terms of size and life span, and the sporophyte phase became more prominent. Heterospory also became more common.

- The evolution of pollen was an important breakthrough in the history of life, because sperm no longer needed to swim to the egg—tiny male gametophytes could be transported through the air via wind or animals.

- Seed plant embryos are dispersed from the parent plant to a new location, encased in a protective housing, and supplied with a store of nutrients.

✔ You should be able to discuss (compare and contrast) the advantages and disadvantages of having spores or seeds serve as the dispersal stage in a plant life cycle.

31.4 Key Lineages of Green Algae and Land Plants

- Green algae include single-celled and multicellular freshwater species that share numerous characteristics with land plants. Single-celled species are symbionts with protists, fungi, and animals.

- Nonvascular plants include mosses and liverworts that are small and have a gametophyte-dominant life cycle.

- Seedless vascular plants include lycopods and ferns that have small, free-living gametophytes and large, longer-lived sporophytes. Dispersal is carried out using spores.

- Seed plants include gymnosperms and angiosperms that dominate the world's floras. Animals are often used to disperse seeds and, in angiosperms, to transport pollen between flowers.

(MB) **MasteringBiology**

1. **MasteringBiology Assignments**

   Tutorials and Activities Fern Life Cycle; Gymnosperms; Highlights of Plant Phylogeny; Moss Life Cycle; Nonvascular Plants; Pine Life Cycle; Plant Evolution and the Phylogenetic Tree; Terrestrial Adaptations of Plants

   Questions Reading Quizzes, Blue-Thread Questions, Test Bank

2. **eText** Read your book online, search, take notes, highlight text, and more.

3. **The Study Area** Practice Test, Cumulative Test, BioFlix® 3-D Animations, Videos, Activities, Audio Glossary, Word Study Tools, Art

You should be able to . . .

✔ TEST YOUR KNOWLEDGE   *Answers are available in Appendix A*

1. Which of the following groups is definitely monophyletic?
   a. nonvascular plants
   b. green algae
   c. green plants
   d. seedless vascular plants

2. The appearance of cuticle and stomata correlated with what event in the evolution of land plants?
   a. the first erect growth forms
   b. the first woody tissues
   c. growth on land
   d. the evolution of the first water-conducting tissues

3. What is a pollen grain?
   a. male gametophyte
   b. female gametophyte
   c. male sporophyte
   d. sperm

4. What do seeds contain?
   a. male gametophyte and nutritive tissue
   b. female gametophyte and nutritive tissue
   c. embryo and nutritive tissue
   d. mature sporophyte and nutritive tissue

5. Land plants may have reproductive structures that (a) protect gametes as they develop, (b) allow sperm to be transported in the absence of water, (c) provide stored nutrients and a protective coat so that offspring can be dispersed by wind away from the parent plant, and (d) provide nutritious tissue around seeds that facilitates dispersal by animals. Name each of these four structures, and state which land plant group or groups have each structure.

6. What does it mean to say that a life cycle is gametophyte dominant versus sporophyte dominant?

✔ TEST YOUR UNDERSTANDING   *Answers are available in Appendix A*

7. In the life cycle of an angiosperm, the egg
   a. forms from meiosis of cells inside the seed.
   b. must be fertilized by a sperm that swims to it with its flagella.
   c. is one of the cells formed by mitosis in the ovule.
   d. is part of a mobile female gametophyte.

8. Soils, water, and the atmosphere are major components of the abiotic (nonliving) environment. Describe how green plants affect the abiotic environment in ways that are advantageous to humans.

9. The evolution of cuticle presented land plants with a challenge that threatened their ability to live on land. Describe this challenge, and

explain why stomata represent a solution. Compare and contrast stomata with the pores found in liverworts.

10. Why was the evolution of lignin-reinforced cell walls significant?

11. Explain the difference between homosporous and heterosporous plants. Where are the microsporangium and megasporangium found in a tulip? What happens to the spores that are produced by these structures?

12. Describe the advantage that flowering plants gain by using animals to transfer pollen from one individual to another.

13. Which of the following statements is *not* true?
   a. Green algae in the lineage called Charophyceae are the closest living relatives of land plants.
   b. The nonvascular plants form a monophyletic group.
   c. The horsetails and the ferns form a distinct lineage. They have vascular tissue but reproduce via spores, not seeds.
   d. According to the fossil record and phylogenetic analyses, angiosperms evolved before the gymnosperms. Angiosperms are the only land plants with vessels.

14. Angiosperms such as grasses, oaks, and maples are wind pollinated. The ancestors of these subgroups were probably pollinated by insects, however. As an adaptive advantage, why might a species "revert" to wind pollination? (Hint: Think about the costs and benefits of being pollinated by insects versus wind.) Why is it logical to observe that wind-pollinated species usually grow in dense stands containing many individuals of the same species? Why is it logical to observe that in wind-pollinated

deciduous trees, flowers form very early in spring—before leaves form?

15. Vessel elements transport water much more efficiently than tracheids, but they are much more susceptible than tracheids to being blocked by air bubbles. Suggest a hypothesis to explain why the vascular tissue of angiosperms consists of a combination of vessel elements and tracheids.

16. You have been hired as a field assistant for a researcher interested in the evolution of flower characteristics in orchids. Design an experiment to determine whether color, size, shape, scent, or amount of nectar is the most important factor in attracting pollinators to a particular species. Assume that you can change any flower's color with a dye and that you can remove petals or nectar stores, add particular scents, add nectar by injection, or switch parts among species by cutting and gluing.

32 Fungi

In this chapter you will learn about

Fungi, the master traders and recyclers in terrestrial ecosystems

by asking

Why and how do biologists study fungi?
32.1, 32.2

by examining

Major themes in their diversification
32.3

by focusing on

Fungi as mutualists

Fungi as decomposers

then, by looking more closely at

Key lineages of fungi   32.4

Variation in reproductive structures and life cycles

Chicken-of-the-woods (*Laetiporus sulphureus*) is a common edible mushroom that tastes somewhat like chicken. Most of its body consists of microscopic filaments that grow into dead trees, decomposing the wood.

Fungi are eukaryotes that grow as single cells or as large, branching networks of multicellular filaments. Familiar fungi include the mushrooms you've encountered in the woods or a grocery store, the molds and mildews in your home, the organism that causes athlete's foot, and the yeasts used in baking and brewing.

Along with the land plants and the animals, the **fungi** are one of three major lineages of large, multicellular eukaryotes that occupy terrestrial environments. When it comes to making a living, the species in these three groups use radically different strategies. Land plants make their own food through photosynthesis. Animals and fungi both feed on plants, protists, fungi, or each other by releasing digestive enzymes and absorbing small molecules, but while many animals carry out digestion in a dedicated digestive tract, fungi release their digestive enzymes to the external environment.

Fungi that absorb nutrients from dead organisms are the world's most important decomposers. Although a few types of organisms are capable of digesting the cellulose in plant cell walls, certain fungi and a handful of bacterial species are the only organisms capable of digesting both the lignin and

✔ When you see this checkmark, stop and test yourself. Answers are available in Appendix A.

cellulose that make up wood. Without fungi, Earth's surface would be piled so high with dead tree trunks and branches that there would be almost no room for animals to move or plants to grow.

Other fungi specialize in absorbing nutrients from living organisms. When fungi absorb these nutrients without providing any benefit in return, they lower the fitness of their host organism and act as parasites. If you've ever had athlete's foot or a vaginal yeast infection, you've hosted a parasitic fungus.

Most of the fungi that live in association with other organisms benefit their hosts, however. In these cases, fungi are not parasites but **mutualists.**

- The roots of most land plants are colonized by an array of mutualistic fungi that provide water and key nutrients such as nitrogen and phosphorus in exchange for sugars that are synthesized by the host plant. The soil around you is alive with an enormous network of fungi that are fertilizing the plants you see aboveground.

- Fungi living inside the shoots of certain plants help ward off herbivores by making toxic compounds.

- Many insects harbor single-celled fungi in their guts that aid their digestion.

- Some insects grow gardens of fungi that they feed with pieces of leaves. The insects then feed off the fungi, which they maintain and continue to cultivate.

In short, fungi are the master traders and recyclers in terrestrial ecosystems. Some fungi release nutrients from dead plants and animals; others obtain nutrients that they then transfer to living plants and animals. Because they recycle key elements such as carbon, nitrogen, and phosphorus and because they transfer key nutrients to plants and animals, fungi have a profound influence on ecosystem productivity and biodiversity. In terms of nutrient cycling, fungi make the world go around.

32.1 Why Do Biologists Study Fungi?

Given their importance to life on land and their intricate relationships to other organisms, it's no surprise that fungi are fascinating to biologists. But there are important practical reasons as well for humans to study fungi. They nourish the plants that nourish us. They affect global climate change, because they are critical to the carbon cycle on land. Unfortunately, many fungi can cause debilitating diseases in humans and crop plants. Let's take a closer look at some of the ways that fungi affect human health and welfare.

Fungi Have Important Economic and Ecological Impacts

In humans, parasitic fungi cause athlete's foot, vaginitis, diaper rash, ringworm, pneumonia, and thrush, among other miseries. But even though these maladies can be serious, in reality no more than approximately 200 species of fungi—out of the hundreds of thousands of existing species—regularly cause illness

in humans. Compared with the frequency of diseases caused by bacteria, viruses, and protists, the incidence of fungal infections in humans is low.

It would be easy to argue, in fact, that fungi have done more to promote human health than degrade it. The first antibiotic that was widely used, penicillin, was isolated from a fungus, and soil-dwelling fungi continue to be the source of many of the most important antibiotics prescribed against bacterial infections.

The major destructive impact that fungi have on people is through the food supply. Fungi known as rusts, smuts, mildews, wilts, and blights cause annual crop losses computed in the billions of dollars. These fungi are particularly troublesome in wheat, corn, barley, and other grain crops (**FIGURE 32.1a**). Saprophytic fungi are also responsible for enormous losses due to spoilage—particularly for fruit and vegetable growers (**FIGURE 32.1b**).

In nature, epidemics caused by fungi have killed 4 billion chestnut trees and tens of millions of American elm trees in North America. The fungal species responsible for these epidemics were accidentally imported on species of chestnut and elm native to other regions of the world. When the fungi arrived in North America and began growing in chestnuts and elms native to North America, the results were catastrophic. The local chestnut and elm populations had virtually no genetic resistance to the pathogens and quickly succumbed. Living chestnut root systems continue to sprout shoots, but these trees rarely live long enough to reproduce.

The chestnut and elm epidemics radically altered the composition of upland and floodplain forests in the eastern United States. Before these fungal epidemics occurred, chestnuts and elms dominated these habitats.

(a) Parasitic fungi infect corn and other crop plants.

(b) Saprophytic fungi rot fruits and vegetables.

FIGURE 32.1 Fungi Cause Problems with Crop Production and Storage. (a) A wide variety of grain crops are parasitized by fungi. Corn smut is a serious disease in sweet corn, although in Mexico the smut fungus is eaten as a delicacy. **(b)** Fungi decompose fruits and vegetables as well as leaves and tree trunks.

Fungi also have important positive impacts on the human food supply:

- Mushrooms are consumed in many cultures; in the industrialized nations they are used in sauces, salads, and pizza.

- The yeast *Saccharomyces cerevisiae* was domesticated thousands of years ago; today it and other fungi are essential to the manufacture of bread, soy sauce, cheese, beer, wine, whiskey, and other products. In most cases, domesticated fungi are used in conditions where the cells grow via fermentation, creating ethanol and by-products like the CO_2 that causes bread to rise and beer and champagne to fizz.

- Chocolate is made from the seeds of *Theobroma cacao*, but it is edible only after the seeds are fermented by several species of fungi.

- Enzymes derived from fungi are used to improve the characteristics of foods ranging from fruit juice and candy to meat.

Fungi Provide Nutrients for Land Plants

Fungi that live in close association with plant roots are said to be **mycorrhizal** (literally, "fungal-root"; see **FIGURE 32.2a**). When biologists first discovered how extensive these fungi–plant associations are, they asked an obvious question: Does plant growth suffer if mycorrhizal fungi are absent?

FIGURE 32.2b shows a result typical of many experiments. In this case, seedlings were grown in the presence and absence of the mycorrhizal fungi normally found on their roots. The photographs document that this species grows three to four times faster in the presence of its normal fungal associates than it does without them.

For farmers, foresters, and ranchers, the presence of normal mycorrhizal fungi can mean the difference between profit and loss. Fungi are critical to the productivity of forests, croplands, and rangelands.

Fungi Accelerate the Carbon Cycle on Land

Fungi that make their living by digesting dead plant material are called **saprophytes.** To understand why saprophytic fungi play a key role in today's terrestrial environments, recall that cells in the vascular tissues of land plants have secondary cell walls containing both lignin and cellulose (see Chapter 31). Wood forms when stems grow in girth by adding layers of lignin-rich vascular tissue.

When trees die, certain fungi are the organisms that break down wood into sugars and other small organic compounds (**FIGURE 32.3**). Fungi use these molecules as food, as do many microorganisms. In addition, when fungi die or are eaten, the molecules are passed along to a wide array of other organisms.

Fungi play an important role in cycling carbon atoms through terrestrial ecosystems. Note that there are two basic components of the **carbon cycle** on land:

1. The fixation of carbon by land plants—meaning that carbon in atmospheric CO_2 is reduced to form sugar, which is then used to synthesize cellulose, lignin, and other complex organic compounds in the bodies of plants.

(a) Mycorrhizal fungi form extensive networks in soil.

Seedling root

Mycorrhiza

Mycorrhizal network

(b) Mycorrhizal fungi increase plant growth.

With mycorrhizal fungi   Without

FIGURE 32.2 Plants Grow Better in the Presence of Mycorrhizal Fungi. (a) Root system of a larch tree seedling; the mycelia from a mycorrhizal fungus are visible. **(b)** Typical experimental results when plants are grown with and without their normal mycorrhizal fungi. (Fungi are not visible in the photo.)

2. The release of CO_2 from nearly all organisms as the result of cellular respiration—meaning the oxidation of glucose and production of the ATP that sustains life.

The fundamental point is that, for many carbon atoms, saprophytic fungi connect the two parts of the carbon cycle.

If fungi had not evolved the ability to digest lignin and cellulose soon after land plants evolved the ability to make these compounds, carbon atoms would have been sequestered in wood for millennia instead of being rapidly recycled into glucose molecules and CO_2. Terrestrial environments would be radically different than they are today, and probably much less productive. On land, fungi make the carbon cycle turn much more rapidly than it would without fungi.

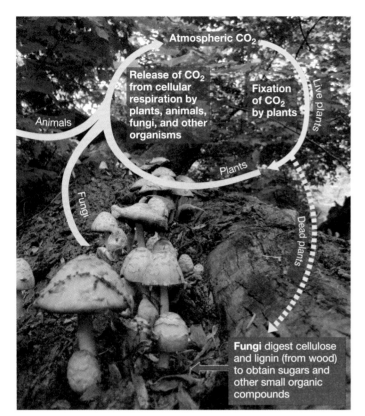

Atmospheric CO₂

Release of CO₂ from cellular respiration by plants, animals, fungi, and other organisms

Fixation of CO₂ by plants

Live plants

Animals

Plants

Fungi

Dead plants

Fungi digest cellulose and lignin (from wood) to obtain sugars and other small organic compounds

FIGURE 32.3 Fungi Speed Up the Carbon Cycle as They Break Down Dead Trees in Terrestrial Ecosystems.

To summarize, biologists study fungi because they affect a wide range of species in nature, including humans. What tools are helping researchers understand the diversity of fungi?

32.2 How Do Biologists Study Fungi?

About 110,000 species of fungi have been described and named to date, and several hundreds more are discovered each year. But the fungi are so poorly studied that the known species are widely regarded as a tiny fraction of the actual total.

To predict the actual number of fungal species alive today, David Hawksworth looked at the ratio of vascular plant species to fungal species in the British Isles—the area where the two groups are the most thoroughly studied. According to Hawksworth's analysis, these islands have an average of six species of fungus for every species of vascular plant. If this ratio holds worldwide, then the estimated total of 275,000 vascular plant species implies that there are perhaps 1.5 million species of fungi.

Although this estimate sounds large, recent data on fungal diversity suggest that it may be an underestimate. Consider what researchers found when they analyzed fungi growing on Barro Colorado Island, Panama: Living on the healthy leaves of just two tropical tree species were a total of 418 distinct morphospecies of fungi. (Recall from Chapter 27 that morphospecies are distinguished from each other by some aspect of morphology.) Because over 310 species of trees and shrubs grow on Barro Colorado, the data suggest that tens of thousands of fungi may be

(a) Single-celled fungi are called yeasts.

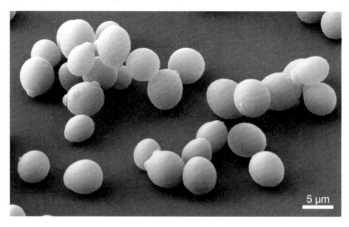

5 μm

(b) Multicellular fungi have weblike bodies called mycelia.

20 mm

FIGURE 32.4 Fungi Have Just Two Growth Forms. Fungi grow **(a)** as single-celled yeasts and/or **(b)** as multicellular mycelia made up of long, thin, highly branched filaments. Both of these scanning electron micrographs are colorized.

native to this island alone. If further work on fungal diversity in the tropics supports these conclusions, there may turn out to be many millions of fungal species.

This viewpoint of fungal diversity was reinforced by an analysis of the fungi living in the guts of beetles. In this study, researchers isolated fungi from 27 species of beetles and characterized them by observing various morphological and physiological traits. In addition, the direct sequencing approach was used to analyze the gene that codes for the RNA molecule in the small subunit of fungal ribosomes (see Chapter 30). The data showed that over 650 fungal species were living in the beetle guts, 200 of which had never been described. Biologists are only beginning to realize the extent of species diversity in fungi.

Let's consider how biologists are working to make sense of all this diversity, beginning with an overview of fungal morphology.

Analyzing Morphological Traits

Compared with animals and land plants, fungi have simple bodies. Only two growth forms occur among them: (**1**) single-celled forms called **yeasts** (**FIGURE 32.4a**), and (**2**) multicellular, filamentous structures called **mycelia** (singular: **mycelium; FIGURE 32.4b**). Many species of fungus grow either as yeasts or as a mycelium, but some regularly adopt both growth forms.

The fossil record for land plants and animals has been extremely useful in understanding their early evolution (see Chapters 25 and 31). In contrast, fossils of fungi are very rare, so biologists rely on morphological and molecular data to explore their past. Because most fungi form mycelia and because this body type is so fundamental to the absorptive mode of life, most studies of fungal morphology have focused on them.

The Nature of the Fungal Mycelium If food sources are plentiful, mycelia can be long lived and grow to be extremely large. Researchers discovered a mycelium of the fungus *Armillaria* growing across 2100 acres (8.8 km^2) in Oregon. This is an area substantially larger than most college campuses. The biologists estimated the individual's weight at hundreds of tons and its age at thousands of years, making it one of the largest and oldest organisms known.

Although most mycelia are much smaller and shorter lived than the individual in Oregon, all mycelia are dynamic. Mycelia constantly grow in the direction of food sources and die back in areas where food is running out. The body shape of a fungus can change almost continuously throughout its life.

The Nature of Hyphae The filaments within a mycelium are called **hyphae** (singular: **hypha**). As **FIGURE 32.5a** shows, hyphae are long, narrow filaments that branch frequently.

In most terrestrial fungi, each filament is divided into cells by cross-walls called **septa** (singular: **septum; FIGURE 32.5b**). Septa do not close off the cells along hyphae completely. Instead, gaps called pores enable a wide variety of materials, even nuclei and other organelles, to flow from one compartment to the next.

Some fungal lineages have hyphae that are **coenocytic** ("common-celled"; pronounced *see-no-SIT-ick*)—meaning that they are not divided into separate cells and thus lack septa (**FIGURE 32.5c**). Coenocytic fungi have hundreds or thousands of nuclei scattered throughout the mycelium. In effect, they are a single, gigantic, multinucleate cell.

Fungi lack complex, long-distance transport systems like those found in plants and animals. One advantage of most fungi having pores in septa is that nutrients can move rapidly from regions of uptake to regions of mycelial growth. Because nutrients and some organelles can flow through the entire mycelium—at least to a degree—the fungal mycelium is intermediate between a multicellular land plant or animal and an enormous single-celled organism.

Mycelia Have a Large Surface Area It's important to appreciate just how thin hyphae are. Plant root tips are typically about 1 mm in diameter, but fungal hyphae are typically less than 10 μm in diameter, or 1/100th the size of a root tip. Fungal mycelia can penetrate tiny fissures in soil and absorb nutrients that are inaccessible to plant roots.

Perhaps the most important aspect of mycelia and hyphae, however, is their shape. Because mycelia are composed of complex, branching networks of extremely thin hyphae, fungi have the highest surface-area-to-volume ratios observed in multicellular organisms and are therefore the best at absorption.

(a) Both the reproductive structure and mycelium are composed of hyphae.

(b) Most hyphae are broken into compartments by septa.

(c) Coenocytic hyphae consist of multinucleate cells.

FIGURE 32.5 Multicellular Fungi Have Unusual Bodies. (a) The feeding structure of a fungus is a mycelium, which is made up of hyphae. In some species, hyphae come together to form multicellular structures such as mushrooms, brackets, or morels that emerge from the ground. **(b)** Hyphae of some terrestrial fungi are divided into cell-like compartments by partitions called septa that contain pores. As a result, the cytoplasm of different compartments is continuous. **(c)** Hyphae of some coenocytic fungi are composed of giant, multinucleate cells.

To drive this point home, consider that the hyphae found in any fist-sized ball of rich soil typically have a surface area equivalent to half a page of this book. Because of their large surface area, fungi are extremely efficient at absorbing nutrients.

The extraordinarily high surface area in a mycelium has a downside, however. The amount of water that evaporates from an organism is a function of its surface area—meaning that these organisms are prone to drying out. As a result, fungi are most abundant in moist habitats.

When conditions dry, the fungal mycelium may die back partially or completely. Reproductive cells called **spores** that are produced by sexual or asexual reproduction are resistant to drying, however. As a result, spores can endure dry periods and then germinate to form a new mycelium when conditions improve. Mycelial growth is dynamic, changing with moisture availability or food supply.

Reproductive Structures Mycelia are an adaptation that supports external digestion and the absorptive lifestyle of fungi. Fungi also produce reproductive organs, which are thick, fleshy structures.

Mushrooms, puffballs, and other dense, multicellular structures that arise from mycelia do not absorb food. Instead, they function in reproduction. Typically they are the only part of a fungus that is exposed to the atmosphere, where drying is a problem. The mass of hyphal filaments on the inside of mushrooms is protected from drying by the densely packed hyphae forming the surface.

In many fungi, including some entire lineages, sexual reproduction has never been observed. In those lineages that do reproduce sexually, important morphological differences among lineages are seen. Most fungal species that undergo sexual reproduction produce one of four types of distinctive reproductive structures.

1. *Swimming gametes and spores*   In certain species that live primarily in water or wet soils, the gametes produced during sexual reproduction have flagella, as do the spores produced during asexual reproduction (**FIGURE 32.6a**). These are the only motile cells known in fungi. Species with swimming gametes are traditionally known as chytrids (pronounced *KYE-trids*).

2. *Zygosporangia*   In some species, haploid hyphae from two individuals meet and become joined, like oxen with a yoke, as shown in **FIGURE 32.6b**. Cells from yoked hyphae fuse to form a distinctive spore-producing structure called a **zygosporangium** (plural: **zygosporangia**; the Greek root *zygo* means to be yoked [see **BioSkills 2** in Appendix B]). Species with a zygosporangium are traditionally known as zygomycetes.

3. *Basidia*   Mushrooms, brackets, and puffballs form specialized club-like cells called **basidia** (singular: **basidium**; "little-club") at the ends of hyphae, each producing four spores (**FIGURE 32.6c**). Species with basidia are traditionally called basidiomycetes or "club fungi."

4. *Asci*   Cups, morels, and some other types of fungi form reproductive structures, specialized sac-like cells called **asci** (singular: **ascus**) at the ends of hyphae, on the surfaces of the fungi, each producing eight spores (**FIGURE 32.6d**). Species with asci are traditionally known as ascomycetes or "sac fungi."

(a) Swimming gametes and spores

Gametes
Spore
Flagella

(b) Zygosporangia: spore-producing structures formed when hyphae are yoked

Zygosporangium
Hypha

(c) Basidia: club-shaped cells where meiosis occurs and 4 spores form

Hypha   Basidium
Spores

(d) Asci: sac-like cells where meiosis occurs and 8 spores form.

Spores
Ascus
Hypha

1 μm   25 μm   20 μm   50 μm

FIGURE 32.6 Four Types of Reproductive Structures Are Observed in Fungi. The dots in the illustrations represent nuclei.

In sum, morphological studies allowed biologists to describe and interpret the mycelial growth habit as an adaptation that makes external digestion and nutrient absorption extremely efficient. Careful analyses of morphological features also allowed researchers to identify four major types of sexual reproductive structures.

Now the question is, within the Fungi, do species that produce swimming gametes and spores, zygosporangia, basidia, and asci each form monophyletic groups—meaning that these distinctive reproductive structures evolved just once? Also, which eukaryotes are most closely related to the fungi?

Evaluating Molecular Phylogenies

Researchers have sequenced and analyzed an array of genes and genomes to establish where fungi fit on the tree of life. In fact, the first complete genome sequence of a eukaryotic species was that of a fungus, the model organism baker's yeast, *Saccharomyces cerevisiae*. The phylogenetic position of fungi as a whole is now well established; the position of lineages within fungi is still the subject of intense research.

Fungi Are Closely Related to Animals   **FIGURE 32.7** shows that fungi are much more closely related to animals than they are to land plants.

The close evolutionary relationship between fungi and animals explains why fungal infections in humans are much more difficult to treat than bacterial infections. Fungi and humans shared a common ancestor relatively recently. As a result, their enzymes and cell components are similar in structure and function. Drugs that disrupt fungal enzymes and cells are also likely to damage humans.

In addition to DNA sequence data, three key morphological traits link animals and fungi:

1. Most animals and fungi synthesize the tough structural material called chitin (see Chapter 5). Chitin is a prominent component of the cell walls of fungi.

2. The flagella that develop in chytrid spores and gametes are similar to those observed in animals: As in animals, the flagella in chytrids are single, are located at the back of reproductive cells, and move in a whiplike manner.

3. Both animals and fungi store food by synthesizing the polysaccharide glycogen. (Green plants, in contrast, synthesize starch as their storage product.)

What Are the Relationships Among the Major Fungal Groups? To understand the relationships among species with swimming gametes and spores, zygosporangia, basidia, and asci, biologists have sequenced a series of genes from an array of fungal species and used the data to estimate the phylogeny of the group. The results, shown in **FIGURE 32.8**, support several important conclusions:

- The single-celled, parasitic eukaryotes called microsporidians are actually fungi.
  Interpretation: They are not a distantly related sister group to fungi, as initially thought. This point is important. Researchers are now testing the hypothesis that **fungicides**—molecules that are lethal to fungi—can cure microsporidian infections in bee colonies, silkworm colonies, and AIDS patients.

- The chytrids and zygomycetes are poorly resolved. The actual order of branching events among lineages with these reproductive structures is still not known (in Figure 32.8, they are collapsed into a **polytomy;** see Chapter 28 and **BioSkills 7** in Appendix B).
  Interpretation: Swimming gametes and the zygosporangium evolved more than once. Or, both structures were present in a common ancestor but then were lost in certain lineages.

- An important group called the Glomeromycota is monophyletic.
  Interpretation: The adaptations that allow these species to live in association with plant roots as mycorrhizae (AMF, discussed in Section 32.3) evolved once.

- The basidiomycetes are monophyletic—they form a lineage called Basidiomycota, or club fungi.
  Interpretation: The basidium evolved once.

- The ascomycetes are monophyletic—they form a lineage called Ascomycota, or sac fungi.
  Interpretation: The ascus evolved once.

- Together, the Basidiomycota and Ascomycota form a monophyletic group.
  Interpretation: Because basidiomycetes and ascomycetes both form septate hyphae and large "fruiting" structures, this growth habit evolved once.

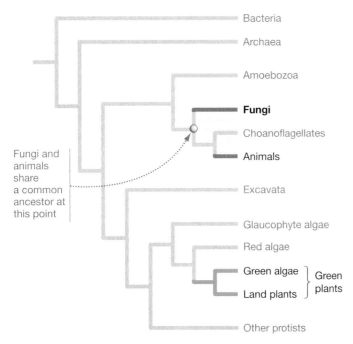

FIGURE 32.7 Fungi Are More Closely Related to Animals than to Land Plants. Phylogenetic tree showing the evolutionary relationships among the green plants, animals, fungi, and some groups of protists. (Choanoflagellates are solitary or colonial protists found in freshwater; see Chapter 33).

FIGURE 32.8 Phylogeny of the Fungi. A phylogenetic tree based on analyses of DNA sequence data. The icons represent the types of sexual reproductive structures observed in each major lineage.

- The sister group to fungi comprises animals plus protists called choanoflagellates.
  Interpretation: Because choanoflagellates and the most ancient groups of animals are aquatic, and because chytrids are aquatic, it is reasonable to hypothesize that the earliest fungi were aquatic and that the switch to terrestrial life occurred early in the evolution of the Fungi.

Although recently advances have been made in understanding the evolutionary history of fungi, the phylogenetic tree in Figure 32.8 is still a work in progress. For example, it is not yet clear where microsporidians are placed relative to several lineages of chytrids and zygomycetes. (Microsporidians lack both swimming gametes and yoked hyphae.) Future work should clarify exactly how fungi diversified.

If you understand that . . .
- The bodies of fungi are either single-celled yeasts or multicellular mycelia.
- During sexual reproduction, different groups of fungi produce distinct reproductive structures.

✓ **You should be able to . . .**

1. Explain why mycelia are interpreted as an adaptation to an absorptive lifestyle.
2. Identify the four types of sexual reproductive structures observed in fungi.

Answers are available in Appendix A.

32.3 What Themes Occur in the Diversification of Fungi?

Why are there so many different species of fungi? This question is particularly puzzling given that fungi share a common attribute: They all make their living by absorbing food directly from their surroundings. In contrast to the diversity of food-getting strategies observed in bacteria, archaea, and protists, all fungi make their living in the same basic way. In this respect, fungi are like plants—virtually all of which make their own food via photosynthesis.

The diversification of land plants was driven not by novel ways of obtaining food, but by adaptations that allowed plants to grow and reproduce in a diverse array of terrestrial habitats (see Chapter 31). Recall that associations with animal pollinators were especially important in the diversification of flowering plants. What drove the diversification of fungi? The answer is the evolution of novel methods for absorbing nutrients from a diverse array of food sources.

This section introduces a few of the ways that fungi go about absorbing nutrients from different food sources, as well as how they produce offspring. Let's explore the diversity of ways that fungi do what they do.

Fungi Participate in Several Types of Symbioses

The first plants in the fossil record are closely associated with fungal fossils; the ability to absorb nutrients from fungi may have been crucial in the early evolution of land plants (see Chapter 31). Close associations between land plants and fungi continue today. Researchers estimate that 90 percent of land plants live in physical contact with fungi. Stated another way, fungi and land plants often have a **symbiotic** ("together-living") relationship.

Although some species of fungi live in association with an array of different land plant species, some documented fungi–plant associations are specific. It is not uncommon for one fungal species to live in only a particular type of tissue, in one plant species.

Scientists categorize these symbiotic relationships as **mutualistic** if they benefit both species, **parasitic** if one species benefits at the expense of the other, or **commensal** if one species benefits while the other is unaffected. However, it is sometimes difficult to know if a species really benefits from the relationship.

Not long after associations between fungi and the roots of land plants were discovered, researchers found that two types of plant–mycorrhizal interactions are particularly common, involving **ectomycorrhizal fungi (EMF)** and **arbuscular mycorrhizal fungi (AMF)**. The two major types of mycorrhizae have distinctive morphologies, geographic distributions, and functions.

- EMF are usually species from the Basidiomycota, though some ascomycetes participate.

- AMF include species from the Glomeromycota.

Mycorrhizae aren't the only type of symbiotic fungi found in plants, however. Researchers have also become interested in fungi that live in close association with the aboveground tissues of land plants—their leaves and stems. Fungi that live in the aboveground parts of plants are called **endophytic** ("inside-plants").

Recent research has shown that endophytic fungi are much more common and diverse than previously suspected. Further, data indicate that at least some species of endophytes are mutualistic.

Ectomycorrhizal Fungi (EMF) EMF like the one shown in **FIGURE 32.9a** are found on many of the tree species in the temperate regions of the world, where warm summers alternate with cold winters (see Chapter 52). In this type of association, hyphae form a dense network that covers a plant's root tips. As Figure 32.9a shows, individual hyphae penetrate between cells in the outer layer of the root, but hyphae do not enter the root cells.

The Greek root *ecto*, which refers to "outer," describes this association accurately: The fungi form an outer sheath on root tips that is often 0.1 mm thick. Hyphae also extend out from the sheath-like portion of the mycelium into the soil.

How do these trees and fungi interact, once they start living together? In the habitats where EMF are abundant, nitrogen atoms tend to remain tied up in dead tissues, in amino

(a) Ectomycorrhizal fungi (EMF) form sheaths around roots and penetrate between root cells.

20 μm

EMF

Hyphae extend outward into the soil

Hyphae extend inward, between cells

Hyphae form a dense, continuous sheath around root

(b) Arbuscular mycorrhizal fungi (AMF) contact plasma membranes of root cells.

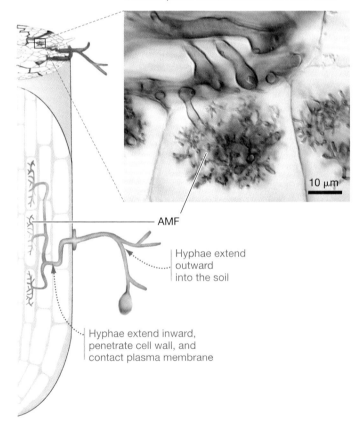

10 μm

AMF

Hyphae extend outward into the soil

Hyphae extend inward, penetrate cell wall, and contact plasma membrane

FIGURE 32.9 Mutualistic Fungi Interact with the Roots of Plants in Two Distinct Ways. (a) Ectomycorrhizal fungi (EMF) form a dense network around the roots of plants. Their hyphae penetrate the intercellular spaces of the root but do not enter the root cells. **(b)** The hyphae of arbuscular mycorrhizal fungi (AMF) penetrate the walls of root cells, where they branch into bushy structures or balloon-like vesicles that are in close contact with the root cell's plasma membrane.

acids and nucleic acids, instead of being available in the soil. The hyphae of EMF penetrate decaying material and release enzymes called peptidases that cleave the peptide bonds between amino acids in dead tissues. The amino acids released by this reaction are absorbed by the hyphae and transported to spaces between the root cells of trees, where they can be absorbed by the plant.

EMF are also able to acquire phosphate ions that are bound to soil particles and transfer the ions to host plants. In return, the fungi receive sugars and other complex carbon compounds from the tree.

Researchers have found that when birch tree seedlings are grown with and without their normal EMF in pots filled with forest soil, only the seedlings with EMF are able to acquire significant quantities of nitrogen and phosphorus. Inspired by such data, one biologist has referred to ectomycorrhiza as the "dominant nutrient-gathering organs in most temperate forest ecosystems."

The hyphae of EMF are like an army of miners that discover, excavate, and deliver precious nuggets of nitrogen to trees. The productivity of the world's most important commercial forests depends on EMF.

Arbuscular Mycorrhizal Fungi (AMF)

In contrast to the hyphae of EMF, the hyphae of arbuscular mycorrhizal fungi (AMF) grow *into* the cells of root tissue. The name *arbuscular* ("little-tree") was inspired by the bushy, highly branched hyphae, shown in **FIGURE 32.9b**, that form between the cell walls and the plasma membrane of root cells. AMF are also called **endomycorrhizal fungi,** because they penetrate the interior of root cell walls.

The key point is that the hyphae of AMF penetrate the cell wall and make direct contact with the plasma membrane of root cells. The highly branched hyphae inside the plant cell wall are thought to be an adaptation that increases the surface area available for exchange of molecules between the fungus and its host. However, AMF do not form a tight sheath around roots, as do EMF. Instead, they form a pipeline extending from inside plant roots into the soil well beyond the root.

AMF are found in a whopping 80 percent of all land plant species. They are particularly common in grasslands and in tropical forests. They are also widespread in temperate climates.

What do AMF do? Plant tissues decompose quickly in the grasslands and tropical forests where AMF flourish because the growing season is long and warm. As a result, nitrogen is often readily available to plants. Phosphorus is usually in short supply, though, because it tends to leach out of soils that experience high rainfall.

To explore the nature of fungi–plant symbioses in detail, researchers have used isotopes as tracers for specific elements (**FIGURE 32.10**, see page 622). For example, to test the hypothesis that fungi obtain food in the form of carbon-containing compounds from their plant associates, biologists have introduced radioactively labeled carbon dioxide into the air surrounding plants that do or do not contain symbiotic fungi. The labeled CO_2 molecules are incorporated into the sugars produced during photosynthesis, and the location of the radioactive atoms can then be followed over time by means of a device that detects radioactivity. If plants feed their fungal symbionts, then labeled carbon compounds should be transferred from the plant to the fungi.

To test the hypothesis that plants are receiving nutrients from their symbiotic fungi in return for sugars, researchers have added radioactive phosphorus atoms or the heavy isotope of nitrogen (^{15}N) to potted plants that do or do not contain symbiotic fungi. If fungi facilitate the transfer of nutrients from soil to plants, then plants grown in the presence of their symbiotic fungi should receive much more of the radioactive phosphorus or heavy nitrogen than do plants grown in the absence of fungi.

Experiments with isotopes used as tracers have shown that sugars and other carbon-containing compounds produced by plants via photosynthesis are transferred to their fungal symbionts. In some cases, as much as 20 percent of the sugars produced by a plant end up in their symbiotic fungi. In exchange, the symbiotic fungi facilitate the transfer of phosphorus or nitrogen—or both—from soil to the plant.

Because phosphorus and nitrogen are in extremely short supply in most environments, the nutrients supplied by symbiotic fungi are critical to the success of the plant. In this way, studies with isotopes have supported the hypothesis that most relationships between fungi and land plants are mutually beneficial.

AMF are also extremely important in soil formation. The cell walls of their hyphae contain large quantities of a glycoprotein called **glomalin.** When the cells die, the glomalin enriches the organic matter in soil and helps bind organic compounds to sand or clay particles. Some recent estimates suggest that over 25 percent of all of the organic matter in soil consists of glomalin.

Are Endophytes Mutualists?

Although endophytic fungi were unknown before the 1940s, they are turning out to be both extremely common and highly diverse.

- Biologists in Brazil are examining tree leaves for the presence of fungi; each time they do, they are discovering several new species of endophytes.

- Recall from Section 32.2 that a study in Panama found hundreds of fungal species living in the leaves of just two tree species. These species are endophytes.

Recent research has shown that some endophytes increase the drought tolerance of their host plants. Endophytes found in some grasses also produce compounds that benefit plants. The compounds deter or even kill herbivores. In exchange for these benefits, endophytes absorb sugars from the plant.

Based on these results, biologists have concluded that the relationship between endophytes and grasses is mutualistic. Similar types of anti-herbivore compounds have recently been documented in an endophyte that lives in morning glories.

In other types of plants, however, researchers have not been able to document benefits for the plant host. The current consensus is that at least some endophytic fungi may be commensals—meaning the fungi and the plants simply coexist with no observable effect, either deleterious or beneficial, on the host plant.

QUESTION: Are mycorrhizal fungi mutualistic?

HYPOTHESIS: Host plants provide mycorrhizal fungi with sugars and other photosynthetic products. Mycorrhizal fungi provide host plants with phosphorus and/or nitrogen from the soil.

NULL HYPOTHESIS: No exchange of food or nutrients occurs between plants and mycorrhizal fungi. The relationship is not mutualistic.

EXPERIMENTAL SETUP:

Labeled carbon treatment:
Labeled CO$_2$
added to air around plant

Mycorrhizal fungi present

Labeled carbon control:
Labeled CO$_2$
added to air around plant

No fungi present

Labeled P or N treatment:

Fungi

Labeled P or N added to soil

Labeled P or N control:

No fungi

Labeled P or N added to soil

PREDICTION FOR LABELED CARBON: A large percentage of the labeled carbon taken up by the plant will be transferred to mycorrhizal fungi. In the control, little labeled carbon will be present in the soil surrounding the roots.

PREDICTION OF NULL HYPOTHESIS, LABELED CARBON: There will be no difference in the localization of carbon in the two treatments.

PREDICTION FOR LABELED P OR N: A large percentage of the labeled P or N taken up by the fungi will be transferred to the plant. In the control, little or no labeled P or N will be taken up by the plant.

PREDICTION OF NULL HYPOTHESIS, LABELED P OR N: There will be no difference between amounts of labeled P or N found in plant in presence or absence of fungi.

RESULTS:

Fungi

Up to 20% of labeled CO$_2$
taken up by plant is transferred
to mycorrhizal fungi

No fungi

Little to no labeled carbon
is found in soil

Large amounts of labeled P or N
are found in host plant

Fungi

Little labeled P or N
is found in host plant

No fungi

CONCLUSION: The relationship between plants and mycorrhizal fungi is mutualistic. Plants provide mycorrhizal fungi with carbohydrates. Mycorrhizal fungi supply host plants with nutrients.

FIGURE 32.10 Experimental Evidence that Mycorrhizal Fungi and Plants Are Mutualistic. Sugars flow from plants to mycorrhizal fungi; key nutrients flow from mycorrhizal fungi to plants.

SOURCES: Bücking, H., and W. Heyser. 2001. Microautoradiographic localization of phosphate and carbohydrates in mycorrhizal roots of *Populus tremula* × *Populus alba* and the implications for transfer processes in ectomycorrhizal associations. *Tree Physiology* 21: 101–107.

✔ **QUESTION** What do these "labeled nutrient" experiments tell you that you didn't already know from experiments like Figure 32.2, where plants are grown with and without mycorrhizae?

FIGURE 32.11 Leaf-cutter Ants Engage in a Symbiosis with Fungi in Underground Nests. The ants provide the fungus with leaf pieces on which to grow and then the ants feed on the fungus. Of the two partners, only the fungus can digest cellulose.

These results support the general realization that most plants are covered with fungi—from the ends of their branches to the tips of their roots. Throughout their lives, many or even most plants are involved in several distinct types of mutualistic relationships with fungi.

Mutualisms with Other Species Do fungi take up residence with species other than land plants? As you'll recall, the answer is yes.

- **Lichens** are a mutualistic partnership usually between a species of ascomycete and either a cyanobacterium or an alga. The nature of this relationship is explored in Section 32.4.

- Many plant-eating insects harbor diverse arrays of yeast species in their guts, where they may aid digestion or detoxification of plant compounds.

- Some ant species actively farm fungi inside their colonies by inoculating leaf pieces with symbiotic fungi. They fertilize and "weed" their fungal gardens and then harvest parts of the fungi for food (**FIGURE 32.11**).

The many unique properties of fungi have made them useful partners in a wide array of symbiotic relationships.

What Adaptations Make Fungi Such Effective Decomposers?

The saprophytic fungi are master recyclers. Although bacteria and archaea are also important decomposers in terrestrial environments, some fungi and a few bacterial species are the only organisms that can digest wood completely. Given enough time, fungi can turn even the hardest, most massive trees into soft soils.

How do fungi do it? You've already been introduced to two key adaptations:

- The large surface area of a mycelium makes nutrient absorption exceptionally efficient.

- Saprophytic fungi can grow toward the dead tissues that supply their food.

What other adaptations help fungi decompose plant tissues?

Extracellular Digestion Large molecules such as starch, lignin, cellulose, proteins, and RNA cannot diffuse across plasma membranes. Only sugars, amino acids, and other small molecules can enter the cytoplasm of cells through membrane carrier proteins or transporters (see Chapter 6). As a result, fungi, like most animals, have to digest their food before they can absorb it.

Fungi perform **extracellular digestion**—digestion that takes place outside the organism. The simple compounds that result from enzymatic action are then absorbed by the hyphae. Most animals also perform extracellular digestion, but it takes place in a digestive tract (see Chapter 44).

As an example of how this process occurs in fungi, consider the enzymes responsible for digesting lignin and cellulose—the two most abundant organic molecules on Earth.

- Lignin is an extremely strong, complex polymer built from monomers that contain six-carbon rings. Recall that most lignin is found in the secondary cell walls of plant vascular tissues, where it furnishes structural support (see Chapter 31).

- Cellulose is a polymer of glucose and is found in the primary and secondary cell walls of all plant cells.

Some basidiomycetes can degrade lignin and digest cellulose completely—to CO_2 and H_2O. Let's take a closer look at how they do it.

Lignin Degradation Biologists have been keenly interested in understanding how basidiomycetes digest lignin. Paper manufacturers are also interested in this process because they need safe, efficient ways to degrade lignin in order to make soft, absorbent paper products.

To find out how lignin-digesting fungi do it, biologists began analyzing the proteins that these species secrete into extracellular space. After purifying these molecules, the investigators tested each protein for the ability to degrade lignin. Using this approach, investigators from two labs independently discovered an enzyme called lignin peroxidase.

Lignin peroxidase catalyzes the removal of a single electron from an atom in the ring structures of lignin. This oxidation step creates a free radical—an atom with an unpaired electron (see Chapter 2). This is an extremely unstable electron configuration, and it leads to a series of uncontrolled and unpredictable reactions that split the polymer into smaller units.

Biologists have referred to this mechanism of lignin degradation as enzymatic combustion. The phrase is apt: The uncontrolled oxidation reactions triggered by lignin peroxidase are analogous to the uncontrolled oxidation reactions that occur when gasoline burns in a car engine.

The unpredictable nature of the reaction is remarkable, because virtually all of the other reactions catalyzed by enzymes are extremely predictable. The lack of predictability here makes sense, however. Unlike proteins, nucleic acids, and most other polymers with a regular and predictable structure, lignin is extremely heterogeneous. Over 10 types of covalent linkages are routinely found between the monomers that make up lignin. But once lignin peroxidase has created a free radical in the ring structure, any of these linkages can be broken.

However, fungi cannot grow with lignin as their sole source of food. The 6-carbon rings in lignin are extremely difficult to metabolize, and as a result they accumulate in soil. If wood-rotting fungi don't use lignin as food, why do they produce enzymes to digest it? The answer is simple. In wood, lignin forms a dense matrix around long strands of cellulose. Degrading the lignin matrix gives hyphae access to huge supplies of energy-rich cellulose. Saprophytic fungi are like miners. But instead of seeking out rare, gem-like nitrogen or phosphorus atoms as do EMF and AMF, the saprophytes use lignin peroxidase to blast away enormous lignin molecules, exposing rich veins of cellulose that can fuel growth and reproduction.

Cellulose Digestion   Once lignin peroxidase has softened wood by stripping away its lignin matrix, the long strands of cellulose that remain can be attacked by enzymes called cellulases.

Like lignin peroxidase, cellulases are secreted into the extracellular environment by fungi. But unlike the uncontrolled chain reaction catalyzed by lignin peroxidase, degradation of cellulose by cellulases is extremely predictable.

Biologists have purified seven different cellulases from the fungus *Trichoderma reesei*. Two of these enzymes catalyze a critical early step in digestion—they cleave long strands of cellulose into a disaccharide called cellobiose. The other cellulases are equally specific and also catalyze hydrolysis reactions. In combination, the suite of seven enzymes in *T. reesei* transforms long strands of cellulose into a simple monomer—glucose—that the fungus can absorb and use as a source of food.

Variation in Reproduction

Recall from Section 32.2 that fungi may produce swimming gametes and spores, yoked hyphae in which nuclei from different individuals fuse to form a zygosporangium inside a protective structure, or specialized spore-producing cells called basidia and asci. As fungi diversified, in addition to adapting to a variety of food sources, they evolved an array of ways to reproduce.

Spores as Key Reproductive Cells   The spore is the most fundamental reproductive cell in fungi. Spores are the dispersal stage in the fungal life cycle and are produced during both asexual and sexual reproduction. (Recall from Chapter 13 that asexual reproduction is based on mitosis, while sexual reproduction is based on meiosis.) Fungi produce spores in such prodigious quantities that it is not unusual for them to outnumber pollen grains in air samples.

If a spore falls on a food source and is able to germinate, a mycelium begins to form. As the fungus expands, hyphae grow in the direction in which food is most abundant. But if food begins to run out, mycelia respond by making spores, which are dispersed by wind or animals.

Why would mycelia reproduce when food is low? The leading hypothesis to answer this question is that spore production allows starving mycelia to disperse offspring to new habitats where more food might be available. Thus, spore production is favored by natural selection when individuals are under nutritional stress.

Multiple Mating Types   In some fungal lineages, hyphae come in many different mating types. Instead of having morphologically distinct males and females that produce sperm and eggs, hyphae of different mating types appear identical.

Hyphae of the same mating type will not combine during sexual reproduction. When zygomycete hyphae grow close to each other, for example, they will not fuse unless the individuals have different alleles of one or more genes involved in mating. If chemical messengers released by two hyphae indicate that they are of different mating types, then fusion and zygosporangium formation follows.

In this way, mating types function as sexes. Instead of having just two sexes, a single fungal species may have tens of thousands of sexes. The basidiomycete *Schizophyllum commune*, for example, is estimated to have 28,000 mating types.

Why so many? The leading hypothesis to explain the existence of mating types is that it helps generate genetic diversity in offspring by increasing the probability of outcrossing. Genetic diversity, in turn, is known to be advantageous in fighting off infections and responding to changes in the environment (see Chapter 13).

How Does Fertilization Occur?   Compared with green plants, protists, and animals, fertilization in fungi has important unique features:

- Only some chytrids produce gametes, which are not considered eggs and sperm. Both male and female gametes are motile, and female gametes are only slightly larger than male gametes.

- In all other lineages of fungi, fertilization occurs in two distinct steps: (1) fusion of hyphae and (2) fusion of nuclei from the fused hyphae. These two steps can be separated by long time spans and even long distances.

In many fungi, the process of sexual reproduction begins when hyphae from two different mating types fuse to form a hybrid cell. When the cytoplasm of two individuals fuses in this way, **plasmogamy** (pronounced *plaz-MAH-ga-mee*) is said to occur (**FIGURE 32.12**). Often, plasmogamy does not immediately lead to nuclear fusion. Rather, the genetically distinct, haploid nuclei persist in the same cell or mycelium. When two or more genetically distinct nuclei exist within a single mycelium, it is **heterokaryotic** ("different kernel"). Most hyphae in heterokaryotic mycelia are **dikaryotic** ("two kernel")—they are divided by septa, and each cell contains two nuclei, one from each mating type.

The distinct nuclei in dikaryotic and heterokaryotic mycelia function independently, even though gene expression must be coordinated for growth and development to occur. For

FIGURE 32.12 Fungi Have Unusual Life Cycles. A generalized fungal life cycle, showing both asexual and sexual reproduction.

✔ **QUESTION** Fungi spend most of their lives feeding. Which is the longest-lived component of this life cycle?

example, in septate fungi the two types of nuclei divide as the hyphae expand, so each compartment that is divided by a septum contains one of each of the two types of nuclei. Biologists are currently investigating how the activities of the two nuclei are coordinated.

In a dikaryotic or heterokaryotic mycelium, one or more pairs of unlike nuclei may eventually fuse to form a diploid zygote. The fusion of nuclei is called **karyogamy** (pronounced *ka-ree-AH-ga-mee*). The nuclei that are produced by karyogamy then divide by meiosis to form haploid spores.

✔ If you understand the relationship between plasmogamy and karyogamy, you should be able to compare and contrast these events with the life cycles of other eukaryotes. For example, which human cells undergo plasmogamy and karyogamy? Are any human cells heterokaryotic?

Asexual Reproduction   As the left side of Figure 32.12 indicates, fungal species can reproduce asexually as well as sexually. In fact, large numbers of ascomycetes have never been observed to reproduce sexually.

During asexual reproduction, spore-forming structures are produced by a haploid mycelium, and spores are generated by mitosis. As a result, offspring are clones—meaning that they are genetically identical to their parent.

Producing microscopic spores that are carried by wind was an adaptation to life on land that allowed terrestrial fungi to disperse widely and to colonize new habitats.

Four Major Types of Life Cycles

Among the sexually reproducing species of fungi, the presence of a heterokaryotic stage and the morphology of the spore-producing structure vary. Morphologies of reproductive structure were introduced in Section 32.2. Let's now take a closer

look at the sexual reproduction portion of each of the four major types of life cycles that have been observed in fungi.

Chytrid Life Cycle   The chytrids are the only type of fungi with species that exhibit alternation of generations. **FIGURE 32.13a** (see page 626) shows how alternation of generations occurs in the well-studied genus *Allomyces*. Here are the key stages in its life cycle:

1. Haploid adults form gametangia, in which male and female swimming gametes are produced by mitosis.

2. Gametes from the same individual or different individuals fuse to form a diploid zygote.

3. The zygote grows into a diploid sporophyte.

The life cycle continues when meiosis occurs in the sporophytic mycelium, inside a structure called a sporangium. The haploid spores produced by meiosis disperse by swimming.

Zygomycete Life Cycle   In zygomycetes, sexual reproduction starts when hyphae from different mating types fuse, as shown in **FIGURE 32.13b**. Plasmogamy forms a spore-forming structure called a zygosporangium that develops a tough, resistant coat. Inside the zygosporangium, nuclei from the mating partners fuse—meaning that karyogamy occurs.

The zygosporangium can persist if conditions become too cold or dry to support growth. When temperature and moisture conditions are again favorable, however, meiosis occurs and the zygosporangium germinates and forms a sporangium. The meiotic products within the sporangium produce haploid spores.

When spores are released and germinate, they grow into new mycelia. These mycelia can reproduce asexually by making sporangia whose haploid spores are produced by mitosis and dispersed by wind.

(a) Chytrids include the only fungi in which alternation of generations occurs. (Species: *Allomyces macrogynus*)

Gametophytic mycelium (*n*)

1 mm

MITOSIS

Gametes form in gametangia

Gametes (*n*)

♂

♀

Gametes and spores (*n*) disperse by water

FUSION

Plasmogamy and karyogamy

Zygote (2*n*)

MITOSIS

Sporophytic mycelium (2*n*)

Spores form in sporangia

1 mm

2*n*

MEIOSIS

n

Spore (*n*)

MITOSIS

Haploid (*n*)

Heterokaryotic (*n* + *n*)

Diploid (2*n*)

(b) Zygomycetes form yoked hyphae that produce a spore-forming structure. (Species: *Rhizopus stolonifer*)

0.1 mm

1 mm

Hyphae of different mating types

Hyphae fuse

PLASMOGAMY

Spores form in zygosporangia

n + *n*

n + *n*

Spores (*n*)

n

MITOSIS

MEIOSIS

Zygote

2*n* KARYOGAMY

FIGURE 32.13 Variation in Sexual Reproduction in Fungi. The sexual part of the life cycle in four major groups of fungi: **(a)** chytrids, **(b)** zygomycetes, **(c)** Basidiomycota, and **(d)** Ascomycota.

✔ **QUESTION** Asexual reproduction, as shown in Figure 32.12, is extremely common in zygomycetes and ascomycetes. What is the difference between mycelia produced via asexual versus sexual reproduction?

Basidiomycota Life Cycle Mushrooms, bracket fungi, and puffballs are sexual reproductive structures produced by members of the Basidiomycota (**FIGURE 32.13c**). Even though their size, shape, and color vary enormously from species to species, all basidiomycete reproductive structures originate from the dikaryotic hyphae of mated individuals.

Inside a mushroom or bracket fungus or puffball, the club-like, spore-producing cells called basidia form at the ends of dikaryotic hyphae. Karyogamy occurs within the basidia. The diploid nucleus that results undergoes meiosis, yielding four haploid spores. The spores are dispersed by wind and then germinate to form a new hyphae.

(c) Basidiomycota have reproductive structures with many spore-producing basidia. (Species: *Amanita muscaria*)

(d) Ascomycota have reproductive structures with many spore-producing asci. (Species: *Cookeina speciosa*)

Spores are eventually ejected from the end of the basidia and are dispersed by the wind. It is not unusual for a single puffball or mushroom to produce a billion spores.

Ascomycota Life Cycle   The mature sexual reproductive structure in Ascomycota is produced by a dikaryotic hypha.

As **FIGURE 32.13d** illustrates, the process usually begins when hyphae or specialized structures from different mating types fuse, forming a heterokaryotic cell containing many independent nuclei. A short dikaryotic hypha with cells containing two nuclei, one from each parent, emerges and eventually grows into a complex reproductive structure whose hyphae have the sac-like, spore-producing structures called asci at their tips.

After karyogamy occurs inside each ascus, meiosis takes place and eight haploid spores are produced. When the ascus matures, the spores inside are forcibly ejected.

If you understand that . . .

- Most fungi live in close association with land plants as mycorrhizae, endophytes, or saprophytes.
- When plants are alive, mycorrhizal fungi associate with their roots, and many fungal species grow as endophytes.
- When plants die, saprophytic fungi degrade their tissues and release nutrients.
- Instead of being based on the fusion of gametes, sexual reproduction in most fungi begins with plasmogamy, or the fusion of hyphae, followed by karyogamy, or nuclear fusion, and meiosis.

✔ You should be able to . . .

1. Describe evidence that mutualism occurs in EMF.
2. Compare and contrast what happens inside a zygosporangium, basidium, and ascus.

Answers are available in Appendix A.

32.4 Key Lineages of Fungi

Based on the current hypothesized phylogeny for the Fungi, it is clear that neither chytrids nor zygomycetes form monophyletic groups and should not be named as single lineages (see Figure 32.8). Researchers continue to collect and analyze additional data—mostly DNA sequences—to identify monophyletic groups and explain the early evolution of fungi.

In addition, the Ascomycota lineage is exceptionally diverse, and the phylogenetic relationships among major subgroups are still being worked out. As a result, the discussion of "key lineages" that follows distinguishes two lifestyles observed in Ascomycota, and not monophyletic groups within the phylum. Research on the phylogenetic relationships within fungi continues.

- Fungi > Microsporidia
- Fungi > Chytrids
- Fungi > Zygomycetes
- Fungi > Glomeromycota
- Fungi > Basidiomycota (Club Fungi)
- Fungi > Ascomycota > Lichen-Formers
- Fungi > Ascomycota > Non-Lichen-Formers

Fungi > Microsporidia

All of the estimated 1500 species of microsporidians are single-celled and parasitic on animal cells. They are distinguished by a unique structure called the polar tube (**FIGURE 32.14**), which allows them to enter the cells they parasitize. The tube shoots out from the microsporidian, penetrates the membrane of the host cell, and acts as a conduit for the contents of the microsporidian cell to enter the host cell. Once inside, the microsporidian replicates and produces a generation of daughter cells, which go on to infect other host cells.

Microsporidians have a dramatically reduced genome. One species has the smallest genome known among eukaryotes—the total number of base pairs present is over 1000 times smaller than the human genome. Unlike other fungi, microsporidians lack functioning mitochondria. As in other fungi, the polysaccharide chitin is a prominent component of their cell wall ✔You should be able to mark the origin of the polar tube and the loss of mitochondria on Figure 32.8.

Absorptive lifestyle Most microsporidians parasitize insects or fish. Because they enter the interior of host cells, they are called intracellular parasites. Their small genome size and lack of functional mitochondria are both adaptations to this lifestyle.

Life cycle Variation in life cycles is extensive. Some species appear to reproduce only asexually. Others produce several different types of sexual or asexual spores, and some must successfully infect several different host species to complete their life cycle.

Human and ecological impacts Species from eight different genera can infect humans. In most cases, however, microsporidians cause serious infections only in AIDS patients and other individuals whose immune systems are not functioning well. Some microsporidians are serious pests in honeybee and silkworm colonies, while others cause severe infections in grasshoppers and are marketed as a biological control agent to reduce crop damage caused by grasshoppers.

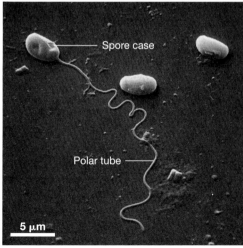

Nosema tractabile

Spore case

Polar tube

5 μm

FIGURE 32.14 Microsporidia Infect Other Cells via a Polar Tube.

Chytrids are largely aquatic and are common in freshwater environments. Some species live in wet soils, and a few have been found in desert soils that are wet only during a rainy season. Species found in dry soils have tough spores that endure harsh conditions. Spores from chytrids have been shown to germinate after a resting period of as long as 31 years. Members of this group are the only fungi that produce motile cells. In each case, the motile gametes use flagella to swim.

Absorptive lifestyle Many species of chytrids have enzymes that allow them to digest cellulose. As a result, these species are important decomposers of plant material in wet soils, ponds, and lakes (**FIGURE 32.15**). However, many of the freshwater species are parasitic, and parasitic chytrids sometimes cause disease epidemics in algae or aquatic insects (including mosquitoes). Other species parasitize mosses, ferns, or flowering plants. Mutualistic chytrids are among the most important of the many organisms living in the guts of deer, cows, elk, and other mammalian herbivores, because the chytrids help these animals digest their food. Chytrids secrete cellulases that degrade the cell walls of grasses and other plants ingested by the animal, releasing sugars that are used by both the chytrids and their hosts.

Life cycle During asexual reproduction, most chytrid species produce spores that swim to new habitats via a flagellum. A few species reproduce sexually as well as asexually and exhibit alternation of generations. (Recall from Section 32.3 that these chytrids are the only fungal species to do so.) In species that reproduce sexually, plasmogamy and karyogamy may occur in a variety of ways: through fusion of hyphae, gamete-forming structures, or gametes.

Human and ecological impacts In addition to the species that serve as decomposers and mutualists, many chytrids are parasites. Biologists are investigating the possibility of using a chytrid that parasitizes mosquito larvae as a biological control agent; potato tubers are sometimes invaded and spoiled by a parasitic chytrid that causes black wart disease. The parasitic chytrid *Batrachochytrium dendrobatidis* is largely responsible for catastrophic declines occurring recently in amphibian populations all over the world.

Allomyces macrogynus

FIGURE 32.15 A Chytrid That is a Common Inhabitant of Soils and Ponds Worldwide.

The zygomycetes ("yoked-fungi") are primarily soil-dwellers. Their hyphae yoke together and fuse during sexual reproduction and then form a durable, thick-walled zygosporangium.

Absorptive lifestyle Many members of zygomycete lineages are saprophytes and live in plant debris. Some parasitize other fungi, however, or are important parasites of insects and spiders.

Life cycle Asexual reproduction is extremely common. Ball-like sporangia are produced at the tips of hyphae that form stalks, and mitosis results in the production of spores that are dispersed by wind. During sexual reproduction, fusion of hyphae occurs only between individuals of different mating types.

Human and ecological impacts The common bread mold *Rhizopus stolonifer* is a frequent household pest—probably the zygomycete that is most familiar to you. Saprophytic and parasitic members of these lineages are responsible for rotting strawberries and other fruits and vegetables and causing large losses in these food industries. The black bread mold pictured in **FIGURE 32.16** is a familiar example of a saprophytic zygomycete. Some species of *Mucor* are used in the production of steroids for medical use.

Species of *Rhizopus* and *Mucor* are used in the commercial production of organic acids, pigments, alcohols, and fermented foods.

Rhizopus nigricans

FIGURE 32.16 Black Bread Molds Are Zygomycetes.

Recent phylogenetic analyses have shown that the arbuscular mycorrhizal fungi (AMF) form a distinct lineage, indicating that Glomeromycota is a major monophyletic group. ✔You should be able to mark the origin of the AMF association on Figure 32.8.

Absorptive lifestyle Recall from Section 32.3 that AMF absorb phosphorus-containing ions or molecules in the soil and transfer them, along with other nutrients and water, into the roots of most of the plants living in grasslands and tropical forests (**FIGURE 32.17**). In exchange, the host plant provides the symbiotic fungi with sugars and other organic compounds.

Life cycle Most species form large spores underground. Glomeromycetes are difficult to grow in the laboratory, so their life cycle is not well known. No one has yet discovered a sexual phase in these species.

Human and ecological impacts Because grasslands and tropical forests occupy large areas of terrestrial habitats on Earth, the AMF are enormously important to both human and natural economies.

50 μm

FIGURE 32.17 AMF Penetrate the Walls of Plant Root Cells.

Fungi > Basidiomycota (Club Fungi)

Although most basidiomycetes form mycelia and produce multicellular reproductive structures, some species have the unicellular growth form. The group is named for basidia, the club-like cells where meiosis and spore formation occur. ✔You should be able to mark the origin of the basidium on Figure 32.8. About 31,000 species of basidiomycetes have already been described, and more are discovered each year.

Absorptive lifestyle Basidiomycetes are important saprophytes. Along with a few soil-dwelling bacteria, some members of this lineage are the only organisms capable of synthesizing lignin peroxidase and are therefore important in wood decomposition. Some basidiomycetes are ectomycorrhizal fungi (EMF) that associate with trees in temperate forests. One subgroup consists entirely of parasitic forms called rusts, including species that cause serious infections in many crop plants. The plant parasites called smut fungi specialize in infecting grasses; a few infect other fungi. Thus, as in other lineages of fungi, the entire array of lifestyles—saprophytic, mutualistic, and parasitic—is found within Basidiomycota.

Life cycle Asexual reproduction through production of spores is common in the Basidiomycota, although not as prevalent as

Astraeus hygrometricus

1 cm

FIGURE 32.18 Puffballs Can Produce Billions of Spores.

in other groups of fungi. Asexual reproduction also occurs through growth and fragmentation of mycelia in the soil or in rotting wood, resulting in genetically identical individuals that are physically independent. During sexual reproduction, all basidiomycetes—even unicellular ones—produce basidia. In the largest subgroup in this lineage, basidia form in large, aboveground reproductive structures called mushrooms, brackets, earthstars, or puffballs (**FIGURE 32.18**). Members of the Basidiomycota and Ascomycota are the only fungi that have heterokaryotic (dikaryotic) mycelia. ✔You should be able to mark the origin of the dikaryotic mycelia on Figure 32.8.

Human and ecological impacts EMF are enormously important in forestry. The temperate forests where these fungi are particularly abundant provide most of the hardwoods and softwoods used in building construction, furniture-making, and papermaking. Throughout the world, mushrooms are cultivated or collected from the wild as a source of food. The white button, crimini, and portabella mushrooms you may have seen in grocery stores are all varieties of the same species, *Agaricus bisporus*. Some of the toxins found in poisonous mushrooms are used in biological research; others are hallucinogens.

About half of the ascomycetes grow in mutualistic association with cyanobacteria and/or single-celled green algae, forming the structures called lichens. Over 15,000 different lichens have been described to date; in most, the fungus involved is an ascomycete (although a few basidiomycetes participate as well). To name a lichen, biologists use the species name assigned to the fungus that participates in the association. Most lichen-formers are found in a single monophyletic group within the Ascomycota. ✔Based on this observation, you should be able to mark the primary origin of the lichen-forming habit on Figure 32.8.

Absorptive lifestyle The fungus in lichens appears to protect the photosynthetic bacterial or algal cells. The fungal hyphae form a dense protective layer that shields the photosynthetic species and reduces water loss. In return, the cyanobacterium or alga provides carbohydrates that the fungus uses as a source of carbon and energy. However, the hyphae of some lichen-forming fungi have been observed to invade algal cells and kill them. This observation suggests a partially parasitic relationship in at least some lichens. The nature of lichen-forming associations is the subject of ongoing research.

Life cycle As **FIGURE 32.19a** shows, many lichens reproduce asexually via the production of small "mini-lichen" structures called soredia that contain both symbionts. Soredia disperse to a new location via wind or water and then develop into a new, mature individual via the growth of the algal or bacterial and fungal symbionts. In addition, the fungal partner may form asci. Spores that are shed from asci germinate to form a small mycelium. If the growing hyphae encounter the appropriate algal or bacterial cells, a new lichen can form.

Human and ecological impacts In abundance and diversity, lichens dominate the Arctic and Antarctic tundras and are extremely common in boreal forests (see Chapter 52). They are the major food of caribou in the winter as well as the most prevalent colonizers of bare rock surfaces throughout the world (**FIGURE 32.19b**). Rock-dwelling lichens are significant because they produce acids that break down the rock surface as they grow—launching the first step in soil formation. Lichens are particularly sensitive to air pollution and are often used as bioindicators of air quality.

(a) Cross section of a lichen, showing three layers

(b) Top view of lichen on a rock

FIGURE 32.19 Lichens Are Associations between a Fungus and a Cyanobacterium or Green Alga. (a) In a lichen, cyanobacteria or green algae live within a dense network of fungal hyphae. **(b)** Lichens often colonize surfaces, such as tree bark or bare rock, where other organisms are rare.

The ascomycetes that do not form lichens are found in virtually every terrestrial habitat as well as some freshwater and marine environments. Although most ascomycetes form mycelia, many are single-celled yeasts. The ascus is a distinguishing characteristic of the Ascomycota. ✔You should be able to mark the origin of the ascus on Figure 32.8.

Absorptive lifestyle Many ascomycetes form mutualistic ecto-mycorrhizal fungi (EMF) associations with tree roots. Ascomycetes are also the most common endophytic fungi on aboveground tissues. Large numbers are saprophytic and are abundant in forest floors and in grassland soils. Parasitic forms are common as well. As in other fungal lineages, the entire array of absorptive lifestyles found in fungi has evolved independently within the Ascomycota.

In addition, about 65 species in one Ascomycota lineage are predatory—primarily on amoebae and other unicellular protists. Some are large enough to capture the microscopic animals called roundworms, however. The predatory ascomycetes trap their prey by means of sticky substances on their cell walls, or they catch prey in snares consisting of looped hyphae (**FIGURE 32.20**). Once a prey individual is captured, hyphae invade its body, digest it, and absorb the nutrients that are released. A predatory ascomycete has been found in fossilized material that is 100 million years old. ✔You should be able to mark the origin of predation on Figure 32.8.

Life cycle The aboveground, asci-bearing reproductive structures of these fungi may be a cup or saucer shape (**FIGURE 32.21**). Spores are also routinely produced asexually, at the ends of specialized hyphae.

Human and ecological impacts Some saprophytic ascomycetes can grow on jet fuel or paint and are used to help clean up contaminated sites. *Penicillium* is an important source of antibiotics, and *Aspergillus* produces citric acid used to flavor soft drinks and candy. Truffles are so highly prized that they can fetch $2000 per pound, and the multibillion-dollar brewing, baking, and wine-making industries would collapse without the yeast *Saccharomyces cerevisiae*. A few parasitic ascomycetes cause infections in humans and other animals. In land plants, parasites from this group cause diseases including Dutch elm disease and chestnut blight.

Arthrobotrys anchonia

Roundworm

Hyphae

50 μm

FIGURE 32.20 Some Ascomycetes Are Predatory.

Cookeina speciosa

1 cm

FIGURE 32.21 Ascomycete Reproductive Structures are Often Cup Shaped.

If you understand . . .

32.1 Why Do Biologists Study Fungi?

- Many fungi species live in close association with land plants. The roots of most plants are infected by mutualistic fungi that provide the plant with nitrogen, phosphorus, and water in exchange for sugars and other products of plant photosynthesis.

- Parasitic fungi are responsible for devastating blights in crops and other plants.

- Once plants die, saprophytic fungi degrade the lignin and cellulose in wood and use nutrients from decaying plant material.

- Because they free up carbon atoms that would otherwise be locked up in wood, fungi speed up the carbon cycle in terrestrial ecosystems.

- Fungi are used widely in the food processing, cheese-making, baking, and brewing industries.

✔ You should be able to predict the results of using fungicides to experimentally exclude fungi from 10-m $\times$ 10-m plots in a forest or grassland near your campus.

32.2 How Do Biologists Study Fungi?

- Morphological analyses of fungi have revealed adaptations that make fungi exceptionally effective at absorbing nutrients from the environment.

- Researchers have determined that fungi exhibit two growth forms—single-celled "yeasts" or mycelia composed of long, filamentous hyphae. Mycelia provide fungi with extremely high surface-area-to-volume ratios. Some fungal bodies are extremely large.

- Septa divide hyphae into cells, but pores in the septa facilitate transport of nutrients and organelles through the hyphae.

- Scientists have determined from morphological studies that fungi possess four known, distinctive types of reproductive structures: (1) chytrids produce motile gametes; (2) zygomycetes make tough zygosporangia; (3) members of the Basidiomycota have club-like, spore-forming structures; and (4) members of the Ascomycota have sac-like, spore-forming structures.

- Recent analyses of DNA sequence data have revealed that fungi are more closely related to animals than they are to plants, and that the Basidiomycota, Ascomycota, and Glomeromycota each form monophyletic groups. Researchers are still trying to understand how the various groups of chytrids and zygomycetes are related to each other and to the other fungal lineages.

✔ You should be able to explain why fungal infections are more difficult to treat than bacterial infections.

32.3 What Themes Occur in the Diversification of Fungi?

- All fungi make their living by absorbing nutrients from living or dead organisms.

- Many fungi live in leaves and stems; in some grasses and other species, these endophytic fungi secrete toxins that discourage herbivores.

- Extracellular digestion, in which enzymes are secreted into food sources, enables fungi to break down extremely large molecules before absorbing them.

- Lignin decomposes through a series of uncontrolled oxidation reactions triggered by the enzyme lignin peroxidase.

- Cellulose digestion occurs in a carefully regulated series of steps, each catalyzed by a specific cellulase.

- Many species of fungi have never been observed to reproduce sexually. Most species can produce haploid spores either sexually or asexually, however.

- Sexual reproduction usually starts when hyphae from genetically different mating types fuse—an event called plasmogamy. If the fusion of nuclei, or karyogamy, does not occur immediately, a heterokaryotic mycelium forms, which in some species is dikaryotic. Heterokaryotic cells may eventually produce spore-forming structures where karyogamy and meiosis take place.

✔ You should be able to explain why most fungi don't need to have gametes to accomplish sexual reproduction.

32.4 Key Lineages of Fungi

- Microsporidia are parasitic, single-celled fungi with reduced genomes and an unusual polar tube used to infect host cells.

- Chytrids are usually aquatic with motile, flagellated spores. Some are terrestrial decomposers and some are parasites.

- Zygomycetes are largely saprophytic on dead plant tissue. During sexual reproduction, they form zygosporangia after hyphae from different mating types are yoked.

- The Glomeromycota is a monophyletic group of soil fungi that form arbuscular mycorrhizae with most land plants.

- The Basidiomycota is a monophyletic group of terrestrial fungi that form four airborne spores on each club-shaped basidium within mushrooms, brackets, and puffballs. Some can degrade wood. Others form ectomycorrhizae with certain trees.

- Ascomycota is a monophyletic group of terrestrial fungi that form eight airborne spores in each ascus. Some form lichens in association with green algae or cyanobacteria.

✔ You should be able to describe some of the characteristics of each major lineage of fungi.

You should be able to . . .

✔ TEST YOUR KNOWLEDGE
Answers are available in Appendix A

1. The mycelial growth habit leads to a body with a high surface-area-to-volume ratio. Why is this important?
   a. Mycelia have a large surface area for absorption.
   b. The hyphae that make up mycelia are long, thin tubes.
   c. Most hyphae are broken up into compartments by walls called septa, although some exist as single, gigantic cells.
   d. Hyphae can infiltrate living or dead tissues.

2. What is plasmogamy?
   a. production of gametes, after fusion of hyphae from different individuals
   b. exchange of nutrients between symbiotic fungi and hosts
   c. fertilization—the fusion of cytoplasm and nuclei from different individuals
   d. fusion of the cytoplasm from different mating types, without nuclear fusion

3. The hyphae of arbuscular mycorrhizal fungi (AMF) form bushy structures after making contact with the plasma membrane of a root cell. Why?
   a. They anchor the fungus inside the root, so the association is more permanent.
   b. They increase the surface area available for the transfer of nutrients.
   c. They produce toxins that protect the plant cells against herbivores.
   d. They break down cellulose and lignin in the plant cell wall.

4. What does it mean to say that a hypha is dikaryotic?
   a. Two nuclei fuse during sexual reproduction to form a zygote.
   b. Two independent nuclei, derived from different individuals, are present in each cell.
   c. The nucleus is diploid or polyploid—not haploid.
   d. It is extremely highly branched, which increases its surface area and thus absorptive capacity.

5. Explain how pores in septa of certain fungi allow mycelia to become larger.

6. Symbiotic partnerships in which one partner benefits and the other is unaffected are called _____.

✔ TEST YOUR UNDERSTANDING
Answers are available in Appendix A

7. The Greek root *ecto* means "outer." Why are ectomycorrhizal fungi, or EMF, aptly named?
   a. Their hyphae form tree-like branching structures inside plant cell walls.
   b. They are mutualistic.
   c. Their hyphae form dense mats that envelop roots but do not penetrate the cell walls.
   d. They transfer nitrogen from outside their plant hosts to the interior.

8. Explain why fungi that degrade dead plant materials are important to the global carbon cycle. Do you accept the text's statement that, without these fungi, "Terrestrial environments would be radically different than they are today and probably much less productive"? Why or why not?

9. Lignin and cellulose provide rigidity to the cell walls of plants. But in most fungi, chitin performs this role. Why is it logical that most fungi don't have lignin or cellulose in their cell walls?

10. Biologists claim that EMF and AMF species are better than plants at acquiring nutrients because they have more surface area and because they are more effective at acquiring phosphorus (P) and/or nitrogen (N). Compare and contrast the surface area of mutualistic fungi and plant roots. Explain why fungi are particularly efficient at acquiring P and N, compared to plants.

11. Using information from Chapters 3 through 6, list two key macromolecules found in plants that contain phosphorus. Because phosphorus is often limiting to plant growth, explain how certain fungi are necessary for rapid plant growth.

12. Compare and contrast the way that fungi degrade lignin with the way that they digest cellulose.

13. Lawns are sometimes fertilized with nitrate that can be washed into neighboring woodlots by rain. If the trees in that woodlot are associated with EMF, what effect might the excess nitrate have on the fungi and/or trees?
    a. Fungal growth may be stimulated, causing the trees to transport more sugars to the fungi.
    b. The fungi may secrete more peptidases to break down the nitrate.
    c. The trees may take up the nitrate directly into their roots and rely less on the fungi.
    d. The excess nitrate may stimulate the fungi to transport more sugar to the trees.

14. The box on chytrids in Section 32.4 mentions that they may be responsible for massive die-offs currently occurring in amphibians. Review Koch's postulates; then design a study showing how you would use Koch's postulates to test the hypothesis that chytrid infections are responsible for the frog deaths (see Chapter 29).

15. Experiments indicate that cellulase genes are transcribed and translated together. If cells are selected to be extremely efficient at digesting cellulose, is this result logical? Would you predict that the gene that codes for lignin peroxidase is transcribed along with the cellulase genes? How would you test your prediction?

16. Many mushrooms are extremely colorful. Fungi do not see, so it is unlikely that colorful mushrooms are communicating with one another. One hypothesis is that the colors serve as a warning to animals that eat mushrooms, much like the bright yellow and black stripes on wasps. Design an experiment capable of testing this hypothesis.

33 An Introduction to Animals

In this chapter you will learn that

Most animal phyla originated suddenly in the Cambrian period and then diversified

by asking

by asking

by asking

What is an animal? **33.1**

What key innovations occurred during animal evolution? **33.2**

What are the themes of animal diversification? **33.3**

looking closer at

Non-bilaterian animals **33.4** —then→ Protostome animals **Ch. 34** —and→ Deuterostome animals **Ch. 35**

Jellyfish such as this hydromedusa are among the most ancient of all animals—they appear in the fossil record over 560 million years ago.

You're an animal, and you have plenty of company—animals are an extraordinarily diverse and species-rich lineage on the tree of life.

The radiation of animals began around 550 million years ago during an event called the **Cambrian explosion** (see Chapter 28). After a slow start, diverse animals suddenly appeared in the fossil record with shells, exoskeletons, internal skeletons, legs, heads, tails, eyes, antennae, jaw-like mandibles, segmented bodies, muscles, and brains. The Cambrian explosion was arguably the most spectacular period of evolutionary change in the history of life.

Today, estimates suggest that there are between 8 million and 50 million animal species alive on our planet, some strikingly similar to their Cambrian ancestors, others quite different. These species range in size and complexity from tiny sponges, which attach to a substrate and contain just a few

✔ When you see this checkmark, stop and test yourself. Answers are available in Appendix A.

cell types, to blue whales, which migrate thousands of kilometers each year in search of food and contain trillions of cells, dozens of distinct tissues, an elaborate skeleton, and highly sophisticated sensory and nervous systems.

Only about 1.4 million animal species have been described and given scientific names to date. As scientists race to discover and describe more animal species, the impact of humans on the planet—ranging from habitat loss to pollution, overharvesting, spread of invasive species, and climate change—is causing the modern rate of animal extinction to accelerate. Some species may go extinct before they are ever described (see Chapter 57).

To analyze the almost overwhelming number and diversity of species in the animal lineage, this chapter presents a broad overview of how these species diversified, along with more detailed information on the characteristics of the lineages that appear first in the fossil record. (See Chapters 34 and 35 to focus on the most species-rich groups.) Let's start by considering what characteristics define animals.

33.1 What Is an Animal?

The ancestors to animals were single-celled protists (Chapter 30). **FIGURE 33.1** reviews how animals fit in the tree of life. Animals occur in a clade, or lineage, called Opisthokonta, along with fungi and the single-celled protists called choanoflagellates. Although scientists cannot be sure exactly what the protist ancestor to all animals looked like, it was likely very similar to today's choanoflagellates.

What characteristics define the animal lineage? Although you may be able to recognize most animals as animals when you see them, some animals look more like plants, fungi, or even protists.

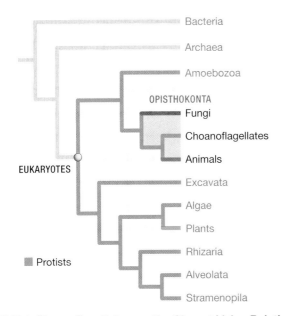

FIGURE 33.1 Choanoflagellates are the Closest Living Relatives to Animals.

Animals are eukaryotes that share key traits:

- *All animals are multicellular,* with cells that lack cell walls but have an extensive **extracellular matrix** (ECM; see Chapter 11). The ECM includes proteins specialized for cell–cell adhesion and communication.

- *All animals are heterotrophs,* meaning that they obtain the carbon compounds they need from other organisms. Most ingest ("eat") their food, rather than absorbing it across their exterior body surfaces.

- *All animals move* under their own power at some point in their life cycle.

- *All animals other than sponges have* **(1)** *neurons* (nerve cells) that transmit electrical signals to other cells; and **(2)** *muscle cells* that can change the shape of the body by contracting.

Multicellular fungi and animals are similar in that they are both multicellular heterotrophs that digest (break down) and absorb nutrients. However, animals are the only multicellular heterotrophs on the tree of life that usually ingest their food first, before they digest it. As a result, digestion in animals typically occurs within the space, or lumen, of the digestive tract rather than in the open exterior, as occurs with fungi.

In most animals, neurons connect to each other, forming a nervous system, and some neurons connect to muscle cells—which may contract in response to electrical signals from neurons. Muscles and neurons are adaptations that allow a large, multicellular body to move efficiently.

What is produced by the combination of multicellularity, heterotrophy, and efficient movement? Eating machines—animals are the largest predators, herbivores, and detritivores on Earth. They are key consumers in virtually every ecosystem, from the deep oceans and tropical forests to alpine ice fields.

Biologists currently recognize 30–35 **phyla**, or major lineages, of animals—the exact number is constantly being debated and revised as additional information comes to light. **TABLE 33.1** (see page 638) lists 29 phyla that are included in most published analyses. How can we make sense of all this diversity? Let's first take a closer look at the key events that occurred during the origin of major animal lineages. Then let's examine the key themes of diversification that occurred *within* lineages.

33.2 What Key Innovations Occurred during the Evolution of Animals?

Biologists who study the evolution of animals consider three types of data:

1. *Fossils* are important because they provide the only direct evidence of what ancient animals looked like, when they existed, and where they lived. However, the fossil record does not represent all animals equally. Fossils are more likely to occur for animals that were abundant, had hard parts, lived in areas where sedimentation was occurring, and/or lived recently (Chapter 28).

TABLE 33.1 An Overview of Major Animal Phyla

Group and Phylum	Common Name or Example Taxa	Number of Described Species
Non-bilaterian Groups		
Porifera	Sponges	8500
Placozoa	Placozoans	1
Ctenophora	Comb jellies	190
Cnidaria	Jellyfish, corals, anemones, hydroids, sea fans	11,500
Acoela	Acoelomate worms	350
Protostomes (lacks typical protostome development)		
Chaetognatha	Arrow worms	120
Protostomes: Lophotrochozoa		
Rotifera	Rotifers	2100
Platyhelminthes	Flatworms	20,000
Nemertea	Ribbon worms	1200
Gastrotricha	Gastrotrichs	400
Acanthocephala	Acanthocephalans	1150
Entoprocta	Entoprocts, kamptozoans	170
Gnathostomulida	Gnathostomulids	100
Annelida	Segmented worms	16,800
Mollusca	Mollusks (clams, snails, octopuses)	85,000
Phoronida	Horseshoe worms	10
Bryozoa	Bryozoa, ectoprocts, moss animals	5700
Brachiopoda	Brachiopods (lamp shells)	550
Protostomes: Ecdysozoa		
Nematoda	Roundworms	25,000
Kinorhyncha	Kinorhynchs	130
Nematomorpha	Hair worms	330
Priapula	Priapulans	16
Onychophora	Velvet worms	165
Tardigrada	Water bears	1045
Arthropoda	Arthropods (spiders, insects, crustaceans)	1,160,000
Deuterostomes		
Echinodermata	Echinoderms (sea stars, sea urchins, sea cucumbers)	7000
Xenoturbellida	Xenoturbillidans	2
Hemichordata	Acorn worms	108
Chordata	Chordates: tunicates, lancelets, sharks, bony fish, amphibians, reptiles (including birds), mammals	65,000

2. *Comparative morphology* provides information about which embryonic, larval, or adult morphological characteristics are common among groups of animals and which are unique to individual lineages (synapomorphies). These data can be used to define the fundamental architecture, or **body plan,** of each animal lineage. In a phylogenetic context, these data can be used to infer which characteristics arose first during the evolution of animals, and which animal groups are more closely related.

3. *Comparative genomics* provides information about the relative similarity of genes or whole genomes of diverse organisms (see Chapter 21). This relatively new source of data is providing dramatic insights into phylogenetic relationships and evolutionary history.

Sometimes these various sources of data on animal evolution suggest the same sequence of evolutionary events. Sometimes they don't. Consider one of the most influential papers on the phylogeny of animals, published in 1997. Using sequences from a gene that codes for RNA in the small subunit of the ribosome, Anna Marie Aguinaldo and colleagues estimated the phylogeny of species from 14 animal phyla. The results were revolutionary, suggesting a different pattern of relationships among animal lineages than had been accepted for years based on morphological data.

The phylogenetic tree in **FIGURE 33.2** is an updated version of the 1997 result, based on further studies of many gene and protein sequences. Some important morphological innovations are mapped onto the tree. Let's explore them by starting at the root of the tree and working toward the branch tips. (See **BioSkills 7** in Appendix B to review how to interpret phylogenetic trees.)

Origin of Multicellularity

Put your finger on the root of the animal tree in Figure 33.2, where the single-celled choanoflagellates outgroup meets the most basal animal branches. What happened at that node?

The first important insight is that animals are a monophyletic group, meaning that all animals share a common ancestor. Data from fossils, comparative morphology, and comparative genomics agree on a single origin of animals.

The second point to note is that sponges (phylum Porifera) include the two most basal, or ancient, lineages of animals. That is, multicellularity appears to have originated in a sponge-like animal. This hypothesis is well supported by diverse evidence. Let's take a closer look.

Fossil Evidence Sponges are the earliest animals to appear in the fossil record, more than 600 million years ago. The early presence of multicellular sponges and absence of fossils of other multicellular organisms are consistent with the basal position of sponges on the phylogeny.

Morphological Evidence Sponges share several key characteristics with the choanoflagellate outgroup:

- Both choanoflagellates and sponges are **sessile,** meaning that adults live permanently attached to a substrate.

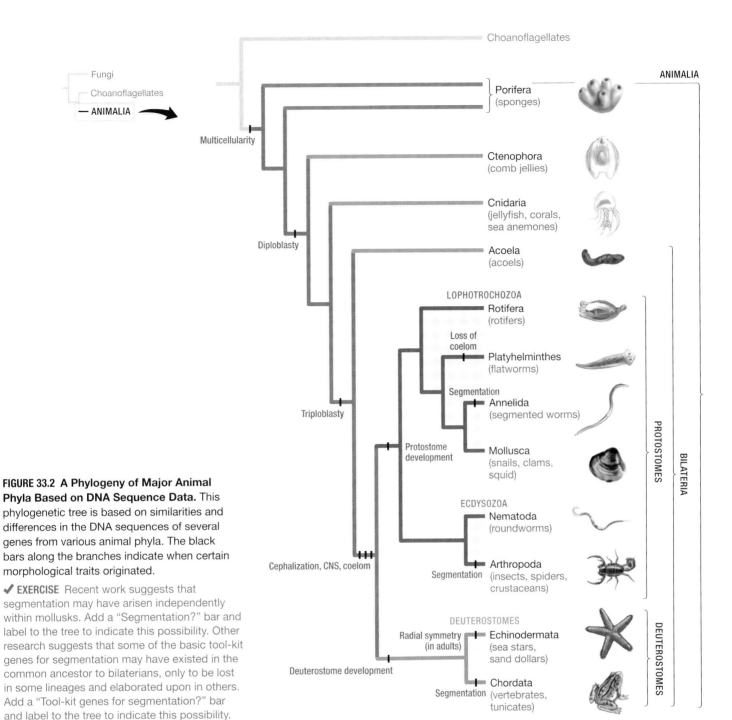

FIGURE 33.2 A Phylogeny of Major Animal Phyla Based on DNA Sequence Data. This phylogenetic tree is based on similarities and differences in the DNA sequences of several genes from various animal phyla. The black bars along the branches indicate when certain morphological traits originated.

✔ **EXERCISE** Recent work suggests that segmentation may have arisen independently within mollusks. Add a "Segmentation?" bar and label to the tree to indicate this possibility. Other research suggests that some of the basic tool-kit genes for segmentation may have existed in the common ancestor to bilaterians, only to be lost in some lineages and elaborated upon in others. Add a "Tool-kit genes for segmentation?" bar and label to the tree to indicate this possibility.

- Both feed in a similar way, using cells with nearly identical morphology. As **FIGURE 33.3** (see page 640) shows, the beating of flagella creates water currents that bring organic debris toward the feeding cells of choanoflagellates and sponges. Sponge feeding cells are called **choanocytes.** Within these choanocytes, food particles are trapped and ingested.

As you can see in the figure, choanoflagellates sometimes form **colonies**—groups of individuals that are attached to each other. Some biologists once considered sponges to be colonies of single-celled protists due to the ability of sponge cells to re-aggregate after being dissociated. But sponges contain many specialized cell types that are dependent on each other, some of which occur in organized layers surrounded by extracellular matrix (ECM). Recent research has shown that some sponges have true **epithelium**—a layer of tightly joined cells that covers the interior and/or exterior surface of the animal. Epithelium is essential to animal form and function (Chapter 42).

Sponges are diverse in size, shape, and composition, distinguished in part by the type of **spicules** they produce—stiff spikes of silica or calcium carbonate ($CaCO_3$) that, along with collagen fibers, provide structural support to the ECM. Despite this morphological diversity, many biologists have described sponges as a

(a) Choanoflagellates are sessile protists; some are colonial.

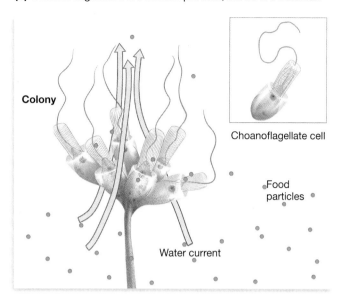

(b) Sponges are multicellular, sessile animals.

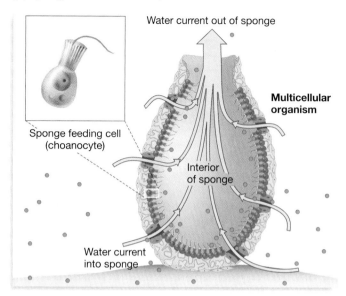

FIGURE 33.3 Choanoflagellates and Sponge Feeding Cells Are Almost Identical in Structure and Function.
(a) Choanoflagellates are suspension feeders. **(b)** A cross section of a simple sponge as it suspension feeds. The beating of flagella produces a water current that brings food into the body of the sponge, where it can be ingested by feeding cells.

monophyletic group based on their unique body plan—systems of tubes and pores that create channels for water currents. Other biologists reject the monophyly of sponges. The recent availability of gene sequence data for diverse sponges has helped to advance the debate, but there is still no consensus.

Molecular Evidence Comparative genomic studies support the hypothesis that sponges are the most basal group of animals, but some reject the hypothesis that the phylum Porifera is monophyletic. According to these studies, sponges are paraphyletic—containing some, but not all, descendants of a common ancestor (see Table 28.2). This concept is indicated on Figure 33.2 by two blue lines, instead of one, representing the sponges.

The paraphyly of sponges is important to the study of animal origins because it provides extra support for the hypothesis that the common ancestor of all animals was sponge-like. If sponges were a monophyletic group, it would be more plausible that their distinguishing characteristics could have evolved later rather than in the common ancestor to all animals (the same way that the most recent common ancestor of humans and chimpanzees was not necessarily chimp-like).

What were the other qualities of the animal ancestor? Comparative genomic studies suggest something surprising. Despite the relative morphological simplicity of sponges, they possess a remarkably complex developmental **tool kit** of genes (Chapter 22). This tool kit contains the genes necessary for all the basic molecular processes required by animals:

- Specialization of cell types
- Regulation of cell cycling and growth
- Adhesion among cells, and between cells and extracellular matrix
- Recognition of self and nonself, thus innate immunity
- Developmental signaling and gene regulation
- Programmed cell death

Thus, a series of important genetic innovations appears to have occurred at the very root of the animal tree along with multicellularity. Subsequent duplication and diversification of these genes contributed to the diversification of animal lineages. ✔If you understand this, you should be able to explain why cancer, which typically involves the mutation of genes in the basic animal genetic tool kit, is thought to be as ancient as multicellularity itself.

Origin of Embryonic Tissue Layers

Sponges possess the basic genetic tool kit for cell–cell adhesion and cell–ECM adhesion, and some sponges even have rudimentary epithelium. However, sponges do not have complex **tissues,** groups of similar cells that are organized into tightly integrated structural and functional units. Animals other than sponges are typically divided into two major groups based on the number of embryonic tissue layers they have. Animals whose embryos have two types of tissue are called **diploblasts** (literally, "two-buds"); animals whose embryos have three types of tissue are called **triploblasts** ("three-buds").

The embryonic tissues are organized in layers, called **germ layers** (Chapter 23). In diploblasts these germ layers are called **ectoderm** and **endoderm** (**FIGURE 33.4**). The Greek roots *ecto* and

Endoderm — — Ectoderm

50 µm

FIGURE 33.4 Diploblastic Animals Have Bodies Built from Ectoderm and Endoderm. This is a cross section through the tube-shaped portion of a hydra's body (phylum Cnidaria). The cells have been stained to make them more visible.

endo refer to outer and inner, respectively; the root *derm* means "skin." (See **BioSkills 2** in Appendix B to review common Latin and Greek roots.) In most cases the outer and inner "skins" of diploblast embryos are connected by a gelatinous material that may contain some cells. In triploblasts, however, there is a germ layer called **mesoderm** between the ectoderm and endoderm. (The Greek root *meso* refers to "middle.")

The embryonic tissues found in animals develop into distinct adult tissues, organs, and organ systems. In triploblasts:

- *Ectoderm* gives rise to skin and the nervous system.

- *Endoderm* gives rise to the lining of the digestive tract.

- *Mesoderm* gives rise to the circulatory system, muscle, and internal structures such as bone and most organs.

In general, then, ectoderm produces the covering of the animal and endoderm generates the digestive tract. Mesoderm gives rise to the tissues in between. The same pattern holds in diploblasts, except that **(1)** muscle is simpler in organization and is derived from ectoderm, and **(2)** reproductive tissues are derived from endoderm.

Look again at the animal phylogeny in Figure 33.2. Traditionally, two groups of animals have been recognized as diploblasts: the Ctenophora (pronounced *ten-AH-for-ah*), or comb jellies, and the Cnidaria (pronounced *ni-DARE-ee-uh*), which include the jellyfish, corals, sea pens, hydra, and anemones. Recent data suggest that some cnidarians have true mesoderm and are triploblasts, complicating the question of exactly when the triploblast body plan evolved. However, the take-home message for the animal phylogeny is still the same—the relatively simple diploblast body plan of ctenophores and cnidarians appears to have originated from ancestral animals later than the sponge lineages but before the other major groups of animals, which are all triploblasts.

The evolution of mesoderm was important because it gave rise to the first complex muscle tissue used in movement. Sponges

lack muscle and are generally sessile, or nonmoving, as adults. Ctenophores have muscle cells that can change the body's shape, but both larvae and adults swim using cilia. Adult cnidarians have muscle cells that can extend and retract the body, such as in sea anemones, and can enable swimming by jet propulsion in jellyfish; but these movements are relatively simple compared to the complex movements performed by many triploblasts.

Origin of Bilateral Symmetry, Cephalization, and the Nervous System

Body symmetry is a key morphological aspect of an animal's body plan. A body is said to be symmetrical if it can be divided by a plane such that the resulting pieces are nearly identical.

Ctenophores, many cnidarians, and some sponges have **radial symmetry** ("spoke symmetry")—meaning that they have at least two planes of symmetry. For example, almost any plane sectioned through the center of the hydra in **FIGURE 33.5a** produces two identical halves. Radial symmetry evolved independently in the phylum Echinodermata (pronounced *ee-KINE-oh-der-ma-ta*)—a group including species such as sea stars, sea urchins, feather stars, and brittle stars.

Organisms with **bilateral symmetry** ("two-sided symmetry") in contrast, have one plane of symmetry and tend to have a long, narrow body. The annelid worm in **FIGURE 33.5b**, for example, has only one plane of symmetry—running lengthwise down its middle.

Based on the topology of branches on the animal phylogeny in Figure 33.2, radial symmetry appears to have arisen earlier in the evolution of animals than bilateral symmetry. Bilateral symmetry occurs in all triploblastic lineages. Where on the tree did bilateral symmetry originate? The answer is more complex than it would seem at first glance. Let's explore why.

(a) Radial symmetry   **(b)** Bilateral symmetry

FIGURE 33.5 Body Symmetry. These colorized SEMs of **(a)** hydra (looking down on oral surface; phylum Cnidaria) and **(b)** a polychaete worm (phylum Annelida) illustrate the difference between radial and bilateral symmetry.

FIGURE 33.6 Morphological Evidence for Bilateral Symmetry in a Sea Anemone. Some corals and sea anemones have long been known to have bilateral symmetry despite the outward appearance of radial symmetry. *Nematostella* is a small sea anemone that burrows in soft marine sediments.

Homology or Convergent Evolution?

Nearly all cnidarians appear radially symmetric at first glance, but a closer inspection of their internal morphology reveals bilateral symmetry in some species, especially in sea anemones (**FIGURE 33.6**). Recall that homology is defined as similarity in traits due to inheritance from a common ancestor (Chapter 25). Is the bilateral symmetry in sea anemones homologous to bilateral symmetry in triploblastic animals, or is it an example of convergent evolution?

Developmental biologists John Finnerty, Mark Martindale, and colleagues used developmental regulatory genes as a tool to address this evolutionary question. In triploblastic, bilaterally symmetric animals, called **bilaterians** for short, bilateral symmetry is achieved by the combination of anterior–posterior ("head–tail") axis formation and dorsal–ventral ("back–belly") axis formation. *Hox* genes are important in the development of the anterior–posterior axis (see Chapter 22), and *decapentaplegic* (*dpp*) genes are important in the development of the dorsal–ventral axis.

Finnerty and Martindale predicted that if bilateral symmetry in the sea anemone *Nematostella vectensis* (an increasingly important model organism in biology) is homologous to bilateral symmetry in bilaterians, the pattern of *Hox* and *dpp* gene expression in the sea anemone would be similar to that found in bilaterians. If convergent evolution had occurred, either *Hox* and *dpp* genes would not be expressed in *Nematostella* or their expression patterns would be unrelated to those in bilaterians.

FIGURE 33.7 shows Finnerty and Martindale's results. During *Nematostella* development, *Hox* genes are expressed along the anterior–posterior axis and *dpp* is expressed asymmetrically about the dorsal–ventral axis, a pattern of gene expression that is parallel to that observed in bilaterians. This evidence supports the hypothesis that bilateral symmetry in this sea anemone is homologous to bilateral symmetry in triploblastic animals—meaning that parts of the genetic tool kit that determine bilateral symmetry arose before the evolutionary split of the cnidarian and bilaterian lineages.

RESEARCH

QUESTION: Is bilateral symmetry in sea anemones homologous to bilateral symmetry in bilaterians?

HYPOTHESIS: Bilateral symmetry is homologous in sea anemones and bilaterians.

NULL HYPOTHESIS: Bilateral symmetry arose independently in sea anemones and bilaterians.

EXPERIMENTAL SETUP:

1. Stain gene products (proteins) of *Hox* gene in developing *Nematostella* embryos and larvae to reveal location of expression.

2. Repeat for other *Hox* and *dpp* gene products.

PREDICTION: The pattern of expression of *Hox* and *dpp* genes in *Nematostella* will be similar to that found in bilaterians.

PREDICTION OF NULL HYPOTHESIS: The pattern of expression of *Hox* and *dpp* genes in *Nematostella* will be unrelated to that found in bilaterians.

RESULTS: Schematic summary of gene expression patterns (longitudinal section):

Hox genes are expressed sequentially along the anterior–posterior axis— *as in bilaterians*

dpp genes are expressed asymmetrically along the dorsal–ventral axis— *as in bilaterians*

CONCLUSION: Bilateral symmetry is homologous in *Nematostella* and bilaterians.

FIGURE 33.7 Genetic Evidence that Bilateral Symmetry Originated in a Common Ancestor of Cnidaria and Bilateria.

SOURCE: Finnerty, J. R., K. Pang, P. Burton, et al. 2004. Origins of bilateral symmetry: *Hox* and *dpp* expression in a sea anemone. *Science* 304: 1335–1337.

✔ **QUESTION** What results would have supported the null hypothesis?

However, *Hox* and *dpp* gene expression in *Nematostella* is not identical to that in bilaterians—and the bilateral symmetry in *Nematostella* is not as extensive as in bilaterians. The take-home message is that the entire genetic tool kit for bilateral symmetry did not evolve all at once in the early Cambrian before the diversification of bilaterians. Some components of the tool kit evolved earlier—in the ancestor to cnidarians and bilaterians—and others evolved later, after the split of these two lineages.

Origin of the Nervous System Over 99 percent of modern animals are bilaterally symmetric. Is this body plan a "key innovation" that led to the diversification of the bilaterians during the Cambrian explosion? One hypothesis is that the evolution of the nervous system and the evolution of the head are tightly linked to bilateral symmetry and that together, these characteristics contributed to the radiation of bilaterians.

How are symmetry and nervous systems related? The function of neurons and nervous systems is to transmit and process information in the form of electrical signals (see Chapter 46).

- Sponges generally lack both nerve cells and symmetry.

- Ctenophores and cnidarians have nerve cells that are organized into a diffuse arrangement called a **nerve net** (**FIGURE 33.8a**). These generally radially symmetric animals either float in water or live attached to a substrate. Radially symmetric organisms are more likely to encounter prey and other aspects of the environment in any direction. As a result, a diffuse nerve net can receive and send signals efficiently.

- All other animals have a **central nervous system,** or CNS. In a CNS, some neurons are clustered into one or more large tracts or cords that project throughout the body; others are clustered into masses called **ganglia** (**FIGURE 33.8b**). Most of the bilaterally symmetric animals living today move through their environment. Bilaterally symmetric organisms tend to encounter prey and other aspects of the environment at the leading end. As a result, it is advantageous to the animal to have many neurons concentrated at that end, with nerve tracts that carry information down the length of the body.

It is intuitive that the evolution of the CNS would coincide with **cephalization:** the evolution of a head, or anterior region, where structures for feeding, sensing the environment, and processing information are concentrated. The large mass of neurons that is located in the head, and that is responsible for processing information to and from the body, is called the cerebral ganglion or **brain.**

To explain the pervasiveness of bilateral symmetry, biologists point out that locating and capturing food is particularly efficient when movement is directed by a distinctive head region and powered by the rest of the body. In combination with the origin of mesoderm, a bilaterally symmetric body plan enabled rapid, directed movement and hunting. Lineages with a triploblastic, bilaterally symmetric, cephalized body had the potential to diversify into an array of formidable eating and moving machines.

Notice in Figure 33.2 that the acoels (phylum Acoela), tiny worms that tend to live among sand grains, are bilaterally symmetric and triploblastic but have a nerve net rather than a CNS—and no head region. Although the position of the acoel branch on the tree is hotly debated, its current position supports the hypothesis that the CNS and cephalization were key to the subsequent radiation of bilaterians.

Acoel worms are named for their lack of a coelom. What is a coelom? What role did it play in the diversification of bilaterians?

Origin of the Coelom

What would you expect an animal body to look like that is triploblastic (formed from ectoderm, mesoderm, and endoderm), bilaterally symmetric, elongated, and cephalized? The basic bilaterian body shape is a *tube within a tube.* The inner tube is the individual's gut with a mouth at one end and an anus at the other, and the outer tube forms the nervous system and skin (**FIGURE 33.9**). The mesoderm in between forms muscles and organs.

(a) Nerve net:
diffuse neurons in hydra

(b) Central nervous system:
clustered neurons in earthworm

Ganglia

FIGURE 33.8 Associations between Body Symmetry and the Nervous System. (a) Radially symmetric animals, like this hydra, have a nerve net. **(b)** Bilaterally symmetric animals, like this earthworm, have a central nervous system.

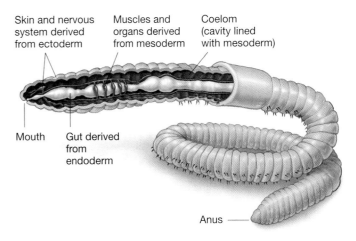

Skin and nervous system derived from ectoderm

Muscles and organs derived from mesoderm

Coelom (cavity lined with mesoderm)

Mouth

Gut derived from endoderm

Anus

FIGURE 33.9 The Tube-within-a-Tube Body Plan Is Common in Animals.

The "tube-within-a-tube" body plan can pose a potential biomechanical and physiological challenge if the inner tube is attached to the outer tube via the mesoderm. One bypass to this physical constraint is a fluid-filled cavity between the inner and outer tubes, called a **coelom** (pronounced *SEE-lum*), as shown in the earthworm in Figure 33.9. The coelom provides a space for the circulation of oxygen and nutrients. It also enables the internal organs to move independently of each other.

Phylogenetic evidence suggests that the coelom arose in the common ancestor of protostomes and deuterostomes (Figure 33.2). The bilaterians that possess a coelom that is completely lined in mesoderm are known as true **coelomates** (**FIGURE 33.10a**). The bilaterians that subsequently lost their coelom, such as the flatworms, are called **acoelomates** ("no-cavity-form"; see **FIGURE 33.10b**). The bilaterians that retained a coelom but lost the mesodermal lining in parts of their coelom, such as nematodes (roundworms), are known as **pseudocoelomates** ("false-cavity-form"; **FIGURE 33.10c**).

The coelom was a critically important innovation during animal evolution, in part because an enclosed, fluid-filled chamber can act as an efficient **hydrostatic skeleton.** Soft-bodied animals with hydrostatic skeletons can move effectively even if they do not have fins or limbs.

(a) Coelomates have an enclosed body cavity completely lined with mesoderm.

(b) Acoelomates have no enclosed body cavity.

(c) Pseudocoelomates have an enclosed body cavity partially lined with mesoderm.

FIGURE 33.10 Animals May or May Not Have a Body Cavity.

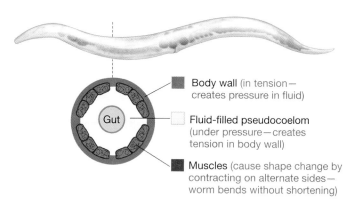

FIGURE 33.11 Hydrostatic Skeletons Allow Limbless Animals to Move.

For example, movement in nematodes is possible because the fluid inside the pseudocoelom stretches the body wall—much like water stretches the wall of a water balloon—putting the wall under tension (**FIGURE 33.11**). In turn, this tension exerts pressure on the fluid inside the coelom. When certain muscles in the body wall contract against the pressurized fluid, the force is transmitted through the fluid, changing the body's shape.

Coordinated contractions and relaxations of different muscles in the body wall enable various shape changes and forms of locomotion (see Chapter 48). The evolution of the coelom improved the ability of bilaterians to move in search of food.

Origin of Protostomes and Deuterostomes

Turn back to Figure 33.2 and put your finger on the node at the origin of the Bilateria, the monophyletic lineage of bilaterians. Now move your finger one node to the right, past the divergence of the Acoels. The common ancestor that existed at this node during the Cambrian was a bilaterally symmetric triploblast with a CNS, cephalization, and a coelom. This ancestor gave rise to a remarkable radiation of diverse animal lineages, only some of which are included in the tree.

What major groups occur within the Bilaterian coelomates? Despite the staggering diversity of adult animals, common themes during embryonic development have enabled biologists to recognize two major subgroups:

1. **protostomes,** in which the mouth develops before the anus, and blocks of mesoderm hollow out to form the coelom; and

2. **deuterostomes,** in which the anus develops before the mouth, and pockets of mesoderm pinch off to form the coelom.

Translated literally, protostome means "first-mouth" and deuterostome means "second-mouth."

To understand the contrasts in how protostome and deuterostome embryos develop, recall that the three embryonic germ layers form during a process called gastrulation (Chapter 23), which begins when cells move from the outside into the center of the embryo. These cell movements create a pore that opens to the outside (**FIGURE 33.12a**). In protostomes, this pore becomes the mouth. The other end of the gut, the anus, forms later. In

(a) Gastrulation (formation of gut and embryonic germ layers)

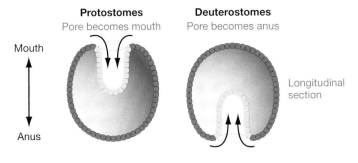

Protostomes
Pore becomes mouth

Deuterostomes
Pore becomes anus

Mouth

Anus

Longitudinal
section

(b) Formation of coelom (body cavity lined with mesoderm)

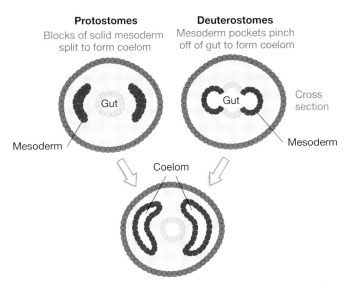

Protostomes
Blocks of solid mesoderm
split to form coelom

Deuterostomes
Mesoderm pockets pinch
off of gut to form coelom

Gut

Gut

Cross
section

Mesoderm

Mesoderm

Coelom

FIGURE 33.12 In Protostomes and Deuterostomes, Events in Early Development Differ. The differences between protostomes and deuterostomes show that there is more than one way to build a bilaterally symmetric, coelomate body plan.

deuterostomes, however, this initial pore becomes the anus; the mouth forms later.

The other developmental difference between the groups arises as gastrulation proceeds and the coelom begins to form. Protostome embryos have two blocks of mesoderm beside the gut. As the left side of **FIGURE 33.12b** indicates, their coelom begins to form when cavities open within each of the two blocks of mesoderm. In contrast, deuterostome embryos have layers of mesodermal cells located on either side of the gut. As the right side of Figure 33.12b shows, their coelom begins to form when these layers bulge out and pinch off to create fluid-filled pockets lined with mesoderm.

In essence, the protostome and deuterostome patterns of development represent two distinct ways of achieving the same end—the construction of a bilaterally symmetric body that contains a cavity lined with mesoderm. The functional or adaptive significance of the differences—if any—is not known.

Two major groups occur within the protostomes: **(1)** the Lophotrochozoa (pronounced *low-foe-tro-ko-ZOH-ah*) includes the mollusks, annelids, flatworms, and rotifers; and **(2)** the

Ecdysozoa (pronounced *eck-die-so-ZOH-ah*) includes the arthropods and the nematodes. **Lophotrochozoans** grow continuously when conditions are good. **Ecdysozoans** grow by shedding their external skeletons or outer coverings and expanding their bodies.

Species in the phylum Platyhelminthes (flatworms) are protostomes but lack a coelom—meaning that the coelom was lost during flatworm evolution. The coelom has also been reduced in an array of lineages including arthropods—which have limbs and thus rely less on hydrostatic skeletons for movement.

The vast majority of animal species are protostomes (see Chapter 34). However, the largest animals and most voracious predators tend to be deuterostomes (see Chapter 35).

Origin of Segmentation

Segmentation is defined by the presence of repeated body structures. A segmented backbone is one of the defining characteristics of **vertebrates,** a monophyletic lineage within the Chordata that includes fish, reptiles (including birds), amphibians, and mammals. Of the animals that are not vertebrates, called **invertebrates** (a paraphyletic group), segmentation occurs in annelids (earthworms and other segmented worms) and arthropods (insects, spiders, and crustaceans)—both very prolific and diverse lineages. Recent evidence suggests that segmentation also occurs to some extent in mollusks (snails, clams, squid).

Before molecular data were available, biologists grouped segmented animals like annelids and arthropods together in the same clade. Then Aguinaldo's landmark 1997 revision of animal relationships overturned this view, suggesting that segmentation arose independently in these groups (see Figure 33.2).

Is segmentation in different phyla an example of convergent evolution? Applying the same logic that Finnerty and Martindale used in their study of bilateral symmetry in sea anemones, several researchers are investigating whether the genes that regulate segmentation in different phyla are homologous. For example, one recent study found that one of the same gene regulatory pathways that regulate segmentation in arthropods, called the hedgehog pathway, is also important in regulating the segmentation in annelid worms.

The conclusion so far is that some of the tool-kit genes for segmentation may have arisen early in animal evolution, meaning that they are homologous in different phyla. However, these genes would subsequently have been lost in some lineages and co-opted in different ways and elaborated upon in others, contributing to the diversity of bilaterian body plans we see today. Research continues to test this hypothesis.

Why have organisms with segmented bodies been so successful in terms of diversity? One leading hypothesis is that segmentation enables specialization. Small changes in the expression of certain tool-kit genes along the length of a body can result in novel numbers, shapes, and sizes of body segments and appendages. *Hox* genes are a good example (see Chapters 22 and 34). Natural selection can then favor variations that have adaptive specializations for life in certain aquatic or terrestrial environments—leading to diversification.

In summing up this tour of the animal phylogeny, it's important to note that although biologists are increasingly confident that most or all of these inferences about animal relationships are correct, the phylogeny of animals is still a work in progress. As more taxa are sampled, larger numbers of characters are analyzed, and methods of phylogenetic analysis are refined, the resolution of the tree of life will continue to improve. Stay tuned.

If you understand that . . .

- The origin of body plans in the major lineages of animals was marked by changes in four fundamental features: the number of embryonic tissues present; the evolution of nervous systems and a bilaterally symmetric, cephalized body; the evolution of a body cavity; and protostome versus deuterostome patterns of development.

✔ **You should be able to . . .**

1. Describe whether your body has a tube-within-a-tube design. Justify your answer.

2. Explain why cephalization was important.

Answers are available in Appendix A.

33.3 What Themes Occur in the Diversification of Animals?

One of the great unresolved mysteries in the study of animal diversity is this: Why did almost all of the phylum-level body plans evolve all at once during the Cambrian, rather than more gradually over time? Animals diversified tremendously *within* phyla after the Cambrian, but no major new lineages arose during the last 500 million years.

One mechanistic hypothesis is that gene regulatory networks are responsible for limiting the origin of new body plans. These networks determine patterns of animal development and thus lay down the body plan. The scientists studying these networks have observed that once the networks are established, they are very resistant to change—that is, changes tend to be fatal. However, a multitude of "switches" and "plug-ins" can be added to networks, providing a mechanism for diversification within lineages after the body plans are established.

As research continues on the origin of body plans, biologists' understanding of the subsequent diversification within lineages is also steadily improving. Several variables have been hypothesized to play a role in animal diversification (Chapter 28):

- *Higher oxygen levels* An increase in oxygen levels may have made the evolution of big, mobile animals possible due to the efficiency of aerobic respiration.

- *The evolution of predation* The earliest animals were sessile and ate organic material. When predators evolved, they exerted selection pressure on other animals for shells, skeletons, rapid movement, and other adaptations for escaping capture.

- *New niches beget more new niches* As animals diversified, they themselves created new niches that could support yet more ecological diversification.

- *New genes, new bodies* As the animal genetic tool kit evolved—such as the duplication and diversification of *Hox* genes—the potential for morphological diversity increased.

It is likely that a combination of all of these variables was important to the diversification of animals. The result? Animals with the same body plan diversified in their strategies for finding food, ingesting food, and reproducing in different ecological niches. Conversely, animals with different body plans have independently evolved similar strategies for feeding and reproducing when they occupy similar ecological niches. What are the major themes of diversification?

Sensory Organs

The evolution of a cephalized body was a major breakthrough in the evolution of animals. Along with a mouth and brain, a concentration of sensory organs in the head region—where the animal initially encounters the environment—is a key aspect of cephalization.

Certain senses are almost universal in animals (see Chapter 47). The common senses include sight, hearing, taste/smell, and touch (see examples in **TABLE 33.2**). At least some ability to sense temperature is also common. But as animals diversified, a wide array of specialized sensory abilities also evolved. For example:

- *Magnetic field* Many birds, sea turtles, sea slugs, and other animals can detect Earth's magnetic field and use it as an aid in navigation.

- *Electric field* Some aquatic predators, such as sharks, are so sensitive to electric fields that they can detect electrical activity in the muscles of passing prey.

- *Barometric pressure* Some birds can sense changes in air pressure, which may aid them in avoiding storms.

Variation in sensory abilities is important: It allows different species of animals to collect information about a wide array of environments, including the presence of food, predators, and mates.

Feeding

To organize the diversity of ways that animals find food, biologists distinguish *what* individuals eat from *how* they eat. Animals within a lineage often pursue different food sources and feeding strategies when they occupy different niches. For example, sea cucumbers and sea stars both have the same echinoderm body plan, but they have different food sources and feeding strategies—the sea cucumber mops up detritus from the seafloor, and the sea star pries open clams and mussels and devours them.

Conversely, animals from *different* lineages often pursue the *same* food sources and feeding strategies when they occupy similar niches. For example, burrowing sea cucumbers and some burrowing polychaete worms (phylum Echinodermata and

SUMMARY TABLE 33.2 Diversification of Sensory Organs

Sense	Example
Sight *Stimulus:* light Flies use their compound eyes to find food and mates, and to escape predators	
Hearing *Stimulus:* sound Bats use their sense of hearing to find prey and to avoid obstacles in the dark	
Taste/smell *Stimulus:* molecules Some male moths have elaborate antennae to detect pheromones in the air	
Touch *Stimulus:* contact, pressure Sea anemones detect and capture prey using their sense of touch	

Other senses	Examples
Stimulus: temperature (thermal energy)	Pit vipers
Stimulus: magnetic field	Sea turtles
Stimulus: electric field	Sharks
Stimulus: barometric pressure	Birds
Stimulus: gravity	Squid

Annelida, respectively) have different body plans but use a similar strategy of consuming detritus on the seafloor.

What Animals Eat: Diversification of Ecological Roles Animals can be classified as (1) **detritivores** that feed on dead organic matter, (2) **herbivores** that feed on plants or algae, or (3) **carnivores** that feed on animals (see **TABLE 33.3**). **Omnivores** eat both plants and animals. These ecological roles have important implications for whole ecosystems because animal feeding moves both energy and nutrients through food webs (Chapter 56).

Another way to categorize the diverse ecological impacts of animal consumption is to consider the impacts on the organisms that animals consume. **Predators** kill and consume all or most of their prey. Predators are usually larger than their prey and kill them quickly using an array of mouthparts and hunting strategies.

SUMMARY TABLE 33.3 Diversification of Ecological Roles

Ecological Role	Example
Detritivores Feed on dead organic matter Millipedes feed on decaying leaves	
Herbivores Feed on plants or algae Pandas eat vast amounts of bamboo	
Carnivores Feed on animals Owls sit and wait for prey	
Omnivores Feed on both plants and animals	*Example:* Humans

Herbivores usually consume plant tissue without killing the whole organism. However, seed predators are an exception because seeds contain an entire—although embryonic—plant.

Like herbivores, **parasites** often harvest nutrients from certain parts of their hosts, but parasites are usually much smaller than their victims. **Endoparasites** live inside their hosts and usually have simple, wormlike bodies. Tapeworms, of the phylum Platyhelminthes (flatworms), are endoparasites with no digestive system. Instead of a mouth, they have hooks or other structures on their head that attach to their host's intestinal wall. Instead of digesting food themselves, they absorb nutrients directly from their surroundings. **Ectoparasites** live on the outside of their hosts. Ectoparasites usually have limbs or mouthparts that allow them to grasp the host, and mouthparts that allow them to pierce their host's skin and suck the nutrient-rich fluids inside. Aphids and ticks are ectoparasites.

Whether animals consume their victims completely or partially, these consumers can drastically affect the fitness of the consumed organisms. Thus, animals are an important agent of natural selection—influencing the evolution of the species they eat (see Chapter 55).

How Animals Feed: Four General Strategies Animal mouthparts vary, and the structure of an animal's mouthparts correlates closely with its method of feeding. Keep this in mind as you review the four general tactics that animals use to obtain food, summarized in **TABLE 33.4**.

Suspension feeders, also known as **filter feeders,** employ a wide array of structures to trap suspended particles—usually detritus or plankton, small organisms that drift in the currents. Sponges, like the individual illustrated in Figure 33.3, are suspension feeders. So are clams and mussels, which pump water through their bodies and trap suspended food on their feathery gills—structures that also function in gas exchange. Baleen whales suspension feed by gulping water, squeezing it out between the horny baleen plates that line their mouths, and trapping shrimp-like organisms called krill inside. Barnacles capture particles with their feathery, jointed legs.

Because particles float in water much more readily than in air, suspension feeding is particularly common in aquatic environments. Many suspension feeders are sessile.

Many **deposit feeders** digest organic matter in the sediments; their food consists of sand- or mud-dwelling bacteria, archaea, protists, and fungi, along with detritus that settles on the surface of the sediments. Earthworms, for example, are annelids that swallow soil as well as leaves and other detritus on the surface of the soil.

The seafloor is also rich in organic matter, which rains down from the surface and collects in food-rich deposits. These deposits are exploited by a wide array of segmented worms (annelids) as well as sea cucumbers.

Unlike suspension feeders, which are diverse in size and shape and use various trapping or filtering systems, deposit feeders tend to be similar in appearance. They usually have simple mouthparts, and their body shape is wormlike. Like suspension feeders, however, deposit feeders occur in a wide variety of lineages.

SUMMARY TABLE 33.4   **Diversification of Feeding Strategies**

Strategy	Example
Suspension feeders (filter feeders) Capture food by filtering out or concentrating particles floating in water or drifting through the air Barnacles use specialized legs to capture plankton	
Deposit feeders Ingest organic material that has been deposited within a substrate or on its surface Sea cucumbers use feeding tentacles to mop up detritus from the seafloor	
Fluid feeders Suck or mop up liquids like nectar, plant sap, blood, or fruit juice Butterflies and moths drink nectar through their extensible, hollow proboscis	
Mass feeders Take chunks of food into their mouths Lions bite chunks of meat off of prey carcasses	

Fluid feeders range from butterflies and moths that feed on nectar with a straw-like proboscis to vampire bats that feed on blood. Fluid feeders are found in a wide array of lineages and often have mouthparts that allow them to pierce seeds, stems, skin, or other structures in order to withdraw the fluids inside.

Mass feeders ingest chunks of food. The structure of their mouthparts correlates with the type of food pieces that they harvest and ingest. Lions, for example, have razor-like teeth for tearing tough flesh into pieces small enough to be swallowed. Snails use a rasp-like structure called a radula to scrape bits of plant tissue or animal flesh into their mouths.

Movement

Animal locomotion provides an array of important functions: finding food, finding mates, escaping from predators, and dispersing to new habitats. Animals move in highly variable ways; they burrow, slither, swim, fly, crawl, walk, or run—mostly powered by muscle (Chapter 48).

Limbs are a prominent feature of species in many phyla. They develop as outpockets of the body wall and tend to be perpendicular to the anterior–posterior and dorsal–ventral body axes. Limbs can take a variety of forms, from the lobe-like limbs of onychophorans (velvet worms) to the jointed limbs of arthropods and vertebrates to the more flexible tube feet of echinoderms and tentacles of mollusks (**TABLE 33.5**).

Traditionally, biologists have hypothesized that the major types of jointed and unjointed animal limbs were not homologous—that is, not descended from a common ancestor. Because the structure of animal appendages is so diverse, it was logical to maintain that at least some appendages evolved independently of each other. As a result, biologists predicted that completely different genes are responsible for each major type of appendage.

Recent results have challenged this view, however. The experiments in question involve a gene called *Distal-less*, or *Dll*, that was originally discovered in fruit flies. (*Distal* means "away from the body.") The *Dll* protein product seems to deliver a simple message as a fruit fly embryo develops: "Grow appendage out this way."

Biologists in Sean Carroll's lab set out to test the hypothesis that *Dll* might be involved in the initial phase of limb or appendage formation in other animals. The team used a fluorescent marker that sticks to the *Dll* gene product to locate tissues where the gene is expressed. When they introduced the fluorescent marker into embryos from annelids, arthropods, echinoderms, and other phyla, they found that it bound to *Dll* in all of them. More important, the highest concentrations of *Dll* gene products were found in cells that form appendages—even in phyla with wormlike bodies that have extremely simple appendages (**FIGURE 33.13**; see page 650). Other experiments have shown that *Dll* is also involved in limb formation in vertebrates.

Based on these findings, biologists are concluding that all animal appendages have some degree of genetic homology and that they are all derived from appendages that were present in a common ancestor. The idea is that simple appendages evolved early in the history of the Bilateria and that, subsequently, evolution by natural selection produced the diversity of limbs observed today.

SUMMARY TABLE 33.5   **Diversification of Limbs**

Type of Limb	Example
Lobe-like limbs	Onychophorans (velvet worms) use lobe-like limbs to crawl
Jointed limbs	The jointed limbs of arthropods (such as this crab) and vertebrates are used for many modes of feeding and locomotion
Parapodia	Polychaete worms use bristled parapodia to crawl and swim
Tube feet	Echinoderms like this sea star use tube feet to crawl
Tentacles	Octopuses use muscular tentacles to crawl and grab prey

Arthropod	Onychophoran	Annelid

Developing legs

Developing segments

FIGURE 33.13 Genetic Evidence that All Animal Appendages Are Homologous. In diverse organisms, *Dll* gene products (stained green and dark brown) are localized in areas of the embryos where appendages are forming. These data suggest that the appendages are related by common ancestry.

Reproduction

An animal may be efficient at moving and eating, but if it does not reproduce, the alleles responsible for its effective locomotion and feeding will not increase in frequency in the population. Evolution by natural selection occurs when individuals with certain alleles produce more surviving offspring than other individuals do (Chapter 25). Organisms live to reproduce.

Given the array of habitats and lifestyles of animals, it's not surprising that they vary tremendously in how they reproduce (**TABLE 33.6**). A few examples here will help illustrate the diversity of animal reproduction (see Chapter 50 for details).

Asexual or Sexual Reproduction? In the phylum Rotifera, an entire lineage called the bdelloids (pronounced *DELL-oyds*) reproduces only asexually, through mitosis, by producing diploid eggs that can mature into adults without being fertilized (a process called parthenogenesis). Even certain fish, lizard, and snail species have never been observed to undergo sexual reproduction.

However, sexual reproduction via meiosis and fusion of gametes is much more common than asexual reproduction in animals; and some animals, like coral polyps, can reproduce either way depending on environmental circumstances. Asexual reproduction tends to be more efficient than sexual reproduction, but sexual reproduction leads to greater genetic diversity, which can be favored by natural selection in variable or unfavorable environments (Chapter 13).

Where Does Fertilization Occur? When sexual reproduction does occur, fertilization may be internal, usually within the body of the female, or external, in the environment.

When internal fertilization takes place, a male typically inserts a sperm-transfer organ into the body of a female. In some animal species, males produce sperm in packets, which females then pick up and insert into their own bodies. But in seahorses, females insert eggs into the male's body, where they are fertilized. (The male is pregnant for a time and then gives birth to live young.)

Reproductive Strategy	Example
Asexual reproduction	
Polyps within a coral colony are genetically identical clones. They are produced by fission (splitting) or budding	
Sexual reproduction External fertilization	
Sea cucumbers reproduce sexually by releasing eggs and sperm into the open water, where fertilization occurs	
Internal fertilization	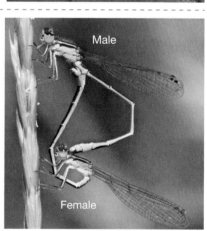 Male Female
When damselflies copulate, the male holds the female behind her head with claspers and the female retrieves sperm from the male's sperm-transfer organ with the tip of her abdomen	

Modes of embryonic development following internal fertilization	Examples
Viviparous species nourish embryos internally and give birth to live young	Most mammals
Oviparous species deposit fertilized eggs; embryos are nourished by yolk	Most insects, birds
Ovoviviparous species retain eggs internally and then give birth to live young	Guppies, garter snakes

External fertilization is extremely common in aquatic species. Females lay eggs onto a substrate or release them into open water. Males shed sperm, which swim, on or near the eggs or into open water.

Where do Embryos Develop? Following internal fertilization, eggs and embryos may be retained in the female's body during development, or fertilized eggs may be laid outside of the body to develop independently of the mother:

- **Viviparous** ("live-bearing") species such as humans and most other mammals nourish embryos inside the body and give birth to live young.

- **Oviparous** ("egg-bearing") species such as chickens and crickets deposit fertilized eggs. The embryos within are nourished by yolk.

- **Ovoviviparous** ("egg-live-bearing") species such as guppies and garter snakes retain eggs inside their body during early development, but the growing embryos are nourished by yolk inside the egg and not by nutrients transferred directly from the mother, as in viviparous species. Ovoviviparous females then give birth to well-developed young.

Most mammals and a few species of sea stars, velvet worms, fish (including sharks), amphibians, and lizards are viviparous; some snails, insects, reptiles, and fishes are ovoviviparous. But the vast majority of animals are oviparous.

Life Cycles

Reproduction is just one component of the diverse life cycles of animals. The vast majority of sexually reproducing animals have diploid-dominant life cycles because the haploid stages—the gametes—are relatively short lived and single celled. But even this general pattern has interesting exceptions, such as haploid males in honeybee colonies. How else do animal life cycles vary?

Perhaps the most spectacular innovation in animal life cycles involves the phenomenon known as **metamorphosis** ("change-form")—a drastic change from one developmental stage to another. The young of some animals, such as zebras, look similar to adults when they are born. Their development is said to be *direct*. However, other animals, such as sea urchins, undergo a dramatic transformation—a metamorphosis—during their life cycle and are said to undergo *indirect* development (**FIGURE 33.14**).

During indirect development, embryogenesis produces **larvae** (singular: **larva**) that look radically different from adults, live in different habitats, and eat different foods. In the case of the sea urchin in Figure 33.14, the larva is bilaterally symmetric and planktonic, and it suspension feeds on single-celled algae; but the adult is radially symmetric and sessile, and it scrapes macroalgae off of rocks.

Through the process of metamorphosis, larvae transform into **juveniles;** they look like adults and live in the same habitats and eat the same foods as adults, but they are still sexually immature. It is only after a period of growth and maturation that juveniles become **adults,** the reproductive stage in the life cycle.

Metamorphosis is extremely common in marine animals as well as in insects (insect metamorphosis is described in Chapter 34)

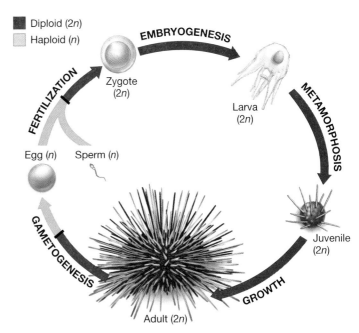

FIGURE 33.14 Many Animal Life Cycles Include Metamorphosis. Most sea urchins have a planktonic, feeding larva that is very different from the sessile adult. Drawings are not to scale.

✔ **EXERCISE** A few sea urchins produce especially large, yolk-rich eggs that develop directly into juveniles and adults, bypassing the feeding larval stage. Circle the part of the life cycle that would be different in this case, and pose a hypothesis to explain why or under what conditions this difference might be adaptive.

and many amphibians. Why would natural selection favor such an indirect path to adulthood? One hypothesis emphasizes dispersal. In marine species that have limited or no movement as adults—such as sponges, corals, sea anemones, clams, and sea urchins—larvae function as a dispersal stage. They are a little like the seeds of many land plants—a life stage that allows individuals to move to new habitats.

Another hypothesis for the occurrence of metamorphosis emphasizes feeding efficiency. Because larvae and adults feed on different foods in different ways, they do not compete with each other and can specialize to take advantage of available food sources. For example, the caterpillar larvae of butterflies can feast on nutritious leaves to acquire the energy needed for growth, while adult butterflies feast on flower nectar to acquire energy needed for finding mates and reproducing.

The diversification of life cycles occurred not only among species in diverse lineages but also within species—such as in organisms that can opt between asexual and sexual reproduction or direct and indirect development, depending on environmental circumstances. The diversification of life cycles, as well as that of sensory structures, ecological roles, feeding strategies, limbs, and reproductive strategies, represent the consequences of evolutionary processes played out in ecological contexts over millions of years.

Let's now examine how the general themes of diversification apply to individual phyla, starting with the most ancient lineages first and then moving on to the extremely diverse protostomes (Chapter 34) and deuterostomes (Chapter 35).

If you understand that . . .

- Most of the animal body plans originated during the Cambrian and subsequently diversified.
- Themes of diversification include evolution of different sensory organs, ecological roles, feeding strategies, type of limbs, type of reproduction, and life cycles.

 You should be able to . . .

1. Explain how the mouthparts of deposit feeders and mass feeders are expected to differ.
2. Suggest why external fertilization is particularly common in aquatic environments, and internal fertilization is particularly common in terrestrial environments.

Answers are available in Appendix A.

33.4 Key Lineages of Animals: Non-Bilaterian Groups

The fossil record indicates that the phyla Porifera (sponges), Ctenophora (comb jellies), and Cnidaria (jellyfish and others) are the most ancient of all major animal lineages. The phylogeny of animals presented in this chapter is consistent with the fossil record—the first sponges, ctenophores, and cnidarians appear at the base of the tree (Figure 33.2). As you consider each group in more detail, try to pick out which characteristics are unique to the group, and which are common to organisms living in similar environments.

- Porifera (Sponges)
- Ctenophora (Comb Jellies)
- Cnidaria (Jellyfish, Corals, Anemones, Hydroids)

Porifera (Sponges)

About 8500 species of sponges have been described to date. Although a few freshwater species are known, most are marine. All sponges are **benthic,** meaning that they live at the bottom of aquatic environments. Sponges are particularly common in rocky, shallow-water habitats of the world's oceans. One sponge species native to the Caribbean can grow to heights of 2 meters (m).

Sponges have commercial and medical value to humans. The dried bodies of certain sponge species are able to hold large amounts of water and thus are prized for use in bathing and washing. In addition, researchers are increasingly interested in the array of toxins that sponges produce to defend themselves against predators and bacterial parasites—possibly for use in cancer chemotherapy.

Feeding Most sponges are suspension feeders. Their cells beat in a coordinated way to produce a water current that flows through small pores in the outer body wall, into chambers inside the body, and out through larger openings (**FIGURE 33.15**). As water passes by feeding cells, organic debris and bacteria, archaea, and small protists are filtered out of the current and then digested. Some deep-sea sponges are predators, however—they capture small crustaceans on hooks that project from the body.

Movement Most adult sponges are sessile, though a few species are reported to move at rates of up to 4 millimeters (mm) per day. Most species produce larvae that swim with the aid of cilia, so they are capable of dispersal. Recent research has confirmed that at least one species can contract its body to expel waste products.

Reproduction Asexual reproduction occurs in a variety of ways, depending on the species. Some sponge cells are totipotent, meaning that when a fragment breaks off, small groups of adult cells have the capacity to develop into complete new adult organisms. Although individuals of most species produce both eggs and sperm, self-fertilization is rare because individuals release their male and female gametes at different times.

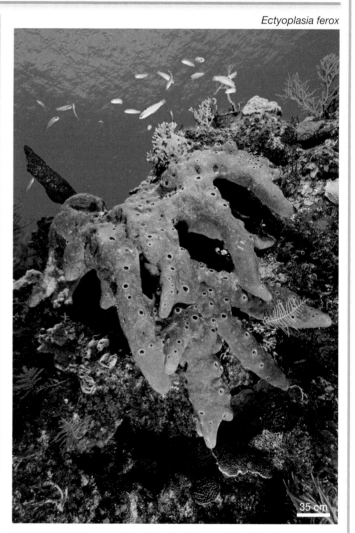

Ectyoplasia ferox

35 cm

FIGURE 33.15 Sponges Are Sessile Suspension Feeders.

Ctenophora (Comb Jellies)

Ctenophores are transparent, ciliated, gelatinous diploblasts that live in marine habitats (**FIGURE 33.16**). Although a few species live on the ocean floor, most are planktonic. Only about 190 species have been described to date, but some are abundant enough to represent a significant fraction of the total planktonic biomass. Accidental introductions of the ctenophore *Mnemiopsis leidyi*, which preys on fish larvae, have devastated fish production in the Black Sea and Caspian Sea. If you have gone swimming in an ocean or bay, you may have bumped up against ctenophores without realizing it since they can be abundant yet nearly invisible in the water. (Don't worry, they are harmless.) Some comb jellies can also be seen at night when they emit pulses of faint phosphorescent light.

Feeding Ctenophores are predators, but lack the stinging cells that are characteristic of Cnidarians. Feeding occurs in several ways, depending on the species. Some comb jellies have long tentacles covered with cells that release an adhesive when they contact prey. These tentacles are periodically wiped across the ctenophore's mouth so that captured prey can be ingested. In other species, prey can stick to mucus on the body and be swept toward the mouth by cilia. Still other species ingest large prey whole.

Movement Adults move via the beating of cilia, which occur in comblike plates. The plates form rows that run the length of the body. Ctenophores are the largest adult animals known to use cilia for locomotion. ✔You should be able to indicate the origin of adult swimming via rows of coordinated cilia on Figure 33.2.

Reproduction Most species have both male and female organs and routinely self-fertilize, though fertilization is external. Larvae are free swimming.

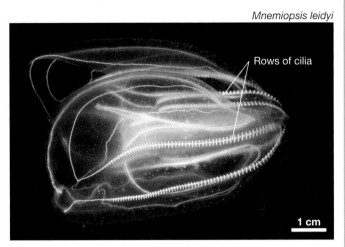

Mnemiopsis leidyi

Rows of cilia

1 cm

FIGURE 33.16 Ctenophores Are Planktonic Predators.

Cnidaria (Jellyfish, Corals, Anemones, Hydroids)

Although a few species of Cnidaria inhabit freshwater, the vast majority of the 11,500 species are marine. They are found in all of the world's oceans, occupying habitats from the surface to the substrate, and are important predators. The phylum comprises four main lineages, as shown in **FIGURE 33.17**.

Many cnidarians are radially symmetric diploblasts consisting of ectoderm and endoderm layers that sandwich gelatinous material known as **mesoglea,** which contains a few scattered ectodermal cells. Recent research indicates that at least some jellyfish are triploblastic and have bilaterally symmetric larvae or are bilaterally symmetric as adults. Cnidarians have a gastrovascular cavity instead of a flow-through gut—meaning there is only one opening to the environment for both ingestion and elimination of wastes (see Chapter 44).

Jellyfish have a life cycle that includes both a sessile **polyp** form that reproduces asexually and a free-floating **medusa** that reproduces sexually, as shown in the case of a hydrozoan in **FIGURE 33.18** (see page 654). Anemones and coral, however, exist only as polyps—never as medusae. Reef-building corals secrete outer skeletons of calcium carbonate that create the physical structure of a coral reef—one of the world's most productive habitats (see Chapter 56).

Feeding A key synapomorphy of the cnidarians is a specialized cell called a **cnidocyte,** which is used in prey capture. When cnidocytes sense a fish or other type of prey, the cells forcibly eject a barbed, spear-like structure, which may contain toxins. The barbs hold the prey, and the toxins subdue it until it can be brought

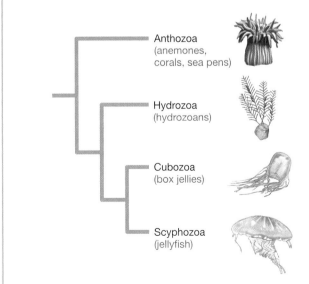

Anthozoa (anemones, corals, sea pens)

Hydrozoa (hydrozoans)

Cubozoa (box jellies)

Scyphozoa (jellyfish)

FIGURE 33.17 There are Four Main Lineages within the Cnidaria.

(Continued on next page)

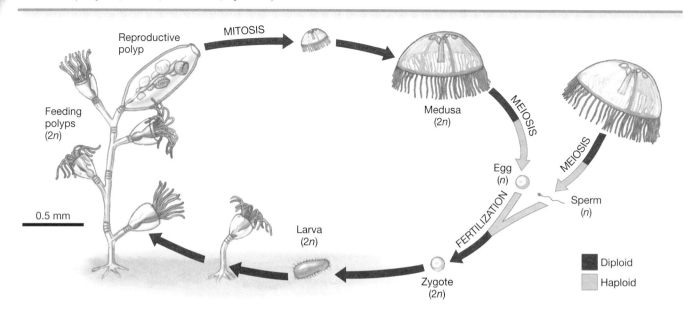

FIGURE 33.18 Cnidarian Life Cycles May Include a Polyp and Medusa Form. This is an example of a hydrozoan called *Obelia*. Colonies are often found attached to seaweed, shells, or rocks and contain hundreds of polyps.

to the mouth and ingested. Cnidocytes are commonly located near the mouths of cnidarians or on elongated structures called tentacles. In a very small number of species, cnidarian toxins can be deadly to humans as well as to prey organisms; in Australia, twice as many people die each year from stings by box jellyfish as from shark attacks (**FIGURE 33.19**). ✔You should be able to indicate the origin of cnidocytes on Figure 33.2.

Besides capturing prey actively, most species of coral, some jellyfish, and many anemones host photosynthetic dinoflagellates. The relationship is mutually beneficial, because the protists supply the cnidarian host with food in exchange for protection.

Movement Both polyps and medusae are capable of movement. In polyps, the gut cavity acts as a hydrostatic skeleton that works with the aid of muscle-like cells to contract or extend the body. Many polyps can also creep along a substrate, using muscle-like cells at their bases. Medusae can contract the muscle-like cells in the rims of their bell-shaped body rhythmically, enabling paddling motions as well as jet propulsion—the forcible flow of water in the opposite direction of movement. Cnidarian larvae swim by means of cilia.

Reproduction Polyps may produce new individuals asexually by (**1**) budding, in which a new organism grows out from the body wall of an existing individual; (**2**) fission, in which an existing adult splits lengthwise to form two individuals; or (**3**) fragmentation, in which parts of an adult regenerate missing pieces to form a complete individual. During sexual reproduction, gametes are usually released from the mouth of a polyp or medusa, and fertilization takes place in the open water. Eggs hatch into larvae that become part of the plankton before settling and developing into polyps.

Chironex fleckeri

FIGURE 33.19 Some Jellyfish Have Long Tentacles, Packed with Cnidocytes That Sting Prey.

If you understand . . .

33.1 What Is an Animal?

- Animals are multicellular, heterotrophic eukaryotes that lack cell walls and ingest their prey.

- All animals move under their own power at some point in their life cycle. All animals other than sponges have specialized nerve and muscle cells that enable complex movements.

- Animals are eating machines, the largest consumers on Earth.

- Animals comprise 30–35 phyla and may number 8 million or more species.

✔ You should be able to predict how the total amount of plant material in a forest would change if all animals were excluded for one year.

33.2 What Key Events Occurred during the Evolution of Animals?

- Animals are a monophyletic group that originated from a protist ancestor that was probably similar to choanoflagellates.

- The basic genetic tool kit for multicellular animals originated at the very root of the animal lineage.

- Sponges lack highly organized, complex tissues other than rudimentary epithelium.

- Most ctenophores and cnidarians have radial symmetry, two embryonic germ layers, and neurons organized into a nerve net. However, some of the regulatory genes responsible for bilateral symmetry arose before the split of cnidarians and bilaterians.

- Most animal species are bilaterians. They have bilateral symmetry, three embryonic tissues, and a coelom—features that gave rise to a tube-within-a-tube body plan.

- Most bilaterians also have a centralized nervous system and cephalized bodies—meaning that a distinctive head region contains the mouth, brain, and sensory organs.

- The bilaterian tube-within-a-tube design is built in one of two fundamental ways—via the protostome or deuterostome patterns of embryonic development.

- Segmentation is a key innovation in annelids, arthropods, and vertebrates.

✔ You should be able to draw a bilaterian that lacks limbs, and label the gut, outer body wall, muscle layers, head region, mouth, anus, brain, and major nerve tracts.

33.3 What Themes Occur in the Diversification of Animals?

- Sensory abilities and sensory structures vary among species and correlate with their habitats and feeding methods.

- Animals fill three different ecological roles depending on their source of food: they are detritivores, herbivores, or carnivores—or omnivores, a combination.

- Animals capture food in four ways: Suspension or filter feeders filter organic material or small organisms from water; deposit feeders swallow sediments; fluid feeders lap or suck up liquids; mass feeders harvest chunks of tissue.

- Although the types of appendages used in animal locomotion range from simple lobe-like limbs to complex jointed legs, the genes that indicate where appendages develop may be homologous.

- Asexual reproduction occurs in many species, but sexual reproduction is predominant. Fertilization may be external or internal, and embryos may complete development inside or outside the mother.

- Development may be direct or indirect. During indirect development, a dramatic morphological transformation takes place from one life stage to another, called metamorphosis.

✔ You should be able to select a phylum and give examples of how the themes of diversification apply within that phylum.

33.4 Key Lineages of Animals: Non-Bilaterian Groups

- Sponges are diverse, benthic suspension feeders.

- Ctenophores are planktonic predators, the largest animals to use cilia for locomotion.

- Cnidarians are diverse and abundant in habitats around the world. The cnidocyte, or stinging cell, is a synapomorphy for this group.

✔ You should be able to recognize sponges, ctenophores, or cnidarians based on their morphology.

MasteringBiology

1. MasteringBiology Assignments

Tutorials and Activities Animal Phylogenetic Tree; Animal Body Plans; Architecture of Animals

Questions Reading Quizzes, Blue-Thread Questions, Test Bank

2. eText Read your book online, search, take notes, highlight text, and more.

3. The Study Area Practice Test, Cumulative Test, BioFlix® 3-D Animations, Videos, Activities, Audio Glossary, Word Study Tools, Art

You should be able to . . .

1. What synapomorphy (shared, derived trait) distinguishes animals as a monophyletic group, distinct from choanoflagellates?
   a. multicellularity
   b. coloniality
   c. heterotrophy
   d. no cell walls

2. True or false? All coelomates are triploblasts.

3. In _____, the mouth develops before the anus; but in _____, the anus develops before the mouth.

4. Why do some researchers maintain that the limbs of all animals are homologous?
   a. Homologous genes, such as *Dll*, are involved in their development.
   b. Their structure—particularly the number and arrangement of elements inside the limb—is the same.
   c. They all function in the same way—in locomotion.
   d. Animal appendages are too complex to have evolved more than once.

5. In a tube-within-a-tube body plan, what is the interior tube?
   a. ectoderm
   b. mesoderm
   c. the coelom
   d. endoderm

6. Which of the following characteristics does not apply to cnidarians?
   a. usually diploblastic
   b. possess a gastrovascular cavity with one opening
   c. undergo metamorphosis
   d. have a central nervous system

7. Explain the difference between a diploblast and a triploblast. How was the evolution of mesoderm associated with the evolution of a coelom?

8. Why is it ecologically significant that animals are heterotrophic *and* multicellular?

9. **QUANTITATIVE** To estimate the relative abundance of the major phyla, calculate how many named species of arthropods, mollusks, and nematode worms exist per named species of chordate (see Table 33.1). Do you think these calculations are likely to be underestimates, or overestimates? Why?

10. Explain how an animal mother nourishes an embryo in oviparous species versus viviparous species.

11. Specific genes are known to play an important role in the segmentation of arthropods. What approach could you use to test the hypothesis that segmentation in vertebrates (phylum Chordata) is homologous to segmentation in arthropods (phylum Arthropoda)?

12. Which of the following is an example of homology (similarity due to common ancestry)?
    a. suspension feeding in sponges and clams
    b. ectoparasite lifestyle in aphids and ticks
    c. cnidocytes (stinging cells) in jellyfish and sea anemones
    d. metamorphosis in sea urchins and insects

13. Suppose that a gene originally identified in nematodes (roundworms) is found to be homologous with a gene that can cause developmental abnormalities in humans. Would it be possible to study this same gene in fruit flies? Explain.

14. Some sea anemones can produce large colonies by reproducing asexually, but they can also produce planktonic larvae by reproducing sexually. Predict the circumstances under which each mode of reproduction would be favored.

15. Radial symmetry evolved in echinoderms (sea stars and sea urchins) from bilaterally symmetric ancestors. Predict whether the echinoderm nervous system is centralized or diffuse. Explain your logic.

16. You have a unique opportunity to explore a pristine coral reef in the South Pacific. You come across a small organism that you do not recognize and that is not in your field guide. Which of the following suites of characteristics would be most helpful in identifying the organism's phylum?
    a. sessile, green, 2 cm tall, abundant
    b. radial symmetry, suspension feeder
    c. nonsegmented, sessile, changes shape
    d. radial symmetry, sessile, two tissue layers

34 Protostome Animals

In this chapter you will learn that

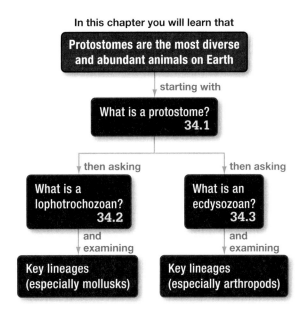

In numbers of individuals and species richness, protostomes are the most abundant and diverse of all animals—and the insects are by far the most abundant and diverse of the protostomes. The insects shown here come from all over the world.

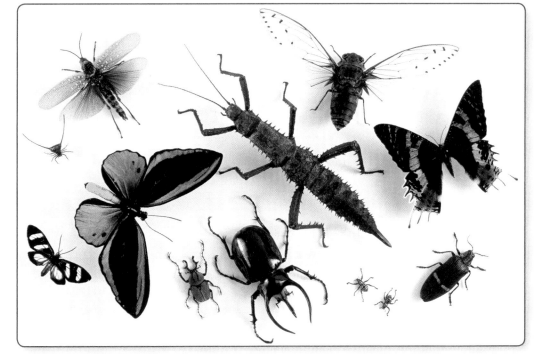

Protostome animals are spectacularly diverse and abundant. To put this in perspective, consider that of the more than 30 animal phyla known to exist, at least 22 are protostomes. Table 33.1 summarized the major protostome phyla. You will likely recognize many of the more familiar ones:

- *Arthropoda*, including the insects, crustaceans (shrimp, lobsters, crabs), chelicerates (spiders and mites), and myriapods (millipedes, centipedes)
- *Mollusca*, including the snails, clams, octopuses, and squid
- *Annelida*, including the segmented worms such as the earthworms, polychaete worms, and leeches

✔ When you see this checkmark, stop and test yourself. Answers are available in Appendix A.

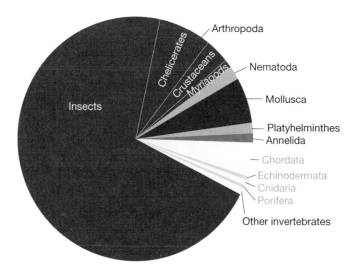

Major non-protostome phyla

Arthropoda

Chelicerates
Crustaceans
Myriapods

Nematoda

Insects

Mollusca

Platyhelminthes
Annelida

— Chordata

Echinodermata
Cnidaria
Porifera

Other invertebrates

FIGURE 34.1 The Relative Diversity of Animal Lineages. Most animals are protostomes. About 70 percent of all known species of animals on Earth are insects, most of them beetles. (Humans and other vertebrates are deuterostomes, in the phylum Chordata.)

DATA: Chapman, A. D. 2009. *Numbers of Living Species in Australia and the World,* 2nd ed. Canberra: Australian Biological Resources Study.

Some protostome phyla have only a few named species, such as the phylum Priapula, which contains 16 named species. On the other end of the spectrum, over 85,000 mollusks have been named thus far, and about 1.2 million arthropod species, of which about a million are insects. Scientists estimate that the actual number of arthropod species may be over 10 million.

The pie chart in **FIGURE 34.1** shows the relative numbers of named species in various animal phyla. Notice that the wedges representing non-protostome phyla, labeled Chordata, Cnidaria, Echinodermata, and Porifera, are small. Almost all animal species are protostomes.

Certain protostomes are not only diverse (in numbers of species), but also extremely abundant (in numbers of individuals). A typical acre of pasture in England is home to almost 18 *million* individual beetles and up to 9 *billion* roundworms. The world population of ants is estimated to be 1 million billion (1,000,000,000,000,000) individuals.

Given the diversity and abundance of protostomes, you should not be surprised that they fill very important ecological roles. As a group, protostomes live in virtually every aquatic and terrestrial habitat in the world and can be detritivores, herbivores, or carnivores. If one of biology's most fundamental goals is to understand the diversity of life on Earth, then protostomes—particularly the mollusks and arthropods—demand our attention.

The ecological importance of protostomes extends to the health and welfare of humans in profound and diverse ways. For example:

- Protostomes, especially shellfish (clams, oysters, scallops, crabs, shrimp, lobsters), are a major direct source of food for humans, which makes them economically important.

- Humans also rely on protostomes indirectly for their ecosystem services. For example, many farmers rely on protostomes like earthworms to prepare the soil, and on bees and other insects to pollinate their crops.

- Some insects have a massive negative impact on food production by damaging crops, while other insects protect crops from pests.

- Some protostomes produce valuable materials such as silk and pearls.

- Many protostomes play an important role in causing or transmitting human diseases, and are common parasites.

- The protostomes include two of the most important model organisms in all of biological science: the fruit fly *Drosophila melanogaster* and the roundworm *Caenorhabditis elegans* (see **BioSkills 13** in Appendix B).

Let's take a closer look at what defines the protostomes and then explore the diversity within the major phyla.

34.1 What Is a Protostome?

There are two major groups of bilaterally symmetric, triploblastic animals: the protostomes and deuterostomes (Chapter 33). The **protostomes** are distinguished from the deuterostomes by two developmental characteristics:

1. During gastrulation, the initial pore that forms in the embryo becomes the mouth—protostome translates to "first-mouth." (In deuterostomes the pore develops into the anus, and the mouth develops later.)

2. If a **coelom** (a body cavity) forms later in development, it forms from cavities that arise within blocks of mesodermal cells (rather than as mesoderm pockets pinching off the gut, as in deuterostomes).

Phylogenetic studies have long supported the hypothesis that protostomes are a monophyletic group, meaning the protostome developmental sequence arose just once. But biologists have found it difficult to sort out the relationships among protostome lineages. During the 500 million years of evolution that followed the diversification of protostomes during the Cambrian explosion, some morphological characteristics evolved independently in different lineages, resulting in convergent evolution. Other characteristics thought to be synapomorphies—shared, derived traits within a lineage—were lost in some groups. As a result, morphological data have led to many alternative hypotheses about phylogenetic relationships.

Recent analyses of DNA sequence data have rejected many previous hypotheses of relationships based on morphological data. As the phylogenetic tree in **FIGURE 34.2** shows, the recent data support two major subgroups within the protostomes. The Lophotrochozoa (pronounced *low-foe-tro-ko-ZOH-ah*) include mollusks and annelid worms. The Ecdysozoa (pronounced *eck-die-so-ZOH-ah*) include the arthropods.

658   UNIT 6   The Diversification of Life

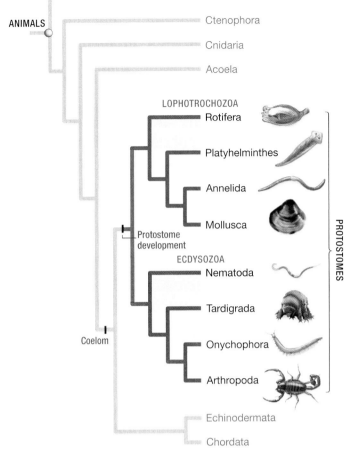

FIGURE 34.2 Protostomes Are a Monophyletic Group Comprising Two Major Lineages. There are 22 phyla of protostomes, but the eight major phyla shown account for about 99.5 percent of the known species.

FIGURE 34.3 The Water-to-Land Transition Occurred Several Times in the Protostomes. The transition occurred only once in the deuterostomes, in phylum Chordata.

✔ **QUESTION** According to this tree, terrestrial living arose seven different times independently in protostomes. Why would you not instead conclude that terrestrial living was ancestral and the aquatic lifestyle arose seven times independently?

Before considering each of these groups in more detail, let's examine the key themes in the evolutionary diversification of the protostomes. Recall that all animal lineages evolved diverse ways of sensing the environment, feeding, moving, reproducing, and developing (see Chapter 33). In addition, diversification in protostomes was triggered by the transition from aquatic to terrestrial habitats, and the modular design of many protostome body plans.

The Water-to-Land Transition

Fossils from the Burgess Shale in Canada, Chengjiang in China, and other sites show that protostome lineages originated in marine environments—that is, in the ocean. Today, protostomes are the most abundant animals in the world's oceans as well as in virtually every freshwater and terrestrial habitat. Like land plants and fungi, protostomes made the transition from aquatic to terrestrial environments.

To help put this achievement in perspective, note that green plants made the move from freshwater to land just once (Chapter 31). It is not yet clear whether fungi moved from aquatic habitats to land once or several times (Chapter 32). Only one lineage among deuterostomes moved onto land (Chapter 35).

But protostomes made a water-to-land transition multiple times as they diversified (**FIGURE 34.3**).

Water-to-land transitions are important because they open up entirely new habitats and new types of resources to exploit. Based on this reasoning, biologists propose that the ability to live in terrestrial environments was a key event in the diversification of several protostome phyla.

Several adaptations were required for protostomes to thrive on land, such as the ability to:

1. exchange gases,

2. avoid drying out, and

3. hold up their bodies under their own weight.

Recall that an **adaptation** is a trait that increases the fitness (reproductive success) of individuals in a particular environment.

Land animals exchange gases with the atmosphere readily as long as they have a large, moist surface area that is exposed to the air (Chapter 45). The bigger challenge is to prevent the gas-exchange surface and other parts of the body from drying out. How do terrestrial protostomes solve this problem?

- Roundworms, earthworms, and other terrestrial protostomes live in humid soils or other moist environments and exchange gases across their body surface. They have a high surface-area-to-volume ratio, which increases the efficiency of gas exchange.

- Some terrestrial arthropods and mollusks have gills or other respiratory structures located *inside* the body, minimizing water loss, such as tracheae in insects (Chapter 43).

- Insects evolved a waxy layer to minimize water loss from the body surface, with openings to respiratory passages that can be closed if the environment dries.

- Insect eggs have a thick membrane that keeps moisture in, and the eggs of slugs and snails have a thin calcium carbonate shell that helps retain water. Desiccation-resistant eggs evolved repeatedly in populations that made the transition to life on land.

Unlike land plants and fungi, many land animals can also move to moister habitats if the area they are in gets too dry.

What are some of the mechanical constraints of living on land? Think for a moment of the weightless feeling you experience in a swimming pool. The upward buoyant force provided by the water counteracts the downward pull on your body mass that is exerted by gravity. Without this buoyant force, animals on land require greater structural support to hold their bodies up and to move. Further, as an animal doubles in length (assuming no change in shape), its weight increases by a factor of 8—so larger organisms feel the effect of gravity more than smaller organisms (Chapter 48). This mechanical constraint has limited the size of many protostomes, including insects, on land.

Modular Body Plans

Morphological and physiological diversification has a genetic basis. Until recent decades, biologists assumed that very different genetic instructions were required to create very different organisms. This reasoning seems logical but turns out to be false. Consider that:

- Multicellular animals have a common tool kit of genes that establish the animal body plan during development. *Hox* genes have been featured in several chapters because they are an important part of this tool kit (see Chapters 22 and 33).

- The genetic tool kit can direct the development of dramatically different types and numbers of structures when the genes are expressed at different times and places during development.

The upshot is that the diversification of animal body plans can occur not just by the generation of new genes over time, but also—and especially—by the changing of the expression pattern of existing genes, in particular regulatory genes. Sean Carroll and colleagues demonstrated this principle clearly when they showed that the *Distal-less* gene is essential to the formation of different types of limbs in diverse lineages, ranging from annelids and arthropods to chordates (see Chapter 33).

When comparing diverse organisms within a lineage, genetically based *modularity* in body plans is evident, meaning that a small set of elements can be reused and rearranged to produce a large variety of outcomes. For example, the change in expression patterns of regulatory genes can account for both the dramatic

increase in the number of vertebrae (back bones) in snakes as well as the loss of limbs in this group (Chapter 22). Thus, a change in expression of preexisting genes generated a dramatically different body plan from that of a shorter, limbed ancestor. Keep the concept of modularity in mind as you read on about the diversity of protostomes.

34.2 What Is a Lophotrochozoan?

The 13 phyla of protostome subgroup Lophotrochozoa include the mollusks, annelids, and flatworms (phylum Platyhelminthes; pronounced *plah-tee-hell-MIN-theez*). Lophotrochozoans are named for two of the three unique morphological traits that define the lineage:

1. a feeding structure called a *lophophore*, which is found in three phyla;

2. a type of larva called a *trochophore*, which is common to many of the phyla; and

3. a spiral pattern of cleavage in embryos.

As **FIGURE 34.4a** shows, a **lophophore** (literally, "tuft-bearer") is a specialized structure that rings the mouth and functions in suspension feeding. Lophophores are found in bryozoans (moss animals), brachiopods (lamp shells), and phoronids (horseshoe worms), together called the lophophorates.

Trochophores are a type of larvae common to marine mollusks, marine annelids, and several other phyla in the Lophotrochozoa. As **FIGURE 34.4b** shows, a **trochophore** (literally, "wheel-bearer") larva has a ring of cilia around its middle. These cilia allow swimming and, in some species, sweep food particles into the mouth.

Trochophore larvae, like other larvae, occur in animals that undergo indirect development—where larvae often look radically different from adults, live in different habitats, and eat different foods (Chapter 33). The process of metamorphosis from larvae to adults in marine species is hypothesized to be an adaptation that allows larvae of sessile or slow-moving adults to disperse to new habitats by floating or swimming in the plankton.

Larvae are not unique to lophotrochozoans—even the earliest known animals, the sponges, had larvae. However, recent analyses suggest that the trochophore larva originated early in the evolution of lophotrochozoans and then evolved into different larval types later in some groups.

Recent phylogenetic analyses of lophotrochozoans have inspired the hypothesis that the spiral pattern of cleavage in embryos is a synapomorphy for this monophyletic group. When cells divide at oblique angles to each other during early embryogenesis, a spiraling pattern of cells results in the blastula (**FIGURE 34.5**). This is in contrast to other patterns of cleavage, such as radial cleavage, where cells divide at right angles to each other. Although spiral cleavage has been highly conserved in some lophotrochozoan phyla, it has been modified or lost in others. ✔ You should be able to indicate the origin of spiral cleavage on Figure 34.2.

(a) Lophophores function in suspension feeding in adults.

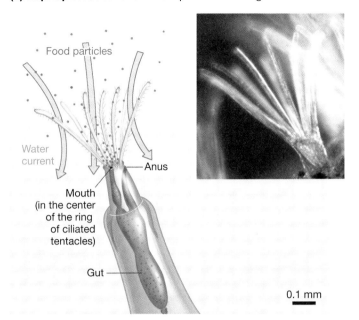

Food particles

Water current

Anus

Mouth (in the center of the ring of ciliated tentacles)

Gut

0.1 mm

(b) Trochophore larvae swim and may feed.

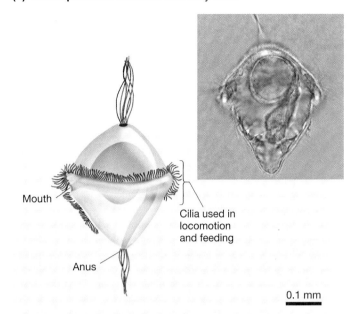

Mouth

Cilia used in locomotion and feeding

Anus

0.1 mm

FIGURE 34.4 Lophotrochozoans Have Distinctive Traits. (a) Three phyla of lophotrochozoans, including the bryozoan shown here, have the feeding structure called a lophophore. **(b)** Many phyla of lophotrochozoans have the type of larva called a trochophore.

The implications of spiral cleavage to adult diversity are becoming evident as modern molecular techniques are applied to studies of comparative embryology. For example, spiral cleavage is clockwise in some snails but counterclockwise in others, resulting in adults with right-handed and left-handed whorls, respectively. It turns out that the cleavage pattern determines the downstream expression of *Nodal* and *Pitx*—tool-kit genes that establish the asymmetry of snails. *Nodal* and *Pitx* are also important in the left-right asymmetries of vertebrates, illustrating how homologous tool-kit genes can have similar but different effects in different lineages.

(a) Spiral cleavage is unique to lophotrochozoans.

(b) Radial cleavage, for comparison

FIGURE 34.5 Spiral Cleavage is a Synapomorphy for the Lophotrochozoans. This pattern of early cell divisions, here contrasted with radial cleavage, has been modified or lost in some lineages. (Some of the cells are stained to help compare the patterns.)

In sum, if you were to observe a lophophore, a trochophore larva, or spiral cleavage in an unfamiliar animal, you could be confident that the animal belongs to the Lophotrochozoa. However, all lophotrochozoans do not possess all three of these characteristics.

Let's briefly consider the diversity of wormlike lophotrochozoans and then look at the most diverse and abundant phylum, the mollusks.

Wormlike Lophotrochozoans

Most protostome phyla include species that have long, thin, tube-like bodies that lack limbs. That is, they are **worms** with a basic tube-within-a-tube design (Chapter 33). The outside tube is the skin, which is derived from ectoderm; the inside tube is the gut, which is derived from endoderm. Muscles and organs derived from mesoderm are located between the two tubes.

Since the coelom evolved in the common ancestor of protostomes and deuterostomes (see Figure 34.2), many lophotrochozoans have a well-developed coelom—not only allowing space for fluids to circulate among organs but also serving as part of a hydrostatic skeleton for movement. The hydrostatic skeleton enables worms to crawl, burrow, and even swim. Specific movements depend on factors such as the organization of muscles within the body, whether the worms are segmented or unsegmented, and the presence or absence of bristles or other structures on the body surface.

As with spiral cleavage and the trochophore larva, however, the coelom was reduced or lost in several lineages. For example, the flatworms—which are mostly small and thin or parasitic—lack a coelom entirely.

(a) Mouthparts for suspension feeding

(b) Mouthparts for deposit feeding

(c) Mouthparts for mass feeding

Feathery mouthparts capture plankton

Long, sticky tentacles reach across substrate and deliver organic particles to mouth (worm body remains hidden)

Eversible proboscis with sharp fangs impales prey

FIGURE 34.6 Many Wormlike Lophotrochozoans Have Specialized Mouthparts. These three examples are polychaete worms, phylum Annelida.

✔ **EXERCISE** It is common to find diverse worms in the same habitat. Suggest a hypothesis to explain why this is possible.

Wormlike lophotrochozoans include suspension, deposit, liquid, and mass feeders (see Chapter 33). Exploiting a diversity of foods is possible in part because worms have such a wide variety of mouthparts for capturing and processing food.

The lophophore feeding structure that is used by lophophorates, such as phoronid worms, is an example of one type of specialized mouthpart (see Figure 34.4a). Let's consider several other examples from the phylum Annelida, the segmented worms.

Segmented worms are a diverse group, especially the marine polychaetes (**FIGURE 34.6**). For example, they may live in tubes and extend their feathery tentacles into the water to suspension feed, burrow in the sediments and reach out with their spaghetti-like tentacles to deposit feed, or spear prey with sharp fangs on the tip of a tubular, extensible feeding appendage called a **proboscis.**

While the diversity of wormlike lophotrochozoans is impressive, you can see from the pie wedges in Figure 34.1 that they do not quite compare to the mollusks in terms of diversity.

What Is a Mollusk?

Mollusks exhibit the lophotrochozoan mode of growth and development. But unlike the lophotrochozoans as a whole, the mollusks can be identified by distinctive characteristics. Mollusks have a specialized body plan based on three major components (**FIGURE 34.7**):

1. the **foot,** a large muscle located at the base of the animal and usually used in movement;

2. the **visceral mass,** the region containing most of the main internal organs and the external gill; and

3. the **mantle,** an outgrowth of the body wall that covers the visceral mass, forming an enclosure called the mantle cavity. In many species, the mantle secretes a hard calcium carbonate shell.

✔You should be able to indicate the origin of the molluscan body plan on Figure 34.2.

Once the molluscan body plan evolved, subsequent diversification was largely driven by adaptations that allowed the mollusks to move, feed, or reproduce in novel ways. The result was a dramatic radiation of mollusks into several lineages, including **bivalves** (clams and mussels), **gastropods** (slugs and snails), **chitons** (mollusks with dorsal shells made of plates), and **cephalopods** (squid and octopuses).

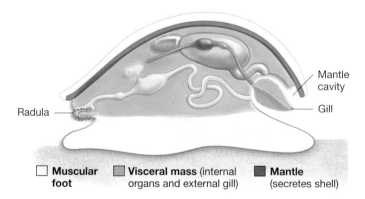

Mantle cavity

Radula

Gill

☐ **Muscular foot**   ☐ **Visceral mass** (internal organs and external gill)   ■ **Mantle** (secretes shell)

FIGURE 34.7 Mollusks Have a Specialized Body Plan. This diagram of a generalized mollusk shows that the mollusk body plan is based on a foot, a visceral mass, and a mantle. Gills are located inside a cavity created by the mantle.

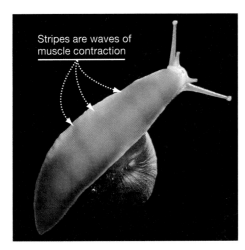

Stripes are waves of muscle contraction

FIGURE 34.8 How Snails Walk on One Foot. Waves of muscle contractions, moving forward or backward along the length of the foot, allow mollusks like this snail to creep along a substrate.

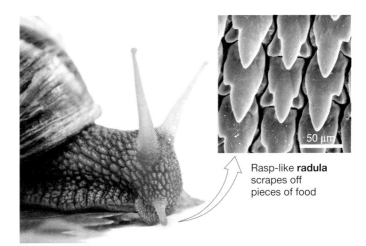

50 µm

Rasp-like **radula** scrapes off pieces of food

FIGURE 34.9 The Radula Is Unique to Mollusks. The size and shape of the radular teeth is highly correlated with the type of food eaten by the species.

The Foot Is a Muscular Hydrostat Snails and chitons have a large, muscular foot at the base of the body that—like your tongue—works as a type of hydrostatic skeleton called a muscular hydrostat. How can an animal move with just one foot? Waves of muscle contractions sweep backward or forward along the length of the foot, allowing individuals to crawl along a surface (**FIGURE 34.8**). Biologists in the field of biomechanics have measured the mechanical properties of snail mucus. They observed that it has the remarkable ability to increase traction for the parts of the foot that are not moving, while lubricating motion for the parts that are advancing.

In bivalves such as clams, the foot is modified as a digging appendage. The foot of cephalopods like squid and octopus is modified to form tentacles for crawling and grasping.

The Visceral Mass Separates Internal Organs from the Hydrostatic Skeleton The visceral mass is the space in all mollusks where organs and surrounding fluids are located. This might not seem like much of an innovation, but compare this body plan to that of worms, where the hydrostatic skeleton and internal organs are inseparable. The separation of the foot and visceral mass in mollusks may have enabled diversification of both.

The coelom itself is highly reduced in most mollusks, functioning mostly in reproduction and excretion of wastes. The organs occupy a different type of body cavity called a **hemocoel** ("blood-hollow"; pronounced *HEE-mah-seal*), where body fluids bathe the organs directly in an open circulatory system (see Chapter 45). A hemocoel is different from a coelom because it is not lined in mesoderm.

At the anterior end of the visceral mass, the mouth has a unique molluscan feeding structure called a **radula**, which functions like a rasp or file. The mollusk moves the radula back and forth over the food source, causing the many sharp plates to scrape material so that it can be ingested (**FIGURE 34.9**). Since it is highly unlikely that such a unique structure would evolve more than once, the radula probably evolved early in molluscan

evolution. It was later lost in the bivalves, which acquire food by suspension feeding.

The Mantle Secretes a Shell In many species the mantle secretes a shell made of calcium carbonate. Some mollusk species have a shell consisting of one, two, or eight parts, or valves; others have no shell at all. The ability to secrete a protective shell may have been an important adaptation during the Cambrian and beyond as the number of predators increased. Note, however, that shells made of calcium carbonate are heavy. Thus, there is a trade-off between protection and mobility. As you might predict, mollusks with thick shells are constrained to aquatic habitats where buoyant forces help support the load. The largest and most agile of all aquatic mollusks, cephalopods such as squid and octopus, have highly reduced shells or none at all. Terrestrial mollusks have thin shells or none at all.

The mantle was also modified in many ways to serve diverse functions other than secreting shells. For example, cephalopods have a mantle lined with muscle. When the cavity surrounded by the mantle fills with water and the mantle muscles contract, a stream of water is forced out of a tube called a **siphon**. The force of the water propels the squid in a form of locomotion called jet propulsion (**FIGURE 34.10**, see page 664). The mantle of bivalves such as clams is modified to form two siphons to control incurrent and excurrent water flow over their gills.

Modular Mollusks How did the novel body plan and lifestyle of mollusks such as squid evolve? Mark Martindale and his colleagues tried to address this question by looking at the role of *Hox* genes in the organization of the squid body plan. They found that:

- Squid have nine *Hox* genes that are homologous to those found in other bilaterians (recall that *Hox* genes are important in establishing the anterior–posterior axis of animals).

- Although *Hox* genes are usually organized in a collinear fashion (see Chapter 22), they were co-opted in a new pattern in

Cavity enclosed by
mantle fills with water

Water is forced out through
siphon; animal moves

FIGURE 34.10 Jet Propulsion in Mollusks. In jet propulsion, muscular contractions of the mantle force the water out through a movable siphon.

✔ **EXERCISE** Circle the part of the squid that is homologous to the snail foot.

squid, suggesting a mechanism by which a change in the regulation of tool-kit genes could result in novel body structures, such as the modification of the molluscan foot into a crown of muscular tentacles.

The take-home message is that mollusks have a modular design. That is, while all mollusks have a foot, visceral mass, and mantle, the sizes, shapes, and functions of these structures have diversified in ways that enable drastically different methods of feeding, moving, and reproducing.

Let's take a closer look at some of the diversity within the main lophotrochozoan phyla.

check your understanding

If you understand that . . .
- Flatworms, annelids, and mollusks are all diverse phyla within the Lophotrochozoa, and all three phyla made the water-to-land transition.
- The mollusk body plan includes a muscular foot, a visceral mass, and a mantle.

✔ **You should be able to . . .**
1. Predict three ways you would expect a snail that lives on land to differ from a snail that lives in the ocean.
2. Design an experiment to test whether the tentacles of an octopus are homologous to the foot of a snail.

Answers are available in Appendix A.

Key Lineages: Lophotrochozoans

The phyla within Lophotrochozoa are united by shared mechanisms of early development and growth, but they are highly diverse in terms of their resulting morphology. You might not even recognize some of them as having a bilaterally symmetrical body plan.

Let's take a closer look at five key lophotrochozoan phyla: **(1)** Bryozoa, **(2)** Rotifera, **(3)** Platyhelminthes, **(4)** Annelida, and **(5)** Mollusca. Because the Mollusca is so species rich and diverse, the detailed descriptions in this section consider the four most familiar lineages of mollusks separately:

- Lophotrochozoans > Bryozoa (Bryozoans)
- Lophotrochozoans > Rotifera (Rotifers)
- Lophotrochozoans > Platyhelminthes (Flatworms)
- Lophotrochozoans > Annelida (Segmented Worms)
- Lophotrochozoans > Mollusca > Bivalvia (Clams, Mussels, Scallops, Oysters)
- Lophotrochozoans > Mollusca > Gastropoda (Snails, Slugs, Nudibranchs)
- Lophotrochozoans > Mollusca > Polyplacophora (Chitons)
- Lophotrochozoans > Mollusca > Cephalopoda (Nautilus, Cuttlefish, Squid, Octopuses)

Bryozoans, sometimes also called ectoprocts or moss animals, are tiny lophophorate animals that live in colonies that look more like a sponge or coral than a bilaterian (**FIGURE 34.11**). The oldest bryozoan fossils date back to the late Cambrian, giving this group the distinction of being the most recent animal phylum to appear in the fossil record. Today, bryozoans live mostly in marine environments, especially in the tropics, but also in some freshwater environments. So far, 5700 species have been described. The colonies range in size, from millimeters to meters, and in shape, from fans to encrusting sheets.

Feeding Bryozoan individuals, or zooids, suspension feed on plankton using a lophophore (see Figure 34.4).

Movement Most bryozoans are sessile. Only the (non-trochophore) larvae are capable of dispersing to new habitats.

Reproduction Bryozoans can increase the size of their colony by asexual reproduction, by budding. They can also reproduce sexually—zooids are hermaphrodites, some having both male and female reproductive organs at the same time, other serving first as males then as females.

Triphyllozoan species

FIGURE 34.11 A Bryozoan Colony. You can see what an individual animal looks like in Figure 34.4.

The 2100 **rotifer** species that have been identified thus far live in damp soils as well as marine and freshwater environments. They are important components of the plankton in freshwater and in brackish (slightly salty) waters. Rotifers have a coelom, and most are less than 1 millimeter (mm) long. Although rotifers do not have a lophophore or a trochophore larval stage, their mode of growth and DNA sequences identify them as a member of the lophotrochozoan lineage.

Feeding Rotifers have a cluster of cilia called a **corona** at their anterior end (**FIGURE 34.12**). In many species, the beating of the cilia in the corona makes suspension feeding possible by creating a current that sweeps microscopic food particles into the mouth. The corona is a synapomorphy of this group. ✔ You should be able to indicate the origin of the corona on Figure 34.2.

Movement Although a few species of rotifers are sessile, most swim via the beating of cilia in the corona.

Reproduction Most reproduction is sexual, but females of some rotifer species can reproduce asexually by a process called **parthenogenesis**—the female produces unfertilized eggs by mitosis; the eggs then hatch into new, asexually produced individuals. In a group of rotifers called the bdelloids, only females have been found and all reproduction appears to be by parthenogenesis. Whether reproduction is sexual or asexual, rotifer development is direct, meaning that fertilized eggs hatch and grow into adults without going through metamorphosis.

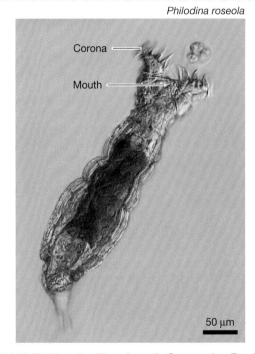

Philodina roseola

Corona

Mouth

50 µm

FIGURE 34.12 Rotifers Are Tiny, Aquatic Suspension Feeders.

The flatworms are a large and diverse phylum. More than 20,000 species have been described in four major lineages:

- *Turbellaria* are free-living flatworms. Although a few turbellarian species are terrestrial, most live on the substrates of freshwater or marine environments. Some marine flatworms are brightly colored and several centimeters long (**FIGURE 34.13a**). However, the freshwater planarians common to biology labs, such as *Dugesia*, are drab and only a few millimeters long.

- *Cestoda* are endoparasitic tapeworms that parasitize fish, mammals, or other vertebrates (**FIGURE 34.13b**).

- *Trematoda* are endoparasitic or ectoparasitic flukes that parasitize vertebrates, arthropods, annelids, and mollusks (**FIGURE 34.13c**). In humans, a fluke is responsible for schistosomiasis—a serious public health issue in many tropical and subtropical nations.

- *Monogenea* are ectoparasites that parasitize fish.

Flatworms are named for the broad, flattened shape of their bodies. (The Greek roots *platy* and *helminth* mean "flat-worm.") Species in the Platyhelminthes are unsegmented and lack a coelom. They also lack structures that are specialized for gas exchange. Further, they do not have blood vessels for circulating oxygen and nutrients to their cells. Based on these observations, biologists interpret the flattened bodies of these animals as an adaptation that gives flatworms an extremely high surface-area-to-volume ratio. This flat body plan allows nutrients and gases to diffuse efficiently to all of the cells inside the animal. The downside is that the body surface has to be moist for gas exchange to take place. Flatworms are restricted to aquatic and moist terrestrial environments.

✔ You should be able to indicate the origin of the flattened, acoelomate body plan on Figure 34.2.

Feeding The Platyhelminthes lack a lophophore and have a digestive tract that is "blind"—meaning it has only one opening for ingestion of food and elimination of wastes.

- Most turbellarians are hunters that prey on protists or small animals; others scavenge dead animals.

- Tapeworms are strictly parasitic and feed on nutrients provided by hosts. They do not have a mouth or a digestive tract, and they obtain nutrients by diffusion across their body wall.

- Flukes are parasites and feed by gulping host tissues and fluids through a mouth.

Movement In general, flatworms do not move much. Some turbellarians can swim by undulating their bodies, and most can creep along substrates with the aid of cilia on their ventral surface. Adult cestodes have hooked attachment structures at their anterior end that permanently attach them to the interior of their host.

Reproduction Turbellarians can reproduce asexually by splitting themselves in half. If they are fragmented as a result of a predator attack, the body parts can regenerate into new individuals. Most turbellarians contain both male and female organs and reproduce sexually by aligning with another individual and engaging in mutual and simultaneous fertilization.

In flukes and tapeworms, individuals reproduce sexually and either cross-fertilize or self-fertilize. The reproductive systems and life cycles of flukes and tapeworms are extremely complex, however. In many cases the life cycle involves two or even three distinct host species; sexual reproduction occurs in the **definitive host,** and asexual reproduction occurs in one or more **intermediate hosts.** For example, humans are the definitive host of the blood fluke *Schistosoma mansoni*, and snails are the intermediate host.

(a) Turbellarians are free living.

Pseudoceros ferrugineus

(b) Cestodes are endoparasitic.

Taenia species

(c) Trematodes are endoparasitic.

Dicrocoelium dendriticum

FIGURE 34.13 Flatworms Have Simple, Flattened Bodies.

Most **annelids** have a segmented body and a coelom. ✔You should be able to indicate the origin of annelid segmentation on Figure 34.2. The 16,800 described species were traditionally divided into groups called Polychaeta (*pol-ee-KEE-ta*) and Clitellata. Recent analyses have supported Clitellata as a monophyletic group, composed of Hirudinea (leeches), and Oligochaeta (*oh-LIG-oh-keet-a*; earthworms and other oligochaetes). However, the Polychaeta are paraphyletic—that is, including some but not all members of a clade (**FIGURE 34.14**). Further, unsegmented worms called Sipunculida and Echiura, which were traditionally thought to be independent phyla, are actually groups within Annelida—they lost their segmentation.

The common ancestor of the annelids had a key synapomorphy in addition to segmentation: numerous, bristle-like extensions called **chaetae** that extend from appendages called **parapodia.** As the name suggests, polychaetes ("many-bristles"; **FIGURE 34.15a**) have more chaetae than oligochaetes ("few-bristles"; **FIGURE 34.15b**). However, highly mobile polychaetes have larger parapodia and more chaetae than more sedentary polychaetes, which have reduced parapodia and chaetae adapted for burrowing in sediments or living in tubes.

Feeding The polychaete lineages have a wide variety of methods for feeding, including suspension feeding, deposit feeding, and mass feeding (Figure 34.6). Virtually all oligochaetes, in contrast, make their living by deposit feeding in soils. The tunnels that they make are critically important in aerating soil, and their feces contribute large amounts of organic matter. About half of the leeches are ectoparasites that attach themselves to fish or other hosts and suck blood and other body fluids (**FIGURE 34.15c**). The nonparasitic leech species are predators or scavengers.

Movement Polychaetes and oligochaetes crawl or burrow with the aid of their hydrostatic skeletons; the parapodia of polychaetes also act as paddles or tiny feet that aid in movement. Many polychaetes and leeches are good swimmers.

Reproduction Asexual reproduction occurs in some polychaetes and oligochaetes by transverse fission or fragmentation—meaning

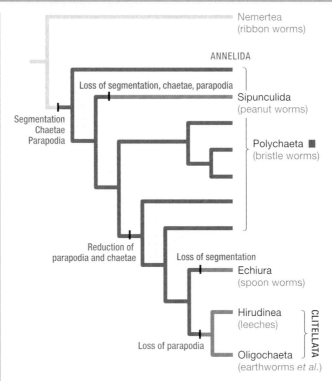

FIGURE 34.14 Phylogeny of Major Groups of Annelids.

that body parts can regenerate a complete individual. Sexual reproduction in polychaetes may begin with internal or external fertilization, depending on the species. Most of the polychaete lineages have separate sexes and usually release their eggs directly into the water. Some species produce eggs that hatch into trochophore larvae, then go on to develop into other larval forms. In oligochaetes and leeches, individuals produce both sperm and eggs and engage in mutual, internal cross-fertilization. Eggs are enclosed in a mucus-rich, cocoon-like structure; after fertilization, offspring develop directly into miniature versions of their parents.

(a) Polychaetes have many bristle-like chaetae.

Hermodice carunculata

(b) Oligochaetes have few or no chaetae.

Chaetogaster species

(c) Some **leeches** are ectoparasites.

Haemadipsa species

FIGURE 34.15 Annelids Are Segmented Worms.

The bivalves are so named because they have a shell with two separate parts, or valves, made of calcium carbonate. The protective shell is hinged, and closes with the aid of muscles attached to the two valves (**FIGURE 34.16**). Because most bivalves live on or under the ocean floor and because they are covered by a hard shell, their bodies are often buried in sediment after death. Bivalves thus fossilize readily and have the most extensive fossil record of any animal, plant, or fungal group.

The smallest bivalves are freshwater clams that are less than 2 mm long; the largest bivalves are giant marine clams that may weigh more than 200 kilograms (kg). All bivalves can sense gravity, touch, and certain chemicals; scallops even have eyes.

Feeding Most bivalves are suspension feeders that take in any type of small animal or protist, or detritus. Suspension feeding is based on a flow of water across gas-exchange structures called **gills** located within the mantle cavity. A single oyster can filter the organic matter from as much as 60 gallons of seawater each day. Bivalves are the only major group of mollusks that lack the feeding structure called a radula.

Movement Clams burrow into sediments with the aid of their muscular foot but are otherwise sedentary. Scallops are able to swim along the surface of soft substrates by clapping their shells together and forcing water out, a form of jet propulsion. Mussels and oysters are sessile, attached to rocks or other solid surfaces.

Chlamys pallium

1 cm

FIGURE 34.16 Bivalves Have Shells with Two Parts.

Reproduction Most bivalves reproduce sexually. Eggs and sperm are shed into the water, and fertilized eggs develop into trochophore larvae. Trochophores then metamorphose into a distinct type of larva called a **veliger,** which continues to feed and swim before settling to the substrate and metamorphosing into an adult form that secretes a shell.

The gastropods ("belly-feet") are named for the large, muscular foot on their ventral side. Marine and terrestrial snails have shells (**FIGURE 34.17**), into which they can retract when they are attacked or when their tissues begin to dry out. Land slugs and nudibranchs (pronounced *NEW-da-branks*; also called sea slugs) lost their shells independently; these gastropods often contain toxins or foul-tasting chemicals to protect them from being eaten. About 70,000 species of gastropods are known, ranging in size from microscopic to almost a meter long. They are the only group of mollusks that made the water-to-land transition.

Feeding The gastropod radula is diverse, varying in size and shape among species (see Figure 34.9). Although most gastropods are herbivores or detritivores, specialized types of teeth allow some gastropods to act as predators. Species called drills, for example, use their radula to bore a hole in the shells of other mollusks.

Movement Waves of contractions along the length of the foot allow gastropods to move by creeping (see Figure 34.8). Sea butterflies are gastropods with a reduced or absent shell but a large, winglike foot that flaps and powers swimming movements.

Reproduction Most reproduction is sexual, but females of some gastropod species can reproduce asexually by producing eggs parthenogenetically. Sexual reproduction in some gastropods begins with internal fertilization. Most marine gastropods produce a veliger larva that may disperse up to several hundred kilometers from the parent. But in some marine species and all terrestrial forms, larvae are not free living. Instead, offspring remain in an egg case while passing through several larval stages and then hatch as miniature versions of the adults.

Maxacteon flammea

5 mm

FIGURE 34.17 Gastropods Have a One-Part Shell or Lack Shells.

The name *Polyplacophora* ("many-plate-bearing") is apt because chitons (pronounced *KITE-uns*) have eight calcium carbonate plates, or valves, along their dorsal surface (**FIGURE 34.18**). The plates form a protective shell. The approximately 1000 species of chitons are exclusively marine, and are found worldwide; the largest reaches 33 centimeters (cm). They are usually found on rocky surfaces in the intertidal zone, where rocks are periodically exposed to air at low tides.

Feeding Chitons have a radula that they use to scrape algae and other organic matter off rocks.

Movement Chitons move by creeping on their broad, muscular foot—as gastropods do.

Reproduction In most chiton species, the sexes are separate and fertilization is external. In some species, however, sperm that are shed into the water enter the female's mantle cavity and fertilize eggs there. Depending on the species, eggs may be enclosed in a membrane and released, or they may be retained until hatching and early development are complete. Most species have trochophore larvae.

Tonicella lineata

1 cm

FIGURE 34.18 Chitons Have Eight Shell Plates.

The cephalopods ("head-feet") have a well-developed "head" that actually consists of the visceral mass, and a foot that is modified to form **tentacles:** long, thin, muscular extensions that aid in movement and prey capture (**FIGURE 34.19**). Except for the nautilus with its chambered shell, the cephalopod species living today have either highly reduced shells or none at all. Most have large brains and image-forming eyes with sophisticated lenses. They also include the largest of all invertebrates living today, the giant squid, which can be more than 10 m long, including the tentacles. The skin of cephalopods contains sophisticated chromatophore ("color-bearing") cells that enable the rapid color changes used for camouflage and communication.

Feeding Cephalopods are highly intelligent predators that hunt by sight and use their tentacles to capture prey—usually fish or crustaceans. They have a radula as well as a structure called a **beak,** which can exert powerful biting forces. Some cuttlefish and octopuses also inject poisons into their prey to subdue them.

Movement Cephalopods can swim by moving their fins to "fly" through the water, or by jet propulsion using the mantle cavity and siphon (see Figure 34.10). Squid are built for speed and hunt small fish by chasing them down. Octopuses, in contrast, crawl along the substrate using their tentacles. They chase down crabs or other crustaceans, or they pry mussels or clams from the substrate and then use their beaks to crush the exoskeletons of their prey. Most cephalopods are able to escape their own predators by releasing a jet of "ink" that disorients their predators while they slip away.

Reproduction Cephalopods have separate sexes, and some species have elaborate courtship rituals that involve color changes and interaction of tentacles. When a male is accepted by a female, he deposits sperm that are encased in a structure called a **spermatophore.** The spermatophore is transferred to the female, and fertilization is internal. Females lay eggs. When they hatch, juveniles develop directly into adults—that is, there is no larval stage.

Octopus bimaculatus

10 cm

FIGURE 34.19 Cephalopods Have Highly Modified Bodies.

34.3 What Is an Ecdysozoan?

Lophotrochozoans grow continuously and incrementally (**FIGURE 34.20a**)—the ancestral manner of growth for animals. In contrast, the Ecdysozoa are defined by a clear synapomorphy: All ecdysozoans grow intermittently by **molting**—that is, by shedding an exoskeleton or external covering.

The Greek root *ecdysis*, which means "to slip out or escape," is appropriate because during a molt, an individual sheds its outer layer, or **cuticle**—called an **exoskeleton** if it is hard—and slips out of it (**FIGURE 34.20b**). Once the old covering is gone, the body is pumped up with fluid to cause it to expand, and a larger cuticle or exoskeleton forms. As ecdysozoans grow and mature, they undergo a succession of molts, sometimes undergoing dramatic morphological transformations along the way.

A stiff body covering is advantageous because it provides an effective structure for muscle attachment and affords protection. However, the trade-off is that during molting, the soft bodies of ecdysozoans are exposed and vulnerable to predators. When a crab or other crustacean has shed an old exoskeleton, its new exoskeleton takes several hours to harden. During this interval, molting individuals hide and do not feed or move about. Experiments have shown that it is much easier for predators to attack and subdue individuals that are not protected by an exoskeleton during the intermolt period.

A hormone called ecdysone is important in regulating the molting cycle (Chapter 49). Is ecdysone a key innovation for the ecdysozoans? No—studies show that this steroid hormone serves a variety of functions in many other animals. Thus, it appears that ecdysone was co-opted in the ecdysozoans to serve a novel function.

The most prominent of the seven ecdysozoan phyla are the roundworms (Nematoda) and arthropods (Arthropoda). Here, let's consider the implications of having an exoskeleton for the most diverse phylum of animals on Earth, the arthropods.

What Is an Arthropod?

In terms of duration in the fossil record, species diversity, and abundance of individuals, arthropods are the most important phylum within the Ecdysozoa. Arthropods appear in the fossil record over 520 million years ago and have long been the most abundant animals observed in both marine and terrestrial environments. Well over a million living species have been described, and biologists estimate that millions or perhaps even tens of millions of arthropod species have yet to be discovered. In terms of species diversity, arthropods are easily the most successful lineage of eukaryotes.

The Arthropod Body Plan In addition to their bilateral symmetry and other protostome characteristics, **arthropods** (literally "joint-foot") are defined by three key features:

1. a *segmented body*, which is organized into prominent regions, or **tagmata,** such as the head, thorax, and abdomen in the grasshopper (**FIGURE 34.21**);

2. an *exoskeleton* made primarily of the polysaccharide chitin (see Chapter 5), strengthened by calcium carbonate ($CaCO_3$) in crustaceans; and

3. *jointed appendages,* which enable the rigid bodies to move.

Instead of being based on muscle contraction within a hydrostatic skeleton, most arthropod locomotion is based on muscles

(a) Lophotrochozoans grow incrementally.

(b) Ecdysozoans grow by molting.

Growth bands

FIGURE 34.20 Lophotrochozoans and Ecdysozoans Differ in Their Mechanism of Growth. (a) Lophotrochozoans do not molt. The growth bands on this clam show periods of slow and rapid incremental growth. **(b)** Ecdysozoans such as this cicada must crawl out of their old exoskeletons and grow new, larger ones.

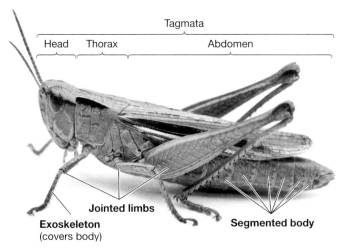

FIGURE 34.21 Arthropods Have a Specialized Body Plan. Arthropods have segmented bodies and jointed limbs, which enable these animals to move despite their hard outer covering, the exoskeleton. Different arthropods have a different number of tagmata, or body regions; insects have three.

that apply force against rigid levers—the exoskeleton—to move legs or wings at the joints. This skeletal architecture enables very rapid and precise movements (Chapter 48).

Arthropods have a highly reduced coelom but, like mollusks, possess an extensive body cavity called a hemocoel. The hemocoel enables movements via a hydrostatic skeleton in caterpillars and other types of arthropod larvae.

The jointed appendages of arthropods are organized in pairs along the segmented body. They have an array of functions: exchanging gases, sensing the environment, feeding, and locomotion via swimming, walking, running, jumping, or flying (Chapter 33). Arthropod appendages provide the ability to sense stimuli and make sophisticated movements in response. ✔ You should be able to indicate the origin of jointed appendages on Figure 34.2.

Why Has the Arthropod Body Plan Been So Successful? You have already learned that mollusk bodies are modular in evolutionary terms. The modularity of arthropod segments, however, takes the prize for evolutionary potential.

Studies on the role of tool-kit genes—such as the *Hox* genes—during development of different arthropods have contributed enormously to our understanding of the process of arthropod diversification. These studies show that small changes in the timing and location of expression of certain tool-kit genes along the length of an arthropod can result in novel shapes and sizes of body segments and appendages.

To understand the evolutionary potential of modularity, consider a hypothetical ancestral arthropod with many uniform body segments, each with a pair of uniform appendages. The redundancy of this body plan can provide a fertile starting place for diversification because some of the segments can continue to function the old way, say for crawling, while other segments can become co-opted for new functions. For example, the appendages at the head end could become specialized as antennae or claws without compromising the ability of the animal to crawl.

Thus, genetic opportunity in the form of variation in the expression of tool-kit genes can combine with ecological opportunity by the process of natural selection to result in the diversification of arthropod body segments and appendages. **FIGURE 34.22** shows an example of how diverse, functionally specialized appendages can result in an animal with a "serialized" body plan, a reflection of its modular structure. The crayfish limbs are said to have serial homology.

Origin of the Wing The process of co-opting old parts for new functions is key to the origin of one of the most remarkable adaptations in the history of life—the insect wing. There are more species of animals on the planet today with insect wings than without. According to data in the fossil record, insects were also the first animals on Earth that had wings and could fly. How did wings evolve?

Some biologists have proposed that wings arose as outgrowths from the thorax, independent from the legs. Others have argued that wings arose from gill-like projections on the branched legs of a wingless ancestor, essentially co-opting the gill-like structures

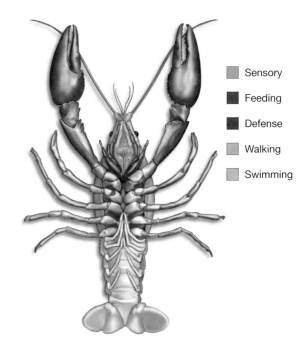

FIGURE 34.22 **Arthropods Have a Modular Body Plan.** This ventral view of a crayfish illustrates how arthropod segments and appendages can be specialized for different functions.

Sensory
Feeding
Defense
Walking
Swimming

(**FIGURE 34.23**). Insects have unbranched appendages, but arthropods such as crustaceans have branched appendages.

Michalis Averof and Stephen Cohen attempted to settle the wing debate using developmental genetics. They probed the developing limbs of brine shrimp (a member of an ancient lineage of crustaceans) for the presence of tool-kit genes *nubbin* and *apterous*. These genes are known to be important in wing development in insects. The results were striking. *Nubbin* and *apterous* are expressed in the gill-like lobe of the crustacean limb, but nowhere else. These results provide strong support for the origin of insect wings from the gill-like branches of ancestral crustacean limbs.

(a) Independent origin hypothesis

(b) Gill co-option hypothesis

FIGURE 34.23 **Alternative Hypotheses for the Origin of Wings in Insects.**

Co-opting one structure to form another may sometimes require thousands or millions of years of evolution within a lineage. It is all the more remarkable, then, that dramatic shape changes can also occur during the lifetime of a single individual. Natural selection operates at all developmental stages in a life cycle.

Arthropod Metamorphosis

Many arthropods undergo a morphological transformation from a larva to a very different adult form. Others do not. Let's use insects as a case study to consider the adaptive implications of metamorphosis in arthropods.

Two Types of Insect Metamorphosis   In insects, the presence or absence of a larval stage defines two distinct types of metamorphosis.

In **hemimetabolous** ("half-change") **metamorphosis**—also called **incomplete metamorphosis**—juveniles called nymphs look like miniature versions of the adult. The aphid nymphs in **FIGURE 34.24a**, for example, shed their external skeletons several times and grow—gradually changing from wingless, sexually immature nymphs to sexually mature adults, some of which can fly. But throughout their life, aphids live in the same habitats and feed on the same food source in the same way: They suck sap.

In **holometabolous** ("whole change") **metamorphosis**—also called **complete metamorphosis**—there is a distinct larval stage. As an example, consider the life cycle of the mosquito, illustrated in **FIGURE 34.24b**:

1. Newly hatched mosquitoes live in quiet bodies of freshwater, where they suspension feed on bacteria, algae, and detritus.

2. When a larva has grown sufficiently, the individual stops feeding and moving and secretes a protective case. The individual is now known as a **pupa** (plural: pupae). During pupation, the pupa's body is completely remodeled into a new, adult form.

3. The adult mosquito flies and feeds—females take blood meals from mammals and males drink nectar from flowers.

What Is the Adaptive Significance of Metamorphosis?   In insects, holometabolous metamorphosis is 10 times more common than hemimetabolous metamorphosis. How can this be explained?

Since most adult insects are mobile, using a larval stage for dispersal does not seem to be a sufficient explanation. The leading hypothesis is based on efficiency in feeding. Because juveniles and adults from holometabolous species feed on different materials in different ways and sometimes even in different habitats, they do not compete with each other.

An alternative hypothesis is based on the advantages of functional specialization. In many moths and butterflies, larvae are specialized for feeding, whereas adults are specialized for mating and feed rarely, if ever. Larvae are largely sessile, whereas adults are highly mobile. If specialization leads to higher efficiency in feeding and reproduction and thus higher fitness, then holometabolous metamorphosis would be advantageous. These hypotheses are not mutually exclusive, however—meaning that both could be correct, depending on the species being considered.

Let's explore the diversity of feeding, moving, and reproducing habits of ecdysozoans in more depth by examining the key lineages separately.

Key Lineages: Ecdysozoans

Seven phyla are currently recognized in the Ecdysozoa (see Table 33.1). Second in diversity to Arthropoda, the phylum Nematoda, or roundworms, is especially noteworthy due to its large number of described species and the mind-boggling abundance of individuals. Of the smaller phyla, Onychophora (*on-ee-KOFF-er-uh*) and Tardigrada are of special interest due to their close relationship to arthropods. They are similar to arthropods

(a) Aphid:
Hemimetabolous metamorphosis
(Incomplete metamorphosis)

Nymphs look like miniature versions of adults and eat the same foods.

(b) Mosquito:
Holometabolous metamorphosis
(Complete metamorphosis)

Larvae look substantially different from adults and eat different foods.

FIGURE 34.24 Insect Metamorphosis Can Be Incomplete or Complete.

(a) Onychophorans are called velvet worms.

Peripatus species

0.5 cm

(b) Tardigrades are called water bears.

Echiniscus species

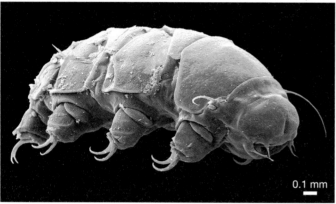

0.1 mm

FIGURE 34.25 Onychophorans and Tardigrades Have Lobe-Shaped Limbs with Claws. The scanning electron micrograph in part (b) has been colorized.

in having a segmented body and limbs. Unlike arthropods, they do not have an exoskeleton and their limbs are not jointed.

The onychophorans, or velvet worms, are small, caterpillar-like organisms that live in moist leaf litter and prey on small invertebrates. Onychophorans have lobe-shaped appendages and segmented bodies with a hemocoel (**FIGURE 34.25a**).

The tardigrades, or water bears, are microscopic animals that live in diverse marine, freshwater, and terrestrial environments, such as in the film of water that covers plants in moist habitats. Tardigrades have a reduced coelom but a prominent hemocoel, and they walk on their clawed, lobe-shaped legs (**FIGURE 34.25b**). Most feed by sucking fluids from plants or animals; others are detritivores, feeding on dead organic matter.

Although the phylogeny of arthropods is still being worked out, most data agree that the phylum as a whole is monophyletic. However, the relationships of lineages within the phylum are currently undergoing a major revision. Insects and myriapods have long been considered sister groups within the arthropods, due in part to their unbranched appendages. However, recent molecular phylogenetic studies—as well as comparisons of new morphological data, developmental data, and evidence in the fossil record—provide strong support for the placement of the insect clade *within* the crustacean lineage, making the crustaceans paraphyletic. It appears that crustaceans successfully invaded land as insects.

In the following exploration of ecdysozoan diversity, insects are described in a separate box, following tradition. However, the revised relationship between insects and crustaceans is then illustrated in the Crustacea box:

- Ecdysozoans > Nematoda (Roundworms)
- Ecdysozoans > Arthropoda > Myriapoda (Millipedes, Centipedes)
- Ecdysozoans > Arthropoda > Insecta (Insects)
- Ecdysozoans > Arthropoda > Crustacea (Shrimp, Lobsters, Crabs, Barnacles, Isopods, Copepods)
- Ecdysozoans > Arthropoda > Chelicerata (Sea Spiders, Horseshoe Crabs, Daddy Longlegs, Mites, Ticks, Scorpions, Spiders)

Species in the phylum Nematoda are commonly called **roundworms** or **nematodes.** Roundworms are unsegmented worms with a pseudocoelom, a tube-within-a-tube body plan, no appendages, and a thick, elastic cuticle that must be molted during growth. Although some nematodes can grow to lengths of several meters, most species are tiny—much less than 1 mm long. They lack specialized systems for exchanging gases and for circulating nutrients. Instead, gas exchange occurs across the body wall, and once ingested, nutrients move by simple diffusion from the gut to other parts of the body.

The nematode *Caenorhabditis elegans* is one of the most thoroughly studied model organisms in biology (see **BioSkills 13** in Appendix B). Species that parasitize humans have also been studied intensively. Pinworms, for example, infect about 40 million people in the United States alone, and *Onchocerca volvulus* causes an eye disease that infects around 20 million people in Africa and Latin America. Advanced infections by the roundworm *Wuchereria bancrofti* result in the blockage of lymphatic vessels, causing fluid accumulation and massive swelling—the condition known as elephantiasis. The potentially fatal disease trichinosis is caused by species in the genus *Trichinella*. The parasites are passed from infected animals to humans by ingesting raw or undercooked meat. An array of symptoms results, including discomfort caused by *Trichinella* larvae encasing themselves in cysts within muscle or other types of tissue (**FIGURE 34.26a**).

A recent phylogenetic analysis of the nematodes has produced an interesting insight: Convergent morphological evolution appears to be extensive. Animal parasitism and plant parasitism both arose several times in different lineages of nematodes. Thus, some nematodes that look very similar and fill a very similar ecological role may be distantly related.

Parasites represent a relatively small fraction of the nematode species that have been described to date, however. The free-living nematodes are important ecologically because they are found in virtually every habitat known (**FIGURE 34.26b**).

While around 25,000 species of roundworms have been named, the actual total may be over a million. Nematodes are also fabulously abundant. Biologists have found 90,000 roundworms in a single rotting apple and have estimated that rich farm soils contain up to 9 billion roundworms per acre. Although they are not the most species-rich animal group, they are among the most abundant. The simple nematode body plan has been extraordinarily successful.

Feeding Roundworms feed on a wide variety of materials, including bacteria, fungi, plant roots, small protists or animals, or detritus. In most cases, the structure of their mouthparts is specialized in a way that increases the efficiency of feeding on a particular type of organism or material.

Movement Roundworms move with the aid of their hydrostatic skeleton. They have no muscles that can change the diameter of the body—their body musculature consists solely of longitudinal muscles that bend the stiff body laterally. ✔ You should be able to indicate the loss of circumferential muscles on Figure 34.2. Most roundworms live in soil or inside a host, so when contractions of their longitudinal muscles cause them to undulate, the movements are resisted by a stiff substrate. As a result, the worm pushes off the substrate, and the body advances.

Reproduction Sexes are separate in most nematode species (but not in *C. elegans*, which is mostly hermaphroditic). Sexual reproduction begins with internal fertilization and culminates in egg laying and the direct development of offspring. Individuals go through a series of four molts over the course of their lifetime.

(a) Parasitic larvae encysted in muscle tissue

Trichinella spiralis

100 µm

(b) Most nematodes are free living.

Caenorhabditis elegans

0.1 mm

FIGURE 34.26 Some Nematodes Are Parasites, but Most Are Free Living.

Ecdysozoans > Arthropoda > Myriapoda (Millipedes, Centipedes)

The **myriapods** have relatively simple bodies, with a head region and a long trunk featuring a series of segments, each bearing one or two pairs of legs (**FIGURE 34.27**). If eyes are present, they consist of a few to many simple structures clustered on the sides of the head. The 16,000 species that have been described to date inhabit terrestrial environments all over the world.

Feeding Millipedes and centipedes have mouthparts that can bite and chew. These organisms live in rotting logs and other types of dead plant material that litters the ground in forests and grasslands. Millipedes are detritivores. Centipedes, in contrast, use a pair of poison-containing fangs just behind the mouth to hunt an array of insects. Large centipedes, with lengths up to 30 cm, can inject enough poison to debilitate a human.

Movement Myriapods walk or run on their many legs; a few species burrow. Some millipedes have over 190 trunk sections, each with two pairs of legs, for a total of over 750 legs. Centipedes usually have fewer than 30 segments, with one pair of legs per segment.

Reproduction Myriapod sexes are separate, and fertilization is internal. Males deposit sperm in packets that are picked up by the female or transferred to her by the male. After females lay eggs in the environment, the eggs hatch into juveniles that develop into adults via a series of molts.

Scolopendra species

1 cm

FIGURE 34.27 Myriapods Have One or Two Pairs of Legs on Each Body Segment.

Ecdysozoans > Arthropoda > Insecta (Insects)

About one million species of insects have been named thus far, but it is certain that many more exist. In terms of species diversity and numbers of individuals, insects dominate terrestrial environments. In addition, the larvae of some species are common in freshwater streams, ponds, and lakes. **TABLE 34.1** (see page 676) describes seven of the most prominent orders of insects. The fruit fly *Drosophila melanogaster*, in the order Diptera, is one of the most thoroughly studied model organisms in biology (see **BioSkills 13** in Appendix B).

Insects are distinguished by having three tagmata—the head, **thorax**, and **abdomen** (see Figure 34.21). Three pairs of walking legs are mounted on the lateral surface of the thorax. Most species have one or two pairs of wings, mounted on the dorsal (back) side of the thorax. Typically the head contains four mouthpart structures, a pair of slender **antennae** that are used to touch and smell, and a pair of compound eyes. A **compound eye** contains many lenses, each associated with a light-sensing, columnar structure (see Chapter 47). (Human and cephalopod eyes are **simple eyes,** meaning they have just one lens.)

Feeding Most insect species have four mouthpart structures (labrum, paired mandibles, paired maxillae, and labium), which have diversified in structure and function in response to natural selection. As a result, insects are able to feed in every conceivable manner and on almost every type of food source available on land. In species with complete (holometabolous) metamorphosis, larvae have wormlike bodies; most tunnel their way through food sources, though some feed by biting leaves. Adults, however, are predators or parasites on plant or animal tissues. Thus, diversification in food sources occurs even within species. Because so many insects make their living by feeding on plant tissues or fluids, the diversification of insects was closely correlated with the diversification of land plants.

Movement Insects use their legs to walk, run, jump, or swim, or they use their wings to fly. When insects walk or run, the sequence of movements usually results in three of their six legs maintaining contact with the ground at all times.

Reproduction Insect sexes are separate, and sexual reproduction is the norm. Mating usually takes place through direct copulation; the male inserts a sperm-transfer organ into the female. Most females lay eggs, but in a few species eggs are retained until hatching. Many species are also capable of reproducing asexually through parthenogenesis, the production of unfertilized eggs via mitosis. Complete metamorphosis is far more common than incomplete metamorphosis.

TABLE 34.1   Prominent Orders of Insects

	Order	Description
	Coleoptera ("sheath-winged") Beetles 360,000 known species	**Key traits:** Hardened forewings, called elytra, protect the membranous hindwings that power flight and act as stabilizers. **Feeding:** Adults are important predators and scavengers. Larvae are called grubs and often have chewing mouthparts. **Life Cycle:** All have complete metamorphosis. **Notes:** The most species-rich lineage on the tree of life—about 25 percent of all described species are beetles.
	Lepidoptera ("scale-winged") Butterflies, moths 180,000 known species	**Key traits:** Wings are covered with tiny, often colorful scales. The forewings and hindwings often hook together and move as a unit. **Feeding:** Larvae usually have chewing mouthparts and either bore through food material or bite leaves; adults often feed on nectar. **Life Cycle:** All have complete metamorphosis. **Notes:** Some species migrate long distances.
	Diptera ("two-winged") Flies (including mosquitoes, gnats, midges) 150,000 known species	**Key traits:** Reduced hindwings, called halteres, act as stabilizers during flight. **Feeding:** Adults are usually liquid feeders; many larvae are maggots, which may be parasitic or feed on rotting material. **Life Cycle:** All have complete metamorphosis. **Notes:** The first flies in fossil record appear 240 million years ago.
	Hymenoptera ("membrane-winged") Ants, bees, wasps 115,000 known species	**Key traits:** Membranous forewings and hindwings lock together. **Feeding:** Most ants feed on plant material; most bees feed on nectar; wasps are predatory and often have parasitic larvae. **Life Cycle:** All have complete metamorphosis. Males are haploid (they hatch from unfertilized eggs) and females are diploid (they hatch from fertilized eggs). Larvae are cared for by adults. **Notes:** Most species live in colonies, and many are eusocial.
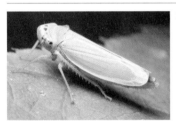	**Hemiptera** ("half-winged") Bugs (including leaf hoppers, aphids, cicadas, scale insects) 85,000 known species	**Key traits:** Some have a thickened forewing with a membranous tip. Mouthparts are modified for piercing and sucking. **Feeding:** Most suck plant juices, but some are predatory. **Life Cycle:** All have incomplete metamorphosis. **Notes:** Called "half-winged" because the forewings of some bugs are hardened near the bases but membranous near the tips.
	Orthoptera ("straight-winged") Grasshoppers, crickets 25,000 known species	**Key traits:** Large, muscular hind legs enable jumping. **Feeding:** Most have chewing mouthparts and are leaf eaters. **Life Cycle:** All have incomplete metamorphosis. **Notes:** In many species, males give distinctive songs to attract mates. They produce sound by rubbing their wings against their legs or against each other, a process called stridulation.
	Odonata ("toothed") Dragonflies, damselflies 6,500 known species	**Key traits:** Four membranous wings and long, slender abdomens. **Feeding:** Larvae and adults are predatory, often on flies. The "odon" refers to strong, toothlike structures on their mandibles. **Life Cycle:** All have complete metamorphosis: Nymphs are aquatic, adults are terrestrial. **Notes:** They hunt by sight and have reduced antennae (used for touch and smell) but enormous eyes with up to 28,000 lenses.

The 47,000 species of **crustaceans** that have been identified to date live primarily in marine and freshwater environments; they are the most diverse arthropods of the sea. A few species of crab and some isopods (known as sowbugs and pillbugs) are terrestrial, however, and recent phylogenetic evidence suggests that the insects represent a highly successful branch of terrestrial crustaceans (**FIGURE 34.28**). (For the purpose of describing diversity here, insects are covered separately on the previous page.)

Crustaceans—especially the planktonic copepods—are common in surface waters around the world, where they are important consumers. Crustaceans are also important grazers and predators in shallow-water benthic environments.

The segmented body of most crustaceans is divided into two distinct regions, or tagmata: (1) the cephalothorax, which combines the head and thorax, and (2) the abdomen. Many crustaceans have a **carapace**—a platelike section of their exoskeleton that covers and protects the cephalothorax (**FIGURE 34.29**). They are the only type of arthropod with two pairs of antennae, and have sophisticated, compound eyes—usually mounted on stalks. Crustacean appendages are usually branched (insect appendages are

unbranched). Other than trilobites—a major arthropod lineage that did not survive the end-Permian mass extinction, 250 million years ago (Chapter 28)—crustaceans were the first arthropods to appear in the fossil record. The tiny ostracods (seed shrimp) have a particularly rich fossil record dating back to the Cambrian.

Feeding Most crustaceans have 4–6 pairs of mouthparts that are derived from jointed appendages, and as a group they use every type of feeding strategy known. Barnacles and many shrimp are suspension feeders that use feathery structures located on head or body appendages to capture prey. Crabs and lobsters are active hunters, herbivores, and scavengers. Typically they have a pair of mouthparts called **mandibles** that bite or chew. Individuals capture and hold their food source with claws or other feeding appendages and then use their mandibles to shred the food into small bits that can be ingested.

Movement Crustaceans typically have many pairs of limbs with diverse forms and functions, including claws, paddle-shaped forms used in swimming, feathery structures used in capturing food that is suspended in water, and slender legs that make sophisticated walking or running movements possible (see Figure 34.22). Barnacles are one of the few types of sessile crustaceans. Adult barnacles cement their heads to a hard substrate, secrete a protective shell of calcium carbonate, and use their feathery legs to capture food particles and transfer them to the mouth.

Reproduction Most individual crustaceans are either male or female, and sexual reproduction is the norm. Fertilization is usually internal, and eggs are typically retained by the female until they hatch. Most crustaceans pass through several distinct larval stages; many species include a larval stage called a **nauplius,** which is usually planktonic. A nauplius has a single eye and appendages that develop into the two pairs of antennae, the mouthparts, and walking/swimming legs of the adult.

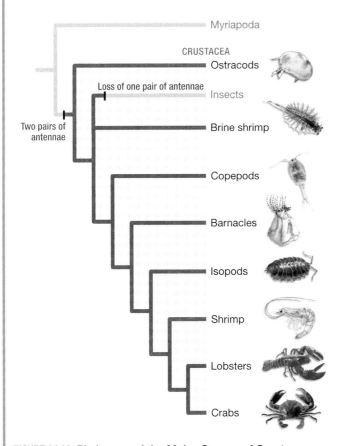

FIGURE 34.28 Phylogeny of the Major Groups of Crustaceans. This phylogeny illustrates the current hypothesis that insects are a terrestrial clade within the crustaceans, overturning traditional classification (which is nonetheless still in widespread use due to its long history).

Cherax quadricarinatus

FIGURE 34.29 Crustaceans Have Two Tagmata: A Cephalothorax and an Abdomen. The carapace protects the cephalothorax (literally "head-chest").

The most prominent lineage within the Chelicerata is the Arachnida, including spiders, scorpions, ticks, mites and daddy longlegs (**FIGURE 34.30**). Most of the 100,000 species of chelicerates are terrestrial, but the horseshoe crabs and sea spiders are marine.

The chelicerate body consists of 2 tagmata, the cephalothorax (prosoma) and abdomen (opisthosoma), and 6 pairs of appendages (**FIGURE 34.31**). The cephalothorax lacks antennae for sensing touch or odor but usually contains eyes. The group is named for a pair of appendages called **chelicerae,** found near the mouth. Depending on the species, the chelicerae are used in feeding, defense, copulation, movement, or sensory reception. In addition to 4 pairs of legs and 1 pair of chelicerae, chelicerates also have a pair of pedipalps, which can be used for diverse functions.

Feeding Spiders, scorpions, and daddy longlegs capture insects or other prey. Although some of these species are active hunters that rove around in search of prey, most spiders are sit-and-wait predators that either pounce on prey that wander near, or that create sticky webs to capture prey. In this case, the spider senses the vibrations of the struggling prey on the web, pounces on it, and administers a toxic bite. Regardless of the mode of capture, spiders generally consume their meals in soup form—they secrete digestive enzymes into the prey to enable digestion to begin externally, then drink up the nutritious liquid. Often all that remains of the prey is an empty exoskeleton.

Mites and ticks are ectoparasitic and use their piercing mouthparts to feed on host animals or, in the case of dust mites, on their hosts' dead skin. Horseshoe crabs eat a variety of animals as well as algae and detritus. Most scorpions feed on insects; the largest scorpion species occasionally eat snakes and lizards. Although the venom of scorpion stings is an effective defense against predators, it is primarily used to kill or paralyze prey.

Movement Like other arthropods, chelicerates move with the aid of muscles attached to an exoskeleton. They walk or crawl on their four pairs of jointed walking legs; some species are also capable of jumping. Horseshoe crabs and some other marine forms can swim slowly. Newly hatched spiders spin long, silken threads that serve as balloons and that can carry them on the wind more than 400 kilometers from the point of hatching.

Reproduction Sexual reproduction is the rule in chelicerates, and fertilization is internal in most groups. Courtship displays are extensive in many chelicerates and may include both visual displays and the release of chemical odorants. In spiders, males may present a dead insect as a gift that the female eats as the pair mates; in some species, the male himself is eaten as sperm is being transferred—the longer the meal lasts, the more successful the sperm transfer. Males use pedipalps to transfer sperm to females. These organs fit into the female reproductive tract in a "lock-and-key" fashion. Differences in the size and shape of male genitalia are often the only way to identify closely related species of spider.

Development in chelicerates is direct, meaning that larval forms and metamorphosis do not occur. In scorpions, females retain fertilized eggs. After the eggs hatch, the young climb on the mother's back. They remain there until they are old enough to hunt for themselves.

FIGURE 34.30 Phylogeny of the Major Groups of Chelicerates.

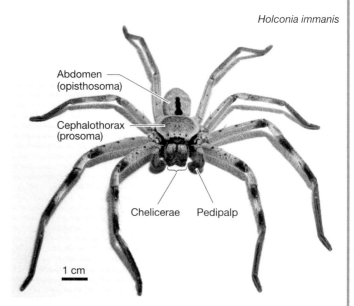

Holconia immanis

FIGURE 34.31 Chelicerates Have Chelicerae and Two Tagmata.

If you understand . . .

34.1 What Is a Protostome?

- Molecular evidence supports the hypothesis that protostomes are a monophyletic group divided into two major subgroups: the Lophotrochozoa and the Ecdysozoa.

- All protostomes are bilaterally symmetric and triploblastic, but the coelom has been reduced or lost in several phyla.

- Protostomes diversified in their methods of sensing their environments, feeding, moving, reproducing, and developing.

- The transition to living on land occurred several times independently during the evolution of protostomes, opening new habitats and driving diversification. The transition was facilitated by watertight shells and exoskeletons, the evolution of specialized structures to minimize water loss and manage gas exchange, and the ability to move to moist locations.

- Many protostome body plans are highly modular due to the underlying expression patterns of developmental tool-kit genes.

✔ *You should be able to explain how both ecological opportunity and genetic opportunity were important in the diversification of protostomes.*

34.2 What Is a Lophotrochozoan?

- Lophotrochozoans grow incrementally. Some phyla have characteristic feeding structures called lophophores, and many have a distinctive type of larvae called trochophores. Spiral cleavage is a synapomorphy for the group but was modified or lost in some phyla.

- Mollusks, annelids, and flatworms are the most diverse phyla within the Lophotrochozoa. All three phyla made the water-to-land transition.

- The mollusk body plan includes a muscular foot, a visceral mass, and a mantle. These components diversified in many ways, enabling mollusks to occupy most aquatic, and some terrestrial, habitats.

✔ You should be able to explain why the absence of a trait known to be found in a clade, such as a trochophore larva, spiral cleavage, or a coelom, does not necessarily exclude a species from that clade.

34.3 What Is an Ecdysozoan?

- Species in the Ecdysozoa grow by molting—meaning that they shed their old external covering and grow a new, larger one.

- Arthropods and nematodes are the most diverse phyla within the Ecdysozoa. A majority of animals on Earth are arthropods.

- The arthropod body plan is segmented and protected by an exoskeleton made primarily of chitin. Appendages are jointed. The diversification of arthropod body plans is based on variation in the number and functional specialization of segments.

- Complete (holometabolous) metamorphosis enables larvae and adults to specialize, creating opportunities for diversification.

- Crustaceans are the most diverse group of arthropods in the sea, and insects are the most diverse group on land. (Recent data suggest that insects are a terrestrial subgroup of Crustacea.)

✔ *You should be able to compare modularity in arthropod body plans with modularity in vertebrates.*

(MB) MasteringBiology

1. MasteringBiology Assignments

  Tutorials and Activities Characteristics of Invertebrates; Ecdysozoans; Protostome Diversity

  Questions Reading Quizzes, Blue-Thread Questions, Test Bank

2. eText Read your book online, search, take notes, highlight text, and more.

3. The Study Area Practice Test, Cumulative Test, BioFlix® 3-D Animations, Videos, Activities, Audio Glossary, Word Study Tools, Art

You should be able to . . .

✔ TEST YOUR KNOWLEDGE
Answers are available in Appendix A

1. **True or false?** The transition from water to land reduced protostome diversity.

2. What is a lophophore?
  a. a specialized filter-feeding structure
  b. the single opening in species with a blind gut
  c. a distinctive type of larva with a band of cilia
  d. a synapomorphy that defines lophotrochozoans

3. What is the function of the arthropod exoskeleton?
  a. Because hard parts fossilize more readily than do soft tissues, the presence of an exoskeleton has given arthropods a good fossil record.
  b. It has no well-established function. (Trilobites had an exoskeleton, and they went extinct.)
  c. It provides protection and functions in locomotion.
  d. It makes growth by molting possible.

4. Why is it logical that Platyhelminthes have flattened bodies?
   a. They have simple bodies and evolved early in the diversification of protostomes.
   b. They lack a coelom, so their body cannot form a rounded tube-within-a-tube design.
   c. A flat body provides a large surface area for gas exchange, which compensates for their lack of gas-exchange organs.
   d. They are sit-and-wait predators that hide from passing prey by flattening themselves against the substrate.

5. One trait that is shared by the Lophotrochozoa and Ecdysozoa is_____.

6. Which protostome phylum is distinguished by having body segments organized into tagmata?
   a. Mollusca
   b. Arthropoda
   c. Annelida
   d. Nematoda

✔ TEST YOUR UNDERSTANDING
Answers are available in Appendix A

7. **QUANTITATIVE** Put the following phyla in order from least diverse (in terms of number of named species) to most diverse: Annelida, Mollusca, Arthropoda, Rotifera.
   a. Annelida, Rotifera, Mollusca, Arthropoda
   b. Mollusca, Rotifera, Annelida, Arthropoda
   c. Rotifera, Annelida, Arthropoda, Mollusca
   d. Rotifera, Annelida, Mollusca, Arthropoda

8. Predict why annelids and arthropods were thought to be closely related, before the phylogenetic analyses in the late 1990s.

9. Pose a hypothesis to explain why a larval stage is adaptive in clams. How could you test your hypothesis?

10. Use your understanding of protostome evolution to predict two adaptations for terrestrial living that occurred in land plants.

11. Pose a hypothesis to explain why the evolution of the wing was such an important event in the evolution of insects.

12. Propose a scenario wherein hemimetabolous metamorphosis would be adaptive.

✔ TEST YOUR PROBLEM-SOLVING SKILLS
Answers are available in Appendix A

13. Recall that the phylum Platyhelminthes includes three major groups: the free-living turbellarians, the parasitic cestodes, and the parasitic trematodes. Because the parasitic forms are so simple morphologically, researchers suggest that they are derived from more complex, free-living forms. Draw a phylogeny of Platyhelminthes that would support the hypotheses that ancestral flatworms were morphologically complex, free-living organisms and that parasitism is a derived condition.

14. Brachiopoda is a phylum within the Lophotrochozoa. Even though they are not closely related to mollusks, brachiopods look and act like bivalve mollusks (clams or mussels). Specifically, brachiopods suspension feed, live inside calcium carbonate shells with two valves that hinge together in some species, and attach to rocks or other hard surfaces on the ocean floor. How is it possible for brachiopods and bivalves to be so similar if they did not share a recent common ancestor?

15. The Mollusca includes a group of about 370 wormlike species called the Aplacophora. Although aplacophorans do not have a shell, their cuticle secretes calcium carbonate scales or spines. They lack a well-developed foot, and some species have a simple radula. They are mostly carnivores and detritivores in the deep sea. The reduced forms of several molluscan synapomorphies suggest that they are ancestral to other mollusks—that is, the earliest group to branch off in the molluscan phylogeny. Evaluate this hypothesis. How could you test it?

16. Examine the figure below summarizing the expression of *Ubx-AbdA Hox* genes (colored red) in different body segments during early development in four crustaceans. The blue appendages represent a novel type of feeding appendage called a maxilliped. Which statement best summarizes the data?

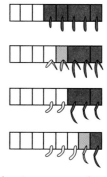

   a. Maxillipeds were lost in segments where *Ubx-AbdA* genes were expressed.
   b. Maxillipeds arose in segments where *Ubx-AbdA* genes were no longer expressed.
   c. Maxillipeds arose in segments where *Ubx-AbdA* genes were expressed.
   d. The presence of maxillipeds does not correlate with the pattern of *Ubx-AbdA* gene expression.

680   UNIT 6   The Diversification of Life

35 Deuterostome Animals

In this chapter you will learn that

Deuterostomes include the largest and most complex animals on Earth

by asking

What is an echinoderm?
35.1

by asking

What is a chordate?
35.2

then asking

What is a vertebrate?
35.3

then examining

The primates and hominins
35.4

In most habitats the "top predators"—animals that prey on other animals and aren't preyed upon themselves—are deuterostomes.

The **deuterostomes** include the largest-bodied and some of the most morphologically complex of all animals. They range from the sea stars that cling to dock pilings to the fish that dart in and out of coral reefs to the wildebeests that migrate across the Serengeti Plains of East Africa.

The deuterostomes were initially grouped together because they all undergo early embryonic development in a similar way. When a humpback whale, sea urchin, or human is just beginning to grow, the gut starts developing from posterior to anterior—the anus forms first and the mouth second. A coelom, if present, develops from outpocketings of mesoderm (see Chapter 33).

✔ When you see this checkmark, stop and test yourself. Answers are available in Appendix A.

More recently, phylogenies based on molecular evidence have confirmed the monophyly of the deuterostomes. Today, most biologists recognize four deuterostome phyla: the Echinodermata, Hemichordata, Xenoturbellida, and Chordata (**FIGURE 35.1**):

- The echinoderms include the sea stars and sea urchins.

- The hemichordates, or "acorn worms," burrow in marine sands or muds and make their living by deposit feeding or suspension feeding.

- A lone genus with two wormlike species, called *Xenoturbella,* was recognized as a distinct phylum in 2006. The phylogenetic placement of these small, ciliated animals has been debated by scientists for years.

- The chordates include the **vertebrates,** or animals with backbones. The vertebrates, in turn, comprise the hagfish, lampreys, sharks and rays, bony fishes, amphibians, mammals, and reptiles (including birds).

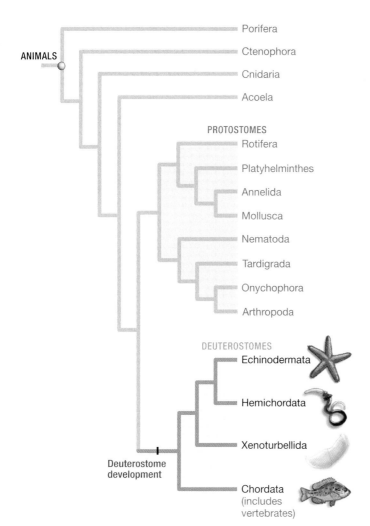

FIGURE 35.1 There Are Four Phyla of Deuterostomes. Vertebrates are in the phylum Chordata.

✔ **QUESTION** Are invertebrates monophyletic or paraphyletic? (Hint: See Table 28.2 if you need help.)

Animals that are not vertebrates are collectively known as **invertebrates.** Over 95 percent of the known animal species are invertebrates—but the vertebrates dominate the deuterostomes.

Although deuterostomes share important features of embryonic development, their adult body plans and their feeding methods, modes of locomotion, and reproductive strategies are highly diverse. This chapter explores deuterostome diversity by introducing the echinoderms and chordates and then taking a more in-depth look at the vertebrates and our own closest ancestors: the hominins.

35.1 What Is an Echinoderm?

The echinoderms (literally, "spiny-skins") were named for the spines or spikes observed in many species. Echinodermata is a large and diverse phylum but is exclusively marine.

In numbers of species and range of habitats occupied, echinoderms are the second-most successful lineage of deuterostomes—next to the vertebrates. To date, biologists have cataloged about 7000 species of echinoderms.

Echinoderms are also abundant. In some deepwater environments, species in this phylum represent 95 percent of the total mass of organisms.

How are these animals put together?

The Echinoderm Body Plan

All deuterostomes are considered bilaterians, because they evolved from an ancestor that was bilaterally symmetric (see Chapter 33). Echinoderm larvae are also bilaterally symmetric (**FIGURE 35.2a**). But a remarkable event occurred early in the evolution of **echinoderms:** the origin of five-sided radial symmetry, called pentaradial symmetry, in adults (**FIGURE 35.2b**).

Radially symmetric animals do not have heads (Chapter 33). As a result, they tend to interact with the environment in all directions at once instead of facing the environment in one direction. Adult echinoderms that are capable of movement tend to move equally well in all directions.

A second noteworthy feature of the echinoderm body is its **endoskeleton:** a hard structure, located just inside a thin layer of epidermal tissue, which provides protection and support (shown without skin in Figure 35.2b). As an individual is developing, cells secrete plates of calcium carbonate inside the skin. Depending on the species, the plates may remain independent and result in a flexible structure or fuse into a rigid case. If the plates do not fuse, they are connected by an unusual tissue that can be reversibly stiff or flexible—depending on conditions.

Another remarkable event in echinoderm evolution was the origin of a unique morphological feature: a series of branching, fluid-filled tubes and chambers called the **water vascular system.** In some groups, one of the tubes is open to the exterior where it meets the body wall, so seawater can flow into and out

(a) Echinoderm larvae are bilaterally symmetric.

50 μm

(b) Adult echinoderms are radially symmetric.

1 cm

FIGURE 35.2 Body Symmetry Differs in Adult and Larval Echinoderms. (a) Bilaterally symmetric sea urchin larva. **(b)** Radially symmetric adult endoskeleton.

of the system. Inside, fluids move via the beating of cilia that line the interior of the tubes and chambers.

FIGURE 35.3a highlights a particularly important part of the water vascular system called tube feet. **Tube feet** are elongated, fluid-filled appendages, each consisting of a balloon-like *ampulla*

on the inside of the body and a tube-like *podium* (literally, "foot"; plural *podia*) projecting on the outside (**FIGURE 35.3b**). The water vascular system forms a specialized hydrostatic skeleton that operates the tube feet. As tube feet extend and contract in a coordinated way along the base of an echinoderm, they alternately grab and release the substrate or food source, enabling movement and feeding. Tube feet can also aid in respiration by increasing the surface area of soft tissue exposed to the aquatic environment.

Radial symmetry in adults, an endoskeleton of calcium carbonate, and the water vascular system are all synapomorphies—traits that identify echinoderms as a monophyletic group. (Turn to Chapter 28 and **BioSkills 7** in Appendix B to review how to read phylogenetic trees.)

How Do Echinoderms Feed?

Depending on the lineage and species in question, echinoderms make their living by mass feeding on algae or other animals, suspension feeding, or deposit feeding. In most cases, an echinoderm's tube feet play a key role in obtaining food.

Many sea stars, for example, prey on bivalves. Clams and mussels respond to sea star attacks by contracting muscles that close the two halves of their shell tight. But by clamping onto each half of the shell with their tube feet, making their endoskeleton rigid, and pulling, sea stars are often able to pry the shell apart a few millimeters (**FIGURE 35.4a**, see page 684). Once a gap exists, the sea star extrudes its stomach from its body and forces the stomach through the opening in the bivalve's shell. Upon contacting the visceral mass of the bivalve, the stomach of the sea star secretes digestive enzymes. It then begins to absorb the small molecules released by enzyme action. Eventually, only the shell of the prey remains.

Tube feet are also used in suspension feeding (**FIGURE 35.4b**). In most cases, tube feet are extended out into the water. When food particles contact them, the tube feet flick the food down to cilia, which sweep the particles toward the animal's mouth.

(a) Echinoderms have a water vascular system.

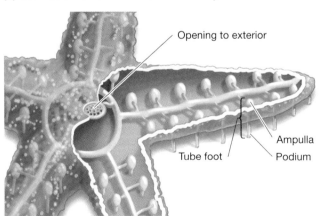

Opening to exterior

Ampulla

Tube foot

Podium

(b) Tube feet project from the underside of the body.

Tube feet

FIGURE 35.3 Echinoderms Have a Water Vascular System. (a) The water vascular system is a series of tubes and chambers that radiates throughout the body, forming a sophisticated hydrostatic skeleton. **(b)** Tube feet aid in movement because they extend from the body and can grab and release the substrate.

(a) Tube feet enable sea stars to get a grip on their prey.

(b) Tube feet enable feather stars to suspension feed.

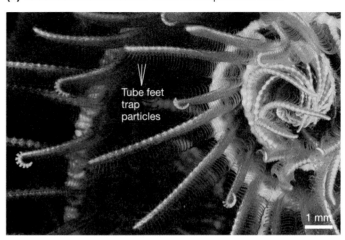

Tube feet
trap
particles

1 mm

FIGURE 35.4 Echinoderms Use Their Tube Feet in Feeding.

In deposit feeders, tube feet secrete mucus that is used to sop up food material on the substrate. The tube feet then roll the food-laden mucus into a ball and move it to the mouth.

Key Lineages: The Echinoderms

The echinoderms living today make up five major lineages: **(1)** feather stars and sea lilies, **(2)** sea stars, **(3)** brittle stars and basket stars, **(4)** sea urchins and sand dollars, and **(5)** sea cucumbers (**FIGURE 35.5**). ✔ You should be able to indicate the origin of pentaradial symmetry in adults, the water vascular system, and the echinoderm endoskeleton on Figure 35.5.

- Most feather stars and sea lilies are sessile suspension feeders.

- Brittle stars and basket stars have five or more long arms that radiate out from a small central disk. They use these arms to suspension feed, deposit feed, or capture small prey animals.

- Sea cucumbers are sausage-shaped animals that suspension feed or deposit feed with the aid of modified tube feet called tentacles that are arranged in a whorl around their mouths.

Sea stars and the sea urchins and sand dollars are described in detail in the two boxes that follow.

- Echinodermata > Asteroidea (Sea Stars)
- Echinodermata > Echinoidea (Sea Urchins and Sand Dollars)

check your understanding

If you understand that . . .

- Echinoderms have a distinctive body plan: pentaradial symmetry in adults, an endoskeleton, and a water vascular system.
- Most echinoderms use their tube feet to move, but they feed in a wide variety of ways—including using their tube feet for mass feeding, suspension feeding, or deposit feeding.

✔ **You should be able to . . .**

Explain how you would determine whether an unidentified animal you encounter along the seashore is an echinoderm.

Answers are available in Appendix A.

FIGURE 35.5 There Are Five Major Lineages of Echinoderms.

The 1900 known species of sea stars have bodies with five or more long arms—in some species, up to 40—radiating from a central region that contains the mouth, stomach, and anus (**FIGURE 35.6**). The sea star's arms are not set off from the central region by clear, joint-like articulations, in contrast to their sister group, the brittle stars. ✔ You should be able to indicate the origin of arms, and the origin of arms that are continuous with the central region, as in sea stars, on Figure 35.5. When fully grown, sea stars can range in size from less than 1 cm in diameter to 1 m across. They live on hard or soft substrates along the coasts of all the world's oceans. Although the spines that are characteristic of some echinoderms are reduced to knobs on the surface of most sea stars, the crown-of-thorns star and a few other species have prominent, upright, movable spines.

Feeding Sea stars are predators or scavengers. Some species pull bivalves apart with their tube feet and feed by everting their stomach into the prey's visceral mass. In some cases, sea star predation on bivalves opens up habitat for other organisms in the community—increasing biodiversity (see Chapter 55). Sponges, barnacles, and snails are also common prey. The crown-of-thorns sea star specializes in feeding on corals and is native to the Indian Ocean and western Pacific Ocean. Its population has skyrocketed recently—possibly because people are harvesting the sea star's major predator, a large snail called the triton, for its shell. Large crown-of-thorns star populations have led to the destruction of large areas of coral reef.

Movement Sea stars crawl with the aid of their tube feet.

Reproduction Sexual reproduction predominates in sea stars, and sexes are separate. At least one sea star arm in each individual is filled with reproductive organs that produce massive amounts of gametes—millions of eggs per female in some species. Some species care for their offspring by holding fertilized eggs on their body until the eggs hatch. Larvae are long-lived and may spend weeks or months in the plankton, possibly dispersing thousands of kilometers. Most sea stars are capable of regenerating arms that are lost in predator attacks or storms. Some species can reproduce asexually by dividing the body in two; each of the two individuals then regenerates the missing half.

Crossaster papposus

5 cm

FIGURE 35.6 Sea Stars May Have Many Arms.

About 1000 species of echinoids are living today; most are sea urchins or sand dollars. Sea urchins have globe-shaped bodies and long spines and crawl along substrates. Sand dollars are flattened and disk-shaped, have short spines, and burrow in soft sediments. ✔ You should be able to indicate the origin of globular or disk-shaped bodies on Figure 35.5.

Feeding Most types of sea urchins are herbivores. In some areas of the world, urchins are extremely important grazers on kelp and other types of algae. In cases when humans have removed predators such as sea otters, and urchin populations have grown in response, the urchins' grazing can destroy kelp forests. The urchins chew away a structure called a holdfast, which anchors kelp to the substrate; the kelp then float away and die. Most echinoids have a unique, jaw-like feeding structure in their mouths, called an Aristotle's lantern, that is made up of five calcium carbonate teeth attached to muscles (**FIGURE 35.7a**, see page 686). In many species, this apparatus can extend and retract as the animal feeds, allowing it to reach more food material. Sand dollars, in contrast, are

suspension feeders. The beating of cilia on their spines creates tiny water currents that bring organic material toward the body. The food particles are then trapped in mucus and transported in long strings to the mouth. In this way, sand dollars harvest nutritious detritus found in sand or in other soft substrates (**FIGURE 35.7b**).

Movement Using their spines, sea urchins crawl and sand dollars burrow. Some species also use their tube feet to aid movement, or just to hold on to the substrate to prevent being washed away by waves and currents.

Reproduction Sexual reproduction predominates in sea urchins and sand dollars (see Figure 33.14). Fertilization is external, and sexes are separate. As in sea stars, larvae are long-lived and may disperse long distances. The larvae of some species undergo asexual reproduction by dividing in two or budding off new individuals, before they undergo metamorphosis.

(Continued on next page)

(a) Sea urchins graze with their teeth.

Lytechinus variegatus

Teeth at center of underside

5 mm

(b) Sand dollars suspension feed.

Echinarachnius parma

1 cm

FIGURE 35.7 Sea Urchins and Sand Dollars Have Different Feeding Strategies.

35.2 What Is a Chordate?

You are a chordate, yet chordates are so diverse that you would probably find it difficult to identify familiar features in some of your fellow chordates—such as the sea squirts, whose small, sessile, globular bodies hardly resemble animals at all. Which characteristics are shared within this phylum?

All **chordates** are defined by the presence of four morphological features at some stage in their life cycles:

1. openings into the throat called **pharyngeal gill slits;**

2. a **dorsal hollow nerve cord,** which runs the length of the body, composed of projections from neurons;

3. a stiff and supportive but flexible rod called a **notochord,** which runs the length of the body; and

4. a muscular, post-anal tail—meaning a tail that contains muscle and extends past the anus.

Electrical signals carried by the dorsal hollow nerve cord coordinate muscle movement, and the notochord stiffens the muscular tail. Together, these traits create a "torpedo"—a long, streamlined animal that can swim rapidly. Fossil evidence suggests that chordates diversified early during the Cambrian radiation of animals, about 550 million years ago.

Three Chordate "Subphyla"

The phylum Chordata is made up of three major lineages: the **(1)** cephalochordates, **(2)** urochordates, and **(3)** vertebrates. Let's examine the four chordate morphological features in each group.

The Cephalochordates   **Cephalochordates** are also called lancelets or amphioxus; they are small, mobile, torpedo-shaped animals with a "fish-like" appearance, and they make their living by suspension feeding (**FIGURE 35.8a**). Adult cephalochordates live in ocean-bottom habitats, where they burrow in sand and suspension feed with the aid of their pharyngeal gill slits.

The dorsal hollow nerve cord in cephalochordates runs parallel to a notochord. Because the notochord stiffens the cephalochordates' bodies, muscle contractions on either side result in fishlike movement when the animals swim during dispersal or mating.

The Urochordates   **Urochordates** are also called tunicates or sea squirts. As **FIGURE 35.8b** shows, pharyngeal gill slits are present in both larvae and adults; these structures function in both feeding and gas exchange.

The dorsal hollow nerve cord, notochord, and tail are present in larvae and in the sexually mature forms of species that move as adults. As in cephalochordates, these features combine to enable swimming movements. As larvae swim or float in the upper water layers of the ocean, they drift to new habitats where food might be more abundant. As a result, larvae function as a dispersal stage in the life cycle.

The Vertebrates   Vertebrates (**FIGURE 35.8c**) include the hagfish, lampreys, sharks and rays, several lineages of bony fishes, amphibians, mammals, and reptiles (including birds). In vertebrates, the dorsal hollow nerve cord is the familiar spinal cord—a bundle of nerve cells that runs from the brain to the posterior of the body. Electrical signals from the spinal cord control movements and the function of organs (see Chapter 46).

Structures called pharyngeal pouches are homologous to pharyngeal gill slits and are present in all vertebrate embryos.

Pharyngeal gill slits
Dorsal hollow nerve cord
Notochord
Muscular, post-anal tail

(a) Cephalochordates

Water flow

Adult

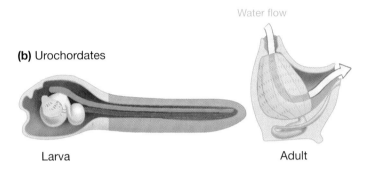

Water flow

(b) Urochordates

Larva

Adult

(c) Vertebrates

Cross section
of embryo

Embryo

Pharyngeal pouches become
gill slits in aquatic vertebrates

FIGURE 35.8 Four Features Distinguish the Three Lineages of Chordates.

In aquatic species, the creases between pouches open into gill slits and develop into part of the main gas-exchange organ—the **gills.** In terrestrial species, however, gill slits do not develop after the pharyngeal pouches form. In these species, the pharyngeal pouches are a vestigial trait (see Chapter 25).

A notochord develops in all vertebrate embryos. It no longer functions in body support and movement, however, for these roles are filled by the vertebral column. Instead, the notochord helps organize the body plan early in development. The notochord secretes proteins that help induce the formation of somites—segmented blocks of tissue that form along the length of the body (Chapter 23). Although the notochord itself disappears, cells in the somites later differentiate into the vertebrae, ribs, and skeletal muscles of the back, body wall, and limbs. In this way, the notochord is instrumental in the development of the trait that gave vertebrates their name.

Key Lineages: The Invertebrate Chordates

There are about 24 species of lancelets, 3000 species of tunicates, and over 62,000 vertebrates. Because they are so species rich and diverse, vertebrate chordates rate their own section in this chapter (Section 35.3). Here let's focus on the tunicates and lancelets. These lineages are sometimes referred to as the invertebrate chordates—meaning, members of the Chordata that lack vertebrae. They are of great interest to evolutionary biologists, who study the morphology, development, and genetics of these animals to better understand the evolutionary history of our own subphylum, the vertebrates.

● Chordata > Cephalochordata (Lancelets)

● Chordata > Urochordata (Tunicates)

Chordata > Cephalochordata (Lancelets)

About two dozen species of cephalochordates have been described to date, all of them found in marine sands. Lancelets—also called amphioxus—have several characteristics that are intermediate between invertebrates and vertebrates. Chief among these is a notochord that is retained in adults, where it functions as an endoskeleton.

Feeding Adult cephalochordates feed by burrowing in sediment until only their heads are sticking out into the water (**FIGURE 35.9**). They take water in through their mouths and trap food particles with the aid of mucus on their pharyngeal gill slits.

Movement Adults have large blocks of muscle arranged in a series along the length of the notochord. Lancelets are efficient swimmers because the flexible, rod-like notochord stiffens the body, making it wriggle when the blocks of muscle contract.

Reproduction Asexual reproduction is unknown, and individuals are either male or female. Gametes are released into the environment, and fertilization is external.

Branchiostoma lanceolatum

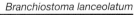

FIGURE 35.9 Lancelets Look Like Fish but Are Not Vertebrates.

The urochordates are also called tunicates; they comprise two major sub-lineages: the sea squirts (or ascidians) and salps. All of the species described to date live in the ocean. Sea squirts live on the ocean floor (**FIGURE 35.10a**); salps live in open water (**FIGURE 35.10b**). The synapomorphies that define the urochordates include an external coat of polysaccharide, called a tunic, that covers and supports the body; a U-shaped gut; and two body openings, called siphons.

Feeding Adult urochordates use their pharyngeal gill slits to suspension feed. A mucus sheet on the inside of the pharynx traps particles present in the water that enters one siphon, passes through the slits, and leaves through the other siphon.

Movement Larvae swim with the aid of the notochord, which functions as a simple endoskeleton. Larvae are a dispersal stage and do not feed. Adult sea squirts are sessile while salps drift in currents.

Reproduction In most species, individuals produce both sperm and eggs. In some species, both sperm and eggs are shed into the water and fertilization is external; in other species, sperm are released into the water but eggs are retained, so that fertilization and early development are internal. Asexual reproduction by budding is also common in some groups, forming large sessile colonies in some sea squirts, and long chains in salps.

(a) Sea squirt

Rhopalaea crassa

1 cm

(b) Salp

Salpa thompsoni

1 cm

FIGURE 35.10 Sea Squirts and Salps Live in Different Habitats.

35.3 What Is a Vertebrate?

The vertebrates are a monophyletic group distinguished by two synapomorphies:

- a column of cartilaginous or bony structures called **vertebrae,** which form along the dorsal side of the body; and

- a **cranium,** a bony, cartilaginous, or fibrous case that encloses the brain. (A skull consists of the cranium in jawless vertebrates, and the cranium and mandible—or jaw—in jawed vertebrates.)

The vertebral column is important because it protects the spinal cord. The cranium is important because it protects the brain along with sensory organs such as eyes. Together, the vertebrae and cranium protect the central nervous system (CNS) and key sensory structures (see Chapter 33).

The vertebrate brain develops as an outgrowth of the most anterior end of the dorsal hollow nerve cord, and it is important to the vertebrate lifestyle. Vertebrates are active predators and herbivores that must sense stimuli in their environments and make rapid, directed movements in response. More complex brains enable more complex interactions with the environment.

In early vertebrates, the brain was divided into three regions with important sensory functions:

1. The **forebrain** housed the sense of smell;

2. the **midbrain** was associated with vision; and

3. the **hindbrain** was responsible for balance and, in some species, hearing.

All vertebrates living today have retained these three regions, but the structures and functions of the regions have evolved over time. For example, in the jawed vertebrates, or **gnathostomes** (pronounced *NATH-oh-stomes*), the hindbrain consists of enlarged regions called the **cerebellum** and **medulla oblongata.** Part of the forebrain also became elaborated into a large structure called the **cerebrum,** especially in birds and mammals. (The structure and functions of the vertebrate brain are analyzed in detail in Chapter 46.) The evolution of a large, three-part brain, protected by a hard cranium, was a key innovation in vertebrate evolution.

An Overview of Vertebrate Evolution

Vertebrates have been the focus of intense research for well over 300 years, in part because they are large and conspicuous and in part because they include humans. All this effort has paid off in an increasingly thorough understanding of how the vertebrates diversified. Let's consider what the data from the fossil record have to say about key events in vertebrate evolution and then examine the current best estimate of the phylogenetic relationships among vertebrates.

The Vertebrate Fossil Record Vertebrate fossils are present in the Chengjiang formation of China, dated to the Cambrian explosion. The detailed vertebrate fossils show that the earliest members of this lineage lived in the ocean about 540 million years ago; they had streamlined, fishlike bodies and a cranium made of cartilage (see an artist's interpretation of an early vertebrate species called *Haikouichthys* in Figure 28.12). **Cartilage** is a strong but flexible tissue—found in your ear and the tip of your nose—that consists of scattered cells in a gel-like matrix of polysaccharides and protein fibers (Chapter 42).

A cartilaginous endoskeleton is a basic vertebrate feature; only in the bony fishes and their descendants, the **tetrapods** (literally, "four-footed"), does the skeleton become composed of bone in the adult. **Bone** is a tissue consisting of cells and blood vessels encased in a matrix made primarily of calcium phosphate, along with a small amount of protein fibers. In species with bony skeletons, growing bones are cartilaginous first and become bony only later in their development.

Following the appearance of vertebrates, the fossil record documents a series of key innovations that occurred as this lineage diversified (**FIGURE 35.11**):

- *Bony exoskeleton* Fossil vertebrates from the early part of the Ordovician period, about 480 million years ago, are the first fossils to have bone. When bone first evolved, it did not occur in the endoskeleton ("inside-skeleton"). Instead, bone was deposited in scalelike plates that formed an **exoskeleton** ("outside-skeleton") (compare with arthropod exoskeletons in Chapter 34). Most of your skull is made up of this type of dermal ("skin") bone—making your skull similar to the "head shields" in Ordovician fishes. Based on the fossils' overall morphology, biologists infer that these animals swam with the aid of a notochord and that they breathed and fed by gulping water and filtering it through their pharyngeal gill slits. Presumably, the bony plates helped provide protection from predators.

- *Jaws* The first fishes with jaws show up early in the Silurian, about 440 million years ago. The evolution of jaws was significant because it improved the ability of fish to use suction feeding to capture prey, and also to bite, meaning that vertebrates were no longer limited to suspension feeding or deposit feeding. Instead, they could make a living as herbivores or predators. Soon after, jawbones with teeth appear in the fossil record. With jaws and teeth, vertebrates became armed and dangerous. The fossil record shows that a spectacular

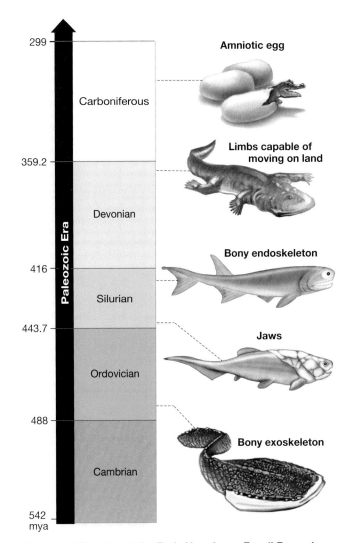

FIGURE 35.11 Timeline of the Early Vertebrate Fossil Record.

radiation of jawed fishes followed, filling marine and freshwater habitats.

- *Bony endoskeleton* In some lineages of fishes from late in the Silurian period, the cartilaginous endoskeleton began to be stiffened by the deposition of bone. Unlike the dermal bone that had evolved earlier, the bony endoskeleton functioned to support movement—rapid swimming.

- *Limbs capable of moving on land* The next great event in the evolution of vertebrates was the transition to living on land. The first animals that had limbs and were capable of moving on land are dated to about 365 million years ago, late in the Devonian. These were the first of the tetrapods—animals with four limbs.

- *Amniotic egg* About 20 million years after the appearance of tetrapods in the fossil record, the first amniotes are present. The Amniota is a lineage of vertebrates that includes all tetrapods other than amphibians. The **amniotes** are named for a signature synapomorphy and adaptation: the amniotic egg. An **amniotic egg** is an egg that has membranes

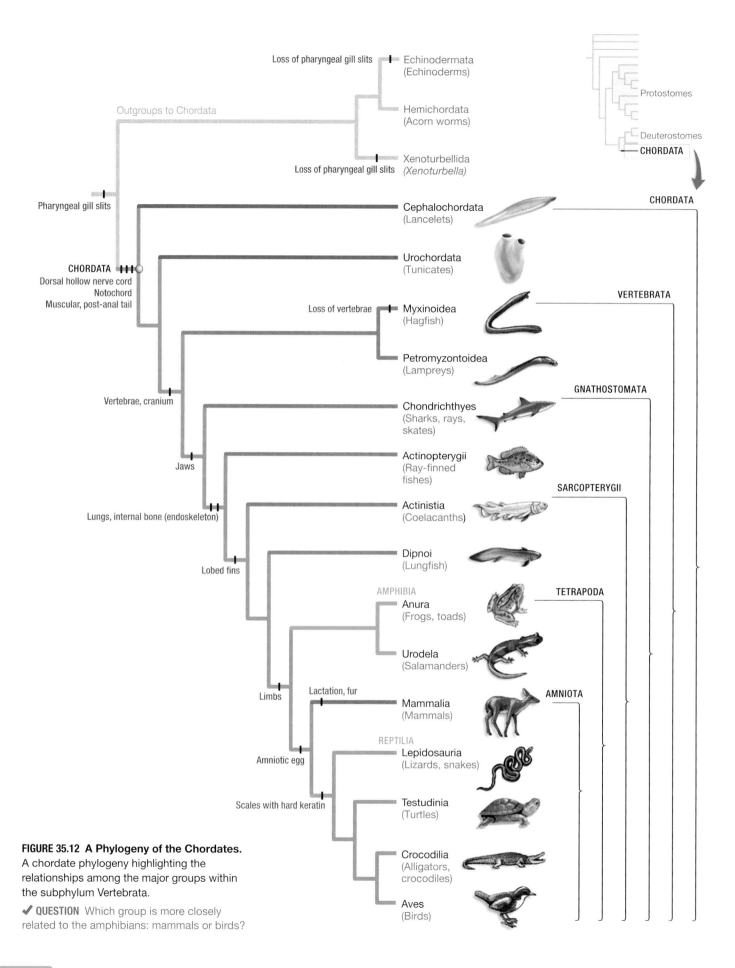

Loss of pharyngeal gill slits — Echinodermata (Echinoderms)

Hemichordata (Acorn worms)

Outgroups to Chordata

Xenoturbellida (*Xenoturbella*)
Loss of pharyngeal gill slits

Pharyngeal gill slits

Cephalochordata (Lancelets)

CHORDATA
Dorsal hollow nerve cord
Notochord
Muscular, post-anal tail

Urochordata (Tunicates)

Loss of vertebrae — Myxinoidea (Hagfish)

Petromyzontoidea (Lampreys)

Vertebrae, cranium

Chondrichthyes (Sharks, rays, skates)

Jaws

Actinopterygii (Ray-finned fishes)

Lungs, internal bone (endoskeleton)

Actinistia (Coelacanths)

Dipnoi (Lungfish)

Lobed fins

AMPHIBIA
Anura (Frogs, toads)

Urodela (Salamanders)

Limbs

Lactation, fur

Mammalia (Mammals)

REPTILIA
Lepidosauria (Lizards, snakes)

Amniotic egg

Testudinia (Turtles)

Scales with hard keratin

Crocodilia (Alligators, crocodiles)

Aves (Birds)

Protostomes

Deuterostomes

CHORDATA

CHORDATA

VERTEBRATA

GNATHOSTOMATA

SARCOPTERYGII

TETRAPODA

AMNIOTA

FIGURE 35.12 A Phylogeny of the Chordates.
A chordate phylogeny highlighting the relationships among the major groups within the subphylum Vertebrata.

✔ **QUESTION** Which group is more closely related to the amphibians: mammals or birds?

surrounding a food supply, a water supply, and a waste repository. These membranes provided support and extra surface area for gas exchange, allowing amniotes to produce larger, better-developed young.

To summarize, the fossil record indicates that vertebrates evolved through a series of major steps, beginning about 540 million years ago with vertebrates whose endoskeleton consisted of a notochord. The earliest vertebrates gave rise to fishes with bony exoskeletons. These fishes gave rise to the jawed vertebrates, including the cartilaginous fishes (sharks and rays) and several lineages of fishes with bony endoskeletons. After the tetrapods emerged and amphibians resembling salamanders began to live on land, the evolution of the amniotic egg paved the way for the diversification of amniotes—specifically, reptiles and the animals that gave rise to mammals. Do phylogenetic trees estimated from analyses of DNA sequence data agree or conflict with these conclusions?

Evaluating Molecular Phylogenies FIGURE 35.12 provides a phylogenetic tree that summarizes the relationships among living chordates, especially vertebrates, based on DNA and other molecular sequence data. The labeled bars on the tree indicate where major morphological innovations occurred as lineages diversified. Although the phylogeny of deuterostomes continues to be a topic of intense research, researchers are increasingly confident that the relationships described in Figure 35.12 are accurate. There are a few important take-home messages from the tree:

- *Outgroup to vertebrates* According to the most recent molecular data, the urochordates are the closest living relatives of the vertebrates.

- *Living outgroup to gnathostomes* The phylogenetic relationships among hagfish, lampreys, and the gnathostomes have been hotly debated. Two hundred years ago, hagfish and lampreys were grouped together as the class Cyclostomata ("round-mouthed"), based on their unique mouthparts. Then, starting in the 1970s, morphologists began to take a closer look and placed hagfish and lampreys in two independent lineages with lampreys as the sister group to gnathostomes. But molecular phylogenies have not supported this view, instead leaning toward the Cyclostomata grouping. So which is correct? Analysis of a new source of data—evolutionarily conserved microRNAs (Chapter 19)—has decided in favor of the Cyclostomata hypothesis. This result is important because it emphasizes that the ancestor to gnathostomes was probably not lamprey-like. Rather, some vertebrate characteristics appear to have been *lost* in the hagfish/lamprey lineages, especially in hagfish.

- *Fossil outgroup to gnathostomes* The diverse lineages of extinct armored fishes are not included in the tree of living vertebrates. If pictured, they would occur as outgroups to the gnathostomes—more closely related to jawed vertebrates than are hagfish and lampreys.

- *"Fish"* No monophyletic group includes all of the living fishlike lineages. Instead, "fishy" organisms are a series of independent monophyletic groups (the living ones are indicated by blue branches in Figure 35.12). Taken together, they form what biologists call a **grade**—a sequence of lineages that are paraphyletic (that include some, but not all, of the descendants of a common ancestor). ✔ If you understand the difference between a paraphyletic and monophyletic group, you should be able to draw what Figure 35.12 would look like if there actually were a monophyletic group called "The Fishes." (See Table 28.2 if you need help.)

- *Mammals* Mammals and reptiles are sister groups and thus are more closely related to each other than either is to the amphibians.

- *Birds* Birds occur within the monophyletic group called reptiles. That is, birds *are* reptiles.

To understand what happened during the diversification of vertebrates, let's explore some of the major morphological innovations involved in feeding, movement, and reproduction in more detail. (For an overview of these themes, see Chapter 33.)

Key Innovations in Vertebrates

Among vertebrates, the most species-rich and ecologically diverse lineages are the ray-finned fishes and the tetrapods. Ray-finned fishes occupy habitats ranging from deepwater environments, which are perpetually dark, to shallow ponds that dry up each year. Tetrapods include the large herbivores and predators in terrestrial environments all over the world. About the same number of species of ray-finned fishes and tetrapods are living today (FIGURE 35.13).

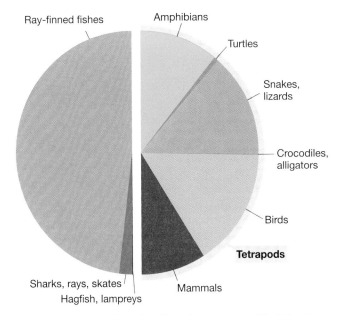

FIGURE 35.13 Relative Species Abundance among Vertebrates. Of the 63,000 species of described vertebrates, about half are ray-finned fishes and about half are tetrapods.

DATA: American Museum of Natural History; birdlife.org; fishbase.org; IUCN red list 2010; Reptile-database.org.

The Vertebrate Jaw The most ancient vertebrates in the fossil record had relatively simple mouthparts that lacked jaws. Until jaws evolved, vertebrates were not able to harvest food by biting.

The leading hypothesis for the origin of the jaw proposes that natural selection acted on developmental regulatory genes that determine the morphology of **gill arches,** which are curved regions of tissue between the gills. The jawless vertebrates have bars of cartilage that stiffen these gill arches. The gill-arch hypothesis proposes that mutation and natural selection increased the size of the most anterior arch and modified its orientation slightly, producing the first working jaw (**FIGURE 35.14**). Four lines of evidence support this hypothesis:

1. Both gill arches and jaws consist of flattened bars of bony or cartilaginous tissue that hinge and bend forward.

2. During development, the same population of cells gives rise to the muscles that move jaws and the muscles that move gill arches.

3. Unlike most other parts of the vertebrate skeleton, both jaws and gill arches are derived from specialized cells in the embryo called neural crest cells.

4. The expression patterns of key developmental regulatory genes, including *Hox* and *Dlx*, are similar in jaws and gill arches.

(a) Jawless vertebrate

Gill arches

Mouth

(b) Intermediate form (basal gnathostomes)

Gill arches

Jaw

(c) Fossil shark

Gill arches

Jaw

FIGURE 35.14 The Gill-Arch Hypothesis for the Evolution of the Jaw. (a) Gill arches support the gills in jawless vertebrates. **(b)** This intermediate form has yet to be found in the fossil record. **(c)** Fossil sharks that appeared later had more elaborate jaws.

Once jaws originated, changes in size and shape enabled a rapid diversification of feeding strategies from suction feeding to biting. Other modifications to the jaw also occurred:

• In most ray-finned fishes, the jaw is protrusible—meaning it can be extended to nip or bite at food.

• Several species-rich lineages of ray-finned fishes have a second specialized jaw called a **pharyngeal** ("throat") **jaw,** consisting of modified gill arches. It is located in the back of the throat and makes food processing particularly efficient. (For more on the structure and function of pharyngeal jaws, see Chapter 44.)

To summarize, the radiation of ray-finned fishes was triggered largely by the evolution and diversification of the jaw, by modifications that made it possible to protrude the jaw, and by the origin of the pharyngeal jaw. The story of tetrapods is different, however. Although jaw structure varies somewhat among tetrapod groups, the adaptation that triggered their initial diversification involved the ability to move and get to food, not to bite it and process it.

The Tetrapod Limb To understand how tetrapods made the water-to-land transition, consider the morphology and behavior of their closest living relatives, the lungfish (**FIGURE 35.15**). Most living species of lungfishes inhabit shallow, oxygen-poor water. To supplement the oxygen taken in by their gills, they have lungs and breathe air. (Note that although lungs originated before the divergence of ray-finned fishes and the Sarcopterygii, they serve primarily as a swim bladder to maintain buoyancy in ray-finned fishes.) Some lungfishes also have fleshy fins supported by bones and are capable of walking short distances along watery mudflats or the bottoms of ponds. In addition, some species can survive extended droughts by burrowing in mud.

Lungfish have a series of adaptations that allow them to survive on land for short periods. Could a lungfish-like organism have evolved into the first land-dwelling vertebrate with limbs?

Fossils provide strong links between the limbs of the ancestors of today's lungfish and the earliest land-dwelling vertebrates. **FIGURE 35.16** shows a phylogeny of the species involved. Note that during the Devonian, the fossil record documents a series of species that indicate a gradual transition from a lobe-like fin to a limb that could support walking on land. The fossil record

Fleshy fins

FIGURE 35.15 Lungfish Have Limb-like Fins. Some lungfishes can walk or crawl short distances on their fleshy fins.

is actually more complete than shown here—the figure is just a sample of the tetrapod and tetrapod-like species known from this interval.

The color coding in Figure 35.16 emphasizes that each fin or limb has a single bony element that is proximal (closest to the body; in blue) and then two bones that are distal (farther from the body) and arranged side by side (these are shown in red), followed by a series of distal elements (in orange). Because the structures are similar, and because no other animal groups have limb bones in this arrangement, the evidence for homology is strong.

The hypothesis that tetrapod limbs evolved from fish fins has also been supported by molecular genetic evidence. Recent work has shown that several regulatory proteins involved in the development of cartilaginous fish fins and ray-finned fish fins and the upper parts of mammal limbs are homologous. For example, the proteins produced by several different *Hox* genes are found at the same times during development and in the same locations in fins and limbs.

These data suggest that these appendages are patterned by the same genes. As a result, the data support the hypothesis that tetrapod limbs evolved from fins. Once the tetrapod limb evolved, natural selection elaborated it into diverse structures that are used for running, gliding, crawling, burrowing, swimming, and flying.

Once tetrapods made the water-to-land transition, they had to survive a number of new terrestrial challenges other than locomotion, such as keeping their eggs from drying out.

The Amniotic Egg Amphibians (frogs and toads, salamanders, and limbless species called caecilians) keep their eggs from drying out by laying their jelly-coated eggs in water. In contrast, reptiles (including birds) and the egg-laying mammals produce an amniotic egg, which has a protective covering that reduces the rate of drying significantly. Species that produce amniotic eggs lay them outside of water.

Amphibian eggs are surrounded by a single membrane, whereas amniotic eggs have four. As **FIGURE 35.17** shows, the outermost membrane of an amniotic egg encloses a supply of water in a protein-rich solution called **albumen**. Albumen cushions the developing embryo and provides nutrients. The three inner membranes surround the embryo itself, the yolk provided by the mother, and the waste from the embryo. The additional membranes are thought to be advantageous because they:

1. provide mechanical support—an important consideration outside of a buoyant aquatic environment; and

2. increase the surface area available for exchange of gases and other materials. Efficient diffusion of molecules is important because it allows females to lay larger eggs that hatch into larger, more independent young.

In addition, amniotic eggs are surrounded by a shell. This layer is leathery in lizards and snakes. It is stiffened by some calcium carbonate deposits in turtles and crocodiles, and by extensive

FIGURE 35.16 Evidence for a Fin-to-Limb Transition. The fossil record documents a series of species that indicate a gradual transition from a lobe-like fin to a limb that could support walking on land.

✔ **QUESTION** This phylogeny was estimated from traits other than the morphology of the limb. Why is this important in addressing how limbs evolved?

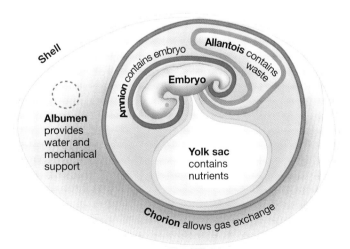

FIGURE 35.17 An Amniotic Egg. Amniotic eggs have four membrane-bound sacs.

calcium carbonate deposits in birds. Lizards, snakes, turtles, and crocodiles often bury their eggs in moist soils, but bird eggs are more watertight and are laid in nests that are exposed to air.

During the evolution of mammals, however, a second major innovation in reproduction occurred—one that eliminated the need for any type of egg laying.

The Placenta   Egg-laying animals are said to be **oviparous,** while species that give birth are termed **viviparous** (Chapter 33). In many viviparous animals, females produce an egg that contains a nutrient-rich yolk. Instead of laying the egg, however, the mother retains it inside her body. In these **ovoviviparous** species, the developing offspring depends on the resources in the egg yolk.

In caecilians, some lizards, and most mammals, however, the eggs that females produce have little yolk. After fertilization occurs and the egg is retained, a combination of maternal and embryonic tissues produces a placenta within the uterus or oviduct. The embryo's contributions are the allantois and chorion—membranes that are involved in gas exchange in an amniotic egg. In the placenta, tissues derived from the allantois and chorion are also involved in the diffusion of gases, nutrients, and wastes.

The **placenta** is an organ that is rich in blood vessels and that facilitates a flow of oxygen and nutrients from the mother to the developing offspring, as well as the removal of nitrogenous wastes and carbon dioxide from the offspring (**FIGURE 35.18**). After a development period called **gestation,** the embryo emerges from the mother's body.

Why did viviparity and the placenta evolve? The leading hypotheses are that retaining the embryo inside the body has several advantages:

1. Offspring develop at a more constant, favorable temperature.

2. Offspring are protected.

3. Offspring are portable—mothers are not tied to a nest.

In effect, a placenta and viviparity are mechanisms for increasing the level of **parental care:** behavior by a parent that improves the ability of its offspring to survive. What other types of parental care evolved in vertebrates?

Parental Care   In species that provide extensive parental care—even if only in the form of exceptionally large amounts of yolk in eggs—mothers participate in a fitness trade-off (see Chapter 25). They can produce fewer offspring, but the offspring that they do produce have a higher probability of survival.

In animals, mechanisms of parental care include incubating eggs to keep them warm during early development, keeping young warm and dry, supplying young with food, and protecting them from danger. In some insect and frog species, mothers carry eggs or newly hatched young on their bodies. In fishes, parents commonly guard eggs during development and fan them with oxygen-rich water.

Mammals and birds provide particularly extensive parental care. In both groups, the mother (and often the father, in birds) continues to feed and care for individuals after birth or hatching—sometimes for many years. Female mammals also **lactate**—meaning that they produce milk and use it to feed their offspring after birth (**FIGURE 35.19**).

Among large animals, the evolution of extensive parental care is hypothesized to be a major reason for the evolutionary success of mammals and birds. In both lineages, mothers produce relatively small numbers of large, high-quality offspring.

Wings and Flight   Following the transition from water to land, the origin of the amniotic egg, and the elaboration of parental care, amniotes made another big transition—into the air. Wings and flight evolved independently in three lineages of tetrapod amniotes: the extinct flying reptiles called pterosaurs (pronounced *TARE-oh-sors*), the bats, and the birds.

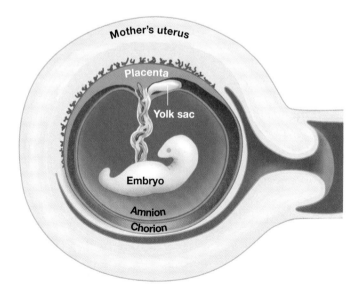

FIGURE 35.18 A Placenta Nourishes a Fetus Internally.

✔ **QUESTION** Compare the relative size of the yolk sac here with that in the amniotic egg (Figure 35.17).

FIGURE 35.19 Parental Care Is Extensive in Mammals. Female mammals, such as this timber wolf, produce highly nutritious milk to nourish their young.

How did flight evolve? The best data sets on this question involve feathered flight in birds. Since the early 2000s, paleontologists have discovered a spectacular series of feathered dinosaur fossils. The newly discovered species and exceptionally well-preserved feather specimens address key questions about the evolution of birds, feathers, and flight:

- **Did birds evolve from dinosaurs?** On the basis of skeletal characteristics, all of these recently discovered fossil species clearly show that birds are part of the monophyletic group called dinosaurs. That is—birds *are* dinosaurs.

- **How did feathers evolve?** The early fossils support a model of feathers evolving in a series of steps, beginning with simple projections from the skin (popularly called "dinofuzz") and culminating in the complex, branched structures familiar today. Although diverse, feathers observed in many non-avian dinosaurs were not strong enough to support gliding or flight. They may have been used for insulation and/or mating displays—recent analysis of pigments in fossil feathers have shown striking color patterns in some dinosaurs (**FIGURE 35.20**).

- **Did birds begin flying from the ground up or from the trees down?** More specifically, did flight evolve with running species that began to jump and glide or make short flights, with the aid of feathers to provide lift? Or did flight evolve from tree-dwelling species that used feathers to glide from tree to tree, much as flying squirrels do today? This question is still unresolved. The observation that many early feathered dinosaurs had "flight" feathers on their legs adds an intriguing piece to the puzzle (see Figure 35.20).

Once dinosaurs evolved feathers and took to the air, the fossil record shows that a series of adaptations made powered, flapping flight increasingly efficient (**FIGURE 35.21**).

- Most dinosaurs have a flat sternum ("breastbone"), but the bird sternum has a projection called the keel, which provides a large surface area to which flight muscles attach.

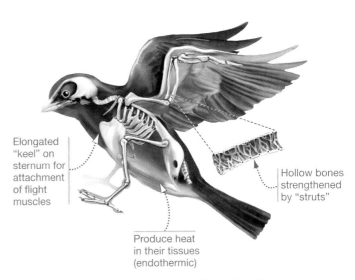

Elongated "keel" on sternum for attachment of flight muscles

Hollow bones strengthened by "struts"

Produce heat in their tissues (endothermic)

FIGURE 35.21 In Addition to Feathers, Birds Have Several Features That Enable Flight.

- Birds are extraordinarily light for their size, primarily because they have a drastically reduced number of bones and because their larger bones are thin walled and hollow—though strengthened by bony "struts."

- Birds are capable of sustained activity year-round because they are **endotherms** ("inside-heated")—they maintain a high body temperature by producing heat in their tissues.

From dinosaurs that jumped or glided from tree to tree, birds have evolved into extraordinary flying machines.

To summarize, the evolution of the jaw gave tetrapods the potential to capture and process a wide array of foods. With limbs, they could move efficiently on land in search of food. The amniotic egg and placenta further enabled vertebrates to diversify into terrestrial niches, and parental care increased the survivorship of their offspring. Finally, some tetrapods were able to defy the pull of gravity, expanding their realm into the air.

FIGURE 35.20 Feathers Evolved in Dinosaurs. This artist's depiction of what *Anchiornis huxleyi,* a feathered dinosaur, might have looked like in life incorporates recent data on fossil feather colors. The colors may have been important in mating displays. Unlike modern birds, many feathered dinosaurs had large feathers on their legs.

check your understanding

If you understand that . . .

- Vertebrates have a distinctive body plan: bilateral symmetry with vertebrae that protect a spinal cord and a cranium that protects the brain.
- An array of key innovations occurred during the evolution of vertebrates: jaws, the tetrapod limb, the amniotic egg, the placenta, extensive parental care, and flight.

✔ **You should be able to . . .**

1. Explain the adaptive significance of each key innovation.

2. Compare adaptations in protostomes (Chapter 34) and deuterostomes for living on land.

Answers are available in Appendix A.

Key Lineages: The Vertebrate Chordates

The fossil record provides an increasingly clear picture of early vertebrate evolution; intensive and continuing research has explored the origins of key vertebrate innovations such as the jaw, the tetrapod limb, and the amniotic egg. Now that you've been introduced to key vertebrate features, let's take a closer look at the key lineages.

"Fish" As discussed earlier, fish are a paraphyletic group consisting of diverse lineages, many of which are extinct. Four of the boxes that follow feature the six major lineages of fishes that are living today:

- Myxinoidea (Hagfish) and Petromyzontoidea (Lampreys)
- Chondrichthyes (Sharks, Rays, Skates)
- Actinopterygii (Ray-Finned Fishes)
- Actinistia (Coelacanths) and Dipnoi (Lungfish)

Amphibia Unlike fish, the **amphibians** form a monophyletic group, the Amphibia. Most amphibians lay their eggs in water and undergo metamorphosis from an aquatic larva to a terrestrial or semiterrestrial adult. The three lineages of amphibians are discussed together in one box:

- Amphibia (Frogs and Toads, Salamanders, Caecilians)

Mammalia **Mammals** are a monophyletic group, today comprising three major lineages:

- Mammalia > Monotremata (Platypuses, Echidnas)
- Mammalia > Marsupiala (Marsupials)
- Mammalia > Eutheria (Placental Mammals)

The relationships among these groups are shown in **FIGURE 35.22**. Monotremes lay eggs; marsupials have a poorly developed placenta

but well-developed pouch for rearing offspring; eutherians have a well-developed placenta and extended pregnancy.

Mammals are endotherms that maintain high body temperatures by oxidizing large amounts of food and generating large amounts of heat. Instead of insulating themselves with feathers, though, mammals retain heat because the body surface is covered with layers of hair or fur. Endothermy evolved independently in mammals and birds. In both groups, endothermy is thought to be an adaptation that enables individuals to maintain high levels of activity—particularly at night or during cold weather.

In addition to being endothermic and having fur, mammals have **mammary glands**—unique structures that make lactation possible. The evolution of mammary glands gave mammals the ability to provide their young with particularly extensive parental care. Mammals are also the only vertebrates with cheek muscles and lips—traits that make suckling milk possible—and the only vertebrates that have a lower jaw formed from a single bone. ✔ You should be able to mark the origin of mammary glands and fur on Figure 35.22.

Mammals evolved when dinosaurs and other reptiles were the dominant large herbivores and predators in terrestrial and aquatic environments. The earliest mammals in the fossil record appear about 195 million years ago; most were small animals that were probably active only at night. Many of the 5500 species of mammals living today have good nocturnal vision and a strong sense of smell, as their ancestors presumably did.

The adaptive radiation that gave rise to today's diversity of mammalian orders did not take place until after the dinosaurs went extinct. Long after the dinosaurs were gone, the mammals diversified into lineages that included large herbivores and large predators—ecological roles that had once been filled by dinosaurs and the ocean-dwelling, extinct reptiles called ichthyosaurs and plesiosaurs.

Reptilia The **reptiles** are a monophyletic group and represent the second major living lineage of amniotes besides mammals (Figure 35.22). One major diagnostic feature distinguishing the reptilian and mammalian lineages is the number and placement of openings in the side of the skull. These skull openings are important: Jaw muscles required for biting and chewing pass through them and attach to bones on the upper part of the skull.

Reptiles have several adaptations for life on land. Their skin is made watertight by a layer of scales made of the protein keratin. Reptiles breathe air through well-developed lungs and lay shelled, amniotic eggs. In snakes and lizards, the egg has a leathery shell; in other reptiles, the shell includes some calcium carbonate.

In many reptiles, the sex of an individual is determined by the environment it experiences during early development. In certain species, for example, high temperatures produce mostly males while low temperatures produce mostly females.

The reptiles include the dinosaurs, ichthyosaurs, plesiosaurs, pterosaurs (flying reptiles), and other extinct lineages that flourished from about 250 million years ago until the mass extinction

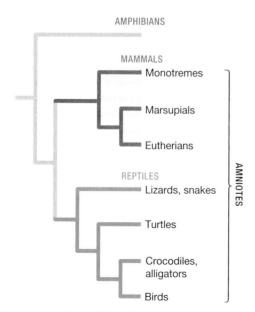

FIGURE 35.22 **Mammals and Reptiles Are Monophyletic Groups.**

at the end of the Cretaceous period, 65 million years ago. Today the Reptilia are represented by four major lineages:

- Reptilia > Lepidosauria (Lizards, Snakes)
- Reptilia > Testudinia (Turtles)
- Reptilia > Crocodilia (Crocodiles, Alligators)
- Reptilia > Aves (Birds)

Except for birds, almost all of the reptiles living today are **ectotherms** ("outside-heated")—meaning that individuals do not use internally generated heat to regulate their body temperature. It would be a mistake, however, to conclude that reptiles other than birds do not regulate their body temperature closely. Reptiles bask in sunlight, seek shade, and perform other behaviors to keep their body temperature at an appropriate level.

Chordata > Vertebrata > Myxinoidea (Hagfish) and Petromyzontoidea (Lampreys)

Hagfish and lampreys are the only living jawless vertebrates, comprising about 120 species. Their relationship to jawed vertebrates has been hotly debated; however, the most recent microRNA evidence suggests that the two groups are more closely related to each other than either is to the jawed vertebrates. These data indicated that hagfish and lampreys, collectively the cyclostomes ("round mouths"), may have *lost* many vertebrate characters over time.

Hagfish and lampreys have long, slender bodies and are aquatic (**FIGURE 35.23**). Hagfish are strictly marine; lampreys include both marine and freshwater forms. Most species are less than a meter long when fully grown. Adult hagfish lack any sort of vertebral column, but adult lampreys have small pieces of cartilage along the length of their dorsal hollow nerve cord. Both hagfish and lampreys have brains protected by a cranium, as do all vertebrates.

Feeding Hagfish are scavengers and predators. They deposit feed on the carcasses of dead fish and whales, and some are thought to burrow through ooze at the bottom of the ocean, feeding on polychaete worms and other buried prey. Lampreys, in contrast, are ectoparasites. They attach to fish or other hosts by suction and then use spines in their mouth and tongue to rasp a hole in the side of their victim. Once the wound is open, they suck blood and other body fluids.

Movement Hagfish and lampreys have a well-developed notochord and swim by making undulating movements. Hagfish are particularly flexible, sometimes forming "knots" with their bodies to rid themselves of the copious slime they create as a defense mechanism when disturbed. Lampreys can move themselves upstream, against the flow of water, by attaching their suckers to rocks and looping the rest of their body forward, like an inchworm. Both groups have tail fins, and lamprey have a dorsal (back) fin; but neither group has the paired lateral (side) appendages found in other vertebrates.

Reproduction Virtually nothing is known about hagfish mating or embryonic development. Some lampreys live in freshwater; others are **anadromous**—meaning they spend their adult life in the ocean but swim up streams to breed. Fertilization is external, and adults die after breeding once. Lamprey eggs hatch into larvae that look and act like lancelets (cephalochordates). The larvae burrow into sediments and suspension feed for several years before metamorphosing into free-swimming adults.

(a) Hagfish

Eptatretus stoutii

Petromyzon marinus

(b) Lamprey

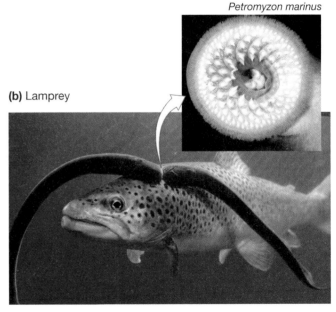

FIGURE 35.23 Hagfish and Lampreys Are Jawless Vertebrates.

Chordata > Vertebrata > Chondrichthyes (Sharks, Rays, Skates)

This lineage is distinguished by a specialized type of reinforced cartilaginous skeleton (*chondrus* is the Greek word for cartilage). Along with jawed vertebrates, or gnathostomes, sharks, rays, and skates also have jaws. They, like the other fish-like gnathostomes, also have paired lateral fins. Paired fins were an important evolutionary innovation because they stabilize the body during rapid swimming—keeping it from pitching up or down, yawing to one side or the other, or rolling. ✔ **You should be able to indicate the origin of paired fins and the chondrichthyan form of cartilage on Figure 35.12.**

Most of the 1200 species of sharks, rays, and skates are marine, though a few species live in freshwater. Sharks have streamlined, torpedo-shaped bodies and an asymmetrical tail—the dorsal portion is longer than the ventral portion (**FIGURE 35.24a**). In contrast, the dorsal–ventral plane of the body in rays and skates is strongly flattened (**FIGURE 35.24b**).

Feeding A few species of sharks and rays suspension feed on plankton, but most species in this lineage are predators. Skates and rays hunt along the ocean floor for snails, clams, and other benthic animals; electric rays capture their prey by stunning them with electric discharges of up to 200 volts. Most sharks, in contrast, are active hunters that chase down prey in open water and bite them. The larger species of sharks feed on large fish or marine mammals. In many marine ecosystems, sharks are referred to as the "top predator" because they are at the top of the food chain—nothing eats them. Yet the largest of all sharks, the whale shark, is a suspension feeder. Whale sharks filter plankton out of water as it passes over their gills.

Movement Rays and skates swim by flapping their greatly enlarged pectoral fins. Sharks swim by undulating their bodies from side to side and beating their large tails.

Reproduction Sharks use internal fertilization, and fertilized eggs may be shed into the water or retained until the young are hatched and well developed. In some viviparous species, embryos are attached to the mother by specialized tissues in a placenta, where the exchange of gases, nutrients, and wastes takes place. (A placenta evolved in certain shark lineages independently of its evolution elsewhere on the tree of life.) Skates are oviparous, but rays are viviparous.

(a) Sharks are torpedo shaped.

Prionace glauca

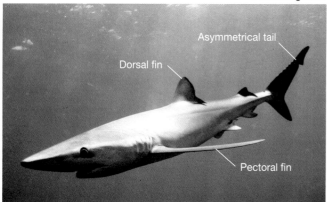

Asymmetrical tail
Dorsal fin
Pectoral fin

(b) Skates and rays are flat.

Dasyatis americana

Pectoral fin

FIGURE 35.24 Sharks and Rays Have Cartilaginous Skeletons.

Chordata > Vertebrata > Actinopterygii (Ray-Finned Fishes)

Actinopterygii (pronounced *ack-tin-op-teh-RIJ-ee-i*) means "ray-finned." These **ray-finned fishes** have fins that are supported by long, bony rods arranged in a ray pattern, in addition to a skeleton made of bone (**FIGURE 35.25**). Their bodies are covered with interlocking scales that provide a stiff but flexible covering, and many have a gas-filled **swim bladder** that evolved from the lungs found in the earliest lineages of fish. The evolution of the swim bladder was an important innovation because it allowed ray-finned fishes to avoid sinking. Most tissues are heavier than water, so the bodies of aquatic organisms tend to sink. Sharks and rays, for example, have to swim to avoid sinking. But in ray-finned fishes, gas is added to the bladder when a fish swims down; gas is removed when the fish swims up. In this way, ray-finned fishes maintain neutral buoyancy in water of various depths and thus various pressures. ✔ **You should be able to indicate the origin of rayed fins and the swim bladder on Figure 35.12.**

The actinopterygians are the most successful vertebrate lineage based on number of species, duration in the fossil record, and extent of habitats occupied. About 30,500 species of ray-finned fishes are known, including the smallest known vertebrate—a species whose adult females average less than 8 mm in length.

(Continued on next page)

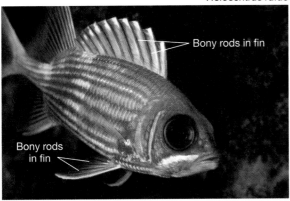

Holocentrus rufus

Bony rods in fin

Bony rods in fin

FIGURE 35.25 In Ray-Finned Fishes, Fins Are Supported by Long, Bony Rods.

The most important major lineage of ray-finned fishes is the Teleostei. About 96 percent of all living fish species, including familiar groups like the tuna, trout, cod, and goldfish, are teleosts. The teleosts underwent an adaptive radiation about the same time that mammals did.

Feeding Teleosts can suck food toward their mouths, grasp it with their protrusible jaws, and then process it with teeth on their jaws and with pharyngeal jaws in their throat. The size and shape of the mouth, the jaw teeth, and the pharyngeal jaw teeth all correlate with the type of food consumed. For example, most predatory teleosts have long, spear-shaped jaws armed with spiky teeth, as well as bladelike teeth on their pharyngeal jaws. Besides being major predators, ray-finned fishes are the most important large herbivores in both marine and freshwater environments.

Movement Like other fish, ray-finned fishes swim by alternately contracting muscles on the left and right sides of their bodies from head to tail, resulting in rapid, side-to-side undulations. Their bodies are streamlined to reduce drag in water. Teleosts have a flexible, symmetrical tail, which reduces the need to use their pectoral (side) fins as steering and stabilizing devices during rapid swimming.

Reproduction Most ray-finned fish species rely on external fertilization and are oviparous; some species have internal fertilization with external development; still others have internal fertilization and are viviparous. Although it is common for fish eggs to be released in the water and left to develop on their own, parental care occurs in some species. Parents may carry fertilized eggs on their fins, in their mouth, or in specialized pouches to guard them until the eggs hatch. In other cases the eggs are laid into a nest that is actively guarded and cared for. In freshwater teleosts, offspring develop directly; but marine species have larvae that undergo metamorphosis into a very different juvenile form, which then grows into an adult.

Chordata > Vertebrata > Actinistia (Coelacanths) and Dipnoi (Lungfish)

Although coelacanths (pronounced *SEEL-uh-kanths*) and lungfish represent independent lineages, they are sometimes grouped together and called **lobe-finned fishes.** Lobe-finned fishes are common and diverse in the fossil record in the Devonian period, about 400 million years ago, but only eight species are living today. They are important, however, because they represent a crucial evolutionary link between the fishes and the tetrapods. Early fishes in the fossil record have fins supported by stiff structures that extend from a base of bone. In ray-finned fishes, the bony elements are reduced. But in lobefins the bony elements extend down the fin, and they branch—similar to the bony elements in the limbs of tetrapods (**FIGURE 35.26**). ✔ You should be able to indicate the origin of extensive fin bones on Figure 35.12.

Coelacanths are marine and occupy habitats 150–700 m below the surface. In contrast, lungfish live in shallow, freshwater ponds and rivers (see Figure 35.15). As their name implies, lungfish have lungs and breathe air when oxygen levels in their habitats drop. Some species burrow in mud and enter a quiescent, sleeplike state when their habitat dries up during each year's dry season.

Feeding Coelacanths prey on fish. Lungfish are omnivorous ("all-eating"), meaning that they eat algae and plant material as well as animals.

Movement Coelacanths and lungfish swim by undulating their bodies; some lungfish can also use their fins to walk along pond bottoms.

Reproduction Sexual reproduction is the rule; fertilization is internal in coelacanths and external in lungfish. Coelacanths are ovoviviparous; lungfish lay eggs. Lungfish eggs hatch into larvae that resemble juvenile salamanders.

Latimeria menadoensis

Fleshy lobes supported by bones

FIGURE 35.26 Coelacanths Are Lobe-Finned Fishes.

Chordata > Vertebrata > Amphibia (Frogs and Toads, Salamanders, Caecilians)

Amphibians are found throughout the world and occupy ponds, lakes, or moist terrestrial environments. Translated literally, their name means "both-sides-living." The name is appropriate because adults of most amphibians feed on land but lay their eggs in water (**FIGURE 35.27**). In many amphibians, gas exchange occurs exclusively or in part across their moist, mucus-covered skin. ✔ You should be able to indicate the origin of "skin-breathing" on Figure 35.12. The 6800 named species of amphibians living today form three distinct clades: (1) frogs and toads, (2) salamanders, and (3) caecilians (pronounced *suh-SILL-ee-uns*). Caecilians lack limbs, and so resemble worms or snakes. They burrow underground in wet tropical regions.

Feeding Adult amphibians are carnivores. Most frogs and toads are sit-and-wait predators that use their long, extensible tongues to capture passing prey. Salamanders also have an extensible tongue, which some species use in feeding. Terrestrial caecilians prey on earthworms and other soil-dwelling animals; aquatic forms eat vertebrates and small fish.

Movement Most amphibians have four well-developed limbs. In water, frogs and toads move by kicking their hind legs to swim; on land, they kick their hind legs out to jump or hop. Salamanders walk on land; in water they undulate their bodies to swim. Caecilians lack limbs; terrestrial forms burrow in moist soils.

Reproduction Reproduction is sexual, breeding occurs in water, and larvae undergo a dramatic metamorphosis into land-dwelling adults. For example, the fishlike tadpoles of frogs and toads develop limbs, and their gills are replaced with lungs. Frogs have external fertilization, but salamanders and caecilians have internal fertilization. Frogs and most salamanders are oviparous, but many caecilians are viviparous.

Rana temporaria

Eggs

FIGURE 35.27 Amphibians Are the Most Ancient Tetrapods.

Chordata > Vertebrata > Mammalia > Monotremata (Platypuses, Echidnas)

The **monotremes** are the most basal lineage of mammals living today and are found in nature only in Australia. Monotremes lay eggs and have metabolic rates that are lower than those of other mammals. Five species exist: one species of platypus (**FIGURE 35.28a**) and four species of echidnas (**FIGURE 35.28b**).

Feeding Monotremes have a leathery beak or bill. The platypus feeds on insect larvae, mollusks, and other small animals in streams. Echidnas feed on ants, termites, and earthworms.

Movement Platypuses swim with the aid of their webbed feet and walk when on land. Echidnas walk on their four legs.

Reproduction Platypuses lay their eggs in a burrow, while echidnas keep their eggs in a pouch on their belly. Young monotremes hatch quickly, and the mother must continue keeping them warm and dry for another four months. Like other mammals, monotremes produce milk and nurse their young. They lack well-defined nipples, however, and instead secrete milk from glands in the skin.

(a) Platypus

Ornithorhynchus anatinus

(b) Echidna

Tachyglossus aculeatus

FIGURE 35.28 Platypuses and Echidnas Lay Eggs.

The 350 known species of **marsupials** live in the Australian region and the Americas and include the familiar opossums (**FIGURE 35.29**), kangaroos, wallabies, and koala.

Feeding Marsupials are herbivores, omnivores, or carnivores. In many cases, convergent evolution has resulted in marsupials that are similar to placental species in morphology and way of life. For example, a recently extinct marsupial called the Tasmanian wolf was a long-legged hunter similar to the wolves of North America and northern Eurasia. A species of marsupial native to Australia specializes in eating ants and looks and acts much like the South American anteater, which is not a marsupial.

Movement Marsupials move by crawling, gliding, walking, running, or hopping.

Reproduction Although females have a placenta that nourishes embryos during development, the young are born after a short embryonic period and are poorly developed. They crawl from the opening of the female's reproductive tract to her nipples, located within a pouch, where they suck milk. They stay attached to their mother until they grow large enough to move independently. ✔ You should be able to indicate the origin of the placenta and viviparity—traits that are also found in eutherian mammals—on Figure 35.22.

Didelphis virginiana

FIGURE 35.29 Marsupials Give Birth after a Short Embryonic Period. Opossums are the only marsupials living in North America.

The 5100 described species of **placental mammals, or eutherians,** are distributed worldwide. They are far and away the most species-rich and morphologically diverse group of mammals.

Biologists group placental mammals into 29 lineages called orders. The six most species-rich orders are the rodents (rats, mice, squirrels; 2255 species), bats (1150 species), insectivores (hedgehogs, moles, shrews; 450 species), artiodactyls (pigs, hippos, whales, deer, sheep, cattle, horses; 329 species), carnivores (dogs, bears, cats, weasels, seals; 285 species), and primates (lemurs, monkeys, apes, humans; 414 species).

Feeding The size and structure of the teeth correlate closely with the diet of placental mammals. Herbivores have large, flat teeth for crushing leaves and other coarse plant material; predators have sharp teeth that are efficient at biting and tearing flesh. The structure of the digestive tract also correlates with the placental mammals' diet. In some plant eaters, for example, the stomach hosts unicellular organisms that digest cellulose and other complex polysaccharides.

Movement In placental mammals, the structure of the limb correlates closely with the type of movement performed. Eutherians fly, glide, run, walk, swim, burrow, or swing from trees. Hindlimbs are reduced or lost in aquatic groups such as whales and dolphins, which swim by undulating their bodies.

Reproduction Eutherians have internal fertilization and are viviparous. An extensive placenta develops from a combination of maternal and fetal tissues; and at birth, young are much better developed than in marsupials—some are able to walk or run minutes after emerging from the mother. ✔ You should be able to indicate the origin of delayed birth (extended development before birth) on Figure 35.22. All eutherians feed their offspring milk until the young have grown large enough to process solid food. A prolonged period of parental care, extending beyond the nursing stage, is common as offspring learn how to escape predators and find food on their own. Some eutherians are highly social (**FIGURE 35.30**).

Suricata suricatta

FIGURE 35.30 Eutherians Are Familiar and Diverse.

Most lizards and snakes are small reptiles with elongated bodies and scaly skin. As these animals grow, they can shed the keratinized surface layer of their skin—all in one piece in snakes, and in flakes in lizards. ✔ You should be able to indicate the origin of scaly skin on Figure 35.22. Most lizards have well-developed jointed legs, but snakes are limbless (**FIGURE 35.31**). The hypothesis that snakes evolved from limbed ancestors is supported by the presence of vestigial hip and leg bones in boas and pythons. About 8900 species of lizards and snakes are alive now.

Feeding Small lizards prey on insects. Although most of the larger lizard species are herbivores, the spectacular 3-meter-long monitor lizard from the island of Komodo is a predator that can eat deer. Snakes are carnivores; some subdue their prey by injecting poison through modified teeth called fangs. Snakes prey primarily on small mammals, amphibians, and invertebrates, which they swallow whole—usually headfirst.

Movement Lizards crawl, climb, or run on their four limbs. Some lizards, such as the water dragons, use their powerful tails for swimming. Snakes and lizards that are limbless burrow through soil, crawl over the ground, or climb trees by undulating their bodies. One genus of snakes in Asia has several adaptations that enable them to glide down from treetops.

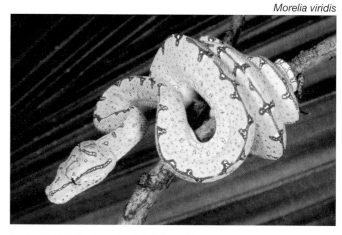

Morelia viridis

FIGURE 35.31 Snakes Are Limbless Predators.

Reproduction Although most lizards and snakes lay eggs, many are ovoviviparous; viviparity has also evolved numerous times in lizards. Most species reproduce sexually, but asexual reproduction, via the production of eggs by mitosis (parthenogenesis), is known to occur in six groups of lizards (including the Komodo dragon) and one snake lineage.

The 320 known species of turtles inhabit freshwater, marine, and terrestrial environments throughout the world. The testudines are distinguished by a shell composed of bony plates that fuse to the vertebrae and ribs (**FIGURE 35.32**). The shell functions in protection from predators—when threatened, most turtles can withdraw their head and legs into it. ✔ You should be able to indicate the origin of the turtle shell on Figure 35.22. The turtles' skulls are highly modified versions of the skulls of other reptiles. Turtles lack teeth, but their jawbone and lower skull form a bony beak. They range in size from the 2-m-long, 900-kg leatherback sea turtle to the 8-cm-long, 140-g speckled padloper tortoise. Tortoises are known for their longevity—some individuals are known to have lived well over 150 years.

Feeding Turtles are either carnivorous—feeding on whatever animals they can capture and swallow—or herbivorous. They may also scavenge dead material. Most marine turtles are carnivorous. Leatherback turtles, for example, feed primarily on jellyfish, and they are only mildly affected by the jellyfish's stinging cnidocytes (see Chapter 33). In contrast, species in the lineage of terrestrial turtles called the tortoises are plant eaters.

Testudo hermanni

FIGURE 35.32 Turtles Have a Shell Consisting of Bony Plates.

Movement Turtles swim, walk, or burrow. Some fully aquatic species have feet that are modified to function as flippers.

Reproduction All turtles are oviparous. Other than digging a nest prior to depositing eggs, parental care is lacking. Some aquatic species navigate long distances to their natal beaches with pinpoint accuracy, to lay eggs.

Chordata > Vertebrata > Reptilia > Crocodilia (Crocodiles, Alligators)

About 24 living species of crocodiles and alligators are known. Most are tropical and live in freshwater or marine environments. They have eyes located on the top of their heads, nostrils located at the top of their long snouts, and other adaptations that allow them to sit semi-submerged in water for long periods of time, breathing air and visually monitoring activity around them (**FIGURE 35.33**).

Feeding Crocodilians are predators. Like other toothed vertebrates, except mammals, their jaws are filled with conical teeth that are continually replaced as they fall out during feeding. Their usual method of killing small prey is by biting through the body wall. Large prey are usually subdued by drowning. Crocodilians eat amphibians, turtles, fish, birds, and mammals; one of the common hunting strategies of crocodiles is to leap out of the water to snatch unwary prey that have come to drink.

Movement Crocodiles and alligators walk on land. In water they swim with the aid of their large, muscular tails.

Reproduction Although crocodilians are oviparous, parental care is extensive. Eggs are laid in earth-covered nests or mounds of vegetation that are guarded by the parents. When young inside the eggs begin to vocalize, parents dig them out and often carry the newly hatched young inside their mouths to nearby water.

Alligator mississippiensis

FIGURE 35.33 Alligators Are Adapted for Aquatic Life.

Crocodilian young can hunt when newly hatched but stay near their mother for up to three years. ✔ You should be able to indicate the origin of extensive parental care in crocodilians—which also occurs in birds (and other dinosaurs)—on Figure 35.22.

Chordata > Vertebrata > Reptilia > Aves (Birds)

The fossil record provides conclusive evidence that birds descended from a lineage of dinosaurs that had **feathers.** Feathers provide insulation, are used for display, and furnish the lift, power, and steering required for flight (**FIGURE 35.34**). Birds have many other adaptations that make flight possible, including large breast muscles used to flap the wings. Bird bodies are lightweight because they have a reduced number of bones and organs and because their hollow bones are filled with air sacs linked to the lungs. Instead of teeth, birds have a horny beak. They are endotherms, meaning that they have a high metabolic rate and use the heat produced, along with the insulation provided by feathers, to maintain a constant body temperature. The 10,000 bird species alive today occupy virtually every habitat, including the open ocean. ✔ You should be able to indicate the origin of feathers, endothermy, and flight on Figure 35.22.

Feeding Plant-eating birds usually feed on nectar or seeds. Some birds are omnivores, although many are predators that capture insects, small mammals, fish, other birds, lizards, mollusks, or crustaceans. The shape of a bird's beak correlates with its diet.

Movement Almost all bird species can fly, although flightlessness has evolved repeatedly in certain groups. The size and shape of birds' wings correlate closely with the type of flying they do. Birds that glide or hover have long, thin wings; species that specialize in explosive takeoffs and short flights have short, stocky wings. Many seabirds are efficient swimmers, using their webbed feet to paddle or flapping their wings to "fly" underwater. Some ground-dwelling birds such as ostrich can run long distances at high speed.

Reproduction Birds are oviparous but provide extensive parental care. In most species, one or both parents build a nest and incubate the eggs. After the eggs hatch, parents feed offspring until they are large enough to fly and find food on their own.

Eudocimus albus

FIGURE 35.34 Birds Are Descendants of Feathered Dinosaurs.

35.4 The Primates and Hominins

Although humans occupy a tiny twig within the mammal branch on the tree of life, there has been a tremendous amount of research on human origins. This section introduces the lineage of mammals called the Primates, the fossil record of human ancestors, and data on the relationships among human populations living today.

The Primates

The lineage known as Primates consists of two main groups: prosimians and anthropoids.

- The **prosimians** ("before-monkeys") typically consist of the lemurs, found in Madagascar, and the lorises, pottos, and tarsiers of Africa and south Asia. However, recent phylogenetic analyses indicate that tarsiers are more closely related to the anthropoids than to the other prosimians, making the prosimians a paraphyletic group (**FIGURE 35.35a**). Most prosimians living today are relatively small in size, reside in trees, and are active at night.

(a) Prosimians

Lemur catta

(b) Anthropoids

Gorilla gorilla

FIGURE 35.35 There Are Two Main Lineages of Primates.

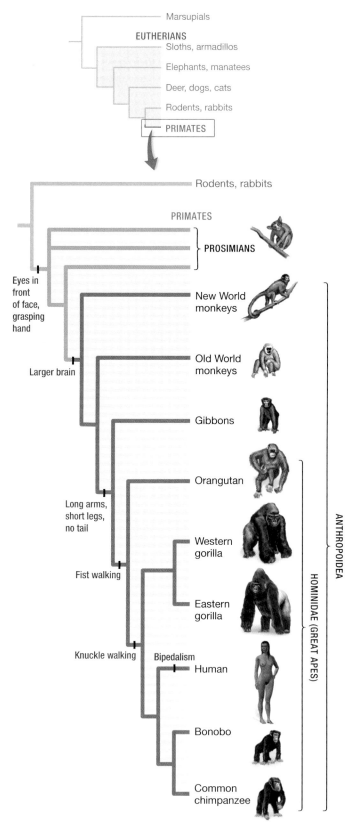

FIGURE 35.36 A Phylogeny of the Primates. Phylogenetic tree estimated from extensive DNA sequence data. According to the fossil record, humans and chimps shared a common ancestor 6 to 7 million years ago.

- The Anthropoidea or **anthropoids** ("human-like") include the New World monkeys found in Central and South America, the Old World monkeys that live in Africa and tropical regions of Asia, the gibbons of the Asian tropics, and the Hominidae, or **great apes**—orangutans, gorillas, chimpanzees, and humans (**FIGURE 35.35b**).

The phylogenetic tree in **FIGURE 35.36** shows the evolutionary relationships among these groups.

What Makes a Primate a Primate?   **Primates** tend to have hands and feet that are efficient at grasping, flattened nails instead of claws on the fingers and toes, brains that are large relative to overall body size, color vision, complex social behavior, and extensive parental care of offspring.

Along with other mammal groups that live in trees or make their living by hunting, primates have eyes located on the front of the face. Eyes that look forward provide better depth perception than do eyes on the sides of the face. The hypothesis here is that good depth perception is important in species that run or swing through trees and/or attack prey.

What Makes a Great Ape a Great Ape?   The great apes are also called **hominids.** Compared with most types of primate, the hominids are relatively large bodied and have long arms, short legs, and no tail.

Although all of the great ape species except for the orangutans live primarily on the ground, they have distinct ways of walking.

- When orangutans come to the ground, they occasionally walk with their knuckles pressed to the ground. More commonly though, they fist walk—that is, they walk with the backs of their hands pressed to the ground.

- Gorillas and chimps only knuckle walk. They also occasionally rise up on two legs—usually in the context of displaying aggression.

- Humans are the only living great ape that is fully **bipedal** ("two-footed")—meaning they walk upright on two legs. Fossil footprints in Tanzania provide direct evidence that bipedalism occurred at least 3.6 million years ago.

Bipedalism is the synapomorphy that defines the hominins. The **hominins** are a monophyletic group comprising *Homo sapiens* and more than twenty extinct, bipedal relatives.

Fossil Humans

From extensive comparisons of DNA sequence data, it is clear that humans are most closely related to the common chimpanzees and bonobos, and that our next nearest living relatives are the gorillas. According to the fossil record, the common ancestor of chimps and humans lived in Africa 6 to 7 million years ago.

The fossil record of hominins, though not nearly as complete as investigators would like, is rapidly improving. More than 20 species have been named to date, and new fossils that inform the debate over the ancestry of humans are discovered every year.

Although naming the hominin species and interpreting their characteristics remain intensely controversial, most researchers agree that the hominins can be organized into four major groups that appeared after the recently characterized *Ardipithecus ramidus*—the oldest hominin known to date. **TABLE 35.1** summarizes key data for

TABLE 35.1 **Characteristics of Selected Hominins**

Genus	Species	Location of Fossils	Estimated Average Braincase Volume (cm³)	Estimated Average Body Size (kg)	Associated with Stone Tools?
■ *Ardipithecus*	A. ramidus	Africa	325	40	no
■ *Australopithecus*	A. afarensis	Africa	460	38	no
	A. africanus	Africa	465	34	no
Paranthropus	P. boisei	Africa	480	41	no?
Early *Homo*	H. habilis	Africa	610	33	**yes**
	H. ergaster	Africa	760	64	**yes**
	H. erectus	Africa, Asia	1000	58	**yes**
■ Recent *Homo*	H. heidelbergensis	Africa, Europe	1200	71	**yes**
	H. neanderthalensis	Middle East, Europe, Asia	1430	72	**yes**
	H. floresiensis	Flores (Indonesia)	380	28	**yes**
	H. sapiens	Middle East, Europe, Asia	1480	64	**yes**

Robson, S. L., and B. Wood. 2008. Hominin life history: Reconstruction and evolution. *Journal of Anatomy* 212: 394–425.

✔ **QUANTITATIVE** How much smaller is the average body size of *H. floresiensis* compared to the average *H. sapiens*? Express your answer as a percentage. How much smaller is the braincase volume of *H. floresiensis* compared to *H. sapiens*? What do these numbers suggest?

(a) Gracile australopithecines (*Australopithecus africanus*)

Vertical posture

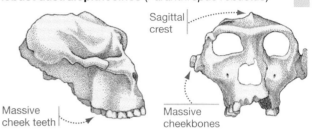

(b) Robust australopithecines (*Paranthropus robustus*)

Sagittal crest

Massive cheek teeth

Massive cheekbones

(c) Early *Homo* (*Homo habilis*)

Flatter face

Larger braincase

(d) Recent *Homo* (*Homo sapiens*)

Flattest face

Largest braincase

FIGURE 35.37 A Timeline of Hominin Evolution. The colored bars indicate the first appearance and last appearance in the fossil record of each species. (Not all named species are shown.)

DATA: Lieberman, D.E. 2009. *Nature* 459: 41–42.

✔ **QUANTITATIVE** Of the species of hominin shown, how many existed 1.8 million years ago (mya) and 100,000 years ago?

FIGURE 35.38 African Hominins Comprise Four Major Groups. The color key links these fossils with Figure 35.37 and Table 35.1.

✔ **QUESTION** The skulls are arranged as they appear in the fossil record, from most ancient (a) to most recent (d). How did the forehead and brow ridge of hominins change through time?

selected species within the four groups. **FIGURE 35.37** provides the time range of each of these species in the fossil record.

Gracile Australopithecines   The small hominins called gracile australopithecines are referred to as *gracile*, or "slender," because they had a slight build. Adult males were about 1.5 meters tall, and their estimated weight was about 36 kg.

The gracile australopithecines are placed in the genus *Australopithecus* ("southern ape"). The name was inspired by the earliest specimens, which came from South Africa. The most recent *Australopithecus* to be named, *A. sediba* in 2010, is also from South Africa.

Several lines of evidence support the hypothesis that the gracile australopithecines were bipedal. The shape of the australopithecine knee and hip are consistent with bipedal locomotion, and the hole in the back of their skulls where the spinal cord connects to the brain is oriented downward (**FIGURE 35.38a**), just as it is in our species, *Homo sapiens*. In chimps, gorillas, and other vertebrates that walk on four feet, this hole is oriented backward.

Robust Australopithecines   The robust australopithecines are grouped in the genus *Paranthropus* ("beside-human"). The name *Paranthropus* was inspired by the hypothesis that the three known species are a monophyletic group that was a side branch during human evolution—an independent lineage that went extinct.

Like the gracile australopithecines, these robust australopithecines were bipedal. They were much stockier than the gracile forms, however—about the same height but an estimated 8–10 kilograms (20 pounds) heavier on average. In addition, their skulls were much broader and more robust.

All three *Paranthropus* species had massive cheek teeth and jaws, very large cheekbones, and a sagittal crest—a flange of bone at the top of the skull (**FIGURE 35.38b**). Because muscles that work the jaw attach to the sagittal crest and cheekbones, researchers conclude that these organisms had tremendous biting power—inspiring the nickname "nutcracker man" for one species. Recent analysis of tooth enamel revealed that the biting power of *Paranthropus* was probably used to crush large quantities of tough grasses.

Early *Homo* Species in the genus *Homo* are called **humans.** As **FIGURE 35.38c** shows, species in this genus have flatter and narrower faces, smaller jaws and teeth, and larger braincases than the earlier hominins do. (The **braincase** is the portion of the skull that encloses the brain.)

The appearance of early members of the genus *Homo* in the fossil record coincides closely with the appearance of tools made of worked stone—most of which are interpreted as handheld choppers or knives. Although the fossil record does not exclude the possibility that *Paranthropus* made tools, many researchers favor the hypothesis that extensive toolmaking was a diagnostic trait of early *Homo*.

Recent *Homo* The recent species of *Homo* date from 1.2 million years ago to the present. As **FIGURE 35.38d** shows, these species have even flatter faces, smaller teeth, and larger braincases than the early *Homo* species do. The 30,000-year-old fossil in the figure, for example, is from a population of *Homo sapiens* (our species) called the **Cro-Magnons.**

The Cro-Magnons were accomplished painters and sculptors who buried their dead in carefully prepared graves. There is also evidence that another species, the **Neanderthal** people (*Homo neanderthalensis*) made art and buried their dead in a ceremonial fashion.

Perhaps the most striking recent *Homo*, though, is *H. floresiensis*. This new species was discovered in 2003 on the island of Flores in Indonesia—also home to a species of dwarfed elephants. *H. floresiensis* consisted of individuals that had braincases smaller than those of gracile australopithecines and that were about a meter tall, inspiring the nickname Hobbit. While some researchers propose that the fossils represent a population of *H. sapiens* stunted by disease, most support full species status. Fossil finds suggest that *H. floresiensis* inhabited Flores from about 100,000 to as recently as 12,000 years ago.

What Can Be Deduced from the Hominin Fossil Record? Although researchers do not have a solid understanding of the phylogenetic relationships among the hominin species, several points are clear from the available fossil data.

1. The shared, derived character that defines the hominins is bipedalism.

2. Several species from the lineage were present simultaneously during most of hominin evolution.

3. Compared with the gracile and robust australopithecines and the other great apes, species in the genus *Homo* have extremely large brains relative to their overall body size.

Why did humans evolve such gigantic brains? The leading hypothesis on this question is that early *Homo* began using symbolic spoken language and tools. The logic here is that increased language use and toolmaking triggered natural selection for the capacity to reason and communicate, which required a larger brain.

To summarize, *Homo sapiens* is the sole survivor of an adaptive radiation of hominins that began in Africa 6–7 million years ago. Why all but one species went extinct is still a mystery, though climate change and competition for food and space may have played a part.

The Out-of-Africa Hypothesis

If hominins originated in Africa and then spread around the world, where did *H. sapiens* originate?

The leading hypothesis for the evolution of *H. sapiens* is called the **out-of-Africa hypothesis.** It contends that *H. sapiens* evolved its distinctive traits in Africa and then dispersed throughout the world. Further, it claims that *H. sapiens* evolved independently of the earlier European and Asian species of *Homo*—meaning there was no interbreeding between *H. sapiens* and Neanderthals, *H. erectus*, or *H. floresiensis*. Do the data support this hypothesis?

Fossil Evidence The first *H. sapiens* fossils appear in East African rocks that date to about 195,000 years ago. For some 130,000 years thereafter, the fossil record indicates that our species occupied Africa while *H. neanderthalensis* resided in Europe and the Middle East. Some evidence suggests that *H. erectus* may still have been present in Asia at that time.

In rocks dated between 60,000 and 30,000 years ago, however, *H. sapiens* fossils are found throughout Europe, Asia, Africa, and Australia. *H. erectus* had disappeared by this time, and *H. neanderthalensis* went extinct after coexisting with *H. sapiens* in Europe for thousands of years.

Taken together, the fossil evidence provides strong support for the African origin of *H. sapiens* and subsequent migration, but it cannot resolve the question of whether *H. sapiens* ever interbred with Neanderthals during their coexistence.

Molecular Evidence Some researchers have taken a molecular approach to understanding the history of our species, providing several major insights.

For example, Luigi Cavali-Sforza and others have created phylogenetic trees based on genetic comparisons of diverse human populations living today. These data agree with the pattern in the fossil record. As **FIGURE 35.39** (see page 708) shows, the first lineages to branch off led to descendant populations that live in Africa today. Later branches gave rise to lineages in central Asia, Europe, East Asia, Polynesia, and the Americas. Based on this observation, it is logical to infer that the ancestral population of modern humans lived in Africa.

Did *H. sapiens* leave Africa in one wave? Sequence data from modern Aboriginal Australians support the hypothesis that

FIGURE 35.39 Phylogeny of Human Populations Living Today.
The phylogeny was estimated from DNA sequence data. As predicted by the out-of-Africa hypothesis, the lineages with the highest genetic diversity and deepest history occur in Africa.

the human migration out of Africa occurred in more than one wave—the first wave moved east and south to Australia, and the second wave moved to Europe and the rest of mainland Asia. While the exact migration routes of *H. sapiens* are hotly debated, **FIGURE 35.40** provides a general summary.

What about the question of interbreeding? There are now two lines of evidence that *H. sapiens* interbred with other species as they migrated out of Africa. First, Svante Pääbo and his colleagues compared gene sequences of extinct Neanderthals and diverse modern humans. They made the remarkable observation that 1 to 4 percent of the genome of indigenous Europeans and Asians—but not Africans—is derived from Neanderthals. These data suggest that *H. sapiens* and Neanderthals interbred in the Middle East after *H. sapiens* left Africa, but before migrating through Europe and Asia.

Second, sequence data from modern Asians and Aboriginal Australians suggest that modern humans interbred with a recently discovered descendant population of *H. erectus*, called the Denisovans, in Asia during their first migration out of Africa. As a result, Aboriginal Australians and their island neighbors—descendants of the first migration—inherited about 5 percent of their nuclear DNA from Denisovans.

In summary, the Neanderthal and Denisovan data reject the strict interpretation of the out-of-Africa hypothesis since a small amount of gene flow occurred between *H. sapiens* and Neanderthals as well as between *H. sapiens* and the Denisovans. However, other fossil and molecular evidence supports the general concept of the out-of-Africa hypothesis. That is, the overall data support an out-of-Africa hypothesis with "leakage."

The study of human evolution is an exciting and dynamic field. And humans continue to evolve.

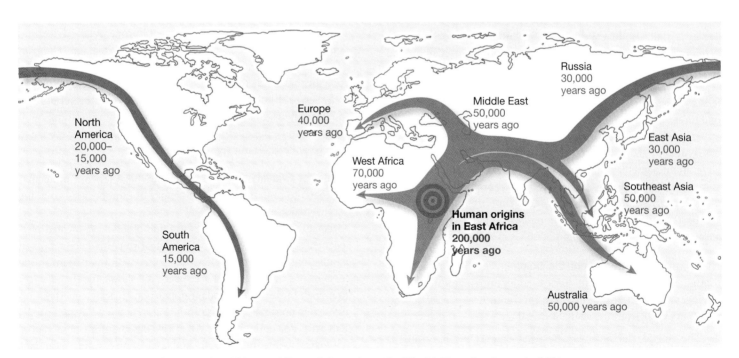

FIGURE 35.40 *Homo sapiens* Originated in Africa and Spread throughout the World. The migration out of Africa probably occurred in two waves—first east and south to Australia, and later to Europe and the rest of Asia and the Americas.

If you understand . . .

35.1 What Is an Echinoderm?

- Echinoderms are among the most important predators and herbivores in marine environments. They include the sea stars, sea urchins, sea cucumbers, and feather stars.

- Echinoderm larvae are bilaterally symmetric but most undergo a metamorphosis into radially symmetric adults.

- Echinoderms have a calcium carbonate endoskeleton.

- The echinoderm water vascular system is composed of fluid-filled tubes and chambers called tube feet. Tube feet may be used for movement, feeding, or respiration.

✔ You should be able to predict what happens when (1) sea stars that prey on mussels and clams are excluded from experimental plots, and (2) sea urchins that graze on kelp are excluded from experimental plots.

35.2 What Is a Chordate?

- Chordates are distinguished by the presence of pharyngeal gill slits, a dorsal hollow nerve cord, a notochord, and a muscular tail that extends past the anus.

- Chordates include cephalochordates, urochordates, and vertebrates.

✔ You should be able to make a simple cartoon of a generic chordate showing its defining features.

35.3 What Is a Vertebrate?

- Vertebrates are distinguished by the presence of a cranium and vertebrae. In some groups of vertebrates, the body plan features an extensive endoskeleton composed of bone.

- The origin of jaws in ray-finned fishes enabled these fishes and their descendants to bite food and process it with teeth.

- The origin of limbs in tetrapods enabled walking and running on land, and flying. Tetrapods include the amphibians, mammals, and reptiles (which include birds).

- The evolution of the amniotic egg allowed tetrapods to lay large eggs on land.

- Parental care was an important adaptation in some groups of ray-finned fishes and tetrapods—particularly mammals.

- Vertebrates are the most important large-bodied predators and herbivores in both marine and terrestrial environments.

✔ You should be able to explain why the tetrapod limb could have evolved in an animal that was aquatic.

35.4 The Primates and Hominins

- Humans are a tiny twig on the tree of life. Chimpanzees and humans diverged from a common ancestor that lived in Africa 6–7 million years ago.

- The fossil record of the past 4.5 million years contains many distinct species of hominins. *Homo sapiens* is the sole surviving representative of an adaptive radiation.

- For most of human evolution, several species were present in Africa or Europe at the same time. Some lineages went extinct without leaving descendant populations.

- The phylogeny of living humans, based on comparisons of DNA sequences, agrees with evidence in the fossil record that *H. sapiens* originated in Africa and later spread throughout Asia, Europe, and the Americas.

✔ You should be able to describe evidence supporting the hypothesis that *Homo sapiens* originated in Africa.

MB MasteringBiology

1. **MasteringBiology Assignments**

   Tutorials and Activities Allometric Growth; Characteristics of Chordates; Chordates; Deuterostome Diversity; Human Evolution; Primate Diversity

   Questions Reading Quizzes, Blue-Thread Questions, Test Bank

2. **eText** Read your book online, search, take notes, highlight text, and more.

3. **The Study Area** Practice Test, Cumulative Test, BioFlix® 3-D Animations, Videos, Activities, Audio Glossary, Word Study Tools, Art

You should be able to . . .

1. True or false? Both echinoderms and vertebrates have endoskeletons.

2. What trait(s) define(s) the vertebrates?
   a. vertebrae and a cranium
   b. jaws and a spinal cord
   c. endoskeleton constructed of bone
   d. endoskeleton constructed of reinforced cartilage

3. Why are the pharyngeal jaws of many ray-finned fishes important?
   a. They allow the main jaw to be protrusible (extendable).
   b. They make it possible for individuals to suck food toward their mouths.
   c. They give rise to teeth that are found on the main jawbones.
   d. They help process food.

4. The two major lineages that make up the living Amniota are reptiles and _____.

5. In chordates, what is the relationship between the dorsal hollow nerve cord and the spinal cord?
   a. The dorsal hollow nerve cord is found only in embryos; the spinal cord is found in adults.
   b. The dorsal hollow nerve cord stiffens the body and functions in movement; the spinal cord relays electrical signals.
   c. The dorsal hollow nerve cord is found in early fossils of chordates; living species have a spinal cord.
   d. The spinal cord is a type of dorsal hollow nerve cord.

6. Most species of hominins are known only from Africa. Which species have been found in other parts of the world as well?
   a. early *Homo*—*H. habilis* and *H. ergaster*
   b. *H. erectus*, *H. neanderthalensis*, and *H. floresiensis*
   c. gracile australopithecines
   d. robust australopithecines

7. Mammals and birds are both endothermic. Which statement is true?
   a. They both inherited this trait from a common ancestor.
   b. Birds are more closely related to mammals than to reptiles.
   c. Endothermy arose independently in these two lineages.
   d. All reptiles are ectothermic.

8. Explain why each of the four key synapomorphies that distinguish chordates is important in feeding or movement. Why is it possible to say that a certain animal is an "invertebrate chordate"?

9. The cells that make up jaws and gill arches are derived from the same population of embryonic cells. Why does this observation support the hypothesis that jaws evolved from gill arches in fish?

10. Explain how the evolution of the placenta and lactation in mammals improved the probability that mammalian offspring would survive, compared to offspring of species without parental care.

11. The text claims that *Homo sapiens* is the sole survivor of an adaptive radiation that took place over the past 6–7 million years. Do you agree with this statement? Why or why not?

12. Describe a phylogeny of modern humans that would reject the out-of-Africa hypothesis.

13. There is some evidence that pharyngeal gill slits occur in certain species of echinoderms that appear early in the fossil record. If confirmed, what do these data suggest?
    a. Echinoderms are chordates.
    b. Pharyngeal gill slits were present in the earliest echinoderms and lost later.
    c. Some lineages of echinoderms are more closely related to chordates than others.
    d. Pharyngeal gill slits should not be used as a trait in phylogenetic analysis.

14. Animals in the species *Xenoturbella* have extremely simple, wormlike bodies. For example, they have a blind gut (only one opening) and no brain. Propose a hypothesis to explain how this simple body plan evolved in this lineage.

15. In arthropods such as crustaceans and insects, changes in the expression patterns of tool-kit genes enabled a dramatic diversification of their segmented appendages and bodies. Design an experiment to test whether this concept applies to vertebrates.

16. QUANTITATIVE Genetic diversity in living human populations is highest in Africa and decreases as a function of distance from Africa. Draw a graph to show this result. Add a label where you would expect to find a data point for the Yanomamö tribe of South America. Why is it important to use indigenous people for this study?

36 Viruses

In this chapter you will learn that

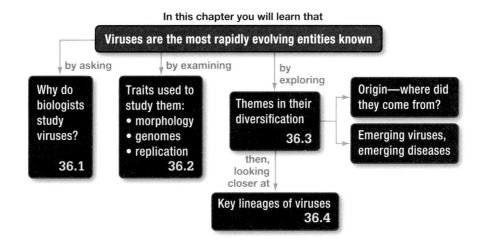

Viruses are the most rapidly evolving entities known

by asking → **Why do biologists study viruses?** 36.1

by examining → **Traits used to study them:**
• morphology
• genomes
• replication
36.2

by exploring → **Themes in their diversification** 36.3

→ **Origin—where did they come from?**

→ **Emerging viruses, emerging diseases**

then, looking closer at → **Key lineages of viruses** 36.4

Photomicrograph created by treating seawater with a fluorescing compound that binds to nucleic acids. The smallest, most abundant dots are viruses. The larger, numerous spots are bacteria and archaea. The largest splotches are protists.

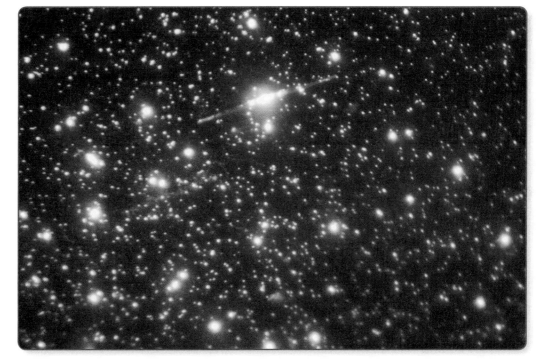

f you have ever been laid low by a high fever, cough, scratchy throat, body ache, and debilitating lack of energy, you may have wondered what hit you. What hit you was probably a **virus:** an obligate, intracellular parasite.

Why obligate? Viral replication is *completely* dependent on host cells. Why intracellular? Viruses must enter a host cell for replication to occur. Why parasite? Viruses reproduce at the expense of their host cells.

Viruses can also be defined by what they are not:

• They are not cells and are not made up of cells, so they are not considered organisms.

• They cannot manufacture their own ATP or amino acids or nucleotides, and they cannot produce proteins on their own.

Viruses enter a **host cell,** take over its biosynthetic machinery, and use that machinery to manufacture a new generation of viruses. Outside of host cells, viruses simply exist.

✔ When you see this checkmark, stop and test yourself. Answers are available in Appendix A.

Characteristics	Viruses	Organisms
Hereditary material	DNA or RNA; can be single stranded or double stranded	DNA; always double stranded
Plasma membrane present?	No	Yes
Can carry out transcription independently?	No—even if a viral polymerase is present, transcription of viral genomes requires use of nucleotides provided by host cell	Yes
Can carry out translation independently?	No	Yes
Metabolic capabilities	Virtually none	Extensive—synthesis of ATP, reduced carbon compounds, vitamins, lipids, nucleic acids, etc.

When you have the flu, influenza viruses enter the cells that line your respiratory tract and use the machinery inside to make copies of themselves. Every time you cough or sneeze, you eject millions or billions of their offspring into the environment. If one of those infectious particles is lucky enough to be breathed in by another person, it may enter his or her respiratory tract cells and start a new infection.

Viruses are not organisms. Most biologists would argue that viruses are not alive, because they are dependent on their host cell to satisfy the five attributes of life (see Chapter 1). Yet viruses have a genome, they are superbly adapted to exploit the metabolic capabilities of their host cells, and they evolve. **TABLE 36.1** summarizes some characteristics of viruses and how they differ from organisms.

The diversity and abundance of viruses almost defy description. Each type of virus infects a specific unicellular species or cell type in a multicellular species, and nearly all organisms examined thus far are parasitized by at least one kind of virus. The surface waters of the world's oceans teem with bacteria and archaea, yet viruses outnumber them in this habitat by a factor of 10 to 1. If you leaned over a boat and filled a wine bottle with seawater, it would contain about 10 billion virus particles—close to one and a half times the world's population of humans.

36.1 Why Do Biologists Study Viruses?

Any study of life's diversity would be incomplete unless it included a look at the acellular parasites that contribute to that diversity. Viruses have directly participated in organismal diversity

by introducing foreign genes into cellular genomes. They can pick up cellular genes and shuttle them from one organism to another. In this way, viruses can promote **lateral gene transfer** (see Chapter 21).

In addition, viruses also contribute their own genetic material to organisms. Researchers estimate that 5–8 percent of the human genome consists of remnants of viral genomes from past infections. Some of these viral genes have even evolved to be part of what makes us human. For example, a protein that is encoded by an abandoned viral gene is necessary for proper development and function of the human placenta.

Viruses are also important from a practical standpoint. To health-care workers, agronomists, and foresters, these parasites are a persistent—and sometimes catastrophic—source of misery and economic loss. Much of the research on viruses is motivated by the desire to minimize the damage they can cause. In the human body, virtually every system, tissue, and cell can be infected by at least one kind of virus, and each of these viruses often infects more than one site (**FIGURE 36.1**).

Recent Viral Epidemics in Humans

Physicians and researchers use the term **epidemic** (literally, "upon-people") to describe a disease that rapidly affects a large number of individuals over a widening area.

Viruses have caused the most devastating epidemics in recent human history. During the eighteenth and nineteenth centuries, it was not unusual for Native American tribes to lose 90 percent of their members over the course of a few years to measles, smallpox, and other viral diseases spread by contact with European settlers. To appreciate the impact of these epidemics, think of 10 close friends and relatives—then remove nine.

An epidemic that is worldwide in scope is called a **pandemic.** The influenza outbreak of 1918–1919, called Spanish flu, qualifies as the most devastating pandemic recorded to date. The strain of influenza virus that emerged in 1918 infected people worldwide and was particularly **virulent**—meaning it tended to cause severe disease. The viral outbreak occurred just as World War I was drawing to a close and killed far more people in eight months than did the five-year conflict itself.

FIGURE 36.2 illustrates the impact of this pandemic in relation to the life expectancy in the United States during the twentieth century. Worldwide, the Spanish flu is thought to have killed up to 50 million people. An estimated 20 million died in the first 8 months of the pandemic.

Current Viral Pandemics in Humans: AIDS

In terms of the total number of people affected, the measles and smallpox epidemics and the 1918 influenza pandemic are almost certain to be surpassed by the incidence of **acquired immune deficiency syndrome (AIDS).** AIDS is an affliction caused by the **human immunodeficiency virus (HIV).**

HIV is now the one of the most intensively studied of all viruses. Since the early 1980s, governments and private corporations from around the world have spent hundreds of millions of dollars on

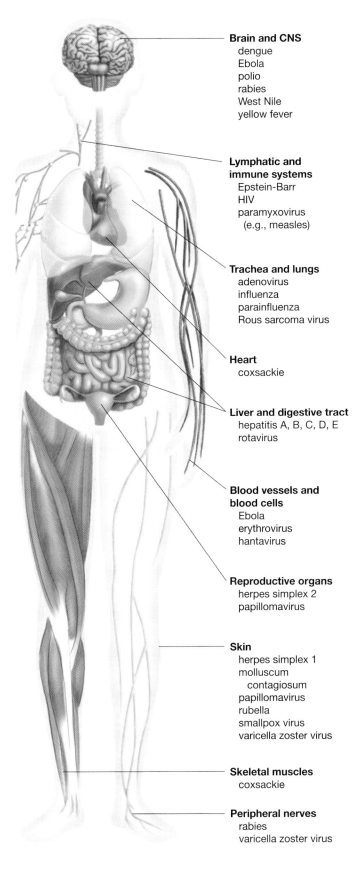

Brain and CNS
dengue
Ebola
polio
rabies
West Nile
yellow fever

Lymphatic and immune systems
Epstein-Barr
HIV
paramyxovirus
  (e.g., measles)

Trachea and lungs
adenovirus
influenza
parainfluenza
Rous sarcoma virus

Heart
coxsackie

Liver and digestive tract
hepatitis A, B, C, D, E
rotavirus

Blood vessels and blood cells
Ebola
erythrovirus
hantavirus

Reproductive organs
herpes simplex 2
papillomavirus

Skin
herpes simplex 1
molluscum
  contagiosum
papillomavirus
rubella
smallpox virus
varicella zoster virus

Skeletal muscles
coxsackie

Peripheral nerves
rabies
varicella zoster virus

FIGURE 36.1 Human Organs and Systems That Are Parasitized by Viruses. Viruses shown in this figure may infect more than one tissue, but for simplicity, usually only one tissue is shown.

(a) Deadly impact of the 1918 influenza pandemic

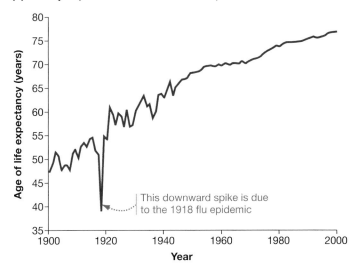

This downward spike is due to the 1918 flu epidemic

(b) Emergency hospital at the start of the 1918 pandemic

FIGURE 36.2 Life Expectancy of Humans over the 20th Century in the United States. (a) This graph illustrates the dramatic effect of the 1918 influenza pandemic in an otherwise promising trend.
(b) The first recorded cases of the 1918 Spanish flu were reported at this U.S. Army emergency hospital (Camp Funston, Kansas).

DATA: Arias, E. 2010. United States life tables, 2006. *National Vital Statistics Reports,* 58 (21): 1–40, Table 10. Hyattsville, MD: National Center for Health Statistics.

HIV research. Given this virus's current and projected impact on human populations around the globe, the investment is justified.

How Does HIV Cause Disease?   Like other viruses, HIV parasitizes specific types of cells—most notably, the helper T cells of the **immune system,** which is the body's defense system against disease. Helper T cells are crucial to the immune system's response to invading pathogens (see Chapter 51).

If an HIV particle succeeds in infecting a helper T cell and reproduces inside, the cell will die as hundreds of new progeny particles are released and infect more cells. Although the body continually replaces helper T cells, the number produced does not keep pace with the number being destroyed by HIV. As

FIGURE 36.3 T-Cell Counts Decline during an HIV Infection.
Graph of changes in the number of helper T cells that are present in the bloodstream over time, based on data from a typical patient infected with HIV. The acute phase may be associated with symptoms such as fever. Few or no disease symptoms occur in the chronic phase. AIDS typically occurs when T-cell counts dip below 200 per mm³ of blood.

DATA: Pantaleo, G., and Fauci, A. S. 1996. *Annual Review of Microbiology* 50: 825-854.

a result, the total number of helper T cells in the bloodstream gradually declines as an HIV infection proceeds (**FIGURE 36.3**).

As the T-cell count drops, the immune system's responses to invading pathogens become less and less effective. Eventually, too few helper T cells are left to fight off pathogens efficiently, allowing bacteria, fungi, protists, or other viruses to multiply unchecked. In almost all cases, one or more of these "opportunistic" infections proves fatal. HIV kills people indirectly—by making them susceptible to diseases that normally do not arise in those with functioning immune systems.

What Is the Scope of the AIDS Pandemic? Researchers with the United Nations AIDS program estimate that AIDS has already killed almost 30 million people worldwide. HIV infection rates have been highest in east and central Africa, where one of the greatest public health crises in history is now occurring. In Botswana, for example, blood-testing programs have confirmed that over 20 percent of the population is HIV positive. Although there may be a lag of as much as 8 to 12 years between the initial infection and the onset of illness, virtually all people who become infected with the virus will die of AIDS.

Currently, the UN estimates the total number of HIV-infected people worldwide at about 34 million. An additional 2.7 million people are infected each year. Most infectious diseases afflict the very young and the very old, but because HIV is primarily a sexually transmitted disease, young adults are most likely to contract the virus and die.

People who become infected with HIV in their late teens or twenties die of AIDS in their twenties or thirties. Tens of millions of people are being lost in the prime of their lives. Physicians, politicians, educators, and aid workers all use the same word to describe the epidemic's impact: staggering.

36.2 How Do Biologists Study Viruses?

Many researchers who study viruses focus on two goals: **(1)** developing vaccines that help hosts fight off disease if they become infected and **(2)** developing antiviral drugs that prevent a virus from replicating efficiently inside the host. Both types of research begin with attempts to isolate the virus in question to learn more about its structure and mode of replication.

Isolating viruses takes researchers into the realm of nanobiology, in which structures are measured in billionths of a meter. (One nanometer, abbreviated nm, is 10^{-9} meter.) Viruses range from about 20 to 300 nm in diameter. They are dwarfed by eukaryotic cells and even by most bacterial cells (**FIGURE 36.4**). Millions of viruses can fit on the period at the end of this sentence.

If virus-infected cells can be grown in culture or harvested from a host individual, researchers can usually isolate the virus by passing solutions of infected cultures or patient samples (e.g., blood, saliva, feces) through a filter. The filters used to study viruses have pores that are large enough for viruses to pass through but are too small to admit cells. If exposing susceptible host cells to the filtrate results in infection, the hypothesis that a virus caused the disease is supported.

In this way, researchers can isolate a virus and confirm that it is the causative agent of infection. These steps are inspired by Koch's postulates, which established the criteria for linking a specific infectious agent with a specific disease (see Chapter 29).

Once biologists have isolated a virus, how do they study and characterize it? The answer is threefold—by analyzing **(1)** the structure of the extracellular infectious particle, referred to as the **virion; (2)** the nature of the genetic material that is transmitted

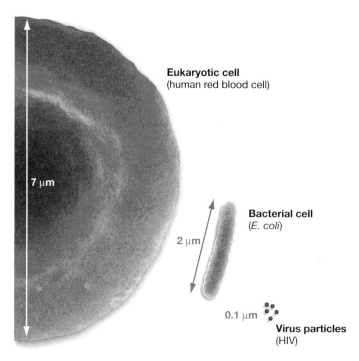

FIGURE 36.4 Viruses Are Tiny.

by virions from one host to another; and (3) variations in how viruses replicate. Let's begin with morphological traits.

Analyzing Morphological Traits

To see the virion of a particular virus, researchers usually rely on transmission electron microscopy (see **BioSkills 11** in Appendix B). Only the very largest viruses, such as the smallpox virus, are visible with a light microscope. Electron microscopy has revealed that viruses come in a wide variety of shapes, and many viruses can be identified by shape alone.

In terms of overall structure, most viruses fall into two general categories: (1) those that are enclosed by just a shell of protein called a **capsid** and (2) those enclosed by both a capsid and a membrane-like **envelope.**

Most viruses produce virions with capsid shapes that are either helical or icosahedral (an icosahedron is a polyhedron with 20 triangular faces) (**FIGURE 36.5a** and **b**). Some viruses, like bacteriophage T4 and the smallpox virus, however, have more complex capsid shapes (**FIGURE 36.5c** and **d**). A virion's capsid serves two functions: it protects the genome while outside the host and is also able to release the genome when infecting a new cell.

Nonenveloped viruses, often called "naked" viruses, use only the capsid shell to protect their genetic material. The naked icosahedral virion illustrated in Figure 36.5b is an adenovirus. You undoubtedly have adenoviruses on your tonsils or in other parts of your upper respiratory passages right now.

Enveloped viruses also have genetic material inside a capsid, or bound to capsid proteins, but the capsid is further enclosed in one or more membranes like the smallpox virus shown in Figure 36.5d. The envelope consists of viral proteins embedded in a phospholipid bilayer derived from a membrane found in a host cell—specifically, the host cell in which the virion was manufactured. Later sections of this chapter will detail how viruses obtain their envelope from an infected host cell.

Analyzing the Genetic Material

In addition to morphology, viruses can be categorized based on the nature of their genome. This is not true for cells. DNA is the hereditary material in all cells, and information flows from DNA to mRNA to proteins (Chapter 16). Although all cells follow this pattern, which is called the central dogma of molecular biology, many viruses break it.

This conclusion traces back to work done in the 1950s, when Heinz Fraenkel-Conrat and colleagues were able to separate the protein and nucleic acid components of a plant virus known as the tobacco mosaic virus, or TMV. Surprisingly, the nucleic acid

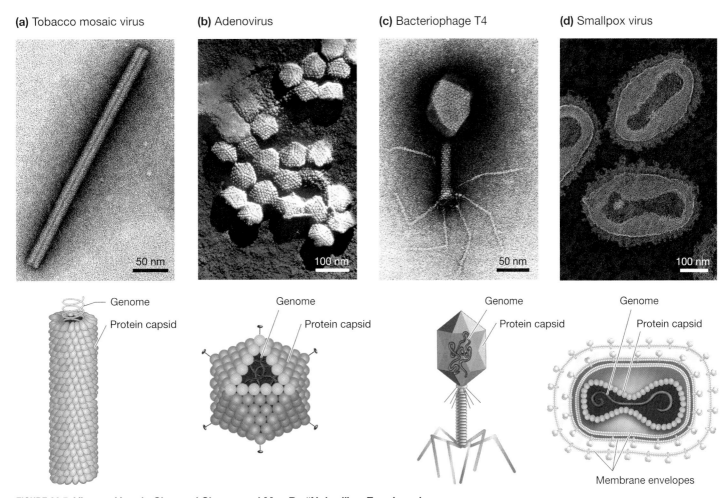

(a) Tobacco mosaic virus   **(b)** Adenovirus   **(c)** Bacteriophage T4   **(d)** Smallpox virus

50 nm   100 nm   50 nm   100 nm

Genome · Protein capsid   Genome · Protein capsid   Genome · Protein capsid   Genome · Protein capsid · Membrane envelopes

FIGURE 36.5 Viruses Vary in Size and Shape, and May Be "Naked" or Enveloped.

portion of this virus consisted of RNA, and the RNA of TMV, by itself, could infect plant tissues and cause disease. This was a confusing result because it showed that the RNA—not DNA—functions as TMV's genetic material.

Subsequent research revealed an amazing diversity of viral genome types. In some groups of viruses, such as the agents that cause measles and flu, the genome consists of RNA. In others, such as the viruses that cause herpes and smallpox, the genome is DNA. Viral genomes may be linear or circular and may consist of a single molecule or be broken up into several different segments.

Further, the RNA and DNA genomes of viruses can be either single stranded or double stranded. The single-stranded RNA genomes can also be classified as "positive sense" or "negative sense" or "ambisense."

- In a **positive-sense** RNA virus, the genome contains the same sequences as the mRNA required to produce viral proteins.

- In a **negative-sense** RNA virus, the base sequences in the genome are complementary to those in viral mRNAs.

- In an **ambisense** RNA virus, the genome has at least one strand that contains two regions: one is positive sense and the other negative sense.

The nature of viruses has been understood only since the 1940s, but they have been the focus of intense research ever since. Although likely millions of types of virus exist, they all appear to infect their host cells in one of two general ways: via **replicative growth,** which produces the next generation of virions and often kills the host cell, or in a dormant manner that suspends production of virions and allows the virus to coexist with the host for a period of time. Let's look first at the more immediately deadly mode.

Analyzing the Phases of Replicative Growth

Six phases are common to replicative growth in virtually all viruses: (**1**) attachment to a host cell and entry into the cytosol;

(**2**) transcription of the viral genome and production of viral proteins; (**3**) replication of the viral genome; (**4**) assembly of a new generation of virions; (**5**) exit from the infected cell; and (**6**) transmission to a new host.

FIGURE 36.6 shows the replicative growth of a bacteriophage. A **bacteriophage** (literally, "bacteria-eater") is a virus that infects bacterial cells. This cycle is referred to as the **lytic cycle,** since it ends with lysis (destruction) of the cell. One thing should stand out as you examine this process: in a single replicative cycle, one virus particle can produce many progeny.

FIGURE 36.7 shows how a bacteriophage's tremendous capacity for replication results in nonlinear growth that proceeds in a stepwise manner. This growth curve is very different from what is observed in the host cells, where each cell produces only two progeny in each generation.

Each type of virus has a particular way of entering a host cell and completing the subsequent phases of the cycle. Let's take a closer look at each phase.

How Do Viruses Enter a Cell? The replicative cycle of a virus begins when a free virion enters a target cell. This is no simple task. All cells are protected by a plasma membrane, and many also have a cell wall. How do viruses breach these defenses, insert themselves into the cytosol inside, and begin an infection?

Most plant viruses are inserted directly into the host-cell cytosol via abrasions or the mouthparts of sucking or biting insects. In contrast, viruses that infect bacterial or animal cells must first attach to a specific molecule on the cell wall or plasma membrane.

After a bacteriophage attaches to its host, it uses an enzyme called lysozyme to degrade part of the cell wall and expose the plasma membrane. The genome of the bacteriophage is then transferred to the cytosol in a process referred to as uncoating, which varies depending on the structure of the virus. The T4 bacteriophage shown in Figure 36.6, for example, uncoats its

PROCESS: REPLICATIVE GROWTH (LYTIC CYCLE)

Virion

Host-cell genome

1. Viral genome enters host cell.

DNA   mRNA   Protein

2. Viral genome is transcribed; viral proteins are produced.

DNA

Protein

3. Viral genome is replicated.

6. Free virions in tissue or environment are transmitted to new host.

5. Particles exit to exterior.

4. Particles assemble inside host.

FIGURE 36.6 The Viral Replicative Cycle. Many viruses follow the same general cycle as this bacteriophage model.

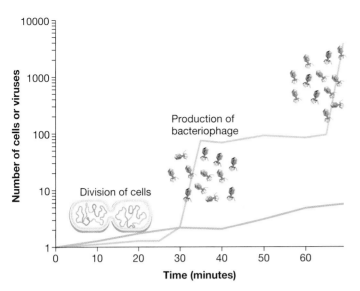

FIGURE 36.7 **Growth Patterns Differ between Viruses and Their Host.** Cells reproduce in an exponential manner, as indicated by a linear pattern when plotted on a logarithmic scale. The production of virions is not exponential; instead, reproduction of bacteriophage like T4 exhibits a stepwise pattern.

DATA: Courtesy of the Undergraduate Biotechnology Laboratory, California Polytechnic State University, San Luis Obispo.

✔ **QUANTITATIVE** Based on the information and trends in this graph, predict the number of extracellular virions that would be produced after 90 minutes if all of those released by 40 minutes were allowed to infect new host cells. Compare this to the number of cells you would predict to be present after 90 minutes.

genome by injecting it through a hollow needle that is stabbed into the bacterial membrane.

Viruses that attack animal cells must first attach to one or more specific molecules in the host cell's plasma membrane. These molecules, often referred to as virus receptors, are typically either membrane proteins or the carbohydrates attached to glycoproteins or glycolipids.

To appreciate how investigators identify virus receptors, consider research on HIV. In 1981—right at the start of the AIDS epidemic—biomedical researchers realized that people with AIDS had few or no T cells possessing a particular membrane protein called **CD4.** These cells, symbolized as $CD4^+$, are called helper T cells because of their key supportive role in the immune response (see Chapter 51).

Two research groups hypothesized that CD4 serves as the receptor for HIV attachment. To test this hypothesis, the researchers systematically blocked different membrane proteins on helper T cells and determined which membrane proteins are involved in HIV attachment. They used **antibodies** to specifically bind to host membrane proteins, including several that bound to CD4 (see **BioSkills 9** in Appendix B to review the activity of antibodies).

The key assumption in their experiments was that antibodies directed against the virus receptor would prevent HIV from entering and infecting the cell. After completing their work, both research teams reached exactly the same result: Antibodies to CD4 inhibited HIV from infecting host cells (**FIGURE 36.8**).

RESEARCH

QUESTION: Does CD4 protein function as the receptor HIV uses to enter host cells?

HYPOTHESIS: CD4 is the membrane protein HIV uses to enter cells.

NULL HYPOTHESIS: CD4 is not the membrane protein HIV uses to enter cells.

EXPERIMENTAL SETUP:

1. Take 160 identical samples of helper T cells from a large population of T cells growing in culture.

×160

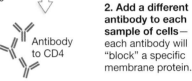

Antibody to protein other than CD4

Antibody to CD4

2. Add a different antibody to each sample of cells—each antibody will "block" a specific membrane protein.

Add HIV   Add HIV

3. Add a constant number of HIV virions to all samples. Incubate cultures under conditions optimal for virus entry.

PREDICTION: HIV will not infect cells with antibody to CD4 but will infect other cells.

PREDICTION OF NULL HYPOTHESIS: HIV will infect cells with antibody to CD4.

RESULTS:

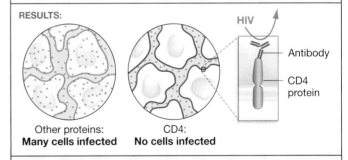

HIV

Antibody

CD4 protein

Other proteins: **Many cells infected**   CD4: **No cells infected**

CONCLUSION: HIV uses CD4 as the receptor to enter helper T cells. Thus, only cells with free, unbound CD4 on their surface can be infected by HIV.

FIGURE 36.8 **Experiments Confirmed that CD4 Is the Receptor Used by HIV to Enter Host Cells.** In this experiment, the antibodies added to each culture bound to a specific protein found on the surface of helper T cells. Antibody binding blocked the membrane protein, so the protein could not be used by HIV to gain entry to the cells.

SOURCE: Dalgleish, A. G., P. C. Beverley, P. R. Clapham, et al. 1984. The CD4 (T4) antigen is an essential component of the receptor for the AIDS retrovirus. *Nature* 312: 763–767. Also Klatzmann, D., E. Champagne, S. Chamaret, et al. 1984. T-lymphocyte T4 molecule behaves as the receptor for human retrovirus LAV. *Nature* 312: 767–768.

✔ **QUESTION** Does this experiment show that CD4 is the only membrane protein required for HIV entry? Explain why or why not.

Although CD4 is necessary for HIV attachment, uncoating the viral genome will not occur unless the virion also binds to a second membrane protein, called a **co-receptor.** In most individuals, membrane proteins called CXCR4 and CCR5 function as co-receptors. Once the virion binds to both CD4 and a co-receptor, the lipid bilayers of the virion's envelope and the plasma membrane of the T cell fuse (**FIGURE 36.9a**). When fusion occurs, HIV has breached the cell boundary. The viral capsid then enters the cytosol and disassembles to release the genomic RNAs, and infection proceeds.

Another common uncoating process is shown in **FIGURE 36.9b**, where the virion is first internalized via endocytosis (see Chapter 7). After a virion, in this case influenza, attaches to a component on the cell surface, it is pulled into the cell in the form of a vesicle called an endosome. When the endosome acidifies—a normal part of endocytosis—the structure of the viral attachment proteins change in a way that promotes fusion of the viral envelope and the endosomal membrane.

Endocytosis is also commonly used by naked viruses to enter animal cells. As with enveloped viruses, acidification of the

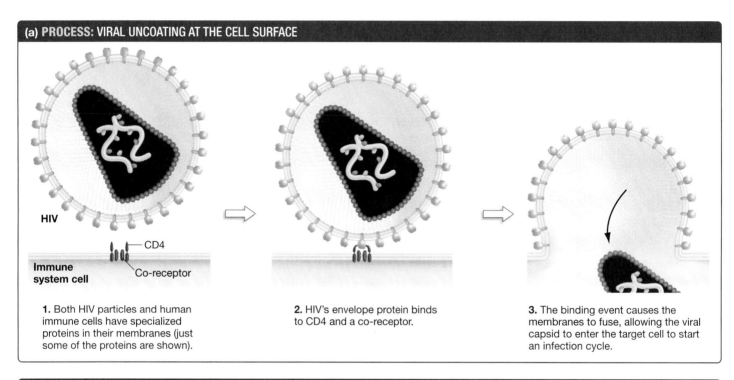

(a) PROCESS: VIRAL UNCOATING AT THE CELL SURFACE

1. Both HIV particles and human immune cells have specialized proteins in their membranes (just some of the proteins are shown).

2. HIV's envelope protein binds to CD4 and a co-receptor.

3. The binding event causes the membranes to fuse, allowing the viral capsid to enter the target cell to start an infection cycle.

(b) PROCESS: VIRAL UNCOATING IN THE ENDOSOME

1. Specialized envelope proteins on influenza attach to a common carbohydrate found on cells, inducing virion uptake via endocytosis.

2. Protons are pumped into the endosome, causing the envelope proteins to change shape.

3. The envelope and endosome membranes are brought together and eventually fuse, releasing the virus genome into the cytoplasm.

FIGURE 36.9 Enveloped Viruses Bind to Host Membranes and Uncoat via Fusion.

endosome serves as a signal for uncoating the genetic material. In this case, low pH alters the structure of the capsid proteins, which disrupts the endosomal membrane so the viral genetic material may enter the cytosol.

The discovery of virus receptors and requirements for uncoating has inspired a search for compounds that block these early events of the replicative cycle. As a result, drugs have been designed to block attachment and uncoating of HIV by targeting the HIV envelope protein and host CCR5 co-receptor. Drugs that interfere with either viral infection or replication of viruses are called **antivirals.**

How Do Viruses Produce Proteins? Production of viral proteins begins soon after the virus uncoats its genome. Depending on the virus, transcription of the viral genome may be accomplished by either host or viral RNA polymerases. In all viruses, however, translation of viral transcripts is entirely dependent on the host. Viruses lack the ribosomes, amino acids, ATP, and most of the other biosynthetic machinery required for translating their own mRNAs into proteins.

Viral proteins are produced and processed in one of two ways, depending on whether the proteins end up in the envelope of a virion or in a capsid.

Viral mRNAs that code for envelope proteins are translated as if they were mRNAs for the cell's own membrane proteins. They are translated by ribosomes attached to the rough endoplasmic reticulum (rough ER), where carbohydrates are added to the protein (see Chapter 7). Depending on the virus, the proteins then may be transported to the Golgi apparatus for further processing. In some viruses, such as HIV, the finished glycoproteins are transported to the plasma membrane, where they are assembled into new virions.

A different route is taken by mRNAs that code for proteins that make up the capsid of a virion. These mRNAs are translated by free ribosomes in the cytosol, just as if they were cellular mRNAs for cytosolic proteins.

In some viruses, long polypeptide sequences called polyproteins are cut into individual proteins by viral enzymes called **proteases.** These enzymes cleave viral polyproteins at specific locations—a critical step in the production of finished viral proteins. In HIV, for example, a polyprotein must be cleaved before it can assemble into new viral capsids.

The discovery that HIV produces a protease triggered research for drugs that would inhibit the enzyme. This effort got a huge boost when researchers identified the three-dimensional structure of HIV's protease, using X-ray crystallographic techniques (see **BioSkills 11** in Appendix B). The enzyme has an opening in its interior where the active site is located (**FIGURE 36.10a**). Viral polyproteins fit into the opening and are cleaved at the active site.

Based on these data, researchers immediately began searching for molecules that could fit into the opening and prevent protease from functioning by blocking the active site (**FIGURE 36.10b**). Several HIV protease inhibitors are currently being used to interfere with viral replication.

How Do Viruses Copy Their Genomes? In addition to transcription and translation, viruses must also copy their genetic material to make a new generation of virions. Viruses depend on the host cell for nucleotide monomers. In addition, some DNA viruses also depend on the host-cell DNA polymerase machinery to replicate their genomes.

In viruses that have an RNA genome, however, the virus must supply its own enzyme to make copies of its genome. Viral enzymes called **RNA replicases** function as *RNA-dependent* RNA polymerases. In other words, RNA replicases synthesize RNA *from an RNA template*, using ribonucleotides provided by the host cell. For example, the RNA replicases of positive-sense single-stranded RNA viruses first convert the genome into double-stranded RNA and then produce multiple positive-sense copies from the negative-sense complementary strand.

(a) HIV's protease enzyme

(b) Could a drug block the active site?

FIGURE 36.10 The Three-Dimensional Structure of HIV's Protease. (a) Ribbon diagram depicting the three-dimensional shape of HIV's protease enzyme. **(b)** Once protease's structure was solved, researchers began synthesizing compounds that were predicted to fit into the active site and prevent the enzyme from working.

First, reverse transcriptase
synthesizes cDNA from RNA

Then, reverse transcriptase
synthesizes double-stranded
DNA from cDNA

FIGURE 36.11 Reverse Transcriptase Catalyzes Synthesis of a Double-Stranded DNA from an RNA Template.
The DNA produced by reverse transcriptase is called a cDNA because its base sequence is complementary to the RNA template.

But not all RNA viruses replicate using RNA replicases. In certain RNA viruses, the genome is first transcribed from RNA to DNA by a viral enzyme called **reverse transcriptase.** This enzyme is an unusual DNA polymerase—one that can make DNA from either an RNA or a DNA template. It first makes **complementary DNA,** or **cDNA,** from a single-stranded RNA template (see Chapter 20). Reverse transcriptase then removes the RNA strand and catalyzes the synthesis of a second, complementary DNA strand, resulting in a double-stranded DNA (**FIGURE 36.11**). The DNA copy is then inserted into the host genome and used as a template for host cellular machinery—including an RNA polymerase—to produce viral mRNAs and genomic RNAs.

Viruses that reverse-transcribe their genome in this way are called **retroviruses** ("backward viruses"). The name is apt because the initial flow of genetic information in this type of virus is RNA → DNA—the opposite of the central dogma (see Chapter 16). HIV is an example of a reverse-transcribing RNA virus.

The first antiviral drugs that were developed to combat HIV act by inhibiting reverse transcriptase. Logically enough, drugs of this type are called reverse transcriptase inhibitors.

How Are New Virions Assembled?

Once viral proteins have been produced and the viral genome has been replicated, assembly of a new generation of virions can take place.

During assembly, viral genomes are packaged into capsids. Some viruses also include copies of non-capsid proteins, often enzymes like polymerases, inside the capsid. In many cases, the details of the assembly process are not well understood. Some viruses assemble the capsid first and then use motor proteins to pull the viral genome inside, while others assemble the capsid around the genomic material.

Enveloped viruses use the host endomembrane system (see Chapter 7) to transport their envelope proteins to the appropriate membrane for assembly. For example, virions of HIV and influenza assemble at the host cell's surface and acquire their envelope from its plasma membrane. Other enveloped viruses may assemble at the surface of internal membranes, such as the rough ER or Golgi apparatus.

In most cases, self-assembly occurs—though some viruses produce proteins that provide scaffolding where new virions are put together. To date, researchers have yet to develop drugs that inhibit the assembly process during a viral infection.

How Do Progeny Virions Exit an Infected Cell?

Most viruses leave a host cell in one of two ways: by budding from cellular membranes or by bursting out of the cell. In general, enveloped viruses bud; nonenveloped viruses burst.

Viruses that bud from one of the host cell's membranes take some of that membrane with them. As a result, their envelope includes host-cell phospholipids along with envelope proteins encoded by the viral genome. In **FIGURE 36.12a**, HIV is shown budding from the plasma membrane. After the budding step, the HIV capsid is further processed by its protease, giving the virion its characteristic cone-like appearance.

Viruses that bud through internal membranes, such as those of the rough ER or Golgi apparatus, are secreted from the cell by being escorted through the endomembrane system (see Chapter 7). Regardless of the membrane used for producing enveloped viruses, exit from the host cell does not require the death of the host.

In contrast, most nonenveloped viruses release their virions from the cell by lysing it—commonly referred to as the burst. For example, bacteriophages produce lytic enzymes that break down the cell wall of the host. Because the cell exerts pressure on the wall, the cell will explode when the wall is damaged—dispersing a new generation of virions into the environment. **FIGURE 36.12b** shows T4 bacteriophages bursting from an infected *E. coli* cell.

How Are Virions Transmitted to New Hosts?

Once the replicative cycle is complete, dozens to several hundred newly assembled virions are in the extracellular space. What happens next?

If the host cell is part of a multicellular organism, the new generation of virions may be transported through the body—often via the bloodstream or lymphatic system. In vertebrates, antibodies produced by the immune system may bind to the virions and mark them for destruction. But if a virion contacts an appropriate host cell before it encounters antibodies, then a new replicative cycle will be started.

The long-term success of the virus, however, is dependent on its ability to be transmitted through the environment from one organism to another. For example, when people cough, sneeze,

(a) Budding of enveloped viruses

(b) Bursting of nonenveloped viruses

FIGURE 36.12 Viruses Leave Infected Cells by Budding or Bursting. (a) In enveloped viruses such as HIV, virions are formed as they bud from membranes of the host cell. **(b)** Naked viruses normally exit the host cell by bursting the cell, as in T4 bacteriophage. Bursting kills the host cell; budding does not require cell death to occur.

✔**QUESTION** Propose a hypothesis to explain why cells often die after extensive budding of enveloped viruses.

wipe a runny nose, or defecate, they help rid their body of viruses and bacteria. But they also expel the pathogens into the environment, sometimes directly onto a new host.

From the virus's point of view, a new host represents an unexploited habitat brimming with resources in the form of target cells. The situation is analogous to that of a multicellular animal dispersing to a new habitat and colonizing it. The alleles carried by these successful colonists increase in frequency in the total population. In this way, natural selection favors alleles that allow viruses to do two things: (**1**) replicate within a host and (**2**) be transmitted to new hosts.

✔ If you understand how different types of viruses produce a new generation of virions, you should be able to compare and contrast the replicative cycles of bacteriophage T4 and HIV.

Analyzing How Viruses Coexist with Host Cells

All viruses undergo replicative growth, but some can arrest the replicative cycle and enter a dormant state. Note that only certain types of viruses are capable of switching between replicative growth and dormant coexistence with the host cell. In bacteriophages, this alternate type of infection is called **lysogeny** (**FIGURE 36.13**, page 722).

The onset of lysogeny is triggered by molecular cues in the host that push the virus out of the replicative cycle. Instead of actively transcribing and replicating the viral DNA, it becomes incorporated into the host's chromosome and the expression of most of the viral genes is shut down.

Although no virions are produced during lysogeny, the host's DNA polymerase replicates the viral DNA each time the cell divides. Copies of the viral genome are passed on to daughter cells

Infection

1. Viral genome enters host cell.

Integration

2. Viral genome integrates into host-cell genome.

Replication of genome

3. Host-cell DNA polymerase copies chromosome.

OR

4. Cell divides. Virus is transmitted to daughter cells.

DNA   mRNA   Protein

At any point after integration, the virus may activate the replicative cycle.

FIGURE 36.13 Some Bacteriophages Can Switch between the Lytic Cycle and Lysogeny. Molecular cues from the host can direct certain bacteriophages toward either lysogeny or the lytic cycle.

just as if it were one of the host's own genes. Some of these introduced genes have been known to significantly alter bacterial hosts. For example, many of the pathogenic strains of *E. coli* that have recently emerged in humans are more virulent due to expression of phage toxins that were introduced via lysogeny.

An integrated bacteriophage genome may be "awakened" by different cues from the host cell, often associated with host stress. Activation leads to excision of the viral DNA from the host genome and reentry into the replicative growth cycle.

In viruses that infect animal cells, the dormant state is called **latency.** As with bacteriophages, only certain animal viruses are capable of arresting their replicative growth cycle. Depending on the virus involved, the genome may or may not be integrated into the host genome.

HIV has a well-characterized latency period that occurs at the start of many new infections. After reverse transcriptase makes a double-stranded DNA version of the viral genome, the viral DNA is inserted into the host-cell genomic DNA. But integration alone is not enough to trigger expression of the viral genes. The helper T cell host must first be activated by the immune response to drive expression of the HIV genome. Until this occurs, HIV remains hidden and silent.

Current antivirals are ineffective against an HIV infection when HIV genes aren't being expressed. For this reason, reducing the likelihood of **transmission**—the spread of pathogens from one individual to another—is presently the most effective way to combat a viral disease like AIDS.

HIV is transmitted from person to person via body fluids such as blood, semen, or vaginal secretions. Faced with decades of disappointing results in drug and vaccine development, public health officials are aggressively promoting preventive medicine, through

- aggressive treatment of venereal diseases—because the lesions caused by chlamydia, genital warts, and gonorrhea encourage

the transmission of HIV-contaminated blood during sexual intercourse;

- condom use; and

- sexual abstinence or monogamy.

The effectiveness of preventive medicine underscores one of this chapter's fundamental messages: Viruses are a fact of life.

check your understanding

C Y U

If you understand that . . .

- Viruses replicate using the energy, substrates, and protein synthesis machinery of their host cells to produce a new generation of virions.
- After infecting a cell, the manner in which viruses proceed through the steps of the replicative cycle will vary depending on the type of virus.

✔ **You should be able to . . .**

Evaluate the claim that the viral replicative cycle more closely resembles the mass production of automobiles than the reproduction of cells.

Answers are available in Appendix A.

36.3 What Themes Occur in the Diversification of Viruses?

The tree of life will never be free of viruses. Mutation and natural selection guarantee that viral genomes will continually adapt to the defenses offered by their hosts, whether those defenses are devised by an immune system or by biomedical researchers.

Two other points are critical to recognize about viral diversity: (1) Biologists do not have a solid understanding of how new viruses originate, but (2) it is certain that viruses will continue to diversify. Because most viral polymerases have high error rates and because viruses lack error repair enzymes, mutation rates are extremely high. Many viruses change constantly—giving them the potential to evolve rapidly.

Where Did Viruses Come From?

No one knows how viruses originated. To address this question, biologists are currently considering three hypotheses to explain where viruses came from.

Origin in Plasmids and Transposable Elements? Like viruses, plasmids and transposable elements are acellular, mobile genetic elements that replicate with the aid of a host cell (see Chapters 20 and 21). Certain viruses are actually indistinguishable from plasmids except for one feature: They encode proteins that form a capsid and allow the genes to exist outside of a cell.

Some biologists hypothesize that simple viruses represent "escaped gene sets." This hypothesis proposes that mobile genetic elements are descended from clusters of genes that physically escaped from prokaryotic or eukaryotic chromosomes long ago.

According to this hypothesis, the escaped gene sets took on a mobile, parasitic existence because they happened to encode the information needed to replicate themselves at the expense of the genomes that once held them. In the case of viruses, the hypothesis is that the escaped genes included the instructions for making a protein capsid and possibly envelope proteins. According to the escaped-genes hypothesis, it is possible that each of the distinct types of viruses represent distinct "escape events."

To support the escaped-genes hypothesis, researchers would need to discover a virus that had so recently derived from intact prokaryotic or eukaryotic genes that the viral genomes still strongly resembled the DNA sequence of those genes.

Origin in Symbiotic Bacteria? Some researchers contend that DNA viruses with large genomes trace their ancestry back to free-living bacteria that once took up residence inside eukaryotic cells. The idea is that these organisms degenerated into viruses by gradually losing the genes required to synthesize ribosomes, ATP, nucleotides, amino acids, and other compounds.

Although this idea sounds speculative, it cannot be dismissed lightly. For example, there is evidence that the mitochondria and chloroplasts of eukaryotic cells originated as intracellular symbionts from ancestors that were independent, free-living cells (see Chapter 30). Investigators contend that, instead of evolving into intracellular symbionts that aid their host cell, DNA viruses became parasites capable of replication and transmission from one host to another.

To support the degeneration hypothesis, researchers have pointed to mimivirus, the largest known virus that infects certain protists. In addition to its large size, the genome of this virus was found to contain some of the genes involved in protein synthesis—genes that are common to cells, but had not previously been observed in viruses. It is still not clear, though, whether these genes are remnants from an ancestral cell or were acquired from a host cell that was infected relatively recently. The discovery of a cell that possesses a genome similar to that of the mimivirus would support the degeneration hypothesis for the origin of this virus.

Origin at the Origin of Life? Recently, some researchers have started to discuss a third alternative to explain the origin of viruses: that they trace their ancestry back to the first, RNA-based forms of life on Earth. If this hypothesis is correct, then the RNA genomes of some viruses are descended from genes found in early inhabitants of the RNA world (see Chapter 4). Some researchers have even suggested that retrovirus-like parasites may have been responsible for transforming the genetic material of cells from RNA to DNA.

To support this hypothesis, advocates point to the ubiquity of viruses—which suggests that they have been evolving along with organisms since life began. The RNA world-origin hypothesis also addresses a problem that weakens the other two hypotheses: Several proteins that are commonly expressed in many viruses are not expressed in any cell examined thus far. If viruses originated at the time of the origin of the cell, then the genes coding for these proteins may have come from an RNA-world pool of genes instead of a cell.

Currently, there is no one widely accepted view of where viruses came from. Because viruses are so diverse, all three hypotheses may be valid.

Emerging Viruses, Emerging Diseases

Although it is not known how the various types of viruses originated, it is certain that viruses will continue to diversify. With alarming regularity, the front pages of newspapers carry accounts of deadly viruses that are infecting humans for the first time.

HIV is an example of a virus responsible for an **emerging disease:** a new illness that suddenly affects significant numbers of individuals in a host population. Another, and perhaps the most infamous of such outbreaks occurred in 1995, when the Ebola virus infected 200 individuals in the Democratic Republic of Congo and had a fatality rate of 80 percent. In both cases, the causative agents were considered emerging viruses because they had switched from their traditional host species to a new host—humans.

Some Emerging Viruses Arise from Genome Reassortment Each year the World Health Organization worries about a new influenza pandemic, one that might resemble the devastating outbreak of 1918. The small changes that arise from the influenza virus's error-prone RNA replicase are primarily responsible for the need for yearly vaccine updates. But influenza can also acquire alleles that are entirely new to the strain. (A virus **strain** consists of populations that have similar characteristics and is the lowest, or most specific, level of taxonomy for viruses.) How do new strains of influenza originate?

Influenza has a single-stranded RNA genome that consists of eight segments, most of which encode only one protein. If two viruses infect the same cell, the replicated genomic segments

PROCESS: GENOMIC REASSORTMENT GENERATES NEW INFLUENZA STRAINS

Influenza

Host cell

Recombinant strain

1. Two different strains of influenza infect the same cell.

2. Replication produces a mix of strain-specific genomic segments in host cytoplasm.

3. Reassortment of genomic segments generates new, recombinant strains.

FIGURE 36.14 Influenza Can Generate New Strains via Genomic Reassortment.

are randomly shuffled in the cytosol. Thus when the progeny are assembled, many have segments from each parent virus (**FIGURE 36.14**).

The ability to randomly mix genomic segments becomes particularly significant when two different strains infect the same cell. In addition to humans, different strains of influenza can infect other animals, such as birds and pigs. While the avian strains do not efficiently infect humans, they can infect pigs—and so can human viral strains. The last point is key— pigs can serve as mixing vessels to produce new recombinant strains. Pandemic strains are thought to emerge from this type of reassortment.

Using Phylogenetic Trees to Understand Emerging Viruses How do researchers know that an emerging virus has "jumped" to a new host? The answer is to analyze the evolutionary history of the virus in question and then estimate a phylogenetic tree that includes its close relatives.

HIV, for example, belongs to a group of viruses called the lentiviruses, which infect a wide range of mammals including house cats, horses, goats, and primates. (*Lenti* is a Latin root that means "slow"; here it refers to the long period observed between the start of an infection by these viruses and the onset of the diseases they cause.) Consider the conclusions that can be drawn from the phylogenetic tree of HIV, shown in **FIGURE 36.15**.

- *There are immunodeficiency viruses.* Many of HIV's closest relatives parasitize cells that are part of the immune system. Several of them cause diseases with symptoms reminiscent of AIDS. These viruses infect monkeys and chimpanzees and are called simian immunodeficiency viruses (SIVs).

- *There are two HIVs.* There are two distinct types of human immunodeficiency viruses, called HIV-1 and HIV-2. Although both can cause AIDS, HIV-1 is far more virulent and is the better studied of the two.

HIV-1's closest known relatives are immunodeficiency viruses isolated from chimpanzees that live in central Africa. In contrast, HIV-2's closest relatives are immunodeficiency viruses that parasitize monkeys called sooty mangabeys.

■ **HIV strains that infect humans**
▨ HIV strains that infect other primates

SIV-Sykes monkey

SIV-sooty mangabey

HIV-2

SIVs from four monkey species

SIV-chimp

HIV-1 strain O ("O" stands for outlier)

SIV-chimp

HIV-1 strain N ("N" stands for new; discovered recently)

HIV-1 strain M ("M" stands for main strain responsible for the AIDS epidemic)

FIGURE 36.15 Phylogeny of HIV Strains and Types. Phylogenetic tree showing the evolutionary relationships among some of the immunodeficiency viruses that infect primates—including chimpanzees, humans, and several species of monkeys.

✔**EXERCISE** On the appropriate branches, indicate where an SIV jumped to humans (draw and label bars across the appropriate branches).

In central Africa, where HIV-1 infection rates first reached epidemic proportions, contact between chimpanzees and humans is extensive. Chimps are hunted for food and kept as pets. Similarly, sooty mangabeys are hunted and kept as pets in western Africa, where HIV-2 infection rates are highest.

- *Multiple "jumps" have occurred.* Several strains of HIV-1 exist. The most important are called O for outlier (meaning, the most distant group relative to other strains), N for new, and M for main. Each of these strains likely represents an independent origin of HIV-1 from a chimp SIV strain.

The last point is particularly important. The existence of distinct strains suggests that HIV-1 has jumped from chimps to humans several times. It may do so again in the future.

Responding to a Virus Outbreak Physicians become alarmed when they see a large number of patients with identical, and unusual, disease symptoms in the same geographic area over a short period of time. The physicians report these cases to public health officials, who take on two urgent tasks: **(1)** identifying the agent that is causing the new illness and **(2)** determining how the disease is being transmitted.

Several strategies can be used to identify the pathogen responsible for an emerging disease. In 1993, an outbreak of an unknown pulmonary syndrome rocked the U.S. Southwest, killing half of the individuals who came down with the disease. Officials recognized strong similarities between symptoms in these cases and symptoms caused by a hantavirus native to northeast Asia, called Hantaan virus. The Hantaan virus rarely causes disease in North America, but it is known to infect rodents.

To determine whether a Hantaan-like virus was responsible for the U.S. outbreak, researchers began capturing mice in the homes and workplaces of afflicted people. About a third of the captured rodents tested positive for the presence of a Hantaan-like virus. Genome sequencing studies confirmed that the virus was a previously undescribed type of hantavirus, and it was aptly named the *sin nombre* virus (Spanish for "no-name virus"). Further, the sequences found in the mice matched those found in infected patients. Based on these results, officials were confident that this rodent-borne hantavirus was causing the wave of human infections.

The next step in the research program, identifying how the agent is being transmitted, is equally critical and takes old-fashioned detective work. By interviewing patients about their activities, researchers called epidemiologists decide how each patient could have acquired the virus. Was there a particular environment or source of food or water shared by those who were infected? Was the illness showing up in health-care workers who were in contact with infected individuals, implying that it was being transmitted from human to human?

If the virus identified as the cause of an illness can be transmitted efficiently from person to person, then the outbreak has the potential to become an epidemic. To date, the sin nombre virus has not been transmitted efficiently from person to person in the United States—only inefficiently from mice to people.

If you understand that . . .

- The origin of viruses is unknown, but three hypotheses have been developed to address it.
- Emerging viruses arise either from genomic reassortment between different strains of virus or via switching host species.

✔ **You should be able to . . .**

1. Compare and contrast the three hypotheses for the origin of viruses.
2. Predict what type of a mutation in an avian flu virus would make it more dangerous to humans, and explain how this change might occur.

Answers are available in Appendix A.

36.4 Key Lineages of Viruses

Because scientists are almost certain that viruses originated more than once throughout the history of life, there is no such thing as the phylogeny of all viruses. Stated another way, unlike the organisms discussed in previous chapters, viruses have no single phylogenetic tree that represents their evolutionary history. Instead, researchers most often focus on comparing base sequences in the genetic material of small, closely related groups of viruses to reconstruct the phylogenies of particular lineages, exemplified by the tree in Figure 36.15.

To organize the genomic diversity of viruses, researchers group them into seven general categories based on the nature of their genetic material and how they replicate. David Baltimore first proposed this approach in 1971; thus, it is called the Baltimore classification. **FIGURE 36.16** (page 726) summarizes the seven classes.

Notice that each of the Baltimore classes has its own unique strategy for transcription and replication of the genetic material. (See Section 36.2 to review how viruses copy their genomes.) All these strategies, however, share the need to produce mRNA, which is then translated by the host cell to make viral proteins.

Within the Baltimore classification system, researchers further distinguish viruses according to **(1)** virion morphology, **(2)** nature of the host species, and **(3)** the type of disease resulting from infection. Using these criteria, biologists identify a total of about 70 virus families.

Although the phylogenetic relationships among viruses is not as clear as what is observed in organisms, those within families are grouped into distinct genera for convenience. Within genera, biologists identify and name types of virus, such as HIV, the measles virus, and smallpox. Within each of these viral types, populations with distinct characteristics may be identified and named as strains.

DNA viruses | RNA viruses | Reverse-transcribing viruses

| Class I | Class II | Class III | Class IV | Class V | Class VI | Class VII |
| dsDNA | ssDNA | dsRNA | (+)ssRNA | (−)ssRNA | (+)ssRNA (RT) | dsDNA (RT) |

Genome replication

To viral mRNA and proteins →

+
mRNA

Proteins

FIGURE 36.16 The Seven Different Strategies for Expression of Viral Genetic Material Converge on the Translation of mRNA. Notice that, although classes I and IV have the same types of genomes as classes VII and VI, they differ in the ways their genomes are replicated.

To get a sense of viral diversity, let's survey some of the most prevalent groups that can be distinguished based on the nature of their genetic material and mode of replication.

- Double-Stranded DNA (dsDNA) Viruses
- Double-Stranded RNA (dsRNA) Viruses
- Positive-Sense Single-Stranded RNA ([+]ssRNA) Viruses
- Negative-Sense Single-Stranded RNA ([−]ssRNA) Viruses
- RNA Reverse-Transcribing Viruses (Retroviruses)

Double-Stranded DNA (dsDNA) Viruses

The double-stranded DNA viruses are a large group, composed of some 21 families and 65 genera. Smallpox is perhaps the most familiar of these viruses (**FIGURE 36.17**). Although smallpox had been responsible for millions of deaths throughout human history, it was eradicated by worldwide vaccination programs. In 1977, smallpox was declared to be extinct in the wild; the only remaining samples of the virus are stored securely in research labs.

Genetic material As their name implies, the genes of these viruses consist of a single molecule of double-stranded DNA. The molecule may be linear or circular.

Host species These viruses parasitize hosts throughout the tree of life, with the notable exception of land plants. They include bacteriophage families called the T-series (T1–T7) and λ, which infect *E. coli*. Other common animal viruses with double-stranded DNA genomes include the human papilloma virus (HPV), herpesviruses, and adenoviruses.

Replicative cycle In most double-stranded DNA viruses that infect eukaryotes, viral genes have to enter the nucleus to be replicated. Often these viruses will replicate their genomes only during S phase, when the host cell's chromosomes are being replicated. Viruses like HPV are able to induce the cells to enter S phase despite the absence of the normal growth signals. This ability may, in fact, be responsible for the link between HPV infections and cervical cancer in women.

FIGURE 36.17 In Humans, Smallpox Can Cause Disease with Easily Recognizable Symptoms.

Double-Stranded RNA (dsRNA) Viruses

There are 7 families of double-stranded RNA viruses and a total of 22 genera. Most of the viruses in this group are nonenveloped.

Genetic material In some families, virions have a genome consisting of 10–12 double-stranded RNA molecules; in other families, the genome is composed of just 1–3 RNA molecules.

Host species A wide variety of organisms, including fungi, land plants, insects, vertebrates, and bacteria, are hosts for viruses with double-stranded RNA genomes. Particularly prominent are viruses that cause disease in rice, corn, sugarcane, and other crops. The bluetongue virus, a double-stranded RNA virus that is transmitted through midges and other biting insects, causes an often fatal disease that has significantly affected the livestock industry (**FIGURE 36.18**). In humans, reovirus and rotavirus infections are the leading cause of infant diarrhea and are responsible for over 110 million cases and a devastating 440,000 infant deaths each year.

Replicative cycle Once inside the cytosol of a host cell, the double-stranded genome of these viruses serves as a template for the synthesis of mRNAs, which are then translated into viral proteins. Some of the proteins form the capsids for a new generation of virions. The genome is copied by a viral enzyme that converts the mRNA transcripts into double-stranded RNA within the newly formed capsids.

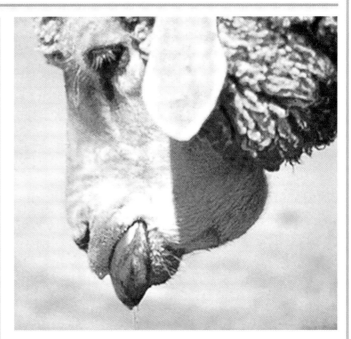

FIGURE 36.18 dsRNA Viruses Parasitize a Wide Array of Organisms, Such as Livestock Infected by the Bluetongue Virus.

Positive-Sense Single-Stranded RNA ([+]ssRNA) Viruses

This is the largest group of viruses known. It has 21 families and 81 genera that include both enveloped and nonenveloped morphologies.

Genetic material The sequence of bases in a positive-sense single-stranded RNA virus is the same as that of a viral mRNA. Stated another way, the genome does not need to be transcribed in order for proteins to be produced. Depending on the species, the genome consists of 1–3 double-stranded RNA molecules.

Host species Most of the commercially important plant viruses belong to this group. Because they kill groups of cells in the host plant and turn patches of leaf or stem white, they are often named mottle viruses, spotted viruses, chlorotic (meaning, lacking chlorophyll) viruses, or mosaic viruses. **FIGURE 36.19** shows a healthy cowpea leaf and a cowpea leaf that has been attacked by a positive-sense single-stranded RNA virus. This group of viruses is also known to infect bacteria, fungi, and animals. Viruses within this group that infect humans include the rhinovirus (common cold), polio, SARS CoV, West Nile virus, and hepatitis A, C, and E.

Replicative cycle When the genome of these viruses enters a host cell, the positive-sense single-stranded RNA is immediately translated into a protein, often a polyprotein that is cleaved to generate functionally distinct proteins. These proteins include proteases that process the polypeptides and polymerases that copy the genome and produce mRNA for translation.

Healthy leaf   Leaf infected with virus

FIGURE 36.19 (+)ssRNA Viruses, Such as the Cowpea Mosaic Virus, Cause Important Plant Diseases.

Negative-Sense Single-Stranded RNA ([−]ssRNA) Viruses

This group has 7 families and 30 genera. Most members of this group are enveloped, but some negative-sense single-stranded RNA viruses lack an envelope.

Genetic material The sequence of bases in a negative-sense single-stranded RNA virus is opposite in polarity to the sequence in a viral mRNA. Stated another way, the virus genome is complementary to the viral mRNA. Depending on the family, the genome may consist of one single-stranded RNA molecule or up to eight separate RNA molecules.

Host species A wide variety of plants and animals are parasitized by viruses that have negative-sense single-stranded RNA genomes. If you have ever suffered from the flu, the mumps, or the measles, then you are painfully familiar with these viruses (**FIGURE 36.20**). The Ebola, sin nombre, and rabies viruses also belong to this group.

Replicative cycle When the genome of a negative-sense single-stranded RNA virus enters a host cell, a viral RNA-dependent RNA polymerase must accompany it to make new viral RNAs. Some of these transcripts serve as mRNA to be translated into viral proteins, while others function as templates for replicating the viral genome.

FIGURE 36.20 (−)ssRNA Viruses, Such as the Measles Virus, Cause Some Common Childhood Diseases.

RNA Reverse-Transcribing Viruses (Retroviruses)

The genomes of the RNA reverse-transcribing viruses are composed of positive-sense single-stranded RNA. This group has only one family, called the retroviruses, with 7 genera.

Genetic material Virions have two copies of a positive-sense single-stranded genome.

Host species Species in this group are known to parasitize only vertebrates—specifically birds, fish, or mammals. The Rous sarcoma virus (chickens), the mouse mammary tumor virus, and the murine (mouse) leukemia virus are well-studied retroviruses that have been shown to contribute to the development of cancer. Of the retroviruses, HIV is the most familiar and, in terms of the human population, by far the deadliest virus in this group (**FIGURE 36.21**).

Replicative cycle Although retroviruses have positive-sense single-stranded RNA, their genome is not translated immediately after it enters the host-cell cytosol. Instead, viral reverse transcriptase, which enters the cell along with the genome, catalyzes the synthesis of a double-stranded DNA version of the viral genome. This DNA is transported into the nucleus and integrated into the host genome by a viral protein called integrase. The virus may remain dormant for a period until induced by host-cell signals, such as the activation of HIV via the immune response. Activation starts up the replicative cycle. Genes are transcribed to mRNA to begin the production of a new generation of virions.

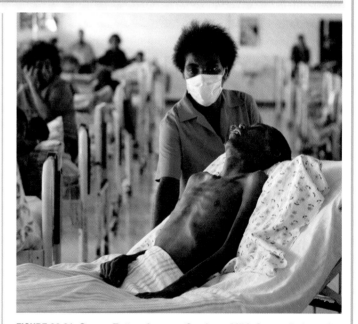

FIGURE 36.21 Some Retroviruses, Such as HIV, Severely Impair the Physiology of the Host.

If you understand . . .

36.1 Why Do Biologists Study Viruses?

- Viruses cause illness and death in organisms from all three domains of life.

- Viruses are specialists—different types of viruses infect particular species and types of cells.

✔ You should be able to explain how HIV causes AIDS and why it takes close to eight years before the clinical signs appear.

36.2 How Do Biologists Study Viruses?

- Most viruses have a capsid made of protein that is either icosahedral or helical in shape. Some viruses are covered by a host-derived membrane called an envelope.

- Viral genomes exhibit considerable diversity. The genetic material may consist of one or more molecules of DNA or RNA that is either double stranded or single stranded.

- Viral entry often depends on specific interactions between viral proteins and molecules on the host-cell surface, called virus receptors. Attachment is followed by uncoating, which is when the viral genome is released into the cytosol.

- The protein-production and genome-replication phases of the replicative cycle depend on biosynthetic machinery, chemical energy, and substrates provided by the host cell.

- Enveloped viruses exit a cell by budding; naked viruses often exit a cell by lysis—killing the host cell in the process.

- Some viruses may enter a dormant phase, when they do not produce virions, but instead coexist with the host cell and transmit genetic material to daughter cells when the host divides.

✔ You should be able to explain how three of the six phases of a viral replicative cycle can be stopped by antivirals, using HIV as an example.

36.3 What Themes Occur in the Diversification of Viruses?

- Three hypotheses have been developed to explain the origin of viruses: (1) They are escaped gene sets from cells; (2) they are the products of degenerate cellular parasites; (3) they coevolved with cells from the RNA world.

- Viruses continue to evolve. Factors influencing rates of viral evolution are errors during genome replication and genomic reassortment.

- Many viruses currently infecting humans are examples of emerging diseases—that is, diseases caused by viruses transmitted from other host species.

✔ You should be able to apply what you know of Darwin's four postulates on natural selection (see Chapter 25) to explain how it is possible for viruses to evolve even though they are not alive.

36.4 Key Lineages of Viruses

- In addition to morphology and the nature of the genetic material, viruses also vary in how the genome is replicated and transcribed to produce mRNA.

✔ You should be able to compare and contrast the transcription and replication of the RNA genomes of class IV and class VI viruses.

(MB) **MasteringBiology**

1. **MasteringBiology Assignments**

   Tutorials and Activities HIV Replicative Cycle; HIV Reproductive Cycle; Phage Lysogenic and Lytic Cycles; Phage Lytic Cycle; Retrovirus (HIV) Reproductive Cycle; Simplified Viral Reproductive Cycle; Viral Replication

   Questions Reading Quizzes, Blue-Thread Questions, Test Bank

2. **eText** Read your book online, search, take notes, highlight text, and more.

3. **The Study Area** Practice Test, Cumulative Test, BioFlix® 3-D Animations, Videos, Activities, Audio Glossary, Word Study Tools, Art

You should be able to . . .

1. What do host cells provide for viruses?
   a. nucleotides and amino acids
   b. ribosomes
   c. ATP
   d. all of the above

2. How do viruses that infect animals enter an animal's cells?
   a. The viruses pass through a wound.
   b. The viruses bind to a membrane component.
   c. The viruses puncture the cell wall.
   d. The viruses lyse the cell.

3. What does reverse transcriptase do?

4. When do most enveloped virions acquire their envelope?
   a. during entry into the host cell
   b. during budding from the host cell
   c. as they burst from the host cell
   d. as they integrate into the host cell's chromosome

5. In the viral replicative cycle, what reaction do viral proteases catalyze?
   a. polymerization of amino acids into peptides
   b. folding of long peptide chains into functional proteins
   c. cutting of polyprotein chains into functional proteins
   d. assembly of virions

6. What features distinguish the seven major categories of viruses?

7. The outer surface of a virion consists of either a membrane-like envelope or a protein capsid. How does the outer surface correlate with a virus's mode of exiting a host cell, and why?

8. Compare and contrast the bacteriophage lytic cycle versus lysogeny by addressing (1) replication of the viral genome, (2) production of virions, and (3) effect on the host cell.

9. Propose a reason why the development of antiviral drugs is more difficult than developing the antibiotics used to treat bacterial infections.

10. How does the diversity of viral genome types support the hypothesis that viruses originated several times independently?

11. What types of data convinced researchers that HIV originated when a simian immunodeficiency virus "jumped" to humans?

12. Of the viruses highlighted in Section 36.4, predict which of the following would be able to make viral proteins by injecting nothing more than its genome into a suitable host cell.
    a. cowpea mosaic ([+]ssRNA) virus
    b. bluetongue (dsRNA) virus
    c. measles ([−]ssRNA) virus
    d. human immunodeficiency (RNA reverse-transcribing) virus

13. Suppose you could isolate a virus that parasitizes the pathogen *Staphylococcus aureus*—a bacterium that causes acne, boils, and a variety of other afflictions in humans. How could you test whether this virus might serve as a safe and effective treatment for a staph infection?

14. If you were in charge of the government's budget devoted to stemming the AIDS epidemic, would you devote most of the resources to drug development or preventive medicine? Defend your answer.

15. Latency is a key adaptation in some animal viruses and likely evolved to prolong the infection by avoiding a strong and effective immune response. Which of the following types of viruses would you expect to require periods of latency?
    a. viruses that have large genomes and require a long time for replication

    b. viruses that require a long time for transmission to new host organisms
    c. viruses that require a long time for assembling into complex structures
    d. viruses that infect cells of the immune system

16. Consider these two contrasting definitions of life: (**1**) An entity is alive if it is capable of replicating itself via the directed chemical transformation of its environment; (**2**) An entity is alive if it is an integrated system for the storage, maintenance, replication, and use of genetic information. According to these definitions, are viruses alive? Explain.

52 An Introduction to Ecology

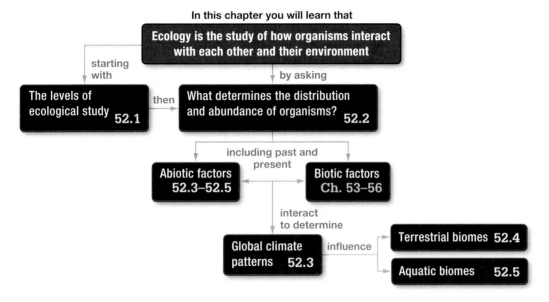

In this chapter you will learn that

Ecology is the study of how organisms interact with each other and their environment

starting with

then

by asking

The levels of ecological study **52.1**

What determines the distribution and abundance of organisms? **52.2**

including past and present

Abiotic factors **52.3–52.5**

Biotic factors **Ch. 53–56**

interact to determine

Global climate patterns **52.3**

influence

Terrestrial biomes **52.4**

Aquatic biomes **52.5**

This heavily oiled bird is being rescued from the Gulf of Mexico following the Deepwater Horizon oil spill in 2010. Increasingly, ecology is the study of how humans are altering the environment in ways that make life difficult for other organisms.

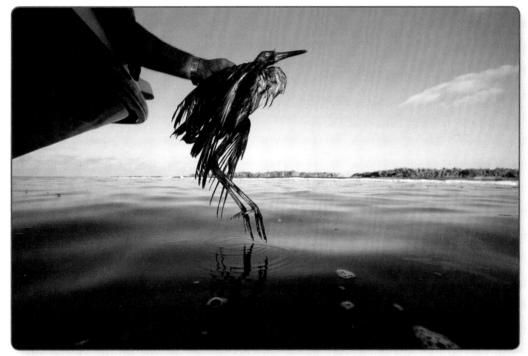

BIG PICTURE

This chapter is part of the Big Picture. See how on pages 1196–1197.

Ecology is the study of how organisms interact with each other and their environment. Except for the episode 65 million years ago, when a mountain-sized asteroid struck Earth, environments are changing more rapidly right now than they have at any time in the past 3.6 billion years. Humans are causing widespread habitat loss and degradation, pollution, introduction of exotic organisms, and climate change—transforming the Earth in unprecedented ways. Efforts to maintain diversity of life, functioning ecological processes, and human health and welfare depend on our ability to cope with these environmental changes. Ecology has become one of the most important and dynamic fields in biological science.

Ecology is also among the most synthetic and integrative fields in biological science. In this unit, you'll be applying what you've learned about biochemistry, genetics, development, physiology,

✔ When you see this checkmark, stop and test yourself. Answers are available in Appendix A.

evolution, and the diversity of life. Ecologists work in subdisciplines ranging from ecological developmental biology to physiological ecology and conservation genetics.

Of these subdisciplines, the intersection of ecology and evolution has been appreciated the longest. The diversity of life (Unit 6) is the result of the evolutionary processes (Unit 5) as they played out in a changing ecological context over 3.6 billion years. Evolution and ecology are intertwined and codependent. It was no coincidence that the German biologist Ernst Haeckel first defined "oekologie" less than a decade after Charles Darwin published *On the Origin of Species by Means of Natural Selection* in 1859 (see Chapter 25).

Ecology's primary goal is to understand the distribution and abundance of organisms. How do interactions with other organisms and the physical environment dictate why certain species live where they do, and how many individuals live there? Some biologists ask why orangutans are restricted to forests in Borneo and Northern Sumatra and why their numbers are declining so rapidly. Other biologists create mathematical models to predict how quickly the human immunodeficiency virus will increase in India, or how long it will take for the current human population of 7 billion to double.

Distribution and abundance are core questions in today's world. Land-use changes and climate change are having dramatic effects on species' distributions, and expanding human populations are causing the abundance of many species to decline, sometimes to zero. Let's delve in.

52.1 Levels of Ecological Study

To understand why organisms live where they do and in what numbers, biologists consider ecology at several levels of analysis. This is a common strategy in biological science. Cell biologists study how cells work at different levels of organization—from individual molecules to cellular structures, cells, and multicellular organisms. Physiologists analyze processes at the level of ions and molecules as well as whole cells, tissues and organs, and complete systems. You can review the levels of organization in The Big Picture of Doing Biology (pages 16–17).

In ecology, researchers work at five main levels: (1) organisms, (2) populations, (3) communities, (4) ecosystems, and (5) the biosphere. The levels aren't rigid or exclusive; researchers routinely draw on multiple levels—including lower levels—to explore a particular question. Let's consider how these levels apply to sockeye salmon, summarized in **TABLE 52.1**.

Organismal Ecology

At the organismal level, researchers explore the morphological, physiological, and behavioral adaptations that allow individuals to live in a particular area (see Chapter 53). For example, after spending four or five years feeding and growing in the ocean, sockeye salmon travel hundreds or thousands of kilometers to return to the stream where they hatched. Females create nests in

SUMMARY TABLE 52.1 **Five Main Levels of Ecological Study**

Level	Sockeye Salmon Example
Organismal ecology How do individuals interact with each other and their physical environment? Salmon migrate from saltwater to freshwater environments to breed	
Population ecology How and why does population size change over space and time? Each female salmon produces thousands of eggs. On average, only a few offspring will survive to return to the same stream to breed	
Community ecology How do species interact, and what are the consequences? Salmon are prey as well as predators	
Ecosystem ecology How do energy and nutrients cycle through the local environment? Salmon die and then decompose, releasing nutrients that are used by other organisms	
Global ecology How is the biosphere affected by global changes in nutrient cycling and climate? Worldwide populations of salmon are affected by climate change	

the gravel stream bottom and lay eggs. Nearby males compete for the chance to fertilize eggs as they are laid. When breeding is finished, all the adults die.

Biologists want to know how these individuals interact with their physical surroundings and with other organisms in and around the stream. How do individuals cope with the transition from living in salt water to living in freshwater? Which females get the best nesting sites and lay the most eggs, and which males are most successful in fertilizing eggs?

Population Ecology

A **population** is a group of individuals of the same species that lives in the same area at the same time. When biologists study population ecology, they focus on how the number and distribution of individuals in a population change over time (see Chapter 54).

For example, researchers use mathematical models to predict the future of salmon populations. Many of these populations have declined due to overharvesting and the loss of habitat quality caused by pollution and dam construction. If the factors that affect population size can be described accurately enough, mathematical models can assess the impact of proposed dams, changes in weather patterns, altered harvest levels, or specific types of protection efforts.

Community Ecology

A biological **community** consists of populations of different species that interact with each other within a particular area. Community ecologists ask questions about the nature of the interactions between species and the consequences of those interactions (see Chapter 55). Research might concentrate on predation, parasitism, and competition or explore how communities respond to fires, floods, and other disturbances.

As an example, consider the interactions among salmon and other species in the marine and stream communities where they live. When they are at sea, salmon eat smaller fish and are themselves hunted and eaten by orcas, sea lions, humans, and other mammals; when they return to freshwater to breed, they are preyed on by bears and bald eagles. In both marine and freshwater habitats, salmon are subject to parasitism and disease. They are also heavily affected by disturbances—particularly changes in their food supply and the quality of their breeding streams.

Ecosystem Ecology

Ecosystem ecology is an extension of organismal, population, and community ecology. An **ecosystem** consists of all the organisms in a particular region along with nonliving components. These physical and chemical, or **abiotic** (literally, "not-living"), components include air, water, and the nonliving parts of soil. At the ecosystem level, biologists study how nutrients and energy move among organisms and through the surrounding atmosphere and soil or water (see Chapter 56).

Salmon are interesting to study at the ecosystem level because they link marine and freshwater ecosystems. They harvest

nutrients in the ocean and then, when they migrate, die, and decompose, they transport those molecules to streams and streamside forests. In this way, salmon transport chemical energy and nutrients from one habitat to another.

Global Ecology

Biologists define the **biosphere** as a thin zone surrounding the Earth where all life exists—the sum of all terrestrial and aquatic ecosystems. It extends to about 5 km below the land surface and 10 km down in the deepest trenches of the sea to over 10 km up into the atmosphere. This may sound like a thick layer, but it represents only 0.002 percent of the diameter of the Earth—and most of life occupies a much thinner layer, comparable to the layer of varnish on a typical classroom globe.

The young field of global ecology is growing rapidly as ecologists scramble to quantify the effects of human impacts on the biosphere. Humans are modifying landscapes and releasing massive amounts of energy and nutrients to the biosphere, causing changes that transcend individual ecosystems.

Since the biosphere is practically a closed system, the actions of people in one part of the world may alter distant ecosystems, creating difficult education and public policy challenges. For example, salmon are sensitive to changes in water temperature. Who is accountable for the impacts of climate change on salmon?

Conservation Biology Applies All Levels of Ecological Study

The five levels of ecological study are synthesized and applied in conservation biology. **Conservation biology** is the effort to study, preserve, and restore threatened genetic diversity in populations, species diversity in communities, and ecosystem function (see Chapter 57).

Conservation biologists are like physicians. But instead of prescribing treatments for sick people, they prescribe remedies for threatened environments to preserve biodiversity, clean air, pure water, and productive soils.

Each chapter in this unit focuses on a different level of ecological study, ending with the study of biodiversity and conservation. Turn to the Big Picture of Ecology (pages 1196–1197) to visualize how the different levels of ecology interact. In this introductory chapter, let's take a global look at factors that determine the distribution and abundance of organisms. In particular, what abiotic factors set the stage for life on Earth?

52.2 What Determines the Distribution and Abundance of Organisms?

When you look around at the plants and animals in your neighborhood, you can tell right away that you are not located deep in the Amazon rain forest—no scarlet macaws, jaguars, anacondas, blue morpho butterflies, or giant kapok trees out there. Different organisms live in different places. Why?

Understanding why organisms are found where they are is one of ecology's most basic tasks. The study of how organisms are distributed geographically is called **biogeography.** Let's consider how both abiotic and biotic factors determine the distribution and abundance of organisms.

Abiotic Factors

The most fundamental observation about the **range,** or geographic distribution, of organisms is simple: No one species can survive the full array of environmental conditions present on Earth.

Temperature is particularly important in determining ranges because temperature has a big impact on the physiology of organisms, and organisms are limited in their ability to regulate their own temperatures (see Chapter 42). *Thermus aquaticus* can thrive in hot springs with temperatures above 70°C because its enzymes do not denature at near-boiling temperatures. But the same cells would die instantly if they were transplanted to the frigid waters near polar ice—even though vast numbers of other microbes are present. The cold-water species have enzymes that work efficiently at near-freezing temperatures.

No organism thrives in both hot springs and cold ocean water. The reason is fitness trade-offs (introduced in Chapter 25). No enzyme can function well in both extremely high temperatures and extremely low temperatures. Because of fitness trade-offs, organisms tend to be adapted to a limited set of physical conditions, or abiotic factors—for example, to a particular temperature and moisture regime on land, or a particular salinity, water depth, and movement regime in water. Every organism has a specific range of tolerance of abiotic conditions.

Adaptation to a limited set of abiotic conditions is just the start of the story explaining why a particular species lives where it does. For example, the temperature and moisture regimes of southern California, in North America, are almost identical to those found in the Mediterranean region of Europe and northern Africa. But the species present at the sites are almost completely different. Why?

Biotic Factors

The ability of a species to persist in a given area is often limited by **biotic** ("living") factors—meaning interactions with other organisms (see Chapters 54 and 55). As an example, consider the distributions of hermit warblers and Townsend's warblers along the Pacific Coast of North America (introduced in Chapter 27).

- Both hermit warblers and Townsend's warblers live in evergreen forests.
- Experiments have shown that male Townsend's warblers directly attack male hermit warblers and evict them from breeding territories.
- The geographic range of Townsend's warblers has been expanding steadily at the expense of hermit warblers.

These observations support the hypothesis that a biotic factor—competition with another species—is limiting the range of hermit warblers. Competition is not the only biotic factor to affect species range, however.

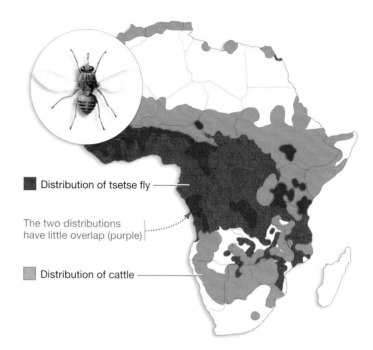

FIGURE 52.1 In Africa, Tsetse Flies Limit the Distribution of Cattle. Tsetse flies carry a disease that is fatal to cattle.

DATA: Food and Agriculture Organization of the United Nations

✔ **QUESTION** What do you think would happen to the range of cattle in Africa if the tsetse fly were eradicated?

- Yucca moths lay their eggs only in the flowers of yucca plants. As a result, this species does not exist outside the range of yucca plants.
- In Africa, the range of domestic cattle is limited by the distribution of tsetse flies (**FIGURE 52.1**). Tsetse flies transmit a parasite that causes the disease trypanosomiasis, which is fatal in cattle (and can be fatal in humans unless treated).
- Most of the songbirds native to Hawaii are limited to alpine habitats—above the range of the mosquitoes that transmit avian malaria.

These examples hint at a key insight: The distribution of organisms is determined not only by conditions present today but also by events in the past. Let's take a closer look at the importance of history.

History Matters: Past Abiotic and Biotic Factors Influence Present Patterns

The abiotic and biotic factors that influence the distribution of organisms are dynamic—they are constantly changing, in part because the Earth itself is changing. The landforms and oceans that may appear static now have been in constant flux for the entire history of life—mountains rise and fall, islands and lakes form and disappear, and whole continents crash together and break apart (see Chapter 28). Associated fluctuations in climate cause glaciers and ice caps to form and melt and sea levels to rise and fall.

These events have an important impact on **dispersal,** the movement of individuals from their place of origin to the location where they live and breed as adults. For example, the formation 3 million years ago of the land bridge from North America to South America, today known as the Isthmus of Panama (see Figure 27.5), created opportunities for dispersal from each continent to the other, introducing new biotic interactions on both continents. Meanwhile, this same geological event separated the Pacific Ocean from the Caribbean Sea, creating barriers to dispersal for marine species.

Alfred Russel Wallace was among the first to document this phenomenon. In addition to codiscovering evolution by natural selection (see Chapter 25), Wallace founded the study of biogeography.

The Wallace Line: Barriers to Dispersal While working in the Malay Archipelago, Wallace realized that the plants and animals native to the more northern and western islands were radically different from the species found on the more southern and eastern islands. Sumatra, to the northwest, has tigers, rhinos, and other species with close relatives in southern Asia. New Guinea, to the southeast, has tree kangaroos and other species with close relatives on Australia. But both islands are tropical wet forests.

This biogeographical demarcation, now known as the **Wallace line,** separates species with Asian and Australian affinities (**FIGURE 52.2**). It exists because a deep trench in the ocean maintained a water barrier to dispersal, even when ocean levels dropped during the most recent glaciations. As a result, landforms on either side of the line remained isolated at a time when most of the other islands became connected.

The Influence of Humans The events Wallace studied occurred before the dispersal of modern humans around the globe from Africa, beginning at least 70,000 years ago (see Chapter 35). Humans have influenced the distribution of species for thousands of years by hunting, clearing forests, and physically moving organisms around. Humans have transported thousands of plants, birds, insects, and other species across physical barriers to new locations—sometimes purposefully and sometimes by accident.

If an **exotic species**—one that is not native—is introduced into a new area, spreads rapidly, and competes successfully with native species, it is said to be an **invasive species.** In North America alone, dozens of invasive species have had devastating effects on native plants and animals. Examples include kudzu (**FIGURE 52.3**, see page 1064), purple loosestrife, reed canary grass, garlic mustard, Russian thistle (also known as tumbleweed), European starlings, African honeybees, red imported fire ants, Argentine ants, and zebra mussels. Similar lists could be compiled for other regions of the world.

The microorganisms that humans introduce to new areas can also dramatically affect species distributions by causing disease. A microscopic fungus introduced to North America in the 1920s swept across the continent, wiping out most Dutch elm trees in its path in a matter of decades.

Humans continue to influence the distribution of species by affecting both biotic and abiotic factors.

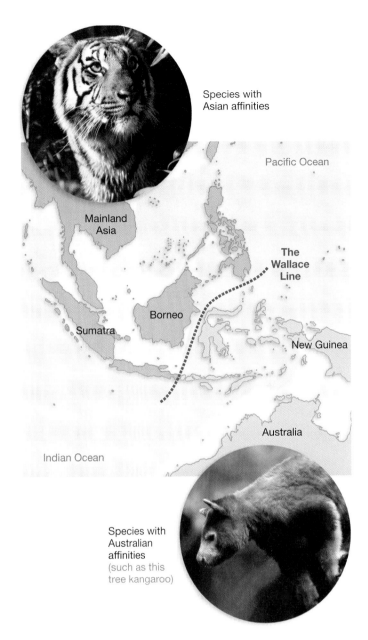

FIGURE 52.2 The "Wallace Line" Divides Organisms with Asian versus Australian Affinities.

Biotic and Abiotic Factors Interact

In many cases, biotic and abiotic factors interact to determine the distribution and abundance of organisms. How can this interaction be measured?

Consider the case of the Argentine ant, *Linepithema humile*, native to South America and rapidly expanding around the world. These small ants form unusually large supercolonies because they mix freely with other Argentine ants while fighting aggressively against native ants. This invasion creates a huge pest-control problem in homes, businesses, and farms, and it also has grave ecological implications. For example, horned lizards are in decline in southern California due to the disappearance of the native ants that they are specialized to eat.

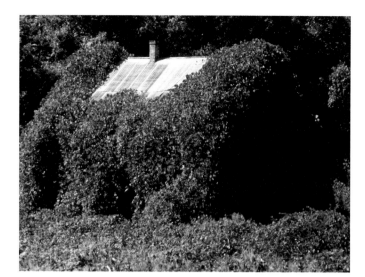

FIGURE 52.3 Introduced Species Can Cause Problems. Kudzu, imported from Asia, has taken over parts of the southeastern U.S.

Argentine ants are dispersed by humans to different continents in food shipments like coffee. But these ants don't thrive everywhere they land—they are most successful in Mediterranean-like ecosystems that have adequate moisture, winters that are not too cold, and summers that are not too hot.

Sean Menke, David Holway, and colleagues took a closer look at factors affecting dispersal of Argentine ants in southern California. They observed that soil moisture was very important in determining the species' range. They also observed that the invasive ants tend to wipe out native ants. But does competition with native ants *interact* with abiotic factors in some way, or do abiotic factors trump biotic factors in this invasive species' success?

To find out, the researchers set up 28 study plots along a contact zone between Argentine ants and native ants at the local edge of the Argentine ant's range—an edge that had been stable for over 10 years. What was keeping the invasive ants from extending their range?

The study plots were randomly assigned to four groups varying in soil moisture and competition, as shown in the "Experimental Setup" section of **FIGURE 52.4**. By counting the number of Argentine ants found in ant baits in each plot at the start of the experiment and again after 12 weeks, the researchers could measure the relative importance of soil moisture and competition to the dispersal of the invasive ants.

As the graph in the "Results" section of Figure 52.4 shows, the Argentine ants did not spread into dry study plots (red bars), even when native ants were not present. They did, however, spread into irrigated plots (blue bars), even when native ants were present—but the native ants slowed them down.

The take-home message is that the interaction of biotic and abiotic factors slowed the spread of invasive ants more than the abiotic factors alone. Argentine ants are at a competitive advantage in moist soils, but at a disadvantage compared to native ants in dry soils. This information can help scientists to predict the dispersal pattern of Argentine ants and design strategies to limit the spread of these invaders.

RESEARCH

QUESTION: Do abiotic and biotic factors interact to determine the distribution of a species?

HYPOTHESIS: Abiotic and biotic factors interact to determine the distribution of invasive Argentine ants.

NULL HYPOTHESIS: Abiotic conditions alone determine the distribution of invasive Argentine ants.

EXPERIMENTAL SETUP:

1. Set up 28 plots (10 × 10 m) in the contact zone between Argentine ants and native ants. Measure invasive ant activity in each plot using ant baits.

Invasive Argentine ants in eucalyptus grove   Contact zone   **Native ants** in the sage scrub

- Dry + native ants present
- Dry + native ants removed
- Irrigation + native ants present
- Irrigation + native ants removed

2. Vary soil moisture and presence of native ants as shown above.

3. After 12 weeks, remeasure invasive ant activity in each plot.

PREDICTION: Invasive ants will displace native ants only in irrigated plots—more so in plots where native ants were removed.

PREDICTION OF NULL HYPOTHESIS: Invasive ants will displace native ants only in irrigated plots—with no difference in plots with native ants present and native ants removed.

RESULTS:

CONCLUSION: Abiotic factors (soil moisture) and biotic factors (competition with native ants) interact in determining the distribution of Argentine ants.

FIGURE 52.4 Experimental Test of the Relative Importance of Biotic and Abiotic Factors in Limiting the Dispersal of a Species. SOURCE: Menke, S. B., et al. 2007. Biotic and abiotic controls of Argentine ant invasion success at local and landscape scales. *Ecology* 88(12): 3164–3173.

✔ **QUANTITATIVE** The bars in the graphs represent the means of several plots. How many replicate plots were used for each treatment? Why were replicates important for this experiment?

The range of every species on Earth is limited by a combination of abiotic and biotic factors that occurred in the past, and that occur in the present. The range of a particular species depends on the capacity to survive climatic conditions, the ability to find food and avoid being eaten, and the ability to disperse.

If a modest amount of irrigation can have a large effect on the dispersal of an ant species, you can imagine how abiotic factors like precipitation can have far-reaching consequences on ecology on a global scale. Let's take a closer look at climate as a major abiotic determinant of the distribution of life on the planet.

52.3 Climate Patterns

Species are adapted to the abiotic conditions present where they live. An array of abiotic factors affect organisms, including the chemistry of water in aquatic habitats, the nature of soils in terrestrial habitats, and **climate**—the prevailing, *long-term* weather conditions found in an area. **Weather** consists of the specific *short-term* atmospheric conditions of temperature, precipitation, sunlight, and wind.

Although your local meteorologist might not always make a good prediction on the night's snowfall, climate occurs in relatively predictable global patterns.

Why Are the Tropics Wet?

One of the most striking climate patterns on Earth involves precipitation. When average annual rainfall is mapped for regions around the globe, it is clear that areas along the equator receive the most moisture. In contrast, locations about 30 degrees latitude north and south of the equator are among the driest on the planet. Why do these patterns occur?

Simple questions often have fascinating answers. It turns out that a major cycle in global air circulation, called a **Hadley cell,** is responsible for making the Amazon River basin wet and the Sahara Desert dry. Hadley cells are named after George Hadley, who in 1735 conceived of the idea of enormous air circulation patterns.

To understand how Hadley cells work, put your finger on the equator in **FIGURE 52.5a** and follow the arrows.

- Air that is heated by the strong sunlight along the equator expands, lowering its pressure and causing it to rise. Note that this warm air holds a great deal of moisture, because warm water molecules tend to stay in vapor form instead of condensing into droplets.

- As air rises above the equator, it radiates heat to space. It also expands into the larger volume of the upper atmosphere, which lowers its density and temperature.

- As rising air cools, its ability to hold water declines. When water vapor cools, water condenses.

The result? High levels of precipitation occur along the equator. The story is just beginning, however.

- As more air is heated along the equator, the cooler, "older" air above Earth's surface is pushed poleward.

- When the air mass has cooled enough, its density increases and it begins to sink.

(a) Circulation cells exist at the equator ...

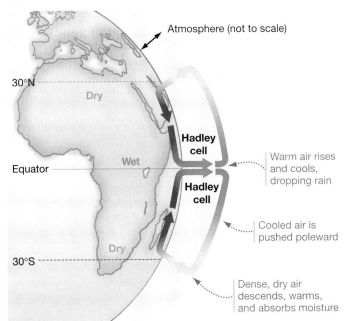

(b) ... and at higher latitudes.

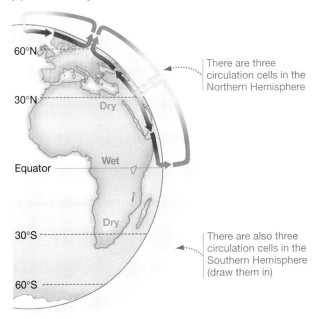

FIGURE 52.5 Global Air Circulation Patterns Affect Rainfall. Hadley cells explain why the tropics are much wetter than regions near 30° latitude north and south. Regions with rising air, such as those at the equator, tend to be wetter than regions with descending air.

- As it sinks, it absorbs more and more solar radiation reflected from Earth's surface and begins to warm while being pushed to the surface by higher-density air above it.

- As the air warms, it also gains water-holding capacity. Thus the air approaching the Earth "holds on" to its water, and little rain occurs where it returns to the surface.

The result? A band of deserts in the vicinity of 30° latitude north and south. These areas are bathed in warm, dry air.

Similar air circulation cells occur between 30° latitude and 60° latitude, and between 60° latitude and the poles, north and south (**FIGURE 52.5b**). ✔ If you understand how Hadley cells work, you should be able to add diagrams to Figure 52.5b showing the three cells that occur in the Southern Hemisphere.

Why Are the Tropics Warm and the Poles Cold?

In general, areas of the world are warm if they receive a large amount of sunlight per unit area; they are cold if they receive a small amount of sunlight per unit area. Over the course of a year, regions at or near the equator receive much more sunlight per unit area—and thus much more energy in the form of heat—than regions that are closer to the poles.

- At the equator, the Sun is often directly overhead. As a result, sunlight strikes the surface at or close to an angle of 90 degrees. At these angles, Earth receives a maximum amount of solar radiation per unit area (**FIGURE 52.6**).

- Because Earth's surface slopes away from the equator, the Sun strikes the surface at lower and lower angles moving toward the poles. When sunlight arrives at a low angle, much less energy is received per unit area.

The pattern of decreasing average temperature with increasing latitude is a result of Earth's spherical shape.

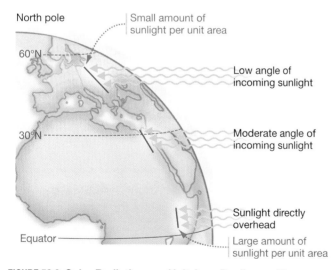

FIGURE 52.6 Solar Radiation per Unit Area Declines with Increasing Latitude. Over the course of a year, the Sun is frequently almost directly overhead at the equator. As a result, equatorial regions receive a large amount of solar radiation per unit area compared to high-latitude regions.

What Causes Seasonality in Weather?

The striking global patterns in temperature and moisture that you've just reviewed are complicated by the phenomenon of seasons. Seasons are defined as regular, annual fluctuations in temperature, precipitation, or both.

As **FIGURE 52.7** shows, seasonality occurs because Earth is tilted on its axis by 23.5 degrees. As a result of this incline, the Northern Hemisphere is tilted toward the Sun in June. In this position, it faces the Sun most directly. In June, the Northern Hemisphere presents its least-acute angle to the Sun and receives the largest amount of solar radiation per unit area.

The Southern Hemisphere, in contrast, is tilted away from the Sun in June. As a result, it presents a steep angle for incoming sunlight and receives its smallest quantity of solar radiation per unit area.

The upshot? In June it is summer in the Northern Hemisphere and winter in the Southern Hemisphere. Conversely, in December it is summer in the Southern Hemisphere and winter in the Northern Hemisphere. In March and September, the equator faces the Sun most directly. During these months, the tropics receive the most solar radiation. If Earth did not tilt on its axis, there would be no seasons.

What Regional Effects Do Mountains and Oceans Have on Climate?

The broad patterns of climate that are dictated by global heating patterns, Hadley cells, and seasonality are modified by regional effects. The most important of these are due to the presence of mountain ranges and proximity to an ocean.

FIGURE 52.8 shows how mountain ranges affect regional climate, using the Cascade Mountain range along the Pacific Coast of North America as an example.

- In this area of the world, the prevailing winds are from the west. These winds bring moisture-laden air from the Pacific Ocean onto the continent.

- As the air masses begin to rise over the mountains, the air cools and releases large volumes of water as rain.

- Once cooled air has passed the crest of a mountain range, the air is relatively dry because much of its moisture content has already been released. Areas that receive this dry air are said to be in a **rain shadow.**

The area along the Pacific Coast from northern California to southeast Alaska hosts some of the only high-latitude rain forests in the world. One area along the Pacific Coast of Washington State gets an average of 340 cm (134 in.) of precipitation each year. Mountain ranges also occur along the Pacific Coast of Southern California and Mexico, but these areas are arid because they are dominated by dry air that is part of the Hadley circulation.

Logically enough, one of the few high-latitude deserts in the world exists to the east of the Cascade Mountains. It is created by a rain shadow and averages less than 25 cm (10 in.) of moisture

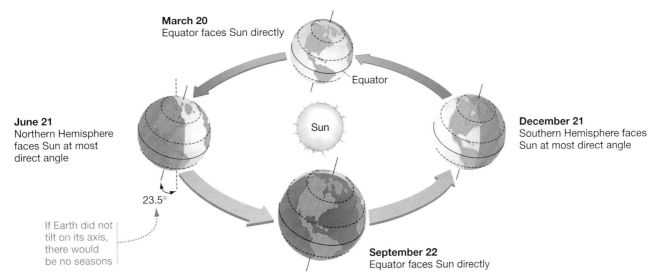

March 20
Equator faces Sun directly

Equator

June 21
Northern Hemisphere
faces Sun at most
direct angle

Sun

December 21
Southern Hemisphere faces
Sun at most direct angle

23.5°

If Earth did not
tilt on its axis,
there would
be no seasons

September 22
Equator faces Sun directly

FIGURE 52.7 Earth's Orbit and Tilt Create Seasons at High Latitudes. Seasons occur due to annual variation in the amount of solar radiation that different parts of the Earth receive.

annually. Similar rain shadows create the Great Basin desert of western North America, the Atacama Desert of South America, and the dry conditions of Asia's Tibetan Plateau.

While the presence of mountain ranges tends to produce extremes in precipitation, the presence of an ocean has a moderating influence on temperature. Water has an extremely high **specific heat,** meaning that it has a large capacity for storing heat energy (Chapter 2). Because water molecules form hydrogen bonds with each other, it takes a great deal of heat energy to boil water or melt ice—in fact, to change the temperature of water at all. The hydrogen bonds have to be broken before the water molecules themselves can begin to absorb heat and move faster, which we measure as increased temperature.

Because water has a high specific heat, it can absorb a great deal of heat from the atmosphere in summer, when the water temperature is cooler than the air temperature. As a result, the ocean moderates summer temperatures on nearby landmasses. This is why the air temperature is cooler near the ocean on a hot summer's day than it is a few miles inland. Similarly, the ocean releases heat to the atmosphere in winter, when the water temperature is warmer than the air temperature. This is why on a calm winter's day the air temperature can be warmer near the

ocean than it is a few miles inland. Because of the large capacity of water for storing heat energy, islands and coastal areas have much more moderate climates than do inland areas.

How do all these climate patterns combine to determine the distribution of organisms on Earth?

check your understanding

If you understand that . . .

• Global climate patterns are dictated by differential heating of Earth's surface, Hadley cells, and seasonality.
• These general patterns are modified at a regional scale by the presence of mountains and oceans.

✔ **You should be able to . . .**

1. Predict whether the North Pole is rainy.
2. Predict where on the globe the highest risk of skin cancer would occur and explain why.

Answers are available in Appendix A.

Air rises over mountains
and cools; rain falls

West
Moisture-laden air blows
onshore from Pacific Ocean

East
Dry air creates
desert conditions

Cascade
Mountains

This area is in
a rain shadow

FIGURE 52.8 Mountain Ranges Create Rain Shadows. When moisture-laden air rises over a mountain range, the air cools enough to condense water vapor. As a result, moisture falls to the ground as precipitation. On the other side of the mountain, little water is left to provide precipitation.

52.4 Types of Terrestrial Biomes

If you could walk from the equator in South America to the North Pole, you would notice startling changes in the organisms around you. Lush tropical forests with broad-leaved evergreen trees would give way to seasonally dry forests and then to deserts. The deserts would yield to the vast grasslands of central North America, which terminate at the boreal forests of the subarctic. If you pressed on, you would reach the end of the trees and the beginning of the most northerly community—the arctic tundra.

If you walked from sea level to the top of the Cascade Mountains of British Columbia, you would experience similar types of changes. Wet forests along the coast would give way to drier forests at higher elevations. Near the peaks, trees would thin and eventually yield to alpine tundras filled with hardy, low-growing plants. Broad-leaved evergreen forests, deserts, grasslands, and tundras are **biomes**: regions characterized by distinct abiotic characteristics and dominant types of vegetation.

Natural Biomes

The global distribution of the most common types of naturally occurring terrestrial biomes is shown in **FIGURE 52.9**—ignoring, for the moment, the effects of humans on the landscape. Notice that different ecosystems may belong to the same biome. For example, the Sonoran Desert in North America and the Sahara

Desert in Africa are both part of the same biome—subtropical desert—but each is a unique ecosystem with its own community of indigenous organisms.

Each of the terrestrial biomes found around the world is associated with a distinctive set of abiotic conditions determined largely by climate:

- *Temperature* is critical because the enzymes that make life possible work at optimal efficiency only in a narrow range of temperatures (see Chapter 8). Temperature also affects the availability of moisture: Water freezes at low temperatures and evaporates rapidly at high temperatures.

- *Moisture* is significant because it is required for life, and because terrestrial organisms constantly lose water to the environment through evaporation or transpiration. To stay alive, they must reduce water loss and replace lost water.

- *Sunlight* is essential because it is required for photosynthesis.

- *Wind* is important because it exacerbates the effects of temperature and moisture. Wind increases heat loss due to evaporation and convection, and it increases water loss due to evaporation and transpiration. Wind also has a direct physical impact on organisms such as birds, flying insects, and plants—it pushes them around.

Of the four components of climate, variation in temperature and moisture are far and away the most important in determining plant distribution and abundance. More specifically, the nature of the terrestrial biome that develops in a particular region

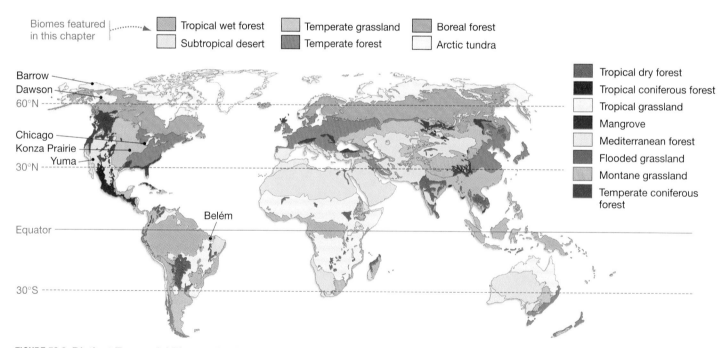

FIGURE 52.9 Distinct Terrestrial Biomes Are Found throughout the World. The six major biomes listed above the map are featured in this chapter, citing data from the labeled locations on the left.

DATA: Olson, D. M., E. Dinerstein, E. D. Wikramanayake, et al. 2001. *BioScience* 51: 933–938.

is governed by **(1)** average annual temperature and precipitation, and **(2)** annual variation in temperature and precipitation. Each biome contains species that are adapted to a particular temperature and moisture regime.

Biologists are particularly concerned with how temperature and moisture influence **net primary productivity (NPP).** NPP is defined as the total amount of carbon that is fixed per year minus the amount that is oxidized during cellular respiration. Fixed carbon that is oxidized in cellular respiration releases energy for the organism but is not used for growth—that is, production of **biomass** (see Chapter 56).

NPP is crucial because it represents the organic matter that is available as food for other organisms. In terrestrial environments, NPP is often estimated by measuring **aboveground biomass**—the total mass of living plants, excluding roots.

Photosynthesis, plant growth, and NPP are maximized on land when temperatures are warm and conditions are wet. Conversely, photosynthesis cannot occur efficiently at low temperatures or under drought stress. Photosynthetic rates are maximized in warm temperatures, when enzymes work efficiently, and in humid weather, when stomata can remain open and CO_2 is readily available.

To help you understand how temperature and precipitation influence biomes, let's take a detailed look at the six terrestrial biomes that represent Earth's most extensive natural vegetation types (see boxes on pages 1070–1072). In each case, you'll be analyzing temperature and precipitation data from a specific location ranging from the wet tropics to the arctic (labeled in Figure 52.9). Data plotted in red indicate the daily average temperature in each biome, graphed throughout the year; data plotted in purple indicate the average monthly precipitation.

- Terrestrial Biomes > Tropical Wet Forest
- Terrestrial Biomes > Subtropical Deserts
- Terrestrial Biomes > Temperate Grasslands
- Terrestrial Biomes > Temperate Forests
- Terrestrial Biomes > Boreal Forests
- Terrestrial Biomes > Arctic Tundra

Anthropogenic Biomes

The map in Figure 52.9 shows the distribution of terrestrial biomes as they would occur naturally. This map is important because it represents the product of 3.6 billion years of interplay between geology, climate, evolution, and ecology.

Now for the rest of the story. The exponential increase in the human population (Chapter 54) is driving land-use changes and other impacts that are so significant that some researchers have called for a new epoch of history, the Anthropocene (see Chapter 28). Consider these data:

- More than 75 percent of Earth's ice-free land shows evidence of direct alteration by humans—for example, farming, logging, and urban development (*not* including the effects of pollution, invasive species, or climate change).

- The remaining regions, or wildlands, account for just 11 percent of terrestrial net primary productivity.

What does the distribution of biomes look like with human impact factored in? Biogeographer Erle Ellis and colleagues have pioneered efforts to map anthropogenic biomes, or "anthromes" worldwide, and with eye-opening results. **FIGURE 52.10** zooms in on one part of the world, eastern North America, to show a close-up of the distribution of anthropogenic biomes. The remaining wildlands (pale gray-green on the map) are so sparse that they are difficult to locate.

As you can see, humans are directly affecting the distribution of terrestrial ecosystems by physically changing the landscape, one forest and suburb at a time. Humans are also changing the global distribution of ecosystems in a more indirect way—by changing the global climate.

Cities, suburbs Villages Rangeland Cropland Semi-natural Wildlands

FIGURE 52.10 Anthropogenic Biomes in Eastern North America.
Several color subcategories have been consolidated in the key for simplicity. For example, cropland can be irrigated or rain-fed, residential, sparsely populated, or remote.

DATA: Ellis, E. C., K. Klein Goldewijk, S. Siebert, et al. 2010. *Global Ecology and Biogeography* 19: 589–606.

✔ QUESTION Where do the remaining wildlands occur?

Terrestrial Biomes > Tropical Wet Forest

Tropical wet forests—also called tropical rain forests—are found in equatorial regions around the world. Plants in this biome have broad leaves and are evergreen. Favorable year-round growing conditions produce riotous growth, leading to extremely high productivity, aboveground biomass, and species diversity. It is not unusual to find over 200 tree species in a single study plot. The diversity of plant sizes and growth forms in wet forest communities produces extraordinary structural diversity, presenting a wide array of habitat types for animals.
✔You should be able to explain why vines and epiphytes increase the productivity of tropical wet forests.

Temperature Compared to other biomes, tropical wet forests show almost no seasonal variation in temperature (**FIGURE 52.11**). This is important because temperatures are high enough to support growth throughout the year.

Precipitation Even the driest month of the year (November in Belém, Brazil) receives over 5 cm (2 in.) of rainfall—considerably more than the *annual* rainfall of many deserts.

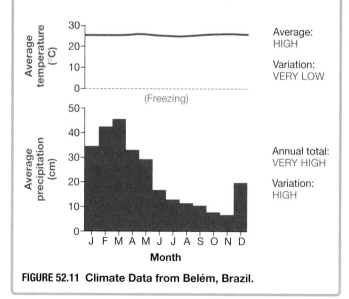

FIGURE 52.11 Climate Data from Belém, Brazil.

Terrestrial Biomes > Subtropical Deserts

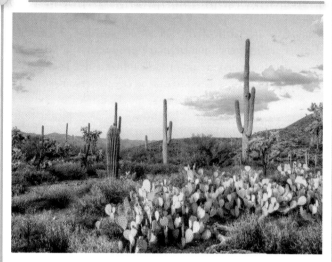

Subtropical deserts are found throughout the world at 30 degrees latitude north and south—a pattern that is explained in Section 52.3. Desert species adapt to the extreme temperatures and aridity by growing at a low rate year-round, or breaking dormancy and growing rapidly in response to any rainfall. Cacti can grow year-round because they have no leaves, or small leaves modified as spines; a thick, waxy coating; and the CAM pathway for photosynthesis (see Chapter 10). The average productivity of desert communities is extremely low.
✔You should be able to explain why most desert plant species have a small surface area.

Temperature Mean monthly temperatures in subtropical deserts vary more than in tropical wet climates (**FIGURE 52.12**). Temperatures fall below freezing in some subtropical deserts.

Precipitation The most striking feature of the subtropical desert climate is low precipitation; average annual precipitation in Yuma, Arizona, is just 7.5 cm (3 in.).

FIGURE 52.12 Climate Data from the Sonoran Desert.

Temperate grasslands are commonly called prairies in North America and steppes in central Eurasia. Grasses are the dominant life-form in temperate grasslands for one of two reasons: (**1**) Conditions are too dry to enable tree growth, or (**2**) encroaching trees are burned out by prairie fires. In contrast to deserts, plant life is extremely dense—virtually every square centimeter is filled, not only on the surface, but also underground where root systems are extensive. Although the productivity of temperate grasslands is generally lower than that of forest communities, grassland soils are often highly fertile. ✔ You should be able to explain why grasslands of North America and Eurasia are the breadbaskets of those continents.

Temperature In the **temperate** zone, plant growth is possible only in spring, summer, and fall months when moisture and warmth are adequate (**FIGURE 52.13**).

Precipitation Conditions are still slightly too hot and dry to support forests; average annual precipitation in Konza Prairie (Kansas) is 83.4 cm (32.5 in.).

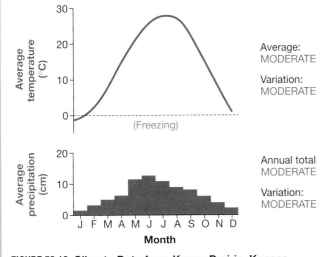

FIGURE 52.13 Climate Data from Konza Prairie, Kansas.

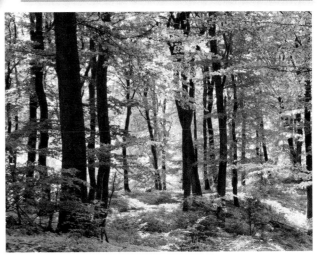

Temperate forests in North America and Europe are dominated by deciduous species, which are leafless in winter and grow new leaves each spring and summer. Needle-leaved evergreens are also common. Most temperate forests have productivity levels that are lower than those of tropical forests yet higher than those of deserts or grasslands. The level of diversity is also moderate. Temperate forests have moderate productivity simply because temperatures do not support photosynthesis year-round. ✔ You should be able to predict the vegetation found in regions with climate characteristics intermediate between those of grasslands and temperate forests.

Temperature Temperate forests experience moderate fluctuations in mean monthly temperatures, sometimes falling below freezing, as in the Chicago area (**FIGURE 52.14**).

Precipitation Precipitation in temperate forests is moderate, and relatively constant throughout the year compared to grassland climates; average annual precipitation in the Chicago area is 85 cm (34 in.), exceeding 5 cm (2 in.) during most months.

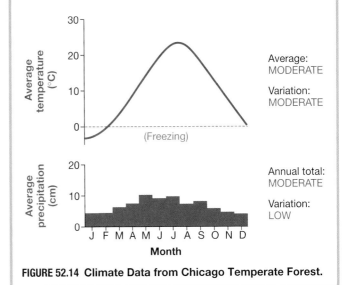

FIGURE 52.14 Climate Data from Chicago Temperate Forest.

Boreal forests, or **taiga,** stretch across most of Canada, Alaska, Russia, and northern Europe and are dominated by highly cold-tolerant conifers, including pines, spruce, fir, and larch trees. Productivity is low, but aboveground biomass is high because slow-growing tree species may be long lived and gradually accumulate large standing biomass. Boreal forests also have exceptionally low species diversity—there are just seven or fewer tree species in Alaska. They also lack most elements of structure seen in tropical or even temperate forests—they often have just a tree layer and a ground layer. ✔ You should be able to predict how the global distribution of boreal forests will change in response to global warming.

Temperature Boreal forests have very cold winters and cool, short summers—an annual fluctuation of over 40°C in the Yukon Territory (**FIGURE 52.15**), and up to 70°C in some areas.

Precipitation Annual precipitation is low, but temperatures are so cold that evaporation is minimal; as a result, moisture is usually abundant enough to support tree growth.

FIGURE 52.15 Climate Data from Yukon Territory.

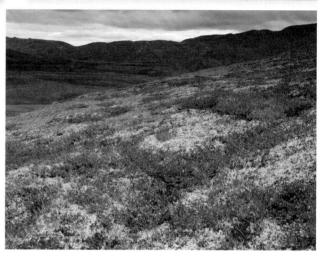

Arctic tundra occurs throughout the arctic regions of the Northern Hemisphere. Tundra has low species diversity, low productivity, and low aboveground biomass—and it is treeless. Most tundra soils are in the perennially frozen state known as **permafrost,** which limits both the release and uptake of nutrients. Unlike desert biomes, however, the ground surface in tundra communities is completely covered with plants or lichens. Animal diversity also tends to be low, although insect abundance—particularly of biting flies—can be staggeringly high. ✔ You should be able to explain why the types of vegetation in arctic tundra are similar to the types found in alpine tundra, which occurs at high elevation.

Temperature The growing season is very short, 10–12 weeks at most in Barrow, Alaska; for the remainder of the year, temperatures are below freezing (**FIGURE 52.16**).

Precipitation Precipitation on the arctic tundra is extremely low, but many arctic soils are saturated year-round due to extremely low evaporation rates.

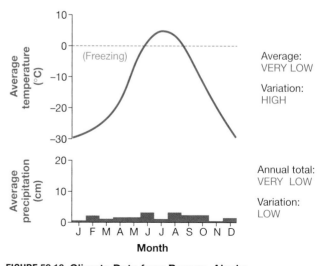

FIGURE 52.16 Climate Data from Barrow, Alaska.

How Will Global Climate Change Affect Terrestrial Biomes?

It is now well established that human-caused emissions of greenhouse gases like CO_2 are causing climate change (see Chapter 56). Climate has changed frequently and radically throughout the history of life, but recent increases in average temperatures around the globe, combined with a projected increase of up to 5.8°C over the next 100 years, represent one of the most rapid periods of climate change since life began.

How will global climate change affect the overall distribution of life on Earth? Biologists use four tools to make predictions:

1. *Simulation studies* are based on computer models of weather patterns in local regions. By increasing average temperatures in these models, researchers can predict how wind and rainfall patterns, storm frequency and intensity, and other aspects of weather and climate will change. In effect, simulation studies predict how temperature and precipitation profiles, like those in this chapter's "Terrestrial Biome" boxes, will change—and how the distribution of organisms might change as a result.

2. *Observational studies* are based on long-term monitoring at fixed sites around the globe. Researchers are documenting changes in key physical variables—temperature and precipitation—as well as changes in the distribution and abundance of organisms, such as the timing of insect emergence and bird migrations. By extrapolation, they can predict how change may continue at these sites.

3. *Historical studies* examine the relationship between CO_2 levels, climate change, and the distribution and abundance of organisms during events that occurred millions of years ago. For example, researchers have been able to measure the pattern of change in the biodiversity of the Amazonian rain forest as a function of past climate fluctuations. These patterns provide valuable data for models designed to predict future changes.

4. *Experiments* are designed to simulate changed climate conditions and to record responses by the organisms present.

Even though the scientific community did not become alert to the reality of global climate change until the 1980s, a large number of multiyear experiments have already been done. Because arctic regions are projected to experience particularly dramatic changes in temperature and precipitation, many of these experiments have been conducted in the tundra biome.

FIGURE 52.17 shows one example of how these experiments are set up. Transparent, open-topped chambers are placed at random locations around a study site, and the characteristics of the vegetation inside the chambers are recorded annually. The same measurements are made in randomly assigned study plots that lack chambers.

Across experiments, the average annual temperature inside the open-topped chambers increases by 1−3°C. This increase is consistent with conservative projections for how temperatures will change over the next century.

FIGURE 52.17 Experimental Increases in Temperature Change Species Composition in Arctic Tundra. The transparent, open-topped chambers shown here are an effective way to increase average annual temperature. Inside the chambers, species composition changes relative to otherwise similar sites nearby.

✔**QUESTION** Why does a clear chamber like this increase average temperature inside?

In an analysis of data from 11 such study sites scattered throughout the arctic, several strong patterns emerged:

- Overall, species diversity decreases.

- Compared with control plots, grasses and shrubs increase inside the chambers, and mosses and lichens decrease.

These experimental results support simulation and observational studies predicting that arctic tundra environments are giving way to boreal forest. Biomes are changing right before our eyes.

Note that not all areas are expected to transition smoothly and incrementally from one biome to another based on changes in temperature and/or rainfall. There is growing evidence that small perturbations in climate in some areas will cause abrupt changes to biomes. For example, scientists predict that decreased rainfall over the Amazon will cause a rapid transition from forest to grassland due to positive feedback loops in the system. Decreases in forest cover will decrease the rate of rain-cloud formation, which will further decrease the rainfall. Decreased rainfall will increase the prevalence of fires, which will further decrease forest cover. The predicted result is a rapid change—a tipping point—from a forested stable state to a grassland stable state, with dramatic effects on biodiversity.

Other simulation and observational studies have converged on a remarkable conclusion: Increases in average global temperature are increasing variability in temperature and precipitation. Stated another way, global climate change is making climates more extreme. Average temperature and average annual precipitation in many regions may not be changing substantially, even though variation is. For example, thus far the frequency of monsoon rains, cyclones, hurricanes, and other types of storm events is not changing. However, the amount of precipitation and the wind speeds associated with these events are increasing. The severity of droughts is also on the rise.

Both human land-use patterns and climate change cause dramatic alterations in the composition and distribution of terrestrial biomes. These factors also affect aquatic biomes. Let's take a closer look.

check your understanding

If you understand that . . .

- In terrestrial environments, distinctive biomes develop based on two aspects of temperature and moisture: average value and variability over the year.
- Human land-use practices are changing terrestrial biomes.
- Global climate change is affecting terrestrial biomes.

✔ **You should be able to . . .**

1. Predict whether tropical dry forests, which have high year-round temperatures but a distinct dry season, are more or less productive than tropical wet forests.

2. Predict how the biome where you live will change over the course of your lifetime.

Answers are available in Appendix A.

52.5 Types of Aquatic Biomes

About 70 percent of the surface of the Earth is covered in water. What types of biomes occur there? And what are the characteristics that distinguish aquatic biomes?

Let's think back to the sockeye salmon in the beginning of the chapter. Most salmon begin their lives in streams, travel to the ocean to feed and grow, then return to their natal streams to mate and die. Thus salmon migrate between a freshwater aquatic biome (a stream) and a marine aquatic biome (the ocean). The major abiotic factors that distinguish streams from oceans include salinity, water depth, water flow, and nutrient availability. Let's consider each factor in turn.

Salinity

Water is an excellent solvent—one reason why it is so essential to life (see Chapter 2). The water in clouds is nearly pure, but as rainwater percolates through soils and flows across rocks, it picks up solutes. The freshwater of ponds, lakes, and rivers has very low solute concentrations. However, as this freshwater steadily flows into the ocean, it continually adds solutes. Over millions of years, this process has resulted in a relatively high concentration of solutes in seawater.

The proportion of solutes dissolved in water determines its **salinity,** generally measured as the number of grams of solute per kilogram of water—a unitless number described as parts per thousand. The salinity of freshwater varies from about 0.06 to 0.3 parts per thousand, whereas the salinity of the open ocean is fairly constant at 35 parts per thousand.

What makes the ocean "salty"? Solutes with a positive charge—like sodium, potassium, magnesium, and calcium—combine with solutes with a negative charge—like chlorine, bromine, and fluorine—to form salts. The concentration of different salts in the ocean is determined by their solubility in water. For example, calcium carbonate has a low solubility in water, so it precipitates as limestone at the bottom of the ocean. But sodium chloride (familiar to you as table salt) has such a high solubility in water that it makes up about 86 percent of sea salt.

Salinity has dramatic effects on osmosis and water balance in organisms (see Chapters 38 and 43); species have physiological adaptations that allow them to cope with a specific range in salinity. Thus salinity is a major determinant of the distribution of organisms in aquatic biomes.

Water Depth

Water absorbs and scatters light, so the amount and types of wavelengths available to organisms change dramatically as water depth increases. Light availability has a major influence on productivity.

At the water surface, all wavelengths of light are equally available. But as **FIGURE 52.18a** shows, ocean water dramatically removes light in the red region of the visible spectrum, resulting in the familiar blue hue of underwater scenes (**FIGURE 52.18b**). This is important because wavelengths in the red region are unavailable for photosynthesis underwater, despite the high absorption peaks for red wavelengths in key photosynthetic pigments like chlorophyll (see Chapter 10).

The quantity of light available to organisms also diminishes rapidly with increasing depth. In pure seawater, the total amount of light available at a depth of 10 m is less than 40 percent of what it is at the surface; virtually no light reaches depths greater than 40 m (131 ft).

Further, seawater is rarely pure and may contain high concentrations of algae, sediments, and other light-reducing components. The **turbidity,** or cloudiness, of water is an important determinant of light penetration. Turbidity is sometimes caused by natural processes such as erosion of river sediments by floodwaters and the erosion of coastal sediments by wave action. Turbidity is also caused by many human activities, such as runoff from agricultural fields and algae blooms caused by nutrient pollution (see Chapter 56).

Water Flow

Water movement is a critical factor in aquatic ecosystems because it presents a physical challenge. It can literally sweep organisms away.

Organisms that live in fast-flowing streams have to cope with the physical force of the water, which constantly threatens to move them downstream. Stream-dwelling organisms like

(a) Red wavelengths are not available underwater.

Note that many organisms require wavelengths of about 680 nm for peak photosynthetic efficiency

(b) Blue wavelengths dominate underwater.

FIGURE 52.18 Light Is Limited Underwater.

DATA: Saffo, M. B. 1987. *BioScience* 37: 654–664.

mayfly larvae have streamlined shapes and behavioral adaptations that help them to maintain their position in the stream. Marine organisms that live in intertidal regions are also exposed to water flow—sometimes to violent wave action during storms—and to tidal currents. A growing field of biomechanics (see Chapter 48) studies the mechanical implications of water flow to the physiology, evolution, and ecology of aquatic organisms.

Water flow can also influence non-mechanical aspects of the abiotic environment, including the availability of oxygen, light, and nutrients. For example, the fast-moving water near the source of streams tends to be high in O_2, clear, and low in nutrients. The slower-flowing water downstream tends to be lower in O_2, more turbid, and more nutrient rich.

Nutrient Availability

In many aquatic ecosystems, nutrients such as nitrogen and phosphorus are in short supply. If water is moving, such as in a small salmon stream, nutrients tend to be washed away. If water is still, nutrients tend to fall to the bottom and collect in the form

PROCESS: OCEAN UPWELLING

1. Winds blow: Along the coast of Peru, the prevailing winds blow north, moving water at the surface.

2. Surface water moves: As the Earth rotates, the moving surface water is forced offshore.

3. Upwelling: As surface water leaves, it is replaced by nutrient-laden water welling up from the bottom.

FIGURE 52.19 Ocean Upwellings Bring Nutrients to the Surface.

of debris. Nutrient levels are important: The scarcity of nutrients limits growth rates in the photosynthetic organisms that provide food for other species.

Let's consider two important events that affect nutrient availability in lakes and oceans. In both cases, they are mechanisms that bring nutrients from the bottom up to the water surface, where they can nourish the growth of photosynthetic species.

Ocean Upwelling In the oceans, nutrients in the sunlit surface waters are constantly lost in the form of dead organisms that rain down into the depths. In certain coastal regions of the world's oceans, however, nutrients are brought up to the surface by currents that cause upwellings.

FIGURE 52.19 shows how this happens off the coast of Peru. Here the prevailing winds blow along the coastline, pushing surface water to the north. Because the Earth rotates constantly, this wind-driven water current is slowly moved offshore. As the surface water moves away from the coast, it is steadily replaced by water moving up from the ocean bottom.

The upwelling water is nutrient rich. In effect, it recycles nutrients that earlier had fallen to the ocean floor. The nutrients support luxuriant growths of photosynthetic cells, which feed an array of small grazing animals, which in turn feed small fish called anchoveta. When currents are favorable and upwelling is steady, the anchoveta fishery is the most productive in the world.

Lake Turnover Because they were scoured by glaciers as recently as 10,000 years ago, the higher latitudes of the world host many lakes. Each year, these bodies of water undergo remarkable changes known as the spring and fall **turnovers.**

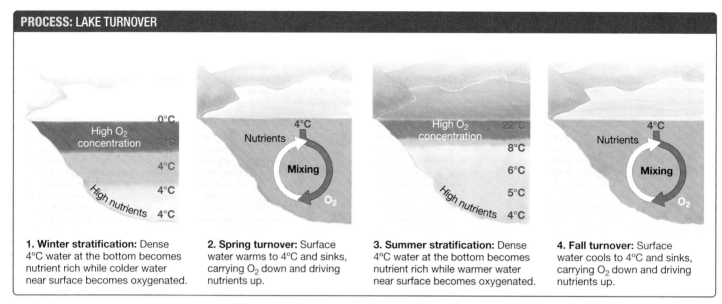

PROCESS: LAKE TURNOVER

High O₂ concentration
0°C
4°C
4°C
4°C
High nutrients

1. Winter stratification: Dense 4°C water at the bottom becomes nutrient rich while colder water near surface becomes oxygenated.

Nutrients
4°C
Mixing
O₂

2. Spring turnover: Surface water warms to 4°C and sinks, carrying O₂ down and driving nutrients up.

High O₂ concentration
22°C
8°C
6°C
5°C
4°C
High nutrients

3. Summer stratification: Dense 4°C water at the bottom becomes nutrient rich while warmer water near surface becomes oxygenated.

Nutrients
4°C
Mixing
O₂

4. Fall turnover: Surface water cools to 4°C and sinks, carrying O₂ down and driving nutrients up.

FIGURE 52.20 In Temperate Regions, Lakes Turn Over Each Spring and Fall.

The spring and fall lake turnovers occur in response to changes in air temperature. To see how this happens, examine the temperature profiles in **FIGURE 52.20**.

Note that in winter, water at the surface of a northern lake is locked up in ice at a temperature of 0°C. The water just under the ice is slightly warmer and is relatively oxygen rich because it was exposed to the atmosphere as winter set in. Because water is densest at 4°C, the water at the bottom of the lake is at 4–5°C. The bottom water is also oxygen poor because organisms that decompose falling organic material use up available oxygen in the process.

A gradient in temperature such as this is called a **thermocline** ("heat-slope"). Thermal stratification is said to occur.

The ice begins to melt, however, when spring arrives. The temperature of the water on the surface rises until it reaches 4°C. The water at the surface of the lake is now heavier than the water below it. As a result, the water on the surface sinks. The water at the bottom of the lake is displaced and comes to the surface, completing the spring turnover.

During the spring turnover, water at the bottom of the lake carries sediments and nutrients from the bottom up to the surface. This flush of nutrients triggers a rapid increase in the growth of photosynthetic algae and bacteria; biologists call it the spring bloom. The lake then stratifies again in the summer.

When temperatures cool in the fall, the water at the surface reaches a temperature of 4°C and sinks, displacing water at the bottom and creating the fall turnover. The fall turnover is important because it brings oxygen-rich water from the surface down to the bottom, and because it again brings nutrients from the bottom up to the sunlit surface waters.

Without the spring and fall turnovers, most freshwater nutrients would remain on the bottom of lakes. These aquatic ecosystems would be much less productive as a result.

The boxes that follow summarize the types of aquatic biomes that are classified based on whether they are freshwater or marine as well as on their water depth, water flow, and nutrient availability.

- Aquatic Biomes > Freshwater > Lakes and Ponds
- Aquatic Biomes > Freshwater > Wetlands
- Aquatic Biomes > Freshwater > Streams
- Aquatic Biomes > Freshwater/Marine > Estuaries
- Aquatic Biomes > Marine > Oceans

How Are Aquatic Biomes Affected by Humans?

Just as on land, humans are affecting aquatic biomes worldwide by causing both direct and indirect effects. For example:

Direct Effects   Humans cause physical changes to aquatic ecosystems by filling or draining wetlands, damming streams, and removing water from ponds and rivers for agricultural irrigation. Humans also cause biological changes by transporting organisms—such as the invasive zebra mussel, which has become a scourge in the Great Lakes of North America and other regions of the world—and overexploiting marine resources, especially top predators such as tuna and sharks, causing cascading effects on food webs (see Chapters 56 and 57).

Indirect Effects   Global climate change influences aquatic ecosystems by the same mechanisms discussed for terrestrial ecosystems. In addition, increased atmospheric CO_2 is causing waterways to become more acidic, limiting the ability of organisms like corals to build skeletons, and having other far-reaching effects (see Chapter 56).

The importance of climate and other abiotic factors to the distribution and abundance of organisms was the main focus of this chapter. Next, let's take a closer look at the biotic factors. How do interactions among organisms of the same species and interactions among different species affect the overall distribution and abundance of organisms?

Lakes and ponds are distinguished from each other by size. Ponds are small; lakes are large enough that the water can be mixed by wind and wave action. Most natural lakes and ponds occur in high latitudes—they formed in depressions created by the scouring action of glaciers thousands of years ago.

Water depth Biologists describe the structure of lakes and ponds by naming zones, some of which overlap (**FIGURE 52.21**).

- The **littoral** ("seashore") **zone** consists of the shallow waters along the shore, where flowering plants are rooted.

- The **limnetic** ("lake") **zone** is offshore and comprises water that receives enough light to support photosynthesis.

- The **benthic** ("depths") **zone** occurs at the substrate.

- Regions of the littoral, limnetic, and benthic zones that receive sunlight are part of the **photic zone.**

- Portions of a lake or pond that do not receive sunlight make up the **aphotic zone.**

Water flow and nutrient availability Water movement is driven by wind and temperature. The littoral and limnetic zones are typically much warmer and better oxygenated than the benthic zone, but the benthic zone is relatively nutrient rich because dead and decomposing bodies sink and accumulate there. The movement of nutrients from the benthic to littoral zones is often dependent on seasonal turnovers.

Organisms Cyanobacteria, algae, and other microscopic organisms, collectively called **plankton,** live in the photic zone, as do the fish and small crustaceans—organisms related to shrimp and crabs—that eat them. In shallow parts of the photic zone, rooted plants are common. Animals (invertebrates and fish) that consume dead organic matter, or **detritus,** are particularly abundant in the benthic zone. ✔You should be able to predict which zone of lakes and ponds produces the most grams of organic material per square meter per year, and to explain your logic.

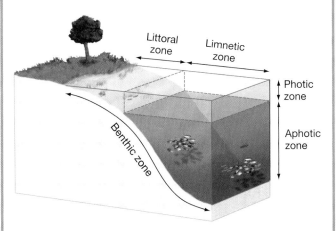

FIGURE 52.21 Lakes Have Distinctive Zones.

Wetlands are shallow-water habitats where the soil is saturated with water for at least part of the year. They are usually distinguished from terrestrial habitats by the presence of "indicator plants" that grow only in saturated soils.

Water depth Wetlands are distinct from lakes and ponds for two reasons: They have only shallow water, and they have **emergent vegetation**—meaning plants that grow above the surface of the water. All or most of the water in wetlands receives sunlight; emergent plants capture sunlight before it strikes the water.

Water flow and nutrient availability Freshwater marshes and swamps are wetland types characterized by a slow but steady flow of water. **Bogs,** in contrast, develop in depressions where water flow is low or nonexistent. If water is stagnant, oxygen is used up during the decomposition of dead organic matter faster than it enters via diffusion from the atmosphere. As a result, bog water is oxygen poor or even anoxic, resulting in slow decomposition rates. Organic acids and other acids build up, lowering the pH of the water. At low pH, nitrogen becomes unavailable to plants. As a result, bogs are nutrient poor. Marshes and swamps, in contrast, are relatively nutrient rich. In terms of total area, bogs are nearly as common as marshes and swamps but occur mostly at northern latitudes.

Organisms The combination of acidity, lack of available nitrogen, and anoxic conditions makes bogs extremely unproductive habitats. Marshes and swamps, in contrast, offer ample supplies of oxygenated water and sunlight, along with nutrients available from rapid decomposition. As a result, they are extraordinarily productive. **Marshes** lack trees and typically feature grasses, reeds, or other nonwoody vegetation; **swamps** are dominated by trees and shrubs. Because their physical environments are so different, there is little overlap in the types of species found in bogs, marshes, and swamps. ✔You should be able to explain why carnivorous plants, which capture and digest insects, are relatively common in bogs but rare in marshes and swamps.

Streams are bodies of water that move constantly in one direction. Creeks are small streams; rivers are large streams.

Water depth Most streams are shallow enough that sunlight reaches the bottom. Availability of sunlight is usually not a limiting factor for organisms, except when turbidity is high.

Water flow and nutrient availability The structure of a typical stream varies along its length. Where it originates at a mountain glacier, lake, or spring, a stream tends to be cold, narrow, and fast. As it descends toward a lake, ocean, or larger river, a stream accepts water from tributaries and becomes larger, warmer, and slower. Oxygen levels tend to be high in fast-moving streams because water droplets are exposed to the atmosphere when moving water splashes over rocks or other obstacles. In contrast, slow-moving streams that lack rapids tend to become relatively oxygen poor. Also, cold water holds more oxygen than warm water does (see Chapter 45). Slow-moving streams tend to be more nutrient rich than fast-moving streams, because decaying matter does not flush away as quickly.

Organisms It is rare to find photosynthetic organisms in small, fast-moving streams; nutrient levels tend to be low, and most of the organic matter present consists of leaves and other materials that fall into the water from outside the stream. Fish, insect larvae, mollusks, and other animals have adaptations that allow them to maintain their positions in the fast-moving portions of streams. As streams widen and slow down, conditions become more favorable for the growth of algae and plants, and the amount of organic matter and nutrients increases. As a result, the same stream often contains completely different types of organisms near its source and near its end, or mouth. ✔You should be able to explain why, in terms of their ability to perform cellular respiration, fish species found in cold, fast-moving streams tend to be much more active than fish species found in warm, slow-moving streams.

Estuaries form where rivers meet the ocean—meaning that freshwater mixes with salt water. In essence, an estuary includes saline marshes (from slightly to highly saline) as well as the body of water that moves in and out of these environments. Salinity varies with **(1)** changes in river flows—it declines when the river floods and increases when the river ebbs—and **(2)** with proximity to the ocean. Salinity has dramatic effects on osmosis and water balance (see Chapters 38 and 43); species that live in estuaries have physiological adaptations that allow them to cope with variations in salinity.

Water depth Most estuaries are shallow enough that sunlight reaches the substrate. Water depth may fluctuate dramatically, however, in response to tides, storms, and floods.

Water flow and nutrient availability Water flow in estuaries fluctuates daily and seasonally due to tides, storms, and floods. The fluctuation is important because it alters salinity, which in turn affects which types of organisms are present. Estuaries are nutrient rich because nutrient-laden sediments are deposited when flowing river water slows as it enters the ocean.

Organisms Because the water is shallow and sunlit, and because nutrients are constantly replenished by incoming river water, estuaries are among the most productive environments on Earth. They serve as a nursery for young fish, which feed on abundant vegetation, benthic invertebrates, and plankton while hiding from predators. Estuaries also tend to have high species diversity—some species may occur throughout the estuary while others occupy specialized zones according to salinity and other abiotic factors. Estuaries are also important feeding grounds for both residential and seasonal bird populations. ✔You should be able to explain why estuaries and freshwater marshes contain few of the same species, even though they are both shallow-water habitats filled with rooted plants.

Oceans form a continuous body of salt water and are remarkably uniform in chemical composition. Regions within an ocean vary markedly in their physical characteristics, however, and have profound effects on the organisms found there.

Water depth Biologists describe the structure of oceans by naming zones (**FIGURE 52.22**).

- The **intertidal** ("between tides") **zone** consists of a rocky, sandy, or muddy beach that is exposed to the air at low tide but submerged at high tide.

- The **neritic zone** extends from the intertidal zone to depths of about 200 m. Its outermost edge is defined by the end of the **continental shelf**—the gently sloping, submerged portion of a continental plate.

- The **oceanic zone** is the "open ocean"—the deepwater region beyond the continental shelf.

- The bottom of the ocean is the **benthic zone.**

- The intertidal and sunlit regions of the neritic, oceanic, and benthic zones make up a **photic zone.**

- Areas that do not receive sunlight are in an **aphotic zone.**

The ocean is indeed very deep (average depth = 3.7 km), but its depth is small compared to its breadth, like a piece of paper.

Water flow and nutrient availability Water movement in the ocean is dominated by different processes at different depths. In the intertidal zone, tides and wave action are the major influences. In the neritic zone, currents that bring nutrient-rich water from the benthic zone of the deep ocean toward shore have a heavy impact. Throughout the ocean, large-scale currents circulate water in the oceanic zone in response to prevailing winds and the Earth's rotation. In general, nutrient availability is dictated by water movement. The neritic and intertidal zones are relatively nutrient rich because they receive nutrients from rivers and upwelling. The oceanic zone, in contrast, constantly loses nutrients due to a steady rain of dead organisms drifting to the benthic zone.

Organisms Each zone in the ocean is populated by distinct species that are adapted to the physical conditions present. Organisms that live in the intertidal zone must be able to withstand physical pounding from waves and desiccation and high temperatures at low tide. Productivity is high, however, due to the availability of sunlight and nutrients contributed by estuaries as well as by currents that sweep in nutrient-laden sediments from offshore areas.

Productivity is also high on the outer edge of the neritic zone, due to nutrients contributed by upwelling at the edge of the continental plate. Almost all of the world's major marine fisheries exploit organisms that live in the neritic zone. In the tropics, shallow portions of the neritic zone may host **coral reefs**—large assemblages of colonial marine corals (related to sea anemones; see Chapter 33) that provide habitat for many other organisms. Mutualistic algae living within the corals are the autotrophs in coral reef ecosystems. Because the water is warm and sunlight penetrates to the ocean floor in these habitats, coral reefs are among the most productive environments in the world (see Chapter 56).

If coral reefs are the rain forests of the ocean, then the oceanic zone is the desert. Sunlight is abundant in the photic zone of the open ocean, but nutrients are extremely scarce. When the photosynthetic organisms and the animals that feed on them die, their bodies drift downward out of the photic zone and are lost. In the open ocean, the benthic zone can be several kilometers below the surface, and there is no mechanism for bringing nutrients back up from the bottom. The aphotic zone of the open ocean is also extremely unproductive because light is absent and photosynthesis is impossible. Most organisms present in the aphotic zone survive on the rain of dead bodies from the photic zone. ✔You should be able to predict whether animals that live in the aphotic zone have functioning eyes.

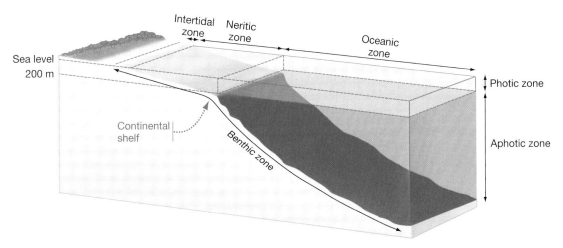

FIGURE 52.22 Oceans Have Distinctive Zones.

Sea level
200 m
Intertidal zone
Neritic zone
Oceanic zone
Photic zone
Aphotic zone
Continental shelf
Benthic zone

If you understand . . .

52.1 Levels of Ecological Study

- Biologists study ecology at five main levels: (1) organisms, (2) populations, (3) communities, (4) ecosystems, and (5) the biosphere.

- The goal of organismal ecology is to understand how individuals interact with other organisms and their physical surroundings.

- Population ecology focuses on how and why populations grow or decline in time and space.

- Community ecology is the study of how different species interact.

- Ecosystem ecology analyzes interactions between organisms and their abiotic environment—particularly the flow of energy and nutrients.

- Global ecology studies processes that transcend ecosystems, such as the effects of climate change.

- To preserve endangered species and communities, conservation biologists apply analytical methods and results from all levels.

✔ You should be able to explain why these five levels are hierarchical.

52.2 What Determines the Distribution and Abundance of Organisms?

- For each species, a unique combination of abiotic and biotic factors determines where individuals live and the size of populations.

- Understanding historical events, such as the movement of entire continents, is important to interpreting current patterns of species distributions.

- Abiotic and biotic factors often interact to produce a different effect on species distributions than either type of factor would have on its own.

✔ You should be able to predict what would happen to the range of Argentine ants in southern California if dry scrubland is converted to irrigated agricultural fields.

52.3 Climate Patterns

- Climate varies around the globe because sunlight is distributed asymmetrically—equatorial regions receive on average more solar radiation than do regions toward the poles.

- Hadley cells create bands of wet and dry habitats, and the tilt of Earth's axis causes seasonality in the amount of sunlight that non-equatorial regions receive.

- The presence of mountains can create local areas of wet or dry habitats, and proximity to an ocean moderates temperatures in nearby terrestrial habitats.

✔ You should be able to explain how climate would be expected to vary around the Earth if the planet were cylindrical and spun on an axis that was at a right angle to the Sun.

52.4 Types of Terrestrial Biomes

- Terrestrial biomes are defined by climatic regimes and distinct types of terrestrial vegetation.

- The major terrestrial biomes include tropical wet forest, subtropical desert, temperate grassland, temperate forest, boreal forest, and arctic tundra.

- Because photosynthesis is most efficient when temperatures are warm and water supplies are ample, the productivity and degree of seasonality in biomes varies with temperature and moisture.

- Humans are changing the terrestrial landscape, creating new anthropogenic biomes—such as cities, suburbs, farms, and ranchlands—in place of natural biomes.

- Global climate change is causing temperature profiles and precipitation patterns to change in ecosystems around the planet.

✔ You should be able to predict how global climate change may affect terrestrial biomes due to increased transpiration from plants.

52.5 Types of Aquatic Biomes

- Salinity, water depth, water flow, and nutrient availability are key factors determining types of aquatic biomes.

- Aquatic biomes include lakes and ponds, wetlands, streams, estuaries, and oceans.

- Humans are affecting aquatic biomes by physically altering them (such as by damming a river), by changing their chemistry (such as by polluting and causing ocean acidification), and by changing species compositions (such as by introducing invasive species).

✔ You should be able to explain why the open ocean is similar to the desert, and why it is not.

(MB) MasteringBiology

1. **MasteringBiology Assignments**

   Tutorials and Activities Adaptations to Biotic and Abiotic Factors; Aquatic Biomes; Tropical Atmospheric Circulation

   Questions Reading Quizzes, Blue-Thread Questions, Test Bank

2. eText Read your book online, search, take notes, highlight text, and more.

3. The Study Area Practice Test, Cumulative Test, BioFlix® 3-D Animations, Videos, Activities, Audio Glossary, Word Study Tools, Art

You should be able to . . .

1. Name the five main levels of study in ecology.

2. Why are certain exotic species considered "invasive"?
   a. They are found in areas where they are not native.
   b. They were introduced by humans—often accidentally.
   c. They spread aggressively and displace native species.
   d. They benefit from being in a new environment.

3. Where do rain shadows exist?
   a. the part of a mountain that receives prevailing winds and heavy rain
   b. the region beyond a mountain range that receives dry air masses
   c. the region along the equator where precipitation is abundant
   d. the region near 30° N and 30° S latitude that receives hot, dry air masses

4. What is the predominant type of vegetation in a tropical wet forest?
   a. shrubs and bunchgrasses
   b. herbs, grasses, and vines
   c. broad-leaved deciduous trees
   d. broad-leaved evergreen trees

5. What is one expected consequence of global climate change?
   a. Average rainfall will increase.
   b. Average rainfall will decrease.
   c. Variability in rainfall will increase.
   d. We cannot make predictions about future rainfall.

6. Typically, where are oxygen levels highest and nutrient levels lowest in a stream?
   a. near its source
   b. near its mouth, or end
   c. where it flows through a swamp or marsh
   d. where it forms an estuary

7. Write questions about the ecology of humans that are relevant to each of the five levels of ecological study.

8. Explain how biotic and abiotic factors determine the distribution of invasive Argentine ants.

9. Why does the Australian Outback receive so little rainfall?
   a. The Australian Outback is in a rain shadow.
   b. A Hadley cell drops warm, dry air in this region.
   c. A Hadley cell pulls warm, moist air up from this region.
   d. The climate in Australia is moderated by the surrounding ocean.

10. Contrast the productivities of the intertidal, neritic, and oceanic zones of marine environments. Explain why large differences in productivity exist.

11. Compare and contrast the terrestrial biomes found at increasing elevation on a mountain with the biomes found at increasing latitude in Figure 52.9.

12. Compare the distribution of the natural terrestrial biomes of eastern North America with the distribution of anthropogenic biomes.

13. Mars has an even more pronounced tilt than Earth does. Does Mars experience seasons? Why or why not?
    a. Yes, the tilt will cause seasons because the Northern and Southern Hemispheres will be tilted toward the Sun, then away, at different times of the year.
    b. Yes, because the tilt will cause the orbit of Mars to be closer to the Sun during part of the year.
    c. No, because the tilt of Mars does not affect the distance of its orbit from the Sun.
    d. No, because the tilt of Mars does not affect the average amount of solar radiation received.

14. Very similar reptile fossils have been discovered on the east coast of South America and the west coast of Africa. Pose a hypothesis to explain this distribution.

15. Scientists are concerned about the future of species that are adapted to high elevations. Do you think their concern is justified?

16. Scientists predict that global climate change will cause a greater increase in average temperature at higher latitudes than in the tropics. This seems like good news since most of the world's biodiversity occurs in the tropics. However, recent data suggest that a smaller temperature change in the tropics will cause a *larger* physiological response there than the larger temperature change will in organisms in temperate areas. Based on what you have learned in this chapter, pose a hypothesis to explain this result.

53 Behavioral Ecology

In this chapter you will learn that

> **Behavioral ecology integrates**
> **genetics, development, anatomy, physiology, evolution, and ecology**

starting with

An introduction: Proximate and ultimate causation and cost–benefit analysis 53.1

then applying these to

What should I eat?	53.2
Who should I mate with?	53.3
Where should I live?	53.4
How should I communicate?	53.5
When should I cooperate?	53.6

This male satin bowerbird is decorating his display area with objects of his preferred color: blue. A female will inspect the bower. She may mate with the male if she finds him, and the display, attractive enough.

BIG PICTURE

This chapter is part of the Big Picture. See how on pages 1196–1197.

To biologists, **behavior** is action—specifically, the response to a stimulus. Pond-dwelling bacteria swim toward drops of blood that fall into the water. Time-lapse movies of growing sunflowers document the steady movement of their flowering heads throughout the day, as they turn to face the shifting Sun. Filaments of the fungus *Arthrobotrys* form loops and then release a molecule that attracts roundworms. When a roundworm touches a loop, the fungal filaments swell—ensnaring the worm. The fungus then grows into the worm's body and digests it.

Among animals, action may be frequent and even spectacular. Peregrine falcons reach speeds of up to 320 km/hr (200 mph) when they dive in pursuit of ducks or other flying prey. Hognose snakes

✔ When you see this checkmark, stop and test yourself. Answers are available in Appendix A.

go belly-up and feign death—mouths agape—when harassed by predators. Deep inside a hive, honeybees communicate about food sources by "dancing."

Ecology is the study of how organisms interact with their physical and biological environments; **behavioral ecology** is the study of how organisms respond to particular stimuli from their environments. What does an *Anolis* lizard do when it gets too hot? How does that same individual respond when a member of the same species approaches it? Or when a house cat approaches? You can see how behavior fits into the Big Picture of Ecology on pages 1196–1197.

Although all organisms respond in some way to signals from their environment, most behavioral research is performed on animals, especially vertebrates, arthropods, or mollusks. With their sophisticated nervous systems and skeletal–muscular systems, these animals can sense, process, and respond rapidly to a wide array of environmental stimuli.

53.1 An Introduction to Behavioral Ecology

Behavior is a particularly integrative field in biological science. While the primary focus of behavioral ecology is at the level of the organism (see Chapter 52 and The Big Picture of Doing Biology on pages 16–17), this field integrates levels from molecules up to ecosystems, depending on the question being asked. To understand why animals and other organisms do what they do, researchers have to ask questions about genetics, development, anatomy, physiology, evolution, and ecology.

To make sense of this diversity, it's helpful to recognize that behavioral ecologists ask questions and test hypotheses at two fundamental levels—proximate and ultimate. Let's first explore these levels and then consider different types of behaviors.

Proximate and Ultimate Causation

The eminent evolutionary biologist Ernst Mayr wrote an influential paper on "cause and effect in biology" in 1961. He made a clear distinction between two types of biological causation:

1. **Proximate** (or mechanistic) **causation** explains *how* actions occur in terms of the genetic, neurological, hormonal, and skeletal–muscular mechanisms involved. For example, does a neural or hormonal signal trigger an animal's courtship display? Does the display vary according to diet, photoperiod, temperature, or other factors?

2. **Ultimate** (or evolutionary) **causation** explains *why* actions occur—based on their evolutionary consequences and history. (Behavior is just like any other phenotype in that it can evolve by natural selection.) Is a particular behavior currently adaptive, meaning that it increases an individual's fitness—its ability to produce viable, fertile offspring? If so, how does the behavior help individuals to produce offspring in a particular environment?

It's important to recognize that efforts to explain behavior at the proximate and ultimate levels are complementary. To understand what an organism is doing, biologists want to know how the behavior happens and why.

To illustrate the proximate and ultimate levels of causation, consider the Argentine ant (*Linepithema humile*) (introduced in Chapter 52). These ants originated in Argentina but have been shipped inadvertently with fruits and other cargoes around the world. Argentine ants are the subject of intense research due to their extreme success as an invasive species.

Argentine Ant Behavior   In their native South America, Argentine ants live in colonies and defend their feeding areas, or **territories,** by fighting with neighboring Argentine ants. However, in places where they have been introduced, such as southern California, Argentine ants from adjacent colonies are "friendly" to each other—workers and queens move freely from nest to nest, forming huge "supercolonies." But when the Argentine ants from supercolonies meet native ants, they are especially aggressive (**FIGURE 53.1**). What's going on?

Proximate Causes   Ants identify each other by smelling waxy hydrocarbon "tags" on each other's exoskeletons. Biochemical research has shown that introduced Argentine ants accept most other members of their species as nest mates because they smell the same as nest mates. That is, they display the same scent tags.

Follow-up research suggests that the introduced ants smell similar because the genes that determine their scent tags are similar; unlike the Argentine ant populations in South America, there is little genetic variation for this trait among colonies in California. Thus the proximate cause for unusually friendly behavior among Argentine ants is the genetically determined scent signal that says "nest mate: don't fight." By contrast, the very unfamiliar scents from native ants signal "foreigner: attack!"

Ultimate Causes   Why is genetic diversity in the supercolonies low? Argentine ants appear to have low genetic variation due to a genetic bottleneck that occurred when the first few ants were introduced to California (see Chapter 26 for a review of genetic bottlenecks).

FIGURE 53.1 Tiny Invasive Argentine Ants Attack a Native Ant.

This chance event happened to be adaptive. Since Argentine ants in supercolonies are not fighting to maintain their feeding territories, they have more time and energy available to produce offspring, thus increasing their fitness. The resulting high density of the supercolonies contributes to their invasiveness. A single, tiny Argentine ant is rarely successful in combat against a single, larger native ant. But when Argentine ants recruit help and fight in mobs, they are a powerful force.

Note, however, that low genetic diversity has its dangers (see Chapter 13). If the supercolony becomes exposed to a virulent pathogen or other environmental change, the entire colony may collapse—a possibility that humans are trying to exploit.

This example illustrates how behavioral ecologists work at the interface of many fields and at both proximate and ultimate levels to address *how* and *why* questions about organisms in their environments. Before considering some of the most prominent questions in behavioral ecology, let's take a closer look at types of behavior.

Types of Behavior: An Overview

Behavior encompasses a diverse array of responses to stimuli. To establish a framework to organize this diversity, let's explore both axes of the graph shown in **FIGURE 53.2**.

Behaviors Vary in Their Flexibility   The horizontal axis of Figure 53.2 maps how flexible a behavior is in terms of its expression within and between individuals of the same species. This axis forms a continuum ranging from fixed, stereotyped behaviors to highly flexible, conditional behaviors.

Consider what happens to a kangaroo rat that is feeding at night and hears the sound of a rattlesnake's rattle. It jumps back—away from the direction of the stimulus. The jump-back behavior is performed the same way every time. Highly inflexible,

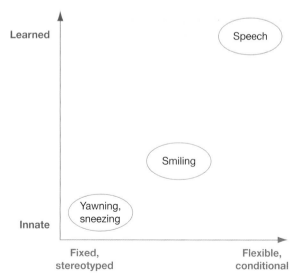

FIGURE 53.2 Behavioral Traits Vary in Flexibility and Dependence on Learning. These two axes represent one way to organize the diversity of behavioral traits. Three human behaviors are mapped as examples.

stereotyped behavior patterns like a kangaroo rat's jump-back behavior are called **fixed action patterns,** or **FAPs.** In humans, yawns and sneezes are familiar FAPs.

At the other end of this axis, behaviors show a great deal of flexibility in response to environmental conditions. For example, human speech is highly flexible and dependent upon factors ranging from inherited sound-making ability to upbringing, education, emotional state, circumstance, and so on. Most behaviors lie between these two extremes of inflexibility and flexibility.

Behaviors Vary in the Extent to Which They Are Learned   The vertical axis of Figure 53.2 plots how much learning is involved in the nature of the behavioral response.

Behaviors that require no learning are said to be innate. **Innate behavior** is inherited—meaning that it is passed genetically from parents to offspring. FAPs are examples of innate behaviors because they are "hardwired"—the jump-back behavior of kangaroo rats and your ability to sneeze are not learned.

When the pioneering behavioral researchers Konrad Lorenz, Niko Tinbergen, and Karl von Frisch first began to explore animal behavior experimentally, they focused on these types of inflexible, innate actions—such as the way goose chicks imprint on their mother and follow her around. This research strategy is common in biological science. If you are addressing a new question—whether it's how DNA is transcribed or how ions cross plasma membranes—it's helpful to start by studying the simplest possible system. Once simple systems are understood, investigators can delve into more complex situations and questions. Lorenz, Tinbergen, and von Frisch won a joint Nobel Prize in 1973 for their behavioral work.

At the other end of the continuum from innate behavior is **learning,** an enduring change in behavior that results from a specific experience in an individual's life. Behaviorist B. F. Skinner is famous for his experiments with rats and mice showing that their behaviors are subject to conditioning. For example, a rat can be trained to pull a lever if food is released as a reward.

The examples plotted in Figure 53.2 indicate that there is a general tendency for innate behaviors to be more fixed and for learned behaviors to be more flexible. For example, in humans, yawning is a relatively fixed, innate behavior while speech is flexible and highly dependent upon learning.

Flexible, Learned Behaviors Often Involve Choice   Most animals have a range of actions that they can perform in response to a situation. Animals take in information from the environment and, based on that information, make decisions about what to do. Animals make choices.

To link condition-dependent behavior to fitness—and thus the ultimate level of explanation—biologists use a framework called **cost–benefit analysis.** Animals appear to weigh the costs and benefits of responding to a particular situation in various ways. Costs and benefits are measured in terms of their impact on fitness—the ability to produce viable, fertile offspring.

Note the phrase "appear to weigh" in the previous paragraph. The decisions made by nonhuman organisms are not—as far as is known—conscious. An Argentine ant does not identify another ant by smell and then consider the pros and cons of reacting as friend or foe. Instead, evolution has shaped the ant's genome to direct a nervous and endocrine system that is capable of (1) taking in information and (2) directing behavior that is likely to pass that genome on to the next generation.

Five Questions in Behavioral Ecology

The following five sections of this chapter introduce some of the most prominent questions in behavioral ecology. In each case, the question is posed as a decision that an animal has to make. The ensuing discussion is divided into research on how the question has been answered at the proximate level and at the ultimate level.

The goals of this chapter are to explore some key topics in behavioral ecology and illustrate the spirit and synthetic nature of the field. Let's begin with an introduction to research on a resource that every animal has to find in its environment: food.

53.2 What Should I Eat?

When animals seek food, they are **foraging.** In some cases, the food source that a species exploits is extremely limited. Giant pandas, for example, subsist almost entirely on the shoots and leaves of bamboo. But individuals must still make decisions about which bamboo plants to harvest and which parts of a particular plant to eat and which to ignore.

In most cases, animals have a relatively wide range of foods that they exploit over the course of their lifetime. To understand why individuals eat the way they do, let's consider (1) research on genetic variation in foraging behavior and (2) the concept of optimal foraging.

Proximate Causes: Foraging Alleles in *Drosophila melanogaster*

When Marla Sokolowski was working as an undergraduate research assistant, she noticed that some of the fruit fly larvae she was studying tended to move after feeding at a particular location, while others tended to stay put. She reared these "rovers" and "sitters" to adulthood, bred them, and found that the offspring of rovers also tended to be rovers, while offspring of sitters tended to be sitters (**FIGURE 53.3**). Thus, this aspect of behavior is inherited. In fruit flies, variation in larval foraging behavior is at least partly due to genetic variation among individuals.

Using information from the *Drosophila melanogaster* genome sequence (see Chapter 21 and **BioSkills 13** in Appendix B) and genetic mapping techniques (Chapter 20), Sokolowski found and cloned a gene associated with the rover–sitter difference. She named it *foraging* (*for*).

Follow-up work showed that the protein product of the *for* gene involved in a signal transduction cascade (see Chapters 11 and 49)—meaning it is involved in the response to a cell–cell

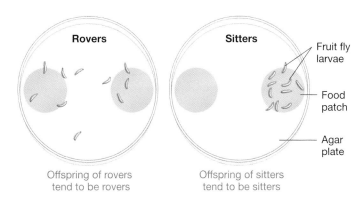

FIGURE 53.3 Foraging Behavior in Fruit Fly Larvae Is Heritable.

signal. Rovers and sitters tend to behave differently when they are foraging because they have different alleles of the *for* gene.

Genes that are homologous to *for* have now been found in honeybees and the roundworm *Caenorhabditis elegans*. Currently, researchers are focused on finding out which signaling pathway uses the *for* gene product and how different alleles change feeding behavior. Presumably, the signaling pathway is associated with feeding or nervous system development. When this aspect of fly feeding is thoroughly understood at the proximate level, biologists will know why variants of the same protein make larvae more likely to move or stay in place after feeding.

Ultimate Causes: Optimal Foraging

By altering population density experimentally and documenting the reproductive success of rovers and sitters, Sokolowski has shown that the rover allele in fruit fly larvae is favored when population density is high and food is in short supply. In contrast, the sitter allele reaches high frequency in low-density populations, where food is abundant. Rovers do better at high density because they are more likely to find unused patches of food; sitters do better at low density because they don't waste energy moving around.

These results are exciting because they link a proximate mechanism with an ultimate outcome. In this case, the presence of certain alleles is responsible for a difference in fitness in specific types of habitats.

In *Drosophila* larvae, foraging is a relatively inflexible and innate behavior—individuals don't "decide" to be rovers or sitters. It is the variability of food supplies that maintains both alleles in the population. But in *Drosophila* adults and in most animal species, foraging behavior is much more flexible and dependent upon learning.

Introduction to Optimal Foraging When biologists set out to study why animals forage in a particular way, they usually start by assuming that individuals make decisions that maximize the amount of usable energy they take in, given the costs of finding and ingesting their food and the risk of being eaten while they're at it. This hypothesis—that animals maximize their feeding efficiency—is called **optimal foraging.**

The specific costs and benefits of foraging depend on the feeding strategy and the environmental conditions of the organism. Different animals have different types of choices and thus different variables subject to optimization. For example:

- Antarctic fur seals forage by diving for small shrimp-like krill. They optimize their foraging by adjusting the duration and location of each dive depending on the depth and density of patches of krill.

- Shorebirds called oystercatchers feed on a variety of prey, including clams, which they have to open to eat. They optimize their foraging by selecting clams of intermediate size. The small clams provide less energy than the large clams, but opening the large clams can take more time and effort, and it can damage the birds' bills.

How do biologists test the optimal foraging hypothesis?

A Test of Optimal Foraging in Desert Gerbils Consider one example from the long-term study of optimal foraging in gerbils. Behavioral ecologists Zvika Abramsky, Michael Rosenzweig, and Aziz Subach posed an interesting question: Do gerbils "weigh" the costs and benefits of foraging? More specifically, if the presence of predators *decreases* the amount of time gerbils spend foraging, and the presence of extra food *increases* foraging time, how much extra food will offset the risk of predation?

To answer this question, Abramsky and his team set up an experiment using two replicate 2-hectare plots in the natural habitat of the gerbil *Gerbillus allenbyi* in the Negev Desert of Israel (**FIGURE 53.4**). Each of these plots had two subplots: a control half and a treatment half, with portals in the fencing to allow gerbils to switch sides. Abramsky started by removing all the gerbils from the plots, setting up seed stations to control the amount of supplementary food that would be available, and setting up sampling stations to measure gerbil track density (number of tracks per given area) during each trial. Earlier research had shown that gerbil track density served as a reliable indicator of the time gerbils spent foraging.

Abramsky and his team then introduced 34 gerbils into each subplot and conducted the following three treatments, replicating their experiments on three different nights:

1. No extra predation risk or seeds in treatment subplots.

2. Increased predation risk—captive barn owls were trained to fly over the treatment subplots three times per hour during the two-hour trials.

3. Increased predation risk plus increased seed availability—varying quantities of supplementary seeds were offered to the treatment subplots during different trials.

What did they find out? The graph at the bottom of Figure 53.4 shows that foraging activity in control and treatment subplots was the same in the absence of owls and extra seed. But when owls were flown over treatment subplots, foraging activity declined dramatically—and some gerbils fled through portals into the control subplots, increasing foraging activity there.

QUESTION: Do gerbils "weigh" the costs and benefits of foraging?

HYPOTHESIS: Gerbils reconcile the risk of predation and the benefits of extra food availability.

NULL HYPOTHESIS: Foraging activity is independent of predation and food availability.

EXPERIMENTAL SETUP:

1. Start with 34 gerbils in each 1-hectare desert subplot in the Negev Desert of Israel.

2. Use the density of gerbil tracks in sampling stations as an index of foraging activity in control versus treatment subplots where:

Treatment 1	Treatment 2	Treatment 3
No owl fly-overs No extra seeds	Owl flyovers No extra seeds	Owl flyovers Extra seeds of varying quantities

3. Replicate experiment on three different nights.

PREDICTION: A certain amount of added seeds will compensate for the decrease in foraging activity due to predation risk.

PREDICTION OF NULL HYPOTHESIS: Gerbil foraging activity will be independent of the presence of owls, extra seeds, or both.

RESULTS:

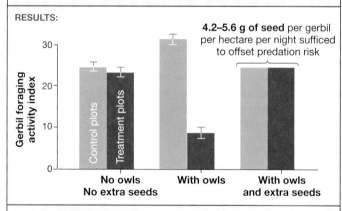

CONCLUSION: Gerbils weigh the relative risk of predation and benefits of extra seeds when they forage.

FIGURE 53.4 Optimal Foraging in Desert Gerbils

SOURCE: Abramsky, Z., M. L. Rosenzweig, and A. Subach. 2002. The costs of apprehensive foraging. *Ecology* 85: 1330–1340.

✔ **QUESTION** Why did the researchers measure foraging activity in the treatment plot before adding owls or seeds?

The researchers also observed a direct relationship between the gerbils' foraging activity and the amount of supplementary seed offered, even with owl flyovers. It turns out that 5 grams of supplementary seeds per individual per hectare per night was sufficient to compensate for the extra risk of predation—that is, to eliminate the difference in foraging between control and treatment subplots. This study was insightful because it enabled predation risk to be measured in energetic terms.

The logic behind optimal foraging is that animals that maximize their feeding efficiency will have more time and energy available for reproduction—thus higher fitness. Animals that forage optimally are expected to be favored by natural selection—thus pass on their heritable behavior patterns to the next generation.

check your understanding

If you understand that . . .
- Feeding behavior can be explored at both proximate and ultimate levels.
- Optimal foraging minimizes fitness costs and maximizes fitness benefits.

✔ **You should be able to . . .**

Predict how the results of the gerbil foraging experiment would change if the owls were permitted to *hunt* in the treatment subplots rather than just flying over, and pose a proximate explanation for the change.

Answers are available in Appendix A.

53.3 Who Should I Mate With?

Biologists sometimes claim that almost all animal behavior revolves around food or sex. Section 53.2 focused on food; here the subject is sex. Let's start by considering the proximate mechanisms responsible for triggering sexual activity in lizards and then go on to explore the ultimate question of why animals choose their mates.

Proximate Causes: How Is Sexual Activity Triggered in *Anolis* Lizards?

Anolis carolinensis lives in the woodlands of the southeastern United States. After spending the winter under a log or rock, males emerge in January and establish feeding and breeding territories (**FIGURE 53.5a**). Females become active a month later, and the breeding season begins in April. By May, females are laying an egg every 10–14 days. By the time the breeding season is complete three months later, the eggs produced by a female will total twice her body mass.

FIGURE 53.5b graphs how sexual condition changes throughout the year, by plotting the size of the testes in males and ovaries of

(a) Display behavior of a male *Anolis* lizard

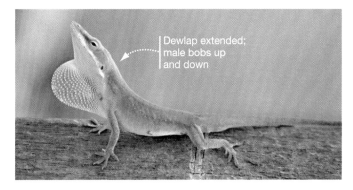

Dewlap extended; male bobs up and down

(b) Changes in sexual organs through the year

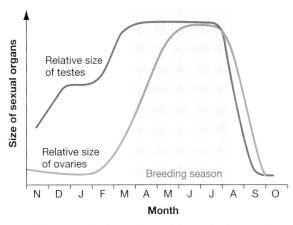

FIGURE 53.5 Sexual Behavior in *Anolis* Lizards Is Seasonal.
DATA: Crews, D. 1975. *Science* 189: 1059–1065.

females—relative to an individual's overall body size. Individuals come into breeding condition in spring, although males experience growth of gonads long before females do. In both sexes, the gonads shrink in size during the fall and winter.

Testosterone and Estradiol What causes the onset of sexual behavior? The proximate answer is sex hormones—testosterone in males and estradiol in females. Testosterone is produced in the testes of males, and estradiol in the ovaries of females (see Chapter 50). The evidence for the effects of sex hormones is direct. Testosterone injections induce courtship activity in castrated males with no prior courtship activity; estradiol injections induce sexual activity in females whose ovaries have been removed.

But a full proximate explanation must also consider environmental and social cues. What environmental cues trigger the production of sex hormones in early spring? How do male and female *Anolis* lizards synchronize their sexual readiness?

Testing the Effects of Light and Social Stimulation To answer these questions, behavioral ecologist David Crews brought a large group of sexually inactive adult lizards into the laboratory during the winter and divided them into five treatment groups. The physical environment was exactly the same in all treatments.

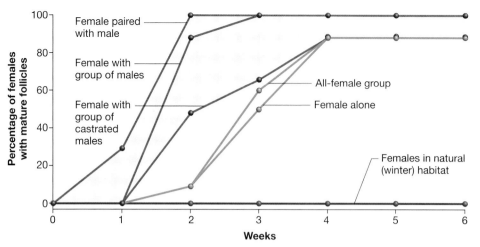

FIGURE 53.6 Exposure to Springlike Conditions and Breeding Males Stimulates Female Lizards to Produce Eggs. The percentage of female lizards with mature follicles, plotted over time. Five treatment groups were exposed to springlike light and temperature in the lab; data are also plotted for females left in a natural (winter) habitat. Each data point represents the average value for a group of 6–10 females.

DATA: Crews, D. 1975. *Science* 189: 1059–1065.

✔**QUESTION** Some critics contend that the females that remained outside do not represent a legitimate control in this experiment. What conditions would represent a better control?

Each lizard received identical food, high "daytime" temperatures, and slightly lower "nighttime" temperatures. Crews also continued to monitor the condition of lizards that remained in natural habitats nearby.

To test the hypothesis that changes in day length signal the arrival of spring and trigger initial changes in sex hormones, Crews exposed the five treatment groups in the laboratory to artificial lighting that simulated the long days and short nights of spring.

To test the hypothesis that social interactions among individuals are responsible for synchronizing sexual behavior, he varied the social setting among the five treatment groups:

1. a single isolated female;

2. a group of females;

3. a single female with a single male;

4. a single female with a group of castrated (nonbreeding) males; and

5. a single female with a group of uncastrated (breeding) males.

Each week, Crews examined the ovaries of females in each treatment. He also monitored the ovaries of females in nearby natural habitats under winter conditions.

As **FIGURE 53.6** shows, the differences in the animals' reproductive systems were dramatic. The data points on this graph represent the average percentage of females with mature ovarian follicles, in each treatment; the points are connected by lines to help you keep track of the different treatments and see the overall patterns. Note two key points:

1. Females that were exposed to springlike conditions began producing eggs; females in the field that were not exposed to springlike conditions did not.

2. Females that were exposed to breeding males began producing eggs much earlier than did the females placed in the other treatment groups.

These results support the hypothesis that two types of stimulation are necessary to produce the hormonal changes that lead to sexual behavior. Females need to experience springlike light and temperatures *and* exposure to breeding males.

Visual Cues from Males Trigger Female Readiness What is it about breeding males that triggers earlier estradiol release and egg production in females?

When males court females, they bob up and down in push-up fashion and extend a brightly colored patch of skin called a dewlap (see Figure 53.5a). To test the hypothesis that this visual stimulation triggers changes in estradiol production, Crews repeated the previous experiment but added a twist: He placed some females with males that had intact dewlaps, and other females with males whose dewlaps had been surgically removed.

The result? Females that were living with dewlap-less males were slow to produce eggs. They were just as slow, in fact, as the females in the first experiment that had been grouped with castrated males. These latter females had not been courted at all.

These data suggest that the dewlap is a key visual signal. The experiments succeeded in identifying the environmental cues—long day length and bobbing dewlaps—that trigger hormone production and the onset of sexual behavior.

Ultimate Causes: Sexual Selection

Why are environmental cues and bobbing dewlaps important to mating lizards? One reason is that lizards that succeed in synchronizing their sexual readiness will likely bear more offspring and have higher fitness than those that do not. *Anolis* lizards are most successful if they reproduce early in the spring when food supplies are increasing but snakes and other predators are not yet hunting to feed their own young. Failing to synchronize their sex or having sex too late would be disastrous for the *Anolis* lizards.

Also, bobbing dewlaps serve as an important signal in mate choice. A specific pattern of selection called **sexual selection** favors individuals possessing traits that increase their ability to obtain mates (introduced in Chapter 26). According to theory and experimental data:

• Females are usually the gender that is pickiest about mate choice because females spend a disproportionate amount of energy on their gametes compared to males.

• Females choose males that contribute good alleles and/or resources to their offspring.

Research on bird species such as zebra finches (see Chapter 26) has shown that individuals with particularly long tails, bright colors, and energetic courtship displays are usually the healthiest and best fed in the population. Because it takes large amounts of energy and resources to produce traits like these, they serve as an honest signal of male quality.

Are bright dewlaps honest signals of male quality? So far, the brightness of lizard dewlaps does not appear correlated with the nutritional health of the lizards. However, lizards with enlarged dewlaps tend to have high bite forces, which are important in fighting with other males to defend their feeding and breeding territories. Males are known to defend their territories fiercely, sometimes with costly consequences. The bobbing dewlap territorial displays serve as honest signals of physical vigor that can help to avert actual confrontation.

There are two types of sexual selection. When females choose males, intersexual ("between-sexes") selection is said to occur. When males compete with one another for mates, intrasexual ("within-sexes") selection is taking place (see Chapter 26). The data on *Anolis* lizards so far suggest that even though female sexual readiness is triggered by male displays, male competition for territories is the main factor determining mate choice.

Data on mate choice in *Anolis* lizards reinforce a central theme of this chapter: Animals usually make decisions in a way that maximizes their fitness.

check your understanding

If you understand that . . .

- The proximate causes of sexual readiness include hormonal, environmental, and social cues.
- At the ultimate level, animals use physical and behavioral cues to choose mates that will increase their fitness.

✔ **You should be able to . . .**

1. Propose one proximate cause of human sexual readiness that is similar to that in *Anolis* lizards.
2. Propose one ultimate cause of human mate choice that is similar to that in *Anolis* lizards.

Answers are available in Appendix A.

53.4 Where Should I Live?

Most animals select the food they eat and the individuals they mate with. In most cases, they also select the habitat where they live. Biologists have explored an array of questions related to habitat selection:

- Should juveniles disperse from the area where they were raised, or should they stay?

- How large an area should be defended against competitors in species that maintain a territory for breeding or feeding, as in *Anolis* lizards?

- Do individuals have higher fitness when they occupy crowded, high-quality habitats or uncrowded, poor-quality habitats?

Let's take a closer look at one of these "lifestyle" decisions: the proximate and ultimate mechanisms responsible for migration. In ecology, **migration** is defined as the long-distance movement of a population associated with a change of seasons. At the proximate level, how do animals know where to go between one seasonal "home" and another? At the ultimate level, what are the fitness benefits of having more than one home range that could overcome the costs of long-distance migrations?

Proximate Causes: How Do Animals Navigate?

To organize research into how animals find their way during migratory movements, biologists distinguish three categories of navigation:

1. **Piloting** is the use of familiar landmarks.

2. **Compass orientation** is movement that is oriented in a specific direction.

3. **True navigation (map orientation)** is the ability to locate a specific place on Earth's surface.

Imagine that you asked someone to blindfold you, drive you to a remote area 100 miles from home, and drop you off by the side of the road without your cell phone. After removing your blindfold, how would you find your way home? Which type of navigation would you use?

Piloting   If you were lucky, you might recognize some landmarks and remember from a previous experience how to return home. The offspring of some migratory birds and mammals navigate in a similar way. They memorize the route when following their parents one or more times—south in the fall and north in the spring—and later are able to pilot the route on their own.

Compass Orientation   If you were dropped off on the side of the road and did not recognize any specific landmarks, which way would you start walking? If you had a compass and knew which direction to head, that would be a big help. Once you got closer to home and found a familiar landmark, you could pilot the rest of the way home.

How do animals perform compass orientation? To date, most research on compass orientation has been done on migratory birds, such as European robins. To determine where north is, these animals appear to use the Sun, the stars, and Earth's magnetic field.

The Sun is difficult to use as a compass reference because its position changes during the day. It rises in the east, is due south at noon (in the Northern Hemisphere), and sets in the west. To use the Sun as a compass reference, then, an animal must have an internal clock that defines morning, noon, and evening. Fortunately, most animals have such a clock. The **circadian clock** that exists in organisms maintains a 24-hour rhythm of chemical activity. The clock is set by the light–dark transitions of day and night. It tells individuals enough about the time of day that they can use the Sun's position to find magnetic north.

The situation is actually simpler on clear nights because migratory birds in the Northern Hemisphere can use the North Star to find magnetic north and select a direction for migration. But what if the weather is cloudy?

Under these conditions, migratory birds appear to orient using Earth's magnetic field. There are two main hypotheses for how animals detect magnetism. One hypothesis contends that animals can detect magnetism by their visual system, through a chemical reaction that involves electron transfer among molecules. An alternative hypothesis maintains that individuals have small particles of magnetic iron—the mineral called magnetite—in their bodies, such as in the upper beaks of birds. Changes in the positions of magnetic particles, in response to Earth's magnetic field, could then be detected and provide reliable information for compass orientation. So far, the data suggest that birds may be capable of both mechanisms.

Although research on mechanisms of compass orientation continues, one important point is clear: Birds and other organisms have multiple mechanisms of finding a compass direction. At least some species can use a Sun compass, a star compass, and a magnetic compass. Which system an individual uses depends on the weather and other circumstances.

True Navigation (Map Orientation) If you were dropped off on the side of the road with a compass *and a map* of your location, finding your way home would be easy.

Do animals have the equivalent of map orientation? Neurobiologist and ecologist Kenneth Lohmann and his colleagues tested this question in green sea turtles. Sea turtles have the remarkable ability to travel around vast areas of the ocean and then return to a specific coastal nesting or feeding site with pinpoint accuracy. How do they do it?

The Earth's magnetic field varies around its surface in predictable gradients of both intensity and angle, or inclination. Lohmann's hypothesis was that the turtles use this magnetic field as a source of precise positional information, much like we use latitude and longitude coordinates to locate ourselves on a map.

To test this hypothesis, Lohmann captured green sea turtles at their coastal feeding grounds in Melbourne Beach, Florida, fitted them with soft harnesses attached to a tracking device, and then let them swim one at a time in a circular pool (**FIGURE 53.7**). A coil system around the pool was used to change the magnetic field to replicate the magnetic signature of locations to the north and south of Melbourne Beach.

The data at the bottom of Figure 53.7 show how the turtles responded. Each dot represents the average swimming direction for one turtle. The arrow represents the average for the whole group of turtles. When the magnetic field in the pool simulated a location 337 km north of Melbourne Beach, the turtles swam south, the direction of "home." When the magnetic field in the pool simulated a location 337 km south of Melbourne Beach, the

RESEARCH

QUESTION: How do sea turtles find their way?

HYPOTHESIS: Green sea turtles use magnetic map orientation to navigate.

NULL HYPOTHESIS: Green sea turtles use compass orientation and piloting to navigate.

EXPERIMENTAL SETUP:

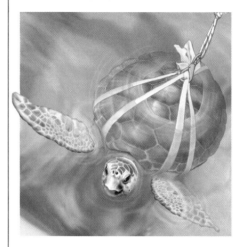

1. Capture green sea turtles in Melbourne Beach, Florida.

2. Place turtles in soft cloth harnesses attached to electronic tracking device in a swimming arena.

3. Monitor direction of swimming in different magnetic fields simulating locations to the north and south of Melbourne Beach.

PREDICTION: Sea turtles will swim south in a simulated north location, and swim north in a simulated south location.

PREDICTION OF NULL HYPOTHESIS: Sea turtles will swim in random directions with respect to the two simulated locations.

RESULTS:

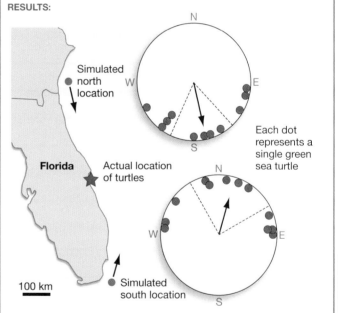

Simulated north location

Florida

Actual location of turtles

100 km

Each dot represents a single green sea turtle

Simulated south location

CONCLUSION: Green sea turtles use magnetic map orientation to navigate.

FIGURE 53.7 Sea Turtles Navigate Using Map Orientation

SOURCE: Lohmann, K. J., C. M. F. Lohmann, L. M. Ehrhart, et al. 2004. Geomagnetic map used in sea-turtle navigation. *Nature* 428: 909–910

✔ **QUESTION** Sketch what the data for the two sites would have to look like to reject Lohmann's hypothesis.

turtles swam north. Note that the direction of magnetic north did not change in either case, so compass orientation alone could not account for the ability of the turtles to navigate. Rather, the turtles were able to locate themselves on the map—they possess map orientation.

Ultimate Causes: Why Do Animals Migrate?

Why do some animals have more than one home range? Migratory movements can be spectacular in their extent and in the navigation challenges they pose:

- Arctic terns nest during summer in tundra in the far north. One population flies south along the coast of Africa to wintering grounds off Antarctica and then flies back north along the eastern coast of South America (**FIGURE 53.8**). Chicks follow their parents south but make the return trip on their own. The journey totals over 32,000 km (20,000 miles).

- Many of the monarch butterflies native to North America spend the winter in the mountains of central Mexico or southwest California. Tagged individuals are known to have flown over 3000 km (1870 miles) to get to a wintering area. In spring, monarchs begin the return trip north. They do not live to complete the trip, however. Instead, mating takes place along the way, and females lay eggs en route. Although adults die on the journey, offspring continue to head north. After several generational cycles of reproduction and death in the original habitats in northern North America, individuals again migrate south to overwinter. Instead of a single generation making the entire migratory cycle, then, the round trip takes several generations.

- Salmon that hatch in rivers along the Pacific Coast of North America and northern Asia migrate to the ocean when they are a few months to several years old, depending on the species. After spending several years feeding and growing in the North Pacific Ocean, they return to the stream where they hatched. There they mate and die (see Chapter 52).

In most cases, at the ultimate level it is relatively straightforward to generate hypotheses explaining why migration exists.

- Arctic terns feed on fish that are available in different parts of the world at different seasons. Thus, individuals that do not migrate might starve; individuals that do migrate should achieve higher reproductive success.

- Monarch butterflies may achieve higher reproductive success by migrating to wintering areas and new breeding areas than by trying to overwinter in northern North America.

- Salmon eggs and young are safer and thrive better in freshwater habitats than they do in the ocean. But adult salmon can find much more food in saltwater habitats than in freshwater and thus grow to bigger size, which confers a benefit in competition for nest sites and mates.

Testing these hypotheses rigorously is more difficult because researchers must compare the fitness of individuals that do and do not migrate. Recently, however, researchers have documented the evolution of sockeye salmon populations that do not migrate to the ocean—in watersheds where migratory sockeye were introduced by humans. The nonmigratory populations are much smaller bodied than the migratory populations—presumably because less food is available in the freshwater habitats.

The fitness of migrant and nonmigrant sockeye salmon appears to be similar in this case, due to fitness trade-offs (see Chapter 25). The larger-bodied migratory salmon are more successful in competing for nesting sites and mates. However, the migratory salmon also experience higher costs in terms of time, energy, and predation risk.

At the ultimate level, researchers are also struggling to understand how a nonmigratory population can evolve into a migratory one that travels to a completely different part of the world. It's likely that habitat changes occurring in response to global climate change (see Chapter 56) will furnish an example. Changes in the range and arrival times of migratory species have already been documented; it may be only a matter of time before a nonmigratory population begins to migrate, to cope with habitat shifts caused by global climate change.

FIGURE 53.8 Arctic Terns Have Spectacular Migrations. The map shows the pathway of 11 arctic terns for a year based on data collected using tiny geolocators.

53.5 How Should I Communicate?

The song of a bird, the bobbing dewlap of a male lizard, and the sentences in this text all have the same overall goal: communication. In biology, **communication** is defined as any process in which a signal from one individual modifies the behavior of a recipient individual. A **signal** is any information-containing behavior or characteristic. Communication is a crucial component of animal behavior. It creates a stimulus that elicits a response.

By definition, communication is a social process. For communication to occur, it is not enough that a signal is sent; the signal must be received and acted on. How does this process play out in honeybees?

Proximate Causes: How Do Honeybees Communicate?

Honeybees are highly social animals that live in hives. Inside the hive, a queen lays eggs that are cared for by workers. Besides caring for young and building and maintaining the hive, workers obtain food for themselves and other members of the colony by gathering nectar and pollen from flowering plants.

Biologists noticed that bees appear to recruit to food sources, meaning that if a new source is discovered by one or a few individuals, many more bees begin showing up over time. This observation inspired the hypothesis that food-finders have some way of communicating the location of food to others. How do bees "talk" to each other?

The Dance Hypothesis In the 1930s Karl von Frisch began studying bee communication by observing bees that built hives inside the glass-walled chambers he had constructed. He found that if he placed a feeder containing sugar water near one of these observation hives, a few of the workers began moving in a circular pattern on the vertical, interior walls of the hive.

Von Frisch called these movements the "round dance" (**FIGURE 53.9a**). Other workers appeared to follow the progress of the dance, first by touching the displaying bee as it danced and then by flying away from the hive in search of the food source.

To investigate the function of these movements further, von Frisch placed feeders containing sugar water at progressively greater distances from the hive. Using this technique, he was able to get bees to visit feeders at a distance of several kilometers from the hive. By catching bees at the feeders and dabbing them with paint, he could individually mark successful food-finders. Follow-up observations at the hive confirmed that marked foragers (**1**) danced when they returned to the hive, and (**2**) returned to the food source with unmarked bees.

To explain these data, von Frisch proposed that the round dance contained information about the location of food. Because bee hives are completely dark, he hypothesized that workers got information from the dance by touching the dancer and following the dancer's movements.

Communicating Directions and Distances When von Frisch began placing feeders at longer distances from the hive, he found that successful food-finders used a new type of display. He named these movements the "waggle dance," because they combined circular movements like those of the round dance with short, straight runs (**FIGURE 53.9b**). During these runs, the dancer vigorously moved her abdomen from side to side.

These observations supported the hypothesis that both the round dance and waggle dance communicate information about food sources. Recent work has shown that the round and waggle dances are actually the same type of behavior—round dances just have a short waggle phase.

Three key observations allowed von Frisch to push our understanding of bee language further:

1. The orientation of the waggle part of the dance varied.

2. The direction of the waggle run correlated with the direction of the food source from the hive.

3. The length of the straight, "waggling" run was proportional to the distance the foragers had to fly to reach the feeder.

(a) The round dance

(b) The waggle dance

Other bee workers follow the progress of the dance by touching the displaying individual

FIGURE 53.9 Honeybees Perform Two Types of Dances. (a) During the round dance, successful food-finders move in a circle. **(b)** During the waggle dance, successful foragers move in a circle but then make straight runs through the circle. During the straight part of the dance, the dancer waggles her abdomen.

By varying the location of the food source and observing the orientation of the waggle dance given by marked workers, von Frisch was able to confirm that dancing bees were communicating the position of the food relative to the current position of the Sun. For example:

- If food is directly away from the Sun's current position, marked bees do the waggle portion of their dance directly downward (**FIGURE 53.10a**).

- If the food is 90 degrees to the right of the Sun, marked bees waggle 90 degrees to the right of vertical (**FIGURE 53.10b**).

These results are nothing short of astonishing. Honeybees do not have large brains, yet they are capable of symbolic language. What's more, they are able to interpret the angle of the waggle dance performed on a vertical surface and to respond by flying horizontally along the corresponding angle.

Further work has confirmed that the dance language of bees includes several modes of communication: tactile information in the movements themselves, sounds made during the dance, and scents that indicate the nature of the food source.

Ultimate Causes: Why Do Honeybees Communicate the Way They Do?

At the ultimate level, it's straightforward to come up with a hypothesis for why honeybees use a complex method of communication to share information about food sources. If more workers find a rich food source quickly, then the total amount of resources harvested as well as the number of offspring produced by the hive should increase. Thus the frequency of bee colonies with successful communication should increase over time.

In general, why do animals use the modes of communication that they do? One of the most general observations about communication is that the type of signal used by an organism correlates with its habitat. Bees, ants, and termites that live in hives or underground rely on olfactory and tactile communication. Similarly, bats, wolves, and other animals that are active at night communicate via sound or scent. In contrast, animals that are active during the day and that live in open or treeless habitats tend to rely on visual communication.

Sound travels much farther than light in aquatic habitats. Based on this observation, it is logical to observe that humpback whales rely on songs for long-distance communication. In some cases, humpback whale songs can travel hundreds of kilometers. Groups of whales use acoustic communication to keep together during migrations, and individual males sing to attract mates and warn rivals.

Each mode of communication has advantages and disadvantages. Songs and calls can carry information over long distances, but are short lived. Thus they have to be repeated to be effective, and frequent repetition requires a large expenditure of time and energy. In addition, acoustic communication attracts predators. It is no surprise that when a hawk or falcon approaches a marsh inhabited by courting red-winged blackbirds, things get very quiet. Communication systems have been honed by natural selection to maximize their benefits and minimize their costs.

(a) Straight runs down the wall of the hive indicate that food is opposite the direction of the Sun.

(b) Straight runs to the right indicate that food is 90° to the right of the Sun.

FIGURE 53.10 The Direction of the Waggle Dance Indicates the Location of Food Relative to the Sun.
Waggle dances are done only when the food is at least 100 m from the hive.

✔**EXERCISE** The length of the straight run in the waggle dance indicates the relative distance of the food. Diagram a waggle dance that indicates food in the direction of the Sun and twice as far away as the food sources indicated by the dances drawn here.

When Is Communication Honest or Deceitful?

At the ultimate level, one of the questions that biologists ask about communication concerns the quality of the information. Is the signal reliable?

As far as is known, honeybees are always honest. Stated another way, honeybee dances consistently "tell the truth" about the location of food sources. This observation is logical, because the honeybees that occupy a hive are closely related and cooperate extensively in the rearing of offspring. As a result, it is advantageous for an individual to convey information accurately. If a food-finder provided inaccurate or misleading information to its hivemates, fewer offspring would be reared and the food-finder's fitness would be reduced.

In many cases, however, natural selection has favored the evolution of deceitful communication. Recent research on deceitful communication has highlighted just how complex interactions between signalers and receivers can be. A few examples will help drive this point home.

Deceiving Individuals of Another Species   FIGURE 53.11 illustrates two of the hundreds of examples of deceitful communication that have been documented among members of different species. Many of the best-studied instances involve prey deceiving predators or predators deceiving prey:

- Hognose snakes are famous for their theatrical ability to play dead (Figure 53.11a). When perturbed by a predator, hognose snakes turn belly up, produce a foul odor, hang their tongues out of their gaping mouths, and may even produce a couple of droplets of blood from their mouths. When turned back onto their bellies, they quickly flip onto their backs again and sustain the charade until the coast is clear.

- Male and female fireflies flash a species-specific signal to each other during courtship. Predatory *Photuris* fireflies can mimic the pattern of flashes given by females of several other species. A *Photuris* female attracts a male of different species with the appropriate set of flashes and then attacks and eats him (Figure 53.11b).

In each of these examples, individuals increase their fitness by providing inaccurate or misleading information to members of a different species. It's important to realize, though, that the signaler is not consciously "lying." Instead, natural selection has simply favored certain behavioral traits that effectively communicate deceitful information.

Deceiving Individuals of the Same Species   In some cases, natural selection has favored the evolution of traits or actions that deceive members of an organism's own species. Perhaps the best-studied type of deceit in nonhuman animals involves the mating system of bluegill sunfish. Male bluegills set up nesting territories in the shallow water along lake edges, fertilize the eggs laid in their nests, care for the developing embryos by fanning them with oxygen-rich water, and protect newly hatched offspring from large predators.

(a) Hognose snakes play dead to avoid being eaten.

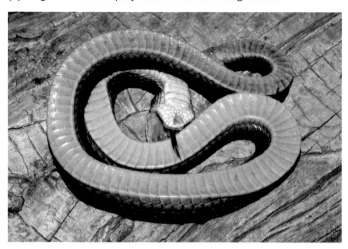

(b) Female *Photuris* fireflies flash the courtship signal of another species and then eat males that respond.

"Femme fatale"

"Victim"

FIGURE 53.11 Deceitful Communication Is Common in Nature.

Some males cheat on this system, however, by mimicking females. To understand how this happens, examine the female, normal male, and female-mimic male in **FIGURE 53.12**. Mimics look like females but have well-developed testes and produce large volumes of sperm. Mimics also act like females during courtship movements with territory-owning males. They even adopt the usual egg-laying posture. When normal females approach the nest and begin courtship, the mimics join in.

The territory-owning male appears to tolerate the mimic, courting the two females at the same time. But when the actual female begins to lay eggs, the mimic responds by releasing sperm—and thus fertilizing some of the eggs—and then darting away. In this way, the mimic fathers offspring but does not help care for them. In addition, mimics do not have to expend time and energy in growing large and making and defending a nest.

Why aren't all males mimics? It turns out that as the number of mimics increases, the reproductive success of mimics declines. This is a common observation in studies of deceitful

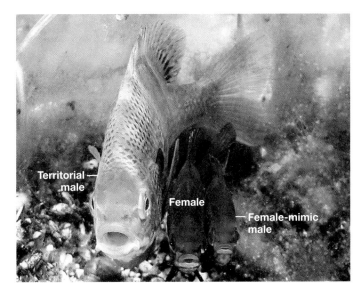

FIGURE 53.12 In Bluegill Sunfish, Some Males Look and Act Like Females.

Territorial male

Female

Female-mimic male

communication—cheating works best within a population when it is relatively rare. The logic behind this hypothesis runs as follows: If deceit becomes extremely common, then natural selection will strongly favor individuals that can detect and avoid or punish liars. But if liars are rare, then natural selection will favor individuals that are occasionally fooled but are more commonly rewarded by responding to signals in a normal way. This is an example of frequency-dependent selection (see Chapter 26).

Further, as the number of mimic males increases in sunfish, competition with other mimics for mating opportunities increases and aggression by territory-owning males on mimic males also increases, lowering the reproductive success of the mimics.

check your understanding

 C
 Y
 U

If you understand that . . .

- Communication is an exchange of information between individuals.
- In most instances, the mode of communication that animals use maximizes the probability that the information will be transferred efficiently.
- Communication can be honest or deceitful, depending on the nature of the information received.

✔ **You should be able to . . .**

1. Explain the costs and benefits of auditory, olfactory, and visual communication.
2. Predict why natural selection favors individuals that can detect deceitful communication and either avoid "liars" or punish them.

Answers are available in Appendix A.

53.6 When Should I Cooperate?

The types of behavior reviewed thus far all have a key common element: They help individuals respond to environmental stimuli in a way that increases their fitness. There is a type of behavior that appears to contradict this pattern, however: altruism.

Altruism is behavior that has a fitness cost to the individual exhibiting the behavior and a fitness benefit to the recipient of the behavior. It is the formal term for self-sacrificing behavior. Altruism decreases an individual's ability to produce offspring but helps others produce more offspring.

In terms of ultimate causation, altruistic behavior appears to be paradoxical. If certain alleles make an individual more likely to be altruistic, those alleles should be selected against (Chapter 25). Meanwhile, individuals with selfish alleles should be more likely to survive and produce more offspring. Over time, selfish alleles should increase in frequency while self-sacrificing alleles should decrease in frequency. Does altruism actually occur? Let's consider four explanations for apparent altruistic behavior.

Kin Selection

There is no question that self-sacrificing behavior occurs in nature. For example, black-tailed prairie dogs perform a behavior called alarm calling (**FIGURE 53.13**). These burrowing mammals live in large communities, called towns, throughout the Great Plains region of North America. When a badger, coyote, hawk, or other predator approaches a town, some prairie dogs give alarm calls that alert other prairie dogs to run to mounds and scan for the threat.

Giving these calls is risky. In several species of ground squirrels and prairie dogs, researchers have shown that alarm-callers draw attention to themselves by calling and are in much greater danger of being attacked than non-callers are.

FIGURE 53.13 Black-Tailed Prairie Dogs Are Highly Social. Prairie dogs live with their immediate and extended family within large groups called towns. The individual with the upright posture has spotted an intruder and may give an alarm call.

The coefficient of relatedness, r, varies between 0.0 and 1.0. If two individuals have no identical alleles that were inherited from the same ancestor, then their r value is 0.0. Because every allele in pairs of identical twins is identical, their coefficient of relatedness is 1.0.

What about other relationships? **FIGURE 53.14a** shows how r is calculated between half-siblings. (To review what the boxes, circles, and lines in a pedigree mean,

see Figure 14.22.) Half-siblings share one parent. Thus, r represents the probability that half-siblings share alleles as a result of inheriting alleles from their common parent. It is critical to realize that in each parent-to-offspring link of descent, the probability of any particular allele being transmitted is 1/2. This is so because meiosis distributes alleles from the parent's diploid genome to their haploid gametes randomly. Thus half the

gametes produced by a parent get one of the alleles present at each gene, and half the gametes produced get the other allele. Half-siblings are connected by two such parent-to-offspring links. The overall probability of two half-siblings sharing the same allele by descent is 1/2 × 1/2 = 1/4. (To review rules for combining probabilities, see **BioSkills 5** in Appendix B.)

To think about this calculation in another way, focus on the red arrows in Figure 53.14a. The left arrow represents the probability that the mother transmits a particular allele to her son. The right arrow represents the probability that the mother transmits the same allele to her daughter. Both probabilities are 1/2. Thus the probability that the mother transmitted the same allele to both her son and daughter is 1/2 × 1/2 = 1/4.

FIGURE 53.14b shows how r is calculated between full siblings. The challenge here is to calculate the probability of two individuals sharing the same allele as a result of inheriting it through their mother or through their father. The probability that full siblings share alleles as a result of inheriting them from one parent is 1/4. Thus the probability that full siblings share alleles inherited from either their mother or their father is 1/4 + 1/4 = 1/2.

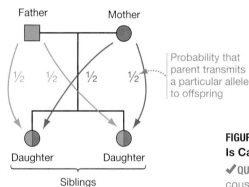

(a) What is the probability that half-siblings inherit the same allele from their common parent?

r between half-siblings:
½ × ½ = ¼

(b) What is the probability that full siblings inherit the same allele from their father or their mother?

Probability that they inherit same allele from **father**:
½ × ½ = ¼

Probability that they inherit same allele from **mother**:
½ × ½ = ¼

r between full siblings:
¼ + ¼ = ½

FIGURE 53.14 The Coefficient of Relatedness (r) Is Calculated from Information in Pedigrees.

✓ **QUANTITATIVE** What is the r between first cousins?

Hamilton's Rule How can natural selection favor the evolution of self-sacrificing behavior? William D. Hamilton answered this question by creating a mathematical model to assess how an allele that contributes to altruistic behavior could increase in frequency in a population.

To model the fate of altruistic alleles, Hamilton represented the fitness cost of the altruistic act to the actor as C and the fitness benefit to the recipient as B. Both C and B are measured in units of offspring produced. His model showed that the allele could spread if

$$Br > C$$

where r is the **coefficient of relatedness.** The coefficient of relatedness is a measure of how closely the actor and beneficiary are related. Specifically, r measures the fraction of alleles in the actor and beneficiary that are identical by descent—that is, inherited from the same ancestor (**Quantitative Methods 53.1**).

The formulation $Br > C$ is called **Hamilton's rule.** It states that altruistic behavior is most likely when three conditions are met:

1. The fitness benefits are high for the recipient.

2. The altruist and recipient are close relatives.

3. The fitness costs to the altruist are low.

When Hamilton's rule holds, alleles associated with altruistic behavior will be favored by natural selection—because close relatives are very likely to have copies of the altruistic alleles. These alleles will increase in frequency in the population.

Inclusive Fitness Hamilton's rule is important because it shows that individuals can pass on their alleles to the next generation not only by having their own offspring but also by helping close relatives produce more offspring. To capture this point, biologists refer to direct fitness and indirect fitness.

- Direct fitness is derived from an individual's own offspring. Parents can increase their direct fitness by spending resources to ensure the welfare of their offspring. This behavior is called parental care (Chapter 35).

- Indirect fitness is derived from helping relatives produce more offspring than they could produce on their own.

The combination of direct and indirect fitness components is called **inclusive fitness.**

Biologists use the term **kin selection** to refer to natural selection that acts through benefits to relatives at the expense of the individual. While kin selection decreases the direct fitness of an individual, it increases the indirect fitness, resulting in an overall increase in an individual's inclusive fitness. ✔ If you understand kin selection, you should be able to explain why it is likely to be common in humans.

Testing Hamilton's Rule Does Hamilton's rule work? Do animals really favor relatives when they act altruistically? To test the kin-selection hypothesis, behavioral ecologist John Hoogland studied which of the inhabitants of a black-tailed prairie dog town were most likely to give alarm calls.

Within a large prairie dog town, individuals live in small groups called coteries that share the same underground burrow. Members of each coterie defend a territory inside the town.

By tagging offspring that were born over several generations, Hoogland identified the genetic relationships among individuals in the town. He determined whether each individual had:

- no close genetic relatives in its coterie;

- no offspring in the coterie but at least one sibling, cousin, uncle, aunt, niece, or nephew; or

- at least one offspring or grandoffspring in the coterie.

The kin-selection hypothesis predicts that individuals who do not have close genetic relatives nearby will rarely give an alarm call. To evaluate this prediction, Hoogland recorded the identity of callers during 698 experiments. In these studies, a stuffed badger was dragged through the colony on a sled.

Were prairie dogs with close relatives nearby more likely to call, or did kinship have nothing to do with the probability of alarm calling? The bar charts in **FIGURE 53.15** show the average proportion of times that individuals in each of the three categories called during an experiment. The data indicate that black-tailed prairie dogs are much more likely to call if they live in a coterie that includes close relatives.

This same pattern—of preferentially dispensing help to kin—has been observed in many other species of social mammals and birds. Most cases of self-sacrificing behavior that have been analyzed to date are consistent with Hamilton's rule and are hypothesized to be the result of kin selection.

Manipulation

Note that altruism refers to a type of behavior with specific fitness consequences to an organism, not to the generosity of an organism. Sometimes altruistic behavior is involuntary.

RESEARCH

QUESTION: Do black-tailed prairie dogs prefer to help relatives when they give an alarm call?

HYPOTHESIS: Individuals give an alarm call only when close relatives are near.

NULL HYPOTHESIS: The presence of relatives has no influence on the probability of alarm calling.

EXPERIMENTAL SETUP:

1. Determine relationships among individuals in prairie dog coterie.

2. Drag stuffed badger across territory of coterie.

3. From observation tower, record which members of coterie give an alarm call.

4. Repeat experiment 698 times. Each prairie dog coterie is tested 6–9 times over a 3-year period.

PREDICTION OF KIN-SELECTION HYPOTHESIS: Individuals in coteries that contain a close genetic relative are more likely to give an alarm call than are individuals in coteries that do not contain a close genetic relative.

PREDICTION OF NULL HYPOTHESIS: The presence of relatives in coteries will not influence the probability of alarm calling.

RESULTS:

CONCLUSION: Alarm calling usually benefits relatives.

FIGURE 53.15 Experimental Evidence That Black-Tailed Prairie Dogs Are More Likely to Give Alarm Calls if Relatives Are Nearby.

SOURCE: Hoogland, J. L. 1983. Nepotism and alarm calling in the black-tailed prairie dog (Cynomys ludovicianus). *Animal Behavior* 31: 472–479.

✔ **QUESTION** What would be an appropriate control in this experiment, and what hypothesis would it test?

For example, one of the most famous applications of Hamilton's rule is in explaining the complex social structure of organisms such as the bees discussed earlier, as well as ants and wasps. In these societies, workers sacrifice most or all of their direct reproduction to help rear the queen's offspring, a phenomenon called **eusociality.**

Why would workers be so altruistic? In insect societies that have a single, diploid queen who mates once with a *haploid* male, all the workers are highly related, with a coefficient of relatedness of 0.75—much higher than the 0.5 of their daughters. These numbers suggest that it would be more adaptive for workers to raise sisters than to raise their own offspring, supporting Hamilton's rule.

However, the queens in some insect societies mate with diploid males, or with multiple males, reducing the coefficient of relatedness of workers. Inclusive fitness is still important, but it is not sufficient to explain why less than 1 in 1000 workers would lay eggs. What else is going on?

One factor is that workers are sometimes manipulated into altruistic behavior. The queen and/or other workers police the nest and kill any eggs laid by workers. The more effective the policing, the less likely the workers are to continue expending energy to lay eggs. In other cases, the queen suppresses reproduction in the workers via pheromones or other chemical signals. Manipulation, or "enforced altruism," makes sense at the ultimate level because efficient colonies will be favored by natural selection, while rampant "cheating" can cause colonies to collapse.

Reciprocal Altruism

During long-term studies of highly social animals such as horses, elephants, dolphins, lions, chimpanzees, and vampire bats, biologists have observed nonrelatives helping each other. Chimps and other primates, for example, may spend considerable time grooming unrelated members of their social group—cleaning their fur and removing ticks and other parasites from their skin. In vampire bats, individuals that have been successful in finding food are known to regurgitate blood meals to non-kin that have not been successful and that are in danger of starving.

How can self-sacrificing behavior like this evolve if kin selection is not acting? The leading hypothesis to explain altruism among nonrelatives is called **reciprocal altruism:** an exchange of fitness benefits that are separated in time. Reciprocal altruists help individuals who have either helped them in the past or are likely to help them in the future.

Data that have been collected so far support the reciprocal altruism hypothesis in some instances.

- Among vervet monkeys, individuals are most likely to groom unrelated individuals that have groomed or helped them in the past.

- Vampire bats are most likely to donate blood meals to non-kin that have previously shared food with them.

Reciprocal altruism is also widely invoked as an explanation for the helpful and cooperative behavior commonly observed among unrelated humans.

Cooperation and Mutualism

Some behaviors may appear altruistic, but actually may be selfish because they increase, rather than decrease, the fitness of the animal exhibiting the behavior. That is, the benefits of assisting others can exceed the costs.

- African wild dogs cannot hunt large prey alone, but they can succeed by hunting as a pack. The fitness of all the dogs increases when they work together. This is an example of cooperation among members of the same species.

- Female horses that spend more time in grooming relationships with other female horses tend to suffer less harassment from males—and to have more foals.

When cooperation occurs among individuals of different species, mutualism is said to occur. The relationship between cleaner shrimp and their host fish is mutualistic because both the shrimp and the fish benefit when the shrimp clears off—and eats—parasites from the fish's gills and jaws (see Chapter 55).

To summarize, organisms are altruistic when doing so either increases their direct fitness or their inclusive fitness, or when they are coerced. As with feeding, mating, finding a place to live, and communicating, helping others has costs and benefits that must be weighed by the organism. This "decision making" occurs at the proximate level through genetic and physiological mechanisms, and at the ultimate level through evolutionary processes.

If you understand . . .

53.1 An Introduction to Behavioral Ecology

- Proximate causation: Experiments and observations focus on understanding how specific gene products, neuron activity, and hormonal signals cause behavior.

- Ultimate causation: Researchers seek to understand the adaptive significance of behavior, or why it enables individuals to survive and reproduce, thus increasing their fitness.

- Behaviors vary in the extent to which they are inflexible or flexible, and in the extent to which they are innate or learned.

- Animals appear to weigh the costs and benefits of responding to a particular situation in various ways. Costs and benefits are measured in terms of their impact on fitness.

✓ You should be able to explain the proximate and ultimate causes for increased sex drive in human teens.

53.2 What Should I Eat?

- Proximate causation: Fruit fly larvae will be sitters or rovers depending on which alleles they carry for a certain gene.

- Ultimate causation: The sitter and rover alleles are maintained in fruit fly populations because the fitness consequences of the alleles depend on fluctuating environmental circumstances.

- Optimal foraging theory predicts that animals will make foraging decisions that maximize energy gain and minimize energy loss, time spent, and predation risk, among other factors. For example, gerbils balance the benefits of foraging for food with the costs of exposure to owls.

✔ You should be able to suggest what variables you would have to consider in testing the principle of optimal foraging in a web-building spider.

53.3 Who Should I Mate With?

- Proximate causation: The onset of sexual behavior in *Anolis* lizards depends on surges of the sex hormones testosterone and estradiol that are triggered by springlike temperatures *and* courtship displays from males.

- Ultimate causation: Sexual selection favors individuals with traits that increase their ability to obtain mates. Females tend to choose mates that provide good alleles and valuable resources. Males compete for the opportunity to mate with females.

- Sexual selection in *Anolis* lizards is determined more by male–male competition than by female choice.

✔ You should be able to generate a hypothesis for why nest decorations are a reliable signal of male quality in bowerbirds (see chapter-opening photo).

53.4 Where Should I Live?

- Many animals migrate between different habitats rather than staying in one place due to the seasonable availability of food and the location of good nesting and mating places.

- Proximate causation: Animals can find their way by following familiar landmarks (piloting); by getting compass information from the Sun, the stars, and Earth's magnetic field; and/or by using the magnetic field for true navigation (map orientation).

- Ultimate causation: The benefits of migrating to food sources and mating and nesting areas must outweigh the energetic costs and danger of the migration.

✔ You should be able to provide an evolutionary hypothesis for the observation that some bird species do not migrate if people supply food for them in feeders.

53.5 How Should I Communicate?

- Honeybees have a symbolic language—their dancing movements communicate accurate information about the direction and distance of food sources.

- Proximate causation: Animals communicate with touch, sounds, scents, and/or sight depending on the environment.

- Ultimate causation: The ability to communicate effectively increases the fitness of animals.

- Communication can be honest or deceitful. Deceitful communication can be adaptive when it is rare.

✔ You should be able to predict the consequences for deceitful communication in bluegill sunfish if most large, territory-owning males were fished out of a lake.

53.6 When Should I Cooperate?

- Proximate causation: Altruistic behavior can be voluntary or coerced.

- Ultimate causation: Altruistic behavior appears paradoxical because it lowers the direct fitness of an individual.

- When altruistic behavior is directed toward close relatives, alleles that lead to self-sacrificing may increase in frequency due to kin selection (increasing the inclusive fitness of the altruist).

- Some animals that live in close-knit social groups engage in reciprocal altruism—meaning they exchange help over time.

- Other animals cooperate because everyone benefits, such as in cooperative foraging. Cooperation among species is called mutualism.

✔ You should be able to describe the characteristics of a species for which helpful behavior is expected to be common.

(MB) MasteringBiology

1. MasteringBiology Assignments

   Tutorials and Activities Animal Behavior and Learning; Homing Behavior in Digger Wasps

   Questions Reading Quizzes, Blue-Thread Questions, Test Bank

2. eText Read your book online, search, take notes, highlight text, and more.

3. The Study Area Practice Test, Cumulative Test, BioFlix® 3-D Animations, Videos, Activities, Audio Glossary, Word Study Tools, Art

You should be able to . . .

1. What do proximate explanations of behavior focus on?
   a. how displays and other types of behavior have changed through time, or evolved
   b. the functional aspect of a behavior, or its "adaptive significance"
   c. genetic, neurological, and hormonal mechanisms of behavior
   d. appropriate experimental methods when studying behavior

2. What do ultimate explanations of behavior focus on?

3. What unit(s) do biologists use when analyzing the costs and benefits of behavior?
   a. proximate versus ultimate
   b. body size (e.g., large versus small)
   c. time
   d. fitness—the ability to survive and produce offspring

4. Which of the following statements about the waggle dance of the honeybee is not correct?
   a. The length of a waggling run is proportional to the distance from the hive to a food source.
   b. Sounds and scents produced by the dancer provide information about the nature of the food source.
   c. The speed of waggling is proportional to the distance from the hive to a food source.
   d. The orientation of the waggling run provides information about the direction of the food from the hive, relative to the Sun's position.

5. Why are biologists convinced that the sex hormone testosterone is required for normal sexual activity in male *Anolis* lizards?
   a. Male *Anolis* lizards with larger testes court females more vigorously than do males with smaller testes.
   b. Testosterone is not found in female *Anolis* lizards.
   c. Male *Anolis* lizards whose gonads had been removed did not develop dewlaps.
   d. Male *Anolis* lizards whose gonads had been removed did not court females.

6. Why does altruism seem paradoxical?
   a. Sometimes altruistic behavior is actually selfish.
   b. Altruism does not actually help others.
   c. Alleles that cause an organism to behave altruistically should be selected against since these alleles should lower the organism's fitness.
   d. Animals behave altruistically to help the species, but sometimes their behavior harms the species.

7. Explain the ultimate causes of honeybee dancing.

8. Is it true that all organisms forage optimally? Why or why not?

9. Why do behaviors like *Anolis* lizard bobbing displays serve as honest signals of male quality?

10. What environmental stimuli cause changes in hormone levels that lead to egg laying in *Anolis* lizards? What biologically relevant information do these stimuli provide?

11. Evaluate this statement: For green sea turtles to return from the open ocean to the beach where they were hatched, they must have both a "map" and a "compass."

12. **QUANTITATIVE** Hamilton's rule states that an altruistic allele could spread in a population if $Br > C$, where B represents the fitness benefit to the recipient, r is the coefficient of relatedness between altruist and recipient, and C represents the fitness cost to the altruist. If $r = 0.5$ between the altruist and the recipient, what would the ratio of costs to benefits have to be for the altruistic allele to spread?
   a. $C/B > 0.5$
   b. $C/B > 0$
   c. $C/B < 0.5$
   b. $C/B < 0$

13. To date, the rover and sitter alleles have been shown to affect foraging behavior only in fruit fly larvae. Design an experiment to test the hypothesis that adults with the rover allele tend to fly farther in search of food sources than adults with the sitter allele.

14. You are invited on a summer internship to study the behavioral ecology of howler monkeys in Costa Rica. Your task will be to follow a troop of howler monkeys during the day and identify what types of fruits and leaves the troop consumes. Which of these general principles will help guide your work? Select all that apply.
   a. condition-dependent behavior
   b. optimal foraging
   c. Hamilton's rule
   d. cost–benefit analysis

15. **QUANTITATIVE** A biologist once remarked that he'd be willing to lay down his life to save two brothers or eight cousins. Explain what he meant.

16. Based on the theory of reciprocal altruism, predict the conditions under which people are expected to donate blood.

54 Population Ecology

In this chapter you will learn that

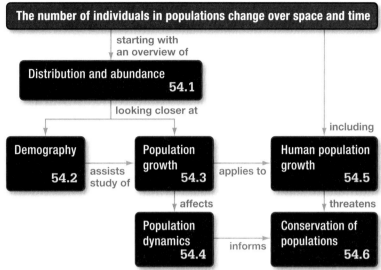

| The number of individuals in populations change over space and time |

starting with
an overview of

Distribution and abundance
54.1

looking closer at

Demography
54.2
— assists study of →
Population growth
54.3
— applies to →
Human population growth
54.5

↓ affects

↓ threatens

Population dynamics
54.4
— informs →
Conservation of populations
54.6

This vivid school of crescent-tail bigeyes off the coast of Papua New Guinea demonstrates how individuals associate together in populations. The distribution and abundance of individuals in populations changes over space and time.

f you asked a biologist to name two of today's most pressing global issues, she might say global climate change and extinction of species. If you were asked to name two of the most pressing issues facing your region, you might say traffic and the price of housing. All four problems have a common cause: recent and dramatic increases in the size of the human population.

In 1940, for example, Mexico City had a population of about 1.6 million. The city grew to 5.4 million in 1960, 13.9 million in 1980, and over 21 million in 2010. How much bigger will it get over the next 50 years? And how large will the entire human population be by the time your kids are in college?

This chapter is part of the Big Picture. See how on pages 1196–1197.

✔ When you see this checkmark, stop and test yourself. Answers are available in Appendix A.

A **population** is a group of individuals of the same species that live in the same area at the same time; **population ecology** is the study of how and why the number of individuals in a population changes over time and space. (You can see how population ecology fits into the Big Picture of Ecology on pages 1196–1197).

With the explosion of human populations across the globe, the massive destruction of natural habitats, changes in climate, and the resulting threats to species throughout the tree of life, population ecology has become a vital field in biological science. The mathematical and analytical tools introduced in this chapter help biologists predict changes in population size and design management strategies to save threatened species.

Let's start by considering some of the basic tools that biologists use to study populations and then follow up with examples of how biologists study changes in the population size of humans and other species over time. The chapter concludes by asking how all of these elements fit together in efforts to limit human population growth and conserve biodiversity.

54.1 Distribution and Abundance

If you were interested in studying the population biology of the blue-ringed octopus, Florida strangler fig, or any other species, you might want to start by asking, "Where do they live, and how many are there?" Distribution and abundance are two fundamental concepts in ecology.

Most species identification guides show the **range,** or geographic distribution, of different species. Two types of factors determine range:

- *Abiotic factors*, such as temperature, rainfall, the presence of geographical structures like mountains and oceans, and large-scale ongoing and historical processes such as continental drift (see Chapter 52);

- *Biotic factors*, such as the past and current presence of other species that provide habitat, food, or competition (discussed in detail in Chapter 55).

Ranges are dynamic—in constant flux as abiotic and biotic factors change over time.

The range of a species can also be considered at different scales. For example, the map in **FIGURE 54.1** shows the vast global range of the common lizard, *Lacerta vivipara* (sometimes called *Zootoca vivipara*), from Ireland in western Europe to Japan in eastern Asia. This map indicates that *L. vivipara* can live in a variety of habitats and is cold adapted—it lives farther north than other reptiles, and only at high altitudes at the southern extent of its range.

Lizards are not distributed uniformly within this range, however. The **population density**—the number of individuals per unit area—of *L. vivipara* varies throughout its range. If you zoom in on one region, such as the British Isles, you can observe that *L. vivipara* is remarkably widespread, yet absent from some areas. Zooming in further still, the distribution of *L. vivipara* is

Global range of *Lacerta vivipara*

FIGURE 54.1 Geographic Range of *Lacerta vivipara*.

clumped according to the distribution of suitable habitats, such as meadows and open woodland.

In general, individual organisms can be arranged in different patterns within populations. The arrangement may be

- *random*, if the position of each individual is independent of the others, as may occur when seeds are dispersed by the wind;

- *clumped*, if the quality of the habitat is patchy or the organisms associate in social groups (such as schools of fish); or

- *uniform*, if negative interactions occur among individuals, such as competition for space, water, or other resources.

Within a certain patch of habitat, the distribution of *L. vivipara* individuals may be relatively uniform due to their maintenance of feeding and mating territories (see Chapter 53). However, as with most species, the distribution of lizards is clumped overall due to varying density of individuals in habitat patches of different quality. The particular dispersion pattern of a population has both proximate causes, in terms of physiological and behavioral mechanisms, and ultimate causes, in terms

To begin a mark–recapture study, researchers catch individuals and mark them with leg bands, ear tags, or some other method of identification. After the marked individuals are released, they are allowed to mix with the unmarked animals in the population for a period of time. Then a second trapping effort is conducted, and the percentage of marked individuals that were captured is recorded.

To estimate the total population from these data, researchers make a key assumption: The percentage of marked and recaptured individuals is equal to the percentage of marked individuals in the entire population. This assumption should be valid if individuals are not moving in and out of the study area, and if no bias exists regarding which individuals are caught in each sample attempt. It

is important that individuals do not learn to avoid or seek out traps after being caught once and that they do not change their behavior or die as a result of being marked or trapped.

The relationship between marked and unmarked individuals can be expressed algebraically:

Number of marked animals in the second sample (the recapture)

Number of marked animals in the first sample

$$\frac{m_2}{n_2} = \frac{m_1}{N}$$

Total number of animals (marked and unmarked) in the second sample

Total population size

(Eq. 54.1)

Having measured m_2, n_2, and m_1, the researcher can estimate N.

✔**QUANTITATIVE** If you understand how to estimate population size in a mark–recapture study, you should be able to do so in this example: Suppose researchers marked 255 animals and later were able to trap a total of 162 individuals in the population, of which 78 were marked. What is the estimate for total population size? Solve for N (give the nearest integer) and check your answer.[1] (Answer is given below.)

$^1 N = 530.$

---

of evolutionary adaptations (see Chapter 53 for an overview of proximate and ultimate causation).

If a species range is small, it may consist of a single population of interbreeding individuals. If a range is large, it may consist of many populations—sometimes contiguous, and sometimes isolated in space. If individuals from a species occupy many small patches of habitat, so that they form many independent populations, they are said to represent a **metapopulation**—a population of populations connected by migration. Because humans are reducing large, contiguous areas of forest and grasslands to isolated patches or reserves, more and more species are being forced into a metapopulation structure.

How can the population size, density, or distribution pattern of individuals be determined? It depends on the species. For sedentary (largely immobile) organisms such as trees and barnacles, researchers sample habitats by counting the individuals that occur along lines of known position and length—called *transects*—or inside rectangular plots—called *quadrats*—set up at random locations in the habitat. These counts can be extrapolated to the entire habitat area to estimate the total population size. In addition, they can be compared to later censuses to document trends over time.

In contrast, counting organisms that are mobile and that do not congregate into herds or flocks or schools is much more challenging. For example, the number of lizards inside sample quadrats or along transects changes constantly as individuals move. Further, it can be difficult to track whether a particular individual has already been counted. But if individuals can be captured and then tagged in some way, the total population size of a mobile species can be estimated by using a research approach called mark–recapture (**Quantitative Methods 54.1**).

Note that individuals within populations vary, and populations are not static in time and space. To get a better understanding of

how researchers quantify population size and structure, let's first examine a field of study in ecology called demography.

54.2 Demography

The number of individuals present in a population depends on four processes: birth, death, immigration, and emigration.

- Populations grow as a result of births—here, meaning any form of reproduction—and **immigration**, which occurs when individuals enter a population by moving from another population.

- Populations decline due to deaths and **emigration**, which occurs when individuals leave a population to join another population.

Analyzing birth rates, death rates, immigration rates, and emigration rates is fundamental to **demography**: the study of factors that determine the size and structure of populations through time.

To predict the future of a population, biologists have to know something about its makeup. They need to know the population's **age structure**—how many individuals of each age are alive. They also need to know how likely individuals of different ages are to survive to the following year, how many offspring are produced by females of different ages, and how many individuals of different ages immigrate and emigrate each **generation**—the average time between a mother's first offspring and her daughter's first offspring.

If a population consists primarily of young individuals with a high survival rate and reproductive rate, the population size should increase over time. But if a population comprises chiefly old individuals with low reproductive rates and low survival rates, then it is almost certain to decline over time. Life tables are useful tools for analyzing this type of information.

Life Tables

A **life table** summarizes the probability that an individual will survive and reproduce in any given time interval over the course of its lifetime. Life tables were invented almost 2000 years ago; ancient Romans used them to predict food needs. In modern times, life insurance companies use life tables to predict the likelihood of a person dying at a given age. Biologists use life tables to study the demographics of endangered species.

***Lacerta vivipara*: A Case Study** To understand how researchers use life tables, consider again the lizard *L. vivipara*. As their name suggests—vivipara means "live birth"—most populations are ovoviviparous and give birth to live young (see Chapter 33).

European biologists Henk Strijbosch and Raymond Creemers set out to construct a life table of a low-elevation population of *L. vivipara* in the Netherlands, with the goal of comparing the results to data that other researchers had collected from *L. vivipara* populations in the mountains of Austria and France and in lowland habitats in Britain and Belgium. They wanted to know whether populations that live in different environments vary in basic demographic features.

To begin the study, Strijbosch, Creemers, or their students visited their study site daily during the seven months that these lizards are active during the year. Each day, the research team captured and marked as many individuals as possible.

Because this program of daily monitoring continued for seven years, the biologists were able to document the number of young produced by each female in each year of her life. If a marked individual was not recaptured in a subsequent year, they assumed that it had died sometime during the previous year. The data allowed researchers to calculate the number of individuals that survived each year in each particular age group as well as how many offspring each female produced. What did the numbers reveal?

TABLE 54.1 shows the data Strijbosch and Creemers collected for females. In most cases, biologists focus on females when calculating life-table data, because the number of males present rarely affects population growth. There are almost always enough males present to fertilize all of the females in breeding condition, so growth rates depend entirely on females. Let's walk through Table 54.1 step by step.

Age Class and Number Alive The first column in Table 54.1 lists the age classes of *L. vivipara*. An **age class** is a group of individuals of a specific age—for example, all female lizards between 4 and 5 years old. These data show that *L. vivipara* females have a maximum life span of about seven years.

But not all females live to be this old. The second column in the table shows how many female lizards survived to reach each age class. Suppose 1000 female *L. vivipara* are born in a particular year. These individuals represent a **cohort**—a group of the same age that can be followed through time. How many individuals would survive to age 1, age 2, and so on? On average, 424 would still be alive one year later, and 308 would still be alive two years later.

Survivorship The third column in Table 54.1 converts the data in the second column to proportions. **Survivorship**—a key component of a life table—is defined as the proportion of offspring produced that survive, on average, to a particular age. For example, the survivorship from birth to age 1 was 0.424 in the Netherlands population. Survivorship from birth to age 2 was 0.308.

Life tables do not usually include data bars—the bars in this table, such as the orange bars for survivorship, are intended to help you visualize the patterns in the numbers in the different columns. Since survivorship varies by several orders of magnitude, in this case from 1.000 to 0.002, biologists typically plot the logarithm of the number of survivors versus age. Using a

TABLE 54.1 Life Table for *Lacerta vivipara* Females in the Netherlands

Age Class (x)	Number of Survivors (N_x)	Survivorship (l_x)	×	Age-Specific Fecundity (m_x)	=	Average Births/Year/Original Female $(l_x m_x)$
0	1000	1.000		0.00		0.00
1	424	0.424		0.04		0.02
2	308	0.308		1.47		0.45
3	158	0.158		2.07		0.33
4	57	0.057		2.44		0.14
5	10	0.010		3.25		0.04
6	7	0.007		3.25		0.03
7	2	0.002		3.25		+ 0.01
						$R_0 = 1.00 =$ Net reproductive rate

SOURCE: H. Strijbosch and R. C. M. Creemers, 1988. Comparative demography of sympatric populations of *Lacerta vivipara and Lacerta agilis*. *Oecologia* 76: 20–26. With kind permission of Springer Science and Business Media.

(a) Three general types of survivorship curves

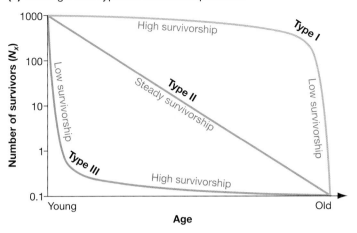

(b) Exercise: Survivorship curve for *Lacerta vivipara*

FIGURE 54.2 Survivorship Curves Identify When Mortality Rates Are Low, Steady, or High.

✔ **QUANTITATIVE** Fill in the graph in part (b) with data on survivorship of *Lacerta vivipara* given in Table 54.1. Compare the shape of the curve to the generalized graphs in part (a).

logarithmic scale on the *y*-axis is a matter of convenience—it makes patterns easier to see (see **BioSkills 6** in Appendix B). The resulting graph is called a **survivorship curve.** Studies on a wide variety of species indicate that three general types of survivorship curves exist (**FIGURE 54.2a**).

1. Humans have what biologists call a type I survivorship curve. In this pattern, survivorship throughout life is high—most individuals approach the species' maximum life span.

2. Type II survivorship curves occur in species where individuals have about the same probability of dying in each year of life. Blackbirds and other songbirds have this type of curve.

3. Many plants have Type III curves—a pattern defined by extremely high death rates for seeds and seedlings but high survival rates later in life.

FIGURE 54.2b provides a graph for you to plot survivorship of *L. vivipara*, using the data reported in Table 54.1. Note the logarithmic scale on the *y*-axis of the graph. What type of survivorship curve does *L. vivipara* have?

Fecundity The fourth column in Table 54.1 shows the data for **fecundity,** the number of female offspring produced by each female in the population. Because the researchers documented the reproductive output of the same *L. vivipara* lizard females year after year, they were able to calculate a quantity called **age-specific fecundity**: the average number of female offspring produced by a female in each age class. The blue data bars in Table 54.1 illustrate the pattern of fecundity in lizards of different ages: older female lizards have a higher age-specific fecundity than younger females.

Reproductive Rate If survivorship is higher in young lizards and fecundity is higher in older lizards, what is the outcome? The purple bars in the fifth column of Table 54.1 show the product of survivorship multiplied by fecundity—most of the offspring in the population are produced by females of intermediate age.

The sum of the purple bars is a measure of **net reproductive rate,** which indicates whether the population is increasing or decreasing (as long as immigration and emigration are insignificant). Since this sum is equal to 1, this population of *L. vivipara* is stable. **Quantitative Methods 54.2** (see page 1106) walks through the calculations in the life table in more detail.

How do the data on the Netherlands populations of *L. vivipara* compare with populations in different types of habitats?

The Role of Life History

The life-table data in Table 54.1 are interesting because they contrast with results from other populations of *L. vivipara*. Consider data on fecundity:

- In the Netherlands, almost no 1-year-old female *L. vivipara* reproduce.

- In Brittany, France, 50 percent of 1-year-old female lizards reproduce.

- In the mountains of Austria, females don't begin breeding until they are 4 years old.

The contrasts are equally stark when looking at survivorship. In the Austrian population, most females live much longer than do individuals in either lowland population—in the Netherlands or France.

In Brittany, fecundity is high but survivorship is low; in Austria, fecundity is low but survivorship is high. The population in the Netherlands is intermediate. In this species, key aspects of the life table vary dramatically among populations.

What Are Fitness Trade-Offs? Why isn't it possible for *L. vivipara* females to have both high fecundity and high survival? The answer is **fitness trade-offs** (see Chapters 25 and 42).

Fitness trade-offs occur because every individual has a restricted amount of time and energy at its disposal—meaning that its resources are limited. If a female lizard devotes a great deal of

If immigration and emigration are not occurring, the data in a life table can be used to calculate a population's growth rate. This approach is logical, because survivorship and fecundity are ways to express death rates and birth rates—the other two factors influencing population size. To see how a population's growth rate can be estimated from life-table data, let's look at each component in a life table more carefully. Recall that life tables usually focus exclusively on females.

Survivorship is symbolized as l_x, where x represents the age class being considered. Survivorship for age class x is calculated by dividing the number of females in that age class (N_x) by the number of females that existed as offspring (N_0):

Survivorship: $l_x = \dfrac{N_x}{N_0}$

Number of females in age class x
Number of original females

(Eq. 54.2)

Age-specific fecundity is symbolized m_x, where x again represents the age class being considered. Age-specific fecundity is calculated as the total number of female offspring produced by females of a particular age, divided by the total number of females of that age class present. It represents the average number of female offspring produced by a female of age x.

Documenting age-specific survivorship and fecundity allows researchers to calculate the average number of female offspring produced by females during each age of life, as $l_x m_x$. For example, if 30 percent of the females in a population live to be 2 years old, and if 2-year-old females have an average of 2 offspring, then the average number of births at age 2, per female born into the population, is $l_x m_x = 0.30 \times 2 = 0.6$.

Summing $l_x m_x$ values over the entire life span of a female gives the net reproductive rate, R_0, of a population:

Sum of all ages — Survivorship at age x — Fecundity at age x

Net reproductive rate: $R_0 = \sum_{i=0}^{x} l_x m_x$

(Eq. 54.3)

Here the Greek letter sigma (Σ) stands for "sum." The expression on the right-hand side of the equals sign is read, "sum the $l_x m_x$ values from each year of life, starting from birth or hatching ($i=0$), through age x."

The net reproductive rate represents the growth rate of a population per generation. The logic behind the equation for R_0 is that the growth rate of a population per generation equals the average number of female offspring that each female produces over the course of her lifetime. A female's average lifetime reproduction, in turn, is a function of survivorship and fecundity at each age class. In Table 54.1, R_0 is the sum of the survivorship multiplied by the fecundity values in the right column.

If R_0 is greater than 1, then the population is increasing in size. If R_0 is less than 1, then the population is declining. ✔QUANTITATIVE If you understand these concepts, you should be able to use the data in Table 54.1 to determine how many female offspring an average *L. vivipara* female produces over the course of her lifetime,[1] and describe whether the population is growing, stable, or declining.[2]

[1]1.00; [2]The population is stable.

energy to producing a large number of offspring, she is not able to devote that same energy to her immune system, growth, nutrient stores, or other traits that increase survival. That is, a female can maximize fecundity, maximize survival, or strike a balance between the two.

An organism's **life history** describes how an individual allocates resources to growth, reproduction, and activities or structures that are related to survival. Traits such as survivorship, age-specific fecundity, age at first reproduction, and growth rate are all aspects of an organism's life history. Understanding variation in life history is all about understanding fitness trade-offs.

In almost all cases, biologists find that life history is shaped by natural selection in a way that maximizes an individual's fitness in its environment. In *L. vivipara* populations, biologists contend that females who live a long time but mature late and have few offspring each year have high fitness in cold, high-elevation habitats, such as Austria. In these habitats, females have to reduce their reproductive output and put more energy into traits that increase survival in a harsh environment. In contrast, females who have short lives but mature early and have large numbers of offspring each year do better in warm, low-elevation habitats, such as Brittany.

✔ If you understand the role that fitness trade-offs play in determining life-history patterns, you should be able to (1) predict how survivorship and fecundity should compare in *L. vivipara* in the warmest versus coldest parts of their range, and (2) comment on why females in these populations lay many eggs instead of giving birth to a relatively small number of live young.

Life-History Patterns across Species The life-history data you've reviewed thus far all involve comparisons of populations within a species. But the same patterns exist when many different species are compared.

In general, individuals from species with high fecundity tend to grow quickly, reach sexual maturity at a young age, and produce many small eggs or seeds. The mustard plant *Arabidopsis thaliana*, for example, germinates and grows to sexual maturity in just 4 to 6 weeks. In this species, individuals usually live only a few months but may produce as many as 10,000 tiny seeds. They live fast and die young.

In contrast, individuals from species with high survivorship tend to grow slowly and invest resources in traits that reduce damage from enemies and increase their own ability to compete for water, sunlight, or food. A coconut palm, for example, may take a decade to mature but live 60–70 years and produce offspring each

Life-history continuum

Low fecundity

High fecundity

Moderate fecundity

Moderate survivorship

High survivorship

Low survivorship

FIGURE 54.3 Life-History Traits Form a Continuum. Every organism can be placed somewhere on this life-history continuum. The placement of a species is most meaningful when it is considered relative to closely related species.

year. Coconut palms invest resources in making stout stems that allow them to grow tall, a relatively extensive root system, large fruits (coconuts), molecules that make herbivores sick, and enzymes that reduce infections from disease-causing fungi. These traits increase survivorship but decrease fecundity.

An *Arabidopsis thaliana* plant and a coconut palm represent two ends of a broad continuum of life-history characteristics (**FIGURE 54.3**). How does a species' location on this continuum affect its population growth rate?

54.3 Population Growth

Life tables provide a snapshot of a population, linking life-history patterns to population growth. Let's take a closer look at changes in population size over time. For conservation biologists, analyzing and predicting changes in population size is fundamental to managing threatened species.

Recall that four processes affect a population's size: Births and immigration add individuals to the population; deaths and emigration remove them. It follows that a population's overall growth rate is a function of birth rates, death rates, immigration rates, and emigration rates.

Quantifying the Growth Rate

A population's growth rate is the change in the number of individuals in the population (ΔN) per unit time (Δt). Here the Greek symbol Δ (delta) means change.

If no immigration or emigration is occurring, then a population's growth rate is equal to the number of individuals (N) in the population times the difference between the birth rate per individual (b) and death rate per individual (d). The difference between the birth rate and death rate per individual is called the per capita rate of increase and is symbolized r. (Per capita means "for each individual.")

If the per capita birth rate is greater than the per capita death rate, then r is positive and the population is growing. But if the per capita death rate begins to exceed the per capita birth rate, then r becomes negative and the population declines. Within populations, r varies through time. Its value can be positive, negative, or 0.

When conditions are optimal for a particular species—meaning birth rates per individual are as high as possible and death rates per individual are as low as possible—then r reaches a maximal value called the **intrinsic rate of increase**, r_{max}. When this occurs, a population's growth rate is expressed as

$$\frac{\Delta N}{\Delta t} = r_{max}N$$

In species such as *Arabidopsis* and fruit flies, which breed at a young age and produce many offspring each year, r_{max} is high. In contrast, r_{max} is low in species such as giant pandas and coconut palms, which take years to mature and produce few offspring each year. Stated another way, r_{max} is a function of a species' life-history traits.

Each species has a characteristic r_{max} that does not change. But at any specific time, a population has an instantaneous growth rate, or per capita rate of increase, symbolized by r. Whereas r_{max} tells you what the maximum growth rate is, r tells you what it is at a particular time. So r is always less than or equal to r_{max}.

The instantaneous growth rate of a population at a particular time is actually likely to be much lower than r_{max}. A population's r is also likely to be different from r values of other populations of the same species (such as in upland and lowland populations of *L. vivipara*), and to change over time (such as in wet years versus dry years). The instantaneous growth rate is dynamic. However, let's consider the simplest case first, where r is constant.

Exponential Growth

The graph in **FIGURE 54.4** (see page 1108) plots changes in population size, for various values of r, under the condition known as exponential growth. **Exponential population growth** occurs when r does not change over time.

The key point about exponential growth is that the growth rate does not depend on the number of individuals in the population. Biologists say that this type of population growth is **density independent.**

It's important to emphasize that exponential growth adds an increasing number of individuals as the total number of individuals, N, gets larger. As an extreme example, an r of 0.02 per year in a population of 1 billion adds over 20 million individuals per year. The same growth rate in a population of 100 adds just

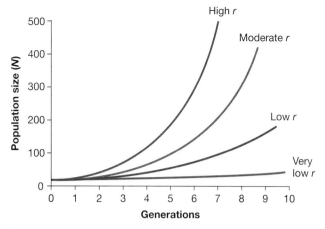

FIGURE 54.4 Exponential Growth Is Independent of Population Size. When the per capita growth rate *r* does not change over time, exponential growth occurs. Population size may increase slowly or rapidly, depending on the size of *r*.

over 2 individuals per year. Even if *r* is constant, the number of individuals added to a population is a function of *N*. The *rate* of increase is the same, but the number of individuals added is not.

In nature, exponential growth is common in two circumstances: **(1)** a few individuals found a new population in a new habitat, or **(2)** a population has been devastated by a storm or some other type of catastrophe and then begins to recover, starting with a few surviving individuals. But it is not possible for exponential growth to continue indefinitely.

If *Arabidopsis* populations grew exponentially for a long period of time, they would eventually fill all available habitat. When population density gets very high, the population's per capita birth rate will decrease and the per capita death rate will increase, causing *r* to decline. Stated another way, growth is often **density dependent.**

Logistic Growth

To analyze what happens when growth is density dependent, biologists use a parameter called carrying capacity. **Carrying capacity, *K*,** is defined as the maximum number of individuals in a population that can be supported in a particular habitat over a sustained period of time.

The carrying capacity of a habitat depends on many factors: food, space, water, soil quality, and resting or nesting sites. Carrying capacity can change from year to year, depending on conditions.

A Logistic Growth Equation If a population of size *N* is below the carrying capacity *K*, then the population should continue to grow. More specifically, a population's growth rate is proportional to $(K - N)/K$:

$$\frac{\Delta N}{\Delta t} = r_{max}N\left[\frac{K-N}{K}\right]$$

This expression is called a logistic growth equation.

In the expression $(K - N)/K$, the numerator defines the number of additional individuals that can be accommodated in a habitat with carrying capacity *K*; dividing $K - N$ by *K* turns this number of individuals into a proportion. Thus $(K - N)/K$ describes the proportion of "unused resources and space" in the habitat. It can be thought of as the environment's resistance to growth.

- When *N* is small, then $(K - N)/K$ is close to 1 and the growth rate should be high. For example: $(1000 - 10)/1000 = 0.99$.

- As *N* gets larger, $(K - N)/K$ gets smaller. For example: $(1000 - 900)/1000 = 0.10$.

- When *N* is at carrying capacity (meaning that $K = N$), then $(K - N)/K$ is equal to 0 and growth stops.

Thus, as a population approaches a habitat's carrying capacity, its growth rate should slow.

The logistic growth equation describes **logistic population growth,** or changes in growth rate that occur as a function of population size. Whereas exponential growth is density independent, logistic growth is density dependent. **Quantitative Methods 54.3** (see pages 1110–1111) explores population growth models in more detail and explains how they relate to the net reproductive rate calculated in the life table.

Graphing Logistic Growth **FIGURE 54.5a** illustrates density-dependent growth in a hypothetical population. The graph plots changes in population size over time, and has three sections.

1. Initially, growth is exponential—meaning that *r* is constant.

2. With time, *N* increases to the point where competition for resources or other density-dependent factors begins to occur. As a result, the growth rate begins to decline.

3. When the population is at the habitat's carrying capacity, the growth rate is 0—the graph of population size versus time is flat.

This is exactly what happened in an experiment on laboratory populations of ciliates, *Paramecium aurelia* and *P. caudatum* (see Chapter 30). Russian biologist Georgii Gause placed 20 individuals from one of the *Paramecium* species into 5 mL of a solution. He created many replicates of these 5-mL environments for each species separately. He kept conditions as constant as possible by adding the same number of bacterial cells every day for food, washing the solution every second day to remove wastes, and maintaining the pH at 8.0.

In the graphs in **FIGURE 54.5b**, each data point represents population size for one of the species—the average of the replicates in the experiment. Note that both species exhibited logistic growth in this environment. The carrying capacity differed in the two species, however. The maximum density of *P. aurelia* averaged 448 individuals per mL, but that for *P. caudatum* averaged just 128 individuals per mL. (When the two species are grown together, one species gets eliminated; see Chapter 55.)

In both of the *Paramecium* species in this experiment, exponential growth could be sustained for only about five days. What factors cause growth rates to change?

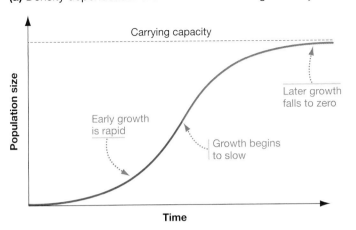

(a) Density dependence: Growth rate slows at high density.

Carrying capacity

Population size

Early growth is rapid

Growth begins to slow

Later growth falls to zero

Time

(b) Logistic growth in ciliates

Population size

Time (days)

Paramecium aurelia alone

Paramecium caudatum alone

FIGURE 54.5 Logistic Growth Is Dependent on Population Size. (a) A curve illustrating the pattern predicted by the logistic growth equation. This pattern often occurs when a small number of individuals colonize an unoccupied habitat. Initially, *r* is high because competition for resources is low to nonexistent. Carrying capacity depends on the quality of the habitat and can vary over time. **(b)** Data from laboratory experiments with two species of *Paramecium*.

DATA: Gause, G. F. 1934. *The Struggle for Existence.* New York: Hafner Press.

What Limits Growth Rates and Population Sizes?

Population sizes change as a result of two general types of factors:

1. *Density-independent factors* alter birth rates and death rates irrespective of the number of individuals in the population, and they usually involve changes in the abiotic environment—variation in weather patterns, or catastrophic events such as cold snaps, hurricanes, volcanic eruptions, or drought.

2. *Density-dependent factors* change in intensity as a function of population size, and they are usually biotic. When trees crowd each other, they have less water, nutrients, and sunlight at their disposal and make fewer seeds.

A Closer Look at Density Dependence To get a better understanding of how biologists study the density-dependent factors that affect population size, consider the data in **FIGURE 54.6**.

The graph in Figure 54.6a presents results of an experimental study of a coral-reef fish called the bridled goby. Each data point represents the average of eight artificial reefs constructed by ecologists Graham Forrester and Mark Steele from pieces of real coral reefs. The initial density, plotted along the *x*-axis, represents the number of bridled gobies released per area at the start of the experiment. The proportion surviving, plotted on the vertical axis, represents the individuals that were still living at that reef 30 days later. The researchers compared the survivorship of

(a) Survival of gobies declines at high population density.

Proportion surviving

Initial density (*N*/m²)

☐ No predators present
■ Predators present

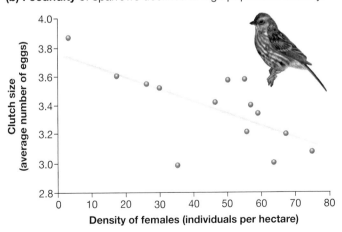

(b) Fecundity of sparrows declines at high population density.

Clutch size (average number of eggs)

Density of females (individuals per hectare)

FIGURE 54.6 Density-Dependent Growth Results from Changes in Survivorship and Fecundity.

DATA: Forrester, G. E. and M. A. Steele 2000. *Ecology* 81: 2416–2427; Arcese, P., and J. N. M. Smith. 1988. *Journal of Animal Ecology* 57: 119–136.

✔ **QUESTION** When song sparrow populations are at high density and extra food is provided to females experimentally, average clutch size is much higher than expected from the data in part (b). Based on these data, state a hypothesis to explain why average clutch size declines at high density.

To explore how biologists model changes in population size in more detail, let's consider data on whooping cranes, which are large, wetland-breeding birds native to North America (**FIGURE 54.7**). Whooping cranes may live more than 20 years in the wild, but it is common for females to be six or seven years old before they breed for the first time. Female cranes lay just one or two eggs per year and usually rear just one chick. Based on these life-history characteristics, you should understand why the growth rate of the whooping crane populations is extremely low.

Conservation biologists monitor this species closely because hunting and habitat destruction reduced the total number of individuals in the world to about 20 in the mid-1940s. Since then, intensive conservation efforts have resulted in a current total population of about 571. That number includes a group of 279 individuals that breed in Wood Buffalo National Park in the Northwest Territories of Canada.

FIGURE 54.7 Whooping Cranes Have Been the Focus of Intensive Conservation Efforts.

In addition to the Wood Buffalo population, two new populations of cranes have been established by releasing offspring raised in captivity. One of these groups lives near the Atlantic coast of Florida year-round; the other group migrates from breeding areas in northern Wisconsin to wintering areas along Florida's Gulf Coast.

According to the biologists who are managing whooping crane recovery, the species will no longer be endangered when the two newly established populations have at least 25 breeding pairs, and when neither population needs to be supplemented with captive-bred young in order to be self-sustaining. How long will this take?

Discrete Growth

Whooping cranes breed once per year, so the simplest way to express a crane population's growth rate is to compare the number of individuals at the start of one breeding season to the number at the start of the following year's breeding season.

To create a general expression for how populations grow over a discrete time interval, biologists use N to symbolize population size. N_0 is the population size at time zero (the starting point), and N_1 is the population size one breeding interval later. In equation form, the growth rate is given as

$$\lambda = \frac{N_1}{N_0}$$

Finite rate of increase:
Population size one breeding interval later
Population size at time zero

(Eq. 54.4)

The parameter λ (lambda) is called the **finite rate of increase.** (In mathematics, a *finite rate* refers to an observed rate over a given period of time. A *parameter* is a variable or constant term that affects the shape

of a function.) The current population size at Wood Buffalo is 279. Suppose that biologists count 300 cranes on the breeding grounds next year. The population growth rate could be calculated as $300/279 = 1.075 = \lambda$. Stated another way, the population will have grown at the rate of 7.5 percent per year. Rearranging the expression in Equation 54.4 gives

$$N_1 = N_0\lambda \qquad \textbf{(Eq. 54.5)}$$

Stated more generally, the size of the population at the end of year t will be given by:

Population size at time zero
Finite rate of increase to the power of t
Population size at end of year t: $N_t = N_0\lambda^t$

(Eq. 54.6)

This equation summarizes how populations grow when breeding takes place seasonally. The size of the population at time t is equal to the starting size, times the finite rate of increase multiplied by itself t times.

In a sense, λ works like the interest rate at a bank. For species that breed once per year, the "interest" on the population is compounded annually. A savings account with a 5 percent annual interest rate increases by a factor of 1.05 per year.

If a population is growing, then its λ is greater than one. The population is stable when λ is 1.0 and declining when λ is less than 1.0.

If a population's age structure is stable, meaning that the proportion of females in each age class is not changing over time, then its finite rate of increase also has a simple relationship to its net reproductive rate (discussed in Quantitative Methods 54.2):

gobies with natural predators present (orange data points), and also with predators excluded from the study areas by cages (red data points). The higher the original density of gobies on the reef, the least likely these fish were to survive. More specifically, a majority of the gobies in the low-density trials survived the study period. In contrast, only about 20 percent of the gobies

in the high-density trials survived. This graph shows a strong density-dependent relationship in *survivorship* both with and without predators.

The graph in Figure 54.6b is from Peter Arcese and James Smith's long-term study of song sparrows on Mandarte Island, British Columbia. Each data point represents a different year.

Finite rate of increase: $\lambda = \dfrac{R_0}{g}$

Net reproductive rate

Generation time

(Eq. 54.7)

In essence, dividing the net reproductive rate by generation time transforms it into a discrete rate.

Continuous Growth

The parameters λ and r (the finite rate of increase and the per capita growth rate, respectively) have a simple mathematical relationship. The best way to understand their relationship is to recall that λ expresses a population's growth rate over a discrete interval of time. In contrast, r gives the population's per capita growth rate at any particular instant. This is why r is also called the instantaneous rate of increase. The relationship between the two parameters is given by

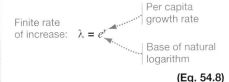

Finite rate of increase: $\lambda = e^r$

Per capita growth rate

Base of natural logarithm

(Eq. 54.8)

where e is about 2.72. (For help with using logarithms, see **BioSkills 6** in Appendix B. Also, note that the relationship between any finite rate and any instantaneous rate is given by finite rate = $e^{\text{instantaneous rate}}$.)

Substituting Equation 54.8 into Equation 54.6 gives

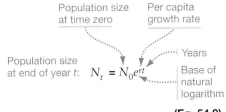

Population size at time zero

Per capita growth rate

Population size at end of year t: $N_t = N_0 e^{rt}$

Years

Base of natural logarithm

(Eq. 54.9)

This expression summarizes how populations grow when they breed continuously, like humans and bacteria do, instead of at defined intervals. For species that breed continuously, the "interest" on the population is compounded continuously. When the growth rates λ and r are equivalent, however, the differences between discrete and continuous growth are negligible.

Because r represents the growth rate at any given time, and because r and λ are so closely related, biologists routinely calculate r for species that breed seasonally. In the whooping crane example, $\lambda = 1.039$ per year = e^r. To solve for r, take the natural logarithm of both sides. (**BioSkills 6** in Appendix B explains how to calculate a natural logarithm using your calculator.) In this case, $r = 0.038$ per year.

The instantaneous rate of increase, r, is also directly related to the net reproductive rate, R_0:

Natural logarithm

Net reproductive rate

Per capita growth rate: $r = \ln \dfrac{R_0}{g}$

Generation time

(Eq. 54.10)

Thus, r can be calculated from life-table data. It is a more useful measure of growth rate than R_0, because r is independent of generation time.

To summarize, biologists have developed several ways of calculating and expressing a population's growth rate:

Finite rate of increase (λ): easiest to understand, but applies only to populations that breed seasonally.

Net reproductive rate (R_0): can be calculated directly from life-table data, but is dependent on generation time.

Instantaneous rate of increase (r): more difficult conceptually, but is the most useful expression for growth rate because it

is independent of generation time and is relevant for species that breed either seasonally or continuously.

Applying the Models

To get a better feel for r and for Equation 54.9, consider the following series of questions about whooping cranes. The key to answering these questions is to realize that Equation 54.9 has just four parameters. Given three of these parameters, you can calculate the fourth. ✔**QUANTITATIVE** If you understand this concept, you should be able to solve these three problems:

1. If 20 individuals were alive in 1941 and 571 existed in 2011, what is r? Here $N_t = 571$, $N_0 = 20$, and $t = 70$ years. Substitute these values into Equation 54.9 and solve for r. Then check your answer at the end of this box.[1]

2. The most recent report issued by the biologists working on the Wood Buffalo crane recovery program estimates that that the flock should be able to sustain an r of 0.048 for the foreseeable future. If the flock currently contains 279 individuals, how long will it take that population to double? Here $N_0 = 279$ and $N_t = 2 \times 279 = 558$. In this case, you solve Equation 54.9 for t. Then check your answer.[2]

3. In 2002, a pair of birds in the flock that lives in Florida year-round successfully raised offspring (nine years after the first cranes were introduced there). In 2003, another pair of cranes in this population bred successfully. If the number of breeding pairs had continued to double each year, how long would it have taken to reach the goal of 25 breeding pairs? (Note that $\lambda = 2.0$ if a population is doubling each year.) Solve for r first (Equation 54.8); then, given that $N_0 = 2$ and $N_t = 25$, solve for t and check your answer.[3]

[1] $r = 0.048$; [2] $t = 14$ years; [3] $t = 3.64$ years.

The density of females, plotted along the horizontal axis, is the number of females that bred on the island; the clutch size, plotted on the vertical axis, is the average number of eggs laid by each female. Females at low density had about 0.5 eggs more per clutch, on average, than females at high density. This graph indicates a strong density-dependent relationship in *fecundity*.

Density-dependent changes in survivorship and fecundity cause logistic population growth. In this way, density-dependent factors define a particular habitat's carrying capacity. If gobies get crowded, they die or emigrate. If song sparrows are crowded, the average number of eggs and offspring that they produce declines.

Carrying Capacity Is Not Fixed It's important to recognize that *K* varies among species and populations. *K* varies because for any particular species, some habitats are better than other habitats due to differences in food availability, space, and other density-dependent factors. Stated another way, *K* varies in space. It also varies with time, because conditions are better in some years than in others.

What's more, the same habitat may have a very different carrying capacity for different species—as the experiment with growing paramecia in 5-mL vials showed. The same area will tend to support many more individuals of a small-bodied species than of a large-bodied species, for example, simply because large individuals demand more space and resources.

These simple observations help explain the variation in total population size that exists among species and among populations of the same species.

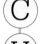

check your understanding

If you understand that . . .

- Populations grow exponentially unless slowed by density dependence.
- Density-dependent factors that influence population growth rates include competition for resources such as food and sunlight.

✔**You should be able to . . .**

Propose a density-dependent factor that limits growth in *Lacerta vivipara* and design an experiment to test this hypothesis.

Answers are available in Appendix A.

54.4 Population Dynamics

The tools introduced in the previous two sections provide a foundation for exploring how biologists study **population dynamics**—changes in populations through time and space. Let's pull together what you have learned about the spatial distribution of populations, demography, and population growth rates—and bring immigration and emigration into the discussion—by considering one of the most important patterns observed: the "blinking on and off" of fragmented metapopulations.

How Do Metapopulations Change through Time?

Glanville fritillaries are an endangered species of butterfly native to the Åland islands off the coast of Finland (**FIGURE 54.8a**). They are a well-studied example of a metapopulation because they occupy isolated patches of habitat within their geographic range. Recent research on Glanville fritillaries by Ilkka Hanski and colleagues illustrates the consequences of dynamic metapopulation structure for endangered species.

(a) Glanville fritillaries live in patches of meadow in Finland.

(b) A metapopulation is made up of small, isolated populations.

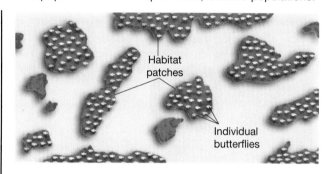

Although some populations go extinct over time...

...migration can restore or establish populations.

FIGURE 54.8 Metapopulation Dynamics Depend on Extinction and Recolonization. The overall size of a metapopulation stays relatively stable even if populations go extinct. These populations may be restored by migration, or unoccupied habitats might be colonized.

Metapopulations Should Be Dynamic Research on Glanville fritillaries began with a survey of meadow habitats on the islands. Because fritillary caterpillars feed on just two types of host plant, *Plantago lanceolata* and *Veronica spicata*, the team was able to pinpoint potential butterfly habitats.

Hanski's group estimated the fritillary population size within each patch of habitat by counting the number of larval webs in the patch. Of the 1502 meadows that contained the host plants, 536 had Glanville fritillaries. The patches ranged from 6 m² to 3 ha in area. Most had only a single breeding pair of adults and one larval group, but the largest population contained hundreds of pairs of breeding adults and 3450 larvae.

FIGURE 54.8b illustrates how a metapopulation like this is expected to change over the years.

- Given enough time, each population within the larger metapopulation is expected to go extinct. The cause could be a catastrophe, such as a storm; it could also be a disease outbreak or a sudden influx of predators.

- Migration from nearby populations can reestablish populations in empty habitat fragments.

In this way, the balance between extinction and recolonization exists within a metapopulation. Even though populations blink on and off over time, the overall metapopulation is maintained at a stable number of individuals.

An Experimental Test Does this metapopulation model correctly describe the population dynamics of fritillaries? To answer this question, Hanski's group conducted a mark–recapture study. They caught and marked many individuals, released them, and revisited fritillary habitats later to recapture as many individuals as possible (see Quantitative Methods 54.1).

Of the 1731 butterflies that the biologists marked and released, 741 were recaptured over the course of the summer study period. Of the recaptured individuals, 9 percent were found in a previously unoccupied patch. This migration rate is high enough to suggest that patches where a population has gone extinct will eventually be recolonized.

Hanski and co-workers repeated the survey two years after their initial census. Just as the metapopulation model predicted, some populations had gone extinct and others had been created. On average, butterflies were lost from 200 patches each year; 114 unoccupied patches were colonized and newly occupied each year. The overall population size was relatively stable even though constituent populations came and went.

To summarize, the history and future of a metapopulation is driven by the birth and death of populations, just as the dynamics of a single population are driven by the birth and death of individuals. Further, migration is an important source of individuals to recolonize patches of vacated habitats. As the final section in this chapter shows, these dynamics have important implications for conserving biodiversity.

Why Do Some Populations Cycle?

Density-dependent factors can be based on *intraspecific* ("within-species") interactions, such as competition among members of a cohort for food, or *interspecific* ("between-species") interactions, such as predation, parasitism, or competition among species for food (discussed in detail in Chapter 55). Analyzing population cycles has been a particularly productive way to understand how intraspecific and interspecific interactions interact. **FIGURE 54.9** shows a classic case of population cycling that involves two species: snowshoe hare and lynx in northern Canada.

The *x*-axis on this graph plots a 50-year interval; the *y*-axes show changes in hare population density (on the left, with a logarithmic scale) and changes in lynx density (on the right). Data for hares are plotted in tan; data for lynx are plotted in blue. Note three key points:

1. The scales on the *y*-axes are different—there are many more hares than lynx. (The *y*-axis for hares is logarithmic to compress the values so they can be compared more easily to changes in the lynx population.)

2. The hare and lynx populations cycle every 10 years on average.

3. Changes in lynx density lag behind changes in hare density by about two years.

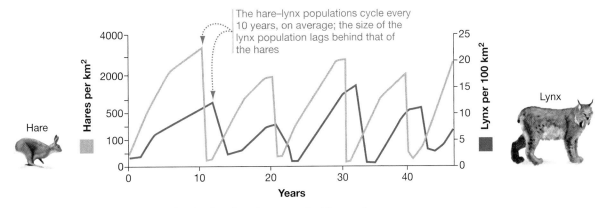

FIGURE 54.9 The Hare and Lynx Population Cycles Are Synchronous, but Lagged.

DATA: MacLulich, D. A. 1937. *University of Toronto Studies Biological Series* 43.

Snowshoe hares are herbivores—they subsist on leaves and stems that grow close to the ground. Lynx are predators, and they subsist mainly on snowshoe hares.

Is It Food or Predation?   To explain the hare–lynx cycle, biologists posed two hypotheses based on density-dependent factors:

1. Hares use up all their food when their populations reach high density and starve; in response, lynx also starve.

2. Lynx populations reach high density in response to increases in hare density. At high density, lynx eat so many hares that the prey population crashes.

Stated another way, hares control lynx population size, or lynx control hare population size. The cycle could be controlled by intraspecific competition for food or interspecific interactions.

A Field Experiment   To test these hypotheses, Charles Krebs and colleagues set up a series of 1-km^2 study plots in boreal forest habitats that were as identical as possible (**FIGURE 54.10**).

• Three plots were left as unmanipulated controls.

• One plot was ringed with an electrified fence made up of a mesh that excluded lynx but allowed hares to pass freely.

• Two plots received additional food for hares year-round (fertilizer was used to increase plant growth).

RESEARCH

QUESTION: What factors control the hare–lynx population cycle?

HYPOTHESIS: Predation, food availability, or a combination of those two factors controls the hare–lynx cycle.

NULL HYPOTHESIS: The hare–lynx cycle isn't driven by predation, food availability, or a combination of those two factors.

EXPERIMENTAL SETUP:

Document hare population in seven study plots (similar boreal forest habitats, each 1 km^2) from 1987 to 1994 (most of a population cycle).

3 plots: Unmanipulated controls

1 plot: Electrified fence excludes lynx but allows free access by hares.

2 plots: Supply extra food for hares.

1 plot: Electrified fence excludes lynx but allows free access by hares; supply extra food for hares.

PREDICTION: Hare populations in at least one type of manipulated plot will be higher than the average population in control plots.

PREDICTION OF NULL HYPOTHESIS: Hare populations in all of the plots will be the same.

RESULTS:

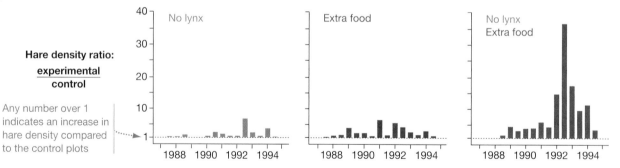

Hare density ratio:

$$\frac{\text{experimental}}{\text{control}}$$

Any number over 1 indicates an increase in hare density compared to the control plots

CONCLUSION: Hare populations are limited by both predation and food availability. When predation and food limitation occur together, they have a greater effect than either factor does independently.

©1995 AAAS

FIGURE 54.10 Experimental Evidence that Predation and Food Availability Drive the Hare–Lynx Cycle.

SOURCE: Krebs, C. J., S. Boutin, R. Boonstra, et al. 1995. Impact of food and predation on the snowshoe hare cycle. *Science* 269: 1112–1115.

✔**QUESTION** Why do you think the extra food did not cause a more dramatic increase in hare density at the *beginning* of the population cycle?

- One plot had a predator-exclusion fence and was also supplemented with food for hares year-round.

The biologists then used the mark–recapture method and radio collars to monitor the size of the hare and lynx populations over an 8-year period.

The graphs in Figure 54.10 plot average hare density in the lynx-exclusion, food-addition, and combination plots relative to controls. A value of 1 on the *y*-axis means that hare density in the experimental and control plots were the same. Each bar represents the average population density over a six-month interval.

The graphs should convince you that plots with predators excluded showed higher hare populations than did control plots. This result supports the hypothesis that predation by lynx reduces hare populations. Plots with supplemental food also had higher populations than controls. But the plot with supplemental food *and* predators excluded had a hare population that became as much as 35 times denser than the population in the control plots.

Note that the effects of the treatments were not even over the course of the study—most of the effects occurred during the decline phase of the cycle, which occurred from 1992 to 1994. As control plots were crashing, the experimental plots retained some or many hares.

These data support the hypothesis that hare populations are limited by availability of food as well as by predation and that food availability and predation intensity interact—meaning the combined effect of food and predation is much larger than their impact in isolation. The leading hypothesis to explain this combined effect is that when hares are at high density, individuals are weakened by nutritional stress and are more susceptible to predation.

Research on population cycles has increased our understanding of density dependence factors in population dynamics. Now let's consider factors that determine dynamics in the human population.

54.5 Human Population Growth

As the life-table data presented in Section 54.1 showed, age dramatically affects the probability that an individual will reproduce and survive to the following year. A population's age structure, in turn, dramatically influences the population's growth over time.

Age Structure in Human Populations

To study age structure in our species, researchers use stacks of horizontal bars to plot the number of males and the number of females in each age cohort—often in five-year intervals between birth (0) and 100. The resulting graph is called a population pyramid or age pyramid.

Age Pyramids In countries where industrial and technological development took place generations ago, like Sweden, survivorship has been high and fecundity low for several decades. An age pyramid like that in **FIGURE 54.11a** results. The most striking pattern in these data is that most of the age classes contain similar numbers of people. The evenness occurs because about the same number of infants are being born each year, and because most of them survive to old age.

In contrast, the age distribution is bottom heavy in less-developed nations like Honduras (**FIGURE 54.11b**). These populations are dominated by the very young, because they are undergoing rapid growth—the number of children being born recently is higher than the number of people existing in older age classes.

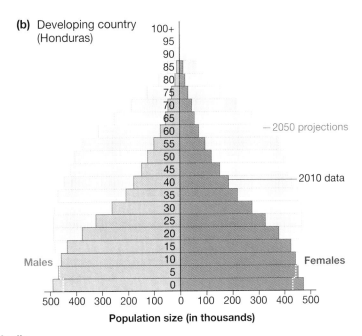

FIGURE 54.11 The Age Structure of Human Populations Varies Dramatically.

DATA: United Nations, *World Population Prospects: The 2010 Revision*.

Analyzing an age pyramid can give biologists important information about a population's history. But studying age distributions can also help researchers predict a population's future.

Look again at Figure 54.11a, and note that the white lines and lighter bars show what the age structure of Sweden is projected to be in 2050. Modest changes in the age distribution will occur because survivorship has increased while fecundity remains the same. But the population is not expected to grow quickly, because only modest numbers of individuals reach reproductive age. The projections highlight a major public policy concern in the industrialized countries: how to care for an increasingly aged population.

Now consider Figure 54.11b again. Thanks to dramatic improvements in health care, most young Hondurans survive to reproductive age. The projected age distribution in 2050 does not continue to flare out, however, because fecundity is declining. The projections illustrate a major public policy concern in less-developed countries: providing education and jobs for an enormous group of young people who will be reaching adulthood during the lifetimes of preceding generations.

Population "Momentum" The data in Figure 54.11b make another important point: Because of recent and rapid population growth in developing countries, overall population size will increase dramatically in these nations over the course of your lifetime. Honduras will have many more people in 2050 than it does now.

A large part of this increase will be due to increased survivorship. But the number of offspring being born each year is also expected to stay high, even though fecundity is predicted to *decline*. This result seems paradoxical, but it is based on a key observation: These populations now contain so many young women that the overall number of births will stay high, even though the average number of children *per female* is much lower than it was a generation ago.

To capture this point, biologists say that these populations have momentum or inertia. Combined with high survivorship, age structures like these make continued increases in the total human population almost inevitable.

Analyzing Change in the Growth Rate of Human Populations

Fossil and molecular evidence indicates that *Homo sapiens* originated in Africa about 200,000 years ago, then migrated into Asia, Europe, Australia, and eventually North and South America (see Chapter 35). The geographic range of our species now consists of most of the land surface of the planet. How did our population size change over that time?

FIGURE 54.12 plots changes in the human population over the past 2000 years. Although the curve's shape looks superficially similar to exponential growth, the growth rate for humans has—until recently—actually *increased* over time, leading to a steeply rising curve.

It is almost impossible to overemphasize just how dramatically the human population has grown recently. It took all of human history to reach a population size of 1 billion in 1804, and only 123 years to reach 2 billion. Most recently, it took only 12 years to add 1 billion people to the population to reach 7 billion.

Consider that until your grandparents' generation, no individual had lived long enough to see the human population double. But many members of the generation born in the early 1920s have lived to see the population *triple*. And to drive home the impact of continued population doubling, consider this: If you were given a penny on January 1, then $0.02 on January 2, $0.04 on January 3, and so on, you would be handed $10,737,418.00 on January 31.

The astonishing growth rate of the human population has raised a sense of alarm that we will overshoot the carrying capacity of our planet—and this concern is not new. For example, Thomas Robert Malthus's book, *An Essay on the Principle of Population*, was widely influential in the early 1800s. Malthus warned that the population cannot continue to increase unchecked—competition for resources will eventually slow population growth, whether by famine, war, or voluntary reduction of family size. Malthus was an important source of inspiration for Charles Darwin in his formulation of the theory of evolution by natural selection (Chapter 25).

FIGURE 54.12 The Human Population Has Been Growing Rapidly. Graph based on estimates from historical and census data.

DATA: Vince, G. 2011. *Science* 334: 32-37.

How Large Is the Current Human Population? As this book goes to press, the world population is estimated at over 7 billion. About 77 million additional people—more than double the population of California—are being added each year.

The consequences of recent and current increases in human population are profound (see Chapter 57). Aside from being the primary cause of habitat loss and species extinctions, overpopulation is linked by researchers to declines in living standards, mass movements of people, political instability, and acute shortages of water, fuel, and other basic resources in many parts of the world.

The one encouraging trend in the data is that the growth rate of the human population has already peaked and begun to decline. The highest growth rates occurred between 1965 and 1970, when populations increased at an average of 2.04 percent per year. Between 1990 and 1995, the overall growth rate in human populations averaged 1.46 percent per year; the current growth rate is 1.2 percent annually.

In humans, r may be undergoing the first long-term decline in history. The question is: Will the human population stabilize or decline in time to prevent global—and potentially irreversible—damage?

Will Human Population Size Peak in Your Lifetime? The United Nations Population Division makes regular projections for how human population size will change between now and 2050. For most readers of this book, these projections describe what the world will look like as you reach your early 60s.

The UN projections are based on three scenarios, which hinge on different values for **fertility** rates—the average number of surviving children that each woman has during her lifetime. Currently, the worldwide average fertility rate is 2.5 children per woman. This represents an enormous reduction in fertility from the 1950s, which averaged 5.0 children per woman. **FIGURE 54.13** shows how total population size is expected to change by 2050

if average fertility continues at 2.5 (high), or declines to 2.0 (medium), or 1.6 (low) children per woman.

The middle number in the projections is statistically the most likely scenario. Its value is close to the replacement rate of the population, 2.1. The **replacement rate** is the average fertility required for each woman to produce exactly enough offspring to replace herself and her offspring's father. (It is slightly above 2.0 to account for the mortality of some women before they bear children). When this fertility rate is sustained for a generation, $r = 0$, and there is **zero population growth.**

A glance at Figure 54.13 should convince you that the three scenarios are starkly different. If average fertility around the world stays at its current level, world population will be closing in on 11 billion about the time you might consider retiring—a 50 percent increase over today's population with no signs of peaking. The low-fertility projection, in contrast, predicts that the total human population in 2050 will be just over 8 billion and will have peaked.

The Role of Fertility Rates The UN's population projections make an important point: The future of the human population hinges on fertility rates—on how many children each woman living today decides to have. Those decisions, in turn, depend on a wide array of factors, including how free women are to choose their family size and how much access women have to education.

Why education? When women are allowed to become educated, they tend to delay having children and have a smaller overall family size. Access to education and reliable birth control methods (see Chapter 50), in addition to overall economic development and access to quality health care, will play a large role in determining how world population changes over your lifetime.

To summarize, the human population may be approaching the end of a period of rapid growth that lasted well over 500 years. Changes in fertility rates will be a major factor in determining how quickly population growth rates decline and how large the population eventually becomes.

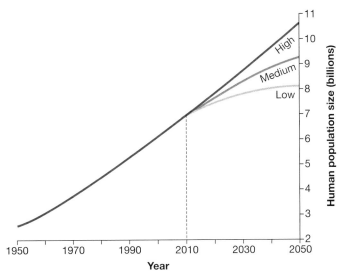

FIGURE 54.13 Projections for Human Population Growth to 2050. The "High," "Medium," and "Low" labels refer to fertility rates.

DATA: United Nations, *World Population Prospects: The 2010 Revision.*

check your understanding

C Y U

If you understand that . . .
- A population's age structure reflects changes in survivorship and fertility rates over time.
- Age structure affects population dynamics because it dictates how many females of reproductive age are in the population.

✔You should be able to . . .
1. Explain why the age structure of human populations differs between developed and developing nations.
2. Explain why changes in fertility rates have such a dramatic impact on projections for the total human population in 2050.

Answers are available in Appendix A.

The human population boom is central to ecology because the growing human population dramatically affects other organisms via habitat destruction, pollution, overharvesting of fish and other resources, and climate change. As the human population rises, the impact on other species increases.

54.6 How Can Population Ecology Help Conserve Biodiversity?

When designing programs to save species threatened with extinction, conservation biologists draw heavily on concepts and techniques from population ecology. As habitat destruction and global climate change push species into decline, the study of population growth rates and population dynamics has taken on an increasingly applied tone.

To illustrate one of the ways that biologists apply the theory and results introduced in this chapter, let's analyze how an understanding of life-table data and geographic structure can help direct conservation action.

Using Life-Table Data

Suppose that you were in charge of reintroducing a population of lizards to a nature reserve, and that earlier research had documented survivorship and fecundity data (**FIGURE 54.14a**).

Your initial plan is to take 1000 newly hatched female lizards from a captive breeding center and release them into the natural habitat, along with enough males for breeding to occur. Is this population likely to become established and grow on its own, or will you need to keep introducing offspring that have been raised in captivity?

Making Population Projections   To answer this question, you need to use the life-table data in Figure 54.14a and calculate two values:

1. How many adults survive to each age class each year?

(a) Life table data from previous study

Age (x)	Survivorship (l_x)	Fecundity (m_x)
0 (birth)	----	0.0
1	0.33	3.0
2	0.20	4.0
3	0.04	5.0

■ 1st-year cohort
■ 2nd-year cohort
■ 3rd-year cohort
□ 4th-year cohort

(b) Predictions: Fate of first-generation females

Year	0 (newborns)	1-year-olds	2-year-olds	3-year-olds	Total population size (N)
1st	1000 (just introduced)				1000
2nd	990 (= 330 × 3.0)	330 (= 1000 × 0.33)			1320 (= 990 + 330)
3rd	800 (= 200 × 4.0)		200 (= 1000 × 0.20)		?
4th	200 (= 40 × 5.0)			40 (= 1000 × 0.04)	?

(c) Predictions: Fate of first- and second-generation females

Year	0 (newborns)	1-year-olds	2-year-olds	3-year-olds	Total population size (N)
1st	1000				1000
2nd	990	330			1320
3rd	800 + 981 (981 = 327 × 3.0)	327 (= 990 × 0.33)	200		?
4th	200 + 792 (792 = 198 × 4.0)		198 (= 990 × 0.20)	40	?

(d) Predictions: Fate of first-, second-, and third-generation females

Year	0 (newborns)	1-year-olds	2-year-olds	3-year-olds	Total population size (N)
4th	200 + 792 + ?	?	198	40	?

FIGURE 54.14 Life-Table Data Can Be Used to Project the Future of a Population.

✔**QUANTITATIVE** Fill in the blanks indicated by the gray question marks. Note that in the bottom row, you must calculate the number of 1-year-olds in the 3rd-year cohort (indicated by blue) and the number of newborns produced by those 1-year-olds (part of the 4th-year cohort, in green).

2. How many offspring are produced by each adult age class—each year over the course of several years?

FIGURE 54.14b starts these calculations for you. The predicted fate of the original 1000 females is indicated in red. Note that in the start of the second year, just 330 of these year-old individuals are expected to survive; only 200 will survive to be 2-year-olds; only 40 will survive to be 3-year-olds.

How many offspring will this first cohort produce? Remember that just 330 of the original 1000 females are expected to remain after one year. As the purple number in Figure 54.14b indicates, these survivors have an average of 3.0 female offspring apiece. Thus, they contribute 990 new female offspring to the population. In the second year, the 200 females that are left from the original cohort have an average of 4.0 female offspring each and contribute 800 juveniles. In their third year, the 40 surviving females average 5.0 offspring and contribute 200 juveniles.

FIGURE 54.14c extends the calculations by predicting what will happen as the offspring of the original females, indicated in purple, begin to breed. ✔QUANTITATIVE By continuing the analysis until the third year, you should be able to predict whether the population of lizards will stay the same, decline, or increase over time. In this way, life-table data can be used to predict the future of populations.

Altering Values for Survivorship and Fecundity Part of the value of a population projection based on life-table data—like the one begun in Figure 54.14—is that it allows biologists to alter values for survivorship and fecundity at particular ages and assess the consequences. For example, suppose that a predatory snake began preying on juvenile lizards. The survivorship of juveniles would decline, decreasing the number of individuals that survive to reproduce each year—with potentially dire consequences for the population size over time.

Analyses like this allow biologists to determine which aspects of survivorship and fecundity are especially sensitive for particular species. The studies done to date support some general conclusions:

- Whooping cranes, sea turtles, spotted owls, and many other endangered species have high juvenile mortality, low adult mortality, and low fecundity. In these species, the fate of a population is extremely sensitive to increases in adult mortality. Based on this insight, conservation biologists have recently begun an intensive campaign to reduce the loss of adult female sea turtles in fishing nets (**FIGURE 54.15**). Previously, most conservation action had focused on protecting eggs and nesting sites.

- Climate change has enormous implications for the life-history traits of populations. For example, the sex of many reptiles is determined by the temperature of the eggs in the nest. Researchers have observed a change in sex ratio of some reptile species due to the warmer climate, and this finding has significant implications for the growth rates of these populations. Direct measurements of life-history traits of populations in the field are essential to mathematical models used to predict ongoing effects of global climate change.

In some or even most cases, the population projections made from life-table data may be too simplistic to be useful. For example, conservationists may need to expand the basic demographic models to account for occasional disturbances such as fires or storms or disease outbreaks. What about species whose overall population dynamics are dictated by a metapopulation structure?

Preserving Metapopulations

Habitat destruction caused by suburbanization and other human activities leaves small populations isolated in pockets of intact habitat (**FIGURE 54.16**; see Chapter 57). Work on Glanville fritillaries and other species has shown that a small, isolated population—even one within a nature preserve—is unlikely to survive over the long term. What can be done about this?

FIGURE 54.15 The mortality of adult sea turtles in fishing nets has a dramatic impact on sea turtle populations. Data from population ecology helps to prioritize conservation action.

FIGURE 54.16 Human activities are turning once intact populations into metapopulations by dividing habitats into fragments.

In Glanville fritillaries, data collected by Hanski's group have identified four attributes of populations that are most likely to persist:

1. They have larger population sizes.

2. They occupy larger geographical ranges.

3. They are closer to neighboring populations (and hence more likely to be colonized).

4. They have higher genetic diversity.

Results like these have important messages for conservation biologists:

- Areas that are being protected for threatened species should be substantial enough in area to maintain large populations that are unlikely to go extinct in the near future. Large populations are more likely than small populations to endure disasters such as storms, and they are much less susceptible to the damaging effects of inbreeding.

- When it is not possible to preserve large tracts of land, an alternative is to establish systems of smaller tracts that are connected by corridors of habitat, so that migration between patches is possible. Constructing "wildlife corridors"

is particularly important when patches of appropriate habitat are separated by highways, subdivisions, rivers, or other obstacles to dispersal.

- If the species that is threatened exists as a metapopulation, it is crucial to preserve at least some patches of unoccupied habitat to provide future homes for immigrants. If a population is lost from a preserve, the habitat should continue to be protected so it can be colonized in the future. This can be difficult to do—nonscientists are often skeptical about preserving scarce land that does not (yet) contain the species of interest.

These results are made more complex by climate change. Data on existing populations show that changes in temperature and precipitation are already changing the ranges of species. For some species with large ranges, such as the lizard *Lacerta vivipara* (see Figure 54.1), local extinctions may occur in some populations, such as in the mountains of Spain, while the range expands northward elsewhere. Other species with smaller ranges or the inability to relocate rapidly are likely to go extinct. Population ecology helps quantify past trends in the distribution and abundance of organisms as well as predict future trends, enabling conservation biologists to prioritize their efforts.

CHAPTER 54 REVIEW

For media, go to MasteringBiology

If you understand . . .

54.1 Distribution and Abundance

- The geographic ranges of species vary in size and patchiness and are determined by a combination of abiotic and biotic factors.

- Individuals in a population have a random, clumped, or uniform distribution depending on the availability of resources, dispersal ability, social behavior, and other factors.

- There are many methods for measuring population size, population density, and distribution patterns—the best method to use depends on the behavior of the species.

✓ You should be able to devise a method for using the pellets (feces) of snowshoe hares to estimate population density.

54.2 Demography

- Life tables summarize how likely it is that individuals of each age class in a population will survive and reproduce.

- Three general types of survivorship curves have been observed: (I) high survivorship throughout life, (II) a constant probability of dying at each year of life, and (III) high mortality in juveniles followed by high survivorship in adults.

- There is a trade-off between survival and reproduction due to the limited availability of resources.

- In analyzing life history, there is a continuum between populations or species with high survivorship and low fecundity and populations or species with low survivorship and high fecundity.

✓ You should be able to predict how the data in the life tables constructed in ancient Rome would differ from a life table for present-day Rome.

54.3 Population Growth

- The growth rate of a population can be calculated from life-table data or from the direct observation of changes in population size over time.

- Exponential growth occurs when the per capita growth rate, r, does not change over time.

- During logistic growth, growing populations approach the carrying capacity, K, of their environment and r declines to 0.

- Density dependence in population growth is due to competition for resources, disease, predation, or other factors that increase in intensity when population size is high. At high density, survivorship and fecundity decrease.

✓ QUANTITATIVE You should be able to draw a logistic growth curve, describe how r changes along the length of the curve, and suggest factors responsible for each change in r.

54.4 Population Dynamics

- In some metapopulations, populations may blink on and off over time even though the overall metapopulation size is stable. Migration can reestablish populations in empty habitat fragments.

- Predator–prey interactions and food availability are density-dependent factors that drive the population cycle of the snowshoe hare and other species.

- ✓ Mosquitoes lay their eggs in water and have an aquatic larval stage. You should be able to predict a pattern of mosquito density over time as a function of rain.

54.5 Human Population Growth

- Age structure has a profound impact on population dynamics. A population with few juveniles and many adults past reproductive age, like the human populations of developed nations, may be declining or stable in size. In contrast, a population with a large proportion of juveniles is likely to increase rapidly in size.

- The total human population has been increasing rapidly and is currently over 7 billion.

- Based on various scenarios for average female fertility, the total human population in 2050 is expected to be between 8 and 11 billion. The future of the human population hinges on fertility rates.

- ✓ You should be able to explain why the human population is projected to continue rapid growth, even though fertility rates are declining.

54.6 How Can Population Ecology Help Conserve Biodiversity?

- Data from population ecology studies help biologists design effective management strategies for endangered species and help prevent other species from becoming endangered.

- ✓ You should be able to explain why a small, isolated population is virtually doomed to extinction, but why population size may be stable in a species consisting of a large number of small, isolated populations.

(MB) MasteringBiology

1. **MasteringBiology Assignments**

   Tutorials and Activities Age Pyramids and Population Growth; Analyzing Age-Structure Pyramids; Human Population Growth; Human Population Growth and Regulation; Investigating Survivorship Curves; Modeling Population Growth; Population Ecology; Population Ecology: Logistic Growth, Techniques for Estimating Population Density and Size

   Questions Reading Quizzes, Blue-Thread Questions, Test Bank

2. **eText** Read your book online, search, take notes, highlight text, and more.

3. **The Study Area** Practice Test, Cumulative Test, BioFlix® 3-D Animations, Videos, Activities, Audio Glossary, Word Study Tools, Art

You should be able to . . .

✓ TEST YOUR KNOWLEDGE

1. What is the defining feature of exponential growth?
   a. The population is growing very quickly.
   b. The growth rate is constant.
   c. The growth rate increases rapidly over time.
   d. The growth rate is very high.

2. List four factors that define population growth.

3. In what populations does exponential growth tend to occur?
   a. populations that colonize new habitats
   b. populations that experience intense competition
   c. populations that experience high rates of predation
   d. populations that have surpassed their carrying capacity

4. True or False? Climate change can influence the population growth of species.

5. If most individuals in a population are young, why is the population likely to grow rapidly in the future?
   a. Death rates will be low.
   b. The population has a skewed age distribution.
   c. Immigration and emigration can be ignored.
   d. Many individuals will begin to reproduce soon.

6. Why have population biologists become particularly interested in the dynamics of metapopulations?
   a. because humans exist as a metapopulation
   b. because whooping cranes exist as a metapopulation
   c. because many populations are becoming restricted to small islands of habitat
   d. because metapopulations explain why populations occupying large, contiguous areas are vulnerable to extinction

✓ TEST YOUR UNDERSTANDING

7. Describe the life-history traits of elephants.
   a. high survivorship, high fecundity
   b. low survivorship, high fecundity
   c. high survivorship, low fecundity
   d. low survivorship, low fecundity

8. Make a table to show what types of traits you would expect to see on the left, middle, and right of the life history continuum (see Figure 54.3). Consider growth habit (herbaceous, shrub, or tree) as well as relative disease- and predator-fighting ability, seed size, seed number, and body size.

9. Offer a hypothesis to explain why humans have undergone near-exponential growth for over 500 years. Why can't exponential growth continue indefinitely? Describe two examples of density-dependent factors that may influence human population growth.

10. Explain why biologists want to maintain (a) "habitat corridors" that connect populations in a metapopulation, and (b) unoccupied habitat that is appropriate for the species in question.

11. Predict which geographic subpopulations of *Lacerta vivipara* may be most vulnerable to the effects of climate change.

12. What is the difference between r (the per capita rate of increase) and r_{max} (the maximum or intrinsic growth rate)?

13. When wild plant and animal populations are logged, fished, or hunted, only the oldest or largest individuals tend to be taken. Many of the commercially important species are long lived and are slow to begin reproducing. If harvesting is not regulated carefully and exploitation is intense, what impact does harvesting have on a population's age structure? How might harvesting affect the population's life table and growth rate?

14. Design a system of nature preserves for an endangered species of beetle whose larvae feed on only one species of sunflower. The sunflowers tend to be found in small patches that are scattered throughout dry grassland habitats.

15. Make a rough sketch of the age distribution in developing versus developed countries, and explain why the shapes of the diagrams are different. How is AIDS, which is a sexually transmitted disease, affecting the age distribution in countries hard hit by the epidemic?

16. In most species the sex ratio is at or near 1.00, meaning that the number of males and females is approximately equal. In China, however, there is a strong preference for male children. According to the 2000 census there, the sex ratio of newborns is almost 1.17, meaning that close to 117 boys are born for every 100 girls. Based on these data, researchers project that China will soon have about 50 million more men than women of marriageable age. Evaluate whether this skewed sex ratio will affect the population growth rate in China.

    a. No, growth rate is not affected by males.
    b. Yes, growth rate will increase.
    c. Yes, growth rate will decrease.
    d. No, the sex ratio affects only fecundity.

55 Community Ecology

In this chapter you will learn that

Communities consist of complex and dynamic interactions among species

starting with → Types of species interactions:
Commensalism + / 0
Competition − / −
Consumption + / −
Mutualism + / +
55.1

important to →

examining → Community structure **55.2**

changes over time causing → Community dynamics: disturbance and succession **55.3**

can vary in → Species richness **55.4**

This chapter explores how species that live in coral reefs and other communities interact with each other, and how communities change over time.

This chapter is part of the Big Picture. See how on pages 1196–1197.

Populations are dynamic. They grow and decline over time and through space (Chapter 54). However, populations do not exist in isolation—individuals within populations also interact with individuals of other species, forming complex assemblages called communities.

A biological **community** consists of all of the species that interact in a certain area. Ecologists want to know how communities work, and how to manage them in a way that will preserve biodiversity, ecosystem function, and an environment that people want to live in. You can see how community ecology fits into the Big Picture of Ecology on pages 1196–1197.

The goal of this chapter is to explore how species interact with each other and to analyze the structure and dynamics of biological communities—how they develop and change over time. Let's delve in.

✔ When you see this checkmark, stop and test yourself. Answers are available in Appendix A.

55.1 Species Interactions

The species in a community interact constantly. Organisms eat each other, exchange nutrients, compete for resources, and provide habitats for each other. In many cases, a population's fate is tightly linked to the other species that share its habitat.

To study species interactions, biologists begin by analyzing the effects of one species on the fitness of another. Recall that the **fitness** of an individual is defined by its ability to survive and produce viable, fertile offspring (Chapter 25).

Does the relationship between two species provide a fitness benefit to members of one species (a "+" interaction) but hurt members of the other species (a "−" interaction)? Or does the association have no effect on the fitness of a participant (a "0" interaction)? There are four general types of interactions:

1. **Commensalism** occurs when one species benefits but the other species is unaffected (+/0).

2. **Competition** occurs when individuals use the same resources—resulting in lower fitness for both (−/−).

3. **Consumption** (including herbivory, predation, and parasitism) occurs when one organism eats or absorbs nutrients from another. The interaction increases the consumer's fitness but decreases the victim's fitness (+/−).

4. **Mutualism** occurs when two species interact in a way that confers fitness benefits to both (+/+).

Let's consider each type of species interaction in more detail. As you analyze each one, watch for three key themes:

1. *Species interactions can affect the distribution and abundance of a particular species.* Species interactions can affect geographic ranges, such as the effect of disease-carrying tsetse flies on the distribution of cattle in Africa (Chapter 52). Species interactions can also have impacts on population size, such as the effects of lynx on hare populations in northern Canada (Chapter 54).

2. *Species act as agents of natural selection when they interact.* Deer are fast and agile in response to natural selection exerted by their major predators, wolves and cougars. The speed and agility of deer, in turn, promote natural selection that favors wolves and cougars that are fast and that have superior eyesight, senses of smell, and hunting strategies. To capture this point, biologists say that **coevolution** is occurring—a pattern of evolution where two species influence each other's adaptations over time.

As species interact over time, it is common to observe a **coevolutionary arms race**—a repeating cycle of reciprocal adaptation. In human civilizations, an arms race occurs when one nation develops a new weapon, which prompts a rival nation to develop a defensive weapon, which pushes the original nation to manufacture an even more powerful weapon, and so on. In biology, coevolutionary arms races occur between predators and prey, parasites and hosts, and other types of interacting species.

3. *The outcome of interactions among species is dynamic and conditional.* Just as two people may alternately help or compete with each other depending on circumstances, the interactions between different organisms may change over time and space.

Commensalism

Of the four general types of interactions, commensalism (+/0) is the least studied. Commensalism can be challenging to demonstrate, because an absence of fitness benefits or costs ("0") can be difficult to quantify. Commensalism is also very conditional.

An example is the relationship between antbirds and the army ants that they follow in the tropics. As the ants march along the forest floor, they hunt insects and small vertebrates. The antbirds follow and pick off prey species that fly or jump up out of the way of the ants (**FIGURE 55.1**). The antbirds benefit, but the ants are not affected.

However, if antbird attacks start to force other insects into the path of the ants, then both ants and birds benefit, and the

(a) Ants stir up insects while hunting ...

(b) ... antbirds tag along and benefit.

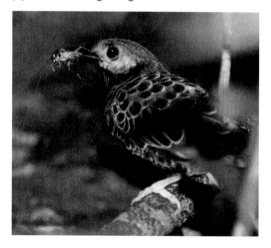

FIGURE 55.1 Commensals Gain a Fitness Advantage but Don't Affect the Species They Depend On. Birds that associate with army ants are commensals. They have no measurable fitness effect on the ants, but gain from the association by capturing insects that try to fly or climb out of the way of the ants.

relationship becomes mutualistic. But if the antbirds begin to steal prey that would otherwise be taken by ants, then the relationship becomes competitive. The outcome of the interaction may depend on the number and types of prey, birds, and ants present and may change over time.

Competition

Competition (−/−) lowers the fitness of the individuals involved for a simple reason: When competitors use resources, those resources are not available to help other individuals survive better and produce more offspring.

Competition that occurs between members of the same species is called **intraspecific** (literally, "within species") **competition.** Intraspecific competition for space, sunlight, food, and other resources intensifies as a population's density increases. As a result, intraspecific competition is a major cause of density-dependent growth (see Chapter 54). **Interspecific** ("between species") **competition** occurs when individuals from different species use the same limiting resources.

Using the Niche Concept to Analyze Interspecific Competition

Early work on interspecific competition focused on the concept of the niche. A **niche** can be thought of as the range of resources that the species is able to use, or the range of conditions it can tolerate. For example, *Anolis* lizards that occupy the crowns of trees on the island of Jamaica occupy a different niche than *Anolis* lizards that live on the ground (Chapter 28).

G. Evelyn Hutchinson proposed that a species' niche could be envisioned by plotting its habitat requirements along a series of axes. **FIGURE 55.2a**, for example, represents one niche axis for a hypothetical species. In this case, the habitat requirement plotted is the size of seeds eaten by members of this population, determined by variables such as mouth or tooth size. Other niche axes could represent other types of foods used or the temperatures, humidity, and other environmental conditions tolerated by the species. In these types of graphs, the *y*-axis represents the number or proportion of resources used, averaged over the population.

Interspecific competition occurs when the niches of two species overlap. The two species plotted in **FIGURE 55.2b**, for example, compete for seeds of intermediate size—assuming that seeds are a limited resource. Individuals that use this "joint" resource are at a disadvantage relative to individuals that use other resources, because they are likely to acquire fewer seeds and have lower fitness as a result. However, if both species experience a similar decrease in fitness due to the overlap of their niches, then **symmetric competition** is said to occur and both species may persist in the area of overlap, even if in reduced numbers.

What Happens when One Species Is a Better Competitor?

When **asymmetric competition** occurs, one species suffers a greater fitness decline than the other species does. The outcome of this interaction depends on the amount of overlap in their niches.

What outcomes are possible? To explore this type of competitive interaction, G. F. Gause conducted a series of experiments with two species of the unicellular pond-dweller *Paramecium*.

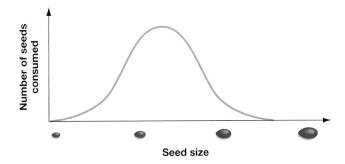

(a) One species eats seeds of a certain size range.

Number of seeds consumed

Seed size (x-axis)

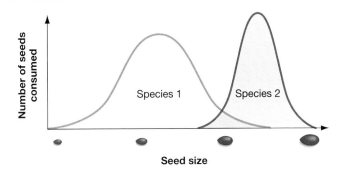

(b) Partial niche overlap: competition for seeds of intermediate size

Number of seeds consumed (y-axis)

Species 1   Species 2

Seed size (x-axis)

FIGURE 55.2 Niche Overlap Leads to Competition. (a) A graph describing one aspect of a species' fundamental niche, meaning the range of resources that it can use or range of conditions it can tolerate. **(b)** Competition occurs when the niches of different species overlap.

- When *P. caudatum* and *P. aurelia* grew in separate laboratory cultures, both species exhibited logistic growth—the number of individuals increased rapidly at first, then leveled off as the population reached its carrying capacity (see Chapter 54).

- When the two species grew in the same culture together, only the *P. aurelia* population exhibited a logistic growth pattern. *P. caudatum*, in contrast, was driven to extinction (**FIGURE 55.3a**, see page 1126).

The stronger competitor drove out the weaker competitor because the two laboratory populations had completely overlapping niches (**FIGURE 55.3b**). Based on these data, Gause proposed that two species that occupy the same niche cannot coexist, a hypothesis called the **competitive exclusion principle.** Some biologists invoke this principle to explain the disappearance of Neanderthals (*Homo neanderthalensis*) after thousands of years of coexistence with our species, *Homo sapiens* (see Chapter 35).

But what happens if the niches of the two species do not overlap completely? The species that is the weaker competitor should be able to retreat to, or persist in, the area of non-overlap. In cases like this, an important distinction arises.

- A **fundamental niche** is the total theoretical range of environmental conditions that a species can tolerate.

(a) Observation: asymmetric competition

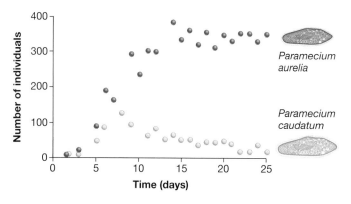

Paramecium aurelia

Paramecium caudatum

(b) Explanation: competitive exclusion due to complete niche overlap

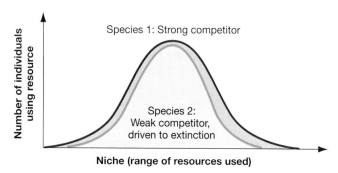

Species 1: Strong competitor

Species 2: Weak competitor, driven to extinction

FIGURE 55.3 Competitive Exclusion in *Paramecium*.

DATA: Gause, G. F. 1934. *The Struggle for Existence*. New York: Hafner Press.

- A **realized niche** is the portion of the fundamental niche that a species actually occupies, given limiting factors such as competition with other species (**FIGURE 55.4**).

Let's take a look at an important field experiment that demonstrated the difference between a fundamental and realized niche.

Experimental Studies of Competition in Nature Joseph Connell began a classic study of competition in the late 1950s, after observing an interesting pattern in an intertidal rocky shore in Scotland. He noticed that there were two species of barnacles

Species 1 (strong competitor)

Species 2 (weak competitor)

Fundamental niche

Realized niche

FIGURE 55.4 Fundamental Niches Are Broader than Realized Niches.

with distinctive distributions. The adults of one species, *Chthamalus stellatus*, occurred in an upper intertidal zone. The adults of the other species, *Semibalanus balanoides*, were restricted to a lower intertidal zone. The upper intertidal zone is a more severe environment for barnacles, because it is exposed to the air and intense heat for longer periods at low tide each day.

Although adult barnacles live attached to rocks, their planktonic larvae are mobile. The young of both species were found together in the lower intertidal zone. Why aren't *Chthamalus* adults found there as well?

To answer this question, Connell considered two hypotheses:

1. Adult *Chthamalus* are outcompeted in the lower intertidal zone by *Semibalanus*.

2. Adult *Chthamalus* are absent from the lower intertidal zone because they do not thrive in the physical conditions there.

Connell tested these hypotheses by performing the experiment shown in **FIGURE 55.5**. He began by removing a number of rocks that had been colonized by *Chthamalus* from the upper intertidal zone and transplanting them into the lower intertidal zone. He screwed the rocks into place and allowed *Semibalanus* larvae to colonize them. Once the spring colonization period was over, Connell divided each rock into two sides. In one half, he removed all *Semibalanus* that were in contact with or next to a *Chthamalus*.

This experimental design allowed Connell to document *Chthamalus* survival in the absence of competition with *Semibalanus*, and compare it with survival during competition. This is a common experimental strategy in competition studies: One of the competitors is removed, and the response by the remaining species is observed.

Connell's results supported the competition hypothesis. On the rocks with both species present, *Semibalanus* killed many of the *Chthamalus* by growing against them and lifting them off the substrate. As the orange bar in the "Results" section of Figure 55.5 shows, *Chthamalus* survival was much higher when all of the *Semibalanus* were removed.

The fundamental niches of these two species partially overlap. Under the conditions tested in this experiment, *Semibalanus* is a superior competitor for space, but *Chthamalus* is able to thrive in a smaller realized niche where *Semibalanus* cannot survive. ✔ If you understand this result, you should be able to sketch and label a graph of the fundamental and realized niches of the two barnacle species as a function of depth in the intertidal zone.

Fitness Trade-Offs in Competition Why haven't *Semibalanus* and other successful competitors taken over the world? Biologists answer this question by invoking the concept of **fitness trade-offs**—inevitable compromises in adaptation (see Chapter 25).

The key insight is that the ability to compete for a particular resource—like space on rocks or edible seeds of a certain size—is only one aspect of an organism's niche. If individuals are extremely good at competing for a particular resource, then they are probably not as good at enduring drought conditions, warding off disease, or preventing predation.

QUESTION: Why is the distribution of adult *Chthamalus* restricted to the upper intertidal zone?

HYPOTHESIS: Adult *Chthamalus* are outcompeted in the lower intertidal zone.

NULL HYPOTHESIS: Adult *Chthamalus* do not thrive in the physical conditions of the lower intertidal zone.

EXPERIMENTAL SETUP:

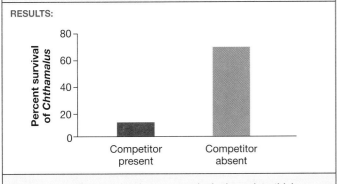

Chthamalus in upper intertidal zone

Mean tide level

Semibalanus in lower intertidal zone

1. Transplant rocks containing young *Chthamalus* to lower intertidal zone.

2. Let *Semibalanus* colonize the rocks.

3. Remove *Semibalanus* from half of each rock. Monitor survival of *Chthamalus* on both sides.

Chthamalus + *Semibalanus*

Chthamalus

PREDICTION: *Chthamalus* will survive better in the absence of *Semibalanus*.

PREDICTION OF NULL HYPOTHESIS: *Chthamalus* survival will be low and the same in the presence or absence of *Semibalanus*.

RESULTS:

Percent survival of *Chthamalus* (y-axis: 0, 20, 40, 60, 80)

Competitor present — approximately 13

Competitor absent — approximately 70

CONCLUSION: *Chthamalus* do not occur in the lower intertidal zone because they are outcompeted by *Semibalanus*.

FIGURE 55.5 Experimental Evidence of Competition.

SOURCE: Connell, J. H. 1961. The influence of interspecific competition and other factors on the distribution of the barnacle *Chthamalus stellatus. Ecology* 42: 710–723.

✔ **QUESTION** Why was it important to carry out both treatments on the same rock? Why not use separate rocks?

In the case of *Semibalanus* and *Chthamalus* growing in the intertidal zone, the fitness trade-off is rapid growth and success in competing for space versus the ability to endure the harsh physical conditions of the upper intertidal zone. *Semibalanus* are fast-growing and large; *Chthamalus* grow slowly but can survive long exposures to the air and to intense sun and heat. Neither species can do both things well.

Also, competitive ability is not an inherent quality of species—it depends on environmental conditions. For example, Argentine ants have unremarkable competitive abilities in their native South America, due to the checks and balances provided by thousands of years of species interactions. But when these ants were introduced to California and the Mediterranean in food shipments, they began to attack aggressively and outcompete native species (Chapter 52). The remarkable competitive success of invasive species poses a threat to communities around the world—and conservation biologists are very concerned about it (see Chapter 57).

Mechanisms of Coexistence: Niche Differentiation   It's important to realize that because competition is a −/− interaction, there is strong natural selection on both species to avoid it. **FIGURE 55.6** shows the predicted outcome of competition over many generations: An evolutionary change in traits reduces the amount of niche overlap, and thus the amount of competition.

An *evolutionary* change in resource use, caused by competition over generations, is called **niche differentiation** or resource partitioning. The evolutionary change that occurs in species' traits, and that enables species to exploit different resources, is called **character displacement.** Character displacement makes niche differentiation possible.

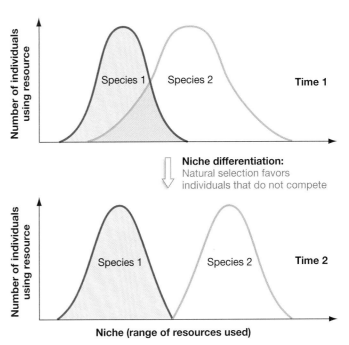

Number of individuals using resource

Species 1   Species 2   **Time 1**

Niche differentiation: Natural selection favors individuals that do not compete

Number of individuals using resource

Species 1   Species 2   **Time 2**

Niche (range of resources used)

FIGURE 55.6 Competition Can Lead to Niche Differentiation. Over evolutionary time, natural selection favors individuals that do not compete.

One spectacular example of character displacement involves the Galápagos finches, famous for their diversity in beak shapes (Chapter 25). Specializations in beak shape are adaptations for harvesting particular food types, such as small or large seeds.

Beak shape varies not only among species of Galápagos finches but also within species. For example, **FIGURE 55.7** shows the frequency of beak depths (in this case, measured from the top half of the beak only) in a population of medium ground finch, *Geospiza fortis*, on the island of Daphne Major in the 1940s. You can see on the *x*-axis that beak depth, an important measure of beak shape, ranged from 3.5 to 5.6 mm. However, on the island of Santa Cruz at around the same time, *G. fortis* lived alongside the small ground finch, *G. fuliginosa*, and had a broader and higher range of beak depths, from 4.0 to 7.5 mm. From these data you could hypothesize either that competition causes character displacement in *G. fortis*, or that the difference between populations is due to other factors such as predation or food availability.

No data were available from the 1940s to test these hypotheses, but Peter and Rosemary Grant observed character displacement in *G. fortis* on Daphne Major more recently. In addition to recording six beak size measurements for hundreds of medium ground finches over three decades starting in 1970, they also made careful observations of feeding behavior, the amount of rainfall each year, the availability of different types of seeds, and the presence of other species.

The Grants observed that the large ground finch, *G. magnirostris*, established a breeding population on Daphne Major in 1982,

living alongside or "sympatric" with *G. fortis*. True to its name (*magnirostris* translates to "large beak"), this finch has a large beak capable of opening large, hard, drought-tolerant seeds. What effect did competition with *G. magnirostris* have on *G. fortis*?

At first the Grants observed a slow but steady decline in average beak size in *G. fortis* after the arrival of *G. magnirostris*. Then in 2004, during a severe drought when competition with *G. magnirostris* was at its most intense, most finches died of starvation. The Grants observed that

- Only *G. fortis* that could eat extremely small seeds efficiently could survive.

- The larger-beaked *G. fortis* had been chased off from the feeding sites with the large seeds by *G. magnirostris*.

- Overall, the average beak size of *G. fortis* declined sharply.

The Grants had observed evolution in action—directional selection on *G. fortis* beak size causing character displacement (reduced average beak size) and niche differentiation (adaptation for different seed sizes). That is, natural selection had favored *G. fortis* individuals that did not compete with *G. magnirostris*. ✔ If you understand niche differentiation, you should be able to sketch and label a graph predicting the fundamental niches of the barnacles *Semibalanus* and *Chthamalus* many years after Connell's experiment, if niche differentiation were to occur.

Consumption

Consumption is a +/− interaction that occurs when one organism eats another. There are three major types of consumption:

1. **Herbivory** takes place when **herbivores** ("plant-eaters") consume plant tissues. Caterpillars chew leaves; marine iguanas feed on algae.

2. **Parasitism** occurs when a **parasite** consumes relatively small amounts of tissue or nutrients from another individual, called the **host.** Parasitism often occurs over a long period. It is not necessarily fatal, and parasites are usually small relative to their host. An array of worms, ticks, mites, and unicellular protists parasitizes humans. In some cases, though, the term parasitism is used more broadly, to denote use of other resources. For example, birds that lay their eggs in other species' nests and induce the other species to raise the young are called social parasites.

3. **Predation** occurs when a predator kills and consumes all or most of another individual. The consumed individual is called the prey. Woodpeckers eat bark beetles; ladybird beetles devour aphids; orcas eat seals. Although many researchers use predation to refer specifically to **carnivores** ("meat-eaters"), predation can also refer to the consumption of plants, especially seeds (which contain entire plant embryos).

To illustrate how biologists analyze the impact of consumption, let's consider a series of questions about herbivores, parasites, predators, and their victims.

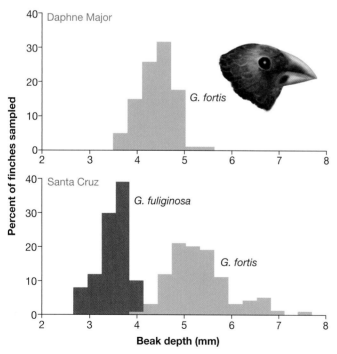

FIGURE 55.7 Niche Differentiation in Galápagos Finches Average beak depths in *Geospiza fortis* were higher on the island with a competitor present.

DATA: Grant, P. 1999. *Ecology and Evolution of Darwin's Finches*. Princeton, NJ: Princeton University Press.

Constitutive Defenses From the standpoint of fitness, consumption is costly for prey and beneficial for consumers. Prey individuals do not passively give up their lives to increase the fitness of their predators, however. Natural selection strongly favors traits that allow individuals to avoid being eaten. Prey species may hide, flee, poison or threaten to poison their predators, school together to confuse their predators, or deploy armor or weapons in defense (**TABLE 55.1**). Traits like these are called **standing or constitutive defenses,** because they are present even in the absence of predators.

The diversity and specificity of constitutive defenses is nothing short of mind-boggling. Consider the discovery made by naturalist Henry Walter Bates as he collected butterflies in the Amazonian rain forest in the mid-1800s. Bates found an astonishing diversity of butterflies in his study area—many more species than in all of Europe—and there was a surprising pattern among the diversity. He noticed pairs of species from different taxonomic groups that were not closely related, yet had very similar wing patterns and occupied the same parts of the forest. One example of this pattern of one species resembling another, called **mimicry,** is shown in **FIGURE 55.8**.

Based on what you have learned about niche differentiation, you might have expected that the traits of species occupying similar niches would become more *different* over time, not more alike. What could explain this seeming paradox in butterflies?

Bates had an insight when he observed that some of the butterflies exuded foul substances when he handled them and that these butterflies were passed over by would-be predators. Concluding that the foul substance thwarted predators, he hypothesized that natural selection favored mimic species that

FIGURE 55.8 Why Do these Butterflies Look So Similar? Henry Walter Bates observed that these two Amazonian butterflies are distantly related, but look more like each other than either does to its closer relatives.

Type of Defense	Example
Cryptic coloration Cryptic coloration helps prey such as this leaf grasshopper escape detection	
Escape behavior Prey may have adaptations to detect predators and run, fly, jump, or swim away	
Toxins and other defense chemicals Prey may lace their tissues with toxic compounds. Many advertise their toxicity with warning coloration	
Schooling/flocking Predators can become confused by groups of prey	
Defense armor and weapons Prey may have protective shells, sharp spines, or other defense structures	

resembled the unpalatable species, a form of mimicry now known as **Batesian mimicry.** Follow-up experimental studies supported this hypothesis—and recent genetic studies have discovered some of the molecular mechanisms responsible for the shared color patterns.

Bates also noticed that some pairs of similar but distantly related butterflies are *both* unpalatable. What was the fitness advantage of this pattern? Several hypotheses were proposed, but German naturalist Fritz Müller hit on the key idea—that the existence of similar-looking unpalatable prey in the same habitat increases the likelihood that predators will learn to avoid them. Thus, co-mimicry boosts the fitness of both species. This type of mimicry, now known as **Müllerian mimicry,** was subsequently supported by field experiments.

Butterflies are not the only organisms to engage in mimicry. Bates and Müller both observed diverse examples, some of them quite shocking, such as a caterpillar with a close resemblance to a venomous snake. **FIGURE 55.9** provides a comparison of Batesian and Müllerian mimics in the case of stinging and nonstinging insects with the familiar yellow and black warning coloration.

One of the take-home messages from studies of mimicry is that the +/− consumption interaction between two species can cause unexpected consequences to other species in the community. That is, natural selection in the form of predator–prey interactions can result in mutualism, commensalism, and other types of interactions among prey species.

Inducible Defenses Although constitutive defenses can be extremely effective, they can also be expensive due to the energy and resources that must be devoted to producing and maintaining them. Based on this observation, it should not be surprising to learn that many prey species have **inducible defenses:** physical, chemical, or behavioral defensive traits that are induced in the prey in response to the presence of a predator.

Induced defenses decline in prey species if predators leave the habitat. Inducible defenses are efficient energetically, but they are slow—it takes time to produce them.

To see how inducible defenses work, consider research on the blue mussels that live in an estuary along the coast of Maine. Biologists had documented that predation on mussels by crabs was high in an area of the estuary with relatively slow tidal currents (a "low-flow" area) but low in an area of the estuary with relatively rapid tidal currents (a "high-flow" area). Mark Bertness and colleagues hypothesized that if blue mussels possess inducible defenses, then heavily defended prey individuals should occur in the low-flow area, where predation pressure is higher, but not in the high-flow area, where water movement reduces the number of crabs present. This is exactly the trend they observed: Blue mussels had thicker shells in the part of the estuary with many crabs.

The correlation between crab density and mussel shell thickness is not enough to demonstrate that predation pressure induced shell thickening. One alternate hypothesis is that crabs eliminated mussels with thin shells from the low-flow areas. The observed differences could also be due to variation in light, temperature, or other abiotic factors that might affect mussel traits but have nothing to do with predation.

To test the inducible-defense hypothesis more rigorously, Bertness and his team carried out the experiment diagrammed in **FIGURE 55.10**. The tank on the right allowed the researchers to measure shell growth in mussels that were "downstream" from crabs, such that they were exposed to the scent of crabs. As predicted by the inducible-defense hypothesis, the mussels exposed to a crab in this way developed significantly thicker shells than did mussels that were not exposed to the scent of a crab. These results suggest that, even without direct contact, mussels can sense the presence of crabs and increase their investment in defenses.

Are Animal Predators Efficient Enough to Reduce Prey Populations?

Prey are typically smaller than predators, have larger litter or clutch sizes, and tend to begin reproducing at a younger age. As a result, they have a much larger intrinsic growth rate. This quantity, symbolized r_{max}, is the maximum growth rate that a population can achieve under ideal conditions (Chapter 54).

If prey reproduce rapidly and are also well defended, can predators kill enough of them to reduce the prey population significantly?

Data from predator-removal programs—in which human hunters actively kill wolves, cougars, coyotes, or other predators—suggest that the answer is yes. For example, when a wolf control program in Alaska during the 1970s decreased predator abundance, the population of moose, on which wolves prey, tripled. This observation suggests that wolves had reduced this moose population far below the carrying capacity—the number that could be supported by the available space and food (see Chapter 54).

Other types of experiments are consistent with this result. For example, the regular population cycles of snowshoe hares appear

(a) Batesian mimics: look dangerous, are **not dangerous**

Hornet moth   Wasp beetle   Hoverfly

(b) Müllerian mimics: look dangerous, are **dangerous**

Paper wasp   Bumblebee   Honeybee

FIGURE 55.9 Prey Species Display Two Basic Forms of Mimicry.

QUESTION: Are mussel defenses induced by the presence of crabs?

HYPOTHESIS: Mussels increase investment in defense in the presence of crabs.

NULL HYPOTHESIS: Mussels do not increase investment in defense in the presence of crabs.

EXPERIMENTAL SETUP:

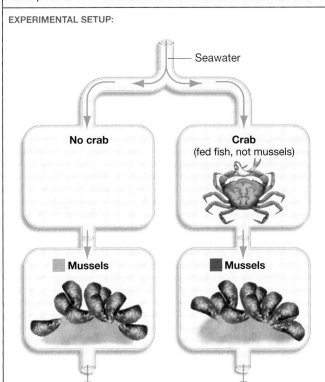

Seawater

No crab

Crab
(fed fish, not mussels)

Mussels

Mussels

PREDICTION: Mussels downstream of the crab tank will have thicker shells than mussels downstream of the empty tank.

PREDICTION OF NULL HYPOTHESIS: Mussels in the two tanks will have shells of equal thickness.

RESULTS:

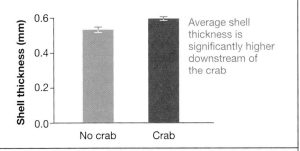

Average shell thickness is significantly higher downstream of the crab

CONCLUSION: Mussels increase investment in defense when they detect crabs. Shell thickness is an inducible defense.

FIGURE 55.10 Experimental Evidence for Inducible Defenses.
SOURCE: Leonard, G. H., M. D. Bertness, and P. O. Yund. 1999. Crab predation, waterborne clues, and inducible defenses in the blue mussel, *Mytilus edulis. Ecology* 80: 1–14.

✔ **QUESTION** Why did the researchers feed the crabs fish instead of mussels?

to be driven, at least in part, by density-dependent increases in predation by lynx (Chapter 54). Taken together, the data indicate that in many instances, predators are efficient enough to reduce prey populations below carrying capacity.

Why Don't Herbivores Eat Everything—Why Is the World Green?

If predators affect the size of populations in prey that can run or fly or swim away, then consumers should have a devastating impact on plants and on mussels, anemones, sponges, and other sessile (nonmoving) animals.

In some cases, this prediction turns out to be correct. For example, consider the results of a recent **meta-analysis**—a study of studies, meaning an analysis of a large number of data sets on a particular question. Biologists who compiled the results of more than 100 studies on herbivory found that the median percentage of mass removed from aquatic algae by herbivores was 79 percent. Herbivores eat the vast majority of algal food available in aquatic biomes. The figure dropped to just 30 percent for aquatic plants, however, and only 18 percent for terrestrial plants.

Why don't herbivores eat more of the food available on land? Stated another way, why is the world green? Biologists routinely consider two hypotheses:

1. ***Top-down control hypothesis: Herbivore populations are limited by predation and disease.*** The "top-down" name is inspired by the food-chain concept (introduced in Chapter 30 and explored in detail in Chapter 56). In a food chain, herbivores are "on top of" plants.

   The logic of top-down control is simple: Predators and parasites remove herbivores that eat plants. As a result, a great deal of plant material remains uneaten.

   In a recent test of this idea, researchers monitored herbivory on islands created by a 1986 dam project in Venezuela. On some small islands in the new lake, predators disappeared. On the predator-free islands, there are now many more herbivores than on similar sites nearby where predators are present. As predicted by the top-down control hypothesis, a much higher percentage of the total plant material is being eaten on the predator-free islands. For example, predator-free islands had just 25 percent of the small trees found on islands that contained predators.

2. ***Bottom-up limitation hypothesis: Plant tissues offer poor nutrition and are well defended.*** The "bottom-up" name reflects the position of plants as the base or bottom of a food chain. The idea here is that herbivore numbers are limited because plant tissues offer poor nutrition, or that plant tissues aren't eaten because they are toxic.

   The poor-nutrition aspect of this hypothesis was inspired by the observation that plant tissues have less than 10 percent of the nitrogen found in animal tissues, by weight. If the growth and reproduction of herbivores are limited by the availability of nitrogen, then their populations will be low and the impact of herbivory on plants relatively slight. Herbivores could eat more plant material to gain nitrogen, but at

Plasmodium cells

HLA-B53

Healthy
liver cell

cp26 protein
from *Plasmodium*

Infected
liver cell

T cell

Infected
liver cell

T cell

Dead
liver cell

1. Healthy liver cell has an HLA-B53 protein on its membrane.

2. HLA-B53 protein displays cp26 protein from *Plasmodium*, indicating that the liver cell is infected.

3. Cytotoxic T cell recognizes HLA-B53 and cp26 complex.

4. Cytotoxic T cell kills liver cell before more *Plasmodium* cells are produced.

FIGURE 55.11 Interactions between the Human Immune System and *Plasmodium*. If HLA-B53 binds to a particular *Plasmodium* protein, then infected cells are recognized and destroyed.

a cost—they would be exposed to predation and expend energy processing the food. Indeed, when nitrogen concentration in plants is increased experimentally by fertilization, it is not unusual to see significant increases in herbivore growth rates.

What's more, most plant tissues are defended by weapons such as thorns, prickles, or hairs, or by potent poisons such as nicotine, caffeine, and cocaine. The punchline? Both top-down control and bottom-up limitations—such as nitrogen limitation and effective defense—are important factors in limiting the impact of herbivory. The mix of factors that keeps the world green varies from species to species and habitat to habitat.

Adaptation and Arms Races Over the long term, how do species that interact via consumption affect each other's evolution? When predators and prey or herbivores and plants interact over time, coevolutionary arms races result. Traits that increase feeding efficiency evolve in predators and herbivores. In response, traits that make prey unpalatable or elusive evolve. In counter-response, selection favors consumers with traits that overcome the prey adaptation. And so on. Coevolution is a major driver of evolutionary change.

To see a coevolutionary arms race in action, consider interactions between humans and the most serious of all human parasites—species in the genus *Plasmodium*. *Plasmodium* species are unicellular protists that cause malaria (Chapter 30). Malaria kills at least 1 million people a year, most of them young children. Recent data suggest that humans and *Plasmodium* are locked in a coevolutionary arms race.

In West Africa, for example, there is a strong association between an allele called *HLA-B53* and protection against malaria. Malaria infections start when *Plasmodium* cells enter a person's bloodstream during a mosquito bite. The *Plasmodium* individuals then infect liver cells, where they begin to multiply.

In liver cells that are infected by *Plasmodium*, HLA-B53 proteins on the surface of the liver cell display a parasite protein called cp26 (**FIGURE 55.11**). The display is a signal that immune system cells can read. It means "I'm an infected cell. Kill me before they kill all of us." Immune system cells destroy the liver cell before the parasite cells inside can multiply.

In this way, people who have at least one copy of the *HLA-B53* allele are better able to beat back malarial infections. People who have the *HLA-B53* allele appear to be winning the arms race against malaria.

Follow-up research has shown that the arms race is far from won, however. *Plasmodium* populations in West Africa now have a variety of alleles for the protein recognized by HLA-B53. Some of these variants bind to HLA-B53 and trigger an immune response in the host, but others escape detection (**TABLE 55.2**). To make sense of these observations, researchers suggest that natural selection has favored the evolution of *Plasmodium* strains with weapons that counter HLA-B53.

Can Parasites Manipulate Their Hosts? To thrive, parasites do not just have to invade tissues and grow while evading defensive responses by their host. They also have to be transmitted to new

TABLE 55.2 Some *Plasmodium* Strains Are Particularly Effective at Infecting Humans

Plasmodium Strain	Infection Rate	Interpretation
cp26	Low	HLA-B53 binds to these proteins. Immune response is effective.
cp29	Low	
cp27	High	HLA-B53 does not bind to these proteins. Immune response is not as effective.
cp28	Average	

hosts. To a parasite, an uninfected host represents uncolonized habitat, teeming with resources. What have biologists learned about how parasites are transmitted to new hosts?

To answer this question, consider species of tree-dwelling ants (**FIGURE 55.12a**) that are parasitized by nematodes (roundworms). When nematodes have matured inside a host individual, they lay eggs in the ant's abdomen. The exoskeleton in the area becomes translucent, and the yellow eggs inside make the structure look reddish instead of the normal black color (**FIGURE 55.12b**). Infected ants also hold the abdomen up in a "flagging" posture, making them look like the berries that grow on the trees occupied by the ants (**FIGURE 55.12c**). Experiments have shown that fruit-eating birds are much more likely to pluck the abdomen from infected ants than uninfected ants.

Why is this interesting? To complete their life cycle, the nematodes have to grow inside birds, before being shed in bird feces that are subsequently eaten by ants.

To interpret their observations, biologists suggest that these nematodes not only change the appearance of the ants, but they also manipulate their behavior. The changes make the parasite much more likely to be transmitted to a new host.

Biologists have now cataloged a large number of case studies like this, from an array of parasites and hosts. Parasite manipulation of hosts is another example of extensive coevolution.

(a) Uninfected ant

(b) Infected ant

(c) Infected ants resemble berries that are eaten by birds.

FIGURE 55.12 A Parasite That Manipulates Host Behavior. When the tropical ant *Cephalotes atratus* becomes infected with parasitic nematodes, its appearance and behavior change.

✔**EXERCISE** Design an experiment to test the hypothesis that infected ants are more likely than uninfected ants to be eaten by birds.

Using Consumers as Biocontrol Agents Research on the dynamics of predator–prey interactions and other aspects of consumption has paid off in practical benefits: In some cases, biologists have been able to control pests by introducing predators or parasites.

The invasive plant purple loosestrife, for example, had succeeded in taking over large expanses of wetlands in North America. Through careful experimentation, however, biologists were able to find two insect species that feed exclusively on purple loosestrife. By introducing the herbivores into wetlands, researchers have been able to reduce purple loosestrife infestations dramatically.

In agriculture and forestry, the use of predators, herbivores, and parasites as biocontrol agents is a key part of **integrated pest management**: strategies to maximize crop and forest productivity while using a minimum of insecticides or other types of potentially harmful compounds. Integrated pest management is a promising and growing area of research. However, ecologists warn that introducing species to communities to limit one species can have unexpected consequences to others, so it is of utmost importance to proceed with extreme caution.

Mutualism

Mutualisms are +/+ interactions that involve a wide variety of organisms and rewards.

- Bees visit flowers to harvest nectar and pollen. The bees benefit by using nectar and pollen as a food source; flowering plants benefit because in the process of visiting flowers, foraging bees carry pollen from one plant to another and accomplish pollination. Adaptations found in flowering plants increase the efficiency of pollination (Chapters 31 and 41).

- One of the most important of all mutualisms occurs between mycorrhizal fungi and plant roots. For example, experimental evidence indicates that mycorrhizal fungi receive sugars and other carbon-containing compounds in exchange for nitrogen or phosphorus acquired by the plant partner (Chapter 32). The plant feeds the fungus; the fungus fertilizes the plant.

- Mutualism occurs between nitrogen-fixing bacteria and the plant species that house them in their tissues (Chapters 29 and 37). The host plants provide sugars and protection to the bacteria; the bacteria supply ammonia or nitrate in return.

- **FIGURE 55.13a** (see page 1134) shows ants in the genus *Crematogaster*, which live in acacia trees native to Africa. The ants live in bulbs at the base of acacia thorns and feed on small structures that grow from leaf tips. The ants protect the tree by attacking and biting herbivores, and by cutting vegetation from the ground below the host tree.

- **FIGURE 55.13b** illustrates cleaner shrimp in action. These shrimp pick external parasites from the jaws and gills of fish. In this mutualism, one species receives dinner while the other obtains medical attention.

The Role of Natural Selection in Mutualism As the examples you've just reviewed show, the rewards from mutualistic interactions range from transporting gametes to acquiring food, housing,

(a) Mutualism between ants and acacia trees

Entrance to ant colony

(b) Mutualism between cleaner shrimp and fish

FIGURE 55.13 Mutualisms Take Many Forms. (a) In certain species of acacia tree, ants in the genus *Crematogaster* live in large bulbs at the base of tree spines and attack herbivores that threaten the tree. The ants eat nutrient-rich tissue produced at the tips of leaves. **(b)** Cleaner shrimp remove and eat parasites that take up residence on the gills of fish.

✔**EXERCISE** Name a possible cost of these mutualistic associations to the acacia tree, the ants, the cleaner shrimp, and the host fish.

medical help, and protection. It is important to note, however, that even though mutualisms benefit both species, the interaction does not involve individuals from different species being altruistic or "nice" to each other.

Expanding on a point that Charles Darwin introduced in 1862, Judith Bronstein described mutualisms as "a kind of reciprocal parasitism; that is, each partner is out to do the best it can by obtaining what it needs from its mutualist at the lowest possible cost to itself." Her point is that the benefits received in a mutualism are a by-product of each individual pursuing its own self-interest by maximizing its fitness—its ability to survive and produce viable offspring.

In this light, it is not surprising that some species "cheat" on mutualistic systems. For example, deceit pollination occurs when certain species of plants produce a showy flower but no nectar

reward. Pollinators receive no reward for making a visit and carrying out pollination. Evolutionary studies show that deceit pollinators evolved from ancestral species that did provide a reward. Over time, a +/+ interaction evolved into a +/− interaction.

Mutualisms Are Dynamic A recent experimental study of mutualism provides another good example of the dynamic nature of these interactions. This study focused on ants and treehoppers. Treehoppers are small, herbivorous insects that feed by sucking sugar out of the phloem of plants. Treehoppers excrete the sugary solution called honeydew from their posteriors. The ants, in turn, harvest the honeydew for food.

It is clear that ants benefit from this association. But do the treehoppers? J. Hall Cushman and Thomas Whitham hypothesized that the ants might protect the treehoppers from their major predator, jumping spiders. These spiders feed heavily on juvenile treehoppers.

To test the hypothesis that ants protect treehoppers, Cushman and Whitham studied ant–treehopper interactions over a three-year period. As **FIGURE 55.14** shows, the researchers marked out a 1000-m² study plot. Each year they removed the ants from one group of the treehopper host plants inside the plot but left the others alone to serve as a control. Then they compared the growth and survival of treehoppers on plants with and without ants.

Recall that this is a common research strategy for studying species interactions. To assess the fitness costs or benefits of the interaction, researchers remove one of the participants experimentally and document the effect on the other participant's survival and reproduction, compared with the survival and reproduction of control individuals that experience a normal interaction.

In both the first and third years of the study, the number of treehopper young on host plants increased in the treatment with ants but showed a significant decline in the treatment with no ants. This result supports the hypothesis that treehoppers benefit from the interaction with ants, because the ants protect the treehoppers from predation by jumping spiders.

In the second year of the study, however, Cushman and Whitham found a very different pattern. There was no difference in offspring survival, adult survival, or overall population size between treehopper populations with ants and those without ants.

Why? The researchers were able to answer this question because they also measured the abundance of spiders that prey on treehoppers in each of the three years. Their census data showed that in the second year of the study, spider populations were very low.

Based on these results, Cushman and Whitham concluded that the benefits of the ant–treehopper interaction depend entirely on predator abundance. Treehoppers benefit from their interaction with ants in years when predators are abundant, but are unaffected in years when predators are scarce. If producing honeydew is costly to treehoppers, then the +/+ mutualism changes to a +/− interaction when spiders are rare.

RESEARCH

QUESTION: Is the relationship between ants and treehoppers mutualistic?

HYPOTHESIS: Ants harvest food from treehoppers. In return, they protect treehoppers from jumping spiders.

NULL HYPOTHESIS: Ants harvest food from treehoppers but are not beneficial to treehopper survival.

EXPERIMENTAL SETUP:

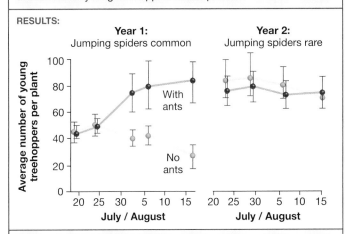

Ant

Treehoppers

Treehoppers

Plants with ants

Plants with ants removed

Study plot, 1000 m²

PREDICTION: Treehopper reproduction will be higher when ants are present than when ants are absent.

PREDICTION OF NULL HYPOTHESIS: There will be no difference in the number of young treehoppers on the plants.

RESULTS:

Year 1: Jumping spiders common

Year 2: Jumping spiders rare

Average number of young treehoppers per plant

With ants

No ants

July / August

July / August

CONCLUSION: Treehoppers benefit from the interaction with ants only when jumping spiders are common.

FIGURE 55.14 Experimental Evidence of Mutualism. The data points on the graphs represent average values from many study plots.

Cushman, J. H., and T. G. Whitham. 1989. Conditional mutualism in a membracid–ant association: temporal, age-specific, and density-dependent effects. *Ecology* 70: 1040–1047.

✔**QUESTION** The researchers assigned the treatments to different plants at random, and they did many replicates of each treatment. Why was this method important?

Mutualism is like parasitism, competition, and other types of species interactions in an important respect: The outcome of the interaction depends on current conditions. Because the costs and benefits of species interactions are fluid, an interaction between the same two species may range from parasitism to mutualism to competition. **TABLE 55.3** (see page 1136) summarizes the fitness effects, short-term impacts on population size, and long-term evolutionary aspects of species interactions.

check your understanding

If you understand that . . .

- Natural selection favors the evolution of traits that decrease competition, resulting in changes in species' niches over time.
- Consumers may reduce the population size of the species they feed on, and they often exert strong natural selection for effective defense mechanisms.
- Mutualisms benefit the species involved and can lead to highly coevolved associations, such as mycorrhizal fungi and symbiotic nitrogen-fixing bacteria.
- Coevolution is a major driver of evolutionary change.

✔ **You should be able to . . .**

1. Explain why niche differentiation (resource partitioning) does not involve a conscious choice by the individuals involved.

2. Explain what a coevolutionary arms race is, and give an example.

Answers are available in Appendix A.

55.2 Community Structure

Up to this point, the focus has been on interactions that occur between two species. But in a community, each species interacts with dozens or hundreds or even thousands of other species within a defined area. The resulting complexity of interactions is mind-boggling. How can biologists begin to characterize communities? In general, a community structure has four key attributes:

1. the total number of species;

2. the relative abundance of those species;

3. the sum of interactions among all species; and

4. the physical attributes of the community, which may include abiotic factors such as geographic size or altitude gradient, and biotic factors such as the physical structure provided by the primary vegetation type, as in a forest or grassland (see Chapter 52).

The processes that shape a particular community can be quite complex. To get a better sense of these processes, let's consider some of the ways that a community's structure can be either predictable or unpredictable.

How Predictable Are Communities?

The first question that biologists asked about woodlands and deserts and shorelines concerned structure: Do biological communities have a tightly prescribed organization and composition, or are they merely loose assemblages of species?

Type of Interaction	Fitness Effects	Short-Term Impact: Distribution and Abundance	Long-Term Impact: Coevolution
Commensalism	+/0	Population size and range of commensal may depend on population size and distribution of host.	Strong selection on commensal to increase fitness benefits in relationship; no selection on host
Competition	−/−	Reduces population size of both species; if competition is asymmetric, the weaker competitor may become locally extinct, or it may retreat to a part of its niche that does not overlap with the stronger competitor.	Niche differentiation via selection to reduce competition
Consumption	+/−	Impact on prey population depends on predator density, prey density, and effectiveness of defenses.	Strong selection on prey for effective defense; strong selection on consumer for traits that overcome defenses
Mutualism	+/+	Population size and range of both species are dependent on each other.	Strong selection on both species to maximize fitness benefits and minimize fitness costs of relationship

The Clements–Gleason Dichotomy Beginning with a paper published in 1936, Frederick Clements hypothesized that biological communities are stable and orderly entities with a highly predictable composition. His reasoning was that species interactions are so extensive—and coevolution so important—that communities have become highly integrated and interdependent units in nature. Stated another way, the species within a community cannot live without each other.

In developing this hypothesis, Clements likened the development of a plant community to the development of an individual organism. He argued that communities develop over time by passing through a series of predictable stages dictated by extensive interactions among species. Eventually, this developmental progression culminates in a final stage, known as a **climax community.** The climax community is stable—it does not change over time.

According to Clements, the nature of a climax community is determined by the area's climate. He predicted that if a fire or other disturbance destroys the climax community, it will reconstitute itself by repeating its predictable developmental stages.

Henry Gleason, in contrast, contended that the community found in a particular area is neither stable nor predictable. He claimed that plant and animal communities are ephemeral associations of species that just happen to share similar climatic requirements.

According to Gleason, it is largely a matter of chance whether a similar community develops in the same area after a disturbance occurs. Gleason downplayed the role of biotic factors, such as species interactions, in structuring communities. Instead, chance historical events—for example, which seeds and juvenile animals happened to arrive after a disturbance—were key elements in determining which species are found at a particular location.

Which viewpoint is more accurate? Let's consider experimental and observational data.

An Experimental Test To explore the predictability of community structure experimentally, ecologists David Jenkins and Arthur Buikema constructed 12 identical ponds (**FIGURE 55.15**). They filled the ponds at the same time with water that contained enough chlorine to kill any preexisting organisms. The Clements and Gleason hypotheses make clear and contrasting predictions:

- If community structure is predictable, then each pond should develop the same community of species once the chlorine has vaporized and made the water habitable.

- If community structure is unpredictable, then each pond should develop a different community.

To test these predictions, the researchers sampled water from the ponds repeatedly for one year. They measured temperature, chemical makeup, and other physical characteristics of the water and recorded the diversity and abundance of each planktonic species by examining the samples under the microscope. Species that make up **plankton** drift in the water and swim little, if at all.

Their results? The rows of colored bars on the graph in Figure 55.15 indicate whether a particular species was found in each of the 12 ponds. Note that the researchers found a total of 61 planktonic species in all of the ponds. However, individual ponds had only 31 to 39 species apiece. This observation is important. Each pond contained just half to two-thirds of the total number of species that lived in the experimental area and that were available for colonization. Several species occurred in most or all 12 of the ponds, but each pond had a unique species assemblage. Why?

To explain their results, Jenkins and Buikema contended that some species are particularly good at dispersing and are likely to colonize all or most of the available habitats. Other species disperse more slowly and tend to reach only one or a few of the

QUESTION: Are communities predictable or unpredictable?

COMMUNITIES-ARE-PREDICTABLE HYPOTHESIS: The group of species present at a particular site is highly predictable.

COMMUNITIES-ARE-UNPREDICTABLE HYPOTHESIS: The group of species present at a particular site is highly unpredictable.

EXPERIMENTAL SETUP:

1. Construct 12 identical ponds. Fill at the same time and sterilize water so that there are no preexisting organisms.

2. Examine water samples from each pond. Identify each plankton species present in each sample.

COMMUNITIES-ARE-PREDICTABLE PREDICTION: Identical plankton communities will develop in all 12 ponds.

COMMUNITIES-ARE-UNPREDICTABLE PREDICTION: Different plankton communities will develop in different ponds.

RESULTS (AFTER 1 YEAR):

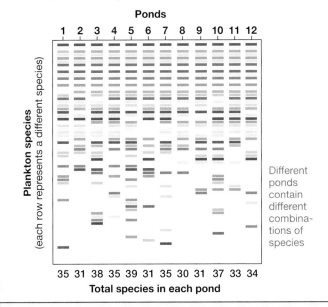

CONCLUSION: Although about half of the species present appear in all or most ponds, each pond has a unique composition. Both hypotheses are partially correct.

FIGURE 55.15 Experimental Evidence that Identical Communities Do Not Develop in Identical Habitats.

SOURCE: Jenkins, D. G., and A. L. Buikema. 1998. Do similar communities develop in similar sites? A test with zooplankton structure and function. *Ecological Monographs* 68: 421–443.

✓**QUESTION** At least 20 species are found in most or all of the ponds. Does this observation suggest that community composition is predictable or unpredictable?

available habitats. Further, the investigators proposed that the arrival of certain competitors or predators early in the colonization process greatly affects which species are able to invade successfully later. As a result, the specific details of community assembly and composition are somewhat contingent and difficult to predict. At least to some degree, communities are a product of chance and history.

Mapping Current and Past Species' Distributions If communities are predictable assemblages, then the ranges of species that make up a particular community should be congruent—meaning that the same group of species should almost always be found living together.

When biologists began documenting the ranges of tree species along elevational gradients, however, they found that species came and went independently of each other. As you go up a mountain, for example, you might find white oak trees growing with hickories and chestnuts. As you continue to gain elevation, chestnuts might disappear, but white oaks and hickories remain. Farther upslope, hickories might drop out of the species mix as white pines and red pines start to appear.

Data on the historical composition of plant communities supported these observations. Studies of fossil pollen documented that the distribution of plant species and communities at specific locations throughout North America has changed radically since the end of the last ice age about 11,000 years ago.

An important pattern emerged from these data: Species do not come and go in the fossil record in tightly integrated units. Instead, the ranges of individual species tend to change independently of one another.

The overall message of research on community structure is that Clements's position was too extreme and that Gleason's view may be closer to being correct. Although both biotic interactions and climate are important in determining which species exist at a certain site, chance and history also play a large role.

How Do Keystone Species Structure Communities?

Even though communities are not predictable assemblages dictated by obligatory species interactions, biotic interactions are still critically important. And not all species are equal. The presence of a particular plant, for example, can have a huge effect on community structure, such as the presence of redwood trees in a redwood forest. Removal of these structural species by harvesting, disease, competition with an invasive species, climate change, or other factors can cause dramatic changes to community structure.

The presence of certain consumers can also have an enormous impact on the species present. In some cases, the structure of an entire community can change dramatically if a single species of predator or herbivore is removed from a community or added to it.

As an example of this research, consider an experiment that Robert Paine conducted in intertidal habitats of the Pacific Northwest of North America. In this environment, the sea star

(a) Predator: *Pisaster ochraceous*

(b) Prey: *Mytilus californianus*

(c) Effect of keystone predator on species richness

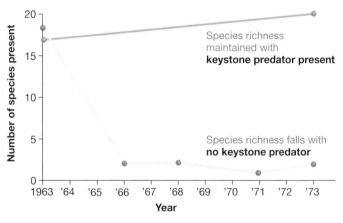

FIGURE 55.16 Keystone Predation Alters Community Structure in a Rocky Intertidal Habitat. Each data point in part (c) represents the average number of species present in several study plots.

DATA: Paine, R. T. 1966. *American Naturalist* 100: 65–75.

Pisaster ochraceous is an important predator (**FIGURE 55.16a**). Its preferred food is the California mussel *Mytilus californianus.*

When Paine removed *P. ochraceous* from experimental areas, what had been diverse communities of algae and invertebrates became overgrown with solid stands of mussels (**FIGURE 55.16b**).

M. californianus is a dominant competitor, but its populations had been held in check by sea star predation.

The graph in **FIGURE 55.16c** shows the number of species present in experimental plots where *P. ochraceous* was present or excluded. When the predator was absent, the species richness of the habitat changed radically.

To capture the effect that a predator such as *P. ochraceous* can have on a community, Paine coined the term keystone species. A **keystone species** has a much greater impact on the distribution and abundance of the surrounding species than its abundance and total biomass would suggest.

An important take-home message of Paine's study is that sea stars are not just affecting their prey. They are also indirectly affecting many other species that interact with their prey. When considering the effects of a keystone species or any other species on the community, it is helpful for biologists to think in terms of the **food web,** the network of exchanges of energy and nutrients among producers, consumers, and decomposers in an ecosystem (see Chapter 56). Food webs focus only on feeding interactions, not all types of species interactions, but are central to the structure of communities.

55.3 Community Dynamics

Once biologists had a basic understanding of how species interact and how communities are structured, they turned to questions about how communities change through time. Like cells, individuals, and species, communities can be described in one word: dynamic.

Disturbance and Change in Ecological Communities

Community composition and structure may change radically in response to changes in abiotic and biotic conditions. Biologists have become particularly interested in how communities respond to disturbance. A **disturbance** is any strong, short-lived disruption to a community that changes the distribution of living and/or nonliving resources.

Forest fires, windstorms, floods, the fall of a large canopy tree, disease epidemics, and short-term explosions in herbivore numbers all qualify as disturbances. These events are important because they alter light levels, nutrients, unoccupied space, or some other aspect of resource availability.

Biologists have come to realize that the impact of disturbance is a function of three factors: (**1**) the type of disturbance, (**2**) its frequency, and (**3**) its severity—for example, the speed and duration of a flood, or the intensity of heat during a fire. Biologists have also come to realize that most communities experience a characteristic type of disturbance. In most cases, disturbances occur with a predictable frequency and severity.

To capture this point, biologists refer to a community's **disturbance regime.** For example, fires kill all or most of the existing trees in a boreal forest every 100 to 300 years, on average.

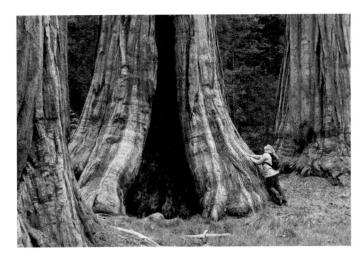

FIGURE 55.17 Giant Sequoia after a Fire.

(a) Fire scars in the growth rings

Fire scars

(b) Reconstructing history from fire scars

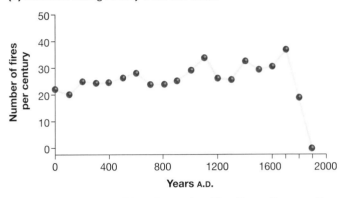

FIGURE 55.18 **History of Disturbance in a Fire-Prone Community.**
Because trees form one ring (light band/dark band) every year,
researchers can count the rings to determine how often fires have
occurred during the last 2000 years in giant sequoia groves.
DATA: Swetnam, T. W. 1993. *Science* 262: 885–889.

In contrast, many small-scale tree falls, usually caused by wind-storms, occur in temperate and tropical forests every few years.

To appreciate why biologists are so interested in understanding disturbance regimes, consider a recent study on the fire history of giant sequoia groves in California (**FIGURE 55.17**). Giant sequoias grow in small, isolated groves on the western side of the Sierra Nevada range. Individuals live more than a thousand years, and many have been scarred repeatedly by fires.

Thomas Swetnam, a specialist on tree-ring research, obtained samples of cross sections through the bases of 90 giant sequoias in five different groves. As **FIGURE 55.18a** shows, the cross sections contained many growth rings that had been scarred by fire. To determine the date of each disturbance, he counted tree rings back from the present.

The graph in **FIGURE 55.18b** plots the average number of fires that occurred each century in the groves, from the year 0 C.E. to 1900. Swetnam found that in most of the groves, 20–40 fires had occurred each century. The data indicated that each tree had been burned an average of 64 times.

This study established that fires had been extremely frequent in these forests for most of recorded history. Because not enough time would pass for large amounts of fuel to accumulate between fires, they were probably of low severity. However, recent human intervention has reduced the frequency of forest fires—with unintended and unfortunate consequences. When fires do occur, the buildup of fuel causes massive conflagrations that are often devastating to the forest community.

Partly because of this work, the biologists responsible for managing sequoia groves now set controlled fires or let low-intensity natural fires burn instead of suppressing them immediately. To maintain communities in good condition, biologists have to ensure that the normal disturbance regime occurs. Otherwise, community composition changes dramatically.

Similarly, studies of disturbance regimes along the Colorado River in southwestern North America inspired land managers to begin releasing a huge pulse of water from the reservoirs behind dams, at intervals. The floods that have resulted were designed to mimic the natural disturbance regime. According to follow-up studies, the artificial floods appear to have benefited the plant and animal communities downstream by depositing nutrients and creating additional sandbar habitat.

Succession: The Development of Communities after Disturbance

Severe disturbances remove all or most of the organisms from an area. The recovery that follows is called **succession.** The name was inspired by the observation that species with certain types of life history strategies tend to succeed others over time.

- **Primary succession** occurs when a disturbance removes the soil and its organisms as well as organisms that live above the surface. Glaciers, floods, volcanic eruptions, and landslides often initiate primary succession.

- **Secondary succession** occurs when a disturbance removes some or all of the organisms from an area but leaves the soil intact. Fire and logging are examples of disturbances that initiate secondary succession.

How does recovery proceed after a severe disturbance? Imagine a landscape before primary succession begins, such as the area surrounding the Kilauea volcano in Hawaii. There are no animals, no plants, *and no soil.* The process of soil building begins with solid rock, which weathers over time to produce gravel, sand, silt, and clay (Chapter 39). As organisms occupy the substrate, they add dead cells and tissues and feces. With time, soil eventually becomes a complex and dynamic mixture of inorganic particles, organic particles, and living organisms that enable plants to take root and add structure to the landscape.

During secondary succession, soil is already present. **FIGURE 55.19** shows a typical sequence of plant communities that develops as succession proceeds. Early successional communities are dominated by species that are short lived, small in stature, and disperse their seeds over long distances. Late successional or climax communities are dominated by species that tend to be long lived, large, and good competitors for resources such as light and nutrients.

The particular sequence of species that appears over time is called a successional pathway. What determines the pattern and rate of species replacement during succession at a certain time and place? To answer this question, biologists focus on three factors:

1. the particular traits of the species involved;

2. how the species interact; and

3. historical and environmental circumstances, such as the size of the area involved and weather conditions.

Let's consider each factor in turn.

The Role of Species Traits Species traits, such as dispersal capability and the ability to withstand extreme dryness, are particularly important early in succession. As common sense would predict, recently disturbed sites tend to be colonized by plants and animals with good dispersal ability. When these organisms arrive, however, they often have to endure harsh environmental conditions.

Pioneering species tend to have "weedy" life histories. A **weed** is a plant that is adapted for growth in disturbed soils. Because soil is often disturbed when biomass is removed, weeds tend to thrive at the start of secondary succession.

Early successional species devote most of their energy to reproduction and little to competitive ability. They are at the "high fecundity, low survivorship" end of the life-history continuum (Chapter 54).

More specifically, pioneering species have small seeds, rapid growth, and a short life span, and they begin reproducing at an early age. As a result, they have a high reproductive rate (high r_{max}—see Chapter 54). But in addition, they can tolerate severe abiotic conditions such as high light levels, poor nutrient availability, and desiccation.

The Role of Species Interactions Once colonization is under way, the course of succession tends to depend less on how species

PROCESS: SECONDARY SUCCESSION

1. Pioneering species
Weedy species become established in disturbed soils.

2. Early successional community
Weedy species are replaced by longer-lived herbaceous species.

3. Mid-successional community
Shrubs and short-lived trees begin to invade.

4. Climax community
Long-lived tree species mature.

FIGURE 55.19 Secondary Succession in Midlatitude Temperate Forests. Succession leads to the development of a temperate forest from a disturbed state (in this case, an abandoned agricultural field).

cope with aspects of the abiotic environment and more on how they interact with other species. This change occurs because plants that grow early in succession change abiotic conditions in a way that makes the conditions less severe.

For example, because plants provide shade, they reduce temperatures and increase humidity. Their dead bodies also add organic material and nutrients to the soil. During succession, existing species can have one of three effects on subsequent species:

1. **Facilitation** takes place when the presence of an early-arriving species makes conditions more favorable for the arrival of certain later species, such as by providing shade or nutrients.

2. **Tolerance** means that existing species do not affect the probability that subsequent species will become established.

3. **Inhibition** occurs when the presence of one species inhibits the establishment or regrowth of another. For example, a plant species that requires high light levels to germinate may be inhibited late in succession by the presence of mature trees that prevent sunlight from reaching the forest floor. Alternately, an established species may produce a chemical that inhibits the growth of other species.

The Role of Chance and History In addition to species traits and species interactions, the pattern and rate of succession depend on their historical and environmental context. For example, the communities that developed after forest fires disturbed

Yellowstone National Park in 1988 depended on the size of the burned patch and how hot the fire had been at that location.

Succession is also affected by the particular weather or climate conditions that occur during the process. Variation in weather and climate causes different successional pathways to occur in the same place at different times.

Analyzing species traits, species interactions, and the historical–environmental context provides a useful structure for understanding why particular successional pathways occur. To see this theoretical framework in action, let's examine data on the course of succession that has occurred in Glacier Bay, Alaska.

A Case Study: Glacier Bay, Alaska An extraordinarily rapid and extensive glacial recession is taking place at Glacier Bay (**FIGURE 55.20**). In just 200 years, glaciers that once filled the bay have retreated approximately 100 km, exposing tracts of barren glacial sediments to colonization. Because of this event, Glacier Bay has become an important site for studying succession.

Figure 55.20 shows the plant communities found in the area.

- Locations that have been ice free for 20 years or less do not have a continuous plant cover. Instead, they host scattered individuals of willow and a small shrub called *Dryas*.

- Areas that have been deglaciated for about 100 years are inhabited by dense thickets of a shrub called Sitka alder and scattered Sitka spruce trees.

- The oldest sites, ranging from 150 to 200 years of soil exposure, are dense forests of Sitka spruce and western hemlock.

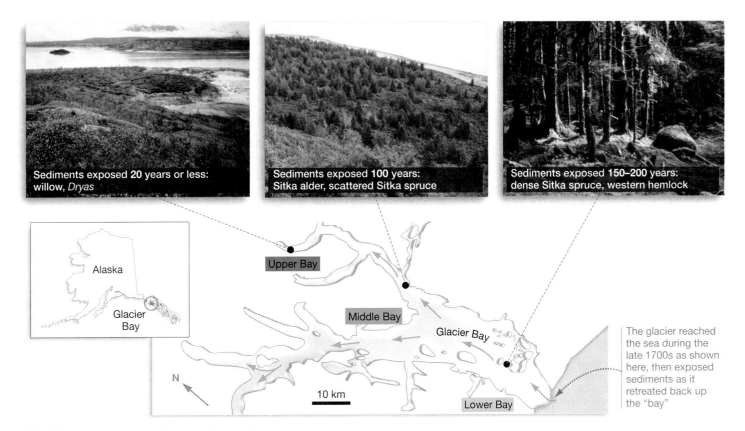

Sediments exposed **20 years or less:** willow, *Dryas*

Sediments exposed **100 years:** Sitka alder, scattered Sitka spruce

Sediments exposed **150–200 years:** dense Sitka spruce, western hemlock

Alaska

Glacier Bay

Upper Bay

Middle Bay

Glacier Bay

Lower Bay

N

10 km

The glacier reached the sea during the late 1700s as shown here, then exposed sediments as it retreated back up the "bay"

FIGURE 55.20 Successional Stages in Glacier Bay.

These observations inspired a hypothesis for the pattern of succession in Glacier Bay: With time, the youngest communities of *Dryas* and willow succeed to alder thickets, which subsequently become dense spruce–hemlock forests. The key claim is that the entire bay is following a single successional pathway.

Research has challenged this hypothesis, however. Biologists who reconstructed the history of each community by studying tree rings found that at least two distinct successional pathways have occurred:

1. In the lower part of the bay, soon after the ice retreated, Sitka spruce began growing and quickly formed dense forests. Western hemlock arrived after spruce and is now common.

2. At middle-aged sites in the upper part of the bay, alder thickets were dominant for several decades, and spruce is just beginning to become common. These forests will probably never be as dense as the ones in the lower bay, however, and there is no sign that western hemlock has begun to establish itself.

In contrast, the youngest sites in the uppermost part of the bay may be following a third pathway. Alder thickets became dominant fairly early, but spruce trees are scarce. Instead, cottonwood trees are abundant.

These data raise an important question: How do species traits, species interactions, and dispersal patterns interact to generate the observed successional pathways?

- *Species traits* Western hemlock is abundant at older sites but largely absent from young ones. This pattern is logical for this species, because its seeds germinate and grow only in soils containing a substantial amount of organic matter. In addition, young hemlock trees can tolerate deep shade but not bright sunlight. Because of these traits, Western hemlock does not tolerate early successional conditions, but thrives later in succession.

- *Species interactions* Sitka alder facilitates the growth of Sitka spruce. The facilitating effect occurs because symbiotic bacteria that live inside nodules on the roots of alder convert atmospheric nitrogen (N_2) to nitrogen-containing molecules that alder use to build proteins and nucleic acids. When alder leaves fall and decay or roots die, the nitrogen becomes available to spruce. Although spruce trees are capable of invading and growing without the presence of alder, they grow faster when alder stands have added nitrogen to the soil. Note that spruce gains from the association. Alder is either unaffected, or negatively affected if it becomes shaded by spruce.

  Competition is another important species interaction. For example, shading by alder trees reduces the growth of spruce trees until spruce trees are tall enough to protrude above the alder thicket. Once the spruce trees breach the alder canopy, however, alder trees die out, unable to compete with spruce trees for light.

- *Historical and environmental context* According to geological evidence, glacial ice was more than 1100 m thick in the upper part of Glacier Bay during the mid-1700s. Because forests grow to an elevation of only 700 m or 800 m in this part of Alaska, the glacier eliminated all of the existing forests. But in the lower part of the bay, the ice was substantially thinner. As a result, some forests remained on the mountain slopes beyond the ice, even at the height of glaciation.

The difference is crucial. In the lower part of the bay, nearby forests provided a ready source of Sitka spruce and western hemlock seed. But in the upper part of the bay, no forests remained to contribute seed. As a result, two dramatically different successional pathways were set in motion.

To summarize, successional pathways are determined by the adaptations that certain species have to their abiotic environment, interactions among species, and the history of the site. Species traits and species interactions tend to make succession predictable; history and chance events contribute a degree of unpredictability to succession.

check your understanding

C Y U

If you understand that . . .

- Disturbance is a normal part of communities.
- The impact of a disturbance depends on its type, frequency, and severity.
- After a disturbance occurs, a succession of species and communities replace those that were lost.
- The exact sequence of species observed is a function of their traits, their interactions, and the history of the site.

✔ You should be able to . . .

1. Explain how early successional species alter the environment in ways that make growing conditions more difficult for themselves.

2. Explain how the presence or absence of a species like alder, where nitrogen fixation occurs, might alter the course of succession.

Answers are available in Appendix A.

55.4 Global Patterns in Species Richness

The diversity of species present is a key feature of biological communities, and it can be quantified in two ways.

1. **Species richness** is a simple count of how many species are present in a given community.

2. **Species diversity** is a weighted measure that incorporates the species' relative abundance.

Although these terms sound similar, they can be very different in communities where a few species dominate. In such a case, species richness might be high but species diversity low (see

Consider the composition of the three hypothetical communities shown in **FIGURE 55.21**. These data can be used to compare two measures of diversity, *species richness* and *species diversity*. Species richness is simply the number of species found in a community. In this case, communities 1 and 2 have equal species richness, and community 3 is lower in richness by one species. It is important to note, however, that communities 2 and 3 have similar relative abundances of each species, or what biologists call high *evenness*. Community 1, in contrast, is highly uneven. Fifty-five percent of the individuals in community 1 belong to species A, and other species are relatively rare. An uneven community has lower effective diversity than its species richness would indicate.

To take evenness into account, other diversity indices have been developed. A simple example is the Shannon index of species diversity, given by the following equation:

$$H' = -\sum_{i=1}^{n} p_i \ln p_i$$

In this equation, p_i is the proportion of individuals in the community that belong to species i. The index is summed over all the species in the study (n).

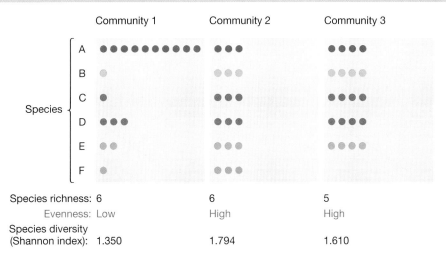

Species richness: 6   6   5
Evenness: Low   High   High
Species diversity
(Shannon index): 1.350   1.794   1.610

FIGURE 55.21 Species Diversity Can Be Quantified.

The Shannon index of species diversity for the hypothetical communities shown in Figure 55.21 was calculated by **(1)** computing the proportion of individuals in each community that belong to each species, **(2)** taking the natural logarithm of each of these proportions (see **BioSkills 6** in Appendix B for a review of logarithms), **(3)** multiplying each natural logarithm times the proportion for each species, and **(4)** summing the total

across the six species in the community and multiplying by −1. ✔ **QUANTITATIVE** If you understand the equation, you should be able to do these calculations and get the same result given in Figure 55.21. Notice that while communities 1 and 2 have the same species richness, community 2 has higher diversity because of its greater evenness. Community 3 has lower species richness than community 1 but higher diversity.

Quantitative Methods 55.1. At scales above that of relatively small study plots, however, it is rare to have data on relative abundance. As a result, species richness is the most popular measure, and biologists sometimes use the terms species richness and species diversity interchangeably.

To introduce research on richness and diversity, let's focus on a simple question: Why are some communities more species rich than others?

Predicting Species Richness: The Theory of Island Biogeography

When researchers first began counting how many species are present in various areas, a strong pattern emerged: Larger patches of habitat contain more species than do smaller patches of habitat. The observation is logical because large areas should contain more types of niches and thus support higher numbers of species.

Early work on species richness highlighted another pattern, however—one that was harder to explain. Islands in the ocean have smaller numbers of species than do areas of the same size on continents. Why?

As it turned out, answering this question has had far-reaching implications in ecology and conservation biology. This situation often happens in biological science: Research on a seemingly esoteric question—often done out of the sheer joy of discovery—turns out to have important implications for completely unrelated issues, or even applications to practical problems that had previously been intractable.

The Role of Immigration and Extinction Robert MacArthur and Edward O. Wilson tackled the question of why islands in the ocean have smaller numbers of species than do areas of the same size on continents. They assumed that speciation occurs so slowly that the number of species on an island is a product of just two processes: immigration and extinction, which should vary with the number of species already present.

- Immigration rates should decline as the number of species on the island increases, because individuals that arrive are more likely to be members of a species that is already there. In addition, competition should prevent new species from becoming established when many species are already present on an island—meaning that immigration should seldom result in establishment of a new population.

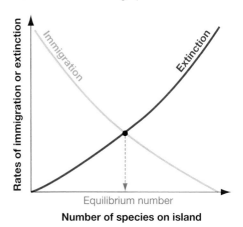

(a) Species richness depends on the number of existing species.

Rates of immigration or extinction

Immigration

Extinction

Equilibrium number

Number of species on island

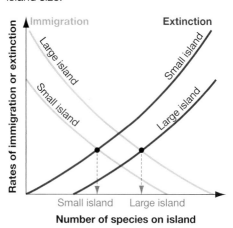

(b) Species richness depends on island size.

Rates of immigration or extinction

Immigration

Extinction

Large island

Small island

Small island

Large island

Small island   Large island

Number of species on island

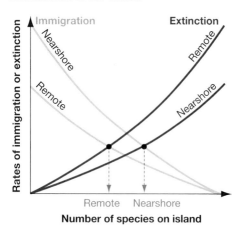

(c) Species richness depends on remoteness of the island.

Rates of immigration or extinction

Immigration

Extinction

Nearshore

Remote

Remote

Nearshore

Remote   Nearshore

Number of species on island

FIGURE 55.22 Theoretical Models Predicting Species Richness as a Function of Island Characteristics. The models illustrate how two key factors affecting species richness—immigration and extinction—depend on the number of existing species, island size, and island remoteness.

✔**QUESTION** Suppose that the mainland habitats closest to an island were wiped out by suburbanization. How would this affect the curves in part (c)?

• Extinction rates should increase as species richness increases, because niche overlap and competition for resources will be more intense.

The curves in **FIGURE 55.22a** illustrate these two predictions. Note that when the two processes interact, the result is an equilibrium—a balance between the arrival of new species and the extinction of existing ones. If species richness changes due to a hurricane or fire or some other disturbance, then continued immigration and extinction should restore the equilibrium value.

The Role of Island Size and Isolation   MacArthur and Wilson also realized that immigration and extinction rates should vary as a function of island size and how far the island is from a continent or other source of immigrants.

• Immigration rates should be higher on large islands that are close to mainlands, because immigrants are more likely to find large islands that are close to shore than small ones that are far from shore.

• Extinction rates should be highest on small islands that are far from shore, because fewer resources are available to support large populations, and because fewer individuals arrive to keep the population going.

Their theory makes two predictions: Species richness should be higher on (**1**) larger islands than smaller islands (**FIGURE 55.22b**), and (**2**) nearshore islands versus remote islands (**FIGURE 55.22c**).

Applying the Theory   The MacArthur–Wilson model, called the theory of island biogeography, is important for several reasons:

• It is relevant to a wide variety of island-like habitats such as alpine meadows, lakes and ponds, and caves. It is also relevant to national parks and other reserves that are surrounded by human development, forming islands of habitat.

• It is relevant to species that have a metapopulation structure (Chapter 54). Immigration and extinction rates in subpopulations should follow the dynamics predicted by the theory.

• It made specific predictions that could be tested. For example, researchers have measured species richness on tiny islands, removed all of the species present, and then measured whether the same number of species recolonized the island and reached an equilibrium number. Predictions about immigration and extinction rates have also been measured by observing the same islands over time.

• It can help inform decisions about the design of natural preserves. In general, the most species-rich reserves should be ones that are (**1**) relatively large, and (**2**) located close to other relatively large habitat areas (Chapters 54 and 57).

Global Patterns in Species Richness

Biologists have long understood that large habitat areas tend to be species rich, and the theory of island biogeography has been successful in framing thinking about how species richness should vary among island-like habitats. But researchers have had a much more difficult time explaining what may be the most striking pattern in species richness.

The Latitudinal Gradient   Biologists who began cataloging the flora and fauna of the tropics in the mid-1800s quickly

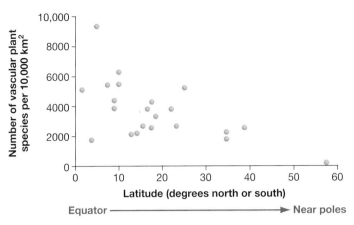

FIGURE 55.23 Species Richness of Vascular Plants Has a Strong Latitudinal Gradient.

DATA: Reid, W. V., and K. R. Miller. 1989. *Keeping Options Alive: The Scientific Basis for Conserving Biodiversity*. Washington, DC: World Resources Institute.

recognized that these habitats contain many more species than do temperate or subarctic environments. Data compiled in the intervening years have confirmed the existence of a strong latitudinal gradient in species diversity—for communities as a whole as well as for many taxonomic groups.

In birds, mammals, fish, reptiles, many aquatic and terrestrial invertebrates, and trees, species diversity declines as latitude increases. **FIGURE 55.23** graphs data on the number of vascular plant species present from the equator toward the poles. This pattern is widespread, although not universal—a number of marine groups, as well as shorebirds, show a *positive* relationship between latitude and diversity. Why does a latitudinal gradient often occur?

To explain why species diversity might decline with increasing latitude, biologists have to consider two fundamental principles:

1. The causal mechanism must be abiotic, because latitude is a physical phenomenon produced by Earth's shape (see Chapter 52). It must be a physical factor that varies predictably with latitude and that could produce changes in species diversity.

2. The species diversity of a particular area is the sum of four processes: speciation, extinction, immigration (colonization), and emigration (dispersal). Thus, the latitudinal gradient must be caused by an abiotic factor that affects the rate of speciation, extinction, immigration, or emigration in a way that would lead to more species in the tropics and fewer near the poles.

Over 30 hypotheses have been proposed to explain the latitudinal gradient. Let's consider three of the most prominent ideas.

The High-Productivity Hypothesis High productivity in the tropics could promote high diversity by increasing speciation rates and decreasing extinction rates. (Recall from Chapter 52 that productivity is the total amount of photosynthesis per unit area per year.) The logic here is that increased biomass production supports more herbivores and thus more predators, parasites, and scavengers. If higher population sizes lead to more intense competition, then speciation rates should increase as niche differentiation occurs within populations of herbivores, predators, parasites, and scavengers.

The evidence for the high-productivity hypothesis is still inconclusive.

- The hypothesis is supported by global patterns in productivity (see Chapter 56). For example, productivity is extremely high in biomes such as tropical rain forests and coral reefs, which have high species diversity.

- It is not supported by experimental addition of fertilizer to aquatic or terrestrial communities. In most of these studies, productivity increases but species diversity decreases. However, critics contend that the temporal and spatial scale of these experiments is too small to be relevant to global patterns.

- It is not supported in some highly productive habitats, such as estuaries (see Chapter 52), where species diversity can be low.

The Energy Hypothesis The energy hypothesis is an extension of the high-productivity hypothesis. Recognizing that some highly productive habitats are species poor, investigators proposed that the key factor can't be productivity alone.

The energy hypothesis claims that high temperatures increase species diversity by increasing both productivity and the likelihood that organisms can tolerate the physical conditions in a region. The idea is that high energy inputs—rather than high nutrient inputs—are the key to high speciation rates and low extinction rates.

Although the energy hypothesis is relatively new, some data support it. For example, biologists who analyzed data on gastropods and other marine invertebrates have documented a strong correlation between the temperature of marine waters and species diversity.

The Area and Age Hypothesis Temperate and arctic latitudes were repeatedly scoured by ice sheets over the last 2 million years, but tropical regions were not. In addition, the land area of the tropics has historically been larger than the land area at high latitudes. Thus, tropical regions have had more time and more space for speciation to occur than other regions have.

Recent data suggest, however, that tropical forests were dramatically reduced in size by widespread drying trends during the ice ages. Existing tropical forests may be much younger than originally thought. If so, then the contrast in the age of northern and southern habitats may not be enough to explain the dramatic difference in species diversity.

What can you conclude from all this? The current consensus is that each of the factors discussed here may influence the species diversity of communities. It is tempting to look for single causes to explain important patterns such as the latitudinal gradient. But in this case, it is probably more realistic to say there are multiple causes.

If you understand . . .

55.1 Species Interactions

- Two interacting species may each experience a net fitness cost (−) or benefit (+) from the interaction. These costs and benefits depend on the conditions that prevail at a particular time and place and may change through time.

- Commensalism (+/0) occurs when one species benefits but the other is unaffected.

- Competition (−/−) occurs when the niches of two species overlap—meaning they use the same resources. Competition may result in the complete exclusion of one species (the weaker competitor). It may also result in niche differentiation, in which competing species evolve traits that allow them to exploit different resources.

- Consumption (+/−) occurs when consumers eat part or all of other individuals, which resist through constitutive defenses or inducible defenses.

- Predators are efficient enough to reduce the size of many prey populations.

- Levels of herbivory are relatively low in terrestrial ecosystems, because predation and disease limit herbivore populations, plants provide little nitrogen, and many plants have defenses.

- Parasites are consumers that generally spend all or part of their life cycle in or on their host (or hosts). They usually have traits that allow them to escape host defenses and even manipulate host behavior to increase their likelihood of transmission to a new host. In turn, hosts have evolved counter-adaptations that help fight off parasites.

- Mutualism (+/+) provides individuals with food, shelter, transport of gametes, defense against predators, or other services. For each species involved, the costs and benefits of mutualism may vary over time and from place to place.

- Interactions among species have two main outcomes: (1) They affect the distribution and abundance of the interacting species, and (2) they are agents of natural selection and thus affect the evolution of the interacting species, by a process called coevolution.

✔ You should be able to give an example of how a mutualistic relationship can evolve into a parasitic one.

55.2 Community Structure

- The attributes of community structure include the number of species present, their relative abundance, the sum of their interactions, and the physical attributes of the community as determined by a combination of abiotic and biotic factors.

- Historical and experimental evidence supports the view that communities are dynamic rather than static, and that their composition is neither entirely predictable nor stable over time.

- The assemblage of species found in a community is primarily a function of climate and chance historical events.

- In some cases, the structure of an entire community can change dramatically if a single species is removed from it or added to it.

✔ You should be able to predict the effects of an invasive species on community structure.

55.3 Community Dynamics

- Each community has a characteristic disturbance regime—meaning a type, severity, and frequency of disturbance that it experiences.

- Three types of factors influence the pattern of succession that occurs after a disturbance: (1) A species' physiological traits influence when it can successfully join a community; (2) interactions among species influence if and when a species can become established; and (3) the historical and environmental context of the site affects which species are present.

✔ You should be able to explain why climate makes the three successional pathways documented in Glacier Bay similar, and why chance historical events make them different.

55.4 Global Patterns in Species Richness

- Species richness is higher in large islands near continents than in small, isolated islands, due to differences in immigration and extinction.

- The theory of island biogeography is relevant to a wide variety of isolated habitats.

- Species richness is higher in the tropics and lower toward the poles, but the mechanism responsible for this pattern is still hotly debated.

✔ You should be able to propose how the theory of island biogeography could be applied by conservation biologists to species richness in a national park.

MB MasteringBiology

1. **MasteringBiology Assignments**

   Tutorials and Activities Experimental Inquiry: Can a Species Niche Be Influenced by Interspecific Competition?; Exploring Island Biogeography; Interspecific Interactions; Life Cycle of a Malaria Parasite; Primary Succession; Species Area Effect and Island Biogeography; Succession

   Questions Reading Quizzes, Blue-Thread Questions, Test Bank

2. **eText** Read your book online, search, take notes, highlight text, and more.

3. **The Study Area** Practice Test, Cumulative Test, BioFlix® 3-D Animations, Videos, Activities, Audio Glossary, Word Study Tools, Art

You should be able to . . .

1. True or false? Species act as agents of natural selection when they interact.

2. What is niche differentiation?
   a. the evolution of traits that reduce niche overlap and competition
   b. interactions that allow species to occupy their fundamental niche
   c. the degree to which the niches of two species overlap
   d. the claim that species with the same niche cannot coexist

3. The relationship between ants and treehoppers in the presence of spiders is an example of
   a. commensalism.
   b. competition.
   c. consumption.
   d. mutualism.

4. What is one advantage of inducible defenses?
   a. They are always present; thus, an individual is always able to defend itself.
   b. They make it impossible for a consumer to launch surprise attacks.
   c. They result from a coevolutionary arms race.
   d. They make efficient use of resources, because they are produced only when needed.

5. Pioneer species tend to have high _____ and lower survivorship.

6. Which of these factors is *not* generally correlated with species diversity?
   a. latitude
   b. productivity
   c. longitude
   d. island size

7. The text claims that species interactions are conditional and dynamic. Do you agree with this statement? Why or why not? Cite specific examples to support your answer.

8. Why does the phrase "coevolutionary arms race" appropriately characterize the long-term effects of species interactions?
   a. Both plants and animals have evolved weapons for defense that are so effective that many plants are not eaten, and predators cannot reduce prey populations to extinction.
   b. Adaptations that give one species a fitness advantage in an interaction are likely to be countered by adaptations in the other species that eliminate this advantage.
   c. In all species interactions except for mutualism, at least one species loses (suffers decreased fitness).
   d. Even mutualistic interactions can become parasitic if conditions change. As a result, interacting species are always "at war."

9. Biologists have tested the hypotheses that communities are highly predictable versus highly unpredictable. State the predictions that these hypotheses make with respect to (a) changes in the distribution of the species in a particular community over time, and (b) the communities that should develop at sites where abiotic conditions are identical. Which hypothesis appears to be more accurate?

10. What is a disturbance? Consider the role of fire in a forest. Compare and contrast the consequences of high-frequency versus low-frequency fire, and high versus low severity of fire.

11. Summarize the life-history attributes of early successional species. Why are these attributes considered adaptations?

12. Explain why high productivity should lead to increased species richness in habitats such as tropical rain forests and coral reefs.

13. Some insects harvest nectar by chewing through the wall of the structure that holds the nectar. As a result, they obtain a nectar reward, but pollination does not occur. Suppose that you observed a certain bee species obtaining nectar in this way from a particular orchid species. Over time, how might you expect the characteristics of the orchid population to change in response to this bee behavior?

14. You are a walking, talking community that includes trillions of bacterial and archaeal cells. Your gut, in particular, contains a complex microbial community. In some circumstances, the use of probiotics, which stimulate the rapid growth of bacteria that are mutalistic or commensal with humans, can eliminate the need to use antibiotics, which can wipe out helpful bacteria along with harmful bacteria. The use of probiotics is an example of which process?
   a. succession
   b. competitive exclusion
   c. parasitism
   d. niche differentiation

15. Suppose that a two-acre lawn on your college's campus is allowed to undergo succession. Describe how species traits, species interactions, and the site's history might affect the community that develops.

16. Design an experiment to test the hypothesis that increasing species richness increases a community's productivity.

56 Ecosystems and Global Ecology

In this chapter you will learn how

Ecosystems are connected by the flow of energy and nutrients

by asking

How does energy flow through ecosystems? 56.1

considering

How do nutrients cycle through ecosystems? 56.2

by asking

including a look at

Global patterns in productivity

affect

Human activities

affect

including a look at

Global biogeochemical cycles

cause

affect

Global climate change 56.3

affect

The reservoir at the Harbeespoort Dam in South Africa is notorious for its spectacular algae blooms caused by nutrient pollution. This chapter explores how humans are altering ecosystems all over the planet.

The data in this chapter carry a simple, but important, message: Some of the most urgent problems facing humans are rooted in ecosystems and global ecology. Acid rain, nitrate pollution, declining freshwater supplies, soil erosion, and global climate change are just a few of the large-scale issues that ecologists are documenting.

The chapter has a simple theme, as well: In nature, everything is connected. This is because the components of an **ecosystem**—the species present in a region, along with abiotic components such as the soil, climate, water, and atmosphere—are linked by the flow of energy and the cycling of **nutrients** needed to sustain life, such as carbon and nitrogen. Further, ecosystems are themselves linked by the global exchange of energy and nutrients throughout the **biosphere**, the thin zone of life surrounding the Earth.

This chapter is part of the Big Picture. See how on pages 1196–1197.

✔ When you see this checkmark, stop and test yourself. Answers are available in Appendix A.

Humans are removing species, changing landscapes, and adding massive amounts of energy and nutrients to ecosystems all over the planet. The biological, chemical, and geological consequences have been so dramatic that scientists have proposed naming a new epoch in the geological timescale—the Anthropocene ("new human epoch") (Chapter 28).

Let's begin our analysis of large-scale ecology by considering the basics of how energy flows through the components of an ecosystem. Then, let's examine the cycling of nutrients. Finally, you will see how the flow of energy and the cycling of nutrients are central to understanding global climate change. You can see how ecosystems fit into the Big Picture of Ecology on pages 1196–1197.

56.1 How Does Energy Flow through Ecosystems?

One of the defining characteristics of life is the use of energy (Chapter 1). Energy enters ecosystems via the primary producers. A **primary producer** is an **autotroph** (literally, "self-feeder")—an organism that can synthesize its own food from inorganic sources. In most ecosystems, primary producers use solar energy to manufacture food via photosynthesis (see Chapter 10). But in deep-sea hydrothermal vents and iron-rich rocks deep below Earth's surface, primary producers use the chemical energy contained in inorganic compounds such as methane (CH_4) or hydrogen sulfide (H_2S) to make food (see Chapter 29).

Primary producers do not *create* energy. Rather, they are *transforming* the energy in sunlight or inorganic compounds into the chemical energy stored in sugars. Recall that the first law of thermodynamics states that energy cannot be created or destroyed, but only transferred and transformed (Chapter 8).

The total amount of chemical energy produced in a given area and time period is called **gross primary productivity (GPP).** The primary producers use this chemical energy in two ways:

1. *Cellular Respiration* It takes energy just to stay alive. Primary producers have to use chemical energy to fuel cellular respiration (or in anaerobic microbes, fermentation) to produce the energy currency, ATP (Chapter 9). ATP, in turn, is used to fuel a diverse array of metabolic processes.

2. *Growth and Reproduction* Chemical energy that isn't used for cellular respiration can be applied to growth and reproduction. Energy that is invested by primary producers in building new tissue or offspring is called **net primary productivity (NPP).**

Because cellular respiration, growth, and reproduction are not perfectly efficient (second law of thermodynamics), some energy is lost during these processes as molecular waste and heat without being used. Linking these concepts together:

$$NPP = GPP - R$$

(where R = energy used in cellular respiration, or lost)

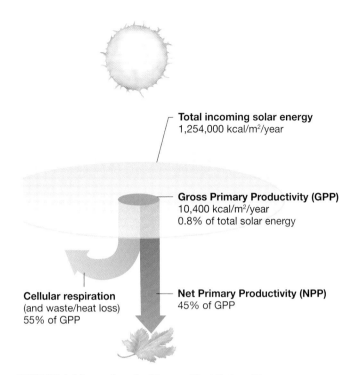

FIGURE 56.1 Measuring the Energy That Enters Ecosystems. These data are from the Hubbard Brook Experimental Forest.

- Total incoming solar energy
  1,254,000 kcal/m²/year
- Gross Primary Productivity (GPP)
  10,400 kcal/m²/year
  0.8% of total solar energy
- Cellular respiration
  (and waste/heat loss)
  55% of GPP
- Net Primary Productivity (NPP)
  45% of GPP

NPP represents the total amount of chemical potential energy that is stored in organic material, or **biomass.** NPP is critically important because it is the amount of energy available via primary producers to the other organisms of an ecosystem.

How Efficient Are Autotrophs at Capturing Solar Energy?

How much of the energy that enters an ecosystem is captured as NPP? To answer this question, let's look at data from the world's most intensively studied ecosystem. Since the early 1960s, ecologists Gene Likens, Herbert Bormann, and colleagues have been studying how energy and nutrients flow through a temperate forest ecosystem in the northeastern United States: the Hubbard Brook Experimental Forest in New Hampshire.

During one year, from June 1, 1969, to May 31, 1970, 1,254,000 kcal of solar radiation per square meter (m²) entered this ecosystem in the form of sunlight. If this amount of energy were available in the form of electricity, it would easily power two 150-watt lightbulbs per square meter of land, burning continuously for one year.

By documenting rates of photosynthesis in a large sample of forest plants, Likens and Bormann calculated that the plants used 10,400 kcal/m² of this energy in photosynthesis—a mere 0.8 percent of the incoming sunlight (**FIGURE 56.1**). Why so inefficient? For comparison, the best solar panels produced today have an efficiency of about 20 percent in normal outdoor conditions.

The answer includes several components.

- Even when conditions are ideal, the pigments that drive photosynthesis can absorb only a fraction of the light wavelengths available and thus a fraction of the total energy received.

- Plants in temperate biomes have drastically reduced photosynthetic rates in winter.
- If conditions get dry during the summer, stomata close to conserve water. Photosynthesis stalls due to a lack of CO_2.
- The efficiency of enzymes is very temperature dependent.

How much of the gross primary production goes to the production of new biomass? By measuring plant growth, Likens and Bormann estimated that only about 45 percent of GPP went to NPP. The other 55 percent either was used for cellular respiration or lost.

Other biologists have done similar studies in other ecosystems. The general pattern is that only a tiny fraction of incoming sunlight is converted to chemical energy, and only a fraction of this gross primary productivity is used to build biomass. What happens to the new biomass once it is formed?

What Happens to the Biomass of Autotrophs?

Graduate student Raymond Lindeman made a key insight about the fate of autotroph biomass in the 1940s, based on his research at Cedar Creek Bog in Minnesota. Lindeman proposed the *energy flow model* for ecosystems, whereby energy could flow from autotrophs through other organisms in the form of biomass.

- **Consumers** eat living organisms. **Primary consumers** eat primary producers; **secondary consumers** eat primary consumers; **tertiary consumers** eat secondary consumers, and so on. You are a primary consumer when you eat a salad, a secondary consumer when you eat beef, and a tertiary consumer when you eat tuna.

- **Decomposers,** or **detritivores,** obtain energy by feeding on the remains of other organisms or waste products. Dead animals and dead plant tissues ("plant litter") are collectively known as **detritus.** Many fungi and bacteria are decomposers, but humans rarely act as decomposers.

Energy flows when one organism eats another. However, a key take-home message is that energy constantly dissipates as it flows through ecosystems and is ultimately "lost" as heat, whereas nutrients continue to cycle (**FIGURE 56.2**). Consider that all the *energy* stored in your tissues right now ultimately will be released into the atmosphere, but all the carbon, nitrogen, and other *nutrients* present in your tissues will persist in other organisms and the abiotic environment for millennia.

Fifteen years after Lindeman's model was published, Howard Odum carried the concept of energy flow to the next level by quantifying energy flow in an aquatic ecosystem. Odum categorized organisms according to their roles. Organisms that obtained energy from the same type of source were said to occupy the same **trophic** ("feeding") **level.**

Trophic Structure: Food Chains and Food Webs
Biologists often begin with a simple model and add complexity. A **food chain** focuses on one possible pathway of energy flow among trophic levels in an ecosystem. **FIGURE 56.3** shows two examples of food chains: a **grazing food chain** and a **decomposer food chain.**

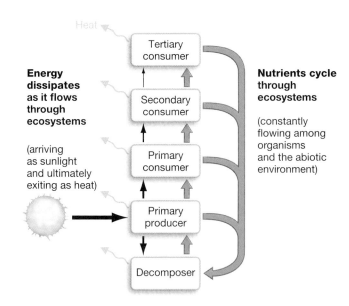

FIGURE 56.2 Energy Eventually Dissipates, but Nutrients Cycle.

At the second trophic level, the decomposer food chain has a **primary decomposer,** which feeds on plant detritus, rather than a primary consumer. Note several key points:

- At higher trophic levels, the grazing and decomposer food chains often merge. For example, the robin in the figure functions as

Trophic level	Feeding strategy	Grazing food chain	Decomposer food chain
5	Quaternary consumer		Cooper's hawk
4	Tertiary consumer	Cooper's hawk	Robin
3	Secondary consumer	Robin	Earthworm
2	Primary decomposer or consumer	Cricket	Bacteria, archaea
1	Primary producer	Live maple leaves	Dead maple leaves

FIGURE 56.3 Trophic Levels Identify Steps in Energy Transfer.

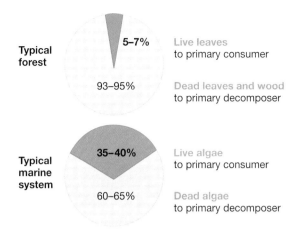

FIGURE 56.5 How Much Biomass Gets Eaten Dead versus Alive?

FIGURE 56.4 Food Webs Describe Trophic Relationships in an Ecosystem. Food webs offer a more comprehensive analysis of feeding relationships than food chains do (compare to Figure 56.3). This food web shows only a fraction of the total feeding relationships and species of a temperate deciduous forest.

a secondary consumer in the grazing food chain and a tertiary consumer in the decomposer food chain.

- It is common for consumers—such as humans and robins—to feed at multiple trophic levels.

Because most consumers feed on a wide array of organisms and at multiple trophic levels, food chains are embedded in more complex **food webs.** The food web shown in **FIGURE 56.4** is a more detailed description of the trophic relationships among the organisms in Figure 56.3. A complete food web would include *all* the organisms interacting in this ecosystem, along with estimates of the amount of energy transferred at each link. Food webs are a way to summarize energy flows and to document the complex trophic interactions that occur in ecosystems.

Energy Flow to Grazers versus Decomposers

Figures 56.3 and 56.4 show that the primary producers—maple leaves in this case—are a source of energy whether alive or dead. How much biomass gets eaten alive versus dead?

The answer to this question varies among habitats. In forest ecosystems, the percentage of NPP consumed alive can be as low as 5–7 percent, because indigestible wood isn't transferred to other organisms until it decays. But in marine habitats, where algae account for most of the primary production, the percentage

of NPP that gets consumed alive is much higher—often 35–40 percent (**FIGURE 56.5**). This is an important point, for two reasons:

1. If you want to understand energy flow in a forest, you have to understand decomposers. Decomposition is less of a factor in marine ecosystems, where primary consumers are key.

2. The decomposer food chain is "leaky." At Hubbard Brook, large amounts of energy leave the forest ecosystem in the form of detritus that washes into streams. In these streams, photosynthesis by aquatic algae and plants introduced only about 10 kcal/m²/yr versus 6039 kcal/m²/yr that washed in from the surrounding forest. In marine ecosystems, energy exits in the form of dead bodies that rain down to the ocean floor.

The "dead versus alive" percentage can vary from season to season or even week to week according to ambient temperatures, light levels, nutrient levels, and species interactions.

Energy Transfer between Trophic Levels

All ecosystems share a characteristic pattern: The total biomass produced each year declines from lower trophic levels up to the higher levels. To understand why, let's go back to the Hubbard Brook Experimental Forest.

Chipmunks are small rodents that are seed predators at Hubbard Brook. On average, these primary consumers harvest 31 kcal/m² of energy each year. Of that total, how much is used to create biomass? Careful measurements revealed that, on average:

- 17.7 percent is unused—meaning that it is not digested and absorbed. Instead, it is excreted.

- 80.7 percent is used for cellular respiration.

- 1.6 percent goes into the production of new chipmunk tissue by growth and reproduction.

The same pattern would be seen in a secondary consumer—a Cooper's hawk, for example, that eats chipmunks and other primary consumers. Consider the energy that a hawk expends to catch prey. This energy cannot be used to make hawk biomass.

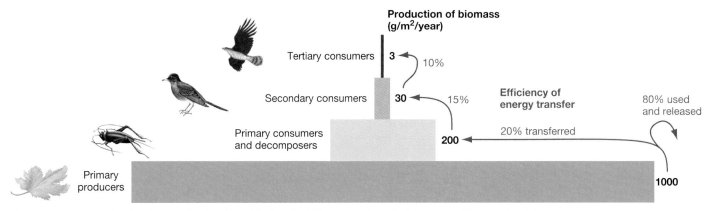

Production of biomass (g/m²/year)

Tertiary consumers **3** — 10%

Secondary consumers **30** — 15% **Efficiency of energy transfer** — 80% used and released

Primary consumers and decomposers **200** — 20% transferred

Primary producers **1000**

FIGURE 56.6 Productivity Declines at Higher Trophic Levels. This pattern is called the pyramid of productivity.

The general point is simple: Because most of the total energy consumed is used for cellular respiration rather than growth and reproduction, the amount of biomass produced at the second trophic level must be less than the amount at the first trophic level; productivity at the third trophic level must be less than that at the second.

The Pyramid of Productivity The data in **FIGURE 56.6** depict the biomass produced at trophic levels 1–4 in a temperate deciduous forest. A separate horizontal bar represents each trophic level. Note that the graph reports two numbers:

1. Productivity is a rate, measured here in units of grams of biomass produced by growth and reproduction per square meter of area each year (g/m²/year).

2. Efficiency is a ratio and thus dimensionless—the units cancel out. (It is reported here as a percentage.) Efficiency is the fraction of biomass transferred from one trophic level to the next.

Biomass production at each trophic level varies widely among ecosystems. But as a rough rule of thumb, the efficiency of biomass transfer from one trophic level to the next is only about 10 percent. ✔ QUANTITATIVE If you understand this concept, you should be able to explain why it takes about 10 times more energy to grow a kilogram of beef than a kilogram of wheat.

Efficiency Varies It's important to recognize that the "10 percent rule" in ecological efficiency masks a great deal of interesting variation. For example:

- Large mammals are more efficient at producing biomass than small mammals, because large animals have a smaller surface-area-to-volume ratio and lose less heat (see Chapter 42).

- Biomass production is much more efficient in ectotherms than in endotherms. Ectotherms rely principally on heat gained from the environment and do not oxidize sugars to keep warm (see Chapter 42). As a result, they devote much less energy to cellular respiration than endotherms do. Chipmunks devote just 1.5 percent of their energy intake to growth; lizards and other ectotherms may devote 20–40 percent.

✔ If you understand this concept, you should be able to explain why it may be more efficient to eat lobster than chicken, even though the two animals feed at the same trophic level.

Biomagnification

The pyramid of productivity helps explain how some pollutants become concentrated in food webs. Some types of molecules do not break down quickly in the environment and are neither digested by organisms nor excreted efficiently. These include heavy metals like mercury, emitted mostly from coal-fired power plants, and organic compounds called POPs—persistent organic pollutants—which are used for herbicides, pesticides, and other industrial applications. Mercury and POPs undergo **biomagnification**—meaning that they increase in concentration at higher levels in a food chain. Why?

Biomagnification begins when persistent atoms or molecules are taken up from air or water by primary producers. A very large quantity of pollutants can be taken up by primary producers—even if the concentration of the pollutant in the environment is low—because of the very large biomass of primary producers at the bottom of the productivity pyramid (Figure 56.6).

Remember that organisms at each trophic level eat about 10 g of tissue to make 1 g of their own tissue. When the primary consumers consume the biomass of the primary producers, they take up all the pollutants but concentrate them in their smaller biomass—because little or none of the pollutant is metabolized or excreted. As a result, the pollutant becomes more concentrated as it moves up each trophic level (**FIGURE 56.7**). The greater the number of trophic levels, the higher the consequences to the top consumers. In many cases, concentrations get high enough in secondary and tertiary consumers to poison them.

One POP, called DDT (dichlorodiphenyltrichloroethane), was made famous by ecologist Rachel Carson in her 1962 book, *Silent Spring*. Carson was alarmed that many birds were dying in areas that were sprayed with DDT to control mosquitoes—the pesticide caused birds to lay thin-shelled eggs that became crushed in the nest. Carson's resulting campaign for responsible chemical use spurred the beginning of the environmental

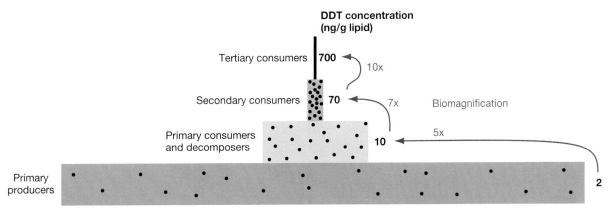

FIGURE 56.7 Biomagnification Can Lead to High Concentrations of Toxic Substances at Higher Trophic Levels. DDT is a pesticide once commonly used to control mosquitoes and other insects.

movement. DDT was banned in the United States in 1972. Populations of top predators like ospreys and bald eagles are back on the rise.

Humans are top predators too, and are vulnerable to biomagnification. ✔ If you understand biomagnification, you should be able to explain how you could get mercury poisoning from eating tuna every day, but not from eating the same quantity of sardines every day.

Top-Down Control and Trophic Cascades

Top predators are not abundant compared to other organisms in the food web (as indicated by the tiny biomass at the peak of the productivity pyramid in Figure 56.6), and usually don't get eaten by other consumers. Are top predators important?

One way to answer this question is to remove top predators from an ecosystem and see what happens. Robert Paine used this approach in his study of the keystone species, *Pisaster ochraceous* (Chapter 55). When Paine removed this sea star from a rocky intertidal habitat on the Pacific Coast, mussels took over—and dramatically changed which other species were present in the community. When a consumer limits a prey population in this way, biologists say that **top-down control** is occurring.

It is rarely practical to remove a top predator from an ecosystem, but sometimes events present an opportunity for a natural experiment. That's exactly what happened in the Greater Yellowstone ecosystem of western North America, where wolves were hunted out in the early 1900s but reintroduced in 1995.

What effect did the absence of wolves have on the Greater Yellowstone ecosystem? During the 1990s, ecologists William Ripple and Robert Beschta had been trying to determine why aspen trees had been declining for most of the last century. Then they made an important connection—the number of aspen trees started to decline at the same time that the wolves had disappeared. Ripple and Beschta hypothesized that there was an indirect link between aspen trees and wolves.

How could they test their hypothesis? One way was to see if the number of aspen trees would begin to increase as the reintroduced wolf population increased. Although the Yellowstone

wolf population has grown to only about 150 individuals since reintroduction, their presence has already led to an increase in aspen—and to many other changes in the food web (**FIGURE 56.8**, see page 1154).

- In Yellowstone, wolves primarily feed on elk. As elk numbers declined and individuals changed their feeding habits due to fear of being in the open, populations of favorite elk foods like aspen, cottonwood, and willow have increased.

- The changes in willow and other plant species triggered an increase in beavers, which compete with elk for these plants.

- Beavers dam streams, forming ponds that provide habitat for frogs, turtles, and an array of other species, whose numbers have increased.

- Wolves do not tolerate coyotes. As coyotes declined, mouse populations increased. The increase in mouse populations led to larger populations of hawks that prey on mice.

When changes in top-down control cause conspicuous effects two or three links away in a food web, biologists say that a **trophic cascade** has occurred. Trophic cascades have been observed in many ecosystems. Atlantic cod and other large predators, for example, have virtually disappeared from large parts of the Atlantic Ocean due to overfishing. Shrimp and crab—species that cod used to eat—have increased in response; the primary producers that shrimp and crab eat have decreased.

The message from these studies is clear: The removal of top predators from ecosystems can have widespread, and sometimes unexpected, effects on the abundance of many species in a community. The underlying cause is energy flow—one of the factors that links components in an ecosystem.

Global Patterns in Productivity

Top predators are not the only players that can profoundly affect the abundance of species in a community. The productivity of primary producers is also very important, and it can have a "bottom-up" effect. Let's consider two big-picture questions about NPP: Where is most of it produced, and what limits its production to those areas?

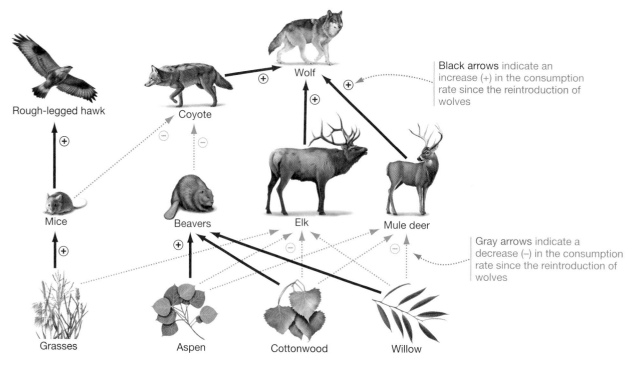

FIGURE 56.8 The Reintroduction of Wolves to Yellowstone Caused a Trophic Cascade. This food web shows only a few of the hundreds of species affected by the reintroduction of wolves.

Keep in mind that productivity is limited by any factor that limits the rate of photosynthesis. The top candidates are temperature and the availabilities of water, sunlight, and nutrients (see Chapter 10 and Unit 7). Let's consider how different limiting factors prevail in different environments.

Is Productivity Higher on the Land or in the Sea? FIGURE 56.9 summarizes data on NPP from around the globe. Note that the highest NPP is indicated in red, yellow, and green, and that these colors occur only on land. In general, NPP on land is much higher than it is in the oceans. Why?

FIGURE 56.9 Net Primary Productivity Varies among Geographic Regions. The terrestrial ecosystems with the highest primary productivity are found in the tropics; tundra and deserts have the lowest. The highest productivity in the oceans occurs in nutrient-rich coastal areas.

DATA: http://earthobservatory.nasa.gov.

The leading hypothesis to explain this pattern is simply that much more light is available to drive photosynthesis on land than in marine environments. Water quickly absorbs many of the light wavelengths that are used in photosynthesis (see Chapter 52). This simple observation also underlines the importance of the evolution of land plants (see Chapter 31). Before green plants invaded the land, NPP in terrestrial environments may have been as low as it is in today's oceans.

Which Terrestrial Ecosystems Are Most Productive? The terrestrial ecosystems with highest productivity are located in the wet tropics. Except in the world's major deserts (such as the large black area representing the Sahara Desert in northern Africa), NPP on land declines from the equator toward the poles. Why? The availability of sunlight and average temperature decreases with latitude. The low productivity in deserts is an exception attributed to the low annual rainfall caused by air circulation patterns (see Chapter 52).

You might recall that the equator-to-poles gradient in NPP is thought to be important in explaining the equator-to-poles gradient in species richness that occurs in most types of plants and animals (see Chapter 55). Areas of high productivity tend to be "hotspots" for high species richness (see Chapter 57).

The global productivity of terrestrial ecosystems is limited by a combination of temperature and availability of water and sunlight. At a local level, however, NPP on land is also limited by nutrient availability. When biologists fertilize a terrestrial ecosystem with nitrogen or phosphorus, NPP almost always increases. Recent research has also shown that NPP in temperate

and boreal forests is increasing due to nitrogen that is blowing in from agricultural fields hundreds or thousands of miles away, where it was originally applied as a fertilizer.

Which Marine Ecosystems Are Most Productive? Productivity patterns in marine ecosystems contrast strongly with those of terrestrial environments. Marine productivity is highest along coastlines, and it can be as high near the poles as in the tropics. Why? The shallow water along coasts receives more nutrients:

- Rivers carry nutrients from terrestrial ecosystems and deposit them on the coasts.

- Nearshore ocean currents bring nutrients from the cold, deep water of the oceanic zone up to the surface, a process called upwelling (Chapter 52).

The coastal areas colored deep blue in Figure 56.9 include most of the world's most important fisheries.

In contrast, the oceanic zones, colored black in Figure 56.9, have extremely low NPP. They lack both river inputs and upwelling. In addition, nutrients in organisms near the surface of the open ocean—where sunlight is abundant—constantly fall to dark, deeper waters in the form of dead cells, where they are no longer available to organisms near the surface. Typically, a square meter (m^2) of open ocean produces a maximum of 35 g (1.2 oz) of organic matter each year. In terms of productivity per m^2, the open ocean is a desert.

Which Biomes Are Most Productive? FIGURE 56.10 presents NPP data a different way—organized by biome instead of by geography

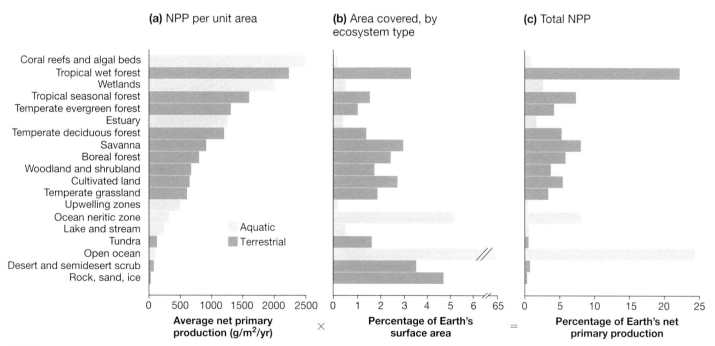

FIGURE 56.10 Net Primary Productivity Varies among Biomes.

DATA: Leith, H., and R. H. Whittaker (eds.). 1975. *Primary Productivity in the Biosphere*. Ecological Studies and Synthesis Vol. 14. New York: Springer-Verlag.

✔QUANTITATIVE Estimate the average NPP per unit area for the open ocean and the total area covered by the open ocean. How do these numbers explain the results in column (c)?

(see Chapter 52). Figure 56.10a provides data on average NPP per square meter per year for each biome; Figure 56.10b documents the total area that is covered by each type of ecosystem; and Figure 56.10c presents the percentage of the world's total productivity—the result of multiplying the data in part (a) by the data in part (b). In each case, data from different types of biomes are plotted as horizontal bars. Note these observations:

- Tropical wet forests and tropical seasonal forests (which have a dry season) cover less than 5 percent of Earth's surface but together account for over 30 percent of total NPP.

- Among aquatic ecosystems, the most productive habitats are algal beds and coral reefs, wetlands, and estuaries.

- Although NPP per square meter is extremely low in the open ocean, this biome is so extensive that its total NPP is high.

One of the reasons biologists are so concerned about the destruction of tropical rain forests and coral reefs is that these are the world's most productive habitats.

The "Human Appropriation" Of the total NPP available on our planet, how much are humans using? A research group led by Helmut Haberl answered this question recently in terms of *potential* NPP—meaning, NPP that would exist without human activity. Using data from a wide array of sources, Haberl's team estimated that humans are currently preventing or appropriating 24 percent of potential NPP. Of this percentage,

- 53 percent is directly harvested and used (timber, food, etc.);

- 40 percent is prevented through land-use changes (such as parking lots and other forms of development); and

- 7 percent is destroyed due to human-induced fires.

This analysis is astonishing. One species—our own—is appropriating almost a quarter of the planet's biomass. Given increases in human population projected to occur over your lifetime (see

check your understanding

If you understand that . . .

- In an ecosystem, energy flows from sunlight to primary producers, and from there to consumers and decomposers.
- Productivity diminishes at each subsequent trophic level in the ecosystem.
- In most ecosystems, NPP is limited by conditions that limit the rate of photosynthesis: temperature and/or the availability of sunlight, water, and nutrients.

✔**You should be able to . . .**

1. Explain why net productivity diminishes with increasing trophic levels.

2. Predict the effect of global warming on NPP.

Answers are available in Appendix A.

Chapter 54), the demand for NPP is likely to reach the physical limit of what can be appropriated.

Humans are affecting the flow of energy through ecosystems as well as the cycling of nutrients within and among ecosystems. Let's take a closer look at nutrient cycling.

Energy is transferred when one organism eats another. But tissues also contain carbon (C), nitrogen (N), phosphorus (P), sulfur (S), calcium (Ca), and other elements that act as nutrients and are essential for normal metabolism, growth, and reproduction. The transfer of both energy and nutrients links the biotic and abiotic components of ecosystems.

Atoms are constantly reused as they move through trophic levels, but they also spend time suspended in air, dissolved in water, or held in soil. The path an element takes as it moves from abiotic systems through producers, consumers, and decomposers and back again is referred to as its **biogeochemical** ("life-Earth-chemical") **cycle.** Because humans are disturbing biogeochemical cycles on a massive scale, research on the topic is exploding.

Nutrient Cycling within Ecosystems

To understand the basic features of a biogeochemical cycle, study the generalized terrestrial nutrient cycle in **FIGURE 56.11**. Start with the "Uptake" arrow on the bottom left, which indicates that nutrients such as nitrogen and phosphorus are taken up from the soil by plants and assimilated into plant tissue; carbon enters primary producers as carbon dioxide from the atmosphere. Note that if the plant tissue is eaten, the nutrients pass to the consumer food web; if the plant tissue dies, the nutrients enter the region labeled "Detritus," which contains wastes and dead organisms.

What happens as detritus decomposes? Nutrients that reside in plant litter, animal excretions, and dead animal bodies are used by bacteria, archaea, roundworms, fungi, and other primary decomposers. These microscopic decomposers and the carbon-containing compounds that they release combine to form what biologists call **soil organic matter**, a complex mixture of partially and completely decomposed detritus. When the detritus becomes completely decayed, it is called **humus** because it is rich in a family of carbon-containing molecules called humic acids. (Chapter 39 described other components of the soil.)

Eventually, decomposition converts the nutrients in soil organic matter to an inorganic form. For example, cellular respiration by soil-dwelling bacteria and archaea converts the nitrogen present in amino acids to ammonium (NH_4^+) or nitrate (NO_3^-) ions, and the carbon to carbon dioxide. Once decomposition is accomplished, the nutrients are available for uptake by plants, completing the cycle of nutrient flow through ecosystems.

What Factors Control the Rate of Nutrient Cycling? Of the many links in a nutrient cycle, the decomposition of detritus most

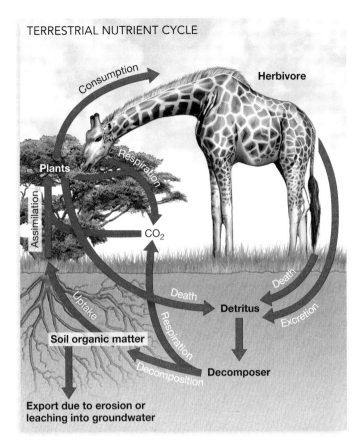

TERRESTRIAL NUTRIENT CYCLE

FIGURE 56.11 Generalized Terrestrial Nutrient Cycle. Nutrients cycle from organism to organism in an ecosystem via assimilation, consumption, and decomposition. Nutrients are exported from ecosystems when water or organisms leave the area.

(a) Boreal forest: Accumulation of detritus and organic matter

(b) Tropical wet forest: Almost no organic accumulation

FIGURE 56.12 Temperature and Moisture Affect Decomposition Rates.

often limits the overall rate at which nutrients move through an ecosystem. Until decomposition occurs, nutrients stay tied up in intact tissues. The decomposition rate, in turn, is influenced by three types of factors:

1. abiotic conditions such as oxygen availability, temperature, and precipitation;

2. the quality of the detritus as a nutrient source for the fungi, bacteria, and archaea that accomplish decomposition; and

3. the abundance and diversity of detritivores present.

To appreciate the importance of abiotic conditions on the decomposition rate, consider the difference in detritus accumulation between boreal forests and tropical wet forests—two moist biomes that differ in temperature. Boreal forests are cool, so the rate of decomposition is slow. Tropical wet forests are warm all year long, so the rate of decomposition is high. **FIGURE 56.12** illustrates typical soils from boreal forests and tropical wet forests. Notice that the uppermost part of the soil in a boreal forest consists of a thick layer of partially decomposed detritus and organic matter. In a tropical forest, this layer is virtually absent.

The contrast occurs because the cold conditions in boreal forests limit the metabolic rates of decomposers. As a result, decomposition fails to keep up with the input of detritus, and organic

matter accumulates. In the tropics, conditions are so favorable for fungi, bacteria, and archaea that decomposition keeps pace with inputs. Nutrients cycle slowly through boreal forests but rapidly through wet tropical forests. As a result, most of the biomass in a tropical forest is living. ✔ If you understand this concept, you should be able to explain why farms created by clearing tropical rain forests tend to perform poorly.

The quality of detritus also exerts a powerful influence on the growth of decomposers, and thus the decomposition rate. Decomposition is inhibited if detritus is **(1)** low in nitrogen, or **(2)** high in lignin. When nitrogen is scarce, such as in acidic bogs, cells have a difficult time making additional proteins and nucleic acids. Lignin—an important constituent of wood—is extremely difficult to digest (see Chapter 32). The presence of lignin is a major reason why wood takes much longer to decompose than leaves do. Only basidiomycete fungi and a few bacteria have the enzymes required to completely degrade lignin.

The presence of oxygen is also critically important. If soils or water become depleted of oxygen, such as in some wetlands, decomposition slows dramatically. Fungi can perform cellular respiration only by using oxygen as the final electron acceptor (see Chapter 9). And even though many bacteria and archaea can thrive in the absence of oxygen and continue decomposition, the growth rates of anaerobic species are a tiny fraction of rates in aerobic bacteria and archaea (to understand why, see Chapter 29).

Sources of Local Nutrient Export and Import Although nutrients are constantly cycled within individual ecosystems, they can also be lost from one ecosystem and exported to another.

• If an herbivore eats a plant and moves out of the ecosystem before excreting the nutrients or dying, the nutrients are exported.

- Nutrients leave ecosystems when flowing water or wind removes particles or inorganic ions and deposits them somewhere else.

Several of the major impacts that humans have on ecosystems—such as farming, logging, burning, and soil erosion—accelerate nutrient export. Farming and logging remove biomass in the form of plants; burning releases nutrients to the atmosphere; soil erosion moves nutrients that are bound to soil particles or dissolved in water. Further, these processes are often interrelated, as when logging leads to soil erosion and runoff.

For an ecosystem to function normally, nutrients that are exported must be replaced by imports. There are three major mechanisms to replace nutrients:

1. Ions that act as nutrients are released as rocks weather (see Chapter 39).

2. Nutrients can blow in on soil particles or arrive dissolved in streams.

3. Nitrogen—a particularly important nutrient in maintaining NPP—is added when nitrogen-fixing bacteria (introduced in Chapter 39) convert molecular nitrogen (N_2) in the atmosphere to usable nitrogen in ammonium or nitrate ions.

In ecosystems that humans manage, nutrients usually have to be actively replaced to maintain productivity. People accomplish this by adding fertilizers or planting nitrogen-fixing crops.

An Experimental Study The ideas presented thus far are logical and provide a strong conceptual foundation for thinking about nutrient cycles. But to explore nutrient movement in more depth, researchers have turned to experiments.

One of the first large-scale experiments on nutrient cycling took place at the Hubbard Brook Experimental Forest, introduced earlier in the chapter. A group of researchers led by Likens and Bormann began by choosing two similar **watersheds**—areas drained by a single stream—for study. Before starting the experiment, they documented that 90 percent of the nutrients in the watersheds was in soil organic matter; 9.5 percent was in plant biomass. The nutrients in the soil were either dissolved in water or attached to small particles of sand or clay (see Chapter 39).

To begin the experimental treatment, the researchers cut all vegetation, including the trees, in one of the two watersheds (**FIGURE 56.13**). In each of the following three years, the clear-cut watershed was treated with an herbicide to prevent vegetation from regrowing. As a result, the experimental watershed was devegetated. Researchers left the other watershed untreated to serve as a control.

The graph in Figure 56.13 documents the amount of total dissolved substances that washed out of the streams in each watershed over the course of the four years of devegetation. Losses from the devegetated site were typically over 10 times as high as they were from the control site.

Why? Plant roots were no longer able to hold soil particles in place, so some of the soil itself washed away. Further, plant roots

RESEARCH

QUESTION: How does the presence of vegetation affect the rate of nutrient export in a temperate-forest ecosystem?

HYPOTHESIS: Presence of vegetation lowers the rate of nutrient export because it increases soil stability and recycling of nutrients.

NULL HYPOTHESIS: Presence of vegetation has no effect on the rate of nutrient export.

EXPERIMENTAL SETUP:

1. Choose two similar watersheds. Document nutrient levels in soil organic matter, plants, and streams.

2. Devegetate one watershed, and leave the other intact.

3. Monitor the amount of dissolved substances in streams.

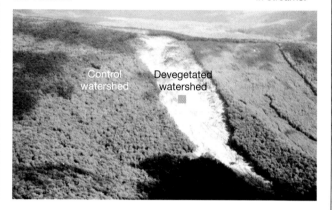

PREDICTION: Amount of dissolved substances in stream in devegetated watershed will be much higher than amount of dissolved substances in stream in control watershed.

PREDICTION OF NULL HYPOTHESIS: No difference will be observed in amount of dissolved substances in the two streams.

RESULTS:

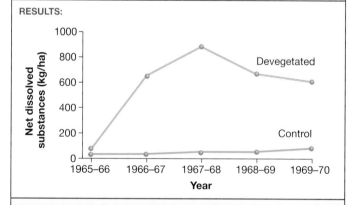

CONCLUSION: Presence of vegetation limits nutrient export. Removing vegetation leads to large increases in nutrient export.

FIGURE 56.13 Experimental Evidence that Deforestation Increases the Rate of Nutrient Export from Ecosystems.

SOURCE: Likens, G. E., F. H. Bormann, N. M. Johnson, et al. 1970. Effects of forest cutting and herbicide treatment on nutrient budgets in the Hubbard Brook watershed-ecosystem. *Ecological Monographs* 40: 23–47.

✔ **QUESTION** Propose a hypothesis to explain why the level of nutrient export from the experimental site appears to peak and then decline.

no longer took up and recycled nutrients that were dissolved in the remaining soil water. Instead, massive quantities of nitrate and other nutrients washed out of the soil and were exported from the ecosystem in runoff water.

Long-term devegetation of this type has occurred in formerly forested areas of the Middle East, North Africa, and elsewhere, due to intensive farming and grazing. Nutrients that used to occur within those soils—and the soils themselves—were exported. Today, the productivity of these regions is a tiny fraction of what it once was. Once soil is lost, it is difficult to regain—soil takes a very long time to form (Chapter 39).

Global Biogeochemical Cycles

When nutrients leave one ecosystem, they enter another. In this way, the movement of ions and molecules among ecosystems links local biogeochemical cycles into one massive global system. As an introduction to how these global biogeochemical cycles work, let's consider the global water, nitrogen, and carbon cycles. These cycles have recently been heavily modified by human activities—with serious ecological consequences.

To understand how biogeochemical cycling works on a global scale, researchers focus on three fundamental questions:

1. What are the nature and size of the reservoirs—areas or "compartments" where elements are stored for a period of time? In the case of carbon, the biomass of living organisms is an important reservoir, as are sediments and soils. Another significant carbon reservoir is buried in the form of coal and oil.

2. How fast does the element move between reservoirs, and what processes are responsible for moving elements from one reservoir to another? The global photosynthetic rate, for example, measures the rate of carbon flow from carbon dioxide (CO_2) in the atmosphere to living biomass. When fossil fuels burn, carbon that was buried in coal or petroleum moves into the atmosphere as CO_2.

3. How does one biogeochemical cycle interact with another biogeochemical cycle? For example, researchers are trying to understand how changes in the nitrogen cycle affect the carbon cycle.

The Global Water Cycle A simplified version of the **global water cycle** appears in **FIGURE 56.14**. The diagram shows the estimated amount of water that moves between major reservoirs over the course of a year, powered by energy from the sun.

To analyze this cycle, begin at the left with evaporation of water out of the ocean, and the subsequent precipitation of water back into the ocean. For the marine component of the cycle, evaporation exceeds precipitation—meaning that, over the oceans, there is a net gain of water to the atmosphere.

When this water vapor moves over the continents, it is joined by a small amount of water that evaporates from lakes and streams and a large volume of water that is transpired (lost) by plants (see Chapter 38). The total volume of water in the atmosphere over land is balanced by the amount of rain and other forms of precipitation that falls on the continents.

The cycle is completed by the water that moves from the land back to the oceans via streams and **groundwater**—water that is found underground. Overall, there is no net gain or loss of water.

Note that 97.5 percent of the Earth's water is salt water (in the ocean). Most of the remaining 2.5 percent of freshwater is locked up in the ice caps and glaciers, leaving about 0.5 percent available for drinking in lakes and rivers, and in groundwater. Most

FIGURE 56.14 The Global Water Cycle.

DATA: Oki, T., and S. Kanae. 2006. *Science* 313: 1068–1072.

✔**QUESTION** Predict how the amount of water evaporated from the ocean is changing in response to global warming. Discuss one possible consequence of this change.

groundwater is stored in **aquifers**—layers of porous rock, sand, or gravel that are saturated with water. Aquifers are said to be closed (contained) when nonporous rock layers overlie them, and open (uncontained) when they can be replenished by water percolating down from the surface (Figure 56.14). Closed aquifers may take thousands of years to recharge.

What effects are humans having on the water cycle? Perhaps the simplest and most direct impacts concern rates of groundwater depletion.

- Asphalt and concrete surfaces reduce the amount of water that percolates from the surface to deep soil and rock layers.

- When grasslands and forests are converted to agricultural fields, root systems that hold water are lost. More water runs off into streams and less percolates into groundwater.

- Irrigated agriculture, along with industrial and household use, is removing massive amounts of water from groundwater storage and bringing it to the surface. The Ogallala–High Plains Aquifer is the largest underground reservoir of freshwater in the United States, underlying 430,000 square kilometers (176,700 square miles) of the Midwest from South Dakota to Texas, and supplying 30 percent of the irrigation water for the country. Billions of gallons are withdrawn each day. The water in this aquifer is estimated to be 10,000 years old, and it is being depleted much faster than it is being recharged (**FIGURE 56.15**), raising concerns about the future supply. Most of the aquifer is closed; only the areas shown in blue have recharged.

The **water table**—the level where soil is saturated with stored water—is dropping on every continent. For example, between 1986 and 2006, the water table north of Beijing, China, experienced a drop of 61 m (200 ft). Throughout India, water tables are falling at a rate of between 1 m and 3 m per year. Similar rates are being documented in Yemen, parts of Mexico, and states in the southern Great Plains region of the United States.

Humans are mining freshwater from aquifers in many parts of the world—meaning that populations have already grown beyond carrying capacity for water. About 40 percent of the world's population has insufficient water resources. Water scarcity tends to lead to strife within and among populations.

The Global Nitrogen Cycle **FIGURE 56.16** illustrates the **global nitrogen cycle.** To follow the cycle, start at the top in the atmosphere. This represents the main reservoir for nitrogen in the form of nitrogen gas (N_2)—a biologically inert molecule that comprises about 78 percent of the air you breathe. Nitrogen is usable by organisms only when it is reduced, or "fixed," meaning that it is converted from unusable atmospheric N_2 to usable ammonium (NH_4^+) or nitrate ions (NO_3^-; see Chapter 39).

The fixation of nitrogen is usually a slow process, so nitrogen often limits primary productivity in terrestrial and aquatic ecosystems. Reduced nitrogen builds up slowly in soils and water over time and then cycles relatively rapidly among soil, microorganisms, fungi, plants, and animals within both terrestrial and aquatic ecosystems. Ultimately, bacteria digest nitrogen-containing molecules and return N_2 to the atmosphere, completing the cycle.

FIGURE 56.15 Declining Water Levels in the Ogallala–High Plains Aquifer from 1950 to 2007.
DATA: U.S. Geological Survey.

Nitrogen fixation occurs naturally due to lightning-driven reactions in the atmosphere and from enzyme-catalyzed reactions in bacteria that live in the soil and oceans. However, the nitrogen cycle has been profoundly altered by human activities. The amount of nitrogen fixation from human sources almost equals the amount of nitrogen fixation from natural sources. As the labels on the right side of the graph in **FIGURE 56.17** indicate, there are three major sources of this human-fixed nitrogen:

1. industrially produced fertilizers;

2. the cultivation of crops, such as soybeans and peas that harbor nitrogen-fixing bacteria; and

3. the burning of fossil fuels, which releases nitric oxide.

Adding nitrogen to terrestrial ecosystems usually increases productivity. But overfertilization with nitrogen has important consequences.

- Nitrogen-laced runoff from agricultural fields causes algae blooms in aquatic ecosystems, such as the lake shown in **FIGURE 56.18**. When the algae die and are decomposed by microbes, oxygen is depleted from the water, causing the formation of oxygen-free "dead zones" (see Chapter 29).

- When nitrogen is added to experimental plots in grasslands in midwestern North America, a few competitively dominant

THE GLOBAL NITROGEN CYCLE

All estimated values in teragrams (10^{12} grams) of nitrogen per year

Major reservoir:
Atmospheric nitrogen (N_2)
(78% of atmosphere)

Nitrogen-fixing cyanobacteria: 140

Bacteria in mud use N-containing molecules as energy sources, excrete N_2: 355

Reservoir: terrestrial organisms: 1200

Industrial fixation: 125

Lightning and rain: 5

Nitrogen-fixing bacteria in roots and soil: 145

Reservoir: Aquatic organisms: 8000

Runoff: 80

Decomposition   Assimilation

Ammonia (NH_3), ammonium (NH_4^+), nitrate (NO_3^-)

Permanent burial: 25

FIGURE 56.16 The Global Nitrogen Cycle. Nitrogen enters ecosystems as ammonia or nitrate via fixation from atmospheric nitrogen. It is exported in runoff and as nitrogen gas given off by bacteria that use nitrogen-containing compounds as an electron acceptor.

DATA: Gruber, N., and J. N. Galloway. 2008. *Nature* 451: 293–296.

species grow rapidly and displace other species that do not respond to nitrogen inputs as strongly. Productivity increases, but species diversity decreases (Chapter 57).

The Global Carbon Cycle The **global carbon cycle** documents the movement of carbon among terrestrial ecosystems, the oceans, and the atmosphere (**FIGURE 56.19**, see page 1162). Of these three reservoirs, the ocean is by far the largest. The atmospheric reservoir is also important despite its relatively small size, because carbon moves into and out of it rapidly—through organisms.

In both terrestrial and aquatic ecosystems, photosynthesis is responsible for taking carbon out of the atmosphere and incorporating it into tissue. Cellular respiration, in contrast, releases carbon that has been incorporated into living organisms to the atmosphere, in the form of carbon dioxide.

How have humans changed the carbon cycle? For thousands of years, humans have altered the global carbon cycle by deforesting land for agriculture and settlements—releasing the carbon stored in trees to the atmosphere.

But the human influence on the carbon cycle increased sharply during the industrial revolution. Burning fossil fuels moves massive amounts of carbon from an inactive geological reservoir, in the form of petroleum or coal (the remains of ancient organisms), to an active reservoir—the atmosphere. When you burn gasoline, you are releasing carbon atoms that have been locked up in petroleum reservoirs for hundreds of millions of years.

Sources of nitrogen fixation

Natural sources: 55%

Lightning

Biological fixation by bacteria

Human sources: 45%

Nitrogen fertilizers

Nitrogen-fixing crops

Burning of fossil fuels

FIGURE 56.17 Humans Are Adding Nitrogen to Ecosystems.

DATA: Canfield, D. E., A. N. Glazer, and P. G. Falkowski. 2010. *Science* 330: 192–196.

FIGURE 56.18 Nutrient Runoff Causes Harmful Algae Blooms.
Blooms, in turn, can lead to oxygen-starved dead zones.

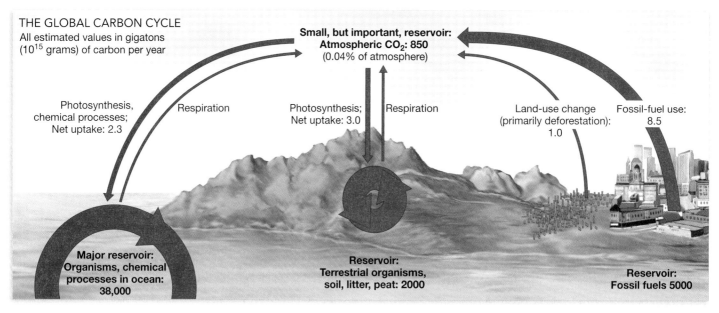

THE GLOBAL CARBON CYCLE
All estimated values in gigatons
(10^{15} grams) of carbon per year

Small, but important, reservoir:
Atmospheric CO_2: 850
(0.04% of atmosphere)

Photosynthesis,
chemical processes;
Net uptake: 2.3

Respiration

Photosynthesis;
Net uptake: 3.0

Respiration

Land-use change
(primarily deforestation):
1.0

Fossil-fuel use:
8.5

Major reservoir:
Organisms, chemical
processes in ocean:
38,000

Reservoir:
Terrestrial organisms,
soil, litter, peat: 2000

Reservoir:
Fossil fuels 5000

FIGURE 56.19 The Global Carbon Cycle. The arrows indicate how carbon moves into and out of ecosystems. Photosynthesis and cellular respiration move carbon in and out of the atmosphere in the form of CO_2. The burning of fossil fuels moves carbon from an inactive geological reservoir, in the form of petroleum or coal, to an active reservoir, the atmosphere.

DATA: Hönisch, B., A. Ridgwell, D. N. Schmidt, et al. 2012. *Science* 335: 1058–1063.

FIGURE 56.20a highlights the dramatic increase in carbon released from fossil-fuel burning over the past century; **FIGURE 56.20b** shows corresponding changes in CO_2 concentrations. Note that CO_2 levels in the atmosphere have now risen to 400 parts per million (ppm), far above the 280 ppm that was typical before 1860. Analysis of CO_2 levels in air bubbles trapped in Antarctic ice reveals that CO_2 fluctuated within a range of about 170 to 300 ppm for the extent of the record—800,000 years. This means that current CO_2 levels are 30 percent higher than the highest level measured over the last 800,000 years.

Why is the increase in atmospheric CO_2 so important? Because it changes the climate.

(a) Increases in fossil-fuel use

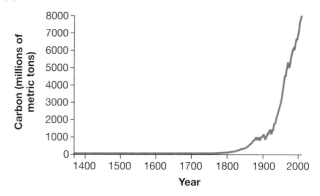

(b) Global changes in atmospheric CO_2 over time

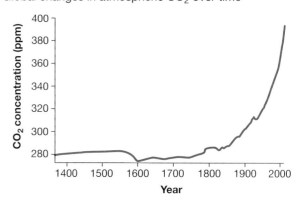

FIGURE 56.20 Humans Are Causing Increases in Atmospheric Carbon Dioxide. (a) Rates of carbon flow from fossil-fuel burning have increased as human populations have increased. **(b)** For centuries, average CO_2 concentrations in the atmosphere were about 280 parts per million (ppm).

DATA: Carbon Dioxide Information Analysis Center.

check your understanding

C Y U

If you understand that . . .
- Biogeochemical cycles trace the movement of water and elements within and among ecosystems.
- Biologists analyze three aspects of these cycles: (1) the nature and size of reservoirs where elements are stored or captured, (2) the processes responsible for moving elements among reservoirs, and (3) interactions among cycles.

✔**You should be able to . . .**

Compare how human activities are altering the global water, nitrogen, and carbon cycles.

Answers are available in Appendix A.

56.3 Global Climate Change

If you are in your late teens or early twenties as you read this text, you are part of a generation that will experience the most dramatic episode of environmental change in human history. The epic event has three main interacting drivers: changes in land use (Chapters 52 and 57), the massive loss of species (see Chapter 57), and global climate change, also called global warming.

Global warming and global climate change refer to the same phenomenon, but with different emphases:

- **Global warming** refers to the increase in the *average* temperature of the planet. The average temperature is very important in determining physical, chemical, and biological processes.

- **Global climate change** refers to the sum of all the changes in local temperature and precipitation patterns that result from global warming—including the frequency and intensity of storms, droughts, and other events. The effects of global warming vary depending on geography, local weather patterns, ocean currents, and other factors. But averaged over the entire planet and over time, the Earth is already much warmer than it was just a few decades ago and is projected to get even warmer.

Why is this warming such a problem? Let's explore the underlying cause of global climate change and then consider the consequences.

What Is the Cause of Global Climate Change?

Climate change is normal. Throughout Earth's history, the average temperature of the atmosphere and local weather patterns have fluctuated. Scientists today are not alarmed because there is change; they are alarmed because the *rate* of change is unprecedented, and because it is caused by human activities. What is going on?

The Greenhouse Effect By the 1950s, scientists were aware that humans were releasing tremendous amounts of CO_2 into the atmosphere. The common reassuring argument at the time was that the vast ocean would be able to absorb this excess CO_2.

However, Roger Revelle, a pioneer in oceanography, and his colleague Hans Suess reported in 1957 that the complex chemistry of the ocean limits how much CO_2 it can absorb. Their data supported a stunning hypothesis posed by the eminent chemist and physicist Svante Arrhenius half a century earlier: The atmospheric CO_2 would rise, which would lead to global warming. Why? Carbon dioxide is a **greenhouse gas:** It traps heat that has been radiated from Earth and keeps it from being lost to space, similar to the way the glass of a greenhouse traps heat (**FIGURE 56.21**).

How could the global warming hypothesis be tested? At the time of Revelle and Suess's research, there were no reliable records of baseline atmospheric CO_2 levels—local CO_2 measurements varied widely due to fluctuating emissions from cities and other variables. So Revelle hired a young geochemist, Charles David

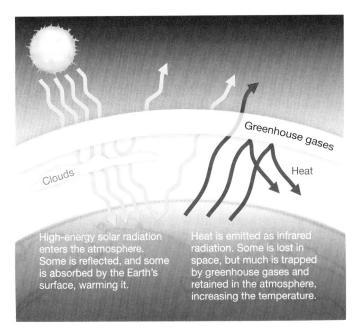

FIGURE 56.21 The Greenhouse Effect. Greenhouse gases in the atmosphere have an effect that is similar to glass in a greenhouse— high-energy radiation in the form of light can easily pass through the gases, but low-energy infrared radiation gets trapped by them.

Keeling, to gather precise measurements of CO_2 concentrations in remote, pristine locations, one of which was the Mauna Loa Observatory in Hawaii. By 1961, Keeling was able to confirm that global CO_2 was rising. Despite many obstacles, he persevered in collecting invaluable records for more than 40 years.

FIGURE 56.22 (see page 1164) shows CO_2 data from Mauna Loa over the last three decades, along with temperature and solar energy data.

- There is no indication that the rate of increase of CO_2 is slowing, despite an array of regulations and incentives to reduce CO_2 emissions.

- The temperature fluctuates, but there is a clear increase in average temperature.

- Even though the amount of solar energy cycles regularly, there is no net increase over time. Thus, fluctuations in solar energy are not sufficient to account for the marked increase in temperature, as one alternative hypothesis had predicted.

Carbon dioxide is only one of several important greenhouse gases in the atmosphere, including methane (CH_4) and nitrous oxide (N_2O). All these gases collectively increase the atmosphere's heat-trapping potential—causing global warming.

Why Is the Climate Changing So Rapidly? The rate of increase of atmospheric CO_2 and temperature are alarming. Why so fast? There are two main reasons. First, the human population has exploded in size. The planet now has over 7 billion people, a 700 percent increase over a mere two centuries—a blink of an eye in Earth's history (see Chapter 54). But an equally important part of the answer involves resource use. The average per capita (per

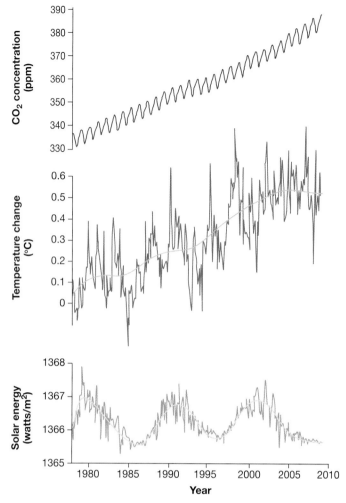

FIGURE 56.22 Recent Changes in Atmospheric CO$_2$, Temperature, and Solar Radiation.

DATA: NOAA; Karl, T. R., J. M. Melillo, T. C. Peterson, and S. J. Hassol (eds.). 2009. *Global Climate Change Impacts in the United States.* Cambridge University Press.

✔**QUESTION** Notice the annual cycles in CO$_2$. Why are atmospheric CO$_2$ concentrations low in the Northern Hemisphere in summer and high in winter?

person) use of fossil fuels is skyrocketing—especially in industrialized countries.

The data in **FIGURE 56.23** illustrate the consequences of population size and resource use. By multiplying the total population size of a country (left graph) by the quantity of CO$_2$ emissions per person in that country (center graph), the total emissions per country can be calculated (right graph). For example, even though the average person in China causes only a quarter of the CO$_2$ emissions that an average person from the United States does, the total emissions from China surpass all others due to its enormous population size.

Conversely, although the United States represents less than 5 percent of the world population, Americans produce an astonishing one-fifth of the global CO$_2$ emissions due to their heavy use of fossil fuels. Looking at the data in another way, the population of the United States was twice that of Nigeria in 2008, but

the average American produced 22 times as much CO$_2$ as the average Nigerian that year.

The bottom line is that the recent exponential increase of the human population, plus the steep increase in per capita fossil fuel use—as well as land-use changes such as deforestation—have caused a steep increase in the emission of CO$_2$ and other greenhouse gasses. These changes, in turn, have caused the recent and dramatic increase in the average temperatures and climate variability around the globe. The rapid rate of climate change is ultimately rooted in human behavior.

How Much Will the Climate Change?

In 1988 an international group composed of hundreds of scientists, called the Intergovernmental Panel on Climate Change (IPCC), was formed to evaluate the consequences of well-documented increases in CO$_2$ and other heat-trapping gases to climate. The group has since produced a series of reports summarizing the state of scientific knowledge on the issue. How has climate already changed, and what changes can we expect in the future?

Temperature The latest IPCC report concludes that the average global temperature increased 0.74°C (1.3°F) during the 20th century, which was double the increase of the 19th century. More recently, four independent records of temperature summarized by NASA concur that the last decade has been the warmest in recorded history.

Predicting the future state of a system as complex and variable as Earth's climate is extremely difficult. Global climate models are the primary tool that scientists use to make these projections. A global climate model is based on a large series of equations that describe how the concentrations of various gases in the atmosphere, solar radiation, transpiration rates, and other parameters interact to affect climate.

The models currently being used suggest that average global temperature will undergo additional increases of 1.1–6.4°C (2.0–11.5°F) by the year 2100. The low number is based on models that assume no further increase in greenhouse gas emissions over present levels; the high number is based on models that assume continued intensive use of fossil fuels and increases in greenhouse gas emissions. To get a sense of what these numbers mean, consider the data in **FIGURE 56.24**. Under a higher-emissions scenario, parts of the southern United States that currently experience about 60 days per year of temperatures above 90°F are expected to experience 150 or more of such days by the end of the century.

Predicting how temperature will change in the future is particularly difficult because ecosystems respond to warming in ways that increase or decrease the concentrations of CO$_2$ and other greenhouse gases and that affect other drivers of global climate, such as the size of the polar ice caps. In essence, warming has a domino effect on other factors that may either further accelerate or reduce warming.

Documenting Positive and Negative Feedback Positive feedback in climate change occurs when changes due to global warming

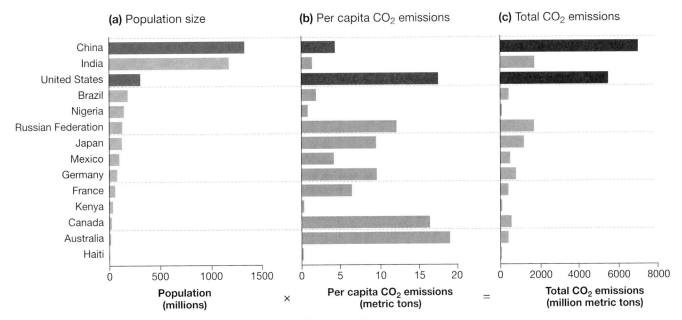

(a) Population size **(b)** Per capita CO_2 emissions **(c)** Total CO_2 emissions

Population (millions) × Per capita CO_2 emissions (metric tons) = Total CO_2 emissions (million metric tons)

FIGURE 56.23 Total Emissions Depend on Population Size and Resource Use.
These data are from 2008. The bars for China and the United States are emphasized for comparison.

DATA: Carbon Dioxide Information Analysis Center (http://cdiac.ornl.gov).

✔**QUANTITATIVE** Estimate what the total emissions in China will be if the average release of CO_2 per person doubles due to increased use of fossil fuels.

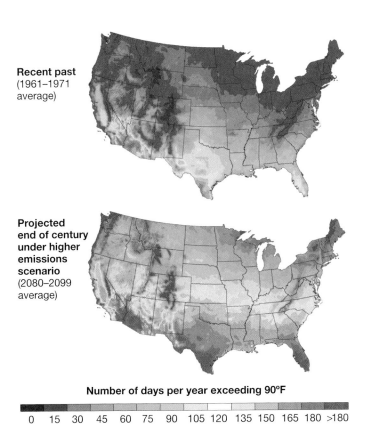

Recent past (1961–1971 average)

Projected end of century under higher emissions scenario (2080–2099 average)

Number of days per year exceeding 90°F

0 15 30 45 60 75 90 105 120 135 150 165 180 >180

FIGURE 56.24 More Extreme Temperatures and Continued High Emissions Are Expected in the Future.

DATA: Karl, T. R., J. M. Melillo, and T.C. Peterson (eds.). 2009. *Global Climate Change Impacts in the United States*. New York: Cambridge University Press.

result in the further acceleration of warming. Many types of positive feedback have been documented. For example:

- Warmer and drier climate conditions have increased the frequency of forest fires in the Rocky Mountains of North America as well as in the boreal forests of Canada. Forest fires release CO_2, which leads to even more warming.

- Traditionally, tundra sequesters carbon in the form of soil organic matter, because decomposition rates are extremely low. During a series of warm summers, however, researchers found that decomposition rates in tundra increased sufficiently to release carbon from stored soil organic matter and to produce a net flow of carbon to the atmosphere.

- Other studies of soil microbes observed that elevated CO_2 levels triggered release of CH_4 and N_2O, greenhouse gases that are less abundant than CO_2 but have a more potent effect on warming.

- Melting of the polar ice caps exposes more open water to the sun. The darker open water absorbs more solar radiation than ice, increasing the rate of warming there.

Negative feedback in climate change occurs when changes due to global warming result in increased uptake and sequestration of CO_2 and other greenhouse gases—meaning that global warming should be reduced. For example, recent experiments have shown that the growth rates of many plant species, including some agricultural crops (but not others), increase in direct response to increasing atmospheric CO_2, thus acting to reduce atmospheric CO_2.

Will ecosystem responses increase or decrease global warming? While the specific predictions of different models vary, the overall trend is the same—it's going to keep getting hotter.

Climate Factors Other than Temperature

Increasing temperatures affect other aspects of climate, and elevated levels of greenhouse gases themselves affect climate, with various negative and positive feedback interactions.

- Temperatures affect the water cycle, changing where, when, and how much water falls as precipitation. Average precipitation has increased about 5 percent in the United States during the past 50 years. Models predict that northern areas may become wetter in the future, but the Southwest is likely to become drier.

- Temperature and precipitation are becoming more variable, and the severity of extreme weather events such as monsoon rains, tornadoes, and hurricanes is increasing, in part because warmer air can hold more moisture (Chapter 52).

- Warmer temperatures have caused the sea level to rise due to expansion of the oceans and the melting of glaciers and polar ice caps.

- The melting of the ice caps affects other aspects of climate such as ocean currents and cloud cover.

Climate is an abiotic component of ecosystems. How does climate change affect organisms?

Consequences to Organisms

Even though global temperatures have risen only slightly in comparison with projections for the next 50–100 years, biologists have already documented dramatic impacts on organisms. The main types of consequences of global climate change to organisms are summarized in **TABLE 56.1**.

Phenology Shifts

The timing of seasonal events, or **phenology,** is changing in many biomes. Spring arrives in the United States 10 days to two weeks earlier than it did just 20 years ago. As a result, many biological activities associated with spring occur earlier. Recent studies have confirmed that migratory bird species are arriving on their breeding grounds in the northeastern United States an average of 13 days earlier than in previous decades, and fruit trees are ripening as much as 18 days earlier.

Note, however, that different species in an ecosystem do not necessarily synchronize changes in their timing. Mismatches can result, disrupting trophic interactions and other types of species interactions (Chapter 55). For example:

- Insectivorous birds like great tits (*Parus major*) rely on the abundance of caterpillars to feed their hatchlings. However, a 20-year study has shown that the peak caterpillar biomass now occurs earlier due to the warmer spring, but the reproductive timing of great tits has not changed. Thus, there is a mismatch that has resulted in fewer and lower-weight great tit fledglings.

- Other researchers observed interactions in a community of plankton in the North Sea for 40 years. Plankton at different trophic levels responded to climate change differently over this period, causing mismatches in trophic relationships and thus a restructuring of the food web.

- Mismatches in interactions due to phenology are not always trophic. The mutualistic interaction between flowers and pollinators is disrupted when the timing of flowering and the emergence of insect pollinators becomes asynchronous.

Geographic Range Shifts

The geographic ranges of many organisms are shifting as their "climate envelopes" of suitable habitat move toward the poles, toward higher altitudes, or toward areas of adequate precipitation.

- Until recently, the range of mountain pine beetles was limited in North America by freezing temperatures. Milder temperatures have triggered massive infestations of beetles in areas where they had not been abundant before (Table 56.1). Researchers fear that an ongoing expansion of these voracious beetles northward and eastward in Canada will threaten the boreal forest ecosystem.

- Copepods are small crustaceans that serve as key primary consumers in marine plankton. Copepods have been studied intensively because they are important commercially: They are eaten by fish that are in turn eaten by other fish that are eaten by humans. A 40-year study in the North Atlantic showed that a southern species of copepod has moved north, while the range of cold-water species has declined.

- When you buy a package of seeds or a potted plant, the planting instructions often indicate which hardiness zones are appropriate for its growth. The U.S. Department of Agriculture released an updated map of hardiness zones in 2012 based on recent temperature and precipitation data. The zones have shifted northward—signifying a range shift of many species.

As with changes in phenology, shifts in the geographic distribution of species can lead to mismatches between species and resources. Mismatches can act as an agent of natural selection and cause evolution, or they can result in the extinction of species.

Evolutionary Adaptation

In some populations, changing climate and species interactions are already causing allele frequencies to change—meaning that some species are evolving in response to global climate change. There are clearly both evolutionary winners and losers, depending on the ecological context and adaptive capacity of the organisms.

For example, ecologists Paul Brakefield and Peter de Jong recently concluded that increasing temperatures in the Netherlands have caused evolution in the two-spot ladybird beetle, *Adalia bipunctata*. There are two heritable color patterns of these beetles: black with red spots and red with black spots (Table 56.1). The frequency of the black morph has declined from 60 percent to 20 in one area in just 30 years. It turns out that the black morphs have lower fitness under warmer conditions because they absorb more solar energy and overheat.

How predictable are evolutionary adaptations to climate change? Luc de Meester and colleagues conducted artificial selection experiments with the water flea, *Daphnia magna*, at different temperatures to try to measure its adaptive capacity. They observed that isolated clones of *Daphnia* evolved in different ways than *Daphnia* living in the presence of competitors, predators, and parasites. Thus, predictive models must consider the complexity of species interactions when making predictions about adaptive capacity—a formidable task.

If changes occur too quickly relative to the life span of the organism, local extinction is likely to occur, rather than adaptation.

Extinctions Biologists are concerned that rapid changes in climate, coupled with other stressors such as land-use changes, will continue to cause high extinction rates (discussed in detail in Chapter 57).

- Behavioral ecologist Barry Sinervo and his team of colleagues found that 12 percent of local lizard populations have gone extinct in Mexico since 1975 because the temperature is changing too rapidly for the lizard populations to adapt. The team used physiological models to discover the mechanism—since lizards were unable to regulate their body temperatures in the heat, they stayed out of the sun during the day, limiting their ability to search for food and thus lowering their fitness.

- As the climate warms, alpine tundra habitats are disappearing. Some plants and animals that have restricted geographic ranges and grow on mountaintops are "running out of habitat" and going extinct.

- Researchers are concerned that habitats may change so rapidly that long-lived trees with limited dispersal will not be able to keep up—their ranges will not be able to change fast enough.

- Polar bears live on sea ice in the Arctic. As sea ice melts, the polar bears are losing their habitat (Table 56.1). If present trends continue, polar bears may be extinct in the wild by 2100.

- Amphibians are in rapid decline globally due to numerous threats—40 percent of amphibian species face imminent extinction. Local changes to temperature and precipitation not only shrink or eliminate the already narrow ranges of many amphibians but also create conditions that favor the chytrid fungus that is causing widespread amphibian mortality.

- Coral reefs are extremely productive in part because corals house symbiotic algae that perform photosynthesis. Warmer water temperatures cause some reef-building corals to expel their photosynthetic algae. When this "coral bleaching" continues due to sustained exposure to warm water, corals begin to die of starvation.

Effects of Ocean Acidification Although the ocean's capacity to absorb CO_2 is not limitless, it is still vast—about one-third of global CO_2 emissions per year is absorbed by the oceans. This absorption has complex consequences for organisms.

When atmospheric CO_2 dissolves, it reacts with water to form bicarbonate (H_2CO_3), which then drops a proton to form

SUMMARY TABLE 56.1 **Effects of Climate Change on Organisms**

Type of Effect	Example
Phenology Shifts The change in the timing of seasonal events such as migration, flowering, hatching, and feeding can result in mismatches between species; such as flower and pollinator.	
Geographic Range Shifts The geographic redistribution of species to more favorable climate envelopes can change ecological interactions. Mountain pine beetles are spreading north and east, devastating pine forests.	
Evolutionary Adaptation Change in allele frequencies can increase fitness in new conditions. The red morph of the ladybird beetle is more adaptive than heat-absorbing black morph in warmer environments.	
Extinction Species that are not able to survive and/or reproduce under new physical or biological conditions go extinct. Polar bears rely on sea ice, which is melting.	
Acidification The ocean becomes more acidic as it absorbs CO_2, causing a wide range of consequences, such as the slowing of skeletal growth in corals, and coral bleaching.	

carbonic acid (HCO_3^-). Increased CO_2 levels in the atmosphere are increasing the rate of this reaction, and the protons that are released are lowering the pH of the oceans. ✔ **If you understand this concept, you should be able to predict what happens to the pH of a glass of water when you blow bubbles into it with a straw.** The acidity of seawater has climbed by 30 percent over the past 150 years—it's now more acidic than it has been for 20 million years. How does acidity affect organisms?

Research is showing that increasing acidity has devastating effects on diverse processes, ranging from the ability of larval fish to smell predators to the ability of sperm to swim. But by far the greatest amount of research has gone into understanding the effects of acidity on skeleton growth. Increased acidity reduces the availability of calcium carbonate ions to organisms that build calcium carbonate skeletons—such as corals, mollusks, and echinoderms. The more acidic the water, the slower the skeletons form. In some conditions, skeletons can dissolve altogether.

Coral reefs are the subject of a tremendous amount of research because they are among the most productive and species-rich ecosystems in the world. Reefs are large, undersea structures built by colonies of corals that secrete calcium carbonate skeletons. A reduction in skeleton growth due to acidity, changes in temperature, or a combination of the two (in addition to bleaching caused by these factors) threatens the ability of corals to create habitat and food for other organisms.

Consequences to Ecosystems

The combination of local changes in climate and corresponding effects on organisms result in changes to whole ecosystems. Understanding the complex impacts of climate change on ecosystems—and predicting future changes—are among the most challenging and important tasks of ecologists today.

As the first two sections of this chapter emphasized, NPP is the basis of both energy flow and nutrient cycling in ecosystems. Thus, ecologists are monitoring effects of climate change on NPP closely.

How Is NPP Changing on Land? A study from 1982 to 1999 showed an overall increase in global terrestrial productivity. However, the most recent data set, from 2000 to 2009—the warmest decade on record—revealed a different trend. The green areas of **FIGURE 56.25** indicate increased productivity, while the blue areas indicate decreased productivity. Overall, the global terrestrial NPP declined.

This result does not seem intuitive at first, because an increase in atmospheric CO_2 would be expected to increase rates of photosynthesis. The researchers of this study, Maosheng Zhao and Steven Running, point out that large-scale droughts in the Southern Hemisphere have decreased NPP in some of the world's most productive forests, counteracting the overall increase in NPP in the Northern Hemisphere.

Other studies have supported this conclusion. For example, a drying trend in the Amazon rain forest, combined with clear-cutting and burning, threatens to turn one of the world's largest carbon sinks into a carbon source. These data illustrate the complex interaction of effects of CO_2, temperature, and precipitation on productivity.

How Is NPP Changing in the Oceans? As on land, there is considerable variation in local productivity in the oceans. For example, productivity has dropped dramatically in large areas of the pelagic zone, but it has increased in a number of other regions. Overall, scientists agree that there is a general correlation between surface water temperatures and decreased productivity—NPP is decreasing in the oceans. What's going on?

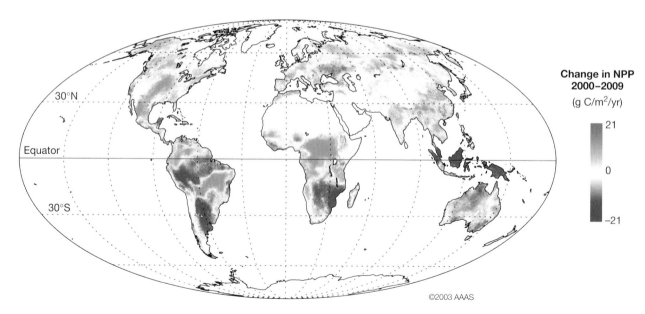

©2003 AAAS

FIGURE 56.25 Recent Changes in Terrestrial NPP. The data shown represent the average annual change in terrestrial NPP from 2000 to 2009.

DATA: Zhao, M., and S. W. Running. 2010. *Science* 329: 940–943.

One of the leading explanations for the drop in marine productivity hinges on changes in water density. To understand the logic behind this hypothesis, recall that lakes can become stratified—that is, have layers of different densities (see Chapter 52). Stratification occurs because water reaches its highest density at 4°C and becomes much less dense at higher temperatures.

Seawater also becomes stratified. This layering is important because the 4°C water in the benthic zone is nutrient rich due to the rain of decomposing organic material from the surface. When warmer surface water becomes lighter, water currents are much less likely to be strong enough to overcome the density difference and bring nutrient-rich, 4°C water all the way up to the surface, where nutrients can spur the growth of photosynthetic bacteria and algae (**FIGURE 56.26**). In this way, the warming global climate is causing surface waters to become more nutrient poor.

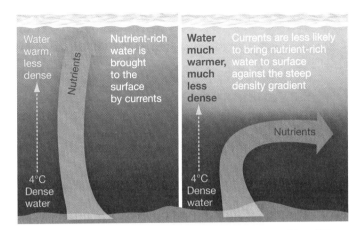

FIGURE 56.26 Temperature Stratification in the Ocean May Affect Productivity.

Local and Global Consequences   What are the consequences of these unprecedented global changes in NPP? There are two main take-home messages:

1. Changes in NPP cause feedback on the rate and pattern of climate change. If terrestrial and marine NPP continue to decline, the rate of absorption of CO_2 will decline, exacerbating climate change with far-reaching effects to humans and other organisms.

2. Changes in NPP combine with other factors, such as changes in precipitation patterns and ocean acidification, to alter food web dynamics—and thus ecosystem function. Maintaining

ecosystem function is critical to the preservation of biodiversity and the ecosystem services that humans rely upon for survival (see Chapter 57).

This chapter began with the theme that everything in nature is connected. As you've seen, humans are very much a part of these connections. Understanding how humans fit into the big picture is key to taking actions to improve ecosystem health and slow global climate change.

CHAPTER 56   REVIEW

For media, go to MasteringBiology (MB)

If you understand . . .

56.1 How Does Energy Flow through Ecosystems?

- Net primary productivity (NPP) is the foundation of ecosystems because it represents the energy, in the form of biomass, available for consumption by other organisms. Humans are reducing and appropriating a large percentage of NPP.

- The feeding relationships among species in a particular ecosystem are described by a food web. Energy is constantly dissipated as it flows from producers to consumers and decomposers in a food web.

- Organisms that acquire energy from the same type of source are said to occupy the same trophic level. Because energy transfer from one trophic level to the next is inefficient, ecosystems have a pyramid of productivity: Biomass production is highest at the lowest trophic level and lower at each higher trophic level.

- Persistent toxins can be magnified in concentration as they are passed up trophic levels in a food web.

- Changes in the species composition of a food web can cause trophic cascades.

- Among terrestrial ecosystems, NPP is highest in tropical wet forests and tropical dry forests. The productivity of terrestrial ecosystems is limited by warmth and moisture.

- Among aquatic ecosystems, productivity is highest in coral reefs, wetlands, and estuaries. Nutrient availability is the key constraint in aquatic ecosystems.

- Although productivity is extremely low in the oceanic zone, the area covered by this ecosystem is so extensive that it accounts for the highest percentage of Earth's overall productivity.

✔ You should be able to explain why it is more energy efficient to eat a pound of tofu (bean curd) than a pound of hamburger.

56.2 How Do Nutrients Cycle through Ecosystems?

- Unlike energy that enters an ecosystem and is ultimately dissipated as heat, nutrients such as carbon, nitrogen, phosphorus, and sulfur move indefinitely through ecosystems in biogeochemical cycles.

- To analyze nutrient cycles, biologists focus on the nature of the reservoirs where elements reside and the processes that move elements between reservoirs.

- The rate of nutrient cycling is strongly affected by the rate of decomposition of detritus.

- Nutrients "leak" from one ecosystem to another when biomass is exported. Experiments have shown that vegetation serves an important role in preventing nutrient export by reducing runoff.

- Humans are harvesting freshwater from reservoirs faster than it can be replenished by natural processes.

- Human activities have doubled the amount of usable nitrogen that enters ecosystems each year, leading to increased productivity and to pollution as well as to loss of biodiversity.

- The burning of fossil fuels by humans releases massive amounts of carbon from an inactive geological reservoir to the atmosphere. Current CO_2 levels are 30 percent higher than the highest level measured over the last 800,000 years.

✔ You should be able to describe conditions under which decomposition rates in terrestrial and marine environments could be extremely low—creating a carbon sink due to a buildup of organic matter.

56.3 Global Climate Change

- Average global temperatures are increasing rapidly because land-use changes and burning of fossil fuels have increased the flow of carbon in the form of CO_2 into the atmosphere. CO_2 is one of the greenhouse gases that trap heat in the atmosphere.

- Both positive and negative feedback are occurring in response to global warming.

- Average temperature is increasing, but local temperatures and precipitation vary. The variability of temperature and rainfall is also increasing. Storm events are becoming more severe.

- Biologists have documented an array of consequences of climate change to organisms, including changes to phenology, geographic ranges, adaptations, extinctions, and acidification.

- Recent data show that the warming associated with climate change is causing a decrease in average NPP both on land and in the oceans, influencing both climate and ecosystem function.

✔ You should be able to propose two mechanisms that humans could adopt to reduce global warming, and explain the logic behind each.

(MB) MasteringBiology

1. MasteringBiology Assignments

Tutorials and Activities Animal Food Production Efficiency and Food Policy; Atmospheric CO_2 and Temperature Changes; Carbon Cycle; Energy Flow and Chemical Cycling; Energy Flow through Ecosystems; What Factors Influence the Loss of Nutrients from a Forest Ecosystem? Food Webs; Global Carbon Cycle; Global Freshwater Resources; Greenhouse Effect; Municipal Solid Waste Trends in the U.S.; Nitrogen Cycle; Prospects for Renewable Energy; Pyramids of Production

Questions Reading Quizzes, Blue-Thread Questions, Test Bank

2. eText Read your book online, search, take notes, highlight text, and more.

3. The Study Area Practice Test, Cumulative Test, BioFlix® 3-D Animations, Videos, Activities, Audio Glossary, Word Study Tools, Art

You should be able to . . .

✔ TEST YOUR KNOWLEDGE

Answers are available in Appendix A

1. True or false? An ecosystem comprises a community and its abiotic environment.

2. Which of the following ecosystems would you expect to have the highest primary production?
   a. subtropical desert
   b. temperate grassland
   c. boreal forest
   d. tropical dry forest

3. Most of the net primary productivity that is consumed is used for what purpose?
   a. respiration by primary consumers
   b. respiration by secondary consumers
   c. growth by primary consumers
   d. growth by secondary consumers

4. What is biomagnification?
   a. biological imaging that provides images or data at fine spatial scales
   b. conversion of an energy input (e.g., sunlight) to chemical energy in the form of sugars, by primary producers
   c. accumulation of certain molecules at high concentration at upper trophic levels
   d. increases in nutrients that lead to pollution

5. Which of the following is normally the longest-lived reservoir for carbon?
   a. atmospheric CO_2
   b. marine plankton (primary producers *and* consumers)
   c. petroleum
   d. wood

6. Devegetation has what effect on ecosystem dynamics?
   a. It increases belowground biomass.
   b. It increases nutrient export.
   c. It increases rates of groundwater recharge (penetration of precipitation to the water table).
   d. It increases the pool of soil organic matter.

7. **QUANTITATIVE** If the GPP of a grassland is 5000 kcal/m^2/year and 55 percent is used up by cellular respiration, what is the NPP?
   a. 2250 kcal/m^2/year
   b. 2750 kcal/m^2/year
   c. 5000 kcal/m^2/year
   d. Not enough information is provided.

8. Predict why the large-scale removal of sharks from the oceans has caused some coral reefs to become overgrown with algae, even though sharks don't eat algae.

9. Explain why decomposition rates in a field in Nebraska would differ from the decomposition rates in a field in the Amazon. How do decomposers regulate nutrient availability in an ecosystem?

10. If you were to select a study organism to measure the effects of climate change on trophic interactions in food webs, would corals be a good choice? Explain your reasoning.

11. Use forests as an example to contrast positive and negative feedbacks on global warming.

12. Why are the open oceans nutrient poor? Why are coastal areas and intertidal habitats relatively nutrient rich?

13. Suppose you had a small set of experimental ponds at your disposal and an array of pond-dwelling algae, plants, and animals. How could you use a radioactive isotope of nitrogen to study the effects of fertilizer runoff in these experimental ecosystems?

14. Suppose that record snows blanket your campus this winter. Your friend says this is proof that global warming isn't really occurring. What is the flaw in your friend's logic?
    a. The average temperature of the Earth is not actually increasing.
    b. Global warming refers to temperatures, but snow is a type of precipitation.
    c. While the average global temperature is increasing, local temperatures and precipitation will vary.
    d. He is confusing global warming and global climate change.

15. Suppose that herbivores were removed from an area within a temperate deciduous forest ecosystem (e.g., by fencing to exclude herbivorous mammals and by spraying insecticides to kill herbivorous insects). Predict what would happen to the rate of nitrogen cycling. Explain the logic behind your prediction.

16. During the Carboniferous period, rates of decomposition slowed even though plant growth was extensive (probably due to the formation of vast, oxygen-poor swamp habitats). As a result, large amounts of biomass accumulated in terrestrial environments (much of this biomass is now coal). The fossil record indicates that atmospheric oxygen increased, atmospheric carbon dioxide decreased, and global temperatures dropped. Explain why.

57 Biodiversity and Conservation Biology

In this chapter you will learn that

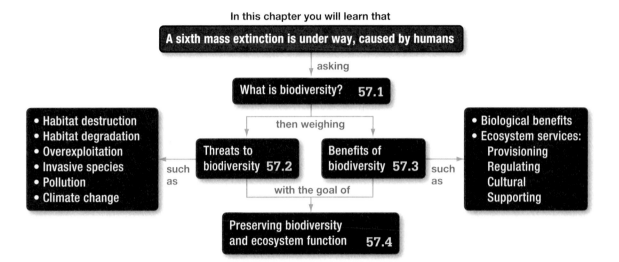

A sixth mass extinction is under way, caused by humans

asking

What is biodiversity? **57.1**

then weighing

• Habitat destruction
• Habitat degradation
• Overexploitation
• Invasive species
• Pollution
• Climate change

such as

Threats to biodiversity **57.2**

Benefits of biodiversity **57.3**

such as

• Biological benefits
• Ecosystem services:
  Provisioning
  Regulating
  Cultural
  Supporting

with the goal of

Preserving biodiversity and ecosystem function **57.4**

The island of Borneo contains some of the most spectacular rain forest in the world, yet it is being cleared at an alarming rate to make way for oil palm plantations. This action threatens both biodiversity and ecosystem function.

When conservation biologist Willie Smits gazed out upon an area near Samboja on the Southeast Asian island of Borneo in 2002, he saw a devastated wasteland:

• almost total extinction of plant and animal life;

• depleted soil fertility and failing crops;

• unusually severe climate: droughts, floods, fires; and

• a destitute human population: 50 percent unemployed, high crime rate, poor nutrition, low life expectancy.

Things were not always this way. Borneo is famous for having some of the most luxuriant rain forests on the planet—15,000 species of plants, 222 species of mammals, and hundreds of birds,

This chapter is part of the Big Picture. See how on pages 1196–1197.

✔ When you see this checkmark, stop and test yourself. Answers are available in Appendix A.

amphibians, and fish. Six thousand of these species occur nowhere else on Earth. What happened in Samboja?

Starting in the 1950s, loggers started cutting down the valuable hardwood trees. Following the eventual clear-cut, the remnants were burned to make room for oil palm plantations to supply the world's demand for cooking oil, products such as lipstick, and biodiesel. This transition to agriculture has been pervasive in Borneo. In just 20 years, Borneo has lost as much forest as would fit in Florida (**FIGURE 57.1a**).

The orangutans have been one major casualty of this land transformation, which is how Smits became interested in Samboja (**FIGURE 57.1b**). For these apes, already threatened by poaching for the pet trade, the massive loss of habitat was the last straw. Smits realized that if he wanted to save the orangutans, he would have to save the rain forest.

The scenario at Samboja is not unique. As the human population soars (Chapter 54) and the available resources plummet, more and more wild areas on Earth are being tamed, often with species losses and unintended consequences for ecosystems and humans. The goal of this chapter is to size up the problem and then explore

(a) Rapid deforestation of Borneo

(b) Orangutans are one casualty of deforestation

FIGURE 57.1 The Deforestation of Borneo (a) Borneo is a large island in Southeast Asia, once completely forested. **(b)** The clearing of lowland rain forest for oil palm plantations has had a devastating effect on species like the orangutan. Conservationist Willie Smits was inspired by the plight of these animals to work toward a solution.

positive steps that can enable a happier outcome. Let's begin by defining a concept that is key to the discussion—biodiversity.

57.1 What Is Biodiversity?

In essence, **biodiversity** means biological diversity. But this concept is more complex than it might seem at first. Perhaps the simplest way to think about biodiversity is in terms of the tree of life—the phylogenetic tree of all organisms (introduced in Chapter 1 and explored in detail in Chapters 29 through 36). If biologists are eventually able to estimate the complete tree of life (using the techniques reviewed in Chapter 28), the branches would represent all of the lineages of organisms living today; the tips would represent all of the species.

When biodiversity increases, branches and tips are added, and the tree of life gets fuller and bushier. When extinctions occur, tips and perhaps branches are removed; the tree of life becomes thinner and sparser (see Chapter 28 and The Big Picture of Evolution on pages 526–527).

Coordinated, multinational efforts are under way to estimate the tree of life—particularly major lineages such as the land plants and vertebrates—and to understand how the tree has changed over time. But even though the tree of life is a compelling way to document biodiversity, it does not tell the entire story.

Biodiversity Can Be Measured and Analyzed at Several Levels

To get a complete understanding of the diversity of life, biologists recognize and analyze biodiversity at the genetic, species, and ecosystem levels. Let's consider each level in turn.

Genetic Diversity **Genetic diversity** is the total genetic information contained within all individuals of a population, species, or group of species. It is measured as the number and relative frequency of all genes (and their alleles) present in a group.

At the genetic level, biodiversity is everywhere around you, from the variety of apples at the grocery store to the variation in singing ability of sparrows in your backyard. Distinct populations of the same species often contain unique alleles that are maintained by genetic drift or have been favored by natural selection as adaptations to local conditions (see Chapter 26).

Recently, efforts to catalog genetic diversity have been advanced by two technical breakthroughs (see Chapter 21):

1. the ability to sequence the entire genome of multiple members of the same species, and

2. the research protocol called environmental sequencing.

In environmental sequencing, researchers take a sample from the soil or water present in a habitat and sequence all or most of the genes present. It's a way to document genetic diversity even if the species present are unknown.

Genetic diversity is important because it represents the adaptive capacity of a population or higher taxonomic group—the ability of that group to persist over time despite changes in the environment.

Species Diversity The diversity of species present is a key feature of biological communities. It is usually quantified in two ways:

1. **Species richness** is a simple count of how many species are present in a given community.

2. **Species diversity** is a weighted measure that incorporates the species' relative abundance.

The key difference is in evenness. In communities that are dominated by a few species, species richness might be high but species diversity low (see Chapter 55). Be aware that since relative abundance data are often not available, biologists sometimes use the terms interchangeably.

Ecologists traditionally measure species diversity by spending hours at their field sites employing transects, traps, quadrats, mark-and-recapture techniques, and other methods to quantify the species present (see Chapter 54). This important work can be painstaking, especially in biodiverse areas such as the mountains of Papua New Guinea where, for example, a lamp can be used to attract hundreds of moth species in a single night.

One new technique for identifying such species is called **bar coding**: the use of a well-characterized gene sequence to identify distinct species, using the phylogenetic species concept (see Chapter 27). The idea is that variation in one gene—such as the gene for cytochrome *c* oxidase subunit 1 in the mitochondria of animals—allows biologists to recognize different species in the same way that variations in bar codes enable grocery store scanners to recognize different products. To date, tens of thousands of species have been bar-coded.

It is important to remember, however, that counting species is not like counting beans in a bowl—all species are not equal. Each species has an evolutionary and ecological context.

In terms of evolution, some species occur in lineages that are extremely species rich—they have what biologists call taxonomic diversity. Prominent examples include the beetles, with 360,000 species, and the orchids, with 35,000 species. Other species occur in lineages that are extremely species poor, such as that of the red panda of Asia, which is represented by a single species (**FIGURE 57.2**). When preservation of taxonomic diversity is important, then the preservation of species from species-poor lineages is a priority. If species like the red panda are wiped out, an entire lineage is lost forever.

In terms of ecology, some species have particularly influential roles in a community: they may be key primary producers, important ecosystem engineers creating habitat for other species, keystone predators, or nitrogen fixers (see Chapter 55). When functional diversity is a priority, preservation of species with unique ecological roles is important.

Ecosystem Diversity Communities vary in the number of species that occur within each trophic level (horizontal diversity) and the number of trophic levels (vertical diversity). **Ecosystem diversity** is a measure of these complex factors plus interactions

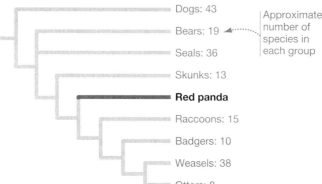

FIGURE 57.2 Phylogenetically Distinct Species May Be High-Priority Targets for Conservation. The red panda has few close relatives and represents a distinct branch on the tree of life. It is a high priority for the preservation of taxonomic diversity.

with the nonliving environment. It is the most difficult level of biodiversity to define and measure because the complexity of ecosystems is so vast.

Intertwined with the concept of ecosystem diversity is **ecosystem function,** which refers to the sum of biological and chemical processes that are characteristic of a given ecosystem—such as primary production, nitrogen cycling, decomposition, and carbon storage. These functions emerge from the sum of the feeding, growing, moving, respiring, excreting, and decomposing processes of the interacting member organisms within their abiotic context. As the field of conservation biology develops, the importance of preserving ecosystem function is gaining more attention.

Note that the three levels of biodiversity are not independent of each other. For example, consider what might happen if the sole species of nitrogen-fixing bacteria disappeared from a grassland community: The genes that code for the nitrogen-fixing enzymes would be lost, followed by loss of the species that depend on that nitrogen as well as loss of important nutrient cycling functions of the ecosystem.

Change through Time Biodiversity can be recognized and quantified on several distinct levels, but it is also dynamic.

- Mutations that create new alleles increase genetic diversity; natural selection, genetic drift, and gene flow may eliminate certain alleles or change their frequency, leading to an increase or decrease in overall genetic diversity (Chapter 26).

- Speciation increases species diversity; extinction decreases it (Chapter 28).
- Changes in climate or other physical conditions can result in the formation of new ecosystems or ecosystem functions, as can the evolution of new species—such as reef-building corals and salt-tolerant marsh grasses. Disturbances such as volcanic eruptions, human activities, and glaciation can destroy ecosystems or ecosystem functions.

Biodiversity is not static. It has been changing since life on Earth began.

How Many Species Are Living Today?

One of the simplest questions about biodiversity is also one of the most difficult to answer: How many species are there on Earth? The answer is not known, but educated guesses range from 10 to 100 million. Given the massive effort that it would take to document every form of life on Earth, the answer will probably never be known.

Biologists are well aware, though, that the approximately 1.5 million species cataloged to date represent a tiny fraction of the number actually present. The issue is even more acute in poorly studied lineages like bacteria and archaea, where researchers routinely discover dozens of new species in each environmental sequencing or direct sequencing study (see Chapters 21 and 29). For example, a recent direct sequencing study estimated that over 500 species of bacteria live in the human mouth, but only about half had been described and named already. Even among well-studied groups like birds and mammals, new species are discovered every year—including new species of primates.

Given that only a fraction of the organisms alive have been discovered to date, how can biologists go about estimating the total number of species on Earth? One approach is based on intensive surveys of species-rich taxonomic groups in defined sites; a second strategy is based on attempts to identify all of the species present in a particular region.

Taxon-Specific Surveys In a taxon-specific survey, biologists conduct an intensive search for a specific taxon in a well-defined area. For example, in their classic study carried out in 1980, Terry Erwin and Janice Scott fogged several trees with insecticide in a Panamanian jungle and recovered 1000 species of beetles. Many of the species were new to science.

More recently, Philippe Bouchet and colleagues conducted a massive survey of marine mollusks in coral-reef habitats along the west coast of New Caledonia, a tropical island in the southwest Pacific Ocean. The team spent more than a year collecting mollusks at 42 sites over a total area of almost 300 km². The survey represented the most thorough sampling effort ever made to determine the species diversity of mollusks, and produced more than 127,000 individuals representing 2738 species. These numbers far exceed the mollusk diversity recorded for any comparable-sized area.

In reporting their findings, Bouchet and his team emphasized that 20 percent of the species found were represented by a single specimen. This observation suggests that many species are exceedingly rare and thus likely to be missed by less intensive sampling efforts. And when the investigators compared their data with a survey in progress at a second site, different in reef structure but only 200 km away from the original site, they found that only 36 percent of species were shared between the two sites. Taken together, the results support the hypothesis that the 85,000 known mollusk species may represent just a third to a half of the actual total.

All-Taxa Surveys As the name suggests, in an all-taxa survey, biologists attempt to find and catalog all of the species present in an area. The most spectacular example of an all-taxa survey so far is the Census of Marine Life, a massive international effort to record the global diversity of life in the oceans. About 2700 scientists from more than 80 countries launched 540 expeditions over a 10-year period ending in 2010 to survey the number and distribution of marine species around the globe—ranging from deep-sea vents to coral reefs to intertidal mud flats.

The participants collected over 30 million records, including 6000 potentially new species. Even a small sediment core only 7 centimeters in diameter and 10 centimeters deep yielded a surprisingly diverse assemblage of deep-sea animals (**FIGURE 57.3**). Many animal phyla are represented in this sample, most of them miniaturized—a possible adaptation to low nutrient availability in the deep sea. The total number of marine species is currently estimated by the census to be 250,000.

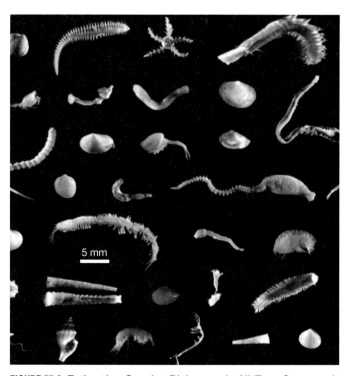

FIGURE 57.3 Estimating Species Richness via All-Taxa Surveys. A sample of the diverse, tiny animals sifted from a small core of deep-sea sediments.

Where Is Biodiversity Highest?

Both taxon-specific surveys and all-taxa surveys provide data on the total number of species as well as on their geographic distributions. What patterns of biodiversity emerge?

Mapping Species Richness and Endemism Patterns of biodiversity are usually measured by dividing geographic areas into a grid of cells and mapping the biodiversity within each cell.

The map in **FIGURE 57.4a**, for example, was constructed by dividing the world's landmasses into a grid of cells 1° latitude by 1° longitude. Using published data, a team of researchers then plotted how many species of birds breed in each cell in the grid, a measure of species richness. The narrow red-and-orange area on the map shows that the number of breeding bird species per area is particularly high in the Andes Mountains of western South America.

Another approach to measuring the distribution of biodiversity is to identify which regions of the world have a high proportion of **endemic species**—meaning, species that are found in a particular area and nowhere else. **FIGURE 57.4b** maps the location of endemic species of breeding birds, based on the same cells and data as the species richness map in Figure 57.4a. Once again, the area of highest species per area occurs in the Andes Mountains. However, the two maps are not identical, so evaluating both species richness and endemism can be more insightful than evaluating either alone.

As you might expect, the areas of highest biodiversity are different for different taxonomic groups. However, some trends do appear.

- The most prominent geographic pattern in the distribution of biodiversity corresponds to latitude: In most, but not all, taxonomic groups, species richness is highest in the tropics and declines toward the poles (see Chapter 55).

- Biodiversity is also higher on land than in the sea. Even though land covers only about 30 percent of Earth's surface, about nine out of 10 species (microbes aside) are terrestrial.

- Areas with greater geographical variation tend to have higher biodiversity than areas that are flat or featureless.

Overall, tropical rain forests are particularly species rich. Even though they represent just 7 percent of Earth's land area, they are thought to contain at least 50 percent of all species present.

Understanding *why* tropical forests are so species rich and *why* the other biodiversity patterns occur is an active research area. The mechanisms that determine patterns of biodiversity are complex because they combine local abiotic factors (solar energy, temperature, precipitation, topography, nutrient availability; Chapters 52 and 56), biotic factors (species interactions; Chapter 55), and historical context (evolutionary history, biogeography; Chapter 52). To complicate matters further, the mechanisms that determine biodiversity differ at the local scale and the regional scale.

(a) Biodiversity distribution in terms of **species richness of birds**

959
720
480
240
1

Number of breeding bird species per cell in grid

FIGURE 57.4 Biodiversity Distribution Maps Show Regions with Extraordinarily High Species Richness and Endemism. Both maps are based on a grid of cells that is 1° longitude by 1° latitude.

DATA: Orme, C. D. L., R. G. Davies, M. Burgess, et al. 2005. *Nature* 436: 1016–1019.

(b) Biodiversity distribution in terms of **endemic species of birds**

89
66
45
22
1

Number of endemic bird species per cell in grid

©2005 Nature Publishing Group

Mapping Biodiversity Hotspots In 1988, conservation biologist Norman Myers pioneered the term **biodiversity hotspot** to characterize regions that are in most urgent need of conservation action—areas where efforts to preserve habitat would have the highest return on investment.

Since 1988, Myers and many collaborators, including organizations such as Conservation International, have identified 34 hotspots that meet two criteria:

1. They contain at least 1500 endemic vascular plant species (more than 0.5 percent of the world's total).

2. They have lost at least 70 percent of their traditional or primary vegetation.

Although these areas represent just 2.3 percent of Earth's land area, they contain over 50 percent of all known plant species and 42 percent of all known terrestrial vertebrate species as endemics.

One criticism of the second criterion of biodiversity hotspots is that it considers cumulative past loss, not current rate of loss. Most notably, hotspots do not include the rain forests found in the Amazon, the Congo River basin, and New Guinea—because those environments have not yet lost 70 percent of their primary vegetation. Effectively preserving these areas along with the hotspots would provide protection for over 65 percent of all land plants in just 5 percent of Earth's land surface.

Further, hotspots focus on species diversity rather than ecosystem diversity—but the importance of ecosystem diversity is becoming more apparent every year. In 2002, scientists in collaboration with the World Wildlife Fund created a map of 238 ecoregions—subsets of biomes (see biome maps in Chapter 52). The ecoregion map, shown in **FIGURE 57.5**, highlights particularly rich areas of biodiversity; it also includes areas with unique communities or abiotic conditions (such as in tundra or desert biomes) or a unique evolutionary history (such as the Galápagos Islands; Chapter 25)—all of which should be priorities for proactive conservation.

Whether global maps of biodiversity hotspots are based on species richness, endemism, level of threat, and/or unique ecological or evolutionary characteristics, they are all useful in their own ways for creating priorities for managing threats to biodiversity.

check your understanding

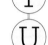

If you understand that . . .

- Biologists document biodiversity at the genetic, species, and ecosystem levels.
- On average, biodiversity tends to be higher in the tropics than near the poles, higher on land than in the sea, and higher in areas with geographic diversity. Tropical rain forests are particularly species rich.
- Biodiversity can be mapped according to species richness, endemism, level of threat, and ecological or evolutionary characteristics.

✔ **You should be able to . . .**

Design a study that would quantify species diversity on your campus.

Answers are available in Appendix A.

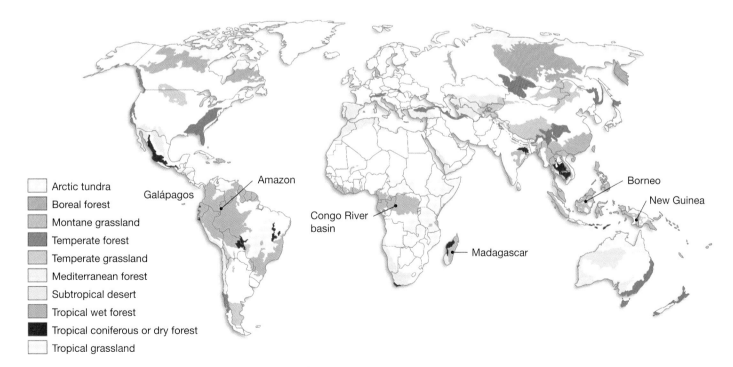

FIGURE 57.5 Ecoregions Designated as High Priority for Conservation. Areas that have the same color belong to the same biome (see Chapter 52).

57.2 Threats to Biodiversity

No species lasts forever. Extinction, like death, is a fact of life. If extinction is natural, why are biologists so concerned about habitat and species conservation? The answer is rate.

Today, species are vanishing faster than at virtually any other time in Earth's history. Modern rates of extinction are 100 to 1000 times greater than the average, or "background," rate recorded in the fossil record over the past 550 million years (see Chapter 28). According to data on the rates of extinction in birds and other well-studied groups, 60 percent of all species are likely to be wiped out within 500 years, a blink of an eye on the evolutionary time scale.

Either directly or indirectly, current extinctions are being caused by the demands of a rapidly growing human population—currently at over 7 billion and increasing by about 80 million people per year (Chapter 54). The island that is planet Earth is far more crowded than it used to be.

How do biologists track extinctions that are under way? Beginning in 1994, an organization called the International Union for the Conservation of Nature (IUCN) began compiling the Red List to quantify the risk of extinction of different species. For example, the data in **FIGURE 57.6** show that an astonishing 13 percent of birds, 25 percent of mammals, and 41 percent of amphibians are threatened—that is, they are either Critically Endangered, Endangered, or Vulnerable. An **endangered species** is a species whose numbers have decreased so drastically that it is almost certain to go extinct unless effective conservation programs are put in place.

Based on the data in hand, the vast majority of biologists agree that the sixth mass extinction in the history of multicellular life is occurring. Between the time you are reading this and the time your great-grandchildren are grown, human impacts on the planet promise to equal or exceed those of the gigantic asteroid that smashed into Earth 65 million years ago (see Chapter 28).

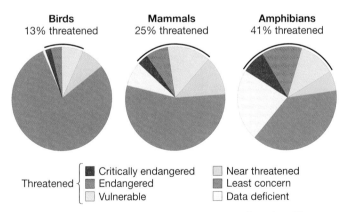

FIGURE 57.6 IUCN Designations of Threatened Species. The total number of threatened species in each group was estimated assuming that the species in the gray (data-deficient) category have the same rate of being threatened as the others. Extinct species are not included.

DATA: IUCN 2012. The IUCN Red List of Threatened Species. Version 2012.2.

Multiple Interacting Threats

What factors are causing so many species to be endangered? Consider a recent study conducted by ecologists Oscar Venter, James Grant, and colleagues. They examined 488 species of endangered plants and animals native to Canada, and they quantified the frequency of different threats based on the categories compiled by the IUCN. The bar graphs in **FIGURE 57.7** show the results. The data reveal several patterns:

- Habitat loss is the single most important factor in the decline of these species. It is a significant issue for over 90 percent of endangered species in terrestrial environments.

- Virtually all of the endangered species are affected by more than one factor. This is important information, because it means that conservation biologists may have to solve more than one problem for any given species to recover.

- Overexploitation is the dominant problem for marine species, while pollution plays a large role for freshwater species.

- Factors beyond human control can be important. These include predation or competition with native species, natural disturbances such as fires or droughts, or the fact that some species have narrow niches and historically are rare. Stated another way, background extinctions will continue to occur.

Similar studies have been conducted for other parts of the world. Do the same conclusions hold up globally? The degree of importance of the different threats varies somewhat according to region, but the take-home message is consistent with the Canadian data. Let's take a closer look at the major threats to species.

Habitat Destruction Humans cause **habitat destruction** (habitat loss) by logging and burning forests, damming rivers, dredging or filling estuaries and wetlands, plowing prairies, grazing livestock, excavating minerals, and building housing

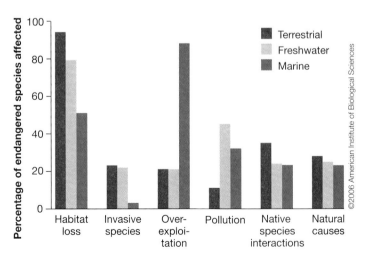

FIGURE 57.7 Endangerment Is Caused by Multiple Factors. These data are from Canada.

✔**QUANTITATIVE** Why do the data for terrestrial, freshwater, and marine species each add up to more than 100 percent?

DATA: Ventner, O. et al. 2006. *BioScience* 56: 903-910.

developments, golf courses, shopping centers, office complexes, airports, and roads. Today, less than a quarter of Earth's land is in its wild state (see map in Figure 52.10), and most of the wild land is near the poles, where biodiversity is lowest.

On a global scale, one of the most important types of habitat destruction is deforestation—especially the conversion of primary forests to agricultural fields and human settlements. Primary forests—which have never been cut—tend to have high biodiversity due to their complex ecological interactions and deep evolutionary history. Research comparing primary forest and reforested areas shows that primary forest is irreplaceable for sustaining biodiversity.

To appreciate the extent of deforestation in some areas, consider satellite images of wet tropical rain forest in the Brazilian state of Mato Grosso that were made in 1985 and in 2000 (**FIGURE 57.8**). Analyses of satellite photos such as these have shown that as many as 3 million hectares were deforested in the Amazon each year during the 1990s. (A hectare, abbreviated ha, is 100 meters by 100 meters—approximately the size of two football fields.)

More recent analyses suggest that the global rate of deforestation has now slowed compared to an average net loss of 16 million ha/year in the 1990s. The latest report from the United Nations Food and Agriculture Organization, released in 2010, states that the world's forests experienced a net loss of about 13 million ha/year between 2000 and 2010.

But conservationists greeted this report with caution. Net losses dropped because of extensive forest regrowth in China, North America, and Europe, even though over 40 million ha of primary forest have been lost in South America, Southeast Asia, and Africa since 2000. If this rate of tropical deforestation continues, over 28 percent of the wet tropical forest that exists today will be gone in your lifetime, including almost half of the Brazilian Amazon.

Forest losses in the tropics are particularly worrisome because these forests contain many biodiversity hotspots, and they moderate the effects of climate change by storing carbon. Further, when forests are cut, soil and nutrients are often lost from the ecosystem due to erosion, inhibiting future recovery (Chapter 56). Meanwhile, the human population continues to increase rapidly in these areas, putting increased pressure on the forests. Addressing the interaction of these scientific and social variables represents one of the most formidable challenges of our time.

Habitat Degradation Aside from destroying natural areas outright, human activities can also cause **habitat degradation**, a reduction in the quality of a habitat. Humans can degrade habitats in many ways, even by using artificial light at night and creating traffic noise. One of the most pervasive patterns of habitat degradation is **habitat fragmentation,** the parsing of contiguous areas of natural habitats into small, isolated fragments. For some species, a narrow road or pipe intruding into their habitat is a sufficient barrier to limit dispersal.

Habitat fragmentation concerns biologists for several reasons. First, like habitat destruction, fragmentation can reduce habitats to a size that is too small to support some species. This situation is especially true for top predators (see Chapters 55 and 56) such as wolves, mountain lions, and grizzly bears, which need vast natural spaces in which to feed, find mates, and reproduce successfully.

When top predators in food webs are lost, trophic cascades may occur, affecting species throughout the food web in complex ways (see Chapter 56). Trophic cascades often have unexpected impacts on many aspects of ecosystem function, ranging from susceptibility to fire, disease, and invasive species to changes in primary productivity and nutrient cycling.

The second concern about habitat fragmentation is that it can force many species into a metapopulation structure (see Chapter 54). The small, isolated populations that make up a metapopulation are much more likely than large populations to be wiped out, for two reasons:

1. Catastrophic events such as storms, disease outbreaks, or fires exterminate small populations more readily than large populations. Biologists refer to episodes like this as stochastic—meaning, chance—events.

FIGURE 57.8 Rates of Deforestation in Tropical Wet Forests Are High. In the satellite image from Mato Grosso, Brazil, at top, the dark green areas are intact forest. In the lower image of the same location 15 years later, the light areas have been burned or logged and converted to agricultural fields and pastures.

2. Small populations suffer from inbreeding depression and random loss of alleles due to genetic drift.

Why do genetic problems occur? Small populations become inbred simply because so few potential mates are available. Over time, virtually all individuals become related. In most populations, inbreeding leads to lowered fitness due to deleterious recessive alleles being homozygous more frequently—a phenomenon known as inbreeding depression (see Chapter 26).

In addition, genetic drift is much more pronounced in small populations than large populations (see Chapter 26). Because drift leads to the random loss or fixation of alleles, it reduces genetic variation in populations. This result is a concern, because if the environment changes due to the evolution of a new disease or global climate change, the population may lack alleles that confer high fitness in the new environment.

Third, fragmentation creates large amounts of degraded "edge" habitat. The edges of intact habitat are subject to invasion by weedy species and are exposed to more intense sunlight and wind, creating difficult conditions for plants.

The degradation of habitat caused by fragmentation is being documented in a long-term experiment in a tropical wet forest. In an area near Manaus, Brazil, that was slated for clear-cutting, ecologists William Laurance, Thomas Lovejoy, and their colleagues set up 66 square, 1-ha experimental plots that remained uncut. Thirty-nine of these study plots were located in fragments designed to contain 1, 10, or 100 ha of intact forest. Twenty-seven of the plots were set up nearby, in continuous wet forest. As the "Experimental setup" section of **FIGURE 57.9** shows, the distribution of the study plots allowed the research team to monitor changes inside forest fragments of different sizes and to compare these changes with conditions in unfragmented forest. Note that many of the study plots were located on forest edges.

When the research group surveyed the plots at least 10 years after the initial cut, they recorded two predominant effects:

1. A startling drop in aboveground **biomass,** or the total amount of fixed carbon, took place in the study plots located near the edges of logged fragments. As the graph in the "Results" section of Figure 57.9 shows, most of the loss occurred within 2–4 years of the fragmentation event.

2. On average, biomass did not decrease in the interior (control) plots. Further, the rate of biomass loss increased as the distance from the plots to the nearest forest edge decreased.

In tropical wet forests, most of the biomass is concentrated in large trees. Due to exposure to high winds and dry conditions, the edges of the experimental plots contained many downed and dying trees. Follow-up work showed that in edge habitats and in the fragmented patches, large-seeded, slow-growing trees typical of undisturbed forests are being replaced by early successional, "weedy" species (see Chapter 55). A large number of bird and understory plant species have disappeared as well.

The take-home message of this experiment and other research is clear: When habitats are fragmented, the quality *and* quantity of habitats decline drastically. In addition to losing over 8 million ha

---

QUESTION: How does fragmentation affect the quality of tropical wet forest habitats?

HYPOTHESIS: Fragmentation reduces the quality of wet forest habitats.

NULL HYPOTHESIS: Fragmentation has no effect on the quality of wet forest habitats.

EXPERIMENTAL SETUP:

PREDICTION: Biomass will decline in forest fragments compared with those of the forest interior, particularly along edges of fragments.

PREDICTION OF NULL HYPOTHESIS: Biomass will be the same inside forest fragments and along edges as in the forest interior.

RESULTS:

CONCLUSION: Biomass declines sharply along edges of forest fragments.

FIGURE 57.9 Experimental Evidence for Edge Effects in Fragmented Forests. Researchers tracked 66 study plots among four 1-hectare (ha) fragments, three 10-ha fragments, and two 100-ha fragments.

SOURCE: Laurance, W. F., S. G. Laurance, L. V. Ferreira, et al. 1997. Biomass collapse in Amazonian forest fragments. *Science* 278: 1117–1118.

✔**QUESTION** The locations of the fragments and study plots were assigned at random. Why was this step important?

of primary forest in South America, Southeast Asia, and Africa each year to clear-cutting, we are losing large amounts of high-quality habitat throughout the world to fragmentation.

Overexploitation Overexploitation refers to any unsustainable removal of wildlife from the natural environment for use by humans. Overexploitation is the dominant threat for marine species.

- Two-thirds of harvestable marine species are now considered fully exploited or depleted, and a recent study of the global fishing industry concluded that 90 percent of individuals in large-bodied fish species from bottom or open-water habitats have already been removed from the world's oceans (see **TABLE 57.1**). The affected species include tuna, swordfish, marlin, cod, halibut, and flounder—and many species of sharks, which are harvested for their fins. Unless stricter fishing limits are established quickly, these popular species will go extinct. Further, removal of these top predators has an immense impact on species at lower trophic levels, causing a cascade of effects on food webs (Chapter 56).

- Overhunting has also emerged recently as a major threat to many mammal populations. Aside from the traditional hunt for game species, the recent growth of the "bushmeat" trade has devastated populations of primates and other mammals in Africa and elsewhere. Adding to the problem, many animals are hunted not for their meat, but for their ivory, skins, horns, or other body parts valued in folk medicines or decorations.

- The capture of exotic animals for the pet trade continues at a rapid rate despite many laws to prevent it. Exotic pets range widely from primates like tamarins and macaques to parrots, iguanas, frogs, corals, and tropical fish.

Overexploitation is a grave concern for marine and terrestrial species worldwide due to the increasing demand and decreasing supply. Ten thousand years ago, only 0.1 percent of Earth's mammalian biomass consisted of humans and their domesticated animals. Today, this number has risen to 90 percent.

Invasive Species   A nonnative species that is introduced into a new area is called an **exotic species.** Exotic species do not necessarily pose problems, and can even be beneficial in some circumstances—such as the arrival of coyotes in the eastern United States (from the west) to fill the niche vacated by wolves and other top predators, which were hunted out.

But some exotic species pose a direct threat to native species by eating them, competing with them, causing disease, or other types of interactions. They can also pose indirect threats by changing the local biotic or abiotic resources. You might recall that if an exotic species is introduced to a new area, grows to large population size, and disrupts species native to the area, it is called an **invasive species** (see Chapter 52).

Global trade and global travel have vastly increased the rate of exchange of plants, animals, fungi, and microorganisms—both purposeful and accidental—sometimes with dire consequences. For example, after brown tree snakes were introduced to the Pacific island of Guam 60 years ago (see Table 57.1), the snake population increased exponentially, causing the extinction of 12 native bird species, decreasing populations of small mammals and lizards, and indirectly affecting plant communities. Islands are particularly vulnerable to the effects of invasive species.

Threat	Example
Habitat destruction and degradation	
The conversion of primary forest to agricultural fields is a particularly devastating cause of habitat destruction.	
Overexploitation	
Overexploitation is the dominant threat for marine species, especially for large predators in top trophic levels, like these endangered bluefin tuna.	
Invasive species	
Invasive species, like this brown tree snake in Guam, are introduced to a new area, multiply rapidly, and threaten native species.	
Pollution	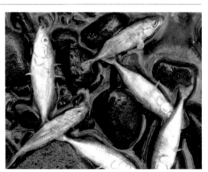
Chemical pollutants have reached every corner of the globe, but are particularly threatening to aquatic species.	
Climate change	
Climate change poses different types of threats to different species. For example, some corals become bleached (lose their symbiotic photosynthetic algae) when water temperature warms.	

CHAPTER 57   Biodiversity and Conservation Biology   **1181**

Pollution Humans change both the biotic environment and the abiotic environment by releasing chemicals into ecosystems. Many types of major pollutants pose both direct and indirect threats to species:

- Industrial pollutants include sulfur (causing acid rain), arsenic, mercury, and lead as well as greenhouse gases such as CO_2 and nitric oxide (NO_2).

- Pharmaceutical pollutants (drugs) like oral contraceptives and ibuprofen are excreted in human urine and pass through wastewater-processing facilities into streams and rivers, where they are still biologically active.

- Pesticides and herbicides are a type of persistent organic pollutant (POP) that have a direct effect where they are used, but also affect species far away—sometimes in extremely remote locations—by dispersing in waterways and undergoing biomagnification in food webs (Chapter 56).

- Nutrients like nitrogen and phosphorus run off of farms and fields, causing eutrophication (Chapter 56).

- Garbage fills landfills, waterways, and ocean gyres, posing both physical and chemical threats.

Many pollutants are persistent. They last in organisms and the environment for a long time and may continue to cause harm long after their initial release—and at great distances from their initial release. Freshwater organisms are particularly vulnerable to chemical pollutants.

Climate Change The acceleration of global climate change due to the burning of fossil fuels, deforestation, and other processes has already had an impact on species, including extinctions (see Chapter 56). Biologists have begun to record how several factors related to climate change influence biodiversity:

- As the global temperature increases, polar and alpine habitats are disappearing, endangering species native to arctic and high-elevation tundra.

- Coastal habitats are threatened by sea-level rise caused by the melting of polar ice caps and glaciers. The most recent analyses project that sea levels will rise an additional 0.8 to 2.0 m by the year 2100, inundating many coastal habitats, including areas with biodiversity hotspots.

- Biomes are becoming redistributed due to changes in both temperature and rainfall, causing mismatches between species distributions and abiotic conditions.

- The species compositions of communities are changing because trees and other slowly dispersing species are unable to keep up with rapid changes in climate.

- Coral reefs are suffering from "bleaching" induced in some species by high temperatures; this change is causing a cascade of effects in those ecosystems.

- Ocean acidification—due to the rise of atmospheric carbon dioxide concentrations and the consequent increase in carbonic acid formation in the oceans—is interfering with many aspects of organismal physiology, including the ability of corals, crustaceans, and other animals to make calcium carbonate skeletons (see Table 56.1).

Although climate change is a natural phenomenon, the climate is currently changing at an extreme rate.

How Will These Threats Affect Future Extinction Rates?

What impact will the different threats have on biodiversity in the future? The formulation of biodiversity projections is a rapidly growing area of research, due to major advances in computer modeling, increased availability of paleontological and current data to input into the models, and increased demand for projections as the world is faced with important decisions about conservation.

The formulation of biodiversity scenarios is a highly integrative field of ecology because it must use information from multiple levels, including

- the effects of evolutionary processes on genetic variation, adaptation, and extinction in populations (Chapter 26);

- the impact of abiotic factors such as temperature and precipitation on the distribution and abundance of organisms (Chapter 52);

- the behavior of organisms in response to changing conditions (Chapter 53);

- the basic demographic characteristics of populations, such as fecundity and survivorship (Chapter 54);

- the consequences of species interactions at the level of communities (Chapter 55) and ecosystems (Chapter 56); and

- the effect of biodiversity threats at all these levels.

You can see how all these levels fit together in the Big Picture of Ecology on pages 1196–1197. Different models integrate different subsets of this list. Let's consider three examples of models ranging from single populations up to global species distributions.

Estimating the Probability that an Individual Species Will Go Extinct A population viability analysis, or PVA, is a model that estimates the likelihood that a population will avoid extinction for a given time period. In most cases, a PVA attempts to combine data on

- age-specific survivorship and fecundity (Chapter 54),

- geographic structure (Chapter 52), and

- rate and severity of habitat disturbance (Chapter 55).

Typically, a population is considered viable if the analysis predicts that it has a 95 percent probability of surviving for at least 100 years. This type of analysis is particularly useful for creating management plans for species like the northern spotted owl and island foxes, which are listed as endangered in the IUCN Red List, the U.S. Endangered Species Act, or other metrics.

Estimating the Effect of Habitat Area on Species Richness One strategy for predicting the future of Earth's biodiversity focuses on

The goal of species–area plots is to quantify the relationship between species richness and habitat area. **FIGURE 57.10** gives an example of a species–area curve—a graph that plots habitat area on the *x*-axis and the number of species present on the *y*-axis. In this case, the habitat areas are islands in the Bismarck Archipelago near New Guinea, in the South Pacific, and the species documented are birds. Each data point represents the number of species found on a different island. Note that the relationship between habitat area and the number of species present is linear when both axes are logarithmic. (For help interpreting logarithms, see **BioSkills 6** in Appendix B.)

The "log–linear" relationship turns out to be typical for other habitats and taxonomic groups as well. When biologists have plotted species–area relationships for plants, butterflies, mammals, or birds from islands or continental habitats around the globe, the relationship is consistently described by a function of the form $S = cA^z$, where:

- S is the number of species.
- A represents the habitat area.
- The c term is a constant that scales the data. Its value is high in species-rich areas, such as coral reefs or tropical wet forests, and low in species-poor areas, such as arctic tundra.
- The exponent z represents the slope of the line on a log–log plot. Thus, z describes how rapidly species numbers change with area. Typically, z is about 0.25.

For example, the solid line drawn through the points in Figure 57.10 is described by the function $S = (18.9)\,A^{0.15}$. This equation can be used to predict the species richness of other areas of a given size, such as an island that has not yet been sampled. If an unsampled island is known to have an area of 8 km², the species richness would be predicted as $S = 18.9 \times 8^{0.15} = 26$.

✔ **QUANTITATIVE** If you understand species–area curves, you should be able to predict the species richness of birds of an island in the Bismarck Archipelago that is 5000 km². Answer: 68

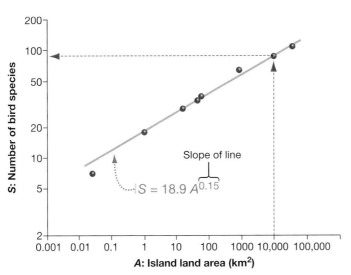

FIGURE 57.10 Species–Area Plots Quantify the Relationship between Species Richness and Habitat Area. The dashed arrows on this graph show the number of bird species that are expected to live on an island with an area of 10,000 km².

DATA: Diamond, J. M. 1975. *Biological Conservation* 7: 129–146.

the consequences of the most urgent problem: habitat loss. Given reasonable projections of how much habitat will be lost over a given time period, biologists can estimate rates of extinction based on **species–area relationships.** The concept here is to use species–area curves (see **Quantitative Methods 57.1**) to extrapolate backward to determine the species richness of a given habitat if its area were to decline due to habitat destruction or climate change.

To understand how biologists use the species–area curve in Quantitative Methods 57.1 to project extinction rates, ask yourself the following question: If sea-level rise and deforestation wiped out 90 percent of the habitats in part of the Bismarck Archipelago—for example, if *A* were reduced from 10,000 km² to 1000 km²—what percentage of species would disappear? According to the graph, the number of bird species present would drop from about 75 to about 53. Thus, the answer is that roughly 30 percent of species would vanish. The slope of the line in species–area curves is species dependent.

The caveat with this method is that using the curve to extrapolate backward can cause overestimation of extinction rates due to assumptions inherent to the model. Thus, the technique must be used with caution and cross-referenced with other projection methods.

Estimating the Effect of Global Climate Change on Species Distributions
How will climate change affect the global distribution of species? Can reliable predictions be made? Bioclimate envelope models, also called niche-based models, use the relationship between current species distributions and abiotic variables such as temperature and precipitation to project the future distributions of species under climate change.

For example, behavioral ecologist Barry Sinervo and an international team of colleagues took a two-step approach. First, they collected extensive data from one part of the globe, measuring extinction rates of 48 species of lizards at 200 sites in Mexico since 1975. The results were shocking: Twelve percent of local populations had already gone extinct because the temperature is changing too rapidly for the lizard populations to adapt (Chapter 56).

Extinction risk: Low [■■■■■■■■■■] High

FIGURE 57.11 Predictions of Global Extinction Rates of Lizards.
The projections are based on current extinction rates in Mexico and projections of global climate change, and were validated using observed extinction rates in regions of four other continents.
DATA: Sinervo, B., F. Mendez-De-La-Cruz, D. B. Miles, et al. 2010. *Science* 328: 894–899.

Sinervo and his team then used their data to predict worldwide responses to global warming given projected climate changes and the relationship between lizard extinctions and climate (**FIGURE 57.11**). They estimated that 4 percent of local lizard populations have already gone extinct worldwide and that lizard species extinctions could reach 20 percent by 2080. These alarming projections were validated with local data, suggesting that climate change is not a problem of the future—it is already a matter of life and death for lizard populations today.

Take-Home Messages about Making Biodiversity Predictions It is important to note that at all levels of study, a great deal of variation exists in projected future biodiversity. And this variation has different sources:

• Projections must make assumptions about future levels of threats to biodiversity, such as the extent of global warming, but these levels are unknown and depend on how rapidly

humans are able to reduce emissions of greenhouse gases and improve land-use practices.

• Variation in some projections is due to knowledge gaps in basic biology. More data collection is greatly needed in these cases for accurate projections.

• Some of the variation is due to differences in how the models are constructed. Scientists address this issue by using multiple models to cross-check projections, and by validating their models with empirical data from the real world.

One of the greatest challenges of making projections is the ability to measure how different factors interact. For example, the precipitous worldwide decline in amphibians is due to a combination of factors ranging from habitat loss, pollution, and climate change to infection by a chytrid fungus. These factors affect each other in complex feedback loops, requiring a systems biology approach that quantifies the network of interactions.

check your understanding

If you understand that . . .

• Species face multiple, interacting threats including habitat destruction and degradation, overexploitation, invasive species, pollution, and climate change.

• Projected extinction rates can be estimated from multiple methods by using a combination of empirical data and mathematical models.

✔ **You should be able to . . .**

1. Explain why fragmentation reduces habitat quality and leads to genetic problems.

2. Describe how you would determine whether each of the threats in Table 57.1 was affecting the orangutans of Borneo.

Answers are available in Appendix A.

57.3 Why Is Biodiversity Important?

One of the questions biologists have to address about biodiversity is one that you may have heard from friends or relatives: Who cares? People are concerned about their grades or job or health or car or love life. Why should they worry about something as abstract as biodiversity?

The answer has two parts, one biological and the other economic and cultural. Let's consider each part in turn.

Biological Benefits of Biodiversity

Do biological communities and ecosystems function better when species diversity is high? The effort to answer this question began in the mid-1980s and has blossomed.

Biodiversity Increases Productivity **FIGURE 57.12** summarizes a classic experiment on how species richness affects one of the most basic aspects of an ecosystem function: the production of biomass. Net primary productivity (NPP) is important because it is the source of chemical energy used by species throughout a food web (see Chapter 56).

As the "Experimental Setup" portion of the figure shows, David Tilman and colleagues divided 32 grassland plant species into five functional categories. The functional groups differ by the timing of their growing season and by whether they allocate most of their resources to manufacturing woody stems or seeds. The researchers planted 13-m by 13-m plots with a mixture of

RESEARCH

QUESTION: Do high species richness and high functional diversity of species increase aspects of ecosystem function such as net primary productivity (NPP)?

HYPOTHESIS: NPP increases with increasing species richness and with increasing functional diversity of species.

NULL HYPOTHESIS:

EXPERIMENTAL SETUP:

Plant a total of 289 experimental plots, each with up to 32 species and up to 5 functional groups:

Cool-season grasses: Grow in spring

Warm-season grasses: Grow in summer

Legumes: Fix nitrogen

Woody plants: Trees, shrubs

Forbs: Lots of seeds

Examples of experimental plots:

13 m

1 species
1 functional group

2 species
1 functional group

6 species
4 functional groups

PREDICTION:

PREDICTION OF NULL HYPOTHESIS:

RESULTS:

CONCLUSION: In this plant community, NPP increases with increasing species richness and increasing functional diversity of plants, at least up to a point.

©1997 AAAS

FIGURE 57.12 Evidence that the Productivity of Ecosystems Depends on the Number and Types of Species Present.

SOURCE: Tilman, D., J. Knops, D. Wedin, et al. 1997. The influence of functional diversity and composition on ecosystem processes. *Science* 277: 1300–1302.

✔**EXERCISE** Write in the null hypothesis and both sets of predictions.

between 0 and 32 randomly chosen species, representing from 0 to 5 functional categories, and measured the amount of above-ground biomass produced in each plot (see **FIGURE 57.13**).

The results, graphed in the "Results" section of Figure 57.12, indicate that both the number and type of species present had important effects on biomass production. Experimental plots with more species and with a wider diversity of functional groups were more productive.

Note that total biomass leveled off as species richness and functional diversity increased, however. This observation suggests that increasing species diversity improves ecosystem function only up to a point. This pattern suggests that at least some species in ecosystems are redundant in terms of their ability to contribute to productivity—a characteristic that is important to resilience.

Follow-up experiments in an array of ecosystems supported the conclusion that species richness has a positive impact on NPP, and they showed that several causal mechanisms may be at work.

- *Resource use efficiency* Diverse assemblages of plant species make more efficient use of the sunlight, water, and nutrients available and thus lead to greater overall productivity. For example, some prairie plants extract water near the soil surface, while others use water available a meter or more below the surface, a result of niche differentiation (Chapter 55). When species diversity is high, more overall water is used and more photosynthesis can occur.

- *Facilitation* Certain species or functional groups facilitate the growth of other species by providing them with nutrients, partial shade, or other benefits. In prairies, the presence of plants in the onion family may discourage herbivores, or the decaying roots, stems, and leaves from nitrogen-fixing species may fertilize other species.

FIGURE 57.13 Tilman's classic biodiversity study is among many conducted at the Cedar Creek Natural History Area in Minnesota.

- *Sampling effects* In many habitats, it is common to observe that one or two species are extremely productive. If the number of species in a study plot is low, it is likely that the "big producer" will be missing and NPP will be low. But if the number of species in a study plot is high, then likely the big producer is present and NPP will be high. Simply due to sampling, then, high-species plots will tend to outproduce low-species plots.

The three mechanisms listed are not mutually exclusive—several can operate at the same time. Further, long-term and large-scale experiments have shown that the reason for the pattern can change over time, that the slope of the curve can continue to rise over time, and that the strength of the diversity–productivity relationship depends on the evolutionary history of the species involved.

Does Biodiversity Lead to Stability? When biologists refer to the stability of a community, they mean its ability to

1. recover to former levels of productivity or species richness after a disturbance;

2. maintain productivity and other aspects of ecosystem function as conditions change over time.

Resistance is a measure of how much a community is affected by a disturbance. **Resilience** is a measure of how quickly a community recovers following a disturbance.

To test the hypothesis that diversity increases resistance and resilience, Tilman's team that did the work highlighted in Figure 57.12 followed up on a natural experiment. A **natural experiment** occurs when comparison groups are created by an unplanned, unmanipulated change in conditions.

In 1987–1988, a severe drought hit their study site. When the drought ended, the team asked whether species richness affected the response to the disturbance. As predicted, drought resistance appeared to be higher in more diverse communities than in less diverse ones.

In addition, the group analyzed the change in biomass in each plot four years after the drought versus the biomass before the drought. The data indicated that most plots containing five or fewer species showed a significant lowering of biomass after the disturbance, whereas all of the plots that contained more than five species regained the same biomass that they had before the disturbance. This is strong evidence that species-rich areas are more resilient following a disturbance.

More recent experiments on California grasslands suggest that species-rich ecosystems also may be less prone to invasion by exotic species. The hypothesis here is that species-rich communities use resources more efficiently—leaving less room for invasives. Taken together, these results from North American grasslands—and similar results from ecosystems around the world—lead biologists to be increasingly confident that species richness has a strongly positive effect on how ecosystems function.

Communities that are more diverse appear to be more productive, more resistant to disturbance and invasion, and more resilient than communities that are less diverse. Increased species richness increases the services provided by ecosystems. By implication, biologists can infer that if ecosystems are simplified by extinctions, then productivity and other attributes might decrease.

Ecosystem Services: Economic and Social Benefits of Biodiversity and Ecosystems

The biological benefits of biodiversity—increased productivity, resistance, and resilience—benefit humans too. But this is just the beginning. The prosperity, health, and happiness of humans are completely dependent upon, and intertwined with, the preservation of biodiversity and functioning ecosystems.

Some species provide obvious direct benefits—bluefin tuna are prized as a food source, and tigers are a cultural icon. Other species offer direct benefits that are more difficult to observe—such as the control of crop-eating insects by bats, a service recently valued at $4 billion/year.

Further, many species indirectly benefit humans simply by serving ecological roles in functioning ecosystems, which in turn provide invaluable services such as soil formation and climate moderation. Considered collectively, all the direct and indirect benefits that humans derive from organisms and the ecosystems they compose are called **ecosystem services.**

In 1997 a team of ecologists and economists led by Robert Costanza estimated the value of ecosystem services at $33 trillion—almost twice the gross national product of all nations combined. Let's take a closer look at some of the material and nonmaterial services that nature provides, summarized in **TABLE 57.2.**

Provisioning Services Wild species have provided the raw material to fuel the development of human societies ever since the first large-scale, highly organized cultures began to develop about 10,000 years ago. People felled vast tracts of forest for building materials and fuel; selectively bred wild plants to yield food and fibers; domesticated animals to provide food, labor, and material goods; harvested the oceans for protein; and processed plants, animals, and fungi as sources of medicines. Today as much as ever, humans rely on other organisms for most of their food and material needs.

The value of provisioning services is relatively easy to quantify because food, fuels, wood, and other materials are traded in the marketplace. For example, mushrooms have an annual food value over $1 billion, the fishing industry has a value of $80–100 billion, and raw materials provided by tropical forests are valued at $300 billion.

Bioprospecting—the exploration of bacteria, archaea, protists, plants, fungi, and animals as novel sources of drugs or ingredients in consumer products—has benefited from the recent explosion of genetic and phylogenetic information. Biologists can now search genomes from a wide array of species to find alleles with desired functions. Recent advances are impressive: researchers have developed a potent painkiller from the paralyzing sting of tropical cone snails and a blood anticoagulant from the saliva of vampire bats.

SUMMARY TABLE 57.2 Ecosystem Services

Type of Service	Example
Provisioning services provide raw materials • Food • Fuel • Fiber and other materials • Medicines • Genetic resources	
Regulating services are part of Earth's life-support system • Climate moderation • Soil formation • Erosion control • O_2 and CO_2 regulation • Water capture • Water purification • Air cleaning • Flood control • Storm mitigation • Waste decomposition • Bioremediation	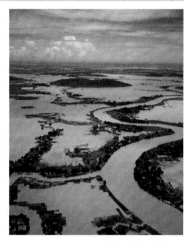
Cultural services enrich quality of life • Aesthetics • Recreation • Education • Spiritual value • Human mental and physical health	
Supporting services enable all the other ecosystem services • Primary productivity • Nutrient cycling • Pollination • Biological control	

Regulating Services Services provided by functioning ecosystems—such as water purification, flood control, and waste decomposition—are part of Earth's life-support system. Yet, these regulating services are often taken for granted until human actions disrupt their function.

This was the case in New York City in the 1990s. New York City is known for its excellent drinking water, piped in from the watersheds of the Catskill Mountains. But the water quality declined as those watersheds became developed—and something had to be done. A cost–benefit analysis revealed that it was far cheaper to restore the ecosystem function of the watersheds ($1 billion) than to filter the water artificially ($6–8 billion up front plus $300 million annually). The "free" water purification services that had been provided by the forested Catskill ecosystem suddenly became quite valuable.

Cultural Services People integrate nature in their lives in diverse cultural ways, ranging from artistic celebrations to recreation, education, and spiritual contemplation. Eminent biologist and author Edward O. Wilson coined the term biophilia to refer to the innate, unconscious bond between humans and other organisms as a result of our coevolutionary history. Wilson's hypothesis is that this bond inspires our deep need for, and appreciation of, nature.

An increasing number of studies are showing that humans who are removed from natural environments—such as people who spend most of their time in cities—often suffer both physical and psychological setbacks. Conversely, people who spend more time outdoors in nature, especially as children, tend to have: reduced stress; reduced rates of obesity, diabetes, depression, and attention deficit and hyperactivity disorder; improved problem-solving skills; enhanced creativity and curiosity; improved sleep patterns; and improved vision, among other traits.

On average, people who spend more time surrounded by nature tend to be healthier and happier. (How much time do you spend outdoors?) Government agencies and businesses are taking note because human health and happiness also have economic value.

Supporting Services Some ecosystem services support all the other categories of services. Primary productivity is a key example because the amount of biomass generated affects the provisioning, regulating, and cultural services that ecosystems can provide.

In a similar way, pollination and pest control are critical to multiple ecosystem services. The value of pollination became clear in a region in China where the excessive use of pesticides led to a particularly sharp decline of bees. Human workers must now pollinate apple, pear, and peach orchards by hand, at great expense (**FIGURE 57.14**). In the United States alone, insect-pollinated crops like almonds, apples, and chocolate produce $40 billion worth of products annually.

The air you breathe, the water you drink, and the soils that grow the food you eat are services that ecosystems provide. The quality of your life is tied to the quality of those ecosystems.

FIGURE 57.14 Hand Pollination of Fruit Trees. Following a decline of bees due to pesticide use, workers in Maoxian County, China, hand pollinate entire orchards.

An Ethical Dimension?

The data reviewed in this section suggest that there are sound economic and biological reasons to preserve species and ecosystem function. People live better, and ecosystems work better, when biodiversity is high.

But in addition, an array of biologists, philosophers, and religious leaders argue that humans have an ethical obligation to preserve species and ecosystems. There are at least three arguments for this position:

1. Organisms have intrinsic worth. Humans diminish the world by extinguishing species and destroying ecosystems.

check your understanding

If you understand that . . .

• Experimental plots with higher species richness tend to have higher productivity and show increased stability in response to disturbance or changed abiotic conditions.
• Biodiversity and functioning ecosystems provide many provisioning, regulating, cultural, and supporting services to humans.

✔ **You should be able to . . .**

Design an experiment to test the hypothesis that the rain forest of Borneo is more effective at building soil, retaining soil nutrients, and minimizing soil erosion than the oil palm plantations are.

Answers are available in Appendix A.

2. Industrialized nations are responsible for most of the environmental harm, yet poor countries are disproportionately affected by the consequences of these actions. It is unethical for rich countries to inflict hardship on poor countries.

3. It is unethical for the current generation to deprive future generations of ecosystem services.

Thus, extinction and disruption of ecosystem function are moral issues as well as biological and economic problems. One of the most important reasons to preserve biodiversity and ecosystem function may simply be that it is the right thing to do.

57.4 Preserving Biodiversity and Ecosystem Function

Extinction is irreversible—and biodiversity loss is not slowing down. But there is reason for hope.

Solving any global problem requires a common goal to be defined. In the case of preserving biodiversity and ecosystem function, the objective is to sustain diverse communities in natural landscapes while supporting the extraction of resources required to maintain the health and well-being of the human population. Given the extreme rate of deterioration of present conditions on Earth, conservation biologists often feel like doctors in a busy emergency room: Triage is necessary to maximize positive outcomes. This effort generates sober acceptance of some inevitable losses but vigorous fighting for important gains.

A number of international organizations and agreements are striving to reduce the threats to biodiversity. Central among them is the Convention on Biological Diversity (CBD), which convened in 1992 to establish conservation targets for 2010—many of which were not met. The CBD convened again in 2010 to establish new goals for 2020. These goals serve as an important organizational framework for the international effort. What's working? And what can you do to help?

Addressing the Ultimate Causes of Loss

To begin, most biologists recognize that it will be impossible to preserve high species diversity and high-functioning ecosystems if two things happen:

1. the human population grows to 10 billion or more—from its current 7 billion—over the next half-century, as projected by the "high-fertility" scenario (see Chapter 54); and

2. the rate of consumption of fossil fuels and other resources does not decline, especially in industrialized nations and growing developing nations (see Chapter 56), because consumption threatens biodiversity not only directly but also indirectly via climate change.

While it is complex and controversial to pinpoint a human carrying capacity for the planet (Chapter 54), it is clear that humans are nearing, or have surpassed, some of the recoverable limits of available resources. The way forward involves both reducing population growth and improving **sustainability**—the managed use of resources at a rate only as fast as the rate at which they are replaced.

You may be familiar with efforts to develop renewable sources of energy, promote agricultural and forestry practices that maintain productivity without the addition of synthetic fertilizers or groundwater, limit fishing catches so that populations do not crash irrecoverably, and recycle metals and other raw materials. In essence, the challenge is to create ways for humans to live off the resources that are being produced continuously on Earth rather than using species to extinction and exhausting resources like petroleum, groundwater, and topsoils that have been stored for centuries or millennia.

It may take decades for human population size and resource use to stabilize and for sustainable development practices to become widespread. In the meantime, thousands of species and untold ecosystem services may be lost. What specific approaches do conservationists use to protect natural resources?

Let's consider several examples, while asking: "What level of biodiversity is this targeting?" and "What threats to biodiversity does this approach address?"

Conservation of Genetic Diversity, Populations, and Species

Despite the discouraging persistence of biodiversity loss, conservation efforts have managed a number of successes in preserving genetic diversity, populations, and species in many locations around the globe. What are these successes?

Management Plans for Invasive Species The costs of eradicating some invasive species are so enormous that they outweigh the benefits. However, global awareness of the problem has slowly led to measures that limit the exchange of species, such as requirements that ships exchange ballast water out at sea (an inhospitable environment for coastal species) rather than in port, that cargoes are inspected at border crossings, and that nurseries stop selling invasive plants. Prevention is more effective than treatment, so education is the key.

Seed Banks Agricultural scientists are preserving diverse strains of crop plants in **seed banks**—long-term storage facilities. In addition, they are using wild relatives of domesticated species in breeding programs aimed at improving crop traits. The goal of both efforts is to avoid the loss of valuable genetic diversity.

Ex Situ Conservation and Reintroduction Some species—including Pere David's deer, Przewalski's horse, and the Arabian oryx—are extinct in the wild and exist only in captivity. These animals are examples of **ex situ conservation**—the preservation of species in zoos, aquaria, wildlife ranches, botanical gardens, or other artificial settings.

Translated literally, ex situ means out of place. The name was inspired by a contrast with in situ ("in place") conservation—the establishment of protected areas where populations can be maintained in their natural habitat. Ex situ conservation is not ideal, but it is preferred over extinction.

In some cases, captive breeding of ex situ individuals and re-introduction into the wild have succeeded in creating wild populations of endangered species. Recall, for example, the successful effort to establish new flocks of whooping cranes by using eggs and young produced in zoos and other breeding centers (see Chapter 54). The peregrine falcon, California condor, and several other species have been rescued from the brink of extinction by captive breeding and reintroduction to the wild.

Genetic Restoration When habitat fragmentation results in small, isolated populations that are vulnerable to the loss of genetic diversity from chance events, genetic restoration can sometimes be accomplished by artificial gene flow.

Consider the case of a remnant population of the endangered Florida panther in the swampland of southern Florida. By the early 1990s, only about 25 adults remained, and they exhibited several recessive disorders and low genetic diversity—both signs of inbreeding. Using a population viability analysis (PVA; see Section 57.2), conservationists gave the population a 95 percent likelihood of extinction in two decades.

In an effort to prevent this outcome, the U.S. Department of the Interior moved eight panthers (locally called "pumas") from Texas to Florida to introduce new alleles into the Florida population. Geneticist Stephen O'Brien and colleagues followed the genetic consequences, some of which are shown in **FIGURE 57.15** (see inbreeding depression data in Chapter 26). The success of the genetic restoration was evident. The number of panthers quadrupled in 10 years, and the offspring of hybridized animals were larger, stronger, and had reduced incidents of defects.

Note, however, that as time went on, the rate of increase in population size leveled off. Also, the panther habitat continues to deteriorate. Ongoing conservation efforts are needed to successfully protect this panther population from extinction.

Wildlife Corridors One longer-term solution to enabling gene flow among isolated populations is the establishment of **wildlife corridors,** strips of undeveloped habitat that connect preserved areas. In some cases, wildlife corridors are as simple as a walkway under a major highway, so that animals such as salamanders can move from one side to the other without being killed. In other cases, a corridor can consist of 1000 km of makeshift paths across private and public lands, such as the corridors used by isolated populations of tigers in southern India.

By facilitating the movement of individuals, the goals of corridors are to

- allow areas to be recolonized if a species or a population was lost;

- encourage gene flow that introduces new alleles—counteracting the deleterious effects of genetic drift and inbreeding.

Do wildlife corridors actually work? To answer this question experimentally, ecologist Ellen Damschen and colleagues established a series of restored natural areas in the middle of a

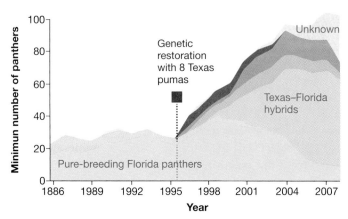

FIGURE 57.15 Experimental Evidence that Gene Flow Can Alleviate Genetic Problems in Small Populations. After eight Texas pumas were introduced to Florida in 1995, the Florida population of panthers quadrupled in size.

DATA: Johnson, W. E., D. P. Onorato, M. E. Roelke, et al. 2010. *Science* 329: 1641–1645.

large, species-poor pine plantation in South Carolina. This project was the equivalent of restoring patches of grassland habitat in the middle of an enormous cornfield. Some of the restoration sites were connected by corridors, while others were isolated. By monitoring the composition of species inside each habitat patch over time, the group was able to show that when compared to unconnected patches, connected patches are steadily gaining more species over time (**FIGURE 57.16**). This is strong evidence that wildlife corridors can increase overall species richness in a metapopulation.

Designing Effective Protected Areas By far the most effective strategy for preserving biodiversity is to protect entire areas of land and sea from exploitation in the first place.

The bad news is that more than two-thirds of the terrestrial biodiversity hotspots lack protection entirely, or have incomplete protection, and less than 0.5 percent of the oceans are protected. Incomplete protection occurs when areas are designated as protected but have insufficient enforcement—such as from poachers.

The good news is that the total amount of protected area is increasing at a steady rate globally. By strategically protecting prime habitat areas, countries worldwide can realize many benefits. For

Unconnected
(winged)

Connected

Unconnected
(rectangular)

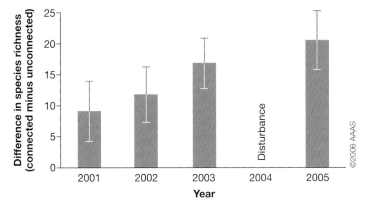

©2006 AAS

Disturbance

FIGURE 57.16 Experimental Evidence that Wildlife Corridors Work. This graph plots the difference in species richness between connected and unconnected plots. Species richness could not be assessed in 2004 because the study plots were burned as part of a program to restore natural habitats.

DATA: Damschen, E. I., N. M. Haddad, J. L. Orrock, et al. 2006. *Science* 313: 1284–1286.

✓**QUANTITATIVE** Draw and label bars showing what this graph would look like if corridors had no effect on species richness.

example, preservation of rain forest in Borneo solves many problems: reducing incidence of droughts, floods, and fires, protecting diverse species, and storing carbon.

Researchers are using a geographic approach called the Gap Analysis Program (GAP) to assess the effectiveness of the current system of protected areas. A GAP tries to identify gaps between geographic areas that are particularly rich in biodiversity and areas that are actually managed for the preservation of biodiversity. Mismatches help to identify areas that are high priorities for conservation efforts.

Conservation of Ecosystem Function

As the biological and economic value of the associated services becomes more apparent, many countries are paying more attention to the conservation and restoration of ecosystem function than ever before. Can degraded ecosystems be restored?

Ecosystem Restoration In many areas of the world, ecosystems are already heavily degraded or lost. The state of Illinois, for example, was estimated to contain 22 million acres of prairie when European settlers arrived; today only about 2000 acres of this

biome remain (0.009 percent). When natural areas are badly degraded or gone, biologists turn to restoration.

Thousands of large-scale and small-scale ecosystem restoration and reforestation projects are now occurring around the globe. Recall the dire scenario in Borneo described at the beginning of this chapter. Willie Smits had been working in Borneo for two decades in a desperate attempt to slow the rate of deforestation and save orangutans from the brink of extinction. Frustrated by his lack of progress, he decided to try a new approach—the restoration of 2000 ha of clear-cut rain forest in Samboja.

While Smits would be the first to agree that primary forest is of greater ecological value than restored forest, his progress in just a few years since 2002 is remarkable:

- Over 1600 tree species in a multilayered canopy
- 137 species of birds (up from 5); 9 species of primates
- rainfall up 25 percent; temperature down 3–5°C
- 3000 people earning income

The cornerstone of Smits's Samboja Lestari ("Everlasting Forest") project is integrating the ecological restoration of the forest along with the economic restoration of the local Dayak tribe—a win–win scenario (**FIGURE 57.17**). Although the size of Samboja Lestari is relatively small and the clear-cutting of rain forest continues in Borneo at a rapid pace, Smits's success is reason for hope. How successful are ecosystem restoration projects in general?

Ecologists Holly Jones and Oswald Schmitz analyzed 240 of such restoration projects to determine the resilience of ecosystems after human disturbances. **FIGURE 57.18** (on page 1192) shows some of their surprising results. Recovery of biomass following disturbances was much slower in forests than in other ecosystems, but unexpectedly fast overall—decades rather than centuries. Similarly, biomass recovery was much slower in agricultural and deforested areas, but more rapid overall than anticipated.

FIGURE 57.17 Ecosystem Restoration Effort in Samboja Lestari, Borneo. Rain forest restoration in Borneo is critical to the survival of orangutans. Indonesian workers prepare rescued orangutans for release into forest sanctuaries like Samboja.

(a) Recovery time in different types of restored ecosystems

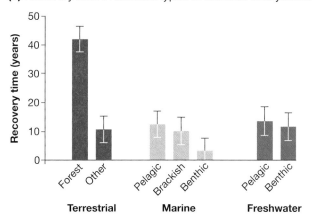

(b) Recovery time following different types of disturbances

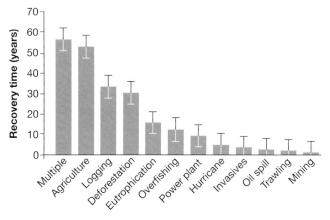

FIGURE 57.18 Average Recovery Times of Ecosystem Biomass.
(a) Recovery of biomass following disturbances was much slower in forests than in other ecosystems, but surprisingly fast overall. **(b)** Recovery was much slower in agricultural and deforested areas, but relatively rapid in other areas.

DATA: Jones, H. P., and O. J. Schmitz. 2009. *PLoS ONE* 4: e5653.

Biomass recovery is not equivalent to biodiversity or ecosystem function, but many of the restoration projects show positive outcomes in terms of species richness and other variables. It is not too late to restore many degraded habitats.

In general, restoration success is positively correlated with the size of the area restored, its connectedness, remoteness, and the intactness of surrounding areas. However, another take-home message from the restoration studies is that an accumulation of many small changes can also have a big effect. If you and all your neighbors planted several native trees, shrubs, or other plants in your yards or neighborhoods, you could have a very positive effect on biodiversity and ecosystem function right where you live.

Quantifying Ecosystem Services Many ecologists are convinced that the traditional approaches to conservation are doomed to fail on their own—nature reserves are too small and isolated to succeed in preserving sufficient biodiversity and ecosystem

function. Many of these scientists advocate for an ecosystem services approach to conservation—a quantification of ecosystem services to make explicit in the marketplace that long-term human prosperity depends on the preservation of these services.

The idea is that the cost of replacing ecosystem services is at least 10 to 100 times higher than the cost of maintaining them—if they are replaceable at all (in which case they are infinitely valuable). Therefore, it is an economic imperative to protect these services before they are lost.

The problem is that quantifying ecosystem services is not an easy task, and incorporating these services into the marketplace requires a complicated collaboration among scientists, governments, nongovernmental organizations (NGOs), businesses, educators, communities, and local leadership. Further, policy mechanisms can be challenging to implement.

Successful experiments are under way, however. For example, the Costa Rican government has made policy changes that allow landowners to be compensated for the ecosystem services provided by their properties, such as carbon storage, biodiversity conservation, and scenic beauty. The payments are funded largely from transportation taxes. The logic here is that the Costa Rican economy will profit more in the long term from **ecotourism**—recreational visits to their wild places (**FIGURE 57.19**)—if their wild places are preserved from logging by private owners in the short term. But the private owners must be compensated.

Numerous other similar programs are in development. A global program called REDD (Reducing Emissions from Deforestation

FIGURE 57.19 Ecotourism in Costa Rica. Ecotourists flock to witness the spectacular biodiversity of the Monteverde Cloud Forest Reserve in Costa Rica.

and forest Degradation) is designed to reduce emissions from de-forestation by paying landowners *not* to clear-cut—thereby slowing climate change. Other than curbing the production of greenhouse gases in the first place, the preservation of forests is currently the single most effective method of slowing climate change—which is otherwise expected to have enormous economic impacts.

In general, although social and political changes are difficult to implement, outcomes that boost biodiversity and ecosystem function are a boon to humans.

Take-Home Message

Your generation is facing the most serious global environmental crisis in the history of our species. The decisions you make, ranging from how many children you have to how much you drive a car, will have far-reaching consequences. Change happens one person at a time. It can be done.

As someone with a background in biology, you have the intellectual tools to make an important, positive impact. In addition to studying life, biologists have an obligation to help save it.

CHAPTER 57 REVIEW

For media, go to MasteringBiology

If you understand . . .

57.1 What Is Biodiversity?

- Biodiversity is quantified at the level of genetic diversity, species diversity, and ecosystem diversity.

- Only about 1.5 million species have been named out of an estimated 10 to 100 million species on Earth.

- On average, biodiversity tends to be higher in the tropics than near the poles, higher on land than in the sea, and higher in areas with geographic diversity.

- Tropical rain forests are particularly species rich, containing at least 50 percent of all species.

- Biodiversity hotspots can be mapped based on species richness, areas of endemism, degree of threat, and/or unique ecological or evolutionary characteristics.

✔ You should be able to explain how you could measure biodiversity in the restored rain forest of Borneo at the genetic, species, and ecosystem levels.

57.2 Threats to Biodiversity

- The sixth mass extinction in the history of life is occurring due to human impacts. Species are vanishing faster than at virtually any other time in Earth's history.

- Habitat destruction and degradation are currently the leading causes of extinction. Conversion of primary forest to agricultural fields and housing developments is particularly devastating.

- Although overexploitation also occurs on land, it is the dominant threat for marine species.

- Invasive species are a long-standing problem on islands, but now they also threaten communities on continents.

- Many types of pollutants are found virtually everywhere on Earth, but they are particularly threatening to aquatic species.

- Climate change is a significant threat to biodiversity due to both direct and indirect impacts.

- To estimate how many species will go extinct in the near future, biologists make calculations based on combinations of variables including current extinction rates, habitat sizes, predicted shifts in climate envelopes, demographics, and species interactions.

✔ You should be able to explain why small, geographically isolated populations of orangutans in isolated islands of forest are at high risk of extinction.

57.3 Why Is Biodiversity Important?

- At the ecosystem level, experiments have shown that high species richness increases aspects of ecosystem function such as productivity, resistance to disturbance, and ability to recover from disturbance.

- Humans gain direct economic benefits from fishing, forestry, agriculture, and other activities that depend on materials provided by ecosystems.

- Humans also benefit from the ability of functioning ecosystems to build and hold soil, moderate local climates, retain and cycle nutrients, retain surface water and recharge groundwater, prevent flooding, and produce oxygen.

- Human happiness and health depend on healthy ecosystems.

✔ You should be able to explain why the high species richness of rain forests leads to high productivity, in terms of the efficiency of resource use, when compared to oil palm plantations.

57.4 Preserving Biodiversity and Ecosystem Function

- The major threats to biodiversity—habitat destruction and climate change—will not lessen until the human population stabilizes and resource use becomes sustainable.

- Solutions to the biodiversity crisis include preserving genetic diversity, protecting key habitats, restoring ecosystems, and integrating conservation with public policy so that protecting ecosystem services is a greater part of both global and local decision-making processes.

- Ecosystem restoration efforts are having a positive impact on biodiversity around the world. Conservation efforts make a difference.

✔ You should be able to evaluate whether the 12 percent of land area protected thus far will be enough to avert a mass extinction.

(MB) MasteringBiology

1. **MasteringBiology Assignments**

   Tutorials and Activities Global Fisheries and Overfishing; Forestation Change; Science, Technology, and Society: DDT; Madagascar and the Biodiversity Crisis; Habitat Fragmentation; Conservation Biology Review; Biodiversity

   Questions Reading Quizzes, Blue-Thread Questions, Test Bank

2. eText Read your book online, search, take notes, highlight text, and more.

3. The Study Area Practice Test, Cumulative Test, BioFlix® 3-D Animations, Videos, Activities, Audio Glossary, Word Study Tools, Art

You should be able to . . .

✔ TEST YOUR KNOWLEDGE
Answers are available in Appendix A

1. What does a species–area plot show?
   a. the overall distribution, or area, occupied by a species
   b. the relationship between the body size of a species and the amount of territory or home range it requires
   c. the number of species found, on average, in tropical versus northern areas
   d. the number of species found, on average, in a habitat of a given size

2. What does species richness refer to?
   a. the number of species in an area
   b. the quality of species in an area
   c. the functional importance of a species
   d. the abundance of individuals within a species

3. What is resilience?
   a. resistance to change during a disturbance
   b. the ability to recover from a disturbance
   c. production of aboveground biomass
   d. total biomass production (above- and belowground)

4. What is a biodiversity "hotspot"?
   a. an area where an all-taxon survey is under way
   b. an area where an environmental sequencing study has been completed
   c. a habitat with high NPP
   d. an area with high species richness and high threat

5. Why do small populations become inbred?
   a. They are often part of a metapopulation structure.
   b. Genetic drift becomes a prominent evolutionary force.
   c. Over time, all individuals become increasingly related.
   d. Natural selection does not operate efficiently in small populations.

6. What is the primary cause of endangerment in marine environments?

✔ TEST YOUR UNDERSTANDING
Answers are available in Appendix A

7. If you were asked to evaluate building materials, what would be most "sustainable" about the use of bamboo as a building material?
   a. Bamboo is an invasive species due to its rapid growth rate.
   b. Bamboo is harvested at the same rate at which it grows.
   c. Bamboo is durable so it does not need to be replaced often.
   d. Bamboo plantations create habitat for other species.

8. Biologists claim that the all-taxa survey now under way at the Great Smoky Mountains National Park in the United States will improve their ability to estimate the total number of species living today. Discuss the benefits and limitations that this data set will provide in understanding the extent of global biodiversity.

9. Some biologists prefer to focus efforts on preserving endangered species while others prefer to focus on preserving ecosystems. What is your opinion?

10. What evidence supports the hypothesis that species richness increases ecosystem functions, such as productivity, resistance to disturbance, and resilience?

11. Explain why the construction of wildlife corridors can help maintain biodiversity in a fragmented landscape.

12. What are ecosystem services? Does anyone own them or pay for them—that is, can you make money from them? How does quantifying ecosystem services affect efforts to protect ecosystems?

13. The population size of mountain pine beetles in the American Northwest has long been held in check by freezing temperatures during the winters. As winters warm, populations of pine beetles are increasing, killing whitebark pine forest. Grizzly bears rely on the nuts of whitebark pine trees. Do you think climate change is affecting the fitness of the grizzly bear population?
 a. No, because climate change is not occurring yet.
 b. No, because climate change is affecting the beetles, not the bears.
 c. Yes, because climate change is decreasing the food supply of the bears, which means they will be less strong.
 d. Yes, because climate change is decreasing the food supply of the bears, which is likely to reduce their ability to survive or reproduce.

14. You are helping to design a series of reserves in a tropical country. List the steps you would recommend for gathering data and creating a plan that would protect a large amount of biodiversity in a small amount of land.

15. The maps to the right chronicle the loss of old-growth forest (more than 200 years old) that occurred in Warwickshire, England, and in the United States. In your opinion, under what conditions is it ethical for conservationists who live in these countries to lobby government officials in Brazil, Indonesia, and other tropical countries to slow the rate of loss of old-growth forest?

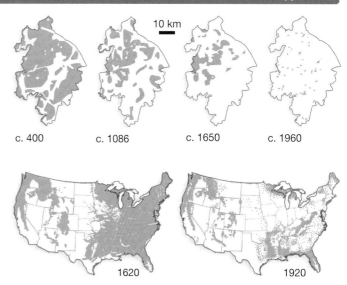

10 km

c. 400   c. 1086   c. 1650   c. 1960

1620   1920

16. Make a list of characteristics that would render a species particularly vulnerable to extinction by humans. Make a list of characteristics that would render a species particularly resistant to pressure from humans. Try to think of an example of each type of species.

BIG PICTURE

The Big Picture

Ecology's primary goal is to understand the distribution and abundance of organisms. To reach this goal, ecologists study how organisms interact with each other and their abiotic environments.

Human activities are causing radical changes in population sizes, climate, soils, water, and atmosphere. As a result, virtually every box and arrow in this concept map is being affected.

In addition to documenting changes, biologists are educating the public and policy makers about the value of biodiversity and healthy ecosystems, and researching the most effective ways to preserve species and keep ecosystems functioning normally.

Note that most boxes in the concept map indicate chapters and sections where you can go for review. Also, be sure to do the blue exercises in the Check Your Understanding box below.

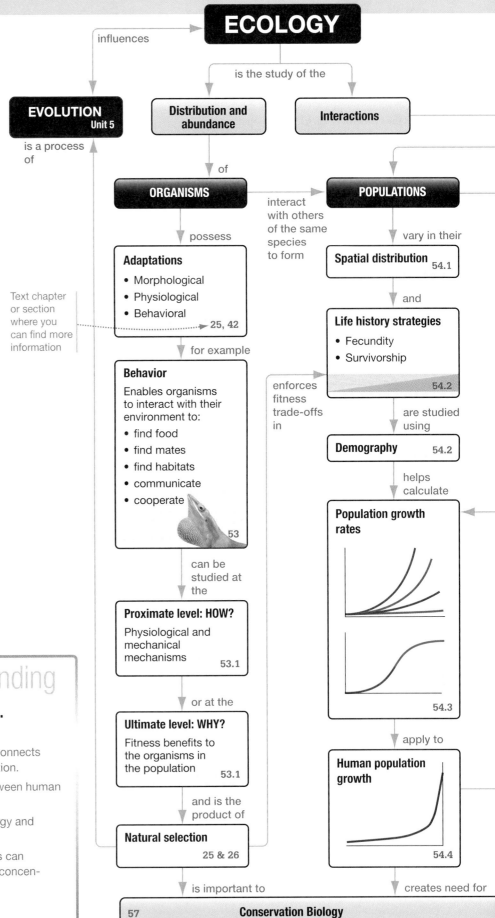

ECOLOGY

influences

is the study of the

EVOLUTION
Unit 5

is a process of

Distribution and abundance

Interactions

of

ORGANISMS

interact with others of the same species to form

POPULATIONS

Text chapter or section where you can find more information

possess

vary in their

Spatial distribution 54.1

Adaptations
• Morphological
• Physiological
• Behavioral
25, 42

and

Life history strategies
• Fecundity
• Survivorship
54.2

for example

enforces fitness trade-offs in

are studied using

Behavior
Enables organisms to interact with their environment to:
• find food
• find mates
• find habitats
• communicate
• cooperate
53

Demography 54.2

helps calculate

Population growth rates
54.3

can be studied at the

Proximate level: HOW?
Physiological and mechanical mechanisms 53.1

or at the

apply to

Ultimate level: WHY?
Fitness benefits to the organisms in the population 53.1

Human population growth
54.4

and is the product of

Natural selection
25 & 26

is important to

creates need for

57   Conservation Biology

check your understanding

 C

 Y

 U

If you understand the big picture . . .

✔ You should be able to . . .

1. Add an arrow and linking phrase that connects species interactions with natural selection.

2. Explain why there is a relationship between human population growth and biodiversity.

3. Compare and contrast the flow of energy and nutrients through ecosystems.

4. Give examples of how human activities can increase or decrease the atmospheric concentration of CO_2.

Answers are available in Appendix A.

1196

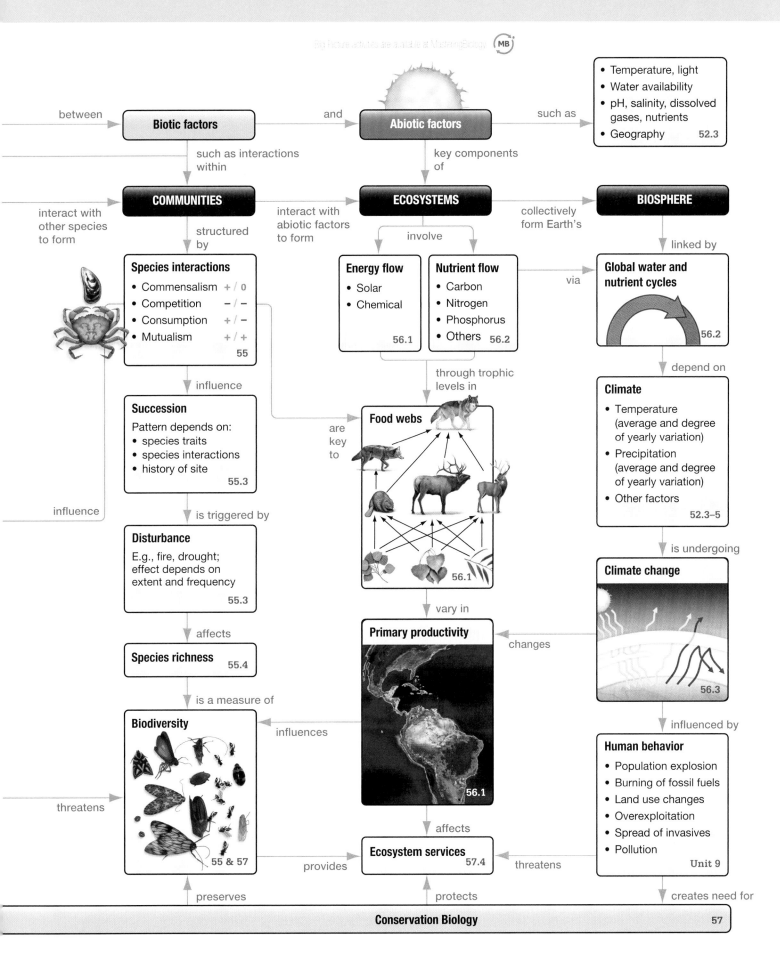

Big Picture activities are available at MasteringBiology **MB**

Biotic factors —between—
Biotic factors —and→ **Abiotic factors** —such as→
- Temperature, light
- Water availability
- pH, salinity, dissolved gases, nutrients
- Geography   52.3

Biotic factors —such as interactions within→ **COMMUNITIES**

Abiotic factors —key components of→ **ECOSYSTEMS**

COMMUNITIES —interact with other species to form→
ECOSYSTEMS —interact with abiotic factors to form→
ECOSYSTEMS —collectively form Earth's→ **BIOSPHERE**

COMMUNITIES —structured by→ **Species interactions**
- Commensalism + / 0
- Competition   − / −
- Consumption   + / −
- Mutualism   + / +
                          55

ECOSYSTEMS —involve→ **Energy flow**
- Solar
- Chemical
                56.1

Nutrient flow
- Carbon
- Nitrogen
- Phosphorus
- Others   56.2

Nutrient flow —via→ **Global water and nutrient cycles**   56.2

BIOSPHERE —linked by→ Global water and nutrient cycles

Global water and nutrient cycles —depend on→ **Climate**
- Temperature (average and degree of yearly variation)
- Precipitation (average and degree of yearly variation)
- Other factors
                          52.3–5

Species interactions —influence→ **Succession**
Pattern depends on:
- species traits
- species interactions
- history of site
                          55.3

Food webs
                          56.1

Energy flow / Nutrient flow —are key to→ Food webs
Nutrient flow —through trophic levels in→ Food webs

Climate —is undergoing→ **Climate change**   56.3

Succession —is triggered by→ **Disturbance**
E.g., fire, drought; effect depends on extent and frequency
                          55.3

Food webs —vary in→ **Primary productivity**   56.1

Climate change —changes→ Primary productivity

Disturbance —affects→ **Species richness**   55.4

Species richness —is a measure of→ **Biodiversity**

Primary productivity —influences→ Biodiversity

Climate change —influenced by→ **Human behavior**
- Population explosion
- Burning of fossil fuels
- Land use changes
- Overexploitation
- Spread of invasives
- Pollution
                          Unit 9

Biodiversity
                          55 & 57

Biotic factors —interact with other species to form / influence→ Biodiversity —threatens→

Primary productivity —affects→ **Ecosystem services**   57.4

Human behavior —threatens→ Ecosystem services

Biodiversity —provides→ Ecosystem services

Biodiversity —preserves→
Ecosystem services —protects→
Human behavior —creates need for→ **Conservation Biology**   57

Answers

CHAPTER 1

IN-TEXT QUESTIONS AND EXERCISES

p. 4 Fig. 1.2 `analyze` If Pasteur had done any of the things listed, he would have had more than one variable in his experiment. This would allow critics to claim that he got different results because of the differences in broth types, heating, or flask types—not the difference in exposure to preexisting cells. The results would not be definitive.

p. 6 `apply` The average kernel protein content would decline, from 11 percent to a much lower value over time.

p. 6 CYU `apply` The data points would all be about 11 percent, indicating no change in average kernel protein content over time.

p. 7 Fig. 1.4 `apply` Molds and other fungi are more closely related to green algae because they differ from plants at two positions (5 and 8 from left) but differ from green algae at only one position (8).

p. 8 Fig. 1.6 `apply` The eukaryotic cell is roughly 10 times the size of the prokaryotic cell.

p. 9 CYU `evaluate` From the sequence data provided, species A and B differ only in one ribonucleotide of the rRNA sequence (position 10 from left). Species C differs from species A and B in four ribonucleotides (positions 1, 2, 9, and 10). A correctly drawn phylogenetic tree would indicate that species A and B appear to be closely related, and species C is more distantly related. See **FIGURE A1.1**.

p. 13 (1) `analyze` You could conclude that the ants weren't navigating normally, because they had been caught and released and transferred to a new channel. **(2)** `analyze` You could conclude that the ants can't navigate normally on their manipulated legs.

p. 13 CYU The key here is to test predation rates during the hottest part of the day (when desert ants actually feed) versus other parts of the day. The experiment would best be done in the field, where natural predators are present. One approach would be to capture a large number of ants, divide the group in two, and measure predation rates (number of ants killed per hour) when they are placed in normal habitat during the hottest part of the day versus an hour before (or after). **(1)** `analyze` The control group here is the normal condition—ants out during the hottest part of the day. If you didn't include a control, a critic could argue that predation did or did not occur because of your experimental setup or manipulation, not because of differences in temperature. **(2)** `analyze` You would need to make sure that there is no difference in body size, walking speed, how they were captured and maintained, or other traits that might make the ants in the two groups more or less susceptible to predators. They should also be put out in the same habitat, so the presence of predators is the same in the two treatments.

p. 12 Fig. 1.9 `analyze` The interpretation of the experiment would not likely change, but your confidence in the conclusions drawn would be reduced if you used just one ant.

IF YOU UNDERSTAND . . .

1.1. `understand` Dead cells cannot regulate the passage of materials between exterior and interior spaces, replicate, use energy, or process information. **1.2** `understand` Observations on thousands of diverse species supported the claim that all organisms consist of cells. The hypothesis that all cells come from preexisting cells was supported when Pasteur showed that new cells do not arise and grow in a boiled liquid unless they are introduced from the air. **1.3** `understand` If seeds with higher protein content leave the most

offspring, then individuals with low protein in their seeds will become rare over time. **1.4** `understand` A newly discovered species can be classified as a member of the Bacteria if the sequence of its rRNA contains some features found only in Bacteria. The same logic applies to classifying a new species in the Archaea or Eukarya. **1.5** `understand` (1) A hypothesis is an explanation of how the world works; a prediction is an outcome you should observe if the hypothesis is correct. (2) Experiments are convincing because they measure predictions from two opposing hypotheses. Both predicted actions cannot occur, so one hypothesis will be supported while the other will not.

YOU SHOULD BE ABLE TO . . .

✔ Test Your Knowledge

1. `remember` d **2.** `understand` d **3.** `remember` populations **4.** `understand` b **5.** `understand` An individual's ability to survive and reproduce **6.** `understand` c

✔ Test Your Understanding

7. `evaluate` That the entity they discovered replicates, processes information, acquires and uses energy, is cellular, and that its populations evolve. **8.** `understand` a **9.** `understand` Over time, traits that increased the fitness of individuals in this habitat became increasingly frequent in the population. **10.** `understand` Individuals with certain traits are selected, in the sense that they produce the most offspring. **11.** `analyze` Yes. If evolution is defined as "change in the characteristics of a population over time," then those organisms that are most closely related should have experienced less change over time. On a phylogenetic tree, species with substantially similar rRNA sequences would be diagrammed with a closer common ancestor—one that had the sequences they inherited—than the ancestors shared between species with dissimilar rRNA sequences. **12.** `understand` A null hypothesis specifies what a researcher should observe when the hypothesis being tested isn't correct.

✔ Test Your Problem-Solving Skills

13. `analyze` A scientific theory is not a guess—it is an idea whose validity can be tested with data. Both the cell theory and the theory of evolution have been validated by large bodies of observational and experimental data. **14.** `apply` If all eukaryotes living today have a nucleus, then it is logical to conclude that the nucleus arose in a common ancestor of all eukaryotes, indicated by the arrow you should have added to the figure. See **FIGURE A1.2**. If it had arisen in a common ancestor of Bacteria or Archaea, then species in those groups would have had to lose the trait—an unlikely event. **15.** `evaluate` The data set was so large and diverse that it was no longer reasonable to argue that noncellular life-forms would be discovered. **16.** `apply` b

BIG PICTURE Doing Biology

p. 16 CYU (1) `understand` Biologists design and carry out a study, either observational or experimental, to test their ideas. As part of this process, they state their ideas as a hypothesis and null hypothesis and make predictions. They analyze and interpret the data they have gathered, and determine whether the data support their ideas. If not, they revisit their ideas and come up with an alternative hypothesis and design another study to test these new predictions. **(2)** `understand` There are many

possible examples. Consider, for example, the experiment on navigation in foraging desert ants (Chapter 1). In addition to testing how the ants use information on stride length and number to calculate how far they are from the nest (multicellular organism and population levels), researchers also could test how the "pedometer" works at the level of cells and molecules. **(3)** `analyze` A hypothesis is a testable statement to explain a specific phenomenon or a set of observations. The word theory refers to proposed explanations for very broad patterns in nature that are supported by a wide body of evidence. A theory serves as a framework for the development of new hypotheses. **(4)** `analyze` The next step is to relate your findings to existing theories and the current scientific literature, and then to communicate your findings to colleagues through informal conversations, presentations at scientific meetings, and eventually publication in peer-reviewed journals.

CHAPTER 2

IN-TEXT QUESTIONS AND EXERCISES

p. 20 Fig. 2.3 `apply` There are 15 electrons in phosphorus, so there must be 15 protons, which is the atomic number. Since the mass number is 31, then the number of neutrons is 16.

p. 23 `understand` *Water:* arrows pointing from hydrogens to oxygen atom; *ammonia:* arrows pointing from hydrogens to nitrogen atom; *methane:* double arrows between carbons and hydrogens; *carbon dioxide:* arrows pointing from carbon to oxygens; *molecular nitrogen:* double arrows between nitrogens.

p. 23 Fig. 2.7 `understand` Oxygen and nitrogen have high electronegativities. They hold shared electrons more tightly than C, H, and many other atoms, resulting in polar bonds.

p. 24 CYU `evaluate` See **FIGURE A2.1**.

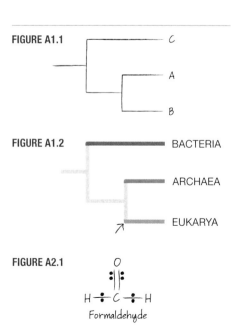

FIGURE A1.1

FIGURE A1.2 — BACTERIA, ARCHAEA, EUKARYA

FIGURE A2.1 — Formaldehyde

p. 25 (1) [evaluate] δ^+H—O$^{\delta-}$—H$^{\delta+}$ (2) [apply] If water were linear, the partial negative charge on oxygen would have partial positive charges on either side. Compared to the actual, bent molecule, the partial negative charge would be much less exposed and less able to participate in hydrogen bonding.

p. 26 Fig. 2.14 [understand] Oils are nonpolar. They have long chains of carbon atoms bonded to hydrogen atoms, which share electrons evenly because their electronegativities are similar. When an oil and water are mixed, the polar water molecules interact with each other via hydrogen bonding rather than with the nonpolar oil molecules, which interact with themselves instead.

p. 28 Table 2.2 [understand] "Cause" (Row 1): electrostatic attractions between partial charges on water and opposite charges on ions; hydrogen bonds; water and other polar molecules. "Biological Consequences" (Row 2): ice to float; freezing solid. "Cause" (Row 4): lots of heat energy; break hydrogen bonds and change water to a gas.

p. 29 [apply] The proton concentration would be 3.2×10^{-9} M.

p. 29 Fig. 2.17 [apply] The concentration of protons would decrease because milk is more basic (pH 6.5) than black coffee (pH 5).

p. 30 [apply] The bicarbonate concentration would increase. The protons (H$^+$) released from carbonic acid would react with the hydroxide ions (OH$^-$) dissociated from NaOH to form H$_2$O, leaving fewer protons free to react with bicarbonate to reform carbonic acid.

p. 32 CYU (1) [apply] The reaction would be spontaneous based on the change in potential energy; the reactants have higher chemical energy than the products. The entropy, however, is not increased based on the number of molecules, although heat given off from the reaction still results in increased entropy. (2) [understand] The electrons are shifted farther from the nuclei of the carbon and hydrogen atoms and closer to the nuclei of the more electronegative oxygen atoms.

p. 32 Fig. 2.19 [remember] See FIGURE A2.2.

p. 33 Fig. 2.21 [analyze] The water-filled flask is the ocean; the gas-filled flask is the atmosphere; the condensed water droplets are rain; the electrical sparks are lightning.

p. 37 Table 2.3 [understand] All the functional groups in Table 2.3, except the sulfhydryl group (—SH), are highly polar. The sulfhydryl group is only very slightly polar.

IF YOU UNDERSTAND . . .

2.1. [understand] The bonds in methane and ammonia are all covalent, but differ in polarity: Methane has nonpolar covalent bonds while ammonia has polar covalent bonds. Sodium chloride does not have covalent bonds; instead, ionic bonds hold the ionized sodium and chloride together. 2.2 [understand] Assuming neutral pH, amino and hydroxyl groups would interact with the partial negative charge on water's oxygen, because they both carry a positive charge (partial for the hydrogen in the hydroxyl). The carboxyl group would interact with the partial positive charges on water's hydrogens, since it would carry a negative charge after losing the proton from its hydroxyl. 2.3 [understand] Like solar radiation, the energy in electricity generates free radicals that would promote the reaction. 2.4 [understand] "Top-down" approach: The reaction responsible for synthesizing acetic acid is observed in cells and can serve as an intermediate for the formation of a more complex molecule (acetyl CoA) that is used by cells throughout the tree of life. "Bottom-up" approach: This reaction can also occur under conditions that mimic the early Earth environment in deep-sea vents. 2.5 [apply] The hydroxyls would increase the solubility of octane by introducing polar covalent bonds, which would make the molecule more hydrophilic. The high electronegativity of oxygen would decrease the potential energy of the modified molecule.

YOU SHOULD BE ABLE TO . . .

✓ Test Your Knowledge

1. [understand] b 2. [remember] a 3. [remember] c 4. [remember] d
5. [remember] Potential energy and entropy. 6. [remember] The prebiotic soup model and the surface metabolism model.

✓ Test Your Understanding

7. [apply] c. Acetic acid has more highly electronegative oxygen atoms than the other molecules. When bonded to carbon or hydrogen, each oxygen will result in a polar covalent bond. 8. [apply] Relative electronegativities would be F > O > H > Na. One bond would form with sodium, and it would be ionic. 9. [understand] When oxygen is covalently bonded to hydrogen, the difference in electronegativities between the atoms causes the electrons to spend more time near the oxygen. In contrast, the atoms in H$_2$ and O$_2$ have the same electronegativities, so they equally share electrons in their covalent bonds. 10. [apply] See FIGURE A2.3. 11. [apply] The dissociation reaction of carbonic acid lowers the pH of the solution by releasing extra H$^+$ into the solution. If additional CO$_2$ is added, the sequence of reactions would be driven to the right, which would make the ocean more acidic. 12. [understand] The carbon framework determines the overall shape of an organic molecule. The functional groups attached to the carbons determine the molecule's chemical behavior, because these groups are likely to interact with other molecules.

✓ Test Your Problem-Solving Skills

13. [analyze] b. 14. [analyze] No, they don't conflict. Shells that are farther from the protons (positive charges) in the nucleus house electrons that have greater potential energy than shells closer to the nucleus. 15. [analyze] One possible concept map relating the structure of water to its properties is shown below (see FIGURE A2.4). 16. [create] In hot weather, water absorbs large amounts of heat due to its high specific heat and high heat of vaporization. In cold weather, water releases the large amount of heat that it has absorbed.

FIGURE A2.2

FIGURE A2.3

FIGURE A2.4

CHAPTER 3

IN-TEXT QUESTIONS AND EXERCISES

p. 43 Fig. 3.2 `understand` The green R-groups contain mostly C and H, which have roughly equal electronegativities. Electrons are evenly shared in C−H bonds and C−S bonds, so the groups are nonpolar. Cysteine has a sulfur that is slightly more electronegative than hydrogen, so it will be less nonpolar than the other green groups. All of the pink R-groups have a highly electronegative oxygen atom with a partial negative charge, making them polar.

p. 44 `apply` From most hydrophilic to most hydrophobic: (1) aspartate, (2) asparagine, (3) tyrosine, (4) valine. The most hydrophilic amino acids will have side chains with full charges (ionized), like aspartate, followed by those with the largest number of highly electronegative atoms, like oxygen or nitrogen. Highly electronegative atoms produce polar covalent bonds with carbon or hydrogen. The most hydrophobic will not have oxygen or nitrogen in their side chains, but instead will have the largest number of C−H bonds, which are nonpolar covalent.

p. 47 CYU `understand` See **FIGURE A3.1**.

p. 52 CYU (1) `understand` Secondary, tertiary, and quaternary structure all depend on bonds and other interactions between amino acids that are linked in a chain in a specific order (primary structure). **(2)** `apply` There would be 20^5 peptides, or 3.2×10^6 different primary sequences.

p. 55 CYU `apply` Amino acid changes would be expected to be in the active site or in regions that affect the folded structure of this site. Either of these changes could result in a different active site that either binds a new substrate or catalyzes a different reaction.

IF YOU UNDERSTAND . . .

3.1. `understand` Look at the R-group of the amino acid. If there is a positive charge, then it is basic. If there is a negative charge, then it is acidic. If there is not a charge, but there is an oxygen atom, then it is polar uncharged. If there is no charge or oxygen, then it is nonpolar.

3.2 `apply` Nonpolar amino acid residues would be found in the interior of a globular protein, grouped with other nonpolar residues due to hydrophobic interactions.

3.3 `analyze` Both calmodulin and infectious prions require some form of induction to achieve their active conformations. Calcium ions are required for calmodulin to fold into its functional structure while prions are induced to change their shape by other, improperly folded prion proteins. **3.4** `analyze` *Catalysis:* Proteins are made of amino acids, which have many reactive functional groups, and can fold into different shapes that allow the formation of active sites. *Defense:* Similar to catalysis, the chemical properties and capacity for different shapes allows proteins to be made that can attach to virtually any type of invading virus or cell. *Signaling:* The flexibility in protein structure allows protein activities to be quickly turned on or off based on binding to signal molecules or ions.

YOU SHOULD BE ABLE TO . . .

✔ Test Your Knowledge

1. `remember` d **2.** `remember` The atoms and functional groups found in the side chains. **3.** `remember` b **4.** `remember` b **5.** `understand` The order and type of amino acids (i.e., the primary structure) contains the information that directs folding. **6.** `understand` a

✔ Test Your Understanding

7. `understand` Because the nonpolar amino acid residues are not able to interact with the water solvent, they are crowded together in the interior of a protein and surrounded by a network of hydrogen-bonded water molecules. This crowding leads to the development of van der Waals interactions that help glue the nonpolar

side chains together. **8.** `understand` No, polymerization is a nonspontaneous reaction because the product molecules have lower entropy than the free form of the reactants and there would be nothing to prevent hydrolysis from reversing the reaction. **9.** `understand` Many possible correct answers, including (1) the presence of an active site in an enzyme that is precisely shaped to fit a substrate or substrates in the correct orientation for a reaction to occur; (2) the doughnut shape of porin that allows certain substances to pass through it; (3) the cable shape of collagen to provide structural support for cells and tissues. **10.** `create` Proteins are highly variable in overall shape and chemical properties due to variation in the composition of R-groups and the array of secondary through quaternary structures that are possible. This variation allows them to fulfill many different roles in the cell. Diversity in the shape and reactivity of active sites also makes them effective catalysts. **11.** `understand` c **12.** `create` In many proteins, especially those involved in cell signaling, their structure is affected by binding to other molecules or ions. Since the shape of the protein is directly involved in its function, the protein's activity is regulated by controlling how it is folded. If proteins were inflexible, this type of control could not occur.

✔ Test Your Problem-Solving Skills

13. `analyze` The side chain of proline is a cyclic structure that is covalently bonded to the nitrogen in the core amino group. This restricts the movement of the side chain relative to the core nitrogen, which further restricts the backbone when the nitrogen participates in a peptide bond with a neighboring amino acid. **14.** `create` See **FIGURE A3.2 15.** `apply` b. Phosphates have a negative charge, so they are most likely to form ionic bonds with the positively-charged side chains of basic amino acid residues. **16.** `apply` The inherited forms likely have some alteration in the primary structure such that the infectious form is spontaneously generated at a higher rate than normal. The amino acid sequence in these prions would likely differ from those transmitted between animals.

CHAPTER 4

IN-TEXT QUESTIONS AND EXERCISES

p. 59 `understand` See **FIGURE A4.1**.

FIGURE A3.1

FIGURE A3.2

FIGURE A4.1

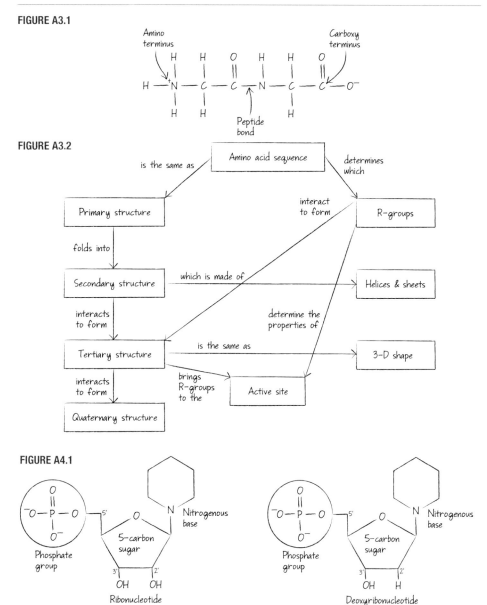

p. 60 Fig. 4.3 `understand` 5′-UAGC-′3
p. 61 CYU `understand` See **FIGURE A4.2**.
p. 63 `understand` If the two strands were parallel, the G-C pairing would align N—H groups together and C═O groups together, which would not allow hydrogen bonds to form.
p. 64 Fig. 4.8 `analyze` It is not spontaneous—energy must be added (as heat) for the reaction to occur.
p. 65 CYU `remember` See **FIGURE A4.3**.

IF YOU UNDERSTAND . . .

4.1. `understand` Cells activate nucleotides by linking additional phosphates to an existing 5′ phosphate. Activation increases the chemical energy in the nucleotides enough to offset the decrease in entropy that will result from the polymerization reaction. **4.2** `understand` C-G pairs involve three hydrogen bonds, so they are more stable than A-T pairs with just two hydrogen bonds. **4.3** `analyze` A single-stranded RNA molecule has unpaired bases that can pair with other bases on the same RNA strand, thereby folding the molecule into stem-and-loop configurations. These secondary structures can further fold on themselves, giving the molecule a tertiary structure. Because DNA molecules are double stranded, with no unpaired bases, further internal folding is not possible. **4.4** `understand` Examples would include (1) the production of nucleotides, and (2) polymerization of RNA. It is thought that nucleotides were scarce during chemical evolution, so their catalyzed synthesis by a ribozyme would have been advantageous. Catalysis by an RNA replicase would have dramatically increased the reproductive rate of RNA molecules.

YOU SHOULD BE ABLE TO . . .

✔ Test Your Knowledge

1. `remember` c **2.** `remember` c **3.** `remember` a **4.** `remember` d **5.** `remember` One end has a free phosphate group on the 5′ carbon; the other end has a free hydroxyl group bonded to the 3′ carbon. **6.** `understand` DNA is a more stable molecule than RNA because it lacks a hydroxyl group on the 2′ carbon and is therefore more resistant to cleavage, and because the two sugar-phosphate backbones are held together by many hydrogen bonds between nitrogenous bases.

✔ Test Your Understanding

7. `apply` In DNA, the secondary structure requires that every guanine pairs with a cytosine and every thymine pairs with an adenine, resulting in consistent ratios between the nucleotides. Chargaff's rules do not apply to RNA, since it is single-stranded and the pairing is not consistent throughout the molecules. **8.** `apply` a; if 30 percent is adenine, then 30 percent would be thymine, since they are base-paired together. This means that 40 percent consists of G-C base pairs, which would be equally divided between the two bases. **9.** `apply` The DNA sequence of the new strand would be 5′-ATCGATATC-3′. The RNA sequence would be the same, except each T would be replaced by a U. **10.** `understand` DNA has limited catalytic ability because it (1) lacks functional groups that can participate in catalysis and (2) has a regular structure that is not conducive to forming shapes required for catalysis. RNA molecules can catalyze some reactions because they (1) have exposed hydroxyl functional groups and (2) can fold into shapes that that can function in catalysis. **11.** `apply` No. Catalytic activity in ribozymes depends on the tertiary structure generated from single-stranded molecules. Double-stranded nucleic acids do not form tertiary structures. **12.** `understand` An RNA replicase would undergo replication and be able to evolve. It would process information in the sense of copying itself, and it would use energy to drive polymerization reactions. It would not be bound by a membrane and considered a cell, however, and it would not be able to acquire energy. It would best be considered as an intermediate step between nonlife and true life (as outlined in Chapter 1).

✔ Test Your Problem-Solving Skills

13. `create` See **FIGURE A4.4 14.** `apply` Yes—if the complementary bases lined up over the entire length of the two strands, they would twist into a double helix analogous to a DNA molecule. The same types of hydrogen bonds and hydrophobic interactions would occur as observed in the "stem" portion of hairpins in single-stranded RNA. **15.** `apply` In a triple helix, the bases are unlikely to align properly for hydrogen bonding to occur, so hydrophobic interactions would probably be more important. **16.** `apply` b; the high temperature would make it more likely that the secondary and tertiary structures would be denatured in the ribozymes. To overcome this effect, you would expect the hairpins to possess more G-C pairs, since they consist of three hydrogen bonds compared to the two found in A-T pairs.

CHAPTER 5

IN-TEXT QUESTIONS AND EXERCISES

p. 73 Fig. 5.2 `understand` See the structure of mannose in **FIGURE A5.1**.
p. 75 CYU `apply` See **FIGURE A5.2**.
p. 78 CYU `remember` They could differ in (1) location of linkages (e.g., 1,4 or 1,6); (2) types of linkages (e.g., α or β); (3) the sequence of the monomers (e.g., two galactose and then two glucose, versus alternating galactose and glucose); and/or (4) whether the four monomers are linked in a line or whether they branch.
p. 79 Fig. 5.6 `apply` The percentage of inhibition would not change for the intact glycoprotein bar. The purified

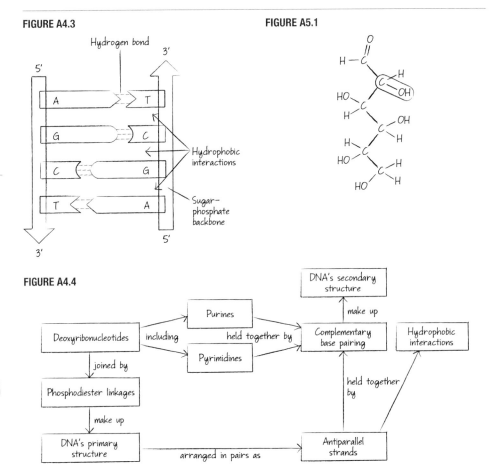

FIGURE A4.2

FIGURE A4.3

FIGURE A4.4

FIGURE A5.1

carbohydrate bar would be at zero inhibition, and the glycoprotein with digested carbohydrate bar would be similar to the intact glycoprotein bar.

p. 80 Fig. 5.7 `understand` All of the C−C and C−H bonds should be circled.

p. 81 CYU (1) `understand` *Aspect 1:* The β-1,4-glycosidic linkages in these molecules result in insoluble fibers that are difficult to degrade. *Aspect 2:* When individual molecules of these carbohydrates align, bonds form between them and produce fibers or sheets that resist pulling and pushing forces. **(2)** `apply` Most are probably being broken down into glucose, some of which in turn is being broken down in reactions that lead to the synthesis of ATP. Some will be resistant to digestion, such as the insoluble cellulose that makes up dietary fiber. This will help retain water and support the digestion and passage of fecal material.

IF YOU UNDERSTAND . . .

5.1. `understand` Molecules have to interact in an extremely specific orientation in order for a reaction to occur. Changing the location of a functional group by even one carbon can mean that the molecule will undergo completely different types of reactions. **5.2** `analyze` Glycosidic linkages can vary more in location and geometry than linkages between amino acids and nucleotides do. This variability increases the structural diversity possible in carbohydrates compared to proteins and nucleic acids. **5.3** `understand` (1) Polysaccharides used for energy storage are formed entirely from glucose monomers joined by α-glycosidic linkages; structural polysaccharides are made up of glucose or other sugars joined by β-glycosidic linkages. (2) The monomers in energy-storage polysaccharides are linked in a helical arrangement; the monomers in structural polysaccharides are linked in a linear arrangement. (3) Energy-storage polysaccharides may branch; structural polysaccharides do not. (4) Individual chains of energy-storage polysaccharides do not associate with each other; adjacent chains of structural polysaccharides are linked by hydrogen bonds or covalent bonds.

YOU SHOULD BE ABLE TO . . .

✔ Test Your Knowledge

1. `remember` d **2.** `remember` Monosaccharides can differ from one another in three ways: (1) the location of their carbonyl group; (2) the number of carbon atoms they contain; and (3) the orientations of their hydroxyl groups. **3.** `remember` a **4.** `remember` c **5.** `remember` a **6.** `understand` The electrons in the C=O bonds of carbon dioxide molecules are held tightly by the highly electronegative oxygen atoms, so they have low potential energy. The electrons in the C−C and C−H bonds of carbohydrates are shared equally, so they have much higher potential energy.

✔ Test Your Understanding

7. `understand` c. **8.** `apply` a; lactose is a disaccharide formed from a β-1,4-glycosidic linkage, so if two glucose molecules were linked with this bond, they would resemble units of cellulose and not be digested by human infants or adults. **9.** `understand` Carbohydrates are ideal for displaying the identity of the cell because they are so diverse structurally. This diversity enables them to serve as very specific identity tags for cells. **10.** `understand` When you compare the glucose monomers in an α-1,4-glycosidic linkage versus in a β-1,4-glycosidic linkage, the linkages are located on opposite sides of the plane of the glucose rings, and the glucose monomers are linked in the same orientation versus having every other glucose flipped in orientation. β-1,4-glycosidic linkages are much more likely to form linear fibers and sheets, so they resist degradation. **11.** `remember` Because (1) no mechanism is

known for the prebiotic polymerization of sugars; (2) no catalytic carbohydrates have been discovered that can perform polymerization reactions; and (3) sugar residues in a polysaccharide are not capable of complementary base pairing. **12.** `understand` Starch and glycogen both consist of glucose monomers joined by α-1,4-glycosidic linkages, and both function as storage carbohydrates. Starch is a mixture of unbranched and branched polysaccharides—called amylose and amylopectin, respectively. All glycogen polysaccharides are branched.

✔ Test Your Problem-Solving Skills

13. `analyze` Carbohydrates are energy-storage molecules, so minimizing their consumption may reduce total energy intake. Lack of available carbohydrate also forces the body to use fats for energy, reducing the amount of fat that is stored. **14.** `apply` d; lactose is a disaccharide of glucose and galactose, which can be cleaved by enzymes expressed in the human gut to release galactose. **15.** `analyze` Amylase breaks down the starch in the cracker into glucose monomers, which stimulate the sweet receptors in your tongue. **16.** `apply` When bacteria contact lysozyme, the peptidoglycan in their cell walls begins to degrade, leading to the death of the bacteria. Lysozyme therefore helps protect humans against bacterial infections.

CHAPTER 6

IN-TEXT QUESTIONS AND EXERCISES

p. 87 `understand` Fatty acids are amphipathic because their hydrocarbon tails are hydrophobic but their carboxyl functional groups are hydrophilic.

p. 87 Fig. 6.4 `apply` At the polar hydroxyl group in cholesterol and the polar head group in phospholipids.

p. 88 CYU (1) `analyze` Fats consist of three fatty acids linked to glycerol; steroids have a distinctive four-ring structure with variable side groups attached; phospholipids have a hydrophilic, phosphate-containing "head" region and a hydrocarbon tail. **(2)** `understand` In cholesterol, the hydrocarbon steroid rings and isoprenoid chain are hydrophobic; the hydroxyl group is hydrophilic. In phospholipids, the phosphate-containing head group is hydrophilic; the hydrocarbon chains are hydrophobic.

p. 89 `apply` Amino acids have amino and carboxyl groups that are ionized in water and nucleotides have negatively charged phosphates. Due to their charge and larger size, both would be placed below the small ions at the bottom of the scale (permeability$<10^{-12}$ cm/sec).

p. 90 CYU `create` See **TABLE A6.1**.

p. 91 Fig. 6.10 `apply` Increasing the number of phospholipids with polyunsaturated tails would increase permeability of the liposomes. Starting from the left, the first line (no cholesterol) would represent liposomes with 50% polyunsaturated phospholipids, the second line would be 20% polyunsaturated phospholipids, and the third line would contain only saturated phospholipids.

p. 92 `apply` If there is a difference in temperature, then there would be a difference in thermal motion. The solute concentration on the side with a higher temperature would decrease because the solute particles would be moving faster and hence be more likely to move to the cooler side of the membrane, where they would slow down.

p. 93 Fig. 6.13 `apply` Higher, because less water would have to move to the right side to achieve equilibrium.

FIGURE A5.2

Start with a monosacchride. This one is a 3-carbon aldose (carbonyl group at end)

Variation 1: 3-carbon ketose (carbonyl group in middle)

Variation 2: 4-carbon aldose

Variation 3: 3-carbon aldose with different arrangement of hydroxyl group

TABLE A6.1

Factor	Effect on permeability	Reason
Temperature	Decreases as temperature decreases.	Lower temperature slows movement of hydrocarbon tails, allowing more interactions (membrane is more dense).
Cholesterol	Decreases as cholesterol content increases.	Cholesterol molecules fill in the spaces between the hydrocarbon tails, making the membrane more tightly packed.
Length of hydrocarbon tails	Decreases as length of hydrocarbon tails increases.	Longer hydrocarbon tails have more interactions (membrane is more dense).
Saturation of hydrocarbon tails	Decreases as degree of saturation increases.	Saturated fatty acids have straight hydrocarbon tails that pack together tightly, leaving few gaps.

p. 94 CYU create See **FIGURE A6.1**.

p. 96 Fig. 6.18 create Repeat the procedure using a lipid bilayer that is free of membrane proteins, such as synthetic liposomes constructed from only phospholipids. If proteins were responsible for the pits and mounds, then this control would not show these structures.

p. 97 apply Your arrow should point out of the cell. There is no concentration gradient for chloride, but the outside has a net positive charge, which favors outward movement of negative ions.

p. 97 Fig. 6.21 analyze No—the 10 replicates where no current was recorded probably represent instances where the CFTR protein was damaged and not functioning properly. (In general, no experimental method works "perfectly.")

p. 101 CYU understand Passive transport does not require an input of energy—it happens as a result of energy already present in existing concentration or electrical gradients. Active transport is active in the sense of requiring an input of energy from, for example, ATP. In cotransport, a second ion or molecule is transported against its concentration gradient along with (i.e., "co") an ion that is transported along its concentration gradient.

p. 101 Fig. 6.26 understand *Diffusion:* description as given; no proteins involved. *Facilitated diffusion:* Passive movement of ions or molecules that cannot cross a phospholipid bilayer readily along a concentration gradient; facilitated by channel or carrier proteins. *Active transport:* Active movement of ions or molecules that will build a gradient; facilitated by pump proteins powered by an energy source such as ATP.

IF YOU UNDERSTAND . . .

6.1. analyze Adding H_2 increases the saturation of the oil by converting C=C bonds into C−C bonds with added hydrogens. Lipids with more C−H bonds tend to be solid at room temperature. **6.2** understand Highly permeable and fluid bilayers possess short, unsaturated hydrocarbon tails while those that are highly impermeable and less fluid contain long, saturated hydrocarbon tails. **6.3** apply (1) The solute will diffuse until both sides are at equal concentrations. (2) Water will diffuse toward the side with the higher solute concentration. **6.4** apply See **FIGURE A6.2**.

YOU SHOULD BE ABLE TO . . .

✔ Test Your Knowledge

1. understand c **2.** remember a **3.** understand b **4.** understand d **5.** understand For osmosis to occur, a concentration gradient and membrane that allows water to pass, but not the solute, must be present. **6.** analyze Channel proteins form pores in the membrane and carrier proteins undergo conformational changes to shuttle molecules or ions across the membrane.

✔ Test Your Understanding

7. apply b **8.** analyze No, because they have no polar end to interact with water. Instead, these lipids would float on the surface of water, or collect in droplets suspended in water, reducing their interaction with water to a minimum. **9.** understand Hydrophilic, phosphate-containing head groups interact with water; hydrophobic hydrocarbon tails associate with each other. A bilayer is more stable than are independent phospholipids in solution. **10.** apply Ethanol's polar hydroxyl group reduces the speed at which it can cross a membrane, but its small size and lack of charge would allow it to slowly cross membranes—between the rates of water and glucose transport. **11.** understand Only nonpolar, hydrophobic amino acid residues would be found in the portion of the protein that crosses the membrane. In the interior of the bilayer, these residues would be hidden from the water solvent and interact with the nonpolar lipid tails. **12.** apply Chloride ions from sodium chloride will move from the left side to the right through the CFTR. Water

will initially move from the right side to the left by osmosis, but as chloride ions move to the right, water will follow. Na^+ and K^+ ions will not move across the membrane.

✔ Test Your Problem-Solving Skills

13. apply c **14.** create Flip-flops should be rare, because they require a polar head group to pass through the hydrophobic portion of the lipid bilayer. To test this prediction, you could monitor the number of dyed phospholipids that transfer from one side of the membrane to the other in a given period of time. **15.** apply Organisms that live in very cold environments are likely to have highly unsaturated phospholipids. The kinks in unsaturated hydrocarbon tails keep membranes fluid and permeable, even at low temperature. Organisms that live in very hot environments would likely have phospholipids with saturated tails, to prevent membranes from becoming too fluid and permeable. **16.** analyze Adding a methyl group makes a drug more hydrophobic and thus more likely to pass through a lipid bilayer. Adding a charged group makes it hydrophilic and reduces its ability to pass through the lipid bilayer. These modifications would help target the drug to either the inside or outside of cells, respectively.

BIG PICTURE Chemistry of Life

p. 104 CYU (1) understand Oxygen is much more electronegative than hydrogen, so within water, the electrons are unequally shared in the O−H covalent bonds. The resulting partial negative charge around the oxygen and partial positive charges around the hydrogen atoms allow for hydrogen bonds to form among water molecules. **(2)** analyze Unlike other macromolecules, nucleic acids can serve as templates for their own replication. RNA is generally single-stranded and can adopt many different

three-dimensional structures. The flexibility in structure, combined with the presence of reactive hydroxyl groups, contribute to the formation of active sites that catalyze chemical reactions. One or more of these catalytic RNA molecules may have evolved the ability to self-replicate. DNA is not likely to have catalyzed its own replication, as it is most often double-stranded, with no clear tertiary structure, and it lacks the reactive hydroxyl groups. **(3)** remember In the amino acid, the nitrogen in the amino (NH_3^+) group and the carbon in the carboxyl (COO⁻) group should be circled. In the nucleotide, the oxygen in the hydroxyl (OH) group and the phosphorus in the phosphate (PO_4^{2-}) group on the nucleotide should be circled. **(4)** understand A line representing a protein should be drawn such that it completely crosses the lipid bilayer at least once. The protein could be involved in a variety of different roles, including transport of substances across the membrane in the form of a channel, carrier, or pump.

CHAPTER 7

IN-TEXT QUESTIONS AND EXERCISES

p. 110 CYU remember (1) The nucleoid compacts the chromosome to fit inside the cell via supercoiling while still keeping it accessible for replication and transmission of information. (2) Photosynthetic membranes increase food production by providing a large surface area to hold the pigments and enzymes required for photosynthesis. (3) Flagella propel cells through liquid, often toward a food source. (4) The layer of thick, strong material stiffens the cell wall and provides protection from mechanical damage.

p. 114 Fig. 7.12 create Storing the toxins in vacuoles prevents the toxins from damaging the plant's own organelles and cells.

FIGURE A6.1

FIGURE A6.2

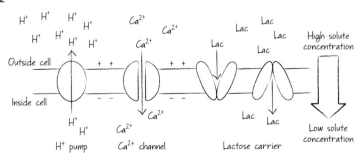

p. 116 CYU (1) `understand` Both organelles contain specific sets of enzymes. Lysosomal enzymes digest macromolecules in the acidic lumen of this organelle, releasing monomers that can be recycled into new macromolecules. Peroxisomes contain catalase and other enzymes that process fatty acids and toxins via oxidation reactions. **(2)** `understand` From top to bottom: administrative/information hub, protein factory, large molecule manufacturing and shipping (protein synthesis and folding center, lipid factory, protein finishing and shipping line, waste processing and recycling center), warehouse, fatty-acid processing and detox center, power station, food-manufacturing facility, support beams, perimeter fencing with secured gates, and leave blank.
p. 118 Fig. 7.16 `remember` See **FIGURE A7.1**.
p. 121 `analyze` **(1)** Nucleotides are small enough that they would diffuse through the nuclear pore complex along their gradients—a passive process that would not require energy. **(2)** Large proteins must be escorted through the nuclear pore complex, which is directional and requires energy, since the protein is concentrated inside the nucleus.
p. 121 Fig. 7.18 `apply` "Prediction": The labeled tail region fragments or the labeled core region fragments of the nucleoplasmin protein will be found in the cell nucleus. "Prediction of null hypothesis": Either both the fragments (no required signal) or neither of them (whole protein signal) will be found in the nucleus of the cell. "Conclusion": The send-to-nucleus signal is in the tail region of the nucleoplasmin protein.
p. 123 `apply` During the chase period, proteins appear to have first entered the Golgi after 7 minutes and then started to move into secretory granules after 37 minutes. This means that in this experiment, it took approximately 30 minutes for the fastest-moving proteins to pass through the Golgi.

p. 126 `apply` In receptor-mediated endocytosis, the conversion of a late endosome to a lysosome is dependent on receiving acid hydrolases from the Golgi. If this receptor is not present, then the enzymes will not be sent and the late endosome will not mature into a lysosome to digest the endocytosed products.
p. 126 CYU (1) `apply` Proteins that enter the nucleus are fully synthesized and have an NLS that interacts with another protein to get it into the organelle. The NLS is not removed. Proteins that enter the ER have a signal sequence that interacts with the SRP during translation. The ribosome is moved to the ER and synthesis continues, moving the protein into the ER. The signal is removed once it enters the organelle. **(2)** `apply` The protein would be in the lysosome. The ER signal would direct the protein into the ER before it is completely synthesized. The M-6-P tag will direct the protein from the Golgi to the late endosome to the lysosome. Thus the complete protein is never free in the cytosol, where the NLS could direct it into the nucleus.
p. 132 CYU `analyze` Actin filaments are made up of two strands of actin monomers, microtubules are made up of tubulin protein dimers that form a tube, and intermediate filaments are made up of a number of different protein subunits. Actin filaments and microtubules exhibit polarity (or directionality), and new subunits are constantly being added or subtracted at either end (but added faster to the plus end). All three elements provide structural support, but only actin filaments and microtubules serve as tracks for motors involved in movement and cell division.
p. 132 Fig. 7.30 `apply` The microtubule doublets of the axoneme would slide past each other completely, but the axoneme would not bend.

IF YOU UNDERSTAND . . .

7.1 `understand` (1) Cells will be unable to synthesize new proteins and will die. (2) In many environments, cells will be unable to resist the osmotic pressure of water entering the cytoplasm and will burst. (3) The cell shape will be different, and cells will not be able to divide.
7.2 `understand` (1) The cell will be unable to produce a sufficient amount of ATP and will die. (2) Reactive molecules, like hydrogen peroxide, will damage the cell, and it will likely die. (3) Nothing will happen, since plants do not have centrioles. **7.3** `analyze` The liver cell would be expected to have more peroxisomes and less rough endoplasmic reticulum than the salivary cells would.
7.4 `create` The addition or removal of phosphates would change the folded structure of the protein, exposing the NLS for nuclear transport. **7.5** `create` As a group, proteins have complex and highly diverse shapes and chemical properties that allow them to recognize a great number of different zip codes in a very specific manner. **7.6** `apply` The Golgi is positioned near the microtubule organizing center, which has microtubules running from the minus end out to the plus end (near the plasma membrane). Kinesin would be used to move these vesicles as it walks toward the plus end.

YOU SHOULD BE ABLE TO . . .
✔ Test Your Knowledge

1. `understand` b **2.** `remember` They have their own small, circular chromosomes; they produce their own ribosomes; and they divide in a manner that is similar to bacterial fission, independent of cellular division. **3.** `remember` c **4.** `remember` b **5.** `remember` a **6.** `understand` The phosphate links to the motor protein and causes it to change shape, which results in the protein moving along the filament.

✔ Test Your Understanding

7. `analyze` All cells are bound by a plasma membrane, are filled with cytoplasm, carry their genetic information (DNA) in chromosomes, and contain ribosomes

FIGURE A7.1

(a) Animal pancreatic cell: Exports digestive enzymes.

Rough ER

Secretory vesicles

(b) Animal testis cell: Exports lipid-soluble signals.

Smooth ER

(c) Plant leaf cell: Manufactures ATP and sugar.

Chloroplasts

Nucleus

Vacuole

(d) Brown fat cells: Burn fat to generate heat in lieu of ATP.

Mitochondria

(the sites of protein synthesis). Some prokaryotes have organelles not found in plants or animals, such as a magnetite-containing structure. Plant cells have chloroplasts, vacuoles, and a cell wall. Animal cells contain lysosomes and lack a cell wall. **8.** analyze a; the endoplasmic reticulum is responsible for synthesizing the membrane proteins required for the transport of solutes across the plasma membrane. **9.** create The NLS will be used to actively import the protein into the nucleus, leaving very little of the protein in the cytoplasm. Diffusion alone would not drive all the protein into the nucleus. **10.** create Ribosome in cytoplasm (signal is synthesized) → Ribosome at rough ER (protein is completed, folded, and glycosylated) → Transport vesicle → Golgi apparatus (protein is processed; has molecular zip code indicating destination) → Transport vesicle → Plasma membrane → Extracellular space. **11.** create This occurs in microfilaments and microtubules because they have ends that differ structurally and functionally—they have different filament growth rates. Intermediate filaments have identical ends, so there is no difference in the rate of assembly between the two ends. **12.** understand Polarized cytoskeletal filaments (microtubules or microfilaments) are present between the organelles. End-directed motor proteins use ATP to move these transport vesicles between them.

✔ Test Your Problem-Solving Skills

13. understand b; fimbriae are involved in bacterial attachment to surfaces and other cells, which would be important in the ability to grow on teeth. **14.** create The proteins must receive a molecular zip code that binds to a receptor on the surface of peroxisomes. They could diffuse randomly to peroxisomes or be transported in a directed way by motor proteins. **15.** create The tails cleaved from nucleoplasmin could be attached to the gold particles that were excluded from crossing the pore complex owing to their size. If these modified particles entered the nucleus, then the tail is not limited to the nucleoplasmin transport alone. **16.** apply The proteins would likely be found in the cytoplasm (e.g., actin and myosin) or imported into the mitochondria. Since there is a high energy demand, you would predict that there are many active mitochondria.

CHAPTER 8

IN-TEXT QUESTIONS AND ANSWERS

p. 139 apply (1) If ΔS is positive (products have more disorder than reactants), then according to the free energy equation, ΔG is more likely to be negative as temperature (T) increases even if ΔH is positive. The increased temperature represents added heat energy that may be used to drive an endothermic reaction to completion, making the reaction spontaneous. (2) Exothermic reactions may be nonspontaneous if they result in a decrease in entropy—meaning that the products are more ordered than the reactants (ΔS is negative).
p. 139 CYU (1) understand Gibbs equation: $\Delta G = \Delta H - T\Delta S$. ΔG symbolizes the change in the Gibbs free energy. ΔH represents the difference in enthalpy (heat, pressure, and volume) between the products and the reactants. T represents the temperature (in degrees Kelvin) at which the reaction is taking place. ΔS symbolizes the change in entropy (amount of disorder). **(2)** understand When ΔH is negative—meaning that the reactants have lower enthalpy than the products—and when ΔS is positive, meaning that the products have higher entropy (are more disordered) than the reactants.
p. 140 Fig. 8.4 understand Each point represents the data from a single test, not an average of many experiments, so it is not possible to calculate the standard error of the average.
p. 144 analyze Redox reactions transfer energy between molecules or atoms via electrons. When oxidized molecules are reduced, their potential energy increases. ATP

hydrolysis is often coupled with the phosphorylation of another molecule. This phosphorylation increases the potential energy of the molecule.
p. 144 CYU (1) understand Electrons in C−H bonds are not held as tightly as electrons in C−O bonds, so they have higher potential energy. **(2)** understand In part, because its three phosphate groups have four negative charges in close proximity. The electrons repulse each other, raising their potential energy.
p. 144 Fig. 8.9 understand The ΔG in the uncoupled reaction would be positive (>0), and each of the steps in the coupled reaction would have a negative (<0) ΔG.
p. 146 Fig. 8.12 understand No—a catalyst affects only the activation energy, not the overall change in free energy.
p. 147 remember (1) binding substrates, (2) transition state, (3) R-groups, (4) structure.
p. 147 Fig. 8.14 analyze See **FIGURE A8.1**.
p. 149 CYU (1) create The rate of the reaction is based primarily on the activity of the enzyme. Once the temperature reaches a level that causes unfolding and inactivation of the enzyme, the rate decreases to the uncatalyzed rate. **(2)** apply The shape change would most likely alter the shape of the active site. If phosphorylation activates catalytic activity, the change to the active site would allow substrates to bind and be brought to their transition state. If phosphorylation inhibits catalytic activity, the shape change to the active site would likely prevent substrates from binding or no longer orient them correctly for the reaction to occur.
p. 150 apply The concentration of A and B would be higher than in the fully functional pathway since they are not depleted to produce C. If D is not being depleted by other reactions, then equilibrium would be established between C and D, resulting in lower concentrations of both.

IF YOU UNDERSTAND . . .

8.1 understand Reactions are spontaneous when the free energy in the products is lower than that of the reactants (ΔG is negative). Enthalpy and entropy are measures used to determine free-energy changes. Enthalpy measures the potential energy of the molecules, and entropy measures the disorder. For exergonic, spontaneous reactions, disorder normally increases and the potential energy stored in the products normally decreases relative to the reactants. **8.2** understand Energetic coupling transfers free energy released from exergonic reactions to drive endergonic reactions. Since endergonic reactions are required for sustaining life, without energetic coupling, life would not exist. **8.3** understand Amino acid R-groups lining the active site interact with the substrates, orienting them in a way that stabilizes the transition state, thereby lowering the activation energy needed for the reaction to proceed. **8.4** analyze Allosteric regulation and phosphorylation cause changes in the conformation of the enzyme that affects its catalytic function. Allosteric regulation involves non-covalent bonding, while phosphorylation is a covalent modification of the enzyme's primary structure. **8.5** apply In the first step of the pathway, the rate would increase as the intermediate, which is the product of the first reaction, is removed. In the last step, the rate would decrease due to the loss of the intermediate, which serves as the substrate for the last reaction.

YOU SHOULD BE ABLE TO . . .

✔ Test Your Knowledge

1. remember c **2.** remember a **3.** remember a **4.** remember The enzyme changes shape, but the change is not permanent. The enzyme shape will return to its original conformation after releasing the products. **5.** remember d **6.** remember When the product of a pathway feeds back to interact with an enzyme early in the same pathway to inhibit its function.

✔ Test Your Understanding

7. understand The shape of reactant molecules (the key) fits into the active site of an enzyme (the lock). Fischer's

original model assumed that enzymes were rigid; in fact, enzymes are flexible and dynamic. **8.** understand d. Energy, such as the thermal energy in fire, must be provided to overcome the activation energy barrier before the reaction can proceed. **9.** understand The phosphorylation reaction is exergonic because the electrons in ADP and the phosphate added to the substrate experience less electrical repulsion, and thus have less potential energy, than they did in ATP. A phosphorylated reactant (i.e., an activated intermediate) gains enough potential energy to shift the free energy change for the reaction from endergonic to exergonic. **10.** apply For the coupled reaction, step 1 has a ΔG of about −3 kcal/mol and step 2 has a ΔG of about −3 kcal/mol. The uncoupled reaction has a ΔG of about +1.3 kcal/mol. **11.** analyze Both are mechanisms that regulate enzymes; the difference is whether the regulatory molecule binds at the active site (competitive inhibition) or away from the active site (allosteric regulation). **12.** apply Catabolic reactions will often have a negative ΔG based on a decrease in enthalpy and increase in entropy. Anabolic reactions are the opposite—a positive ΔG that is based on an increase in enthalpy and decrease in entropy.

✔ Test Your Problem-Solving Skills

13. create See **FIGURE A8.2**. **14.** apply Without the coenzyme, the free-radical-containing transition state would not be stabilized and the reaction rate would drop dramatically. **15.** analyze The data suggest that the enzyme and substrate form a transition state that requires a change in the shape of the active site, and that each movement corresponds to one reaction. **16.** apply b. The sugar likely functions as an allosteric regulator to activate the enzyme.

CHAPTER 9

IN-TEXT QUESTIONS AND EXERCISES

p. 156 Fig. 9.2 remember *Glycolysis:* "What goes in" = glucose, NAD^+, ADP, inorganic phosphate; "What comes out" = pyruvate, NADH, ATP. *Pyruvate processing:* "What goes in" = pyruvate, NAD^+; "What comes out" = NADH, CO_2, acetyl CoA. *Citric acid cycle:* "What goes in" = acetyl CoA, NAD^+, FAD, GDP or ADP, inorganic phosphate; "What comes out" = NADH, $FADH_2$, ATP or GTP, CO_2. *Electron transport and oxidative phosphorylation:* "What goes in" = NADH, $FADH_2$, O_2, ADP, inorganic phosphate; "What comes out" = ATP, H_2O, NAD^+, FAD.

FIGURE A8.1

Maximum speed of reaction

Rate of product formation

Most or all active sites are occupied

Reaction is most sensitive to changes in substrate concentration

Substrate concentration

FIGURE A8.2

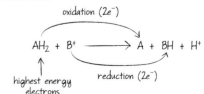

oxidation (2e⁻)

$$AH_2 + B^+ \longrightarrow A + BH + H^+$$

highest energy electrons

reduction (2e⁻)

p. 161 apply If the regulatory site had a higher affinity for ATP than the active site, then ATP would always be bound at the regulatory site, and glycolysis would always proceed at a very slow rate.

p. 161 Fig. 9.9 remember "Positive control": AMP, NAD$^+$, CoA (reaction substrates). "Negative control by feedback inhibition": acetyl CoA, NADH, ATP (reaction products).

p. 165 CYU remember (1) and (2) are combined with the answer to p. 171 CYU (3) Start with 12 triangles on glucose. (These triangles represent the 12 pairs of electrons that will be moved to electron carriers during redox reactions throughout glycolysis and the citric acid cycle.) Move two triangles to the NADH circle generated by glycolysis and the other 10 triangles to the pyruvate circle. Then move these 10 triangles through the pyruvate dehydrogenase square, placing two of them in the NADH circle next to pyruvate dehydrogenase. Add the remaining eight triangles in the acetyl CoA circle. Next move the eight triangles in the acetyl CoA circle through the citric acid cycle, placing six of them in the NADH circle and two in the FADH$_2$ circle generated during the citric acid cycle. (4) These boxes are marked with stars in the diagram.

p. 165 Fig. 9.13 apply NADH would be expected to have the highest amount of chemical energy since its production is correlated with the largest drop in free energy in the graph.

p. 167 Fig. 9.15 understand The proton gradient arrow should start above in the inner membrane space and point down across the membrane into the mitochondrial matrix. *Complex I:* "What goes in" = NADH; "What comes out" = NAD$^+$, e$^-$, transported H$^+$. *Complex II:* "What goes in" = FADH$_2$; "What comes out" = FAD, e$^-$, H$^+$. *Complex III:* "What goes in" = e$^-$, H$^+$; "What comes out" = e$^-$, transported H$^+$. *Complex IV:* "What goes in" = e$^-$, H$^+$, O$_2$; "What comes out" = H$_2$O, transported H$^+$.

p. 169 explain "Indirect" is accurate because most of the energy released during glucose oxidation is not used to produce ATP directly. Instead, this energy is stored in reduced electron carriers that are used by the ETC to generate a proton gradient across a membrane. These protons then diffuse down their concentration gradient across the inner membrane through ATP synthase, which drives ATP synthesis.

p. 169 Fig. 9.17 create They could have placed the vesicles in an acidic solution that has a pH below that of the solution in the vesicle. This would set up a proton gradient across the membrane to test for ATP synthesis.

p. 171 CYU understand See **FIGURE A9.1**. To illustrate the chemiosmotic mechanism, take the triangles (electrons) piled on the NADH and FADH$_2$ circles and move them through the ETC. While moving these triangles, also move dimes from the mitochondrial matrix to the intermembrane space. As the triangles exit the ETC, add them to the oxygen to water circle. Once all the dimes have been pumped by the ETC into the intermembrane space, move them through ATP synthase back into the mitochondrial matrix to fuel the formation of ATP.

p. 173 CYU understand Electron acceptors such as oxygen have a much higher electronegativity than pyruvate. Donating an electron to O$_2$ causes a greater drop in potential energy, making it possible to generate much more ATP per molecule of glucose.

IF YOU UNDERSTAND . . .

9.1 understand The radioactive carbons in glucose can be fully oxidized by the central pathways to generate CO$_2$, which would be radiolabeled. Other molecules, like lipids and amino acids, would also be expected to be radiolabeled since they are made using intermediates from the central pathways in other anabolic pathways. **9.2** apply See **FIGURE A9.2**. **9.3** understand Pyruvate dehydrogenase accomplishes three different tasks that would be expected to require multiple enzymes and active sites: CO$_2$ release, NADH production, and linking of an acetyl group to CoA. **9.4** apply NADH would decrease if a drug poisoned the acetyl CoA and oxaloacetate-to-citrate enzyme, since the citric acid cycle would no longer be able to produce NADH in the steps following this reaction in the pathway. **9.5** apply The ATP synthase allows protons to reenter the mitochondrial matrix after they have been pumped out by the ETC. By blocking ATP synthase, you would expect the pH of the matrix to increase (decreased proton concentration). **9.6** understand Organisms that produce ATP by fermentation would be expected to grow more slowly than those that produce ATP via cellular respiration simply because fermentation produces fewer ATP molecules per glucose molecule than cellular respiration does.

YOU SHOULD BE ABLE TO . . .

✔ Test Your Knowledge

1. understand Glycolysis → Pyruvate processing → citric acid cycle → ETC and chemiosmosis. The first three steps are responsible for glucose oxidation; the final step produces the most ATP. **2.** remember b **3.** understand d **4.** understand Most of the energy is stored in the form of NADH. **5.** remember c **6.** remember a

✔ Test Your Understanding

7. understand Stored carbohydrates can be broken down into glucose that enters the glycolytic pathway. If carbohydrates are absent, products from fat and protein catabolism can be used to fuel cellular respiration or fermentation. If ATP is plentiful, anabolic reactions use

---

FIGURE A9.1

Eukaryotic mitochondrion

FIGURE A9.2

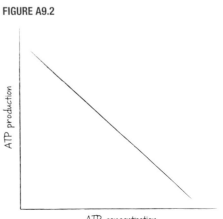

intermediates of the glycolytic pathway and the citric acid cycle to synthesize carbohydrates, fats, and proteins. **8.** analyze Both processes produce ATP from ADP and P_i, but substrate-level phosphorylation occurs when enzymes remove a "high-energy" phosphate from a substrate and directly transfer it to ADP, while oxidative phosphorylation occurs when electrons move through an ETC and produce a proton-motive force that drives ATP synthase. **9.** understand Aerobic respiration is much more productive because oxygen has extremely high electronegativity compared with other electron acceptors, resulting in a greater release of energy during electron transport and more proton pumping. **10.** apply b **11.** analyze Both phosphofructokinase and isocitrate dehydrogenase are regulated by feedback inhibition, where the product of the reaction or series of reactions inhibits the enzyme activity. They differ in that phosphofructokinase is regulated by allosteric inhibition while isocitrate dehydrogenase is controlled by competitive inhibition. **12.** understand Oxidative phosphorylation is possible via a proton gradient that is established by redox reactions in the ETC. ATP synthase consists of a membrane-associated F_o unit and a F_1 unit joined by a rotor shaft. When protons flow through the F_o unit, it spins the rotor shaft within the fixed F_1 unit. This spinning shaft causes structural changes in the F_1 that drives the synthesis of ATP from ADP and P_i.

✔ Test Your Problem-Solving Skills

13. create When complex IV is blocked, electrons can no longer be transferred to oxygen, the final acceptor, and cellular respiration stops. Fermentation could keep glycolysis going, but it is inefficient and unlikely to fuel a cell's energy needs over the long term. Cells that lack the enzymes required for fermentation would die first. **14.** apply Because mitochondria with few cristae would have fewer electron transport chains and ATP synthase molecules, they would produce much less ATP than mitochondria with numerous cristae. **15.** apply For each glucose molecule, two ATP are produced in glycolysis and two ATP are produced in the citric acid cycle via substrate-level phosphorylation. A total of 10 NADH and 2 $FADH_2$ molecules are produced from glycolysis, pyruvate oxidation, and the citric acid cycle. If each NADH were to yield 3 ATP, and each $FADH_2$ were to yield 2 ATP, then a total of 34 ATP would be produced via oxidative phosphorylation. Adding these totals would result in 38 ATP. A cell will not produce this much ATP, because the proton-motive force is used in other transport steps and because of other issues that may reduce the overall efficiency. **16.** apply b

CHAPTER 10

IN-TEXT QUESTIONS AND EXERCISES

p. 177 understand See **FIGURE A10.1**. This reaction is endergonic because there are more high-energy chemical bonds in the products compared with the reactants, and there is a decrease in entropy.
p. 177 Fig. 10.1 apply See **FIGURE A10.1**.
p. 180 Fig. 10.6 apply See **FIGURE A10.2**.

p. 182 Fig. 10.9 apply The energy state corresponding to a photon of green light would be located between the energy states corresponding to red and blue photons.
p. 183 CYU apply The outer pigments would be more likely to absorb blue photons (short wavelength, high energy), and interior pigments would absorb red photons (long wavelength, low energy). This establishes a pathway to direct photon energy toward the reaction center since resonance energy is transferred from higher to lower energy levels.
p. 184 Fig. 10.11 understand Yes—otherwise, changes in the production of oxygen could be due to differences in the number of chloroplasts, not differences in the rate of photosynthesis.
p. 186 analyze Light → Antenna complex → Reaction center → Pheophytin → ETC → Proton gradient → ATP synthase. Electrons from water are donated to the reaction center to replace those that were transferred to pheophytin.
p. 188 remember (1) Plastocyanin transfers electrons that move through the cytochrome complex in the ETC to the reaction center of photosystem (PS) I. (2) After they are excited by a photon and donated to the initial electron acceptor.
p. 190 CYU analyze In mitochondria, high-energy electrons are donated by NADH or $FADH_2$ (primary donors) and passed through an ETC to generate a proton-motive force. The low-energy electrons at the end of the chain are accepted by O_2 (terminal acceptor) to form water. In chloroplasts, low-energy electrons are donated by H_2O (primary donor), energized by photons or resonance energy, and passed through an ETC to generate a proton-motive force. These electrons are then excited a second time by photons or resonance energy, and the high-energy electrons are accepted by $NADP^+$ (terminal acceptor) to form NADPH.
p. 190 Fig. 10.18 understand The researchers didn't have any basis on which to predict these intermediates. They needed to perform the experiment to identify them.
p. 192 apply Each complete cycle requires 3 ATP and 2 NADPH molecules. To complete 6 runs through the cycle, a total of 18 ATP and 12 NADPH molecules are needed. By following the number of carbons, it is apparent that only three RuBP molecules are required, since they are fully regenerated every 3 cycles: 3 RuBP (15 carbons) fix and reduce 3 CO_2 to generate 6 G3P (18 carbons), yielding 1 G3P (3 carbons); the other 5 are used to regenerate 3 RuBP (15 carbons). The regeneration of RuBP means that only three would be required for continued runs through the Calvin cycle.
p. 194 Fig. 10.24 apply The morning would have the highest concentration of organic acids in the vacuoles of CAM plants, since these acids are made during the night and used up during the day.
p. 195 CYU (1) understand (a) C_4 plants use PEP carboxylase to fix CO_2 into organic acids in mesophyll cells. These organic acids are then transported into bundle-sheath cells, where they release carbon dioxide to rubisco. (b) CAM plants take in CO_2 at night and have enzymes that fix it into organic acids stored in the central vacuoles of photosynthesizing cells. During the day, the organic acids are processed to release CO_2 to rubisco.

(c) By diffusion through a plant's stomata when they are open. **(2)** apply The concentration of starch would be highest at the end of the day and lowest at the start of the day. Starch is made and stored in the chloroplasts of leaves during periods of high photosynthetic activity during the day. At night, it is broken down to make sucrose, which is transported throughout the plant to drive cellular respiration. (Cellular respiration also occurs during the day, but the impact is minimized due to the photosynthetic production of sugar.)

IF YOU UNDERSTAND . . .

10.1 understand The Calvin cycle depends on the ATP and NADPH produced by the light-capturing reactions, so it is not independent of light. **10.2** understand Most of the energy captured by pigments in chloroplasts is converted into chemical energy by reducing electron acceptors in ETCs. When pigments are extracted, the antenna complexes, reaction centers, and ETCs have been disassembled, so the energy is given off as fluorescence and heat. **10.3** understand Oxygen is produced by a critical step in photosynthesis: splitting water to provide electrons to PS II. If oxygen production increases, it means that more electrons are moving through the photosystems. **10.4** apply Each CO_2 that is fixed and reduced by the Calvin cycle requires 2 NADPH, which means that 12 NADPH molecules are required for a 6-carbon glucose. Each NADPH is made when two high-energy electrons reduce $NADP^+$. Each of these high-energy electrons originates from H_2O only after being excited by 2 photons (one in PS II and one in PS I). This means that 48 photons are required to produce 24 high-energy electrons to reduce 12 $NADP^+$ molecules for the fixation and reduction of 6 CO_2 to make glucose. Photorespiration would increase the number of photons required, since some of the CO_2 that is fixed would be released.

YOU SHOULD BE ABLE TO . . .
✔ Test Your Knowledge

1. remember d **2.** understand c **3.** remember c **4.** understand b **5.** remember The conversion of light energy to chemical energy occurs when electrons are transferred from excited pigments to an electron carrier in the photosystems. **6.** remember The electron transport chain that accepts electrons from PS II. Plastocyanin is the molecule that transfers electrons from this chain to the PS I reaction center.

✔ Test Your Understanding

7. understand The electrons taken from water in PS II are excited twice by either photons or resonance energy. When excited in PS II, the electrons are transferred to PQ and used to build a proton-motive force that makes ATP. After reaching PS I, they are excited a second time and will either be used to reduce $NADP^+$ to make NADPH (noncyclic) or be transported back to PQ to produce more ATP (cyclic). **8.** analyze c **9.** understand The fixation phase is when CO_2 is fixed to RuBP by rubisco to form 3-phosphoglycerate. The reduction phase uses ATP to phosphorylate the carbons and NADPH to

---

FIGURE A10.1

FIGURE A10.2

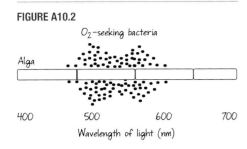

reduce them with high-energy electrons to form G3P. The regeneration phase uses more ATP to convert some of the G3P to RuBP to continue the cycle. **10.** [understand] Photorespiration occurs when levels of CO_2 are low and O_2 are high. Less sugar is produced because (1) CO_2 doesn't participate in the initial reaction catalyzed by rubisco and (2) when rubisco catalyzes the reaction with O_2 instead, one of the products is eventually broken down to CO_2 in a process that uses ATP. **11.** [understand] In both C_4 and CAM plants, atmospheric CO_2 is brought in through stomata and first captured by fixing it to a 3-carbon molecule by PEP carboxylase. The C_4 pathway and CAM differ in the timing of this first fixation step—it occurs during the day in C_4 plants and during the night in CAM plants. They also differ in the location of the Calvin cycle with respect to this first fixation step. In C_4 plants, the two processes occur in different cells, while in CAM plants they occur in the same cell, but at different times (Calvin cycle during the day). **12.** [analyze] Photosynthesis in chloroplasts produces sugar, which is used as a source of carbon for building organic molecules and energy for cellular respiration. Mitochondria harvest the energy stored in sugar to produce ATP, which is used to drive many cellular activities.

✔ Test Your Problem-Solving Skills

13. [apply] (1) O_2, ATP, and NADPH would be formed by noncyclic electron flow. (2) No O_2 or NADPH would be formed, but ATP may be made by cyclic electron flow. (3) Initially, O_2 and NADPH would be formed by noncyclic electron flow, but no ATP would be made. Without ATP, the Calvin cycle would halt and, once all the NADP$^+$ is used up, noncyclic electron flow would switch to cyclic electron flow. **14.** [evaluate] Because rubisco evolved in a high CO_2, low O_2 environment, which would minimize the impact of photorespiration, the hypothesis is credible. But once O_2 levels increased, any change in rubisco that minimized photorespiration would give individuals a huge advantage over organisms with "old" forms of rubisco. There has been plenty of time for such changes to occur, making the "holdover" hypothesis less credible. **15.** [analyze] b; the wavelength of light could excite PS I, but not PS II, resulting in cyclic electron flow since no electrons could be harvested from water by PS II. **16.** [create] No—they are unlikely to have the same complement of photosynthetic pigments. Different wavelengths of light are available in various layers of a forest and water depths. It is logical to predict that plants and algae have pigments that absorb the available wavelengths efficiently. One way to test this hypothesis would be to isolate pigments from species in different locations and test the absorbance spectra of each.

BIG PICTURE Energy

p. 198 CYU (1) [understand] Photosynthesis uses H_2O as a substrate and releases O_2 as a by-product; cellular respiration uses O_2 as a substrate and releases H_2O as a by-product. **(2)** [understand] Photosynthesis uses CO_2 as a substrate; cellular respiration releases CO_2 as a by-product. **(3)** [analyze] CO_2 fixation would essentially stop; CO_2 would continue to be released by cellular respiration. CO_2

levels in the atmosphere would increase rapidly, and production of new plant tissue would cease—meaning that most animals would quickly starve to death. **(4)** [analyze] ATP "is used by" the Calvin cycle; photosystem I "yields" NADPH.

CHAPTER 11

IN-TEXT QUESTIONS AND EXERCISES

p. 204 CYU [analyze] Plant cell walls and animal ECMs are both fiber composites. In plant cell walls the fiber component consists of cross-linked cellulose fibers, and the ground substance is pectin. In animal ECMs the fiber component consists of collagen fibrils, and the ground substance is proteoglycan.
p. 207 [apply] Developing muscle cells could not adhere normally, and muscle tissue would not form properly. The embryo would die.
p. 207 Fig. 11.9 [apply] "Prediction": Cells treated with an antibody that blocks membrane proteins involved in adhesion will not adhere. "Prediction of null hypothesis": All cells will adhere normally.
p. 209 CYU (1) [analyze] The three structures differ in composition, but their function is similar. The middle lamella in plants is composed of pectins that glue adjacent cells together. Tight junctions are made up of membrane proteins that line up and "stitch" adjacent cells together. Desmosomes are "rivet-like" structures composed of proteins that link the cytoskeletons of adjacent cells. **(2)** [understand] The plasma membranes of adjacent plant cells are continuous at plasmodesmata and share portions of the smooth endoplasmic reticulum. Gap junctions connect adjacent animal cells by forming protein-lined pores. Both structures result in openings between the cells that allow cytosol, including ions and small molecules, to be shared.
p. 211 Fig. 11.12 [create] The steroid hormone likely changes the structure of the receptor such that it now exposes a nuclear localization signal, which is required for the protein to be transported into the nucleus.
p. 213 [analyze] The spy is the signaling molecule that arrives at the cell surface (the castle gate). The guard is the G-protein-coupled receptor in the plasma membrane, and the queen is the G protein. The commander of the guard is the enzyme that is activated by the G protein to produce second messengers (the soldiers).
p. 214 [apply] **(1)** The red dominos (RTK components) would be the first two dominos in the chain, followed by the black domino (Ras). This black domino would start two or more new branches, each one represented by a single domino of one color (e.g., green) to represent the activation of one type of kinase. Each of these green dominos would then again branch out, knocking down two or more branches consisting of single dominos of a different color (blue). The same branching pattern would result from each blue domino knocking down two or more yellow dominos. **(2)** Each single black domino (Ras) would require 10 green dominos, 100 blue dominos, and 1000 yellow dominos.
p. 215 CYU (1) [understand] Each cell–cell signaling molecule binds to a specific receptor protein. A cell can respond to a signal only if it has the appropriate receptor. Only certain cell types will have the appropriate receptor for a given signaling molecule. **(2)** [understand] Signals are amplified if one or more steps in a signal transduction pathway, involving either second messengers or a phosphorylation cascade, result in the activation of multiple downstream molecules.
p. 215 Fig. 11.16 [apply] **(1)** cell responses A and C **(2)** cell responses A, B, and C **(3)** cell responses B and C.

IF YOU UNDERSTAND . . .

11.1 [apply] (1) Cells without functional integrin molecules would likely die as a result of not being able to send the

appropriate anchorage-dependent survival signals. (2) Cells would be more sensitive to pulling or shearing forces; both cells and tissues would be weaker and more susceptible to damage. **11.2** [apply] Cells would not be co-ordinated in their activity, so the heart tissue would not contract in unison and the heart would not beat. **11.3** [analyze] Adrenalin binds to both heart and liver cells, but the activated receptors trigger different signal transduction pathways and lead to different cell responses. **11.4** [analyze] The signal transduction pathways are similarly organized in both unicellular and multicellular organisms—consisting of signaling molecules, receptors, and second messengers. There is more variety in the means of transmitting the signal between cells in multicellular organisms compared to unicellular organisms. For example, there are no direct connections such as gap junctions or plasmodesmata in unicellular organisms.

YOU SHOULD BE ABLE TO . . .

✔ Test Your Knowledge

1. [remember] Fiber composites consist of cross-linked fiber components that withstand tension and a ground substance that withstands compression. The cellulose microfibrils in plants and collagen fibrils in animals functionally resemble the steel rods in reinforced concrete. The pectin in plants and proteoglycan in animals functionally resemble the concrete ground substance. **2.** [remember] b **3.** [remember] a **4.** [understand] b **5.** [remember] d **6.** [remember] Responses that affect which proteins are produced and those that affect the activity of existing proteins.

✔ Test Your Understanding

7. [understand] b **8.** [understand] If each enzyme in the cascade phosphorylates many copies of the enzyme in the next step of the cascade, the initial signal will be amplified many times over. **9.** [analyze] All three are made up of membrane-spanning proteins that directly interact between adjacent cells. Tight junctions seal adjacent animal cells together; gap junctions allow a flow of material from the cytosol of one to the other. Desmosomes firmly secure adjacent cells to one another but do not affect the movement of substances between cells or into the cells. **10.** [understand] When dissociated cells from two sponge species were mixed, the cells sorted themselves into distinct aggregates that contained only cells of the same species. By blocking membrane proteins with antibodies and isolating cells that would not adhere, researchers found that specialized groups of proteins, including cadherins, are responsible for selective adhesion. **11.** [create] Signaling molecule crosses plasma membrane and binds to intracellular receptor (reception) → Receptor changes conformation, and the activated receptor complex moves to target site (processing) → Activated receptor complex binds to a target molecule (e.g., a gene or membrane pump), which changes its activity (response) → Signaling molecule falls off receptor or is destroyed; receptor changes to inactive conformation (deactivation). **12.** [understand] Information from different signals may conflict or be reinforcing. "Crosstalk" between signaling pathways allows cells to integrate information from many signals at the same time instead of responding to each signal in isolation.

✔ Test Your Problem-Solving Skills

13. [apply] d **14.** [analyze] (a) The response would have to be extremely local—the activated receptor complex would have to affect nearby proteins. (b) No amplification could occur, because the number of signaling molecules dictates the amount of the response. (c) The only way to regulate the response would be to block the receptor or make it more responsive to the signaling molecule. **15.** [create] Chitin forms chains that can cross-link with one another. This fungal cell wall would likely need to be either relaxed or destroyed, and new cell wall synthesis

would be coordinated with the directional growth.
16. `create` Antibody binding to the two parts of the receptor may be causing them to dimerize. Since dimerization normally results after the signaling molecule binds to the receptor, the antibodies could be activating the receptor by mimicking this interaction. The result would be signal transduction even in the absence of the signaling molecule.

CHAPTER 12

IN-TEXT QUESTIONS AND EXERCISES

p. 223 `understand` **(1)** Genes are segments of chromosomes that code for RNAs and proteins. **(2)** Chromosomes are made of chromatin. **(3)** Sister chromatids are identical copies of the same chromosome, joined together.
p. 224 Fig. 12.5 `apply` (1) prophase cells would have 4 chromosomes with $2x$ DNA. (2) Anaphase cells would

have 8 chromosomes and $2x$ DNA. (3) Each daughter cell will have 4 chromosomes with x DNA.
p. 226 `apply` See **TABLE A12.1**.
p. 227 Fig. 12.6 `apply` The chromosome and black bar would move at the same rate toward the spindle pole.
p. 229 CYU (1) `apply` See **FIGURE A12.1**. **(2)** `apply` Loss of the motors would result in two problems: (1) It would reduce the ability of chromosomes to attach to microtubules via their kinetochores; (2) cytokinesis would be inhibited since the Golgi-derived vesicles would not be moved to the center of the spindle to build the cell plate.
p. 230 Fig. 12.10 `create` Inject cytoplasm from an M-phase frog egg into a somatic cell that is in interphase. If the somatic cell starts mitosis, then the meiotic factor is not limited to gametes.
p. 230 Fig. 12.11 `analyze` In effect, MPF turns itself off after it is activated. If this didn't happen, the cell might undergo mitosis again before the cell has replicated its DNA.

p. 231 `understand` MPF activates proteins that get mitosis under way. MPF consists of a cyclin and a Cdk, and it is turned on by phosphorylation at the activating site and dephosphorylation at the inhibitory site. Enzymes that degrade cyclin reduce MPF levels.
p. 232 CYU (1) `remember` $G_1 \rightarrow S \rightarrow G_2 \rightarrow M$. Checkpoints occur at the end of G_1 and G_2 and during M phase. **(2)** `understand` Cdk levels are fairly constant throughout the cycle, but cyclin increases during interphase and peaks in M phase. This accumulation of cyclin is a prerequisite for MPF activity, which turns on at the end of G_2, initiating M phase, and declines at the end of M phase.

IF YOU UNDERSTAND . . .

12.1 `understand` The G_1 and G_2 phases give the cell time to replicate organelles and grow before division as well as perform the normal functions required to stay alive. Chromosomes replicate during S phase and are separated from one another during M phase. Cytokinesis also occurs during M phase, when the parent cell divides into two daughter cells. **12.2** `apply` In cells that do not dissolve the nuclear envelope, the spindle must be constructed inside the nucleus to attach to the chromosomes and separate them. **12.3** `apply` Cells would prematurely enter M phase, shortening the length of G_2 and resulting in the daughter cells' being smaller than normal. **12.4** `analyze` The absence of growth factors in normal cells would cause them to arrest in the G_1 phase—eventually all the cells in the culture would be in G_1. The cancerous cells are not likely to be dependent on these growth factors, so the cells would not arrest and would continue through the cell cycle.

YOU SHOULD BE ABLE TO . . .

✔ Test Your Knowledge

1. `remember` b **2.** `remember` d **3.** `remember` d **4.** `remember` c
5. `understand` Daughter chromosomes were observed to move toward the pole faster than do the marked regions of fluorescently labeled kinetochore microtubules.
6. `remember` The cycle would arrest in M phase, and cytokinesis would not occur.

✔ Test Your Understanding

7. `apply` For daughter cells to have identical complements of chromosomes, all the chromosomes must be replicated during the S phase, the spindle apparatus must connect with the kinetochores of each sister chromatid in prometaphase, and the sister chromatids of each replicated chromosome must be partitioned in anaphase and fully separated into daughter cells by cytokinesis. **8.** `create` One possible concept map is shown in **FIGURE A12.2**. **9.** `understand` Microinjection experiments suggested that something in the cytoplasm of M-phase cells activated the transition from interphase to M phase. The control for this experiment was to inject cytoplasm from a G_2-arrested oocyte into another G_2-arrested oocyte. **10.** `apply` Protein kinases phosphorylate proteins. Phosphorylation changes a protein's shape, altering its function (activating or inactivating it). As a result, protein kinases regulate the function of proteins. **11.** `understand` Cyclin concentrations change during the cell cycle. At high concentration, cyclins bind to a specific cyclin-dependent kinase (or Cdk), forming a dimer. This dimer becomes active MPF by changing its shape through the phosphorylation (activating site) and dephosphorylation (inhibitory site) of Cdk. **12.** `analyze` b.

✔ Test Your Problem-Solving Skills

13. `analyze` a; adding up each phase allows you to determine that the cell cycle is 8.5 hours long. After 9 hours, the radiolabeled cells would have passed through a full cycle and be in either S or G_2—none would have entered M phase. **14.** `apply` The embryo passes through multiple rounds of the cell cycle, but cytokinesis does

FIGURE A12.1

FIGURE A12.2

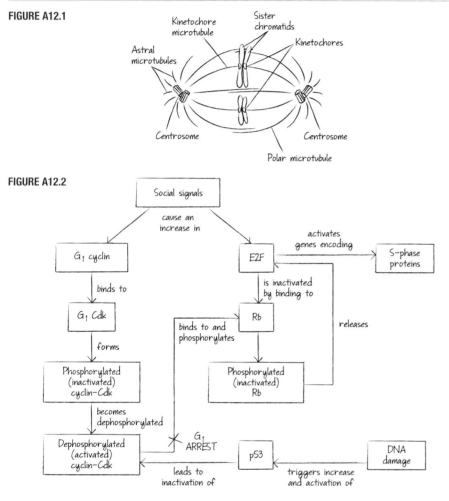

TABLE A12.1

	Prophase	Prometaphase	Metaphase	Anaphase	Telophase
Spindle apparatus	Starts to form	Contacts and moves chromosomes	Anchors poles to membrane and produces tension at kinetochores	Pulls chromatids apart	Defines site of cytokinesis
Nuclear envelope	Present	Disintegrates	Nonexistent	Nonexistent	Re-forms
Chromosomes	Condense	Attach to microtubules	Held at metaphase plate	Sister chromatids separate into daughter chromosomes	Collect at opposite poles

not occur during M phases. **15.** apply Early detection of cancers leads to a greater likelihood of survival. The widespread implementation of breast and prostate exams allows for the identification and removal of benign tumors before they become malignant. **16.** analyze Cancer requires many defects. Older cells have had more time to accumulate defects. Individuals with a genetic predisposition to cancer start out with some cancer-related defects, but this does not mean that the additional defects required for cancer to occur will develop.

CHAPTER 13

IN-TEXT QUESTIONS AND EXERCISES

p. 239 apply $n = 4$; the organism is diploid with $2n = 8$.

p. 240 remember See **FIGURE A13.1**. Because the two sister chromatids are identical and attached, it is sensible to consider them as parts of a single chromosome.

p. 242 apply There will be four DNA molecules in each gamete because the 8 replicated chromosomes in a diploid cell are reduced to 4 replicated chromosomes per cell at the end of meiosis I. In meiosis II, the sister chromatids of each replicated chromosome are separated. Each cell now contains 4 unreplicated chromosomes, each with a single molecule of DNA.

p. 245 apply Crossing over would not occur and the daughter cells produced by meiosis would be diploid, not haploid. There would be no reduction division.

p. 246 CYU (1) understand Use four long and four short pipe cleaners (or pieces of cooked spaghetti) to represent the chromatids of two replicated homologous chromosomes (four total chromosomes). Mark two long and two short ones with a colored marker pen to distinguish maternal and paternal copies of these chromosomes. Twist identical pipe cleaners (e.g., the two long colored ones) together to simulate replicated chromosomes. Arrange the pipe cleaners to depict the different phases of meiosis I as follows: *Early prophase I:* Align sister chromatids of each homologous pair to form two tetrads. *Late prophase I:* Form one or more crossovers between non-sister chromatids in each tetrad. (This is hard to simulate with pipe cleaners—you'll have to imagine that each chromatid now contains both maternal and paternal segments.) *Metaphase I:* Line up homologous pairs (the two pairs of short pipe cleaners and the two pairs of long pipe cleaners) at the metaphase plate. *Anaphase I:* Separate homologs. Each homolog still consists of sister chromatids joined at the centromere. *Telophase I and cytokinesis:* Move homologs apart to depict formation of two haploid cells, each containing a single replicated copy of two different chromosomes. **(2)** understand During anaphase I, homologs (not sister chromatids, as in mitosis) are separated, making the cell products of meiosis I haploid. **(3)** understand The pairing of homologs in metaphase I and their separation in anaphase I so that one goes to one daughter cell and the other to the other daughter cell means that each daughter cell obtains precisely one copy of each type of chromosome.

p. 248 apply Each gamete would inherit either all maternal or all paternal chromosomes. This would limit genetic variation in the offspring by precluding the many possible gametes containing various combinations of maternal and paternal chromosomes.

p. 248 CYU (1) apply See **FIGURE A13.2**. Maternal chromosomes are white and paternal chromosomes are black. Daughter cells with other possible combinations of chromosomes than shown could result from meiosis of this parent cell. **(2)** understand Crossing over would increase the genetic diversity of these gametes by creating many different combinations of maternal and paternal alleles along each of the chromosomes. **(3)** analyze Asexual reproduction generates no appreciable genetic diversity. Self-fertilization is preceded by meiosis so it generates gametes, through crossing over and independent

assortment, that have combinations of alleles not present in the parent. Outcrossing generates the most genetic diversity among offspring because it produces new combinations of alleles from two different individuals.

p. 249 Fig. 13.11 apply See **FIGURE A13.3**.

p. 251 Fig. 13.14 apply Asexually: 64 (16 individuals from generation three produce 4 offspring per individual). Sexually: 16 (8 individuals from generation three form 4 couples; each couple produces 4 offspring).

p. 252 Fig. 13.15 apply The rate of outcrossing is predicted to rise initially, as the pathogen selects for resistant roundworms, and then to fall as the roundworms in the population gain resistance and take advantage of the increased numbers of offspring offered by having more hermaphrodites that can reproduce by self-fertilization.

IF YOU UNDERSTAND . . .

13.1 remember See the right panel of Figure 13.7 and note how the cells transition from diploid to haploid in meiosis I. Also note that each chromosome contains two sister chromatids before and after meiosis I.

13.2 understand (a) See Figure 13.10 and note how the two different ways of aligning two homologous pairs of chromosomes at metaphase I of meiosis can create four different combinations of maternal and paternal chromosomes in daughter cells. (b) See Figure 13.11. Note that for each homologous pair with one crossover, two chromatids are recombinant and two are unaltered. Since each chromatid will produce a chromosome at the end of meiosis II, your drawing should show that two out of four of these chromosomes are recombinant.

13.3 apply (a) Your model should show events similar to those of Figure 13.12. (b) Your model should show two cells with one of each type of chromosome (n), one cell with an extra copy of one chromosome ($n + 1$), and one gamete without any copies of one chromosome ($n - 1$).

13.4 remember Sexual reproduction will likely occur during times when conditions are changing rapidly, because genetically diverse offspring may have an advantage in the new conditions.

YOU SHOULD BE ABLE TO . . .

✔ **Test Your Knowledge**

1. remember b **2.** remember a **3.** remember b **4.** understand b
5. remember 1/2 **6.** remember mitosis.

✔ **Test Your Understanding**

7. remember Homologous chromosomes are similar in size, shape, and gene content, and originate from different parents. Sister chromatids are exact copies of a chromosome that are generated when chromosomes are replicated (S phase of the cell cycle). **8.** understand Refer to Figure 13.7 as a guide for this exercise. The four pens represent the chromatids in one replicated homologous pair; the four pencils, the chromatids in a different homologous pair. To simulate meiosis II, make two "haploid cells"—each with a pair of pens and a pair of pencils representing two replicated chromosomes (one of each type in this species). Line them up in the middle of the cell; then separate the two pens and the two pencils in each cell such that one pen and one pencil go to each of four daughter cells. **9.** understand Meiosis I is a reduction division because homologs separate—daughter cells have just one of each type of chromosome instead of two. Meiosis II is not a reduction division because sister chromatids separate— daughter cells have unreplicated chromosomes instead of replicated chromosomes, but still just one of each type. **10.** apply b. **11.** apply Tetraploids produce diploid gametes, which combine with a haploid gamete from a diploid individual to form a triploid offspring. Mitosis proceeds normally in triploid cells because mitosis doesn't require forming pairs of chromosomes. But during meiosis in a triploid, homologous chromosomes can't pair up correctly. The third set of chromosomes does not have a homologous partner to pair with. **12.** understand Asexually produced individuals are genetically identical, so if one is susceptible to a new disease, all are. Sexually produced individuals are genetically unique, so if a new disease strain evolves, at least some plants are likely to be resistant.

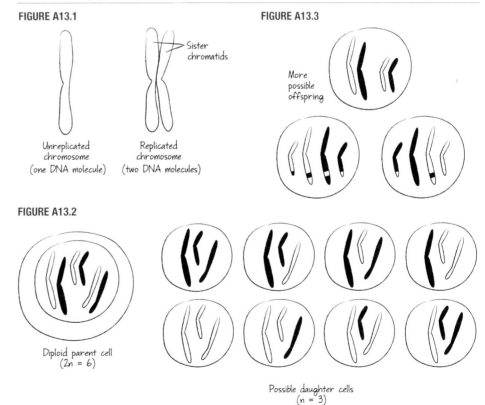

FIGURE A13.1

Sister chromatids

Unreplicated chromosome (one DNA molecule)

Replicated chromosome (two DNA molecules)

FIGURE A13.2

Diploid parent cell ($2n = 6$)

Possible daughter cells ($n = 3$)

FIGURE A13.3

More possible offspring

13. apply The gibbon would have 22 chromosomes in each gamete, and the siamang would have 25. Each somatic cell of the offspring would have 47 chromosomes. The offspring should be sterile because it has some chromosomes that would not form homologous pairs at prophase I of meiosis. **14.** apply c **15.** apply Aneuploidy is the major cause of spontaneous abortion. If spontaneous abortion is rare in older women, it would result in a higher incidence of aneuploid conditions such as Down syndrome in older women, as recorded in the Figure 13.13. **16. (a)** create Such a study might be done in the laboratory, controlling conditions in identical dishes. A population of rotifers infected with fungus could be established in each dish. One dish of rotifers would be kept moist; the other dishes of rotifers would be allowed to dry out. After various periods of time, water would be added to each dish and then the rotifers would be observed to see if fungal infections reappeared. **(b)** create Wind disperses the rotifer to new and often pathogen-free environments. In this case, the ticket to a sex-free existence is not genetic diversity but the evolution of an alternative means of evading pathogens made possible by fungus-infected rotifers ridding themselves of the pathogen when they dry.

CHAPTER 14

IN-TEXT QUESTIONS AND EXERCISES

p. 260 Fig. 14.3 evaluate An experiment is a failure if you didn't learn anything from it. That is not the case here.
p. 261 Table 14.2 analyze Row 6: 3.14:1; Row 7: 266.
p. 262 apply Filling in the top and side of a Punnett square requires writing out the types and ratios of gametes. For a cross involving one gene (monohybrid cross), this amounts to applying the principle of segregation as one allele is segregated from another. The phenotype ratios are 1:1 round:wrinkled; the genotype ratios are 1:1 Rr:rr.
p. 262 Fig. 14.4 understand No—the outcome (the expected offspring genotypes that the Punnett square generates) will be the same.

p. 263 CYU (1) apply See answer to Problem 13 in Test Your Problem-Solving Skills, below. **(2)** apply See answer to Problem 15 in Test Your Problem-Solving Skills.
p. 265 apply $AABb \rightarrow AB$ and Ab. $PpRr \rightarrow PR$, Pr, pR, and pr. $AaPpRr \rightarrow APR$, APr, ApR, Apr, aPR, apR, apr, and aPr.
p. 265 CYU (1) apply See answer to Problem 14 in Test Your Problem-Solving Skills. **(2)** apply See answer to Problem 16 in Test Your Problem-Solving Skills.
p. 269 CYU (1) understand See **FIGURE A14.1**. Segregation of alleles occurs when homologs that carry those alleles are separated during anaphase I. One allele ends up in each daughter cell. **(2)** understand See **FIGURE A14.2**. Independent assortment occurs because homologous pairs line up randomly at the metaphase plate during metaphase I. The figure shows two alternative arrangements of homologs in metaphase I. As a result, it is equally possible for a gamete to receive the following four combinations of alleles: YR, Yr, yR, yr.
p. 270 Fig. 14.12 apply XWY, XWy, XwY, Xwy.

p. 271 Fig. 14.13 analyze Random chance (or perhaps red-eyed, gray-bodied males don't survive well).
p. 273 Fig. 14.16 analyze The gene colored orange is *ruby*; the gene colored blue is *miniature wings*.
p. 275 Fig. 14.19 understand In this case of gene-by-gene interaction, the 9:3:3:1 ratio comes from four different forms of one trait (comb shape), whereas in a standard dihybrid cross, the four different phenotypes come from two different phenotypes for each of two genes.
p. 276 CYU (1) apply The comb phenotype results from interactions between alleles at two different genes, not a single gene. Matings between rose- and pea-comb chickens produce F_2 offspring that may have a new combination of alleles and thus new phenotypes. **(2)** apply Kernel color in wheat is influenced by alleles at many different genes, not a single gene. F_2 offspring have a normal distribution of phenotypes, not a 3:1 ratio.
p. 276 Fig. 14.21 understand Because there are many different genotypes that can produce intermediate coloration and fewer that can produce the extremes of coloration.

FIGURE A14.1

FIGURE A14.2

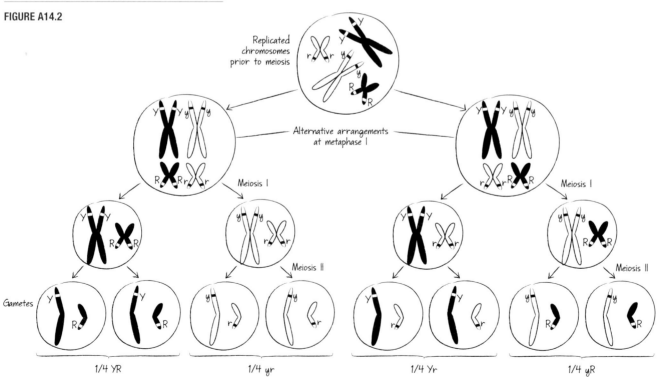

p. 279 Fig. 14.23 (apply) A heterozygous female and a color-blind male. 1 color-blind male : 1 color-blind female.

IF YOU UNDERSTAND . . .

14.1 (understand) Because crosses within a pure line never produce individuals with a different phenotype, this indicates that there must be only one allele in pure-line individuals. **14.2** (apply) B and b. **14.3** (apply) BR, Br, bR, and br, in equal proportions. The B and b alleles are located on different but homologous chromosomes, which separate into different daughter cells during meiosis I. The $BbRr$ notation indicates that the B and R genes are on different chromosomes. As a result, the chromosomes line up independently of each other in metaphase of meiosis I. The B allele is equally likely to go to a daughter cell with R as with r; likewise, the b allele is equally likely to go to a daughter cell with R as with r. **14.4** (apply) There are many ways to show this. If eye color in *Drosophila* is chosen as an example, then a pair of Punnett squares like those in Figure 14.11 illustrate how reciprocal crosses involving an X-linked recessive gene give different results. **14.5** (understand) Linkage refers to the physical connection of two alleles on the same chromosome. Crossing over breaks this linkage between particular alleles as segments of maternal and paternal homologs are exchanged. **14.6** (apply) See **FIGURE A14.3** for the pedigree. Note that the birth order of daughters and sons is arbitrary; other birth orders are equally valid. A Punnett square will show that all the sons are predicted to be X^+Y and all the daughters X^+X^c, where X^+ shows the dominant, X-linked allele for normal color vision and X^c shows the X-linked recessive color blindness allele.

YOU SHOULD BE ABLE TO . . .

✔ Test Your Understanding

1. (understand) d 2. (understand) b 3. (understand) a 4. (understand) d
5. (understand) a 6. (understand) a 7. (evaluate) d 8. (remember) b 9. (apply) b
10. (understand) d

✔ Test Your Problem-Solving Skills

11. (apply) **Example Solution** Here you're given offspring phenotypes and you're asked to infer parental genotypes. To do this you have to propose hypothetical parental genotypes to test, make a Punnett square to predict the offspring genotypes, and then see if the predicted offspring phenotypes match the data. In this case, coming up with a hypothesis for the parental genotypes is relatively straightforward because the problem states that the trait is due to one gene and two alleles. No information on sex is given, so assume the gene is autosomal (the simplest case). Now look at the second entry in the chart. It shows a 3 : 1 ratio of offspring from a winged individual that self-fertilizes. This result is consistent with the hypothesis that W^+ is dominant and W^- recessive and that this parent's genotype is W^+W^-. Now let's look at the first cross in the chart. If W^+ is dominant, then a wingless parent must be W^-W^-. When you do the Punnett square to predict offspring genotypes from selfing, you find that all the offspring will produce wingless fruits, consistent with the data. In the third cross, all the offspring make winged fruits even though one of the parents produces wingless

FIGURE A14.3

fruits and thus is W^-W^-. This would happen only if the winged parent is W^+W^+. (If this reasoning isn't immediately clear to you, work the Punnett square.) In the fourth cross, you could get offspring that all make winged fruits if the parents were W^+W^+ and W^+W^-, or if the parents were W^+W^+ and W^-W^-. Either answer is correct. Again, you can write out the Punnett squares to see that this statement is correct.

12. (apply) **Example Solution** Here you are given parental and offspring phenotypes and are asked to infer the parental genotypes. As a starting point, assume that the coat colors are due to the simplest genetic system possible: one autosomal gene with two alleles, where one allele is dominant and the other recessive. Because female II produces only black offspring, it's logical to suppose that black is dominant to brown. Let's use B for black and b for brown. Then the male parent is bb. To produce offspring with a 1 : 1 ratio of black : brown coats, female I must be Bb. But to produce all black offspring, female II must be BB. This model explains the data, so you can accept it as correct.

13. (apply) 3/4; 1/256 (see BioSkills 5 in Appendix B); 1/2 (the probabilities of transmitting the alleles or having sons does not change over time). **14.** (apply) Your answer to the first three parts should conform to the F_1 and F_2 crosses diagrammed in Figure 14.5b, except that different alleles and traits are being analyzed. The recombinant gametes would be Yi and yI. Yes—there would be some individuals with yellow seeds and constricted pods and with green seeds and inflated pods. **15.** (apply) Cross 1: non-crested (Cc) × non-crested (Cc) = 22 non-crested ($C\_$); 7 crested (cc). Cross 2: crested (cc) × crested (cc) = 20 crested (cc). Cross 3: non-crested (Cc) × crested (cc) = 7 non-crested (Cc); 6 crested (cc). Non-crested (C) is the dominant allele. **16.** (apply) This is a dihybrid cross that yields progeny phenotypes in a 9 : 3 : 3 : 1 ratio. Let O stand for the allele for orange petals and o the allele for yellow petals; let S stand for the allele for spotted petals and s the allele for unspotted petals. Start with the hypothesis that O is dominant to o, that S is dominant to s, that the two genes are found on different chromosomes so they assort independently, and that the parent individual's genotype is $OoSs$. If you do a Punnett square for the $OoSs \times OoSs$ mating, you'll find that progeny phenotypes should be in the observed 9 : 3 : 3 : 1 proportions. **17.** (analyze) Let D stand for the normal allele and d for the allele responsible for Duchenne-type muscular dystrophy. The woman's family has no history of the disease, so her genotype is almost certainly DD. The man is not afflicted, so he must be DY. (The trait is X-linked, so he has only one allele; the "Y" stands for the Y chromosome.) Their children are not at risk. The man's sister could be a carrier, however—meaning she has the genotype Dd. If so, then half of the second couple's male children are likely to be affected. **18.** (understand) Your stages of meiosis should look like a simplified version of Figure 13.7, except with $2n = 4$ instead of $2n = 6$. The A and a alleles could be on the red and blue versions of the longest chromosome, and the B and b alleles could be on the red and blue versions of the smallest chromosome, similar to the way the hair- and eye-color genes are shown in Figure 13.10. The places you draw them are the locations of the A and B genes, but each chromosome has only one allele. Each pair of red and blue chromosomes is a homologous pair. Sister chromatids bear the same allele (e.g., both sister chromatids of the long blue chromosomes might bear the a allele). Chromatids from the longest and shortest chromosomes are not homologous. To identify the events that result in the principles of segregation and independent assortment, see Figures 14.7 and 14.8 and substitute A, a, and B, b for R, r and Y, y. **19.** (apply) Half their offspring should have the genotype iI^A and the type A blood phenotype. The other half of their offspring should have the genotype iI^B and the type B blood

phenotype. Second case: the genotype and phenotype ratios would be 1 : 1 : 1 : 1 I^AI^B (type AB): I^Ai (type A): I^Bi (type B): ii (type O). **20.** (apply) Because the children of Tukan and Valco had no eyes and smooth skin, you can conclude that the allele for eyelessness is dominant to eyes and the allele for smooth skin is dominant to hooked skin. E = eyeless, e = two eye sets, S = smooth skin, s = hooked skin. Tukan is $eeSS$; Valco is $EEss$. The children are all $EeSs$. Grandchildren with eyes and smooth skin are eeS-. Assuming that the genes are on different chromosomes, one-fourth of the children's gametes are ee and three-fourths are S-. So 1/4 ee × 3/4 S- × 32 = 6 children would be expected to have two sets of eyes and smooth skin. **21.** (create) Although the mothers were treated as children by a reduction of dietary phenylalanine, they would have accumulated phenylalanine and its derivatives once they went off the low-phenylalanine diet as young adults. Children born of such mothers were therefore exposed to high levels of phenylalanine during pregnancy. For this reason, a low-phenylalanine diet is recommended for such mothers throughout the pregnancy. **22.** (apply) According to Mendel's model, palomino individuals should be heterozygous at the locus for coat color. If you mated palomino individuals, you would expect to see a combination of chestnut, palomino, and cremello offspring. If blending inheritance occurred, however, all the offspring would be palomino. **23.** (apply) Because this is an X-linked trait, the father who has hemophilia could not have passed the trait on to his son. Thus, the mother in couple 1 must be a carrier and must have passed the recessive allele on to her son, who is XY and affected. To educate a jury about the situation, you should draw what happens to the X and Y during meiosis and then make a drawing showing the chromosomes in couple 1 and couple 2, with a Punnett square showing how these chromosomes are passed to the affected and unaffected children. **24.** (apply) The curved-wing allele is autosomal recessive; the lozenge-eyed allele is sex-linked (specifically, X-linked) recessive. Let L be the allele for long wings and l be the allele for curved wings; let X^R be the allele for red eyes and X^r the allele for lozenge eyes. The female parent is LlX^RX^r; the male parent is LlX^RY. **25.** (apply) Albinism indicates the absence of pigment, so let b stand for an allele that gives the absence of blue and y for an allele that gives the absence of yellow pigment. If blue and yellow pigment blend to give green, then both green parents are $BbYy$. The green phenotype is found in $BBYY$, $BBYy$, $BbYY$, and $BbYy$ offspring. The blue phenotype is found in $BByy$ or $Bbyy$ offspring. The yellow phenotype is observed in $bbYY$ or $bbYy$ offspring. Albino offspring are $bbyy$. The phenotypes of the offspring should be in the ratio 9 : 3 : 3 : 1 as green : blue : yellow : albino. Two types of crosses yield $BbYy$ F_1 offspring: $BByy \times bbYY$ (blue × yellow) and $BBYY \times bbyy$ (green × albino). **26.** (apply) The chance that their first child will have hemophilia is 1/2. This is because all sons will have the disease and there is a 1/2 chance of having a firstborn son. The chance of having a carrier as their first child is also 1/2. This is because all daughters and none of the sons will be carriers and there is a 1/2 chance of having a firstborn daughter. (Recall that males cannot carry an X-linked recessive trait—with only one X chromosome, males either have the trait or not.) **27.** (apply) Autosomal dominant.

CHAPTER 15

IN-TEXT QUESTIONS AND EXERCISES

p. 286 Fig. 15.2 (analyze) The lack of radioactive protein in the pellet (after centrifugation) is strong evidence; they could also make micrographs of infected bacterial cells before and after agitation.

p. 287 Fig. 15.3 (apply) 5'-TAG-3'.

p. 288 Fig. 15.5 (apply) The same two bands should appear, but the upper band (DNA containing only ^{14}N) should get bigger and darker and the lower band (hybrid DNA) should get smaller and lighter in color since each succeeding generation has relatively less heavy DNA.

p. 290 apply See **FIGURE A15.1**. The new strands grow in opposite directions, each in the $5' \rightarrow 3'$ direction.

p. 292 apply Helicase, topoisomerase, single-strand DNA-binding proteins, primase, and DNA polymerase are all required for leading-strand synthesis. If any one of these proteins is nonfunctional, DNA replication will not occur.

p. 293 apply See **FIGURE A15.2**. If DNA ligase were defective, then the leading strand would be continuous, and the lagging strand would have gaps in it where the Okazaki fragments had not been joined.

p. 294 CYU (1) understand DNA polymerase adds nucleotides only to the free $3' -OH$ on a strand. Primase synthesizes a short RNA sequence that provides the free $3'$ end necessary for DNA polymerase to start working. **(2)** understand The need to begin DNA synthesis many times on the lagging strand requires many new primers. Since DNA polymerase I is needed to remove primers, it is required predominantly on the primer-rich lagging strand.

p. 296 Fig. 15.13 apply As long as the RNA template could bind to the "overhanging" section of single-stranded DNA, any sequence could produce a longer strand. For example, 5'-CCCAUUCCC-3' would work just as well.

p. 297 CYU (1) understand This is because telomerase is needed only to replicate one end of a linear DNA and bacterial DNAs lack ends because they are circular. **(2)** understand Since telomerase works by extending one strand of DNA without any external template and because DNA synthesis requires a template, telomerase must contain an internal template to allow it to extend a DNA chain.

p. 299 Fig. 15.16 analyze They are lower in energy and not absorbed effectively by the DNA bases.

p. 300 CYU (1) apply The mutation rate would be predicted to rise because differences in base-pair stability and shape make it possible for DNA polymerase to distinguish correct from incorrect base pairs during DNA replication. **(2)** apply The mutation rate should increase because without a way to distinguish which strand to use as a template for repair, about half of mismatches on average would be repaired using the incorrect strand as a template. **(3)** remember The enzyme that removes the dimer and surrounding DNA is specific to nucleotide excision repair. DNA polymerase and DNA ligase work in both nucleotide excision repair and normal DNA synthesis.

p. 300 Fig. 15.18 apply Exposure to UV radiation can cause formation of thymine dimers. If thymine dimers are not repaired, they represent mutations. If such mutations occur in genes controlling the cell cycle, cells can grow abnormally, resulting in cancers.

IF YOU UNDERSTAND . . .

15.1 apply These results would not allow distinguishing whether DNA or protein was the genetic material.

15.2 understand The bases added during DNA replication are shown in red type.

    Original DNA: **CAATTACGGA**
                    GTTAATGCCT
   Replicated DNA: **CAATTACGGA**
                    GTTAATGCCT
                    CAATTACGGA
                    GTTAATGCCT

15.3 understand See **FIGURE A15.3**. **15.4** understand Because cancer cells divide nearly without limit, it's important for these cells to have active telomerase so that chromosomes don't shorten to the point where cell division becomes impossible. **15.5** understand If errors in DNA aren't corrected, they represent mutations. When DNA repair systems fail, the mutation rate increases. As the mutation rate increases, the chance that one or more cell-cycle genes will be mutated increases. Mutations in these genes often result in uncontrolled cell division, ultimately leading to cancer.

YOU SHOULD BE ABLE TO . . .

✔ Test Your Knowledge

1. remember d **2.** understand a **3.** remember topoisomerase **4.** remember b **5.** remember c **6.** remember telomerase

✔ Test Your Understanding

7. remember Labeling DNA or labeling proteins. **8.** understand DNA is constantly damaged, and many pathways have evolved to repair this onslaught of damage. If a DNA repair pathway is inactivated by mutation, damage is inefficiently repaired. Consequently mutation rates increase, and the increased number of mutations increases the probability that cancer-causing mutations will occur. **9.** understand On the lagging strand, DNA polymerase moves away from the replication fork. When helicase unwinds a new section of DNA, primase must build a new primer on the template for the lagging strand (closer to the fork) and another polymerase molecule must begin synthesis at this point. This makes the lagging-strand synthesis discontinuous. On the leading strand, DNA polymerase moves in the same direction as helicase, so synthesis can continue, without interruption, from a single primer (at the origin of replication). **10.** understand Telomerase binds to the overhang at the end of a chromosome. Once bound, it begins catalyzing the addition of deoxyribonucleotides to the overhang in the $5' \rightarrow 3'$ direction, lengthening the overhang. This allows primase, DNA polymerase, and ligase to catalyze the addition of deoxyribonucleotides to the lagging strand in the $5' \rightarrow 3'$ direction, restoring the lagging strand to its original length. **11.** apply a (Because the ability to distinguish which strand contains the incorrect base would be lost). **12.** analyze d (The regularity of DNA's structure allows enzymes to recognize any type of damage that distorts this regular structure.)

✔ Test Your Problem-Solving Skills

13. analyze c (If DNA polymerase could synthesize DNA $3' \rightarrow 5'$ as well as the normal $5' \rightarrow 3'$, then both newly synthesized DNA strands could be extended to follow the replication fork.) **14.** apply (a) In **FIGURE A15.4**, the gray lines represent DNA strands containing radioactivity. (b) After one round of replication in radioactive solution, one double-stranded DNA would be radioactive in both strands and the other would not be radioactive in either strand. After another round of DNA synthesis, this time in nonradioactive solution, one of the four DNA molecules would be radioactive in both strands and the other three DNA molecules would contain no radioactivity in any strand. **15.** analyze (a) The double mutant of both *uvrA* and *recA* is most sensitive to UV light; the single mutants are in between; and the wild type is

---

FIGURE A15.1

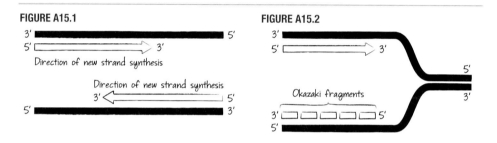

Direction of new strand synthesis

Direction of new strand synthesis

FIGURE A15.2

Okazaki fragments

FIGURE A15.3

FIGURE A15.4

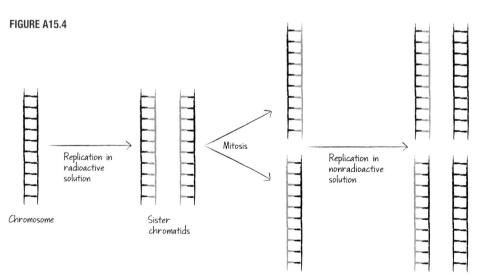

Chromosome

Replication in radioactive solution

Sister chromatids

Mitosis

Replication in nonradioactive solution

least sensitive. (b) The *recA* gene contributes more to UV repair through most of the UV dose levels. But at very high UV doses, the *uvrA* gene is somewhat more important than the *recA* gene. **16.** apply About 4600 seconds or 77 minutes. This answer comes from knowing that replication proceeds bidirectionally, so replication from each fork is predicted to replicate half the chromosome. This is 4.6 million base pairs/2 = 2.3 million base pairs. At 500 base pairs per second, this requires 2.3 million base pairs/500 base pairs per second = 4600 seconds. To obtain the time in minutes, divide 4600 seconds by 60 seconds per minute.

CHAPTER 16

IN-TEXT QUESTIONS AND EXERCISES

p. 305 Fig. 16.1 apply No, it could not make citrulline from ornithine without enzyme 2. Yes, it would no longer need enzyme 2 to make citrulline.

p. 306 Fig. 16.2 create Many possibilities: strain of fungi used, exact method for creating mutants and harvesting spores to grow, exact growing conditions (temperature, light, recipe for growth medium—including concentrations of supplements), objective criteria for determining growth or no growth.

p. 309 CYU remember Change in DNA sequence, change in sequence of transcribed mRNA, potential change in amino acid sequence of protein, likely altered protein function (if amino acid sequence was altered), likely change in phenotype.

p. 312 analyze (1) The codons in Figure 16.4 are translated correctly. (2) See **FIGURE A16.1**. (3) There are many possibilities (just pick alternative codons for one or more of the amino acids); one is an mRNA sequence (running 5′ → 3′): 5′ GCG-AAC-GAU-UUC-CAG 3′. To get the corresponding DNA sequence, write this sequence but substitute Ts for Us: 5′ GCG-AAC-GAT-TTC-CAG 3′. Now write the complementary bases, which will be in the 3′ → 5′ direction: 3′ CGC-TTG-CTA-AAG-GTC 5′. When this second strand is transcribed by RNA polymerase, it will produce the mRNA given with the proper 5′ → 3′ orientation.

p. 312 CYU (1) apply Note the 3′ → 5′ polarity of the DNA sequences in the accompanying table, and in the subsequent answer. This means that the complementary mRNA codon will read 5′ → 3′. U (rather than T) is the base transcribed from A.

DNA	mRNA Codon	Amino Acid
ATA	UAU	Tyrosine
ATG	UAC	Tyrosine
ATT	UAA	Stop
GCA	UGC	Cysteine

(2) understand The ATA → ATG mutation would have no effect on the protein. The ATA → ATT mutation introduces a stop codon, so the resulting polypeptide would be shortened. This would result in synthesis of a mutant protein much shorter than the original protein. The ATA → GCA mutation might have a profound effect on the protein's conformation because cysteine's structure is different from tyrosine's.

p. 312 Fig. 16.7 apply See **FIGURE A16.1**.

p. 314 Fig. 16.9 analyze Chromosomes 2, 3, 6, 10, 13, 14, 15, 18, 19, 21, 22, and the X chromosome show aneuploidy. Virtually every chromosome has structural rearrangements, and translocations are the most obvious. These are seen when two or more different colors occur on the same chromosome.

IF YOU UNDERSTAND . . .

16.1 understand Ornithine, citrulline or arginine could be added to allow growth. As Figure 16.1 shows, these compounds are made after the steps catalyzed by the enzymes 1, 2, and 3 that are needed to produce arginine.

16.2 understand An inhibitor of RNA synthesis will eventually prevent the synthesis of new proteins because newly synthesized mRNA is needed for translation. **16.3** apply AUG UGG AAA/AAG CAA/CAG **16.4** understand Since redundancy is having more than one codon specify a particular amino acid, redundancy makes it possible for there to be a point mutation without altering the amino acid. This is a silent mutation. A silent mutation is likely to be neutral because there is no change in amino acid sequence.

YOU SHOULD BE ABLE TO . . .

✔ Test Your Knowledge

1. remember d **2.** understand a **3.** understand d **4.** remember Because there is no chemical complementarity between nucleotides and amino acids; and because in eukaryotes, DNA is in the nucleus but translation occurs in the cytoplasm. **5.** understand d **6.** remember A codon that signals the end of translation.

✔ Test Your Understanding

7. understand Because the Morse code and genetic code both use simple elements (dots and dashes; 4 different bases) in different orders to encode complex information (words; amino acid sequences).

8.
Substrate 1 →A Substrate 2 →B Substrate 3 →C
Substrate 4 →D Substrate 5 →E Biological Sciazine
apply Substrate 3 would accumulate. Hypothesis: The individuals have a mutation in the gene for enzyme D.

9. understand They supported an important prediction of the hypothesis: Losing a gene (via mutation) resulted in loss of an enzyme. **10.** understand c **11.** understand In a triplet code, addition or deletion of 1–2 bases disrupts the reading frame "downstream" of the mutation site(s), resulting in a dysfunctional protein. But addition or deletion of 3 bases restores the reading frame—the normal sequence is disrupted only between the first and third mutation. The resulting protein is altered but may still be able to function normally. Only a triplet code would show these patterns. **12.** understand A point mutation changes the nucleotide sequence of an existing allele, creating a new one, so it always changes the genotype. But because the genetic code is redundant, and because point mutations can occur in DNA sequences that do not code for amino acids, these point mutations do not change the protein product and therefore do not change the phenotype.

✔ Test Your Problem-Solving Skills

13. apply See **FIGURE A16.2**. **14.** analyze Every copying error would result in a mutation that would change the amino acid sequence of the protein and would likely affect its function. **15.** analyze Before the central dogma was understood, DNA was known to be the hereditary material, but no one knew how particular sequences of bases resulted in the production of RNA and protein products. The central dogma clarified how genotypes produce phenotypes. **16.** analyze c

CHAPTER 17

IN-TEXT QUESTIONS AND EXERCISES

p. 318 Fig. 17.1 remember RNA is synthesized in the 5′ → 3′ direction; the DNA template is "read" 3′ → 5′.

p. 321 CYU (1) apply Transcription would be reduced or absent because the missing nucleotides are in the −10 region, one of the two critical parts of the promoter. **(2)** understand NTPs are required because the three phosphate groups raise the monomer's potential energy enough to make the polymerization reaction exergonic.

p. 322 Fig. 17.5 apply There would be no loops—the molecules would match up exactly.

p. 323 CYU (1) understand The subunits contain both RNA (the *ribonucleo*– in the name) and proteins. **(2)** understand The cap and tail protect mRNAs from degradation and facilitate translation.

p. 326 Fig. 17.11 analyze If the amino acids stayed attached to the tRNAs, the gray line in the graph would stay high and the green line low. If the amino acids were transferred to some other cell component, the gray line would decline but the green line would be low.

p. 327 understand (1) The amino acid attaches on the top right of the L-shaped structure. (2) The anticodon is antiparallel in orientation to the mRNA codon, and it contains the complementary bases.

p. 332 CYU understand E is for exit—the site where uncharged tRNAs are ejected; P is for peptidyl (or peptide bond)—the site where peptide bond formation takes place; A is for aminoacyl—the site where aminoacyl tRNAs enter.

IF YOU UNDERSTAND . . .

17.1 apply Transcription would continue past the normal point because the insertion of nucleotides would disrupt the structure of the RNA hairpin that functions as a terminator. **17.2** apply The protein-coding segment of the gene is predicted to be longer in eukaryotes because of the presence of introns. **17.3** understand The tRNAs act as adaptors because they couple the information contained in the nucleotides of mRNA to that contained in the amino acid sequence of proteins. **17.4** apply An incorrect amino acid would appear often in proteins. This is because the altered synthetase would sometimes add the correct amino acid for a particular tRNA and at other

---

FIGURE A16.1

mRNA sequence:
5′ AUG-CUG-GAG-GGG-GUU-AGA-CAU 3′

Amino acid sequence:
Met-Leu-Glu-Gly-Val-Arg-His

FIGURE A16.2

Bottom DNA strand:
3′ AACTT-TAC(start)-GGG-CAA-ACC-TCT-AGC-CCA-ATG-TCG-ATC(stop)-AGTTTC 5′

mRNA sequence:
5′ AUG-CCC-GUU-UGG-AGA-UCG-GGU-UAC-AGC-UAG 3′

Amino acid sequence:
Met-Pro-Val-Trp-Arg-Ser-Gly-Tyr-Ser

times add the incorrect amino acid. **17.5** create One possible concept map is shown in **FIGURE A17.1**.

YOU SHOULD BE ABLE TO . . .

✔ Test Your Knowledge

1. understand c **2.** remember c **3.** understand To speed the correct folding of newly synthesized proteins **4.** remember d **5.** remember At the 3′ end **6.** apply b

✔ Test Your Understanding

7. understand Basal transcription factors bind to promoter sequences in eukaryotic DNA and facilitate the binding of RNA polymerase. As part of the RNA polymerase holoenzyme, sigma binds to a promoter sequence in bacterial DNA and to allow RNA polymerase to initiate at the start of genes. **8.** analyze The wobble rules allow a single tRNA to pair with more than one type of mRNA codon. This is distinct from redundancy, in which more than one codon can specify a single amino acid. If the wobble rules did not exist, there would need to be one tRNA for each amino-acid-specifying codon in the redundant genetic code. **9.** apply b **10.** apply After a peptide bond forms between the polypeptide and the amino acid held by the tRNA in the A site, the ribosome moves down the mRNA. As it does, an uncharged tRNA leaves the E site. The now-uncharged tRNA that was in the P site enters the E site; the tRNA holding the polypeptide chain moves from the A site to the P site, and a new aminoacyl tRNA enters the A site. **11.** apply The ribosome's active site is made up of RNA, not protein. **12.** understand The separation allows the aminoacyl tRNA to place the amino acid into the ribosome's active site while reaching to the distant codon on the mRNA.

✔ Test Your Problem-Solving Skills

13. create Ribonucleases degrade mRNAs that are no longer needed by the cell. If an mRNA for a hormone that increased heart rate were never degraded, the hormone would be produced continuously and heart rate would stay elevated—a dangerous situation. **14.** apply d **15.** analyze The regions most crucial to the ribosome's function should be the most highly conserved: the active site, the E, P, and A sites, and the site where mRNAs initially bind. **16.** analyze The most likely locations are one of the grooves or channels where RNA, DNA, and ribonucleotides move through the enzyme—plugging one of them would prevent transcription.

CHAPTER 18

IN-TEXT QUESTIONS AND EXERCISES

p. 338 understand See **TABLE A18.1**.
p. 338 Fig. 18.1 analyze Write "Slowest response, most efficient resource use" next to the transcriptional control label. Write "Fastest response, least efficient resource use" next to the post-translational control label.
p. 339 Fig. 18.2 apply Plates from all three treatments must be identical and contain identical growth medium, except for the presence of the sugars labeled in the figure. Also, all plates must be grown under the same physical conditions (temperature, light) for the same time.
p. 340 Fig. 18.3 apply Use a medium with all 20 amino acids when producing a master plate of mutagenized *E. coli* colonies; then use a replica plate that contains all the amino acids except tryptophan. Choose cells from the master plate that did *not* grow on the replica plate.
p. 341 understand *lacZ* codes for the β-galactosidase enzyme, which breaks the disaccharide lactose into glucose and galactose. *lacY* codes for the galactoside permease enzyme, which transports lactose into the bacterial cell. *lacI* codes for a protein that shuts down production of the other *lac* products. When lactose is absent, the *lacI* product prevents transcription. This is logical because there is no reason for the cell to make β-galactosidase

and galactoside permease if there is no lactose to metabolize. But when lactose is present, it interacts with *lacI* in some way so that *lacZ* and *lacY* are induced (their transcription can occur). When lactose is present, the enzymes that metabolize it are expressed.
p. 343 apply A mutation that prevents lactose binding to repressor is predicted to prevent transcription of the *lac* operon under any condition. This is because the repressor would never come off the operator. A mutation in the operator that prevents repressor binding is predicted to lead to constitutive expression of the *lac* operon.
p. 343 Fig. 18.8 understand Put the "Repressor protein" on the operator. No transcription will take place. Then put the "RNA polymerase" on the promoter. No transcription will take place. Finally, put "lactose" on the repressor protein and then remove the resulting lactose–repressor complex from the operon. Transcription will begin.
p. 344 CYU (1) understand It is logical that the genes for metabolizing lactose should be expressed only when lactose is available. **(2)** understand See **FIGURE A18.1**.
p. 344 apply This mutation is predicted to lower rates of transcription initiation because AraC's binding to

RNA polymerase is essential for AraC to work as an activator.

IF YOU UNDERSTAND . . .

18.1 understand Production of β-galactosidase and galactosidase permease are under transcriptional control—transcription depends on the action of regulatory proteins. The activity of the repressor is under post-translational control. **18.2** apply Treat *E. coli* cells with a mutagen, and create a master plate that is grown at 33°C. Replica-plate this master plate and grow the replica plate at 42°C. Look for colonies that are on the master plate at 33°C but not on the replica plate at 42°C. **18.3** understand The operator is the parking brake; the repressor locks it in place, and the inducer releases it. **18.4** apply *ara* initiator mutations are likely to affect positive and negative control because AraC must bind to the *ara* initiator sequence for both forms of control. **18.5** create Mutations that create operators for the SOS regulon repressor protein would put new genes under control of the repressor and incorporate them into the regulon.

FIGURE A17.1

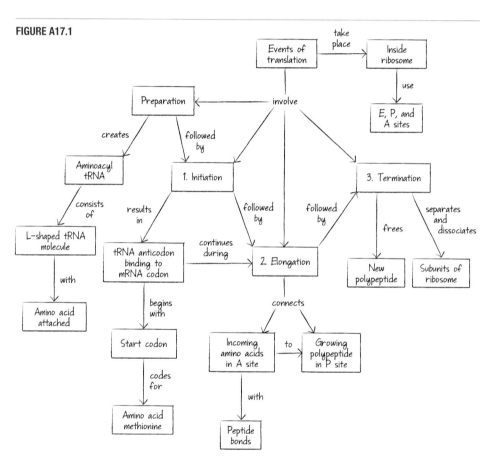

TABLE A18.1

Name	Function
lacZ	Gene for β-galactosidase
lacY	Gene for galactoside permease
Operator	Binding site for repressor
Promoter	Binding site for RNA polymerase
Repressor	Shuts down transcription
Lactose	Binds to repressor and stimulates transcription (removes negative control)
Glucose	At low concentration, increases transcription

YOU SHOULD BE ABLE TO . . .

✔ Test Your Knowledge

1. `remember` b **2.** `understand` d **3.** `apply` b **4.** `remember` DNA; RNA; **5.** `understand` When glucose and another sugar are present in the environment, inducer exclusion prevents the use of the other sugar and allows only use of glucose. **6.** `remember` a

✔ Test Your Understanding

7. `understand` The glycolytic enzymes are always needed in the cell because they are required to produce ATP, and ATP is always needed. **8.** `understand` Positive control means that a regulatory protein, when present, causes transcription to increase. Negative control means that a regulatory protein, when present, prevents transcription. **9.** `apply` b. **10.** `apply` Regulation of the *lac* operon should be normal. The location of the *lacI* gene isn't important, because the gene produces a protein that diffuses within the cell to the operator. **11.** `analyze` b (since the activator needs to bind to regulatory sequences to activate gene expression, preventing DNA binding would cripple the regulon and prevent cholera).**12.** `analyze` The *lac* operon would be strongly induced. Once inside the cell, the IPTG will bind to the repressor, causing it to release from DNA. IPTG cannot be broken down, so its concentration will remain high. Finally, since glucose is absent, there will be no inducer exclusion to inhibit IPTG transport through the galactoside permease transporter.

✔ Test Your Problem-Solving Skills

13. `create` Set up cultures with individuals that all come from the same colony of toluene-tolerating bacteria. Half the cultures should have toluene as the only source of carbon; half should have glucose or another common source of carbon. The glucose-containing medium serves as a control to ensure that cells can be grown in the lab. Cells will grow in both cultures if they are able to use toluene as a source of carbon; they will grow only in glucose-containing medium if toluene cannot be used as a carbon source. **14.** `apply` At a rate of 1×10^{-4} mutants per cell, you would on average find one mutant in every 10,000 (1×10^4) cells. Therefore, you should screen a bit more (~ 2–3 times more) than 10,000 cells to be reasonably sure of finding at least one mutant. **15.** `apply` b **16.** `analyze` Cells with functioning β-galactosidase will produce blue colonies; cells with *lacZ* mutations or *lacY* mutations will not produce β-galactosidase and will produce white colonies. The *lac* promoter could be mutated so that RNA polymerase cannot bind.

CHAPTER 19

IN-TEXT QUESTIONS AND EXERCISES

p. 352 Fig. 19.4 `analyze` Acetylation of histones decondenses chromatin and allows transcription to begin, so HATs are elements in positive control. Deacetylation condenses chromatin and inactivates transcription, so HDACs are elements in negative control.
p. 353 CYU (1) `apply` Many more genes than normal are predicted to be expressed because the inability to methylate DNA would lead to more decondensed chromatin.
(2) `understand` Addition of acetyl groups to histones or methyl groups to DNA can cause chromatin to decondense or condense, respectively. Different patterns of acetylation or methylation will determine which genes in muscle cells versus brain cells can be transcribed and which are not available for transcription.
p. 353 Fig. 19.5 `create` They could do something to change histone modifications to see how this affects gene transcription instead of just making the observation that certain histone modifications and low rates of transcription go together.
p. 355 Fig. 19.6 `analyze` A typical eukaryotic gene usually contains introns and is regulated by multiple enhancers. Bacterial operons lack introns and enhancers. The

promoter-proximal element found in some eukaryotic genes is comparable to the *araC* binding site in the *ara* operon of bacteria. Bacterial operons have a single promoter but code for more than one protein; eukaryotic genes code for a single product.
p. 356 `analyze` The basal transcription factors found in muscle and nerve cells are similar or identical; the regulatory transcription factors found in each cell type are different.
p. 356 `understand` DNA forms loops when distant regulatory regions, such as silencers and enhancers, are brought close to the promoter through binding of regulatory transcription factors to mediator.
p. 356 CYU (1) `analyze` Bacterial regulatory sequences are found close to the promoter; eukaryotic regulatory sequences can be close to the promoter or far from it. Bacterial regulatory proteins interact directly with RNA polymerase to initiate or prevent transcription; eukaryotic regulatory proteins influence transcription by altering chromatin structure or binding to the basal transcription complex through mediator proteins.
(2) `understand` Certain regulatory proteins decondense chromatin at muscle- or brain-specific genes and then activate or repress the transcription of cell-type-specific genes. Muscle-specific genes are expressed only if muscle-specific regulatory proteins are produced and activated.
p. 358 `understand` Alternative splicing does not occur in bacteria because bacterial genes do not contain introns—in bacteria, each gene codes for a single product. Alternative splicing is part of step 3 in Figure 19.1.

p. 359 `remember` Step 4 and step 5. RNA interference either **(1)** decreases the life span of mRNAs or **(2)** inhibits translation.
p. 360 CYU (1) `understand` It became clear that a single gene can code for multiple products instead of a single one.
(2) `understand` The miRNAs interfere with mRNAs by targeting them for destruction or preventing them from being translated.
p. 362 CYU (1) `understand` Many different types of mutations can disrupt control of the cell cycle and initiate cancer. These mutations can affect any of the six levels of control over gene regulation outlined in Figure 19.1.
(2) `understand` The p53 protein is responsible for shutting down the cell cycle in cells with damaged DNA. If the protein does not function, then cells with damaged DNA—and thus many mutations—continue to divide. If these cells have mutations in genes that regulate the cell cycle, then they may continue to divide in an uncontrolled fashion.

IF YOU UNDERSTAND . . .

19.1 `understand` Because eukaryotic RNAs are not translated as soon as transcription occurs, it is possible for RNA processing to occur, which creates variation in the mRNAs produced from a primary RNA transcript and in their life spans. **19.2** `understand` The default state of eukaryotic genes is "off," because the highly condensed state of the chromatin makes DNA unavailable to RNA polymerase. **19.3** `understand` See **FIGURE A19.1**. **19.4** `understand` Alternative splicing makes it possible for a single gene to

---

FIGURE A18.1

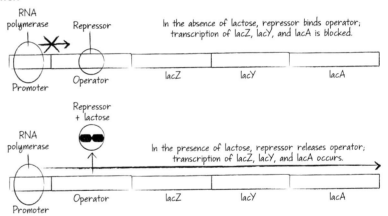

In the absence of lactose, repressor binds operator; transcription of lacZ, lacY, and lacA is blocked.

In the presence of lactose, repressor releases operator; transcription of lacZ, lacY, and lacA occurs.

FIGURE A19.1

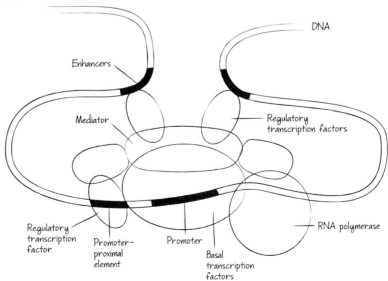

code for multiple products. **19.5** analyze One difference is that, in bacteria, genes that need to be turned on together are often clustered together in operons. Eukaryotes do not use operons. A similarity is in the way eukaryotes turn on many genes together and the strategy of bacterial regulons. Here, genes that are in many different locations share a DNA regulatory sequence and are activated when a regulatory transcription factor binds to the regulatory sequence. **19.6** understand In mutant cells that lack a form of p53 that can bind to enhancers, DNA damage cannot arrest the cell cycle, the cell cannot kill itself, and damaged DNA is replicated. This leads to mutations that can move the cell farther down the road to cancer.

YOU SHOULD BE ABLE TO . . .

✔ Test Your Knowledge

1. remember c **2.** remember A tumor suppressor is a gene or protein that holds cell division in check unless conditions are right for the cell to divide. **3.** understand d **4.** remember The set of regulatory transcription factors present in a particular cell, not differences in DNA sequence, are largely responsible for which genes are expressed. **5.** remember b **6.** remember d

✔ Test Your Understanding

7. analyze (a) Enhancers and the *araC* site are similar because both are sites in DNA where regulatory proteins bind. They are different because enhancers generally are located at great distances from the promoter, whereas the *araC* site is located nearer the promoter. (b) Promoter-proximal elements and the *lac* operon operator are both regulatory sites in DNA located close to the promoter. (c) Basal transcription factors and sigma are proteins that must bind to the promoter before RNA polymerase can initiate transcription. They differ because sigma is part of the RNA polymerase holoenzyme, while the basal transcription complex recruits RNA polymerase to the promoter. **8.** understand If changes in the environment cause changes in how spliceosomes function, then the RNAs and proteins produced from a particular gene could change in a way that helps the cell cope with the new environmental conditions. **9.** analyze (a) Enhancers and silencers are both regulatory sequences located at a distance from the promoter. Enhancers bind regulatory transcription factors that activate transcription; silencers bind regulatory proteins that shut down transcription. (b) Promoter-proximal elements and enhancers are both regulatory sequences that bind positive regulatory transcription factors. Promoter-proximal elements are located close to the promoter; enhancers are far from the promoter. (c) Transcription factors bind to regulatory sites in DNA; mediator does not bind to DNA but instead forms a bridge between regulatory transcription factors and

basal transcription factors. **10.** apply Inhibition of a histone deacetylase is predicted to leave acetyl groups on histones longer than normal. This is predicted to keep gene transcription going longer than normal, leading to higher levels of particular proteins and a change in the phenotype of the cell. **11.** analyze c; this is because there are many more modifications in the histone code relative to the genetic code. **12.** apply The cell is predicted to arrest in the cell cycle and most likely will commit suicide through activation of apoptosis genes because of the continually active *p53*.

✔ Test Your Problem-Solving Skills

13. apply c; This is because promoters and enhancers are brought into close physical proximity when transcription begins (see Figure 19.8). Since rats of malnourished mothers initiate *Hnf4a* gene transcription infrequently, the promoter and enhancer will be together less often in these animals compared with rats born to well-nourished mothers. **14.** understand See **FIGURE A19.2**. The value for the normal diet should be shown as 1.0 and the value for the low-protein diet should be shown as 0.64 (0.64 comes from the ratio of the cpm of the low-protein diet divided by the cpm of the normal diet, or 7368/11,478). **15.** create You could treat a culture of DNA-damaged cells with a drug that stops transcription and then compare them with untreated DNA-damaged cells. If transcriptional control regulates p53 levels, then the p53 level would be lower in the treated cells versus control cells. If control of p53 levels is post-translational, then in both cultures the p53 level would be the same. *Other approaches:* Add labeled NTPs to damaged cells and see if they are incorporated into mRNAs for p53; or add labeled amino acids to damaged cells and see if they are incorporated into completed p53 proteins; or add labeled phosphate groups and see if they are added to p53 proteins. **16.** create The double-stranded RNA could be cut by the same enzyme that creates double-stranded miRNAs from miRNA precursors. This double-stranded RNA may be incorporated into RISC and converted into a single-stranded RNA. The single-stranded RNA held by RISC would work in RNA interference to trigger the destruction of the complementary viral mRNA.

BIG PICTURE Genetic Information

p. 366 CYU (1) remember Star = DNA, mRNA, proteins. **(2)** understand RNA "is reverse transcribed by" reverse transcriptase "to form" DNA. **(3)** analyze E = splicing, etc.; E = meiosis and sexual reproduction (along with their links). **(4)** analyze Chromatin "makes up" chromosomes; independent assortment and recombination "contribute to" high genetic diversity.

CHAPTER 20

IN-TEXT QUESTIONS AND EXERCISES

p. 372 understand Isolate the DNA and cut it into small fragments with EcoRI, which leaves sticky ends. Cut copies of a plasmid or other vector with EcoRI. Mix the fragments and plasmids under conditions that promote complementary base pairing by sticky ends of fragments and plasmids. Use DNA ligase to catalyze formation of phosphodiester bonds and seal the sequences.
p. 372 Fig. 20.4 apply No—many times, because many copies of each type of mRNA were present in the pituitary cells, and many pituitary cells were used to prepare the library.
p. 373 apply The probe must be single stranded so that it will bind by complementary base pairing to the target DNA, and it must be labeled so that it can be detected. The probe will base-pair only with fragments that include a sequence complementary to the probe's sequence. A probe with the sequence 5′ AATCG 3′ will

bind to the region of the target DNA that has the sequence 5′ CGATT 3′ as shown here:

5′ AATCG 3′
3′ TCCGGTTAGCATTACCATTTT 5′

p. 374 CYU (1) understand When the endonuclease makes a staggered cut in a palindromic sequence and the strands separate, the single-stranded bases will bind ("stick") to the single-stranded bases left where the endonuclease cut the same palindrome at a different location. **(2)** understand The word probe means to examine thoroughly. A DNA probe "examines" a large set of sequences thoroughly and binds to one—the one that has a complementary base sequence.
p. 375 Fig. 20.7 analyze The polymerase will begin at the 3′ end of each primer. On the top strand in part (b), it will move to the left; on the bottom strand, it will move to the right. As always, synthesis is in the 5′-to-3′ direction.
p. 376 CYU (1) understand Denaturation makes DNA single stranded so the primer can bind to the sequence during the annealing step. Once the primer is in place, *Taq* polymerase can synthesize the rest of the strand during the extension step. It is a "chain reaction" because the products of each reaction cycle are used in the next reaction cycle—this is why the number of copies doubles in each cycle. **(2)** apply One of many possible answers is shown in **FIGURE A20.1**.
p. 378 CYU understand If ddNTPs were present at high concentration, they would almost always be incorporated—meaning that only fragments from the first complementary base in the sequence would be produced.
p. 379 understand Using a map with many markers makes it more likely that there will be one marker that is very tightly linked to the gene of interest—meaning that a form of the marker will almost always be associated with the phenotype you are tracking.
p. 382 CYU create Start with a genetic map with as many polymorphic markers as possible. Determine the genotype at these markers for a large number of individuals who have the same type of alcoholism—one that is thought to have a genetic component—as well as a large number of unaffected individuals. Look for particular versions of a marker that is almost always found in affected individuals. Genes that contribute to a predisposition to alcoholism will be near that marker.
p. 383 Fig. 20.11 apply The insertion will probably disrupt the gene and have serious consequences for the cell and potentially the individual.

IF YOU UNDERSTAND . . .

20.1 understand Special features of DNA are needed to allow replication in a cell. It is unlikely that any DNA fragment generated by cutting the DNA of one species with a restriction enzyme would replicate when inserted into a bacterial cell. By placing DNA fragments within plasmids that normally replicate in a bacterial cell, the inserted DNA can be replicated along with the plasmid. **20.2** analyze Advantages of cloning in cells: No knowledge of the sequence is required. Disadvantages of cloning in cells: It is slower and technically more difficult than PCR. PCR advantages: It is fast and easy, and it can amplify a DNA sequence that is rare in the sample. PCR disadvantage: It requires knowledge of sequences on either side of the target gene, so primers can be designed. **20.3** understand The length of each fragment is dictated by where a ddNTP was incorporated into the growing strand, and each ddNTP corresponds to a base on the template strand. Thus the sequence of fragment sizes corresponds to the sequence on the template DNA. **20.4** understand In this case, both affected and unaffected individuals would have the same marker, so there would be no way to associate a particular form of the marker with the gene and phenotype you are interested in. **20.5** understand An advantage of using retrovirus vectors is that any foreign genes they carry are integrated into

FIGURE A19.2

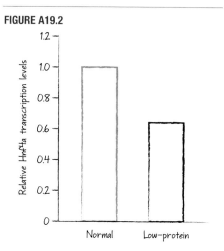

human chromosomes, so delivered genes become a permanent feature of the cell's genome. The concern is that they may integrate at sites that alter gene function in ways that harm cell function, or even worse, lead to cancer. **20.6** understand Genes can be inserted into the Ti plasmids carried by *Agrobacterium*, and the recombinant plasmids transferred to plant cells infected by the bacterium.

YOU SHOULD BE ABLE TO . . .

✔ Test Your Knowledge

1. understand They cut DNA at specific sites, known as recognition sites, to produce DNA fragments useful for cloning. **2.** remember b **3.** understand ddNTPs lack the −OH (hydroxyl) group on the 3′ carbon of deoxyribose sugar that is required to extend the DNA chain during synthesis. **4.** understand a **5.** remember b **6.** understand d

✔ Test Your Understanding

7. understand When a restriction endonuclease cuts a "foreign gene" sequence and a plasmid, the same sticky ends are created on the excised foreign gene and the cut plasmid. After the sticky ends on the foreign gene and the plasmid anneal, DNA ligase catalyzes closure of the DNA backbone, sealing the foreign gene into the plasmid DNA (see **FIGURE A20.2**). **8.** apply If the DNA at each cycle steadily doubles, it is predicted to yield a 33.6 million-fold increase (a 2^{25}-fold increase). **9.** understand A cDNA library is a collection of complementary DNAs made from all the mRNAs present in a certain group of cells. A cDNA library from a human nerve cell would be different from one made from a human muscle cell, because nerve cells and muscle cells express many different genes that are specific to their cell type. **10.** remember Genetic markers are genes or other loci

that have known locations in the genome. When these locations are diagrammed, they represent the physical relationships between landmarks—in other words, they form a map. **11.** understand d; these regulatory sequences promote expression of introduced genes in the endosperm—the rice grain eaten as food. **12.** analyze PCR and cellular DNA synthesis are similar in the sense of producing copies of a template DNA. Both rely on primers and DNA polymerase. The major difference between the two is that PCR copies only a specific target sequence, but the entire genome is copied during cellular DNA synthesis.

✔ Test Your Problem-Solving Skills

13. create You could use a computer program to identify likely promoter sequences in the sequence data and then look for sequences just downstream that have an AUG start codon and codons that could be part of the protein-coding exons of a potential protein about 500 amino acids (500 codons or 1500 bases) long. **14.** analyze False. Since somatic-cell modifications cannot be passed on to future generations, but germ-line modifications may be passed on to offspring, germ-line modification opens a new set of ethical questions. **15.** analyze In both techniques, researchers use an indicator to identify either a gene of interest or a colony of bacteria with a particular trait. The problem is the same—picking one particular thing (a certain gene or a cell with a particular mutation) out of a large collection. **16.** analyze (a) Primer 1b binds to the top right strand and would allow DNA polymerase to synthesize the top strand across the target gene. Primer 1a, however, binds to the upper left strand and would allow DNA polymerase to synthesize the upper strand *away* from the target gene. Primer 2a binds to the bottom left strand and would allow DNA polymerase to synthesize the bottom strand across the target gene. Primer 2b, however, binds

to the bottom right strand and would allow DNA polymerase to synthesize away from the target gene. (b) She could use primer 1b with primer 2a.

CHAPTER 21

IN-TEXT QUESTIONS AND EXERCISES

p. 391 understand If no overlap occurred, there would be no way of ordering the fragments correctly. You would be able to put fragments only in random order—not the correct order.
p. 391 Fig. 21.2 understand Shotgun sequencing is based on fragmenting the genome into many small pieces.
p. 393 Fig. 21.3 understand Because the mRNA codons that match up with each strand are oriented in the 5′-to-3′ direction.
p. 395 CYU (1) understand Parasites don't need genes that code for enzymes required to synthesize molecules they acquire from their hosts. **(2)** understand If two closely related species inherited the same gene from a common ancestor, the genes should be similar. But if one species acquired the same gene from a distantly related species via lateral gene transfer, then the genes should be much less similar.
p. 399 Fig. 21.9 understand
Chromosome 1:
ψβ2-ε-Gγ-Aγ-ψβ1-δ-ψβ2-ε-Gγ-Aγ-ψβ1-δ-β
Chromosome 2: β
p. 400 CYU (1) understand Eukaryotic genomes have vastly different numbers of transposable elements and other repeated sequences that don't directly contribute to phenotype. So genome size can vary widely without changing the number of coding genes or the complexity of the organism. **(2)** apply Because most of the genome is transcribed, this number is very close to the percentage of the genome that codes for exons. In humans, the ratio would be roughly 2%.
p. 401 understand Start with a microarray containing exons from a large number of human genes. Isolate mRNAs from brain tissue and liver tissue, and make labeled cDNAs from each. Probe the microarray with cDNA made from each type of tissue, and record where binding occurs. Binding events identify exons that are transcribed in each type of tissue. Compare the results to identify genes that are expressed in brain but not liver, or in liver but not brain.
p. 402 understand Cell replication is an emergent property because it is due to the interaction of proteins working in a network, yet is a property that could not be predicted from the analysis of any one of these proteins.

IF YOU UNDERSTAND . . .

21.1 understand If a search of human gene sequence databases revealed a gene that was similar in base sequence, and if follow-up work confirmed that the mouse and human genes were similar in their pattern of exons and introns and regulatory sequences, then the researchers could claim that they are homologous. **21.2** understand The genome of a parasite is predicted to be smaller and to have fewer genes, particularly for metabolism, than the genome of a nonparasite. **21.3** understand These features include variable numbers of transposable elements; noncoding repeated sequences; and noncoding, non-repetitive DNA. These add to genome size without adding genes that directly influence phenotype. Additionally, mechanisms such as alternative splicing and the possibility of many noncoding regulatory RNAs can create situations in which few protein-coding genes but many different proteins are expressed in intricately regulated ways—features that would increase morphological complexity without increasing gene number.
21.4 understand Color-coded clusters shown in Figure 21.13 represent networks of interacting proteins that work together to carry out a particular cellular function. But it's clear from the figure that smaller, clusters are also connected into larger networks.

FIGURE A20.1

```
5′ CATGACTATTACGTATCGGGTACTATGCTATCGATCTAGCTACGCTAGCT 3′
3′ GTACTGATAATGCATAGCCCATGATACGATAGCTAGATCGATGCGATCGA 5′
```

Primer #1, which will anneal to the 3′ end of the top strand:

```
5′ AGCTAGCGTAGCTAGATCGAT 3′
```

Primer #2, which will anneal to the 3′ end of the bottom strand:

```
5′ CATGACTATTACGTATCGGGT 3′
```

The primers will bind to the separated strands of the parent DNA sequence as follows:

```
5′ CATGACTATTACGTATCGGGTACTATGCTATCGATCTAGCTACGCTAGCT 3′
                               3′ TAGCTAGATCGATGCGATCGA 5′
```

```
5′ CATGACTATTACGTATCGGGT 3′
3′ GTACTGATAATGCATAGCCCATGATACGATAGCTAGATCGATGCGATCGA 5′
```

FIGURE A20.2

Sticky ends on cut plasmid and gene to be inserted are complementary and can base pair

DNA ligase forms four phosphodiester bonds between gene and plasmid DNA at the points indicated by arrows

YOU SHOULD BE ABLE TO . . .

✔ Test Your Knowledge

1. `remember` c 2. `understand` a 3. `remember` d 4. `remember` b 5. `understand` Finding two or more genes of similar sequence in the genome. 6. `remember` Having a sequence that is clearly related to a functional gene but with a crippling mutation such as a stop codon or deletion.

✔ Test Your Understanding

7. `analyze` Computer programs are used to scan sequences in both directions to find ATG start codons, a gene-sized logical sequence with recognizable codons, and then a stop codon. One can also look for characteristic promoter, operator, and other regulatory sites. It is more difficult to identify open reading frames in eukaryotes because their genomes are so much larger and because of the presence of introns and repeated sequences. 8. `evaluate` c 9. `apply` Water flea gene density is about 31,000 genes/200 Mbp = 155 genes/Mbp. In humans, gene density is about 21,000 genes/3000 Mbp = 7 genes/Mbp. The relative gene density of water flea/human = 155/7 = 22. 10. `analyze` Because with many different alleles, the chance that a match is coincidental is low relative to the chance that they match by identity (they come from the same person). 11. `understand` A DNA microarray experiment identifies which genes are being expressed in a particular cell at a particular time. If a series of experiments shows that different genes are expressed in cells at different times or under different conditions, it implies that expression was turned on or off in response to changes in age or changes in conditions. 12. `understand` Homology is a similarity among different species that is due to their inheritance from a common ancestor. If a newly sequenced gene is found to be homologous with a known gene of a different species, it is assumed that the gene products have similar function.

✔ Test Your Problem-Solving Skills

13. `analyze` If "gene A" is not necessary for existence, it can be lost by an event like unequal crossing over (on the chromosome with deleted segments) with no ill effects on the organism. In fact, individuals who have lost unnecessary genes are probably at a competitive advantage, because they no longer have to spend time and energy copying and repairing unused genes. 14. `apply` If the grave were authentic, it might include two very different parental patterns along with five children whose patterns each represented a mix between the two parents. The other unrelated individuals would have patterns not shared by anyone else in the grave. 15. `apply` d; mutation of such a central gene is likely to influence the phenotypes associated with all the genes it interacts with. If these interacting genes are involved in many different functions, the effects of mutation of a central gene are likely to be widespread, or pleiotropic. 16. `apply` You would expect that the livers and blood of chimps and humans would function similarly, but that strong differences occur in brain function. The microarray data support this prediction and suggest chimp and human brains are different because certain genes are turned on or off at different times and expressed in different amounts.

CHAPTER 22

IN-TEXT QUESTIONS AND EXERCISES

p. 410 CYU `understand` Biologists have been able to grow entire plants from a single differentiated cell taken from an adult. They have also succeeded in producing animals by transferring the nucleus of a fully differentiated cell to an egg whose nucleus has been removed.
p. 411 Fig. 22.4 `apply` Anterior structures are predicted to be shifted farther back in the embryo, since higher-than-normal levels of the anterior-specifying Bicoid protein would be present in posterior regions.
p. 412 Fig. 22.6 `apply` Since Bicoid is located in the anterior cytoplasm and works to turn on genes that produce anterior structures, a prediction is that this manipulation would create a concentration gradient of Bicoid with its highest point in the middle of the mutant embryo. This would lead to the most anterior-like structures in the center of the embryo and more posterior structures forming on either side.
p. 414 Fig. 22.10 `create` Genes are usually considered homologous if they have similar DNA sequence and exon–intron structure.
p. 415 CYU (1) `understand` Bicoid is a transcription factor, and cells that experience a high versus medium versus low concentrations of Bicoid express Bicoid-regulated genes at different levels. **(2)** `understand` Bicoid proteins control the expression of gap genes, which—in effect—tell cells which third of the embryo they are in along the anterior–posterior axis. Gap genes control the expression of pair-rule genes, which organize the embryo into individual segments. Pair-rule genes control expression of segment polarity genes, which establish an anterior–posterior polarity to each individual segment. Segment polarity genes turn on homeotic genes, which then trigger genes for producing segment-specific structures like wings or legs.

IF YOU UNDERSTAND . . .

22.1 `apply` The seedling would be misshapen because controlling the plane of cleavage is essential for formation of normal structures in plants. **22.2.** `understand` Cells in the stem contain a complete set of genes, so if they can be de-differentiated and receive appropriate signals from other cells, they could differentiate into any cell type and cooperate to form any structure of the plant, including roots. **22.3** `analyze` Because segmentation genes work in a cascade, mutation of a pair-rule gene is expected to influence the expression of genes in the next level, segment polarity genes, but not affect the expression of earlier acting genes, such as gap genes. **22.4** `apply` This procedure is predicted to lead to the loss of limbs because the regulatory sequence replacement will cause the regions of expression of *Hoxc8* and *Hoxc6* to overlap.

YOU SHOULD BE ABLE TO . . .

✔ Test Your Knowledge

1. `remember` b 2. `remember` stem cells 3. `understand` a 4. `remember` True 5. `remember` a 6. `remember` a conserved gene that can be expressed at different times and places during development to produce different types of structures in different organisms.

✔ Test Your Understanding

7. `understand` They occur in both plants and animals, and are responsible for the changes that occur as an embryo develops. 8. `understand` If an adult cell used for cloning had some permanent change or loss of genetic information during development, it should not be able to direct the development of a viable adult. But if the adult cell is genetically equivalent to a fertilized egg, then it should be capable of directing the development of a new individual. 9. `understand` The researchers exposed adult flies to treatments that induce mutations and then looked for embryos with defects in the anterior–posterior body axis or body segmentation. The embryos had mutations in genes required for body axis formation and segmentation. 10. `understand` Development—specifically differentiation—depends on changes in gene expression. Changes in gene expression depend on differences in regulatory transcription factors. 11. `create` Differentiation is triggered by the production of transcription factors, which induce other transcription factors, and so on—a sequence that constitutes a regulatory cascade—as

development progresses. At each step in the cascade, a new subset of genes is activated—resulting in a step-by-step progression from undifferentiated to fully differentiated cells. 12. `analyze` c. This result shows that a mouse tool-kit gene is similar enough (conserved) to a fly gene to be able to take over its function.

✔ Test Your Problem-Solving Skills

13. `apply` There would be more Bicoid protein farther toward the posterior and less in the anterior—meaning that the anterior segments would be "less anterior" in their characteristics and the posterior segments would be "more anterior." 14. `analyze` The result means that the human gene was expressed in worms and that its protein product has performed the same function in worms as in humans. This is strong evidence that the genes are homologous and that their function has been evolutionarily conserved. 15. `apply` c. Since the morphogen concentration is halved every 100 μm away from the posterior pole, dropping to 1/16 the highest concentration would require 4 "halvings" of the initial concentration ($1/2 \times 1/2 \times 1/2 \times 1/2 = 1/16$), and this would take place in 4 steps of 100 μm each—or 400 μm from the morphogen source. This is where the leg is predicted to be formed. 16. `create` Tool-kit genes—such as *Hox* genes—are responsible for triggering the production of structures like wings or legs in particular segments. One hypothesis to explain the variation is that changes in gene expression of tool-kit genes led to different numbers of segments that express leg-forming *Hox* genes.

CHAPTER 23

IN-TEXT QUESTIONS AND EXERCISES

p. 421 `understand` If the bindin-like protein were blocked, it would not be able to bind to the egg-cell receptor for sperm. Fertilization would not take place.
p. 422 CYU (1) `apply` This change is predicted to allow the sperm of species A to bind to—and likely fertilize—the eggs of species B. **(2)** `understand` The entry of a sperm triggers a wave of calcium ion release around the egg-cell membrane. Thus, calcium ions act as a signal that indicates "A sperm has entered the egg." In response to the increase in calcium ion concentration, cortical granules fuse with the egg-cell membrane and release their contents to the exterior, causing the destruction of the sperm receptor and, in sea urchins, the formation of a fertilization envelope.
p. 424 CYU `understand` If cytoplasmic determinants could not be localized, they would be distributed to all blastomeres. This would specify the same fate for all blastomeres and their descendants instead of different fates.
p. 426 CYU (1) `understand` Ectoderm and endoderm, because these are the inside and outside layers. **(2)** `apply` The future anterior–posterior and dorsal–ventral region of the embryo could first be identified as the blastopore forms and cells begin migrating into the embryo. The blastopore marks the future posterior of the embryo, and the region opposite the blastopore is where the anterior will form. Cells that first migrate into the blastopore come from the dorsal side, so this is the future dorsal region; the ventral region is on the opposite side.
p. 427 Fig. 23.9 `understand` Because neural tube formation depends on rearrangements of the cytoskeleton, a drug that locked the cytoskeleton in place is predicted to prevent formation of the neural tube.
p. 428 Fig. 23.12 `analyze` It made gene expression possible in non-muscle cells. A "general purpose" promoter can lead to transcription of an inserted cDNA in any type of cell, including fibroblasts.

IF YOU UNDERSTAND . . .

23.1 `understand` Bindin and the egg-cell receptor for sperm bind to each other. For this to happen, their tertiary structures must fit together somewhat like the structures of a lock and key. **23.2** `understand` If the cytoplasmic determinant is localized to a certain part of the egg cytoplasm, blastomeres will either contain or not contain the determinant, because the egg's cytoplasm becomes divided between blastomeres. **23.3** `analyze` Both cells would contain cell–cell signals and transcription factors specific to mesodermal cells, but the anterior cell would contain anterior-specific signals and regulatory transcription factors while the posterior cell would contain posterior-specific signals and regulators. **23.4** `understand` Cell–cell signals from the notochord direct the formation of the neural tube; subsequently, cell–cell signals from the notochord, neural tube, and ectoderm direct the differentiation of somite cells. Cells expand during neural tube formation. Somite cells move to new positions, proliferate, and differentiate.

YOU SHOULD BE ABLE TO . . .

✔ Test Your Knowledge

1. `remember` Bindin is a protein on the sperm-cell head that binds to a receptor on the egg plasma membrane to allow species-specific attachment of sperm and egg. **2.** `understand` c **3.** `understand` c **4.** `understand` d **5.** `remember` ectoderm; **6.** `remember` a

✔ Test Your Understanding

7. `evaluate` As mammals, mice form a placenta that allows the embryo to obtain nutrients from the mother. There's no need for lots of nutrient reserves to support development. Frogs develop outside the mother's body, and the embryo is on its own. There is a need for nutrient reserves, and the large egg provides this. Eggs contain stores of nutrients. **8.** `apply` This treatment is predicted to promote polyspermy, because there would be no release of Ca^{2+} inside the cell to trigger a block to polyspermy. **9.** `understand` They contain different types and/or concentrations of cytoplasmic determinants. **10.** `analyze` a; this is because ectodermal cells have to constrict on their dorsal margins and expand on their ventral margin to fold ectoderm into a neural tube. **11.** `understand` When transplanted early in development, somite cells become the cell type associated with their new location. But when transplanted later in development, somite cells become the cell type associated with the original location. These observations indicate that the same cell is not committed to a particular fate until later in development. **12.** `apply` More than 50,000 fibroblasts would need to be screened. This is because if 2×10^{-5} of all mRNAs and cDNAs coded for MyoD, then 50,000 cells would need to be screened on average to find one with a MyoD cDNA (50,000 comes from the quotient of $1/2 \times 10^{-5}$ MyoD cDNAs per cDNA). Since this is an average, in some groups of 50,000 cells, there may not be a MyoD cDNA. Therefore, to be reasonably sure of finding at least one MyoD cDNA, more than 50,000 cells should be screened.

✔ Test Your Problem-Solving Skills

13. `analyze` The interactions that allow proteins to bind to each other are extremely specific—certain amino acids have to line up in certain positions with specific charges and/or chemical groups exposed. Sperm–egg binding is species specific because each form of bindin binds to a different egg-cell receptor for sperm. **14.** `evaluate` Translation is not needed for cleavage to occur normally—meaning that all the proteins needed for early cleavage are present in the egg. **15.** `create` The yellow pigment may be a cytoplasmic determinant, or it may be only a molecule that is found in the same cells as the cytoplasmic determinant. Without isolating the pigment and adding it to cells to see if it acts as a cytoplasmic determinant, these two possibilities cannot be distinguished. **16.** `apply` a (See Figure 23.10 and note how the notochord is closest

to the somite region that forms bone. If an additional notochord were transplanted near the top of the neural tube, signals from the notochord would be expected to convert the nearby somite into bone-producing somite, adding to the amount of bone.)

CHAPTER 24

IN-TEXT QUESTIONS AND EXERCISES

p. 434 Fig. 24.2 `remember` See **FIGURE A24.1**.
p. 436 CYU (1) `understand` Cells located on the outside of the embryo become epidermal tissue; cells in the interior become vascular tissue; cells in between become ground tissue. **(2)** `apply` The *MONOPTEROS* gene would be over-expressed, producing higher levels of the MONOPTEROUS transcription factor. The embryonic cells would get information indicating that they are in the shoot apex, where MONOPTEROS levels are high. As a result, the root would not develop normally.
p. 437 Fig. 24.6 `analyze` Many animals have three body axes analogous to the three axes of leaves: anterior–posterior (like the leaf's proximal–distal), dorsal–ventral (like the leaf's adaxial–abaxial), and a left–right axis (like the leaf's mediolateral axis). The main plant body, in contrast, has only two axes: apical–basal and radial.
p. 438 CYU (1) `create` Simply planting a cut stem will often produce a new root meristem. This occurs whenever the cutting produces a new plant. **(2)** `create` If the plant could be genetically engineered to express higher levels of the PHAN transcription factor in leaves, the likely result would be a shift in leaf type from simple to compound.
p. 440 `apply` Mutual inhibition between the *A* and *C* genes helps create separate regions of gene expression. If *C* gene expression is lost, then the inhibition of *A* gene expression is also lost, and *A* genes are expressed in their normal location and where they are normally inhibited by *C* gene expression.
p. 441 CYU (1) `analyze` Plants generate many reproductive structures—flowers—many times and in many places during the plant's life. Plant reproductive cells are created from vegetative cells, not cells that are set aside early in development as reproductive cells. In contrast, animals form reproductive structures (ovaries and testes) in set locations, produce reproductive structures only once during development and set aside specialized reproductive cells (the germ line) early in their development. **(2)** `analyze` Like *Hox* genes, *MADS-box* genes code for DNA-binding proteins that function as transcriptional regulators. Both Hox proteins and MADS-domain proteins are involved in regulatory transcription cascades that lead to development of specific structures in specific locations. Although they perform similar functions, *Hox* genes and *MADS-box* genes evolved independently and differ in their base sequences. Thus, these genes and their protein products are not homologous.

IF YOU UNDERSTAND . . .

24.1 `apply` If there were only one master regulatory gene for patterning, the predicted result of loss of that gene's function by mutation would be disorganization of the entire apical–basal axis, not just removal of parts of it as seen in the actual mutants. **24.2** `apply` Many things could be imagined, but all should point in the direction of many repeated body parts (for example, 21 wings in a variety of shapes and positions on the body) and a very large overall size (if the fly is old) due to continuous growth. **24.3** `apply` The prediction is that sepals would develop from every whorl. This is because when the *C* gene product is absent, *A* gene expression spreads across the flower. Since there is no *B* gene expression, only sepals would form—the structure associated with only *A* gene expression.

YOU SHOULD BE ABLE TO . . .

✔ Test Your Knowledge

1. `understand` This is because mechanisms for development evolved independently in plants and animals. **2.** `understand` b **3.** `remember` b **4.** `remember` d **5.** `remember` c **6.** `understand` a.

✔ Test Your Understanding

7. `understand` c. **8.** `understand` Animal cells that lead to sperm or egg formation are "sequestered" early in development, so that they are involved only in reproduction and undergo few rounds of mitosis. Because mutations occur in every round of mitosis, plant eggs and sperm should contain many more mutations than animal eggs and sperm. **9.** `understand` Meristem cells are undifferentiated and divide to produce a cell that remains in the meristem and a cell that can commit to a particular pathway of differentiation. These are properties of stem cells. **10.** `understand` Just like the three tissue layers in plant embryos, the tissues produced in the SAMs and RAMs of a 300-year-old oak tree can differentiate into all of the specialized cell types found in a mature plant. **11.** `evaluate` The ABC model predicted that the *A*, *B*, and *C* genes would each be expressed in specific whorls in the developing flower. The experimental data showed that the prediction was correct. **12.** `understand` Cell proliferation is responsible for expansion of meristems and growth in size. Asymmetric cell division along with cell expansion is responsible for giving the adult plant a particular shape. Differentiation is responsible for generating functional tissues and organs in the adult plant. Cell–cell signals are responsible for setting up the apical–basal axis of the embryo and for guiding the pollen tube to the egg.

✔ Test Your Problem-Solving Skills

13. `create` Four. This is because there are four states of gene activity (in any cell, *D* can be on or off and *E* can be on or off) and therefore four structures that could be specified using this code. Sets of on-off switches are known as Boolean systems and are the basis for information storage in computers and (sometimes) in cells. **14.** `analyze` Because vegetative growth is continuous and occurs in all directions (unrestricted), it could be considered indeterminate. In contrast, reproductive growth produces mature reproductive organs and stops when those organs are complete. Because its duration is limited, it could be considered determinate. **15.** `apply` a; see Figure 24.10 to understand the reasoning. *A* and *B* together create petals, and *B* and *C* together create stamens. **16.** `analyze` One hundred times more likely. This is because 100 times more cell divisions were required to create the oak egg, and the rate of mutation per cell division is constant. (You don't need to consider the absolute rate of 1×10^{-6}/division, since the question asks about relative chances only).

FIGURE A24.1

CHAPTER 25

IN-TEXT QUESTIONS AND ANSWERS

p. 448 Fig. 25.4 `analyze` Under special creation, fossils with transitional features would be explained as separate types, unrelated to other organisms and just coincidentally having intermediate features.

p. 448 Fig. 25.5 `analyze` If vestigial traits result from inheritance of acquired characteristics, some individuals must have lost the traits during their own lifetimes and passed the reduced traits on to their offspring. For example, a certain monkey's long tail might have been bitten off by a predator. The new traits would then somehow have passed to the individual's eggs or sperm, resulting in shorter-tailed offspring, until humans with a coccyx resulted.

p. 453 CYU (1) `evaluate` If birds evolved from dinosaurs you would expect to find transitional fossil dinosaurs with feathers. Such fossils have been found. **(2)** `analyze` The DNA sequences of chimpanzees and humans are so similar because we share a recent common ancestor.

p. 455 (1) `understand` Relapse occurred because the few bacteria remaining after drug therapy were not eliminated by the patient's weakened immune system and began to reproduce quickly. **(2)** `understand` No—almost all of the cells present at the start of step 3 would have been resistant to the drug.

p. 456 `understand` When antibiotics are overused, some bacteria will be killed, including the bacteria that are harmless or beneficial to us. However, bacteria that are resistant to these antibiotics will flourish and multiply, reducing the likelihood that antibiotic treatment will be effective in the future.

p. 457 Fig. 25.16 `analyze` "Prediction": Beak measurements were different before and after the drought. "Prediction of null hypothesis": No difference was found in beak measurements before and after the drought.

p. 459 CYU (1) `understand` *Postulate 1:* Traits vary within a population. *Postulate 2:* Some of the trait variation is heritable. *Postulate 3:* There is variation in reproductive success (some individuals produce more offspring than others). *Postulate 4:* Individuals with certain heritable traits produce the most offspring. The first two postulates describe heritable variation; the second two describe differential reproductive success. **(2)** `understand` Beak size and shape and body size vary among individual finches, in part because of differences in their genotypes (some alleles lead to deeper or shallower beaks, for example). When a drought hit, individuals with deep beaks survived better and produced more offspring than individuals with shallow beaks.

p. 460 `analyze` In biology, an adaptation is any heritable trait that increases an individual's ability to produce offspring in a particular environment. In everyday English, adaptation is often used to refer to an individual's nonheritable adjustment to meet an environmental challenge, a phenomenon that biologists call acclimatization. The phenotypic changes resulting from acclimatization are not passed on to offspring.

IF YOU UNDERSTAND . . .

25.1 `analyze` Typological thinking is based on the idea that species are unchanging types and that variations within species are unimportant or even misleading. Population thinking recognizes the variation that occurs within a species as critically important. **25.2** `remember` Many answers are possible. For example, the evolution of an antibiotic-resistant TB bacterium demonstrates that species change through time. The homologous sequences in human and fruit fly genes demonstrate that these species are related by common ancestry. **25.3** `analyze` In biology, fitness is the ability of an individual to produce viable, fertile offspring, relative to that ability in other individuals in the population. In everyday English, fitness is a physical attribute

that is acquired as a result of practice or exercise. **25.4** `apply` Under special creation, changes in *Mycobacterium tuberculosis* populations would be explained as individual creative events governed by an intelligent creator. Under the theory of evolution by inheritance of acquired characters, changes in *M. tuberculosis* populations would be explained by the cells trying to transcribe genes in the presence of the drug, and their *rpoB* gene becoming altered as a result. **25.5** `understand` Brain size in *H. sapiens* might be constrained by the need for babies to pass through the mother's birth canal, by the energy required to maintain a large brain as an adult, or by lack of genetic variation for even larger brain size. Flying speed in falcons might be constrained by loss of maneuverability (and thus less success in hunting), the energy demands of extremely rapid flight, or lack of genetic variation for even faster flight.

YOU SHOULD BE ABLE TO . . .

✓ Test Your Knowledge

1. `remember` b **2.** `remember` False **3.** `remember` a **4.** `understand` d **5.** `remember` homologous **6.** `remember` c

✓ Test Your Understanding

1. `understand` Mutation produces new genetic variations, at random, without any forethought as to which variations might prove adaptive in the future. Individuals with mutations that are disadvantageous won't produce many offspring, but individuals with beneficial mutations will produce many offspring. The beneficial mutations will thus increase in frequency through selection. **2.** `analyze` Artificial selection is determined by human choice and is goal directed, whereas natural selection is the unplanned differential reproductive success of individuals that vary in their heritable traits. **3.** `analyze` b (In some environments, being big and strong lowers fitness.) **4.** `analyze` There were 751 finches before the drought and 90 finches afterward, so only $90/751 = 0.12$, or 12 percent, survived. There were more finches with deep beaks before the drought only because the population was much larger, but the *average* beak depth increased after the drought. **5.** `analyze` The evidence for within-patient evolution is that DNA sequences at the start and end of treatment were identical except for a single nucleotide change in the *rpoB* gene. If rifampin were banned, *rpoB* mutant strains likely would have had lower fitness in the drug-free environment and would not continue to increase in frequency in *M. tuberculosis* populations. **6.** `apply` (1) Traits vary within rabbit populations, such as fur color and hearing ability. (2) Some of this variation is heritable. (3) More rabbits are produced than can survive. (4) Rabbits with certain heritable traits, such as the ability to camouflage in their environment or hear the approach of foxes, produce more offspring.

✓ Test Your Problem-Solving Skills

1. `analyze` The theory of evolution by natural selection predicts that white and brown deer mice are descendants of an ancestral population that varied in color. The white mice had higher fitness in the beach environment, where they were more likely to escape the notice of predators. Likewise, the brown mice had a higher fitness in the shady forest environment. Over time, the average color of each population changed. Special creation claims that the two colors of mice do not change and were created by divine intervention. Evolution by inheritance of acquired characteristic predicts that the mice in the different environments needed to change color, so they did so and then passed their traits onto their offspring. **2.** `analyze` b (No, because changes in height due to nutrition and reduced incidence of disease are not heritable.) **3.** `evaluate` Evidence to support a common ancestor of humans and chimpanzees about 6–7 million years ago includes the following: fossils from Africa that date from 6–7 million years ago with primate characteristics that

are common to chimps and humans, and DNA sequence comparisons showing homologies between chimps and humans (the rate of change of some genes can be used to estimate time of divergence). **4.** `evaluate` The theory of evolution fits the six criteria as follows: (1) and (2): It provides a common underlying mechanism responsible for puzzling observations such as homology, geographic proximity of similar species, the law of succession in the fossil record, vestigial traits, and extinctions. (3) and (4): It suggests new lines of research to test predictions about the outcome of changing environmental conditions in populations, about the presence of transitional forms in the fossil record, and so on. (5): It is a simple idea that explains the tremendous diversity of living and fossil organisms and why species continue to change today. (6): The realization that all organisms are related by common descent and that none are higher or lower than others was a surprise.

CHAPTER 26

IN-TEXT QUESTIONS AND EXERCISES

p. 467 (1) `apply` The frequencies of the three genotypes are shown by $p^2 + 2pq + q^2 = 1$.

$$\text{Freq}(A_1A_1) = p^2 = (0.6)^2 = 0.36.$$
$$\text{Freq}(A_1A_2) = 2pq = 2(0.6)(0.4) = 0.48.$$
$$\text{Freq}(A_2A_2) = q^2 = (0.4)^2 = 0.16.$$

(2) `apply` Allele frequencies in the offspring gene pool:

$$\text{Freq}(A_1) = 0.36 + 1/2(0.48) = 0.60.$$
$$\text{Freq}(A_2) = 1/2(0.48) + 0.16 = 0.40.$$

(3) `apply` Evolution has not occurred, because the allele frequencies have not changed.

p. 469 Table 26.1 `apply` The observed allele frequencies, calculated from the observed genotype frequencies, are 0.43 for *M* and 0.57 for *N*. The expected genotypes, calculated from the observed allele frequencies under the Hardy–Weinberg principle, are 0.185 for *MM*, 0.49 for *MN*, 0.325 for *NN*.

p. 470 CYU `apply` Given the observed genotype frequencies, the observed allele frequencies are

$$\text{Freq}(A_1) = 0.574 + 1/2(0.339) = 0.744.$$
$$\text{Freq}(A_2) = 1/2(0.339) + 0.087 = 0.256.$$

Given these allele frequencies, the genotype frequencies expected under the Hardy–Weinberg principle are A_1A_1: $0.744^2 = 0.554$; A_1A_2: $2(0.744 \times 0.256) = 0.381$; A_2A_2: $0.256^2 = 0.066$. There are 4 percent too few heterozygotes observed, relative to the expected proportion. One of the assumptions of the Hardy–Weinberg principle is not met at this gene in this population, at this time.

p. 470 `apply` The proportions of homozygotes should increase, and the proportions of heterozygotes should decrease, but allele frequencies will not change if no selection is occurring at this locus.

p. 474 `apply` There will be an excess of observed genotypes containing the favored allele compared to the proportion expected under Hardy–Weinberg proportions.

p. 478 (1) `understand` Sperm are small and hence relatively cheap to produce, whereas eggs are large and require a large investment of resources to produce. **(2)** `understand` Sperm are inexpensive to produce, so reproductive success for males depends on their ability to find mates—not on their ability to find resources to produce sperm. The opposite pattern holds for females. Sexual selection is based on variation in ability to find mates, so is more intense in males—leading to more exaggerated traits in males.

p. 479 `apply` If allele frequencies are changing due to drift, the populations in Table 26.1 would behave like the simulated populations in Figure 26.12—frequencies would drift up and down over time, and diverge.

p. 479 Fig. 26.12 `analyze` See FIGURE A26.1.

p. 480 Fig. 26.13 `analyze` The smaller the population, the greater the chance that genetic drift will cause alleles to be fixed or lost in the population in a short time. Starting with a smaller population makes the experiment easier.

p. 481 Fig. 26.14 (analyze) Original population: freq(A_1) = $(9 + 9 + 11)/54 = 0.54$; New population: freq(A_1) = $(2 + 2 + 1)/6 = 0.83$. The frequency of A_1 has increased dramatically.

p. 482 CYU (1) (understand) When allele frequencies fluctuate randomly up and down, sooner or later the frequency of an allele will hit 0. That allele thus is lost from the population, and the other allele at that locus is fixed. **(2)** (understand) In small populations, sampling error is large. For example, the accidental death of a few individuals would have a large impact on allele frequencies.

IF YOU UNDERSTAND . . .

26.1 (apply) The frequencies of two alleles are shown by $p + q = 1$. The frequencies of the three genotypes are shown by $p^2 + 2pq + q^2 = 1$.

$$\text{If freq}(A_1) = 0.2, \text{ then freq}(A_2) = 0.8.$$
$$\text{Freq}(A_1A_1) = p^2 = (0.2)^2 = 0.04.$$
$$\text{Freq}(A_1A_2) = 2pq = 2(0.2)(0.8) = 0.32.$$
$$\text{Freq}(A_2A_2) = q^2 = (0.8)^2 = 0.64.$$

26.2 (apply) It increased homozygosity of recessive alleles associated with diseases such as hemophilia. As a result, the royal families were plagued by genetic diseases. **26.3** (understand) The Hardy–Weinberg principle predicts that allele frequencies will stay the same over time. Natural selection favors some alleles over others, causing the frequency of those alleles to increase. **26.4** (apply) Genetic drift will rapidly reduce genetic variation in small populations. Thus, it is a high priority to keep the populations of endangered animals as large as possible. **26.5** (apply) In some cases, gene flow could be used to increase genetic variation and reduce inbreeding depression. However, gene flow can also be deleterious if the new alleles reduce the fitness of the endangered population. **26.6** (apply) Mutation rates themselves are too slow to affect an endangered population in a short time—unless a deleterious or beneficial mutation exposes the population to selection. (In instances where chemicals cause mutations in organisms, such as pesticides causing abnormal development in frogs, actions could be taken to remove the source of chemicals.)

YOU SHOULD BE ABLE TO . . .

✔ Test Your Knowledge

1. (understand) It defines what genotype and allele frequencies should be expected if evolutionary processes and nonrandom mating are *not* occurring. **2.** (remember) b **3.** (understand) a **4.** (remember) b **5.** (remember) True **6.** (understand) c

✔ Test Your Understanding

1. (apply) d; freq(A_1) = 0.9, so freq(A_2) = 0.1. Freq(A_2A_2) = $q^2 = (0.1)^2 = 0.01$. The number of babies with cystic

FIGURE A26.1

fibrosis = $2500(0.01) = 25$. **2.** (apply) Inbreeding causes an increase in the number of homozygotes. Individuals who are homozygous for recessive deleterious alleles are exposed to—and eliminated by—directional natural selection. **3.** (evaluate) This statement is false. Mutations do not occur because an organism wants or needs them. They just happen by accident and can be beneficial, neutral, or deleterious. **4.** (understand) Sexual selection is most intense on the sex that makes the least investment in the offspring, so that sex tends to have exaggerated traits that make individuals successful in competition for mates. In this way, sexual dimorphism evolves. **5.** (analyze) An allele in males, because a successful male can have as many as 100 offspring—10 times more than a successful female. **6.** (create) Your concept map should have linking verbs that relate the following information: Selection may decrease genetic variation or maintain it (e.g., if heterozygotes are favored). Genetic drift reduces it. Gene flow may increase or decrease it (depending on whether immigrants bring new alleles or emigrants remove alleles). Mutation increases it.

✔ Test Your Problem-Solving Skills

1. (apply) If we let p stand for the frequency of the loss-of-function allele, we know that $p^2 = 0.0001$; therefore $p = \sqrt{0.0001} = 0.01$. By subtraction, the frequency of normal alleles is 0.99. Under the Hardy–Weinberg principle, the frequency of heterozygotes is $2pq$, or $2 \times 0.01 \times 0.99$, which is 0.0198. **2.** (apply) c **3.** (analyze) If males never lift a finger to help females raise children, the fundamental asymmetry of sex is pronounced and sexual dimorphism should be high. If males invest a great deal in raising offspring, then the fundamental asymmetry of sex is small and sexual dimorphism should be low. **4.** (apply) Captive-bred young, transferred adults, and habitat corridors could be introduced to counteract the damaging effects of drift and inbreeding in the small, isolated populations. It is important, though, to introduce individuals only from similar habitats and connect patches of similar habitat with corridors—to avoid introducing alleles that lower fitness.

CHAPTER 27

IN-TEXT QUESTIONS AND EXERCISES

p. 493 CYU (1) (understand) Biological species concept: Reproductive isolation means that no gene flow is occurring. Morphological species concept: If populations are evolving independently, they may have evolved morphological differences. Phylogenetic species concept: If populations are evolving independently, they should have synapomorphies that identify them as independent twigs on the tree of life. **(2)** (understand) Biological species concept: It cannot be used to evaluate fossils, species that reproduce asexually, or species that do not occur in the same area and therefore never have the opportunity to mate. Morphospecies concept: It cannot identify species that differ in traits other than morphology and is subjective—it can lead to differences of opinion that cannot be resolved by data. Phylogenetic species concept: Reliable phylogenetic information exists only for a small number of organisms.

p. 494 Fig. 27.4 (analyze) The disruption of the salt marsh was a vicariance event that split the range of seaside sparrows into two areas (Gulf and Atlantic coasts), limiting gene flow and eventually resulting in two monophyletic groups.

p. 496 Fig. 27.7 (analyze) It creates genetic isolation—the precondition for divergence—by separating apple maggot flies from hawthorn flies, which mate on hawthorn fruits.

p. 497 (create) The difference in timing of ripening of apples and hawthorns causes divergence by separating apple flies and hawthorn flies not only in space (apple fruits versus hawthorn fruits) but also in time. The

earlier ripening of apples means that apple maggots develop in relatively warm conditions in the fall compared to hawthorn maggots. Natural selection may favor traits in apple maggots that make them adapted to warmer temperatures and traits in hawthorn maggots that make them adapted to cooler temperatures.

p. 498 (apply) A diploid grape plant experiences a defect in meiosis resulting in the formation of diploid gametes. The individual self-fertilizes, producing tetraploid offspring. The tetraploid grape plants self-fertilize or mate with other tetraploid individuals, producing a tetraploid population.

p. 499 (apply) A tetraploid ($4n$) species (in this case, wheat) gives rise to diploid ($2n$) gametes, and a diploid wheat species gives rise to haploid (n) gametes. When a haploid gamete (n) fertilizes a diploid gamete ($2n$), a sterile triploid offspring results. If an error in mitosis occurs and chromosome number doubles before meiosis, then hexaploid wheat is formed.

p. 499 CYU (understand) Use one of the cases given in the chapter to illustrate dispersal (Galápagos finches), vicariance (snapping shrimp), or habitat specialization (apple maggot flies). In each case, drift will cause allele frequencies to change randomly. Differences in the habitats occupied by the isolated populations will cause allele frequencies to change under natural selection. Any mutation that occurs in one population will not occur in the other due to the lack, or low level, of gene flow. Or use polyploidization in maidenhair fern or *Tragopogon* to illustrate how a massive mutation can itself create immediate genetic isolation and divergence.

p. 501 Fig. 27.12 (evaluate) If the same processes were at work in the past that are at work in an experiment with living organisms, then the outcome of the experiment should be a valid replication of the outcome that occurred in the past.

p. 502 (analyze) Gene regions in the experimental hybrid would not align with those of *H. annuus* and *H. petiolaris*, meaning that the gray bar representing the DNA of the experimental hybrid could have had no orange or red regions corresponding to *H. annuus* and *H. petiolaris*.

IF YOU UNDERSTAND THAT . . .

27.1 (apply) Human populations would not be considered separate species under the biological concept, because human populations can successfully interbreed. They would not be considered separate species under the morphological species concept, because all human populations have the same basic morphology. Although human races differ in minor superficial attributes such as skin color and hair texture, they have virtually identical anatomy and physiology in all other regards. Nor would human populations be separate species under the phylogenetic species concept, because they all arose from a very recent common ancestor. (DNA comparisons have revealed that human races are remarkably similar genetically and do not differ enough genetically to qualify even for subspecies status.) **27.2** (create) (Many possibilities) Collect a group of fruit flies that are living in your kitchen, separate the individuals into different cages, and expose the two new populations to dramatically different environmental conditions—heat, lighting, food sources. In essence: Create genetic isolation and then conditions for divergence due to drift, mutation, and selection. **27.3** (evaluate) The sample experiment described above is an example of vicariance, because the experimenter fragmented the habitat into isolated cages. (A different mechanism of speciation may have been involved in your example.) **27.4** (apply) Hybrid individuals will increase in frequency, and natural selection will favor parents that hybridize. The hybrid zone will increase. The two parent populations may go extinct if hybrids live in the same environment.

YOU SHOULD BE ABLE TO . . .

✔ Test Your Knowledge

1. `remember` a 2. `remember` b 3. `understand` a 4. `understand` a
5. `understand` d 6. `remember` False

✔ Test Your Understanding

1. `analyze` Direct observation: Galápagos finch coloniza-tion, apple maggot flies, *Tragopogon* allopolyploids, maidenhair ferns. Indirect studies: snapping shrimp, sunflowers. 2. `understand` a 3. `analyze` Based on biological and morphospecies criteria (breeding range and morpholog-ical traits, respectively), one species with six subspecies of seaside sparrows were recognized. However, based on phylogenetic criteria (comparison of gene sequences), biologists concluded that seaside sparrows represent only two monophyletic groups. 4. `evaluate` Some flies breed on apple fruits, others breed on hawthorn fruits. This reduces gene flow. Because apple fruits and hawthorn fruits have different scents and other characteristics, selection is causing the populations to diverge. 5. `apply` Decreasing. Gene flow tends to equalize allele frequen-cies among populations. 6. `apply` c

✔ Test Your Problem-Solving Skills

1. `analyze` b 2. `create` Sexual selection can promote diver-gence because it affects gene flow directly. For example, if blond women began mating only with blond men, and dark-haired women began mating only with dark-haired men, gene flow would decrease between these two popu-lations even though they live in the same geographic area. Given enough time, these two populations would diverge due to mutation, genetic drift, and natural selec-tion. 3. `evaluate` Two things: (1) Some flies happen to have alleles that allow them to respond to apple scents; others happen to have alleles that allow them to respond to hawthorn scents. There is no "need" involved. (2) The alleles for scent response exist; they are not acquired by spending time on the fruit. 4. `analyze` If the isolated populations and habitat fragments are small enough, the species is likely to dwindle to extinction due to in-breeding and loss of genetic variation or catastrophes like a severe storm or a disease outbreak. If the isolated populations survive, they are likely to diverge into new species because they are genetically isolated and because the habitats may differ.

CHAPTER 28

IN-TEXT QUESTIONS AND EXERCISES

p. 507 Fig. 28.1 `apply` Many trees are possible. See **FIGURE A28.1**.
p. 510 `understand` Because whales and hippos share SINES 4–7 and no other species has these four, they are synapo-morphies that define them as a monophyletic group. The similarity is unlikely to be due to homoplasy because the chance of four SINEs inserting in exactly the same place in two different species, independently, is very small.
p. 510 CYU `analyze` Hair and limb structures in humans and whales are examples of homology because they are traits that can be traced to a common ancestor. All mammals have hair and similar limb bone structure. However, extensive hair loss and advanced social behav-ior in whales and humans are examples of homoplasy. These traits are not common to all mammalian species and likely arose independently during the evolution of specific mammalian lineages.
p. 511 Fig. 28.4 `apply` See **FIGURE A28.2**. SINEs group (19, 20) identifies peccaries and pigs as a monophyletic group.
p. 513 `apply` Since some organisms fossilize better than others—such as mollusks with hard shells—a paleontol-ogist would be mistaken to conclude that mollusks were more abundant than other organisms at a certain time just because mollusks are more abundant in the fossil

record. Similarly, a paleontologist would be mistaken to conclude that no organisms occupied an ancient desert based on the lack of fossil evidence; organisms in deserts do not fossilize well. Several other examples are possible.
p. 517 `create` In habitats on the mainland, complex com-munities have been established for a long time, so tar-weeds in California and *Anolis* lizards on the mainland experience greater competition for resources and are limited to specific niches.
p. 520 CYU `understand` In terms of ecological opportunity, many resources were available because no other animals (or other types of organisms) existed to exploit them. Also, the evolution of new species in new niches made new niches available for predators. Morphological in-novations like limbs and complex mouthparts were im-portant because they made it possible for animals to live in habitats other than the benthic area, and to consume new types of food.
p. 520 Fig. 28.14 `analyze` The dinosaurs went extinct during the end-Cretaceous. About 15 percent of families went extinct.

IF YOU UNDERSTAND . . .

28.1 `apply` Humans are the only mammal that walks up-right on two legs, so this character is a synapomorphy that defines the hominins. **28.2** `apply` Snail shells are hard structures, and snails often live in marine environments that fossilize readily, so snail fossils are relatively abun-dant. Abundant fossils are more helpful than rare fossils for estimating changes in relative abundance over time. **28.3** `analyze` They could eat food that was available off the substrate. And once animals lived off (and away from) the bottom, it created an opportunity for other animals who could eat them to evolve. **28.4** `evaluate` This statement is false. Human-induced extinctions today are not due to poor adaptation to the normal environment. Humans are changing habitats rapidly and in unusual ways (such as clear-cutting forests), so most species cannot adapt to these changes rapidly enough to avoid going extinct. In this way, human-induced extinctions are more similar to mass extinctions than to background extinctions.

YOU SHOULD BE ABLE TO . . .

✔ Test Your Knowledge

1. `remember` c 2. `understand` a 3. `understand` b 4. `understand` False
5. `understand` d 6. `remember` True

✔ Test Your Understanding

1. `evaluate` The fossil record is biased because recent, abun-dant organisms with hard parts that live underground or in environments where sediments are being depos-ited are most likely to fossilize. Even so, it is the only data available on what organisms that lived in the past looked like, and where they lived. 2. `create` An enormous amount of animal diversity appeared in a time frame that was short compared to the sweep of earlier Earth history. The most common analogy is to an explosion, referring to the sudden and rapid radiation of lineages. Another analogy is your hand, where one line (your arm) comes into a node (your palm) and several lineages (your fingers) come out. Many other answers are possible. 3. `understand` Homoplasy is rare relative to homology, so phylogenetic trees that minimize the total change required are usually more accurate. In the case of the artiodactyl astragalus, parsimony was misleading because the astragalus was lost when whales evolved limblessness—creating two changes (a loss following a gain) instead of one (a gain). 4. `create` (Many answers are possible.) Adaptive radiation of *Anolis* lizards after they colonized new islands in the Caribbean. *Hypothesis:* After lizards arrived on each new island, where there were no predators or competitors, they rapidly diversi-fied to occupy four distinct types of habitats on each island. Adaptive radiation during the Cambrian period

following a morphological innovation. *Hypothesis:* Ad-ditional *Hox* genes made it possible to organize a large, complex body; the evolution of complex mouthparts and limbs made it possible for animals to move and find food in new ways. 5. `apply` About 90 percent of spe-cies went extinct during the end-Permian, so about 10 percent survived, written as 0.1 in decimals. $1880 \times 0.1 = 188$ would make it to graduation. 6. `apply` d

✔ Test Your Problem-Solving Skills

1. `understand` b 2. `evaluate` Many trees are possible. See **FIGURE A28.3**. Sponges are the outgroup. The most par-simonious tree has the fewest number of changes (five in this case). Adding more characters to the matrix can improve the reliability of the results. 3. `understand` It would be helpful to have (1) a large crater or other physical evidence from a major impact dated at 251 mya, (2) shocked quartz and microtektites in rock layers dating to the end-Permian era, and (3) high levels of iridium or other elements common in space rocks and rare on Earth, dated to the time of the extinctions. 4. `create` Many concept map topologies are possible. Maps should follow this logic: Oxygen is an effective final electron acceptor in cellular respiration because of its high electronegativ-ity. Organisms that use it as a final electron acceptor can produce more usable energy than organisms that do not use oxygen, but only if it is available. With more available energy, aerobic organisms can grow larger and move faster.

BIG **PICTURE** Evolution

p. 526 CYU (1) `understand` Circle = inbreeding, sexual selec-tion, nonrandom mating, natural selection, genetic drift,

---

FIGURE A28.1

```
                                          Outgroup
                                          Species 1
                                          Species 2
              A → C        C → A          Species 3
                           C → G          Species 4
```

FIGURE A28.2

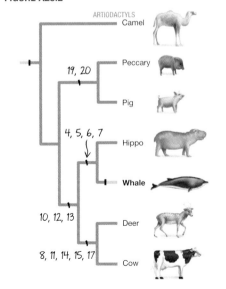

ARTIODACTYLS
Camel
Peccary
Pig
19, 20
4, 5, 6, 7
Hippo
Whale
10, 12, 13
Deer
8, 11, 14, 15, 17
Cow

mutation, and gene flow. **(2)** analyze Adaptation "increases" fitness; synapomorphies "identify branches on" the tree of life. **(3)** analyze Several answers possible, e.g., adaptive radiations and mass extinctions → can be studied using → the fossil record. **(4)** analyze Genetic drift, mutation, and gene flow "are random with respect to" fitness.

CHAPTER 29

IN-TEXT QUESTIONS AND EXERCISES

p. 529 Fig. 29.1 analyze Prokaryotic—it would require just one evolutionary change, the origin of the nuclear envelope in Eukarya. If it were eukaryotic, it would require that the nuclear envelope was lost in both Bacteria and Archaea—two changes are less parsimonious (see Chapter 28).

p. 529 Table 29.1 apply See **FIGURE A29.2**.

p. 534 Fig. 29.5 analyze Weak—different culture conditions may have revealed different species.

p. 535 CYU (1) create Conditions should mimic a spill—sand or stones with a layer of crude oil, or seawater with oil floating on top. Add samples, from sites contaminated with oil, that might contain cells capable of using molecules in oil as electron donors or electron acceptors. Other conditions (temperature, pH, etc.) should be realistic. **(2)** create Use metagenomic analysis. After isolating DNA from a soil sample, fragment and sequence the DNA. Compare the sequences to those from known organisms and use the data to place the species on the tree of life.

p. 538 apply cyanobacteria—photoautotrophs; *Clostridium aceticum*—chemoorganoautotroph; *Nitrosopumilus* sp.—chemolithoautotrophs; helicobacteria— photoheterotrophs; *Escherichia coli*—chemoorganoheterotroph; *Beggitoa*—chemolithotrophic heterotrophs.

p. 539 Fig. 29.10 apply Table 29.5 contains the answers. For example, for organisms called sulfate reducers you would have H_2 as the electron donor, SO_4^{2-} as the electron acceptor, and H_2S as the reduced by-product. For humans the electron donor is glucose ($C_6H_{12}O_6$), the electron acceptor is O_2, and the reduced by-product is water (H_2O).

p. 541 understand Different species of bacteria that have different forms of bacterial chlorophyll will not absorb the same wavelengths of light and will therefore not compete.

p. 542 Fig. 29.12 apply Aerobic respiration. More free energy is released when oxygen is the final electron acceptor than when any other molecule is used, so more ATP can be produced and used for growth.

p. 543 apply If bacteria and archaea did not exist, then (1) the atmosphere would have little or no oxygen, and (2) almost all nitrogen would exist in molecular form (the gas N_2).

p. 543 Fig. 29.13 apply Add a label for "animals" in the upper right quarter of the circle, and draw arrows leading from "Organic compounds with amino groups" to "animals" and from "animals" to "NH_3."

p. 544 CYU evaluate Eukaryotes can only (1) fix carbon via the Calvin–Benson pathway, (2) use aerobic respiration with organic compounds as electron donors, and (3) perform oxygenic photosynthesis. Among bacteria and archaea, there is much more diversity in pathways for carbon fixation, respiration, and photosynthesis, along with many more fermentation pathways.

p. 545a, p. 545b, p. 546a, p. 546b, p. 547 apply See **FIGURE A29.1**.

p. 548 Fig. 29.21 apply About 600,000 *N. maritimus* cells would fit end to end on a meter stick.

IF YOU UNDERSTAND . . .

29.1 analyze Eukaryotes have a nuclear envelope that encloses their chromosomes; bacteria and archaea do not. Bacteria have cell walls that contain peptidoglycan, and archaea have phospholipids containing isoprene subunits in their plasma membranes. Thus, the exteriors of a bacterium and archaeon are radically different. Archaea and eukaryotes also have similar machinery for processing genetic information. **29.2** apply After extracting and sequencing DNA from bacteria that live in an insect gut, analyze the resulting sequences for genes that are similar to known genes involved in nitrogen fixation. **29.3** understand They are compatible, and thrive in each others' presence—one species' waste product is the other species' food.

YOU SHOULD BE ABLE TO . . .

✔ Test Your Knowledge

1. understand c **2.** remember b **3.** remember d **4.** remember d **5.** remember False—the enzyme nitrogenase is poisoned by oxygen. **6.** remember peptidoglycan

✔ Test Your Understanding

1. understand c **2.** understand An electron donor provides the potential energy required to produce ATP. **3.** evaluate Yes. The array of substances that bacteria and archaea can use as electron donors, electron acceptors, and fermentation substrates, along with the diversity of ways that they can fix carbon and perform photosynthesis, allows them to live just about anywhere. **4.** apply They should get energy from reduced organic compounds "stolen" from their hosts. **5.** understand Large amounts of potential energy are released and ATP produced when oxygen is the electron acceptor, because oxygen is so electronegative.

FIGURE A28.3

FIGURE A29.1

FIGURE A29.2

Large body size and high growth rates are not possible without large amounts of ATP. **6.** understand They are paraphyletic because the prokaryotes include some (bacteria and archaea) but not all (eukaryotes) groups derived from the common ancestor of all organisms living today.

✓ Test Your Problem-Solving Skills

1. understand c (It is pathogenic on a wide variety of organisms). **2.** analyze This result supports their hypothesis because the drug poisons the enzymes of the electron transport chain and prevents electron transfer to Fe^{3-}, which is required to drive magnetite synthesis. If magnetite had still formed, another explanation would have been needed. **3.** create Hypothesis: A high rate of tooth cavities in Western children is due to an excess of sucrose in the diet, which is absent from the diets of East African children. To test this hypothesis, switch a group of East African children to a diet that contains sucrose, switch a group of Western children to a diet lacking sucrose, and monitor the presence of *S. mutans* in both groups of children. **4.** create Look in waters or soil polluted with benzene-containing compounds. Put samples from these environments in culture tubes where benzene is the only source of carbon. Monitor the cultures and study the cells that grow efficiently.

CHAPTER 30

IN-TEXT QUESTIONS AND EXERCISES

p. 529 CYU (1) understand Opisthokonts have a flagellum at the base or back of the cell; alveolate cells contain unique support structures called alveoli; stramenopiles—meaning "straw hairs"—have straw-like hairs on their flagella. **(2)** understand In direct sequencing, DNA is isolated directly from the environment and analyzed to place species on the tree of life. It is not necessary to actually see the species being studied.

p. 557 CYU (1) understand *Plasmodium* species are transmitted to humans by mosquitoes. If mosquitoes can be prevented from biting people, they cannot spread the disease. **(2)** apply Iron added → primary producers (photosynthetic protists and bacteria) bloom → more carbon dioxide taken up from atmosphere during photosynthesis → consumers bloom, eat primary producers → bodies of primary producers and consumers fall to bottom of ocean → large deposits of carbon-containing compounds form on ocean floor.

p. 560 apply See **FIGURE A30.1**.

p. 560 Fig. 30.8 analyze Two—one derived from the original bacterium and one derived from the eukaryotic cell that engulfed the bacterium.

p. 561 apply A photosynthetic bacterium (e.g., a cyanobacterium) could have been engulfed by a larger eukaryotic cell. If it was not digested, it could continue to photosynthesize and supply sugars to the host cell.

p. 562 apply Membrane infoldings observed in bacterial species today support the hypothesis's plausibility—they confirm that the initial steps actually could have happened. The continuity of the nuclear envelope and ER are consistent with the hypothesis, which predicts that the two structures are derived from the same source (infolded membranes).

p. 564 apply The chloroplast genes label should come off of the branch that leads to cyanobacteria. (If you had a phylogeny of just the cyanobacteria, the chloroplast branch would be located somewhere inside.)

p. 565 analyze The acquisition of the mitochondrion and the chloroplast represent the transfer of entire genomes, and not just single genes, to a new organism.

p. 566 apply Yes—when food is scarce or population density is high, the environment is changing rapidly (deteriorating). Offspring that are genetically unlike their parents may be better able to cope with the new and challenging environment.

p. 569a understand (1) Alternation of generations refers to a life cycle in which there are multicellular haploid phases and multicellular diploid phases. A gametophyte is the multicellular haploid phase that produces gametes; the sporophyte is the multicellular diploid phase that produces spores. A spore is a cell that grows into a multicellular individual but is not produced by fusion of two cells. A zygote is a cell that grows into a multicellular individual but *is* produced by fusion of two cells (gametes). Gametes are haploid cells that fuse to form a zygote. (2) See **FIGURE A30.2**.

p. 569b apply Most likely, the two types of amoebae evolved independently. The alternative hypothesis is that the common ancestor of alveolates, stramenopiles, rhizarians, plants, opisthokonts, and amoebozoa were amoeboid and that this growth form was lost many times. See **FIGURE A30.1**.

p. 569c, p. 569d, p. 569e, p. 570a, p. 570b, p. 570c apply See **FIGURE A30.1**.

p. 571 analyze If euglenids could take in food via phagocytosis (ingestive feeding), then that would have provided a mechanism by which a smaller photosynthetic protist could have been engulfed and incorporated into the cell via secondary endosymbiosis.

p. 572a, p. 572b, p. 573, p. 574. apply See **FIGURE A30.1**.

IF YOU UNDERSTAND . . .

30.1 understand Photosynthetic protists use CO_2 and light to produce sugars and other organic compounds, so they furnish the first or primary source of organic material in an ecosystem. **30.2** remember Eukaryotic cells contain many organelles, including a nucleus, endomembrane system, and an extensive cytoskeleton. Most prokaryotic cells contain few or no organelles, and no nucleus or endomembrane system and lack a complex cytoskeleton. **30.3** understand (1) Outside membrane was from host eukaryote; inside from engulfed cyanobacterium. (2) From the outside in, the four membranes are derived from the eukaryote that engulfed a chloroplast-containing eukaryote, the plasma membrane of the eukaryote that was engulfed, the outer membrane of the engulfed cell's chloroplast, and the inner membrane of its chloroplast.

YOU SHOULD BE ABLE TO . . .

✓ Test Your Knowledge

1. understand b **2.** remember b **3.** remember a **4.** remember b **5.** remember False. **6.** remember diatoms.

✓ Test Your Understanding

1. understand a **2.** understand Because all eukaryotes living today have cells with a nuclear envelope, it is valid to infer that their common ancestor also had a nuclear envelope. Because bacteria and archaea do not have a nuclear envelope, it is valid to infer that the trait arose in the common ancestor of eukaryotes. **3.** understand The host cell provided a protected environment and carbon compounds for the endosymbiont; the endosymbiont provided increased ATP from the carbon compounds **4.** understand It confirmed a fundamental prediction made by the hypothesis and could not be explained by any alternative hypothesis. **5.** understand All alveolates have alveoli, which are unique structures that function in supporting the cell. Among alveolates there are species that are (1) ingestive feeders, photosynthetic, or parasitic, and that (2) move using cilia, flagella, or a type of amoeboid movement.

✓ Test Your Problem-Solving Skills

1. analyze d **2.** analyze If the apicoplast that is found in *Plasmodium* (the organism that causes malaria) is genetically similar to chloroplasts, and if glyphosate poisons chloroplasts, it is reasonable to hypothesize that glyphosate will poison the apicoplast and potentially kill the *Plasmodium*. This would be a good treatment strategy

for malaria because humans have no chloroplasts, provided that the glyphosate produces no other effects that would be detrimental to humans. **3.** evaluate Primary producers usually grow faster when CO_2 concentration increases, but to date they have not grown fast enough to make CO_2 levels drop—CO_2 levels have been increasing steadily over decades. **4.** analyze Given that lateral gene transfer can occur at different points in a phylogenetic history, specific genes can become part of a lineage by a different route from that taken by other genes of the organism. In the case of chlorophyll *a*, its history traces back to a bacterium being engulfed by a protist and forming a chloroplast.

FIGURE A30.1

FIGURE A30.2

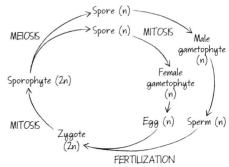

CHAPTER 31

IN-TEXT QUESTIONS AND EXERCISES

p. 582 CYU `understand` (1) Green algae and land plants share an array of morphological traits that are synapomorphies, including the chlorophylls they contain; (2) green algae appear before land plants in the fossil record; and (3) on phylogenetic trees estimated from DNA sequence data, green algae and land plants share a most recent common ancestor—green algae are the initial groups to diverge, and land plants diverge subsequently.

p. 583 Fig. 31.6 `remember` If you find the common ancestor of all the green algae (at the base of the Ulvophyceae), the lineages that are collectively called green algae don't include that common ancestor and all of its descendants—only some of its descendants. The same is true for non-vascular plants (the common ancestor here is at the base of the Hepaticophyta) and seedless vascular plants (the common ancestor here is at the base of the Lycophyta).

p. 586 Fig. 31.8 `understand` Water flows more easily through a short, wide pipe than through a long, skinny one because there is less resistance from the walls of the pipe.

p. 587 Fig. 31.9 `understand` To map an innovation on a phylogenetic tree, biologists determine the location(s) on the tree that is (are) consistent with all descendants from that point having the innovation (unless, in rare cases, the innovation was lost in a few descendants).

p. 589 `understand` Alternation of generations occurs when there are multicellular haploid individuals and multicellular diploid individuals in a life cycle. One hypothesis is that this life cycle evolved because the plants with a large sporophyte were more successful at using the wind to disperse many spores, while a small gametophyte produces gametes that can swim.

p. 590 `apply` In the hornwort photo, the sporophyte is the spike-like green and yellow structure; the gametophyte is the leafy-looking structure underneath; the horsetail gametophyte is the microscopic individual on the left; the sporophyte is the much larger individual on the right.

p. 590 Fig. 31.14 `understand` There are multicellular haploid stages and multicellular diploid stages in these plants.

p. 592 Fig. 31.18 `analyze` Gymnosperm gametophytes are microscopic, so they are even smaller than fern gametophytes. The gymnosperm gametophyte is completely dependent on the sporophyte for nutrition, while fern gametophytes are not. Fern gametophytes are photosynthetic and even supply nutrition to the young sporophytes.

p. 593 Fig. 31.19 `analyze` Consistent—the fossil data suggest that gymnosperms evolved earlier, and gymnosperms have larger gametophytes than angiosperms.

p. 595 Fig. 31.21 `analyze` To eliminate all other variables except for the one being tested, such as the possibility of pollination by other insects in the field. Scales differ because the two insect species visited flowers at very different frequencies.

p. 596 CYU (1) `understand` Cuticle prevents water loss from the plant; UV-absorbing compounds allow plants to be exposed to high light intensities without damage to their DNA; vascular tissue moves water up from the soil and moves photosynthetic products down to the roots.
(2) `apply` See **FIGURE A31.1**.
p. 600a, p. 600b, p. 601, p. 603a, p. 603b, p. 604, p. 605, p. 606 `apply` See **FIGURE A31.1**.
p. 596 Fig. 31.23 `apply` See **FIGURE A31.1**.

IF YOU UNDERSTAND . . .

31.1 `apply` Rates of soil formation will decline; rates of soil loss will increase. **31.2** `understand` All green algae and land plants contain chlorophyll *a* and *b*, have similar arrangements of thylakoid membranes, and store starch.
31.3 `analyze` Both spores and seeds have a tough, protective coat, so they can survive while being dispersed to a new location. Seeds have the advantage of carrying a store of nutrients with them—when a spore germinates, it must make its own food via photosynthesis right away.

YOU SHOULD BE ABLE TO . . .

✔ Test Your Knowledge

1. `remember` c **2.** `remember` c **3.** `remember` a **4.** `remember` c **5.** `remember` (a) Gametangia are found in all land plant groups except angiosperms; (b) pollen is found in gymnosperms and angiosperms; (c) seeds are found in gymnosperms and angiosperms, (d) fruit is found in angiosperms. **6.** `understand` In a gametophyte-dominant life cycle, the gametophyte is larger and longer lived than the sporophyte and produces most of the nutrition. In a sporophyte-dominant life cycle, the sporophyte generation is the larger, longer-lived, and photosynthetic phase of the life cycle.

✔ Test Your Understanding

1. `remember` c **2.** `understand` Plants build and hold soils required for human agriculture and forestry, and they increase water supplies that humans can use for drinking, irrigation, or industrial use. Plants release oxygen that we breathe. **3.** `analyze` Cuticle prevents water loss from leaves but also prevents entry of CO_2 required for photosynthesis. Stomata allow CO_2 to diffuse but can close to minimize water loss. Liverwort pores allow gas exchange but cannot be closed if conditions become dry. **4.** `understand` They provided the support needed for plants to grow upright and not fall over in response to wind or gravity. Erect growth allowed plants to compete for light. **5.** `apply` Homosporous plants produce a single type of spore that develops into a gametophyte that produces both egg and sperm. Heterosporous plants produce two different types of spores that develop into two different gametophytes that produce either egg or sperm. In a tulip, the microsporangium is found within the stamen, and the megasporangium is found within the ovule. Microspores divide by mitosis to form male gametophytes (pollen grains); megaspores divide by mitosis to form the female gametophyte. **6.** `understand` Animal pollinators increase the efficiency of pollination so plants can save energy by making less pollen.

✔ Test Your Problem-Solving Skills

1. `understand` d **2.** `analyze` A "reversion" to wind pollination might be favored by natural selection because it is costly to produce a flower that can attract animal pollinators. Because wind-pollinated species grow in dense clusters, they can maximize the chance that the wind will carry pollen from one individual to another (less likely if the individuals are far apart). Wind-pollinated deciduous trees flower in early spring before their developing leaves begin to block the wind. **3.** `create` The combination represents a compromise between efficiency and safety. Tracheids can still transport water if vessels become blocked by air bubbles. **4.** `create` Alter one characteristic of a flower, and present the flower to the normal pollinator. As a control, present the normal (unaltered) flower to the normal pollinator. Record the amount of time the pollinator spends in the flower, the amount of pollen removed, or some other measure of pollination success. Repeat for other altered characteristics. Analyze the data to determine which altered characteristic affects pollination success the most.

CHAPTER 32

IN-TEXT QUESTIONS AND EXERCISES

p. 619 CYU (1) `understand` Because they are made up of a network of thin, branching hyphae, mycelia have a large surface area, which makes absorption efficient. **(2)** `remember` Swimming spores and gametes, zygosporangia, basidia, asci.

FIGURE A31.1

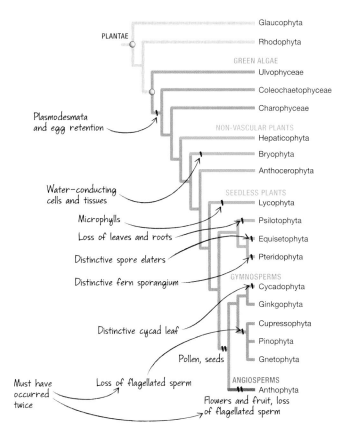

p. 622 Fig. 32.10 **analyze** Labeled nutrient experiments identify which nutrients are exchanged and in which direction. They explain *why* plants do better in the presence of mycorrhizae, and why the fungus also benefits.

p. 625 **analyze** Human sperm and egg undergo plasmogamy followed by karyogamy during fertilization, but heterokaryotic cells do not occur in humans or other eukaryotes besides fungi.

p. 625 Fig. 32.12 **apply** The haploid mycelium.

p. 626 Fig. 32.13 **analyze** Mycelia produced by asexual reproduction are genetically identical to their parent; mycelia produced by sexual reproduction are genetically different from both parents (each spore has a unique genotype).

p. 628 CYU (1) **understand** When birch tree seedlings are grown in the presence and absence of EMF, individuals denied their normal EMF are not able to acquire sufficient nitrogen and phosphorus. (2) **analyze** Each type of fungus involves meiosis and the production of haploid spores. In a zygosporangium, meiosis occurs in a multinucleate cell; in basidia and asci, this cell has a single diploid nucleus. In asci, meiosis is followed by one round of mitosis.

p. 628, p. 630a, p. 630b, p. 630c, p. 631, p. 632a, p. 632b **apply** See FIGURE A32.1.

IF YOU UNDERSTAND . . .

32.1 **apply** Loss of mycorrhizal fungi would decrease nutrient delivery to plants and reduce their growth inside the experimental plots, compared to control plots with intact fungi. Also, lack of fungi would slow decay of dead plant material, causing a dramatic buildup of dead organic material. 32.2 **understand** Like humans, fungi are composed of eukaryotic cells. Bacteria are prokaryotic and thus have many more unique targets for antibiotics. 32.3 **understand** Haploid hyphae fuse, forming a heterokaryotic mycelium. When karyogamy occurs, a diploid nucleus forms—just as when gametes fuse. 32.4 **understand** See the diversity boxes.

YOU SHOULD BE ABLE TO . . .

✔ Test Your Knowledge

1. **remember** a 2. **remember** d 3. **remember** b 4. **remember** b 5. **understand** Pores in septa allow nutrients to move from regions of acquisition to regions of mycelial growth; 6. **remember** commensal.

✔ Test Your Understanding

7. **understand** c 8. **evaluate** Along with a few bacteria, fungi are the only organisms that can digest wood completely. If the wood is not digested, carbon remains trapped in wood. Without fungi, CO_2 would be tied up and unavailable for photosynthesis, and the presence of undecayed organic matter would reduce the space available for plants to grow. 9. **analyze** Fungi produce enzymes that degrade cellulose and lignin. 10. **analyze** Plant roots have much smaller surface area than EMF or AMF. Hyphae are much smaller than the smallest portions of plant roots so can penetrate dead material more efficiently. Extracellular digestion—which plant roots cannot do—allows fungi to break large molecules into small compounds that can be absorbed. 11. **apply** DNA and RNA contain phosphorus. Mycorrhizal fungi effectively increase the volume of soil mined by root systems and can make adequate amounts of phosphorus available for supporting rapid cell division. 12. **analyze** Both compounds are processed via extracellular digestion. Different enzymes are involved, however. The degradation of lignin is uncontrolled and does not yield useful products; digestion of cellulose is controlled and produces useful glucose molecules.

✔ Test Your Problem-Solving Skills

13. **apply** c 14. **create** (1) Confirm that the chytrid fungus is found only in sick frogs and not healthy frogs.

(2) Isolate the chytrid fungus and grow it in a pure culture. (3) Expose healthy frogs to the cultured fungus and see if they become sick. (4) Isolate the fungus from the experimental frogs, grow it in culture, and test whether it is the same as the original fungus. 15. **create** Each of the different cellulase enzymes attacks cellulose in a different way, so producing all the enzymes together increases the efficiency of the fungus in breaking down cellulose completely. It is likely that lignin peroxidase is produced along with cellulases, so they can act in concert to degrade wood. To test this idea, you could harvest enzymes secreted from a mycelium before and after it contacts wood in a culture dish, and see if the cellulases and lignin peroxidase appear together once the mycelium begins growing on the wood. 16. **create** Collect a large array of colorful mushrooms that are poisonous; also capture mushroom-eating animals, such as squirrels. Present a hungry squirrel with a choice of mushrooms that have been dyed or painted a drab color versus others treated with a solution that is identical to the dye or paint used but uncolored. Record which mushrooms the squirrel eats. Repeat the test with many squirrels and many mushrooms.

CHAPTER 33

IN-TEXT QUESTIONS AND EXERCISES

p. 639 Fig. 33.2 **apply** The label and bar for "Segmentation?" should go on the branch to the left of the clam shell and "Mollusca" label. The label and bar for "Toolkit genes for segmentation?" should go on the same branch as cephalization, CNS, and coelom.

p. 640 **apply** The genetic tool kit required for multicellularity includes genes that regulate the cell cycle, and thus cell division. When something goes wrong in this process, uncontrolled growth called cancer can result. Thus, cancer may be as ancient as multicellularity itself.

p. 642 Fig. 33.7 **analyze** Stained *Hox* and *dpp* gene products would either not be found in *Nematostella* at all or would not occur in the same anterior–posterior and dorsal–ventral pattern as observed in bilaterians.

p. 646 CYU (1) **apply** Yes, your body has a tube-within-a-tube design. Humans are vertebrates, vertebrates are chordates, chordates are bilaterians, and all bilaterians have a tube-within-a-tube design, consisting of a bilaterally symmetrical, slender body with an inner tube (gut), an outer tube (mostly skin), muscles and organs derived from mesoderm in between, and one end (head) cephalized with sensory organs and brain. (2) **understand** When an animal moves through an environment in one direction, its ability to acquire food and perceive and respond to threats is greater if its feeding, sensing, and information-processing structures are at the leading end.

p. 651 Fig. 33.14 **evaluate** The larva and metamorphosis arrow should be circled in the life cycle. It might be adaptive to skip the feeding larval stage if more food resources are available for the mother (who must expend energy to produce the yolk) than for the larva, if predation pressure on larvae is high, or if other environmental factors for larvae are unfavorable, resulting in high mortality of the larvae.

p. 652 CYU (1) **understand** The mouthparts of deposit feeders are relatively simple because these animals simply gulp relatively soft material. The mouthparts of mass feeders are more complex because they have to tear off and process chunks of relatively hard material. (2) **create** Gametes that are shed into aquatic environments can float or swim. This cannot happen on land, so internal fertilization is more common there. Also, gametes are more at risk of drying out on land, and internal fertilization eliminates that risk.

p. 653 **apply** The origin of cilia-powered swimming should be marked on the Ctenophora branch.

p. 654 **apply** The origin of cnidocytes should be marked on the Cnidaria branch.

IF YOU UNDERSTAND . . .

33.1 **apply** It should increase dramatically—animals would no longer be consuming plant material. 33.2 **understand** Your drawing should show the tube-within-a-tube design of a worm, with one end labeled head and containing the mouth and brain. From the brain, one or two major nerve tracts should run the length of the body. The gut should go from the mouth to the other end, ending in the anus. In between the gut and outer body wall, there should be blocks or layers of muscle. 33.3 **apply** There are many possible answers. For example, within the phylum Arthropoda: The sensory systems have diversified (e.g., eyes are important for sight in flies, antennae are important for smelling in moths); feeding strategies have diversified (e.g., butterflies are fluid feeders but caterpillars are mass feeders); ecological roles have diversified (e.g., butterflies are herbivores but spiders are carnivores); limbs have diversified (e.g., butterflies have legs and wings but caterpillars have lobe-like limbs); and some insects, like grasshoppers, have direct development whereas others, like mosquitoes, have indirect development. 33.4 **remember** If you see a

FIGURE A32.1

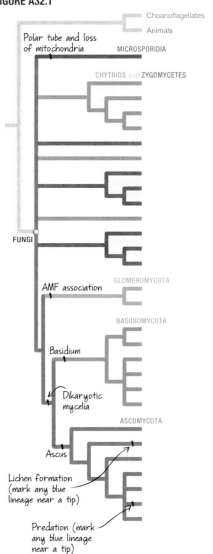

Choanoflagellates
Animals

Polar tube and loss of mitochondria — MICROSPORIDIA

CHYTRIDS and ZYGOMYCETES

FUNGI

AMF association — GLOMEROMYCOTA

BASIDIOMYCOTA

Basidium

Dikaryotic mycelia

ASCOMYCOTA

Ascus

Lichen formation (mark any blue lineage near a tip)

Predation (mark any blue lineage near a tip)

sessile, multicellular animal with no symmetry or radial symmetry, with outcurrent holes, and choanocytes, it is a sponge. If you see a radially symmetric, transparent, planktonic organism moving via rows of cilia, it is a ctenophore. If you see a diploblastic, outwardly radially symmetric animal that is either a single polyp or colony of multiple polyps, or it is a medusa, then the animal is a cnidarian.

YOU SHOULD BE ABLE TO . . .

✔ Test Your Knowledge

1. `remember` a **2.** `remember` True **3.** `remember` protostomes, deuterostomes **4.** `understand` a **5.** `remember` d **6.** `remember` d

✔ Test Your Understanding

1. `understand` Diploblasts have two types of embryonic tissue (ectoderm, endoderm); triploblasts have three (ectoderm, mesoderm, endoderm). Mesoderm made it possible for an enclosed, mesoderm-lined cavity to develop, creating a coelom. **2.** `analyze` Many unicellular organisms are heterotrophic, but they can consume only small bits of food. Animals are multicellular, so they are larger and can consume larger amounts of food—making them important consumers in food webs. **3.** `apply` There are about 18.4 named arthropods for every named species of chordate (1,160,000/63,000), about 1.3 named mollusks for every named species of chordate (85,000/63,000), and about 0.4 named species of nematode worm for every named species of chordate (25,000/63,000). These numbers are likely to be underestimates because there are probably many more arthropods, mollusks, and nematode worms than have been described, yet relatively few of the larger, more familiar chordates (including vertebrates) left to be described. **4.** `understand` In oviparous species, the mother adds nutrient-rich yolk to the egg that nourishes the developing embryo. In viviparous species, the mother transfers nutrients directly from her body to the growing embryo. **5.** `evaluate` You could stain the gene products of the segmentation genes in chordates (such as a mouse embryo) and observe whether the expression pattern correlates with the expression pattern observed in arthropods (such as a fly embryo). **6.** `apply` c

✔ Test Your Problem-Solving Skills

1. `analyze` Yes—if the same gene is found in nematodes and humans, it was likely found in the common ancestor of protostomes and deuterostomes. If so, then fruit flies should also have this gene. **2.** `apply` Asexual reproduction is likely to be favored when the sea anemones have plenty of resources such as space and food. Sexual reproduction is likely to be favored when the sea anemones are stressed due to lack of resources, predation, infection, or other factors. In this case, it is adaptive to reproduce in a way that enables dispersal to a new site and/or genetically diverse offspring, some of which will be more likely to survive selection than a clone of identical animals would. **3.** `apply` Since echinoderms evolved from bilaterally symmetric ancestors with nerve cords, they should have nerve cords too. However, the organization of the nerve cords should be more diffuse, because echinoderms need to take in and process information from multiple directions—not just one. **4.** `apply` d

CHAPTER 34

IN-TEXT QUESTIONS AND EXERCISES

p. 659 Fig. 34.3 `apply` We know that aquatic living was the ancestral trait because the outgroups (e.g., Cnidaria, Porifera) were aquatic. We also have fossil evidence that the aquatic lifestyle was ancestral.
p. 660, p. 662, p. 665, p. 666, p. 667, p. 671, p. 674 `apply` See **FIGURE A34.1.**

p. 662 Fig. 34.6 `evaluate` Different phyla of worms have different feeding strategies and mouthparts, and they eat different foods, so they are not all in direct competition for food.
p. 664 CYU (1) `apply` Snails on land would be expected to have gills or lungs inside the body or otherwise protected from drying out, thinner shells so that they are not as heavy, and desiccation-resistant eggs. **(2)** `evaluate` The tentacles of the octopus could be probed during development for *Hox* genes known to be important in establishing the foot of snails. If the expression pattern is similar, homology would be supported.
p. 664 Fig. 34.10 `apply` You should have drawn a circle around all the tentacles.

IF YOU UNDERSTAND . . .

34.1 `remember` Ecological opportunity was important in the diversification of protostomes because many lineages made the transition from water to land, radiating into new niches. Genetic opportunity was important because changes in the expression of developmental tool-kit genes enabled diverse modifications of body plans based on existing genes. **34.2** `understand` One source of diversification is the loss of ancestral traits; just because a species does not possess a trait does not mean that its ancestor didn't. **34.3** `analyze` Both arthropods and vertebrates have segmented bodies, which are especially amenable to modularity since a change in the expression pattern of developmental tool-kit genes can change the number, size, and type of segments, exposing morphological diversity to the adaptive forces of natural selection.

YOU SHOULD BE ABLE TO . . .

✔ Test Your Knowledge

1. `remember` False **2.** `remember` a **3.** `remember` c **4.** `understand` c **5.** `remember` Both groups are bilaterally symmetric triploblasts with the protostome pattern of development. **6.** `remember` b

✔ Test Your Understanding

1. `remember` d **2.** `apply` Both annelids and arthropods have segmented bodies, unlike other protostomes such as flatworms and nematodes. **3.** `evaluate` Clams are relatively sessile as adults. Because larvae can swim, they provide a mechanism to disperse to new locations. To test this idea, you could measure the distance traveled by clam larvae and measure their success in recruiting to new habitats compared to larvae that are artificially limited to stay near their parents. **4.** `apply` Land plants would require cuticles or other mechanisms to keep their tissues from drying out, a mechanism of keeping gametes from

drying out, and structural support to retain body shape despite the effects of gravity. **5.** `evaluate` The ability to fly allowed insects to disperse to new habitats and find new food sources efficiently. **6.** `analyze` It might be adaptive for juveniles of hemimetabolous individuals to use the same food source as the adults when the food source is plentiful.

✔ Test Your Problem-Solving Skills

1. `analyze` See **FIGURE A34.2. 2.** `analyze` If the ancestors of brachiopods and mollusks lived in similar habitats and experienced natural selection that favored similar traits, then they would have evolved to have similar forms and habitats. This is called convergent evolution (see Chapter 28). **3.** `evaluate` This hypothesis is reasonable because simple characteristics often occur before more elaborate ones; this could be true for the radula and foot in aplacophorans. An alternate hypothesis would be that these structures' simple forms are derived characteristics that are adaptations for parasitism, but the evidence suggests that aplacophorans are carnivores and detritivores, not parasites. The position of the aplacophorans in the phylogeny could be tested by using DNA sequence data rather than morphological characters in a phylogenetic analysis. **4.** `apply` b

CHAPTER 35

IN-TEXT QUESTIONS AND EXERCISES

p. 682 Fig. 35.1 `apply` Invertebrates are a paraphyletic group. The group does not include all the descendants of the common ancestor because vertebrates are excluded.
p. 684 CYU `understand` If it's an echinoderm, it should have five-part radial symmetry, a calcium carbonate endoskeleton just underneath the skin, and a water vascular system (e.g., visible tube feet).
p. 684, p. 685a, p. 685b `apply` See **FIGURE A35.1.**

FIGURE A34.2

FIGURE A34.1

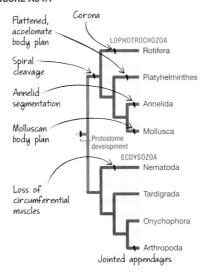

FIGURE A35.1

p. 690 Fig. 35.12 apply Mammals and birds are equally related to amphibians, because birds and mammals share a common ancestor, and this ancestor shares a common ancestor with amphibians.

p. 691 apply See FIGURE A35.2

p. 693 Fig. 35.16 analyze If this phylogeny were estimated on the basis of limb traits, it would be based on the *assumption* that the limb evolved from fish fins in a series of steps. But because the phylogeny was estimated from other types of data, it is legitimate to use it to analyze the evolution of the limb without making a circular argument.

p. 694 Fig. 35.18 analyze The yolk sac is smaller here—their function in an amniotic egg has been taken over by the placenta.

p. 695 CYU (1) understand Jaws allow animals to capture food efficiently and process it by crushing or tearing. The ability to process food more efficiently increased the importance of jawed vertebrates as herbivores and carnivores. The tetrapod limb enables vertebrates to locomote on land. The amniotic egg enables vertebrates to lay eggs on land; amniotic eggs are resistant to drying out. The placenta protects the growing fetus and enables the mother to be mobile. Parental care reduces the number of offspring that can be raised, but increases their survivorship. Flight enabled pterosaurs, bats, and birds to expand into new habitats. **(2)** analyze Protostomes and deuterostomes

living on land both have mechanisms for preventing their eggs from drying out (e.g., thick membranes in snail and insect eggs; amniotic egg in tetrapods), preventing their skin from drying out (e.g., waxy cuticle in insects, scaly skin in reptiles), and keeping their respiratory surfaces from drying out (e.g., lungs in land snails and tracheal system in insects; lungs in tetrapods).

p. 696 apply See FIGURE A35.3.

p. 698a, p. 698b, p. 699, p. 700 apply See FIGURE A35.4.

p. 701a, p. 701b, p. 702a, p. 702b, p. 703a, p. 703b apply See FIGURE A35.3.

p. 705 Table 35.1 analyze 44 percent, 26 percent; relative to body size, *H. floresiensis* has a smaller brain than *H. sapiens*.

p. 706 Fig. 35.37 apply Five hominin species existed 1.8 mya, and three existed 100,000 years ago.

p. 706 Fig. 35.38 analyze The forehead became much larger, and the face became "flatter"; the brow ridges are less prominent in later skulls than in earlier skulls.

IF YOU UNDERSTAND . . .

35.1 apply (1) Mussels and clams will increase dramatically; (2) kelp density will increase.

35.2 understand Your cartoon should have a dorsal hollow nerve chord along the top, dorsal edge; another line representing the notochord below the nerve cord; a tail

labeled at one end; and a series of lines or bumps representing pharyngeal gill slits near the other end.

35.3 understand Like today's lungfish, the earliest tetrapods could have used their limbs for pulling themselves along the substrate in shallow-water habitats. **35.4** understand The earliest fossils of *H. sapiens* are found in Africa, and phylogenetic analyses of living human populations indicate that the most basal groups are all African.

YOU SHOULD BE ABLE TO . . .

✔ **Test Your Knowledge**

1. remember True **2.** remember a **3.** understand d **4.** remember mammals **5.** understand d **6.** remember b

✔ **Test Your Understanding**

1. understand c **2.** understand Pharyngeal gill slits function in suspension feeding. The notochord furnishes a simple endoskeleton that stiffens the body; electrical signals that coordinate movement are carried by the dorsal hollow nerve cord to the muscles in the post-anal tail, which beats back and forth to make swimming possible. Cephalochordates and Urochordates are chordates, but they are not vertebrates. **3.** analyze If jaws are derived forms of gill arches, then the same genes and the same cells should be involved in the development of the jaw and the gill arches. **4.** understand Increased parental care allows

FIGURE A35.2

FIGURE A35.3

FIGURE A35.4

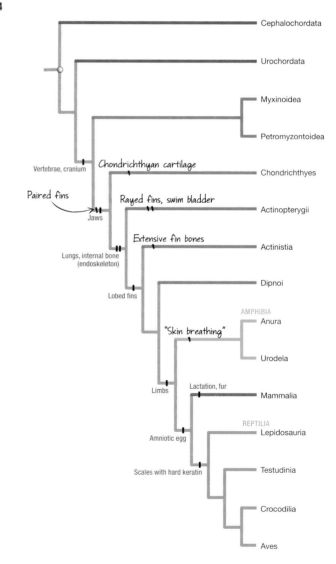

offspring to be better developed, and thus have increased chances of survival, before they have to live on their own. **5.** evaluate Yes: Over a short time interval, many species that occupy an array of foods and habitats evolved. Changes in tooth and jaw structure, tool use, and body size suggest that different hominin species exploited different types of food. **6.** evaluate A phylogeny of modern humans would reject the out-of-Africa hypothesis if the most ancestral lineages in the phylogeny were from any other region but Africa, such as Asia or Europe.

✓ Test Your Problem-Solving Skills

1. apply b **2.** evaluate Xenoturbellidans either retain traits that were present in the common ancestor of all deuterostomes, or they have lost many complex morphological characteristics. **3.** evaluate Vertebrate limbs are diverse in form and function (they can serve as arms, flippers, wings, etc.) but have homologous bones within. Molecular probes could be used to visualize the expression patterns of tool-kit genes like *Hox* in the embryonic limbs of different vertebrate species to show how changes in the expression of homologous genes can cause changes to the structure and function of limbs. **4.** analyze See **FIGURE A35.5**. It is important to use indigenous people for this study to minimize the amount of gene flow (transfer of alleles from one population to another) that occurred during the history of the study populations due to immigration and interbreeding.

CHAPTER 36

IN-TEXT QUESTIONS AND EXERCISES

p. 717 Fig. 36.7 apply Starting with 1 virion, approximately 80 virions were produced after 40 minutes. If all of these virions infected new cells, you expect each of the 80 infected cells to produce 80 virions by 90 minutes, or a total of 6400 virions. The cells appear to replicate once every 30 minutes, which means that they will have replicated 3 times during this period. The total number of cells at 90 minutes would be approximately 2^3, or 8 cells. There would be 800 times as many virions as cells.
p. 717 Fig. 36.8 analyze No—it only shows that CD4 is required. (Subsequent work showed that other proteins are involved as well.)
p. 721 analyze In T4, the viral genome is injected into the host while HIV inserts its genome via membrane fusion. HIV must first convert its genome into double-stranded DNA before replication, which is not necessary for T4, which has a DNA genome. Both viruses use double-stranded DNA to make mRNA that is translated into viral proteins. T4 virions are made in the cytosol of the cell and are released when the cell bursts. HIV virions bud from the cell surface, which does not require cell death.
p. 721 Fig. 36.12 create Budding may disrupt the integrity of the host-cell plasma membrane enough to kill the cell.
p. 722 CYU analyze Like cars in the assembly line of an automobile plant, new virions are assembled from premanufactured parts, and large numbers of progeny are produced per generation. Cells, on the other hand, reproduce by the division of a single integrated unit, resulting in only two progeny per generation.
p. 724 Fig. 36.15 apply You should have the following bars and labels: on branch to HIV-2 (sooty mangabey to human); on branch to HIV-1 strain O (chimp to human); on branch to HIV-1 strain N (chimp to human); on branch to HIV-1 strain M (chimp to human).

p. 725 CYU (1) analyze The escaped-genes and degenerate-cell hypotheses state that viruses originated from cells, while the RNA-world hypothesis suggests that viruses originated in parallel with, maybe even influenced, the origin of cells. The escaped-gene and RNA-world hypotheses state that the viruses originated from parasitic molecules. The degenerate-cell hypothesis states that viruses originated from parasitic organisms. **(2)** apply A mutation that made transmission between humans more efficient would make it more dangerous. Such mutation could occur via genomic reassortment in pigs, where an avian virus and a human virus could co-infect cells and produce recombinants.

IF YOU UNDERSTAND . . .

36.1 understand HIV infects helper T cells. These cells play a central role in the human immune response. As latently infected cells are activated by the immune response, HIV replication causes cell death. Over several years, steady decline in helper T cells leads to a compromised immune response that allows other infections to eventually kill the infected person. **36.2** understand (1) Fusion inhibitors block viral envelope proteins or host cell receptors. (2) Protease inhibitors prevent processing/assembly of viral proteins. (3) Reverse transcriptase inhibitors block reverse transcriptase, preventing replication of the genome. **36.3** apply There is heritable variation among virions, due to random changes that occur as their genomes are copied. There is also differential reproductive success among virions in their ability to successfully infect host cells. This differential success is due to the presence of certain heritable traits. **36.4** analyze Both classes have positive-sense single-stranded RNA genomes. Class IV viruses use an RNA replicase to make a negative-sense copy of the genome, which is then used as a template for producing mRNA and genomic RNAs. Class VI viruses use reverse transcriptase to convert their RNA genome into dsDNA, which is then integrated into the host genome. There it serves as a template to for transcribing viral mRNA and genomic RNAs by host-cell RNA polymerase.

YOU SHOULD BE ABLE TO . . .

✓ Test Your Knowledge

1. remember d **2.** remember b **3.** understand In class VI and VII viruses, reverse transcriptase converts positive-sense single-stranded RNA to double-stranded DNA.
4. remember b **5.** remember c **6.** remember The unique type of genetic material included in the virion (classes I–V) in addition to the manner in which the genome is replicated (compare classes VI and VII with classes IV and I, respectively).

✓ Test Your Understanding

7. analyze A virus with an envelope exits host cells by budding. A virus that lacks an envelope exits host cells by lysis (or other mechanisms that don't involve budding). **8.** analyze (1) Rate of viral genome replication is much higher via the lytic cycle. In lysogeny, the viral genome can replicate only when the host cell replicates. (2) Only the lytic cycle produces virions. (3) The lytic

cycle results in host cell death, while the host cell continues to survive during lysogeny. **9.** create Viruses rely on host-cell enzymes to replicate, whereas bacteria do not. Therefore, many drugs designed to disrupt the virus life cycle cannot be used, because they would kill host cells as well. Only viral-specific proteins are good targets for drug design. **10.** evaluate Each major hypothesis to explain the origin of viruses is associated with a different type of genome. Escaped genes: single- and double-stranded DNA or possibly single-stranded RNA. Degenerate cells: double-stranded DNA. RNA world: single- or double-stranded RNA. **11.** understand The phylogenies of SIVs and HIVs show that the two shared common ancestors, but that SIVs are ancestral to the HIVs. Also, there are plausible mechanisms for SIVs to be transmitted to humans through butchering or contact with pets, but fewer or no plausible mechanisms for HIVs to be transmitted to monkeys or chimps. **12.** apply a; the single-stranded positive-sense RNA genome of this virus can serve as an mRNA to produce viral proteins, while all of the other possibilities require viral enzymes to transcribe the genome into mRNA.

✓ Test Your Problem-Solving Skills

13. create Culture the *Staphylococcus* strain outside the human host and then add the virus to determine whether the virus kills the bacterium efficiently. Then test the virus on cultured human cells to determine whether the virus harms human cells. Then test the virus on monkeys or other animals to determine if it is safe. Finally, you could test the virus on human volunteers.
14. evaluate Prevention is currently the most cost-effective program, but it does not help people who are already infected. Treatment with effective drugs not only prolongs lives but also reduces virus loads in infected people, so that they have less chance of infecting others. **15.** apply b; by prolonging the infection in this manner, these viruses are more likely to be transmitted to a new host.
16. evaluate Viruses cannot be considered to be alive by the definition given in (1), because viruses are not capable of replicating by themselves. By the definition given in (2), it can be argued that viruses are alive because they store, maintain, replicate, and use genetic information—although they cannot perform all these tasks on their own.

CHAPTER 37

IN-TEXT QUESTIONS AND EXERCISES

p. 732 Fig. 37.1 apply New branches and leaves should develop to the right (the plant will also lean that way); new lateral roots will develop to the left.
p. 734 Fig. 37.3 analyze Lawn grasses are shallow rooted, so when the top part of soil dries out, the plants cannot grow.
p. 735 Table 37.1 understand For aerobic respiration, which generates ATP needed to keep cells alive.
p. 736 Fig. 37.5 evaluate The individuals are genetically identical—thus, any differences in size or shape in the different habitats are due to phenotypic plasticity and not genetic differences.
p. 738 Fig. 37.6 analyze Large surface area provides more area to capture photons, but also more area from which to lose water and be exposed to potentially damaging tearing forces from wind.

FIGURE A35.5

p. 739 CYU (1) remember See **FIGURE A37.1** **(2)** understand The generalized body of a plant has a taproot with many lateral roots and broad leaves. Examples of deviations from this generalized body structure follow. Modified roots: Unlike taproots, *fibrous roots* do not have one central root, and *adventitious roots* arise from stems. Modified stems: *Stolons* grow along the soil and grow roots and leaves at each node, and *rhizomes* grow horizontally underground. Both of these modified stems function in asexual reproduction. Modified leaves: The *needle-shaped leaves* of cacti don't lose water to transpiration compared with typical broad leaves, and they protect the plant from herbivory. *Tendrils* on climbing plants are modified leaves that do not photosynthesize, but wrap around trees or other substrates to facilitate climbing.
p. 745 Fig. 37.16 evaluate Many possible answers; for example, in leaf cells, genes encoding proteins involved in photosynthesis would be expressed. These genes wouldn't be expressed in the root cells.
p. 747 CYU (1) understand The apical meristem gives rise to the three primary meristems. The apical meristem is a single mass of cells localized at the tip of a root or shoot; the primary meristems are localized in distinctive locations behind the apical meristem. **(2)** understand Epidermal cells are flattened and lack chloroplasts; they secrete the cuticle (in shoots) or extend water and nutrient-absorbing root hairs (in roots) throughout the plant. Parenchyma cells are "metabolic workhorse cells" found throughout the plant body; they perform photosynthesis and synthesize, store, and/or transport materials. Tracheids are long, slim cells with pits in their secondary cell walls; they are found in xylem and are dead when mature, and they conduct water and solutes up the plant. Sieve-tube elements are long, thin cells found in phloem; they are alive when mature and conduct sugars and other solutes up and down the plant.
p. 750 understand Secondary growth adds cells only to the inside and outside of the vascular cambium, so the nail remains the same distance above the ground as the tree grows.

p. 750 Fig. 37.24 remember You should have labeled a light band as early wood, and the dark band to its right as late wood; both bands together comprise one growth ring.
p. 751 CYU (1) apply The rings are small on the shaded side and large on the sunny side. **(2)** remember See **FIGURE A37.2**.

IF YOU UNDERSTAND . . .

37.1 analyze Phenotypic plasticity is more important in (1) environments where conditions vary because it gives individuals the ability to change the growth pattern of their roots, shoots, and stems to access sunlight, water, and other nutrients as the environment changes; and (2) in long-lived species because it gives individuals a mechanism to change their growth pattern as the environment changes throughout their lifetime.
37.2 analyze The zone of cellular maturation is where root hairs emerge. Root hairs are the site where most of the water and ions are absorbed in the root system. **37.3** apply (1) In both cases, the structure would stop growing; (2) the shoot would lack epidermal cells and would die.
37.4 apply (1) The plant would not produce secondary xylem and phloem, so its girth and transport ability would be reduced, though it would still produce bark. (2) The trunk would have much wider tree rings on one side than the other side.

YOU SHOULD BE ABLE TO . . .

✔ Test Your Knowledge

1. remember b **2.** understand a **3.** remember c **4.** remember a **5.** remember Cactus spines are modified leaves; thorns are modified stems. **6.** analyze Collenchyma and sclerenchyma cells both have thickened cell walls and function in providing structural support. Unlike collenchyma cells, sclerenchyma cells have lignified cell walls and are dead at maturity.

✔ Test Your Understanding

7. remember b **8.** evaluate The general function of both systems is to acquire resources: The shoot system captures light and carbon dioxide; the root system absorbs water and nutrients. Vascular tissue is continuous throughout both the shoot and root systems. Diversity in roots and shoots enables plants of different species to live together in the same environment without directly competing for resources. **9.** analyze Continuous growth enhances phenotypic plasticity because it allows plants to grow and respond to changes or challenges in their environment (such as changes in light and water availability). **10.** apply Cuticle reduces water loss; stomata facilitate gas exchange. Plants from wet habitats should have a relatively large number of stomata and thin cuticle. Plants

living in dry habitats should have relatively few stomata and thick cuticle. **11.** analyze Parenchyma cells in the ground tissue perform photosynthesis and/or synthesize and store materials. In vascular tissue, parenchyma cells in rays conduct water and solutes across the stem; other parenchyma cells differentiate into sieve-tube elements and companion cells in phloem. **12.** understand Cells produced to the inside of the vascular cambium differentiate into secondary xylem; cells produced to the outside of the vascular cambium differentiate into secondary phloem.

✔ Test Your Problem-Solving Skills

13. apply c **14.** remember Asparagus—stem; Brussels sprouts—lateral buds; celery—petiole; spinach—leaf (petiole and blade); carrots—taproot; potato—modified stem. **15.** create They grow continuously, so do not have alternating groups of large and small cells that form rings. **16.** analyze Girdling disrupts transport of solutes in secondary phloem. The tree's root system starves.

CHAPTER 38

IN-TEXT QUESTIONS AND EXERCISES

p. 756 apply (1) Add a solute (e.g., salt or sugar) to the left side of the U-tube, so that solute concentration is higher than on the right side. (2) Water would move from the right side of the U-tube into the left side.
p. 757 apply Pull the plunger up.
p. 757 Fig. 38.2 apply It increases (becomes less negative), which reduces the solute potential gradient between the two sides.
p. 758 CYU (1) apply When a cell is in equilibrium with pure water, the negative solute potential of the cell will be balanced by the positive pressure potential (turgor pressure). **(2)** analyze Irrigation water contains low concentrations of salts so the solute potential will be near zero. As water evaporates, salts remain in the soil, lowering the solute potential.
p. 760 Fig. 38.7 understand See **FIGURE A38.1**.
p. 761 apply The individual would not be able to exclude toxic solutes from the xylem. If these solutes were transported throughout the plant in high enough concentrations, they could damage tissues. The mutant could also not maintain ions in the xylem and so could not generate root pressure.
p. 764 Fig. 38.12 analyze The line on the graph would go up, and large jumps would occur each time the light level was turned up, because increased transpiration rates would increase negative pressure (tension) at the leaf surface and thus increase the water-potential gradient.
p. 765 CYU (1) understand At the air–water interface under a stoma, water molecules cannot hydrogen-bond in all directions—they can bond only with water molecules below them. This creates tension at the surface. When transpiration occurs, the curvature of the meniscus increases—increasing tension. **(2)** apply When a nearby stoma closes, the humidity inside the air space increases, transpiration decreases, and surface tension on menisci decreases. The same changes occur when a rain shower starts. When air outside the leaf dries, though, the opposite occurs: The humidity inside the air space decreases, transpiration increases, and surface tension on menisci increases.
p. 768 Fig. 38.17 understand See **FIGURE A38.2**.
p. 771 CYU (1) understand In early spring, growing leaves are sinks and phloem sap moves toward them. In midsummer, leaves are sources and phloem sap moves away from them. **(2)** create The movement of water and solutes in sieve elements is by bulk flow, so stationary organelles would provide resistance to the flow.
p. 771 Fig. 38.22 create With time the plant will metabolize the radioactive sucrose, and levels of radioactivity will decline.

FIGURE A37.1

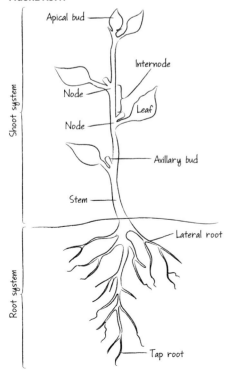

Apical bud
Internode
Shoot system
Node
Leaf
Node
Axillary bud
Stem
Lateral root
Root system
Tap root

FIGURE A37.2

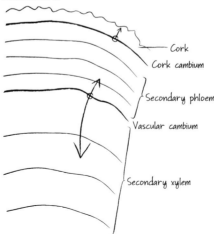

Cork
Cork cambium
Secondary phloem
Vascular cambium
Secondary xylem

IF YOU UNDERSTAND . . .

38.1 analyze Because xylem cells are dead at maturity, there are no plasma membranes for water to cross and thus no solute potential. The only significant force acting on water in xylem is pressure. **38.2** evaluate It drops, because transpiration rates from stomata decrease, reducing the water potential gradient between roots and leaves. **38.3** understand CAM plants open their stomata only at night, and thus they transpire when temperatures are cooler. **38.4** analyze To create high pressure in phloem near sources and low pressure near sinks, water has to move from xylem to phloem near sources and from phloem back to xylem near sinks. If xylem were not close to phloem, this water movement could not occur, and the pressure gradient in phloem wouldn't exist.

YOU SHOULD BE ABLE TO . . .

✓ Test Your Knowledge

1. understand d **2.** apply b **3.** understand a **4.** remember c **5.** remember Endodermal cells. **6.** remember Companion cells.

✓ Test Your Understanding

7. understand c **8.** apply See FIGURE A38.3. **9.** analyze The force responsible for root pressure is the high solute potential of root cells at night, when they continue to accumulate ions from soil and water follows by osmosis. The mechanism is active, in the sense that ions are imported

against their concentration gradient. Capillarity is driven by adhesion of water molecules to the sides of xylem cells that creates a pull upward, and by cohesion with water molecules below. Cohesion-tension, generated by transpiration, is driven by the loss of water molecules from menisci in leaves, which creates a large negative pressure (tension). Both capillarity and cohesion-tension are passive processes. **10.** understand Water moves up xylem because of transpiration-induced tension created at the air–water interface under stomata, which is communicated down to the roots by the cohesion of water molecules. Sap flows in phloem because of a pressure gradient that exists between source and sink cells, driven by differences in sucrose concentration and flows of water into or out of nearby xylem. **11.** understand By pumping protons out, companion cells create a strong electrochemical gradient favoring entry of protons. A cotransport protein (a symporter) uses this proton gradient to import sucrose molecules *against* their concentration gradient. **12.** apply When it germinates—sucrose stored inside the seed is released to the growing embryo, which cannot yet make enough sucrose to feed itself.

✓ Test Your Problem-Solving Skills

13. apply b **14.** create Plants would not grow as quickly or as tall, because the taller they grew, the more energy they would need to use to transport water and thus the less they would have available for growth. **15.** analyze Because

they cannot readily replace water that is lost to transpiration, they have to close stomata. During a heat wave, transpiration cannot cool the plants' tissues. The trees bake to death. **16.** create Closing aquaporins will slow or stop movement of water from cells into the xylem and out of the plant via transpiration. Because water does not leave the cells, they are able to maintain turgor (normal solute potentials).

CHAPTER 39

IN-TEXT QUESTIONS AND EXERCISES

p. 776 Fig. 39.1 evaluate To test the hypothesis that most of a plant's mass comes from CO_2 in the atmosphere, design an experiment where plants are grown in the presence or absence of CO_2 and compare growth, and/or grow plants in the presence of labeled CO_2 and document the presence of labeled carbon in plant tissues.

p. 777 Fig. 39.3 evaluate Because soil is so complex, it would be difficult to defend the assumption that the soils with and without added copper are identical except for the difference in copper concentrations. Also, it would be better to compare treatments that had no copper versus normal amounts of copper, instead of comparing with and without added copper.

p. 783 Fig. 39.10 analyze The proton gradient arrow in (b) should begin above the membrane and cross the

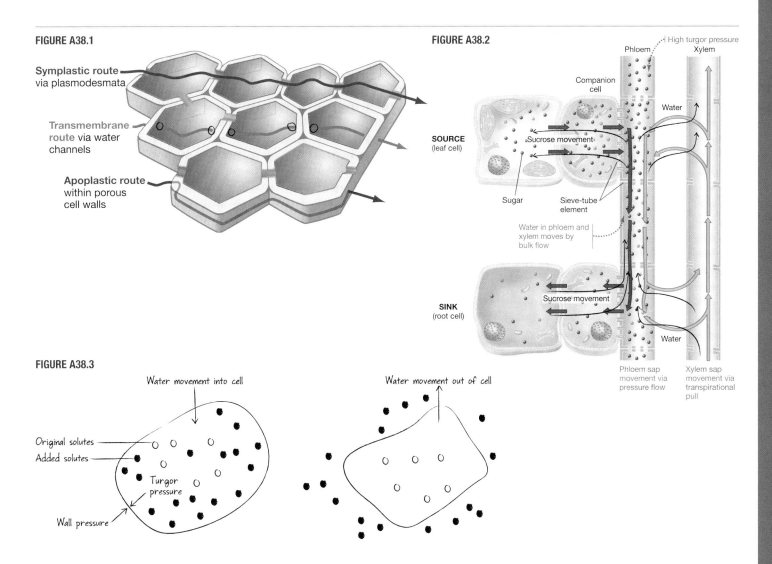

FIGURE A38.1

Symplastic route via plasmodesmata

Transmembrane route via water channels

Apoplastic route within porous cell walls

FIGURE A38.2

High turgor pressure
Phloem   Xylem

Companion cell

Water

SOURCE (leaf cell)

Sucrose movement

Sugar

Sieve-tube element

Water in phloem and xylem moves by bulk flow

SINK (root cell)

Sucrose movement

Water

Phloem sap movement via pressure flow

Xylem sap movement via transpirational pull

FIGURE A38.3

Water movement into cell

Water movement out of cell

Original solutes

Added solutes

Turgor pressure

Wall pressure

membrane pointing down. The electrical gradient arrow in (c) should also begin above the membrane (marked positive) and cross the membrane pointing down (toward negative).

p. 784a apply (1) If there is no voltage across the root-hair membrane, there is no electrical gradient favoring entry of cations, and absorption stops. (2) There is no route for cations to cross the root hair membrane, so absorption stops.

p. 784b apply (1) If there is no proton gradient across the root-hair membrane, there is no gradient favoring entry of protons, so symport of anions stops. (2) There is no route for anions to cross the root-hair membrane, so absorption stops.

p. 786 CYU (1) understand Proton pumps establish an electrical gradient across root-hair membranes, and the inside of the membrane is much more negative than the outside. Cations follow this electrical gradient into the cell. Anions are transported with protons via symporters. **(2)** create See **FIGURE A39.1**.

p. 788 CYU (1) analyze Many possible answers, including the following: If N-containing ions are abundant in the soil, the energetic cost of maintaining nitrogen-fixing bacteria may outweigh the benefits in terms of increased nitrogen availability. If mycorrhizal fungi provide a plant with nitrogen-containing ions, the energetic cost of maintaining nitrogen-fixing fungi may outweigh the benefits in terms of increased nitrogen availability. Plants that grow near species with N-fixing bacteria might gain nitrogen by absorbing nitrogen-containing ions after root nodules die, or by "stealing it" (roots of different species will often grow together). There could be a genetic constraint (see Chapter 25): Plant species may simply lack the alleles required to manage the relationship. Note: To be considered correct, your hypothesis should be plausible (based on the underlying biology) and testable. **(2)** analyze Many possible answers, including: Because the bacterial cell enters root cells, it is important for the plant to have reliable signals that it is not a parasitic bacterium. It is advantageous for the plant to be able to reject nitrogen-fixing bacteria if usable nitrogen is already abundant in the soil.

p. 790 Fig. 39.17 analyze Mistletoe is parasitic (it extracts water and certain nutrients from host trees); bromeliads are not (they are epiphytes and don't affect the fitness of their host plants).

IF YOU UNDERSTAND . . .

39.1 evaluate Add nitrogen to an experimental plot and compare growth within the plot to growth of a similar plot nearby. Do many replicates of the comparison, to convince yourself and others that the results are not due to unusual circumstances in one or a few plots.

39.2 understand Loam is a mixture of sand, clay, and organic matter so it can retain water and nutrients, and provide aeration, all properties that benefit plants.

39.3 apply A higher membrane voltage would allow cations in soil to cross root-hair membranes more readily through membrane channels and would allow anions to enter more readily via symporters. Nutrient absorption should increase. **39.4** evaluate Grow pea plants in the presence of rhizobia, exposed to air with (1) N_2 containing the heavy isotope of nitrogen and (2) radioactive carbon dioxide. Allow the plant to grow and then analyze the rhizobia and plant tissues. If the rhizobia–pea plant interaction is mutualistic, then ^{15}N-containing compounds and radioactive-carbon-containing compounds should be observed in both rhizobia and plant tissues. As a control, grow pea plants in the presence of the labeled compounds but without rhizobia. If the mutualism hypothesis is correct, the plant should contain labeled carbon but no heavy nitrogen. **39.5** analyze Despite the abundance of rain in tropical rain forests, many epiphytes there have limited access to water because their roots don't reach the soil. Like desert plants, these epiphytes are adapted to conserve water.

✔ Test Your Knowledge

1. remember b **2.** understand d **3.** remember d **4.** understand c **5.** remember Mycorrhizal fungi obtain most of their nitrogen from proteins in soil organic matter. Rhizobia obtain nitrogen from N_2 gas. **6.** remember Plant

✔ Test Your Understanding

7. remember b **8.** evaluate (1) Grow a large sample of corn plants with a solution containing all essential nutrients needed for normal growth and reproduction. Grow another set of genetically identical (or similar) corn plants in a solution containing all essential nutrients except iron. Compare growth, color, and seed production (and/or other attributes) in the two treatments. (2) Grow corn plants in solutions containing all essential nutrients in normal amounts, but with varying concentrations of iron. Determine the lowest iron concentration at which plants exhibit normal growth rates. **9.** understand If nutrients in soil are scarce, there is intense natural selection favoring alternative ways of obtaining nutrients—for example, by digesting insects or stealing nutrients. **10.** apply Higher in soils with both sand and clay. Clay fills spaces between sand grains and slows (1) passage of water through soil and away from roots, and (2) leaching of anions in soil water. Clay particles also provide negative charges that retain cations so they are available to plants. The presence of sand keeps soil loose enough to allow roots to penetrate. **11.** understand Some metal ions are poisonous to plants, and high levels of other ions are toxic. Passive mechanisms of ion exclusion—such as a lack of ion channels that allow passage into root cells—do not require an expenditure of ATP. Active mechanisms of exclusion—such as production of metallothioneins—require an expenditure of ATP. **12.** analyze (1) The amounts required and the mechanisms of absorption are different. Much larger amounts of C, H, and O are needed than mineral nutrients. CO_2 enters

plants via stomata, and water is absorbed passively along a water potential gradient; mineral nutrients enter via active uptake in root hairs or via mycorrhizae. (2) Macronutrients are required in relatively large quantities and usually function as components of macromolecules; micronutrients are required in relatively small quantities and usually function as enzyme cofactors.

✔ Test Your Problem-Solving Skills

13. apply The soil lost 60 g while the tree gained 74,000 g. Dividing 60 by 74,000 and converting to a percentage equals 0.08%, far less than 4%. The water that van Helmont used may have contained many of the macro- and micronutrients that were incorporated into plant mass. To test this hypothesis, conduct an experiment of the same design but have one treatment with pure water and one treatment with water containing solutes ("hard water"). Compare the percentage of soil mass incorporated into the willow tree in both treatments. Another explanation: Willow may be unusual. To test this hypothesis, repeat the experiment with willow and several other species under identical conditions, and compare the percentage of soil mass incorporated into the plants. A third hypothesis: Van Helmont's measurements were inaccurate. Repeat the experiment. **14.** analyze Acid rain inundates the soil with protons, which bind with the negatively charged clay particles—displacing the cations that are normally bound there. Cations may then be leached away. **15.** evaluate If the hypothesis is correct, vanadate should inhibit phosphorus uptake. To test this hypothesis, grow plants in solution with radioactively labeled phosphorus without vanadate and in solution with vanadate. Measure the amount of labeled phosphorus in root cells with and without vanadate. **16.** apply Alder should decompose much faster, because it provides more nitrogen to the fungi and bacteria that are responsible for decomposition—meaning that they can grow faster.

FIGURE A39.1

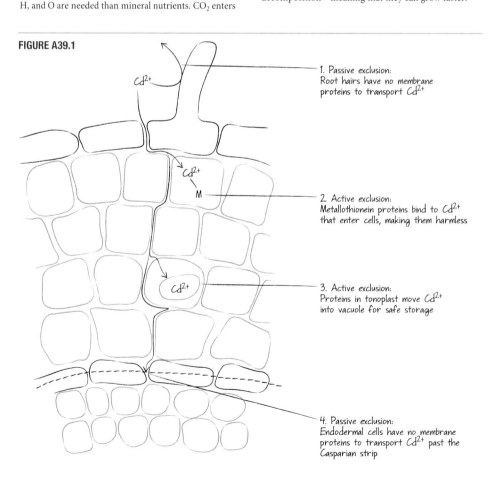

Cd^{2+}

Cd^{2+}
M

Cd^{2+}

1. Passive exclusion:
Root hairs have no membrane proteins to transport Cd^{2+}

2. Active exclusion:
Metallothionein proteins bind to Cd^{2+} that enter cells, making them harmless

3. Active exclusion:
Proteins in tonoplast move Cd^{2+} into vacuole for safe storage

4. Passive exclusion:
Endodermal cells have no membrane proteins to transport Cd^{2+} past the Casparian strip

CHAPTER 40

IN-TEXT QUESTIONS AND EXERCISES

p. 796 CYU (1) [understand] The only cells that can respond to a specific environmental signal or hormone are cells that have an appropriate receptor for that signal or hormone. **(2)** [remember] Once a receptor is activated, it triggers production of many second messengers or the activation of many proteins in a phosphorylation cascade.

p. 797 Fig. 40.4 [evaluate] Wild-type plants already have a normal *PHOT1* gene and can respond to directional blue light. Inserting a defective gene into a wild-type plant would have no effect.

p. 798 Fig. 40.5 [evaluate] For "Tip removed": Cut the tip and replace it, to control for the hypothesis that the wound itself influences bending (not the loss of the tip). For "Tip covered": Cover the tip with a transparent cover instead of opaque one, to control for the hypothesis that the cover itself affects bending (not excluding light).

p. 799 CYU (1) [apply] Using a syringe or other device, apply auxin to the west side of each stem, just below the tip. **(2)** [apply] Stems would become very long and would not respond to directional light.

p. 801 Table 40.1 [analyze] The average germination rate of lettuce seeds last exposed to red light is about 99 percent. The average for seeds last exposed to far-red light is about 50 percent. Both of these values are much higher than the germination rate of buried seeds that receive no light at all; their germination rate is only 9 percent.

p. 803 CYU (1) [understand] Lettuce grows best in full sunlight. If a seed receives red light, it indicates that the seed is in full sunlight—meaning that conditions for growth are good. But if a seed receives far-red light, it indicates that the seed is in shade—meaning that conditions for growth are poor. **(2)** [understand] Night length changes very predictably throughout the year, so it is a reliable signal for plants to perceive.

p. 803 Fig. 40.10 [apply] An appropriate control treatment would be to graft a leaf from a plant that had never been exposed to a short-night photoperiod onto a new plant that also had never been exposed to the correct conditions for flowering. If the hypothesis is correct, the grafted leaf should not induce flowering.

p. 804 Fig. 40.12 [evaluate] (One possibility) Direct pressure from a statolith changes the shape of the receptor protein. The shape change triggers a signal transduction cascade, leading to a gravitropic cell response.

p. 805 CYU (1) [analyze] In both roots and shoots, auxin is redistributed asymmetrically in response to a signal. In phototropism, auxin is redistributed to the shaded side of the shoot tip in response to blue light. In gravitropism, auxin is redistributed to the lower part of the root or shoot in response to gravity. **(2)** [understand] In root cells, auxin concentrations indicate the direction of gravity. The gravitropic response is triggered by changes in the distribution of auxin in root-tip cells.

p. 809 Fig. 40.19 [analyze] The dwarfed individuals can put more of their available energy into reproduction because they are using less energy for growth than taller individuals.

p. 810 Fig. 40.21 [analyze] Without these data, a critic could argue that the stomata closed in response to water potentials in the leaf, not a signal from the roots.

p. 813 CYU (1) [analyze] ABA: Stomata should close and photosynthesis should stop compared to individuals that do not receive extra ABA. GA: Their stems should elongate rapidly, compared to plants that do not receive extra GA. **(2)** [understand] The rotten apple produces ethylene, which, being a gas, stimulates all of the other apples in the bushel basket to ripen too fast.

p. 816 [analyze] Sacrificing a small number of cells due to the hypersensitive response is far less costly for the plant than death.

p. 819 CYU (1) [analyze] Because specific receptor proteins recognize and match specific proteins produced by pathogens, having a wide array of these proteins allows individuals to recognize and respond to a wide array of pathogens. **(2)** [understand] If herbivores are not present, synthesizing large quantities of proteinase inhibitors wastes resources (ATP and substrates) that could be used for growth and reproduction.

IF YOU UNDERSTAND . . .

40.1 [analyze] An external signal arrives at a receptor cell/protein (a person sees a barn on fire); the person calls (the receptor transduces the signal); the dispatcher rings a siren (a second messenger is produced or phosphorylation cascade occurs); the fire department members get to the pumper truck (gene expression or some other cell activity changes). **40.2** [apply] Sun-adapted (high-light) plants are likely to have a significant positive phototropic response (i.e., growth toward light), whereas shade-adapted (low-light) plants are less likely to respond phototropically. **40.3** [evaluate] They do not perform photosynthesis and therefore do not need to switch behavior based on being located in full sunlight versus shade. **40.4** [understand] Starch is denser than water, so it falls quickly to the bottom of cells. **40.5** [understand] A reduction in plant height in a windy environment might keep them upright. **40.6** [evaluate] Those hormones need to be transported throughout the plant, because cells throughout the body need to respond to the signal—not in a single direction. **40.7** [understand] In the hypersensitive response, a few cells are sacrificed to prevent a pathogen from spreading throughout the plant.

YOU SHOULD BE ABLE TO . . .

✔ Test Your Knowledge

1. [remember] d **2.** [remember] b **3.** [remember] d **4.** [understand] b
5. [remember] hypersensitive response **6.** [remember] plasma membrane

✔ Test Your Understanding

7. [remember] a **8..** [remember] The P_{fr}–P_r switch in phytochromes lets plants know whether they are in shade or sunlight, and triggers responses that allow sun-adapted plants to avoid shade. Phototropins are activated by blue light, which indicates full sunlight, and trigger responses that allow plants to photosynthesize at full capacity. **9.** [analyze] Transduce means to convert energy from one form to another. There are many examples: A touch may be converted to an electrochemical potential, light energy or pressure into a shape change in a protein, and so on. **10.** [understand] (Many possibilities) Cells or tissue involved: Auxin directs gravitropism in roots but phototropism in shoots. Developmental stage or age: GA breaks dormancy in seeds but extends stems in older individuals. Concentration: Large concentrations of auxin in stems promote cell elongation; small concentrations do not. Other hormones: The concentration of GA relative to ABA influences whether germination proceeds. **11.** [apply] Ethylene promotes senescence; increased ethylene sensitivity in leaves (combined with reduced auxin concentrations) causes abscission; fruits exposed to ethylene ripen faster than fruits that are not exposed to ethylene. Produce dealers can maximize profits by delaying ripening during transit of fruit by removing ethylene and then adding ethylene just before fruits are displayed. **12.** [apply] When one leaf on the mutant plant is chewed on by a herbivore, the damaged leaf would not be able to signal to the rest of the plant, so it would not be able to fend off future herbivore attacks.

✔ Test Your Problem-Solving Skills

13. [apply] a **14.** [evaluate] Small-seeded plants need to perform photosynthesis early in seedling development or they will starve. Therefore, they need to germinate in direct sunlight. The food reserves in large-seeded plants can support seedling growth for a relatively long time without photosynthesis. Therefore, they do not need to germinate in direct sunlight. **15.** [evaluate] In these seeds, (a) little or no ABA is present, or (b) ABA is easily leached from the seeds, so its inhibitory effects are eliminated. Test hypothesis (a) by determining whether ABA is present in the seeds. Test hypothesis (b) by comparing the amount of ABA present before and after running water—enough to mimic the amount of rain that falls in a tropical rain forest over a few days or weeks—over the seeds. **16.** [apply] New types of plant resistance provide a selective pressure for pathogens to evolve new mechanisms to evade the resistance.

CHAPTER 41

IN-TEXT QUESTIONS AND EXERCISES

p. 825 [understand] (1) The microsporocyte divides by meiosis to generate male spores; the megasporocyte divides by meiosis to generate female spores. (2) In angiosperms, the female gametophyte stays within the ovary at the base of the flower even after it matures and produces an egg cell. Fertilization and seed development take place in the same location.

p. 829 CYU (1) [analyze] Both are diploid cells that divide by meiosis to produce spores. The megasporocyte produces a megaspore (female spore); a microsporocyte produces a microspore (male spore). **(2)** [analyze] Both are multicellular individuals that produce gametes by mitosis. The female gametophyte is larger than the male gametophyte and produces an egg; the male gametophyte produces sperm.

p. 829 Fig. 41.8 [understand] A gametophyte is the multicellular individual that produces gametes by mitosis. The pollen grain is a gametophyte because it is multicellular and produces sperm by mitosis.

p. 832 CYU (1) [understand] Insects feed on nectar and/or pollen in flowers. Flowers provide food rewards to insects to encourage visitation. The individuals that attract the most pollinators produce the most offspring. **(2)** [understand] One product of double fertilization is the diploid zygote, which will eventually grow by mitosis into a mature sporophyte. The other product is the triploid endosperm nucleus, which will grow by mitosis to form a source of nutrients for the embryo.

p. 833 [remember] The endosperm nucleus in the central cell is triploid; the zygote is diploid; the synergid and other cells remaining from the female gametophyte are haploid.

p. 834 [analyze] A cotyledon and a radicle develop into organs with distinct functions, but both are composed of the same tissues: Epidermis is derived from protoderm, ground tissue is derived from ground meristem, and vascular tissue is derived from the procambium.

p. 835 Fig. 41.14 [evaluate] Hypothesis: Fruit changes color when it ripens as a signal to fruit eaters. The color change is advantageous because seeds are not mature in unripe fruit and are unlikely to survive. Fruit eaters disperse mature seeds in ripe fruit, which are more likely to survive.

p. 836 Fig. 41.16 [evaluate] Hackberries served as a control to demonstrate that animals that avoided chilies were hungry.

p. 837 Fig. 41.17 [understand] Beans—their cotyledons are aboveground.

IF YOU UNDERSTAND . . .

41.1 [remember] The sporophyte is diploid, the gametophyte is haploid, spores and gametes are haploid, and zygotes are diploid. **41.2** [apply] There would be four embryo sacs in each ovule, and thus four eggs. **41.3** [apply] Endosperm tissue is generated when two haploid, female nuclei fuse with one haploid sperm nucleus, forming a cell with three sets of chromosomes. In contrast, a zygote is formed by the fusion of one haploid sperm nucleus and one haploid egg, forming a cell with two sets of chromosomes. **41.4** `analyze` A cold requirement for germination ensures that seeds from plants native to habitats that experience cold winters will not germinate in the fall and fail to become established before winter.

YOU SHOULD BE ABLE TO . . .

✔ Test Your Knowledge

1. `remember` b **2.** `understand` c **3.** `remember` a **4.** `understand` d
5. `remember` Megasporocytes (female) and microsporocytes (male) undergo meiosis to generate megaspores and microspores, respectively. These divide mitotically to give rise to female and male gametophytes—the embryo sac and pollen grain, respectively. **6.** `remember` Pollination is the transfer of pollen from an anther to a stigma. Fertilization is the fusion of haploid gametes.

✔ Test Your Understanding

7. `understand` c **8.** `analyze` Asexual reproduction is a quick, efficient way for a plant to produce large numbers of offspring. The disadvantage is that offspring are genetically identical to the parent and thus are vulnerable to the same diseases and able to thrive only in habitats similar to those inhabited by the parent. **9.** `analyze` Sepals are the outermost structures in a flower and usually protect the structures within; petals are located just inside the sepals and usually function to advertise the flower to pollinators. In wind- and bee-pollinated flowers, sepals probably do not vary much in overall function or general structure. Petals are often lacking in wind-pollinated species; they tend to be broad and flat (to provide a landing site) and colored purple, blue, or yellow in bee-pollinated species (often with ultraviolet markings). **10.** `analyze` Outcrossing increases genetic diversity among offspring, making them more likely to thrive if environmental conditions change from the parental generation. However, outcrossing requires cross-pollination to be successful. Self-fertilization results in relatively low genetic diversity among offspring but ensures that pollination succeeds. **11.** `remember` The female gametophyte develops inside the ovule; the ovary develops into the pericarp of the fruit after fertilization. A mature ovule contains both gametophyte and sporophyte tissue; the ovary and carpel are all sporophyte tissue. **12.** `analyze` The endosperm of a corn seed and the cotyledons of a bean seed both contain nutrients needed for the seed to germinate. The bean cotyledons have absorbed the nutrients of the endosperm.

✔ Test Your Problem-Solving Skills

13. `remember` c **14.** `apply` Subtracting the number of used pollen grains per plant (600) from the number of pollen grains generated per plant (10 million) divided by 10 million reveals that 99.994% of the pollen is wasted.
15. `evaluate` In response to "cheater" plants, insects should be under intense selection pressure to detect and avoid species that do not offer a food reward. Individuals would have to be able to distinguish scents, colors, or other traits that identify cheaters. In response to cheater insects, plants possessing unusually thick or toxic petals or sepals that prevented insects from bypassing the anthers on their way to the nectaries would be more successful at producing seed in the next generation.
16. `evaluate` Acorns have a large, edible mass and are usually animal dispersed (e.g., by squirrels that store them and forget some). Cherries have an edible fruit and are animal dispersed. Burrs stick to animals and are dispersed as the animals move around. Dandelion seeds float in wind. To estimate the distance that each type of seed is dispersed from the parent, (1) set up "seed traps" to capture seeds at various distances from the parent plant; (2) sample locations at various distances from the parent and analyze young individuals—using techniques introduced in Chapter 21 to determine if they are offspring from the parent being studied; (3) mark seeds, if possible, and find them again after dispersal.

ᴮᴵᴳ ᴘᴵᴄᴛᴜʀᴇ Plant and Animal Form and Function

p. 840 CYU (1) `understand` Plants and animals may accumulate compounds in the epidermis and skin (such as waxes and other lipids) that reduce water loss. Both plants and animals evolved closable openings in this barrier to allow regulated gas exchange. Plants and animals may also change the amount of the body exposed to the sun (in plants by moving leaves, in animals by changing posture or taking shelter) to reduce evaporative water loss.
(2) `understand` Many plants and animals keep their eggs inside their bodies and depend on sperm being deposited near them rather than releasing both to the environment where they could dry out. **(3)** `understand` Very large animals and plants need to build strong supporting structures such as thick bones or trunks to withstand the force of gravity. They also need elaborate internal transport systems to bring nutrients to each cell and carry wastes away.
(4) `analyze` Cellular respiration fits into one of the key functions for survival—Nutrition. Both plants and animals depend on sugar (produced by plants via photosynthesis) as their energy source to make ATP via cellular respiration.

CHAPTER 42

IN-TEXT QUESTIONS AND EXERCISES

p. 844 Fig. 42.2 `create` There would be no difference between experimental and control males.
p. 847 Fig. 42.4 `analyze` Based on the projections called axons that go to specific cells and tissues, neurons transmit signals to specific tissues rather than broadly throughout the body.
p. 850 CYU `analyze` *Connective tissues* consist of cells embedded in a matrix. The density of the matrix determines the rigidity of the connective tissue and its function in padding/protection, structural support, or transport. *Nervous tissue* is composed of neurons and support cells. Neurons have long projections that function as "biological wires" for transmitting electrical signals. *Muscle tissue* has cells that can contract and functions in movement. *Epithelial tissue* has tightly packed layers of cells with distinct apical and basal surfaces. They protect underlying tissues and regulate the entry and exit of materials into these tissues.
p. 850 `apply` A newborn.
p. 851 Fig. 42.10 `apply` The puppy has to eat relatively more because it has a higher mass-specific metabolic rate than an adult dog.
p. 852 Fig. 42.11 `apply` There should be no oxygen uptake on either side of the chamber.
p. 853 CYU (1) `understand` As three-dimensional size increases, volume increases more quickly than does surface area, so the surface area/volume ratio decreases.
(2) `analyze` The squirrel has a higher mass-specific metabolic rate because it is a smaller animal. It therefore needs more energy per unit body mass than a bear.
p. 858 CYU (1) `analyze` Endotherms can remain active during the winter and at night and sustain high levels of aerobic activities such as running or flying, but they require large amounts of food energy. Ectotherms need much less food and can devote a larger proportion of their food intake to reproduction, but they have a hard time maintaining high activity levels at night or in cold weather. **(2)** `apply` The whale's tongue would become colder because not as much heat would be transferred from warm arterial blood to cool venous blood with concurrent flow as with countercurrent flow, and as a result more heat from the arterial blood would be lost to the environment.

IF YOU UNDERSTAND . . .

42.1 `understand` Some adaptations to cold environments in mammals include the evolution of thick coats of hair,

countercurrent heat exchangers in the extremities, and dark coloration to absorb radiation. Acclimatization may include further thickening and darkening of hair upon exposure to cold. **42.2** `apply` Loosening tight junctions would compromise the barrier between the external and internal environments. It would allow water and other substances (including toxins) to pass more readily across the epithelial boundary. **42.3** `create` King Kong is endothermic, and his huge mass would generate a great deal of heat. His relatively small surface area, especially as compared to a normal-sized gorilla, would not be able to dissipate the heat—especially if it were covered with fur. **42.4** `create` Individuals could acclimatize by normal homeostatic mechanisms (increased sweating, seeking shade, etc.). Individuals with alleles that allowed them to function better at higher temperatures would produce more offspring than individuals without those alleles. Over time, this would lead to adaptation via natural selection.
42.5 `create` *Apatosaurus* was likely a homeotherm because of its huge size. Because of its low surface area/volume ratio, metabolically produced heat would be trapped inside the dinosaur's body. Even if exposed to cold or warm temperatures, the dinosaur's temperature would not change much because it takes a long time for it to lose or gain heat across such a relatively small surface area.

YOU SHOULD BE ABLE TO . . .

✔ Test Your Knowledge

1. `understand` True **2.** `remember` Epithelial **3.** `understand` b
4. `understand` d **5.** `understand` a **6.** `analyze` a

✔ Test Your Understanding

7. `understand` It is where environmental changes are sensed first. As an effector, it has a large surface area for losing or gaining heat. **8.** `analyze` A warm frog can move, breathe, and digest much faster than a cold frog, because enzymes work faster (rates of chemical reactions increase) at 35°C than at 5°C. A warm frog needs much more food, though, to support this high metabolic rate. **9.** `understand` *Absorptive sections* have many folds and projections that increase their surface area for absorption. *Capillaries* have a high surface area because they are thin and highly branched, making exchange of fluids and gases efficient. *Beaks of Galápagos finches* have sizes and shapes that correlate with the type of food each species eats. Large beaks are used to crack large seeds; long, thin beaks are used to pick insects off surfaces; etc. *Fish gills* are thin, flattened structures with a large surface area, which facilitates the efficient exchange of gases and wastes. **10.** `analyze` A larger sphere has a relatively smaller surface area for its interior volume than does a smaller sphere, because surface area increases with size at a lower rate than volume does. Diffusion will be more efficient in the smaller sphere.
11. `apply` See **FIGURE A42.1**. The graph shows that as size (length) increases, the volume increases more rapidly than the surface area. **12.** `understand` c. This is an example of negative feedback because insulin is released in response to an increase in a parameter (blood sugar concentration)

FIGURE A42.1

and acts to reduce that parameter so that homeostasis is maintained. In the other examples, homeostasis is not maintained.

✔ Test Your Problem-Solving Skills

13. `create` Large individuals require more food, so are more susceptible to starvation if food sources can't be defended (e.g., when seeds are scattered around). They may also be slower and/or more visible to predators. **14.** `create` You would have to show changes in the frequencies of alleles whose products are affected by temperature—for example, increases in the frequencies of alleles for enzymes that operate best at higher temperature. **15.** `create` The heat-dissipating system should have a very high surface area/volume ratio. Based on biological structures, it should be flattened (thin) and highly folded and branched, and/or contain tubelike projections. **16.** `understand` a. Turtles are ectothermic, so they do not expend much energy to produce body heat. Mice are endothermic, and expend energy to regulate body temperature. Mice will therefore have a higher basal metabolic rate and consume more food.

CHAPTER 43

IN-TEXT QUESTIONS AND EXERCISES

p. 863 Fig. 43.1 `understand` The black and white molecules are moving down their concentration gradients, and the concentration of red molecules does not affect the concentration of black or white molecules.
p. 867 Fig. 43.6 `understand` The sodium–potassium ATPase performs primary active transport; the sodium–chloride–potassium cotransporter performs secondary active transport; the chloride and potassium channels are involved in passive transport.
p. 869 CYU `apply` The osmolarity of the fish's tissue would decrease because the pumps would no longer be able to import sodium.
p. 869 Fig. 43.8 `analyze` The underside of the abdomen is shaded by the grasshopper's body. This location is relatively cool and should reduce the loss of water during respiration.
p. 871 CYU (1) `understand` Uric acid is insoluble in water and can be excreted without much water loss. **(2)** `understand` When electrolytes are reabsorbed in the hindgut epithelia, water follows along an osmotic gradient.
p. 873 `understand` Blood contains cells and large molecules as well as ions, nutrients, and wastes; the filtrate contains only ions, nutrients, and wastes.
p. 876 `apply` (1) Less water reabsorption will occur because the osmotic gradient is not as steep. (2) Filtrate osmolarity will be lower because less water has been reabsorbed. (3) Salt reabsorption will be reduced because the concentration of NaCl in the filtrate is lower.
p. 878 `apply` Ethanol consumption inhibits water reabsorption, leading to a larger volume of less concentrated urine. Nicotine consumption increases water reabsorption, leading to a smaller volume of more concentrated urine.
p. 878 CYU `apply` Water intake increases blood pressure and filtration rate in the renal corpuscle. It leads to lowered electrolyte concentration in the blood and filtrate and the production of large volumes of dilute urine. Eating large amounts of salt results in concentrated, hyperosmotic urine. Water deprivation triggers ADH release and the production of concentrated, hyperosmotic urine.
p. 879 Fig. 43.19 `create` Carrying a large volume of water in the bladder may be energetically costly to the animal. The animal also might not be able to run as quickly and may have an increased predation risk.

IF YOU UNDERSTAND . . .

43.1 `understand` A sodium gradient establishes an osmotic gradient that moves water. It also sets up an electrochemical gradient that can move ions against their electrochemical gradient by secondary active transport through a cotransporter, or with their electrochemical gradient by passive diffusion through a channel.
43.2 `analyze` Cartilaginous fishes such as sharks maintain a high concentration of urea in their blood, which raises their osmolarity to about the same as seawater. Therefore, they lose very little water to the environment. Bony fishes drink seawater and maintain their osmolarity lower than that of seawater by actively transporting excess ions out of the gills. Sharks actively excrete sodium chloride through a rectal gland. **43.3** `understand` When comparing chloride cells in gills of salmon in the ocean versus freshwater, the cells are in different locations on the gills, different forms of Na^+/K^+-ATPase exist, and $Na^+K^+Cl^-$ transporters are located on the basolateral side of the membrane in ocean salmon and on the apical side of the membrane in freshwater salmon. **43.4** `apply` Aquatic insects produce more urea, and especially more ammonia, than terrestrial insects because these nitrogenous wastes are energetically less expensive to produce than uric acid. Ammonia and urea require more water to be excreted, but aquatic insects have access to plenty of water. **43.5** `create` The longer the loop of Henle, the steeper the osmotic gradient in the interstitial fluid in the medulla. Steep osmotic gradients allow a great deal of water to be reabsorbed as urine passes through the collecting duct, resulting in highly concentrated urine.

YOU SHOULD BE ABLE TO . . .

✔ Test Your Knowledge

1. `remember` True **2.** `understand` c **3.** `apply` a **4.** `remember` d **5.** `understand` c **6.** `remember` bladder

✔ Test Your Understanding

7. `create` b. Fishes excrete nitrogenous wastes in the form of ammonia. Although the ammonia can easily diffuse across the fish's gills into the water, freshwater ponds are often small and ammonia can build up to potentially toxic levels. **8.** `analyze` Ammonia is toxic and must be diluted with large amounts of water to be excreted safely. Urea and uric acid are safer and do not have to be excreted with large amounts of water but are more expensive to produce in terms of energy expenditure. Fish excrete ammonia; mammals excrete urea; insects excrete uric acid. You would expect the embryos inside terrestrial eggs to excrete uric acid, since it is the least toxic and is insoluble in water. **9.** `understand` Mitochondria are needed to produce ATP. A key component of salt regulation in fish is the Na^+/K^+-ATPase, which requires ATP to function. Because ATP fuels establishment of the ion gradients necessary for the cotransport mechanisms used by chloride cells as well as other epithelial cells, an abundance of mitochondria would be expected in a chloride cell. **10.** `create` The wax layer should be thicker in insects inhabiting dry habitats, because the animals there are under more osmotic stress (more evaporation due to low humidity; less available drinking water). Insects that live in extremely humid habitats may lack the ability to close their spiracles; animals that live in humid habitats may excrete primarily ammonia or urea and have relatively short loops of Henle. In each case, you could test these predictions by comparing individuals of closely related species that live in dry versus humid habitats. **11.** `create` Without a loop of Henle, there would be no concentration gradient in the medullary interstitial fluid, so water would not be absorbed from the pre-urine in the collecting duct; concentrated urine could not be formed. **12.** `create` Freshwater has such a low osmolarity that if invertebrates osmoconformed to it, they would lack sufficient quantities of ions to conduct electrical currents in their excitable tissues.

✔ Test Your Problem-Solving Skills

13. `apply` c. Without aquaporins in the collecting duct, water cannot be reabsorbed, which would result in increased urine volume and decreased urine osmolarity. **14.** `create` The mammal's urine volume would increase dramatically. Ouabain would poison Na^+/K^+-ATPase pumps in the nephron, which would prevent reabsorption of ions and nutrients. Without the osmotic gradient established by reabsorption of ions and nutrients, the water would remain in the nephron and exit as urine. **15.** `analyze` The observation that the Na^+ concentration decreased by 30 percent, even as volume also decreased, indicates that Na^+ ions were reabsorbed to a greater degree than water. The observation that the urea concentration increased by 50 percent indicates that urea was not reabsorbed. **16.** `analyze` The hypothesis that the mussels are osmoconformers is supported because their hemolymph is very close in osmolarity to that of the wide range of water osmolarities into which they are placed.

CHAPTER 44

IN-TEXT QUESTIONS AND EXERCISES

p. 891 `understand` Nutrient absorption occurs in the small intestine, but not in the esophagus or stomach, and the rate of absorption increases with available surface area.
p. 894 Fig. 44.14 `understand` Frog eggs don't normally make the sodium-glucose cotransporter protein, so the researchers could be confident that if the protein appeared, it was from the injected RNA. They would not be confident of this if the RNAs were injected into rabbit epithelial cells, where the protein was probably already present.
p. 896 `apply` (1) Decreased energy yield from foods due to reduced glucose absorption, (2) increased fecal glucose content due to the passing of unabsorbed glucose into the colon, (3) watery feces due to decreased water reabsorption in the large intestine (lower osmotic gradient). (An additional effect would be increased flatulence due to the metabolism of unabsorbed glucose by bacteria that produce gases as waste products.)
p. 896 CYU (1) `understand` *Mouth:* Food is taken in; teeth physically break down food into smaller particles; salivary amylase begins to break down carbohydrates; lingual lipase initiates the digestion of fats. *Esophagus:* Food is moved to the stomach via peristaltic contractions. *Stomach:* HCl denatures proteins; pepsin begins to digest them. *Small intestine:* Pancreatic enzymes complete the digestion of carbohydrates, proteins, lipids, and nucleic acids. Most of the water and all of the nutrients are absorbed here. *Large intestine:* Water is reabsorbed and feces are formed. *Anus:* Feces accumulate in the rectum and are expelled out the anus. **(2)** `apply` If the release of bile salts is inhibited, fats would not be digested and absorbed efficiently, and they would pass into the large intestine. The individual would likely produce fatty feces and lose weight over time. Trypsin inactivation would inhibit protein digestion and reduce absorption of amino acids, likely leading to weight loss and muscle atrophy due to protein deprivation. If the Na^+-glucose cotransporter is blocked, then glucose would not be absorbed into the bloodstream. The individual would likely become sluggish from lack of energy.
p. 897 Fig. 44.16 `understand` In type 2 diabetes mellitus, the top, black arrow from glucose (in the blood) to glycogen (inside liver and muscle cells) is disrupted. In type 1 diabetes mellitus, the green arrow on the top left, indicating insulin produced by the pancreas, is disrupted.
p. 898 `understand` Individuals with type 1 diabetes mellitus have normal insulin receptors on their liver cells but do not produce and release insulin from the pancreas. Individuals with type 2 diabetes mellitus produce and release insulin, but they have fewer insulin receptors and/or these receptors do not respond as well to insulin. In both types of diseased individuals, glucose concentrations in the blood remain high. In individuals without disease, both insulin production and insulin receptors

are normal, so liver cells respond to insulin by taking up glucose from the blood and storing it.

p. 898 Fig. 44.17 analyze The data do not prove that obesity causes type 2 diabetes mellitus. The study provides correlational data and cannot directly assign causation. However, the study provides strong evidence of an association between obesity and type 2 diabetes.

p. 899 CYU (1) understand Type 1 diabetes mellitus is an autoimmune disease, meaning that the body's immune system mistakenly kills the insulin-producing cells of the pancreas. Type 2 diabetes mellitus is caused by a combination of factors, including genetic predisposition, obesity, and poor diet. **(2)** apply When the blood sugar of individuals with type 1 diabetes mellitus gets too high, they should inject themselves with insulin, which will trigger the absorption and storage of glucose by their cells. When their blood glucose gets too low, they should eat something with a high concentration of sugar (such as orange juice or a candy bar) to increase their blood glucose levels quickly. (If glucose is dangerously low, a glucagon shot may be administered.)

IF YOU UNDERSTAND . . .

44.1 analyze One serving of skim milk has approximately 80 kcal, while whole milk has about 152 kcal. **44.2** apply Its mouth or tongue would be modified to make probing of flowers and sucking nectar efficient. It would not require teeth, and its digestive tract would be relatively simple. For example, the stomach would not have to pulverize food or digest large amounts of protein, and the large intestine would not have to process and store large amounts of bulky waste. **44.3** analyze Amylase will work best at a neutral or slightly alkaline pH (typical of mouth and small intestine), whereas pepsin will work best at a low pH (typical of the stomach). **44.4** understand Type 1 diabetes mellitus leads to high glucose because insulin is not produced, and therefore glucose cannot be moved from blood into cells. Type 2 diabetes mellitus leads to high blood glucose because insulin does not bind in sufficient amounts to its receptor, so glucose cannot be transported from blood into cells.

YOU SHOULD BE ABLE TO . . .

✔ Test Your Knowledge

1. remember a **2.** understand d **3.** remember d **4.** remember b **5.** remember insulin, glucagon **6.** remember False

✔ Test Your Understanding

7. understand The bird crop is an enlarged sac that can hold quickly ingested food; in leaf-eating species it is filled with symbiotic organisms and functions as a fermentation vessel. The cow rumen is an enlarged portion of the stomach; the elephant cecum is enlarged relative to other species. Both structures are filled with symbiotic organisms and function as fermentation vessels. **8.** analyze Digestive enzymes break down macromolecules (proteins, carbohydrates, nucleic acids, lipids). If they weren't produced in an inactive form, they would destroy the cells that produce and secrete them. **9.** understand b **10.** understand Pima living in the United States are genetically similar to Pima living in Mexico, but they have higher rates of type 2 diabetes. This suggests that the difference in diabetes rates is due to environmental factors such as diet and exercise. **11.** understand Humans have sharp canines and incisors, which may help to bite and tear meat. Humans also have flat molars, which help in grinding grains and other plant material. **12.** apply In a type 1 diabetic who took too much insulin, the blood glucose would become very low. The person would become sluggish and disoriented, and might slip into a diabetic coma.

✔ Test Your Problem-Solving Skills

13. create A proton pump inhibitor reduces activity of the proton pumps in parietal cells, reducing the amount of HCl produced. The contents of the stomach would therefore become less acidic, helping to prevent acid reflux. **14.** analyze b **15.** evaluate The result is still valid because the injection caused secretion—if it is also true that lack of injection led to no secretion. The criticism is also somewhat valid, however, because the researchers couldn't rule out the hypothesis that signaling from nerves plays some sort of role, too. **16.** create Terrestrial animals are exposed to increased risk of water loss, and the large intestine is where water reabsorption occurs. In most cases, fish do not need to reabsorb large amounts of water from their feces.

CHAPTER 45

IN-TEXT QUESTIONS AND EXERCISES

p. 905 CYU (1) analyze Oxygen partial pressure is high in mountain streams because the water is cold, mixes constantly, and has a high surface area (due to white water). Oxygen partial pressure is low at the ocean bottom because that is far from the surface, where gas exchange takes place, and there is relatively little mixing. **(2)** create *Warm-water species:* Large amount of air, because the oxygen-carrying capacity of warm water is low. *Vigorous algal growth:* Small amount of air, because algae contribute oxygen to the water through photosynthesis. *Sedentary animals:* Small amount of air, because sedentary animals require relatively little oxygen.

p. 906 Fig. 45.4 analyze External gills are ventilated passively and are efficient because they are in direct contact with water. They are exposed to predators and mechanical damage, however. Internal gills are protected but have to be ventilated by some type of active mechanism for water flow.

p. 907 Fig. 45.6 apply If "concurrent flow" occurred, oxygen transfer would stop, because the partial pressure gradient driving diffusion would fall to zero partway along the length of the capillary.

p. 909 Fig. 45.7 analyze Using many animals increases confidence that the results are true for most or all members of the population and are not due to one or a few unusual individuals or circumstances.

p. 912 CYU (1) analyze Common features include large surface area, short diffusion distance (a thin gas exchange membrane), and a mechanism that keeps fresh air or water moving over the gas exchange surface. Only fish gills use a countercurrent exchange mechanism; only tracheae deliver oxygen directly to cells without using a circulatory system; only lungs (mammalian) contain "dead space"—areas that are not involved in gas exchange. **(2)** apply The P_{O_2} would go down as oxygen is used up, the P_{CO_2} would go up as CO_2 diffuses into the blood but cannot be exhaled, and pH would drop as the CO_2 dissolves in blood to form bicarbonate and H^+ ions.

p. 913 create There would be an even larger change in the oxygen saturation of hemoglobin in response to an even smaller change in the partial pressure of oxygen.

p. 914 Fig. 45.16 analyze According to the data in the figure, when the tissue P_{O_2} is 30 mm Hg, the oxygen saturation of hemoglobin is about 15 percent for blood at pH 7.2 and about 25 percent at pH 7.4. Therefore, about 85 percent of the oxygen is released from hemoglobin at pH 7.2, but only about 75 percent of the oxygen is released at pH 7.4.

p. 915 Fig. 45.18 analyze In the lungs, a strong partial pressure gradient favors diffusion of dissolved CO_2 from blood into the alveoli. As the partial pressure of CO_2 in the blood declines, hydrogen ions leave hemoglobin and react with bicarbonate to form more CO_2, which then diffuses into the alveoli and is exhaled from the lungs.

p. 916 CYU analyze The curve for Tibetans should be shifted to the left relative to people adapted to sea level—meaning that their hemoglobin has a higher affinity for oxygen at all partial pressures.

p. 921 Fig. 45.24 analyze Air from the alveoli mixes with air in the "dead space" in the bronchi and trachea on its way out of the body. This dead space air is from the previous inhalation (P_{O_2} = 160 mm Hg; P_{CO_2} = 0.3), so when the alveolar air mixes with the dead space air, the partial pressures in the exhaled air achieve levels intermediate between that of inhaled and alveolar air.

p. 922 apply Without the delay at the AV node, the ventricles would not have the chance to fully fill with blood from the atria. As a consequence, the volume of blood ejected from the ventricles would decline.

p. 924 CYU (1) understand Blood plasma that leaks out of the capillaries and is not reabsorbed into capillaries enters lymphatic vessels that eventually merge with blood vessels. **(2)** understand See **FIGURE A45.1**.

IF YOU UNDERSTAND . . .

45.1 understand Cellular respiration. **45.2** create It is harder to extract oxygen from water than it is to extract it from air. This limits the metabolic rate of water breathers. **45.3** analyze Large animals have a relatively small body surface area relative to their volume. If they had to rely solely on gas exchange across their skin, the skin surface area would not be large enough to exchange the volume of oxygen needed to meet their metabolic needs. **45.4** create Since O_2 cannot compete as well as CO for binding sites in hemoglobin, O_2 transport decreases. As oxygen levels in the blood drop, tissues (most crucially in the brain) become deprived of oxygen, and suffocation occurs. **45.5** apply Their circulatory systems should have relatively high pressures—achieved by independent systemic and pulmonary circulations powered by a four-chambered heart—to maximize the delivery of oxygenated blood to metabolically active tissues.

YOU SHOULD BE ABLE TO . . .

✔ Test Your Knowledge

1. understand b **2.** remember c **3.** understand d **4.** understand d **5.** remember c **6.** understand An open circulatory system is less efficient than a closed circulatory system because it is not capable of directing the hemolymph toward specific organs, and because the pressure is lower, resulting in slower flow of hemolymph.

✔ Test Your Understanding

7. understand During exercise, P_{O_2} decreases in the tissues and P_{CO_2} increases. The increase in P_{CO_2} lowers tissue pH. The drop in P_{O_2} and pH causes more oxygen to be released from hemoglobin. **8.** understand In blood, CO_2 is converted to carbonic acid, which dissociates into a proton and a bicarbonate ion. Because the protons bind to deoxygenated hemoglobin, they do not cause a dramatic drop in blood pH (hemoglobin acts as a buffer). Also, small decreases in blood pH trigger a homeostatic response that increases breathing rate and expulsion of CO_2. **9.** understand Because airflow through bird lungs is unidirectional, the gas exchange surfaces are

---

FIGURE A45.1

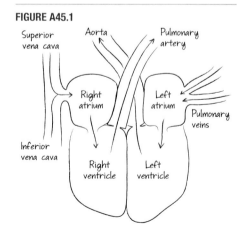

continuously ventilated with fresh, oxygenated air. The trachea and bronchi in a mammalian lung do not have a gas exchange surface, and bidirectional airflow in these species means that "stale" air has to be expelled before "fresh" air can be inhaled. As a result, the alveoli are not ventilated continuously. **10.** *understand* Lungs increase the temperature of the air and are moist to allow greater solubility of gases (increasing k); alveoli present a large surface area (large A); the epithelium of alveoli is thin (small D); and constant delivery of deoxygenated blood to alveoli maintains a steep partial pressure gradient, favoring diffusion of oxygen into the body ($P_2 - P_1$ is high). **11.** *analyze* Yes—the trait compensates for the small P_{O_2} gradient between stagnant water and the blood of the carp. **12.** *analyze* c

✔ Test Your Problem-Solving Skills

13. *create* Cold water carries more oxygen than warm water does, so icefish blood can carry enough oxygen to supply the tissues with oxygen even in the absence of hemoglobin. The oxygen and carbon dioxide are simply dissolved in the blood. **14.** *analyze* The insect tracheal system delivers O_2 directly to respiring cells; in humans, O_2 is first taken up by the blood of the circulatory system, which then delivers it to the cells. The human respiratory system also has specialized muscles devoted to ventilating the lungs. The open circulatory system of the insect is a low-pressure system, which makes use of pumps (hearts) and body movements to circulate hemolymph. The closed circulatory system of humans is a high-pressure system, which can respond to rapid changes in O_2 demand by tissues. **15.** *understand* If the pulmonary circulation were under pressure as high as that found in the systemic circulation, large amounts of fluid would be forced out of capillaries in the lungs. There is a conflict between the thin surface required for efficient gas exchange and the thickness needed in blood vessels to withstand high pressure. **16.** *analyze* b

CHAPTER 46

IN-TEXT QUESTIONS AND EXERCISES

p. 931 Fig. 46.3 *apply* No—as K^+ leaves the cell along its concentration gradient, the interior of the cell becomes more negative. As a result, an electrical gradient favoring movement of K^+ into the cell begins to counteract the concentration gradient favoring movement of K^+ out of the cell. Eventually, the two opposing forces balance out, and there is no net movement of K^+.
p. 932a *apply* $E_{Ca^{2+}} = 58$ mV $\times \log 1$ mM$/0.0001$ mM$)/$ $(+2) = 116$ mV.
p. 932b *apply* The membrane potential would be near zero because the Na^+ and K^+ would diffuse across the membrane along their concentration gradients.
p. 934 CYU (1) *apply* The resting potential would fall (be much less negative) because K^+ could no longer leak out of the cell. **(2)** *understand* The size of an action potential from a particular neuron does not vary, so it cannot contain information. Only the frequency of action potentials from the same neuron varies.
p. 934 *understand* See **FIGURE A46.1**.
p. 935 (1) *understand* Na^+ channels open when the cell membrane depolarizes, and as more channels open, more Na^+ flows into the cell, further depolarizing the cell, causing more voltage-gated Na^+ channels to open. **(2)** *apply* Sodium would flow inward as potassium flowed outward, and their charges would cancel each other out. The membrane would not depolarize.
p. 938 CYU (1) *understand* Once the threshold level of depolarization is attained, the probability that the voltage-gated sodium channels will open approaches 100 percent. But below threshold, the massive opening of Na^+ channels does not occur. This is why the action potential is all or none. **(2)** *apply* The membrane would depolarize and stay depolarized.

p. 938 Fig. 46.10 *understand* Get a solution taken from the synapse between the heart muscle and the vagus nerve *without* the nerve being stimulated. Expose a second heart to this solution. There should be no change in heart rate.
p. 940 *analyze* The inside of the cell becomes more positive (depolarized). This shifts the membrane potential closer to the threshold, making it more likely that the postsynaptic cell will fire an action potential.
p. 942 CYU *apply* The EPSP would not be as strong if an enzyme that broke down excitatory neurotransmitter increased in concentration.
p. 943 Fig. 46.16 *analyze* In the rest-and-digest mode, pupils take in less light stimulation, the heartbeat slows to conserve energy, liver conserves glucose, and digestion is promoted by release of gallbladder products. In the fight-or-flight mode, the pupils open to take in more light, the heartbeat increases to support muscle activity, and the liver releases glucose into the blood to fuel the fight or the flight (for example, from a predator).
p. 944 Fig. 46.18 *understand* Point to your forehead and top of your head for the frontal lobe, the top and top rear of your head for the parietal lobe, the back of your head for the occipital lobe, and the sides of your head (just above your ear openings) for the temporal lobes.
p. 946 Fig. 46.21 *analyze* No—for example, part (b) indicates that the size of the brain area devoted to the trunk is no bigger than the size of the brain area devoted to the thumb.

p. 948 CYU (1) *understand* One method is to study individuals with known brain lesions and correlate the location of the defect with a deficit in mental or physical function. Another is to directly stimulate brain areas in conscious patients during brain surgery and record the response. **(2)** *remember* Learning and memory involve changes in the number, sensitivity, and placement of synapses and neurons.

IF YOU UNDERSTAND . . .

46.1 *understand* See **FIGURE A46.2**. **46.2** *understand* Because the behavior of voltage-gated Na^+ and K^+ channels does not change. Every time a membrane depolarizes past the threshold and all of the voltage-gated Na^+ channels are open, the same amount of Na^+ will enter the cell, and the cell will depolarize to the same extent. In addition, K^+ channels open at the same time and allow the same amount of K^+ to leave the cell, so that it repolarizes to the same extent. **46.3** *understand* Summation means that EPSPs and IPSPs produced at various sites combine to determine the response or lack of response in a postsynaptic cell. Thus the behavior of a postsynaptic cell depends on the integration of input from many synapses. **46.4** The animal would likely die because its basic bodily functions (circulation, breathing, digestion, etc.) would be disrupted.

FIGURE A46.1

FIGURE A46.2

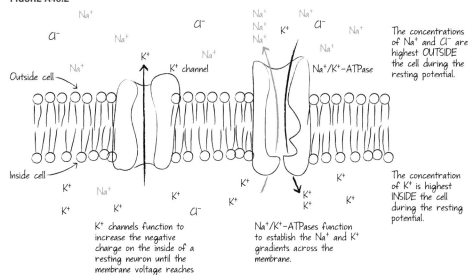

K⁺ channels function to increase the negative charge on the inside of a resting neuron until the membrane voltage reaches the equilibrium potential.

Na⁺/K⁺–ATPases function to establish the Na⁺ and K⁺ gradients across the membrane.

The concentrations of Na⁺ and Cl⁻ are highest OUTSIDE the cell during the resting potential.

The concentration of K⁺ is highest INSIDE the cell during the resting potential.

✔ **Test Your Knowledge**

1. `remember` b 2. `remember` d 3. `understand` c 4. `remember` a 5. `remember`
b 6. `understand` See **FIGURE A46.3**.

✔ **Test Your Understanding**

7. `understand` The Na^+/K^+-ATPase pumps 3 Na^+ out of
the cell for every 2 K^+ it brings in. Since more positive
charges leave the cell than enter it, there is a difference
in charge on the two sides of the membrane and thus
a voltage. 8. `analyze` The ligand-gated channel opens or
closes in response to binding by a small molecule (for
example, a neurotransmitter); the voltage-gated channel
opens or closes in response to changes in membrane
potential. 9. `understand` EPSPs and IPSPs are produced by
flow of ions through channels in the membrane, which
change the membrane potential in the postsynaptic
cell. If a flow of ions at one point depolarizes the cell
but a nearby flow of ions hyperpolarizes the cell, then
in combination the flows of ions cancel each other out.
But if two adjacent flows of ions depolarize the cell, then
their effects are additive and sum to produce a total that
is more likely to reach threshold. 10. `analyze` The somatic
system responds to external stimuli and controls voluntary skeletal muscle activity, such as movement of
arms and legs. The autonomic system controls internal
involuntary activities, such as digestion, heart rate,
and gland activities. 11. `apply` A drug that activates the
sympathetic nervous system would increase heart rate.
12. `understand` a

✔ **Test Your Problem-Solving Skills**

13. `apply` The neurotransmitters will stay in the synaptic
cleft longer, so the amount of binding to ligand-gated
channels will increase. Ion flows into the postsynaptic
cell will increase dramatically, affecting its membrane
potential and likelihood of firing action potentials.
14. `analyze` The benefit of these approaches is that they
allow the specific role of the brain region to be assessed
in a living person. However, diseased or damaged
brains may not respond like healthy and undamaged
brains. Also, the extent of lesions and the exact location
of electrical stimulation may be difficult to determine,
making correlations between the regions affected and
the response imprecise. 15. `analyze` c (Because the current is reduced from normal levels, but not eliminated
completely, there must be more than one type of potassium channel present. The venom probably knocks out
just one specific type of channel.) 16. `create` A reasonable
hypothesis is that the new neurons arise from stem cells
in the brain.

CHAPTER 47

IN-TEXT QUESTIONS AND EXERCISES

p. 959 CYU (1) `apply` A punctured eardrum wouldn't
vibrate correctly and would result in hearing loss at all
frequencies in the affected ear. If the stereocilia are too
short to come into contact with the tectorial membrane,
vibration of the basilar membrane will not cause them
to bend, and sound will not be detected. A loss in basilar
membrane flexibility would result in the inability to
hear lower-pitched sounds, such as human speech.
(2) `understand` The lateral line system senses pressure waves
in the water, allowing fishes to detect the presence of
such objects as prey, predators, or potential mates.
p. 959 Fig. 47.9 `create` Vision could potentially be used to
detect prey in diurnal catfish. To control for this, catfish
could be blinded or "blindfolded" with an opaque material over their eyes to permit study of the lateral line
system.
p. 961 Fig. 47.12 `create` The change in retinal's shape
would induce a change in opsin's shape. A protein's

function correlates with its shape, so opsin's function is
likely to change as well.
p. 962 `analyze` cGMP is a second messenger in the sense
that it signals the sodium channel that transducing is inactive, and it is also a ligand that opens sodium channels.
Transducin is like a G protein because it switches from
"off" to "on" in response to a receptor protein, activating
a key protein as a result. Closing Na^+ channels stops the
entry of positive charges into the cell, making the inside
of the cell more negative relative to the outside.
p. 962 Fig. 47.13 `understand` This ion flow occurs when the
photoreceptor cell is not receiving light—when the cell
is in the dark.
p. 963 Fig. 47.15 `create` The S, M, and L opsin proteins are
each different in structure. These structural differences
affect the ability of retinal to absorb specific frequencies
of light and change shape.
p. 964 CYU (1) `understand` Retinal acts like an on–off switch
that indicates whether light has fallen on a rod cell.
When retinal absorbs light, it changes shape. The shape
change triggers events that result in a change in action
potentials, signaling that light has been absorbed.
(2) `apply` Retina detachment from a jarring car accident
would separate the retina from the optic nerve, resulting
in blindness. The mutation would produce blue–purple
color blindness, because the S opsin responds to wavelengths in that region of the spectrum. A clouded lens
would reduce the amount of light that reaches the retina,
reducing visual sensitivity.
p. 967 CYU (1) `understand` Extremely hot food damages taste
receptors, so those proteins would no longer be able to
respond to their chemical triggers. **(2)** `create` One hypothesis to explain why the vomeronasal organ is vestigial
in most primates is that primates have evolved other
complex methods of communication, including gestures,
facial expressions, and language, that render sensation
with the vomeronasal organ less important.
p. 967 `apply` (1) A rattlesnake with its eyes covered will
strike effectively because its pits will sense the prey.
(2) A rattlesnake with its eyes and pits covered with cotton cloth will strike effectively. The cloth will block vision
but the heat from the animal will penetrate the cloth to
stimulate the thermoreceptors in the pits. (3) A rattlesnake with its eyes and pits covered with an opaque heat-
blocking material will not be able to strike effectively,
because the material blocks vision and heat energy.

p. 969 CYU (1) `create` Electrogenic fishes may have a survival advantage over other fishes in murky water, where
vision would be compromised. **(2)** `create` Female sea turtles likely use magnetic cues to locate their natal beaches.

IF YOU UNDERSTAND . . .

47.1 `create` The sensory neurons that used to serve that
limb are cut, but may still fire action potentials that
stimulate brain areas associated with awareness of pain
in the missing limb. **47.2** `analyze` Each of these senses involves ion channels that open in response to application
of pressure to cell membranes. **47.3.** `understand` Because
different animal species have different sets of opsin
molecules that respond to different wavelengths of light
absorbed by retinal they see different colors. **47.4** `create`
Smells are part of the sensation of flavor, and people
with nasal congestion cannot smell. In addition to smell
being important, the small number of taste receptors can
send many different frequencies of action potentials to
the brain, and the taste receptors can be stimulated in
many different combinations. **47.5** `understand` Capsaicin
binds to thermoreceptors that are also activated at high
temperatures, leading to the perception of heat. It can
also bind to some nociceptors.

YOU SHOULD BE ABLE TO . . .

✔ **Test Your Knowledge**

1. `understand` a 2. `remember` d 3. `understand` c 4. `remember` b 5. `remember`
false 6. `remember` magnetoreception

✔ **Test Your Understanding**

7. `understand` b 8. `understand` The many possible examples
include several blue opsins allowing coelacanths to distinguish the blue wavelengths present in their deep-sea
habitat, infrasound hearing allowing elephants to hear
over long distances, ultrasonic hearing allowing bats to
hunt by echolocation and moths to avoid bat predators,
red and yellow opsins allowing fruit-eating mammals
to distinguish ripe from unripe fruit. 9. `analyze` The data
provide evidence that a pheromone acted to synchronize
the women's menstrual cycles because the average time
between onset of a woman's cycle and that of the rest
of the women declined over time following move-in to
the dorms in fall. It declined most rapidly after they first
moved in together. The leveling of the graph over time

FIGURE A46.3

DEPOLARIZATION:
Na^+ channels open, allowing
influx of Na^+ into cell.

REPOLARIZATION:
K^+ channels are open, allowing K^+ to exit
the cell; Na^+ channels are inactivated.

HYPERPOLARIZATION:
K^+ channels have remained open
and are closing slowly, temporarily
hyperpolarizing the cell.

Resting potential

Resting potential:
K^+, Na^+, and Cl^- are at
equilibrium with respect to
their electrochemical
gradients.

Resting potential
is reestablished.

Membrane potential (mV)

40

0

-65

Time (msec)

0   2   4

shows that the onset dates became more synchronized. **10.** `understand` Ion channels in hair cells are thought to open in response to physical distortion caused by the bending of stereocilia. Ion channels in taste receptors, in contrast, change shape and open when certain bitter-tasting molecules bind to them. **11.** `evaluate` Dalton's hypothesis was reasonable because blue fluid absorbs red light and would prevent it from reaching the retina, and this could be tested by dissecting his eyes. Although the hypothesis was rejected, it inspired a rigorous test and required researchers to think of alternative explanations. **12.** `create` To determine whether electric eels are electro-genic, place an eel in a tank and measure the electricity of water to determine if the eel is emitting an electric field. To determine whether they are electroreceptive, place an eel in a tank in the dark and add an object the size and shape of a prey item that emits no scent or taste. Block the lateral line system with a chemical as in Figure 47.9. If the eel locates the object, it must do so us-ing electroreception.

✔ Test Your Problem-Solving Skills

13. `understand` a; vomeronasal organs evolved in tetrapods, presumably as an adaptation for communication in ter-restrial habitats. **14.** `analyze` Both the lateral line system and electroreceptors use hair cells to sense changes in the water. Hair cells in the lateral line system contain mechanoreceptors that are depolarized in response to changes in water pressure. In contrast, electroreceptors cause hair cells to depolarize in response to changes in the electric field. **15.** `evaluate` It is unlikely that a perfume contains human pheromones because they have never actually been characterized, although their presence is suspected. **16.** `create` Block all but a few hundred om-matidia in the center of each compound eye of many dragonflies with an opaque material, and observe any differences that might occur in their ability to detect and pursue prey compared to individuals whose ommatidia had been covered by a transparent material.

CHAPTER 48

IN-TEXT QUESTIONS AND EXERCISES

p. 974 `understand` The trucks are the Z discs, the ropes are the thin filaments, and the burly weightlifters are the thick filaments.

p. 974 Fig. 48.2 `analyze` The dark band includes thin fila-ments and a dense concentration of bulbous structures extending from the thick filament; the light band con-sists of thin filaments only.

p. 977 CYU (1) `understand` In a sarcomere, thick myosin filaments are sandwiched between thin actin filaments. When the heads on myosin contact actin and change conformation, they pull the actin filaments toward one another, shortening the whole sarcomere. **(2)** `apply` Increased acetylcholine release would result in an in-creased rate of muscle-cell contraction. Preventing con-formational changes in troponin would prevent muscle contraction. Blocking the uptake of calcium ions into the sarcoplasmic reticulum would lead to sustained muscle contraction.

p. 980 `apply` The breast meat of pigeons is composed of dark meat because these muscles are specialized for en-durance. Their dark color is due to a high concentration of myoglobin.

p. 981 `create` When earthworms shorten their longitu-dinal muscles, the segment containing these muscles shortens and squeezes the internal fluid, which becomes pressurized and pushes in all directions, expanding the circumference of the segment and pushing laterally against the burrow.

p. 983 CYU (1) `analyze` Exoskeletons are made of chitin, proteins, and other substances like calcium carbonate. Endoskeletons are made of calcium phosphate, calcium carbonate, and proteins. Exoskeletons occur on the outsides of animals, whereas endoskeletons occur on the insides. Both types of skeletons are composed of rigid levers separated by joints such that motions are a result of changes in joint angles. Both types of skeletons attach to skeletal muscle and serve to transmit muscle forces. Hydrostatic skeletons also transmit muscle forces but are made of soft tissues that vary from animal to animal. Shape changes in hydrostatic skeletons occur not from changes in joint angles, but from shape changes of the bodies themselves. **(2)** `apply` No movement of the arm would occur, because for these antagonistic muscles to produce movement, one of them must be contracted while the other is relaxed.

p. 986 Fig. 48.16 `apply` Galloping at 3.5 m/s would cost about 75 percent more energy.

p. 987 Fig. 48.18 `apply` About 10 times more costly.

IF YOU UNDERSTAND . . .

48.1 `apply` Paralysis, because acetylcholine has to bind to its receptors on the membrane of postsynaptic muscle fibers for action potentials to propagate in the muscle and cause contraction. **48.2** `analyze` Skeletal muscle is multinucleate, unbranched, striated, and voluntary. Cardiac muscle is branched, striated, and involuntary. Smooth muscle is smooth and involuntary. **48.3** `analyze` When muscles contract in hydrostatic skeletons, the body deforms by becoming longer and thinner, or shorter and narrower, or by bending side to side. In vertebrates and arthropods, muscle contractions cause changes in the joint angles between rigid segments rather than shape changes in the segments themselves. **48.4** `understand` On land, an animal is constrained by its weight; larger animals cannot move as freely and are in greater danger of breaking than are smaller animals. In water, animals are constrained by their shapes. Some shapes would cause high drag, which would inhibit locomotion. In air, animals are constrained by their weight. Larger animals must create more lift to over-come gravity.

YOU SHOULD BE ABLE TO . . .

✔ Test Your Knowledge

1. `remember` a **2.** `remember` c **3.** `understand` d **4.** `remember` True **5.** `understand` a **6.** `understand` False.

✔ Test Your Understanding

7. `create` The key observation was that the banding pat-tern of sarcomeres changed during contraction. Even though the entire unit became shorter, only some por-tions moved relative to each other. This observation suggested that some portions of the structure slid past other portions. Muscle fibers have to have many mi-tochondria because large amounts of ATP are needed to power myosin heads to move along actin filaments; large amounts of calcium stored in smooth ER are needed to initiate contraction by binding to troponin. **8.** `apply` c **9.** `analyze` Acetylcholine reduces the rate and force of contraction of cardiac muscle, whereas it stimulates skeletal muscle to contract. **10.** `create` ATP is no longer produced after death, so calcium cannot be actively transported from the cytosol into the sarco-plasmic reticulum to enable the binding of myosin and actin. The proteins do not unbind, because this process requires ATP. **11.** `apply` The oxygen consumption of the runner would increase because the arches of his or her feet would no longer store as much elastic energy, which normally reduces the energetic cost of running. **12.** `create` Because diverse animals experience the same physical constraints where they live, natural selection favors the same kind of adaptations to those con-straints, resulting in convergent evolution. For example,

dolphins and ichthyosaurs are distantly related but have a similar body shape due to their similar ability to swim rapidly through water.

✔ Test Your Problem-Solving Skills

13. `apply` In cardiac muscle, the binding of acetylcholine to its receptors causes the heart rate to slow. Ingestion of nightshade would increase heart rate because it blocks acetylcholine receptors. **14.** `apply` If the sarcomere is in a stretched state before stimulation, the initial force pro-duction would be reduced because less overlap would occur between actin and myosin; thus, fewer myosin heads could engage in the pull. **15.** `apply` c:

$$0.5 = v^2/(9.8 \text{ m/s}^2 \times 0.9 \text{ m})$$
$$0.5 = v^2/8.8 \text{ m}^2/\text{s}^2$$
$$v^2 = 0.5 \times 8.8 \text{ m}^2/\text{s}^2$$
$$v = \text{square root of } 4.4 \text{ m}^2/\text{s}^2$$
$$v = 2.1 \text{ m/s}$$

16. `evaluate` The bones of *T. rex* are available and can be analyzed to determine their structural properties, such as the mechanical advantage of the leg joints and the ability of the bones to withstand the different forces that the skel-eton would experience at different speeds based on data from living animals. A very close estimate could be made, but the exact speed would be hard to determine without observing the dinosaur in action, since behavior affects speed.

CHAPTER 49

IN-TEXT QUESTIONS AND EXERCISES

p. 994 `apply` The current temperature is analogous to the sensory input; the thermostat, to the CNS; the thermo-stat signal, to the cell-to-cell signal; and the furnace, to the effector. If feedback inhibition fails, hormone pro-duction will continually increase.

p. 996 Fig. 49.4 `understand` Injecting the dog with a liquid extract from a different part of the body, or with a solu-tion that is similar in chemical composition to the ex-tract from the pancreas (e.g., in pH or ions present) but lacking molecules produced by the pancreas.

p. 997 CYU (1) `understand` It is difficult to differentiate be-tween the nervous and endocrine systems because some neurons secrete hormones and some endocrine glands respond to neural signals. **(2)** `apply` Thyroid hormone receptors are intracellular because thyroid hormones are lipid soluble.

p. 998 `create` The cells could differ in receptors, signal transduction systems, or response systems (genes or proteins that can be activated).

p. 1001 `understand` Stressed individuals would be less able to heal or to fight pathogens.

p. 1001 Fig. 49.9 `understand` The saline injection controlled for any stress induced by the injection procedure and for introducing additional fluids.

p. 1002 Fig. 49.11 `apply` Same—a defect in both signal and receptor has the same effect—no response—as a defec-tive signal alone or a defective receptor alone.

p. 1003 CYU (1) `analyze` In amphibians, an increase in thyroid hormones stimulates metamorphosis from a tadpole to an adult without a resting stage. In insects, a decrease in juvenile hormone and increase in ecdysone stimulate metamorphosis during a resting (pupal) stage. **(2)** `understand` By reducing immune system function, pro-moting release of fatty acids from storage cells and use of amino acids from muscles for energy production, and preventing release of glucose in response to signals from insulin, cortisol conserves glucose supplies for use by the brain.

p. 1004 Fig. 49.12 `create` Inject ACTH and monitor cortisol levels. If cortisol does not increase, adrenal failure is likely.

p. 1006 CYU (1) [understand] ACTH triggers the release of cortisol, but cortisol inhibits ACTH release by blocking the release of CRH from the hypothalamus and suppressing ACTH production. High levels of cortisol tend to lower cortisol levels in the future. **(2)** [understand] Processing centers in the brain are responsible for synthesizing a wide array of sensory input. To start a response to this sensory input, they stimulate neurosecretory cells in the hypothalamus. Neurohormones from these cells travel to the anterior pituitary, where they trigger the production and release of other hormones.

p. 1010 CYU (1) [apply] The steroid-hormone–receptor complex would probably fail to bind to the hormone-response element. Then gene expression would not change in response to the hormone—the arrival of the hormone would have little or no effect on the target cell. **(2)** [analyze] The mode of action would be similar to that of a peptide—hormone binding would have to trigger a signal transduction event that resulted in production of a second messenger or a phosphorylation cascade.

IF YOU UNDERSTAND . . .

49.1 [understand] The production and release of hormones are often controlled, directly or indirectly, by electrical or chemical signals from the nervous system. A good example is the hypothalamic–pituitary axis, where neuroendocrine signals regulate endocrine signals. **49.2** [analyze] Electrical signals are short lived and localized. All of the processes listed in the question are long-term changes in the organism that require responses by tissues and organs throughout the body. **49.3** [apply] The hypothalamus would continue to release CRH, stimulating the pituitary to release ACTH, which in turn would stimulate continued release of cortisol from the adrenal cortex. Sustained, high circulating levels of cortisol have negative effects. **49.4** [apply] A cell can respond to more than one hormone at a time, because cells produce different receptors for each hormone.

YOU SHOULD BE ABLE TO . . .

✔ Test Your Knowledge

1. [understand] c **2.** [understand] a **3.** [remember] a **4.** [remember] c **5.** [remember] Mammals—eye and pineal gland; lizards—pineal gland; birds—diffuse in brain **6.** [remember] True

✔ Test Your Understanding

7. [understand] b **8.** [create] This is one reason that the same hormone can trigger different effects in different tissues. For example, epinephrine binds to four different types of receptors in different tissues—eliciting a different response from each. **9.** [analyze] The posterior pituitary is an extension of the hypothalamus; the anterior pituitary is independent—it communicates with the hypothalamus via chemical signals in blood vessels. The posterior pituitary is a storage area for hypothalamic hormones; the anterior pituitary synthesizes and releases an array of hormones in response to releasing hormones from the hypothalamus. **10.** [analyze] Steroid hormones act directly, and alter gene expression. They usually bind to receptors inside the cell, forming a complex that binds to DNA and activates transcription. Nonsteroid hormones act indirectly, and activate proteins. They bind to receptors on the cell surface and trigger production of a second messenger or a phosphorylation cascade, ending in activation of proteins already present in the cell. **11.** [create] Subject volunteers to travel and jet lag, measure ACTH and cortisol levels, and correlate these results with their perceived level of jet lag symptoms. **12.** [create] Studies have shown that treatment with leptin is largely ineffective in reducing food intake in obese humans.

✔ Test Your Problem-Solving Skills

13. [apply] c (dehydration would result, because ADH would not be present) **14.** [create] There are two basic strategies: (1) remove the structure from some individuals and compare their behavior and condition to untreated individuals in the same environment, or (2) make a liquid extract from the structure, inject it into some individuals, and compare their behavior and condition to individuals in the same environment that were injected with water or a saline solution. **15.** [evaluate] The researcher's hypothesis initially sounds unreasonable, because peptide hormones cannot enter cells to initiate gene transcription. However, a second messenger activated by the hormone could possibly lead to changes in gene transcription. **16.** [evaluate] The graph shows that atrazine treatment reduces the testosterone concentrations in male frogs to near the levels of females. This result supports the hypothesis that atrazine is an endocrine disruptor that feminizes male amphibians.

CHAPTER 50

IN-TEXT QUESTIONS AND EXERCISES

p. 1015 Fig. 50.3 [create] Isolate and identify the molecules found in "crowded" water. Test each molecule by adding it, at the same concentration found in crowded water, to clean water occupied by a single *Daphnia* and recording whether the female produces a male-containing brood. Repeat with many test females, and for each molecule identified. As a control, record the number of male-containing broods produced in clean water.

p. 1016 [apply] Asexual reproduction would be expected in environments where conditions change little over time.

p. 1019 Fig. 50.7 [evaluate] No—the data are consistent with the displacement hypothesis, but there is no direct evidence for it. There are other plausible explanations for the data.

p. 1021 CYU (1) [analyze] Oviparity usually requires less energy input from the mother after egg laying, and mothers do not have to carry eggs around as long—meaning that they can lay more eggs and be more mobile. However, before egg laying, mothers have to produce all the nutrition the embryo will require in the egg, and eggs may not be well protected after laying. Viviparity usually increases the likelihood that the developing offspring will survive until birth, but it limits the number of young that can be produced to the space available in the mother's reproductive tract. If viviparous young can be nourished longer than oviparous young, then they may be larger and more capable of fending for themselves. **(2)** [create] Divide a population of sperm into two groups. Subject one group of sperm to spermkillerene at concentrations observed in the female reproductive tract (the experimental group) but not the other group (the control). Document the number of sperm that are still alive over time in both groups.

p. 1025 Fig. 50.13 [create] (Note: There is more than one possible answer—this is an example.) Having two gonads is an "insurance policy" against loss or damage. To test this idea, surgically remove one gonad from a large number of male and female rats. Do a similar operation on a large number of similar male and female rats, but do not remove either gonad. Once the animals have recovered, place them in a barn or other "natural" setting and let them breed. Compare reproductive success of individuals with paired versus unpaired gonads.

p. 1026 Fig. 50.15 [analyze] It is similar. In both cases, the hypothalamus produces a releasing factor (GnRH or CRH) that acts on the pituitary. The releasing factor stimulates release of regulatory hormones from the anterior pituitary (LH and FSH or ACTH). These hormones travel via the bloodstream and act on the gonads or adrenals to induce the release of hormones from these glands. In both cases, the hormones are involved in negative feedback control of the regulatory hormones from the pituitary.

p. 1029 [apply] (1) There should be an LH spike early in the cycle, followed by early ovulation. (This might not actually occur, though, if the follicle is not yet ready to rupture.) (2) LH levels should remain low—no mid-cycle spike and no ovulation.

p. 1030 CYU (1) [understand] FSH triggers maturation of an ovarian follicle. Its level rises at the end of a cycle because it is no longer inhibited by progesterone, which stops being produced at high levels when the corpus luteum degenerates. **(2)** [apply] The drug would keep FSH levels low, meaning that follicles would not mature and would not begin producing estradiol and progesterone. The uterine lining would not thicken.

p. 1032 [analyze] (1) The uterine lining may not be maintained adequately—if it degenerates, a miscarriage is likely. (2) Pregnancy tests detect hCG in urine. By the third trimester, hCG is no longer being produced by the placenta.

p. 1034 Fig. 50.23 [apply] The birth will be more difficult, as limbs or other body parts can get caught.

p. 1034 Fig. 50.24 [apply] A mother's chance of dying in childbirth in 1760 was a little over 1000 in 100,000 live births, or close to 1.0 percent. If she gave birth 10 times, she would have a 10 percent chance of dying in childbirth.

IF YOU UNDERSTAND . . .

50.1 [create] Parasites could wipe out a population of asexually produced animals because the animals do not vary genetically. However, a population of sexually produced animals would have enough genetic variability that some individuals could present better resistance to the parasites. **50.2** [create] Parental care demands resources (time, nutrients) that cannot be used to produce more eggs. Sperm competition is not likely when external fertilization occurs, and most fishes use external fertilization. **50.3** [create] The most common methods are surgical ligation (tying off) of the fallopian tubes in a woman to stop the delivery of the egg to the uterus and cutting the vas deferens (vasectomy) in a man to stop the delivery of sperm into the semen. **50.4** [apply] FSH and LH levels would increase following ovariectomy, because sex steroids would no longer be present to exert negative feedback upon the anterior pituitary. **50.5** [understand] The placenta is highly vascularized to provide a large surface area for transfer of oxygen and nutrients from the mother's circulation to the fetus's, and of wastes from the fetus's circulation to the mother's.

YOU SHOULD BE ABLE TO . . .

✔ Test Your Knowledge

1. [remember] a **2.** [understand] b **3.** [remember] a **4.** [remember] b **5.** [remember] Luteinizing hormone (LH) and follicle-stimulating hormone (FSH) **6.** [remember] True

✔ Test Your Understanding

7. [remember] Every offspring that is produced sexually is genetically unique; offspring that are produced asexually have DNA from their mother only and often are genetically identical to her. **8.** [create] *Daphnia* females produced males only when they were exposed to short day lengths *and* water from crowded populations *and* low food levels. These conditions are likely to occur in the fall. Sexual reproduction could be adaptive in fall if genetically variable offspring are better able to thrive in conditions that occur the following spring. **9.** [analyze] Spermatogenesis generates four haploid sperm cells from each primary spermatocyte; oogenesis produces only one haploid egg cell from each primary oocyte. Egg cells are much larger than sperm cells because they contain more cytoplasm. In males, the second meiotic division occurs right after the first meiotic division, but in females it is delayed until fertilization. **10.** [understand] LH triggers release of estradiol, but at low levels estradiol inhibits further release of LH. LH and FSH trigger release of progesterone, but progesterone inhibits further release of LH and FSH. High levels of estradiol trigger release of more LH. The follicle can produce high levels of estradiol only if it has grown and matured—meaning that it is ready for ovulation to occur. **11.** analyze c **12.** apply Females should be choosier about their second mate because the sperm of the second male has an advantage in sperm competition.

✔ Test Your Problem-Solving Skills

13. create If all sheep were genetically identical, it is less likely that any individuals could survive a major adverse event (e.g., a disease outbreak or other environmental challenge) than if the population were composed of genetically diverse individuals, which are likely to vary in their ability to cope with the new conditions. **14.** apply Oviparous populations should produce larger eggs than viviparous populations. Because oviparous species deposit eggs in the environment, the eggs must contain all the nutrients and water required for the entire period of embryonic development. But because embryos in viviparous species receive nourishment directly from the mother, it is likely that their eggs are smaller. **15.** create One hypothesis might be that the shell membrane is a vestigial trait (see Chapter 25)—specifically, an "evolutionary holdover" from an ancestor of today's marsupials that laid eggs. **16.** analyze b

CHAPTER 51

IN-TEXT QUESTIONS AND EXERCISES

p. 1040 apply See **FIGURE A51.1**. Each type of Toll protein is a pattern-recognition receptor that responds to a different group of pathogens. Since bacteria and fungi are very different pathogens, you would not expect a receptor that recognizes fungi to be required for a response against bacteria, so flies lacking Toll would survive as well as the wild-type control.

p. 1041 CYU analyze (1) Applying direct pressure constricts blood vessels, mimicking the effect of histamine released from mast cells. (2) Cleaning the wound to remove dirt and debris mimics the phagocytic activity of macrophages and neutrophils to eliminate foreign cells and material. (3) Applying bandages impregnated with platelet-recruiting compounds mimics the effect of chemokines released from injured tissues and macrophages, which attract circulating leukocytes and platelets to facilitate blood clotting and wound repair.

p. 1041 Fig. 51.3 understand Rubor (reddening) is due to increased blood flow to the infected or injured area. Mast cells trigger dilation of vessels by releasing signaling molecules. Calor (heat) is the result of fever, which is activated by the release of cytokines from macrophages that are in the area of infection. Tumor (swelling) occurs due to dilation of blood vessels triggered by mast cells and macrophages.

p. 1046 apply (1) In a human, 200 different light chains are possible (see text). The recombination of the V, D, and J segments results in $51 \times 27 \times 6 = 8262$ possible heavy chains. Since these chains combine to produce BCRs, a total of $200 \times 8262 = 1,652,400$ are possible within a human. (2) Only one type of antibody can be produced from a single human B cell.

p. 1047 CYU (1) understand The production of BCRs and TCRs via DNA recombination is required for an effective response. Selection against self-reactive lymphocytes is required for a safe response. **(2)** understand The cell is able to make a larger number of differing variable regions per unit of DNA when the variable region is segmented. For example, 3 segments of V and 3 segments of J can randomly assemble into 9 different variable light chains, whereas only 6 light chains would be possible with the same amount of DNA if the V regions were preassembled.

p. 1048 understand "Clonal" refers to the cloning—producing many exact copies—of cells that are "selected" by the binding of their receptor to a specific antigen (or MHC–peptide complex, which is introduced later).

p. 1050 understand See **TABLE A51.1**.

p. 1051 CYU create Individuals who are heterozygous for MHC genes have greater variability in their MHC proteins than do individuals who are homozygous for these genes. The increased variability in MHC proteins allows a greater variety of peptides to be presented to T cells, which thus would be able to recognize, attack, and eliminate a wider range of pathogens compared with T cells in homozygotes. As a result, heterozygous individuals would likely suffer fewer infections than homozygous individuals.

p. 1054 Fig. 51.17 apply See **FIGURE A51.2**. The response to the new antigen would resemble the primary response that occurred after the first injection of the original antigen.

p. 1055 CYU analyze Attenuated viruses are able to infect the cells of a vaccinated host, but not well enough to cause disease. Inactivated viruses are not able to infect cells at all, and instead remain as extracellular particles. Since attenuated viruses can infect cells, their antigens will be processed and presented by class I MHC complexes, which will promote a stronger cell-mediated response than the inactivated viruses.

IF YOU UNDERSTAND . . .

51.1 understand Toll-like receptors recognize molecular patterns of broad groups of pathogens, such as viruses, fungi, and bacteria. Upon binding to these molecules, the receptors cause signals to be transduced that activate cellular responses directed at eliminating these types of pathogens. **51.2** analyze The BCR locus would be smaller in mature B cells compared to immature cells, since recombination between the gene segments resulted in removing sections of DNA. **51.3** understand Antigen-presenting cells activate the CD4$^+$ T cells that are required for the full activation of B cells to differentiate into plasma cells. **51.4** understand Extracellular viruses are agglutinated and inactivated by antibodies produced in the humoral response. Intracellular viruses are eliminated by the destruction of their host via cytotoxic T cells in the cell-mediated response. **51.5** apply The IgE antibodies generate hypersensitive responses. When these are severe, they can lead to anaphylactic shock. If IgE antibodies were generated against self molecules, you would expect a strong response that will lead to either anaphylactic shock or chronic inflammation.

YOU SHOULD BE ABLE TO . . .

✔ Test Your Knowledge

1. remember c **2.** understand a **3.** remember b **4.** understand They are identical except that a B-cell receptor has a transmembrane domain that allows it to be located in the B-cell

---

TABLE A51.1

MHC type	Origin of Peptide	Type of T cell that binds	Activity stimulated
Class I MHC	Host cell cytosol	CD8+	Activates T cell to secrete molecules that kill infected cell
Class II MHC	Extracellular environment	CD4+	Activates T cell to secrete cytokines that support the immune response of other cells

FIGURE A51.2

FIGURE A51.1

plasma membrane. **5.** `remember` c **6.** `understand` The hygiene hypothesis claims that immune disorders such as allergies and autoimmunity have increased in frequency due to the implementation of hygienic practices that decrease common bacterial, viral, and parasitic infections.

✔ Test Your Understanding

7. `understand` B cells are first activated to divide via interactions between the BCRs and their complementary antigens. They are then fully activated to differentiate and continue to divide when their class II MCH–peptide complex interacts with an activated CD4⁺ T cell that has a complementary TCR. **8.** `analyze` Both the BCR and TCR interact with epitopes of antigens via binding sites located in the variable regions of their polypeptide chains. A TCR is like one "arm" of a BCR. The BCR can interact with an antigen directly, while the TCR interacts with its epitope in the context of an MHC protein presented by another cell. **9.** `analyze` Vaccines have to contain an antigen that can stimulate an appropriate primary immune response. Vaccines have not worked for HIV, because the antigens on this virus are constantly changed through mutation, rendering it unrecognizable to memory cells generated following vaccination. **10.** `analyze` b **11.** `analyze` Pattern-recognition receptors on leukocytes bind to surface molecules that are present on many pathogens. BCRs and TCRs bind to particular epitopes of antigens that are pathogen specific. Pattern-recognition-receptor binding directly activates the cell, while BCRs and TCRs often require additional signals to become fully activated. **12.** `understand` The mixing and matching of different combinations of gene segments from the variable and joining regions of the light-chain immunoglobulin gene, along with diversity regions of the heavy-chain gene, gives lymphocytes a unique sequence in both chains of the BCR. The variable regions in the alpha and beta chains of the TCR are assembled from segments in a similar manner.

✔ Test Your Problem-Solving Skills

13. `apply` d **14.** `create` If B cells have receptors that bind to self molecules during maturation, they would become activated during the maturation process. Instead of the signal causing the cells to divide, these immature B cells might respond to the signal by becoming permanently inactivated or might be induced to undergo apoptosis. **15.** `analyze` Natural selection favors individuals that can create a large array of antibodies, because the high mutation rates and rapid evolution observed in pathogens means that they will constantly present the immune system with new antigens. If these antigens were not recognized by the immune system, the pathogens would multiply freely and kill the individual. **16.** `analyze` Tissue from the patient's own body is marked with the major histocompatibility (MHC) proteins that the patient's immune system recognizes as self. This results in the preservation, rather than the destruction, of the transplanted tissue. Tissue from a different person is marked with MHC proteins that are unique to that individual. The body of the transplant recipient will recognize the MHC proteins (and other molecules) of the donor as foreign, resulting in a full immune response and rejection of the grafted tissue.

CHAPTER 52

IN-TEXT QUESTIONS AND EXERCISES

p. 1062 Fig. 52.1 `apply` It would increase, because cattle would no longer succumb to the disease carried by tsetse flies. (In Africa, cattle are limited more by biotic conditions than abiotic conditions.)
p. 1064 Fig. 52.4 `apply` Since there were 28 plots and four treatments, there were probably about seven replicate plots per treatment. This is important because the abiotic and biotic characteristics of individual plots are likely to vary in natural landscapes.
p. 1066 `apply` See **FIGURE A52.1**.
p. 1067 CYU (1) `apply` There is little rain at the North Pole because dense, dry air descends and absorbs moisture. **(2)** `apply` The highest risk of skin cancer is in the tropics where the Sun is often directly overhead, resulting in a large amount of solar radiation per unit area.
p. 1069 Fig. 52.10 `analyze` Most of the remaining wildlands occur in Canada, at the top of the image.
p. 1070a `understand` Vines and epiphytes increase productivity because they are photosynthetic organisms that fill space between small trees and large trees—they capture light and use nutrients that might not be used in a forest that lacked vines and epiphytes.
p. 1070b `understand` Most leaves have a large surface area to capture light. Light is abundant in deserts, however, and plants with a large surface area would be susceptible to high water loss and/or overheating.
p. 1071a `understand` Crop grains like wheat and corn are grasses, so they thrive in the grassland biome.
p. 1071b `apply` There is a continuous grass cover and scattered trees. (A biome like this is called a savanna.)
p. 1072a `apply` Boreal forests should move north.
p. 1072b `understand` High elevations present an abiotic environment (precipitation and temperature) that is similar to the conditions in arctic tundra. As a result, the plant species present will have similar adaptations to cope with these physical conditions. These similarities are the result of convergent evolution.
p. 1073 Fig. 52.17 `analyze` Visible wavelengths enter the chamber from the top and through the glass, warming the plants and ground inside the chamber. Some of this heat energy is retained within the mini-greenhouse.
p. 1074 CYU (1) `analyze` Tropical dry forests are probably less productive than tropical wet forests because they have less water available to support photosynthesis during some periods of the year. **(2)** `apply` Your answer will depend on where you live. For example, if global warming continues at the current projected rates, several effects can be expected. If you live in a coastal area, you can expect water levels to rise. In general, plant communities will change because average temperature and moisture—and variation in temperature and moisture—will change. Also, development is likely to increase in most areas, transforming the landscape.
p. 1077a `apply` The littoral zone, because light is abundant and nutrients are available from the substrate.
p. 1077b `apply` Bogs are nitrogen poor, so plants that are able to capture and digest insects have a large advantage. This advantage does not exist in marshes and swamps, where nutrients are more readily available.
p. 1078a `apply` Cold water contains more oxygen than warm water, so it can support a higher rate of cellular respiration.
p. 1078b `apply` Species that live in estuaries must be able to tolerate variable salinity; freshwater marsh species do not. The abiotic environment (salt concentration) is so different that few species grow well in both habitats.
p. 1079 `apply` No—the aphotic zone is lightless, so natural selection favors individuals that do not invest energy in developing and maintaining eyes.

IF YOU UNDERSTAND . . .

52.1 `understand` Populations are made up of individual organisms; communities are made up of populations of different species; ecosystems are made up of groups of communities along with the abiotic environment; the biome is made up of all the world's ecosystems. **52.2** `apply` The Argentine ants would likely increase their range into the moist agricultural fields, displacing native ants. **52.3** `apply` All latitudes would get equal amounts of sunlight all year round. There would be no seasons and no changes in climate with latitude. **52.4** `apply` If precipitation does not increase, then increased transpiration rates will put plants under water stress and reduce productivity. **52.5** `apply` Like deserts, open oceans have plenty of sunlight but very low productivity. Unlike deserts, which are water limited, open oceans are nutrient limited.

YOU SHOULD BE ABLE TO . . .

✔ Test Your Knowledge

1. `remember` Organismal, population, community, ecosystem, and biosphere/global **2.** `remember` c **3.** `remember` b **4.** `remember` d **5.** `remember` c **6.** `remember` a

✔ Test Your Understanding

7. `create` Examples of possible questions: *Organismal:* How do humans cope with extremely hot (or cold) weather conditions? *Population:* How large will the total human population be in 50 years? *Community:* How are humans affecting ocean species such as cod and tuna? *Ecosystem:* How does nitrogen runoff from agricultural fields affect the availability of oxygen in the Gulf of Mexico? *Global:* How will human-induced changes in global temperature affect sea level? **8.** `analyze` *Biotic:* These ants compete with native ants. *Abiotic:* Argentine ants require moist conditions, not too cold, not too hot. These factors interact; Argentine ants outcompete native ants in moist environments, but competition slows down their dispersal. **9.** `apply` b **10.** `analyze` Productivity in intertidal and neritic zones is high because sunlight is readily available and because nutrients are available from estuaries and deep-ocean currents. Productivity in the oceanic zone is extremely low—even though light is available at the surface—because nutrients are scarce. The deepest part of the oceanic zone may have nutrients available from the substrate but lacks light to support photosynthesis. **11.** `analyze` As elevation increases, biomes often change in a similar way as increasing in latitude (e.g., you might go from temperate forest to a boreal-type forest to tundra). **12.** `analyze` The natural biome of Eastern North America is primarily temperate forest. However, most of this area is now occupied by human-influenced biomes such as cities, suburbs, and farmland.

FIGURE A52.1

✔ Test Your Problem-Solving Skills

13. `apply` a **14.** `create` A single population of lizards was split when South America split from Africa due to continental drift. Fossils are too old to represent animals dispersed by human travel. **15.** `evaluate` Yes. As average global temperatures increase, some organisms may be able to adapt by moving their range. As mountaintops warm, however, organisms adapted to cold conditions will have nowhere to go. Ecologists have good reason to be concerned about these possible extinctions. **16.** `create` Temperature is relatively constant in the tropics all year, whereas temperature fluctuates dramatically at higher latitudes. Organisms that are physiologically adapted to constant temperatures may not have as much genetic variation for traits that enable adaptation to higher temperatures, compared to organisms at higher latitudes, which are adapted to tolerate different temperatures at different times of the year.

CHAPTER 53

IN-TEXT QUESTIONS AND EXERCISES

p. 1086 Fig. 53.4 `analyze` Even though half of each plot serves as a control, the natural landscape varies within plots, so there is a chance that foraging in the two subplots could be different due to factors not measured in the experiment. By measuring foraging activity before adding owls and seeds, the researchers established an additional control and a baseline.
p. 1087 CYU `apply` If owls were allowed to hunt within the treatment subplots, the threat of predation would increase. A larger amount of supplementary seed would be required to compensate for the increased risk of foraging. Gerbils do not consciously calculate costs and benefits. One possible proximate explanation is that the presence of hunting owls would increase the stress level in the gerbils (e.g., the hormone cortisol), which could suppress the hunger of the gerbils or reduce their drive to find food.
p. 1088 Fig. 53.6 `analyze` To form a better control, bring females into the lab and give them the same food and housing conditions as the treatment groups, but expose them to artificial lighting that simulates the short daylight conditions of winter.
p. 1089 CYU (1) `analyze` As in lizards, a rise in sex hormones in humans is a proximate cause of sexual readiness. **(2)** `analyze` Humans, like lizards, tend to choose a mate who will increase their fitness. For example, women often prefer men who will be able to help provide for offspring.
p. 1090 Fig. 53.7 `apply` The data would either be distributed randomly around the circles, showing no preference for direction, or the turtles would swim the same direction in both cases, suggesting that they are using compass orientation rather than map orientation.
p. 1093 Fig. 53.10 `apply` See FIGURE A53.1.
p. 1095 CYU (1) `analyze` Auditory communication allows individuals to communicate over long distances but can be heard by predators. Olfactory communication is effective in the dark, and scents can continue to carry information long after the signaler has left, but scents do not carry long distances. Visual communication is effective during the day but can be seen by predators. **(2)** `apply` The ability to detect deceit protects the individual from fitness costs (e.g., being eaten, having a mate's eggs fertilized by someone else); avoiding or punishing "liars" should also lower the frequency of deceit (because it becomes less successful). Alleles associated with detecting and avoiding or punishing liars should be favored by natural selection and increase in frequency.
p. 1096 Fig. 53.14 `apply` Between first cousins, $r = 1/2n \times 1/2n \times 1/2n = 1/8$.

p. 1097 `apply` Humans often live near kin, are good at recognizing kin, and in many cases can give kin resources for protection that increase fitness.
p. 1097 Fig. 53.15 `create` The control would be to drag an object of similar size through the colony, at similar times of day. This would test the hypothesis that prairie dogs are reacting to the presence of the experimenter and the disturbance—not to a predator.

IF YOU UNDERSTAND . . .

53.1 `apply` At the proximate level of causation, human teens experience an increase in the sex hormones testosterone and estrogen, which increases sex drive. At the ultimate level of causation, teens with a higher sex drive will, on average, have more offspring, which will increase their fitness and pass along their alleles at a higher rate than individuals with a lower sex drive. **53.2** `apply` You would have to consider variables that provide costs and benefits to the web-building spiders. For example, spinning silk and building webs uses up energy, so the size of the web, the amount of material used to build the web, and the frequency of building webs should be considered, as well as the effect of web-building on vulnerability to predation. But large, well-built webs can increase prey capture, so the rate of capture of prey of various types of webs should be considered. **53.3** `create` Only a bowerbird with plenty of resources could afford the time and energy to build a highly decorated bower. Thus, the decoration behavior is likely an honest signal of male fitness. **53.4** `create` Migration is condition-dependent in some birds. If enough food is available locally, the costs of migration outweigh the benefits, so the birds' fitness will be higher if they stay put. **53.5** `apply` Deceitful communication should decrease. If large males are gone, few nests are available and more female-mimic males would compete at them. Most female-mimic males would have no (or fewer) eggs to fertilize and no male to take care of the eggs they did fertilize. There would be less natural selection pressure favoring deceitful communication. **53.6** `understand` (1) Long-lived, so that extensive interactions with kin and nonrelatives are possible; (2) good memory, to record reciprocal interactions; (3) kin nearby, to make inclusive fitness gains possible; (4) ability to share fitness benefits (e.g., food) between individuals.

YOU SHOULD BE ABLE TO . . .

✔ Test Your Knowledge

1. `understand` c **2.** `understand` Adaptive significance of behavior; how behavior affects fitness **3.** `remember` d **4.** `understand` c **5.** `understand` d **6.** `understand` c

✔ Test Your Understanding

7. `understand` The fitness of honeybees depends on how much energy they can bring into the hive by foraging. Hives with honeybees that successfully communicate locations of food resources will be favored by natural

---

FIGURE A53.1

The straight run of the upward waggle dance is twice as long as in Fig. 53.10

---

selection such that the frequency of alleles that determine the communication behavior will increase in the population. **8.** `evaluate` Evolution favors organisms that forage optimally—maximizing benefits and minimizing costs—increasing the frequency of optimal foraging in populations over time. However, individuals in populations vary and environmental circumstances vary. Some individuals will forage suboptimally and suffer reduced fitness. **9.** `apply` Only males that are in good shape (well nourished and free of disease) have the resources required to perform their displays. **10.** `understand` Longer day lengths indicate the arrival of spring, when renewed plant growth and insect activity make more food available. Courtship displays from males indicate the presence of males that can fertilize eggs. **11.** `analyze` This statement is true; sea turtles need both a map and a compass. A map provides information about the spatial relationships of places on a landscape; a compass allows sea turtles to orient the map—so that it aligns with the actual landscape. **12.** `apply` c

✔ Test Your Problem-Solving Skills

13. `create` One possibility would be to set up a testing arena with food supplies at various distances from the food source where adult flies are originally released. Identify rover and sitter adults by genotype and test them individually; then record how far each travels during a set time interval. If the rover and sitter alleles affect foraging movements in adults, then rovers should be more likely to move (and to move farther) than sitters. **14.** `analyze` a, b, c, d; the howler monkeys make condition-dependent decisions about what to eat and are likely to prefer high-energy foods, when available, or foods with nutrients or compounds that would increase their fitness. Hamilton's rule could be considered if some monkeys give the highest-quality foods to close relatives rather than consuming it themselves. **15.** `analyze` Individuals have an r of 1/2 with full siblings and an r of 1/8 with first cousins. If the biologist will lose his life, he needs to save two siblings, or eight first cousins, to keep the "lost copy" of his altruism alleles in the population.
16. `apply` People would be expected to donate blood under two conditions: (1) They either received blood before or expect to have a transfusion in the future, and/or (2) they receive some other benefit in return, such as a good reputation among people who might be able to help them in other ways.

CHAPTER 54

IN-TEXT QUESTIONS AND EXERCISES

p. 1105 Fig. 54.2 `apply` See FIGURE A54.1 (see A:48)
p. 1106 (1) `apply` Compared to northern populations, fecundity should be high and survivorship low in southern populations. **(2)** `apply` Fecundity can be much higher if females lay eggs instead of retaining them in their bodies and giving birth to live young.
p. 1109 Fig. 54.6 `create` At high population density, competition for food limits the amount of energy available to female song sparrows for egg production.
p. 1112 CYU `create` One possibility is competition for food. To test this idea, compare carrying capacity in identical fenced-in areas that differ only in the amount of food added. (You could also test a hypothesis of space limitation by doubling the size of the enclosure but keeping the amount of food the same.)
p. 1114 Fig. 54.10 `apply` When the population of hares was low, the rate of increase was not density dependent on food. That is, plenty of food was available for the few rabbits, so competition for this resource was low.
p. 1117 CYU (1) `understand` Developed nations have roughly equal numbers of individuals in each age class because the fertility rate has been constant and survivorship high for many years. Developing countries have many more

children and young people than older people, because the fertility rate and survivorship have been increasing. **(2)** understand Because survivorship is high in most human populations, changes in overall growth rate depend almost entirely on fertility rates.

p. 1118 Fig. 54.14 apply The total population size of first-generation females in the third year = 1000; in the fourth year = 240. The total population size of first- and second-generation females in the third year = 2308; in the fourth year = 1230. The number of 1-year-olds created by third-generation females (blue) is $981 \times 0.33 = 324$. The number of newborns (green) created by these 1-year-olds is $324 \times 4.0 = 1296$. Thus, the total population size at the end of the fourth year is expected to be $200 + 792 + 1296 + 324 + 198 + 40 = 2850$.

p. 1119 apply The population appears to be increasing over time: From the original 1000 females, there are 2308 females in the third year.

IF YOU UNDERSTAND . . .

54.1 create First you would need to observe snowshoe hares to determine how many pellets one individual can produce in one day, on average. Then you could use a quadrat to measure the number of fresh pellets produced in a known area of habitat in one day and use your pellets/hare number to convert this value to hares/area. **54.2** analyze In ancient Rome, survivorship was probably low and fecundity high—women started reproducing at a young age and did not live long, on average. In Rome today, the population has high survivorship and low fecundity. **54.3** apply See **FIGURE A54.2 54.4** apply The population density of mosquitoes would not be constant over time. Ignoring the number of predators that might be present, you would expect high densities of these short-lived insects after rains, followed by low densities during extended dry periods. **54.5** understand Fewer children are being born per female, but there are many more females of reproductive age due to high fecundity rates in the previous generation. **54.6** apply Small, isolated

populations are likely to be wiped out by bad weather, a disease outbreak, or changes in the habitat. But in a metapopulation, migration between the individual small populations helps reestablish populations and maintain the overall size of the metapopulation.

YOU SHOULD BE ABLE TO . . .

✔ Test Your Knowledge

1. remember b **2.** remember births, deaths, immigration, emigration **3.** remember a **4.** remember True **5.** understand d **6.** understand c

✔ Test Your Understanding

7. remember c **8.** apply See **TABLE A54.1 9.** create The population has undergone near-exponential growth recently because advances in nutrition, sanitation, and medicine have allowed humans to live at high density without suffering from decreased survivorship and fecundity. Eventually, however, growth rates must slow as density-dependent effects such as disease and famine increase death rates and lower birth rates. **10.** understand (a) Corridors allow individuals to move between populations, increasing gene flow and making it possible to recolonize habitats where populations have been lost. (b) Maintaining unoccupied habitat makes it possible for the habitat to be recolonized. **11.** apply *Lacerta vivipara* is most vulnerable at the southern edge of its ranges, such as in the mountains of Spain. These lizards are cold adapted and may become locally extinct in areas that become too warm **12.** analyze r_{max} gives the population growth rate in the absence of density-dependent limitation; r is the actual growth rate, which is usually affected by density-dependent factors.

✔ Test Your Problem-Solving Skills

13. apply Fewer older individuals will be left in the population; there will be relatively more young individuals. If too many older individuals are taken, population growth rate may decline sharply as reproduction stops or slows. (But if relatively few older individuals are taken, more

resources are available to younger individuals and their survivorship and fecundity, and the population's overall growth rate, may increase.) **14.** create The sunflowers and beetles are both metapopulations. To preserve them, you must preserve as many of the sunflower patches as possible (or plant more) and maintain corridors (that may be smaller sunflower patches) along which the beetles can migrate between the patches. **15.** apply As a sexually transmitted disease, AIDS will reduce the number of sexually active adults. If the epidemic continues unabated, the numbers of both reproductive-age adults and children will decline, causing a top-heavy age distribution dominated by older adults and the elderly. See **FIGURE A54.3**. **16.** analyze c

CHAPTER 55

IN-TEXT QUESTIONS AND EXERCISES

p. 1126 apply Your graph should look like Figure 55.4, with *Semibalanus* on the left, *Chthamalus* is on the right, and the *x*-axis indicating height in the intertidal zone. The realized niche of *Chthamalus* is in the upper intertidal zone.

p. 1127 Fig. 55.5 apply If Connell had done the treatments on different rocks, a critic could argue that differences in survival were due to differences in the nature of the rocks—not differences in competition.

p. 1128 apply Your graph should look like the bottom graph in Figure 55.6, showing that the niches of *Semibalanus* and *Chthamalus* are nonoverlapping with respect to height in the intertidal zone.

p. 1131 Fig. 55.10 apply This experimental design tested the hypothesis that mussels can sense the presence of crabs. If the crabs had been fed mussels, a critic could argue that the mussels detect the presence of damaged mussels—not crabs. (As it turns out, the mussels can do both. The experimenters did the same experiment with broken mussel shells in the chamber instead of a fish-fed crab.)

FIGURE A54.1

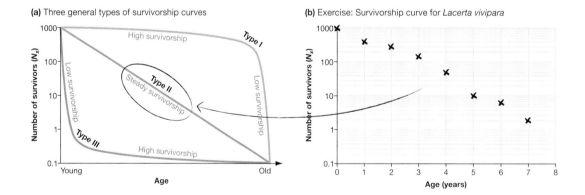

(a) Three general types of survivorship curves

(b) Exercise: Survivorship curve for *Lacerta vivipara*

TABLE A54.1

Trait	Life–History Continuum		
	Left	Middle	Right
	High fecundity Low survivorship	Intermediate	Low fecundity High survivorship
Growth habit	Herbaceous	Shrub	Tree
Disease- and predator–fighting ability	Low	Medium	High
Seed size	Small	Medium	Large
Seed number	Many	Moderate	Few
Body size	Small	Medium	Large

FIGURE A54.2

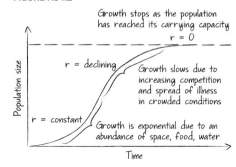

Growth stops as the population has reached its carrying capacity $r = 0$

$r =$ declining

Growth slows due to increasing competition and spread of illness in crowded conditions

$r =$ constant

Growth is exponential due to an abundance of space, food, water

p. 1133 Fig. 55.12 **create** Put equal numbers of infected and uninfected ants in a pen that includes a bird predator. Count how many of each type are eaten over time.

p. 1134 Fig. 55.13 **apply** Many possible answers. For example, acacia trees may expend energy in producing the large bulbs the ants live in and the food they eat. The *Crematogaster* ants may expend energy and risk injury or death in repelling herbivores. The cleaner shrimp may occasionally get injured or eaten by the fish they are cleaning, or their diets may be somewhat restricted by the association. The host fish may miss meals or be at greater risk of predation when they are undergoing cleaning by the shrimp. In both examples of mutualism, however, the overall associations are positive for both parties.

p. 1135 CYU (1) **understand** The individuals do not choose or try to have traits that reduce competition—they simply have those traits (or not). Resource partitioning just happens, because individuals with traits that allow them to exploit different resources produce more offspring, which also have those traits. (Recall from Chapter 25 that natural selection occurs on individuals, but adaptive responses such as resource partitioning are properties of populations.) (2) **understand** When species interact via consumption, a trait that gives one species an advantage will exert natural selection on individuals of the other species who have traits that reduce that advantage. This reciprocal adaptation will continue indefinitely. An example is the interaction of *Plasmodium* with the human immune system: The human immune system has evolved the ability to detect proteins from the *Plasmodium* and kill infected cells; in response, *Plasmodium* has evolved different proteins that the immune system does not detect.

p. 1135 Fig. 55.14 **apply** These steps controlled for the alternative hypothesis that differences in treehopper survival were due to differences in the plants they occupied—not the presence or absence of ants.

p. 1137 Fig. 55.15 **analyze** Predictable, at least to a degree.

p. 1142 CYU (1) **understand** The shade provided by early successional species increases humidity, and decomposition of their tissues adds nutrients and organic material to the soil. These conditions favor growth by later successional species, which can outcompete the early successional species. (2) **understand** The presence or absence of a plant species where nitrogen fixation occurs would dramatically alter nutrient conditions, and thus the speed of succession and the types of species that could become established. For example, species that require high nitrogen would be favored on sites where alder grew, and species that can tolerate low nitrogen would thrive on sites where alder is absent.

p. 1142 Fig. 55.22 **apply** The effect would be to make the island more remote, which would lower the rate of immigration and move the whole immigration curve downward. The rate of extinction would increase, shifting the curve upward. The overall effect would be to decrease the number of species, because the island effectively would have become more remote.

p. 1143 **analyze** The calculations for community 1 are as follows. (Note that your final answer is sensitive to the number of digits you use in your calculations; see BioSkills 1 in Appendix B).

Species A	Species B
$p_A = 10/18 = 0.555$	$p_B = 1/18 = 0.0555$
$\ln 0.555 = -0.589$	$\ln 0.0555 = -2.891$
$-0.589 \times 0.555 = -0.327$	$-2.891 \times 0.0555 = -0.160$

Species C	Species D
$p_C = 1/18 = 0.0555$	$p_D = 3/18 = 0.167$
$\ln 0.0555 = -2.891$	$\ln 0.167 = -1.790$
$-2.891 \times 0.0555 = -0.160$	$-1.790 \times 0.167 = -0.299$

Species E	Species F
$p_E = 2/18 = 0.111$	$p_F = 1/18 = 0.0555$
$\ln 0.111 = -2.198$	$\ln 0.0555 = -2.891$
$-2.198 \times 0.111 = -0.244$	$-2.891 \times 0.0555 = -0.160$

Summing the values of $p \times \ln p$ for each species and multiplying by -1 gives the Shannon index of species diversity for community 1:

$$(-1)[(-0.327) + (-0.160) + (-0.160) + (-0.299) + (-0.244) + (-0.160)] = 1.350$$

Similar calculations would give species diversity values of 1.794 for community 2 and 1.610 for community 3.

IF YOU UNDERSTAND . . .

55.1 understand A mutualistic relationship becomes a parasitic one if one of the species stops receiving a benefit. The treehopper–ant mutualism becomes parasitic in years when spiders are rare, because the treehoppers no longer derive a benefit but pay a fitness cost (producing honeydew that the ants eat). **55.2 apply** An invasive species could have a large impact on community structure by replacing the dominant plant species that creates physical structure, which in turn affects other members of the community; by replacing a keystone species that has an indirect effect on many other species; or by outcompeting many species, causing a direct reduction in species richness. **55.3 apply** All species at Glacier Bay must be able to survive in a cold climate with the local amount of precipitation. The species earliest in the succession must also be able to grow on rock exposed as the glacier melts. But chance, historical differences in seed sources, and presence or absence of alder (and nitrogen fixation) created differences in the species present. **55.4 apply** Many national parks are surrounded by altered habitats, so they are functionally similar to islands. Species richness in the park is more likely to be preserved if it is large and located nearby other wilderness areas.

YOU SHOULD BE ABLE TO . . .

✔ Test Your Knowledge

1. remember True **2. remember** a **3. remember** d **4. understand** d **5. remember** fecundity **6. remember** c

✔ Test Your Understanding

7. evaluate Yes—the treehopper–ant mutualism is parasitic or mutualistic, depending on conditions; competition can evolve into no competition over time if niche differentiation occurs; arms races mean that the outcome of host–parasite interactions can change over time, and so forth (many other examples). **8. understand** b **9. evaluate** (a) If community composition is predictable, then the species present should not change over time. But if composition is not predictable, then the species should undergo significant changes over time. (b) If community composition is predictable, then two sites with identical abiotic factors should develop identical communities. If community composition is not predictable, then sites with identical abiotic factors should develop variable communities. In most tests, the data best match the predictions of the "not predictable" hypothesis, though communities show elements of both. **10. analyze** Disturbance is any short-lived event that changes the distribution of resources. Compared to low-frequency fires, high-frequency fires would tend to be less severe (less fuel builds up) and would tend to exert more intense natural selection for adaptations to resist the effects of fire. Compared to low-severity fires, high-severity fires would open up more space for pioneering species and would tend to exert more intense natural selection for adaptations to resist the effects of fire. **11. apply** Early successional species are adapted to disperse to new environments (small seeds) and grow and reproduce quickly (reproduce at an early age, grow quickly). They can tolerate severe abiotic conditions (high temperature, low humidity, low nutrient availability) but have little competitive ability. These species are able to enter a new environment (with no competitors) and thrive. These attributes are considered adaptations because they increase the fitness of these species. **12. analyze** The idea is that high productivity will lead to high population density of consumers, leading to competition and intense natural selection favoring niche differentiation that leads to speciation.

✔ Test Your Problem-Solving Skills

13. analyze Natural selection will favor orchid individuals that have traits that resist bee attack: thicker flower walls, nectar storage in a different position, a toxin in the flower walls, and so on. Individuals could also be favored if their anthers were in a position that accomplished pollination even if bees eat through the walls of the nectar-storage structure. **14. apply** b **15. apply** The exact answer will depend on the location of the campus. The first species to appear must possess good dispersal ability, rapid growth, quick reproductive periods, and tolerance for very harsh and severe conditions. The two-acre plot is likely to be colonized first by pioneer species that have very "weedy" characteristics. But once colonization is under way, the course of succession will depend more on how the various species interact with each other. The presence of one species can inhibit or facilitate the arrival and establishment of another. For example, an early-arriving species might provide the shade and nutrients required by a late-arriving species. The site's history and nearby ecosystems may influence which species appear at each stage; for instance, an undisturbed ecosystem nearby could be a source for native species. The pattern and rate of this succession is also influenced by the overall environmental conditions affecting it. Only species with traits appropriate to the local climate are likely to colonize the site. **16. create** One reasonable experiment would involve constructing artificial ponds and introducing different numbers of plankton species

FIGURE A54.3

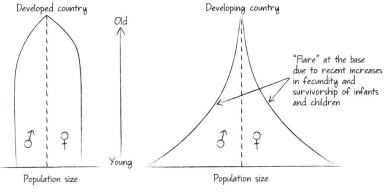

Developed country

Similar numbers of people in most age classes due to same number of infants born each year and high survivorship

♂ ♀

Population size

Developing country

Old ↑

"Flare" at the base due to recent increases in fecundity and survivorship of infants and children

♂ ♀

Young

Population size

to different ponds, but the same total number of individuals. (Any natural immigration to the ponds would have to be prevented.) After a period of time, remove all of the plankton and measure the biomass present. Make a graph with number of species on the x-axis and total biomass on the y-axis. If the hypothesis is correct, the line of best fit through the data should have a positive slope.

CHAPTER 56

IN-TEXT QUESTIONS AND EXERCISES

p. 1152 `apply` To grow a kilogram of beef, first you have to grow 10 kg of grain or grass and feed it to the cow. Only 10 percent of this 10 kg will be used for growth and reproduction—the other 9 kg is used for cellular respiration or lost as heat.

p. 1152 `apply` Crustaceans and fish are ectothermic, so they are much more efficient at converting primary production into the biomass in their bodies than are endothermic birds and mammals.

p. 1153 `apply` Tuna are top predators in marine food webs. Mercury occurs in low concentrations in the water, but accumulates in higher concentrations at each trophic level. A person who eats a lot of tuna will accumulate mercury over time, potentially causing mercury poisoning. Sardines are primary consumers, a low trophic level, and therefore are not in danger of biomagnification of mercury.

p. 1155 Fig. 56.10 `apply` Average NPP = 125 g/m²/year. Area = 65%. Although NPP in the ocean is very small, the ocean is so vast that it ends up being the largest contributor to total NPP.

p. 1156 CYU (1) `understand` At each trophic level, most of the energy that is consumed is lost to cellular respiration and ultimately heat, metabolism, or other maintenance activities, which leaves only a small percentage of energy for biomass production (growth and reproduction).
(2) `apply` Since NPP depends on temperature, it is logical to predict that NPP will increase with global warming. (However, Section 56.3 explains that global warming changes precipitation patterns, ocean stratification, and other factors that can limit NPP).

p. 1157 `understand` When the trees of tropical rain forests are hauled away and the remnants of the forest are burned, the nutrients are exported from the site, leaving only poor soil to nourish the crops.

p. 1158 Fig. 56.13 `create` One logical hypothesis is that the total amount exported increases as tree roots and other belowground organic material decay and begin to wash into the stream; the amount exported should begin to decline as nitrate reserves become exhausted—eventually, there is no more nitrate to wash away.

p. 1159 Fig. 56.14 `apply` Much more water should evaporate. One possibility is that more water vapor will be blown over land and increase precipitation over land.

p. 1162 CYU `apply` **Water**: perhaps the most direct impacts concern rates of groundwater replenishment. Converting biomes into farms or suburbs decreases groundwater recharge and increases runoff. Irrigation pumps groundwater to the surface, where much of it runs off into streams. **Nitrogen**: Humans introduce huge quantities of nitrogen into ecosystems as fertilizer. **Carbon**: Humans are releasing huge quantities of stored carbon, in the form of fossil fuels, into the atmosphere as CO_2. Whereas humans are depleting supplies of freshwater below natural levels, we are increasing quantities of N and C above natural levels.

p. 1164 Fig. 56.22 `analyze` Photosynthesis increases in summer, resulting in removal of CO_2 from the atmosphere.

p. 1165 Fig. 56.23 `apply` $2 \times 7000 = 14{,}000$ million metric tons of CO_2, double the current total.

p. 1168 `apply` The pH of the water will fall (acidity will increase) because CO_2 from your breath reacts with water to form carbonic acid.

IF YOU UNDERSTAND . . .

56.1 `apply` Tofu is a plant, a primary producer at the base of the productivity pyramid. Cows are primary consumers, one level up. It would be about 10 times more efficient to eat tofu, because 10 times the quantity of plants would be necessary to provide one pound of beef.
56.2 `apply` Decomposition is slowest in cold, wet temperatures on land and in anoxic conditions in the ocean (or in freshwater). Stagnant water often becomes anoxic as decomposers use up available oxygen that is not replenished by diffusion from the atmosphere. **56.3** `apply` Many possible answers, including (1) reducing carbon dioxide emissions by finding alternatives to fossil fuels should decrease the amount of carbon dioxide released into the atmosphere, (2) recycling programs decrease carbon dioxide emissions from manufacturing plants and power plants, and (3) reforestation can tie up carbon dioxide in wood.

YOU SHOULD BE ABLE TO . . .

✔ Test Your Knowledge
1. `remember` True **2.** `understand` d **3.** `understand` a **4.** `remember` c
5. `remember` c **6.** `remember` b

✔ Test Your Understanding
7. `apply` a **8.** `analyze` Sharks normally eat fish that eat the herbivorous fish that keep the growth of algae on corals in check. When sharks are absent, more of the intermediate fish survive, which decreases the number of herbivorous fish, causing algae to thrive on the coral reef. **9.** `analyze` Warmer, wetter climates speed decomposition; cool temperatures slow it. Decomposition regulates nutrient availability because it releases nutrients from detritus and allows them to reenter the food web.
10. `evaluate` Yes, corals would be a good choice. Corals are important ecologically because they have a high NPP, providing energy to diverse animals in reef food webs. Corals are also vulnerable to climate change from the increase in water temperature, which can cause bleaching (loss of photosynthetic algae), as well as from the effects of ocean acidification, which slows the rate of coral growth. **11.** `analyze` Forest fires are an example of positive feedback; drier, hotter summers increase the risk of fires, and fires release CO_2 to the atmosphere, increasing global warming and the risk of fires. Forest growth is an example of negative feedback; as CO_2 increases in the atmosphere, the rate of forest growth in some areas may increase (due to increased availability of CO_2, increased temperature, and/or increased precipitation), reducing the concentration of CO_2 in the atmosphere. **12.** `analyze` The open ocean has almost no nutrient input from the land and has little upwelling to supply nutrients from the deep ocean. In contrast, intertidal and coastal areas receive large inputs of nutrients from rivers as well as from upwellings from ocean depths.

✔ Test Your Problem-Solving Skills
13. `create` One possibility is to add radioactive nitrogen to the water of the ponds in natural concentrations (control ponds) and elevated nitrogen concentrations (treatment ponds) and then follow this isotope through the primary producers, primary consumers, and secondary consumers in the system by measuring the amount of radioisotope in the tissues of organisms, and quantity of biomass, at each trophic level in the treatment versus control ponds **14.** `evaluate` c **15.** `apply` Without herbivores, there is no link in the nitrogen cycle between primary producers and secondary consumers. All of the plant nitrogen would go to the primary decomposers and back

into the soil. If decomposition is rapid enough, nitrogen would cycle quickly between primary producers, decomposers, the soil, and back to primary producers.
16. `apply` Atmospheric oxygen would increase due to extensive photosynthesis, but carbon dioxide levels would decrease because little decomposition was occurring. The temperature would drop because fewer greenhouse gases would be trapping heat reflected from the Earth's surface.

CHAPTER 57

IN-TEXT QUESTIONS AND EXERCISES

p. 1177 CYU `create` Do an all-taxon survey: organize experts and volunteers to collect, examine, and identify all species present. This could include direct sequencing studies (see Chapter 29) to document the bacteria and archaea present in different habitats on campus.
p. 1178 Fig. 57.7 `apply` Nearly all endangered species are threatened by more than one factor and therefore appear on the graph more than once, leading to species totals greater than 100 percent.
p. 1180 Fig. 57.9 `analyze` Because the treatments and study areas were assigned randomly, there is no bias involved in picking certain areas that are unusual in terms of their biomass or species diversity. They should represent a random sample of biomass and species diversity in fragments of various sizes versus intact forest.
p. 1183 Using the equation $S = (18.9)A^{0.15}$, plug in 5000 km² for A and solve for S such that $S = 18.9 \times 5000^{0.15} = 68$ species.
p. 1184 CYU (1) `understand` Fragmentation reduces habitat quality by creating edges that are susceptible to invasion and loss of species, due to changed abiotic conditions. Genetic problems occur inside fragments as species become inbred and/or lose genetic diversity via genetic drift. **(2)** `create` Habitat destruction and fragmentation in Borneo are obvious from the clear-cut areas. Overexploitation could be assessed by interviewing local citizens and law enforcement officers to the history of poaching orangutans for the pet trade or for other uses. Effects from invasive species could be assessed by published or local accounts of community structure over time. Pollution could be assessed by direct measurement of soil, water, air, and possibly tissue samples (in dead orangutans, feces, or blood samples). The effects of climate change could be assessed based on local weather records.
p. 1185 Fig. 57.12 `analyze` *Null hypothesis*: NPP is not affected by species richness or functional diversity of species. *Prediction*: NPP will be greater in plots with more species and more functional groups. *Prediction of null*: There will be no difference in NPP based on species richness or number of functional groups.
p. 1188 CYU `create` Establish several study plots in an area of Borneo where some of the forest has been converted to palm plantation but the adjacent forest (of similar elevation and topography) has rain forest still intact—half of the plots in the plantation and half in the rain forest. Monitor soil depth, soil nutrient levels, precipitation, and stream sediment levels regularly, and compare after several years.
p. 1191 Fig. 57.16 `apply` See **FIGURE A57.1**.

IF YOU UNDERSTAND . . .

57.1 `apply` At the genetic level, you could measure the genetic diversity of the orangutan population by comparing DNA sequences obtained from blood or feces samples. This information could help you to determine if a genetic bottleneck had occurred during the clear-cutting period. At the species level, you could monitor the increase in species richness of trees, birds, mammals, and other groups as the restoration progressed. At the ecosystem level, you could measure the ability of the growing forest to retain nutrients in the soil, prevent erosion, trap carbon, increase the prevalence of rain, and other functions **57.2** (apply) Small, geographically isolated populations of orangutans have low genetic diversity due to lack of gene flow and genetic drift and can suffer from inbreeding depression. Low genetic diversity makes them vulnerable to changes in their habitat, and small population size makes them susceptible to catastrophic events such as storms, disease outbreaks, or fires. **57.3** (apply) If many species are present, and each uses the available resources in a unique way, then a larger proportion of all available resources should be used—leading to higher biomass production. **57.4** (evaluate) Twelve percent of land area is already protected, but a mass extinction event is still going on. This is because 12 percent is not sufficient to save most species, and because many or most protected areas are not located in biodiversity hotspots.

YOU SHOULD BE ABLE TO . . .

✔ Test Your Knowledge

1. (remember) d **2.** (remember) a **3.** (remember) b **4.** (remember) d
5. (understand) c **6.** (understand) overexploitation

✔ Test Your Understanding

7. (analyze) b **8.** (analyze) By comparing the number of species estimated to reside in the park before and after the survey, biologists will have an idea of how much actual species diversity is being underestimated in other parts of the world—where research was at the level of Great Smoky Mountains National Park before the survey. The limitation is that it may be difficult to extrapolate from the data—no one knows if the situation at this park is typical of other habitats. **9.** (evaluate) No one correct answer; there are good arguments for both. In a case like the orangutans of Borneo, it is not possible to save the orangutans without saving their ecosystem. On the other hand, it is much easier to attract money and volunteers to help save a charismatic primate like an orangutan than to save a patch of forest. Both types of efforts are needed, their relative importance depends on the circumstance. **10.** (evaluate) In experiments with species native to North American grasslands, study plots that have more species produce more biomass, change less during a disturbance (drought), and recover from a disturbance faster than study plots with fewer species. **11.** (understand) Wildlife corridors facilitate the movement of individuals. Corridors allow areas to be recolonized if a species is lost and enable the introduction of new alleles that can counteract genetic drift and inbreeding in small, isolated

populations. **12.** (apply) Ecosystem services are beneficial effects that ecosystems have on humans. It is rare to own or pay for these services, so no one has a vested interest in maintaining them. This is a primary reason that ecosystems are destroyed, even though the services they offer are valuable. The exception would be landowners who are paid *not* to log their forests, to maintain ecosystem functions like carbon storage.

✔ Test Your Problem-Solving Skills

13. (apply) d **14.** (create) Catalog existing biodiversity at the genetic, species, and ecosystem levels. Using these data, find areas that have the highest concentration of biodiversity at each level. Protect as many of these areas as possible, and connect them with corridors of habitat. **15.** (evaluate) There is no "correct" answer. One argument is that conservationists could lobby officials in Brazil and Indonesia to learn from the mistakes made in developed nations, and preserve enough forests to maintain biodiversity and keep ecosystem services intact—avoiding the expenditures that developed countries had to make to clean up pollution and restore ecosystems and endangered species. **16.** (analyze) Species with specialized food or habitat requirements, large size (and thus large requirements for land area), small population size, and low reproductive rate are vulnerable to extinction. A good example is the koala bear. Species that are particularly resistant to pressure from humans possess the opposite of many of these traits. A good example of a resistant species is the Norway rat.

BIG PICTURE Ecology

p. 1196 CYU (1) (analyze) Species interactions "are agents of" natural selection. **(2)** (analyze) As the human population climbs, more resources are needed to support the population. The harvesting of resources affects other species both directly (e.g. overfishing) and indirectly (e.g. climate change), resulting in decreased biodiversity. **(3)** (analyze) Both energy and nutrients flow through trophic levels in food webs. However, energy flows in one direction, ultimately dissipating as heat, whereas nutrients continue to cycle through ecosystems. **(4)** (apply) Human behavior "increases emissions of" CO_2 "causes" greenhouse effect "causes" climate change.

Bioskills

BIOSKILLS 1; p. B:2 CYU (1) (apply) 3.1 miles **(2)** (apply) 37°C
(3) (apply) Multiply your weight in pounds by 1/2.2 (0.45).

(4) (apply) 4 **(5)** (apply) Two significant figures. When you multiply, the answer can have no more significant figures than the least accurate measurement—in this case, 1.6.

BIOSKILLS 2; p. B:3 CYU (1) (apply) "different-yoked-together" **(2)** (apply) "sugary-loosened" **(3)** (apply) "study-of-form" **(4)** (apply) "three-bodies"

BIOSKILLS 3; p. B:6 CYU (1) (apply) about 18% **(2)** (apply) a dramatic drop (almost 10%) **(3)** (understand) No—the order of presentation in a bar chart does not matter (though it's convenient to arrange the bars in a way that reinforces the overall message). **(4)** (apply) 11 **(5)** (apply) 68 inches

BIOSKILLS 4; p. B:7 CYU (apply) Test 2, the estimate based on the larger sample—the more replicates or observations you have, the more precise your estimate of the average should be.

BIOSKILLS 5; p. B:8 CYU (1) (apply) $1/2 \times 1/2 \times 1/2 \times 1/2 = 1/16$ **(2)** (apply) $1/6 + 1/6 + 1/6 = 1/2$

BIOSKILLS 6; p. B:9 CYU (1) (understand) exponential **(2)** (apply) $\ln N_t = \ln N_0 + rt$

BIOSKILLS 7; p. B:9 Fig. B7.1. (analyze) See **FIGURE BA.1**.
p. B:10 Fig. B7.2. (analyze) See **FIGURE BA.2**.
p. B:11 Fig. B7.3. (analyze) Figure B7.3d is different.

FIGURE BA.1

FIGURE BA.2

FIGURE A57.1

BIOSKILLS 8; p. B:12 Fig. B8.1. (apply) See **FIGURE BA.3**.

BIOSKILLS 9; p. B:13 Fig. B9.1 (understand) DNA and RNA are acids that tend to drop a proton in solution, giving them a negative charge.
p. B:15 CYU (analyze) The lane with no band comes from a sample where RNA X is not present. The same size RNA X is present in the next two lanes, but the faint band has very few copies while the dark band has many. In the fourth lane, the band is formed by a smaller version of RNA X, and relatively few copies are present.

BIOSKILLS 10; p. B:17 CYU (1) (understand) size and/or density. **(2)** (apply) Mitochondria, because they are larger in size compared with ribosomes.

BIOSKILLS 11; p. B:19 CYU (1) (analyze) No—it's just that no mitochondria happened to be present in this section

sliced through the cell. **(2)** (explain) Understanding a molecule's structure is often critical to understanding how it functions in cells.

BIOSKILLS 12; p. B:21 CYU (1) (analyze) It may not be clear that the results are relevant to noncancerous cells that are not growing in cell culture—that is, that the artificial conditions mimic natural conditions. **(2)** (analyze) It may not be clear that the results are relevant to individuals that developed normally, from an embryo—that is, that the artificial conditions mimic natural conditions.

BIOSKILLS 13; p. B:23 Fig. B13.1 (analyze) This is human body temperature—the natural habitat of *E. coli*.
p. B:24 CYU (1) (analyze) *Caenorhabditis elegans* would be a good possibility, because the cells that normally die have already been identified. You could find mutant individuals that lacked normal cell death; you could compare

the resulting embryos to normal embryos and be able to identify exactly which cells change as a result. **(2)** (analyze) Any of the multicellular organisms in the list would be a candidate, but *Dictyostelium discoideum* might be particularly interesting because cells stick to each other only during certain points in the life cycle. **(3)** (analyze) *Mus musculus*—as the only mammal in the list, it is the organism most likely to have a gene similar to the one you want to study.

BIOSKILLS 14; p. B:26 CYU (synthesize) Many examples are possible. See Figure 1.9 in Chapter 1, as an example of the format to use for your Research Box.

BIOSKILLS 15; p. B:27 CYU (1) (analyze) See **FIGURE BA.4**.
(2) (analyze) See **FIGURE BA.4**.

FIGURE BA.3

Molecular formula: CO_2

Structural formula: $O = C = O$

Ball-and-stick model:

Space-filling model:

FIGURE BA.4

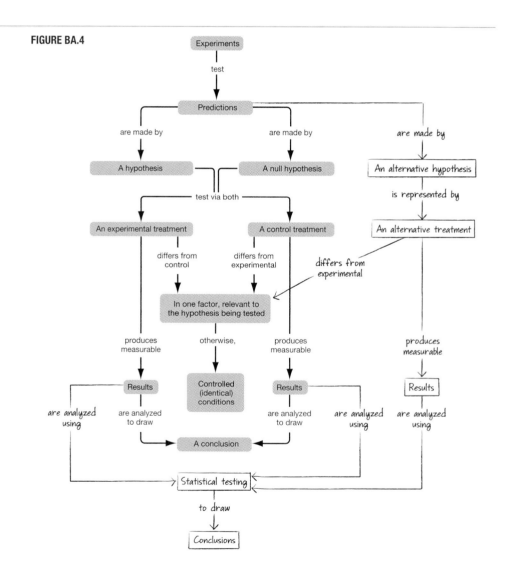

BioSkills

BIOSKILL 1 the metric system and significant figures

Scientists ask questions that can be answered by observing or measuring things—by collecting data. What units are used to make measurements? When measurements are reported, how can you tell how reliable the data are?

The Metric System

The metric system is the system of units of measure used in every country of the world but three (Liberia, Myanmar, and the United States). It is also the basis of the SI system—the International System of Units (abbreviated from the French, *Système international d'unités*)—used in scientific publications.

The popularity of the metric system is based on its consistency and ease of use. These attributes, in turn, arise from the

system's use of the base 10. For example, each unit of length in the system is related to all other measures of length in the system by a multiple of 10. There are 10 millimeters in a centimeter; 100 centimeters in a meter; 1000 meters in a kilometer.

Measures of length in the English system, in contrast, do not relate to each other in a regular way. Inches are routinely divided into 16ths; there are 12 inches in a foot; 3 feet in a yard; 5280 feet (or 1760 yards) in a mile.

If you have grown up in the United States and are accustomed to using the English system, it is extremely important to begin developing a working familiarity with metric units and values. **Tables B1.1** and **B1.2** (see B:2) should help you get started with this process.

As an example, consider the following question: An American football field is 120 yards long, while rugby fields are 144 meters

TABLE B1.1 **Metric System Units and Conversions**

Measurement	Unit of Measurement and Abbreviation	Metric System Equivalent	Converting Metric Units to English Units
Length	kilometer (km)	1 km = 1000 m = 10^3 m	1 km = 0.62 mile
	meter (m)	1 m = 100 cm	1 m = 1.09 yards = 3.28 feet = 39.37 inches
	centimeter (cm)	1 cm = 0.01 m = 10^{-2} m	1 cm = 0.3937 inch
	millimeter (mm)	1 mm = 0.001 m = 10^{-3} m	1 mm = 0.039 inch
	micrometer (μm)	1 μm = 10^{-6} m = 10^{-3} mm	
	nanometer (nm)	1 nm = 10^{-9} m = 10^{-3} μm	
Area	hectare (ha)	1 ha = 10,000 m^2	1 ha = 2.47 acres
	square meter (m^2)	1 m^2 = 10,000 cm^2	1 m^2 = 1.196 square yards
	square centimeter (cm^2)	1 cm^2 = 100 mm^2 = 10^{-4} m^2	1 cm^2 = 0.155 square inch
Volume	liter (L)	1 L = 1000 mL	1 L = 1.06 quarts
	milliliter (mL)	1 mL = 1000 μL = 10^{-3} L	1 mL = 0.034 fluid ounce
	microliter (μL)	1 μL = 10^{-6} L	
Mass	kilogram (kg)	1 kg = 1000 g	1 kg = 2.20 pounds
	gram (g)	1 g = 1000 mg	1 g = 0.035 ounce
	milligram (mg)	1 mg = 1000 μg = 10^{-3} g	
	microgram (μg)	1 μg = 10^{-6} g	
Temperature	Kelvin (K)*		K = °C + 273.15
	degrees Celsius (°C)		°C = $\frac{5}{9}$ (°F − 32)
	degrees Fahrenheit (°F)		°F = $\frac{9}{5}$°C + 32

*Absolute zero is −273.15 °C = 0 K.

TABLE B1.2 Prefixes Used in the Metric System

Prefix	Abbreviation	Definition
nano–	n	$0.000\,000\,001 = 10^{-9}$
micro–	μ	$0.000\,001 = 10^{-6}$
milli–	m	$0.001 = 10^{-3}$
centi–	c	$0.01 = 10^{-2}$
deci–	d	$0.1 = 10^{-1}$
–	–	$1 = 10^{0}$
kilo–	k	$1000 = 10^{3}$
mega–	M	$1\,000\,000 = 10^{6}$
giga–	G	$1\,000\,000\,000 = 10^{9}$

long. In yards, how much longer is a rugby field than an American football field? To solve this problem, first convert meters to yards: $144\ \text{m} \times 1.09\ \text{yards/m} = 157$ yards (note that the unit "m" cancels out). The difference in yards is thus: $157 - 120 = 37$ yards. If you did these calculations on a calculator, you might have come up with 36.96 yards. Why has the number of yards been rounded off? The answer lies in significant figures. Let's take a closer look.

Significant Figures

Significant figures or "sig figs"—the number of digits used to report the measurement—are critical when reporting scientific data. The number of significant figures in a measurement, such as 3.524, is the number of digits that are known with some degree of confidence (3, 5, and 2) plus the last digit (4), which is an estimate or approximation. How do scientists know how many digits to report?

Rules for Working with Significant Figures

The rules for counting significant figures are summarized here:

- All nonzero numbers are always significant.
- Leading zeros are never significant; these zeros do nothing but set the decimal point.
- Embedded zeros are always significant.

- Trailing zeros are significant *only* if the decimal point is specified (Hint: Change the number to scientific notation. It is easier to see the "trailing" zeros.)

Table B1.3 provides examples of how to apply these rules. The bottom line is that significant figures indicate the precision of measurements.

Precision versus Accuracy

If biologists count the number of bird eggs in a nest, they report the data as an exact number—say, 3 eggs. But if the same biologists are measuring the diameter of the eggs, the numbers will be inexact. Just how inexact they are depends on the equipment used to make the measurements. For example, if you measure the width of your textbook with a ruler several times, you'll get essentially the same measurement again and again. Precision refers to how closely individual measurements agree with each other. So, you have determined the length with precision, but how do you know if the ruler was accurate to begin with?

Accuracy refers to how closely a measured value agrees with the correct value. You don't know the accuracy of a measuring device unless you calibrate it, by comparing it against a ruler that is known to be accurate. As the sensitivity of equipment used to

check your understanding

If you understand BioSkill 1

✔ **You should be able to . . .**

1. **QUANTITATIVE** Calculate how many miles a runner completes in a 5.0-kilometer run.
2. **QUANTITATIVE** Calculate your normal body temperature in degrees Celsius (Normal body temperature is 98.6°F.).
3. **QUANTITATIVE** Calculate your current weight in kilograms.
4. **QUANTITATIVE** Calculate how many liters of milk you would need to buy to get approximately the same volume as a gallon of milk.
5. **QUANTITATIVE** Multiply the measurements 2.8723 and 1.6. How many significant figures does your answer have? Why?

TABLE B1.3 Rules for Working with Significant Figures

Example	Number of Significant Figures	Scientific Notation	Rule
35,200	5	3.52×10^{4}	All nonzero numbers are always significant
0.00352	3	3.52×10^{-3}	Leading zeros are not significant
1.035	4	$1.035\ (\times 10^{0})$	Imbedded zeros are always significant
200	1	2×10^{2}	Trailing zeros are significant only if the decimal point is specified
200.0	4	2.000×10^{2}	Trailing zeros are significant only if the decimal point is specified

make a measurement increases, the number of significant figures increases. For example, if you used a kitchen scale to weigh out some sodium chloride, it might be accurate to 3 ± 1 g (1 significant figure); but an analytical balance in the lab might be accurate to 3.524 ± 0.001 g (4 significant figures).

In science, only the numbers that have significance—that are obtained from measurement—are reported. It is important to follow the "sig fig rules" when reporting a measurement, so that data do not appear to be more accurate than the equipment allows.

Combining Measurements

How do you deal with combining measurements with different degrees of accuracy and precision? The simple rule to follow is that the accuracy of the final answer can be no greater than the least accurate measurement. So, when you multiply or divide measurements, the answer can have no more significant figures than the least accurate measurement. When you add or subtract measurements, the answer can have no more decimal places than the least accurate measurement.

As an example, consider that you are adding the following measurements: 5.9522, 2.065, and 1.06. If you plug these numbers into your calculator, the answer your calculator will give you is 9.0772. However, this is incorrect—you must round your answer off to the nearest value, 9.08, to the least number of decimal places in your data.

It is important to nail down the concept of significant figures and to practice working with metric units and values. The Check Your Understanding questions in this BioSkill should help you get started with this process.

BIOSKILL 2 some common Latin and Greek roots used in biology

Greek or Latin Root	English Translation	Example Term
a, an	not	anaerobic
aero	air	aerobic
allo	other	allopatric
amphi	on both sides	amphipathic
anti	against	antibody
auto	self	autotroph
bi	two	bilateral symmetry
bio	life, living	bioinformatics
blast	bud, sprout	blastula
co	with	cofactor
cyto	cell	cytoplasm
di	two	diploid
ecto	outer	ectoparasite
endo	inner, within	endoparasite
epi	outer, upon	epidermis
exo	outside	exothermic
glyco	sugary	glycolysis
hetero	different	heterozygous
homo	alike	homozygous
hydro	water	hydrolysis
hyper	over, more than	hypertonic
hypo	under, less than	hypotonic
inter	between	interspecific
intra	within	intraspecific
iso	same	isotonic
logo, logy	study of	morphology
lyse, lysis	loosen, burst	glycolysis
macro	large	macromolecule

Greek or Latin Root	English Translation	Example Term
meta	change, turning point	metamorphosis
micro	small	microfilament
morph	form	morphology
oligo	few	oligopeptide
para	beside	parathyroid gland
photo	light	photosynthesis
poly	many	polymer
soma	body	somatic cells
sym, syn	together	symbiotic, synapsis
trans	across	translation
tri	three	trisomy
zygo	yoked together	zygote

check your understanding

If you understand BioSkill 2

✔ **You should be able to . . .**

Provide literal translations of the following terms:

1. heterozygote

2. glycolysis

3. morphology
4. trisomy

Graphs are the most common way to report data, for a simple reason. Compared to reading raw numerical values in a table or list, a graph makes it much easier to understand what the data mean.

Learning how to read and interpret graphs is one of the most basic skills you'll need to acquire as a biology student. As when learning piano or soccer or anything else, you need to understand a few key ideas to get started and then have a chance to practice—a lot—with some guidance and feedback.

Getting Started

To start reading a graph, you need to do three things: read the axes, figure out what the data points represent—that is, where they came from—and think about the overall message of the data. Let's consider each step in turn.

What Do the Axes Represent?

Graphs have two axes: one horizontal and one vertical. The horizontal axis of a graph is also called the x-axis or the abscissa. The vertical axis of a graph is also called the y-axis or the ordinate. Each axis represents a variable that takes on a range of values. These values are indicated by the ticks and labels on the axis. Note that each axis should *always* be clearly labeled with the unit or treatment it represents.

FIGURE B3.1 shows a scatterplot—a type of graph where continuous data are graphed on each axis. Continuous data can take an array of values over a range. In contrast, discrete data can take only a restricted set of values. If you were graphing the average height of men and women in your class, height is a continuous variable, but gender is a discrete variable.

For the example in this figure, the x-axis represents time in units of generations of maize; the y-axis represents the average percentage of the dry weight of a maize kernel that is protein.

To create a graph, researchers plot the independent variable on the x-axis and the dependent variable on the y-axis (Figure B3.1a). The terms independent and dependent are used because the values on the y-axis depend on the x-axis values. In our example, the researchers wanted to show how the protein content of maize kernels in a study population changed over time. Thus, the protein concentration plotted on the y-axis depended on the year (generation) plotted on the x-axis. The value on the y-axis always depends on the value on the x-axis, but not vice versa.

In many graphs in biology, the independent variable is either time or the various treatments used in an experiment. In these cases, the y-axis records how some quantity changes as a function of time or as the outcome of the treatments applied to the experimental cells or organisms.

(a) Read the axes—what is being plotted?

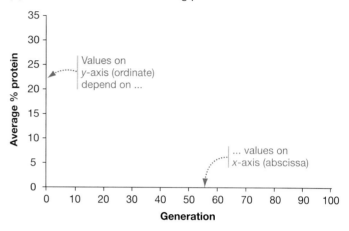

(b) Look at the bars or data points—what do they represent?

(c) What's the punchline?

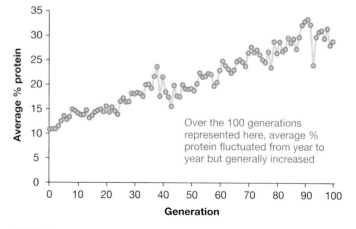

FIGURE B3.1 Scatterplots Are Used to Graph Continuous Data.

What Do the Data Points Represent?

Once you've read the axes, you need to figure out what each data point is. In our maize kernel example, the data point in Figure B3.1b represents the average percentage of protein found in a sample of kernels from a study population in a particular generation.

If it's difficult to figure out what the data points are, ask yourself where they came from—meaning, how the researchers got them. You can do this by understanding how the study was done and by understanding what is being plotted on each axis. The *y*-axis will tell you what they measured; the *x*-axis will usually tell you when they measured it or what group was measured. In some cases—for example, in a plot of average body size versus average brain size in primates—the *x*-axis will report a second variable that was measured.

In other cases, a data point on a graph may represent a relative or arbitrary unit of measurement. The data point shows the ratio of the amount of a substance, intensity, or other quantities, relative to a predetermined reference measurement. For example, the *y*-axis might show the percentage of relative activity of an enzyme—the rate of the enzyme-catalyzed reaction, scaled to the highest rate of activity observed (100 percent)—in experiments conducted under conditions that are identical except for one variable, such as pH or temperature (see Figure 8.15).

What Is the Overall Trend or Message?

Look at the data as a whole, and figure out what they mean. Figure B3.1c suggests an interpretation of the maize kernel example. If the graph shows how some quantity changes over time, ask yourself if that quantity is increasing, decreasing, fluctuating up and down, or staying the same. Then ask whether the pattern is the same over time or whether it changes over time.

When you're interpreting a graph, it's extremely important to limit your conclusions to the data presented. Don't extrapolate beyond the data, unless you are explicitly making a prediction based on the assumption that present trends will continue. For example, you can't say that the average percentage of protein content was increasing in the population before the experiment started, or that it will continue to increase in the future. You can say only what the data tell you.

Types of Graphs

Many of the graphs in this text are scatterplots like the one shown in Figure B3.1c, where individual data points are plotted. But you will also come across other types of graphs in this text.

Scatterplots, Lines, and Curves

Scatterplots sometimes have data points that are by themselves, but at other times data points will be connected by dot-to-dot lines to help make the overall trend clearer, as in Figure B3.1c, or may have a smooth line through them.

A *smooth line* through data points—sometimes straight, sometimes curved—is a mathematical "line of best fit." A line of best fit represents a mathematical function that summarizes the relationship between the *x* and *y* variables. It is "best" in the sense of fitting the data points most precisely. The line may pass through some of the points, none of the points, or all of the points.

Curved lines often take on characteristic shapes depending on the relationships between the *x* and *y* variable. For example, a bell-shaped curve depicts a normal distribution in which most data points are clumped near the middle, while a sigmoid or S-shaped curve exhibits small changes at first, which then accelerate and approach maximal value over time. Data from studies on population growth (see Chapter 54), enzyme kinetics (see Chapter 8), and oxygen–hemoglobin dissociation (see Chapter 45) typically fall on a curved line.

Bar Charts, Histograms, and Box-and-Whisker Plots

Scatterplots, or line-of-best-fit graphs, are the most appropriate type of graph when the data have a continuous range of values and you want to show individual data points. But other types of graphs are used to represent different types of distributions:

- *Bar charts* plot data that have discrete or categorical values instead of a continuous range of values. In many cases the bars might represent different treatment groups in an experiment, as in **FIGURE B3.2a** (see B:6). In this graph, the height of the bar indicates the average value. Statistical tests can be used to determine whether a difference between treatment groups is significant (see **BIOSKILLS 4**).

- *Histograms* illustrate frequency data and can be plotted as numbers or percentages. **FIGURE B3.2b** shows an example where height is plotted on the *x*-axis, and the number of students in a population is plotted on the *y*-axis. Each rectangle indicates the number of individuals in each interval of height, which reflects the relative frequency, in this population, of people whose heights are in that interval. The measurements could also be recalculated so that the *y*-axis would report the proportion of people in each interval. Then the sum of all the bars would equal 100 percent. Note that if you were to draw a smooth curve connecting the top of the bars on this histogram, the smooth curve would represent the shape of a bell.

- *Box-and-whisker plots* allow you to easily see where most of the data fall. Each box indicates where half of the data numbers are. The whiskers indicate the lower extreme and the upper extreme of the data. The vertical line inside each box indicates the median—meaning that half of the data are above this value and half are below (see Figure 1.9 for an example).

When you are looking at a bar chart that plots values from different treatments in an experiment, ask yourself if these values are the same or different. If the bar chart reports averages over discrete ranges of values, ask what trend is implied—as you would for a scatterplot.

(a) Bar chart

(b) Histogram

FIGURE B3.2 Bar Charts and Histograms. (a) Bar charts are used to graph data that are discontinuous or categorical. **(b)** Histograms show the distribution of frequencies or values in a population.

When you are looking at a histogram, ask whether there is a "hump" in the data—indicating a group of values that are more frequent than others. Is the hump in the center of the distribution of values, toward the left, or toward the right? If so, what does it mean?

Similarly, when you are looking at a box-and-whisker plot, ask yourself what information the graph gives you. What is the

range of values for the data? Where are half the data points? Below what value is three quarters of the data?

Getting Practice

Working with this text will give you lots of practice with reading graphs—they appear in almost every chapter. In many cases we've inserted an arrow to represent your instructor's hand at the whiteboard, with a label that suggests an interpretation or draws your attention to an important point on the graph. In other cases, you should be able to figure out what the data mean on your own or with the help of other students or your instructor.

check your understanding

If you understand BioSkill 3

✔ **You should be able to . . .**

1. **QUANTITATIVE** Determine the total change in average percentage of protein in maize kernels, from the start of the experiment until the end.

2. **QUANTITATIVE** Determine the trend in average percentage of protein in maize kernels between generation 37 and generation 42.

3. Explain whether the conclusions from the bar chart in Figure B3.2a would be different if the data and label for Treatment 3 were put on the far left and the data and label for Treatment 1 on the far right.

4. **QUANTITATIVE** Determine approximately how many students in this class are 70 inches tall, by using Figure B3.2b.

5. **QUANTITATIVE** Determine the most common height in the class graphed in Figure B3.2b.

BIOSKILL 4 using statistical tests and interpreting standard error bars

When biologists do an experiment, they collect data on individuals in a treatment group and a control group, or several such comparison groups. Then they want to know whether the individuals in the two (or more) groups are different. For example, in one experiment student researchers measured how fast a product formed when they set up a reaction with three different concentrations of reactants (introduced in Chapter 8). Each treatment—meaning, each combination of reactant concentrations—was replicated many times.

FIGURE B4.1 graphs the average reaction rate for each of the three treatments in the experiment. Note that Treatments 1, 2, and 3 represent increasing concentrations of reactants. The thin

"I-beams" on each bar indicate the standard error of each average. The standard error is a quantity that indicates the uncertainty in the calculation of an average.

For example, if two trials with the same concentration of reactants had a reaction rate of 0.075 and two trials had a reaction rate of 0.025, then the average reaction rate would be 0.050. In this case, the standard error would be large. But if two trials had a reaction rate of 0.051 and two had a reaction rate of 0.049, the average would still be 0.050, but the standard error would be small.

In effect, the standard error quantifies how confident you are that the average you've calculated is the average you'd observe if

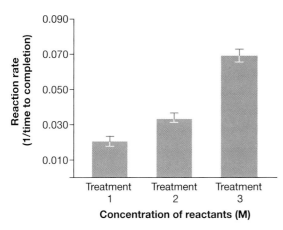

FIGURE B4.1 Standard Error Bars Indicate the Uncertainty in an Average.

you did the experiment under the same conditions an extremely large number of times. It is a measure of precision (see **BIOSKILLS 1**).

Once they had calculated these averages and standard errors, the students wanted to answer a question: Does reaction rate increase when reactant concentration increases?

After looking at the data, you might conclude that the answer is yes. But how could you come to a conclusion like this objectively, instead of subjectively?

The answer is to use a statistical test. This can be thought of as a three-step process.

1. Specify the null hypothesis, which is that reactant concentration has no effect on reaction rate.

2. Calculate a test statistic, which is a number that characterizes the size of the difference among the treatments. In this case, the test statistic compares the actual differences in reaction rates among treatments to the difference predicted by the null hypothesis. The null hypothesis predicts that there should be no difference.

3. The third step is to determine the probability of getting a test statistic at least as large as the one calculated just by chance. The answer comes from a reference distribution—a mathematical function that specifies the probability of getting various values of the test statistic if the null hypothesis is correct. (If you take a statistics course, you'll learn which test

statistics and reference distributions are relevant to different types of data.)

You are very likely to see small differences among treatment groups just by chance—even if no differences actually exist. If you flipped a coin 10 times, for example, you are unlikely to get exactly five heads and five tails, even if the coin is fair. A reference distribution tells you how likely you are to get each of the possible outcomes of the 10 flips if the coin is fair, just by chance.

In this case, the reference distribution indicated that if the null hypothesis of no actual difference in reaction rates is correct, you would see differences at least as large as those observed only 0.01 percent of the time just by chance. By convention, biologists consider a difference among treatment groups to be statistically significant if there is less than a 5 percent probability of observing it just by chance. Based on this convention, the student researchers were able to claim that the null hypothesis is not correct for reactant concentration. According to their data, the reaction they studied really does happen faster when reactant concentration increases.

You'll likely be doing actual statistical tests early in your undergraduate career. To use this text, though, you only need to know what statistical testing does. And you should take care to inspect the standard error bars on graphs in this book. As a *very* rough rule of thumb, averages often turn out to be significantly different, according to an appropriate statistical test, if there is no overlap between two times the standard errors.

check your understanding

If you understand BioSkill 4
✔ **You should be able to . . .**

QUANTITATIVE Determine which of the following tests used to estimate the average height of individuals in a class is likely to have the smallest standard error, and why.

- Measuring the height of two individuals chosen at random to estimate the average.

- Measuring the height of every student who showed up for class on a particular day to estimate the average.

In several cases in this text, you'll need to combine probabilities from different events in order to solve a problem. One of the most common applications is in genetics problems. For example, Punnett squares work because they are based on two fundamental rules of probability. Each rule pertains to a distinct situation.

The Both-And Rule

The both-and rule—also known as the product rule or multiplication rule—applies when you want to know the probability that two or more independent events occur together. Let's use the rolling of two dice as an example. What is the probability of rolling two sixes? These two events are independent, because the probability of rolling a six on one die has no effect on the probability of rolling a six on the other die. (In the same way, the probability of getting a gamete with allele R from one parent has no effect on the probability of getting a gamete with allele R from the other parent. Gametes fuse randomly.)

The probability of rolling a six on the first die is 1/6. The probability of rolling a six on the second die is also 1/6. The probability of rolling a six on *both* dice, then, is $1/6 \times 1/6 = 1/36$. In other words, if you rolled two dice 36 times, on average you would expect to roll two sixes once.

In the case of a cross between two parents heterozygous at the R gene, the probability of getting allele R from the father is 1/2 and the probability of getting R from the mother is 1/2. Thus, the probability of getting both alleles and creating an offspring with genotype RR is $1/2 \times 1/2 = 1/4$.

The Either-Or Rule

The either-or rule—also known as the sum rule or addition rule—applies when you want to know the probability of an event happening when there are several different ways for the same event or outcome to occur. In this case, the probability that the event will occur is the sum of the probabilities of each way that it can occur.

For example, suppose you wanted to know the probability of rolling either a one or a six when you toss a die. The probability of drawing each is 1/6, so the probability of getting one or the other is $1/6 + 1/6 = 1/3$. If you rolled a die three times, on average you'd expect to get a one or a six once.

In the case of a cross between two parents heterozygous at the R gene, the probability of getting an R allele from the father and an r allele from the mother is $1/2 \times 1/2 = 1/4$. Similarly, the probability of getting an r allele from the father and an R allele from the mother is $1/2 \times 1/2 = 1/4$. Thus, the combined probability of getting the Rr genotype in either of the two ways is $1/4 + 1/4 = 1/2$.

check your understanding

If you understand BioSkill 5

✔ **You should be able to . . .**

1. **QUANTITATIVE** Calculate the probability of getting four "tails" if four students each toss a coin.

2. **QUANTITATIVE** Calculate the probability of getting a two, a three, or a six after a single roll of a die.

You have probably been introduced to logarithms and logarithmic notation in algebra courses, and you will encounter logarithms at several points in this course. Logarithms are a way of working with powers—meaning, numbers that are multiplied by themselves one or more times.

Scientists use exponential notation to represent powers. For example,

$$a^x = y$$

means that if you multiply a by itself x times, you get y. In exponential notation, a is called the base and x is called the exponent. The entire expression is called an exponential function.

What if you know y and a, and you want to know x? This is where logarithms come in. You can solve for exponents using logarithms. For example,

$$x = \log_a y$$

This equation reads, x is equal to the logarithm of y to the base a. Logarithms are a way of working with exponential functions. They are important because so many processes in biology (and chemistry and physics, for that matter) are exponential. To understand what's going on, you have to describe the process with an exponential function and then use logarithms to work with that function.

Although a base can be any number, most scientists use just two bases when they employ logarithmic notation: 10 and e (sometimes called Euler's number after Swiss mathematician Leonhard Euler). What is e? It is a rate of exponential growth shared by many natural processes, where e is the limit of $(1 + \frac{1}{n})^n$ (as n tends to infinity). Mathematicians have shown that the base e is an irrational number (like π) that is approximately equal to 2.718. Like 10, e is just a number; $10^0 = 1$ and, likewise, $e^0 = 1$. But both 10 and e have qualities that make them convenient to use in biology (as well as chemistry and physics).

Logarithms to the base 10 are so common that they are usually symbolized in the form $\log y$ instead of $\log_{10} y$. A logarithm to the base e is called a natural logarithm and is symbolized ln (pronounced *EL-EN*) instead of log. You write "the natural logarithm of y" as $\ln y$.

Most scientific calculators have keys that allow you to solve problems involving base 10 and base e. For example, if you know y, they'll tell you what $\log y$ or $\ln y$ are—meaning that they'll solve for x in our first example equation. They'll also allow you to find a number when you know its logarithm to base 10 or base e. Stated another way, they'll tell you what y is if you know x, and y is equal to e^x or 10^x. This is called taking an antilog. In most cases, you'll use the inverse or second function button on your calculator to find an antilog (above the log or ln key).

To get some practice with your calculator, consider this equation:

$$10^2 = 100$$

If you enter 100 in your calculator and then press the log key, the screen should say 2. The logarithm tells you what the exponent is. Now press the antilog key while 2 is on the screen. The calculator screen should return to 100. The antilog solves the exponential function, given the base and the exponent.

If your background in algebra isn't strong, you'll want to get more practice working with logarithms—you'll see them frequently during your undergraduate career. Remember that once you understand the basic notation, there's nothing mysterious about logarithms. They are simply a way of working with exponential functions, which describe what happens when something is multiplied by itself a number of times—like cells that divide and then divide again and then again.

Using logarithms will also come up when you are studying something that can have a large range of values, like the concentration of hydrogen ions in a solution or the intensity of sound that the human ear can detect. In cases like this, it's convenient to express the numbers involved as exponents. Using exponents makes a large range of numbers smaller and more manageable. For example, instead of saying that hydrogen ion concentration in a solution can range from 1 to 10^{-14}, the pH scale allows you to simply say that it ranges from 1 to 14. Instead of giving the actual value, you're expressing it as an exponent. It just simplifies things.

check your understanding

If you understand BioSkill 6

✔ **You should be able to . . .**

Use the equation $N_t = N_0 e^{rt}$ (see Chapter 54).

1. Explain what type of function this equation describes.

2. **QUANTITATIVE** Determine how you would write the equation, after taking the natural logarithm of both sides.

Phylogenetic trees show the evolutionary relationships among species, just as a genealogy shows the relationships among people in your family. They are unusual diagrams, however, and it can take practice to interpret them correctly.

To understand how evolutionary trees work, consider **FIGURE B7.1**. Notice that a phylogenetic tree consists of a root (the most ancestral branch in the tree), branches, nodes, and tips.

- Branches represent populations through time. In this text, branches are drawn as horizontal lines. In most cases the length of the branch is arbitrary and has no meaning, but in some cases branch lengths are proportional to time or the extent of genetic difference among populations (if so, there will be a scale at the bottom of the tree). The vertical lines on the tree represent splitting events, where one group broke into two independent groups. Their length is arbitrary—chosen simply to make the tree more readable.

- Nodes (also called forks) occur where an ancestral group splits into two or more descendant groups (see point A in Figure B7.1). Thus, each node represents the most recent common ancestor of the two or more descendant populations that emerge from it. If more than two descendant groups emerge from a node, the node is called a polytomy (see node C). A polytomy usually means that the populations split from one another so quickly that it is not possible to tell which split off earlier or later.

- Tips (also called terminal nodes) are the tree's endpoints, which represent groups living today or a dead end—a branch

ending in extinction. The names at the tips can represent species or larger groups such as mammals or conifers.

Recall that a taxon (plural: taxa) is any named group of organisms (see Chapter 1). A taxon could be a single species, such as *Homo sapiens*, or a large group of species, such as Primates. Tips connected by a single node on a tree are called sister taxa.

The phylogenetic trees used in this text are all rooted. This means that the first, or most basal, node on the tree—the one on the far left in this book—is the most ancient. To determine where the root on a tree occurs, biologists include one or more outgroup species when they are collecting data to estimate a particular phylogeny. An outgroup is a taxonomic group that is known to have diverged before the rest of the taxa in the study. Outgroups are used to establish whether a trait is ancestral or derived. An ancestral trait is a characteristic that existed in an ancestor; a derived trait is a characteristic that is a modified form of the ancestral trait, found in a descendant.

In Figure B7.1, Taxon 1 is an outgroup to the monophyletic group consisting of taxa 2–6. A monophyletic group consists of an ancestral species and all of its descendants. The root of a tree is placed between the outgroup and the monophyletic group being studied. This position in Figure B7.1 is node A. Note that black hash marks are used to indicate a derived trait that is shared among the red branches, and another derived trait that is shared among the orange branches.

Understanding monophyletic groups is fundamental to reading and estimating phylogenetic trees. Monophyletic groups may also be called lineages or clades and can be identified using the "one-snip test": If you cut any branch on a phylogenetic tree, all of the branches and tips that fall off represent a monophyletic group. Using the one-snip test, you should be able to convince yourself that the monophyletic groups on a tree are nested. In Figure B7.1, for example, the monophyletic group comprising node A and taxa 1–6 contains a monophyletic group consisting of node B and taxa 2–6, which includes the monophyletic group represented by node C and taxa 4–6.

To put all these new terms and concepts to work, consider the phylogenetic tree in **FIGURE B7.2**, which shows the relationships between common chimpanzees and six human and humanlike species that lived over the past 5–6 million years. Chimps functioned as an outgroup in the analysis that led to this tree, so the root was placed at node A. The branches marked in red identify a monophyletic group called the hominins.

To practice how to read a tree, put your finger at the tree's root, at the far left, and work your way to the right. At node A, the ancestral population split into two descendant populations. One of these populations eventually evolved into today's chimps; the other gave rise to the six species of hominins pictured. Now

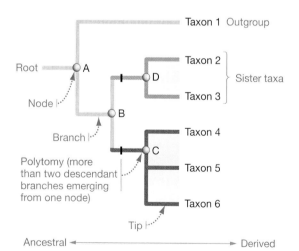

FIGURE B7.1 Phylogenetic Trees Have Roots, Branches, Nodes, and Tips.

✔**EXERCISE** Circle all four monophyletic groups present.

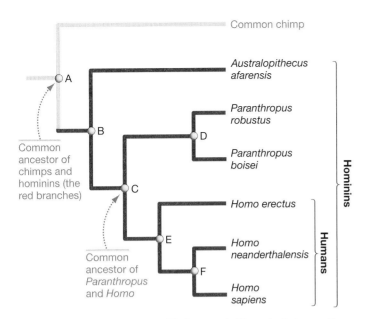

FIGURE B7.2 An Example of a Phylogenetic Tree. A phylogenetic tree showing the relationships of species in the monophyletic group called hominins.

✔**EXERCISE** All of the hominins walked on two legs—unlike chimps and all of the other primates. Add a mark on the phylogeny to show where upright posture evolved, and label it "origin of walking on two legs." Circle and label a pair of sister species. Label an outgroup to the monophyletic group called humans (species in the genus *Homo*).

continue moving your finger toward the tips of the tree until you hit node C. It should make sense to you that at this splitting event, one descendant population eventually gave rise to two *Paranthropus* species, while the other became the ancestor of humans—species in the genus *Homo*. As you study Figure B7.2, consider these two important points:

1. There are many equivalent ways of drawing this tree. For example, this version shows *Homo sapiens* on the bottom. But the tree would be identical if the two branches emerging from node E were rotated 180°, so that the species appeared in the order *Homo sapiens*, *Homo neanderthalensis*, *Homo erectus*. Trees are read from root to tips, not from top to bottom or bottom to top.

2. No species on any tree is any higher or lower than any other. Chimps and *Homo sapiens* have been evolving exactly the same amount of time since their divergence from a common ancestor—neither species is higher or lower than the other. It is legitimate to say that more ancient groups like *Australopithecus afarensis* have traits that are ancestral or more basal—meaning, that appeared earlier in evolution—compared to traits that appear in *Homo sapiens*, which are referred to as more derived.

FIGURE B7.3 presents a chance to test your tree-reading ability. Five of the six trees shown in this diagram are identical in terms of the evolutionary relationships they represent. One differs. The key to understanding the difference is to recognize that the ordering of tips does not matter in a tree—only the ordering of nodes (branch points) matters. You can think of a tree as being like a mobile: The tips can rotate without changing the underlying relationships.

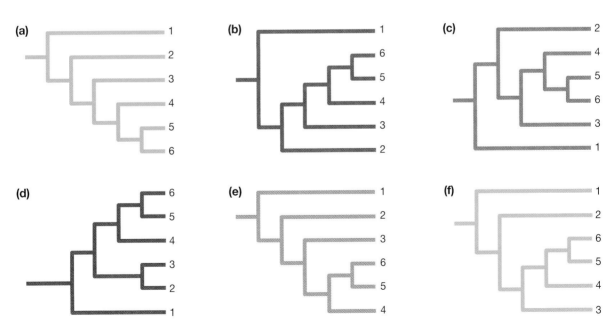

FIGURE B7.3 Alternative Ways of Drawing the Same Tree.

✔**QUESTION** Five of these six trees describe exactly the same relationships among taxa 1 through 6. Identify the tree that is different from the other five.

If you haven't had much chemistry yet, learning basic biological chemistry can be a challenge. One stumbling block is simply being able to read chemical structures efficiently and understand what they mean. This skill will come much easier once you have a little notation under your belt and you understand some basic symbols.

Atoms are the basic building blocks of everything in the universe, just as cells are the basic building blocks of your body. Every atom has a one- or two-letter symbol. **Table B8.1** shows the symbols for most of the atoms you'll encounter in this book. You should memorize these. The table also offers details on how the atoms form bonds as well as how they are represented in some visual models.

When atoms attach to each other by covalent bonding, a molecule forms. Biologists have a couple of different ways of representing molecules—you'll see each of these in the book and in class.

- Molecular formulas like those in **FIGURE B8.1a** simply list the atoms present in a molecule. Subscripts indicate how many of each atom are present. If the formula has no subscript, only one atom of that type is present. A methane (natural gas) molecule, for example, can be written as CH_4. It consists of one carbon atom and four hydrogen atoms.

TABLE B8.1 **Some Attributes of Atoms Found in Organisms**

Atom	Symbol	Number of Bonds It Can Form	Standard Color Code*
Hydrogen	H	1	white
Carbon	C	4	black
Nitrogen	N	3	blue
Oxygen	O	2	red
Sodium	Na	1	—
Magnesium	Mg	2	—
Phosphorus	P	5	orange or purple
Sulfur	S	2	yellow
Chlorine	Cl	1	—
Potassium	K	1	—
Calcium	Ca	2	—

*In ball-and-stick or space-filling models.

- Structural formulas like those in **FIGURE B8.1b** show which atoms in the molecule are bonded to each other. Each bond is indicated by a dash. The structural formula for methane in-

	Methane	Ammonia	Water	Oxygen
(a) Molecular formulas:	CH_4	NH_3	H_2O	O_2
(b) Structural formulas:				
(c) Ball-and-stick models:				
(d) Space-filling models:				

FIGURE B8.1 Molecules Can Be Represented in Several Different Ways.

✔**EXERCISE** Carbon dioxide consists of a carbon atom that forms a double bond with each of two oxygen atoms, for a total of four bonds. It is a linear molecule. Write carbon dioxide's molecular formula and then draw its structural formula, a ball-and-stick model, and a space-filling model.

dicates that each of the four hydrogen atoms forms one covalent bond with carbon, and that carbon makes a total of four covalent bonds. Single covalent bonds are symbolized by a single dash; double bonds are indicated by two dashes.

Even simple molecules have distinctive shapes, because different atoms make covalent bonds at different angles. Ball-and-stick and space-filling models show the geometry of the bonds accurately.

- In a ball-and-stick model, a stick is used to represent each covalent bond (see **FIGURE B8.1c**).

- In space-filling models, the atoms are simply stuck onto each other in their proper places (see **FIGURE B8.1d**).

To learn more about a molecule when you look at a chemical structure, ask yourself three questions:

1. *Is the molecule polar—meaning that some parts are more negatively or positively charged than others?* Molecules that contain nitrogen or oxygen atoms are often polar, because these atoms have such high electronegativity (see Chapter 2). This trait is important because polar molecules dissolve in water.

2. *Does the structural formula show atoms that might participate in chemical reactions?* For example, are there charged atoms or amino or carboxyl $(-COOH)$ groups that might act as a base or an acid?

3. *In ball-and-stick and especially space-filling models of large molecules, are there interesting aspects of overall shape?* For example, is there a groove where a protein might bind to DNA, or a cleft where a substrate might undergo a reaction in an enzyme?

BIOSKILL 9 separating and visualizing molecules

To study a molecule, you have to be able to isolate it. Isolating a molecule is a two-step process: the molecule has to be separated from other molecules in a mixture and then physically picked out or located in a purified form. **BIOSKILLS 9** focuses on the techniques that biologists use to separate nucleic acids and proteins and then find the particular one they are interested in.

Using Electrophoresis to Separate Molecules

In molecular biology, the standard technique for separating proteins and nucleic acids is called gel electrophoresis or, simply, electrophoresis (literally, "electricity-moving"). You may be using electrophoresis in a lab for this course, and you will certainly be analyzing data derived from electrophoresis in this text.

The principle behind electrophoresis is simple. Proteins (when denatured and coated with a special detergent) and nucleic acids carry a charge. As a result, these molecules move when placed in an electric field. Negatively charged molecules move toward the positive electrode; positively charged molecules move toward the negative electrode.

To separate a mixture of macromolecules so that each one can be isolated and analyzed, researchers place the sample in a gelatinous substance. More specifically, the sample is placed in a "well"—a slot in a sheet or slab of the gelatinous substance. The "gel" itself consists of long molecules that form a matrix of fibers. The gelatinous matrix has pores that act like a sieve through which the molecules can pass.

When an electrical field is applied across the gel, the molecules in the well move through the gel toward an electrode. Molecules that are smaller or more highly charged for their size move faster than do larger or less highly charged molecules. As they move, then, the molecules separate by size and by charge. Small and highly charged molecules end up at the bottom of the gel; large, less-charged molecules remain near the top.

An Example "Run"

FIGURE B9.1 (see B:14) shows the electrophoresis setup used in an experiment investigating how RNA molecules polymerize. In this case, the investigators wanted to document how long RNA molecules became over time, when ribonucleoside triphosphates were present in a particular type of solution.

Step 1 shows how they loaded samples of macromolecules, taken on different days during the experiment, into wells at the top of the gel slab. This is a general observation: Each well holds a different sample. In this and many other cases, the researchers also filled a well with a sample containing fragments of known size, called a size standard or "ladder."

In step 2, the researchers immersed the gel in a solution that conducts electricity and applied a voltage across the gel. The molecules in each well started to run down the gel, forming a lane. After several hours of allowing the molecules to move, the researchers removed the electric field (step 3). By then, molecules of different size and charge had separated from one another. In this case, small RNA molecules had reached the bottom of the gel. Above them were larger RNA molecules, which had run more slowly.

PROCESS: GEL ELECTROPHORESIS

Samples of macromolecules collected on different days

2 4 6 8 14

Fragments of known size

Wells —

Gel —

Power supply

−

+

Molecules that are smaller and more highly charged run farther than molecules that are larger and less highly charged

1. Load cavities ("wells") in gel with samples.

2. Hook up power supply and run gel. Molecules separate over time as some migrate faster than others.

3. Remove gel after samples have run its length.

FIGURE B9.1 Macromolecules Can Be Separated via Gel Electrophoresis.

✔**QUESTION** DNA and RNA run toward the positive electrode. Why are these molecules negatively charged?

Why Do Separated Molecules Form Bands?

When researchers visualize a particular molecule on a gel, using techniques described in this section, the image that results consists of bands: shallow lines that are as wide as a lane in the gel. Why?

To understand the answer, study **FIGURE B9.2**. The left panel shows the original mixture of molecules. In this cartoon, the size of each dot represents the size of each molecule. The key is to realize that the original sample contains many copies of each specific molecule, and that these copies run down the length of the gel together—meaning, at the same rate—because they have the same size and charge.

It's that simple: Molecules that are alike form a band because they stay together.

Using Thin Layer Chromatography to Separate Molecules

Gel electrophoresis is not the only way to separate molecules. Researchers also use a method called thin layer chromatography. This method was developed in the early 1900s by botanists who were analyzing the different-colored pigments from leaves of a plant (see Chapter 10), hence the name chromatography from the Greek words khroma for "color" and graphein, "to write."

In this method, rather than loading the sample into the well of a gel, the samples are deposited or "spotted" near the bottom of a stiff support, either glass or plastic, that is coated with a thin layer of silica gel, cellulose, or a similar porous material. The coated support is placed in a solvent solution. As the solvent

PROCESS: FORMATION OF BANDS ON GELS

Well

1. Start with a mixture of molecules in a well.

2. As electrophoresis starts, molecules begin to separate by size and charge.

3. As electrophoresis continues, separation increases. Molecules with the same size and charge "run" at the same rate.

4. If each molecule is visualized, the result is a set of bands.

FIGURE B9.2 On a Gel, Molecules That Are Alike Form Bands.

wicks upward through the coating by capillary action, it carries the molecules in the mixture with it. Molecules are carried at different rates, based on their size and solubility in the solvent.

Visualizing Molecules

Once molecules have been separated using electrophoresis or thin layer chromatography, they have to be detected. Unfortunately, although plant pigments are colored, proteins and nucleic acids are invisible unless they are tagged in some way. Let's first look at two of the most common tagging systems and then consider how researchers can tag and visualize specific molecules of interest and not others.

Using Radioactive Isotopes and Autoradiography

When molecular biology was getting under way, the first types of tags in common use were radioactive isotopes—forms of atoms that are unstable and release energy in the form of radiation.

In the polymerization experiment diagrammed in Figure B9.1, for example, the researchers had attached a radioactive phosphorus atom to the monomers—ribonucleoside triphosphates—used in the original reaction mix. Once polymers formed, they contained radioactive atoms. When electrophoresis was complete, the investigators visualized the polymers by laying X-ray film over the gel. Because radioactive emissions expose film, a black dot appears wherever a radioactive atom is located in the gel. So many black dots occur so close together that the collection forms a dark band.

This technique for visualizing macromolecules is called autoradiography. The autoradiograph that resulted from the polymerization experiment is shown in **FIGURE B9.3**. The samples, taken on days 2, 4, 6, 8, and 14 of the experiment, are noted along the bottom. The far right lane contains macromolecules of known size; this lane is used to estimate the size of the molecules in the experimental samples. The bands that appear in each sample lane represent the different polymers that had formed.

Reading a Gel

One of the keys to interpreting or "reading" a gel, or the corresponding autoradiograph, is to realize that darker bands contain more radioactive markers, indicating the presence of many radioactive molecules. Lighter bands contain fewer molecules.

To read a gel, then, you look for (1) the presence or absence of bands in some lanes—meaning, some experimental samples—versus others, and (2) contrasts in the darkness of the bands—meaning, differences in the number of molecules present.

For example, several conclusions can be drawn from the data in Figure B9.3. First, a variety of polymers formed at each stage. After the second day, for example, polymers from 12 to 18 monomers long had formed. Second, the overall length of polymers produced increased with time. At the end of the fourteenth day, most of the RNA molecules were between 20 and 40 monomers long.

FIGURE B9.3 Autoradiography Is a Technique for Visualizing Macromolecules. The molecules in a gel can be visualized in a number of ways. In this case, the RNA molecules in the gel exposed an X-ray film because they had radioactive atoms attached. When developed, the film is called an autoradiograph.

Starting in the late 1990s and early 2000s, it became much more common to tag nucleic acids with fluorescent tags. Once electrophoresis is complete, fluorescence can be detected by exposing the gel to an appropriate wavelength of light; the fluorescent tag fluoresces or glows in response (fluorescence is explained in Chapter 10).

Fluorescent tags have important advantages over radioactive isotopes: (1) They are safer to handle. (2) They are faster—you don't have to wait hours or days for the radioactive isotope to expose a film. (3) They come in multiple colors, so you can tag several different molecules in the same experiment and detect them independently.

Using Nucleic Acid Probes

In many cases, researchers want to find one specific molecule—a certain DNA sequence, for example—in the collection of molecules on a gel. How is this possible? The answer hinges on using a particular molecule as a probe.

You'll learn in more detail about how probes work in this text (Chapter 20). Here it's enough to get the general idea: A probe is a marked molecule that binds specifically to your molecule of interest. The "mark" is often a radioactive atom, a fluorescent tag, or an enzyme that catalyzes a color-forming or light-emitting reaction.

If you are looking for a particular DNA or RNA sequence on a gel, for example, you can expose the gel to a single-stranded probe that binds to the target sequence by complementary base pairing. Once it has bound, you can detect the band through autoradiography or fluorescence.

- **Southern blotting** is a technique for making DNA fragments that have been run out on a gel single stranded, transferring them from the gel to a nylon membrane, and then probing them to identify segments of interest. The technique was named after its inventor, Edwin Southern.

- **Northern blotting** is a technique for transferring RNA fragments from a gel to a nylon membrane and then probing them to detect target segments. The name is a lighthearted play on Southern blotting—the protocol from which it was derived.

Using Antibody Probes

How can researchers find a particular protein out of a large collection of different proteins? The answer is to use an antibody. An antibody is a protein that binds specifically to a section of a different protein (see Chapter 51 if you want more detail on the structure of antibodies and their function in the immune system).

To use an antibody as a probe, investigators attach a tag molecule—often an enzyme that catalyzes a color-forming reaction—to the antibody and allow it to react with proteins in a mixture. The antibody will stick to the specific protein that it binds to and then can be visualized thanks to the tag it carries.

If the proteins in question have been separated by gel electrophoresis and transferred to a membrane, the result is called a western blot. The name western is an extension of the Southern and northern patterns.

Using Radioimmunoassay and ELISA to Measure Amounts of Molecules

Another important method that makes use of antibodies is called a radioimmunoassay. This method is used when investigators want to measure tiny amounts of a molecule, such as a hormone in the blood. In this case, a known quantity of a hormone is labeled with a radioactive tag. This tagged hormone is then mixed with a known amount of antibody, and the two bind to one another. Next, a sample of blood, containing an unknown quantity of that same hormone, is added. The hormone from the blood and the radiolabeled hormone compete for antibody binding sites. As the concentration of unlabeled hormone increases, more of it binds to the antibody, displacing more of the radiolabeled hormone. The amount of unbound radiolabeled hormone is then measured. Using known standards as a reference, the amount of hormone in the blood can be determined.

Another commonly used technique based on similar principles is called ELISA (enzyme-linked immunosorbent assay). In this case, the amount of a particular molecule is measured using colorimetric signals instead of a radioactive signal.

check your understanding

If you understand BioSkill 9

✓**You should be able to . . .**

Interpret a gel that has been stained for "RNA X." One lane contains no bands. Two lanes have a band in the same location, even though one of the bands is barely visible and the other is extremely dark. The fourth lane has a faint band located below the bands in the other lanes.

Biologists use a technique called differential centrifugation to isolate specific cell components. Differential centrifugation is based on breaking cells apart to create a complex mixture and then separating components in a centrifuge. A centrifuge accomplishes this task by spinning cells in a solution that allows molecules and other cell components to separate according to their density or size and shape. The individual parts of the cell can then be purified and studied in detail, in isolation from other parts of the cell.

The first step in preparing a cell sample for centrifugation is to release the cell components by breaking the cells apart. This can be done by putting them in a hypotonic solution, by exposing them to high-frequency vibration, by treating cells with a detergent, or by grinding them up. Each of these methods breaks apart plasma membranes and releases the contents of the cells.

The resulting pieces of plasma membrane quickly reseal to form small vesicles, often trapping cell components inside. The solution that results from the homogenization step is a mixture of these vesicles, free-floating macromolecules released from the cells, and organelles. A solution like this is called a cell extract or cell homogenate.

When a cell homogenate is placed in a centrifuge tube and spun at high speed, the components that are in solution tend to move outward, along the red arrow in **FIGURE B10.1a**. The effect is similar to a merry-go-round, which seems to push you outward in a straight line away from the spinning platform. In response to this outward-directed force, the cell homogenate exerts a centripetal (literally, "center-seeking") force that pushes the homogenate away from the bottom of the tube. Larger, denser molecules or particles resist this inward force more readily than do smaller, less dense ones and so reach the bottom of the centrifuge tube faster.

To separate the components of a cell extract, researchers often perform a series of centrifuge runs. Steps 1 and 2 of

(a) How a centrifuge works

When the centrifuge spins, the macromolecules tend to move toward the bottom of the centrifuge tube (red arrow)

The solution in the tube exerts a centripetal force, which resists movement of the molecules to the bottom of the tube (blue arrow)

Very large or dense molecules overcome the centripetal force more readily than smaller, less dense ones. As a result, larger, denser molecules move toward the bottom of the tube faster.

(b) PROCESS: DIFFERENTIAL CENTRIFUGATION

1. Start with uniform cell homogenate in centrifuge tube.

2. After low-speed spin, pellet contains large components. Transfer supernatant to new tube.

3. After medium-speed spin, pellet contains medium components. Transfer supernatant to new tube.

4. After high-speed spin, pellet contains small components.

(c) PROCESS: SUCROSE DENSITY–GRADIENT CENTRIFUGATION

1. Add sample to tube of variable-density solution.

2. Run centrifuge. Cell components are separated into distinct bands based on size and/or density.

3. To extract specific cell components for analysis, puncture bottom of tube with needle and collect drops from specific bands.

FIGURE B10.1 Cell Components Can Be Separated by Centrifugation. (a) The forces inside a centrifuge tube allow cell components to be separated. **(b)** Through a series of centrifuge runs made at increasingly higher speeds, an investigator can separate fractions of a cell homogenate by size via differential centrifugation. **(c)** A high-speed centrifuge run can achieve extremely fine separation among cell components by sucrose density–gradient centrifugation.

FIGURE B10.1b illustrate how an initial treatment at low speed causes larger, heavier parts of the homogenate to move below smaller, lighter parts. The material that collects at the bottom of the tube is called the pellet, and the solution and solutes left behind form the supernatant ("above-swimming"). The supernatant is placed in a fresh tube and centrifuged at increasingly higher speeds and longer durations. Each centrifuge run continues to separate cell components based on their size and density.

To accomplish separation of macromolecules or organelles, researchers frequently follow up with centrifugation at extremely high speeds. One strategy is based on filling the centrifuge tube with a series of sucrose solutions of increasing density (**FIGURE B10.1c**). The density gradient allows cell components to separate on the basis of small differences in size, shape, and density. When the centrifuge run is complete, each cell component occupies a distinct band of material in the tube, based on how quickly each component moves through the increasingly dense gradient of sucrose solution during the centrifuge run. A researcher can then collect the material in each band for further study.

check your understanding

If you understand BioSkill 10

✓ **You should be able to ...**

1. Explain the physical basis for separating molecules or cell components via centrifugation given that electrophoresis separates molecules by charge and size.

2. Determine which cell component—ribosomes or mitochondria—you would expect to find closer to the bottom of the tube if you centrifuge a cell homogenate by sucrose density–gradient centrifugation.

BIOSKILL — biological imaging: microscopy and x-ray crystallography

A lot of biology happens at levels that can't be detected with the naked eye. Biologists use an array of microscopes to study small multicellular organisms, individual cells, and the contents of cells. And to understand what individual macromolecules or macromolecular machines like ribosomes look like, researchers use data from a technique called X-ray crystallography.

You'll probably use dissecting microscopes and compound light microscopes to view specimens during your labs for this course, and throughout this text you'll be seeing images generated from other types of microscopy and from X-ray crystallographic data. Among the fundamental skills you'll be acquiring as an introductory student, then, is a basic understanding of how these techniques work. The key is to recognize that each approach for visualizing microscopic structures has strengths and weaknesses. As a result, each technique is appropriate for studying certain types or aspects of cells or molecules.

Light and Fluorescence Microscopy

If you use a dissecting microscope during labs, you'll recognize that it works by magnifying light that bounces off a whole specimen—often a live organism. You'll be able to view the specimen in three dimensions, which is why these instruments are sometimes called stereomicroscopes, but the maximum magnification possible is only about 20 to 40 times normal size (20✕ to 40✕).

To view smaller objects, you'll probably use a compound microscope. Compound microscopes magnify light that is passed *through* a specimen. The instruments used in introductory labs are usually capable of 400✕ magnifications; the most sophisticated compound microscopes available can achieve magnifications of about 2000✕. This is enough to view individual bacterial or eukaryotic cells and see large structures inside cells, like condensed chromosomes (see Chapter 12). To prepare a specimen for viewing under a compound light microscope, the tissues or cells are usually sliced to create a section thin enough for light to pass through efficiently. The section is then dyed to increase contrast and make structures visible. In many cases, different types of dyes are used to highlight different types of structures.

To visualize specific proteins, researchers use a technique called immunostaining. After preparing tissues or cells for viewing, the specimen is stained with fluorescently tagged antibodies. In this case, the cells are viewed under a fluorescence microscope. Ultraviolet or other wavelengths of light are passed through the specimen. The fluorescing tag emits visible light in response. The result? Beautiful cells that glow green, red, or blue.

Electron Microscopy

Until the 1950s, the compound microscope was the biologist's only tool for viewing cells directly. But the invention of the electron microscope provided a new way to view specimens. Two basic types of electron microscopy are now available: one that allows researchers to examine cross sections of cells at extremely high magnification, and one that offers a view of surfaces at somewhat lower magnification.

Transmission Electron Microscopy

The transmission electron microscope (TEM) is an extraordinarily effective tool for viewing cell structure at high magnification. TEM forms an image from electrons that pass through a specimen, just as a light microscope forms an image from light rays that pass through a specimen.

Biologists who want to view a cell under a transmission electron microscope begin by "fixing" the cell, meaning that they treat it with a chemical agent that stabilizes the cell's structure and contents while disturbing them as little as possible. Then the researcher permeates the cell with an epoxy plastic that stiffens the structure. Once this epoxy hardens, the cell can be cut into extremely thin sections with a glass or diamond knife. Finally, the sectioned specimens are impregnated with a metal—often lead. (The reason for this last step is explained shortly.)

FIGURE B11.1a outlines how the transmission electron microscope works. A beam of electrons is produced by a tungsten filament at the top of a column and directed downward. (All of the air is pumped out of the column, so that the electron beam isn't scattered by collisions with air molecules.) The electron beam passes through a series of lenses and through the specimen. The lenses are actually electromagnets, which alter the path of the beam much like a glass lens in a dissecting or compound microscope bends light. The electromagnet lenses magnify and focus the image on a screen at the bottom of the column. There the electrons strike a coating of fluorescent crystals, which emit visible light in response—just like a television screen. When the microscopist moves the screen out of the way and allows the electrons to expose a sheet of black-and-white film or to be detected by a digital camera, the result is a micrograph—a photograph of an image produced by microscopy.

The image itself is created by electrons that pass through the specimen. If no specimen were in place, all the electrons would pass through and the screen (and micrograph) would be uniformly bright. Unfortunately, cell materials by themselves would also appear fairly uniform and bright. This is because an atom's ability to deflect an electron depends on its mass. In turn, an atom's mass is a function of its atomic number. The hydrogen, carbon, oxygen, and nitrogen atoms that dominate biological molecules have low atomic numbers. This is why cell biologists must saturate cell sections with lead solutions. Lead has a high atomic number and scatters electrons effectively. Different macromolecules take up lead atoms in different amounts, so the metal acts as a "stain" that produces contrast. With TEM, areas of dense metal scatter the electron beam most, producing dark areas in micrographs.

The advantage of TEM is that it can magnify objects up to 250,000×—meaning that intracellular structures are clearly visible. The downsides are that researchers are restricted to observing dead, sectioned material, and they must take care that the preparation process does not distort the specimen.

Scanning Electron Microscopy

The scanning electron microscope (SEM) is the most useful tool biologists have for looking at the surfaces of structures. Materials are prepared for scanning electron microscopy by coating their surfaces with a layer of metal atoms. To create an image of this surface, the microscope scans the surface with a narrow beam of electrons. Electrons that are reflected back from the surface or that are emitted by the metal atoms in response to the beam then strike a detector. The signal from the detector controls a second electron beam, which scans a TV-like screen and forms an image magnified up to 50,000 times the object's size.

Because SEM records shadows and highlights, it provides images with a three-dimensional appearance (**FIGURE B11.1b**). It cannot magnify objects nearly as much as TEM can, however.

(a) Transmission electron microscopy: High magnification of cross sections

Tungsten filament (source of electrons)

Condenser lens

Specimen

Objective lens

Projector lens

Image on fluorescent screen

Cross section of *E. coli* bacterium

0.2 µm

(b) Scanning electron microscopy: Lower magnification of surfaces

Surface view of *E. coli* bacteria

1 µm

FIGURE B11.1 There Are Two Basic Types of Electron Microscopy.

Studying Live Cells and Real-Time Processes

Until the 1960s, biologists were unable to get clear, high-magnification images of living cells. But a series of innovations over the past 50 years has made it possible to observe organelles and subcellular structures in action.

The development of video microscopy, where the image from a light microscope is captured by a video camera instead of by an eye or a film camera, proved revolutionary. It allowed specimens to be viewed at higher magnification, because video cameras are more sensitive to small differences in contrast than are the human eye or still cameras. It also made it easier to keep live specimens functioning normally, because the increased light sensitivity of video cameras allows them to be used with low illumination, so specimens don't overheat. And when it became possible to digitize video images, researchers began using computers to remove out-of-focus background material and increase image clarity.

A more recent innovation was the use of a fluorescent molecule called green fluorescent protein, or GFP, which allows researchers to tag specific molecules or structures and follow their movement in live cells over time. This was a major advance over immunostaining, in which cells have to be fixed. GFP is naturally synthesized in jellyfish that fluoresce, or emit light. By affixing GFP to another protein (using genetic engineering techniques described in Chapter 20) and then inserting it into a live cell, investigators can follow the protein's fate over time and even videotape its movement. For example, researchers have videotaped GFP-tagged proteins being transported from the rough ER through the Golgi apparatus and out to the plasma membrane. This is cell biology: the movie.

GFP's influence has been so profound that the researchers who developed its use in microscopy were awarded the 2008 Nobel Prize in Chemistry.

Visualizing Structures in 3-D

The world is three-dimensional. To understand how microscopic structures and macromolecules work, it is essential to understand their shape and spatial relationships. Consider three techniques currently being used to reconstruct the 3-D structure of cells, organelles, and macromolecules.

- **Confocal microscopy** is carried out by mounting cells that have been treated with one or more fluorescing tags on a microscope slide and then focusing a beam of ultraviolet or other wavelengths of light at a specific depth within the specimen. The fluorescing tag emits visible light in response. A detector for this light is then set up at exactly the position where the emitted light comes into focus. The result is a sharp image of a precise plane in the cell being studied (**FIGURE B11.2a**). Note that if you viewed the same specimen under a conventional fluorescence microscope, the image would be blurry because it results from light emitted by the entire cell (**FIGURE B11.2b**). By altering the focal plane, a researcher can record images from

(a) Confocal fluorescence image of single cell

(b) Conventional fluorescence image of same cell

25 µm

FIGURE B11.2 Confocal Microscopy Provides Sharp Images of Living Cells. (a) The confocal image of this mouse intestinal cell is sharp, because it results from light emitted at a single plane inside the cell. **(b)** The conventional image of this same cell is blurred, because it results from light emitted by the entire cell.

an array of depths in the specimen; a computer can then be used to generate a 3-D image of the cell.

- **Electron tomography** uses a transmission electron microscope to generate a 3-D image of an organelle or other subcellular structure. The specimen is rotated around a single axis while the researcher takes many "snapshots." The individual images are then pieced together with a computer. This technique has provided a much more accurate view of mitochondrial structure than was possible using traditional TEM (see Chapter 7).

- **X-ray crystallography, or X-ray diffraction analysis**, is the most widely used technique for reconstructing the 3-D structure of molecules. As its name implies, the procedure is based on bombarding crystals of a molecule with X-rays. X-rays are scattered in precise ways when they interact with the electrons surrounding the atoms in a crystal, producing a diffraction pattern that can be recorded on X-ray film or other types of detectors (**FIGURE B11.3**). By varying the orientation of the X-ray beam as it strikes a crystal and documenting the

check your understanding

 If you understand BioSkill 11

 ✔ **You should be able to . . .**

1. Interpret whether the absence of mitochondria in a transmission electron micrograph of a cancerous human liver means that the cell lacks mitochondria.

 2. Explain why the effort to understand the structure of biological molecules is worthwhile even though X-ray crystallography is time consuming and technically difficult. What's the payoff?

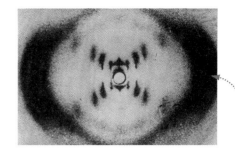

The patterns are determined by the structure of the molecules within the crystal

FIGURE B11.3 X-Ray Crystallography. When crystallized molecules are bombarded with X-rays, the radiation is scattered in distinctive patterns. The photograph at the right shows an X-ray film that recorded the pattern of scattered radiation from DNA molecules.

diffraction patterns that result, researchers can construct a map representing the density of electrons in the crystal. By relating these electron-density maps to information about the primary structure of the nucleic acid or protein, a 3-D model of the molecule can be built. Virtually all of the molecular models used in this book were built from X-ray crystallographic data.

BIOSKILL 12 cell and tissue culture methods

For researchers, there are important advantages to growing plant and animal cells and tissues outside the organism itself. Cell and tissue cultures provide large populations of a single type of cell or tissue and the opportunity to control experimental conditions precisely.

Animal Cell Culture

The first successful attempt to culture animal cells occurred in 1907, when a researcher cultivated amphibian nerve cells in a drop of fluid from the spinal cord. But it was not until the 1950s and 1960s that biologists could routinely culture plant and animal cells in the laboratory. The long lag time was due to the difficulty of re-creating conditions that exist in the intact organism precisely enough for cells to grow normally.

To grow in culture, animal cells must be provided with a liquid mixture containing the nutrients, vitamins, and hormones that stimulate growth. Initially, this mixture was serum, the liquid portion of blood; now, serum-free media are available for certain cell types. Serum-free media are preferred because they are much more precisely defined chemically than serum.

In addition, many types of animal cells will not grow in culture unless they are provided with a solid surface that mimics the types of surfaces that enable cells in the intact organisms to adhere. As a result, cells are typically cultured in flasks (**FIGURE B12.1a**, left; see B:22).

Even under optimal conditions, though, normal cells display a finite life span in culture. In contrast, many cultured cancerous cells grow indefinitely. This characteristic correlates with a key feature of cancerous cells in organisms: Their growth is continuous and uncontrolled.

Because of their immortality and relative ease of growth, cultured cancer cells are commonly used in research on basic aspects of cell structure and function. For example, the first human cell type to be grown in culture was isolated in 1951 from a malignant tumor of the uterine cervix. These cells are called HeLa cells in honor of their donor, Henrietta Lacks, who died soon thereafter from her cancer. HeLa cells continue to grow in laboratories around the world (Figure B12.1a, right).

Plant Tissue Culture

Certain cells found in plants are totipotent—meaning that they retain the ability to divide and differentiate into a complete, mature plant, including new types of tissue. These cells, called parenchyma cells, are important in wound healing and asexual reproduction. But they also allow researchers to grow complete adult plants in the laboratory, starting with a small number of parenchyma cells.

Biologists who grow plants in tissue culture begin by placing parenchyma cells in a liquid or solid medium containing all the

(a) Animal cell culture: immortal HeLa cancer cells

(b) Plant tissue culture: tobacco callus

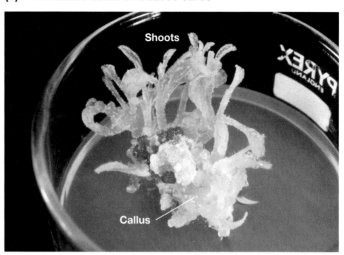

FIGURE B12.1 Animal and Plant Cells Can Be Grown in the Lab.

nutrients required for cell maintenance and growth. In the early days of plant tissue culture, investigators found not only that specific growth signals called hormones were required for successful growth and differentiation but also that the relative abundance of hormones present was critical to success.

The earliest experiments on hormone interactions in tissue cultures were done with tobacco cells in the 1950s by Folke Skoog and co-workers. These researchers found that when the hormone called auxin was added to the culture by itself, the cells enlarged but did not divide. But if the team added roughly equal amounts of auxin and another growth signal called cytokinin to the cells, the cells began to divide and eventually formed a callus, or an undifferentiated mass of parenchyma cells.

By varying the proportion of auxin to cytokinins in different parts of the callus and through time, the team could stimulate the growth and differentiation of root and shoot systems and produce whole new plants (**FIGURE B12.1b**). A high ratio of auxin to cytokinin led to the differentiation of a root system, while a high ratio of cytokinin to auxin led to the development of a shoot system. Eventually Skoog's team was able to produce a complete plant from just one parenchyma cell.

The ability to grow whole new plants in tissue culture from just one cell has been instrumental in the development of genetic engineering (see Chapter 20). Researchers insert recombinant genes into target cells, test the cells to identify those that successfully express the recombinant genes, and then use tissue culture techniques to grow those cells into adult individuals with a novel genotype and phenotype.

check your understanding

 If you understand BioSkill 12

✔ **You should be able to . . .**

 1. Identify a limitation of how experiments on HeLa cells are interpreted.

 2. State a disadvantage of doing experiments on plants that have been propagated from single cells growing in tissue culture.

Research in biological science starts with a question. In most cases, the question is inspired by an observation about a cell or an organism. To answer it, biologists have to study a particular species. Study organisms are often called model organisms, because investigators hope that they serve as a model for what is going on in a wide array of species.

Model organisms are chosen because they are convenient to study and because they have attributes that make them appropriate for the particular research proposed. They tend to have some common characteristics:

- **Short generation time and rapid reproduction** This trait is important because it makes it possible to produce offspring quickly and perform many experiments in a short amount of time—you don't have to wait long for individuals to grow.

- **Large numbers of offspring** This trait is particularly important in genetics, where many offspring phenotypes and genotypes need to be assessed to get a large sample size.

- **Small size, simple feeding and habitat requirements** These attributes make it relatively cheap and easy to maintain individuals in the lab.

The following notes highlight just a few model organisms supporting current work in biological science.

Escherichia coli

Of all model organisms in biology, perhaps none has been more important than the bacterium *Escherichia coli*—a common inhabitant of the human gut. The strain that is most commonly worked on today, called K-12 (**FIGURE B13.1a**; see B:24), was originally isolated from a hospital patient in 1922.

During the last half of the twentieth century, key results in molecular biology originated in studies of *E. coli*. These results include the discovery of enzymes such as DNA polymerase, RNA polymerase, DNA repair enzymes, and restriction endonucleases; the elucidation of ribosome structure and function; and the initial characterization of promoters, regulatory transcription factors, regulatory sites in DNA, and operons. In many cases, initial discoveries made in *E. coli* allowed researchers to confirm that homologous enzymes and processes existed in an array of organisms, often ranging from other bacteria to yeast, mice, and humans.

The success of *E. coli* as a model for other species inspired Jacques Monod's claim that "Once we understand the biology of *Escherichia coli*, we will understand the biology of an elephant." The genome of *E. coli* K-12 was sequenced in 1997, and the strain continues to be a workhorse in studies of gene function, biochemistry, and particularly biotechnology. Much remains to be learned, however. Despite over 60 years of intensive study, the function of about a third of the *E. coli* genome is still unknown.

In the lab, *E. coli* is usually grown in suspension culture, where cells are introduced to a liquid nutrient medium, or on plates containing agar—a gelatinous mix of polysaccharides. Under optimal growing conditions—meaning before cells begin to get crowded and compete for space and nutrients—a cell takes just 30 minutes on average to grow and divide. At this rate, a single cell can produce a population of over a million descendants in just 10 hours. Except for new mutations, all of the descendant cells are genetically identical.

Dictyostelium discoideum

The cellular slime mold *Dictyostelium discoideum* is not always slimy, and it is not a mold—meaning a type of fungus. Instead, it is an amoeba. Amoeba is a general term that biologists use to characterize a unicellular eukaryote that lacks a cell wall and is extremely flexible in shape. *Dictyostelium* has long fascinated biologists because it is a social organism. Independent cells sometimes aggregate to form a multicellular structure.

Under most conditions, *Dictyostelium* cells are haploid (*n*) and move about in decaying vegetation on forest floors or other habitats. They feed on bacteria by engulfing them whole. When these cells reproduce, they can do so sexually by fusing with another cell then undergoing meiosis, or asexually by mitosis, which is more common. If food begins to run out, the cells begin to aggregate. In many cases, tens of thousands of cells cohere to form a 2-mm-long mass called a slug (**FIGURE B13.1b**). (This is not the slug that is related to snails.)

After migrating to a sunlit location, the slug stops and individual cells differentiate according to their position in the slug. Some form a stalk; others form a mass of spores at the tip of the stalk. (A spore is a single cell that develops into an adult organism, but it is not formed from gamete fusion like a zygote is.) The entire structure, stalk plus mass of spores, is called a fruiting body. Cells that form spores secrete a tough coat and represent a durable resting stage. The fruiting body eventually dries out, and the wind disperses the spores to new locations, where more food might be available.

Dictyostelium has been an important model organism for investigating questions about eukaryotes:

- Cells in a slug are initially identical in morphology but then differentiate into distinctive stalk cells and spores. Studying this process helped biologists better understand how cells in plant and animal embryos differentiate into distinct cell types.

(a) Bacterium *Escherichia coli* (strain K-12)

0.5 μm

(b) Slime mold *Dictyostelium discoideum*

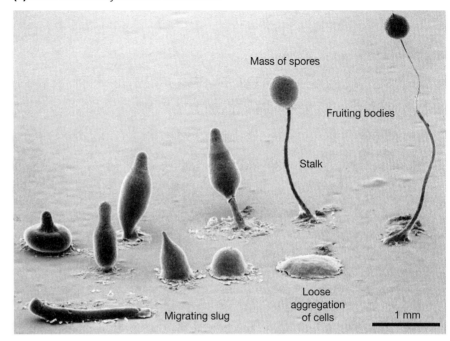

Mass of spores

Fruiting bodies

Stalk

Migrating slug

Loose aggregation of cells

1 mm

(c) Thale cress *Arabidopsis thaliana*

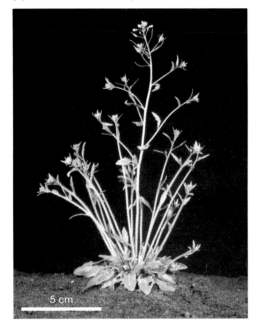

5 cm

(e) Fruit fly *Drosophila melanogaster*

0.5 mm

(f) Roundworm *Caenorhabditis elegans*

0.1 mm

(d) Yeast *Saccharomyces cerevisiae*

5 μm

(g) Mouse *Mus musculus*

1 cm

FIGURE B13.1 Model Organisms.

✔ **QUESTION** *E. coli* is grown at a temperature of 37°C. Why?

- The process of slug formation has helped biologists study how animal cells move and how they aggregate as they form specific types of tissues.

- When *Dictyostelium* cells aggregate to form a slug, they stick to each other. The discovery of membrane proteins responsible for cell–cell adhesion helped biologists understand some of the general principles of multicellular life (highlighted in Chapter 11).

Arabidopsis thaliana

In the early days of biology, the best-studied plants were agricultural varieties such as maize (corn), rice, and garden peas. When biologists began to unravel the mechanisms responsible for oxygenic photosynthesis in the early to mid-1900s, they relied on green algae that were relatively easy to grow and manipulate in the lab—often the unicellular species *Chlamydomonas reinhardii*—as an experimental subject.

Although crop plants and green algae continue to be the subject of considerable research, a new model organism emerged in the 1980s and now serves as the preeminent experimental subject in plant biology. That organism is *Arabidopsis thaliana*, commonly known as thale cress or wall cress (**FIGURE B13.1c**).

Arabidopsis is a member of the mustard family, or Brassicaceae, so it is closely related to radishes and broccoli. In nature it is a weed—meaning a species that is adapted to thrive in habitats where soils have been disturbed.

One of the most attractive aspects of working with *Arabidopsis* is that individuals can grow from a seed into a mature, seed-producing plant in just four to six weeks. Several other attributes make it an effective subject for study: It has just five chromosomes, has a relatively small genome with limited numbers of repetitive sequences, can self-fertilize as well as undergo cross-fertilization, can be grown in a relatively small amount of space and with a minimum of care in the greenhouse, and produces up to 10,000 seeds per individual per generation.

Arabidopsis has been instrumental in a variety of studies in plant molecular genetics and development, and it is increasingly popular in ecological and evolutionary studies. In addition, the entire genome of the species has now been sequenced, and studies have benefited from the development of an international "*Arabidopsis* community"—a combination of informal and formal associations of investigators who work on *Arabidopsis* and use regular meetings, e-mail, and the Internet to share data, techniques, and seed stocks.

Saccharomyces cerevisiae

When biologists want to answer basic questions about how eukaryotic cells work, they often turn to the yeast *Saccharomyces cerevisiae*.

S. cerevisiae is unicellular and relatively easy to culture and manipulate in the lab (**FIGURE B13.1d**). In good conditions, yeast cells grow and divide almost as rapidly as bacteria. As a result,

the species has become the organism of choice for experiments on control of the cell cycle and regulation of gene expression in eukaryotes. For example, research has confirmed that several of the genes controlling cell division and DNA repair in yeast have homologs in humans; and when mutated, these genes contribute to cancer. Strains of yeast that carry these mutations are now being used to test drugs that might be effective against cancer.

S. cerevisiae has become even more important in efforts to interpret the genomes of organisms like rice, mice, zebrafish, and humans. It is much easier to investigate the function of particular genes in *S. cerevisiae* by creating mutants or transferring specific alleles among individuals than it is to do the same experiments in mice or zebrafish. Once the function of a gene has been established in yeast, biologists can look for the homologous gene in other eukaryotes. If such a gene exists, they can usually infer that it has a function similar to its role in *S. cerevisiae*. It was also the first eukaryote with a completely sequenced genome.

Drosophila melanogaster

If you walk into a biology building on any university campus around the world, you are almost certain to find at least one lab where the fruit fly *Drosophila melanogaster* is being studied (**FIGURE B13.1e**).

Drosophila has been a key experimental subject in genetics since the early 1900s. It was initially chosen as a focus for study by T. H. Morgan, because it can be reared in the laboratory easily and inexpensively, matings can be arranged, the life cycle is completed in less than two weeks, and females lay a large number of eggs. These traits made fruit flies valuable subjects for breeding experiments designed to test hypotheses about how traits are transmitted from parents to offspring (see Chapter 14).

More recently, *Drosophila* has also become a key model organism in the field of developmental biology. The use of flies in developmental studies was inspired largely by the work of Christianne Nüsslein-Volhard and Eric Wieschaus, who in the 1980s isolated flies with genetic defects in early embryonic development. By investigating the nature of these defects, researchers have gained valuable insights into how various gene products influence the development of eukaryotes (see Chapter 22). The complete genome sequence of *Drosophila* has been available to investigators since the year 2000.

Caenorhabditis elegans

The roundworm *Caenorhabditis elegans* emerged as a model organism in developmental biology in the 1970s, due largely to work by Sydney Brenner and colleagues. (*Caenorhabditis* is pronounced *see-no-rab-DIE-tiss*.)

C. elegans was chosen for three reasons: (**1**) Its cuticle (soft outer layer) is transparent, making individual cells relatively easy to observe (**FIGURE B13.1f**); (**2**) adults have exactly 959 nonreproductive cells; and, most important, (**3**) the fate of each cell in an embryo can be predicted because cell fates are invariant among individuals. For example, when researchers examine a 33-cell *C. elegans* embryo, they know exactly which of the 959 cells in the adult will be derived from each of those 33 embryonic cells.

In addition, *C. elegans* are small (less than 1 mm long), are able to self-fertilize or cross-fertilize, and undergo early development in just 16 hours. The entire genome of *C. elegans* has now been sequenced.

Mus musculus

The house mouse *Mus musculus* is the most important model organism among mammals. For this reason, it is especially prominent in biomedical research, where researchers need to work on individuals with strong genetic and developmental similarities to humans.

The house mouse was an intelligent choice of model organism in mammals because it is small and thus relatively inexpensive to maintain in captivity, and because it breeds rapidly. A litter can contain 10 offspring, and generation time is only 12 weeks—meaning that several generations can be produced in a year. Descendants of wild house mice have been selected for docility and other traits that make them easy to handle and rear; these populations are referred to as laboratory mice (**FIGURE B13.1g**).

Some of the most valuable laboratory mice are strains with distinctive, well-characterized genotypes. Inbred strains are

virtually homogenous genetically (see Chapter 26) and are useful in experiments where gene-by-gene or gene-by-environment interactions have to be controlled. Other populations carry mutations that knock out genes and cause diseases similar to those observed in humans. These individuals are useful for identifying the cause of genetic diseases and testing drugs or other types of therapies.

BIOSKILL primary literature and peer review

As part of the process of doing science, biologists communicate their results to the scientific community through publications in scientific journals that report on their original research discoveries (see Chapter 1). These published reports are referred to, interchangeably, as the primary literature, research papers, or primary research articles.

What Is the Primary Literature?

Scientists publish "peer-reviewed" papers. This means that several experts in the field have carefully read the paper and considered its strengths and weaknesses. Reviewers write a critique of the paper and make a recommendation to the journal editor as to whether the paper should be published. Often reviewers will suggest additional experiments that need to be completed before a paper is considered acceptable for publication. The peer review process means that research discoveries are carefully vetted before they go to press.

A primary research paper can be distinguished from secondary sources—such as review articles, textbooks, and magazine articles—by looking for key characteristics. A primary research paper includes a detailed description of methods and results,

written by the researchers who did the work. A typical paper contains a Title, Abstract, Introduction, Materials and Methods (or Experimental Design), Results and Discussion (**Table B14.1**), although the order and name of the sections varies among journals.

Getting Started

At first, trying to read the primary literature may seem like a daunting task. A paper may be peppered with unfamiliar terms and acronyms. If you tried to read a research paper from start to finish, like you might read a chapter in this textbook, it would be a frustrating experience. But, with practice, the scientific literature becomes approachable, and it is well worth the effort. The primary literature is the cutting edge, the place to read firsthand about the process of doing science. Becoming skilled at reading and evaluating scientific reports is a powerful way to learn how to think critically—to think like a biologist.

To get started, try breaking down reading the primary research article into a series of steps:

1. Read the authors' names. Where are they from? Are they working as a team or alone? After delving into the literature,

TABLE B14.1 Sections of a Primary Research Paper

Section	Characteristics
Title	Short, succinct, eye-catching
Abstract	Summary of Methods, Results, Discussion. Explains why the research was done and why the results are significant.
Introduction	Background information (what past work was done, why the work was important). States the objectives and hypotheses of the study and explains why the study is important.
Materials and Methods	Explains how the work was done and where it was done.
Results	Explains what the data show.
Discussion	Explains why the data show what they show, how the analysis relates to the objectives from the Introduction, and the significance of findings and how they advance the field.

certain familiar names will crop up again and again. You'll begin to recognize the experts in a particular field.

2. Read the title. It should summarize the key finding of the paper and tell you what you can expect to learn from the paper.

3. Read the abstract. The abstract summarizes the entire paper in a short paragraph. At this point, it might be tempting to stop reading. But sometimes authors understate or overstate the significance and conclusions of their work. You should never cite an article as a reference after having read only the abstract.

4. Read the Introduction. The first couple of paragraphs should make it clear what the objectives or hypotheses of the paper are; the remaining paragraphs will give you the background information you need to understand the point of the paper.

5. Flip through the article and look at the figures, illustrations, and tables, including reading the legends.

6. Read the Results section carefully. Ask yourself these questions: Do the results accurately describe the data presented in the paper? Were all the appropriate controls carried out in an experiment? Are there additional experiments that you think should have been performed? Are the figures and tables clearly labeled?

7. Consult the Materials and Methods section to help understand the research design and the techniques used.

8. Read the Discussion. The first and last paragraphs usually summarize the key findings and state their significance. The Discussion is the part of the paper where the results are explained in the context of the scientific literature. The authors should explain what their results mean.

Getting Practice

The best way to get practice is to start reading the scientific literature as often as possible. You could begin by reading some of the references cited in this textbook. You can get an electronic copy of most articles through online databases such as PubMed, Science-Direct, or Google Scholar, or through your institution's library.

After reading a primary research paper, you should be able to paraphrase the significance of the paper in a few sentences, free of technical jargon. You should also be able to both praise and criticize several points of the paper. As you become more familiar with reading the scientific literature, you're likely to start thinking about what questions remain to be answered. And, you may even come up with "the next experiment."

check your understanding

If you understand BioSkill 14

✔ **You should be able to . . .**

After choosing a primary research paper on a topic in biology that you would like to know more about, select one figure in the Results section that reports on the experiment and construct a Research box (like the ones in this textbook) that depicts this experiment.

A concept map is a graphical device for organizing and expressing what you know about a topic. It has two main elements: **(1)** concepts that are identified by words or short phrases and placed in a box or circle, and **(2)** labeled arrows that physically link two concepts and explain the relationship between them. The concepts are arranged hierarchically on a page with the most general concepts at the top and the most specific ideas at the bottom.

The combination of a concept, a linking word, and a second concept is called a proposition. Good concept maps also have cross-links—meaning, labeled arrows that connect different elements in the hierarchy as you read down the page.

Concept maps, initially developed by Joseph Novak in the early 1970s, have proven to be an effective studying and learning tool. They can be particularly valuable if constructed by a group, or when different individuals exchange and critique concept maps they have created independently. Although concept maps vary widely in quality and can be graded using objective criteria, there are many equally valid ways of making a high-quality concept map on a particular topic.

When you are asked to make a concept map in this text, you will usually be given at least a partial list of concepts to use. As an example, suppose you were asked to create a concept map on experimental design and were given the following concepts: results, predictions, control treatment, experimental treatment, controlled (identical) conditions, conclusions, experiment, hypothesis to be tested, null hypothesis. One possible concept map is shown in **FIGURE B15.1**.

Good concept maps have four qualities:

1. They exhibit an organized hierarchy, indicating how each concept on the map relates to larger and smaller concepts.

2. The concept words are specific—not vague.

3. The propositions are accurate.

4. There is cross-linking between different elements in the hierarchy of concepts.

As you practice making concept maps, go through these criteria and use them to evaluate your own work as well as the work of fellow students.

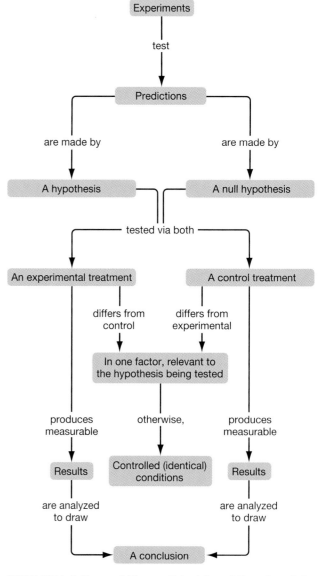

FIGURE B15.1 A Concept Map on Principles of Experimental Design.

check your understanding

If you understand BioSkill 15

✔ **You should be able to . . .**

1. Add an "Alternative hypothesis" concept to the map in Figure B15.1, along with other concepts and labeled linking arrows needed to indicate its relationship to other information on the map. (Hint: Recall that investigators often contrast a hypothesis being tested with an alternative hypothesis that does not qualify as a null hypothesis.)

2. Add a box for the concept "Statistical testing" (see **BIOSKILLS 4**) along with appropriately labeled linking arrows.

Most students have at one time or another wondered why a particular question on an exam seemed so hard, while others seemed easy. The explanation lies in the type of cognitive skills required to answer the question. Let's take a closer look.

Categories of Human Cognition

Bloom's Taxonomy is a classification system that instructors use to identify the cognitive skill levels at which they are asking students to work, particularly on practice problems and exams. Bloom's Taxonomy is also a very useful tool for students to know—it can help you to figure out the appropriate level at which you should be studying to succeed in a course.

Bloom's Taxonomy distinguishes six different categories of human thinking: remember, understand, apply, analyze, evaluate, and create. One of the most useful distinctions lies not in the differences among the six categories, but rather in the difference between high-order cognitive (HOC) and low-order cognitive (LOC) skills. **FIGURE B16.1** shows how the different levels of the taxonomy can be broken into HOC and LOC skills.

Skills that hallmark LOCs include recall, explanation, or application of knowledge in the exact way that you have before (remember, understand, apply), while skills that typify HOCs include the application of knowledge in a new way, as well as the breakdown, critique, or creation of information (analyze, evaluate, and create). Most college instructors will assume students are proficient at solving LOC questions and will expect you to frequently work at the HOC levels. HOC problems usually require use of basic vocabulary and applying knowledge—working at this level helps students to master the LOC levels.

Six Study Steps to Success

Bloom's Taxonomy provides a useful guide for preparing for an exam, using the following six steps:

1. *Answer in-chapter questions while reading the chapter.* All questions in this book have been assigned Bloom's levels, so you can review the question answers and the Bloom's level while you study.

2. *Identify the Bloom's level(s) of the questions that you are having greatest difficulty answering.* While working through

FIGURE B16.1 Bloom's Taxonomy.

the text, take note of the content and Bloom's level(s) that you find the most challenging.

3. *Use the Bloom's Taxonomy Study Guide* (**Table B16.1;** see **B:30**) *to focus your study efforts at the appropriate Bloom's level.* Table B16.1 lists specific study methods that can help you practice your understanding of the material at both the LOC and HOC levels, whether you are studying alone or with a study group.

4. *Complete the end-of-chapter questions as if you're taking an exam, without looking for the answers.* If you look at the chapter text or jump to the answers, then you really aren't testing your ability to work with the content and have reduced the questions to the lowest Bloom's level of remember.

5. *Grade your answers and note the Bloom's level of the questions you got wrong.* At what level of Bloom's Taxonomy were the questions you missed?

6. *Use the Bloom's Taxonomy Study Guide to focus your study efforts at the appropriate Bloom's level.* If you missed a lot of questions, then spend more time studying the material and find other resources for quizzing yourself.

By following these steps and studying at the HOC levels, you should succeed in answering questions on in-class exams.

	Individual Study Activities	Group Study Activities
Create (HOC) Generate something new	• Generate a hypothesis or design an experiment based on information you are studying • Create a model based on a given data set • Create summary sheets that show how facts and concepts relate to each other • Create questions at each level of Bloom's Taxonomy as a practice test and then take the test	• Each student puts forward a hypothesis about biological process and designs an experiment to test it. Peers critique the hypotheses and experiments • Create a new model/summary sheet/concept map that integrates each group member's ideas
Evaluate (HOC) Defend or judge a concept or idea	• Provide a written assessment of the strengths and weaknesses of your peers' work or understanding of a given concept based on previously determined criteria	• Provide a verbal assessment of the strengths and weaknesses of your peers' work or understanding of a given concept based on previously described criteria, and have your peers critique your assessment
Analyze (HOC) Distinguish parts and make inferences	• Analyze and interpret data in primary literature or a textbook without reading the author's interpretation and then compare the authors' interpretation with your own • Analyze a situation and then identify the assumptions and principles of the argument • Compare and contrast two ideas or concepts • Construct a map of the main concepts by defining the relationships of the concepts using one- or two-way arrows	• Work together to analyze and interpret data in primary literature or a textbook without reading the author's interpretation, and defend your analysis to your peers • Work together to identify all of the concepts in a paper or textbook chapter, construct individual maps linking the concepts together with arrows and words that relate the concepts, and then grade each other's concept maps
Apply (HOC or LOC) Use information or concepts in new ways (HOC) or in the same ways (LOC)	• Review each process you have learned and then ask yourself: What would happen if you increase or decrease a component in the system, or what would happen if you alter the activity of a component in the system? • If possible, graph a biological process and create scenarios that change the shape or slope of the graph	• Practice writing out answers to old exam questions on the board, and have your peers check to make sure you don't have too much or too little information in your answer • Take turns teaching your peers a biological process while the group critiques the content
Understand (LOC) Explain information or concepts	• Describe a biological process in your own words without copying it from a book or another source • Provide examples of a process • Write a sentence using the word • Give examples of a process	• Discuss content with peers • Take turns quizzing each other about definitions, and have your peers check your answer
Remember (LOC) Recall information	• Practice labeling diagrams • List characteristics • Identify biological objects or components from flash cards • Quiz yourself with flash cards • Take a self-made quiz on vocabulary • Draw, classify, select, or match items • Write out the textbook definitions	• Check a drawing that another student labeled • Create lists of concepts and processes that your peers can match • Place flash cards in a bag and take turns selecting one for which you must define a term • Do the preceding activities, and have peers check your answers

APPENDIX C Periodic Table of Elements

Legend:
- 1 — Atomic number
- H — Atomic symbol
- Hydrogen — Name
- 1.008 — Atomic weight*

- Most common elements in living things
- Other major elements (minerals) found in living things
- Important trace elements (minerals) found in living things
- Elements mostly found in non-living things
- Elements not found in nature (synthesized by scientists)

1	2	3	4	5	6	7	8	9	10	11	12	13	14	15	16	17	18
1 H Hydrogen 1.008																	**2 He** Helium 4.003
3 Li Lithium 6.941	**4 Be** Beryllium 9.012											**5 B** Boron 10.81	**6 C** Carbon 12.01	**7 N** Nitrogen 14.01	**8 O** Oxygen 16.00	**9 F** Fluorine 19.00	**10 Ne** Neon 20.18
11 Na Sodium 22.99	**12 Mg** Magnesium 24.31											**13 Al** Aluminum 26.98	**14 Si** Silicon 28.09	**15 P** Phosphorus 30.97	**16 S** Sulfur 32.07	**17 Cl** Chlorine 35.45	**18 Ar** Argon 39.95
19 K Potassium 39.10	**20 Ca** Calcium 40.08	**21 Sc** Scandium 44.96	**22 Ti** Titanium 47.87	**23 V** Vanadium 50.94	**24 Cr** Chromium 52.00	**25 Mn** Manganese 54.94	**26 Fe** Iron 55.85	**27 Co** Cobalt 58.93	**28 Ni** Nickel 58.69	**29 Cu** Copper 63.55	**30 Zn** Zinc 65.38	**31 Ga** Gallium 69.72	**32 Ge** Germanium 72.64	**33 As** Arsenic 74.92	**34 Se** Selenium 78.96	**35 Br** Bromine 79.90	**36 Kr** Krypton 83.80
37 Rb Rubidium 85.47	**38 Sr** Strontium 87.61	**39 Y** Yttrium 88.91	**40 Zr** Zirconium 91.22	**41 Nb** Niobium 92.91	**42 Mo** Molybdenum 95.96	**43 Tc** Technetium [98]†	**44 Ru** Ruthenium 101.1	**45 Rh** Rhodium 102.9	**46 Pd** Palladium 106.4	**47 Ag** Silver 107.9	**48 Cd** Cadmium 112.4	**49 In** Indium 114.8	**50 Sn** Tin 118.7	**51 Sb** Antimony 121.8	**52 Te** Tellurium 127.6	**53 I** Iodine 126.9	**54 Xe** Xenon 131.3
55 Cs Cesium 132.9	**56 Ba** Barium 137.3	*	**72 Hf** Hafnium 178.5	**73 Ta** Tantalum 180.9	**74 W** Tungsten 183.8	**75 Re** Rhenium 186.2	**76 Os** Osmium 190.2	**77 Ir** Iridium 192.2	**78 Pt** Platinum 195.1	**79 Au** Gold 197.0	**80 Hg** Mercury 200.6	**81 Tl** Thallium 204.4	**82 Pb** Lead 207.2	**83 Bi** Bismuth 209.0	**84 Po** Polonium [209]	**85 At** Astatine [210]	**86 Rn** Radon [222]
87 Fr Francium [223]	**88 Ra** Radium [226]	**	**104 Rf** Rutherfordium [265]	**105 Db** Dubnium [268]	**106 Sg** Seaborgium [271]	**107 Bh** Bohrium [270]	**108 Hs** Hassium [277]	**109 Mt** Meitnerium [276]	**110 Ds** Darmstadtium [281]	**111 Rg** Roentgenium [280]	**112 Cn** Copernicium [285]	**113 Uut** Ununtrium [284]	**114 Fl** Flerovium [289]	**115 Uup** Ununpentium [288]	**116 Lv** Livermorium [293]	**117 UUs** Ununseptium [294]	**118 UUo** Ununoctium [294]

*Lanthanides

57 La Lanthanum 138.9	58 Ce Cerium 140.1	59 Pr Praseodymium 140.9	60 Nd Neodymium 144.2	61 Pm Promethium [145]	62 Sm Samarium 150.4	63 Eu Europium 152.0	64 Gd Gadolinium 157.3	65 Tb Terbium 158.9	66 Dy Dysprosium 162.5	67 Ho Holmium 164.9	68 Er Erbium 167.3	69 Tm Thulium 168.9	70 Yb Ytterbium 173.1	71 Lu Lutetium 175.0

**Actinides

89 Ac Actinium [227]	90 Th Thorium 232.0	91 Pa Protactinium 231.0	92 U Uranium 238.0	93 Np Neptunium [237]	94 Pu Plutonium [244]	95 Am Americium [243]	96 Cm Curium [247]	97 Bk Berkelium [247]	98 Cf Californium [251]	99 Es Einsteinium [252]	100 Fm Fermium [257]	101 Md Mendelevium [258]	102 No Nobelium [259]	103 Lr Lawrencium [262]

DATA: Wieser, M. E., and M. Berglund. 2009. *Pure and Applied Chemistry* 81: 2131–2156.

*Atomic weights are reported to four significant figures.

†For elements with a variable number of protons and/or neutrons, the mass number of the longest-lived isotope of the element is reported in brackets.

Glossary

5′ cap A modified guanine (G) nucleotide (7-methylguanylate) added to the 5′ end of eukaryotic mRNAs. Helps protect the mRNA from being degraded and promotes efficient initiation of translation.

abdomen A region of the body; in insects, one of the three prominent body regions called tagmata.

abiotic Not alive (e.g., air, water, and some components of soil). Compare with **biotic**.

aboveground biomass The total mass of living plants in an area, excluding roots.

abscisic acid (ABA) A plant hormone that inhibits growth; it stimulates stomatal closure and triggers dormancy.

abscission In plants, the normal (often seasonal) shedding of leaves, fruits, or flowers.

abscission zone The region at the base of a petiole where cell wall degradation occurs; results in the dropping of leaves.

absorption In animals, the uptake of ions and small molecules, derived from food, across the lining of the intestine and into the bloodstream.

absorption spectrum The amount of light of different wavelengths absorbed by a pigment. Usually depicted as a graph of light absorbed versus wavelength. Compare with **action spectrum**.

acclimation A change in a study organism's phenotype that occurs in response to laboratory conditions.

acclimatization A change in an individual's phenotype that occurs in response to a change in natural environmental conditions.

acetyl CoA A molecule produced by oxidation of pyruvate (the final product of glycolysis) in a reaction catalyzed by pyruvate dehydrogenase. Can enter the citric acid cycle and is used as a carbon source in the synthesis of fatty acids, steroids, and other compounds.

acetylation Addition of an acetyl group (−COCH₃) to a molecule. Acetylation of histone proteins is important in controlling chromatin condensation.

acetylcholine (Ach) A neurotransmitter, released by nerve cells at neuromuscular junctions, that triggers contraction of skeletal muscle cells but slows the rate of contraction in cardiac muscle cells. Also used as a neurotransmitter between neurons.

acid Any compound that gives up protons or accepts electrons during a chemical reaction or that releases hydrogen ions when dissolved in water.

acid-growth hypothesis The hypothesis that auxin triggers elongation of plant cells by increasing the activity of proton pumps, making the cell wall more acidic and leading to expansion of the cell wall and an influx of water.

acoelomate A bilaterian animal that lacks an internal body cavity (coelom). Compare with **coelomate** and **pseudocoelomate**.

acquired immune deficiency syndrome (AIDS) A human disease characterized by death of immune system cells (in particular helper T cells) and subsequent vulnerability to other infections. Caused by the human immunodeficiency virus (HIV).

acrosome A caplike structure, located on the head of a sperm cell, that contains enzymes capable of dissolving the outer coverings of an egg.

ACTH See **adrenocorticotropic hormone**.

actin A globular protein that can be polymerized to form filaments. Actin filaments are part of the cytoskeleton and constitute the thin filaments in skeletal muscle cells.

actin filament A long fiber, about 7 nm in diameter, composed of two intertwined strands of polymerized actin protein; one of the three types of cytoskeletal fibers. Involved in cell movement. Also called a *microfilament*. Compare with **intermediate filament** and **microtubule**.

action potential A rapid, temporary change in electrical potential across a membrane, from negative to positive and back to negative. Occurs in cells, such as neurons and muscle cells, that have an excitable membrane.

action spectrum The relative effectiveness of different wavelengths of light in driving a light-dependent process such as photosynthesis. Usually depicted as a graph of some measure of the process versus wavelength. Compare with **absorption spectrum**.

activation energy The amount of energy required to initiate a chemical reaction; specifically, the energy required to reach the transition state.

activator A protein that binds to a DNA regulatory sequence to increase the frequency of transcription initiation by RNA polymerase.

active site The location in an enzyme molecule where substrates (reactant molecules) bind and react.

active transport The movement of ions or molecules across a membrane against an electrochemical gradient. Requires energy (e.g., from hydrolysis of ATP) and assistance of a transport protein (e.g., pump).

adaptation Any heritable trait that increases the fitness of an individual with that trait, compared with individuals without that trait, in a particular environment.

adaptive immune response See **adaptive immunity**.

adaptive immunity Immunity to a particular pathogen or other antigen conferred by activated B and T cells in vertebrates. Characterized by specificity, diversity, memory, and self–nonself recognition. Also called *adaptive immune response*. Compare with **innate immunity**.

adaptive radiation Rapid evolutionary diversification within one lineage, producing many descendant species with a wide range of adaptive forms.

adenosine triphosphate (ATP) A molecule consisting of an adenine base, a sugar, and three phosphate groups that can be hydrolyzed to release energy. Universally used by cells to store and transfer energy.

adhesion The tendency of certain dissimilar molecules to cling together due to attractive forces. Compare with **cohesion**.

adipocyte A fat cell.

adrenal gland Either of two small endocrine glands, one above each kidney. The outer portion (cortex) secretes several steroid hormones; the inner portion (medulla) secretes epinephrine and norepinephrine.

adrenaline See **epinephrine**.

adrenocorticotropic hormone (ACTH) A peptide hormone, produced and secreted by the anterior pituitary, that stimulates release of steroid hormones (e.g., cortisol, aldosterone) from the adrenal cortex.

adult A sexually mature individual.

adventitious root A root that develops from a plant's shoot system instead of from the plant's root system.

aerobic Referring to any metabolic process, cell, or organism that uses oxygen as an electron acceptor. Compare with **anaerobic**.

afferent division The part of the nervous system that transmits information about the internal and external environment to the central nervous system. Consists mainly of sensory neurons. Compare with **efferent division**.

age class All the individuals of a specific age in a population.

age-specific fecundity The average number of female offspring produced by a female in a certain age class.

age structure The proportion of individuals in a population that are of each possible age.

agglutination Clumping together of cells or viruses by antibodies or other cross-linking molecules.

aggregate fruit A fruit (e.g., raspberry) that develops from a single flower that has many separate carpels. Compare with **multiple** and **simple fruit**.

AIDS See **acquired immune deficiency syndrome**.

albumen A solution of water and protein (particularly albumins), found in amniotic eggs, that nourishes the growing embryo. Also called *egg white*.

alcohol fermentation Catabolic pathway in which pyruvate produced by glycolysis is converted to ethanol in the absence of oxygen.

aldosterone A hormone that stimulates the kidney to conserve salt and water and promotes retention of sodium; produced in the adrenal cortex.

allele A particular version of a gene.

allergen Any molecule (antigen) that triggers an allergic response (an allergy).

allergy An IgE-mediated abnormal response to an antigen, usually characterized by dilation of blood vessels, contraction of smooth muscle cells, and increased activity of mucus-secreting cells.

allopatric speciation Speciation that occurs when populations of the same species become geographically isolated, often due to dispersal or vicariance. Compare with **sympatric speciation**.

allopatry Condition in which two or more populations live in different geographic areas. Compare with **sympatry**.

allopolyploidy (adjective: allopolyploid) The state of having more than two full sets of chromosomes (polyploidy) due to hybridization between different species. Compare with **autopolyploidy**.

allosteric regulation Regulation of a protein's function by binding of a regulatory molecule, usually to a specific site distinct from the active site, that causes a change in the protein's shape.

α-amylase See **amylase**.

α-helix (alpha-helix) A protein secondary structure in which the polypeptide backbone coils into a spiral shape stabilized by hydrogen bonding.

alternation of generations A life cycle involving alternation of a multicellular haploid stage (gametophyte) with a multicellular diploid stage (sporophyte). Occurs in most plants and some protists.

alternative splicing In eukaryotes, the splicing of primary RNA transcripts from a single gene in different ways to produce different mature mRNAs and thus different polypeptides.

altruism Any behavior that has a fitness cost to the individual (lowered survival and/or reproduction) and a fitness benefit to the recipient. See **reciprocal altruism**.

alveolus (plural: alveoli) Any of the tiny air-filled sacs of a mammalian lung.

ambisense RNA virus A ssRNA virus whose genome consists of at least one strand that contains both positive-sense and negative-sense regions.

aminoacyl tRNA A transfer RNA molecule that is covalently bound to an amino acid.

aminoacyl-tRNA synthetase An enzyme that catalyzes the addition of a particular amino acid to its corresponding tRNA molecule.

ammonia (NH_3) A small molecule, produced by the breakdown of proteins and nucleic acids, that is very toxic to cells. Is a strong base that gains a proton to form the ammonium ion (NH_4^+).

amnion The innermost of the membranes surrounding the embryo in an amniotic egg.

amniotes A major lineage of vertebrates (Amniota) that reproduce with amniotic eggs. Includes all reptiles (including birds) and mammals—all tetrapods except amphibians.

amniotic egg An egg that has a watertight shell or case enclosing a membrane-bound water supply (the amnion), food supply (yolk sac), and waste sac (allantois).

amoeboid motion See **cell crawling**.

amphibians A lineage of vertebrates, many of which breathe through their skin and feed on land but lay their eggs in water. Represent the earliest tetrapods; include frogs, salamanders, and caecilians.

amphipathic Containing hydrophilic and hydrophobic regions.

ampullae of Lorenzini Structures on the heads of sharks that contain cells with electroreceptors.

amylase Any enzyme that can break down starch by catalyzing hydrolysis of the glycosidic linkages between the glucose residues.

amyloplasts Starch-storing organelles (plastids) in plants. In root cap cells, they settle to the bottom of the cell and may be used as gravity detectors.

anabolic pathway Any set of chemical reactions that synthesizes large molecules from smaller ones. Generally requires an input of energy. Compare with **catabolic pathway**.

anadromous Having a life cycle in which adults live in the ocean (or large lakes) but migrate up freshwater streams to breed and lay eggs.

anaerobic Referring to any metabolic process, cell, or organism that uses an electron acceptor other than oxygen, including fermentation or anaerobic respiration. Compare with **aerobic**.

anaphase A stage in mitosis or meiosis during which chromosomes are moved to opposite poles of the spindle apparatus.

anatomy The study of the physical structure of organisms.

ancestral trait A trait found in the ancestors of a particular group.

aneuploidy (adjective: aneuploid) The state of having an abnormal number of copies of a certain chromosome.

angiosperm A flowering vascular plant that produces seeds within mature ovaries (fruits). The angiosperms form a single lineage. Compare with **gymnosperm**.

animal A member of a major lineage of eukaryotes (Animalia) whose members typically have a complex, large, multicellular body; eat other organisms; and are mobile.

animal model Any disease that occurs in a non-human animal and has parallels to a similar disease of humans. Studied by medical researchers in hopes that findings may apply to human disease.

anion A negatively charged ion.

annelids Members of the phylum Annelida (segmented worms). Distinguished by a segmented body and a coelom that functions as a hydrostatic skeleton. Annelids belong to the lophotrochozoan branch of the protostomes.

annual Referring to a plant whose life cycle normally lasts only one growing season—less than one year. Compare with **perennial**.

anoxygenic Referring to any process or reaction that does not produce oxygen. Photosynthesis in purple sulfur and purple nonsulfur bacteria, which does not involve photosystem II, is anoxygenic. Compare with **oxygenic**.

antagonistic muscle group A set of two or more muscles that reextend one another by transmitting their forces via the skeleton.

antenna (plural: antennae) A long appendage of the head that is used to touch or smell.

antenna complex Part of a photosystem, containing an array of chlorophyll molecules and accessory pigments, that receives energy from light and directs the energy to a central reaction center during photosynthesis.

anterior Toward an animal's head and away from its tail. The opposite of posterior.

anterior pituitary The part of the pituitary gland containing endocrine cells that produce and release a variety of peptide hormones in response to other hormones from the hypothalamus. Compare with **posterior pituitary**.

anther The pollen-producing structure at the end of a stamen in flowering plants (angiosperms).

antheridium (plural: antheridia) The sperm-producing structure in most land plants except angiosperms.

anthropoids One of the two major lineages of primates, including apes, humans, and all monkeys. Compare with **prosimians**.

antibiotic Any substance, such as penicillin, that can kill or inhibit the growth of bacteria.

antibody A protein produced by B cells that can bind to a specific part of an antigen, tagging it for removal by the immune system. All monomeric forms of antibodies consist of two light chains and two heavy chains, which vary between different antibodies. Also called *immunoglobulin*.

anticodon The sequence of three bases (a triplet) in a transfer RNA molecule that can bind to an mRNA codon with a complementary sequence.

antidiuretic hormone (ADH) A peptide hormone, secreted from the posterior pituitary gland, that stimulates water retention by the kidney. Also called *vasopressin*.

antigen Any foreign molecule, often a protein, that can stimulate an innate or adaptive response by the immune system.

antigen presentation Process by which small peptides, derived from ingested particulate antigens (e.g., bacteria) or intracellular antigens (e.g., viruses in infected cell) are complexed with MHC proteins and transported to the cell surface, where they are displayed and can be recognized by T cells.

antiparallel Describing the opposite orientation of nucleic acid strands that are hydrogen bonded to one another, with one strand running in the $5' \rightarrow 3'$ direction and the other in the $3' \rightarrow 5'$ direction.

antiporter A carrier protein that allows an ion to diffuse down an electrochemical gradient, using the energy of that process to transport a different substance in the opposite direction *against* its concentration gradient. Compare with **symporter**.

antiviral Any drug or other agent that can interfere with the transmission or replication of viruses.

aorta In terrestrial vertebrates, the major artery carrying oxygenated blood away from the heart.

aphotic zone Deep water receiving no sunlight. Compare with **photic zone**.

apical Toward the top. In plants, at the tip of a branch. In animals, on the side of an epithelial layer that faces the environment and not other body tissues. Compare with **basal**.

apical–basal axis The shoot-to-root axis of a plant.

apical bud A bud at the tip of a stem or branch, where growth occurs to lengthen the stem or branch.

apical dominance Inhibition of lateral bud growth by the apical meristem at the tip of a plant branch.

apical meristem A group of undifferentiated plant cells, at the tip of a shoot or root, that is responsible for primary growth. Compare with **cambium**.

apodeme Any of the chitinous ingrowths of the exoskeleton to which muscles attach.

apomixis The formation of mature seeds without fertilization occurring; a type of asexual reproduction.

apoplast In plants, the region outside plasma membranes consisting of the porous cell walls and the intervening extracellular air space. Compare with **symplast**.

apoptosis Series of tightly controlled changes that lead to the self-destruction of a cell. Occurs frequently during embryological development and as part of the immune response to remove infected or cancerous cells. Also called *programmed cell death*.

appendix A blind sac (having only one opening) that extends from the cecum in some mammals.

aquaporin A type of channel protein through which water can move by osmosis across a plasma membrane.

aquifer An underground layer of porous rock, sand, or gravel that is saturated with water.

***ara* operon** A set of three genes in *E. coli* that are transcribed into a single mRNA and required for metabolism of the sugar arabinose. Transcription of the *ara* operon is controlled by the AraC regulatory protein.

araC The regulatory gene (written as *araC*) or regulatory protein (when written as AraC) of the *E. coli ara* operon.

arbuscular mycorrhizal fungi (AMF) Fungi from the Glomeromycota lineage whose hyphae enter the root cells of their host plants. Also called *endomycorrhizal fungi*.

Archaea One of the three taxonomic domains of life, consisting of unicellular prokaryotes distinguished by cell walls made of certain polysaccharides not found in bacterial or eukaryotic cell walls, plasma membranes composed of unique isoprene-containing phospholipids, and ribosomes and RNA polymerase similar to those of eukaryotes. Compare with **Bacteria** and **Eukarya**.

archegonium (plural: archegonia) The egg-producing structure in most land plants except angiosperms.

arteriole Any of the many tiny vessels that carry blood from arteries to capillaries.

arteriosclerosis Hardening and loss of elasticity of arteries.

artery Any thick-walled blood vessel that carries blood (oxygenated or not) under relatively high pressure away from the heart to organs throughout the body. Compare with **vein**.

arthropods Members of the phylum Arthropoda. Distinguished by a segmented body; a hard, jointed exoskeleton; paired, jointed appendages; and an extensive body cavity called a hemocoel. Arthropods belong to the ecdysozoan branch of the protostomes.

articulation A movable point of contact between two rigid components of a skeleton, such as between bones of a vertebrate endoskeleton or between segments of cuticle in an arthropod exoskeleton. See **joint**.

artificial selection Deliberate manipulation by humans, as in animal and plant breeding, of the genetic composition of a population by allowing only individuals with desirable traits to reproduce.

ascus (plural: asci) Specialized spore-producing cell found at the ends of hyphae in "sac fungi" (Ascomycota).

asexual reproduction Any form of reproduction where offspring inherit DNA from only one parent. Includes binary fission, budding, and parthenogenesis. Compare with **sexual reproduction**.

astral microtubules Mitotic and meiotic microtubules that have arisen from the two spindle poles and interact with proteins on the plasma membrane.

asymmetric competition Ecological competition between two species in which one species suffers a much greater fitness decline than the other. Compare with **symmetric competition**.

atomic number The number of protons in the nucleus of an atom, giving the atom its identity as a particular chemical element.

atomic weight The average mass of an element that is based on the relative proportions of all the naturally occurring isotopes.

ATP synthase A large membrane-bound protein complex that uses the energy of protons flowing through it to synthesize ATP.

ATP See **adenosine triphosphate**.

atrioventricular (AV) node A region of the heart between the right atrium and right ventricle where electrical signals from the atrium are slowed briefly before spreading to the ventricle. This delay allows the ventricle to fill with blood before contracting. Compare with **sinoatrial (SA) node**.

atrium (plural: atria) A thin-walled chamber of the heart that receives blood from veins and pumps it to a neighboring chamber (the ventricle).

autocrine Relating to a chemical signal that affects the same cell that produced and released it.

autoimmunity A pathological condition in which the immune system attacks self cells or tissues of an individual's own body.

autonomic nervous system The part of the peripheral nervous system that controls internal organs and involuntary processes, such as stomach contraction, hormone release, and heart rate. Includes parasympathetic and sympathetic nerves. Compare with **somatic nervous system**.

autophagy The process by which damaged organelles are surrounded by a membrane and delivered to a lysosome to be recycled.

autopolyploidy (adjective: autopolyploid) The state of having more than two full sets of chromosomes (polyploidy) due to a mutation that doubled the chromosome number. All the chromosomes come from the same species. Compare with **allopolyploidy**.

autosomal inheritance The inheritance patterns that occur when genes are located on autosomes rather than on sex chromosomes.

autosome Any chromosome other than a sex chromosome (i.e,. any chromosome other than the X or Y in mammals).

autotroph Any organism that can synthesize reduced organic compounds from simple inorganic sources such as CO_2 or CH_4. Most plants and some bacteria and archaea are autotrophs. Also called *primary producer*. Compare with **heterotroph**.

auxin Indoleacetic acid (IAA), a plant hormone that stimulates phototropism and cell elongation.

axillary bud A bud that forms at a node and may develop into a lateral (side) branch. Also called *lateral bud*.

axon A long projection of a neuron that can propagate an action potential.

axon hillock The site in a neuron where an axon joins the cell body and where action potentials are first triggered.

axoneme A structure found in eukaryotic cilia and flagella and responsible for their motion; composed of two central microtubules surrounded by nine doublet microtubules (9 + 2 arrangement).

B cell A type of lymphocyte that matures in the bone marrow and, with T cells, is responsible for adaptive immunity. Produces antibodies and also functions in antigen presentation. Also called *B lymphocyte*.

B-cell receptor (BCR) An immunoglobulin protein embedded in the plasma membrane of mature B cells and to which antigens bind. Apart from the transmembrane domain, it is identical in structure to antibodies.

BAC library A collection of all the sequences found in the genome of a species, inserted into **bacterial artificial chromosomes (BACs)**.

background extinction The average rate of low-level extinction that has occurred continuously throughout much of evolutionary history. Compare with **mass extinction**.

Bacteria One of the three taxonomic domains of life, consisting of unicellular prokaryotes distinguished by cell walls composed largely of peptidoglycan, plasma membranes similar to those of eukaryotic cells, and ribosomes and RNA polymerase that differ from those in archaeans or eukaryotes. Compare with **Archaea** and **Eukarya**.

bacterial artificial chromosome (BAC) An artificial version of a bacterial chromosome that can be used as a cloning vector to produce many copies of large DNA fragments.

bacteriophage Any virus that infects bacteria.

baculum A bone inside the penis; usually present in mammals with a penis that lacks erectile tissue.

balancing selection A mode of natural selection in which no single allele is favored in all populations of a species at all times. Instead, there is a balance among alleles in terms of fitness and frequency.

ball-and-stick model A representation of a molecule where atoms are shown as balls—colored and scaled to indicate the atom's identity—and covalent bonds are shown as rods or sticks connecting the balls in the correct geometry.

bar coding The use of well-characterized gene sequences to identify species.

bark The protective outer layer of woody plants, composed of cork cells, cork cambium, and secondary phloem.

baroreceptors Specialized nerve cells in the walls of the heart and certain major arteries that detect changes in blood pressure and trigger appropriate responses by the brain.

basal Toward the base. In plants, toward the root or at the base of a branch where it joins the stem. In animals, on the side of an epithelial layer that abuts underlying body tissues. Compare with **apical**.

basal body The microtubule organizing center for cilia and flagella in eukaryotic cells. Consists of nine triplets of microtubules arranged in a circle and establishes the structure of axonemes. Structurally identical with a centriole.

basal lamina A thick, collagen-rich extracellular matrix that underlies most epithelial tissues in animals and connects it to connective tissue.

basal metabolic rate (BMR) The total energy consumption by an organism at rest in a comfortable environment. For aerobes, often measured as the amount of oxygen consumed per hour.

basal transcription factor Proteins, present in all eukaryotic cells, that bind to promoters and help initiate transcription. Compare with **regulatory transcription factor**.

base Any compound that acquires protons or gives up electrons during a chemical reaction or accepts hydrogen ions when dissolved in water.

basidium (plural: basidia) Specialized spore-producing cell at the ends of hyphae in club fungi, members of the Basidiomycota.

basilar membrane The membrane on which the bottom portion of hair cells sits in the vertebrate cochlea.

basolateral Toward the bottom and sides. In animals, the side of an epithelial layer that faces other body tissues and not the environment.

Batesian mimicry A type of mimicry in which a harmless or palatable species resembles a dangerous or poisonous species. Compare with **Müllerian mimicry**.

beak A structure that exerts biting forces and is associated with the mouth; found in birds, cephalopods, and some insects.

behavior Any action by an organism, often in response to a stimulus.

behavioral ecology The study of how organisms respond to particular abiotic and biotic stimuli from their environment.

beneficial In genetics, referring to any mutation, allele, or trait that increases an individual's fitness.

benign tumor A mass of abnormal tissue that appears due to unregulated growth but does not spread to other organs. Benign tumors are not cancers. Compare with **malignant tumor**.

benthic Living at the bottom of an aquatic environment.

benthic zone The area along the bottom of an aquatic environment.

β-pleated sheet (beta-pleated sheet) A protein secondary structure in which the polypeptide backbone folds into a sheetlike shape stabilized by hydrogen bonding.

bilateral symmetry An animal body pattern in which one plane of symmetry divides the body into a left side and a right side. Typically, the body is long and narrow, with a distinct head end and tail end. Compare with **radial symmetry**.

bilaterian A member of a major lineage of animals (Bilateria) that are bilaterally symmetrical at some point in their life cycle, have three embryonic germ layers, and have a coelom. All protostomes and deuterostomes are bilaterians.

bile A complex solution produced by the liver, stored in the gallbladder, and secreted into the intestine. Contains steroid derivatives called bile salts that are responsible for emulsification of fats during digestion.

binary fission The process of cell division used for asexual reproduction of many prokaryotic cells. The genetic material is replicated and partitioned to opposite sides of a growing cell, which is then divided in half to create two genetically identical cells.

biodiversity The diversity of life considered at three levels: genetic diversity (variety of alleles and/or genes in a population, species, or group of species); species diversity (variety and relative abundance of species present in a certain area); and ecosystem diversity (variety of communities and abiotic components in a region).

biodiversity hotspot A region that is extraordinarily rich in species.

biofilm A complex community of bacteria enmeshed in a polysaccharide-rich, extracellular matrix that allows them to attach to a surface.

biogeochemical cycle The pattern of circulation of an element or molecule among living organisms and the environment.

biogeography The study of how species and populations are distributed geographically.

bioinformatics The field of study concerned with managing, analyzing, and interpreting biological information, particularly DNA sequences.

biological species concept The definition of a species as a population or group of populations that are reproductively isolated from other groups. Members of a species have the potential to interbreed in nature to produce viable, fertile offspring but cannot interbreed successfully with members of other species. Compare with **morphospecies** and **phylogenetic species concepts**.

bioluminescence The emission of light by a living organism via an enzyme-catalyzed reaction.

biomagnification In animal tissues, an increase in the concentration of particular molecules that may occur as those molecules are passed up a food chain.

biomass The total mass of all organisms in a given population or geographical area; usually expressed as total dry weight.

biome A large terrestrial or marine region characterized by distinct abiotic characteristics and dominant types of vegetation.

biomechanics A field of biology that applies the principles of physics and engineering to analyze the mechanical structure and function of organisms.

bioprospecting The effort to find commercially useful compounds by studying organisms—especially species that are poorly studied to date.

bioremediation The use of living organisms, usually bacteria or archaea, to degrade environmental pollutants.

biosphere The thin zone surrounding the Earth where all life exists; the sum of all terrestrial and aquatic ecosystems.

biotechnology The application of biological techniques and discoveries to medicine, industry, and agriculture.

biotic Living, or produced by a living organism. Compare with **abiotic**.

bipedal Walking primarily on two legs; characteristic of hominins.

bipolar cell A type of cell in the vertebrate retina that receives information from one or more photoreceptors and passes it to other bipolar cells or ganglion cells.

bivalent The structure formed by synapsed homologous chromosomes during prophase of meiosis I. Also known as a *tetrad*.

bivalves A lineage of mollusks that have shells made of two parts, or valves, such as clams and mussels.

bladder A mammalian organ that holds urine until it can be excreted.

blade The wide, flat part of a plant leaf.

blastocoel Fluid-filled cavity in the blastula of many animal species.

blastocyst The mammalian blastula. A roughly spherical structure composed of trophoblast cells on the exterior and a cluster of cells (the inner cell mass) that fills part of the interior space.

blastomere A cell created by cleavage divisions in early animal embryos.

blastopore An opening (pore) in the surface of some early embryos, through which cells move during gastrulation.

blastula In vertebrate development, a ball of cells (blastomere cells) typically surrounding a fluid-filled cavity (the blastocoel). The blastula is formed by cleavage of a zygote and undergoes gastrulation. See **blastocyst**.

blood A type of connective tissue consisting of red blood cells and leukocytes suspended in a fluid portion, an extracellular matrix called plasma. Transports materials throughout the vertebrate body.

body mass index (BMI) A mathematical relationship of weight and height used to assess obesity in humans. Calculated as weight (kg) divided by the square of height (m^2).

body plan The basic architecture of an animal's body, including the number and arrangement of limbs, body segments, and major tissue layers.

bog A freshwater wetland that has no or almost no water flow, resulting in very low oxygen levels and acidic conditions.

Bohr shift The rightward shift of the oxygen–hemoglobin equilibrium curve that occurs with decreasing pH. It results in hemoglobin being more likely to release oxygen in the acidic environment of exercising muscle.

bone A type of vertebrate connective tissue consisting of living cells and blood vessels within a hard extracellular matrix composed of calcium phosphate ($CaPO_4$) and small amounts of calcium carbonate ($CaCO_3$) and protein fibers.

bone marrow The soft tissue filling the inside of large bones; contains stem cells that develop into red blood cells and leukocytes throughout life.

Bowman's capsule The hollow, double-walled, cup-shaped portion of a nephron that surrounds a glomerulus in the vertebrate kidney.

brain A large mass of neurons, located in the head region of an animal, that is involved in information processing; may also be called the cerebral ganglion.

brain stem The most posterior portion of the vertebrate brain, connecting to the spinal cord and responsible for autonomic body functions such as heart rate, respiration, and digestion.

braincase See **cranium**.

branch (1) A part of a phylogenetic tree that represents populations through time. (2) Any extension of a plant's shoot system.

brassinosteroids A family of steroid hormones found in plants; stimulate growth.

bronchiole Any of the small tubes in mammalian lungs that carry air from the bronchi to the alveoli.

bronchus (plural: bronchi) In mammals, one of a pair of large tubes that lead from the trachea to each lung.

bryophyte See **non-vascular plants**.

budding Asexual reproduction in which an outgrowth from the parent breaks free as an independent individual; occurs in yeasts and some invertebrates.

buffer A substance that, in solution, acts to minimize changes in the pH of that solution when acid or base is added.

bulbourethral gland In male mammals, either of a small pair of glands at the base of the urethra that secrete an alkaline mucus (part of semen), which lubricates the tip of the penis and neutralizes acids in the urethra during copulation. In humans, also called *Cowper's glands*.

bulk flow The directional mass movement of a fluid due to pressure differences, such as movement of water through plant xylem and phloem, and movement of blood in animals.

bulk-phase endocytosis Nonspecific uptake of extracellular fluid by pinching off the plasma membrane to form small membrane-bound vesicles; considered to be a means of retrieving membrane

from the surface following exocytosis. Compare with **receptor-mediated endocytosis**.

bundle-sheath cell A type of cell found around the vascular tissue (veins) of plant leaves.

C_3 pathway The most common form of photosynthesis in which atmospheric CO_2 is fixed by rubisco to form 3-phosphoglycerate, a three-carbon molecule. Used in first phase of the Calvin cycle.

C_4 pathway A variant type of photosynthesis in which atmospheric CO_2 is first fixed by PEP carboxylase into four-carbon acids, rather than the three-carbon molecules of the classic C_3 pathway. Used to concentrate CO_2 to reduce photorespiration in the Calvin cycle while stomata are closed.

C_4 photosynthesis A variant type of photosynthesis in which atmospheric CO_2 is first fixed into four-carbon sugars, rather than the three-carbon sugars of classic C_3 photosynthesis. Enhances photosynthetic efficiency in hot, dry environments by reducing loss of oxygen due to photorespiration.

cadherin Any of a class of cell-surface proteins involved in selective cell–cell adhesion. Important for coordinating movements of cells and the establishment of tissues during embryological development.

callus In plants, a mass of undifferentiated cells that can generate roots and other tissues necessary to create a mature plant.

Calvin cycle In photosynthesis, the set of reactions that use NADPH and ATP formed in the light-dependent reactions to drive the fixation of CO_2, reduction of the fixed carbon to produce sugar, and the regeneration of the substrate used to fix CO_2. Also called *light-independent reactions*.

calyx All of the sepals of a flower.

CAM See **crassulacean acid metabolism**.

cambium (plural: cambia) In woody plants, tissue that consists of two types of cylindrical meristems that increase the width of roots and shoots through the process of secondary growth. See **vascular cambium** and **cork cambium**.

Cambrian explosion The rapid diversification of animal body types and lineages that occured between the species present in the Doushantuo faunas (around 570 mya), Ediacaran faunas (565–542 mya), and the Early Cambrian faunas (525–515 mya).

cancer General term for any tumor whose cells grow in an uncontrolled fashion, invade nearby tissues, and spread to other sites in the body.

capillarity The tendency of water to move up a narrow tube due to adhesion, cohesion, and surface tension (also called capillary action).

capillary Any of the many small, thin-walled blood vessels that permeate all tissues and organs, and allow exchange of gases and other molecules between blood and body cells.

capillary bed A thick network of capillaries.

capsid A shell of protein enclosing the genome of a virus particle.

carapace In crustaceans, a large platelike section of the exoskeleton that covers and protects the cephalothorax (e.g., a crab's "shell").

carbohydrate Any of a class of molecules that contain a carbonyl group, several hydroxyl groups, and

several to many carbon-hydrogen bonds. See **monosaccharide** and **polysaccharide**.

carbon cycle, global The worldwide movement of carbon among terrestrial ecosystems, the oceans, and the atmosphere.

carbon fixation The process of converting gaseous carbon dioxide into an organic molecule, often associated with photosynthesis. See also **PEP carboxylase** and **rubisco**.

carbonic anhydrase An enzyme that catalyzes the formation of carbonic acid (H_2CO_3) from carbon dioxide and water.

carboxylic acids Organic acids with the form R-COOH (a carboxyl group).

cardiac cycle One complete heartbeat cycle, including systole and diastole.

cardiac muscle The muscle tissue of the vertebrate heart; responsible for pumping blood. Consists of long, branched fibers that are electrically connected and that initiate their own contractions; not under voluntary control. Compare with **skeletal** and **smooth muscle**.

cardiovascular disease A group of diseases of the heart and blood vessels caused by poor diet, obesity, inactivity, genetics, tobacco use, age, and other factors.

carnivore (adjective: carnivorous) An animal whose diet consists predominantly of meat, or other animals. Most members of the mammalian taxon Carnivora are carnivores. Some plants are carnivorous, trapping and killing small animals and then absorbing nutrients from the prey's body. Compare with **herbivore** and **omnivore**.

carotenoid Any of a class of accessory pigments, found in chloroplasts, that absorb wavelengths of light not absorbed by chlorophyll; typically appear yellow, orange, or red. Includes carotenes and xanthophylls.

carpel The female reproductive organ in a flower. Consists of the stigma, to which pollen grains adhere; the style, through which the pollen tube grows; and the ovary, which houses the ovule. Compare with **stamen**.

carrier protein A membrane protein that facilitates diffusion of a small molecule (e.g., glucose) across a membrane by a process involving a reversible change in the shape of the protein. Also called *carrier* or *transporter*. Compare with **channel protein**.

carrier A heterozygous individual carrying a normal allele and a recessive allele for an inherited trait; does not display the phenotype of the recessive trait but can pass the recessive allele to offspring.

carrying capacity (*K*) The maximum population size of a certain species that a given habitat can support.

cartilage A type of vertebrate connective tissue that consists of relatively few cells scattered in a stiff matrix of polysaccharides and protein fibers. Provides structural support.

Casparian strip In plant roots, a waxy layer containing suberin, a water-repellent substance that prevents movement of water through the walls of endodermal cells, thus blocking the apoplastic pathway of water and ion movement into the vascular tissue.

cast A type of fossil, formed when the decay of a body part leaves a void that is then filled with minerals that later harden.

catabolic pathway Any set of chemical reactions that breaks down large, complex molecules into smaller ones, releasing energy in the process. Compare with **anabolic pathway**.

catalysis (verb: catalyze) Acceleration of the rate of a chemical reaction due to a decrease in the free energy of the transition state, called the activation energy.

catalyst Any substance that increases the rate of a chemical reaction without itself undergoing any permanent chemical change.

catecholamines A class of small compounds, derived from the amino acid tyrosine, that are used as hormones or neurotransmitters. Include epinephrine, norepinephrine, and dopamine.

cation A positively charged ion.

cation exchange The release (displacement) of cations, such as magnesium and calcium from soil particles, by protons in acidic soil water. The released cations are available for uptake by plants.

CD4 A membrane protein on the surface of some T cells in humans. $CD4^+$ T cells can give rise to helper T cells.

CD8 A membrane protein on the surface of some T cells in humans. $CD8^+$ T cells can give rise to cytotoxic T cells.

Cdk See **cyclin-dependent kinase**.

cDNA See **complementary DNA**.

cDNA library A set of cDNAs from a particular cell type or stage of development. Each cDNA is carried by a plasmid or other cloning vector and can be separated from other cDNAs. Compare with **genomic library**.

cecum A blind sac between the small intestine and the colon. Is enlarged in some species (e.g., rabbits) that use it as a fermentation vat for digestion of cellulose.

cell A highly organized compartment bounded by a thin, flexible structure (plasma membrane) and containing concentrated chemicals in an aqueous (watery) solution. The basic structural and functional unit of all organisms.

cell body The part of a neuron that contains the nucleus; where incoming signals are integrated. Also called the *soma*.

cell crawling A form of cellular movement involving actin filaments in which the cell produces bulges (pseudopodia) that stick to the substrate and pull the cell forward. Also called *amoeboid motion*.

cell cycle Ordered sequence of events in which a eukaryotic cell replicates its chromosomes, evenly partitions the chromosomes to two daughter cells, and then undergoes division of the cytoplasm.

cell-cycle checkpoint Any of several points in the cell cycle at which progression of a cell through the cycle can be regulated.

cell division Creation of new cells by division of preexisting cells.

cell-mediated (immune) response The type of immune response that involves generation of cytotoxic T cells from $CD8^+$ T cells. Defends against pathogen-infected cells, cancer cells, and transplanted cells. Compare with **humoral (immune) response**.

cell membrane See **plasma membrane**.

cell plate A flattened sac-like structure formed in the middle of a dividing plant cell from Golgi-derived vesicles containing cell wall material; ultimately divides the cytoplasm into two separate cells.

cell sap An aqueous solution found in the vacuoles of plant cells.

cell theory The theory that all organisms are made of cells and that all cells come from preexisting cells.

cell wall A fibrous layer found outside the plasma membrane of most bacteria and archaea and many eukaryotes.

cellular respiration A common pathway for production of ATP, involving transfer of electrons from compounds with high potential energy through an electron transport chain and ultimately to an electron acceptor (often oxygen).

cellulase An enzyme that can break down cellulose by catalyzing hydrolysis of the glycosidic linkages between the glucose residues.

cellulose A structural polysaccharide composed of glucose monomers joined by β-1,4-glycosidic linkages. Found in the cell wall of algae, plants, and some bacteria and fungi.

Cenozoic era The most recent interval of geologic time, beginning 65.5 million years ago, during which mammals became the dominant vertebrates and angiosperms became the dominant plants.

central dogma The scheme for information flow in the cell: DNA → RNA → protein.

central nervous system (CNS) Large numbers of neurons aggregated into clusters called ganglia in bilaterian animals. In vertebrates, the central nervous system consists of the brain and spinal cord. Compare with **nerve net** and **peripheral nervous system (PNS)**.

centriole One of two small cylindrical structures found together within the centrosome near the nucleus of a eukaryotic cell (not found in plants). Consists of microtubule triplets and is structurally identical with a basal body.

centromere Constricted region of a replicated chromosome where the two sister chromatids are joined and the kinetochore is located.

centrosome Structure in animal and fungal cells, containing two centrioles, that serves as a microtubule organizing center for the cell's cytoskeleton and for the spindle apparatus during cell division.

cephalization The formation in animals of a distinct anterior region (the head) where sense organs and a mouth are clustered.

cephalochordates One of the three major chordate lineages (Cephalochordata), comprising small, mobile organisms that live in marine sands and suspension feed; also called *lancelets* or *amphioxus*. Compare with **urochordates** and **vertebrates**.

cephalopods A lineage of mollusks including the squid, octopuses, and nautiluses. Distinguished by large brains, excellent vision, tentacles, and a reduced or absent shell.

cerebellum Posterior section of the vertebrate brain; involved in coordination of complex muscle movements, such as those required for locomotion and maintaining balance.

cerebrum The most anterior section of the vertebrate brain. Divided into left and right hemispheres and four lobes: frontal lobe, involved in complex decision making (in humans); occipital lobe, receives and interprets visual information; parietal lobe, involved in integrating sensory and motor functions; and temporal lobe, functions in memory, speech (in humans), and interpreting auditory information.

cervix The bottom portion of the uterus, containing a canal that leads to the vagina.

chaetae (singular: chaeta) Bristle-like extensions found in some annelids.

channel protein A protein that forms a pore in a cell membrane. The structure of most channels allows them to admit just one or a few types of ions or molecules. Compare with **carrier protein**.

character Any genetic, morphological, physiological, or behavioral characteristic of an organism to be studied.

character displacement The evolutionary tendency for the traits of similar species that occupy overlapping ranges to change in a way that reduces interspecific competition.

chelicerae A pair of clawlike appendages found around the mouth of certain arthropods called chelicerates (spiders, mites, and allies).

chemical bond An attractive force binding two atoms together. Covalent bonds, ionic bonds, and hydrogen bonds are types of chemical bonds.

chemical energy The potential energy stored in covalent bonds between atoms.

chemical equilibrium A dynamic but stable state of a reversible chemical reaction in which the forward reaction and reverse reactions proceed at the same rate, so that the concentrations of reactants and products remain constant.

chemical evolution A The theory that simple chemical compounds in the early atmosphere and ocean combined via chemical reactions to form larger, more complex substances, eventually leading to the origin of life and the start of biological evolution.

chemical reaction Any process in which one compound or element is combined with others or is broken down; involves the making and/or breaking of chemical bonds.

chemiosmosis An energetic coupling mechanism whereby energy stored in an electrochemical proton gradient is used to drive an energy-requiring process such as production of ATP.

chemokine Any of a subset of cytokines that acts as a chemical signal attracting leukocytes to a site of tissue injury or infection.

chemolithotroph An organism (bacteria or archaea) that produces ATP by oxidizing inorganic molecules with high potential energy such as ammonia (NH_3) or methane (CH_4). Also called *lithotroph*. Compare with **chemoorganotroph**.

chemoorganotroph An organism that produces ATP by oxidizing organic molecules with high potential energy such as sugars. Also called *organotroph*. Compare with **chemolithotroph**.

chemoreception A sensory system in which receptors are activated in response to the binding of chemicals.

chemoreceptor A sensory cell or organ specialized for detection of specific molecules or classes of molecules.

chiasma (plural: chiasmata) The X-shaped structure formed during meiosis by crossing over between non-sister chromatids in a pair of homologous chromosomes.

chitin A structural polysaccharide composed of *N*-acetyl-glucosamine (NAG) monomers joined end to end by β-1,4-glycosidic linkages. Found in cell walls of fungi and many algae, and in external skeletons of insects and crustaceans.

chitons A lineage of marine mollusks that have a protective shell formed of eight calcium carbonate plates.

chlorophyll Any of several closely related green pigments, found in chloroplasts, that absorb light during photosynthesis.

chloroplast A chlorophyll-containing organelle, bounded by a double membrane, in which photosynthesis occurs; found in plants and photosynthetic protists. Also the location of starch, amino acid, fatty acid, purine, and pyrimidine synthesis.

choanocyte A specialized, flagellated feeding cell found in choanoflagellates (protists that are the closest living relatives of animals) and sponges (the most ancient animal phylum).

cholecystokinin A peptide hormone secreted by cells in the lining of the small intestine. Stimulates the secretion of digestive enzymes from the pancreas and of bile from the liver and gallbladder.

chordate Any member of the phylum Chordata. Chordates are deuterostomes distinguished by a dorsal hollow nerve cord, pharyngeal gill slits, a notochord, and a post-anal tail. Includes vertebrates, cephalochordata, and urochordata.

chromatid One of the two identical double-stranded DNAs composing a replicated chromosome that is connected at the centromere to the other strand.

chromatin The complex of DNA and proteins, mainly histones, that compose eukaryotic chromosomes. Can be highly compact (heterochromatin) or loosely coiled (euchromatin).

chromatin remodeling The process by which the DNA in chromatin is unwound from its associated proteins to allow transcription or replication. May involve chemical modification of histone proteins or reshaping of the chromatin by large multiprotein complexes in an ATP-requiring process.

chromatin remodeling complexes Sets of enzymes that use energy from ATP hydrolysis shift nucleosomes on DNA to expose regulatory sequences to transcription factors.

chromosome theory of inheritance The principle that genes are located on chromosomes and that patterns of inheritance are determined by the behavior of chromosomes during meiosis.

chromosome Gene-carrying structure consisting of a single long molecule of double-stranded DNA and

associated proteins (e.g., histones). Most prokaryotic cells contain a single, circular chromosome; eukaryotic cells contain multiple noncircular (linear) chromosomes located in the nucleus.

cilium (plural: cilia) One of many short, filamentous projections of some eukaryotic cells, containing a core of microtubules. Used to move the cell as well as to circulate fluid or particles around the surface of a stationary cell. See **axoneme**.

circadian clock An internal mechanism found in most organisms that regulates many body processes (sleep–wake cycles, hormonal patterns, etc.) in a roughly 24-hour cycle.

circulatory system The system responsible for moving oxygen, carbon dioxide, and other materials (hormones, nutrients, wastes) within the animal body.

cisternae (singular: cisterna) Flattened, membrane-bound compartments that make up the Golgi apparatus.

cisternal maturation The process of cargo movement through the Golgi apparatus by residing in cisternae that mature from *cis* to *trans* via the import and export of different Golgi enzymes.

citric acid cycle A series of eight chemical reactions that start with citrate (deprotonated citric acid) and ends with oxaloacetate, which reacts with acetyl CoA to form citrate—forming a cycle that is part of the pathway that oxidizes glucose to CO_2. Also known as the *Krebs cycle* or *tricarboxylic acid* (*TCA*) *cycle*.

clade See **monophyletic group**.

cladistic approach A method for constructing a phylogenetic tree that is based on identifying the unique traits (shared, derived characters, called synapomorphies) of each monophyletic group.

Class I MHC protein A type of MHC protein that is present on the plasma membrane of virtually all nucleated cells and functions in presenting antigen to CD8$^+$ T cells.

Class II MHC protein A type of MHC protein that is present only on the plasma membrane of certain cells in the immune response, such as dendritic cells, macrophages, and B cells. It functions in presenting epitopes of antigens to CD4$^+$ T cells.

cleavage In animal development, the series of rapid mitotic cell divisions, with little cell growth, that produces successively smaller cells (blastomeres) and transforms a zygote into a multicellular blastula.

cleavage furrow A pinching in of the plasma membrane that occurs as cytokinesis begins in animal cells and deepens until the cytoplasm is divided into two daughter cells.

climate The prevailing, long-term weather conditions in a particular region.

climax community The stable, final community that develops from ecological succession.

clitoris A rod of erectile tissue in the external genitalia of female mammals. Is formed from the same embryonic tissue as the male penis and has a similar function in sexual arousal.

cloaca In a few mammals and many nonmammalian vertebrates, a body cavity opening to the outside and used by both the excretory and reproductive systems.

clonal selection theory The dominant explanation of the generation of an adaptive immune response. According to the theory, the immune system retains a vast pool of inactive lymphocytes, each with a unique receptor for a unique epitope. Lymphocytes that encounter their complementary epitopes are stimulated to divide (selected and cloned), producing daughter cells that combat infection and confer immunity.

clone (1) An individual that is genetically identical to another individual. (2) A lineage of genetically identical individuals or cells. (3) As a verb, to make one or more genetic replicas of a cell or individual.

cloning vector A plasmid or other agent used to transfer recombinant genes into cultured host cells. Also called simply *vector*.

closed circulatory system A circulatory system in which the circulating fluid (blood) is confined to blood vessels and flows in a continuous circuit. Compare with **open circulatory system**.

cnidocyte A specialized stinging cell found in cnidarians (e.g., jellyfish, corals, and anemones) and used in capturing prey.

co-receptor Any membrane protein that acts with some other membrane protein in a cell interaction or cell response.

cochlea The organ of hearing in the inner ear of mammals, birds, and crocodilians. A coiled, fluid-filled tube containing specialized pressure-sensing neurons (hair cells) that detect sounds of different pitches.

coding strand See **non-template strand**.

codominance An inheritance pattern in which heterozygotes exhibit both of the traits seen in each type of homozygous individual.

codon A sequence of three nucleotides in DNA or RNA that codes for an amino acid or a start or stop signal for protein synthesis.

coefficient of relatedness (*r*) A measure of how closely two individuals are related. Calculated as the probability that an allele in two individuals is inherited from the same ancestor.

coelom An internal, usually fluid-filled body cavity that is completely or partially lined with mesoderm.

coelomate An animal that has a true coelom, completely lined with mesoderm. Compare with **acoelomate** and **pseudocoelomate**.

coenocytic Containing many nuclei and a continuous cytoplasm through a filamentous body, without the body being divided into distinct cells. Some fungi are coenocytic.

coenzyme A small organic molecule that is a required cofactor for an enzyme-catalyzed reaction. Often donates or receives electrons or functional groups during the reaction.

coenzyme A (CoA) A molecule that is required for many cellular reactions and that is often transiently linked to other molecules, such as acetyl groups (see **acetyl CoA**).

coenzyme Q A nonprotein molecule that shuttles electrons between membrane-bound complexes in the mitochondrial electron transport chain. Also called **ubiquinone** or *Q*.

coevolution A pattern of evolution in which two interacting species reciprocally influence each other's adaptations over time.

coevolutionary arms race A series of adaptations and counter-adaptations observed in species that interact closely over time and affect each other's fitness.

cofactor An inorganic ion, such as a metal ion, that is required for an enzyme to function normally. May be bound tightly to an enzyme or associate with it transiently during catalysis.

cohesion The tendency of certain like molecules (e.g., water molecules) to cling together due to attractive forces. Compare with **adhesion**.

cohesion-tension theory The theory that water movement upward through plant vascular tissues is due to loss of water from leaves (transpiration), which pulls a cohesive column of water upward.

cohort A group of individuals that are the same age and can be followed through time.

coleoptile A modified leaf that covers and protects the stems and leaves of grass seedlings.

collagen A fibrous, pliable, cable-like glycoprotein that is a major component of the extracellular matrix of animal cells. Various subtypes differ in their tissue distribution, some of which are assembled into large fibrils in the extracellular space.

collecting duct In the vertebrate kidney, a large straight tube that receives filtrate from the distal tubules of several nephrons. Involved in the regulated reabsorption of water.

collenchyma cell In plants, an elongated cell with cell walls thickened at the corners that provides support to growing plant parts; usually found in strands along leaf veins and stalks. Compare with **parenchyma cell** and **sclerenchyma cell**.

colon The portion of the large intestine where feces are formed by compaction of wastes and reabsorption of water.

colony An assemblage of individuals. May refer to an assemblage of semi-independent cells or to a breeding population of multicellular organisms.

commensalism (adjective: commensal) A symbiotic relationship in which one organism (the commensal) benefits and the other (the host) is not harmed. Compare with **mutualism** and **parasitism**.

communication In ecology, any process in which a signal from one individual modifies the behavior of another individual.

community All of the species that interact with each other in a certain area.

companion cell In plants, a cell in the phloem that is connected via many plasmodesmata to adjacent sieve-tube elements. Companion cells provide materials to maintain sieve-tube elements and function in the loading and unloading of sugars into sieve-tube elements.

compass orientation A type of navigation in which movement occurs in a specific direction.

competition In ecology, the interaction of two species or two individuals trying to use the same limited resource (e.g., water, food, living space). May occur between individuals of the same species (intraspecific competition) or different species (interspecific competition).

competitive exclusion principle The principle that two species cannot coexist in the same ecological niche in the same area because one species will outcompete the other.

competitive inhibition Inhibition of an enzyme's ability to catalyze a chemical reaction via binding of a nonreactant molecule that competes with the substrate(s) for access to the active site.

complement system A set of proteins that circulate in the bloodstream and can destroy bacteria by forming holes in the bacterial plasma membrane.

complementary base pairing The association between specific nitrogenous bases of nucleic acids stabilized by hydrogen bonding. Adenine pairs only with thymine (in DNA) or uracil (in RNA), and guanine pairs only with cytosine.

complementary DNA (cDNA) DNA produced in the laboratory using an RNA transcript as a template and reverse transcriptase; corresponds to a gene but lacks introns. Also produced naturally by retroviruses.

complementary strand A newly synthesized strand of RNA or DNA that has a base sequence complementary to that of the template strand.

complete digestive tract A digestive tract with two openings, usually called a mouth and an anus.

complete metamorphosis See **holometabolous metamorphosis**.

compound eye An eye formed of many independent light-sensing columns (ommatidia); occurs in arthropods. Compare with **simple eye**.

concentration gradient Difference across space (e.g., across a membrane) in the concentration of a dissolved substance.

condensation reaction A chemical reaction in which two molecules are joined covalently with the removal of an −OH from one and an −H from another to form water. Also called a *dehydration reaction*. Compare with **hydrolysis**.

conduction (1) Direct transfer of heat between two objects that are in physical contact. Compare with **convection**. (2) Transmission of an electrical impulse along the axon of a nerve cell.

cone cell A photoreceptor cell with a cone-shaped outer portion that is particularly sensitive to bright light of a certain color. Also called simply *cone*. Compare with **rod cell**.

connective tissue An animal tissue consisting of scattered cells in a liquid, jellylike, or solid extracellular matrix. Includes bone, cartilage, tendons, ligaments, and blood.

conservation biology The effort to study, preserve, and restore threatened genetic diversity, populations, communities, and ecosystems.

constant (C) region The invariant amino acid sequence in polypeptides that are used to make antibodies, B-cell receptors, and T-cell receptors. Apart from antibody class types (IgG, IgM, etc.), this region remains constant within an individual. Compare with **variable (V) region**.

constitutive Always occurring; always present. Commonly used to describe enzymes and other proteins that are synthesized continuously or mutants in which one or more genetic loci are constantly expressed due to defects in gene control.

constitutive defense A defensive trait that is always manifested even in the absence of a predator or pathogen. Also called *standing defense*. Compare with **inducible defenses**.

constitutive mutant An abnormal (mutated) strain that produces a product at all times, instead of under certain conditions only.

consumer See **heterotroph**.

consumption Predation or herbivory.

continental shelf The portion of a geologic plate that extends from a continent under seawater.

continuous strand See **leading strand**.

contraception Any of several methods to prevent pregnancy.

control In a scientific experiment, a group of organisms or samples that do not receive the experimental treatment but are otherwise identical to the group that does.

convection Transfer of heat by movement of large volumes of a gas or liquid. Compare with **conduction**.

convergent evolution The independent evolution of similar traits in distantly related organisms due to adaptation to similar environments and a similar way of life.

cooperative binding The tendency of the protein subunits of hemoglobin to affect each other's oxygen binding such that each bound oxygen molecule increases the likelihood of further oxygen binding.

coprophagy The eating of feces.

coral reef A large assemblage of colonial marine corals that usually serves as shallow-water, sunlit habitat for many other species as well.

co-receptor Any membrane protein that acts with some other membrane protein in a cell interaction or cell response.

core enzyme The enzyme responsible for catalysis in a multipart holoenzyme.

cork cambium One of two types of cylindrical meristem, consisting of a ring of undifferentiated plant cells found just under the cork layer of woody stems and roots; produces new cork cells on its outer side. Compare with **vascular cambium**.

cork cell A cell in the protective outermost layer of a woody stem and root that produces and accumulates waxes that make the cell less permeable to water and gases.

corm A rounded, thick underground stem that can produce new plants via asexual reproduction.

cornea The transparent sheet of connective tissue at the very front of the eye in vertebrates and some other animals. Protects the eye and helps focus light.

corolla All of the petals of a flower.

corona The cluster of cilia at the anterior end of a rotifer; in many species it facilitates suspension feeding.

corpus callosum A thick band of neurons that connects the two hemispheres of the cerebrum in the mammalian brain.

corpus luteum A yellowish structure that secretes progesterone in an ovary. Is formed from a follicle that has recently ovulated.

cortex (1) In animals, the outermost region of an organ, such as the kidney or adrenal gland. (2) In plants, a layer of ground tissue found outside the vascular bundles of roots and outside the pith of a stem.

corticotropin-releasing hormone (CRH) A peptide hormone, produced and secreted by the hypothalamus, that stimulates the anterior pituitary to release ACTH.

cortisol A steroid hormone, produced and secreted by the adrenal cortex, that increases blood glucose and prepares the body for stress. The major glucocorticoid hormone in some mammals. Also called *hydrocortisone*.

cost–benefit analysis Decisions or analyses that weigh the fitness costs and benefits of a particular action.

cotransporter A transmembrane protein that facilitates diffusion of an ion down its previously established electrochemical gradient and uses the energy of that process to transport some other substance, in the same or opposite direction, *against* its concentration gradient. Also called *secondary active transporter*. See **antiporter** and **symporter**.

cotyledon The first leaf, or seed leaf, of a plant embryo. Used for storing and digesting nutrients and/or for early photosynthesis.

countercurrent exchanger In animals, any anatomical arrangement that allows the maximum transfer of heat or a soluble substance from one fluid to another. The two fluids must be flowing in opposite directions and have a heat or concentration gradient between them.

covalent bond A type of chemical bond in which two atoms share one or more pairs of electrons. Compare with **hydrogen bond** and **ionic bond**.

cranium A bony, cartilaginous, or fibrous case that encloses and protects the brain of vertebrates. Forms part of the skull. Also called *braincase*.

crassulacean acid metabolism (CAM) A variant type of photosynthesis in which CO_2 is fixed and stored in organic acids at night when stomata are open and then released to feed the Calvin cycle during the day when stomata are closed. Helps reduce water loss and CO_2 loss by photorespiration.

cristae (singular: crista) Sac-like invaginations of the inner membrane of a mitochondrion. Location of the electron transport chain and ATP synthase.

Cro-Magnon A prehistoric European population of modern humans (*Homo sapiens*) known from fossils, paintings, sculptures, and other artifacts.

crop A storage organ in the digestive system of certain vertebrates.

cross A mating between two individuals that is used for genetic analysis.

cross-pollination Pollination of a flower by pollen from another individual, rather than by self-fertilization. Also called *crossing*.

cross-talk Interactions among signaling pathways, triggered by different signals, that modify a cellular response.

crossing over The exchange of segments of non-sister chromatids between a pair of homologous chromosomes that occurs during meiosis I.

crosstalk Interactions among signaling pathways that modify a cellular response.

crustaceans A lineage of arthropods that includes shrimp, lobster, and crabs. Many have a carapace (a platelike portion of the exoskeleton covering the cephalothorax) and mandibles for biting or chewing.

cryptic species A species that cannot be distinguished from similar species by easily identifiable morphological traits.

culture In cell biology, a collection of cells or a tissue growing under controlled conditions, usually in suspension or on the surface of a dish of solid growth medium.

Cushing's disease A human endocrine disorder caused by loss of feedback inhibition of cortisol on ACTH secretion. Characterized by high ACTH and cortisol levels and wasting of body protein reserves.

cuticle A protective coating secreted by the outermost layer of cells of an animal or a plant; often functioning to reduce evaporative water loss.

cyanobacteria A lineage of photosynthetic bacteria formerly known as blue-green algae. Likely the first life-forms to carry out oxygenic photosynthesis.

cyclic AMP (cAMP) Cyclic adenosine monophosphate; a small molecule, derived from ATP, that is widely used by cells in signal transduction and transcriptional control.

cyclic electron flow Path of electrons in which excited electrons of photosystem I are transferred back to plastoquinone (PQ), the start of the electron transport chain normally associated with photosystem II. Instead of reducing $NADP^+$ to make NADPH, the electron energy is used to make ATP via photophosphorylation. Compare with **noncyclic electron flow**.

cyclin One of several regulatory proteins whose concentrations fluctuate cyclically throughout the cell cycle. Involved in the control of the cell cycle via cyclin-dependent kinases.

cyclin-dependent kinase (Cdk) Any of several related protein kinases that are functional only when bound to a cyclin and are activated by other modifications. Involved in control of the cell cycle.

cytochrome *c* (cyt *c*) A soluble protein that shuttles electrons between membrane-bound complexes in the mitochondrial electron transport chain.

cytokine Any of a diverse group of signaling proteins, secreted largely by cells of the immune system, whose effects include stimulating leukocyte production, recruiting cells to the site of infection, tissue repair, and fever. Generally function to regulate the type, intensity, and duration of an immune response.

cytokinesis Division of the cytoplasm to form two daughter cells. Typically occurs immediately after division of the nucleus by mitosis or meiosis.

cytokinins A class of plant hormones that stimulate cell division and retard aging.

cytoplasm All the contents of a cell, excluding the nucleus, bounded by the plasma membrane.

cytoplasmic determinant A regulatory transcription factor or signaling molecule that is distributed unevenly in the cytoplasm of the egg and that directs early pattern formation in an embryo.

cytoplasmic streaming The directed flow of cytosol and organelles that facilitates distribution of materials within some large plant and fungal cells. Occurs along actin filaments and is powered by myosin.

cytoskeleton In eukaryotic cells, a network of protein fibers in the cytoplasm that are involved in cell shape, support, locomotion, and transport of materials within the cell. Prokaryotic cells have a similar but much less extensive network of fibers.

cytosol The fluid portion of the cytoplasm, excluding the contents of membrane-enclosed organelles.

cytotoxic T cell A type of $CD8^+$ effector T cell that induces apoptosis in infected and cancerous cells. Recognizes target cells via interactions with complementary class I MHC–peptide complexes. Also called *cytotoxic T lymphocyte (CTL)* and *killer T cell*. Compare with **helper T cell**.

dalton (Da) A unit of mass equal to 1/12 the mass of one carbon-12 atom; about the mass of 1 proton or 1 neutron.

daughter strand The strand of DNA that is newly replicated from an existing template strand of DNA.

day-neutral plant A plant whose flowering time is not affected by the relative length of day and night (the photoperiod). Compare with **long-day** and **short-day plant**.

dead space Air passages that are not involved in gas exchange with the blood; examples are the trachea and bronchi.

deciduous Describing a plant that sheds leaves or other structures at regular intervals (e.g., each autumn).

decomposer See **detritivore**.

decomposer food chain An ecological network of detritus, decomposers that eat detritus, and predators and parasites of the decomposers.

deep sequencing A method to learn the types of mRNAs or DNA sequences present in cells, and their relative amounts, involving the preparation and sequencing of cDNA libraries.

definitive host The host species in which a parasite reproduces sexually. Compare with **intermediate host**.

dehydration reaction See **condensation reaction**.

deleterious In genetics, referring to any mutation, allele, or trait that reduces an individual's fitness.

deletion In genetics, refers to the loss of part of a chromosome.

demography The study of factors that determine the size and structure of populations through time.

dendrite A short extension from a neuron's cell body that receives signals from other neurons.

dendritic cell A type of leukocyte that ingests and digests foreign antigens, moves to a lymph node, and presents the antigens' epitopes, in the context of MHC proteins on its membrane, to $CD4^+$ and $CD8^+$ T cells.

dense connective tissue A type of connective tissue, distinguished by having an extracellular matrix dominated by collagen fibers. Found in tendons and ligaments.

density dependent In population ecology, referring to any characteristic that varies depending on population density.

density independent In population ecology, referring to any characteristic that does not vary with population density.

deoxyribonucleic acid (DNA) A nucleic acid composed of deoxyribonucleotides that carries the genetic information of a cell. Generally occurs as two intertwined strands, but these can be separated. See also **double helix**.

deoxyribonucleoside triphosphate (dNTP) A monomer used by DNA polymerase to polymerize DNA. Consists of the sugar deoxyribose, a base (A, T, G, or C), and three phosphate groups.

deoxyribonucleotide See **nucleotide**.

depolarization A change in membrane potential from its resting negative state to a less negative or a positive state; a normal phase in an action potential. Compare with **hyperpolarization**.

depolarization Change in membrane potential from its resting negative state to a less negative or to a positive state; a normal phase in an action potential. Compare with **hyperpolarization**.

deposit feeder An animal that eats its way through a food-containing substrate.

derived trait A trait that is clearly homologous with a trait found in an ancestor of a particular group, but that has a new form.

dermal tissue system The tissue forming the outer layer of a plant; also called *epidermis*.

descent with modification The phrase used by Darwin to describe his hypothesis of evolution by natural selection.

desmosome A type of cell–cell attachment structure, consisting of cadherin proteins, that is anchored to intermediate filaments. Serves to link the cytoskeletons of adjacent animal cells and form strong cell–cell attachments throughout a tissue. Compare with **gap junction** and **tight junction**.

detergent A type of small amphipathic molecule used to solubilize hydrophobic molecules in aqueous solution.

determination In development, the commitment of a cell to a particular differentiated fate. Once a cell is fully determined, it can differentiate only into a particular cell type (e.g., liver cell, brain cell).

detritivore An organism whose diet consists mainly of dead organic matter (detritus). Various bacteria, fungi, protists, and animals are detritivores. Also called *decomposer*.

detritus A layer of dead organic matter that accumulates at ground level or on seafloors and lake bottoms.

deuterostomes A major lineage of bilaterian animals that share a pattern of embryological development, including formation of the anus earlier than the mouth, and formation of the coelom by pinching off of layers of mesoderm from the gut. Includes echinoderms and chordates. Compare with **protostomes**.

developmental homology A similarity in embryonic form, or in the fate of embryonic tissues, that is due to inheritance from a common ancestor.

diabetes mellitus A disease caused by defects in insulin production (type 1) or in the response of cells to insulin (type 2). Characterized by abnormally high blood glucose levels and huge amounts of glucose-containing urine.

diaphragm An elastic, sheetlike structure. In mammals, the muscular sheet of tissue that separates the chest and abdominal cavities. It contracts and moves downward during inhalation, expanding the chest cavity.

diastole The portion of the cardiac cycle during which the atria or ventricles of the heart are relaxed. Compare with **systole**.

diastolic blood pressure The force exerted by blood against artery walls during relaxation of the heart's left ventricle. Compare with **systolic blood pressure**.

dicot Any flowering plant (angiosperm) that has two cotyledons (embryonic leaves) upon germination. The dicots do not form a monophyletic group. Also called *dicotyledonous plant*. Compare with **eudicot** and **monocot**.

dideoxy sequencing A laboratory technique for determining the exact nucleotide sequence of DNA. Relies on the use of dideoxynucleoside triphosphates (ddNTPs), which terminate DNA replication.

diencephalon The part of the mammalian brain that relays sensory information to the cerebellum and functions in maintaining homeostasis.

differential centrifugation Procedure for separating cellular components according to their size and density by spinning a cell homogenate in a series of centrifuge runs. After each run, the supernatant is removed from the deposited material (pellet) and spun again at progressively higher speeds.

differential gene expression Expression of different sets of genes in cells with the same genome. Responsible for creating different cell types.

differentiation The process by which any unspecialized cell becomes a distinct specialized cell type (e.g., liver cell, brain cell), usually by changes in gene expression. Also called *cell differentiation*.

diffusion Spontaneous movement of a substance from one region to another, often with a net movement from a region of high concentration to one of low concentration (i.e., down a concentration gradient).

digestion The physical and chemical breakdown of food into molecules that can be absorbed into the body of an animal.

digestive tract The long tube that begins at the mouth and ends at the anus. Also called *alimentary canal* or *gastrointestinal (GI) tract*.

dihybrid cross A mating between two parents that are heterozygous for two different genes.

dikaryotic Describing a cell or fungal mycelium having two haploid nuclei that are genetically distinct.

dimer An association of two molecules that may be identical (homodimer) or different (heterodimer).

dioecious Describing an angiosperm species that has male and female reproductive structures on separate plants. Compare with **monoecious**.

diploblast (adjective: diploblastic) An animal whose body develops from two basic embryonic cell layers

or tissues—ectoderm and endoderm. Compare with **triploblast**.

diploid (1) Having two sets of chromosomes (2*n*). (2) A cell or an individual organism with two sets of chromosomes, one set inherited from the mother and one set from the father. Compare with **haploid**.

direct sequencing A technique for identifying and studying microorganisms that cannot be grown in culture. Involves detecting and amplifying copies of certain specific genes in the microorganisms' DNA, sequencing these genes, and then comparing the sequences with the known sequences from other organisms.

directional selection A mode of natural selection that favors one extreme phenotype with the result that the average phenotype of a population changes in one direction. Generally reduces overall genetic variation in a population. Compare with **disruptive selection** and **stabilizing selection**.

disaccharide A carbohydrate consisting of two monosaccharides (sugar residues) linked together.

discontinuous strand See **lagging strand**.

discrete trait An inherited trait that exhibits distinct phenotypes rather than the continuous variation characteristic of a quantitative trait such as body height.

dispersal The movement of individuals from their place of origin (birth, hatching) to a new location.

disruptive selection A mode of natural selection that favors extreme phenotypes at both ends of the range of phenotypic variation. Maintains overall genetic variation in a population. Compare with **stabilizing selection** and **directional selection**.

distal tubule In the vertebrate kidney, the convoluted portion of a nephron into which filtrate moves from the loop of Henle. Involved in the regulated reabsorption of sodium and water. Compare with **proximal tubule**.

disturbance In ecology, any strong, short-lived disruption to a community that changes the distribution of living and/or nonliving resources.

disturbance regime The characteristic disturbances that affect a given ecological community.

disulfide bond A covalent bond between two sulfur atoms, typically in the side chains of certain amino acids (e.g., cysteine). Often contributes to tertiary and quaternary levels of protein structure.

DNA See **deoxyribonucleic acid**.

DNA cloning Any of several techniques for producing many identical copies of a particular gene or other DNA sequence.

DNA fingerprinting Any of several methods for identifying individuals by unique features of their genomes. Commonly involves using PCR to produce many copies of certain short tandem repeats (microsatellites) and then analyzing their lengths.

DNA helicase An enzyme that breaks hydrogen bonds between nucleotides of DNA, "unzipping" a double-stranded DNA molecule.

DNA library See **cDNA library** and **genomic library**.

DNA ligase An enzyme that joins pieces of DNA by catalyzing the formation of a phosphodiester bond between the pieces.

DNA methylation The addition of a methyl group ($-CH_3$) to a DNA molecule.

DNA methyltransferase A class of eukaryotic enzymes that add a methyl group to cytosines in DNA. Methylation of DNA leads to chromatin condensation and is an important means of regulating gene expression in eukaryotes.

DNA microarray A set of single-stranded DNA fragments, representing thousands of different genes that are permanently fixed to a small glass slide. Can be used to determine which genes are expressed in different cell types, under different conditions, or at different developmental stages.

DNA polymerase Any enzyme that catalyzes synthesis of DNA from deoxyribonucleotide triphosphates (dNTPs).

domain (1) A taxonomic category, based on similarities in basic cellular biochemistry, above the kingdom level. The three recognized domains are Bacteria, Archaea, and Eukarya. (2) A section of a protein that has a distinctive tertiary structure and function.

dominant Referring to an allele that determines the same phenotype when it is present in homozygous or heterozygous form.. Compare with **recessive**.

dormancy A temporary state of greatly reduced metabolic activity and growth in plants or plant parts (e.g., seeds, spores, bulbs, and buds).

dorsal Toward an animal's back and away from its belly. The opposite of ventral.

dorsal hollow nerve chord See **nerve chord**.

double fertilization An unusual form of reproduction seen in flowering plants, in which one sperm cell fuses with an egg to form a zygote and the other sperm cell fuses with two polar nuclei to form the triploid endosperm.

double helix The secondary structure of DNA, consisting of two antiparallel DNA strands wound around each other.

Down syndrome A human developmental disorder caused by trisomy of chromosome 21.

downstream In genetics, the direction in which RNA polymerase moves along a DNA strand. Compare with **upstream**.

duplication In genetics, refers to an additional copy of part of a chromosome.

dynein A class of motor proteins that uses the chemical energy of ATP to "walk" toward the minus end of a microtubule. Dyneins are responsible for bending of cilia and flagella, play a role in chromosome movement during mitosis, and can transport vesicles and organelles.

early endosome A small transient organelle that is formed by the accumulation of vesicles from receptor-mediated endocytosis and is an early stage in the formation of a lysosome.

ecdysone An insect hormone that triggers either molting (to a larger larval form) or metamorphosis (to the adult form), depending on the level of juvenile hormone.

ecdysozoans A major lineage of protostomes (Ecdysozoa) that grow by shedding their external skeletons (molting) and expanding their bodies. Includes

arthropods, nematodes, and other groups. Compare with **lophotrochozoans**.

echinoderms A major lineage of deuterostomes (Echinodermata) distinguished by adult bodies with five-sided radial symmetry, a water vascular system, and tube feet. Includes sea urchins, sand dollars, and sea stars.

echolocation The use of echoes from vocalizations to obtain information about locations of objects in the environment.

ecological selection Also known as environmental selection. A type of natural selection that favors individuals with heritable traits that enhance their ability to survive and reproduce in a certain physical and/or biological environment, excluding their ability to obtain a mate. Compare with **sexual selection**.

ecology The study of how organisms interact with each other and with their surrounding environment.

ecosystem All the organisms that live in a geographic area, together with the nonliving (abiotic) components that affect or exchange materials with the organisms; a community and its physical environment.

ecosystem diversity The variety of biotic components in a region along with abiotic components, such as soil, water, and nutrients.

ecosystem function The sum of biological and chemical processes that are characteristic of a given ecosystem—such as primary production, nitrogen cycling, and carbon storage.

ecosystem services All of the benefits that humans derive, directly or indirectly, from ecosystem functions.

ecotourism Tourism that is based on observing wildlife or experiencing other aspects of natural areas.

ectoderm The outermost of the three basic cell layers (germ layers) in most animal embryos; gives rise to the outer covering and nervous system. Compare with **endoderm** and **mesoderm**.

ectomycorrhizal fungi (EMF) Fungi whose hyphae form a dense network that covers their host plant's roots but do not enter the root cells.

ectoparasite A parasite that lives on the outer surface of the host's body.

ectotherm An animal that gains most of its body heat from external sources as opposed to metabolic processes. Compare with **endotherm**.

effector Any cell, organ, or structure with which an animal can respond to external or internal stimuli. Usually functions, along with a sensor and integrator, as part of a homeostatic system.

efferent division The part of the nervous system that carries commands from the central nervous system to the body. Consists primarily of motor neurons.

egg A mature female gamete and any associated external layers (such as a shell). Larger and less mobile than the male gamete. In animals, also called *ovum*.

ejaculation The release of semen from the copulatory organ of a male animal.

ejaculatory duct A short duct through which sperm move during ejaculation; connects the vas deferens to the urethra.

electric current A flow of electrical charge past a point. Also called *current*.

electrical potential Potential energy created by a separation of electrical charges between two points. Also called *voltage*.

electrocardiogram (EKG) A recording of the electrical activity of the heart, as measured through electrodes on the skin.

electrochemical gradient The combined effect of an ion's concentration gradient and electrical (charge) gradient across a membrane that affects the diffusion of ions across the membrane.

electrogenic fish Any of various kinds of fishes having specialized electric organs that emit a current into the water to detect objects.

electrolyte Any compound that dissociates into ions when dissolved in water. In nutrition, any of the major ions necessary for normal cell function.

electromagnetic spectrum The entire range of wavelengths of radiation extending from short wavelengths (high energy) to long wavelengths (low energy). Includes gamma rays, X-rays, ultraviolet, visible light, infrared, microwaves, and radio waves (from short to long wavelengths).

electron acceptor A reactant that gains an electron and is reduced in a reduction–oxidation reaction.

electron carrier Any molecule that readily accepts electrons from and donates electrons to other molecules. Protons may be transferred with the electrons in the form of hydrogen atoms.

electron donor A reactant that loses an electron and is oxidized in a reduction–oxidation reaction.

electron shell A group of orbitals of electrons with similar energies. Electron shells are arranged in roughly concentric layers around the nucleus of an atom, and electrons in outer shells have more energy than those in inner shells. Electrons in the outermost shell, the valence shell, often are involved in chemical bonding.

electron transport chain (ETC) Any set of membrane-bound protein complexes and mobile electron carriers involved in a coordinated series of redox reactions in which the potential energy of electrons is successively decreased and used to pump protons from one side of a membrane to the other.

electronegativity A measure of the ability of an atom to attract electrons toward itself from an atom to which it is bonded.

electroreception A sensory system in which receptors are activated by electric fields.

electroreceptor A sensory cell or organ specialized to detect electric fields.

element A substance, consisting of atoms with a specific number of protons. Elements preserve their identity in chemical reactions.

elongation (1) The process by which RNA lengthens during transcription. (2) The process by which a polypeptide chain lengthens during translation.

elongation factors Proteins involved in the elongation phase of translation, assisting ribosomes in the synthesis of the growing peptide chain.

embryo A young, developing organism; the stage after fertilization and zygote formation.

embryo sac The female gametophyte in flowering plants.

embryogenesis The production of an embryo from a zygote. Embryogenesis is an early event in development of animals and plants.

Embryophyta An increasingly popular name for the lineage called land plants; reflects their retention of a fertilized egg.

embryophyte A plant that nourishes its embryos inside its own body. All land plants are embryophytes.

emergent property A property that stems from the interaction of simpler elements and that is impossible to predict from the study of individual elements.

emergent vegetation Any plants in an aquatic habitat that extend above the surface of the water.

emerging disease Any infectious disease, often a viral disease, that suddenly afflicts significant numbers of humans for the first time; often due to changes in the host species for a pathogen or to radical changes in the genetic material of the pathogen.

emigration The migration of individuals away from one population to other populations. Compare with **immigration**.

emulsification (verb: emulsify) The dispersion of fat into an aqueous solution. Usually requires the aid of an amphipathic substance such as a detergent or bile salts, which can break large fat globules into microscopic fat droplets.

endangered species A species whose numbers have decreased so much that it is in danger of extinction throughout all or part of its range.

endemic species A species that lives in one geographic area and nowhere else.

endergonic Referring to a chemical reaction that requires an input of energy to occur and for which the Gibbs free-energy change (ΔG) is greater than zero. Compare with **exergonic**.

endocrine Relating to a chemical signal (hormone) that is released into the bloodstream by a producing cell and acts on a distant target cell.

endocrine disruptor An exogenous chemical that interferes with normal hormonal signaling.

endocrine gland A gland that secretes hormones directly into the bloodstream or interstitial fluid instead of into ducts. Compare with **exocrine gland**.

endocrine system All of the glands and tissues that produce and secrete hormones into the bloodstream.

endocytosis General term for any pinching off of the plasma membrane that results in the uptake of material from outside the cell. Includes phagocytosis, pinocytosis, and receptor-mediated endocytosis. Compare with **exocytosis**.

endoderm The innermost of the three basic cell layers (germ layers) in most animal embryos; gives rise to the digestive tract and organs that connect to it (liver, lungs, etc.). Compare with **ectoderm** and **mesoderm**.

endodermis In plant roots, a cylindrical layer of cells that separates the cortex from the vascular tissue and location of the Casparian strip.

endomembrane system A system of organelles in eukaryotic cells that synthesizes, processes, transports, and recycles proteins and lipids. Includes the endoplasmic reticulum (ER), Golgi apparatus, and lysosomes.

endomycorrhizal fungi See **arbuscular mycorrhizal fungi (AMF)**.

endoparasite A parasite that lives inside the host's body.

endophyte (adjective: endophytic) A fungus that lives inside the aboveground parts of a plant in a symbiotic relationship. Compare with **epiphyte**.

endoplasmic reticulum (ER) A network of interconnected membranous sacs and tubules found inside eukaryotic cells. See **rough** and **smooth endoplasmic reticulum**.

endoskeleton Bony and/or cartilaginous structures within the body that provide support. Examples are the spicules of sponges, the plates in echinoderms, and the bony skeleton of vertebrates. Compare with **exoskeleton**.

endosperm A triploid ($3n$) tissue in the seed of a flowering plant (angiosperm) that serves as food for the plant embryo. Functionally analogous to the yolk in some animal eggs.

endosymbiont An organism that lives in a symbiotic relationship inside the body of its host.

endosymbiosis An association between organisms of two different species in which one lives inside the cell or cells of the other.

endosymbiosis theory The theory that mitochondria and chloroplasts evolved from prokaryotes that were engulfed by host cells and took up a symbiotic existence within those cells, a process termed primary endosymbiosis. In some eukaryotes, chloroplasts may have originated by secondary endosymbiosis; that is, when a cell engulfed a chloroplast-containing protist and retained its chloroplasts.

endotherm An animal whose primary source of body heat is internally generated. Compare with **ectotherm**.

endothermic Referring to a chemical reaction that absorbs heat. Compare with **exothermic**.

energetic coupling In cellular metabolism, the mechanism by which energy released from an exergonic reaction (commonly, hydrolysis of ATP) is used to drive an endergonic reaction.

energy The capacity to do work or to supply heat. May be stored (potential energy) or available in the form of motion (kinetic energy).

enhancer A regulatory sequence in eukaryotic DNA that may be located far from the gene it controls or within introns of the gene. Binding of specific proteins to an enhancer enhances the transcription of certain genes.

enrichment culture A method of detecting and obtaining cells with specific characteristics by placing a sample, containing many types of cells, under a specific set of conditions (e.g., temperature, salt concentration, available nutrients) and isolating those cells that grow rapidly in response.

enthalpy (H) A quantitative measure of the amount of potential energy, or heat content, of a system plus the pressure and volume it exerts on its surroundings.

entropy (S) A quantitative measure of the amount of disorder of any system, such as a group of molecules.

envelope (viral) A membrane that encloses the capsids of some viruses. Normally includes specialized proteins that attach to host-cell surfaces.

environmental sequencing See **metagenomics**.

enzyme A protein catalyst used by living organisms to speed up and control biological reactions.

epicotyl In some embryonic plants, a portion of the embryonic stem that extends above the cotyledons.

epidemic The spread of an infectious disease throughout a population in a short time period. Compare with **pandemic**.

epidermis The outermost layer of cells of any multicellular organism.

epididymis A coiled tube wrapped around each testis in reptiles, birds, and mammals. The site of the final stages of sperm maturation.

epigenetic inheritance Pattern of inheritance involving differences in phenotype that are not due to differences in the nucleotide sequence of genes.

epinephrine A catecholamine hormone, produced and secreted by the adrenal medulla, that triggers rapid responses related to the fight-or-flight response. Also called *adrenaline*.

epiphyte (adjective: epiphytic) A nonparasitic plant that grows on the trunks or branches of other plants and is not rooted in soil.

epithelial tissues See **epithelium**.

epithelium (plural: epithelia) An animal tissue consisting of sheetlike layers of tightly packed cells that line an organ, a gland, a duct, or a body surface. Also called *epithelial tissue*.

epitope A small region of a particular antigen to which an antibody, B-cell receptor, or T-cell receptor binds.

equilibrium potential The membrane potential at which there is no net movement of a particular ion into or out of a cell.

ER signal sequence A short amino acid sequence that marks a polypeptide for transport to the endoplasmic reticulum, where synthesis of the polypeptide chain is completed and the signal sequence removed. See **signal recognition particle**.

erythropoietin (EPO) A peptide hormone, released by the kidney in response to low blood-oxygen levels, that stimulates the bone marrow to produce more red blood cells.

esophagus The muscular tube that connects the mouth to the stomach.

essential amino acid Any amino acid that an animal cannot synthesize and must obtain from the diet. May refer specifically to one of the eight essential amino acids of adult humans: isoleucine, leucine, lysine, methionine, phenylalanine, threonine, tryptophan, and valine.

essential nutrient Any chemical element, ion, or compound that is required for normal growth, reproduction, and maintenance of a living organism and that cannot be synthesized by the organism.

EST See **expressed sequence tag**.

ester linkage The covalent bond formed by a condensation reaction between a carboxyl group and a hydroxyl group. Ester linkages join fatty acids to glycerol to form a fat or phospholipid.

estradiol The major estrogen produced by the ovaries of female mammals and many other vertebrates. Stimulates development of the female reproductive tract, growth of ovarian follicles, and growth of breast tissue in mammals.

estrogens A class of steroid hormones, including estradiol, estrone, and estriol, that generally promote female-like traits. Secreted by the gonads, fat tissue, and some other organs.

estrous cycle A female reproductive cycle in which the uterine lining is reabsorbed rather than shed in the absence of pregnancy and the female is sexually receptive only briefly during mid-cycle (estrus). It is seen in all mammals except Old World monkeys and apes (including humans). Compare with **menstrual cycle**.

ethylene A gaseous plant hormone associated with senescence that induces fruits to ripen and flowers to fade.

eudicot A member of a monophyletic group (lineage) of angiosperms that includes complex flowering plants and trees (e.g., roses, daisies, maples). All eudicots have two cotyledons, but not all dicots are members of this lineage. Compare with **dicot** and **monocot**.

Eukarya One of the three taxonomic domains of life, consisting of unicellular organisms (most protists, yeasts) and multicellular organisms (fungi, plants, animals) distinguished by a membrane-bound cell nucleus, numerous organelles, and an extensive cytoskeleton. Compare with **Archaea** and **Bacteria**.

eukaryote A member of the domain Eukarya; an organism whose cells contain a nucleus, numerous membrane-bound organelles, and an extensive cytoskeleton. May be unicellular or multicellular. Compare with **prokaryote**.

eusociality A complex social structure in which workers sacrifice most or all of their direct reproduction to help rear the queen's offspring. Common in insects such as ants, bees, wasps, and termites.

eutherians A lineage of mammals (Eutheria) whose young develop in the uterus and are not housed in an abdominal pouch. Also called *placental mammals*.

evaporation The energy-absorbing phase change from a liquid state to a gaseous state. Many organisms evaporate water as a means of heat loss.

evo-devo Popular term for evolutionary developmental biology, a research field focused on how changes in developmentally important genes have led to the evolution of new phenotypes.

evolution (1) The theory that all organisms on Earth are related by common ancestry and that they have changed over time, and continue to change, via natural selection and other processes. (2) Any change in the genetic characteristics of a population over time, especially, a change in allele frequencies.

ex situ conservation Preserving species outside of natural areas (e.g., in zoos, aquaria, or botanical gardens).

excitable membrane A plasma membrane that is capable of generating an action potential. Neurons, muscle cells, and some other cells have excitable membranes.

excitatory postsynaptic potential (EPSP) A change in membrane potential, usually depolarization, at a neuron dendrite that makes an action potential more likely.

exergonic Referring to a chemical reaction that can occur spontaneously, releasing heat and/or increasing entropy, and for which the Gibbs free-energy change (ΔG) is less than zero. Compare with **endergonic**.

exocrine gland A gland that secretes some substance through a duct into a space other than the circulatory system, such as the digestive tract or the skin surface. Compare with **endocrine gland**.

exocytosis Secretion of intracellular molecules (e.g., hormones, collagen), contained within membrane-bound vesicles, to the outside of the cell by fusion of vesicles to the plasma membrane. Compare with **endocytosis**.

exon A transcribed region of a eukaryotic gene or region of a primary transcript that is retained in the mature RNA. Except for 5′ and 3′ UTRs, mRNA exons code for amino acids. Compare with **intron**.

exoskeleton A hard covering secreted on the outside of the body, used for body support, protection, and muscle attachment. Prominent in arthropods. Compare with **endoskeleton**.

exothermic Referring to a chemical reaction that releases heat. Compare with **endothermic**.

exotic species A nonnative species that is introduced into a new area. Exotic species often are competitors, pathogens, or predators of native species.

expansins A class of plant proteins that break hydrogen bonds between components in the primary cell wall to allow it to expand for cell growth.

exponential population growth The accelerating increase in the size of a population that occurs when the growth rate is constant and density independent. Compare with **logistic population growth**.

expressed sequence tag (EST) A portion of a transcribed gene (synthesized from an mRNA in a cell), used to find the physical location of that gene in the genome.

extant species A species that is living today.

extensor A muscle that pulls two bones farther apart from each other, increasing the angle of the joint, as in the extension of a limb or the spine. Compare with **flexor**.

extinct species A species that has died out.

extracellular digestion Digestion that takes place outside of an organism, as occurs in many fungi that make and secrete digestive enzymes.

extracellular matrix (ECM) A complex meshwork in which animal cells are embedded, consisting of proteins (e.g., collagen, proteoglycan, laminin) and polysaccharides produced by the cells.

extremophile A bacterium or archaean that thrives in an "extreme" environment (e.g., high-salt, high-temperature, low-temperature, or low-pressure).

F_1 generation First filial generation. The first generation of offspring produced from a mating (i.e., the offspring of the parental generation).

facilitated diffusion Passive movement of a substance across a membrane with the assistance of transmembrane carrier proteins or channel proteins.

facilitation In ecological succession, the phenomenon in which early-arriving species make conditions more favorable for later-arriving species. Compare with **inhibition** and **tolerance**.

facultative anaerobe Any organism that can survive and reproduce by performing aerobic respiration when oxygen is available or fermentation when it is not.

FAD/FADH₂ Oxidized and reduced forms, respectively, of flavin adenine dinucleotide. A nonprotein electron carrier that functions in the citric acid cycle and oxidative phosphorylation.

fallopian tube A narrow tube connecting the uterus to the ovary in humans, through which the egg travels after ovulation. Site of fertilization and cleavage. In nonhuman animals, called *oviduct*.

fast muscle fiber Type of skeletal muscle fiber that is white in color, generates ATP by glycolysis, and contracts rapidly but fatigues easily. Also called *fast glycolytic*, or *Type IIb*, *fiber*.

fat A lipid consisting of three fatty acid molecules joined by ester linkages to a glycerol molecule. Also called *triacylglycerol* or *triglyceride*.

fatty acid A lipid consisting of a hydrocarbon chain bonded at one end to a carboxyl group. Used by many organisms to store chemical energy; a major component of animal and plant fats and phospholipids.

fauna All the animal species characteristic of a particular region, period, or environment.

feather A specialized skin outgrowth, composed of β-keratin, present in all birds as well as in some non-avian dinosaurs. Used for flight, insulation, display, and other purposes.

feces The waste products of digestion.

fecundity The average number of female offspring produced by a single female in the course of her lifetime.

feedback inhibition A type of control in which high concentrations of the product of a metabolic pathway inhibit one of the enzymes early in the pathway. A form of negative feedback.

fermentation Any of several metabolic pathways that regenerate oxidizing agents, such as NAD⁺, by transferring electrons to a final electron acceptor in the absence of an electron transport chain. Allows pathways such as glycolysis to continue to make ATP.

ferredoxin In photosynthetic organisms, an iron- and sulfur-containing protein in the electron transport chain of photosystem I. Can transfer electrons to the enzyme NADP⁺ reductase, which catalyzes formation of NADPH.

fertility The average number of surviving children that each woman has during her lifetime.

fertilization Fusion of the nuclei of two haploid gametes to form a zygote with a diploid nucleus.

fertilization envelope A physical barrier that forms around a fertilized egg in many animals. The fertilization envelope prevents fertilization by more than one sperm (polyspermy).

fetal alcohol syndrome (FAS) A condition marked by hyperactivity, severe learning disabilities, and depression. Thought to be caused by exposure of an individual to high blood alcohol concentrations during embryonic development.

fetus In live-bearing animals, the unborn offspring after the embryonic stage. It usually is developed sufficiently to be recognizable as belonging to a certain species.

fiber In plants, a type of elongated sclerenchyma cell that provides support to vascular tissue. Compare with **sclereid**.

Fick's law of diffusion A mathematical relationship that describes the rates of diffusion of gases.

fight-or-flight response Rapid physiological changes that prepare the body for emergencies. Includes increased heart rate, increased blood pressure, and decreased digestion.

filament Any thin, threadlike structure, particularly (1) the threadlike extensions of a fish's gills or (2) part of a stamen: the slender stalk that bears the anthers in a flower.

filter feeder See **suspension feeder**.

filtrate Any fluid produced by filtration, in particular the fluid ("pre-urine") in the Malpighian tubules of insects and the nephrons of vertebrate kidneys.

filtration A process of removing large components from a fluid by forcing it through a filter. Occurs in a renal corpuscle of the vertebrate kidney, allowing water and small solutes to pass from the blood into the nephron.

fimbria (plural: fimbriae) A long, needlelike projection from the cell membrane of prokaryotes that is involved in attachment to nonliving surfaces or other cells.

finite rate of increase (λ) The rate of increase of a population over a given period of time. Calculated as the ending population size divided by the starting population size. Compare with **intrinsic rate of increase**.

first law of thermodynamics The principle of physics that energy is conserved in any process. Energy can be transferred and converted into different forms, but it cannot be created or destroyed.

fission (1) A form of asexual reproduction in which a prokaryotic cell divides to produce two genetically similar daughter cells by a process similar to mitosis of eukaryotic cells. Also called *binary fission*. (2) A form of asexual reproduction in which an animal splits into two or more individuals of approximately equal size; common among invertebrates.

fitness The ability of an individual to produce viable offspring relative to others of the same species.

fitness trade-off See **trade-off**.

fixed action pattern (FAP) Highly stereotyped behavior pattern that occurs in a certain invariant way in a certain species. A form of innate behavior.

flaccid Limp as a result of low internal (turgor) pressure (e.g., a wilted plant leaf). Compare with **turgid**.

flagellum (plural: flagella) A long, cellular projection that undulates (in eukaryotes) or rotates (in prokaryotes) to move the cell through an aqueous environment. See **axoneme**.

flatworms Members of the phylum Platyhelminthes. Distinguished by a broad, flat, unsegmented body that lacks a coelom. Flatworms belong to the lophotrochozoan branch of the protostomes.

flavin adenine dinucleotide See **FAD/FADH$_2$**.

flexor A muscle that pulls two bones closer together, decreasing the joint angle, as in the flexing of a limb or the spine. Compare with **extensor**.

floral meristem A group of undifferentiated plant stem cells that can give rise to the four organs making up a flower.

florigen In plants, a protein hormone that is synthesized in leaves and transported to the shoot apical meristem, where it stimulates flowering.

flower In angiosperms, the part of a plant that contains reproductive structures. Typically includes a calyx, a corolla, and one or more stamens and/or carpels. See **perfect** and **imperfect flower**.

fluid connective tissue A type of connective tissue, distinguished by having a liquid extracellular matrix; includes blood.

fluid feeder Any animal that feeds by sucking or mopping up liquids such as nectar, plant sap, or blood.

fluid-mosaic model The widely accepted hypothesis that the plasma membrane and organelle membranes consist of proteins embedded in a fluid phospholipid bilayer.

fluorescence The spontaneous emission of light from an excited electron falling back to its normal (ground) state.

follicle In a mammalian ovary, a sac of supportive cells containing an egg cell.

follicle-stimulating hormone (FSH) A peptide hormone, produced and secreted by the anterior pituitary; it stimulates (in females) growth of eggs and follicles in the ovaries or (in males) sperm production in the testes.

follicular phase In a menstrual cycle, the first major phase, during which follicles grow and estradiol levels increase; ends with ovulation.

food Any nutrient-containing material that can be consumed and digested by animals.

food chain A relatively simple pathway of energy flow through a few species, each at a different trophic level, in an ecosystem. Might include, for example, a primary producer, a primary consumer, a secondary consumer, and a decomposer. A subset of a **food web**.

food web The complex network of interactions among species in an ecosystem formed by the transfer of energy and nutrients among trophic levels. Consists of many food chains.

foot One of the three main parts of the mollusk body; a muscular appendage, used for movements such as crawling and/or burrowing into sediment.

foraging Searching for food.

forebrain One of the three main regions of the vertebrate brain; includes the cerebrum, thalamus, and hypothalamus. Compare with **hindbrain** and **midbrain**.

fossil Any physical trace of an organism that existed in the past. Includes tracks, burrows, fossilized bones, casts, and so on.

fossil record All of the fossils that have been found anywhere on Earth and that have been formally described in the scientific literature.

founder effect A change in allele frequencies that often occurs when a new population is established from a small group of individuals (founder event) due to sampling error (i.e., the small group is not a representative sample of the source population).

fovea In the vertebrate eye, a portion of the retina where incoming light is focused; contains a high proportion of cone cells.

frameshift mutation The addition or deletion of a nucleotide in a coding sequence that shifts the reading frame of the mRNA.

free energy The energy of a system that can be converted into work. It may be measured only through the change in free energy in a reaction. See **Gibbs free-energy change**.

free radicals Any substance containing one or more atoms with an unpaired electron. Unstable and highly reactive.

frequency The number of wave crests per second traveling past a stationary point. Determines the pitch of sound and the color of light.

frequency-dependent selection A pattern of selection in which certain alleles are favored only when they are rare; a form of balancing selection.

fronds The large leaves of ferns.

frontal lobe In the vertebrate brain, one of the four major areas in the cerebrum.

fruit In flowering plants (angiosperms), a mature, ripened plant ovary (or group of ovaries), along with the seeds it contains and any adjacent fused parts; often functions in seed dispersal. See **aggregate**, **multiple**, and **simple fruit**.

fruiting body A structure formed in some prokaryotes, fungi, and protists for spore dispersal; usually consists of a base, a stalk, and a mass of spores at the top.

functional genomics The study of how a genome works; that is, when and where specific genes are expressed and how their products interact to produce a functional organism.

functional group A small group of atoms bonded together in a precise configuration and exhibiting particular chemical properties that it imparts to any organic molecule in which it occurs.

fundamental niche The total theoretical range of environmental conditions that a species can tolerate. Compare with **realized niche**.

fungi A lineage of eukaryotes that typically have a filamentous body (mycelium) and obtain nutrients by absorption.

fungicide Any substance that can kill fungi or slow their growth.

G protein Any of various proteins that are activated by binding to guanosine triphosphate (GTP) and inactivated when GTP is hydrolyzed to GDP. In G-protein-coupled receptors, signal binding directly triggers the activation of a G protein, leading to production of a second messenger or initiation of a phosphorylation cascade.

G₁ phase The phase of the cell cycle that constitutes the first part of interphase before DNA synthesis (S phase).

G₂ phase The phase of the cell cycle between synthesis of DNA (S phase) and mitosis (M phase); the last part of interphase.

gallbladder A small pouch that stores bile from the liver and releases it as needed into the small intestine during digestion of fats.

gametangium (plural: gametangia) (1) The gamete-forming structure found in all land plants except angiosperms. Contains a sperm-producing antheridium and an egg-producing archegonium. (2) The gamete-forming structure of some chytrid fungi.

gamete A haploid reproductive cell that can fuse with another haploid cell to form a zygote. Most multicellular eukaryotes have two distinct forms of gametes: egg cells (ova) and sperm cells.

gametogenesis The production of gametes (eggs or sperm).

gametophyte In organisms undergoing alternation of generations, the multicellular haploid form that arises from a single haploid spore and produces gametes. Compare with **sporophyte**.

ganglion (plural: ganglia) A mass of neurons in a centralized nervous system.

ganglion cell In the retina, a type of neuron whose axons form the optic nerves.

gap junction A type of cell–cell attachment structure that directly connects the cytosolic components of adjacent animal cells, allowing passage of water, ions, and small molecules between the cells. Compare with **desmosome** and **tight junction**.

gastrin A hormone produced by cells in the stomach lining in response to the arrival of food or to a neural signal from the brain. Stimulates other stomach cells to release hydrochloric acid.

gastropods A lineage of mollusks distinguished by a large, muscular foot and a unique feeding structure, the radula. Include slugs and snails.

gastrulation The process of coordinated cell movements, including the moving of some cells from the outer surface of the embryo to the interior, that results in the formation of three germ layers (endoderm, mesoderm, and ectoderm) and the axes of the embryo.

gated channel A channel protein that opens and closes in response to a specific stimulus, such as the binding of a particular molecule or a change in voltage across the membrane.

gemma (plural: gemmae) A small reproductive structure that is produced asexually in some plants during the gametophyte phase and can grow into a mature gametophyte; most common in non-vascular plants, particularly liverworts and club mosses, and in ferns

gene A section of DNA (or RNA, for some viruses) that encodes information for building one or more related polypeptides or functional RNA molecules along with the regulatory sequences required for its transcription.

gene duplication The formation of an additional copy of a gene, typically by misalignment of chromosomes during crossing over. Thought to be an important evolutionary process in creating new genes.

gene expression The set of processes, including transcription and translation, that convert information in DNA into aproduct of a gene, most commonly a protein.

gene family A set of genetic loci whose DNA sequences are extremely similar. Thought to have arisen by duplication of a single ancestral gene and subsequent mutations in the duplicated sequences.

gene flow The movement of alleles between populations; occurs when individuals leave one population, join another, and breed.

gene pool All the alleles of all the genes in a certain population.

gene therapy The treatment of an inherited disease by introducing a normal form of the gene.

generation The average time between a mother's first offspring and her daughter's first offspring.

genetic bottleneck A reduction in allelic diversity resulting from a sudden reduction in the size of a large population (population bottleneck) due to a random event.

genetic code The set of all codons and their meaning.

genetic correlation A type of evolutionary constraint in which selection on one trait causes a change in another trait as well; may occur when the same gene(s) affect both traits.

genetic diversity The diversity of alleles or genes in a population, species, or group of species.

genetic drift Any change in allele frequencies due to random events. Causes allele frequencies to drift up and down randomly over time, and eventually can lead to the fixation or loss of alleles.

genetic equivalence Having all different cell types of a multicellular individual possess the same genome.

genetic homology Similarity in DNA nucleotide sequences, RNA nucleotide sequences, or amino acid sequences due to inheritance from a common ancestor.

genetic map A list of genes on a chromosome that indicates their position and relative distances from one another. Also called a *linkage map*. Compare with **physical map**.

genetic marker A genetic locus that can be identified and traced in populations by laboratory techniques or by a distinctive visible phenotype.

genetic recombination A change in the combination of alleles on a given chromosome or in a given individual. Also called *recombination*.

genetic screen Any technique that identifies individuals with a particular type of mutation.

genetic variation (1) The number and relative frequency of alleles present in a particular population. (2) The proportion of phenotypic variation in a trait that is due to genetic rather than environmental influences in a certain population in a certain environment.

genetics The study of the inheritance of traits.

genital (plural: genitalia) Any external copulatory organ.

genome All the hereditary information in an organism, including not only genes but also stretches of DNA that do not contain genes.

genome annotation The process of analyzing a genome sequence to identify key features such as genes, regulatory sequences, and splice sites.

genomic library A set of DNA segments representing the entire genome of a particular organism. Each segment is carried by a plasmid or other cloning vector and can be separated from other segments. Compare with **cDNA library**.

genomics The field of study concerned with sequencing, interpreting, and comparing whole genomes from different organisms.

genotype All the alleles of every gene present in a given individual. Often specified only for the alleles of a particular set of genes under study. Compare with **phenotype**.

genus (plural: genera) In Linnaeus' system, a taxonomic category of closely related species. Always italicized and capitalized to indicate that it is a recognized scientific genus.

geologic time scale The sequence of eons, eras, and periods used to describe the geologic history of Earth.

germ cell In animals, any cell that can potentially give rise to gametes. Also called *germ-line cells*.

germ layer In animals, one of the three basic types of tissue formed during gastrulation; gives rise to all other tissues. See **endoderm**, **mesoderm**, and **ectoderm**.

germ line In animals, any of the cells that are capable of giving rise to reproductive cells (sperm or egg). Compare with **germ cell**.

germ theory of disease The theory that infectious diseases are caused by bacteria, viruses, and other microorganisms.

germination The process by which a seed becomes a young plant.

gestation The period of development inside the mother, from implantation to birth, in those species that have live birth.

gibberellins A class of hormones, found in plants and fungi, that stimulate growth. Gibberellic acid (GA) is one of the major gibberellins.

Gibbs free-energy change (ΔG) A measure of the change in enthalpy and entropy that occurs in a given chemical reaction. ΔG is less than 0 for spontaneous reactions and greater than 0 for nonspontaneous reactions.

gill Any organ in aquatic animals that exchanges gases and other dissolved substances between the blood and the surrounding water. Typically, a filamentous outgrowth of a body surface.

gill arch In aquatic vertebrates, a curved region of tissue between the gills. Gills are suspended from the gill arches.

gill filament In fish, any of the many long, thin structures that extend from gill arches into the water and across which gas exchange occurs.

gill lamella (plural: gill lamellae) Any of hundreds to thousands of sheetlike structures, each containing a capillary bed, that make up a gill filament.

gland An organ whose primary function is to secrete some substance, either into the blood (endocrine gland) or into some other space such as the gut or skin (exocrine gland).

glia Collective term for several types of cells in nervous tissue that are not neurons and do not conduct electrical signals but perform other functions, such as providing support, nourishment, or electrical insulation. Also called *glial cells*.

global carbon cycle See **carbon cycle, global**.

global climate change The global sum of all the local changes in temperature and precipitation patterns that accompany global warming (or in some past events, global cooling).

global gene regulation The regulation of multiple bacterial genes that are not part of one operon.

global nitrogen cycle See **nitrogen cycle, global**.

global warming A sustained increase in Earth's average surface temperature.

global water cycle See **water cycle, global**.

glomalin A glycoprotein that is abundant in the hyphae of arbuscular mycorrhizal fungi; when hyphae decay, it is an important component of soil.

glomerulus (plural: glomeruli) (1) In the vertebrate kidney, a ball-like cluster of capillaries, surrounded by Bowman's capsule, at the beginning of a nephron. (2) In the brain, a ball-shaped cluster of neurons in the olfactory bulb.

glucagon A peptide hormone produced by the pancreas in response to low blood glucose. Raises blood glucose by triggering breakdown of glycogen and stimulating gluconeogenesis. Compare with **insulin**.

glucocorticoids A class of steroid hormones, produced and secreted by the adrenal cortex, that increase blood glucose and prepare the body for stress. Include cortisol and corticosterone. Compare with **mineralocorticoids**.

gluconeogenesis Synthesis of glucose, often from non-carbohydrate sources (e.g., proteins and fatty acids). In plants, used to produce glucose from products of the Calvin cycle. In animals, occurs in the liver in response to low insulin levels and high glucagon levels.

glucose Six-carbon monosaccharide whose oxidation in cellular respiration is the major source of ATP in animal cells.

glyceraldehyde-3-phosphate (G3P) The phosphorylated three-carbon compound formed as the result of carbon fixation in the first step of the Calvin cycle.

glycerol A three-carbon molecule that forms the "backbone" of phospholipids and most fats.

glycogen A highly branched storage polysaccharide composed of glucose monomers joined by α-1,4- and α-1,6-glycosidic linkages. The major form of stored carbohydrate in animals.

glycolipid Any lipid molecule that is covalently bonded to a carbohydrate group.

glycolysis A series of 10 chemical reactions that oxidize glucose to produce pyruvate, NADH, and ATP.

Used by organisms as part of fermentation or cellular respiration.

glycoprotein Any protein with one or more covalently bonded carbohydrates, typically oligosaccharides.

glycosidic linkage The covalent linkage formed by a condensation reaction between two sugar monomers; joins the residues of a polysaccharide.

glycosylation Addition of a carbohydrate group to a molecule.

glyoxysome Specialized type of peroxisome found in plant cells and packed with enzymes for processing the products of photosynthesis.

gnathostomes Animals with jaws. Most vertebrates are gnathostomes.

Golgi apparatus A eukaryotic organelle, consisting of stacks of flattened membranous sacs (cisternae), that functions in processing and sorting proteins and lipids destined to be secreted or directed to other organelles. Also called *Golgi complex*.

gonad An organ, such as a testis or an ovary, that produces reproductive cells.

gonadotropin-releasing hormone (GnRH) A peptide hormone, produced and secreted by the hypothalamus, that stimulates release of follicle-stimulating hormone (FSH) and luteinizing hormone (LH) from the anterior pituitary.

grade In taxonomy, a group of species that share some, but not all, of the descendants of a common ancestor. Also called a *paraphyletic group*.

Gram-negative Describing bacteria that look pink when treated with a Gram stain. These bacteria have a cell wall composed of a thin layer of peptidoglycan and an outer phospholipid layer. Compare with **Gram-positive**.

Gram-positive Describing bacteria that look purple when treated with a Gram stain. These bacteria have cell walls composed of a thick layer of peptidoglycan and no outer phospholipid later. Compare with **Gram-negative**.

Gram stain A dye that distinguishes the two general types of cell walls found in bacteria. Used to routinely classify bacteria as Gram-negative or Gram-positive.

granum (plural: grana) In chloroplasts, a stack of flattened, membrane-bound thylakoid discs where the light reactions of photosynthesis occur.

gravitropism The growth or movement of a plant in a particular direction in response to gravity.

grazing food chain The ecological network of herbivores and the predators and parasites that consume them.

great apes See **hominids**.

green algae A paraphyletic group of photosynthetic organisms that contain chloroplasts similar to those in green plants. Often classified as protists, green algae are the closest living relatives of land plants and form a monophyletic group with them.

greenhouse gas An atmospheric gas that absorbs and reflects infrared radiation, so that heat radiated from Earth is retained in the atmosphere instead of being lost to space.

gross primary productivity In an ecosystem, the total amount of carbon fixed by photosynthesis (or more

rarely, chemosynthesis), including that used for cellular respiration, over a given time period. Compare with **net primary productivity**.

ground meristem The middle layer of a young plant embryo. Gives rise to the ground tissue system.

ground tissue An embryonic tissue layer that gives rise to parenchyma, collenchyma, and sclerenchyma—tissues other than the epidermis and vascular tissue. Also called *ground tissue system*.

groundwater Any water below the land surface.

growth factor Any of a large number of signaling molecules that are secreted by certain cells and that stimulate other cells to grow, divide, or differentiate.

growth hormone (GH) A peptide hormone, produced and secreted by the mammalian anterior pituitary, that promotes lengthening of the long bones in children and muscle growth, tissue repair, and lactation in adults. Also called *somatotropin*.

guanosine triphosphate (GTP) A nucleotide consisting of guanine, a ribose sugar, and three phosphate groups. Can be hydrolyzed to release free energy. Commonly used in RNA synthesis and also functions in signal transduction in association with G proteins.

guard cell One of two specialized, crescent-shaped cells forming the border of a plant stoma. Guard cells can change shape to open or close the stoma. See also **stoma**.

gustation The perception of taste.

guttation Excretion of water droplets from plant leaves; visible in the early morning. Caused by root pressure.

gymnosperm A vascular plant that makes seeds but does not produce flowers. The gymnosperms include five lineages of green plants (cycads, ginkgoes, conifers, redwoods, and gnetophytes). Compare with **angiosperm**.

H$^+$-ATPase See **proton pump**.

habitat degradation The reduction of the quality of a habitat.

habitat destruction Human-caused destruction of a natural habitat, replaced by an urban, suburban, or agricultural landscape.

habitat fragmentation The breakup of a large region of a habitat into many smaller regions, separated from others by a different type of habitat.

Hadley cell An atmospheric cycle of large-scale air movement in which warm equatorial air rises, moves north or south, and then descends at approximately 30° N or 30° S latitude.

hair cell A pressure-detecting sensory cell that has tiny "hairs" (stereocilia) jutting from its surface. Found in the inner ear, lateral line system, and ampullae of Lorenzini.

hairpin A secondary structure in RNA consisting of a stable loop formed by hydrogen bonding between purine and pyrimidine bases on the same strand.

Hamilton's rule The proposition that an allele for altruistic behavior will be favored by natural selection only if Br C, where B = the fitness benefit to the recipient, C = the fitness cost to the actor, and r = the coefficient of relatedness between recipient and actor.

haploid (1) Having one set of chromosomes (1*n* or *n* for short). (2) A cell or an individual organism with one set of chromosomes. Compare with **diploid**.

haploid number The number of different types of chromosomes in a cell. Symbolized as *n*.

Hardy–Weinberg principle A principle of population genetics stating that genotype frequencies in a large population do not change from generation to generation in the absence of evolutionary processes (e.g., mutation, gene flow, genetic drift, and selection), and nonrandom mating.

haustorium (plural: haustoria) Highly modified stem or root of a parasitic plant. The haustorium penetrates the tissues of a host and absorbs nutrients and water.

hearing The sensation of the wavelike changes in air pressure called sound.

heart A muscular pump that circulates blood throughout the body.

heart murmur A distinctive sound caused by backflow of blood through a defective heart valve.

heartwood The older xylem in the center of an older stem or root, containing protective compounds and no longer functioning in water transport.

heat Thermal energy that is transferred from an object at higher temperature to one at lower temperature.

heat of vaporization The energy required to vaporize 1 gram of a liquid into a gas.

heat-shock proteins Proteins that facilitate refolding of proteins that have been denatured by heat or other agents.

heavy chain The larger of the two types of polypeptide chains in an antibody or B-cell receptor molecule; composed of a variable (*V*) region, which contributes to the antigen-binding site, and a constant (*C*) region. Differences in heavy-chain constant regions determine the different classes of immunoglobulins (IgA, IgE, etc.). Compare with **light chain**.

helper T cell A CD4$^+$ effector T cell that secretes cytokines and in other ways promotes the activation of other lymphocytes. Activated by interacting with complementary class II MHC–peptide complexes on the surface of antigen-presenting cells such as dendritic cells.

heme A small molecule that binds to each of the four polypeptides that combine to form hemoglobin; contains an iron ion that can bind oxygen.

hemimetabolous metamorphosis A type of metamorphosis in which the animal increases in size from one stage to the next, but does not dramatically change its body form. Also called *incomplete metamorphosis*.

hemocoel A body cavity, present in arthropods and some mollusks, containing a pool of circulatory fluid (hemolymph) bathing the internal organs. Unlike a coelom, a hemocoel is not lined in mesoderm.

hemoglobin An oxygen-binding protein consisting of four polypeptide subunits, each containing an oxygen-binding heme group. The major oxygen carrier in mammalian blood.

hemolymph The circulatory fluid of animals with open circulatory systems (e.g., insects) in which the fluid is not confined to blood vessels.

herbaceous Referring to a plant that is not woody.

herbivore (adjective: herbivorous) An animal that eats primarily plants and rarely or never eats meat. Compare with **carnivore** and **omnivore**.

herbivory The practice of eating plant tissues.

heredity The transmission of traits from parents to offspring via genetic information.

heritable Referring to traits that can be transmitted from one generation to the next.

hermaphrodite An organism that produces both male and female gametes.

heterokaryotic Describing a cell or fungal mycelium containing two or more haploid nuclei that are genetically distinct.

heterospory (adjective: heterosporous) In seed plants, the production of two distinct types of spores: microspores, which become the male gametophyte, and megaspores, which become the female gametophyte. Compare with **homospory**.

heterotherm An animal whose body temperature varies markedly. Compare with **homeotherm**.

heterotroph Any organism that cannot synthesize reduced organic compounds from inorganic sources and that must obtain them from other organisms. Some bacteria, some archaea, and virtually all fungi and animals are heterotrophs. Also called *consumer*. Compare with **autotroph**.

heterozygote advantage A pattern of natural selection that favors heterozygous individuals compared with homozygotes. Tends to maintain genetic variation in a population, thus is a form of balancing selection.

heterozygous Having two different alleles of a gene.

hexose A monosaccharide (simple sugar) containing six carbon atoms.

hibernation An energy-conserving physiological state, marked by a decrease in metabolic rate, body temperature, and activity, that lasts for a prolonged period (weeks to months). Occurs in some animals in response to winter cold and scarcity of food. Compare with **torpor**.

hindbrain One of the three main regions of the vertebrate brain, responsible for balance and sometimes hearing; includes the cerebellum and medulla oblongata. Compare with **forebrain** and **midbrain**.

histamine A molecule released from mast cells during an inflammatory response that, at high concentrations, causes blood vessels to constrict to reduce blood loss from tissue damage.

histone One of several positively charged (basic) proteins associated with DNA in the chromatin of eukaryotic cells.

histone acetyltransferases (HATs) A class of eukaryotic enzymes that loosen chromatin structure by adding acetyl groups to histone proteins.

histone code The hypothesis that specific combinations of chemical modifications of histone proteins contain information that influences chromatin condensation and gene expression.

histone deacetylases (HDACs) A class of eukaryotic enzymes that condense chromatin by removing acetyl groups from histone proteins.

holoenzyme A multipart enzyme consisting of a core enzyme (containing the active site for catalysis) along with other required proteins.

holometabolous metamorphosis A type of metamorphosis in which the animal completely changes its form; includes a distinct larval stage. Also called *complete metamorphosis*.

homeobox A DNA sequence of about 180 base pairs that codes for a DNA binding motif called the homeodomain in the resulting protein. Genes containing a homeobox usually play a role in controlling development of organisms from fruit flies to humans.

homeostasis (adjective: homeostatic) The array of relatively stable chemical and physical conditions in an animal's cells, tissues, and organs. May be achieved by the body's passively matching the conditions of a stable external environment (conformational homeostasis) or by active physiological processes (regulatory homeostasis) triggered by variations in the external or internal environment.

homeotherm An animal that has a constant or relatively constant body temperature. Compare with **heterotherm**.

homeotic mutation A mutation that causes one body part to be substituted for another.

hominids Members of the family Hominidae, which includes humans and extinct related forms, chimpanzees, gorillas, and orangutans. Distinguished by large body size, no tail, and an exceptionally large brain. Also called *great apes*.

hominins Any extinct or living species of bipedal apes, such as *Australopithecus africanus*, *Homo erectus*, and *Homo sapiens*.

homologous See **homology**.

homologous chromosomes In a diploid organism, chromosomes that are similar in size, shape, and gene content. Also called *homologs*.

homology (adjective: homologous) Similarity among organisms of different species due to their inheritance from a common ancestor. Features that exhibit such similarity (e.g., DNA sequences, proteins, body parts) are said to be homologous. Compare with **homoplasy**.

homoplasy (adjective: homoplastic) Similarity among organisms of different species due to reasons other than common ancestry, such as convergent evolution. Features that exhibit such similarity (e.g., the wings of birds and bats) are said to be homoplastic, or convergent. Compare with **homology**.

homospory (adjective: homosporous) In seedless vascular plants, the production of just one type of spore. Compare with **heterospory**.

homozygous Having two identical alleles of a gene.

hormone Any of many different signaling molecules that circulate throughout the plant or animal body and can trigger characteristic responses in distant target cells at very low concentrations.

hormone-response element A specific sequence in DNA to which a steroid hormone–receptor complex can bind and affect gene transcription.

host An individual that has been invaded by an organism such as a parasite or a virus, or that provides habitat or resources to a commensal organism.

host cell A cell that has been invaded by a parasitic organism or a virus and provides an environment that is conducive to the pathogen's growth and reproduction.

Hox genes A class of genes found in several animal phyla, including vertebrates, that are expressed in a distinctive pattern along the anterior–posterior axis in early embryos and control formation of specific structures. *Hox* genes code for transcription factors with a DNA-binding sequence called a homeobox.

human Any member of the genus *Homo*, which includes modern humans (*Homo sapiens*) and several extinct species.

human chorionic gonadotropin (hCG) A glycoprotein hormone produced by a human embryo and placenta from about week 3 to week 14 of pregnancy. Maintains the corpus luteum, which produces hormones that preserve the uterine lining.

Human Genome Project The multinational research project that sequenced the human genome.

human immunodeficiency virus (HIV) A retrovirus that causes acquired immune deficiency syndrome (AIDS) in humans.

humoral (immune) response The type of immune response that is mediated through the production and secretion of antibodies, complement proteins, and other soluble factors that eliminate extracellular pathogens. Compare with **cell-mediated (immune) response**.

humus The decayed organic matter in soils.

hybrid The offspring of parents from two different strains, populations, or species.

hybrid zone A geographic area where interbreeding occurs between two species, sometimes producing fertile hybrid offspring.

hydrocarbon An organic molecule that contains only hydrogen and carbon atoms.

hydrogen bond A weak interaction between two molecules or different parts of the same molecule resulting from the attraction between a hydrogen atom with a partial positive charge and another atom (usually O or N) with a partial negative charge. Compare with **covalent bond** and **ionic bond**.

hydrogen ion (H$^+$) A single proton with a charge of 1+; typically, one that is dissolved in solution or that is being transferred from one atom to another in a chemical reaction.

hydrolysis A chemical reaction in which a molecule is split into smaller molecules by reacting with water. In biology, most hydrolysis reactions involve the splitting of polymers into monomers. Compare with **condensation reaction**.

hydrophilic Interacting readily with water. Hydrophilic compounds are typically polar compounds containing partially or fully charged atoms. Compare with **hydrophobic**.

hydrophobic Not readily interacting with water. Hydrophobic compounds are typically nonpolar compounds that lack partially or fully charged atoms. Compare with **hydrophilic**.

hydrophobic interactions Very weak interactions between nonpolar molecules, or nonpolar regions of the same molecule, when exposed to an aqueous solvent. The surrounding water molecules support these interactions by interacting with one another and encapsulating the nonpolar molecules.

hydroponic growth Growth of plants in liquid cultures instead of soil.

hydrostatic skeleton A system of body support involving a body wall in tension surrounding a fluid or soft tissue under compression.

hydroxide ion (OH$^-$) An oxygen atom and a hydrogen atom joined by a single covalent bond and carrying a negative charge; formed by dissociation of water.

hygiene hypothesis The claim that immune disorders arise in individuals less likely to have been exposed to pathogens and parasites, especially in early childhood. Provides an explanation for the increased risk of allergies and autoimmune disease in countries with high levels of sanitation.

hyperosmotic Comparative term designating a solution that has a greater solute concentration, and therefore a lower water concentration, than another solution. Compare with **hyposmotic** and **isosmotic**.

hyperpolarization A change in membrane potential from its resting negative state to an even more negative state; a normal phase in an action potential. Compare with **depolarization**.

hypersensitive reaction An intense allergic response by cells that have been sensitized by previous exposure to an allergen.

hypersensitive response In plants, the rapid death of a cell that has been infected by a pathogen, thereby reducing the potential for infection to spread throughout a plant. Compare with **systemic acquired resistance**.

hypertension Abnormally high blood pressure.

hypertonic Comparative term designating a solution that, if inside a cell or vesicle, results in the uptake of water and swelling or even bursting of the membrane-bound structure. This solution has a greater solute concentration than the solution on the other side of the membrane. Used when the solute is unable to pass through the membrane. Compare with **hypotonic** and **isotonic**.

hypha (plural: hyphae) One of the long, branching strands of a fungal mycelium (the mesh-like body of a fungus). Also found in some protists.

hypocotyl The stem of a very young plant; the region between the cotyledon (embryonic leaf) and the radicle (embryonic root).

hyposmotic Comparative term designating a solution that has a lower solute concentration, and therefore a higher water concentration, than another solution. Compare with **hyperosmotic** and **isosmotic**.

hypothalamic–pituitary axis The functional interaction of the hypothalamus and the anterior pituitary gland, which are anatomically distinct but work together to regulate most of the other endocrine glands in the body.

hypothalamus A part of the brain that functions in maintaining the body's internal physiological state by regulating the autonomic nervous system, endocrine system, body temperature, water balance, and appetite.

hypothesis A testable statement that explains a phenomenon or a set of observations.

hypotonic Comparative term designating a solution that, if inside a cell or vesicle, results in the loss of water and shrinkage of the membrane-bound structure. This solution has a lower solute concentration than the solution on the other side of the membrane. Used when the solute is unable to pass through the membrane. Compare with **hypertonic** and **isotonic**.

immigration The migration of individuals into a particular population from other populations. Compare with **emigration**.

immune system The system whose primary function is to defend the host organism against pathogens. Includes several types of cells (e.g., leukocytes). In vertebrates, several organs are also involved where specialized cells develop or reside (e.g., lymph nodes and thymus).

immunity (adjective: immune) State of being protected against infection by disease-causing pathogens.

immunization The conferring of immunity to a particular disease by artificial means.

immunoglobulin (Ig) Any of the class of proteins that are structurally related to antibodies.

immunological memory The ability of the immune system to "remember" an antigen and mount a rapid, effective adaptive immune response to a pathogen encountered years or decades earlier. Based on the formation of memory lymphocytes.

impact hypothesis The hypothesis that a collision between the Earth and an asteroid caused the mass extinction at the K–P boundary, 65 million years ago.

imperfect flower A flower that contains male parts (stamens) *or* female parts (carpels) but not both. Compare with **perfect flower**.

implantation The process by which an embryo buries itself in the uterine or oviductal wall and forms a placenta. Occurs in mammals and some other viviparous vertebrates.

in situ hybridization A technique for detecting specific DNAs and mRNAs in cells and tissues by use of labeled complementary probes. Can be used to determine where and when particular genes are expressed in embryos.

inbreeding Mating between closely related individuals. Increases homozygosity of a population and often leads to a decline in the average fitness via selection (inbreeding depression).

inbreeding depression In inbred offspring, fitness declines due to deleterious recessive alleles that are homozygous, thus exposed to selection.

inclusive fitness The combination of (1) direct production of offspring (direct fitness) and (2) extra production of offspring by relatives in response to help provided by the individual in question (indirect fitness).

incomplete digestive tract A digestive tract that has just one opening.

incomplete dominance An inheritance pattern in which the heterozygote phenotype is in between the homozygote phenotypes.

incomplete metamorphosis See **hemimetabolous metamorphosis**.

independent assortment, principle of The concept that each pair of hereditary elements (alleles of the same gene) segregates (separates) independently of alleles of other genes during meiosis. One of Mendel's two principles of genetics.

indeterminate growth A pattern of growth in which an individual continues to increase its overall body size throughout its life.

induced fit Change in the shape of the active site of an enzyme, as the result of initial weak binding of a substrate, so that it binds substrate more tightly.

inducer A small molecule that triggers transcription of a specific gene, often by binding to and inactivating a repressor protein.

inducible defense A defensive trait that is manifested only in response to the presence of a consumer (predator or herbivore) or pathogen. Compare with **constitutive defense**.

infection thread An invagination of the plasma membrane of a root hair through which beneficial nitrogen-fixing bacteria enter the roots of their host plants (legumes).

inflammatory response An aspect of the innate immune response, seen in most cases of infection or tissue injury, in which the affected tissue becomes swollen, red, warm, and painful.

ingestion The act of bringing food into the digestive tract.

inhibition In ecological succession, the phenomenon in which early-arriving species make conditions less favorable for the establishment of certain later-arriving species. Compare with **facilitation** and **tolerance**.

inhibitory postsynaptic potential (IPSP) A change in membrane potential, usually hyperpolarization, at a neuron dendrite that makes an action potential less likely.

initiation (1) In an enzyme-catalyzed reaction, the stage during which enzymes orient reactants precisely as they bind at specific locations within the enzyme's active site. (2) In DNA transcription, the stage during which RNA polymerase and other proteins assemble at the promoter sequence and open the strands of DNA to start transcription. (3) In translation, the stage during which a complex consisting of initiation factor proteins, a ribosome, an mRNA, and an aminoacyl tRNA corresponding to the start codon is formed.

initiation factors A class of proteins that assist ribosomes in binding to a messenger RNA molecule to begin translation.

innate behavior Behavior that is inherited genetically, does not have to be learned, and is typical of a species.

innate immune response See **innate immunity**.

innate immunity A set of barriers to infection and generic defenses against broad types of pathogens. Produces an immediate response that involves many different leukocytes, which often activate an inflammatory response. Compare with **acquired immunity**.

inner cell mass (ICM) A cluster of cells, in the interior of a mammalian blastocyst, that eventually develop into the embryo. Contrast with **trophoblast**.

inner ear The innermost portion of the mammalian ear, consisting of a fluid-filled system of tubes that includes the cochlea (which receives sound vibrations from the middle ear) and the semicircular canals (which function in balance).

insects A terrestrial lineage of arthropods distinguished by three tagmata (head, thorax, abdomen), a single pair of antennae, and unbranched appendages.

insulin A peptide hormone produced by the pancreas in response to high levels of glucose (or amino acids) in blood. Enables cells to absorb glucose and coordinates synthesis of fats, proteins, and glycogen. Compare with **glucagon**.

integral membrane protein Any membrane protein that spans the entire lipid bilayer. Also called *transmembrane protein*. Compare with **peripheral membrane protein**.

integrated pest management In agriculture or forestry, systems for managing insects or other pests that include carefully controlled applications of toxins, introduction of species that prey on pests, planting schemes that reduce the chance of a severe pest outbreak, and other techniques.

integrator A component of an animal's nervous system that functions as part of a homeostatic system by evaluating sensory information and triggering appropriate responses. See **effector** and **sensor**.

integrin Any of a class of cell-surface proteins that bind to laminins and other proteins in the extracellular matrix, thus holding cells in place.

intercalated disc A type of specialized connection between adjacent heart muscle cells that contains gap junctions, allowing electrical signals to pass between the cells.

intermediate filament A long fiber, about 10 nm in diameter, composed of one of various proteins (e.g., keratins, lamins); one of the three types of cytoskeletal fibers. Used to form networks that help maintain cell shape and hold the nucleus in place. Compare with **actin filament** and **microtubule**.

intermediate host The host species in which a parasite reproduces asexually. Compare with **definitive host**.

intermediate muscle fiber Type of skeletal muscle fiber that is pink in color, generates ATP by both glycolysis and aerobic respiration, and has contractile properties that are intermediate between slow fibers and fast fibers. Also called fast oxidative fiber.

interneuron A neuron that passes signals from one neuron to another. Compare with **motor neuron** and **sensory neuron**.

internode The section of a plant stem between two nodes (sites where leaves attach).

interphase The portion of the cell cycle between one M phase and the next. Includes the G_1 phase, S phase, and G_2 phase.

intersexual selection The sexual selection of an individual of one gender for mating by an individual of the other gender (usually by female choice).

interspecific competition Competition between members of different species for the same limited resource. Compare with **intraspecific competition**.

interstitial fluid The plasma-like fluid found in the region (interstitial space) between cells.

intertidal zone The region between the low-tide and high-tide marks on a seashore.

intrasexual selection Competition among members of one gender for an opportunity to mate (usually male–male competition).

intraspecific competition Competition between members of the same species for the same limited resource. Compare with **interspecific competition**.

intrinsic rate of increase (r_{max}) The rate at which a population will grow under optimal conditions (i.e., when birthrates are as high as possible and death rates are as low as possible). Compare with **finite rate of increase**.

intron A region of a eukaryotic gene that is transcribed into RNA but is later removed. Compare with **exon**.

invasive species An exotic (nonnative) species that, upon introduction to a new area, spreads rapidly and competes successfully with native species.

inversion A mutation in which a segment of a chromosome breaks from the rest of the chromosome, flips, and rejoins in reversed orientation.

invertebrates A paraphyletic group composed of animals without a backbone; includes about 95 percent of all animal species. Compare with **vertebrates**.

involuntary muscle Muscle that contracts in response to stimulation by involuntary (parasympathetic or sympathetic), but not voluntary (somatic), neural stimulation.

ion An atom or a molecule that has lost or gained electrons and thus carries an electric charge, either positive (cation) or negative (anion), respectively.

ion channel A type of channel protein that allows certain ions to diffuse across a plasma membrane down an electrochemical gradient.

ionic bond A chemical bond that is formed when an electron is completely transferred from one atom to another so that the atoms remain associated due to their opposite electric charges. Compare with **covalent bond** and **hydrogen bond**.

iris A ring of pigmented muscle just behind the cornea in the vertebrate eye that contracts or expands to control the amount of light entering the eye through the pupil.

isosmotic Comparative term designating a solution that has the same solute concentration and water concentration as another solution. Compare with **hyperosmotic** and **hyposmotic**.

isotonic Comparative term designating a solution that, if inside a cell or vesicle, results in no net uptake or loss of water and thus no effect on the volume of the membrane-bound structure. This solution has the same solute concentration as the solution on the other side of the membrane. Compare with **hypertonic** and **hypotonic**.

isotope Any of several forms of an element that have the same number of protons but differ in the number of neutrons.

joint A place where two components (bones, cartilages, etc.) of a skeleton meet. May be movable (an articulated joint) or immovable (e.g., skull sutures).

juvenile An individual that has adult-like morphology but is not sexually mature.

juvenile hormone An insect hormone that prevents larvae from metamorphosing into adults.

karyogamy Fusion of two haploid nuclei to form a diploid nucleus. Occurs in many fungi, and in animals and plants during fertilization of gametes.

karyotype The distinctive appearance of all of the chromosomes in an individual, including the number of chromosomes and their length and banding patterns (after staining with dyes).

keystone species A species that has an exceptionally great impact on the other species in its ecosystem relative to its abundance.

kidney In terrestrial vertebrates, one of a paired organ situated at the back of the abdominal cavity that filters the blood, produces urine, and secretes several hormones.

kilocalorie (kcal) A unit of energy often used to measure the energy content of food. A kcal of energy raises 1 kg of water 1°C.

kin selection A form of natural selection that favors traits that increase survival or reproduction of an individual's kin at the expense of the individual.

kinesin A class of motor proteins that uses the chemical energy of ATP to "walk" toward the plus end of a microtubule. Used to transport vesicles, particles, organelles and chromosomes.

kinetic energy The energy of motion. Compare with **potential energy**.

kinetochore A protein complex at the centromere where microtubules attach to the chromosome. Contains motor proteins and microtubule-binding proteins that are involved in chromosome segregation during M phase.

kinetochore microtubules Microtubules in the spindle formed during mitosis and meiosis that are attached to the kinetochore on a chromosome.

kinocilium (plural: kinocilia) A single cilium that juts from the surface of many hair cells and functions in detection of sound or pressure.

knock-out allele A mutant allele that does not produce a functional product. Also called a *null allele* or *loss-of-function allele*.

Koch's postulates Four criteria used to determine whether a suspected infectious agent causes a particular disease.

labium majus (plural: labia majora) One of two outer folds of skin that surround the labia minora, clitoris, and vaginal opening of female mammals.

labium minus (plural: labia minora) One of two folds of skin inside the labia majora and surrounding the opening of the urethra and vagina.

labor The strong muscular contractions of the uterus that expel the fetus during birth.

***lac* operon** A set of three genes in *E. coli* that are transcribed into a single mRNA and required for lactose metabolism. Studies of the *lac* operon revealed many insights about gene regulation.

lactation (verb: lactate) Production of milk to feed offspring, from mammary glands of mammals.

lacteal A small lymphatic vessel extending into the center of a villus in the small intestine. Receives chylomicrons containing fat absorbed from food.

lactic acid fermentation Catabolic pathway in which pyruvate produced by glycolysis is converted to lactic acid in the absence of oxygen.

lagging strand In DNA replication, the strand of new DNA that is synthesized discontinuously in a series of short pieces that are later joined. Also called *discontinuous strand*. Compare with **leading strand**.

laminins An abundant protein in the extracellular matrix that binds to other ECM components and to integrins in plasma membranes; helps anchor cells in place. Predominantly found in the basal lamina; many subtypes function in different tissues.

large intestine The distal portion of the digestive tract, consisting of the cecum, colon, and rectum. Its primary function is to compact the wastes delivered from the small intestine and absorb enough water to form feces.

larva (plural: larvae) An immature stage of an animal species in which the immature and adult stages have different body forms.

late endosome A membrane-bound vesicle that arises from an early endosome, accepts lysosomal enzymes from the Golgi, and matures into a lysosome.

latency In viruses that infect animals, the ability to coexist with the host cell in a dormant state without producing new virions. The viral genetic material is replicated as the host cell replicates. Genetic material may or may not be integrated in the host genome, depending on the virus.

lateral bud A bud that forms at the nodes of a stem and may develop into a lateral (side) branch. Also called *axillary bud*.

lateral gene transfer Transfer of DNA between two different species.

lateral line system A pressure-sensitive sensory organ found in many aquatic vertebrates.

lateral root A plant root that extends horizontally from another root.

leaching Loss of nutrients from soil via percolating water.

leading strand In DNA replication, the strand of new DNA that is synthesized in one continuous piece. Also called *continuous strand*. Compare with **lagging strand**.

leaf The main photosynthetic organ of vascular plants.

leak channel Ion channel that allows ions to leak across the membrane of a neuron in its resting state.

learning An enduring change in an individual's behavior that results from specific experience(s).

leghemoglobin An iron-containing protein similar to hemoglobin. Found in infected cells of legume root nodules where it binds oxygen, preventing it from poisoning a bacterial enzyme needed for nitrogen fixation.

legumes Members of the pea plant family. Many form symbiotic associations with nitrogen-fixing bacteria in their roots.

lens A transparent structure that focuses incoming light onto a retina or other light-sensing apparatus of an eye.

lenticel Spongy segment in bark that allows gas exchange between cells in a woody stem and the atmosphere.

leptin A hormone produced and secreted by fat cells (adipocytes) that acts to stabilize fat-tissue mass partly by inhibiting appetite and increasing energy expenditure.

leukocyte Any of several types of blood cells, including neutrophils, macrophages, and lymphocytes, that reside in tissues and circulate in blood and lymph. Functions in tissue repair and defense against pathogens. Also called *white blood cell*.

lichen A symbiotic association of a fungus, often in the Ascomycota lineage, and a photosynthetic alga or cyanobacterium.

life cycle The sequence of developmental events and phases that occurs during the life span of an organism, from fertilization to offspring production.

life history The sequence of events in an individual's life from birth to reproduction to death, including how an individual allocates resources to growth, reproduction, and activities or structures that are related to survival.

life table A data set that summarizes the probability that an individual in a certain population will survive and reproduce in any given year over the course of its lifetime.

ligament Connective tissue that joins bones of an endoskeleton together.

ligand Any molecule that binds to a specific site on a receptor molecule.

ligand-gated channel An ion channel that opens or closes in response to binding by a certain molecule. Compare with **voltage-gated channel**.

light chain The smaller of the two types of polypeptide chains in an antibody or B-cell receptor molecule; composed of a variable (*V*) region, which contributes to the antigen-binding site, and a constant (*C*) region. Compare with **heavy chain**.

lignin A substance, found in the secondary cell walls of some plants, that is exceptionally stiff and strong; a complex polymer built from six-carbon rings. Most abundant in woody plant parts.

limiting nutrient Any essential nutrient whose scarcity in the environment significantly reduces growth and reproduction of organisms.

limnetic zone Open water (not near shore) that receives enough sunlight to support photosynthesis.

lineage See **monophyletic group**.

LINEs (long interspersed nuclear elements) The most abundant class of transposable elements in human genomes; can create copies of itself and insert them elsewhere in the genome. Compare with **SINEs**.

lingual lipase An enzyme produced by glands in the tongue. It breaks down fat molecules into fatty acids and monoglycerides.

linkage In genetics, a physical association between two genes because they are on the same chromosome; the inheritance patterns resulting from this association.

linkage map See **genetic map**.

lipid Any organic substance that does not dissolve in water, but dissolves well in nonpolar organic solvents. Lipids include fatty acids, fats, oils, waxes, steroids, and phospholipids.

lipid bilayer The basic structural element of all cellular membranes; consists of a two-layer sheet of phospholipid molecules with their hydrophobic tails oriented toward the inside and their hydrophilic heads toward the outside. Also called *phospholipid bilayer*.

littoral zone Shallow water near shore that receives enough sunlight to support photosynthesis. May be marine or freshwater; often flowering plants are present.

liver A large, complex organ of vertebrates that performs many functions, including storage of glycogen, processing and conversion of food and wastes, and production of bile.

lobe-finned fish Fish with fins supported by bony elements that extend down the length of the structure.

locomotion Movement of an organism under its own power.

locus (plural: loci) A gene's physical location on a chromosome.

logistic population growth The density-dependent decrease in growth rate as population size approaches the carrying capacity. Compare with **exponential population growth**.

long interspersed nuclear elements See **LINEs**.

long-day plant A plant that blooms in response to short nights (usually in late spring or early summer in the Northern Hemisphere). Compare with **day-neutral** and **short-day plant**.

loop of Henle In the vertebrate kidney, a long U-shaped loop in a nephron that extends into the medulla. Functions as a countercurrent exchanger to set up an osmotic gradient that allows reabsorption of water from the collecting duct.

loose connective tissue A type of connective tissue consisting of fibrous proteins in a soft matrix. Often functions as padding for organs.

lophophore A specialized feeding structure found in some lophotrochozoans and used in suspension (filter) feeding.

lophotrochozoans A major lineage of protostomes (Lophotrochozoa) that grow by extending the size of their skeletons rather than by molting. Many phyla have a specialized feeding structure (lophophore) and/or ciliated larvae (trochophore). Includes rotifers, flatworms, segmented worms, and mollusks. Compare with **ecdysozoans**.

loss-of-function allele See **knock-out allele**.

LUCA The *l*ast *u*niversal *c*ommon *a*ncestor of cells. This theoretical entity is proposed to be the product of chemical evolution and provided characteristics of life that are shared by all living organisms on Earth today.

lumen The interior space of any hollow structure (e.g., the rough ER) or organ (e.g., the stomach).

lung Any respiratory organ used for gas exchange between blood and air.

luteal phase The second major phase of a menstrual cycle, after ovulation, when the progesterone levels are high and the body is preparing for a possible pregnancy.

luteinizing hormone (LH) A peptide hormone, produced and secreted by the anterior pituitary, that stimulates estrogen production, ovulation, and formation of the corpus luteum in females and testosterone production in males.

lymph The mixture of fluid and white blood cells that circulates through the ducts and lymph nodes of the lymphatic system in vertebrates.

lymph node Any of many small, oval structures that lymph moves through in the lymphatic system. Filters the lymph and screens it for pathogens and other antigens. Major sites of lymphocyte activation.

lymphatic system In vertebrates, a body-wide network of thin-walled ducts (or vessels) and lymph nodes, separate from the circulatory system. Collects excess fluid from body tissues and returns it to the blood; also functions as part of the immune system.

lymphocyte A cell that circulates through the bloodstream and lymphatic system and is responsible for the development of adaptive immunity. In most cases belongs to one type of leukocyte—either B cells or T cells.

lysogeny In viruses that infect bacteria (bacteriophages), the ability to coexist with the host cell in a dormant state without producing new virions. The viral genetic material is integrated in the host chromosome and is replicated as the host cell replicates. Compare with **lytic cycle**.

lysosome A small, acidified organelle in an animal cell containing enzymes that catalyze hydrolysis reactions and can digest large molecules. Compare with **vacuole**.

lysozyme An enzyme that functions in innate immunity by digesting bacterial cell walls. Occurs in lysosomes of phagocytes and is secreted in saliva, tears, mucus, and egg white.

lytic cycle A type of viral replicative growth in which the production and release of virions kills the host cell. Compare with **lysogeny**.

M phase The phase of the cell cycle during which cell division occurs. Includes mitosis or meiosis and often cytokinesis.

M-phase-promoting factor (MPF) A complex of a cyclin and cyclin-dependent kinase that, when activated, phosphorylates a number of specific proteins needed to initiate mitosis in eukaryotic cells.

macromolecular machine A group of proteins that assemble to carry out a particular function.

macromolecule Any very large organic molecule, usually made up of smaller molecules (monomers) joined together into a polymer. The main biological macromolecules are proteins, nucleic acids, and polysaccharides.

macronutrient Any element (e.g., nitrogen) that is required in large quantities for normal growth, reproduction, and maintenance of a living organism. Compare with **micronutrient**.

macrophage A type of leukocyte in the innate immune system that participates in the inflammatory response by secreting cytokines and phagocytizing invading pathogens and apoptotic cells. Also serves as an antigen-presenting cell to activate lymphocytes.

MADS box A DNA sequence that codes for a DNA-binding motif in proteins; present in floral organ identity genes in plants. Functionally similar sequences are found in some fungal and animal genes.

magnetoreception A sensory system in which receptors are activated in response to magnetic fields.

magnetoreceptor A sensory cell or organ specialized for detecting magnetic fields.

major histocompatibility protein See **MHC protein**.

maladaptive Describing a trait that lowers fitness.

malaria A human disease caused by five species of the protist *Plasmodium* and passed to humans by mosquitoes.

malignant tumor A tumor that is actively growing and disrupting local tissues or is spreading to other organs. Cancer consists of one or more malignant tumors. Compare with **benign tumor**.

Malpighian tubules A major excretory organ of insects, consisting of blind-ended tubes that extend from the gut into the hemocoel. Filter hemolymph to form "pre-urine" and then send it to the hindgut for further processing.

mammals One of the two lineages of amniotes (vertebrates that produce amniotic eggs) distinguished by hair (or fur) and mammary glands. Includes the monotremes (platypuses), marsupials, and eutherians (placental mammals).

mammary glands Specialized exocrine glands that produce and secrete milk for nursing offspring. A diagnostic feature of mammals.

mandibles Any mouthpart used in chewing. In vertebrates, the lower jaw. In insects, crustaceans, and myriapods, the first pair of mouthparts.

mantle One of the three main parts of the mollusk body; the thick outer tissue that protects the visceral mass and may secrete a calcium carbonate shell.

marsh A wetland dominated by grasses and other nonwoody plants.

marsupials A lineage of mammals (Marsupiala) that nourish their young in an abdominal pouch after a very short period of development in the uterus.

mass extinction The extinction of a large number of diverse evolutionary groups during a relatively short period of geologic time (about 1 million years). May occur due to sudden and extraordinary environmental changes. Compare with **background extinction**.

mass feeder An animal that ingests chunks of food.

mass number The total number of protons and neutrons in an atom.

mast cell A type of leukocyte that is stationary (embedded in tissue) and helps trigger the inflammatory response, including secretion of histamine, to infection or injury. Particularly important in allergic responses and defense against parasites.

maternal chromosome A chromosome inherited from the mother.

mechanical advantage The ratio of force exerted on a load to the muscle force of the effort. A measure of the force efficiency of a mechanical system.

mechanoreception A sensory system in which receptors are activated in response to changes in pressure.

mechanoreceptor A sensory cell or organ specialized for detecting distortions caused by touch or pressure. One example is hair cells in the cochlea.

mediator Regulatory proteins in eukaryotes that form a physical link between regulatory transcription factors that are bound to DNA, the basal transcription complex, and RNA polymerase.

medium A liquid or solid that supports the growth of cells.

medulla The innermost part of an organ (e.g., kidney or adrenal gland).

medulla oblongata In vertebrates, a region of the brain stem that along with the cerebellum forms the hindbrain.

medusa (plural: medusae) The free-floating stage in the life cycle of some cnidarians (e.g., jellyfish). Compare with **polyp**.

megapascal (MPa) A unit of pressure (force per unit area) equivalent to 1 million pascals (Pa).

megasporangium (plural: megasporangia) In heterosporous species of plants, a spore-producing structure that produces megaspores, which go on to develop into female gametophytes.

megaspore In seed plants, a haploid (n) spore that is produced in a megasporangium by meiosis of a diploid ($2n$) megasporocyte; develops into a female gametophyte. Compare with **microspore**.

meiosis In sexually reproducing organisms, a special two-stage type of cell division in which one diploid ($2n$) parent cell produces haploid (n) cells (gametes); results in halving of the chromosome number. Also called *reduction division*.

meiosis I The first cell division of meiosis, in which synapsis and crossing over occur and homologous chromosomes are separated from each other, producing daughter cells with half as many chromosomes (each composed of two sister chromatids) as the parent cell.

meiosis II The second cell division of meiosis, in which sister chromatids are separated from each other. Similar to mitosis.

melatonin A hormone, produced by the pineal gland, that regulates sleep–wake cycles and seasonal reproduction in vertebrates.

membrane potential A difference in electric charge across a cell membrane; a form of potential energy. Also called *membrane voltage*.

memory Retention of learned information.

memory cells A type of lymphocyte responsible for maintaining immunity for years or decades after an infection. Descended from a B cell or T cell activated during a previous infection or vaccination.

meniscus (plural: menisci) The concave boundary layer formed at most air–water interfaces due to adhesion and surface tension.

menstrual cycle A female reproductive cycle seen in Old World monkeys and apes (including humans) in which the uterine lining is shed (menstruation) if no pregnancy occurs. Compare with **estrous cycle**.

menstruation The periodic shedding of the uterine lining through the vagina that occurs in females of Old World monkeys and apes, including humans.

meristem (adjective: meristematic) In plants, a group of undifferentiated cells, including stem cells, which can divide and develop into various adult tissues throughout the life of a plant. See also **apical meristem** and **ground meristem**.

mesoderm The middle of the three basic cell layers (germ layers) in most animal embryos; gives rise to muscles, bones, blood, and some internal organs (kidney, spleen, etc.). Compare with **ectoderm** and **endoderm**.

mesoglea A gelatinous material, containing scattered ectodermal cells, that is located between the ectoderm and endoderm of cnidarians (e.g., jellyfish, corals, and anemones).

mesophyll cell A type of cell, found near the surfaces of plant leaves, that is specialized for the light-dependent reactions of photosynthesis.

Mesozoic era The interval of geologic time, from 251 million to 65.5 million years ago, during which gymnosperms were the dominant plants and dinosaurs the dominant vertebrates. Ended with extinction of the dinosaurs (except birds).

messenger RNA (mRNA) An RNA molecule transcribed from DNA that carries information (in codons) that specifies the amino acid sequence of a polypeptide.

meta-analysis A comparative analysis of the results of many smaller, previously published studies.

metabolic pathway A linked series of biochemical reactions that build up or break down a particular molecule; the product of one reaction is the substrate of the next reaction.

metabolic rate The total energy use by all the cells of an individual. For aerobic organisms, often measured as the amount of oxygen consumed per hour.

metabolic water The water that is produced as a by-product of cellular respiration.

metagenomics The inventory of all the genes in a community or ecosystem by sequencing, analyzing, and comparing the genomes of the component organisms. Also called *environmental sequencing*.

metagenomics The inventory of all the genes in a community or ecosystem by sequencing, analyzing, and comparing the genomes of the component organisms. Sequencing of all or most of the genes present in an environment directly (also called *environmental sequencing*).

metallothioneins Small plant proteins that bind to and prevent excess metal ions from acting as toxins.

metamorphosis Transition from one developmental stage to another, such as from the larval to the adult form of an animal.

metaphase A stage in mitosis or meiosis during which chromosomes line up in the middle of the cell.

metaphase plate The plane along which chromosomes line up in the middle of the spindle during metaphase of mitosis or meiosis; not an actual structure.

metapopulation A population made up of many small, physically isolated populations connected by migration.

metastasis The spread of cancerous cells from their site of origin to distant sites in the body where they may establish additional tumors.

methanogen A prokaryote that produces methane (CH_4) as a by-product of cellular respiration.

methanotroph An organism (bacteria or archaea) that uses methane (CH_4) as its primary electron donor and source of carbon.

methyl salicylate (MeSA) A molecule that is hypothesized to function as a signal, transported among tissues, that triggers systematic acquired resistance in plants—a response to pathogen attack.

MHC protein Any of a large set of mammalian cell-surface glycoproteins involved in marking cells as self and in antigen presentation to T cells. Also called *MHC molecule*. Compare with **Class I** and **Class II MHC protein**.

microbe Any microscopic organism, including bacteria, archaea, and various tiny eukaryotes.

microbiology The field of study concerned with microscopic organisms.

microfibril Bundled strands of cellulose that serve as the fibrous component in plant cell walls.

microfilament See **actin filament**.

micronutrient Any element (e.g., iron, molybdenum, magnesium) that is required in very small quantities for normal growth, reproduction, and maintenance of a living organism. Compare with **macronutrient**.

micropyle The tiny pore in a plant ovule through which the pollen tube reaches the embryo sac.

microRNA (miRNA) A small, single-stranded RNA associated with proteins in an RNA-induced silencing complex (RISC). Processed from a longer premiRNA gene transcript. Can bind to complementary sequences in mRNA molecules, allowing the associated proteins of RISC to degrade the bound mRNA or inhibit its translation. See **RNA interference**.

microsatellite A noncoding stretch of eukaryotic DNA consisting of a repeating sequence 2 to 6 base pairs long. Also called *short tandem repeat* or *simple sequence repeat*.

microsporangium (plural: microsporangia) In heterosporous species of plants, a spore-producing structure that produces microspores, which go on to develop into male gametophytes.

microspore In seed plants, a haploid (n) spore that is produced in a microsporangium by meiosis of a diploid ($2n$) microsporocyte; develops into a male gametophyte. Compare with **megaspore**.

microtubule A long, tubular fiber, about 25 nm in diameter, formed by polymerization of tubulin protein dimers; one of the three types of cytoskeletal fibers. Involved in cell movement and transport of materials within the cell. Compare with **actin filament** and **intermediate filament**.

microtubule organizing center (MTOC) General term for any structure (e.g., centrosome and basal body) that organizes microtubules in cells.

microvilli (singular: microvillus) Tiny protrusions from the surface of an epithelial cell that increase the surface area for absorption of substances.

midbrain One of the three main regions of the vertebrate brain; includes sensory integrating and relay centers. Compare with **forebrain** and **hindbrain**.

middle ear The air-filled middle portion of the mammalian ear, which contains three small bones (ossicles) that transmit and amplify sound from the tympanic membrane to the inner ear. Is connected to the throat via the eustachian tube.

migration (1) In ecology, a seasonal movement of large numbers of organisms from one geographic location or habitat to another. (2) In population genetics, movement of individuals from one population to another.

millivolt (mV) A unit of voltage equal to 1/1000 of a volt.

mimicry A phenomenon in which one species has evolved (or learns) to look or sound like another species. See **Batesian mimicry** and **Müllerian mimicry**.

mineral One of various inorganic substances that are important components of enzyme cofactors or of structural materials in an organism.

mineralocorticoids A class of steroid hormones, produced and secreted by the adrenal cortex, that regulate electrolyte levels and the overall volume of body fluids. Aldosterone is the principal one in humans. Compare with **glucocorticoids**.

minisatellite A noncoding stretch of eukaryotic DNA consisting of a repeating sequence that is 6 to 100 base pairs long. Also called *variable number tandem repeat (VNTR)*.

mismatch repair The process by which mismatched base pairs in DNA are fixed.

missense mutation A point mutation (change in a single base pair) that changes one amino acid for another within the sequence of a protein.

mitochondrial matrix Central compartment of a mitochondrion, which is lined by the inner membrane; contains mitochondrial DNA, ribosomes, and the enzymes for pyruvate processing and the citric acid cycle.

mitochondrion (plural: mitochondria) A eukaryotic organelle that is bounded by a double membrane and is the site of aerobic respiration and ATP synthesis.

mitogen-activated protein kinases (MAPK) Enzymes that are involved in signal transduction pathways that often lead to the induction of cell replication. Different types are organized in a series, where one kinase activates another via phosphorylation. See also **phosphorylation cascade**.

mitosis In eukaryotic cells, the process of nuclear division that results in two daughter nuclei genetically identical with the parent nucleus. Subsequent cytokinesis (division of the cytoplasm) yields two daughter cells.

mode of transmission The type of inheritance observed as a trait is passed from parent to offspring. Some common types are autosomal recessive, autosomal dominant, and X-linked recessive.

model organism An organism selected for intensive scientific study based on features that make it easy to work with (e.g., body size, life span), in the hope that findings will apply to other species.

molarity A common unit of solute concentration equal to the number of moles of a dissolved solute in 1 liter of solution.

mole The amount of a substance that contains 6.022×10^{23} of its elemental entities (e.g., atoms, ions, or molecules). This number of molecules of a compound will have a mass equal to the molecular weight of that compound expressed in grams.

molecular chaperone A protein that facilitates the folding of newly synthesized proteins into their correct three-dimensional shape. Usually works by an ATP-dependent mechanism.

molecular formula A notation that indicates only the numbers and types of atoms in a molecule, such as H_2O for the water molecule. Compare with **structural formula**.

molecular weight The sum of the atomic weights of all of the atoms in a molecule; roughly, the total number of protons and neutrons in the molecule.

molecule A combination of two or more atoms held together by covalent bonds.

molting A method of body growth, used by ecdysozoans, that involves the shedding of an external protective cuticle or skeleton, expansion of the soft body, and growth of a new external layer.

monocot Any flowering plant (angiosperm) that has a single cotyledon (embryonic leaf) upon germination. Monocots form a monophyletic group. Also called a monocotyledonous plant. Compare with **dicot**.

monoecious Describing an angiosperm species that has both male and female reproductive structures on each plant. Compare with **dioecious**.

monohybrid cross A mating between two parents that are both heterozygous for one given gene.

monomer A small molecule that can covalently bind to other similar molecules to form a larger macromolecule. Compare with **polymer**.

monophyletic group An evolutionary unit that includes an ancestral population and all of its descendants but no others. Also called a *clade* or *lineage*. Compare with **paraphyletic group** and **polyphyletic group**.

monosaccharide A molecule that has the molecular formula $(CH_2O)_n$ and cannot be hydrolyzed to form any smaller carbohydrates. Also called *simple sugar*. Compare with **oligosaccharide** and **polysaccharide**.

monosomy The state of having only one copy of a particular type of chromosome in an otherwise diploid cell.

monotremes A lineage of mammals (Monotremata) that lay eggs and then nourish the young with milk. Includes just five living species: the platypus and four species of echidna, all with leathery beaks or bills.

morphogen A molecule that exists in a concentration gradient and provides spatial information to embryonic cells.

morphology The shape and appearance of an organism's body and its component parts.

morphospecies concept The definition of a species as a population or group of populations that have measurably different anatomical features from other groups. Also called *morphological species concept*. Compare with **biological species concept** and **phylogenetic species concept**.

motor neuron A nerve cell that carries signals from the central nervous system (brain and spinal cord) to an effector, such as a muscle or gland. Compare with **interneuron** and **sensory neuron**.

motor protein A class of proteins whose major function is to convert the chemical energy of ATP into motion. Includes dynein, kinesin, and myosin.

MPF See **M-phase-promoting factor**.

mRNA See **messenger RNA**.

mucosal-associated lymphoid tissue (MALT) Collective term for lymphocytes and other leukocytes associated with skin cells and mucus-secreting epithelial tissues in the gut and respiratory tract. Plays an important role in preventing entry of pathogens into the body.

mucous cell A type of cell found in the epithelial layer of the stomach; responsible for secreting mucus into the stomach.

mucus (adjective: mucous) A slimy mixture of glycoproteins (called mucins) and water that is secreted in many animal organs for lubrication. Serves as a barrier to protect surfaces from infection.

Müllerian inhibitory substance A peptide hormone, secreted by the embryonic testis, that causes regression (withering away) of the female reproductive ducts.

Müllerian mimicry A type of mimicry in which two (or more) harmful species resemble each other. Compare with **Batesian mimicry**.

multicellularity The state of being composed of many cells that adhere to each other and do not all express the same genes, with the result that some cells have specialized functions.

multiple allelism The existence of more than two alleles of the same gene.

multiple fruit A fruit (e.g., pineapple) that develops from many separate flowers and thus many carpels. Compare with **aggregate** and **simple fruit**.

multiple sclerosis (MS) A human autoimmune disease in which the immune system attacks the myelin sheaths that insulate axons of neurons.

muscle fiber A single muscle cell.

muscle tissue An animal tissue consisting of bundles of long, thin, contractile cells (muscle fibers). Functions primarily in movement.

mutagen Any physical or chemical agent that increases the rate of mutation.

mutant An individual that carries a mutation, particularly a new or rare mutation.

mutation Any change in the hereditary material of an organism (DNA in most organisms, RNA in some viruses). The only source of new alleles in populations.

mutualism (adjective: mutualistic) A symbiotic relationship between two organisms (mutualists) that benefits both. Compare with **commensalism** and **parasitism**.

mutualist Organism that is a participant and partner in a mutualistic relationship. See **mutualism**.

mycelium (plural: mycelia) A mass of underground filaments (hyphae) that form the body of a fungus. Also found in some protists and bacteria.

mycorrhiza (plural: mycorrhizae) A mutualistic association between certain fungi and the roots of most vascular plants, sometimes visible as nodules or nets in or around plant roots.

myelin sheath Multiple layers of myelin, derived from the cell membranes of certain glial cells, wrapped around the axon of a neuron and providing electrical insulation.

myocardial infarction Death of cardiac muscle cells when deprived of oxygen.

myoD A transcription factor that is a master regulator of muscle cell differentiation (short for "*myo*blast *determination*").

myofibril Long, slender structure composed of contractile proteins organized into repeating units (sarcomeres) in vertebrate heart muscle and skeletal muscle.

myosin Any one of a class of motor proteins that use the chemical energy of ATP to move along actin filaments in muscle contraction, cytokinesis, and vesicle transport.

myriapods A lineage of arthropods with long segmented trunks, each segment bearing one or two pairs of legs. Includes millipedes and centipedes.

NAD⁺/NADH Oxidized and reduced forms, respectively, of nicotinamide adenine dinucleotide. A nonprotein electron carrier that functions in many of the redox reactions of metabolism.

NADP⁺/NADPH Oxidized and reduced forms, respectively, of nicotinamide adenine dinucleotide phosphate. A nonprotein electron carrier that is reduced during the light-dependent reactions in photosynthesis and extensively used in biosynthetic reactions.

natural experiment A situation in which a natural change in conditions enables comparisons of groups, rather than a manipulation of conditions by researchers.

natural selection The process by which individuals with certain heritable traits tend to produce more surviving offspring than do individuals without those traits, often leading to a change in the genetic makeup of the population. A major mechanism of evolution.

nauplius A distinct planktonic larval stage seen in many crustaceans.

Neanderthal A recently extinct European species of hominin, *Homo neanderthalensis*, closely related to but distinct from modern humans.

nectar The sugary fluid produced by flowers to attract and reward pollinating animals.

nectary A nectar-producing structure in a flower.

negative control Of genes, when a regulatory protein shuts down expression by binding to DNA on or near the gene.

negative feedback A self-limiting, corrective response in which a deviation in some variable (e.g., concentration of some compound) triggers responses aimed at returning the variable to normal. Represents a means of maintaining homeostasis. Compare with **positive feedback**.

negative pressure ventilation Ventilation of the lungs by expanding the rib cage so as to "pull" air into the lungs. Compare with **positive pressure ventilation**.

negative-sense RNA virus An ssRNA virus whose genome contains sequences complementary to those in the mRNA required to produce viral proteins. Compare with **ambisense virus** and **positive-sense virus**.

nematodes See **roundworms**.

nephron One of many tiny tubules inside the kidney that function in the formation of urine.

neritic zone Shallow marine waters beyond the intertidal zone, extending down to about 200 meters, where the continental shelf ends.

nerve A long, tough strand of nervous tissue, typically containing thousands of axons, wrapped in connective tissue; carries impulses between the central nervous system and some other part of the body.

nerve cord In chordate animals, a hollow bundle of nerves extending from the brain along the dorsal (back) side of the animal, with cerebrospinal fluid inside a central channel. One of the defining features of chordates.

nerve net A nervous system in which neurons are diffuse instead of being clustered into large ganglia or tracts; found in cnidarians and ctenophores.

nervous tissue An animal tissue consisting of nerve cells (neurons) and various supporting cells.

net primary productivity (NPP) In an ecosystem, the total amount of carbon fixed by photosynthesis over a given time period minus the amount oxidized during cellular respiration. Compare with **gross primary productivity**.

net reproductive rate (R_0) The growth rate of a population per generation; equivalent to the average number of female offspring that each female produces over her lifetime.

neural tube A folded tube of ectoderm that forms along the dorsal side of a young vertebrate embryo; gives rise to the brain and spinal cord.

neuroendocrine Referring to nerve cells (neurons) that release hormones into the blood or to such hormones themselves.

neurogenesis The birth of new neurons from central nervous system stem cells.

neurohormones Hormones produced by neurons.

neuron A cell that is specialized for the transmission of nerve impulses. Typically has dendrites, a cell body, and a long axon that forms synapses with other neurons. Also called *nerve cell*.

neurosecretory cell A nerve cell (neuron) that produces and secretes hormones into the bloodstream. Principally found in the hypothalamus. Also called *neuroendocrine cell*.

neurotoxin Any substance that specifically destroys or blocks the normal functioning of neurons.

neurotransmitter A molecule that transmits signals from one neuron to another or from a neuron to a muscle or gland. Examples are acetylcholine, dopamine, serotonin, and norepinephrine.

neutral In genetics, referring to any mutation or mutant allele that has no effect on an individual's fitness.

neutrophil A type of leukocyte, capable of moving through body tissues, that engulfs and digests pathogens and other foreign particles; also secretes various compounds that attack bacteria and fungi.

niche The range of resources that a species can use and the range of conditions that it can tolerate. More broadly, the role that species plays in its ecosystem.

niche differentiation The evolutionary change in resource use by competing species that occurs as the result of character displacement.

nicotinamide adenine dinucleotide See **NAD⁺/NADH**.

nitrogen cycle, global The movement of nitrogen among terrestrial ecosystems, the oceans, and the atmosphere.

nitrogen fixation The incorporation of atmospheric nitrogen (N_2) into ammonia (NH_3), which can be used to make many organic compounds. Occurs in only a few lineages of bacteria and archaea.

nociceptor A sensory cell or organ specialized to detect tissue damage, usually producing the sensation of pain.

Nod factors Molecules produced by nitrogen-fixing bacteria that help them recognize and bind to roots of legumes.

node (1) In animals, any small thickening (e.g., a lymph node). (2) In plants, the part of a stem where leaves or leaf buds are attached. (3) In a phylogenetic tree, the point where two branches diverge, representing the point in time when an ancestral group split into two or more descendant groups. Also called *fork*.

node of Ranvier One of the unmyelinated sections that occurs periodically along a neuron's axon and serves as a site where an action potential can be regenerated.

nodule Globular structure on roots of legume plants that contain symbiotic nitrogen-fixing bacteria.

noncyclic electron flow Path of electron flow in which electrons pass from photosystem II, through an electron transport chain, to photosystem I, and ultimately to NADP⁺ during the light-dependent reactions of photosynthesis. See also **Z scheme**.

nondisjunction An error that can occur during meiosis or mitosis, in which one daughter cell receives two copies of a particular chromosome and the other daughter cell receives none.

nonpolar covalent bond A covalent bond in which electrons are equally shared between two atoms of the same or similar electronegativity. Compare with **polar covalent bond**.

nonsense mutation A point mutation (change in a single base pair) that converts an amino-acid-specifying codon into a stop codon.

non-sister chromatids The chromatids of a particular type of chromosome (after replication) with respect to the chromatids of its homologous chromosome. Crossing over occurs between non-sister chromatids. Compare with **sister chromatids**.

non-template strand The strand of DNA that is not transcribed during synthesis of RNA. Its sequence corresponds to that of the mRNA produced from the other strand. Also called *coding strand*.

non-vascular plants A paraphyletic group of land plants that lack vascular tissue and reproduce using spores. The non-vascular plants include three lineages of green plants (liverworts, mosses, and hornworts). These lineages are sometimes called *bryophytes*.

norepinephrine A catecholamine used as a neurotransmitter in the sympathetic nervous system. Also is produced by the adrenal medulla and functions as a hormone that triggers rapid responses relating to the fight-or-flight response.

notochord A supportive but flexible rod that occurs in the back of a chordate embryo, ventral to the developing spinal cord. Replaced by vertebrae in most adult vertebrates. A defining feature of chordates.

nuclear envelope The double-layered membrane enclosing the nucleus of a eukaryotic cell.

nuclear lamina A lattice-like sheet of fibrous nuclear lamins, which are one type of intermediate filament. Lines the inner membrane of the nuclear envelope, stiffening the envelope and helping to organize the chromosomes.

nuclear lamins Intermediate filaments that make up the nuclear lamina layer—a lattice-like layer inside the nuclear envelope that stiffens the structure.

nuclear localization signal (NLS) A short amino acid sequence that marks a protein for delivery to the nucleus.

nuclear pore An opening in the nuclear envelope that connects the inside of the nucleus with the cytoplasm and through which molecules such as mRNA and some proteins can pass.

nuclear pore complex A large complex of dozens of proteins lining a nuclear pore, defining its shape and regulating transport through the pore.

nuclease Any enzyme that can break down RNA or DNA molecules.

nucleic acid A macromolecule composed of nucleotide monomers. Generally used by cells to store or transmit hereditary information. Includes ribonucleic acid and deoxyribonucleic acid.

nucleoid In prokaryotic cells, a dense, centrally located region that contains DNA but is not surrounded by a membrane.

nucleolus In eukaryotic cells, a specialized structure in the nucleus where ribosomal RNA processing occurs and ribosomal subunits are assembled.

nucleosome A repeating, bead-like unit of eukaryotic chromatin, consisting of about 200 nucleotides of DNA wrapped twice around eight histone proteins.

nucleotide excision repair The process of removing a damaged region in one strand of DNA and replacing it with the correct sequence using the undamaged strand as a template.

nucleotide A molecule consisting of a five-carbon sugar (ribose or deoxyribose), a phosphate group, and one of several nitrogen-containing bases. DNA and RNA are polymers of nucleotides containing deoxyribose (deoxyribonucleotides) and ribose (ribonucleotides), respectively. Equivalent to a nucleoside plus one phosphate group.

nucleus (1) The center of an atom, containing protons and neutrons. (2) In eukaryotic cells, the large organelle containing the chromosomes and surrounded by a double membrane. (3) A discrete clump of neuron cell bodies in the brain, usually sharing a distinct function.

null allele See **knock-out allele**.

null hypothesis A hypothesis that specifies what the results of an experiment will be if the main hypothesis being tested is wrong. Often states that there will be no difference between experimental groups.

nutrient Any substance that an organism requires for normal growth, maintenance, or reproduction.

occipital lobe In the vertebrate brain, one of the four major areas in the cerebrum.

oceanic zone The waters of the open ocean beyond the continental shelf.

odorant Any volatile molecule that conveys information about food or the environment.

oil An unsaturated fat that is liquid at room temperature.

Okazaki fragment Short segment of DNA produced during replication of the lagging strand template. Many Okazaki fragments make up the lagging strand in newly synthesized DNA.

olfaction The perception of odors.

olfactory bulb A bulb-shaped projection of the brain just above the nose. Receives and interprets odor information from the nose.

oligodendrocyte A type of glial cell that wraps around axons of some neurons in the central nervous system, forming a myelin sheath that provides electrical insulation. Compare with **Schwann cell**.

oligopeptide A chain composed of fewer than 50 amino acids linked together by peptide bonds. Often referred to simply as *peptide*.

oligosaccharide A linear or branched polymer consisting of less than 50 monosaccharides joined by glycosidic linkages. Compare with **monosaccharide** and **polysaccharide**.

ommatidium (plural: ommatidia) A light-sensing column in an arthropod's compound eye.

omnivore (adjective: omnivorous) An animal whose diet regularly includes both meat and plants. Compare with **carnivore** and **herbivore**.

oncogene Any gene whose protein product stimulates cell division at all times and thus promotes cancer development. Often is a mutated form of a gene involved in regulating the cell cycle. See **proto-oncogene**.

one-gene, one-enzyme hypothesis The hypothesis that each gene is responsible for making one enzyme. This hypothesis has expanded to include genes that produce proteins other than enzymes or that produce RNAs as final products.

oogenesis The production of egg cells (ova).

oogonium (plural: oogonia) In an ovary, any of the diploid cells that can divide by mitosis to create primary oocytes (which can undergo meiosis) and more oogonia.

open circulatory system A circulatory system in which the circulating fluid (hemolymph) is not confined to blood vessels. Compare with **closed circulatory system**.

open reading frame (ORF) Any DNA sequence, ranging in length from several hundred to thousands of base pairs long, that is flanked by a start codon and a stop codon. ORFs identified by computer analysis of DNA may be functional genes, especially if they have other features characteristic of genes (e.g., promoter sequence).

operator In prokaryotic DNA, a binding site for a repressor protein; located near the start of an operon.

operculum The stiff flap of tissue that covers the gills of teleost fishes.

operon A region of prokaryotic DNA that codes for a series of functionally related genes and is transcribed from a single promoter into one mRNA.

opsin A transmembrane protein that is covalently linked to retinal, the light-detecting pigment in rod and cone cells.

optic nerve A bundle of neurons that runs from the eye to the brain.

optimal foraging The concept that animals forage in a way that maximizes the amount of usable energy they take in, given the costs of finding and ingesting their food and the risk of being eaten while they're at it.

orbital The region of space around an atomic nucleus in which an electron is present most of the time.

organ A group of tissues organized into a functional and structural unit.

organ system Groups of tissues and organs that work together to perform a function.

organelle Any discrete, membrane-bound structure within a cell (e.g., mitochondrion) that has a characteristic structure and function.

organic For a compound, containing carbon and hydrogen and usually containing carbon–carbon bonds. Organic compounds are widely used by living organisms.

organism Any living entity that contains one or more cells.

organogenesis A stage of embryonic development that follows gastrulation and that creates organs from the three germ layers.

origin of replication The site on a chromosome at which DNA replication begins.

osmoconformer An animal that does not actively regulate the osmolarity of its tissues but conforms to the osmolarity of the surrounding environment.

osmolarity The concentration of dissolved substances in a solution, measured in osmoles per liter.

osmoregulation The process by which a living organism controls the concentration of water and salts in its body.

osmoregulator An animal that actively regulates the osmolarity of its tissues.

osmosis Diffusion of water across a selectively permeable membrane from a region of low solute concentration (high water concentration) to a region of high solute concentration (low water concentration).

ossicles, ear In mammals, three bones found in the middle ear that function in transferring and amplifying sound from the outer ear to the inner ear.

ouabain A plant toxin that poisons the sodium–potassium pumps of animals.

out-of-Africa hypothesis The hypothesis that modern humans (*Homo sapiens*) evolved in Africa and spread to other continents, replacing other *Homo* species without interbreeding with them.

outcrossing Reproduction by fusion of the gametes of different individuals, rather than by self-fertilization.

outer ear The outermost portion of the mammalian ear, consisting of the pinna (ear flap) and the ear canal. Funnels sound to the tympanic membrane.

outgroup A taxon that is closely related to a particular monophyletic group but is not part of it.

oval window A membrane separating the fluid-filled cochlea from the air-filled middle ear; sound vibrations pass through it from the middle ear to the inner ear in mammals.

ovary The egg-producing organ of a female animal, or the fruit- and seed-producing structure in the female part of a flower.

overexploitation Unsustainable removal of wildlife from the natural environment for use by humans.

oviduct See **fallopian tube**.

oviparous In animals, producing eggs that are laid outside the body where they develop and hatch. Compare with **ovoviviparous** and **viviparous**.

ovoviviparous In animals, producing eggs that are retained inside the body until they are ready to hatch. Compare with **oviparous** and **viviparous**.

ovulation The release of an egg from an ovary of a female vertebrate. In humans, an ovarian follicle releases an egg at the end of the follicular phase of the menstrual cycle.

ovule In flowering plants, the structure inside an ovary that contains the female gametophyte and eventually (if fertilized) becomes a seed.

ovum (plural: ova) See **egg**.

oxidation The loss of electrons from an atom or molecule during a redox reaction, either by donation of an electron to another atom or molecule, or by the shared electrons in covalent bonds moving farther from the atomic nucleus.

oxidative phosphorylation Production of ATP molecules by ATP synthase using the proton gradient established via redox reactions of an electron transport chain.

oxygen–hemoglobin equilibrium curve The graphed depiction of the percentage of hemoglobin in the blood that is bound to oxygen at various partial pressures of oxygen.

oxygenic Referring to any process or reaction that produces oxygen. Photosynthesis in plants, algae, and cyanobacteria, which involves photosystem II, is oxygenic. Compare with **anoxygenic**.

oxytocin A peptide hormone, secreted by the posterior pituitary, that triggers labor and milk production in females and that stimulates pair bonding, parental care, and affiliative behavior in both sexes.

p53 A tumor-suppressor protein (molecular weight of 53 kilodaltons) that responds to DNA damage by stopping the cell cycle, turning on DNA repair

machinery, and, if necessary, triggering apoptosis. Encoded by the *p53* gene.

pacemaker cell Any of a group of specialized cardiac muscle cells in the sinoatrial (SA) node of the vertebrate heart that have an inherent rhythm and can generate an electrical impulse that spreads to other heart cells.

paleontologists Scientists who study the fossil record and the history of life.

Paleozoic era The interval of geologic time, from 542 million to 251 million years ago, during which fungi, land plants, and animals first appeared and diversified. Began with the Cambrian explosion and ended with the extinction of many invertebrates and vertebrates at the end of the Permian period.

pancreas A large gland in vertebrates that has both exocrine and endocrine functions. Secretes digestive enzymes into a duct connected to the intestine and secretes several hormones (notably, insulin and glucagon) into the bloodstream.

pancreatic amylase An enzyme produced by the pancreas that breaks down glucose chains by catalyzing hydrolysis of the glycosidic linkages between the glucose residues.

pancreatic lipase An enzyme that is produced in the pancreas and acts in the small intestine to break bonds in complex fats, releasing small lipids.

pandemic The spread of an infectious disease in a short time period over a wide geographic area and affecting a very high proportion of the population. Compare with **epidemic**.

parabiosis An experimental technique for determining whether a certain physiological phenomenon is regulated by a hormone; consists of surgically uniting two individuals so that hormones can pass between them.

paracrine Relating to a chemical signal that is released by one cell and affects neighboring cells.

paraphyletic group A group that includes an ancestral population and *some* but not all of its descendants. Compare with **monophyletic group**.

parapodia (singular: parapodium) Appendages found in some annelids from which bristle-like structures (chaetae) extend.

parasite An organism that lives on a host species (ectoparasite) or in a host species (endoparasite) and that damages its host.

parasitism (adjective: parasitic) A symbiotic relationship between two organisms that is beneficial to one organism (the parasite) but detrimental to the other (the host). Compare with **commensalism** and **mutualism**.

parasitoid An organism that has a parasitic larval stage and a free-living adult stage. Most parasitoids are insects that lay eggs in the bodies of other insects.

parasympathetic nervous system The part of the autonomic nervous system that stimulates responses for conserving or restoring energy, such as reduced heart rate and increased digestion. Compare with **sympathetic nervous system**.

parenchyma cell In plants, a general type of cell with a relatively thin primary cell wall. These cells, found in leaves, the centers of stems and roots, and fruits, are involved in photosynthesis, storage, and

transport. Compare with **collenchyma cell** and **sclerenchyma cell**.

parental care Any action by which an animal expends energy or assumes risks to benefit its offspring (e.g., nest building, feeding of young, defense).

parental generation The adults used in the first experimental cross of a breeding experiment.

parental strand A strand of DNA that is used as a template during DNA synthesis.

parietal cell A cell in the stomach lining that secretes hydrochloric acid.

parietal lobe In the vertebrate brain, one of the four major areas in the cerebrum.

parsimony The logical principle that the most likely explanation of a phenomenon is the most economical or simplest. When applied to comparison of alternative phylogenetic trees, it suggests that the one requiring the fewest evolutionary changes is most likely to be correct.

parthenogenesis Development of offspring from unfertilized eggs; a type of asexual reproduction.

partial pressure The pressure of one particular gas in a mixture of gases; the contribution of that gas to the overall pressure.

particulate inheritance The observation that genes from two parents do not blend together in offspring, but instead remain separate or particle-like.

pascal (Pa) A unit of pressure (force per unit area).

passive transport Diffusion of a substance across a membrane. When this event occurs with the assistance of membrane proteins, it is called *facilitated diffusion*.

patch clamping A technique for studying the electrical currents that flow through individual ion channels by sucking a tiny patch of membrane to the hollow tip of a microelectrode.

paternal chromosome A chromosome inherited from the father.

pathogen (adjective: pathogenic) Any entity capable of causing disease, such as a microbe, virus, or prion.

pattern formation The series of events that determines the spatial organization of an entire embryo or parts of an embryo, for example, setting the major body axes early in development.

pattern-recognition receptor On leukocytes, a class of membrane proteins that bind to molecules commonly associated with foreign cells and viruses and signal responses against broad types of pathogens. Part of the innate immune response.

peat Semi-decayed organic matter that accumulates in moist, low-oxygen environments such as *Sphagnum* (moss) bogs.

pectin A gelatinous polysaccharide found in the primary cell wall of plant cells. Attracts and holds water, forming a gel that resists compression forces and helps keep the cell wall moist.

pedigree A family tree of parents and offspring, showing inheritance of particular traits of interest.

penis The copulatory organ of male mammals, used to insert sperm into a female.

pentose A monosaccharide (simple sugar) containing five carbon atoms.

PEP carboxylase An enzyme that catalyzes addition of CO_2 to phosphoenolpyruvate, a three-carbon compound, forming a four-carbon organic acid. See also **C₄ pathway** and **crassulacean acid metabolism (CAM)**.

pepsin A protein-digesting enzyme present in the stomach.

peptide See **oligopeptide**.

peptide bond The covalent bond formed by a condensation reaction between two amino acids; links the residues in peptides and proteins.

peptidoglycan A complex structural polysaccharide found in bacterial cell walls.

perennial Describing a plant whose life cycle normally lasts for more than one year. Compare with **annual**.

perfect flower A flower that contains both male parts (stamens) and female parts (carpels). Compare with **imperfect flower**.

perforation In plants, a small hole in the primary and secondary cell walls of vessel elements that allows passage of water.

pericarp The part of a fruit, formed from the ovary wall, that surrounds the seeds and protects them. Corresponds to the flesh of most edible fruits and the hard shells of most nuts.

pericycle In plant roots, a layer of cells just inside the endodermis that give rise to lateral roots.

peripheral membrane protein Any membrane protein that does not span the entire lipid bilayer and associates with only one side of the bilayer. Compare with **integral membrane protein**.

peripheral nervous system (PNS) All the components of the nervous system that are outside the central nervous system (the brain and spinal cord). Includes the somatic nervous system and the autonomic nervous system.

peristalsis Rhythmic waves of muscular contraction. In the digestive tract, pushes food along. In animals with hydrostatic skeletons, enables crawling.

permafrost A permanently frozen layer of icy soil found in most tundra and some taiga.

permeability The tendency of a structure, such as a membrane, to allow a given substance to diffuse across it.

peroxisome An organelle found in most eukaryotic cells that contains enzymes for oxidizing fatty acids and other compounds, including many toxins, rendering them harmless. See **glyoxysome**.

petal Any of the leaflike organs arranged around the reproductive organs of a flower. Often colored and scented to attract pollinators.

petiole The stalk of a leaf.

pH A measure of the concentration of protons in a solution and thus of acidity or alkalinity. Defined as the negative of the base-10 logarithm of the proton concentration: $pH = -\log[H^+]$.

phagocytosis Uptake by a cell of small particles or cells by invagination and pinching off of the plasma membrane to form small, membrane-bound vesicles; one type of **endocytosis**.

pharyngeal gill slits A set of parallel openings from the throat to the outside that function in both feeding and gas exchange. A diagnostic trait of chordates.

pharyngeal jaw A secondary jaw in the back of the throat; found in some fishes, it aids in food processing. Derived from modified gill arches.

phenology The timing of events during the year, in environments where seasonal changes occur.

phenotype The detectable traits of an individual. Compare with **genotype**.

phenotypic plasticity Within-species variation in phenotype that is due to differences in environmental conditions. Occurs more commonly in plants than animals.

pheophytin The molecule in photosystem II that accepts excited electrons from the reaction center chlorophyll and passes them to an electron transport chain.

pheromone A chemical signal, released by an individual into the external environment, that can trigger changes in the behavior or physiology or both of another member of the same species.

phloem A plant vascular tissue that conducts sugars between roots and shoots; contains sieve-tube elements and companion cells. Primary phloem develops from the procambium of apical meristems; secondary phloem, from the vascular cambium. Compare with **xylem**.

phosphatase An enzyme that removes phosphate groups from proteins or other molecules. Phosphatases are often used in the inactivation of signaling pathways that involve the phosphorylation and activation of proteins.

phosphodiester linkage Chemical linkage between adjacent nucleotide residues in DNA and RNA. Forms when the phosphate group of one nucleotide condenses with the hydroxyl group on the sugar of another nucleotide. Also known as *phosphodiester bond*.

phosphofructokinase The enzyme that catalyzes synthesis of fructose-1,6-bisphosphate from fructose-6-phosphate, a key reaction in glycolysis (step 3). Also called *6-phosphofructokinase*.

phospholipid A class of lipid having a hydrophilic head (including a phosphate group) and a hydrophobic tail (consisting of two hydrocarbon chains). Major components of the plasma membrane and organelle membranes.

phosphorylase An enzyme that breaks down glycogen by catalyzing hydrolysis of the α-glycosidic linkages between the glucose residues.

phosphorylation (verb: phosphorylate) The addition of a phosphate group to a molecule.

phosphorylation cascade A series of enzyme-catalyzed phosphorylation reactions commonly used in signal transduction pathways to amplify and convey a signal inward from the plasma membrane.

photic zone In an aquatic habitat, water that is shallow enough to receive some sunlight (whether or not it is enough to support photosynthesis). Compare with **aphotic zone**.

photon A discrete packet of light energy; a particle of light.

photoperiod The amount of time per day (usually in hours) that an organism is exposed to light.

photoperiodism Any response by an organism to the relative lengths of day and night (i.e., photoperiod).

photophosphorylation Production of ATP molecules by ATP synthase using the proton-motive force generated as light-excited electrons flow through an electron transport chain during photosynthesis.

photoreception A sensory system in which receptors are activated by light.

photoreceptor A molecule, a cell, or an organ that is specialized to detect light.

photorespiration A series of light-driven chemical reactions that consumes oxygen and releases carbon dioxide, basically undoing photosynthesis. Usually occurs when there are high O_2 and low CO_2 concentrations inside plant cells; often occurs when stomata must be kept closed to prevent dehydration.

photoreversibility A change in conformation that occurs in certain plant pigments when they are exposed to the particular wavelengths of light that they absorb; triggers responses by the plant.

photosynthesis The complex biological process that converts the energy of light into chemical energy stored in glucose and other organic molecules. Occurs in most plants, algae, and some bacteria.

photosystem One of two types of units, consisting of a central reaction center surrounded by antenna complexes, that is responsible for the light-dependent reactions of photosynthesis.

photosystem I A photosystem that contains a pair of P700 chlorophyll molecules and uses absorbed light energy to reduce $NADP^+$ to NADPH.

photosystem II A photosystem that contains a pair of P680 chlorophyll molecules and uses absorbed light energy to produce a proton-motive force for the synthesis of ATP. Oxygen is produced as a by-product when water is split to obtain electrons.

phototroph An organism (most plants, algae, and some bacteria) that produces ATP through photosynthesis.

phototropins A class of plant photoreceptors that detect blue light and initiate various responses.

phototropism Growth or movement of an organism in a particular direction in response to light.

phylogenetic species concept The definition of a species as the smallest monophyletic group in a phylogenetic tree. Compare with **biological species concept** and **morphospecies concept**.

phylogenetic tree A branching diagram that depicts the evolutionary relationships among species or other taxa.

phylogeny The evolutionary history of a group of organisms.

phylum (plural: phyla) In Linnaeus' system, a taxonomic category above the class level and below the kingdom level. In plants, sometimes called a *division*.

physical map A map of a chromosome that shows the number of base pairs between various genetic markers. Compare with **genetic map**.

physiology The study of how an organism's body functions.

phytochrome A specialized plant photoreceptor that exists in two shapes depending on the ratio of red to far-red light and is involved in the timing of certain physiological processes, such as flowering, stem elongation, and germination.

pigment Any molecule that absorbs certain wavelengths of visible light and reflects or transmits other wavelengths.

piloting A type of navigation in which animals use familiar landmarks to find their way.

pineal gland An endocrine gland, located in the brain, that secretes the hormone melatonin.

pioneering species Those species that appear first in recently disturbed areas.

pit In plants, a small hole in the secondary cell walls of tracheids and vessel elements that allows passage of water.

pitch The sensation produced by a particular frequency of sound. Low frequencies are perceived as low pitches; high frequencies, as high pitches.

pith In the shoot systems of plants, ground tissue located to the inside of the vascular bundles.

pituitary gland A small gland located directly under the brain and physically and functionally connected to the hypothalamus. Produces and secretes an array of hormones that affect many other glands and organs.

placenta A structure that forms in the pregnant uterus from maternal and fetal tissues. Delivers oxygen to the fetus, exchanges nutrients and wastes between mother and fetus, anchors the fetus to the uterine wall, and produces some hormones. Occurs in most mammals and in a few other vertebrates.

placental mammals See **eutherians**.

plankton Drifting organisms (animals, plants, archaea, or bacteria) in aquatic environments.

Plantae The monophyletic group that includes red, green, and glaucophyte algae, and land plants.

plasma The non-cellular portion of blood.

plasma cell A B cell that produces large quantities of antibodies after being activated by interacting with antigen and a CD4+ T cell via peptide presentation. Also called an *effector B cell.*

plasma membrane A membrane that surrounds a cell, separating it from the external environment and selectively regulating passage of molecules and ions into and out of the cell. Also called *cell membrane.*

plasmid A small, usually circular, supercoiled DNA molecule independent of the cell's main chromosome(s) in prokaryotes and some eukaryotes.

plasmodesmata (singular: plasmodesma) Physical connections between two plant cells, consisting of membrane-lined gaps in the cell walls through which the two cells' plasma membranes, cytoplasm, and smooth ER can connect directly. Functionally similar to gap junctions in animal cells.

plasmogamy Fusion of the cytoplasm of two individuals. Occurs in many fungi.

plastocyanin A small protein that shuttles electrons originating from photosystem II to the reaction center of photosystem I during photosynthesis.

plastoquinone (PQ) A nonprotein electron carrier in the chloroplast electron transport chain. Receives excited electrons from photosystem II (noncyclic) or photosystem I (cyclic) and passes them to more electronegative molecules in the chain. Also transports protons from the stroma to the thylakoid lumen, generating a proton-motive force.

platelet A small membrane-bound cell fragment in vertebrate blood that functions in blood clotting. Derived from large cells in the bone marrow.

pleiotropy (adjective: pleiotropic) The ability of a single gene to affect more than one trait.

ploidy The number of complete chromosome sets present. *Haploid* refers to a ploidy of 1; *diploid,* a ploidy of 2; *triploid,* a ploidy of 3; and *tetraploid,* a ploidy of 4.

point mutation A mutation that results in a change in a single base pair in DNA.

polar (1) Asymmetrical or unidirectional. (2) Carrying a partial positive charge on one side of a molecule and a partial negative charge on the other. Polar molecules are generally hydrophilic.

polar body Any of the tiny, nonfunctional cells that are made as a by-product during meiosis of a primary oocyte, due to most of the cytoplasm going to the ovum.

polar covalent bond A covalent bond in which electrons are shared unequally between atoms differing in electronegativity, resulting in the more electronegative atom having a partial negative charge and the other atom, a partial positive charge. Compare with **nonpolar covalent bond**.

polar microtubules Mitotic and meiotic microtubules that have arisen from the two spindle poles and overlap with each other in the middle of the spindle apparatus.

polar nuclei In flowering plants, the nuclei in the female gametophyte that fuse with one sperm nucleus to produce the endosperm. Most species have two.

pollen grain In seed plants, a male gametophyte enclosed within a protective coat of sporopollenin.

pollen tube In flowering plants, a structure that grows out of a pollen grain after it reaches the stigma, extends down the style, and through which two sperm cells are delivered to the ovule.

pollination The process by which pollen reaches the carpel of a flower (in flowering plants), transferred from anther to stigma, or reaches the ovule directly (in conifers and their relatives).

pollination syndrome Suites of flower characters that are associated with certain types of pollinators and that have evolved through natural selection imposed by the interaction between flowers and pollinators.

poly(A) signal In eukaryotes, a short sequence of nucleotides near the 3′ end of pre-mRNAs that signals cleavage of the RNA and addition of the poly(A) tail.

poly(A) tail In eukaryotes, a sequence of about 100–250 adenine nucleotides added to the 3′ end of newly transcribed messenger RNA molecules.

polygenic inheritance Having many genes influence one trait.

polymer Any long molecule composed of small repeating units (monomers) bonded together. The main biological polymers are proteins, nucleic acids, and polysaccharides.

polymerase chain reaction (PCR) A laboratory technique for rapidly generating millions of identical copies of a specific stretch of DNA. Works by incubating the original DNA sequence of interest with primers, nucleotides, and DNA polymerase.

polymerization (verb: polymerize) The process by which many identical or similar small molecules (monomers) are covalently bonded to form a large molecule (polymer).

polymorphic species A species that has two or more distinct phenotypes in the same interbreeding population at the same time.

polymorphism (adjective: polymorphic) (1) The occurrence of more than one allele at a genetic locus in a population. (2) The occurrence of more than two distinct phenotypes of a trait in a population.

polyp The immotile (sessile) stage in the life cycle of some cnidarians (e.g., jellyfish). Compare with **medusa**.

polypeptide A chain of 50 or more amino acids linked together by peptide bonds. Compare with **oligopeptide** and **protein**.

polyphyletic group An unnatural group based on convergent (homoplastic) characteristics that are not present in a common ancestor. Compare with **monophyletic group**.

polyploidy (adjective: polyploid) The state of having more than two full sets of chromosomes, either from the same species (autopolyploidy) or from different species (allopolyploidy).

polyribosome A messenger RNA molecule along with more than one attached ribosome and their growing peptide strands.

polysaccharide A linear or branched polymer consisting of many monosaccharides joined by glycosidic linkages. Compare with **monosaccharide** and **oligosaccharide**.

polytomy A node in a phylogenetic tree that depicts an ancestral branch dividing into three or more descendant branches; usually indicates that insufficient data were available to resolve which taxa are more closely related.

population A group of individuals of the same species living in the same geographic area at the same time.

population density The number of individuals of a population per unit area.

population dynamics Changes in the size and other characteristics of populations through time and space.

population ecology The study of how and why the number of individuals in a population changes over time and space.

population thinking The ability to analyze trait frequencies, event probabilities, and other attributes of populations of molecules, cells, or organisms.

pore In land plants, an opening in the epithelium that allows gas exchange. See also **stoma**.

positive control Of genes, when a regulatory protein triggers expression by binding to DNA on or near the gene.

positive feedback A physiological mechanism in which a change in some variable stimulates a response that increases the change. Relatively rare in organisms but is important in generation of the action potential. Compare with **negative feedback**.

positive pressure ventilation Ventilation of the lungs by using positive pressure in the mouth to "push" air into the lungs. Compare with **negative pressure ventilation**.

positive-sense RNA virus An ssRNA virus whose genome contains the same sequences as the mRNA required to produce viral proteins. Compare with **ambisense virus** and **negative-sense virus**.

posterior Toward an animal's tail and away from its head. The opposite of anterior.

posterior pituitary The part of the pituitary gland that contains the ends of hypothalamic neurosecretory cells and from which oxytocin and antidiuretic hormone are secreted. Compare with **anterior pituitary**.

postsynaptic neuron A neuron that receives signals, usually via neurotransmitters, from another neuron at a synapse. Compare with **presynaptic neuron**.

post-translational control Regulation of gene expression by modification of proteins (e.g., addition of a phosphate group or sugar residues) after translation.

postzygotic isolation Reproductive isolation resulting from mechanisms that operate after mating of individuals of two different species occurs. The most common mechanisms are the death of hybrid embryos or reduced fitness of hybrids.

potential energy Energy stored in matter as a result of its position or molecular arrangement. Compare with **kinetic energy**.

prebiotic soup model Hypothetical explanation for chemical evolution whereby small molecules reacted with one another in a mixture of organic molecules condensed into a body of water, typically in reference to the early oceans.

Precambrian The interval between the formation of the Earth, about 4.6 billion years ago, and the appearance of most animal groups about 542 million years ago. Unicellular organisms were dominant for most of this era, and oxygen was virtually absent for the first 2 billion years.

pre-mRNA In eukaryotes, the primary transcript of protein-coding genes. Pre-mRNA is processed to form mRNA.

predation The killing and eating of one organism (the prey) by another (the predator).

predator Any organism that kills other organisms for food.

prediction A measurable or observable result of an experiment based on a particular hypothesis. A correct prediction provides support for the hypothesis being tested.

pressure-flow hypothesis The hypothesis that sugar movement through phloem tissue is due to differences in the turgor pressure of phloem sap.

pressure potential (ψ_P) A component of the potential energy of water caused by physical pressures on a solution. It can be positive or negative. Compare with **solute potential** (ψ_S).

presynaptic neuron A neuron that transmits signals, usually by releasing neurotransmitters, to another neuron or to an effector cell at a synapse.

prezygotic isolation Reproductive isolation resulting from any one of several mechanisms that prevent individuals of two different species from mating.

primary active transport A form of active transport in which a source of energy like ATP is directly used to move ions against their electrochemical gradients.

primary cell wall The outermost layer of a plant cell wall, made of cellulose fibers and gelatinous polysaccharides, that defines the shape of the cell and withstands the turgor pressure of the plasma membrane.

primary consumer An herbivore; an organism that eats plants, algae, or other primary producers. Compare with **secondary consumer**.

primary decomposer A decomposer (detritivore) that consumes detritus from plants.

primary growth In plants, an increase in the length of stems and roots due to the activity of apical meristems. Compare with **secondary growth**.

primary immune response An adaptive immune response to a pathogen that the immune system has not encountered before. Compare with **secondary immune response**.

primary meristem In plants, three types of partially differentiated cells that are produced by apical meristems, including protoderm, ground meristem, and procambium. Compare with **apical meristem** and **cambium**.

primary oocyte Any of the large diploid cells in an ovarian follicle that can initiate meiosis to produce a haploid secondary oocyte and a polar body.

primary producer Any organism that creates its own food by photosynthesis or from reduced inorganic compounds and that is a food source for other species in its ecosystem. Also called *autotroph*.

primary spermatocyte Any of the diploid cells in the testis that can initiate meiosis I to produce two secondary spermatocytes.

primary structure The sequence of amino acid residues in a peptide or protein; also the sequence of nucleotides in a nucleic acid. Compare with **secondary**, **tertiary**, and **quaternary structure**.

primary succession The gradual colonization of a habitat of bare rock or gravel, usually after an environmental disturbance that removes all soil and previous organisms. Compare with **secondary succession**.

primary transcript In eukaryotes, a newly transcribed RNA molecule that has not yet been processed to a mature RNA. Called *pre-mRNA* when the final product is a protein.

primase An enzyme that synthesizes a short stretch of RNA to use as a primer during DNA replication.

primates The lineage of mammals that includes prosimians (lemurs, lorises, etc.), monkeys, and great apes (including humans).

primer A short, single-stranded RNA molecule that base-pairs with a DNA template strand and is elongated by DNA polymerase during DNA replication.

probe A radioactively or chemically labeled single-stranded fragment of a known DNA or RNA sequence that can bind to and detect its complementary sequence in a sample containing many different sequences.

proboscis A tubular, often extensible feeding appendage with which food can be obtained.

procambium A primary meristem tissue that gives rise to the vascular tissue.

product Any of the final materials formed in a chemical reaction.

progesterone A steroid hormone produced and secreted by the corpus luteum in the ovaries after ovulation and by the placenta during gestation; protects the uterine lining.

programmed cell death Regulated cell death that is used in development, tissue maintainance, and destruction of infected cells. Can occur in different ways; apoptosis is the best-known mechanism.

prokaryote A member of the domain Bacteria or Archaea; a unicellular organism lacking a nucleus and containing relatively few organelles or cytoskeletal components. Compare with **eukaryote**.

prolactin A peptide hormone, produced and secreted by the anterior pituitary, that promotes milk production in female mammals and has a variety of effects on parental behavior and seasonal reproduction in other vertebrates.

prometaphase A stage in mitosis or meiosis during which the nuclear envelope breaks down and microtubules attach to kinetochores.

promoter A short nucleotide sequence in DNA that binds a sigma factor (in bacteria) or basal transcription factors (in eukaryotes) to enable RNA polymerase to begin transcription. In bacteria, several contiguous genes are often transcribed from a single promoter. In eukaryotes, each gene generally has its own promoter.

promoter-proximal element In eukaryotes, regulatory sequences in DNA that are close to a promoter and that can bind regulatory transcription factors.

proofreading The process by which a DNA polymerase recognizes and removes a wrong base added during DNA replication and then continues synthesis.

prophase The first stage in mitosis or meiosis during which chromosomes become visible and the spindle apparatus forms. Synapsis and crossing over occur during prophase of meiosis I.

prosimians One of the two major lineages of primates, a paraphyletic group including lemurs, pottos, and lorises. Compare with **anthropoids**.

prostate gland A gland in male mammals that surrounds the base of the urethra and secretes a fluid that is a component of semen.

prosthetic group A non-amino acid atom or molecule that is permanently attached to an enzyme or other protein and is required for its function.

protease An enzyme that can break up proteins by cleaving the peptide bonds between amino acid residues.

proteasome A macromolecular machine that destroys proteins that have been marked by the addition of ubiquitin.

protein A macromolecule consisting of one or more polypeptide chains composed of 50 or more amino acids linked together. Each protein has a unique sequence of amino acids and generally possesses a characteristic three-dimensional shape.

protein kinase An enzyme that catalyzes the addition of a phosphate group to another protein, typically activating or inactivating the substrate protein.

proteinase inhibitors Defense compounds, produced by plants, that induce illness in herbivores by inhibiting digestive enzymes.

proteoglycan A type of highly glycosylated protein found in the extracellular matrix of animal cells that attracts and holds water, forming a gel that resists compression forces.

proteome The complete set of proteins produced by a particular cell type.

proteomics The systematic study of the interactions, localization, functions, regulation, and other features of the full protein set (proteome) in a particular cell type.

protist Any eukaryote that is not a green plant, animal, or fungus. Protists are a diverse paraphyletic group. Most are unicellular, but some are multicellular or form aggregations called colonies.

protocell A hypothetical pre-cell structure consisting of a membrane compartment that encloses replicating macromolecules, such as ribozymes.

protoderm The exterior layer of a young plant embryo that gives rise to the epidermis.

proto-oncogene Any gene that encourages cell division in a regulated manner, typically by triggering specific phases in the cell cycle. Mutation may convert it into an oncogene. See **oncogene**.

proton pump A membrane protein that can hydrolyze ATP to power active transport of protons (H^+ ions) across a membrane against an electrochemical gradient. Also called H^+-ATPase.

proton-motive force The combined effect of a proton gradient and an electric potential gradient across a membrane, which can drive protons across the membrane. Used by mitochondria and chloroplasts to power ATP synthesis via the mechanism of chemiosmosis.

protostomes A major lineage of animals that share a pattern of embryological development, including formation of the mouth earlier than the anus, and formation of the coelom by splitting of a block of mesoderm. Includes arthropods, mollusks, and annelids. Compare with **deuterostomes**.

proximal tubule In the vertebrate kidney, the convoluted section of a nephron into which filtrate moves from Bowman's capsule. Involved in the largely unregulated reabsorption of electrolytes, nutrients, and water. Compare with **distal tubule**.

proximate causation In biology, the immediate, mechanistic cause of a phenomenon (how it happens), as opposed to why it evolved. Also called *proximate explanation*. Compare with **ultimate causation**.

pseudocoelomate An animal that has a coelom that is only partially lined with mesoderm. Compare with **acoelomate** and **coelomate**.

pseudogene A DNA sequence that closely resembles a functional gene but is not transcribed. Thought to have arisen by duplication of the functional gene followed by inactivation due to a mutation.

pseudopodium (plural: pseudopodia) A temporary bulge-like extension of certain protist cells used in cell crawling and ingestion of food.

puberty The various physical and emotional changes that an immature human undergoes in reaching reproductive maturity. Also the period when such changes occur.

pulmonary artery A short, thick-walled artery that carries oxygen-poor blood from the heart to the lungs.

pulmonary circulation The part of the circulatory system that sends oxygen-poor blood to the lungs. It is separate from the rest of the circulatory system (the systemic circulation) in mammals and birds.

pulmonary vein A short, thin-walled vein that carries oxygen-rich blood from the lungs to the heart. Humans have four such veins.

pulse–chase experiment A type of experiment in which a population of cells or molecules at a particular moment in time is marked by means of a labeled molecule (pulse) and then their fate is followed over time (chase).

pump Any membrane protein that can hydrolyze ATP and change shape to power active transport of a specific ion or small molecule across a plasma membrane against its electrochemical gradient. See **proton pump**.

pupa (plural: pupae) A metamorphosing insect that is enclosed in a protective case.

pupil The hole in the center of the iris through which light enters a vertebrate or cephalopod eye.

pure line In animal or plant breeding, a strain that produces offspring identical with themselves when self-fertilized or crossed to another member of the same population. Pure lines are homozygous for most, if not all, genetic loci.

purifying selection Selection that lowers the frequency of or even eliminates deleterious alleles.

purines A class of small, nitrogen-containing, double-ringed bases (guanine, adenine) found in nucleotides. Compare with **pyrimidines**.

pyrimidines A class of small, nitrogen-containing, single-ringed bases (cytosine, uracil, thymine) found in nucleotides. Compare with **purines**.

pyruvate dehydrogenase A large enzyme complex, located in the mitochondrial matix, that is responsible for converting pyruvate to acetyl CoA during cellular respiration.

quantitative trait A trait that exhibits continuous phenotypic variation (e.g., human height), rather than the distinct forms characteristic of discrete traits.

quaternary structure In proteins, the overall three-dimensional shape formed from two or more polypeptide chains (subunits); determined by the number, relative positions, and interactions of the subunits. In single stranded nucleic acids, the hydrogen bonding between two or more distinct strands will form this level of structure through hydrophobic interactions between complementary bases. Compare with **primary**, **secondary**, and **tertiary structures**.

quorum sensing Cell–cell signaling in unicellular organisms, in which cells of the same species communicate via chemical signals. It is often observed that cell activity changes dramatically when the population reaches a threshold size, or quorum.

radial symmetry An animal body pattern that has at least two planes of symmetry. Typically, the body is in the form of a cylinder or disk, and the body parts radiate from a central hub. Compare with **bilateral symmetry**.

radiation Transfer of heat between two bodies that are not in direct physical contact. More generally, the emission of electromagnetic energy of any wavelength.

radicle The root of a plant embryo.

radioactive isotope A version of an element that has an unstable nucleus, which will release radiation energy as it decays to a more stable form. Decay often results in the radioisotope becoming a different element.

radioimmunoassay A competitive binding assay in which the quantity of hormone in a sample can be estimated. Uses radioactively labeled hormones that compete with the unknown hormone to bind with an antibody.

radula A rasping feeding appendage in mollusks such as gastropods (snails, slugs).

rain shadow The dry region on the side of a mountain range away from the prevailing wind.

range The geographic distribution of a species.

Ras protein A type of G protein that is activated by enzyme-linked cell-surface receptors, including receptor tyrosine kinases. Activated Ras then initiates a phosphorylation cascade, culminating in a cell response.

ray-finned fishes Members of the Actinopterygii, a diverse group of fishes with fins supported by bony rods arranged in a ray pattern.

rays In plant shoot systems with secondary growth, a lateral row of parenchyma cells produced by vascular cambium. Transport water and nutrients laterally across the stem.

Rb protein A tumor suppressor protein that helps regulate progression of a cell from the G_1 phase to the S phase of the cell cycle. Defects in Rb protein are found in many types of cancer.

reactant Any of the starting materials in a chemical reaction.

reaction center Centrally located component of a photosystem containing proteins and a pair of specialized chlorophyll molecules. It is surrounded by antenna complexes that transmit resonance energy to excite the reaction center pigments.

reading frame A series of non-overlapping, three-base-long sequences (potential codons) in DNA or RNA. The reading frame for a polypeptide is set by the start codon.

realized niche The portion of the fundamental niche that a species actually occupies given limiting factors such as competition with other species. Compare with **fundamental niche**.

receptor-mediated endocytosis Uptake by a cell of certain extracellular macromolecules, bound to specific receptors in the plasma membrane, by pinching off the membrane to form small membrane-bound vesicles.

receptor tyrosine kinase (RTK) Any of a class of enzyme-linked cell-surface signal receptors that undergo phosphorylation after binding a signaling molecule. The activated, phosphorylated receptor then triggers a signal transduction pathway inside the cell.

recessive Referring to an allele whose phenotypic effect is observed only in homozygous individuals. Compare with **dominant**.

reciprocal altruism Altruistic behavior that is exchanged between a pair of individuals at different times (i.e., sometimes individual A helps individual B, and sometimes B helps A).

reciprocal cross A cross in which the mother's and father's phenotypes are the reverse of that examined in a previous cross.

recombinant Possessing a new combination of alleles. May refer to a single chromosome or DNA molecule, or to an entire organism.

recombinant DNA technology A variety of techniques for isolating specific DNA fragments and introducing them into different regions of DNA or a different host organism.

rectal gland A salt-excreting gland in the digestive system of sharks, skates, and rays.

rectum The last portion of the digestive tract. It is where feces are held until they are expelled.

red blood cell A hemoglobin-containing cell that circulates in the blood and delivers oxygen from the lungs to the tissues.

redox reaction Any chemical reaction that involves either the complete transfer of one or more electrons from one reactant to another, or a reciprocal shift in the position of shared electrons within one or more of the covalent bonds of two reactants. Also called *reduction–oxidation reaction.*

reduction The gain of electrons by an atom or molecule during a redox reaction, either by acceptance of an electron from another atom or molecule, or by the shared electrons in covalent bonds moving closer to the atomic nucleus.

reduction–oxidation reaction See **redox reaction**.

reflex An involuntary response to environmental stimulation. May involve the brain (conditioned reflex) or not (spinal reflex).

refractory No longer responding to stimuli that previously elicited a response. An example is the tendency of voltage-gated sodium channels to remain closed immediately after an action potential.

regulatory sequence Any segment of DNA that is involved in controlling transcription of a specific gene by binding a regulatory transcription factor protein.

regulatory transcription factor General term for proteins that bind to DNA regulatory sequences (eukaryotic enhancers, silencers, and promoter-proximal elements), but not to the promoter itself, leading to an increase or decrease in transcription of specific genes. Compare with **basal transcription factor**.

regulon A large set of genes in bacteria that are controlled by a single type of regulatory molecule. Regulon genes are transcribed in response to environmental cues and allow cells to respond to changing environments.

reinforcement In evolutionary biology, the natural selection for traits that prevent interbreeding between recently diverged species.

release factors Proteins that trigger termination of translation when a ribosome reaches a stop codon.

renal corpuscle In the vertebrate kidney, the ball-like structure at the beginning of a nephron, consisting of a glomerulus and the surrounding Bowman's capsule. Acts as a filtration device.

replacement rate The number of offspring each female must produce over her entire life to "replace" herself and her mate, resulting in zero population growth. The actual number is slightly more than 2 because some offspring die before reproducing.

replica plating A method of identifying bacterial colonies that have certain mutations by transferring cells from each colony on a master plate to a second (replica) plate and observing their growth when exposed to different conditions.

replication fork The Y-shaped site at which a double-stranded molecule of DNA is separated into two single strands for replication.

replicative growth The process by which viruses produce new virions.

replisome The macromolecular machine that copies DNA; includes DNA polymerase, helicase, primase, and other enzymes.

repolarization Return to a resting potential after a membrane potential has changed; a normal phase in an action potential.

repressor (1) In bacteria, a protein that binds to an operator sequence in DNA to prevent transcription when an inducer is not present and that comes off DNA to allow transcription when an inducer binds to the repressor protein. (2) In eukaryotes, a protein that binds to a silencer sequence in DNA to prevent or reduce gene transcription.

reproductive development The phase of plant development that involves development of the flower and reproductive cells. Follows vegetative development and occurs when a shoot apical meristem (SAM) transitions to a flower-producing meristem.

reptiles One of the two lineages of amniotes (vertebrates that produce amniotic eggs) distinguished by adaptations for life and reproduction on land. Living reptiles include turtles, snakes and lizards, crocodiles and alligators, and birds. Except for birds, all are ectotherms.

resilience, community A measure of how quickly a community recovers following a disturbance.

resistance, community A measure of how much a community is affected by a disturbance.

respiratory system The collection of cells, tissues, and organs responsible for gas exchange between an animal and its environment.

resting potential The membrane potential of a cell in its resting, or normal, state.

restriction endonucleases Bacterial enzymes that cut DNA at a specific base-pair sequence (restriction site). Also called *restriction enzymes.*

retina A thin layer of light-sensitive cells (rods and cones) and neurons at the back of a simple eye, such as that of cephalopods and vertebrates.

retinal A light-absorbing pigment that is linked to the protein opsin in rods and cones of the vertebrate eye.

retrovirus A virus with an RNA genome that reverse-transcribes its RNA into a double-stranded DNA sequence, which is then inserted into the host's genome as part of its replicative cycle.

reverse transcriptase An enzyme that can synthesize double-stranded DNA from a single-stranded RNA template.

rhizobia (singular: rhizobium) Members of the bacterial genus *Rhizobium;* nitrogen-fixing bacteria that live in root nodules of members of the pea family (legumes).

rhizoid The hairlike structure that anchors a non-vascular plant to the substrate.

rhizome A modified stem that runs horizontally underground and produces new plants at the nodes (a form of asexual reproduction). Compare with **stolon**.

rhodopsin A transmembrane complex that is instrumental in detection of light by rods of the vertebrate eye. Is composed of the transmembrane protein opsin covalently linked to retinal, a light-absorbing pigment.

ribonucleic acid (RNA) A nucleic acid composed of ribonucleotides that usually is single stranded. Functions include structural components of ribosomes (rRNA), transporters of amino acids (tRNA), and messages of the DNA code required for protein synthesis (mRNA), among others.

ribonucleotide See **nucleotide**.

ribosomal RNA (rRNA) An RNA molecule that forms part of the ribosome.

ribosome A large macromolecular machine that synthesizes proteins by using the genetic information encoded in messenger RNA. Consists of two subunits, each composed of ribosomal RNA and proteins.

ribosome binding site In a bacterial mRNA molecule, the sequence just upstream of the start codon to which a ribosome binds to initiate translation. Also called the *Shine–Dalgarno sequence.*

ribozyme Any RNA molecule that can act as a catalyst, that is, speed up a chemical reaction.

ribulose bisphosphate (RuBP) A five-carbon compound that combines with CO_2 in the first step of the Calvin cycle during photosynthesis.

RNA See **ribonucleic acid**.

RNA interference (RNAi) Degradation of an RNA molecule or inhibition of its translation following its binding by a short RNA (microRNA) whose sequence is complementary to a portion of the mRNA.

RNA polymerase An enzyme that catalyzes the synthesis of RNA from ribonucleotides using a DNA template.

RNA processing In eukaryotes, the changes that a primary RNA transcript undergoes to become a mature RNA molecule. For pre-mRNA it includes the addition of a 5′ cap and poly(A) tail and splicing to remove introns.

RNA replicase A viral enzyme that can synthesize RNA from an RNA template. Also called an *RNA-dependent RNA polymerase.*

RNA world hypothesis Proposal that chemical evolution produced RNAs that could catalyze key reactions involved in their own replication and basic metabolism, which led to the evolution of proteins and the first life-form.

rod cell A photoreceptor cell with a rod-shaped outer portion that is particularly sensitive to dim light but not used to distinguish colors. Also called simply *rod*. Compare with **cone cell**.

root (1) An underground part of a plant that anchors the plant and absorbs water and nutrients. (2) The most ancestral branch in a phylogenetic tree.

root apical meristem (RAM) A group of undifferentiated plant stem cells at the tip of a plant root that can differentiate into mature root tissue.

root cap A small group of cells that covers and protects the root apical meristem. Senses gravity and determines the direction of root growth.

root hair A long, thin outgrowth of the epidermal cells of plant roots, providing increased surface area for absorption of water and nutrients.

root pressure Positive pressure of xylem sap in the vascular tissue of roots. Generated during the night as a result of the accumulation of ions from the soil and subsequent osmotic movement of water into the xylem.

root system The belowground part of a plant.

rotifer Member of the phylum Rotifera. Distinguished by a cluster of cilia, called a corona, used in suspension feeding in marine and freshwater environments. Rotifers belong to the lophotrochozoan branch of the protostomes.

rough endoplasmic reticulum (rough ER) The portion of the endoplasmic reticulum that is dotted with ribosomes. Involved in synthesis of plasma membrane proteins, secreted proteins, and proteins localized to the ER, Golgi apparatus, and lysosomes. Compare with **smooth endoplasmic reticulum**.

roundworms Members of the phylum Nematoda. Distinguished by an unsegmented body with a pseudocoelom and no appendages. Roundworms belong to the ecdysozoan branch of the protostomes. Also called *nematodes*.

rubisco The enzyme that catalyzes the first step of the Calvin cycle during photosynthesis: the addition of a molecule of CO_2 to ribulose bisphosphate. See also **carbon fixation**.

ruminant Member of a group of hoofed mammals (e.g., cattle, sheep, deer) that have a four-chambered stomach specialized for digestion of plant cellulose. Ruminants regurgitate cud, a mixture of partially digested food and cellulose-digesting bacteria, from the largest chamber (the rumen) for further chewing.

salinity The proportion of solutes dissolved in water in natural environments, often designated in grams of solute per kilogram of water (cited as parts per thousand).

salivary amylase An enzyme that is produced by the salivary glands and that can break down starch by catalyzing hydrolysis of the glycosidic linkages between the glucose residues.

salivary gland A type of vertebrate gland that secretes saliva (a mixture of water, mucus-forming glycoproteins, and digestive enzymes) into the mouth.

sampling error The selection of a nonrepresentative sample from some larger population, due to chance.

saprophyte An organism that feeds primarily on dead plant material.

sapwood The younger xylem in the outer layer of wood of a stem or root, functioning primarily in water transport.

sarcomere The repeating contractile unit of a skeletal muscle cell; the portion of a myofibril located between adjacent Z disks.

sarcoplasmic reticulum Sheets of smooth endoplasmic reticulum in a muscle cell. Contains high concentrations of calcium, which can be released into the cytoplasm to trigger contraction.

saturated Referring to lipids in which all the carbon-carbon bonds are single bonds. Such compounds have relatively high melting points. Compare with **unsaturated**.

scanning electron microscope (SEM) A microscope that produces images of the surfaces of objects by reflecting electrons from a specimen coated with a layer of metal atoms. Compare with **transmission electron microscope**.

scarify To scrape, rasp, cut, or otherwise damage the coat of a seed. Necessary in some species to trigger germination.

Schwann cell A type of glial cell that wraps around axons of some neurons outside the brain and spinal cord, forming a myelin sheath that provides electrical insulation. Compare with **oligodendrocyte**.

scientific name The unique, two-part name given to each species, with a genus name followed by a species name—as in *Homo sapiens*. Scientific names are always italicized, and are also known as Latin names.

sclereid In plants, a relatively short type of sclerenchyma cell that usually functions in protection, such as in seed coats and nutshells. Compare with **fiber**.

sclerenchyma cell In plants, a cell that has a thick secondary cell wall and provides support; typically contains the tough structural polymer lignin and usually is dead at maturity. Includes fibers and sclereids. Compare with **collenchyma cell** and **parenchyma cell**.

scrotum A sac of skin containing the testes and suspended just outside the abdominal body cavity of many male mammals.

second law of thermodynamics The principle of physics that the entropy of the universe or any closed system always increases.

second-male advantage The reproductive advantage, in some species, of a male who mates with a female last, after other males have mated with her.

second messenger A nonprotein signaling molecule produced or activated inside a cell in response to stimulation at the cell surface. Commonly used to relay the message of a hormone or other extracellular signaling molecule.

secondary active transport Transport of an ion or molecule in a defined direction that is often against its electrochemical gradient, in company with an ion or molecule being transported along its electrochemical gradient. Also called *cotransport*.

secondary cell wall The thickened inner layer of a plant cell wall formed by certain cells as they mature and have stopped growing; in water-conducting cells, contains lignin. Provides support or protection.

secondary consumer A carnivore; an organism that eats herbivores. Compare with **primary consumer**.

secondary growth In plants, an increase in the width of stems and roots due to the activity of cambium. Compare with **primary growth**.

secondary immune response The adaptive immune response to a pathogen that the immune system has encountered before. Normally much faster and more efficient than the primary response, due to immunological memory. Compare with **primary immune response**.

secondary metabolites Molecules that are closely related to compounds in key synthetic pathways and that often function in defense.

secondary oocyte A cell produced by meiosis I of a primary oocyte in the ovary. If fertilized, will complete meiosis II to produce an ootid (which develops into an ovum) and a polar body.

secondary spermatocyte A cell produced by meiosis I of a primary spermatocyte in the testis. Can undergo meiosis II to produce spermatids.

secondary structure In proteins, localized folding of a polypeptide chain into regular structures (i.e., alpha-helix and beta-pleated sheet) stabilized by hydrogen bonding between atoms of the peptide backbone. In nucleic acids, elements of structure (e.g., helices and hairpins) stabilized by hydrogen bonding and hydrophobic interactions between complementary bases. Compare with **primary**, **tertiary**, and **quaternary structures**.

secondary succession Gradual colonization of a habitat after an environmental disturbance (e.g., fire, windstorm, logging) that removes some or all previous organisms but leaves the soil intact. Compare with **primary succession**.

secretin A peptide hormone produced by cells in the small intestine in response to the arrival of food from the stomach. Stimulates secretion of bicarbonate (HCO_3^-) from the pancreas.

sedimentary rock A type of rock formed by gradual accumulation of sediment, particularly sand and mud, as in riverbeds and on the ocean floor. Most fossils are found in sedimentary rocks.

seed A plant reproductive structure consisting of an embryo, associated nutritive tissue (endosperm), and an outer protective layer (seed coat). In angiosperms, develops from the fertilized ovule of a flower.

seed bank A repository where seeds, representing many different varieties of domestic crops or other species, are preserved.

seed coat A protective layer around a seed that encases both the embryo and the endosperm.

segment A well-defined, repeated region of the body along the anterior–posterior body axis, containing structures similar to other nearby segments.

segmentation Division of the body or a part of it into a series of similar structures; exemplified by the body segments of insects and worms and by the somites of vertebrates.

segmentation genes A group of genes that control the formation and patterning of body segmentation in embryonic development. Includes maternal genes, gap genes, pair-rule genes, and segment polarity genes.

segregation, principle of The concept that each pair of hereditary elements (alleles of the same gene) separate from each other during meiosis. One of Mendel's two principles of genetics.

selective adhesion The tendency of cells of one tissue type to adhere to other cells of the same type.

selective permeability The property of a membrane that allows some substances to diffuse across it much more readily than other substances.

selectively permeable membrane Any membrane across which some solutes can move more readily than others.

self-fertilization The fusion of two gametes from the same individual to form offspring. Also called *selfing*.

self molecule A molecule that is synthesized by an organism and is a normal part of its cells and/or body; as opposed to nonself, or foreign, molecules.

semen The combination of sperm and accessory fluids that is released by male mammals and reptiles during ejaculation.

semiconservative replication The way DNA replicates, in which each strand of an existing DNA molecule serves as a template to create a new complementary DNA strand. It is called semiconservative because each newly replicated DNA molecule conserves one of the parental strands and contains another, newly replicated strand.

seminal vesicle In male mammals, either of a pair of reproductive glands that secrete a sugar-containing fluid into semen, which provides energy for sperm movement.

senescence The genetically programmed, active process of aging.

sensor Any cell, organ, or structure with which an animal can sense some aspect of the external or internal environment. Usually functions, along with an integrator and effector, as part of a homeostatic system.

sensory neuron A nerve cell that carries signals from sensory receptors to the central nervous system. Compare with **interneuron** and **motor neuron**.

sepal One of the protective leaflike organs enclosing a flower bud and later part of the outermost portion of the flower.

septum (plural: septa) Any wall-like structure. In fungi, septa divide the filaments (hyphae) of mycelia into cell-like compartments.

serotonin A neurotransmitter involved in many brain functions, including sleep, pleasure, and mood.

serum The liquid that remains when cells and clot material are removed from clotted blood. Contains water, dissolved gases, growth factors, nutrients, and other soluble substances. Compare with **plasma**.

sessile Permanently attached to a substrate; not capable of moving to another location.

set point A normal or target value for a regulated internal variable, such as body heat or blood pH.

sex chromosome Chromosomes that differ in shape or in number in males and females. For example, the X and Y chromosomes of many animals. Compare with **autosome**.

sex-linked inheritance Inheritance patterns observed in genes carried on sex chromosomes. In this case, females and males have different numbers of alleles of a gene. Often creates situations in which a trait appears more often in one sex. Also called *sex-linkage*.

sexual dimorphism Any trait that differs between males and females.

sexual reproduction Any form of reproduction in which genes from two parents are combined via fusion of gametes, producing offspring that are genetically distinct from both parents. Compare with **asexual reproduction**.

sexual selection A type of natural selection that favors individuals with traits that increase their ability to obtain mates. Acts more strongly on males than females. (Compare with **ecological selection**.)

shell A hard, protective outer structure.

Shine–Dalgarno sequence See **ribosome binding sequence**.

shoot In a plant embryo, the combination of hypocotyl and cotyledons, which will become the aboveground portions of the plant.

shoot apical meristem (SAM) A group of undifferentiated plant stem cells at the tip of a plant stem that can differentiate into mature shoot tissues.

shoot system The aboveground part of a plant comprising stems, leaves, and flowers (in angiosperms).

short-day plant A plant that blooms in response to long nights (usually in late summer or fall in the Northern Hemisphere). Compare with **day-neutral** and **long-day plant**.

short tandem repeats (STRs) Relatively short DNA sequences that are repeated, one after another, down the length of a chromosome. See **microsatellite**.

shotgun sequencing A method of sequencing genomes that is based on breaking the genome into small pieces, sequencing each piece separately, and then figuring out how the pieces are connected.

sieve plate In plants, a pore-containing structure at each end of a sieve-tube element in phloem.

sieve-tube element In plants, an elongated sugar-conducting cell in phloem that lacks nuclei and has sieve plates at both ends, allowing sap to flow to adjacent cells.

sigma A bacterial protein that associates with the core RNA polymerase to allow recognition of promoters.

signal In behavioral ecology, any information-containing behavior or characteristic.

signal receptor Any cellular protein that binds to a particular signaling molecule (e.g., a hormone or neurotransmitter) and triggers a response by the cell. Receptors for lipid-insoluble signals are transmembrane proteins in the plasma membrane; those for many lipid-soluble signals (e.g., steroid hormones) are located inside the cell.

signal recognition particle (SRP) An RNA–protein complex that binds to the ER signal sequence in a polypeptide as it emerges from a ribosome and transports the ribosome–polypeptide complex to the ER membrane, where synthesis of the polypeptide is completed.

signal transduction The process by which a stimulus (e.g., a hormone, a neurotransmitter, or sensory information) outside a cell is converted into an intracellular signal required for a cellular response. Usually involves a specific sequence of molecular events, or signal transduction pathway, that may lead to amplification of the signal.

signal transduction cascade See **phosphorylation cascade**.

silencer A regulatory sequence in eukaryotic DNA to which repressor proteins can bind, inhibiting gene transcription.

silent mutation A point mutation that changes the sequence of a codon without changing the amino acid that is specified.

simple eye An eye with only one light-collecting apparatus (e.g., one lens), as in vertebrates and cephalopods. Compare with **compound eye**.

simple fruit A fruit (e.g., apricot) that develops from a single flower that has a single carpel or several fused carpels. Compare with **aggregate** and **multiple fruit**.

simple sequence repeat See **microsatellite**.

SINEs (short interspersed nuclear elements) The second most abundant class of transposable elements in human genomes; can create copies of itself and insert them elsewhere in the genome. Compare with **LINEs**.

single nucleotide polymorphism (SNP) A site on a chromosome where individuals in a population have different nucleotides. Can be used as a genetic marker to help track the inheritance of nearby genes.

single-strand DNA-binding proteins (SSBP) A protein that attaches to separated strands of DNA during replication or transcription, preventing them from re-forming a double helix.

sink Any tissue, site, or location where an element or a molecule is consumed or taken out of circulation (e.g., in plants, a tissue where sugar exits the phloem). Compare with **source**.

sinoatrial (SA) node In the right atrium of the vertebrate heart, a cluster of cardiac muscle cells that initiates the heartbeat and determines the heart rate. Compare with **atrioventricular (AV) node**.

siphon A tubelike appendage of many mollusks, often used for feeding or propulsion.

sister chromatids The paired strands of a recently replicated chromosome, which are connected at the centromere and eventually separate during anaphase of mitosis and meiosis II. Compare with **non-sister chromatids**.

sister species Closely related species that occupy adjacent branches in a phylogenetic tree.

skeletal muscle The muscle tissue attached to the bones of the vertebrate skeleton. Consists of long, unbranched muscle fibers with a characteristic striped (striated) appearance; controlled voluntarily. Compare with **cardiac** and **smooth muscle**.

sliding-filament model The hypothesis that thin (actin) filaments and thick (myosin) filaments slide past each other, thereby shortening the sarcomere. Shortening of all the sarcomeres in a myofibril results in contraction of the entire myofibril.

slow muscle fiber Type of skeletal muscle fiber that is red in color due to the abundance of myoglobin, generates ATP by oxidative phosphorylation, and contracts slowly but does not fatigue easily. Also called *slow oxidative*, or *Type I, fiber*.

small intestine The portion of the digestive tract between the stomach and the large intestine. The site of the final stages of digestion and of most nutrient absorption.

small nuclear ribonucleoproteins See **snRNPs**.

smooth endoplasmic reticulum (smooth ER) The portion of the endoplasmic reticulum that does not have ribosomes attached to it. Involved in synthesis and secretion of lipids. Compare with **rough endoplasmic reticulum**.

smooth muscle The unstriated muscle tissue that lines the intestine, blood vessels, and some other organs. Consists of tapered, unbranched cells that can sustain long contractions. Not voluntarily controlled. Compare with **cardiac** and **skeletal muscle**.

snRNPs (small nuclear ribonucleoproteins) Complexes of proteins and small RNA molecules that function as components of spliceosomes during splicing (removal of introns from pre-mRNAs).

sodium–potassium pump A transmembrane protein that uses the energy of ATP to move sodium ions out of the cell and potassium ions in. Also called *Na^+/K^+-ATPase*.

soil organic matter Organic (carbon-containing) compounds found in soil.

solute Any substance that is dissolved in a liquid.

solute potential (ψ_S) A component of the potential energy of water caused by a difference in solute concentrations at two locations. Can be zero (pure water) or negative. Compare with **pressure potential (ψ_P)**.

solution A liquid containing one or more dissolved solids or gases in a homogeneous mixture.

solvent Any liquid in which one or more solids or gases can dissolve.

soma See **cell body**.

somatic cell Any type of cell in a multicellular organism except eggs, sperm, and their precursor cells. Also called *body cells*.

somatic hypermutation Mutation that occurs in the variable regions of immunoglobulin genes when B cells are first activated and in memory cells, resulting in novel variation in the receptors that bind to antigens.

somatic nervous system The part of the peripheral nervous system (outside the brain and spinal cord) that controls skeletal muscles and is under voluntary control. Compare with **autonomic nervous system**.

somatostatin A hormone secreted by the pancreas and hypothalamus that inhibits the release of several other hormones.

somite A block of mesoderm that occurs in pairs along both sides of the developing neural tube in a vertebrate embryo. Gives rise to muscle, vertebrae, ribs, and the dermis of the skin.

sori In ferns, a cluster of spore-producing structures (sporangia) usually found on the underside of fronds.

source Any tissue, site, or location where a substance is produced or enters circulation (e.g., in plants, the tissue where sugar enters the phloem). Compare with **sink**.

space-filling model A representation of a molecule where atoms are shown as balls—color-coded and scaled to indicate the atom's identify—attached to each other in the correct geometry.

speciation The evolution of two or more distinct species from a single ancestral species.

species An evolutionarily independent population or group of populations. Generally distinct from other species in appearance, behavior, habitat, ecology, genetic characteristics, and so on.

species–area relationship The mathematical relationship between the area of a certain habitat and the number of species that it can support.

species diversity The variety and relative abundance of the species present in a given ecological community.

species richness The number of species present in a given ecological community.

specific heat The amount of energy required to raise the temperature of 1 gram of a substance by 1°C; a measure of the capacity of a substance to absorb energy.

sperm A mature male gamete; smaller and more mobile than the female gamete.

sperm competition Competition between the sperm of different males to fertilize eggs inside the same female.

spermatid An immature sperm cell.

spermatogenesis The production of sperm. Occurs continuously in a testis.

spermatogonium (plural: spermatogonia) Any of the diploid cells in a testis that can give rise to primary spermatocytes.

spermatophore A gelatinous package containing sperm cells that is produced by males of species that have internal fertilization without copulation.

sphincter A muscular valve that can close off a tube, as in a blood vessel or a part of the digestive tract.

spicule Stiff spike of silica or calcium carbonate that provides structural support in the body of many sponges.

spindle apparatus The array of microtubules responsible for moving chromosomes during mitosis and meiosis; includes kinetochore microtubules, polar microtubules, and astral microtubules.

spines In plants, modified leaves that are stiff and sharp and that function in defense.

spiracle In insects, a small opening that connects air-filled tracheae to the external environment, allowing for gas exchange.

spleen A dark red organ, found near the stomach of most vertebrates, that filters blood, stores extra red blood cells in case of emergency, and plays a role in immunity.

spliceosome In eukaryotes, a large, complex assembly of snRNPs (small nuclear ribonucleoproteins) that catalyzes removal of introns from primary RNA transcripts.

splicing The process by which introns are removed from primary RNA transcripts and the remaining exons are connected together.

stigma The sticky tip at the end of a flower carpel to which pollen grains adhere.

sporangium (plural: sporangia) A spore-producing structure found in seed plants, some protists, and some fungi (e.g., chytrids).

spore (1) In bacteria, a dormant form that generally is resistant to extreme conditions. (2) In eukaryotes, a single haploid cell produced by mitosis or meiosis (not by fusion of gametes) that is capable of developing into an adult organism.

sporophyte In organisms undergoing alternation of generations, the multicellular diploid form that arises from two fused gametes and produces haploid spores. Compare with **gametophyte**.

sporopollenin A watertight material that encases spores and pollen of modern land plants.

stabilizing selection A mode of natural selection that favors phenotypes near the middle of the range of phenotypic variation. Reduces overall genetic variation in a population. Compare with **disruptive selection** and **directional selection**.

stamen The male reproductive structure of a flower. Consists of an anther, in which pollen grains are produced, and a filament, which supports the anther. Compare with **carpel**.

standing defense See **constitutive defense**.

stapes The last of three small bones (ossicles) in the middle ear of vertebrates. Receives vibrations from the tympanic membrane and by vibrating against the oval window passes them to the cochlea.

starch A mixture of two storage polysaccharides, amylose and amylopectin, both formed from α-glucose monomers. Amylopectin is branched, and amylose is unbranched. The major form of stored carbohydrate in plants.

start codon The AUG triplet in mRNA at which protein synthesis begins; codes for the amino acid methionine.

statocyst A sensory organ of many arthropods that detects the animal's orientation in space (e.g., whether the animal is flipped upside down).

statolith A tiny stone or dense particle found in specialized gravity-sensing organs in some animals such as lobsters, and in gravity-sensing tissues of plants.

statolith hypothesis The hypothesis that amyloplasts (dense, starch-storing plant organelles) serve as statoliths in gravity detection by plants.

stem cell Any relatively undifferentiated cell that can divide to produce a daughter cell that remains a stem cell and a daughter cell that can differentiate into specific cell types.

stems Vertical, aboveground structures that make up the shoot system of plants.

stereocilium (plural: stereocilia) One of many stiff outgrowths from the surface of a hair cell that are involved in detection of sound by terrestrial vertebrates or of waterborne vibrations by fishes.

steroid A class of lipid with a characteristic four-ring hydrocarbon structure.

sticky ends The short, single-stranded ends of a DNA molecule cut by a restriction endonuclease. Tend to form hydrogen bonds with other sticky ends that have complementary sequences.

stolon A modified stem that runs horizontally over the soil surface and produces new plants at the nodes (a form of asexual reproduction). Compare with **rhizome**.

stoma (plural: stomata) Generally, a pore or opening. In plants, a microscopic pore, surrounded by specialized cells that open the pore, on the surface of a leaf or stem through which gas exchange occurs. See also **guard cells**.

stomach A tough, muscular pouch in the vertebrate digestive tract between the esophagus and small intestine. Physically breaks up food and begins digestion of proteins.

stop codon Any of three mRNA triplets (UAG, UGA, or UAA) that cause termination of protein synthesis. Also called a *termination codon*.

strain The lowest, most specific level of taxonomy that refers to a population of individuals that are genetically very similar or identical.

striated muscle Muscle tissue containing protein filaments organized into repeating structures that give the cells and tissues a banded appearance.

stroma The fluid matrix of a chloroplast in which the thylakoids are embedded. Site where the Calvin cycle reactions occur.

structural formula A two-dimensional notation in which the chemical symbols for the constituent atoms are joined by straight lines representing single (–), double (=), or triple (≡) covalent bonds. Compare with **molecular formula**.

structural homology Similarities in adult organismal structures (e.g., limbs, shells, flowers) that are due to inheritance from a common ancestor.

style The slender stalk of a flower carpel connecting the stigma and the ovary.

suberin Waxy substance found in the cell walls of cork tissue and in the Casparian strip of endodermal cells.

subspecies A population that has distinctive traits and some genetic differences relative to other populations of the same species but that is not distinct enough to be classified as a separate species.

substrate (1) A reactant that interacts with a catalyst, such as an enzyme or ribozyme, in a chemical reaction. (2) A surface on which a cell or organism sits.

substrate-level phosphorylation Production of ATP or GTP by the transfer of a phosphate group from an intermediate substrate directly to ADP or GDP. Occurs in glycolysis and in the citric acid cycle.

succession In ecology, the gradual colonization of a habitat after an environmental disturbance (e.g., fire, flood), usually by a series of species. See **primary** and **secondary succession**.

sucrose A disaccharide formed from glucose and fructose. One of the two main products of photosynthesis.

sugar Synonymous with carbohydrate, though usually used in an informal sense to refer to small carbohydrates (monosaccharides and disaccharides).

summation The additive effect of different postsynaptic potentials on a nerve or muscle cell, such that several subthreshold stimulations can cause or inhibit an action potential.

supporting connective tissue A type of connective tissue distinguished by having a firm extracellular matrix. Includes bone and cartilage.

surface metabolism model Hypothetical explanation for chemical evolution whereby small molecules reacted with one another through catalytic activity associated with a surface, such as the mineral deposits found in deep-sea hydrothermal vents.

surface tension The cohesive force that causes molecules at the surface of a liquid to stick together, thereby resisting deformation of the liquid's surface and minimizing its surface area.

survivorship On average, the proportion of offspring that survive to a particular age.

survivorship curve A graph depicting the percentage of a population that survives to different ages.

suspension feeder Any organism that obtains food by filtering small particles or small organisms out of water or air. Also called *filter feeder*.

sustainability The planned use of environmental resources at a rate no faster than the rate at which they are naturally replaced.

sustainable agriculture Agricultural techniques that are designed to maintain long-term soil quality and productivity.

swamp A wetland that has a steady rate of water flow and is dominated by trees and shrubs.

swim bladder A gas-filled organ of many ray-finned fishes that regulates buoyancy.

symbiosis (adjective: symbiotic) Any close and prolonged physical relationship between individuals of two different species. See **commensalism**, **mutualism**, and **parasitism**.

symmetric competition Ecological competition between two species in which both suffer similar declines in fitness. Compare with **asymmetric competition**.

sympathetic nervous system The part of the autonomic nervous system that stimulates fight-or-flight responses, such as increased heart rate, increased blood pressure, and decreased digestion. Compare with **parasympathetic nervous system**.

sympatric speciation The divergence of populations living within the same geographic area into different species as the result of their genetic (not physical) isolation. Compare with **allopatric speciation**.

sympatry Condition in which two or more populations live in the same geographic area, or close enough to permit interbreeding. Compare with **allopatry**.

symplast In plants, the space inside the plasma membranes. The symplast of adjacent cells is often connected through plasmodesmata. Compare with **apoplast**.

symporter A cotransport protein that allows an ion to diffuse down an electrochemical gradient, using the energy of that process to transport a different substance in the same direction *against* its concentration gradient. Compare with **antiporter**.

synapomorphy A shared, derived trait found in two or more taxa that is present in their most recent common ancestor but is missing in more

distant ancestors. Useful for inferring evolutionary relationships.

synapse The interface between two neurons or between a neuron and an effector cell.

synapsis The physical pairing of two homologous chromosomes during prophase I of meiosis. Crossing over is observed during synapsis.

synaptic cleft The space between two communicating nerve cells (or between a neuron and effector cell) at a synapse, across which neurotransmitters diffuse.

synaptic plasticity Long-term changes in the responsiveness or physical structure of a synapse that can occur after particular stimulation patterns. Thought to be the basis of learning and memory.

synaptic vesicle A small neurotransmitter-containing vesicle at the end of an axon that releases neurotransmitter into the synaptic cleft by exocytosis.

synaptonemal complex A network of proteins that holds non-sister chromatids together during synapsis in meiosis I.

synthesis (S) phase The phase of the cell cycle during which DNA is synthesized and chromosomes are replicated.

system A defined set of interacting chemical components under observation.

systemic acquired resistance (SAR) A slow, widespread response of plants to a localized infection that protects healthy tissue from invasion by pathogens. Compare with **hypersensitive response**.

systemic circulation The part of the circulatory system that sends oxygen-rich blood from the lungs out to the rest of the body. It is separate from the pulmonary circulation in mammals and birds.

systemin A peptide hormone, produced by plant cells damaged by herbivores, that initiates a protective response in undamaged cells.

systems biology The study of the structure of networks and how interactions between individual network components such as genes or proteins can lead to emergent biological properties.

systole The portion of the cardiac cycle during which the heart muscles are contracting. Compare with **diastole**.

systolic blood pressure The force exerted by blood against artery walls during contraction of the heart's left ventricle. Compare with **diastolic blood pressure**.

T cell A type of lymphocyte that matures in the thymus and, with B cells, is responsible for adaptive immunity. Involved in activation of B cells (CD4$^+$ helper T cells) and destruction of infected cells (CD8$^+$ cytotoxic T cells). Also called *T lymphocytes*.

T-cell receptor (TCR) A type of transmembrane protein found on T cells that can bind to antigens displayed on the surfaces of other cells. Composed of two polypeptides, called the alpha chain and beta chain, that consist of variable and constant regions. See **antigen presentation**.

T tubule Any of the membranous tubes that extend into the interior of muscle cells, propagating action potentials throughout the cell and triggering release of calcium from the sarcoplasmic reticulum.

tagmata (singular: tagma) Prominent body regions in arthropods, such as the head, thorax, and abdomen in insects.

taiga A vast forest biome throughout subarctic regions, consisting primarily of short coniferous trees. Characterized by intensely cold winters, short summers, and high annual variation in temperature.

taproot A large, vertical main root of a plant's root system.

taste bud A sensory structure, found chiefly in the mammalian tongue, containing spindle-shaped cells that respond to chemical stimuli.

TATA-binding protein (TBP) A protein that binds to the TATA box in eukaryotic promoters and is a component of the basal transcription complex.

TATA box A short DNA sequence in many eukaryotic promoters about 30 base pairs upstream from the transcription start site.

taxon (plural: taxa) Any named group of organisms at any level of a classification system.

taxonomy The branch of biology concerned with the classification and naming of organisms.

tectorial membrane A membrane, located in the vertebrate cochlea, that takes part in the transduction of sound by bending the stereocilia of hair cells in response to sonic vibrations.

telomerase An enzyme that adds DNA to the ends of chromosomes (telomeres) by catalyzing DNA synthesis from an RNA template that is part of the enzyme.

telomere The end of a linear chromosome that contains a repeated sequence of DNA.

telophase The final stage in mitosis or meiosis during which daughter chromosomes (homologous chromosomes in meiosis I) have separated and new nuclear envelopes begin to form around each set of chromosomes.

temperate Having a climate with pronounced annual fluctuations in temperature (i.e., warm summers and cold winters) but typically neither as hot as the tropics nor as cold as the poles.

temperature A measurement of thermal energy present in an object or substance, reflecting how much the constituent molecules are moving.

template strand An original nucleic acid strand used to make a new, complementary copy based on hydrogen bonding between nitrogenous bases.

temporal lobe In the vertebrate brain, one of the four major areas in the cerebrum.

tendon A band of tough, fibrous connective tissue that connects a muscle to a bone.

tentacle A long, thin, muscular appendage typically used for feeling and feeding. Occurs in different forms in diverse animals such as cephalopod mollusks and sea anemones.

termination (1) In enzyme-catalyzed reactions, the final stage in which the enzyme returns to its original conformation and products are released. (2) In transcription, the dissociation of RNA polymerase from DNA. (3) In translation, the dissociation of a ribosome from mRNA when it reaches a stop codon.

territory An area that is actively defended by an animal from others of its species and that provides exclusive or semi-exclusive use of its resources by the owner.

tertiary consumers In a food chain or food web, organisms that feed on secondary consumers. Compare with **primary consumer** and **secondary consumer**.

tertiary structure The overall three-dimensional shape of a single polypeptide chain, resulting from multiple interactions among the amino acid side chains and the peptide backbone. In single-stranded nucleic acids, the three-dimensional shape is formed by hydrogen bonding and hydrophobic interactions between complementary bases. Compare with **primary**, **secondary**, and **quaternary structure**.

testcross The breeding of an individual that expresses a dominant phenotype but has an unknown genotype with an individual having only recessive alleles for the traits of interest. Used to order to infer the unknown genotype from observation of the phenotypes seen in offspring.

testis (plural: testes) The sperm-producing organ of a male animal.

testosterone A steroid hormone, produced and secreted by the testes, that stimulates sperm production and various male traits and reproductive behaviors.

tetrad The structure formed by synapsed homologous chromosomes during prophase of meiosis I. Also known as a *bivalent*.

tetrapod Any member of the lineage that includes all vertebrates with two pairs of limbs (amphibians, mammals, and reptiles, including birds).

texture A quality of soil, resulting from the relative abundance of different-sized particles.

theory An explanation for a broad class of phenomena that is supported by a wide body of evidence. A theory serves as a framework for the development of new hypotheses.

thermal energy The kinetic energy of molecular motion.

thermocline A steep gradient (cline) in environmental temperature, such as occurs in a thermally stratified lake or ocean.

thermophile A bacterium or archaean that thrives in very hot environments.

thermoreception A sensory system in which receptors are activated by changes in heat energy.

thermoreceptor A sensory cell or an organ specialized for detection of changes in temperature.

thermoregulation Regulation of body temperature.

thick filament A filament composed of bundles of the motor protein myosin; anchored to the center of the sarcomere. Compare with **thin filament**.

thigmotropism Growth or movement of an organism in response to contact with a solid object.

thin filament A filament composed of two coiled chains of actin and associated regulatory proteins; anchored at the Z disk of the sarcomere. Compare with **thick filament**.

thorax A region of the body; in insects, one of the three prominent body regions, along with the head and abdomen, called tagmata.

thorn A modified plant stem shaped as a sharp, protective structure. Helps protect a plant against feeding by herbivores.

threshold potential The membrane potential that will trigger an action potential in a neuron or other excitable cell. Also called simply *threshold*.

thylakoid A membrane-bound network of flattened sac-like structures inside a plant chloroplast that functions in converting light energy to chemical energy. A stack of thylakoid discs is a granum.

thymus An organ, located in the anterior chest or neck of vertebrates, in which immature T cells that originated in the bone marrow undergo maturation.

thyroid gland A gland in the neck that releases thyroid hormone (which increases metabolic rate) and calcitonin (which lowers blood calcium).

thyroid hormones Either of two hormones, triiodothyronine (T_3) or thyroxine (T_4), produced by the thyroid gland. See **triiodothyronine** and **thyroxine**.

thyroid-stimulating hormone (TSH) A peptide hormone, produced and secreted by the anterior pituitary, that stimulates release of thyroid hormones from the thyroid gland.

thyroxine (T_4) A lipid-soluble hormone, derived from the amino acid tyrosine, containing four iodine atoms and produced and secreted by the thyroid gland. Acts primarily to increase cellular metabolism. In mammals, T_4 is converted to the more active hormone triiodothyronine (T_3) in the liver.

Ti plasmid A plasmid carried by *Agrobacterium* (a bacterium that infects plants) that can integrate into a plant cell's chromosomes and induce formation of a gall.

tight junction A type of cell–cell attachment structure that links the plasma membranes of adjacent animal cells, forming a barrier that restricts movement of substances in the space between the cells. Most abundant in epithelia (e.g., the intestinal lining). Compare with **desmosome** and **gap junction**.

tip The end of a branch on a phylogenetic tree. Represents a specific species or larger taxon that has not (yet) produced descendants—either a group living today or a group that ended in extinction. Also called *terminal node*.

tissue A group of cells that function as a unit, such as muscle tissue in an animal or xylem tissue in a plant.

tolerance In ecological succession, the phenomenon in which early-arriving species do not affect the probability that subsequent species will become established. Compare with **facilitation** and **inhibition**.

tonoplast The membrane surrounding a plant vacuole.

tool-kit genes A set of key developmental genes that establishes the body plan of animals and plants; present at the origin of the multicellular lineages and elaborated upon over evolutionary time by a process of duplication and divergence. Includes *Hox* genes.

top-down control The hypothesis that population size is limited by predators or herbivores (consumers).

topoisomerase An enzyme that prevents the twisting of DNA ahead of the advancing replication fork by cutting the DNA, allowing it to unwind, and rejoining it.

torpor An energy-conserving physiological state, marked by a decrease in metabolic rate, body temperature, and activity, that lasts for a short period (overnight to a few days or weeks). Occurs in some small mammals when the ambient temperature drops significantly. Compare with **hibernation**.

totipotent Capable of dividing and developing to form a complete, mature organism.

toxin A poison produced by a living organism, such as a plant, animal, or microorganism.

trachea (plural: tracheae) (1) In insects, any of the small air-filled tubes that extend throughout the body and function in gas exchange. (2) In terrestrial vertebrates, the airway connecting the larynx to the bronchi. Also called *windpipe*.

tracheid In vascular plants, a long, thin, water-conducting cell that has pits where its lignin-containing secondary cell wall is absent, allowing water movement between adjacent cells. Compare with **vessel element**.

trade-off In evolutionary biology, an inescapable compromise between two traits that cannot be optimized simultaneously. Also called *fitness trade-off*.

trait Any observable characteristic of an individual.

transcription The process that uses a DNA template to produce a complementary RNA.

transcription factor General term for a protein that binds to a DNA regulatory sequence to influence transcription. It includes both regulatory and basal transcription factors.

transcriptional activator A eukaryotic regulatory transcription factor that binds to regulatory DNA sequences in enhancers or promoter-proximal elements to promote the initiation of transcription.

transcriptional control Regulation of gene expression by various mechanisms that change the rate at which genes are transcribed to form messenger RNA. In negative control, binding of a regulatory protein to DNA represses transcription; in positive control, binding of a regulatory protein to DNA promotes transcription.

transcriptome The complete set of genes transcribed in a particular cell.

transduction The conversion of information from one mode to another. For example, the process by which a stimulus outside a cell is converted into a response by the cell.

transfer RNA (tRNA) An RNA molecule that has an anticodon at one end and an amino acid attachment site at the other. Each tRNA carries a specific amino acid and binds to the corresponding codon in messenger RNA during translation.

transformation (1) Incorporation of external DNA into a cell. Occurs naturally in some bacteria; can be induced in the laboratory. (2) Conversion of a normal mammalian cell to one that divides uncontrollably.

transgenic A plant or animal whose genome contains DNA introduced from another individual, often from a different species.

transition state A high-energy intermediate state of the reactants during a chemical reaction that must be achieved for the reaction to proceed. Compare with **activation energy**.

transitional feature A trait that is intermediate between a condition observed in ancestral (older) species and the condition observed in derived (younger) species.

translation The process by which a polypeptide (a string of amino acids joined by peptide bonds) is synthesized from information in codons of messenger RNA.

translational control Regulation of gene expression by various mechanisms that alter the life span of messenger RNA or the efficiency of translation.

translocation (1) In plants, the movement of sugars and other organic nutrients through the phloem by bulk flow. (2) A type of mutation in which a piece of a chromosome moves to a nonhomologous chromosome. (3) The movement of a ribosome down a messenger RNA during translation.

transmembrane protein See **integral membrane protein**.

transmission The passage or transfer of (1) a disease from one individual to another or (2) electrical impulses from one neuron to another.

transpiration Loss of water vapor from aboveground plant parts. Occurs primarily through stomata.

transporter See **carrier protein**.

transposable elements Any of several kinds of DNA sequences that are capable of moving themselves, or copies of themselves, to other locations in the genome. Include LINEs and SINEs.

tree of life The phylogenetic tree that includes all organisms.

trichome A hairlike appendage that grows from epidermal cells in the shoot system of some plants. Trichomes exhibit a variety of shapes, sizes, and functions depending on species.

triiodothyronine (T_3) A lipid-soluble hormone, derived from the amino acid tyrosine, containing three iodine atoms and produced and secreted by the thyroid gland. Acts primarily to increase cellular metabolism. In mammals, T_3 has a stronger effect than does the related hormone thyroxine (T_4).

triose A monosaccharide (simple sugar) containing three carbon atoms.

triplet code A code in which a "word" of three letters encodes one piece of information. The genetic code is a triplet code because a codon is three nucleotides long and encodes one amino acid.

triploblast (adjective: triploblastic) An animal whose body develops from three basic embryonic cell layers or tissues: ectoderm, mesoderm, and endoderm. Compare with **diploblast**.

trisomy The state of having three copies of one particular type of chromosome in an otherwise diploid cell.

tRNA See **transfer RNA**.

trochophore A larva with a ring of cilia around its middle that is found in some lophotrochozoans.

trophic cascade A series of changes in the abundance of species in a food web, usually caused by the addition or removal of a key predator.

trophic level A feeding level in an ecosystem.

trophoblast The exterior of a blastocyst (the structure that results from cleavage in embryonic development of mammals).

tropomyosin A regulatory protein present in thin (actin) filaments that blocks the myosin-binding sites on these filaments in resting muscles, thereby preventing muscle contraction.

troponin A regulatory protein, present in thin (actin) filaments, that can move tropomyosin off the myosin-binding sites on these filaments, thereby triggering muscle contraction. Activated by high intracellular calcium.

true navigation The type of navigation by which an animal can reach a specific point on Earth's surface. Also called *map orientation*.

trypsin A protein-digesting enzyme present in the small intestine that activates several other protein-digesting enzymes.

tube foot One of the many small, mobile, fluid-filled extensions of the water vascular system of echinoderms; the part extending outside the body is called a podium, while the bulb within the body is the ampulla. Used in locomotion, feeding, and respiration.

tuber A modified plant rhizome that functions in storage of carbohydrates.

tuberculosis A disease of the lungs caused by infection with the bacterium *Mycobacterium tuberculosis*.

tumor A mass of cells formed by uncontrolled cell division. Can be benign or malignant.

tumor suppressor A protein (e.g., p53 or Rb) that prevents cell division, such as when the cell has DNA damage. Mutant genes that code for tumor suppressors are associated with cancer.

tundra The treeless biome in polar and alpine regions, characterized by short, slow-growing vegetation, permafrost, and a climate of long, intensely cold winters and very short summers.

turbidity Cloudiness of water caused by sediments and/or microscopic organisms.

turgid Swollen and firm as a result of high internal pressure (e.g., a plant cell containing enough water for the cytoplasm to press against the cell wall). Compare with **flaccid**.

turgor pressure The outward pressure exerted by the fluid contents of a living plant cell against its cell wall.

turnover In lake ecology, the complete mixing of upper and lower layers of water of different temperatures; occurs each spring and fall in temperate-zone lakes.

tympanic membrane The membrane separating the middle ear from the outer ear in terrestrial vertebrates, or similar structures in insects. Also called the *eardrum*.

ubiquinone See **coenzyme Q**.

ulcer A hole in an epithelial layer, exposing the underlying tissues to damage.

ultimate causation In biology, the reason that a trait or phenomenon is thought to have evolved; the adaptive advantage of that trait. Also called *ultimate explanation*. Compare with **proximate causation**.

umami The taste of glutamate, responsible for the "meaty" taste of most proteins and of monosodium glutamate.

umbilical cord The cord that connects a developing mammalian embryo or fetus to the placenta and through which the embryo or fetus receives oxygen and nutrients.

unequal crossover An error in crossing over during meiosis I in which the two non-sister chromatids match up at different sites. Results in gene duplication in one chromatid and gene loss in the other.

unsaturated Referring to lipids in which at least one carbon-carbon bond is a double bond. Double bonds produce kinks in hydrocarbon chains and decrease the compound's melting point. Compare with **saturated**.

upstream In genetics, opposite to the direction in which RNA polymerase moves along a DNA strand. Compare with **downstream**.

urea The major nitrogenous waste of mammals, adult amphibains, and cartilaginous fishes. Compare with **ammonia** and **uric acid**.

ureter In vertebrates, a tube that transports urine from one kidney to the bladder.

urethra The tube that drains urine from the bladder to the outside environment. In male vertebrates, also used for passage of semen during ejaculation.

uric acid A whitish excretory product of birds, reptiles, and terrestrial arthropods. Used to remove from the body excess nitrogen derived from the breakdown of amino acids. Compare with **ammonia** and **urea**.

urochordates One of the three major chordate lineages (Urochordata), comprising sessile or floating, filter-feeding animals that have a polysaccharide covering (tunic) and two siphons through which water enters and exits; include tunicates and salps. Compare with **cephalochordates** and **vertebrates**.

uterus The organ in which developing embryos are housed in mammals and some other viviparous vertebrates.

vaccination Artificially producing immunological memory against a pathogen by using isolated antigens or altered versions of the pathogen to stimulate an adaptive immune response in the absence of disease.

vaccine A preparation designed to stimulate an immune response against a particular pathogen without causing illness. Vaccines consist of inactivated (killed) pathogens, live but weakened (attenuated) pathogens, or parts of a pathogen (subunit vaccine).

vacuole A large organelle in plant and fungal cells that usually is used for bulk storage of water, pigments, oils, or other substances. Some vacuoles contain enzymes and have a digestive function similar to lysosomes in animal cells.

vagina The birth canal of female mammals; a muscular tube that extends from the uterus through the pelvis to the exterior.

valence The number of unpaired electrons in the outermost electron shell of an atom; often determines how many covalent bonds the atom can form.

valence electron An electron in the outermost electron shell, the valence shell, of an atom. Valence electrons tend to be involved in chemical bonding.

valence shell The outermost electron shell of an atom.

valve In circulatory systems, any of the flaps of tissue that prevent backward flow of blood, particularly in veins and in the heart.

van der Waals interactions A weak electrical attraction between two nonpolar molecules that have been brought together through hydrophobic interactions. Often contributes to tertiary and quaternary structures in proteins.

variable number tandem repeat See **minisatellite**.

variable (V) region The amino acid sequence that changes in polypeptides used to make antibodies, B-cell receptors, and T-cell receptors, This portion of the protein is highly variable within an individual and forms the epitope-binding site. Compare with **constant (C) region**.

vas deferens (plural: vasa deferentia) A muscular tube that stores and transports semen from the epididymis to the ejaculatory duct. Also called the *ductus deferens*.

vasa recta In the vertebrate kidney, a network of blood vessels that runs alongside the loop of Henle of a nephron. Functions in reabsorption of water and solutes from the filtrate.

vascular bundle In a plant stem, a cluster of xylem and phloem strands that run the length of the stem.

vascular cambium One of two types of cylindrical meristem, consisting of a ring of undifferentiated plant cells in the stem and root of woody plants; produces secondary xylem (wood) and secondary phloem. Compare with **cork cambium**.

vascular tissue In plants, tissue that transports water, nutrients, and sugars. Made up of the complex tissues xylem and phloem, each of which contains several cell types. Also called *vascular tissue system*.

vector A biting insect or other organism that transfers pathogens from one species to another. See also **cloning vector**.

vegetative development The phase of plant development that involves growth and the the production of all plant structures except the flower.

vein Any blood vessel that carries blood (oxygenated or not) under relatively low pressure from the tissues toward the heart. Compare with **artery**.

veliger A distinctive type of larva, found in mollusks.

vena cava (plural: venae cavae) Either of two large veins that return oxygen-poor blood to the heart.

ventral Toward an animal's belly and away from its back. The opposite of dorsal.

ventricle (1) A thick-walled chamber of the heart that receives blood from an atrium and pumps it to the body or to the lungs. (2) Any of several small fluid-filled chambers in the vertebrate brain.

venule Any of the body's many small veins (blood vessels that return blood to the heart).

vertebrae (singular: vertebra) The cartilaginous or bony elements that form the backbones of vertebrate animals.

vertebrates One of the three major chordate lineages (Vertebrata), comprising animals with a dorsal column of cartilaginous or bony structures (vertebrae) and a skull enclosing the brain. Includes fishes, amphibians, mammals, and reptiles (including birds). Compare with **cephalochordates** and **urochordates**.

vessel element In vascular plants, a short, wide, water-conducting cell that has gaps through both the primary and secondary cell walls, allowing unimpeded passage of water between adjacent cells. Compare with **tracheid**.

vestigial trait A reduced or incompletely developed structure that has no function or reduced function, but is clearly similar to functioning organs or structures in closely related species.

vicariance The physical splitting of a population into smaller, isolated populations by a geographic barrier.

villi (singular: villus) Small, fingerlike projections (1) of the lining of the small intestine or (2) of the fetal portion of the placenta adjacent to maternal arteries. Function to increase the surface area available for absorption of nutrients and gas exchange.

virion The infectious extracellular particle that is produced from a viral infection; used for transmitting the virus between hosts. It consists of a DNA or RNA genome enclosed within a protein shell (capsid) that may be further enveloped in a phospholipid bilayer. Compare with **virus**.

virulence (adjective: virulent) Referring to the ability of pathogens to cause severe disease in susceptible hosts.

virus An obligate, intracellular parasite that is acellular, but uses host-cell biosynthetic machinery to replicate. Compare with **virion**.

visceral mass One of the three main parts of the mollusk body; contains most of the internal organs and external gill.

visible light The range of wavelengths of electromagnetic radiation that humans can see, from about 400 to 700 nanometers.

vitamin Any of various organic micronutrients that usually function as coenzymes.

vitelline envelope A fibrous sheet of glycoproteins that surrounds mature egg cells in many vertebrates. Surrounded by a thick gelatinous matrix (the jelly layer) in some aquatic species. In mammals, called the *zona pellucida*.

viviparous Producing live young (instead of eggs) that develop within the body of the mother before birth. Compare with **oviparous** and **ovoviviparous**.

volt (V) A unit of electrical potential (voltage).

voltage Potential energy created by a separation of electric charges between two points. Also called *electrical potential*.

voltage clamping A technique for imposing a constant membrane potential on a cell. Widely used to investigate ion channels.

voltage-gated channel An ion channel that opens or closes in response to changes in membrane voltage. Compare with **ligand-gated channel**.

voluntary muscle Muscle that contracts in response to stimulation by voluntary (somatic), but not involuntary (parasympathetic or sympathetic), neural stimulation.

vomeronasal organ A paired sensory organ, located in the nasal region, containing chemoreceptors that bind odorants and pheromones.

wall pressure The inward pressure exerted by a cell wall against the fluid contents of a living plant cell.

Wallace line A line in the Indonesian region that demarcates two areas, each of which is characterized by a distinct set of animal species.

water potential (ψ) The potential energy of water in a certain environment compared with the potential energy of pure water at room temperature and atmospheric pressure. In living organisms, ψ equals the solute potential (ψ_S) plus the pressure potential (ψ_P).

water-potential gradient A difference in water potential in one region compared with that in another region. Determines the direction that water moves, always from regions of higher water potential to regions of lower water potential.

water table The upper limit of the underground layer of soil that is saturated with water.

water vascular system In echinoderms, a system of fluid-filled tubes and chambers that functions as a hydrostatic skeleton.

watershed The area drained by a single stream or river.

Watson–Crick pairing See **complementary base pairing**.

wavelength The distance between two successive crests in any regular wave, such as light waves, sound waves, or waves in water.

wax A class of lipid with extremely long, saturated hydrocarbon tails. Harder and less greasy than fats.

weather The specific short-term atmospheric conditions of temperature, moisture, sunlight, and wind in a certain area.

weathering The gradual wearing down of large rocks by rain, running water, temperature changes, and wind; one of the processes that transform rocks into soil.

weed Any plant that is adapted for growth in disturbed soils.

white blood cell Any of several types of blood cells, including neutrophils, macrophages, and lymphocytes, that circulate in blood and lymph and function in defense against pathogens.

wild type The most common phenotype seen in a wild population.

wildlife corridor Strips of wildlife habitat connecting populations that otherwise would be isolated by human-made development.

wilt To lose turgor pressure in a plant tissue.

wobble hypothesis The hypothesis that some tRNA molecules can pair with more than one mRNA codon by tolerating some non-standard base pairing in the third base, so long as the first and second bases are correctly matched.

wood Xylem resulting from secondary growth; forms strong supporting material. Also called *secondary xylem*.

worm An animal with a long, thin, tubelike body lacking limbs.

Woronin body A dense organelle in certain fungi that plugs pores in damaged septa to prevent leakage of cytoplasm.

X-linked inheritance Inheritance patterns for genes located on the mammalian X chromosome. Also called *X-linkage*.

X-ray crystallography A technique for determining the three-dimensional structure of large molecules, including proteins and nucleic acids, by analysis of the diffraction patterns produced by X-rays beamed at crystals of the molecule.

xenoestrogens Foreign chemicals that bind to estrogen receptors or otherwise induce estrogen-like effects.

xeroderma pigmentosum (XP) A human disease characterized by extreme sensitivity to ultraviolet light. Caused by an autosomal recessive allele that inactivates the nucleotide excision DNA repair system.

xylem A plant vascular tissue that conducts water and ions; contains tracheids and/or vessel elements.

Primary xylem develops from the procambium of apical meristems; secondary xylem, or wood, from the vascular cambium. Compare with **phloem**.

Y-linked inheritance Inheritance patterns for genes located on the mammalian Y chromosome. Also called *Y-linkage*.

yeast Any fungus growing as a single-celled form. Also, a specific lineage of Ascomycota.

yolk The nutrient-rich cytoplasm inside an egg cell; used as food for the growing embryo.

Z disk The structure that forms each end of a sarcomere. Contains a protein that binds tightly to actin, thereby anchoring thin filaments.

Z scheme Model for changes in the potential energy of electrons as they pass from photosystem II to photosystem I and ultimately to NADP$^+$ during the light-dependent reactions of photosynthesis. See also **noncyclic electron flow**.

zero population growth (ZPG) A state of stable population size due to fertility staying at the replacement rate for at least one generation.

zona pellucida The gelatinous layer around a mammalian egg cell. In other vertebrates, called the *vitelline envelope*.

zone of (cellular) division In plant roots, a group of apical meristematic cells just behind the root cap where cells are actively dividing.

zone of (cellular) elongation In plant roots, a group of young cells, derived from primary meristem tissues and located behind the apical meristem, that are increasing in length.

zone of (cellular) maturation In plant roots, a group of plant cells, located several millimeters behind the root cap, that are differentiating into mature tissues.

zygosporangium (plural: zygosporangia) The distinctive spore-producing structure in fungi that are members of the Zygomycota.

zygote The cell formed by the union of two gametes; a fertilized egg.

Credits

Photo Credits

Frontmatter Title page Eric Isselée/Fotolia **p. iii** IFE, URI-IAO, UW, Lost City Science Party; NOAA/OAR/OER; The Lost City 2005 Expedition **p. v** R. B. Taylor/Photo Researchers, Inc. **p. vi** Brian Johnston **p. viii** Anthony Bannister/Photo Researchers, Inc. **p. x** Tim Laman/National Geographic Stock **p. xiii** Lee W. Wilcox **p. xiv** Steve Winter/National Geographic/Getty Images **p. xvi** Associated Press/Gerald Herbert **p. xviii (bottom left)** Natalie B. Fobes Photography

Chapter 1 Opener Thierry Montford/Biosphoto/Photo Researchers Inc. **1.1a** Biophoto Associates/Photo Researchers, Inc. **1.1b** Brian J. Ford **1.1c** The Print Collector/Alamy **1.6a** Steve Gschmeissner/Photo Researchers, Inc. **1.6b** Kwangshin Kim/Photo Researchers, Inc. **1.7** Corbis Premium RF/Alamy **1.10** Reproduced with permission from H. Wolf. 2011. Odometry and Insect Navigation. *Journal of Experimental Biology* 214: 1629–1641.

Chapter 2 Opener IFE, URI-IAO, UW, Lost City Science Party; NOAA/OAR/OER; The Lost City 2005 Expedition **2.1** Dragan Trifunovic/Shutterstock **2.6c** Zedcor Wholly Owned/Photos.com **2.11** LianeM/Alamy **2.15b** Dietmar Nill/Getty Images **2.15c** John Sylvester/First Light/AGE fotostock **2.29** picture-alliance/Judaica-Samml/Newscom

Chapter 3 3.9ab Janice Carr, CDC

Chapter 4 Opener SSPL/The Image Works **4.5** National Cancer Institute

Chapter 5 Opener Peter Arnold/Alamy

Chapter 6 6.2a Multiart/iStockphoto **6.2b** lepas2004/iStockphoto **6.2c** ilker canikligil/Shutterstock **6.6L** Dr. rer. nat. Markus Drechsler **6.15** Jack W. Szostak **6.18** Don W. Fawcett/Photo Researchers, Inc.

Chapter 7 Opener Dr. Torsten Wittmann/SPL/Photo Researchers, Inc. **7.1** SPL/Photo Researchers, Inc. **7.2a** Gopal Murti/Photo Researchers, Inc. **7.2b** Biology Pics/Photo Researchers, Inc. **7.3** Eye of Science/Photo Researchers, Inc. **7.4** Eye of Science/Photo Researchers, Inc. **7.7** Don W. Fawcett/Photo Researchers, Inc. **7.8** Omikron/Photo Researchers, Inc. **7.9LR** Don W. Fawcett/Photo Researchers, Inc. **7.10** Biophoto Associates/Photo Researchers, Inc. **7.11** Don W. Fawcett/Photo Researchers, Inc. **7.12** E.H. Newcomb & W.P Wergin/Biological Photo Service **7.13** Fawcett/Friend/Photo Researchers, Inc. **7.14** Don W. Fawcett/Photo Researchers, Inc. **7.15** E.H. Newcomb & W.P. Wergin/Biological Photo Service **7.16a** Dr. Don Fawcett/S. Ito & A. Like/Photo Researchers, Inc. **7.16b** Don W. Fawcett/Photo Researchers, Inc. **7.16c** Biophoto Associates/Photo Researchers, Inc. **7.16d** From "*Caveolin-1* expression is essential for proper nonshivering thermogenesis in brown adipose tissue." Cohen, A. W., Schubert, W., Brasaemle, D. L., Scherer, P. E., Lisanti, M. P. *Diabetes.* 2005 Mar; 54(3): 679–86, Fig. 6. **7.17** Don W. Fawcett/Photo Researchers, Inc. **7.26** Micrograph by Dr. Conly L. Rieder, Wadsworth Center, Albany, New York 12201-0509 **7.27ab** Reproduced by permission of the American Society for Cell Biology from *Molecular Biology of the Cell* 9 (12), December 1998, cover. ©1999 by the American Society for Cell Biology. Image courtesy of Bruce J. Schnapp, Oregon Health Sciences University. **7.28** John E. Heuser, M.D., Washington University School of Medicine, St. Louis, Missouri **7.29a** SPL/Photo Researchers, Inc. **7.29b** Biomedical Imaging Unit, Southampton General Hospital/Photo Researchers, Inc. **7.30** Don W. Fawcett/Photo Researchers, Inc.

Chapter 8 Opener Richard Megna/Fundamental Photographs **8.3** Thomas Eisner

Chapter 9 Opener Raiden32/Used under the Creative Commons license. http://creativecommons.org/licenses/by-sa/3.0/deed.en **9.1a** Oliver Hoffmann/Shutterstock **9.1b** Gergo Orban/Shutterstock **9.8** Terry Frey **9.16** Michael Delannoy

Chapter 10 Opener R.B. Taylor/Photo Researchers, Inc. **10.3T** John Durham/Science Photo Library/Photo Researchers, Inc. **10.3MB** Barbara Erienborn/Electron micrograph by Wm. P. Wergin, courtesy of E.H. Newcomb, University of Wisconsin **10.5b** Sinclair Stammers/Photo Researchers, Inc. **10.18LR** James A. Bassham, Lawrence Berkeley Laboratory, UCB (retired) **10.21ab** Dr. Jeremy Burgess/Science Photo Library/Photo Researchers, Inc.

Chapter 11 Opener Roger J. Bick & Brian J. Poindexter/UT-Houston Medical School/Photo Researchers, Inc. **11.1** Jochen Tack/imagebroker/Alamy **11.2** Biophoto Associates/Photo Researchers, Inc. **11.3b** ©ROCKWATER, Inc./*Journal of Bone and Joint Surgery*, 1997, 79, Instructional Course Lectures, The American Academy of Orthopaedic Surgeons—Articular Cartilage. Part I: Tissue Design and Chondrocyte-Matrix Interactions, Buckwalter, 600–611 **11.5** Biophoto Associates/Photo Researchers, Inc. **11.7aLR** Don W. Fawcett/Photo Researchers, Inc. **11.8** Don W. Fawcett/Photo Researchers, Inc. **11.10a** Dr. Don Fawcett/Photo Researcher's, Inc. **11.10b** E. H. Newcomb & W.P. Wergin/Biological Photo Service **11.17a** A. Nern and R. A. Arkowitz. 2000. "G proteins mediate changes in cell shape by stabilizing the axis of polarity." *Molecular Cell* 5(5): 853–864, Fig. 2C. **11.17b** R. H. Kessin. 2003. "Cell motility: Making streams." *Nature* 422: 481–482. ©2003 Nature Publishing Group. Used with permission.

Chapter 12 Opener G. Gimenz-Martin/SPL/Photo Researchers, Inc. **12.1T** Gopal Murti/Photo Researchers, Inc. **12.1B** Biophoto Associates/Photo Researchers, Inc. **12.5(1–7)** Micrographs by Conly L. Rieder, Department of Biology, Rensselaer Polytechnic Institute, Troy, New York. **12.8a** Ed Reschke/Getty **12.8b** Michael Danilchik

Chapter 13 Opener David Phillips/The Population Council/Photo Researchers, Inc. **13.1** Look at Sciences/Photo Researchers, Inc. **13.6** Jupiterimages/Photos.com **13.7(1–9)** Warren Rosenberg/Fundamental Photographs **13.15** Dr. Rebecca Shulte

Chapter 14 Opener Brian Johnston **14.9a** Benjamin Prud'homme/Nicolas Gompel **14.9LR** From "Learning to Fly: Phenotypic Markers in Drosophila." A poster of common phenotypic markers used in Drosophila genetics. Jennifer Childress, Richard Behringer, and Georg Halder. 2005. *Genesis* 43(1). Cover illustration. **14.16** Carla Fernanda Reis

Chapter 15 Opener Dr. Gopal Murti/Science Photo Library/Photo Researchers, Inc. **15.1** Eye of Science/Photo Researchers, Inc. **15.5** Reproduced by permission of Matthew S. Meselson, Harvard University, from M. Meselson and F.W. Stahl, "The replication of DNA in Escherichia coli." *PNAS* 44(7): 671–682 (July 1958), p. 675, Fig. 4. **15.7** Dr. Gopal Murti/Science Photo Library/Photo Researchers, Inc.

Chapter 16 Opener J. Craig Venter Institute **16.4a** Rod Williams/Nature Picture Library **16.4b** Clint Cook & Janet P. Crossland, The Peromyscu Genetic Stock Center at the U. of South Carolina **16.9** Courtesy of Hesed M. Padilla-Nash and Thomas Ried. Affiliation is Section of Cancer Genomics, Genetics Branch, Center for Research, National Cancer Institute, National Institutes of Health, Bethesda, MD 20892

Chapter 17 Opener Oscar Miller/SPL/Photo Researchers, Inc. **17.5** Bert W. O'Malley, M.D., Baylor College of Medicine **17.8** From Hamkalo and Miller, "Electronmicroscopy of Genetic Material," Figure 6a, p. 379. Reproduced with permission, from *Annual Review of Biochemistry*, Volume 42. ©1973 by Annual Reviews, Inc. **17.9** Dr. Elena Kiseleva/Photo Researchers, Inc.

Chapter 18 Opener Stephanie Schuller/Photo Researchers, Inc. **18.UN1** EDVOTEK, The Biotechnology Education Company (www.edvotek.com)/Jeff Chirikjian

Chapter 19 19.2 Ada Olins/Don Fawcett/Photo Researchers, Inc. **19.4TB** ©2002 from *Molecular Biology of the Cell* 4/e Fig. 4-23 by Bruce Alberts et al. Reproduced by Permission of Garland Science Taylor & Francis Books, Inc., and Dr. Barbara Hamkalo.

Chapter 20 Opener International Rice Research Institute (IRRI) used under Creative Commons 2.0 license. http://creativecommons.org/licenses/by-nc-sa/2.0/deed.en **20.1** Brady-Handy Photograph Collection, Library of Congress. Reproduction number: LC-DIG-cwpbh-02977. **20.12** Science Source/Photo Researchers, Inc. **20.15BLR** Golden Rice

Chapter 21 Opener Martin Krzywinski/Science Photo Library/Alamy **21.8BLR** Test results provided by GENDIA (www.gendia.net) **21.11** Agilent Technologies, Inc. **21.12** Camilla M. Kao and Patrick O. Brown, Stanford University.

Chapter 22 Opener Anthony Bannister/Photo Researchers, Inc. **22.1a** From "The BMP antagonist Gremlin regulates outgrowth, chondrogenesis and programmed cell death in the developing limb." R. Merino, J. Rodriguez-Leon, D. Macias, Y. Ganan, A.N. Economides and J. M. Hurle. *Development* 126: 5515-5522 (1999), Fig. 6 E & F / Juan M. Hurle. **22.1b** Reproduced by permission of Elsevier Science from K. Kuida et al., "Reduced apoptosis and cytochrome c–mediated caspase activation in mice lacking caspase 9. *Cell* 94:325-337 (1998), Figs. 2E and 2F. ©1998 by Elsevier Science Ltd. Image courtesy of Keisuke Kuida, Vertex Pharmaceuticals. **22.2** John Chadwick/Agence France Presse/Newscom **22.3a** Richard Hutchings/Photo Researchers, Inc. **22.3b** From Drs. Adamson and Walls, "Embryonic and Neonatal Phenotyping of Gentically Engineered Mice." *ILAR Journal* 47(2), Fig. 12A, Optical projection tomographhy (OPT) image of a day 12.5 mouse embryo. (The rotation of the embryo and the art overlays do not appear in original paper.) **22.5** Christiane Nüsslein-Volhard, Max Planck Institute of Developmental Biology, Tubingen, Germany **22.6** Wolfgang Driever, University of Freiburg, Germany **22.7ac** Jim Langeland, Stephen Paddock, and Sean Carroll, University of Wisconsin at Madison **22.7b** Stephen J. Small, New York University **22.8L** Thomas Deerinck, NCMIR/Photo Researchers, Inc. **22.8M** David Scharf/Science Source/Photo Researchers, Inc. **22.8R** Eye of Science/Photo Researchers, Inc. **22.11ab** Reprinted from "The Origin of Form" by Sean B. Carroll, *Natural History*, November 2005. Burke, A. C. 2000. *Hox Genes and the Global Patterning of the Somitic Mesoderm*. In *Somitogenesis*. C. Ordahl (ed.) *Current Topics in Developmental Biology*, Vol. 47. Academic Press.

Chapter 23 Opener Dr. Yorgos Nikas/Photo Researchers, Inc. **23.2a** MIA Tegner/Science Photo Library/Photo Researchers, Inc. **23.3a(1–3)** Michael Whitaker/Science Photo Library/Photo Researchers, Inc. **23.3b(1–3)** Victor D. Vacquier, Scripps Institution of Oceanography, University of California at San Diego **23.6** David Shook, University of Virginia **23.8ab** Kathryn W. Tosney, University of Miami

Chapter 24 Opener Bill Stevenson/AGE Fotostock **24.5** Ed Reschke/Peter Arnold Images/Getty Images **24.7ab** Reprinted by permission from *Nature*, Vol. 424, July 24, 2003, p. 439, by Kim et al. ©2004 Macmillan Publishers Ltd. Photograph supplied by Dr. Neelima Sinah. **24.9a–d** John L. Bowman, University of California at Davis

Chapter 25 Opener R.B. Taylor/Photo Researchers, Inc. **25.2L** Gary Crabbe/AGE Fotostock **25.2RTMB** Michael Quinn, Grand Canyon NPS **25.3a** Gerd Weitbrecht for Landesbildungsserver Baden-Wurttemberg (www.schule-bw.de)/Gerd Weitbrecht **25.3b** angelhell/iStockphoto **25.5L** Januario Dolores **25.5R** BSIP/Photo Researchers, Inc. **25.6a1** AGE fotostock/SuperStock **25.6a2** George D. Lepp/Photo Researchers, Inc. **25.6a3** David Hosking/FLPA **25.6a4** Stefan Huwiler/AGE fotostock **25.8a** Photo by Michael K. Richardson. Reproduced by permission from *Anatomy and Embryology* 305, Fig. 7. Copyright © Springer-Verlag GmbH & Co KG, Heidelberg, Germany. **25.8b** Photo by Professor R. O'Rahilly. Reproduced by permission from *Anatomy and Embryology* 305, Fig.8. Copyright © Springer-Verlag GmbH & Co KG, Heidelberg, Germany. **25.8c** From: Richardson, M. K., et al. *Science*. Vol. 280, p. 983c, Issue #5366, May 15, 1998. Embryo from Professor R. O'Rahilly, National Museum of Health and Medicine/Armed Forces Institute of Pathology. **25.10** Walter J. Gehring **25.12LT** PetStockBoys/Alamy **25.12LB** Arco Images GmbH/Alamy **25.12RT** Arco Images GmbH/Alamy **25.12RB** Juniors Bildarchiv/Alamy **25.15** Alison Wright/ National Geographic Stock **25.15inset** D. & S. Tollerton/AGE fotostock **25.16** Kim Quillin **25.18a** American Association for the Advancement of Science **25.18b** The Royal Society of London **25.T2-1** dibrova/Shutterstock **25.T2-2** James H. Robinson/Photo Researchers, Inc. **25.T2-3** Erlend Haarberg/NHPA/Photoshot **25.T2-4** D. & S. Tollerton/AGE fotostock

Chapter 26 Opener REUTERS/Chaiwat Subprasom **26.8** Cyril Laubscher/Dorling Kindersley **26.9** Otorohanga Kiwi House, New Zealand **26.10** Robert Rattner **26.11a** Steve Mackay Photography/Flickr/Getty Images **26.11b** Phil Savoie/Nature Picture Library **26.13** Andre Karwath **26.16** Mark Conlin/V&W/Image Quest Marine **26.18** Alasdair Thomson/iStockphoto

Chapter 27 Opener Jason Rick/Loren H. Rieseberg, Indiana University **27.10B** Cubo Images/SuperStock **27.10TL** Michael Lustbader/Photo Researchers, Inc. **27.10TR** Stephen Lynn Clancy **27.11TMB** H. Douglas Pratt/National Geographic Image Collection

Chapter 28 Opener Indiana9 fossils and L & M Collectibles **28.5a** GEUS **28.5b** imv/iStockphoto **28.5c** Mike Viney/The Virtual Petrified Wood Museum **28.5d** Mike Viney/The Virtual Petrified Wood Museum **28.7a–f** Ron Blakey, Colorado Plateau Geosystems, Inc. **28.9TL** Dr. Gerald D. Carr, PhD **28.9BL** Forest & Kim Starr **28.9R** Forest & Kim Starr **28.10a** Eladio Fernandez **28.10b** Jonathan B. Losos, Washington University in St. Louis **28.11bBL** Verisimilus, Used under a Creative Commons license. http://creativecommons.org/licenses/by-sa/3.0/ **28.11bBR** James Gehling/South Australian Museum **28.11bTL** Alan Sirulnikoff/Science Source **28.11bTR** Ed Reschke/Getty **28.12** Jun-Yuan Chen **28.T4a** National Park Service **28.T4b** Glen A. Izett, USGS-NASA Langley **28.T4c** Peter Schulte/American Association for the Advancement of Science **28.T4d** NASA

Chapter 29 Opener John Hammond **29.2** Doc Searls **29.2inset** University of Maryland Biotechnology Institute **29.3** Rodney M. Donlan and Janice Carr, CDC **29.4** Accent Alaska/Alamy **29.5** Reprinted with permission from S.V. Liu, et al., "Thermophilic Fe(III)-reducing bacteria from the deep subsurface." *Science* 277: 1106–1109 (1997), Fig. 2. ©1997 AAAS. **29.8aT** Tom Deerinck **29.8aB** Reprinted with permission from H.N. Schulz, et al., "Dense populations of a giant sulfur bacterium in Namibian shelf sediments." *Science* 284: 493–495, Fig. 1b, August 4, 1999. ©1999 AAAS. **29.8bTB** SciMAT/Photo Researchers, Inc. **29.8cT** Linda M. Stannard, University of Cape Town/Photo Researchers, Inc. **29.8cB** Dr. Ralf Wagner **29.9a** R. Hubert, Microbiology Program, Dept. of Animal Science, Iowa State University **29.11** Bruce J. Russell/Biomedia Associates **29.15** Andrew Syred/Photo Researchers, Inc. **29.16** David Dalton **29.17** S. Amano, S. Miyadoh and T. Shomura/The Society for Actinomycetes Japan **29.18** Bill Schwartz, Public Health Image Library (PHIL), CDC **29.19** Jordan L. Cocchiaro, Yadunanda Kumar, Elizabeth R.

Fischer, Ted Hackstadt, and Raphael H. Valdivia, *PNAS*. **29.20a** Yves V. Brun **29.20b** Dr. George L. Barron **29.21** Martin Koenneke, MARUM, University of Bremen **29.22** Dr. Kenneth M. Stedman/Portland State University **29.22inset** Eye of Science/Photo Researchers, Inc. **29.23inset** Prof. Dr. Karl O. Stetter and Prof. Dr. Reinhard Rachel, University of Regensburg, Germany. **29.23** MARUM, University of Bremen

Chapter 30 Opener TIM LAMAN/National Geographic Stock **30.2L** Scenics & Science/Alamy **30.2M** Mirko Zanni/WaterFrame/AGE fotostock **30.2R** Francois Gohier/Photo Researchers, Inc. **30.4** Pete Atkinson/NHPA/Photoshot **30.4inset** Woods Hole Oceanographic Institution Biology **30.6** Biophoto Associates/Photo Researchers, Inc. **30.11a** Steve Gschmeissner/Photo Researchers, Inc. **30.11b** Biophoto Associates/Photo Researchers **30.11c** Andrew Syred/Photo Researchers, Inc. **30.12a** Biophoto Associates/Photo Researchers, Inc. **30.12b** Photoshot Holdings Ltd./Bruce Coleman **30.15** Eric V. Grave/Photo Researchers, Inc. **30.19** Jeffrey Lepore/Photo Researchers, Inc. **30.20** Eye of Science/Photo Researchers, Inc. **30.21** Tai-Soon Young/Chungbuk National University College of Medicine **30.22** David J. Patterson **30.23** Francis Abbott/Nature Picture Library **30.24** @Colomban de Vargas, SBRoscoff **30.25** David J. Patterson **30.26** Inra **30.27** Vern B. Carruthers, Dept. of Microbiology & Immunology, U. of Michigan Medical School **30.28** Andrew Syred/SPL/Photo Researchers, Inc. **30.29** Linden Gledhill Photography **30.30** Kelvin Aitken/V&W/Image Quest Marine

Chapter 31 Opener RF Company/Alamy **31.1** USDA/NRCS **31.2** Hugh Iltis, The Doebley Lab, University of Wisconsin-Madison **31.3L** Linda Graham, University of Wisconsin-Madison **31.3R** Lee W. Wilcox **31.4a** Lee W. Wilcox **31.4b** sacco/iStockphoto **31.4c** Zinatova Olga/Shutterstock **31.7** Lee W. Wilcox **31.10b** BlueRidgeKitties, used under a creative commons license: http://creativecommons.org/licenses/by-nc-sa/2.0/deed.en **31.10LR** Jon Monroe, James Madison University **31.15a** Lee W. Wilcox **31.15bLR** Lee W. Wilcox **31.17** Patrick J. Lynch/Photo Researchers, Inc. **31.20a** Larry Deack **31.20b** Quest786/Shutterstock **31.20c** John Severns, Severnjc **31.22a** Maxim Khytra/Shutterstock **31.22b** Andreas Nilsson/Shutterstock **31.24T.1&2** Dr. Keith Wheeler/Photo Researchers, Inc. **31.24T.3** jonphoto/iStockphoto **31.24T.4** Noodle Snacks **31.24B.1** Garry DeLong/Photo Researchers, Inc. **31.24B.2** Dr. Keith Wheeler/Photo Researchers, Inc. **31.24B.3** ATELIER Simon/iStockphoto **31.24B.4** Shutterstock **31.26a** Doug Allan/Nature Picture Library **31.26b** Roland Birke/Peter Arnold Images/Getty **31.26c** Lee W. Wilcox **31.27** Andrew J. Martinez/Photo Researchers, Inc. **31.28** Linda Graham **31.29** Center for Aquatic and Invasive Plants **31.30** Adrian Davies/Nature Picture Library **31.31a** Wildlife GmbH/Alamy **31.31b** Photononstop/SuperStock **31.32** Lee W. Wilcox **31.33** Lee W. Wilcox **31.34** Gordon Dickson **31.35a** Greg Vaughn/Alamy **31.35b** Lee W. Wilcox **31.36L** F Hecker/AGE fotostock **31.36R** SPL/Photo Researchers, Inc. **31.37a** Martin Land/Science Photo Library/Photo Researchers, Inc. **31.37b** brigg/iStockphoto **31.38** Maxfocus/iStockphoto **31.39** Peter Chadwick/Dorling Kindersley **31.40ab** Lee W. Wilcox **31.41** Geoff Renner/Robert Harding **31.42a** Brian Johnston **31.42b1&2** Lee W. Wilcox

Chapter 32 Opener Michael P. Gadomski/Photo Researchers, Inc. **32.1a** Inga Spence/Photo Researchers, Inc. **32.1b** EGON/Alamy **32.2a** David Read, Department of Animal and Plant Sciences, University of Sheffield **32.2b** Mycorrhizal Applications, Inc. **32.3** Paul W. Romaine **32.4a** Eye of Science/Photo Researchers, Inc. **32.4b** Emmanuel LATTES/Alamy **32.5aR** Susumu Nishinaga/Photo Researchers, Inc. **32.5bL** Dr. George L. Barron **32.5bR** Biophoto Associates/Photo Researchers, Inc. **32.5c** Dr. Patrick C. Hickey, NIPHT Limited & Prof. Nick D. Read, University of Edinburgh **32.6a** Melvin Fuller/The Mycological Society of America **32.6b** Ed Reschke/ Peter Arnold/Getty **32.6c** Biophoto Associates/Photo Researchers, Inc. **32.6d** Nino Santamaria **32.9ab** Mark Brundrett **32.11** Scott Camazine/Photo Researchers, Inc. **32.14** Courtesy of Dr. J.I. Ronny Larsson, Professor of Zoology, and *The European Journal of Protistology*, "The Continuation of Protistologica." 17, 1981, 407–422. **32.15** Thomas J. Volk **32.16** Gregory G. Dimijian/Photo Researchers, Inc. **32.17** Dr. Jim Deacon **32.18** David Slater/DanitaDelimont.com "Danita Delimont Photography"/Newscom **32.19b** Daniel Mosquin **32.20** N. Allen and G. Barron, University of Guelph, Guelph, Ontario **32.21** Adrian Hepworth/NHPA/Photoshot

Chapter 33 Opener Bill Curtsinger/National Geographic Image Collection **33.4** Walker England/Photo Researchers, Inc. **33.5a** Steve Gschmeissner/Photo Researchers, Inc. **33.5b** Susumu Nishinaga/Photo Researchers, Inc. **33.6** www.deepseawaters.com **33.7** From "Origins of bilateral symmetry: *Hox* and *dpp* expression in a sea anemone." JR Finnerty, et al. *Science*. 2004 May 28; 304(5675): 1335–1337. Epub 2004 May 6. Fig. 2i. **33.T2-1** Perennou Nuridsany/Photo Researchers, Inc. **33.T2-2** Barry Mansell/Nature Picture Library **33.T2-3** Pascal Goetgheluck/Photo Researchers, Inc. **33.T2-4** Joze Maucec/Shutterstock **33.T3-1** Juniors Bildarchiv/Alamy **33.T3-2** Olga Khoroshunova/Fotolia **33.T3-3** Len Rue, Jr./Photo Researchers, Inc. **33.T4-1** Sanamyan/Alamy **33.T4-2** Todd Husman **33.T4-3** Heinz-jürgen Landshoe/AGE Fotostock **33.T4-4** Four Oaks/Shutterstock **33.T5-1** Francesco Tomasinelli/Photo Researchers, Inc. **33.T5-2** Michael Zysman/Shutterstock **33.T5-3** D P Wilson/Frank Lane Picture Agency **33.T5-4** Lawson Wood/Specialist Stock **33.T5-5** Jeff Rotman/Photo Researchers, Inc.

46.23 Genevieve Anderson **46.24b** Reproduced by permission of Elsevier Science from K.C. Martin, et al. "Synapse-specific, long-term facilitation of *Aplysia* sensory to motor synapses: A function for local protein synthesis in memory storage." *Cell* 91: 927–938 (1997), Fig. 2a. **46.25a** Nigel J. Dennis/Photo Researchers, Inc.

Chapter 47 Opener Neil Hardwick/Alamy Images **47.3** Carole Hackney **47.7** Stephen Dalton/Nature Picture Library **47.8** Piet Spaans Viridiflavus/Used under the Creative Commons licence. http://creativecommons.org/licenses/by-sa/2.5/deed.en **47.9** Kirsten Pohlmann, Leibniz-Institut für Gewässerökologie und Binnenfischerei, and Thomas Breithaupt, University of Hull **47.10** Andrew Syred/Photo Researchers, Inc. **47.12** Dr. Don Fawcett/T. Kwwabara/Photo Researchers, Inc. **47.1** Dmitry Smirnov **47.14LR** David Quillin **47.18a** Corbis/Alamy Images **47.18b** Ted Kinsman/Photo Researchers, Inc. **47.19** Dr. Steven Campana, Bedford Institute of Oceanography, Canada ©2003 Canadian Shark Research Lab. **47.20** Christopher Zarnick/Animal Graphics

Chapter 48 Opener Bence Mate/Nature Picture Library **48.1** Don W. Fawcett/Photo Researchers, Inc. **48.T1L** Biophoto Associates/Photo Researchers, Inc. **48.T1M** Manfred Kage/Peter Arnold/Photolibrary/Getty Images **48.T1R** Nina Zanetti/Pearson Education **48.T2** National Library of Medicine **48.12TB** Eadweard Muybridge Collection/Kingston Museum/Photo Researchers, Inc. **48.15** Anders Hedenström, Lund University, Sweden

Chapter 49 Opener Anthony Mercieca/Photo Researchers, Inc. **49.5L** Stephen Dalton/Photo Researchers, Inc. **49.5M** Duncan McEwan/Nature Picture Library **49.5R** Juniors/SuperStock **49.6L** Joël Héras/Biosphoto/Photo Researchers, Inc. **49.6M** Andrew Darrington/Alamy **49.6R** Perennou Nuridsany/Photo Researchers, Inc. **49.8** Kenneth M. Highfill/Photo Researchers, Inc. **49.10** The Jackson Laboratory, Bar Harbor, Maine

Chapter 50 Opener Jurgen & Christine Sohns/FLPA **50.1a** LUTRA/NHPA/Photoshot **50.1b** Andrew J. Martinez/Photo Researchers, Inc. **50.1c** Charles J. Cole **50.2** Bruce J. Russell/Biomedia Associates **50.5** Privatdozent Dr. med. habil. Holger Jastrow **50.7TB** Reprinted by permission from Catherine S.C. Price et al., "Sperm competition between *Drosophila* males involves both development and incapacitation." *Nature* 400: 449–452 (1999), Figs. 2 and 3. ©1999 Macmillan Magazines Ltd. Image courtesy of Jerry A. Coyne, University of Chicago. **50.8** Premaphotos/Alamy **50.10a** Johanna Liljestrand Rönn **50.10inset** Fleur Champion de Crespigny **50.19ab** John Cancalosi/Nature Picture Library **50.19c** John Cancalosi/Alamy **50.20a** Claude Edelmann/Petit Format/Photo Researchers, Inc. **50.20b** Lennart Nilsson/Scanpix Sweden AB **50.20c** Petit Format/Nestle/Science Source/Photo Researchers, Inc.

Chapter 51 Opener Eye of Science/Photo Researchers, Inc. **51.1** Susumu Nishinaga/Photo Researchers, Inc. **51.2** From "The Dorsoventral Regulatory Gene Cassette *spatzle/Toll/cactus* Controls the Potent Antifungal Response in *Drosophila* Adults." Lemaitre, B.; Nicolas, E.; Michaut, L., Reichhart, J-M.; Hoffmann, J. A. *Cell* 86(6): pp.973–983. **51.13a** Steve Gschmeissner/Photo Researchers, Inc. **51.13b** Steve Gschmeissner/SPL/Photo Researchers, Inc.

Chapter 52 Opener Associated Press/Gerald Herbert **52.T1-1** Paul Nicklen/National Geographic/Getty Images **52.T1-2** Sergey Gorshkov/Nature Picture Library **52.T1-3** Juniors Bildarchiv GmbH/Alamy Images **52.T1-4** Lynn M. Stone/Nature Picture Library **52.T1-5** Russ Heinl/All Canada Photos/SuperStock **52.2T** Purestock/Getty Images **52.2B** Comstock/Getty Images **52.3** Bruce Coleman/Alamy Images **52.UN1** stillfx/AGE fotostock **52.UN2** Anton Foltin/Shutterstock **52.UN3** David Hartnett **52.UN4** Inga Nielsen/Shutterstock **52.UN5** Bill Brooks/Alamy **52.UN6** Pixtal Images/Photolibrary **52.17** Fairfax Photos/Sandy Scheltema/Courtesy of the AGE/The AGE Photo Sales **52.18b** Gerald Nowak/WaterFrame/AGE fotostock **52.UN7** imagebroker.net/SuperStock **52.UN8** Yuriy Kulyk/Shutterstock **52.UN9** John Warburton Lee/SuperStock

Chapter 53 Opener Tim Laman/National Geographic/Getty Images **53.1** Alex Wild Photography **53.5a** walkingmoon/iStockphoto **53.11a** Suzanne L. & Joseph T. Collins/Photo Researchers, Inc. **53.11b** Robert L. Dunne/Photoshot **53.12** Dr. Bryan D. Neff **53.13** Tom McHugh/Photo Researchers, Inc.

Chapter 54 Opener Reinhard Dirscherl/AGE fotostock **54.1inset** Hecker/Sauer/AGE fotostock **54.7** USDA/NRCS/Natural Resources Conservation Service **54.8a** JC Schou/Biopix.dk (www.Biopix.dk) **54.12R** Janine Wiedel Photolibrary/Alamy **54.15** Jeff Rotman/Oxford Scientific/Getty Images **54.16** Robert Thompson/Nature Picture Library

Chapter 55 Opener Scott Tuason/Image Quest Marine **55.1a** PREMAPHOTOS/Nature Picture Library **55.1b** David Tipling/Alamy Images **55.8** Henry Walter Bates 1862. "Contributions to an insect fauna of the Amazon Valley." *Lepidoptera: Heliconidae. Trans. Linn. Soc.* 23: 495–566. **55.T1-1** Francesco Tomasinelli/Photo Researchers, Inc. **55.T1-2** Stephen Dalton/Photo Researchers, Inc. **55.T1-3** Dirk Ercken/Shutterstock **55.T1-4** DJ Mattaar/Shutterstock **55.T1-5** NatureDiver/Shutterstock **55.12a–c** Stephen P. Yanoviak **55.13a** Raul Gonzalez Perez/Photo Researchers, Inc. **55.13b** Reinhard Dirscherl/Alamy Images **55.16a** Image Quest Marine **55.16b** Richard Wong/Alamy **55.17** Neale Clark/Robert Harding World Imagery/Alamy Images **55.18** Tony C. Caprio **55.19(1)** Elenamiv/Shutterstock **55.19(2–4)** Michael P. Gadomski/Photo Researchers, Inc. **55.20L** Glacier Bay National Park Photo **55.20M** Christopher L. Fastie, Middlebury College **55.20R** Glacier Bay National Park Photo

Chapter 56 Opener Eric Nathan/AGE fotostock **56.12a** Richard Hartnup **56.12b** Randall J. Schaetzl **56.13** John Campbell, USDA Forest Service, Northern Research **56.T1-1** avorym/Fotolia **56.T1-2** Environmental Stock Images by Streeter Photography/Alamy Images **56.T1-3** Robert Pickett/Papilio/Alamy **56.T1-4** Arterra Picture Library/Alamy Images **56.T1-5** Mark Conlin/Image Quest Marine

Chapter 57 Opener Art Wolfe/Newscom **57.1** Sergey Uryadnikov/AGE fotostock **57.2** DLILLC/Corbis **57.3** Craig McClain **57.8TB** Planet Observer/Science Source/Photo Researchers, Inc. **57.T1-1** David Woodfall/Photo Researchers, Inc. **57.T1-2** Jeff Rotman/Photo Researchers, Inc. **57.T1-3** Theo Allofs/Getty Images **57.T1-4** Simon Booth/Photo Researchers, Inc. **57.T1-5** Peter Scoones/Photo Researchers, Inc. **57.13** From D. Tilman, 2005. "Causes, consequences and ethics of biodiversity." *Nature* 405: 208–211, Fig. 1. ©2000 Macmillan Magazines Ltd. **57.T2-1** Fedor Kondratenko/Shutterstock **57.T2-2** Brian Brake/Photo Researchers, Inc. **57.T2-3** Specta/Shutterstock **57.T2-4** koh sze kiat/Shutterstock **57.14** TPG/Associated Press **57.15T** Mark Conlin/Alamy **57.16T** Ellen Damschen **57.17** STERN/Black Star/Newscom **57.19** Brian Bailey/Taxi/Getty Images

Appendix A: Answers A7.1ab Don W. Fawcett/Photo Researchers, Inc. **A7.1c** Biophoto Associates/Photo Researchers, Inc. **A7.1d** From "Caveolin-1 expression is essential for proper nonshivering thermogenesis in brown adipose tissue." Cohen, A. W., Schubert, W., Brasaemle, D. L., Scherer, P. E., Lisanti, M. P. *Diabetes.* 2005 Mar; 54(3): 679–86, Fig. 6.

Appendix B: BioSkills B9.3 Reproduced by permission from J. P. Ferris, et al. "Synthesis of long prebiotic oligomers on mineral surfaces." *Nature* 381: 59–61 (1996), Fig. 2. ©1996 Macmillan Magazines Ltd. Image courtesy of James P. Ferris, Rensselaer Polytechnic Institute. **B11.1a** Biology Media/Photo Researchers, Inc. **B11.1b** Janice Carr/Centers for Disease Control and Prevention (CDC) **B11.2ab** Michael W. Davidson/Molecular Expressions **B11.3** Rosalind Franklin/Photo Researchers, Inc. **B12.1aL** National Cancer Institute **B12.1aR** E.S. Anderson/Photo Researchers, Inc. **B12.1b** Sinclair Stammers/Photo Researchers, Inc. **B13.1a** Kwangshin Kim/Photo Researchers, Inc. **B13.1b** Richard L. (Larry) Blanton, Ph.D. **B13.1c** Holt Studios International/Photo Researchers, Inc. **B13.1d** Custom Medical Stock Photo, Inc. **B13.1e** Graphic Science/Alamy Images **B13.1f** Sinclair Stammers/Photo Researchers, Inc. **B13.1g** dra_schwartz/iStockphoto

Illustration and Text Credits

Chapter 1 1.3 Based on S. P. Moose, J. W. Dudley, and T. R. Rocheford. 2004. Maize selection passes the century mark: A unique resource for 21st century genomics. *Trends in Plant Science* 9 (7): 358–364, Fig. 1a; and the Illinois long-term selection experiment for oil and protein in corn (University of Illinois at Urbana–Champaign). **1.7a** Based on T. P. Young and L. A. Isbell. 1991. Sex differences in giraffe feeding ecology: Energetic and social constraints. *Ethology* 87: 79–80, Figs. 5a, 6a. **1.9** Adapted by permission of AAAS and the author from M. Wittlinger, R. Wehner, and H. Wolf. 2006. The ant odometer: Stepping on stilts and stumps. *Science* 312: 1965–1967, Figs. 1, 2, 3. (http://www.sciencemag.org/content/312/5782/1965.short).

Chapter 2 T2-1 Data: D. R. Lide (editor). 2008. Standard thermodynamic properties of chemical substances. In *CRC Handbook of Physics and Chemistry.* 89th ed. Boca Raton, FL: CRC Press. **2.21** After S. L. Miller. 1953. A production of amino acids under possible primitive Earth conditions. *Science* 117 (3046): 528–529.

Chapter 3 Opener PDB ID: 2DN2. S. Y. Park, T. Yokoyama, N. Shibayama, et al. 2006. 1.25 Å resolution crystal structures of human hemoglobin in the oxy, deoxy and carbonmonoxy forms. *Journal of Molecular Biology* 360: 690–701. **3.8a** PDB ID: 1TGH. Z. S. Juo, T. K. Chiu, P. M. Leiberman, et al. 1996. How proteins recognize the TATA box. *Journal of Molecular Biology* 261: 239–254. **3.8b** PDB ID: 2X9K. F. Korkmaz-Ozkan, S. Koster, W. Kuhlbrandt, et al. 2010. Correlation between the OmpG secondary structure and its pH-dependent alterations monitored by FTIR. *Journal of Molecular Biology* 401: 56–67. **3.8c** PDB ID: 2PTC. M. Marquart, J. Walter, J. Deisenhofer, et al. 1983. The geometry of the reactive site and of the peptide groups in trypsin, trypsinogen and its complexes with inhibitors. *Acta Crystallographica* Section B39: 480–490. **3.8d** PDB ID: 1CLG. J. M. Chen, C. E. King, S. H. Feairheller, et al. 1991. An energetic evaluation of a "Smith" collagen microfibril model. *Journal of Protein Chemistry* 10: 535–552. **3.11b, left** PDB ID: 2MHR. S. Sheriff, W. A. Hendrickson, and J. L. Smith. 1987. Structure of myohemerythrin in the azidomet state at 1.7/1.3 resolution. *Journal of Molecular Biology* 197: 273–296. **3.11b, middle** PDB ID: 1FTP. N. H. Haunerland, B. L. Jacobson, G. Wesenberg, et al. 1994. Three-dimensional structure of the muscle fatty-acid-binding protein isolated from the desert locust *Schistocerca gregaria. Biochemistry* 33: 12378–12385. **3.11b, right** PDB ID: 1IXA. M. Baron, D. G. Norman, T. S. Harvey, et al. 1992.

The three-dimensional structure of the first EGF-like module of human factor IX: Comparison with EGF and TGF-alpha. *Protein Science* 1: 81–90. **3.12a** PDB ID: 1D1L. P. B. Rupert, A. K. Mollah, M. C. Mossing, et al. 2000. The structural basis for enhanced stability and reduced DNA binding seen in engineered second-generation Cro monomers and dimers. *Journal of Molecular Biology* 296: 1079–1090. **3.12b** PDB ID: 2DN2. S.Y. Park, T. Yokoyama, N. Shibayama, et al. 2006. 1.25 Å resolution crystal structures of human hemoglobin in the oxy, deoxy and carbonmonoxy forms. *Journal of Molecular Biology* 360: 690–701. **3.14a** PDB ID: 1DMO. M. Zhang, T. Tanaka, and M. Ikura. 1995. Calcium-induced conformational transition revealed by the solution structure of apo calmodulin. *Nature Structural Biology* 2: 758–767. **3.14b** PDB ID: 3CLN. Y. S. Babu, C. E. Bugg, and W. J. Cook. 1988. Structure of calmodulin refined at 2.2 Å resolution. *Journal of Molecular Biology* 204: 191–204. **3.16** PDB ID: 2PTC. M. Marquart, J. Walter, J. Deisenhofer, et al. 1983. The geometry of the reactive site and of the peptide groups in trypsin, trypsinogen and its complexes with inhibitors. *Acta Crystallographica* Section B39: 480–490.

Chapter 4 **T4-1** PDB ID: 1EHZ. H. Shi and P. B. Moore. 2000. The crystal structure of yeast phenylalanine tRNA at 1.93 Å resolution: A classic structure revisited. *RNA* 6: 1091–1105. **4.11** PDB ID: 1X8W. F. Guo, A. R. Gooding, and T. R. Cech. 2004. Structure of the *Tetrahymena* ribozyme: Base triple sandwich and metal ion at the active site. *Molecular Cell* 16: 351–362.

Chapter 5 **5.6** H. M. Florman, K. B. Bechtol, and P. M. Wassarman. 1984. Enzymatic dissection of the functions of the mouse egg's receptor for sperm. *Developmental Biology* 106: 243–255. Also H. M. Florman and P. M. Wassarman. 1985. O-linked oligosaccharides of mouse egg ZP3 account for its sperm receptor activity. *Cell* 41: 313–324; J. D. Bleil and P. M. Wassarman. 1988. Galactose at the nonreducing terminus of O-linked oligosaccharides of mouse egg zona pellucida glycoprotein ZP3 is essential for the glycoprotein's sperm receptor activity. *PNAS* 85: 6778–6782.

Chapter 6 **Opener** CHARMM-GUI Archive—Library of Pure Lipid Bilayer (www.charmm-gui.org/?doc=archive&lib=lipid_pure), POPE Bilayer Library (pope_n256.pdb). Reference: S. Jo, T. Kim, and W. Im. 2007. Automated builder and database of protein/membrane complexes for molecular dynamics simulations. *PLoS ONE* 2 (9): e880. **6.10** Data: J. de Gier, J. G. Mandersloot, and L. L. van Deenen. 1968. Lipid composition and permeability of liposomes. *Biochimica et Biophysica Acta* 150: 666–675. **6.21** Data: C. E. Bear, F. Duguay, A. L. Naismith, et al. 1992. Purification and functional reconstitution of the cystic fibrosis transmembrane conductance regulator (CFTR). *Cell* 68: 809–818. **6.22** PDB ID: 2ZZ9. K. Tani, T. Mitsuma, Y. Hiroaki, et al. 2009. Mechanism of aquaporin-4's fast and highly selective water conduction and proton exclusion. *Journal of Molecular Biology* 389: 694–706. **6.23** PDB ID: 1K4C. Y. Zhou, J. H. Morais-Cabral, A. Kaufman, et al. 2001. Chemistry of ion coordination and hydration revealed by a K⁺ channel–Fab complex at 2.0 Å resolution. *Nature* 414: 43–48. **6.23** PDB ID: 3FB7. L. G. Cuello, V. Jogini, D. M. Cortes, et al. Open KcsA potassium channel in the presence of Rb⁺ ion. (To be published.) **7.18** A. D. Mills, R. A. Laskey, P. Black, et al. 1980. An acidic protein which assembles nucleosomes in vitro is the most abundant protein in *Xenopus* oocyte nuclei. *Journal of Molecular Biology* 139: 561–568. Also C. Dingwall, S. V. Sharnick, and R. A. Laskey. 1982. A polypeptide domain that specifies migration of nucleoplasmin into the nucleus. *Cell* 30: 449–458.

Chapter 7 **7.5** Reprinted with kind permission from Springer Science+Business Media B.V. from David S. Goodsell, *The Machinery of Life*. 2nd ed., 2009.

Chapter 8 **8.8** PDB ID: 1Q18. V. V. Lunin, Y. Li, J. D. Schrag, et al. 2004. Crystal structures of *Escherichia coli* ATP-dependent glucokinase and its complex with glucose. *Journal of Bacteriology* 186: 6915–6927. **8.8** PDB ID: 2Q2R. A. T. Cordeiro, A. J. Caceres, D. Vertommen, et al. 2007. The crystal structure of *Trypanosoma cruzi* glucokinase reveals features determining oligomerization and anomer specificity of hexose-phosphorylating enzymes. *Journal of Molecular Biology* 372: 1215–1226. **8.14a&b** Data: N. N. Nawani, B. P. Kapadnis, A. D. Das, et al. 2002. Purification and characterization of a thermophilic and acidophilic chitinase from Microbispora sp. V2. *Journal of Applied Microbiology* 93: 865–975, Figs. 7, 8a. Also N. N. Nawani and B. P. Kapadnis. 2001. One-step purification of chitinase from *Serratia marcescens* NK1, a soil isolate. *Journal of Applied Microbiology* 90: 803–808, Figs. 3, 4. **8.15** PDB ID: 2ERK. B. J. Canagarajah, A. Khokhlatchev, M. H. Cobb, et al. 1997. Activation mechanism of the MAP kinase ERK2 by dual phosphorylation. *Cell* (Cambridge, MA) 90: 859–869. **8.15** PDB ID 3ERK. Z. Wang, B. J. Canagarajah, J. C. Boehm, et al. 1998. Structural basis of inhibitor selectivity in MAP kinases. *Structure* 6: 1117–1128.

Chapter 9 **9.7** PDB ID: 4PFK. P. R. Evans and P. J. Hudson. 1981. Phosphofructokinase: Structure and control. *Philosophical Transactions R. Soc. Lond. B: Biol. Sci.* 293: 53–62. **9.13** Data: X. Li, R. K. Dash, R. K. Pradhan, et al. 2010. A database of thermodynamic quantities for the reactions of glycolysis and the tricarboxylic acid cycle. *Journal of Physical Chemistry B* 114: 16068–16082, Table 4. **9.14** Data: D.F. Wilson, M. Erecinska, and P. L. Dutton. 1974. Thermodynamic relationships in mitochondrial oxidative phosphorylation. *Annual Review of Biophysics and Bioengineering* 3: 203–230, Tables 1, 3. Also V. D. Sled, N. I. Rudnitzky, Y. Hatefi, et al. 1994. Thermodynamic analysis of flavin in mitochondrial NADH: Ubiquinone oxidoreductase (complex I). *Biochemistry* 33: 10069–10075.

9.17 E. Racker and W. Stoeckenius. 1974. Reconstitution of purple membrane vesicles catalyzing light-driven proton uptake and adenosine triphosphate formation. *Journal of Biological Chemistry* 249: 662–663.

Chapter 10 **10.6** T. W. Engelmann. 1882. Oxygen excretion from plant cells in a microspectrum. *Botanische Zeitung* 40: 419–426. **10.7** Reprinted with kind permission from Springer Science+Business Media B.V. from The Photosynthetic Process. In *Concepts in Photobiology: Photosynthesis and Photomorphogenesis*, edited by G. S. Singhal, G. Renger, S. K. Sopory et al. Dordrecht: Kluwer Academic; co-published with Narosa Publishing House (New Delhi), pp. 11–51, Fig. 5. **10.11** R. Govindjee, Govindjee, and G. Hoch. 1964. Emerson enhancement effect in chloroplast reactions. *Plant Physiology* 39: 10–14. **10.18** A. A. Benson, J. A. Bassham, M. Calvin, et al. 1950. The path of carbon in photosynthesis. V. Paper chromatography and radioautography of the products. *Journal of the American Chemistry Society* 72: 1710–1718. **10.20a** PDB ID: 1RCX. T. C. Taylor and I. Andersson. 1997. The structure of the complex between rubisco and its natural substrate ribulose 1,5-bisphosphate. *Journal of Molecular Biology* 265: 432–444.

Chapter 11 **11.4** Reprinted with kind permission from Springer Science+Business Media B.V. from David S. Goodsell, *The Machinery of Life*. 2nd ed., 2009. **11.7b** B. Alberts, A. Johnson, J. Lewis, et al. 2002. *Molecular Biology of the Cell*. 4th ed., Fig. 19.5, p. 1069. **11.9** K. Hatta and M. Takeichi. 1986. Expression of N-cadherin adhesion molecules associated with early morphogenetic events in chick development. *Nature* 320: 447–449. Also M. Takeichi. 1988. The cadherins: Cell–cell adhesion molecules controlling animal morphogenesis. *Development* 102: 639–655.

Chapter 12 **12.6** G. J. Gorbsky, P. J. Sammack, and G. G. Borisey. 1987. Chromosomes move poleward during anaphase along stationary microtubules that coordinately dissemble from their kinetochore ends. *Journal of Cell Biology* 104: 9–18. **12.9** Based on J. L. Ptacin, S. F. Lee, E. C. Garner, et al. 2010. A spindle-like apparatus guides bacterial chromosome segregation. *Nature Cell Biology* 12: 791–798, Fig. 5. **12.10** Y. Masui and C. L. Markert. 1971. Cytoplasmic control of nuclear behavior during meiotic maturation of frog oocytes. *Journal of Experimental Zoology* 177: 129–145. **12.13** Data: the website of the National Cancer Institute (www.cancer.gov), Common Cancer Types, November 2010.

Chapter 13 **13.13** Data: National Down Syndrome Society, 2012. **13.15** L. T. Morran, O. G. Schmidt, I. A. Gelarden, et al. 2011. Running with the red queen: Host-parasite coevolution selects for biparental sex. *Science* 333: 216–218.

Chapter 14 **14.3** G. Mendel. 1866. Versuche über Pflanzen-hybriden. *Verhandlungen des naturforschenden Vereines in Brünn* 4: 3–47. English translation available from Electronic Scholarly Publishing (www.esp.org). **T14.2** Data: G. Mendel. 1866. Versuche über Pflanzen-hybriden. *Verhandlungen des naturforschenden Vereines in Brünn* 4: 3–47. **14.13** Data: T. H. Morgan. 1911. An attempt to analyze the constitution of the chromosomes on the basis of sex-limited inheritance in *Drosophila*. *Journal of Experimental Zoology* 11: 365–414. **14.20** Data: Pan American Health Organization/WHO. 2004. *Epidemiological Bulletin* 25: 9–13, Graph 1.

Chapter 15 **15.2** A. D. Hershey and M. Chase. 1952. Independent functions of viral protein and nucleic acid in growth of bacteriophage. *Journal of General Physiology* 36: 39–56. **15.5** Adapted by permission of Dr. Matthew Meselson after M. Meselson and F. W. Stahl. 1958. The replication of DNA in *Escherichia coli*. *PNAS* 44: 671–682, Fig. 6. **15.14** Data: R. C. Allsopp, H. Vaziri, C. Patterson, et al. 1992. Telomere length predicts replicative capacity of human fibroblasts. *PNAS* 89: 10114–10118. **15.18a** Data: J. E. Cleaver. 1970. DNA repair and radiation sensitivity in human (xeroderma pigmentosum) cells. *International Journal of Radiation Biology* 18: 557–565, Fig. 3. **15.18b** Data: J .E. Cleaver. 1972. Xeroderma pigmentosum: Variants with normal DNA repair and normal sensitivity to ultraviolet light. *Journal of Investigative Dermatology* 58: 124–128, Fig. 1. **15.UN2** Graph adapted by permission of the Radiation Research Society from P. Howard-Flanders and R. P. Boyce. 1966. DNA repair and genetic recombination: Studies on mutants of *Escherichia coli* defective in these processes. *Radiation Research Supplement* 6: 156–184, Fig. 8.

Chapter 16 **16.2** Data: A. M. Srb and N. H. Horowitz. 1944. The ornithine cycle in *Neurospora* and its genetic control. *Journal of Biological Chemistry* 154: 129–139.

Chapter 17 **17.2a** PDB ID: 3IYD. B. P. Hudson, J. Quispe, S. Lara-Gonzalez, et al. 2009. Three-dimensional EM structure of an intact activator-dependent transcription initiation complex. *PNAS* 106: 19830–19835. **17.11** Based on M. B. Hoagland, M. L. Stephenson, J. F. Scott, et al. 1958. A soluble ribonucleic acid intermediate in protein synthesis. *Journal of Biological Chemistry* 231: 241–257, Fig. 6. **17.13** PDB ID: 1ZJW: I. Gruic-Sovulj, N. Uter, T. Bullock, et al. 2005. tRNA-dependent aminoacyl-adenylate hydrolysis by a nonediting class I aminoacyl-tRNA synthetase. *Journal of Biological Chemistry* 280: 23978–23986. **17.14b** PDB IDs: 3FIK, 3FIH. E. Villa, J. Sengupta, L. G. Trabuco, et al. 2009. Ribosome-induced changes in elongation factor Tu conformation control GTP hydrolysis. *PNAS* 106: 1063–1068.

Chapter 18 **18.2** Data: A. B. Pardee, F. Jacob, and J. Monod. 1959. The genetic control and cytoplasmic expression of "inducibility" in the synthesis of ß-galactosidase by *E. coli*. *Journal of Molecular Biology* 1: 165–178.

Chapter 19 **Opener** PDB ID: 1ZBB. T. Schalch, S. Duda, D. F. Sargent, et al. 2005. X-ray structure of a tetranucleosome and its implications for the chromatin fibre. *Nature* 436: 138–141. **19.5** I. Sandovici, N. H. Smith, and M. D. Nitert. 2011. Maternal diet and aging alter the epigenetic control of a promoter–enhancer interaction at the *Hnf4a* gene in rat pancreatic islets. *PNAS* 108: 5449–5454. **19.7b** PDB ID: 1MDY. P. C. Ma, M. A. Rould, H. Weintraub, et al. 1994. Crystal structure of MyoD bHLH domain–DNA complex: Perspectives on DNA recognition and implications for transcriptional activation. *Cell* (Cambridge, MA) 77: 451–459.

Chapter 21 **21.1** Data: European Nucleotide Archive/EMBL–Bank Release Notes. Release 110, December 2011 (www.ebi.ac.uk/embl/). **21.4a&b** Data: (a) Y. Hou and S. Lin. 2009. *PLoS ONE* 4 (9): e6978, Supplemental Table S1. (b) KEGG: *Kyoto Encyclopedia of Genes and Genomes*, KEGG Organisms: Complete genomes (www.genome.jp/kegg/). **21.8** Reproduced by permission of GENDIA from www.paternity.be/information_EN.html#identitytest, Examples 1 and 2. **21.10** Based on G. T. Ryan. 2005. Synergy between sequence and size in large-scale genomics. *Nature Reviews Genetics* 6, Box 3, p. 702. **21.13** V. Pancaldi, O. S. Sarac, C. Rallis, et al. 2012. Predicting the fission yeast protein interaction network. *G3: Genes, Genomes, Genetics* 2 (4): 453–567, Fig. 5.

Chapter 22 **22.10a&b** Reproduced by permission from Macmillan Publishers Ltd after J. C. Pearson, D. Lemons, and W. McGinnis. 2005. Modulating *Hox* gene functions during animal body patterning. *Nature Reviews Genetics* 6: 893–904, Fig. 1.

Chapter 23 **23.12** H. Weintraub, S. J. Tapscott, R. L. Davis, et al. 1989. Activation of muscle-specific genes in pigment, nerve, fat, liver, and fibroblast cell lines by forced expression of MyoD. *PNAS* 86: 5434–5438.

Chapter 25 **25.4** Based on E. B. Daeschler et al. 2006. A Devonian tetrapod-like fish and the evolution of the tetrapod body plan. *Nature* 440: 757–763, Fig. 6; P. E. Ahlberg and J. A. Clack. 2006. A firm step from water to land. *Nature* 440: 747–749, Fig. 1; N. H. Shubin et al. 2006. The pectoral fin of *Tiktaalik roseae* and the origin of the tetrapod limb. *Nature* 440: 764–771, Fig. 4; M. Hildebrand and G. Goslow. 2001. *Analysis of Vertebrate Structures*, 5th ed. John Wiley and Sons, Inc. **25.11** *Indohyus* reproduced by permission from Macmillan Publishers Ltd after J. G. M. Thewissen, L. N. Cooper, M. T. Clementz, et al. 2007. Whales originated from aquatic artiodactyls in the Eocene epoch of India. *Nature* 450: 1190–1194, Fig. 5. *Rhodhocetus* reproduced by permission of AAAS after P. D. Gingerich, M. ul Haq, I. S. Zalmout, et al. 2001. Origin of whales from early Artiodactyls: Hands and feet of Eocene Protocetidae from Pakistan. *Science* 293: 2239–2242, Fig. 3. (http://www.sciencemag.org/content/293/5538/2239.abstract). *Dorudon* reproduced by permission after P. D. Gingerich, M. ul Haq, W. Von Koenigswald, et al. 2009. New protocetid whale from the middle Eocene of Pakistan: Birth on land, precocial development, and sexual dimorphism. *PLoS ONE* 4 (2): e4366, Fig. 1B. New data for Dorudon from M. D. Uhen. 2004. Form, function, and anatomy of *Dorudon atrox* (Mammalia, Cetacea): An Archaeocete from the Middle to Late Eocene of Egypt. University of Michigan Papesrs on Paleontology No. 34. *Delphinapterus* reproduced by permission of Skulls Unlimited International, Inc. (www.SkullsUnlimited.com). **25.14** Adapted by permission from D. P. Genereux and C. T. Bergstrom. 2005. Evolution in action: Understanding antibiotic resistance. In *Evolutionary Science and Society: Educating a New Generation*, edited by J. Cracraft and R. W. Bybee. Colorado Springs, CO: BSCS, 145–153, Fig 3. Data: Centers for Disease Control, 2004. **25.16** Data: P. T. Boag and P. R. Grant. 1981. Intense natural selection in a population of Darwin's finches (*Geospizinae*) in the Galápagos. *Science New Series* 214 (4516): 82–85, Table 1. **25.17** Body size and beak shape graph reproduced by permission of AAAS from P. R. Grant and B. R. Grant. 2002. *Science* 296:707–711, Fig. 1. Beak size graph reprinted by permission of AAAS from P. R. Grant and B. R. Grant. 2006. Evolution of character displacement in Darwin's finches. *Science* 313: 224–226, Fig. 2. (http://www.sciencemag.org/content/296/5568/707.abstract; http://www.sciencemag.org/content/313/5784/224.abstract).

Chapter 26 **T26.1** Data: W. C. Boyd. 1950. *Genetics and the Races of Man*. Boston: Little, Brown. Data are from Chapter 8, Blood Groups: Table 27, p. 238. **T26.2** Data: T. Markow, P. H. Hedrick, K. Zuerlein, et al. 1993. HLA polymorphism in the Havasupai: Evidence for balancing selection. *American Journal of Human Genetics* 53: 943–952, Table 3. **26.4** Data: W. E. Johnson, E. P. Onorato, M. E. Roelke, et al. 2010. Genetic restoration of the Florida panther. *Science* 329: 1641–1645. **26.5b** Data: C. R. Brown and M. B. Brown. 1998. Intense natural selection on body size and wing and tail asymmetry in cliff swallows during severe weather. *Evolution* 52: 1461–1475. **26.6b** Data: M. N. Karn, H. Lang–Brown, H. J. MacKenzie, et al. 1951. Birth weight, gestation time and survival in sibs. *Annals of Eugenics* 15: 306–322. **26.7b** Data: T. B. Smith. 1987. Bill size polymorphism and intraspecific niche utilization in an African finch. *Nature* 329: 717–719. **26.8b** Data: J. D. Blount, N. B. Metcalf, T. R. Birkhead, et al. 2003. Carotenoid modulation of immune function and sexual attractiveness in zebra finches. *Science* 300: 125–127, Fig. 1b. **26.10 b&c** Data: B. J. Le Boeuf and R. S. Peterson. 1969. Social status and mating activity in elephant seals. *Science* 163: 91–93. **26.12** Reproduced by permission of Pearson Education, Inc., from S. Freeman and J. Herron. 2004. *Evolutionary Analysis*. 3rd ed., Figs. 6.15a, 6.15c. ©2004. **26.13** W. E. Kerr and S. Wright. 1954. Experimental studies

of the distribution of gene frequencies in very small populations of *Drosophilia melanogaster*: I. Forked. *Evolution* 8: 172–177. **26.16b** Data: H. Araki, B. Cooper, and M. S. Blouin. 2009. Carry-over effect of captive breeding reduces reproductive fitness of wild-born descendants in the wild. *Biology Letters* 5 (5): 621–624, Fig. 2b. **26.17** Data: S. F. Elena, V. S. Cooper, and R. E. Lenski. 1996. Punctuated evolution caused by selection of rare beneficial mutations. *Science New Series* 272: 1802–1804. **26.18b** N. Moran and T. Jarvik. 2010. Lateral transfer of genes from fungi underlies carotenoid production in aphids. *Science* 328: 624–627. **27.3a** Based on J. C. Avise and W. S. Nelson. 1989. Molecular genetic relationships of the extinct dusky seaside sparrow. *Science* 243 (4891): 646–648, Fig. 2.

Chapter 27 **27.3b** J. C. Avise and W. S. Nelson. 1989. Molecular genetic relationships of the extinct dusky seaside sparrow. *Science* 243 (4891): 646–648. **27.5b** Data: N. Knowlton, L. A. Weigt, L. A. Solorzano, et al. 1993. Divergence in proteins, mitochondrial DNA, and reproductive compatibility across the Isthmus of Panama. *Science* 260 (5114): 1629–1632, Fig. 1. **27.7** Adapted by permission of Wiley-Blackwell from H. R. Dambroski, C. Linn Jr., S. H. Berlocher, et al. 2005. The genetic basis for fruit odor discrimination in *Rhagoletis* flies and its significance for sympatric host shifts. *Evolution* 59 (9): 1953–1964, Figs. 1A, 1B. **27.11b** Data: S. E. Rohwer, E. Bermingham, and C. Wood. 2001. Plumage and mitochondrial DNA haplotype variation across a moving hybrid zone. *Evolution* 55: 405–422. **27.12** Data: L. H. Rieseberg, B. Sinervo, C. R. Linder, et al. 1996. Role of gene interactions in hybrid speciation: Evidence from ancient and experimental hybrids. *Science* 272 (5262): 741–745.

Chapter 28 **28.4c** Data adapted by permission of the publisher from M. A. Nikaido, P. Rooney, and N. Okada. 1999. Phylogenetic relationships among cetartiodactyls based on insertions of short and long interspersed elements: Hippopotamuses are the closest extant relatives of whales. *PNAS* 96: 10261–10266, Figs. 1, 7. ©1999 National Academy of Sciences, USA. **28.6, 28.7, 28.8** Data: International Commission on Stratigraphy, "International Stratigraphic Chart, 2009" (www.stratigraphy.org/column.php?id=Chart/TimeScale). The data on this site are modified from F. M. Gradstein and J. C. Ogg (eds.). 2004. *A Geologic Time Scale 2004*. Cambridge, UK: Cambridge University Press; and J. G. Ogg, G. Ogg, and F. M. Gradstein. 2008. *The Concise Geologic Time Scale*. Cambridge, UK: Cambridge University Press. **28.9** B. G. Baldwin, D. W. Kyhos, and J. Dvorak. 1990. Chloroplast DNA evolution and adaptive radiation in the Hawaiian Silversword Alliance (*Asteraceae–Madiinae*). *Annals of the Missouri Botanical Garden* 77 (1): 96–109, Fig. 2. **28.10c** J. B Losos, K. I. Warheit, and T. W. Schoener. 1997. Adaptive differentiation following experimental island colonization in Anolis lizards. *Nature* 387: 70–73. **28.13** Data: J. W. Valentine, D. H. Erwin, and D. Jablonski. 1996. Developmental evolution of metazoan body plans: The fossil evidence. *Developmental Biology* 173: 373–381, Fig. 3. Also R. de Rosa et al. 1999. *Hox* genes in brachiopods and priapulids and protostome evolution. *Nature* 399: 772–776; D. Chourrout, F. Delsuc, P. Chourrout, et al. 2006. Minimal *ProtoHox* cluster inferred from bilaterian and cnidarian *Hox* complements. *Nature* 442: 684–687. **28.14** Data: M. J. Benton. 1995. Diversification and extinction in the history of life. *Science* 268: 52–58.

Chapter 29 **29.12** Data: D. F. Wilson, M. Erecinska, and P. L. Dutton. 1974. Thermodynamic relationships in mitochondrial oxidative phosphorylation. *Annual Review of Biophysics and Bioengineering* 3: 203–230, Tables 1, 3. **30.1** S. M. Adl, A. G. Simpson, M. A. Farmer, et al. 2005. The new higher level classification of eukaryotes with emphasis on the taxonomy of protists. *Journal of Eukaryotic Microbiology* 52: 399–451. Also N. Arisue, M. Hasegawa, and T. Hashimoto. 2005. Root of the eukaryota tree as inferred from combined maximum likelihood analyses of multiple molecular sequence data. *Society for Molecular Biology and Evolution, Molecular Biology and Evolution* 22: 409–420; V. Hampl, L. Hug, J. W. Leigh, et al. 2009. Phylogenomic analyses support the monophyly of Excavata and resolve relationships among eukaryotic "supergroups." *PNAS* 106 (10): 3859–3864, Figs. 1, 2, 3; J. D. Hackett, H. S. Yoon, S. Li, et al. 2007. Phylogenomic analysis supports the monophyly of cryptophytes and haptophytes and the association of Rhizaria with chromalveolates. *Molecular Biology and Evolution* 24: 1702–1713, Fig. 1; P. Schaap, T. Winckler, M. Nelson, et al. 2006. Molecular phylogeny and evolution of morphology in the social amoebas. *Science* 314: 661–663.

Chapter 30 **30.1, 30.7** S. M. Adl, A. G. Simpson, M. A. Farmer, et al. 2005. The new higher level classification of eukaryotes with emphasis on the taxonomy of protists. *Journal of Eukaryotic Microbiology* 52: 399–451. Also N. Arisue, M. Hasegawa, and T. Hashimoto. 2005. Root of the eukaryota tree as inferred from combined maximum likelihood analyses of multiple molecular sequence data. *Society for Molecular Biology and Evolution, Molecular Biology and Evolution* 22: 409–420; V. Hampl, L. Hug, J. W. Leigh, et al. 2009. Phylogenomic analyses support the monophyly of Excavata and resolve relationships among eukaryotic "supergroups." *PNAS* 106 (10): 3859–3864, Figs. 1, 2, 3; J. D. Hackett, H. S. Yoon, S. Li, et al. 2007. Phylogenomic analysis supports the monophyly of cryptophytes and haptophytes and the association of Rhizaria with chromalveolates. *Molecular Biology and Evolution* 24: 1702–1713, Fig. 1; P. Schaap, T. Winckler, M. Nelson, et al. 2006. Molecular phylogeny and evolution of morphology in the social amoebas. *Science* 314: 661–663.

Chapter 31 31.6, 31.9, 31.23 Y.-L. Qiu, L. Li, B. Wang, et al. 2006. The deepest divergences in land plants inferred from phylogenomic evidence. *PNAS* 103 (42): 15511–15516, Fig. 1. Also K. S. Renzaglia, S. Schuette, R. J. Duff, et al. 2007. Bryophyte phylogeny: Advancing the molecular and morphological frontiers. American Bryological and Lichenological Society, *The Bryologist* 110 (2): 179–213; J.-F. Pombert, C. Otis, C. Lemieux, et al. 2005. The chloroplast genome sequence of the green alga *Pseudendoclonium akinetum* (*Ulvophyceae*) reveals unusual structural features and new insights into the branching order of chlorophyte lineages. *Molecular Biology and Evolution* 22: 1903–1918. **31.21** Based on M. E. Hoballah, T. Gubitz, J. Stuurman, et al. 2007. Single gene-mediated shift in pollinator attraction in Petunia. *Plant Cell* 19: 779–790, Fig. 6g, 6h. **31.25** P. S. Soltis and D. E. Soltis. 2004. The origin and diversification of angiosperms. *American Journal of Botany* 91 (10): 1614–1626, Figs. 1, 2, 3.

Chapter 32 32.7 S. M. Adl, A. G. Simpson, M. A. Farmer, et al. 2005. The new higher level classification of eukaryotes with emphasis on the taxonomy of protists. *Journal of Eukaryotic Microbiology* 52: 399–451. Also N. Arisue, M. Hasegawa, and T. Hashimoto. 2005. Root of the Eukaryota tree as inferred from combined maximum likelihood analyses of multiple molecular sequence data. *Molecular Biology and Evolution* 22: 409–420; V. Hampl, L. Hug, J. W. Leigh, et al. 2009. Phylogenomic analyses support the monophyly of Excavata and resolve relationships among eukaryotic "supergroups." *PNAS* 106: 3859–3864, Figs. 1, 2, 3; J. D. Hackett, H. S. Yoon, S. Li, et al. 2007. Phylogenomic analysis supports the monophyly of cryptophytes and haptophytes and the association of Rhizaria with chromalveolates. *Molecular Biology and Evolution* 24: 1702–1713, Fig 1; P. Schaap, T. Winckler, M. Nelson, et al. 2006. Molecular phylogeny and evolution of morphology in the social amoebas. *Science* 314: 661–663. **32.8** T. Y. James, F. Kauff, C. L. Schoch, et. al. 2006. Reconstructing the early evolution of fungi using a six-gene phylogeny. *Nature* 443: 818–822, Fig. 1. Also D. J. McLaughlin, D. S. Hibbett, F. Lutzoni, et al. 2009. The search for the fungal tree of life. *Trends in Microbiology* 17: 488–497; D. S. Hibbett, M. Binder, J. F. Bischoff, et al. 2007. A higher-level phylogenetic classification of the Fungi. *Mycological Research* 111: 509–547; H. Wang, X. Zhao, L. Gao, et al. 2009. A fungal phylogeny based on 82 complete genomes using the composition vector method. *BMC Evolutionary Biology* 9: 195. **32.10** After H. Bücking and W. Heyser. 2001. Microautoradiographic localization of phosphate and carbohydrates in mycorrhizal roots of *Populus tremula* x *Populus alba* and the implications for transfer processes in ectmycorrhizal associations. *Tree Physiology* 21: 101–107.

Chapter 33 33.2 A. M. A. Aguinaldo, J. M. Turbeville, L. S. Linford, et al. 1997. Evidence for a clade of nematodes, arthropods, and other moulting animals. *Nature* 387: 489–493, Figs. 1, 2, 3. Also C. W. Dunn, A. Hejnol, D. Q. Matus, et al. 2008. Broad phylogenomic sampling improves resolution of the animal tree of life. *Nature* 452: 745–750, Figs. 1, 2. **33.7** Based on J. R. Finnerty, K. Pang, P. Burton, et al. 2004. Origins of bilateral symmetry: *Hox* and *Dpp* expression in a sea anemone. *Science* 304: 1335–1337.

Chapter 34 34.1 A. D. Chapman. 2009. *Numbers of Living Species in Australia and the World.* 2nd ed. Canberra: Australian Biological Resources Study. **34.2** A. M. A. Aguinaldo, J. M. Turbeville, L. S. Linford, et al. 1997. Evidence for a clade of nematodes, arthropods, and other moulting animals. *Nature* 387: 489–493, Figs. 1, 2, 3. Also C. W. Dunn, A. Hejnol, D. Q. Matus, et al. 2008. Broad phylogenomic sampling improves resolution of the animal tree of life. *Nature* 452: 745–750, Figs. 1, 2. **34.28, 34.30** Based on data in J. Regier, J. W. Schultz, A. Zwick, et al. 2010. Arthropod relationships revealed by phylogenomic analysis of nuclear protein-coding sequences. *Nature* 463: 1079–1083. **34.UN1** Based on M. Averof and N. H. Patel. 1997. Crustacean appendage evolution associated with changes in *Hox* gene expression. *Nature* 388: 682–686, Fig. 4.

Chapter 35 35.1 After C. W. Dunn, A. Hejnol, D. Q. Matus, et al. 2008. Broad phylogenomic sampling improves resolution of the animal tree of life. *Nature* 452: 745–750, Figs. 1, 2. Also F. Delsuc, H. Brinkmann, D. Chourrout, et al. 2006. Tunicates and not cephalochordates are the closest living relatives of vertebrates. *Nature* 439: 965–968, Fig. 1. **35.12** J. E. Blair and S. B. Hedges. 2005. Molecular phylogeny and divergence times of deuterostome animals. *Molecular Biology and Evolution* 22: 2275–2284, Figs. 1, 3, 4. **35.13** Data: American Museum of Natural History; BirdLife International (birdlife.org); FishBase (www.fishbase.org); IUCN 2010. *The IUCN Red List of Threatened Species* (http://www.iucnredlist.org); The Reptile Database (reptile-database.org). **35.16** Based on E. B. Daeschler, N. H. Shubin, F. A. Jenkins, et al. 2006. A Devonian tetrapod-like fish and the evolution of the tetrapod body plan. *Nature* 440: 757–763, Fig. 6. Also N. H. Shubin, E. B. Daeschler, F. A. Jenkins, et al. 2006. The pectoral fin of *Tiktaalik roseae* and the origin of the tetrapod limb. *Nature* 440:764–771, Fig. 4. **35.20** Color plate by Michael DiGiorgio. From Q. Li, K.-Q. Gao, J. Vinther, et al. 2010. Plumage color patterns of an extinct dinosaur. *Science* 327, Fig. 4, p. 1371. Reprinted by permission of the artist, Michael DiGiorgio. **35.22** J. E. Blair and S. B. Hedges. 2005. Molecular phylogeny and divergence times of deuterostome animals. *Molecular Biology and Evolution,* 22: 2275–2284, Figs. 1, 3, 4. Also F. R. Liu, M. M. Miyamoto, N. P. Freire, et al. 2001. Molecular and morphological supertrees for eutherian (placental) mammals. *Science*

291: 1786–1789; A. B. Prasad, M. W. Allard, E. D. Green, et al. 2008. Confirming the phylogeny of mammals by use of large comparative sequence data sets. Oxford University Press, *Molecular Biology and Evolution* 25: 1795–1808. **35.36** A. B. Prasad, M. W. Allard, E. D. Green, et al. 2008. Confirming the phylogeny of mammals by use of large comparative sequence data sets. Oxford University Press, *Molecular Biology and Evolution* 25 (9):1795–1808, Figs. 1, 2, 3. **35.37** Data: D. E. Lieberman. 2009. Palaeoanthropology: *Homo floresiensis* from head to toe. *Nature* 459: 41–42. **T35.1** S. L. Robson and B. Wood. 2008. Hominin life history: Reconstruction and evolution. *Journal of Anatomy* 212: 394–425. **35.39** J. Z. Li, D. M. Absher, H. Tang, et al. 2008. Worldwide human relationships inferred from genome-wide patterns of variation. *Science* 319: 1100–1104, Fig. 1. **35.40** After L. L. Cavalli-Sforza and M. W. Feldman. 2003. The application of molecular genetic approaches to the study of human evolution. *Nature Genetics Supplement* 33: 266–275, Fig. 3; Data: Genographic Project at National Geographic: https://genographic.nationalgeographic.com/genographic/lan/en/atlas.html; dates from August 2009 summary map: http://ngm.nationalgeographic.com/big-idea/02/queens-genes; M. Rasmussen, X. Guo, Y. Wang, et al. 2011. An Aboriginal Australian genome reveals separate human dispersals into Asia. *Science* 334: 94–98.

Chapter 36 36.2a Data: E. Arias. 2010. United States life tables, 2006. *National Vital Statistics Reports* 58 (21): 1–40, Table 10. Hyattsville, MD: National Center for Health Statistics. **36.3** Adapted with permission of Annual Reviews from G. Pantaleo and A. S. Fauci. 1996. Immunopathogensis of HIV infection. *Annual Review of Microbiology* 50: 825–854. Fig. 1a, p. 829. **36.7** Data courtesy of the Undergraduate Biotechnology Laboratory, California Polytechnic State University, San Luis Obispo. **36.8** A. G. Dalgleish, P. C. Beverley, P. R. Clapham, et al. 1984. The CD4 (T4) antigen is an essential component of the receptor for the AIDS retrovirus. *Nature* 312: 763–767. Also D. Klatzmann, E. Champagne, S. Chamaret, et al. 1984. T-lymphocyte T4 molecule behaves as the receptor for human retrovirus LAV. *Nature* 312: 767–768. **36.15** F. Gao, E. Bailes, D. L. Robertson, et al. 1999. Origin of HIV-1 in the chimpanzee *Pan troglodytes troglodytes*. *Nature* 397: 436–441, Fig. 2.

Chapter 37 37.3 Illustration adapted by permission from "Root Systems of Prairie Plants," by Heidi Natura. ©1995 Conservation Research Institute. **37.5** After J. Clausen, D. D. Keck, and M. Hiesey. 1945. Experimental studies on the nature of species II: Plant evolution through amphiploidy and autoploidy, with examples from the *Madiinae*. Publication No. 564. Washington, DC: Carnegie Institute of Washington.

Chapter 38 38.3 Data: R. G. Cline and W. S. Campbell. 1976. Seasonal and diurnal water relations of selected forest species. *Ecology* 57: 367–373, Fig. 1. **38.12** Adapted by permission of the American Society of Plant Biologists from C. M. Wei, M. T. Tyree, and E. Steudle. 1999. Direct measurement of xylem pressure in leaves of intact maize plants. A test of the cohesion-tension theory taking hydraulic architecture into consideration. *Plant Physiology* 121: 1191–1205, Fig. 6. ©1999 by American Society of Plant Biologists. **38.22** J. R. Gottwald, P. J. Krysan, J. C. Young, et al. 2000. Genetic evidence for the in planta role of phloem-specific plasma membrane sucrose transporters. *PNAS* 97: 13979–13984.

Chapter 39 39.3 D. I. Arnon and P. R. Stout. 1939. The essentiality of certain elements in minute quantity for plants with special reference to copper. *Plant Physiology* 14: 371–375.

Chapter 40 40.4 Based on S. Inoue, T. Kinoshita, M. Matsumoto, et al. 2008. Blue light-induced autophosphorylation of phototropin is a primary step for signaling. *PNAS* 105: 5626–5631. **40.5** After C. Darwin and F. Darwin. 1897. *The Power of Movement in Plants*. New York: D. Appleton. **40.10** After A. Lang, M. K. Chailakhyan, and I. A. Frolova. 1977. Promotion and inhibition of flower formation in a day-neutral plant in grafts with a short-day plant and a long-day plant. *PNAS* 74: 2412–2416. **T40.1** H. A. Borthwick, S. B. Hendricks, M. W. Parker, et al. 1952. A reversible photoreaction controlling seed germination. *PNAS* 38: 662–666, Table 1. **40.21** After P. G. Blackman and W. J. Davies. 1985. Root to shoot communication in maize plants of the effect of soil drying. *Journal of Experimental Botany* 36: 39–48.

Chapter 41 41.10 Y.-L. Qiu, L. Li, B. Wang, et al. 2006. The deepest divergences in land plants inferred from phylogenomic evidence. *PNAS* 103 (42): 15511–15516, Fig. 1. Also K. S. Renzaglia, S. Schuette, R. J. Duff, et al. 2007. Bryophyte phylogeny: Advancing the molecular and morphological frontiers. American Bryological and Lichenological Society, *The Bryologist* 110 (2): 179–213; J.-F. Pombert, C. Otis, C. Lemieux, et al. 2005. The chloroplast genome sequence of the green alga *Pseudendoclonium akinetum* (*Ulvophyceae*) reveals unusual structural features and new insights into the branching order of chlorophyte lineages. *Molecular Biology and Evolution* 22: 1903–1918. **41.17** J. J. Tewksbury and G. P. Nabhan. 2001. Directed deterrence by capsaicin in chilies. *Nature* 412: 403–404.

Chapter 42 42.2 A. M. Kerr, S. N. Gershman, and S. K. Sakaluk. 2010. Experimentally induced spermatophore production and immune responses reveal a trade-off in crickets. *Behavioral Ecology* 21: 647–654. **42.10** Data: K. Schmidt-Nielsen. 1984. *Scaling: Why Is Animal Size So Important?* Cambridge, UK: Cambridge University Press. **42.11** Adapted by permission of The Company of Biologists from

P. R. Wells and A. W. Pinder. 1996. The respiratory development of Atlantic salmon. II. Partitioning of oxygen uptake among gills, yolk sac and body surfaces. *Journal of Experimental Biology* 199: 2737–2744, Figs. 1A, 3.

Chapter 43 **43.13** Anterolateral view reproduced by permission from Macmillan Publishers Ltd after S. E. Pierce, J. A. Clack, and J. R. Hutchinson. 2012. Three-dimensional limb joint mobility in the early tetrapod *Ichthyostega. Nature* 486: Fig. 1, p. 523. **43.14** Adapted by permission of Elsevier from K. J. Ullrich, K. Kramer, and J. W. Boyer. 1961. Present knowledge of the counter-current system in the mammalian kidney. *Progress in Cardiovascular Diseases* 3: 395–431, Figs. 5, 7. ©1961. **43.19** J. R. Davis and D. F. DeNardo. 2007. The urinary bladder as a physiological reservoir that moderates dehydration in a large desert lizard, the Gila monster *Heloderma suspectum. Journal of Experimental Biology* 210 (8): 1472–1480. Reprinted with permission from The Company of Biologists.

Chapter 44 **44.14** After E. M. Wright. 1993. The intestinal Na$^+$/glucose cotransporter. *Annual Review of Physiology* 55: 575–589. Also K. Takata, T. Kasahara, M. Kasahar, et al. 1992. Immunohistochemical localization of Na$^+$-dependent glucose transporter in rat jejunum. *Cell and Tissue Research* 267: 3–9. **44.17a** Graph reproduced with permission from The American Diabetes Association from L. O. Schulz, P. H. Bennett, E. Ravussin, et al. 2006. Effects of traditional and Western environments on prevalence of type 2 diabetes in Pima Indians in Mexico and the U.S. *Diabetes Care* 29: 1866–1871, Fig. 1. ©2006 American Diabetes Association. **44.17b** Data: L. O. Schulz, P. H. Bennett, E. Ravussin, et al. 2006. Effects of traditional and Western environments on prevalence of type 2 diabetes in Pima Indians in Mexico and the U.S. *Diabetes Care* 29: 1866–1871, Table 2.

Chapter 45 **45.7** Based on Y. Komai. 1998. Augmented respiration in a flying insect. *Journal of Experimental Biology* 201: 2359–2366, Fig. 7. **45.9** R. A. Berner. 1999. Atmospheric oxygen over Phanerozoic time. *PNAS* 96: 10955–10957. Fig. 2. Copyright ©1999 National Academy of Sciences, USA. Reprinted with permission. **45.12** W. Bretz and K. Schmidt Nielsen. 1971. Bird respiration: Flow patterns in the duck lung. *Journal of Experimental Biology* 54: 103–118. **45.21** Data: E. H. Starling. 1896. On the absorption of fluids from the connective tissue spaces. *Journal of Physiology* 19: 312–326. **45.23** J. E. Blair and S. B. Hedges. 2005. Molecular phylogeny and divergence times of deuterostome animals. *Molecular Biology and Evolution* 22: 2275–2284, Figs. 1, 3, 4.

Chapter 46 **46.10** Data: O. Loewi. 1921. Über humorale Übertragbarkeit der Herznervenwirkung. *Pflügers Archiv European Journal of Physiology* 189: 239–242. **46.22** Data: R. Q. Quiroga, L. Reddy, G. Kreiman, et al. 2005. Invariant visual representation by single neurons in the human brain. *Nature* 435: 1102–1107, Fig. 1B. **46.24a&c** Based on K. C. Martin, A. Casadio, H. Zhu, et al. 1997. Synapse-specific, long-term facilitation of *Aplysia* sensory to motor synapses: A function for local protein synthesis in memory storage. *Cell* 91: 927–938.

Chapter 47 **47.2a&b** Data: J. E. Rose, J. E. Hind, D. J. Anderson, et al. 1971. Some effects of stimulus intensity on response of auditory nerve fibers in the squirrel monkey. *Journal of Neurophysiology* 34: 685–699. **47.9** K. Pohlmann, J. Atema, and T. Breithaupt. 2004. The importance of the lateral line in nocturnal predation of piscivorous catfish. *Journal of Experimental Biology* 207: 2971–2978. **47.15** Based on D. M. Hunt, K. S. Dulai, J. K. Bowmaker, et al. 1995. The chemistry of John Dalton's color blindness. *Science* 267: 984–988, Fig. 3. **47.UN1** Reprinted by permission from Macmillan Publishers Ltd. from M. K. McClintock. 1971. Menstrual synchrony and suppression. *Nature* 229: 244–245, Fig. 1.

Chapter 48 **48.3** PDB ID: 1KWO. D. M. Himmel, S. Gourinath, L. Reshetnikova, et al. 2002. Crystallographic findings on the internally uncoupled and near-rigor states of myosin: Further insights into the mechanics of the motor. *PNAS* 99: 12645–12650. **48.7** After Exercise Information—Interactive Muscle Map (www.askthetrainer.com/exercise-information/). **48.15** Reprinted by permission from Macmillan Publishers Ltd from D. F. Hoyt and C. R. Taylor. 1981. Gait and the energetics of locomotion in horses. *Nature* 292: 239–240, Fig. 2, p. 240. **48.16** S. Vogel. 2003. *Comparative Biomechanics: Life's Physical World.* 2nd ed. Princeton, NJ: Princeton University Press, Fig. 3.5, p. 60. Reprinted by permission of Princeton University Press. **48.17** Reprinted with permission from AAAS from K. Schmidt-Nielsen. 1972. Locomotion: Energy cost of swimming, flying, and running. *Science* 177: 222–228. (http://www.sciencemag.org/content/177/4045/222.extract).

Chapter 49 **49.4** W. M. Bayliss and E. H. Starling. 1902. The mechanism of pancreatic secretion. *Journal of Physiology* 28: 325–353. **49.7** Graph from J. Vanecek. 1998. Cellular mechanisms of melatonin action. *Psychological Reviews* 78: 687–721, Fig. 1, p. 688. Original Data: H. Illnerová, K. Hoffmann, J. Vanácek, et al. 1984. Adjustment of pineal melatonin and N-acetyltransferase rhythms to change from long to short photoperiod in the Djungarian hamster Phodopus sungorus. *Neuroendocrinology* 38: 226–231. **49.9** Data: B. D. King, L. Sokoloff, and R. L. Wechsler. 1952. The effects of l-epinephrine and l-nor-epinephrine upon cerebral circulation and metabolism in man. *Journal of Clinical Investigation* 31: 273–279. Also P. S. Mueller and D. Horwitz. 1962. Plasma free fatty acid and blood glucose

responses to analogues of norepinephrine in man. *Journal of Lipid Research* 3: 251–255. **49.11** Reprinted by permission from Macmillan Publishers Ltd. from D. L. Coleman. 2010. A historical perspective on Leptin. *Nature Medicine* 16: 1097–1099, Fig. 2. **49.12a&b** Data: G. V. Upton, A. Corbin, C. C. Mabry, et al. 1973. Evidence for the internal feedback phenomenon in human subjects: Effects of ACTH on plasma CRF. *Acta Endocrinologica* 73: 437–443. **48.13** Anterolateral view reproduced by permission from Macmillan Publishers Ltd after S. E. Pierce, J. A. Clack, and J. R. Hutchinson. 2012. Three-dimensional limb joint mobility in the early tetrapod Ichthyostega. *Nature* 486: Fig. 1, p. 523. **49.14** Data: D. Toft and J. Gorski. 1966. A receptor molecule for estrogens: Isolation from the rat uterus and preliminary characterization. *PNAS* 55: 1574–1581. **49.16b** Data: T. W. Rall, E. W. Sutherland, and J. Berthet. 1957. The relationship of epinephrine and glucagon to liver phosphorylase. *Journal of Biological Chemistry* 224: 463–475. **49.17** Reproduced from E. W. Sutherland. 1971. Studies on the mechanism of hormone action. (A lecture delivered in Stockholm, December 11, 1971.) ©1971 Nobel Foundation; E. W. Sutherland. 1972. Studies on the mechanism of hormone action. *Science* 177: 401–408, Fig. 6. Also G. A. Robison, R. W. Butcher, I. Oye, et al. 1965. The effect of epinephrine on adenosine 3',5'-phosphate levels in the isolated perfused rat heart. *Molecular Pharmacology* 1: 168–177. **49.UN1** Data: T. B. Hayes, A. Collins, M. Lee, et al. 2002. Hermaphroditic, demasculinized frogs after exposure to the herbicide atrazine at low ecologically relevant doses. *PNAS* 99: 5476–5480.

Chapter 50 **50.3a** Based on R. G. Stross and J. C. Hill. 1965. Diapause induction in *Daphnia* requires two stimuli. *Science* 150: 1462–1464, Fig. 3. **50.3b** Data: O. T. Kleiven, P. Larsson, and A. Hobæk. 1992. Sexual reproduction in *Daphnia magna* requires three stimuli. *Oikos* 65: 197–206, Table 4. **50.7** Based on C. S. C. Price, K. A. Dyer, and J. A. Coyne. 1999. Sperm competition between *Drosophila* males involves both displacement and incapacitation. *Nature* 400: 449–452, Fig. 3c. **50.9** R. Shine and M. S. Y. Lee. 1999. A reanalysis of the evolution of viviparity and egg-guarding in squamate reptiles. The Herpetologists' League, Inc., *Herpetologica* 55 (4): 538–549, Figs. 1, 2, 3. **50.10b** Reproduced by permission of Elsevier from C. Hotzy and G. Arnqvist. 2009. Sperm competition favors harmful males in seed beetles. *Current Biology* 19: 404–407, Fig. 2. ©2009. **50.17, 50.18** Data: R. Stricker, R. Eberhart, M.-C. Chevallier, et al. 2006. Establishment of detailed reference values for luteinizing hormone, follicle stimulating hormone, estradiol, and progesterone during different phases of the menstrual cycle on the Abbott ARCHITECT* analyzer. *Clinical Chemistry and Laboratory Medicine* 44: 883–887, Tables 1A, 1B. **50.22b** Data: V. W. Swayze II., V. P. Johnson, J. W. Hanson, et al. 1997. Magnetic resonance imaging of brain anomalies in fetal alcohol syndrome. *Pediatrics* 99: 232–240. **50.24** Adapted by permission of the publisher and author from U. Högberg and I. Joelsson. 1985. The decline in maternal mortality in Sweden, 1931–1980. *Acta Obstetricia et Gynecologica Scandinavica* 64: 583–592, Fig. 1. Taylor & Francis Group, http://www.informaworld.com.

Chapter 51 **51.2** Based on B. Lemaitre, E. Nicolas, L. Michaut, et al. 1996. The dorsoventral regulatory gene cassette *spätzle/Toll/cactus* controls the potent antifungal response in *Drosophila* adults. *Cell* 86: 973–983. **51.5a** PDB ID: 1IGT. L. J. Harris, S. B. Larson, K. W. Hasel, et al. 1997. Refined structure of an intact IgG2a monoclonal antibody. *Biochemistry* 36: 1581–1597. **51.5b** PDB ID: 1TCR. K. C. Garcia, M. Degano, R. L. Stanfield, et al. 1996. An aß T cell receptor structure at 2.5 Å and its orientation in the TCR–MHC complex. *Science* 274: 209–219. **51.6** PDB ID: 1HMG. M. Knossow, R. S. Daniels, A. R. Douglas, et al. 1984. Three-dimensional structure of an antigenic mutant of the influenza virus haemagglutinin. *Nature* 311 (5987): 678–680. **51.10** PDB ID: 1DLH. L. J. Stern, J. H. Brown, T. S. Jardetzky, et al. 1994. Crystal structure of the human class II MHC protein HLA–DR1 complexed with an influenza virus peptide. *Nature* 368: 215–221. **51.17** Based on B. Alberts, A. Johnson, J. Lewis, et al. 2002. *Molecular Biology of the Cell.* 4th ed. New York: Garland Science, Taylor & Francis Group. Modified from Fig. 24-10. **52.1** Data: Food and Agricultural Organization of the United Nations.

Chapter 52 **52.4** S. B. Menke, R. N. Fisher, W. Jetz, et al. 2007. Biotic and abiotic controls of Argentine ant invasion success at local and landscape scales. *Ecology* 88 (12): 3164–3173. **52.9** D. M. Olson, E. Dinerstein, E. D. Wikramanayake, et al. 2001. Terrestrial ecoregions of the world: A new map of life on Earth. *BioScience* 51: 933–938. **52.11, 52.12, 52.13, 52.14, 52.15, 52.16** Data: E. Aguado and J. E. Burt. 1999. *Understanding Weather and Climate.* Upper Saddle River, NJ: Prentice Hall. **52.18a** Based on M. B. Saffo. 1987. New light on seaweeds. *BioScience* 37 (9): 654–664, Fig. 1. ©1987 American Institute of Biological Sciences.

Chapter 53 **53.3** After M. Sokolowski. 2001. *Drosophila*: Genetics meets behavior. *Nature Reviews Genetics* 2: 879–890, Fig. 2. **53.4** Z. Abramsky, M. L. Rosenzweig, and A. Subach. 2002. The costs of apprehensive foraging. *Ecology* 85: 1330–1340. **53.5b** Adapted by permission of the author from D. Crews. 1975. Psychobiology of reptilian reproduction. *Science* 189: 1059–1065, Fig. 1. **53.6** Adapted by permission of the author from D. Crews. 1975. Psychobiology of reptilian reproduction. *Science* 189: 1059–1065, Fig. 2. **53.7** Reprinted by permission from Macmillan Publishers Ltd from K. J. Lohmann, C. M. F. Lohmann, L. M. Ehrhart, et al. 2004. Geomagnetic map used in sea-turtle navigation. *Nature* 428: 909. **53.8** C. Egevang, et al. 2010. Tracking of

Arctic terns *Sterna paradisaea* reveals longest animal migration. *PNAS* 107(5): 2078–2081, Fig. 1B. ©2010 National Academy of Sciences, USA. Reprinted with permission. **53.15** Data: J. L. Hoogland. 1983. Nepotism and alarm calling in the black-tailed prairie dog (*Cynomys ludovicianus*). *Animal Behavior* 31: 472–479.

Chapter 54 **T54.1** Data reproduced from H. Strijbosch and R. C. M. Creemers. 1988. Comparative demography of sympatric populations of *Lacerta vivipara* and *Lacerta agilis*. *Oecologia* 76: 20–26. **54.5b** Data: G. F. Gause. 1934. *The Struggle for Existence*. Baltimore, MD: Williams & Wilkins. **54.6a** G. E. Forrester and M. A. Steele. 2000. Variation in the presence and cause of density-dependent mortality in three species of reef fishes. *Ecology* 81 (9): 2416–2427, Fig. 5. Copyright © 2000 by the Ecological Society of America. Reprinted with permission. **54.6b** Data: P. Arcese and J. N. M. Smith. 1988. Effects of population density and supplemental food on reproduction in song sparrows. *Journal of Animal Ecology* 57: 119–136, Fig. 2. **54.9** Data: D. A. MacLulich. 1937. Fluctuations in the numbers of the varying hare (*Lepus americanus*). *University of Toronto Studies Biological Series* 43. **54.10** Graphs reproduced by permission of AAAS from C. J. Krebs, S. Boutin, R. Boonstra, et al. 1995. Impact of food and predation on the snowshoe hare cycle. *Science* 269: 1112–1115, Fig. 3. (http://www.sciencemag.org/content/269/5227/1112). **54.11a** Data: United Nations, Department of Economic and Social Affairs, Population Division. 2011. *World Population Prospects: The 2010 Revision*. **54.11b** Data: United Nations, Department of Economic and Social Affairs, Population Division. 2011. *World Population Prospects: The 2010 Revision*. **54.13** United Nations, Department of Economic and Social Affairs, Population Division. 2011. *World Population Prospects: The 2010 Revision*.

Chapter 55 **55.2a&b, 55.3a&b** G. F. Gause. 1934. *The Struggle for Existence*. Baltimore. MD: Williams & Wilkins. **55.5** Data: J. H. Connell. 1961. The influence of interspecific competition and other factors on the distribution of the barnacle *Chthamalus stellatus*. *Ecology* 42: 710–723. **55.7** P. R. Grant. 1999. *Ecology and Evolution of Darwin's Finches*. ©1986 Princeton University Press. Fig. 15, p. 78. Reprinted by permission of Princeton University Press. **55.10** Adapted by permission of Ecological Society of America from G. H. Leonard, M. D. Bertness, and P. O. Yund. 1999. Crab predation, waterborne cues, and inducible defenses in the blue mussel, *Mytilus edulis*. *Ecology* 80: 1–14. **55.14** Based on J. H. Cushman and T. G. Whitham. 1989. Conditional mutualism in a membracid–ant association: temporal, age-specific, and density-dependent effects. *Ecology* 70 (4): 1040–1047. **55.15** Based on D. G. Jenkins and A. L. Buikema. 1998. Do similar communities develop in similar sites? A test with zooplankton structure and function. *Ecological Monographs* 68: 421–443. **55.16** Adapted with kind permission from Springer Science+Business Media B.V. from R. T. Paine. 1974. Intertidal community structure. *Oecologia* 15: 93–120. ©1974 Springer-Verlag. **55.18b** Data: T. W. Swetnam. 1993. Fire history and climate change in giant sequoia groves. *Science* 262: 885–889. **55.21a** Adapted by permission of Wiley-Blackwell from R. H. MacArthur and E. O. Wilson. 1963. An equilibrium theory of insular zoogeography. *Evolution* 17 (4): 373–387, Fig. 4. **55.21 b & c** Data: R. H. MacArthur and E. O. Wilson. 1963. An equilibrium theory of insular zoogeography. *Evolution* 17 (4): 373–387, Fig. 5. **55.23** Data: W. V. Reid and K. R. Miller. 1989. *Keeping Options Alive: The Scientific Basis for Conserving Biodiversity*. Washington, DC: World Resources Institute. **56.10** Data: H. Leith and R. H. Whittaker (eds.). 1975. *Primary Productivity in the Biosphere. Ecological Studies and Synthesis*, Vol. 14. New York: Springer-Verlag.

Chapter 56 **56.5** Based on D. E. Canfield, A. N. Glazer, and P. G. Falkowski. 2010. The evolution and future of Earth's nitrogen cycle. *Science* 330: 192–196. **56.13** Data: G. E. Likens, F. H. Bormann, N. M. Johnson, et al. 1970. Effects of forest cutting and herbicide treatment on nutrient budgets in the Hubbard Brook watershed-ecosystem. *Ecological Monographs* 40: 23–47. **56.14** T. Oki and S. Kanae. 2006. Global hydrological cycles and world water resources. *Science* 313: 1068–1072. **56.15** U.S Geological Survey: Changes in water levels and storage in the High Plains Aquifer, predevelopment to 2009. Fact Sheet, July 2011, Fig. 1. **56.16** N. Gruber and J. N. Galloway. 2008. An Earth-system perspective of the global nitrogen cycle. *Nature* 451: 293–296. **56.17** Based on D. E. Canfield, A. N. Glazer, and P. G. Falkowski. 2010. The evolution and future of Earth's nitrogen cycle. *Science* 330: 192–196.

56.19 B. Hönisch, A. Ridgwell, D. N. Schmidt, et al. 2012. The geological record of ocean acidification. *Science* 335: 1058–1063. **56.20** Carbon Dioxide Information Analysis Center (http://cdiac.ornl.gov/). **56.22, top** NOAA. **56.22, middle & bottom** T. R. Karl, J. M. Melillo, and T. C. Peterson (eds.). 2009. *Global Climate Change Impacts in the United States*. New York: Cambridge University Press, p. 6. **56.23** Carbon Dioxide Information Analysis Center (http://cdiac.ornl.gov/). **56.24** T. R. Karl, J. M. Melillo, and T. C. Peterson (eds.). 2009. *Global Climate Change Impacts in the United States*. New York: Cambridge University Press, p. 34. **56.25** Reproduced by permission of AAAS from M. Zhao and S. W. Running. 2010. Drought-induced reduction in global terrestrial net primary production from 2000 through 2009. *Science* 329: 940–943, Fig. 2. (http://www.sciencemag.org/content/329/5994/940.abstract).

Chapter 57 **57.2** F.-R. Liu, M. M. Miyamoto, N. P. Freire, et al. 2001. Molecular and morphological supertrees for eutherian (placental) mammals. *Science* 291: 1786–1789, Fig. 1. Also A. B. Prasad, M. W. Mallard, E. D. Green, et al. 2008. Confirming the phylogeny of mammals by use of large comparative sequence data sets. Oxford University Press, *Molecular Biology and Evolution* 25 (9):1795–1808; U. Arnason, A. Gullberg, and A. Janke. 2004. Mitogenomic analyses provide new insights into cetacean origin and evolution. Elsevier, *Gene* 333: 27–34, Fig. 1. **57.4a&b** Adapted by permission of Macmillan Publishers Ltd after C. D. L. Orme, R. G. Davies, M. Burgess, et al. 2005. Global hotspots of species richness are not congruent with endemism or threat. *Nature* 436: 1016–1019, Figs. 1a, 1c. **57.5** D. M. Olson and E. Einerstein. 2002. The Global 200: Priority ecoregions for global conservation. *Annals of the Missouri Botanical Garden* 89: 199–224. Fig. 2, p. 201. Reprinted by permission of the Missouri Botanical Garden Press. **57.6** Data: IUCN 2012. *The IUCN Red List of Threatened Species. Version 2012.2*. (http://www.iucnredlist.org). **57.7** Reproduced by permission of the University of California Press Journals and the author from O. Venter et al. 2006. Threats to endangered species in Canada. *BioScience* 56 (11): 903–910, Fig 2. ©2006 American Institute of Biological Sciences. **57.9** Data: W. F. Laurance, S. G. Laurance, L. V. Ferreira, et al. 2000. Conservation: Rainforest fragmentation kills big trees. *Nature* 404: 836. **57.9** Based on W. F. Laurance, S. G. Laurance, L.V . Ferreira, et al. 1997. Biomass collapse in Amazonian forest fragments. *Science* 278: 1117–1118, Fig. 2. **57.10** Adapted by permission of Elsevier from J. M. Diamond. 1975. The island dilemma: Lessons of modern biogeographic studies for the design of natural reserves. *Biological Conservation* 7: 129–146. ©1975. **57.11** Reproduced by permission of AAAS from B. Sinervo, F. Mendez de la Cruz, D. B. Miles, et al. 2010. Erosion of lizard diversity by climate change and altered thermal niches. *Science* 328: 894–899, Fig. 3. (http://www.sciencemag.org/content/328/5980/894.abstract). **57.12** Graphs reproduced by permission of AAAS from D. Tilman, J. Knops, D. Wedin, et al. 1997. The influence of functional diversity and composition on ecosystem processes. *Science* 277: 1300–1302, Fig. 1. (http://www.sciencemag.org/content/277/5330/1300.short). **57.14** Data: W. E. Johnson, D. P. Onorato, M. E. Roelke, et al. 2010. Genetic restoration of the Florida panther. *Science* 329: 1641–1645. **57.15** Reproduced by permission of AAAS from E. I. Damschen, N. M. Haddad, J. L. Orrock, et al. 2006. Corridors increase plant species richness at large scales. *Science* 313: 1284–1286, Fig. 2B. (http://www.sciencemag.org/content/313/5791/1284.abstract). **57.17** Data: H. P. Jones and O. J. Schmitz. 2009. Rapid recovery of damaged ecosystems. *PLoS One* 4 (5): e5653, Fig. 4. **57.UN1, top** Data: A. P. Dobson. 1996. *Conservation and Biodiversity*. Scientific American Library. New York: Freeman. Original source of England maps: Reproduced by permission of the author from D. S. Wilcove, C. H. McLellan, and A. P. Dobson. 1986. Habitat fragmentation in the temperate zone. In *Conservation Biology: The Science of Scarcity and Diversity*, edited by M. E. Soule. Sunderland, MA: Sinauer Associates, pp. 237–256, Fig. 1. **57.UN1, bottom** Data: A. P. Dobson. 1996. *Conservation and Biodiversity*. Scientific American Library. New York: Freeman. Original source of U.S. Maps: W. B. Greeley, Chief, U.S. Forest Service. 1925. Relation of geography to timber supply. *Economic Geography* 1: 1–11.

Appendix B: BioSkills **TB16.1** A. Crowe, C. Dicks, and M. P. Wenderoth. 2008. Biology in bloom: Implementing Bloom's Taxonomy to enhance student learning in Biology. *CBE–Life Sciences Education* 7: 368–381, Table 3.

Index

Boldface page numbers indicate a glossary entry; page numbers followed by an *f* indicate a figure; those followed by *t* indicate a table.

I:2   INDEX

Animal form and function, 842–60
  adaptations of, 843–45
  anatomy and physiology in, 843, 849*f*
  body size effects on physiology in,
    850–53
  body temperature regulation in,
    854–58
  dry habitats and, 842–43
  fitness trade-offs in, 843–45
  homeostasis in, 853–54
  organs and organ systems in, 849–50
  structure-function relationships in,
    845–46
  tissues in, 845–49
Animal models, disease, **382**
Animal movement
  bipedalism, 705–7
  diversification of limbs for, 649–50*f*
  ecdysozoan, 674–78
  echinoderm, 685–86
  evolution of muscle-generated, 972–73
  hydrostatic skeletons and, 644
  invertebrate, 687–88
  locomotion in, 983–88
  lophotrochozoan, 665–69
  mollusk, 663–64*f*, 669
  muscle contraction in, 973–77
  muscle tissue and, 847–48, 977–80
  neurons and muscle cells of, 637
  skeletal systems in, 980–83
  types of, 973
  vertebrate, 697–703
Animal nutrition, 882–901
  cellulose as dietary fiber for, 78
  digestion and absorption of nutrients
    in, 886–97
  fetal, during pregnancy, 1032–33
  food capture and mouthparts in,
    884–86
  food types in, 892–93
  four steps of, 883*f*
  glucose and nutritional homeostasis
    in, 897–99
  glycogen and starch hydrolysis in, 76,
    80–81
  herbivore limitation by poor, 1131–32
  human digestive tract, 887*f*
  human mineral and electrolyte
    requirements, 885*t*
  human vitamin requirements, 884*t*
  mammalian digestive enzymes in, 895*t*
  nutritional homeostasis in, 897–99
  nutritional requirements, 883–84
  waste elimination in, 896–97
Animal reproduction, 1013–36
  asexual vs. sexual, 1014
  deceitful communication in, 1094–95
  diversity in, 650–51
  ecdysozoan, 674–78
  echinoderm, 685–86
  in energy flow, 1149
  fertilization and egg development in,
    1018–21
  fitness trade-offs in, 843–45
  gametogenesis in, 1016–18*f*
  hormones and, 998–99
  invertebrate, 687–88
  life cycles of, 651–52
  life tables and, 1104–5, 1106*f*
  lophotrochozoan, 665–69
  mammalian pregnancy and birth in,
    1030–34
  mammalian reproductive systems in,
    1021–25
  mammalian sex hormones in, 1025–30
  mating behaviors and (*see* Mating)

  migration and, 1091
  net reproductive rate in, 1105
  parthenogenesis, 665
  sexual selection and, 475–78
  switching modes of, between asexual
    and sexual, 1014–16
  vertebrate, 693–94, 697–703
Anion(s), **22***f*, 781–83
Annelids, 638*t*, 657–58, 667
Aniston, Jennifer, 946, 947*f*
Annual growth rings, tree, 751, 1139
Annuals, **599**
Anolis lizards, 516, 1087–89, 1125
Anoxic oceans, 521
Anoxygenic photosynthesis, **186**, 539*t*,
  541, 545
Antagonistic muscle groups, **981**
Antagonistic-pair feedback systems, 857
Antarctic fur seals, 1086
Antbirds, 1124–25
Antenna complex, photosystem, **182**–83
Antennae, insect, **675**
Anterior body axis, **410**
Anterior pituitary gland, **1004**, 1006
Antheridia, **587**
Anthers, **827**
Anthocerophyta, 602
Anthophyta (angiosperms), 608–9*f*. *See
  also* Angiosperms
Anthrax, 532
Anthropocene epoch, 515
Anthropogenic biomes, 1069
Anthropoids, 704*f*–**5**
Antibacterial compounds, 815
Antibiotic resistance, 395, 454–56
Antibiotics, **532**, **1039**
  actinobacteria and, 546
  bacterial ribosomes and, 529
  fungi and, 613, 632
  in human semen, 1023*t*
  lysozyme as, 1039
  pathogenic bacteria and, 532–33
Antibodies, **206**, **717**, **1042**
  adaptive immune response and,
    1042 (*see also* Adaptive immune
    response)
  B-cell activation and secretion of,
    1050–51
  binding of, to epitopes, 1045
  cell adhesion proteins and, 206–7
  five classes of immunoglobulins and,
    1045*t* (*see also* Immunoglobulins)
  in fluorescent microscopy, B:18
  in HIV research, 717–18
  humoral response and, 1051–52
  as probes, B:16
  proteins as, 54
  specificity and diversity of, 1045–46
Anticodons, **327**, 328–29*f*
Antidiuretic hormone (ADH), **877**–78,
  1003, 1005–6
Antifungal compounds, 815
Antigen presentation, **1045**, 1048–49
Antigens, **1040**, 1044–50
Antilogs, logarithms and, B:9
Antiparallel DNA strands, **62**, 286–87
Antiporters, **770**, 786–**87**, **865**
Antiviral drugs, 714
Ants, 11–14, 623, 1063–64, 1083–85,
  1124–25, 1133–35
Anus, 644–45, 658, 886, 887*f*
Aorta, **917**
Aphids, 485, 672*f*, 768–69*f*
Aphotic zone, lake and pond, **1077**
Aphotic zone, ocean, **1079**
Apical-basal axis, **434**, 435–36

Apical buds, **735**
Apical cells, **434**
Apical dominance, **806**–7
Apical meristems, 436, 437*f*, **740**–42, 748
Apical mutants, 435
Apical side, epithelium, **849**
Apicomplexans, 573
Apodemes, **983**
Apomixis, **825**
Apoplast, 208–9*f*, **760**
Apoplastic pathway, 760–61, 785
Apoptosis, **232**, **407**, **1053**
  in cancer, 362–63
  in cell-mediated response, 1053–54
  in Huntington's disease, 382
  in organogenesis, 426
  as programmed cell death, 406*t*, 407–8
  tumor suppressors and, 232
Appendix, 887*f*, **896**
Apple maggot flies, 496–97
Apply (Bloom's taxonomy skill),
  B:29*f*–B:30*t*
Aquaporins, **98**, **865**, **896**
  large intestine water absorption and, 896
  in osmoregulation, 865, 877–78
  in plant water movement, 760
  selectivity and structure of, 98
Aquatic biomes
  animal locomotion in, 984, 985, 988
  animal transition to terrestrial
    environments from, 659–60
  behavior of oxygen and carbon dioxide
    in, 904–5
  direct and indirect human impacts
    on, 1076
  estuaries as freshwater and
    marine, 1078
  external fertilization in, 1018
  freshwater lakes, ponds, wetlands, and
    streams, 1077–78
  gas exchange by gills in, 906–7
  global net primary productivity of,
    1154–56
  global water cycle and, 1159–60
  nitrogen pollution in, 1160
  nutrient availability in, 1075–76
  oceans as marine, 1079
  osmoregulation in, 862–64, 866–68
  plant transition to terrestrial
    environments from, 584–95
  protists in, 553, 556
  salinity in, 1074
  water depth and water flow in,
    1074–75
Aqueous solutions
  acid-base reactions in, 28–30
  chemical evolution in, 24
  molarity of, 29
  properties of water and, 25–28
Aquifers, 1159–**60**
ara operon, 344, 345*f*
Arabian oryx, 842–43, 855, 864
Arabidopsis thaliana. *See* Mustard plant
Arabinose, 344, 345*f*
araC gene, **344**
AraC protein, 344, 345*f*
Arachnids, 678
Arbuscular mycorrhizal fungi (AMF),
  620–22, 630
Arcese, Peter, 1111–12
Archaea, **529**. *See also* Archaea domain
  aerobic vs. anaerobic respiration
    in, 171
  ammonia-oxidizing, 539*t*
  bacteria vs., as prokaryotes, 528–29
    (*see also* Bacteria; Prokaryotes)

citric acid cycle in, 162
human diseases and, 531
phospholipids in plasma membranes
  of, 529
phylogeny of, 536*f*
as prokaryotes, 107
Archaea domain. *See also* Archaea
  characteristics of, 529*t*
  key lineages of, 544, 548–49
  phospholipids in, 87
  phylogeny of, 7*f*–8, 536*f*
  as prokaryotic, 107, 528–29 (*see also*
    Prokaryotes)
Archaeopteryx, 512
Archegonia, **587**
Arctic terns, 1091
Arctic tundra, **1072**, 1073*f*, 1165, 1167
Ardipithecus genus, 705*f*
Area, metric units and conversions
  for, B:1*t*
Area and age hypothesis on species
  richness, 1145
Argentine ant, 1063–64, 1083–84, 1085.
  See also Ants
Arginine, 305–6
Aristotle, 445
Armor, defensive, 1129*f*
Arms races, coevolutionary, 1124, 1132
Arrangement, leaf, 738
Arrhenius, Svante, 1163
Arteries, 871, **917**–18, 921
Arterioles, **917**–18, 924
Arteriosclerosis, **924**
Arthrobacter, 546
Arthropods, **670**
  body plan of, 670–71
  as ecdysozoans, 673
  exoskeletons of, 983
  key lineages of, 638*t*, 675–78
  metamorphosis in, 672
  origin of wings in evolution of, 671–72
  as protostomes, 657–58
Articulations (joints), **981**–82
Artificial membranes, 88–89
Artificial selection, 5–6, 257, **453**, **579**.
  See also Crosses
Artiodactyls, 510–11*f*
Ascaris (roundworms), 238. *See also*
  Roundworms
Asci, **617**–18
Ascidians, 688
Ascomycetes, 617–18, 627, 631–32
Ascorbic acid, 884*t*
Asexual reproduction, **220**, **825**, **1014**.
  See also Animal reproduction
  animal, 650, 1014
  by fragmentation, 601, 603
  fungal, 617, 625
  mitosis and, 220 (*see also* Mitosis)
  plant, 825–26
  prokaryotic vs. eukaryotic, 566
  sexual reproduction vs., 247, 566, 1014
    (*see also* Sexual reproduction)
  switching between sexual and,
    1014–16
Asteroidea, 685
Asteroid impact hypothesis, 521–23
Asthma, 1056
Astragalus, 510
Astral microtubules, **225**
Astrobiologists, 531
Asymmetric cell division, plant, 433–34
Asymmetric competition, **1125**–26
Athletes
  EPO abuse by, 1003
  human growth hormone and, 374

Atlantic cod, 1153
Atlantic salmon, 483
Atmosphere. *See also* Air
  behavior of oxygen and carbon dioxide
    in, 903–4
  Cambrian explosion and oxygen in,
    519–20
  chemical evolution conditions in, 30,
    33–35
  end-Permian mass extinction and
    carbon dioxide in, 521
  greenhouse effect from carbon dioxide
    in, 1163
  plant nitrogen fixation of nitrogen
    from, 787–89
  prokaryotic oxygen revolution and
    oxygen in, 541–42
  water potential in, 758–59
Atomic level, anatomy and physiology
    at, 849*f*
Atomic number, **19**
Atomic weight, **20**, 29
Atoms
  chemical bonding of, in molecules,
    21–24
  in chemical structures, B:12–B:13
  energy transformation in, 138*f*
  linking carbon, into organic
    molecules, 36
  radioactive decay of, 446–47
  structure of, 19–21
ATP (adenosine triphosphate), **60**, **143**
  actin filaments and, 128–29
  in active transport by pumps, 99–100
  in attachment of amino acids to
    tRNAs, 327
  in Calvin cycle, 191–92
  cell usage of, 119
  in cytokinesis, 228
  in endergonic reactions, 143–44
  as energy source for nucleic acid
    polymerization, 60
  energy transfer from glucose to, 81
  in enzyme catalysis, 145
  in glycolysis, 158*f*–61
  GTP vs., 162, 325
  microtubules, vesicle transport, and,
    130–31
  in muscle contraction, 974–76, 979
  in photosynthesis, 178
  in photosystems, 185–89
  prokaryotic metabolic diversity in
    producing, 538–41
  synthesis of, by ATP synthase and
    chemiosmosis, 168–70
  synthesis of, from glucose oxidation in
    cellular respiration, 154–56
  synthesis of, in fermentation, 172–73
  synthesis of, in mitochondria, 115–16
  yield of, in cellular respiration, 170–72
ATP synthase, 168–**69**, 170–71,
    185–86, 189
Atria, **919**, 921
Atrioventricular (AV) node, **922**
Atrioventricular valves, 919
Attachments, intercellular. *See* Cell–cell
    attachments
Attenuated virus vaccines, 1055
Australia, 701
Australian redback spiders, 1020
Australopithecus genus, 705*t*, 706, B:11
Autocrine signals, **992**

Autoimmune diseases, 1056
Autoimmunity, **1056**
Autonomic nervous system, **942**–43,
    977–78
Autophagy, **126**
Autopolyploids, **497**–98
Autoradiography, 948–49, B:15–B:16
Autosomal genes, 258*t*, 268
Autosomal inheritance, **269**, 277–78, 299
Autosomal traits
  Huntington's disease as, 378
  pituitary dwarfism as, 369
  recessive and dominant, 277–78
Autosomes, **238**, 240*t*
Autotrophs, **176**–77, **538**, **1149**–50
Auxin, **798**
  in apical dominance, 806–7
  as blue-light phototropic hormone,
    797–800
  in cell and tissue cultures, B:22
  as gravitropic signal, 804–5
  as morphogen in plant
    development, 412
  as plant growth regulator, 813*t*
  in setup of body axes in
    embryogenesis, 435–36
  structure and function of, 210*t*
Averages, standard error and, B:6–B:7
Averof, Michalis, 671
Aves, 703. *See also* Birds
Avian malaria, 1062
Axel, Richard, 966
Axes, body, 425, 435–36
Axes, graph, B:4–B:5
Axes, leaf, 436–38
Axillary buds, **735**
Axon hillock, **942**
Axons, **847**, **930**
  action potential propagation down,
    936–38
  measuring membrane potentials of,
    932–33
  in neuron anatomy, 847, 930
  squid, 130
Azoneme, **132**

B

B-cell receptors (BCRs), **1043**–47, 1048,
    1050–51
B-cell tumors, 1045–46
B cells, **1043**
  activation and function of, 1054*t*
  activation of, 1050–51
  antigen receptors of, 1044–47
  as lymphocytes, 1042–43
  memory cells and, 1053–55
Baboons, 1013*f*
BAC libraries, **390**–91
Bacillus anthracis, 532
Bacillus thuringiensis, 385, 545
Backbones
  nucleic acid, 59–63, 65
  peptide, 46, 48–50
Background extinctions, 520–21. *See also*
    Extinctions
Bacteria, **529**. *See also* Bacteria domain;
    Prokaryotes
  aerobic vs. anaerobic respiration
    in, 171
  archaea vs., 528–29
  in bioremediation, 151

cell division by binary fission in,
    228–29
in cell theory experiment, 3–5
citric acid cycle in, 162
DNA of, 107–8
DNA synthesis in, 291*f*
evolution of antibiotic resistance in,
    454–56
flagella of, 131
gene regulation in (*see* Gene
    regulation, bacterial)
introducing recombinant plasmids into
    cells of, for transformation, 372
isolating, with enrichment
    cultures, 534*f*
as model organism (*see Escherichia coli*
    [*E. coli*])
nitrogen-fixing, 542–43, 787–89, 1133,
    1158, 1160
optimal temperature and pH for
    enzyme function in, 148
origin of viruses in symbiotic, 723
pathogenic, 531–33
peptidoglycan as structural
    polysaccharide in, 76, 78, 529
phospholipids in, 87
photosynthesis in, 177, 184,
    186–87, 515
proteins as defense against, 54
RNA polymerase in, 318
transcription and translation in,
    318–20*f*, 324*f*–25, 329–30
in transgenic plant development,
    385–86
Bacteria domain. *See also* Bacteria
  characteristics of, 529*t*
  key lineages of, 544, 545–47
  phylogeny of, 7*f*, 8, 536*f*
  as prokaryotic, 107, 528–29 (*see also*
    Prokaryotes)
Bacterial artificial chromosomes (BACs),
    390–91
Bacteriophages, 50, 715, **716**
Bacteriorhodopsin, 169, 541
Baculum, **1022**–23
Baker's yeast. *See Saccharomyces
    cerevisiae*
Balancing selection, **474**–75*t*
Ball-and-stick models, **24**, B:12*f*–B:13
Baltimore, David, 725
Bands, molecular, B:14–B:16
Banting, Frederick, 897
Baobab trees, 731*f*
Bar charts, B:5–B:7
Bar coding, species, **1174**
Bark, **750**, 750
Barnacles, 677, 1126, 1127*f*
Barometric pressure, animal sensing of,
    646, 647*t*
Baroreceptors, **924**
Barriers
  to conception, 1030*t*
  to dispersal, 1063
  to pathogens, 848–49, 1038–39
  plant cuticle as, 743, 815
  to polyspermy, 421–23
Bartel, David, 68–69
Basal body, **132**
Basal cells, **434**
Basal lamina, epithelium, **849**
Basal metabolic rate (BMR), **851**–52
Basal mutants, 435–36
Basal transcription factors, **320**,
    325*t*, **356**
Basalts, 521
Base, exponential notation, B:9

Base(s), **28**–30, 38. *See also* Nitrogenous
    bases, nucleotide
Base *e*, exponential notation, B:9
Base 10
  exponential notation and, B:9
  metric system scientific notation and,
    B:1–B:2
Basic-odor hypothesis, 966
Basic side chains, 43*f*, 44
Basidia, **617**–18, 626–27
Basidiomycetes, 617–18, 626–27, 630
Basilar membrane, **956**–57
Basilisk lizards, 972*f*
Basket stars, 684
Basolateral side, epithelium, **849**
Bateman, A. J., 475
Bateman-Trivers theory, 475
Bates, Henry Walter, 1129–30
Batesian mimicry, 1129–**30**
Bateson, William, 273–75
Bats, 694, 952–53, 957–58
Bayliss, William, 893
Bdelloids, 650, 665
Bdellovibrios, 547
Beadle, George, 305
Beagle naval ship, 449–50
Beaks, cepalopod, **669**
Beaks, finch, 494–95, 1128
Beaumont, William, 890
Bedbugs, 1020
Bees, 594, 826–27, 856, 1092–93, 1097–
    98, 1133
Beeswax, 86*f*
Beetles, 138, 139*f*, 268, 1166*t*, 1167
Behavior, **1082**–85, 1129*f*. *See also*
    Behavioral ecology
Behavioral contraception
    methods, 1030*t*
Behavioral ecology, **1083**
  altruism in, 1095–98
  behaviors in, 1082–83, 1084–85
  communication in, 1091–95
  five questions of, 1085
  foraging in, 1085–87
  mating in, 1087–89
  migration in, 1089–91
  proximate and ultimate causation in,
    1083–84
Behavioral prezygotic isolation, 491*t*
Bendall, Faye, 184
Beneficial alleles, **484**–85
Beneficial mutations, 313
Benign tumors, **233**
Benthic habitats, 519
Benthic organisms, 651, **652**
Benthic zone, lake and pond, **1077**
Benthic zone, ocean, **1079**
Benzene, 27*t*
Benzo[α]pyrene, 299
Beriberi, 147
Bermingham, Eldredge, 500–501
Bertness, Mark, 1130, 1131*f*
Beschta, Robert, 1153
Best, Charles, 897
Beta-blockers, 210
β-carotene, 180–81*f*, 385
β-galactosidase, 338–43
β-globin genes, 351, 399
β-glucose, 74, 75*f*, 76
β-glycosidic linkages, 75*f*–76
β-pleated sheet (beta-pleated sheet),
    48–49
β-tubulin, 129
Beutler, Bruce, 1040
Biewener, Andrew, 987–88
Bikonta, 558

Boldface page numbers indicate a glossary entry; page numbers followed by an
f indicate a figure; those followed by *t* indicate a table.

Bilateral symmetry, **641**–44, 682–83*f*
Bilaterians, 638*t*, 641–**42**
Bile, **894**–96
Binary fission, **228**–29
Bindin, 421
Biochemical pathways, plant, 766
Bioclimate envelope models, 1183
Biocontrol agents, consumers as, 1133
Biodiversity, **1173**
  biological benefits of, 1184–87
  changes in, through time, 1174–75
  economic and social benefits of
    ecosystem services and, 1187–88
  estimating, using surveys, 1175
  ethical issues on, 1188–89
  genetic, species, and ecosystem levels
    of, 1173–74
  global climate change and, 1073
  loss of, in Borneo, 1172–73
  mapping, 1176–77
  preserving ecosystem function and,
    1189–93 (*see also* Conservation
    biology)
  threats to, 1178–84 (*see also* Threats,
    biodiversity)
  tree of life and, 1173
  ultimate causes of losses in, 1189
Biodiversity hotspots, **1177**, 1190–91
Biofilms, **532**
Biogeochemical cycles, **1156**, 1159–62.
  See also Nutrient cycling
Biogeography, **494**, **1062**. *See also*
  Geographic distribution and
  abundance of organisms
Bioindicators, lichens as, 631
Bioinformatics, 391–**92**. *See also*
  Genomics
Biological benefits, biodiversity, 1184–87
Biological control agents, 628
Biological evolution, chemical evolution
  vs., 19, 57–58. *See also* Chemical
  evolution; Evolution
Biological imaging techniques, 119
Biological impact, prokaryotic, 530
Biological species concept, **490**, 492*t*
Biology
  astrobiology, 531
  biogeography, 494
  biomechanics, 973, 984–88
  cell theory and cells in, 2–5 (*see also*
    Cell[s])
  characteristics of living organisms in, 2
    (*see also* Life; Organisms)
  chemical evolution and (*see* Chemical
    evolution)
  chemical reactions in (*see* Chemical
    reactions)
  conservation biology and
    bioremediation, 533 (*see also*
    Conservation biology)
  developmental, 405–6 (*see also*
    Development)
  evo-devo (evolutionary-
    developmental), 416
  forensic, 376
  genetic engineering in (*see* Genetic
    engineering)
  microbiology, 530
  nanobiology, 714
  science and methods of, 1, 9–14 (*see
    also* Quantitative Methods boxes;
    Science; Bioskills; Process boxes)
  speciation, phylogenetic tree of
    life, and taxonomy in, 6–9 (*see
    also* Phylogenies; Speciation;
    Taxonomy; Tree of life)

systems biology, 402
  theory of evolution by natural
    selection in, 5–6 (*see also*
    Evolution; Natural selection)
Bioluminescence, **573**
Biomagnification, **1152**–53
Biomass, **1069**, **1149**, **1180**
  loss of, by habitat fragmentation and
    degradation and, 1180
  net primary production and, 1149–50
    (*see also* Net primary productivity
    [NPP])
  plants and, 776
  recovery of, after disturbances,
    1191–92
  in terrestrial biomes, 1069
  trophic structure and (*see* Trophic
    structure)
Biomechanics, **973**, 984–88
Biomes, **1068**. *See also* Aquatic biomes;
  Terrestrial biomes
  effects of global climate change
    on, 1182
  most productive, 1154–56
Biophilia, 1188
Bioprospecting, **1187**
Bioremediation, **151**, **533**, 632. *See also*
  Conservation biology
Bioskills, B:1–B:30
  biological imaging with microscopy
    and x-ray crystallography,
    B18–B:21
  cell and tissue culture methods,
    B:21–B:22
  combining probabilities, B:8
  Latin and Greek word roots, B:3
  making concept maps, B:28
  metric system and significant figures,
    B:1–B:3
  model organisms, B:23–B:26
  primary literature and peer review,
    B:26–B:27
  reading chemical structures, B:12–B:13
  reading graphs, B:4–B:6
  reading phylogenetic trees, B:10–B:11
  separating and visualizing molecules,
    B:13–B:16
  separating cell components by
    centrifugation, B:17–B:18
  using Bloom's taxonomy, B:29–B:30*t*
  using logarithms, B:9
  using statistical tests and interpreting
    standard error bars, B:6–B:7
Biosphere, 1060*t*, **1061**, 1148–49. *See
  also* Ecosystem ecology; Global
  ecology
Biotechnology, **374**
  agricultural, 385–86 (*see also*
    Transgenic organisms)
  recombinant DNA technology in,
    374 (*see also* Recombinant DNA
    technology)
Biotic factors, **1062**
  in community structure, 1135
  in density-dependent population
    growth, 1108–12
  in geographic distribution and
    abundance of organisms,
    1062–65, 1102
  of natural biomes, 1068–69
Bipedal organisms, **705**, 706–7
Bipolar cells, **960**, 961*f*
Birds
  adaptive radiations of, 517
  allopatric speciation in, 494–95
  biotic factors in distribution of, 1062

crop as modified esophagus in, 889
disruptive selection of, 473–74
evolution of wings and flight in,
  694–95
female choice for good alleles in, 475–76
female reproductive tract in, 1024
form-function correlation in, 845–46*f*
fossils of, 512
gizzards as modified stomachs in,
  891–92
impacts of global climate change on,
  1166–67
loss of species of, from invasive
  species, 1181
magnetoreception by, 968–69
mapping species richness and
  endemism of, 1176
mating behaviors of, 1020, 1082*f*
maturation of eggs of, 1024*f*
migration of, 1089–91
neurogenesis in, 948–49
nitrogenous wastes of, 865
parental care in, 694
as reptiles, 696–97, 703
seed dispersal by, 835–36
sexual dimorphism in, 478*f*
survivorship curves for, 1105
ventilation of lungs of, 910–11*f*
in vertebrate phylogeny, 691
Birth, human, 473, 1034
Birth control, human, 1029–30
Birth rates, 1103–7, 1108, 1109
Bisphenol A, 1000
Bis prefix, 158
Bitterness taste receptors, 965
Bivalents, 240*t*, **242**, 244, 246*f*
Bivalves, **662**, 668
Black-bellied seedcrackers, 474
Blackburn, Elizabeth, 296
Black-tailed prairie dogs, 1095–97
Bladders, **872**, 879
Blade, leaf, **737**
BLAST program, 392
Blastocoel, **425**
Blastocysts, **424**
Blastomeres, **423**, 425
Blastopores, **425**
Blastula, **423**
Bleaching, coral, 1167–68
Blending inheritance hypothesis, 257
Blind spot, 960
Blindness, 385, 547
Blobel, Günter, 123–24
Blood, **847**. *See also* Leukocytes (white
  blood cells)
  acclimatization vs. adaptation for
    oxygen content of, 459–60
  alveoli and, in mammal lungs, 909, 916
  in bird lungs, 910
  blood vessels and, 917–18
  bones, bone marrow, and, 982–83
  carbon dioxide transport in, 915–16
  chromatin remodeling in, 351
  circulation of, through hearts, 919–21
  connective tissue of, 846*f*
  glucagon and sugar levels in, 54
  glucose carrier protein for cells of
    human, 99
  hemoglobin and oxygen in, 54 (*see also*
    Hemoglobin)
  hemophilia and human, 278
  homeostasis for oxygen in, 911–12
  human ABO types of, 272, 468–69
  immune systems and transfusions
    of, 1053
  insulin and sugar levels in, 47–48

measuring hormone amounts in, using
  radioimmunoassay, B:16
microscopic images of cells of
  human, 3*f*
oxygen transport in hemoglobin of,
  912–15
patterns in pressure and flow of,
  923–25
regulation of oxygen levels in, 1003
serum of, in cell cultures, B:21
sickle-cell disease and human, 48
Blood cells, 912, 982–83, **1039**–41, 1042*t*
Blood pressure, 923–24
Blood serum, 233
Blood types, human, 272, 468–69
Blood vessels
  cancer and, 233*f*
  in closed circulatory systems, 917–18
  kidney, 871, 873, 876, 877*f*
Blooms, algal, 555–56, 1148*f*, 1160
Bloom's taxonomy, B:29–B:30*t*
Blotting techniques, B:16
Blue light, 179–81*f*
Blue light phototropic responses, plant,
  796–800, 818*f*
Blue mussels, 1130, 1131*f*
Blue opsin, 963–64
Bluegill sunfish, 1094–95
Bluetongue virus, 727
Bmp4 gene, 458–59*f*
Body. *See also* Body plans
  animal locomotion and size of, 986–88
  animal regulation of temperature of
    (*see* Thermoregulation)
  barriers to entry of pathogens into,
    1038–39
  brassinosteroids in size of plant, 812
  Cambrian explosion and new, 519–20
  convergent evolution of streamlined,
    509–10
  definition of axes of, in
    gastrulation, 425
  differential gene expression in
    development of, 410–15 (*see also*
    Differential gene expression)
  directional selection on size of cliff
    swallow, 473
  fungal, 615–18
  gas exchange through diffusion across,
    905–6, 916
  genes and proteins in setup of axes of,
    in plant embryogenesis, 435–36
  organogenesis and, 426–29
  physiological effects of size of animal,
    850–53
  plant (*see* Primary growth, plant;
    Secondary growth, plant)
  plant nutrition and plant, 775
  skeletal systems in maintenance of
    posture of, 980
Body mass index, **899**
Body odor, genotypes and, 470
Body plans, **638**. *See also* Bodies
  arthropod, 670–71
  bilateral symmetry in animal, 641–44
  chelicerate, 678
  chordate, 686–87
  comparative morphology and animal,
    638 (*see also* Evolutionary
    innovations, animal)
  crustacean, 677
  echinoderm, 682–83
  insect, 675
  modular protostome, 660
  mollusk, 662–64
  myriapod, 675

Bogs, **1077**
Bohr shift, **914**, 915
Boluses, 884
Bombadier beetles, 138, 139*f*
Bond saturation, lipid, 85–86, 90
Bonds, B:12–B:13. *See also* Covalent bonds; Hydrogen bonds; Ionic bonds
Bone, 427–29, **689**, 846*f*, **847**, **981**–82
Bone marrow, 982–83, **1043**
Bone marrow transplants, 384
Bony fishes, osmoregulation by, 862–63
Boreal forests, **1072**, 1157
Bormann, Frank, 1149–50
Borneo, 1172–73, 1191
Borthwick, H. A., 800–801
Both-and rule, probability, B:8
Bottom-up limitation hypothesis, herbivore population, 1131–32
Bottom-up research approach, chemical evolution, 32–33
Bouchet Philippe, 1175
Boveri, Theodor, 257, 266–67
Bowerbird, 1082*f*
Bowman's capsule, **873**, 878*t*
Box-and-whisker plots, B:5–B:6
Bracket fungi, 617, 626–27
Brain stem, **944**
Braincase, **707**
Brains, **643**
anatomy of, 944–46
central nervous system and, 929
diffusion spectrum imaging of neurons in, 928*f*
in hormonal control of urine formation, 877
Huntington's disease and degeneration of, 278, 378, 382
in initiation of muscle contraction, 976–77
neural tubes and, 427
origin of, in animal evolution, 643
primate, 705
prions and diseases of, 53, 369–70
sensory organ transmission of information to, 954
in thermoregulation, 856–57
vertebrate, 688
Brakefield, Paul, 1167
Branches, phylogenetic tree, **506**, B:10–B:11
Branches, plant, **735**
Branching adaptations, animal surface-area, 852–53
Brassinosteroids, 210*t*, **812**, 813*t*
Bread mold, 305, 629. *See also Neurospora crassa*
Bread yeast. *See Saccharomyces cerevisiae*
Breathing. *See* Ventilation
Breeding. *See also* Artificial selection
captive, 482–83, 1189–90
selective, 257 (*see also* Crosses)
Brenner, Sydney, 310–11, B:25
Brewer's yeast. *See Saccharomyces cerevisiae*
Bridled goby fish, 1109–11
Briggs, Winslow, 799
Brine shrimp, 671
Bristlecone pine trees, 432*f*, 607, 732
Brittle stars, 684
Britton, Roy, 324

Broca, Paul, 944–45
Bronchi, **909**
Bronchioles, **909**
Bronstein, Judith, 1134
Brown algae, 7, 552*f*, 557, 568–69, 574. *See also* Algae
Brown fat cells, 118
Brown kiwis, 476
Brown tree snake, 1181
Brushtail possums, 1031*f*
Bryophytes (mosses), 580–81, 601–2, 824*f*–25
Bryozoans, 638*t*, 665
Buchner, Hans and Edward, 158
Buck, Linda, 966
Budding, animal, 654, 685, 688, **1014**
Budding, viral, 720, 721*f*
Buds, axillary and apical, 735
Buds, flower, 826
Buffers, pH, **30**, **916**
Buikema, Arthur, 1136–37
Building materials, 579–80
Bulbourethral gland, **1022**–23*f*
Bulbs, 739
Bulk flow, **762**
Bulk-phase endocytosis, **127**
Bundle-sheath cells, **193**–94, **766**
Burials, fossils and, 512
Bursa-dependent lymphocytes, 1043. *See also* B cells
Bursting, viral, 720, 721*f*
Butter, 86*f*
Buttercup root, 72*f*
Butterflies, 1091, 1112–13, 1129–30
Bypass vessels, 920–21

C

C-terminus, 46, 328
C$_3$ pathway, **193**
C$_4$ pathway, **193**–94, **766**
Cacti, 194, 736–37
Cactus mice, 835–36
Cadherins, **207**
Caecilians, 700
Caenorhabditis elegans. See Roundworms
Caffeine, 815
Calcium. *See also* Calcium ions
animal bones and, 982–83
in ecosystem nutrient cycles, 1156
as human nutrient, 885*t*
as plant nutrient, 780*t*
Calcium carbonate, 556–57, 572, 600, 981
Calcium ions
as electrolytes, 862
in polyspermy prevention, 421–23
protein folding and, 53
as second messengers, 213*t*
Calcium phosphate, 981
Calculators, scientific, B:9
Calibration, accuracy and, B:2–B:3
California mussel, 1137–38
Callus, **744**, B:22
Calmodulin, 53, 458
Calvin, Melvin, 178, 190–91
Calvin cycle, **178**, 190–95, 541, 766. *See also* Photosynthesis
Calyx, **826**
CAM (crassulacean acid metabolism), **194**–95, 766, **766**, 1070
Cambium, **748**–50

Cambrian explosion, **518**–20, **636**–37
cAMP (cyclic adenosine monophosphate; cyclic AMP), 213*t* **1009**–10
Cancer, **232**
anticancer drugs, 607
cell cultures of, B:21, B:22*f*
chromosome mutations and, 314–15
defective Ras proteins and, 214
as family of diseases, 234
from gene therapy for severe combined immunodeficiency (SCID), 384–85
genetic basis of uncontrolled cell growth in, 361–63
genetic testing for, 382
homology and, 451
loss of cell-cycle control and, 233–34
mutations and, 298–301
properties of cancer cells, 233
retroviruses and, 728
telomerase and, 297
types and severity of, 232*t*
types of cell division defects in, 232–33
Cannabis sativa, 828*f*
Cannibalization, mating and, 1020
Capillaries, **852**–53*f*, 873, **917**–18, 923
Capillarity, **761**
Capillary action, plant, 759, 761–62
Capillary beds, **917**–18
Caps, mRNA, 323–24
Capsaicin, 835–36
Capsids, 285–86, **715**, 720
Captive breeding, 482–83, 1189–90
Carapace, **677**
Carbohydrates, **72**
animal digestion and transportation of, 883, 888, 893–94
cell characteristics and, 107
cell identity functions of, 78–80
energy storage functions of, 76, 80–81
fermentation of, 540
glucose production from, 155
glycosylation of, to form glycoproteins, 124
in metabolic pathways, 157–58
monosaccharides, 73–75
photosynthetic production of, 176–78 (*see also* Photosynthesis)
polysaccharide types and structures, 77*t*
polysaccharides, 75–78
processing of, in photosynthesis, 195
reduction of carbon dioxide to produce (*see* Calvin cycle)
structural functions of, 76, 78, 117
types of, 72–73
Carbon
in amino acid structure, 42
Calvin cycle fixation of, during photosynthesis (*see* Calvin cycle)
in citric acid cycle, 162
in ecosystem nutrient cycles, 1156
electronegativity of, 21
functional groups and atoms of, 37*t*
lipids, hydrocarbons, and, 85–86
in living organisms, 19
in metabolic pathways, 157
in net primary productivity, 1069
in organic molecules, 36–38
prokaryotic strategies for obtaining and fixing, 538–41
in protein secondary structure, 48–49
in redox reactions, 141–42
simple molecules from, 23
Carbon–carbon bonds, 36

Carbon cycle, **614**
global, 1161–62
plants and, 579
saprophytic fungi and, 614–15
Sphagnum moss and, 602
Carbon dioxide
animal gas exchange and, 902–3
atmospheric, in global carbon cycle, 1161
behavior of, in air, 903–4
behavior of, in water, 904–5
carbonic anhydrase protein and, 54
covalent bonds of, 23
end-Permian atmospheric, 521
global climate change and, 556–57, 579, 1073
land plants and, 584
mechanisms for increasing concentration of, 193–95
in origin-of-life experiments, 34–36
oxidation of acetyl CoA to, in citric acid cycle, 162–65
passage of, through stomata of leaves, 192–93
photosynthetic carbohydrate production from, 177–78 (*see also* Photosynthesis)
plant pathways to increase, in dry habitats, 766
plant water loss and requirements for, 754–55
in prokaryotic metabolism, 539*t*, 541
in pyruvate processing, 161–62
reduction of, in Calvin cycle (*see* Calvin cycle)
transport of, in blood, 915–16
in volcanic gases, 30
Carbon fixation, **190**, 541. *See also* Calvin cycle
Carbonic acid, 30
Carbonic anhydrase, 54, **890**, 915–16
Carboniferous period, 582
Carbonyl functional group, 37*t*–38, 45–46, 48–49*f*, 73–74
Carboxy-terminus, 46, 328
Carboxyl functional group, 37*t*–38, 42, 45–46, 85–87, 162
Carboxylic acids, 37*t*, **162**
Carboxypeptidase, 895*t*
Cardiac cycle, **923**. *See also* Hearts
Cardiac muscle, 847*f*–**48**, **978**
Cardiac tissue, 200*f*
Cardiovascular disease, **924**
Carnauba palms, 743
Carnivores, 579, **647**–48, **1128**
Carnivorous plants, **789**–90
Carotenes, 180–82
Carotenoids, **180**. *See also* Pigments, photosynthetic
in photosynthesis, 180–82
in sexual selection, 475–76
synthesis of, by aphids, 485
Carpellate flowers, 827(*footnote*)
Carpels, **439**, **592**, **827**–28
Carrier proteins, **99**, **769**, **864**
facilitated diffusion via, 99
in osmoregulation, 864–65
in translocation, 769–72
Carriers, recessive trait, **277**–78
Carrion flowers, 594
Carroll, Sean, 419, 649–50*f*, 660
Carrying capacity, **1108**–9*f*, 1112, 1130–31, 1189
Carson, Rachel, 999–1000, 1152–53
Cartilage, **689**, 846*f*, **847**, **981**–82
Cascades, regulatory gene, 413–14

Chemical equilibrium, **30**
Chemical evolution, **18**–40
  atoms, ions, and molecules as building blocks of, 19–24
  biological evolution vs., 57–58
  chemical reactions, chemical energy, and, 30–32
  deep-sea hydrothermal vents and, 18*f*
  importance of carbon and organic molecules from, 36–38
  monosaccharides in, 73, 74
  nucleic acid formation by, 60–61
  nucleotide production by, 59
  Oparin–Haldane theory of, 41–42
  pattern and process components of theory of, 18–19
  polysaccharides and monosaccharides in, 78
  properties of water and early oceans as environments for, 25–30
  protein polymerization in, 44–45, 55
  proton-motive force in, 170
  research approaches and model systems for, 32–36
  RNA as first self-replicating molecule in, 57–58, 68–69
  switch from, to biological evolution, 19
Chemical modifications, post-translational, 332
Chemical reactions, **28**, 136–53
  acid-base reactions, 28–30
  chemical evolution conditions for, 30
  chemical structures and, B:13
  dynamics of, 30–31
  effects of temperature and concentration on reaction rates of, 139–40*f*
  energetic coupling of exergonic and endergonic reactions in, 141–44
  energy transformations in, 31, 137–39
  enzyme catalysis of, 54, 144–47, 150–52
  eukaryotic organelles and, 110
  factors affecting enzyme function in, 148–50
  life, energy, enzymes, and, 136–37
  in origin-of-life experiments, 33–36
  plasma membranes and, 84
  rate of, in cells, 119
  spontaneous, 31–32
Chemical signaling
  in animals, 992–93 (*see also* Hormones, animal)
  in cell–cell interactions of development, 406*t*, 407
  in differential gene expression and body pattern formation, 410–15
Chemical structures, reading, B:12–B:13
Chemicals
  animal sensing of (*see* Chemoreception)
  biomagnification of, 1152–53
  as constitutive defenses, 1129*f*
  endocrine disruptors, 999–1000
  transfer of toxic, from human mother to fetus, 1033
Chemiosmosis, **169**–70, 186
Chemokines, **1040**–41, 1042*t*
Chemolithotrophs, **538**
Chemoorganotrophs, **538**–41
Chemoreception, **964**–67
  gustation (taste) as, 964–65
  olfaction (smell) as, 964, 965–67

Chemoreceptors, **953**
Chestnut blight, 613, 632, 815
Chewing, 888
Chiasmata (chiasma), **243**
Chicken-of-the-woods fungus, 612*f*
Childbirth, 1034
Chimpanzees, 705, 724–25, 1098, B:10–B:11
Chitin, **76**, 77*t*, 78, 618
Chitons, **662**, 669, 983
Chlamydiae, 544, 547
Chlamydia trachomatis, 547
Chlamydomonas reinhardtii, B:25
Chloride cells, 866–67, 868
Chloride ions, 97–98, 862
Chlorophyll *a*, 180–81*f*
Chlorophyll *b*, 180–81*f*
Chlorophylls, 178, **180**–82, 796. *See also* Pigments, photosynthetic
Chloroplasts, **116**, **178**, **743**
  Calvin cycle in, 191
  in green plants, 580
  in moss cells, 176*f*
  origin of, in protists, 563–64
  photosynthesis in, 116, 117*t*, 118, 178–79*f* (*see also* Photosynthesis)
  phototropins and, 797
Choanocytes, **639**
Choanoflagellates
  animals and, 638–39
  fungi and, 619
Chocolate, fungi and, 614
Choice, behavioral, 1084–85
Cholecystokinin, **893**
Cholesterol, 87, 90, 91*f*, 995
Cholodny, N. O., 799
Cholodny–Went hypothesis, 799
Chondrichthyes, 698
Chordates, **426**, **686**
  body plans of, 686
  brains and spinal chords in, 427
  as deuterostomes, 682
  key lineages of invertebrate, 687–88
  notochords in, 426
  phylogenetic tree of, 638*t*, 690*f*
  subphyla of, 686–87
  vertebrate (*see* Vertebrates)
Chorion membrane, 694
Chromatids, **220**, 226*t*
Chromatin, **223**–26, 349–53
Chromatin remodeling, **349**–53
  bacterial vs. eukaryotic gene regulation and, 360–61*f*
  chromatin's basic structure and, 350–51
  in eukaryotic transcription initiation model, 356, 357*f*
  evidence for, in active genes, 351
  inheritance of chromatin modifications and, 352–53
  mechanisms of, 351–52
Chromatin-remodeling complexes, **352**, 356, 357*f*
Chromatography, 180, 190–91, B:14–B:15
Chromosome theory of inheritance, **266**–79. *See also* Genetics; Mendelian genetics
  allele codominance and incomplete dominance in, 272, 273*f*
  exceptions and extensions to Mendelian rules in, 269–77*t*
  human inheritance in, 277–79

importance of, 256–57
  linkage in, 270–71, 274*f*
  meiosis as explanation for Mendelian principles in, 266–67
  multiple allelism in, 271–72
  phenotypes vs. genotypes in, 273–75
  pleiotropic genes in, 272–73
  quantitative traits and polygenic inheritance in, 275–76
  sex linkage in, 268–69
  testing of, 267–69
Chromosomes, **107**, **220**. *See also* Chromosome theory of inheritance; DNA (deoxyribonucleic acid); Genes
  in characteristics of domains of life, 529*t*
  in chloroplasts, 116
  discovery of, 220
  eukaryotic, 112
  gene loci on, 266
  genetic maps of (*see* Genetic maps)
  Hox genes on, 414–15
  linkage and linked genes of, 270–71, 274*f*
  in meiosis, 238
  in mitosis, 223–27
  mutations of, 314–15, 483–84
  nuclear lamina and, 119
  number of, found in some organisms, 223
  number of human, 238, 239
  prokaryotic, in nucleoids, 107–8
  sex (*see* Sex chromosomes)
  sister chromatids of, 239–40
  sympatric speciation by polyploidy in, 497–99
  synthesis of telomeres of, 295–97
  types of, 238
  vocabulary for cell makeup of, 240*t*
Chrysanthemum, 815
Chymotrypsin, 895*t*
Chytrids, 617, 618, 625, 626*f*, 629
Cichlid jaws, 884–86
Cigarette smoke, 298
Cilia, **131**
  cell movement by, 132–33
  flagella vs., 131–32
  protist, 565–66
  rotifer, 665
Ciliates, 573
Circadian clock, **1089**–90
Circular chromosomes, 529*t*
Circulatory systems, **903**, 912–25
  blood pressure and blood flow in, 923–25
  closed, 917–19
  hearts in, 919–23
  open, 916–17
  oxygen and carbon dioxide transport in blood of, 912–16
  respiratory systems and, in gas exchange and circulation, 902–3 (*see also* Respiratory systems)
Cis surface, Golgi appartus, 113*f*–14, 124–25
Cisternae, 113–14, 124–25
Cisternal maturation, **125**
Citrate, 162
Citric acid, 632, 1023*t*
Citric acid cycle, 156, **162**–65
Citrulline, 305
Clades, 491–93, **507**, 536, B:10–B:11. *See also* Lineages; Monophyletic groups
Cladistic approach (analysis), **507**–8
Cladograms, 508

Clams, 662, 668, 916–17
Class I and class II MHC proteins, **1048**–51
Classification, viral, 725–26
Clausen, Jens, 736
Clay, 781–82
Cleaner shrimp, 1133, 1134*f*
Cleavage, **423**–24, 660–61
Cleavage furrow, **228**
Cleaver, James, 299–300
Clefts, chemical structure, B:13
Clements, Frederick, 1136–37
Clements–Gleason dichotomy, 1136–37
Cliff swallows, 473
Climate, **1065**
  global change of (*see* Global climate change)
  global patterns in, 1065–67
  green plants, water, and moderation of, 578
  tree growth rings and research on, 751
  weather and, 1065, 1066
Climax community, **1136**
Clinton, Bill, 399
Clitoris, **1024**–25*f*
Cloaca, **878**–79, **897**, **1019**
Clonal selection theory, **1048**
Clones, **247**, **409**, **825**, **1014**
  asexual reproduction and, 247, 825–26, 1014
  in clonal selection theory, 1048
  fungal, 625
  nuclear transfer techniques for creating, 409
Cloning vectors, **370**. *See also* Plasmids
Closed circulatory systems, **917**–19
Clostridium acetium, 539*t*, 540
Clotting, blood, 278
Club fungi, 617–18, 630. *See also* Basidiomycetes
Club mosses, 603
Clumped distribution, 1102
CNGB3 gene, 481–82
Cnidaria, 638*t*, 641, 653–54
Cnidocytes, **653**–54
Coal, 579, 582, 603
Coat color, mouse, 308, 309*f*, 313
Cocaine, 815
Cochlea, **956**–57
Coconut palms, 1106–7
Coding strands, DNA, 318–20*f*
Codominance, **272**, 277*t*
Codons, **310**
  in genetic code, 310–12
  in genome annotation, 392–93
  in mRNA specification of amino acids, 325–26
  in point mutations, 313–14
  in ribosomal translation, 328–31
  in wobble hypothesis on types of tRNAs, 327–28
Coefficient of relatedness, **1096**
Coelacanths, 699, 964
Coelom, **644**, **658**
  formation of, in embryonic development, 644–45
  lophotrochozoan, 661
  origin of, in animal evolution, 643–44
  protostome vs. deuterostome, 658
Coelomates, **644**
Coenocytic fungi, **616**
Coenzyme A (CoA), **161**–62
Coenzyme Q (ubiquinone), **166**–68
Coenzymes, **147**
Coevolution, 815, **830**, 831*f*, **1124**, 1132
Coevolutionary arms race, **1124**, 1132

Coexistence, niche differentiation and, 1127–28
Cofactors, **147**, 166
Cognitive skills, Bloom's taxonomy of, B:29–B:30*t*
Cohen, Stephen, 671
Cohesins, 246
Cohesion, **26**, **762**
Cohesion-tension theory, 759, **762**–64
Cohorts, **1104**
Cold virus, 727
Coleochaetes, 600
Coleoptiles, **796**, 798–99
Collagen, 47, **202**–3
Collecting duct, 872–73, **877**–78
Collenchyma cells, **744**–45, 748*t*
Colon, **896**
Colonies, **639**, 665
Colonization. *See* Dispersal; Migration
Color, flower, 593–94, 595*f*, 826–27
Color blindness, 278–79, 481–82, 963–64
Colorimetric signals, B:16
Color variations, mutations and, 485
Coloration, cryptic, 1129*f*
Colorized scanning electron micrographs, 1037*f*
Comb jellies, 653
Commensalism, **1124**–25, 1136*t*
Commensal relationships, **620**
Commercial applications, extremophiles and, 531
Common ancestry, 4–5, 8, 446, 449–51. *See also* Ancestral traits
Communication, animal, **1091**–95
Communication, intercellular. *See* Cell–cell gaps; Cell–cell signaling
Communication, plant, 817–18
Communities, **1061**, **1123**
  biodiversity and stability of, 1186–87
  community ecology and, 1060*t*, 1061, 1123 (*see also* Community ecology)
  disturbance regimes in, 1138–39
  effects of global climate change on, 1182
  environmental sequencing (metagenomics) of, 395
  genetic diversity of, 1173–74
  key attributes of, 1135
  keystone species in, 1137–38
  post-disturbance succession in, 1139–42
  predictability of, 1135–37
  structure of, 1135–38
Community ecology, 1123–47
  communities in, 1060*t*, 1061, 1123 (*see also* Communities)
  community dynamics in, 1138–42
  community structure in, 1135–38
  global patterns in species richness in, 1142–45
  species interactions in, 1124–35, 1136*t*
Companion cells, **747**, **767**
Comparative genomics, 638, 640
Comparative morphology, 638–39
Compartmentalization, eukaryotic cell, 110
Compass orientation, migration and, **1089**–90
Competition, **1124**, 1125–28
  in ant behavior, 1083–84
  as biotic factor in geographic distribution and abundance of organisms, 1062
  competitive exclusion principle of, 1125–26

experimental studies of, 1126
fitness and impacts of, 1136*t*
fitness trade-offs in, 1126–27
giraffe necks and food vs. sexual, 9–11
intraspecific vs. interspecific, 1125
male-male, 476–77
niche differentiation and mechanisms of coexistence vs., 1127–28
niches and, 1125
as threat to biodiversity, 1178
Competitive exclusion principle, **1125**–26
Competitive inhibition, **149**
Complement system proteins, 54, **1052**
Complementary base pairing, **63**, **287**. *See also* Nitrogenous bases, nucleotide
  DNA double helix model and, 63
  in mistakes in DNA synthesis, 298–99
  polysaccharides and, 78
  RNA secondary structure and, 65–66
  transcription factors and, 355–56
  in wobble hypothesis on types of tRNAs, 327–28
Complementary DNAs (cDNAs), **370**, 372, 401–2, **720**
Complementary strand, DNA, **64**
Complete digestive tracts, **886**–87
Complete metamorphosis, **672**
Complex tissues, plant, 746
Complexes I–IV (of electron transport chain), 167*f*–68*t*
Complexity, evolution and, 460
Compound eyes, **675**, **960**
Compound leaves, 737–38*f*
Compound light microscopes, B:18
Compression fossils, 512
Compression-resisting elements, 201
Computer programs
  asteroid impact models, 522
  genetic drift simulations, 479
  global climate change models, 1073
  in parsimony analysis, 508
Concentration
  hormone, 995–96
  reactant, 139–40*f*
Concentration gradients, **92**, **862**
  in diffusion and osmosis across lipid bilayers, 91–93
  electrochemical gradients as, 97 (*see also* Electrochemical gradients)
  morphogen, 411–12
Concept maps, B:28
Concurrent exchange, 907*f*
Condensation, chromosome, 246
Condensation reactions, **45**, 59–61, 75–76. *See also* Dehydration reactions
Condensed replicated chromosomes, 220, 221*f*, 223
Condoms, 722, 1030*t*
Conducting cells, plant, 760. *See also* Phloem tissue; Xylem tissue
Conduction, heat exchange and, **855**
Cones, eye, **960**–64
Cones, gymnosperm, 592, 607
Confocal microscopy, B:20
Conformational homeostasis, 853
Connections, intercellular. *See* Intercellular connections
Connective tissue, **847**, 848*f*
Connell, Joseph, 1126, 1127*f*
Conservation, soil, 779–81
Conservation biology, **1061**. *See also* Biodiversity
  addressing ultimate causes of biodiversity losses in, 1189

bioremediation in, 151, 533, 632
conservation of ecosystem function in, 1191–93
conservation of genetic diversity, populations, and species in, 1189–91
ecology and, 1061
estimating biodiversity using surveys in, 1175
genetic drift in, 479
mapping of biodiversity in, 1176–77
phylogenetic species concept and, 492–93
population ecology in preservation of endangered species by, 1118–20
population growth equations in, 1110*f*–11*f*
predator-removal programs in, 1130–31
predicting extinctions in, 1182–84
preservation of phylogenetically distinct species by, 1174
Conservation International, 1177
Conservative, genetic code as, 312
Conservative replication, DNA, 287–89
Constant (C) regions, **1045**–46
Constitutive defenses, **1129**–30
Constitutive mutants, **340**–41
Constitutive transcription, **338**
Constraints, natural selection, 461–62
Consumers, **1150**
  animals as, 637, 647–49
  as biocontrol agents, 1133
  biomagnification in, 1152–53
  top-down control by, and trophic cascades, 1153, 1154*f*
  in trophic structure, 1150–52
Consumption, **1124**, 1128–33
  adaptation and arms races in, 1132
  constitutive defenses against, 1129–30
  consumers as biocontrol agents and, 1133
  efficiency of predators at reducing prey populations by, 1130–31
  fitness and impacts of, 1136*t*
  human, of resources, 1164, 1189
  inducible defenses against, 1130, 1131*f*
  limitations on herbivore, 1131–32
  parasite manipulation of hosts in, 1132–33
  types of, 1128
Continental shelf, **1079**
Continents, in time line of life, 515–16
Continuous data, graphing, B:4–B:6
Continuous population growth, 1110*f*–11*f*
Continuous strands, **292**
Contraception, human, **1029**–30
Contractile proteins, 54
Contraction, muscle. *See* Muscle contraction
Control groups, **13**, B:6–B:7
Convection, heat exchange and, **855**
Convention on Biological Diversity (CBD), 1189
Convergent evolution, **508**–10, 642–43, 986
Conversions, metric unit to English unit, B:1*t*
Cooperation, 1098. *See also* Altruism; Mutualism(s)
Cooperative binding, 912, **913**–14
Copepods, 677, 1167
Copper, as plant nutrient, 777–78, 785
Coprophagy, **896**
Copulation, 1019. *See also* Mating

Copulatory plugs, 1022
Copying
  DNA, 64, 287–89 (*see also* DNA synthesis)
  RNA, 67
Coral polyps, 650
Coral-reef fish, 1109–11
Coral reefs, **1079**
  communities of, 1123*f*
  impacts of global climate change on, 1167–68, 1182
  in ocean biomes, 1079
  red algae and, 572
  sea stars and, 685
Corals, 653–54, 1166*t*
Co-receptors, viral, **718**
Core enzymes, 319
Cork cambium, **748**, 750
Cork cells, **750**
Corms, **825**
Corn, 5–6, 385, 392, 579, 828*f*
Cornea, **960**, 961*f*
Corolla, flower, **827**
Corona, **665**
Corpus callosum, **944**
Corpus luteum, **1027**
Corridors, wildlife, 1120, 1190, 1191*f*
Cortex, kidney, **872**
Cortex, plant root, **759**–61
Cortex, plant shoot, **742**
Cortical granules, 422
Corticosteroids, 1055
Corticotropin-releasing hormone (CRH), **1004**
Cortisol, 995, **1000**–1001
Costanza, Robert, 1187
Costa Rica, ecotourism in, 1192
Cost, of animal locomotion, 986, 987*f*, 988
Cost-benefit analysis, **1084**–87, 1093, 1134–35
Cotransport. *See* Secondary active transport (cotransport)
Cotransporters, **770**, **864**
  in animal glucose transport, 893–94
  in osmoregulation, 864–65, 867–68
  in plant nutrient uptake, 784
  resting potential and, 931
Cotton, 579
Cotyledons, **434**–35, **596**, **833**–34, 837
Countercurrent exchangers, 857–58, 875, 907
Covalent bonds, **21**
  in chemical structures, B:12–B:13
  nonpolar and polar, of molecules, 21–23*f*
  in organic molecules, 36, 38
  protein tertiary structure and, 49–50*f*
  single, double, and triple, 23
Covalent modifications, enzyme regulation via, 149–50
Cowpox, 1054–55
Coyne, Jerry, 500
Crabs, 677, 1130, 1131*f*
Crane, Peter, 585
Cranium, **688**
Crassulacean acid metabolism (CAM), **194**–95, 766, **766**, 1070
Create (Bloom's taxonomy skill), B:29*f*–B:30*t*
Creemers, Raymond, 1104–5
Crenarchaeota, 544, 548
Creosote bushes, 825
Crescent-tail bigeyes, 1101*f*
Cress. *See* Mustard plant
Cretaceous mass extinction, 521–23

Crews, David, 1087–88
Crick, Francis
    adaptor molecule hypothesis of, for
        mRNA specification of amino
        acids, 325–26
    articulation of central dogma of
        molecular biology by, 308
    discovery of double helix secondary
        structure of DNA by, 57f, 61–63,
        286–87
    genetic code hypothesis of, 307
    triplet code hypothesis experiments
        of, 310–11
    wobble hypothesis of, on types of
        tRNAs, 327–28
Crickets, 843–45
Cristae, **115**, **161**, 166–68
Cro protein, 50–51f
Crocodiles, 703, 882f
Cro-Magnons, **707**
Crop, bird, **889**
Crops, transgenic. *See* Transgenic plants
Cross-links, concept map, B:28
Cross-pollination, **830**
Crosses (cross fertilization), 257–**58**,
    B:25–B:26. *See also* Dihybrid
    crosses; Monohybrid crosses;
    Reciprocal crosses; Testcrosses
Crossing over, **243**
    calculating frequencies of, for genetic
        maps, 274f
    chromosome exchange by, 243
    genetic variation from, 248, 249f
    role of, in chromosome theory of
        inheritance, 270–71
    unequal, 397–98
Crosstalk, **215**, 814–15
Crustaceans, **677**, 917, 983
Cryoelectron tomography, of
    mitochondrion, 161f
Cryptic coloration, 1129f
Cryptic female choice, 1020
Cryptic species, **491**
Crystal lattice, ice, 27f
Ctenophora, 638t, 641, 653
Cultural services, ecosystem, 1187t, 1188
Cultures, cell and tissue, **122**
    animal and plant, B:21–B:22
    in anthrax research, 532
    in cytokinin research, 807
    discovery of gap phases with, 221–22
    in DNA nucleotide excision repair
        research, 299–300
    genetic equivalence in differentiated
        plant cells and, 408
    for model organisms, B:23
    in prokaryotic research, 534, 540
    in pulse–chase experiments, 122
Cupressophyta, 607
Curve-billed thrashers, 835–36
Curved lines, scatterplot, B:5
Cushing's disease, **1004**
Cushman, J. Hall, 1134–35
Cuticle, **670**, **743**, **870**
    as defense barrier, 815, 1038
    fossilized, 581–82
    water loss prevention by, 584–85,
        765–66, 869–71
Cuticle
    of roundworm as model organism, B:25
Cuttings, plant, 438, 736f, 744
Cuttlefish, 669

Cuvier, Baron George, 447
Cyanobacteria, **542**
    as bacterial lineage, 544, 545
    chloroplasts and photosynthesis of,
        564–65
    in lichens, 598, 623, 631
    metabolic diversity and, 539t
    photosystems in, 184
    prokaryotic oxygen revolution and,
        541–42
Cycads, 606
Cyclic adenosine monophosphate (cyclic
    AMP; cAMP), 213t, **1009**–10
Cyclic electron flow, Z-scheme model, **189**
Cyclic guanosine monophosphate
    (cGMP), 213t, 214, 962
Cyclic photophosphorylation, 189
Cyclin-dependent kinase (Cdk), **230**, 234
Cyclins, **230**, 234, 360
Cyclostomata hypothesis, 691
Cysteine, 49, 308, 324
Cystic fibrosis (CF), 97–98, 867–68
Cystic fibrosis transmembrane regulator
    (CFTR), 867–68
Cytochrome c, **167**, 168t
Cytochrome complex, 185, 189
Cytochromes, 166–67
Cytokines, **992**, **1040**–41, 1042t, B:22
Cytokinesis, **129**, **220**, **243**
    as animal cell division, 128f–29
    daughter cells resulting from, 227–29
    mitosis, meiosis, and, 220, 223, 225f,
        243–44
Cytokinins, **807**–8, 813t
Cytoplasm, **109**. *See also* Cytosol
    cleavage of, 423–24
    cytokinesis as division of, 220, 223,
        227–29 (*see also* Cytokinesis)
    prokaryotic, 109
    protein synthesis in, 119–20
Cytoplasmic determinants, **423**
Cytoplasmic streaming, **129**
Cytosine, 58–59, 61–63, 65–66, 287
Cytoskeletons, **109**
    eukaryotic, 116, 117t, 127–32, 204
    prokaryotic, 109
    prokaryotic vs. eukaryotic, 110t
    protist, 562
Cytosol, **110**, 129, 159, 164f. *See also*
    Cytoplasm
Cytotoxic T cells, **1050**–53, 1054t

D

Daddy longlegs, 678
Dalton, **20**
Dalton, John, 963–64
Daltonism, 963(*footnote*)
Damaged DNA
    in cancer, 362–63
    cell-cycle checkpoints and, 231–32
    repair of, 299–301
Damschen, Ellen, 1190
Dance hypothesis, honeybee language,
    1092–93
Danielli, James, 94–96f
Daphnia, 1014–16, 1167
Darwin, Charles
    experiments of, on gravitropic
        response, 803–4
    experiments of, on phototropic
        response, 796, 797–98

on pollination, 830
    postulates of, on natural selection,
        453–54
    on sexual selection, 475
    on speciation, 489
    testing of postulates of, 456
    theory of evolution by natural
        selection by, 5, 444, 446
Darwin, Francis, 796, 797–98, 803–4
Data, graphing, B:4–B:6
Data, scientific, 9–10
Data matrix, phylogenetic, 506–8
Data points, graph, B:4–B:5
Databases, genome, 390, 392
Databases, online scientific literature, B:27
Dating, geological, 572
Dating, radiometric, 446–47, 453, 513
Daughter cells
    in meiosis, 240–45
    in mitosis, 223–26
Daughter strands, DNA, **287**–89
Davson, Hugh, 94–96f
Day length, mating and, 1087–88
Day-neutral plants, **802**
DDT (dichlorodiphenyltrichloroethane),
    1152–53
Deactivation, cell–cell signal, 214
Deactivation, M phase-promoting factor,
    231
Dead space, **910**
Dead zones, 544, 1160
Death, cell. *See* Apoptosis; Programmed
    cell death
Death, plant, 812–13
Death rates, animal, 1103–7, 1108, 1109,
    1118, 1119f. *See also* Survivorship
Death rates, human, 473, 1034
Deceitful communication, animal,
    1094–95
Deciduous plants, **606**, 1071
Decisions, behavioral, 1084–85
Decomposer food chains, **1150**
Decomposers, 564, **1150**
    fungal, 612–15, 623–24, 629
    nutrient cycling rates and, 1156–57
    plasmodial slime molds as, 570
    protist, 574
    in trophic structure, 1150–51
Decomposition rates, 1156–57
De-differentiation, cell, 407, 438
Deep-sea hydrothermal vents. *See also*
    Chemical evolution; Origin-of-life
    research
    chemical evolution conditions at, 18f,
        30, 74
    nucleic acid polymerization at, 60
    photosynthesis at, 541
    proton-motive force and chemical
        evolution at, 170
    ribose problem and, 59
    in surface metabolism model of
        chemical evolution, 33, 35–36
Deep sequencing, 400–**402**
Deepwater Horizon oil spill, 1059f
Deer, 478f
Defense, protein functions in, 54
Defense responses, plant
    epidermis as barrier, 743
    to herbivore attacks, 817–19
    noxious compounds in vacuoles and,
        115
    overview of, 818t
    to pathogens, 815–17
    poor herbivore nutrition as, 1131–32
Defenses, constitutive (or standing),
    1129–30

Definition, species, 490–92
Definitive hosts, **666**
Deforestation
    erosion from, 779–81
    and global carbon cycle, 1160
    habitat destruction by, 1179
    loss of biodiversity from, 1172–73
    nutrient cycling and, 1158–59
    reducing emissions from, 1192–93
Degeneration hypothesis, 723
Dehydration, human, 843
Dehydration reactions, **45**, 75–76, 86–87
De Jong, Peter, 1167
Deleterious alleles, **471**–72, 484
Deleterious mutations, 313
Deletions, chromosome, **314**
De Meester, Luc, 1167
Demography, **1103**–7
Denaturation, DNA, 375–76
Denatured proteins, **52**
Dendrites, **847**, **930**
Dendritic cells, **1048**–50
Dense connective tissue, 846t, **847**
Density, water, 26–27, 28t
Density-dependent population growth,
    1108–12
Density-gradient centrifugation, 289,
    1007
Density-independent population growth,
    1107–8, 1109
Dental disease, 531
Deoxyribonucleic acid (DNA). *See* DNA
    (deoxyribonucleic acid)
Deoxyribonucleoside triphosphates
    (dNTPs), **290**
Deoxyribonucleotides, **58**–59, 285–87.
    See also Nucleotides
Deoxyribonucleotide triphosphates
    (dNTPs), 377–78
Deoxyribose, 58–60, 78
Dependent assortment, 264f–65
Dependent variables, graphing, B:4–B:6
Dephosphorylation, 150, 332
Depolarization, **933**–35
Depolarized membranes, sensory
    transduction and, **953**–54
Deposit feeders, 648, 662f, **884**
Derived traits, **506**, B:10–B:11. *See also*
    Synapomorphies
Dermal tissue systems, plant, **740**
    embryogenesis and, 833
    primary plant body and tissues of, 740,
        742, 748t
    protection of surfaces by epidermal
        cells in, 743
    regulation of gas exchange and water
        loss by stomata in, 743–44
    trichomes in, 744
Descartes, René, 973
Descendent groups, phylogenetic trees
    and, B:10–B:11
Descending limb, loop of Henle, 876,
    878t
Descent from common ancestors. *See*
    Common ancestry
Descent with modification, 5, **446**
Desert gerbils, 1086–87
Desert locust, 869
Deserts
    rain shadows and, 1066–67
    subtropical, 1070
Design, experimental, 11–14, B:28
Desmosomes, 205f, **206**
Detergents, **96**
Determination, cell, 427–**29**, 436–38
Detritivores, **647**–48, **1150**

Boldface page numbers indicate a glossary entry; page numbers followed by an
f indicate a figure; those followed by *t* indicate a table.

I:10   INDEX

DNA (deoxyribonucleic acid)
    (continued)
  discovery of double helix secondary
      structure of, 61–63, 286–87
  electron micrograph of, 317f
  eukaryotic, 112
  expression of (see Gene expression)
  as genetic information–containing
      molecule, 64
  metabolic pathways and, 157
  methylation of, 351
  mitochondrial, 115–16
  model of condensed eukaryotic, 348f
  mutations in, 483–84
  next-generation technologies for
      sequencing of, 378
  in nucleus, 119–20
  physical model of, 57f, 62f
  polymerization of
      deoxyribonucleotides to
      form, 59–61
  as poor catalytic molecule, 64–65
  primary structure and secondary
      structure of, 286f
  prokaryotic, 107–8
  as probe(s), 372–73, 381t, 401–2,
      B:15–B:16
  prokaryotic vs. eukaryotic, 110t
  repair of damaged, 297–99
  RNA structure vs. structure of, 66t
  using restriction endonucleases and
      DNA ligase to cut and paste,
      370–72
  visualizing, B:15–B:16
DNA cloning, 370, 374–76
DNA fingerprinting, 397–98f
DNA helicase, 290–91, 294t
DNA libraries, 372–73, 381t
DNA ligases, 293–94t, 370, 371–72, 380t
DNA methylation, 351
DNA methyl-transferases, 351
DNA microarrays, 400–402
DNA polymerases, 289
  in DNA synthesis, 289–94t, 295
  in genetic engineering, 380t
  in polymerase chain reaction, 374–76
  proofreading by, 297
  RNA polymerases vs., 318
  in telomere synthesis, 295–96
DNA sequences
  cancer and regulatory, 363
  in genetic code, 310–12
  genetic homology and, 450–51
  phylogenies based on, 510, 511f,
      558–59
  regulatory, in transcription initiation,
      353–54
  silent mutations in, as nonadaptive
      traits, 461
  transcription factors and specific,
      355–56
DNA sequencing, 376–78. See also
      Whole-genome sequencing
DNA synthesis, 284–303
  comprehensive model of, 289–95
  correcting mistakes in, 298–99
  in dideoxy sequencing, 377–78
  discovery of DNA as genetic material
      in genes and, 284–86
  discovery of secondary structure of
      DNA and, 286–87
  DNA polymerase in, 289–90

electron micrograph of, 284f
opening and stabiliizing of double
    helix in, 290–91
process of, 64
proteins required for bacterial, 294t
repairing damaged DNA and
    correcting mistakes in, 297–301
starting of replication in, 290
synthesis of lagging strand in, 292–95
synthesis of leading strand in, 291–92
telomere replication in, 295–97
testing of early hypotheses about DNA
    replication and, 287–89
xeroderma pigmentosum (XP) case
    study of damaged DNA, 299–301
DNA viruses, 715–16, 726, 726f
DNase enzyme, 351
Dogs, life cycle of, 241–42
Dolly (cloned sheep), 409
Dolphins, 452–53, 509–10, 985, 986
Domains, taxonomic, 7f, 8, 107, 528–29
Dominant alleles, 258t, 272, 273f, 277t
Dominant traits, 259. See also Recessive
    traits
  codominance and incomplete
      dominance of, 272, 273f, 277t
  human, 277–79
  Huntington's disease as, 378
  in Mendelian genetics, 259–60
Dopamine, 941t
Dormancy, plant, 751
Dormancy, seed, 808, 809–10, 836–37
Dormancy, viral, 721–22
Dorsal body axis, 410
Dorsal hollow nerve cord, 686
Dot-to-dot lines, scatterplot, B:5
Double bonds, 23
Double fertilization, plant, 593, 832–33
Double helix secondary structure, DNA,
    62, 286
  discovery of, 61–63, 286–87
  electron micrograph of, 284f
  opening and stabiliizing of, in DNA
      synthesis, 290–91, 294t
  transcription factors and, 355–56
Double-stranded DNA (dsDNA)
    viruses, 726
Double-stranded RNA (dsRNA)
    viruses, 727
Double-stranded viruses, 716, 726–27
Doubly compound leaves, 737–38f
Doughnut-shaped proteins, 47f
Doushantuo microfossils, 518
Downstream DNA, 319
Drag, animal locomotion and, 985
Drosophila bifurca. See Fruit flies
Drosophila melanogaster. See Fruit flies
Drought, natural selection in Galápagos
    finches and, 456–59f
Drugs. See also Diseases and disorders,
    human; Medicine
  anti-AIDS, 309
  anticancer, 607
  anti-HIV, 719, 720
  antiviral, 714, 719, 722
  beta-blocker, 210
  derived from green plants, 580
  derived from land plants, 580t
  ephedrine, 608
  evolution of bacterial resistance to,
      454–56
  fungal infections and fungicides, 618

fungi in production of, 629, 632
homology and, 451
plants and, 580
resistance to, 532–33
Dryer, W. J., 1046
Dry habitats
  carbon dioxide concentration
      mechanisms for, 193–95
  gas exchange in, 905
  plant adaptations to, 754–55, 757–58,
      765–66
Drying, seed maturation and, 834
Dscam gene, 358
Duero river hydroelectric dam, 154f
Duplications, chromosome, 314
Dusky seaside sparrow, 492–93
Dutch elm disease, 632, 1063
Dwarf plants, defective gibberellin in,
    808–9
Dwarfism, pituitary. See Pituitary
    dwarfism
Dyes, microscopy, B:18
Dynamite tree, 834–35
Dynein, 133
Dysentery, 555t
Dystrophin, 200f

E

E site, ribosome, 328–29f
E2F protein, 234
Eardrums, 956
Early endosome, 126
Early prophase I, 242, 244
Ear ossicles, 956
Ears, mammalian, 955–58
Earthworms, 658, 659
Ebola virus, 723, 728
Ecdysone, 670, 998
Ecdysozoans, 638t, 645, 658, 659f, 670–
    78. See also Protostomes
Echidnas, 700
Echinoderms, 638t, 641, 682–86f
Echinoidea, 685–86f
Echolocation, 957–58
Ecological selection, 478
Ecology, 1059–81. See also Community
    ecology; Conservation biology;
    Ecosystem ecology; Global
    ecology; Population ecology
  aquatic biomes in, 1074–79
  biodiversity scenarios in, 1182–84
  climate patterns in, 1065–67
  geographic distribution and
      abundance of organisms in,
      1061–65
  levels of study in, 1060–61
  organismal, 1060–61
  pyramid of production in, 1152
  summary of levels of study of, 1060t
  terrestrial biomes in, 1068–74
Economic benefits
  of biodiversity, 1187–88
  of fungi, 613–15
Ecosystem diversity, 1174–75
Ecosystem ecology, 1148–59
  in ecology, 1060t, 1061
  ecosystem energy flow in, 1149–56
  ecosystem nutrient cycling in, 1156–59
  ecosystems and biosphere in, 1148–49
      (see also Global ecology)
Ecosystem function, 1174
  in biodiversity, 1174–75
  conservation of, 1191–93
Ecosystem services, 578, 1187
  benefits of, 1187–88

of green plants, 578–79
protostome, 658
quantifying, 1192–93
Ecosystems, 578, 1061, 1148
  animal locomotion in, 986
  animals in, 647–48
  biosphere and, 1148–49 (see also
      Global ecology)
  ecosystem ecology and, 1060t, 1061
      (see also Ecosystem ecology)
  energy flow in (see Energy flow,
      ecosystem)
  fungi in, 613–15
  human impacts on, 1149
  impacts of global climate change on,
      1168–69
  most productive, 1154–56
  nutrient cycling in, 1156–59
  prokaryotes in, 541–44
  protists in, 556–57
  protostomes in, 658
  restoration of, 1191–93
  services of (see Ecosystem services)
Ecotourism, 1192
Ectoderm, 424f–25, 640–41
Ectomoss, 665
Ectomycorrhizal fungi (EMF), 620–21,
    622f, 632
Ectoparasites, 648, 666, 667
Ectotherms, 697, 855–56, 1152
Edge effects, habitat fragmentation
    and, 1180
Ediacaran fauna fossils, 518
Education, human population growth
    and, 1117
Effector B cells, 1050–51
Effector cells, hormone, 993–94, 1001
Effector genes, 414f
Effector T cells, 1050–51
Effectors, 854
Efferent division, PNS, 942
Eggs, 237, 823, 1014
  amniotic, 689–91, 693–94
  bird, 1024f
  cell-cycle regulation and frog, 229–30f
  fertilization of, 5, 420–21, 433,
      1018–20
  gametogenesis of, 241
  glycoproteins in fertilization of, 79–80
  human, 237f
  insect, 659
  land plant, 587–88
  mammalian, 1023–25
  meiosis and, 219, 237–38
  of monotreme mammals, 701
  plant, 438, 823, 827–28
  polyspermy prevention and, 421–23
  reptile and bird, 696
  in sexual reproduction, 1014–15
  sexual selection and, 475
  structure and function of, 1017–18f
  variations in development of, 1020–21
Either-or rule, probability, B:8
Ejaculation, 1022
Ejaculatory duct, 1022–23t
Elastase, 895t
Electric current, 931
Electrical activation, heart, 921–22
Electrical energy, chemical evolution
    and, 31, 33–34
Electrical fields
  electroreception of, 646, 647t, 967–68
  in gel electrophoresis, B:13–B:14
  generation of, by electrogenic fishes,
      968, 969f
Electrical potential, 930

Enzymes (*continued*)
  hydrolysis of carbohydrates by, to
      release glucose, 80–81
  nucleic acid polymerization and,
      60–61
  in one-gene, one-enzyme hypothesis
      on gene expression, 305–7
  origin-of-life research on, as first living
      molecules, 55
  in peroxisomes, 115
  post-translational control by, 360
  prokaryotic, 108
  in pyruvate processing, 161–62
  in regulation and evolution of
      metabolic pathways, 150–52
  regulation of, 149–50
  in replisomes, 294–95
  ribozymes vs., 330–31
  RNA ribozymes as, 67–69
  in RNA synthesis, 307–8
  smooth ER and, 114
  in telomere synthesis, 296–97
  in visualizing molecules, B:15–B:16
Ephedrine, 608
Epicotyl, **833**
Epidemics, **712**
  diabetes mellitus, 898–99
  fungal, 613
  viral, 712
Epidermal cells, plant, 743, 748*t*
Epidermal tissue, plant, 435, 437*f*
Epidermis, plant, **435**, 437*f*, **740**,
      759, 833
Epididymis, **1022**–23*f*
Epigenetic inheritance, **352**–53
Epinephrine (adrenaline), **977**, **1000**
  adrenal glands and, 995
  beta-blockers and, 210
  binding of, to receptor, 1008–10
  control of, by sympathetic nerves, 1006
  fight-or-flight response and, 922, 1000
  model for action of, 1009*f*
  smooth muscle and, 977–78
Epiphyseal plate, 999
Epiphytes, **603**, **789**
Epithelia, **204**, **639**, **848**–49, 854, 873–74
Epitopes, **1045**
Epochs, 515
Equilibrium, 30, 91–93
Equilibrium potentials, **931**, 932*f*
Equisetophyta (horsetails), 605
Equivalence, genetic, 408–9
ER lumen, 123–24
ER signal sequence, **123**–24
Erosion
  devegetation and, 578
  soil composition and, 778
  soil conservation and, 779–81
Error, standard, B:6–B:7
Errors, DNA synthesis, 297–301, 313–14
Erwin, Terry, 1175
Erythropoietin (EPO), **1003**
Escape behavior, 1129*f*
Escaped-genes hypothesis, 723
Escherichia coli (*E. coli*)
  BAC libraries and, 390
  cell structures of, 107–9
  discovery of DNA in genes of, in
      Hershey–Chase experiment,
      285–86
  DNA synthesis mistakes and
      mutations of, 297–99

food poisoning and, 532
gene expression in, 336–37
genetic engineering of recombinant
    human growth hormone using,
    370–74
genome of, 393–94
lactose metabolism of, as
    model, 338–44
in Meselson–Stahl DNA replication
    experiment, 288*f*, 289
metabolic diversity and, 539*t*
as model organism, B:23, B:24*f*
mutations and fitness in, 484–85
as proteobacteria, 547
scanning electron micrograph of, 336*f*
types of lactose metabolism mutants
    in, 340*t*
Esophagus, 848*f*, 887*f*, **889**
Essay on the Principle of Population, An
    (book), 453, 1116
Essential amino acids, **883**
Essential nutrients, animal and human,
    883–84, 885*t*
Essential nutrients, plant, **776**–78, 780*t*
Ester linkages, 86–**87**
Estradiol, **999**, **1025**
  binding of, to intracellular receptors,
      1006–8*f*
  in mammalian reproduction, 1025–26
  in mating, 1087–88
  ovaries and, 999
  in puberty, 1026–27
Estrogens, **999**
  binding of, to intracellular receptors,
      1006–8*f*
  structure and function of, 210*t*
Estrous cycle, **1027**
Estrus, 11
Estuaries, **1078**
Ethanol, 27*t*, 115, 172–73*f*, 540
Ethical issues
  on biodiversity, 1188–89
  on gene therapy, 384–85
  on genetic testing, 382
  on recombinant human growth
      hormone, 374
Ethylene, 210*t*, **812**–13
Eudicots, **596**, 742*f*, 834, 837
Euglenids, 571
Eukarya domain, **552**. *See also*
    Eukaryotes
  characteristics of, 529*t*
  phospholipids in, 87
  in phylogenetic tree of life, 7*f*–8
  protists in, 552–53
Eukaryotes, **8**. *See also* Eukarya domain
  aerobic respiration in, 170–72
  cell–cell attachments in, 204–7
  cells of (*see* Eukaryotic cells)
  characteristics of, 552–53
  discovery of split genes and introns in,
      321–22
  DNA synthesis in, 291*f* (*see also* DNA
      synthesis)
  embryos of, 219 (*see also* Embryos)
  genomes of (*see* Eukaryotic genomes)
  initiating translation in, 330
  major lineages of protists and, 558*t*
  model of condensed DNA of, 348*f*
  model organisms for (*see* Model
      organisms)
  phylogenetic tree of, 559*f*

prokaryotes vs., 8, 552–53
regulation of gene expression in (*see*
    Gene regulation, eukaryotic)
RNA polymerases for, 318
RNA processing in, 321–24, 325*t*
transcription and translation in,
    320–21*f*, 325
Eukaryotic cells, 110–33. *See also* Cell(s)
  animal and plant, 111*f* (*see also*
      Animal cells; Plant cells)
  cytoskeletons of, 127–33
  endomembrane system of, 121–27
  images of, 106*f*, 111*f*, 118*f*
  nuclear transport in, 119–21
  organelles in, 110–17*t*
  prokaryotic cells vs., 8, 107, 110
  whole-cell dynamism of, 118–19
Eukaryotic genomes
  challenges in sequencing of, 395–96
  gene families in, 398–99
  identifying genes in, 393
  insights from Human Genome Project
      and human, 399–400
  number of genes in, 394*f*
  prokaryotic genomes vs., 393–96
  transposable elements and repeated
      sequences in genomes of, 396–98*f*
  whole-genome sequencing of, 392
Euler, Leonhard, B:9
Euler's number, B:9
European robins, 968–69
Euryarchaeota, 544, 549
Eusociality, **1097**–98
Eutherians, 696, **701**, 1031–34
Evaluate (Bloom's taxonomy skill),
    B:29*f*–B:30*t*
Evaporation, **855**
  gas exchange and, 907
  in global water cycle, 1159
  heat exchange and, 855
  insects and, 869–70
  plants and (*see* Transpiration)
Evergreen trees, 1071
Evidence, evolutionary, 446–53
Evo-devo (evolutionary-developmental
    biology), **415**, 790
Evolution, **5**, **445**. *See also* Adaptation(s);
    Natural selection
  of aerobic respiration, 542
  of animal innovations (*see*
      Evolutionary innovations, animal)
  of animal locomotion, 986
  Aristotle's great chain of being and, 445
  arms races in, 815, 1124, 1133
  behavior causation in (*see* Ultimate
      causation)
  cell theory and theory of, 6
  changes in developmental gene
      expression and, 416
  common misconceptions about
      natural selection, adaptation, and,
      459–62
  of contemporary populations, 449
  convergent, 508–10
  current examples of, 448
  evidence for, as pattern component of
      theory of, 446–53
  evolutionary conservation of chemical
      signals and regulatory genes,
      414–15
  of evolutionary theories, 445–46
  experimental, 484–85
  of giraffe necks, 9–11
  global climate change and, 1166*t*, 1167
  human, 705–8
  of introns, 323

lateral gene transfer and, 394–95
mapping land plants, on phylogenetic
    trees, 586, 587*f*
of metabolic pathways, 150–52
of mouthparts, 884–86
of multichambered hearts with
    multiple circulations, 919–21
of muscle-generated animal
    movement, 972–73
mutations and, 485
natural selection as process component
    of theory of, 453–54 (*see also*
    Natural selection)
of neurons and muscle cells, 929
niche differentiation in, 1127–28
nitrogenous wastes and, 865
as not goal oriented, 460–61
of oviparous, viviparous, and
    ovoviviparous species, 1020–21
of oxygenic photosynthesis, 189
phylogenetic trees of (*see* Phylogenetic
    trees)
Plato's typological thinking and,
    445–46
point mutations and, 313
of pollination, 830–31
of populations, 2
of predation, 519–20
processes of (*see* Evolutionary
    processes)
prokaryotic impact on, 541–44
religious faith vs. theory of, 9
RNA world hypothesis and, as
    characteristic of life, 69
selfish vs. self-sacrificing behavior
    and, 461
speciation and, 489–90 (*see also*
    Speciation)
summary of common misconceptions
    about, with corrections, 460*t*
summary of evidence for, 452*t*
switch from chemical, to biological, 19
    (*see also* Chemical evolution)
theory of, by Charles Darwin and
    Alfred Russel Wallace, 5–6,
    444–45, 446
theory of, by Jean-Baptiste de
    Lamarck, 445–46
vertebrate, 689–91
of whales, 510–11*f*
whole-genome sequencing and study
    of, 392
Evolutionary-developmental (evo-devo)
    biology, **415**, 790
Evolutionary innovations, animal,
    637–46. *See also* Animal(s)
  data on, 637–38
  origin of bilateral symmetry, 641–44
  origin of coelom, 643–44
  origin of embryonic tissue layers,
      640–41
  origin of multicellularity, 638–40
  origin of nervous system and
      cephalization, 643
  origin of protostomes and
      deuterostomes, 644–45
  origin of segmentation, 645–46
Evolutionary processes, 465–88. *See also*
    Evolution
  effects on evolution of four, 465–66
  gene flow, 466, 482–83
  genetic drift, 465, 478–82
  Hardy–Weinberg principle in
      analyzing change in allele
      frequencies, 466–70
  mutation, 466, 483–86

Fishes (continued)
  nitrogenous wastes of, 865
  osmoregulation by, 862–64, 866–69
  overfishing and, 1153, 1181
  populations of, 1101f
  in vertebrate phylogeny, 691
Fission, 654, 667, **1014**
Fitness, **6, 454, 1124**
  altruism and, 1095–98
  effects of evolutionary processes on,
    465–66, 486t
  gene flow and, 482–83
  genetic drift and, 479
  hemoglobins and, 915
  heterozygous individuals and, 470
  inbreeding depression and, 471–72
  maladaptive traits and, 192
  mutations and, 484–85
  natural selection, adaptation, and
    biological, 6, 454
  point mutations and, 313
  sexual selection and, 475
  species interactions and, 1124
Fitness trade-offs, **462, 1105, 1126**
  in animal form and function, 843–45
  in competition, 1126–27
  as constraint on natural selection, 462
  endothermy vs. ectothermy, 856
  in life histories, 1105–6
  of mutualisms, 1134–35
  nitrogenous wastes and, 865
  parental care and, 694
  between skeletal protection and
    mobility, 980
Fixation phase, Calvin cycle, 191
Fixed action patterns (FAPs), **1084**
Flaccid cells, **756**
Flagella, **109, 557, 1017**
  bacterial vs. eukaryotic, 131
  cell movement by, 132–33
  in characteristics of domains of life, 529t
  fungal, 617, 618
  prokaryotic, 109
  protist, 557, 565–66
  of sperm, 1017
Flagellin, 131
Flattening adaptations, animal surface-
    area, 852–53
Flatworms, 638t, 666
Flavin adenine dinucleotide (FAD), **142**,
    163f–65
Flavin-containing protein (FMN), 166
Flavins, 166
Flemming, Walther, 220
Flexibility, behavioral, 1084–85
Flexibility, polypeptide, 46
Flexors, **982–83**
Flies. See Fruit flies
Flight
  animal locomotion in, 984
  bird, 703
  evolution of vertebrate, 694–95
  insect, 671–72
Flightless birds, 703
Flocking, 1129f
Flood basalts, 521
Floods, succession after, 1139
Floral meristems, 438, **439–41**
Floral mimics, 739f
Florida panthers, 471–72, 1190
Florigen, **803**
Flour beetle, 869

FLOWERING LOCUS T (FT) gene, 803
Flowers, **592, 823**
  angiosperm radiation and, 596
  buds and, 735
  fading of, 812
  in female gametophyte production,
    827–28
  floral meristems and formation of,
    438–41
  light reception and flowering of, 802–3
  in male gametophyte production,
    828–29
  as morphological innovations, 517
  in plant reproduction, 592–93, 823
  pollination of, 593–94, 608–9f
  as reproductive structures, 826–29
Flu virus. See Influenza viruses
Fluid connective tissue, 846t, **847**
Fluid feeders, 648t, **649, 884**
Fluid-mosaic model, 94–**95**, 96f
Fluidity, plasma membrane, 90, 91f
Fluids, human seminal, 1023t
Flukes, 666
Fluorescence, 182–84, 377, B:15–B:16
Fluorescence microscopy, B:18, B:20
Folate, 884t
Folding
  animal surface-area adaptations in,
    852–53
  protein, 48–49, 51–53, 331–32, 359–60
  small intestine, 892
Follicle-stimulating hormone (FSH),
    210t, **1006, 1026**, 1027–30
Follicles, **1026**
Follicular phase, **1027**
Food, **883**
  angiosperms as human, 608
  animal communication and, 1092–93
  animal mouthparts and, 884–86
  animal nutrition and, 882–83
  animal sources of, 646–48
  disruptive selection and, 474
  as ecosystem provisioning service, 1187
  foraging behaviors and (see Foraging)
  fungi and human, 613–14, 630
  giraffe necks and competition for, 9–11
  green plants as human, 579
  migration and, 1091
  model organism, B:23
  plant pathogens and human, 815
  plant reproduction and human, 823
  in population cycles, 1114–15
  protostomes as human, 658
  transgenic crops as, 385–86
Food and Drug Administration
    (FDA), 374
Food and Nutrition Board, U.S. National
    Academy of Sciences, 883
Food chains, **556, 1150–51**
Food poisoning, 532
Food webs, **1138**, 1150–**51**, 1169
Foot, mollusk, **662–63**
Foraging, 11–14, **1085–87**
Foraminiferans, 572
Forces, animal locomotion, 985–86
Forebrain, **688**
Foreign cells, immune system and,
    1046–47, 1053
Forensic biology, 376, 397–98f
Forestry
  green plants in, 579–80
  sustainable, 781

Forests
  boreal, 1072
  decomposition rates in, 1157
  fires in, 1165
  net primary production of, 1155, 1156
  temperate, 1071
  tropical wet, 1070
Forks, phylogenetic tree, B:10–B:11
Form-function correlations, 19, 118,
    845–46, 884–85, 984–85
Formaldehyde, 34, 73, 74
Formulas, molecular and structural, 24,
    B:12–B:13
Forrester, Graham, 1109–11
Fossil fuels
  in biodiversity loss, 1189
  in global carbon cycle, 1161–62
  global climate change and, 1163
  in global nitrogen cycle, 1160
Fossil record, **446, 511**
  animals in, 636, 637, 638
  bryozoans in, 665
  of Cambrian explosion, 518–19f
  cetaceans in, 452–53
  DNA in, 64–65
  evidence for evolutionary change in,
    446–49
  feathered dinosaurs in, 694–95
  fish in, 692–93
  fossil formation and, 511–12
  fungi in, 616
  hominins and humans in, 705–8
  life's time line in, 513–16
  limitations of, 512–13
  mass extinctions in, 520–23
  pollen grains in, and ranges of
    species, 1137
  prokaryotes in, 530
  studying green plants using, 581–82
  studying history of life using, 511 (see
    also History of life)
  vertebrates in, 689–91
Fossilization, 511–12
Fossils, **446, 511**
  amplification of DNA in, with
    polymerase chain reaction, 375–76
  formation of, 511–12
  ginkgoes as living, 606
  spores, 581–82
  trilobytes, 505f
  vascular tissue in, 585
Founder effect, **481**, 495
Four-o'-clocks, 272, 273f
Fovea, 961f–**62**
Fox, Arthur, 965
Fraenkel-Conrat, Heinz, 715–16
Fragmentation, 601, 603, 652, 654, 667
Frameshift mutations, **313**–14t
Franklin, Rosalind, 61–63
Free energy, **137**–39, 145–47, 165f
Free radicals, 34, 182, 298
Free ribosomes, 123–24
Freeze-fracture electron microscopy,
    95–96f
Frequencies, genetic maps and
    recombination, 274f
Frequency, sound, **955**–58
Frequency-dependent selection, **474**
Frequency distributions, histograms of,
    275–76, B:5–B:6
Freshwater biomes
  characteristics of, 1076–78
  fish osmoregulation in, 863–64, 868
  global water cycle and, 1159–60
Frogs, 229–30f, 317f, 424f–25, 700, 909,
    997–98

Fronds, **604**
Frontal lobe, **944**, 945
Fructose, 157, 195, 1023t
Fructose-1,6-bisphosphate, 158f–59, 160f
Fructose-6-phosphate, 158f–59, 160f
Fructose bisphosphate, 158
Fruit flies
  bicoid gene of, 411–12
  chromosomes of, 238
  foraging alleles in, 1085
  genetic drift in, 479–81
  genetic map of, 274f
  genome of, 392
  giant sperm of, 1020
  Hox genes of, 412–16, 508–9
  innate immunity in, 1039–40
  in linked gene experiments, 270–71
  as model organism, 675, B:24f, B:25
  mRNA sequences from one gene
    in, 358
  number of chromosomes in, 238
  as protostome, 658
  reinforcement of divergence in, 500
  second-male advantage in, 1019–20
  in testing of chromosome theory of
    inheritance, 267–69
Fruiting bodies, **547, 563**, B:23–B:25
Fruits, **594, 823**
  angiosperm radiation and, 596
  animal color vision and, 964
  auxin and, 807
  development of, for seed dispersal,
    833, 834–36
  ethylene and ripening of, 812–13f
  in plant reproduction, 594–95, 823
Fuel, green plants as, 579
Function-form correlations, 19, 118,
    845–46, 884–85, 984–85
Functional genomics, **389**–90, 400–402.
    See also Genomics
Functional groups, 37–38, 44
Functions, exponential, B:9
Fundamental niches, **1125**
Fungi, 612–35
  adaptations of, as decomposers,
    612–13, 623–24
  appearance of, in Paleozoic era, 515
  biological methods for studying,
    615–19
  biological reasons for studying, 613–15
  in cell theory experiment, 3–5
  chitin as structural polysaccharide in,
    76, 78
  diversification themes of, 582, 619–28
  economic and ecological impacts of,
    613–14
  in ecosystems, 612–13
  key lineages of, 618–19, 628–32
  lichens as green algae and, 598
  life cycles of, 625–28
  molecular phylogenies of, 618–19
  morphological traits of, 615–18
  mycorrhizal, 614, 784, 1133
  phylogenetic relationship of animals
    to, 8, 618
  reproductive variation in, 624–25
  saprophytic, and carbon cycle on land,
    614–15
  in symbiotic relationships, 613,
    619–23
  vacuoles in cells of, 114–15
Fungicides, **618**
Fusion
  fungal, 624–25
  gamete, 241, 246–47, 420–21
  interbreeding and, 499, 502t

Boldface page numbers indicate a glossary entry; page numbers followed by an
f indicate a figure; those followed by *t* indicate a table.

I:16 INDEX

G

G-protein-coupled receptors, 211–14, 216
G proteins, **212**
G_0 state, 229, 233–34
G_1 phase, 221–**22**, 223f, 229, 231–34
G_2 phase, 221–**22**, 223f, 232
G3P. *See* Glyceraldehyde-3-phosphate (G3P)
Gage, Phineas, 945
Gaits, 986, 987f
Galactose, 73–74, 354
Galactoside permease, 340
Galápagos finches, 449–50, 456–59f, 494–95, 845–46f, 1127
Galen, 973
Galileo, 987
Gallbladder, 887f, **894**
Galls, 385–86
Galvani, Luigi, 973
Gametangia, **587**
Gametes, **219**, **238**, **823**. *See also* Eggs; Sperm
  in external fertilization, 1018
  in fertilization, 237–38, 241–42, 420–21
  fungal, 617, 624
  gametogenesis of plant, 433, 438, 588–89
  meiosis and, 219
  in plant reproduction, 823–25
  polyspermy prevention and, 421–23
Gametic barriers and prezygotic isolation, 491t
Gametogenesis, **241**, **1016**. *See also* Oogenesis; Spermatogenesis
  animal, 1016–18f
  in external fertilization, 1018
  meiosis and, 241
  plant, 433, 438, 588–89
Gametophytes, **567**, **588**, **823**
  in alternation of generations, 588–89
  land plant evolution away from life cycles dominated by, 589–90
  in land plant life cycles, 823–25
  production of, 827–29
  protist, 568–69
  in seed evolution, 592
Gamma-aminobutyric acid (GABA), 941t
Gamow, George, 310
Ganglia, **643**
Ganglion cells, **960**, 961f
Gap Analysis Program (GAP), 1191
Gap genes, 412, 413f, 414f
Gap junctions, 205f, **207**–8
Gap phases, 221–**22**, 223f
Gaps, cell–cell. *See* Cell–cell gaps
Garden peas, Mendelian genetics and, 257–66. *See also* Mendelian genetics
Gas exchange
  amniotic egg, 693
  amphibian, 700
  animal, 902–3 (*see also* Circulatory systems; Respiratory systems)
  diffusion and skin-breathing in, 852
  fish lungs in, 692–93
  gills in (*see* Gills)
  during human pregnancy, 1032–33
  insect, 869
  in osmoregulation, 862–63
  plant stomata in, 743–44
  pneumatophore roots in, 734, 735f
  by protostomes, 659–60
  by salmon, 851–52

terrestrial adaptations for, 659–60
tree bark and, 750
Gases
  chemical evolution and volcanic, 24, 30, 34–35
  greenhouse, 1163, 1192–93
Gastric juice, 890–91
Gastrin, **893**
Gastropods, **662**, 668
Gastrovascular cavities, 886–87f
Gastrula, 424
Gastrulation, **407**, **424**
  in animal embryonic development, 644–45
  cell movement in, 407, 424–25
  definition of body axes in, 425
  formation of germ layers in, 425
  in frog embryo, 424f
  protostome vs. deuterostome, 658
Gated channels, **98**–99. *See also* Membrane proteins
Gause, G. F., 1125–26
GDP (guanosine diphosphate), 211, 214
Gel electrophoresis, 378, 894, B:13–B:14
Gemmae, **601**, 603
Gene-by-environment interactions, 273, 277t
Gene-by-gene interactions, 273–75, 277t
Gene duplication, **398**–99, 483, 485
Gene expression, **305**–16, **337**
  bacterial gene regulation and, 336–37 (*see also* Gene regulation, bacterial)
  bacterial vs. eukaryotic, 360–61
  in cell determination, 429
  in cell differentiation, 409–10 (*see also* Differential gene expression)
  central dogma of molecular biology on, 307–9
  eukaryotic gene regulation and (*see* Gene regulation, eukaryotic)
  evolutionary change from changes in, 416
  functional genomics and, 400–402
  genetic code and, 307, 310–12
  linking genotypes and phenotypes in, 308
  molecular biology, DNA, and, 304–5
  mutations of genes and chromosomes and, 313–15
  one-gene, one-enzyme hypothesis on, 305–7
  plant cell–cell signals and, 795–96
  point mutations and, 483
  protein synthesis and, 317–18 (*see also* Protein synthesis)
  RNA as intermediary between genes and proteins in, 307–8
  steroids and, 1007–8f
  transcription and translation in, 308
Gene families, **398**–99, 508
Gene flow, **466**, **482**
  in allopatric speciation, 494–95
  effects of, on fitness, 483
  as evolutionary process, 466, 486t
  genetic diversity and, 483
  genetic restoration with artificial, 1190
  Hardy–Weinberg principle and, 468
  measuring, 482–83
  reproductive isolation and, 490
  speciation and, 489
  in sympatric speciation, 495
Gene pools, **466**–67
Gene recombination, immune system, 1046, 1047f. *See also* Recombinant DNA technology

Gene regulation, bacterial, 336–47
  eukaryotic vs., 348–49, 360–61
  of gene expression, 336–37 (*see also* Gene expression)
  global, 344–45
  identifying regulated genes for, 339–41
  information flow and overview of, 337–39f
  negative control of transcription in, 341–44
  positive control of transcription in, 344, 345f
Gene regulation, eukaryotic, 348–65
  bacterial gene regulation vs., 348–49, 360–61
  cancer and defects in, 361–63
  chromatin remodeling in, 349–53
  environment, differential gene expression, and, 348–49
  overview of levels of, 349
  post-transcriptional, 356–60
  in transcription initiation, 353–56
  translational and post-translational, 359–60
Gene therapy, 381, **383**–85
Generations, **1103**
Genes, **107**, **238**, **261**. *See also* Chromosomes; DNA (deoxyribonucleic acid); Genomes
  alleles of, 238, 239f, 247, 261 (*see also* Alleles)
  in animal diversification, 646
  Cambrian explosion and new, 519–20
  cancer and tumor-suppressor, 361–63
  case study of *HLA* genes in humans, 469–70
  chromosomes, DNA, and, 220
  cloned, in DNA libraries, 372
  for color vision, 964
  control of flower structures by, 439–41
  in control of plant body axes, 435–36
  definitions of, 392
  in determination of leaf shape, 436–38
  developing phylogenetic tree of life from rRNA sequences in, 6–8
  discovery of split eukaryotic, 321–22
  dwarf plants and defective, 808–9
  effects on phenotypes of interactions between, 273–75, 277t
  eukaryotic cell nucleus and, 112
  exceptions and extensions to Mendelian rules on, 277t
  expression of (*see* Gene expression)
  finding disease-causing, 378–82
  gene families of, 398–99
  in genetic constraints on natural selection, 461–62
  genetic information in, 2 (*see also* Genetics)
  genetic maps of (*see* Genetic maps)
  Hershey–Chase experiment on DNA as genetic material in, 285–86
  hormones and, in leptin and energy homeostasis, 1001–3
  identifying, in genome sequences, 392–93
  identifying regulated, with replica plating, 339–41
  inserting, into human cells, 383
  inserting, into plasmids, 370, 371f
  lactose metabolism, 340–44
  lateral gene transfer of, 394–95, 399
  linkage and linked, 270–71, 274f, 277t

loci of, on chromosomes, 266
  in Mendelian genetics, 258t
  multiple traits and pleiotropic, 272–73
  mutations of (*see* Mutation[s])
  natural selection of, in Galápagos finches, 458–59f
  number of, in genomes, 394–96, 399–400
  in one-gene, one-enzyme hypothesis on gene expression, 305–7
  in particulate inheritance, 261–63
  plant pathogen resistance, 815
  polygenic inheritance and, 275–76
  prokaryotic, 107–8
  redefinition of, 358
  regulation of bacterial (*see* Gene regulation, bacterial)
  regulatory (*see* Regulatory genes)
  RNA as intermediary between proteins and, 307–8
  for taste receptors, 965
  tumor suppressors, 232–33
Genes-in-pieces hypothesis, 321–22
Genetic bottlenecks, **481**–82
Genetic code, **310**
  code words in, 310–11
  cracking of, 311–12
  hypothesis of, by Francis Crick, 307
  in mRNA specification of amino acids, 325–26
  in transfer of amino acids to proteins by tRNAs, 326–28
Genetic constraints, natural selection and, 461–62
Genetic correlation, 461–62
Genetic counselors, 376
Genetic disorders, human, 97–98, 277–79, 383–85. *See also* Diseases and disorders, human
Genetic divergence
  in allopatric speciation, 494f
  interbreeding and reinforcement of, 499–500
  speciation and, 490
  in sympatric speciation, 496f
Genetic diversity, **1173**. *See also* Genetic variation
  behavior and, 1083–84
  biodiversity and, 1173–74
  conservation of, 1189–90
  directional selection and, 473
  fungal, 615
  genetic drift, gene flow, mutations, and, 483–84
  habitat fragmentation and loss of, 1182
  in preservation of metapopulations, 1120
  switching of reproductive modes for, 1016
Genetic drift, **465**, **478**–82
  in allopatric speciation, 495
  causes of, in natural populations, 481–82
  effects of, on laboratory populations, 480f
  as evolutionary process, 465, 486t
  experimental studies of, 479–81
  founder effect and, 481
  genetic bottlenecks and, 481–82
  genetic diversity and, 483
  Hardy–Weinberg principle and, 468
  key points about fitness, genetic variation, and, 479
  loss of biodiversity from, 1180
  simulation studies of, 478–79

Boldface page numbers indicate a glossary entry; page numbers followed by an
f indicate a figure; those followed by *t* indicate a table.

Boldface page numbers indicate a glossary entry; page numbers followed by an
f indicate a figure; those followed by *t* indicate a table.

Boldface page numbers indicate a glossary entry; page numbers followed by an *f* indicate a figure; those followed by *t* indicate a table.

Boldface page numbers indicate a glossary entry; page numbers followed by an *f* indicate a figure; those followed by *t* indicate a table.

Boldface page numbers indicate a glossary entry; page numbers followed by an *f* indicate a figure; those followed by *t* indicate a table.

I:26   INDEX

Boldface page numbers indicate a glossary entry; page numbers followed by an
f indicate a figure; those followed by *t* indicate a table.

Boldface page numbers indicate a glossary entry; page numbers followed by an *f* indicate a figure; those followed by *t* indicate a table.

Boldface page numbers indicate a glossary entry; page numbers followed by an *f* indicate a figure; those followed by *t* indicate a table.

Random distribution, 1102
Random mating, 468–70
Random mutations, 312
Ranges, species, **1062**, **1102**. *See also* Geographic distribution and abundance of organisms
  for human population, 1116
  impacts of global climate change on, 1166t, 1167
  mapping, 1137
  in population ecology, 1102–3
  in preservation of metapopulations, 1120
Ras protein, **213**, 214, 232–33
Ray-finned fishes, 692, **698**–99
Rayment, Ivan, 974–76
Rays, cambium, **749**
Rb protein, **234**
Reabsorption, animal, 870–71, 873–74, 877–78
Reactants, chemical reaction, 30–32, 139–40f, 143–47
Reaction center, photosystem, **183**–84
Reaction rates, chemical reaction, 139–40f, 145–47
Reactivity, amino acid side chain, 44
Reading frame, DNA sequence, **310**–11, 313, 392–93
Realized niches, **1126**
Receptacle, plant flower, 826
Receptor-mediated endocytosis, **126**
Receptor tyrosine kinases (RTKs), **213**–14
Receptors. *See also* B-cell receptors (BCRs); Enzyme-linked receptors; G-protein-coupled receptors; Signal receptors; T-cell receptors (TCRs)
  animal hormone, 1006–10
  animal sensory, 856–57, 953
  cell–cell signal, 210
  plant sensory system, 794–97, 800–803
Recessive alleles, 258t, 471
Recessive traits, **259**–60, 277–79, 299, 369. *See also* Dominant traits
Reciprocal altruism, **1098**
Reciprocal crosses, 258t, **260**–61, 268–69
Recognition phase, adaptive immune response, 1041–47
Recolonization, metapopulation, 1112f–13
Recombinant DNA technology, **369**–74. *See also* Genetic engineering
  agricultural biotechnology, transgenic crops, and, 385–86
  creating animal models of disease using, 382
  discovery of gene recombination, 1046, 1047f
  engineering human growth hormone using, 370–74
  ethical concerns about, 374
  gene therapy and, 383–85
  in genetic engineering, 368–69, 381t
  pituitary dwarfism, human growth hormone, and, 369–70
  plant cell and tissue cultures in, B:22
Recombinant organisms, **270**–71, 274f
Recommended Dietary Allowances (RDAs), 883
Rectal glands, shark, **866**–88
Rectum, **896**
Red algae, 569, 572. *See also* Algae
Red/far-red light responses, plant, 179–81f, 800–803, 818t, 837

Red-green color blindness, 278–79, 963–64
REDD (Reducing Emissions from Deforestation and Forest Degradation), 1192–93
Re-differentiation, cell, 407
Redox (reduction-oxidation) reactions
  in Calvin cycle, 190–92
  in electron transport chain, 166–67
  peroxisomes and, 115
  in photosystems, 183–85
  prokaryotic nitrogen fixation and, 542–43
  transfer of energy via electrons in, 141–43
Reduction, **141**, 191, 241–42. *See also* Redox (reduction-oxidation) reactions
Re-extension, muscle, 980
Reference distributions, statistical tests and, B:7
Reflex, spinal, **944**
Refractory state, **936**
Regeneration phase, Calvin cycle, 191
Regulating services, ecosystem, 1187t, 1188
Regulation
  of animal body temperature, 854–58
  of animal homeostasis, 853–54
  of animal hormone production, 1003–6
  of blood pH, 915–16
  of blood pressure and blood flow, 924
  cancer as loss of cell-cycle, 233–34
  of cell cycle, 229–31
  of citric acid cycle, 162–63f
  of enzymes, 149–50
  of flower structures, 439–41
  of gene expression (*see* Gene regulation, bacterial; Gene regulation, eukaryotic)
  of glycolysis, 159–61
  homeostatic, of ventilation, 911–12
  kidney, 872, 877–78
  of mammalian menstrual cycle, 1027–30
  of mammalian puberty, 1026–27
  of metabolic pathways, 150
  of pancreatic enzymes, 893
  of photosynthesis, 195
  of plant body axes, 435–36
  of plant leaf shape, 436–38
  of protein folding, 53
  of pyruvate processing, 161–62
  of telomerase, 297
Regulatory cascades, 413–14, 436–38, 439–41
Regulatory genes
  evolutionary change and, 416
  evolutionary conservation of, 414–15
  Hox genes as, for developmental positional information, 412–14
  morphogens, 410–12
  tool-kit genes as, 415
  transcription factors, 410, 411–15 (*see also* Transcription factors)
Regulatory homeostasis, 853–58
Regulatory hormones, regulation of, 1026–27
Regulatory molecules, cell-cycle, 229–31
Regulatory proteins
  in bacterial gene regulation, 337
  in control of plant body axes, 435–36
  in determination of leaf shape, 436–38
  in differentiation of muscle cells, 428f–29

  in eukaryotic transcription initiation, 353–56, 357f
  in negative control, 341–44
  in positive control, 341, 344, 345f
  regulatory DNA sequences and, 353–54, 355f
  as transcription factors in differential gene expression, 354–56
Regulatory sequences, 353–**54**, 355f, 356, 357f, 363
Regulatory transcription factors, **410**. *See also* Transcription factors
Regulons, **345**, 360
Reinforcement, **499**–500
Reintroduction programs, species, 1189–90
Relative dating, 446–47, 453
Release factors, **331**, 332f
Religious faith vs. science, 9
Remember (Bloom's taxonomy skill), B:29f–B:30t
Remodeling, chromatin. *See* Chromatin remodeling
Renal blood vessels, 871
Renal corpuscle, 872, **873**, 878t
Renewable energy, 1189
Repair, DNA, 297–301
Repetition, scientific experiments and, 13
Replacement rate, **1117**
Replica plating, **339**–41
Replicated chromosomes, 239–40
Replication, chromosome, 220–21
Replication, DNA, 287–90. *See also* DNA synthesis
Replication fork, **290**–95
Replicative growth, viral, **716**–21
Replisomes, **294**–95
Repolarization, **933**–34
Reports, scientific, B:26–B:27
Repressors, **341**–44, 345f, **354**
Reproduction
  animal (*see* Animal reproduction)
  cell cycle and, 219–20 (*see also* Cell cycle)
  evolution by natural selection and, 5–6, 454, 456
  evolution of giraffe necks and, 9–11
  fungal, 616f, 617–18, 624–25
  as goal of living organisms, 2
  mitosis and, 220 (*see also* Mitosis)
  in model organisms, B:23
  plant (*see* Plant reproduction)
  protist, 566
Reproductive development, plant, **433**, 438–41
Reproductive isolation
  allopatric speciation and, 494–95
  biological species concept and, 490
  ecological niches and, 496–97
  mechanisms of, 491t
  sympatric speciation and, 495–99
Reptiles, **696**
  birds as, 694–95
  lineages of, 696–97, 702–3
  nitrogenous wastes of, 865
  urine formation in, 878–79
  in vertebrate phylogeny, 691
Research papers, B:26–B:27
Reservoirs, biogeochemical, 1159
Residues, amino acids as, 45–46
Resilience, community, **1186**–87, 1191–92
Resistance, community, **1186**–87
Resource partitioning, 1127–28
Resources
  as ecosystem provisioning services, 1187

  human consumption of, 1164, 1189
  species richness and efficient use of, 1186
Respiratory systems, **903**–16
  air and water as respiratory media in, 903–5
  circulatory systems and, in gas exchange and circulation, 902–3 (*see also* Circulatory systems)
  Fick's law of diffusion and, 905–6
  fish gills in, 906–7
  homeostatic control of ventilation in, 911–12
  insect trachaea in, 907–9
  organs of, 905–12
  pulmonary circulation and, 920f
  vertebrate lungs in, 909–11f
Responder cells, plant, 794–96
Response, cell–cell signal, 214
Resting potentials, **931**–32
Restoration, ecosystem, 1191–92
Restriction endonucleases, **370**–72, 379, 380t
Results, primary literature, B:27t
Reticulum, 891
Retina, eye, **960**, 961f
Retinal, 147, 961f–**62**
Retinoblastoma, 234
Retro-evolution hypothesis, 150–51f
Retroviruses, **383**–85, 720, 726f, 728
Revelle, Roger, 1163
Reverse transcriptase, **309**, **720**
  in gene therapy, 383
  in genetic engineering, 380t
  inhibitors, 720
  in long interspersed nuclear elements (LINEs), 396, 397f
  producing cDNAs with, 370
Rheumatoid arthritis, 1056
Rhinovirus, 727
Rhizaria, 558t, 569, 572
Rhizobia, **787**–89
Rhizoids, **598**
Rhizomes, **603**, **737**, **825**
Rhodopsin, 395, 961f–**62**
Rhynie Chert, 585
Ribbon diagrams, 48–49f
Ribonuclease, 52
Ribonucleic acid (RNA). *See* RNA (ribonucleic acid)
Ribonucleotides, 6–7, **58**–59, 93–94. *See also* Nucleotides
Ribonucleotide triphosphate (NTP), 318–19
Ribose, 58–60, 65, 68–69, 73, 78
Ribosomal RNAs (rRNAs), **119**, **328**
  developing tree of life from sequences of, 6–8
  in gene families, 398
  phylogenies based on, 535–36
  in ribosomes, 328, 330
Ribosome binding sites, **329**–30
Ribosomes, **108**, **324**
  in cracking of genetic code, 311
  elongation of polypeptides during translation in, 330–31
  eukaryotic, 112, 113
  initiation of translation in, 329–30
  mitochondria and, 115–16
  polypeptide synthesis by, 119
  post-translational modifications of proteins in, 331–32
  prokaryotic, 108
  protein synthesis by free, 123–24
  ribosomal RNA and structure of, 328–29f

Boldface page numbers indicate a glossary entry; page numbers followed by an
f indicate a figure; those followed by *t* indicate a table.

I:34　INDEX

Silent mutations, **313**–14*t*, 479
Silent Spring (book), 999–1000, 1152–53
Silk moths, 966–67
Silurian-Devonian explosion, 582
Simian immunodeficiency viruses
    (SIVs), 724
Simmons, Robert, 10–11
Simple eyes, **675**, 960–64
Simple fruits, **834**, 835*f*
Simple leaves, 737–38*f*
Simple molecules, 23–24
Simple sequence repeats, 396–98*f*
Simulation studies, genetic drift, 478–79
Simulation studies, global climate
    change, 1073
Sinervo, Barry, 1167, 1183–84
SINEs (short interspersed nuclear
    elements), **510**, 511*f*
Singer, S. Jon, 95–96*f*
Single bonds, 23
Single nucleotide polymorphisms
    (SNPs), 379, 380*t*
Single-strand DNA-binding proteins
    (SSBPs), **290**–91, 294*t*
Single-stranded RNA viruses, 716
Single-trait experiments, Gregor
    Mendel's, 259–63
Sinks, plant sugar, **766**
    connections between sources and,
        766–67
    low pressure near, 768
    phloem unloading into, 771–72
Sinoatrial (SA) node, **921**–22
Siphon, **663**, 688
Sister chromatids, **220**, **239**
    in meiosis, 239–46
    in mitosis, 220, 221*f*, 223–26
Sister species, **495**
Sister taxa, B:10–B:11
Sit-and-wait predators, amphibian, 700
Sixth mass extinction possibility, 523
Size
    animal locomotion and body, 986–88
    cell-cycle checkpoints and cell, 231
    directional selection on cliff swallow
        body, 473
    disruptive selection on black-bellied
        seedcracker beak, 474
    genome, 390
    human population, 1117
    island biogeography and island, 1144
    largest fungal mycelium, 616
    male penis, 1021–22
    microsporidia as smallest eukaryotic
        genome, 628
    model organism, B:23
    organic molecule, 38
    physiological effects of animal body,
        850–53
    population, 1108–12, 1120
    of prokaryotic and eukaryotic
        cells, 110
    prokaryotic diversity in, 536, 537*f*
    prokaryotic vs. eukaryotic genomes,
        393–94, 395–96
    scrotum, 1021
    sperm and egg, 420
    testes, 1020
    virus, 714
Size standards (or "ladders"), gel
    electrophoresis, B:13
Skeletal muscle, **847**, **978**–80

Skeletal systems, 980–83. *See also*
    Animal movement
    animal locomotion and size of, 987–88
    endoskeletons, 981–83
    exoskeletons, 983
    functions and types of, 980–81
    hydrostatic skeletons, 981
    skeletal muscle and, 978–80
Skeletons, animal
    cartilaginous, 698
    hydrostatic (*see* Hydrostatic skeletons)
    notochords in organogenesis of, 426
    somites as precursors to bone cells of,
        in organogenesis, 427–29
    vertebrate bony, 689
Skills, Bloom's taxonomy of, B:29–B:30*t*
Skin
    amphibian, 700
    as barrier to pathogens, 1038–39
    cancer of human, 300–301
    gas exchange via, 852
    lymphocytes and, 1042
    somites as precursors to cells of, in
        organogenesis, 427–29
    in thermoregulation, 857–58
Skoog, Folke, B:22
Skulls
    reptile, 696
    vertebrate, 688
Slack, Roger, 194
Sleeping sickness, 555*t*
Sliding clamp, 292, 294*t*
Sliding-filament model, 973–**74**
Slime molds, 216, B:23–B:25
Slow (slow oxidative) muscle fibers,
    979–80
Slugs, 662, 668, 1020
Slugs, cellular slime mold, B:23–B:25
Small intestines, **892**–96
    carbohydrate digestion and
        transportation in, 893–94
    chemical digestion in, 888
    digestion of lipids in, 895*f*
    fermentation in, 172
    folding and projections that increase
        surface area of, 892
    hormones and, 996–97
    human, 848*f*, 887*f*
    lipid digestion and bile transport in,
        894–96
    pancreatic enzyme regulation in, 893
    protein processing by pancreatic
        enzymes in, 892–93
    water absorption in, 896
Small nuclear rebonucleoproteins
    (snRNPs), **322**–23
Smallpox virus, 715, 726, 1054–55
Smell, 646, 647*t*, 965–67
Smith, James, 1111–12
Smits, Willie, 1172–73, 1191
Smoke, seed germination and, 837
Smooth ER (smooth endoplasmic
    reticulum), 112–**13**, 117*t*, 208
Smooth lines, scatterplot, B:5
Smooth muscle, 847*f*–**48**, **977**
Smut, 630
Snails, 662, 668, 1020
Snakes, 416, 702, 884, 885*f*, 967,
    968*f*, 1181
Snapping shrimp, 495
Snowshoe hare, 1113–15
Social controls, cell-cycle, 231, 233–34

Social stimulation, mating and, 1087–88
Sockeye salmon, 1060–61, 1074, 1091
Sodium channels, 934, 935
Sodium chloride (table salt), 22–23*f*, 1074
Sodium-glucose cotransporter, 893–94
Sodium hydroxide, 28–29
Sodium ions, electrolytes as, 862
Sodium poisoning, plant, 785
Sodium-potassium pump, **99**, **864**
    active transport and, 99–101*f*
    in animal glucose absorption, 894
    in fish osmoregulation, 867–68
    hormonal control of, 877
    in insect osmoregulation, 871
    in mammalian osmoregulation, 874
    in neuron membranes, 933*f*
    in osmoregulation, 864–65
    resting potentials and, 931–32
Soil, 778–82
    arbuscular mycorrhizal fungi (AMF)
        and, 621
    effects of composition of, on properties
        of, 782*t*
    fertilization of, 1155
    formation of, and textures of, 778–79*f*
    green plant building and holding
        of, 578
    importance of conservation of, 779–81
    lichens and, 631
    nutrient availability in, 781–82
    organic matter and humus in, 1156–57
    plant adaptations to dry and salty,
        757–58
    plant nutrients from, 776–78, 780*t*
    plant nutrition and, 775
    Sphagnum moss and, 602
    succession and, 1140
    water potential in, 757–59
Soil conservation, 779–81
Soil organic matter, **1156**–57
Solar energy. *See* Light energy
Solubility
    amino acid, 44
    of gas in water, 904
    of lipids, 85, 86–87
    lipid-soluble vs. lipid-insoluble
        hormones, 209, 210–11
Solute potential, **756**
    in plant adaptations to dry habitations,
        757–58
    water potential and, 755–57
Solutes, **25**, **92**, **755**, **862**
    in aquatic salinity, 1074
    concentration of, and solubility of
        gases in water, 904
    diffusion and, 91
    diffusion and osmosis of, across lipid
        bilayers, 91–94
    movement of water and, in plants (*see*
        Translocation; Water transport,
        plant)
    transport of, across plasma membranes
        by membrane proteins (*see*
        Membrane proteins)
Solutions, **25**, **755**
    acid-base reactions in, 28–30
    aqueous, and properties of water,
        25–28
    chemical evolution in aqueous, 24
    molarity of, 29
    oxygen and carbon dioxide amounts
        in, 904
    pH buffers in, 30
    pH scale of acidity and alkalinity of, 29
Solvent, **25**–26, 28*t*
Soma, **930**

Somatic cells, **219**–20, **297**
Somatic hypermutation, **1051**, 1054
Somatic motor neurons, 979
Somatic nervous system, **942**
Somatostatin, **992**
Somites, **426**–29
Songbirds, 948–49, 1109*f*, 1111–12
Sooty mangabeys, 724–25
Sori, **604**
Sound energy, 137. *See also*
    Mechanoreception
Sources, plant sugar, **766**
    connections between sinks and, 766–67
    high pressure near, 768
    sugar concentration in sieve-tube
        members at, 770–71
Sourness taste reception, 965
Southern, Edwin, B:16
Southern blotting, B:16
Soybeans, transgenic, 385
Space-filling models, **24**, B:12*f*–B:13
Spanish flu, 712, 713*f*. *See also* Influenza
    viruses
Spanish moss, 789
Spark-discharge experiment, 33–34
Sparrows, 492–93, 1109*f*, 1111–12
Spatial avoidance mechanisms, plant
    self-fertilization, 830
Spawning, 1018
Special creation, theory of, 445, 446,
    447, 453
Speciation, **6**, **490**
    adaptive radiations and (*see* Adaptive
        radiations)
    allopatric, 494–95
    contemporary, 451
    disruptive selection and, 474
    evolutionary processes and, 489–90
    outcomes of contact between isolated
        populations and, 499–502
    pollination by animals and, 831–32
    species concepts and, 490–93
    species diversity and, 1175
    species richness and rates of, 1145
    summary of outcomes of contact
        between isolated populations
        and, 502*t*
    sympatric, 495–99
    tree of life and, 6
Species, **490**. *See also* Population(s)
    adaptive radiations of, 516–20
    in Aristotle's theory, 445
    biological species concept of, 490, 491*t*
    Cambrian explosion diversification
        of, 518–20
    with closed circulatory systems, 917
    concepts of, 490–93
    conservation of, 1189–91
    deceitful communication among and
        between, 1094–95
    disruptive selection and formation of
        new, 472
    diversity of (*see* Biodiversity; Species
        diversity; Species richness)
    dusky seaside sparrow as species
        definition case, 492–93
    endangered (*see* Endangered species)
    endemic, 1176
    estimating probablilites of extinction
        of individual, 1182
    evolution and common ancestry of,
        446, 449–51
    evolution and extinct, 447
    evolution of, by natural selection,
        5–6 (*see also* Evolution; Natural
        selection)

Boldface page numbers indicate a glossary entry; page numbers followed by an
f indicate a figure; those followed by *t* indicate a table.

evolution of oviparous, viviparous, and ovoviviparous, 1020–21
exotic and invasive, 1063, 1064*f*
gamete recognition of same, in fertilization, 421
geographic distribution and abundance of (*see* Geographic distribution and abundance of organisms)
global warming and loss of, 1163
Hox genes of multiple, 414–15
impacts of global climate change on, 1166–68
interactions of (*see* Species interactions)
keystone, 1137–38
lateral gene transfer between, 394–95
life-history patterns across, 1106–7
mass extinctions of, 520–23
mechanisms of reproductive isolation of, 491*t*
morphospecies concept of, 490–91
number of animal, 636–37
number of fungal, 615
origination of, 6 (*see also* Speciation)
outgroup, B:10–B11
phylogenetic species concept of, 491–92
population cycles and interactions within and between, 1113–15
protostome, 657–58
reintroduction of, 1153, 1154*f*
relationships of fungal, 618–19
relative abundance of vertebrate, 691*f*
selecting, for whole-genome sequencing, 392
successional, 1140–42
summary of concepts of, 492*t*
taxonomy and scientific names for, 8–9
variations in nitrogenous wastes of, 865–66
without circulatory systems, 916
Species-area plots, 1183*f*
Species-area relationships, 1182–**83**
Species diversity, **1142**, **1174**
biodiversity, species richness, and, 1174
in biodiversity hotspots, 1177
global climate change and, 1073
latitudinal gradient in global, 1144–45
measuring, 1143*f*
of protists, 555
speciation, extinction, and, 1175
species richness and, 1142–43 (*see also* Species richness)
Species interactions, 1123–35, 1136*t*
commensalism in, 1124–25
competition in, 1125–28
consumption in, 1128–33
four types of, 1123, 1136*t*
mutualisms in, 1133–35
species richness and, 1186
in succession, 1140–41, 1142
three themes of, 1124
Species richness, **1142–45**, **1174**
biodiversity, species diversity, and, 1174
estimates of, using surveys, 1175
estimating, 1182–83
hypotheses for latitudinal gradient in global, 1144–45
mapping, 1176
measuring species diversity and, 1143*f*
net primary productivity and, 1185–86
of protostomes, 657–58
species diversity and, 1142–43

theory of island biogeography and predicting, 1143–44
vertebrate, 691*f*
Specific heat, **27**, 28*t*, **1067**
Specificity
adaptive immune response, 1042
antibody, 1045–46
Spectrophotometer, 845
Sperm, **237**, **823**, **1014**
competition by, 1019–20, 1022
fertilization by, 420–21
gametogenesis of, 241
glycoproteins in fertilization by, 79–80
human, 237*f*
meiosis and, 219, 237–38
microscopic images of cells of human, 3*f*
plant, 827, 829
plant development of, 438
in plant fertilization, 433
in plant reproduction, 823
polyspermy prevention and, 421–23
in sexual reproduction, 1014–15
sexual selection and, 475
spermatogenesis of, 1016–17, 1022
structure and function of, 1017
testes size and, 1020
transport and delivery of, by male reproductive systems, 1022
Spermatids, **1016**
Spermatogenesis, **1016**–17, 1022
Spermatogonia, **1016**
Spermatophores, **669**, **843–45**, **1019**
Sperm competition, **1019–20**, 1022
Sphagnum moss, 601–2
Sphenophyta (horsetails), 605
Sphincters, stomach, 889
Spicules, **639**
Spiders, 678, 1020
Spinal cords, 427, 686, 688, 943–44
Spindle apparatus, **224**, **241**
in meiosis, 241
in mitosis, 224–27
Spine, human, 843
Spines, cactus, **740**
Spiracles, **869**, **907**
Spiral cleavage, 660–61
Spirochetes, 544, 546
Spleen, **1043**
Spliceosomes, **323**, **357**
Splicing, RNA, 322–23. *See also* Alternative splicing, RNA
Sponges, 206, 638–41, 652, 981
Spongiform encephalopathies, 53
Spontaneous abortions, 470
Spontaneous chemical reactions. *See also* Exergonic reactions
characteristics of, 31–32
energetic coupling of endergonic reactions and, 141–44
in formation of micelles and lipid bilayers, 88
Gibbs free-energy change and, 138–39
Spontaneous generation hypothesis, 3–4, 18–19, 445–46
Sporangia, **582**, **823–25**
Spores, **567**, **617**, **823**
cellular slime mold, B:23–B:25
fossilized, 581–82
fungal, 617, 624
land plant production of, 588–89
in plant reproduction, 823–25
protist sporophytes and, 568
Sporophytes, **567**, **588**, **823**
in alternation of generations, 588–89

land plant evolution toward life cycles dominated by, 589–90
in land plant life cycles, 823–25
protist, 568–69
seeds and, 833
Sporopollenin, **582**, 587, 829, **829**
Spotted owls, 1118
Spruces, 607
Squid, 130, 662, 669, 932–33
Srb, Adrian, 305–7
Stability, community, 1186–87
Stability, DNA, 64, 65
Stabilizing selection, **473**, 475*t*
Stahl, Franklin, 288*f*, 289
Staining, cell, B:18–B:19
Stalked bacteria, 547
Stamens, **439**, **592**, **827**, 828–29
Staminate flowers, 827(*footnote*)
Standard error bars, B:6–B:7
Standing defenses, **1129**–30
Stapes, **956**
Staphylococcus aureus, 456
Starch, **76**, **195**
digestion of, 889
hydrolysis of, to release glucose, 80–81
in metabolic pathways, 157
as plant storage polysaccharide, 72*f*, 76, 78
processing of, in photosynthesis, 195
structure of, 77*t*
Starling, Ernest, 893, 918–19
Start codon, **311**, 329–30
Statistical tests, B:5, B:7
Statocyst, **954**
Statolith hypothesis, **804**
Statoliths, **804**
Steelhead trout, 482–83
Stem-and-loop structure, RNA, 66
Stem cells, **407**
in cell proliferation, 407
in gene therapy for severe combined immunodeficiency (SCID), 384
in plant embryogenesis, 434–35
Stems, **735**
elongation of, 799–802
eudicot vs. monocot, 742*f*
modified, 736–37
in shoot systems, 735 (*see also* Shoot systems)
Stereocilia, **955**
Stereomicroscopes, B:18
Sterilization, heat, 3–4
Steroids, **87**, **812**. *See also* Sex hormones
brassinosteroids as plant, 812
fungi in production of, 629
as hormones, 995–96, 999
as lipids, 87
receptors for, 1006–8*f*
testis cells and, 118
Steudle, Ernst, 764
Stevens, Nettie Maria, 238–39, 268
Sticky ends, 371–72
Stigma, **827**
Stochastic events, biodiversity loss and, 1179
Stolons, **737**
Stomach acid, 28
Stomachs, **889**–92
avian gizzard as modified, 891–92
chemical digestion in, 888
discovery of digestion in, 890
echinoderm, 683
human, 848*f*, 887*f*
protein digestion in, 890

in ruminants, 891
stomach acid production by parietal cells of, 890–91
ulcers as infectious disease of, 891
Stomata, **584**, **743**, **810**
abscisic acid in closing of guard cells of, 810–12
passage of oxygen and carbon dioxide through, 192–93
phototropins and, 797
regulation of gas exchange and water loss by, 743–44
water loss prevention by, 584–85, 754, 765–66
Stoneworts, 600
Stop codons, **311**, 313, 331, 332*f*
Storage polysaccharides, 72*f*, 76, 77*t*, 80–81, 155
Strains, viral, **723**–26
Stramenopila, 557, 558*t*, 570, 574
Strands
DNA, 61–63
nucleic acid, 59–60
Stratton, Charles, 369*f*
Streamlined bodies, convergent evolution of, 509–10
Streams, **1078**
Stress, hormones in, 1000–1001
Striated muscle, **977**, 978
Strijbosch, Henk, 1104–5
Stroma, **116**, **178**–79*f*, 189, 191
Structural formulas, 24, B:12–B:13
Structural homologies, **450**–51*f*
Structural polysaccharides, 72*f*, 76, 77*t*, 78–80
Structural proteins, 54
Structure-function correlations. *See* Form-function correlations
Struggle for existence hypothesis, 453
Study guide, Bloom's taxonomy, B:29–B:30*t*
Sturtevant, A. H., 274*f*
Style, carpel, **827**
Subach, Aziz, 1086–87
Subspecies, **492–93**
Substrate-level phosphorylation, **159**, 162–63*f*, 167. *See also* Phosphorylation
Substrates, **54**, **143**
concentration of, and catalysis rate, 147
in endergonic reactions, 143–44
enzyme catalysis and, 54, 144–45
eukaryotic organelles and, 110
in evolution of metabolic pathways, 150–51
globular shape of, 47*f*
in nucleic acid polymerization, 60
Subtropical deserts, **1070**
Subunit virus vaccines, 1055
Subunits, polypeptide, 50
Suburbanization, 1161
Succession, **1139**–42
Succession, law of, 447
Succulent leaves, 739
Sucrose, **195**, **767**
in glycolysis experiments, 158–59
plant transport of (*see* Translocation)
processing of, in photosynthesis, 195
Sucrose density-gradient centrifugation, B:17*f*–B:18
Suess, Hans, 1163
Sugar beets, 771–72
Sugar-phosphate backbone, nucleic acid, 59–63, 65

Sugars, **72**
  as carbohydrates, 72–73 (*see also* Carbohydrates)
  chemical evolution and nucleotide production from, 59
  dry seeds and, 834
  glucagon and blood levels of, 54
  insulin and blood levels of, 47–48
  in metabolic pathways, 157
  in nucleotides, 58–60
  photosynthetic production of, 176–78 (*see also* Photosynthesis)
  plant transport of (*see* Translocation)
  in RNA (*see* Ribose)
Sulfate, 171
Sulfhydryl functional group, 37*t*–38
Sulfur, 38, 49, 177, 324
  as human nutrient, 885*t*
  as plant nutrient, 780*t*
  in proteins vs. in DNA, 285–86
Sum rule, probability, B:8
Summation, 941–**42**
Sun, compass orientation by, 1089–90
Sun leaves, 738
Sundews, 790
Sunfish, 1094–95
Sunflowers, 489*f*, 501–2, 1082
Sunlight. *See* Light energy
Supercoiled bacterial DNA, 107–8
Supernatants, centrifugation, B:18
Support structures, protist, 562
Supporting connective tissue, 846*t*, **847**
Supporting services, ecosystem, 1187*t*, 1188
Surface area
  animal adaptations that increase, 852–53
  animal gas exchange and, 905, 916
  fungal mycelia, 616–17
Surface/area relationships, plant, 732–33
Surface area/volume, animal body, 850–53
Surface metabolism model, chemical evolution, 33, 35–36, 45, 60
Surface tension, **26**, 762
Surgery, cancer, 233
Survival, evolution and, 5–6, 9–11
Survivorship, **1104**
  in density-dependent logistic population growth, 1109–12
  in human population growth, 1115–18
  in life histories, 1105–7
  in life tables, 1104–5, 1106*f*
  in preservation of endangered species, 1118, 1119*f*
Survivorship curves, 1104–**5**
Suspension feeders, **648**, 662*f*, **884**
Suspensor, plant, 434
Sussman, Michael, 770
Sustainability, **1189**
Sustainable agriculture, **781**
Sustainable forestry, 781
Sutton, Walter, 257, 266–67
Swallowing, 889
Swammerdam, Jan, 973
Swamps, **1077**
Sweating, 864
Sweden
  age structure of human population in, 1115–16
  maternal mortality rates in, 1034
Sweet potato hawkmoth, 908

Sweetness taste receptors, 965
Swetnam, Thomas, 1139
Swim bladder, **698**
Swimming, 109, 132–33, 617, 984, 985, 988
Symbiosis, **560**, 619–23, 723, **891**. *See also* Mutualism(s)
Symbiotic relationships, **619**, **784**, 787–89. *See also* Mutualism(s)
Symbols, chemical structure, B:12–B:13
Symmetric competition, **1125**
Symmetry, body
  animal bilateral, 641–44
  echinoderm, 682–83*f*
Sympathetic nervous system, 924, **943**, 1006
Sympathetic neurons, 977
Sympatric speciation, 495–99
Sympatry, **495**, 500
Symplast, 208–9*f*, **760**
Symplastic route, 760–61
Symporters, 770, **865**
Synapomorphies, 491–**92**, **507–8**, **553**
Synapses, **939**. *See also* Electrical signaling, animal; Nervous systems
  memory, learning, and changes in, 946–48
  neurotransmitter functions and, 938–39, 940
  neurotransmitter release and structure of, 939–40
  postsynaptic potentials and, 940–42
Synapsis, **242**
Synaptic cleft, **939**
Synaptic plasticity, 946, **947**–48
Synaptic vesicles, **939**
Synaptonemal complex, **246**
Synergids, 832
Synthesis (S) phase. *See* S (synthesis) phase
Syphilis, 546
Systemic acquired resistance (SAR), **816**
Systemic circulation, 919–21
Systemin, 210*t*, **817**
Systems, 30–32
Systems biology, 390, **402**. *See also* Genomics
Systole, **923**
Systolic blood pressure, **923**
Szilard, Leo, 342
Szostak, Jack, 94, 296

T

T-cell receptors (TCRs), 1044*f*, **1045**–47, 1048
T cells, **1043**
  activation and function of, 1054*t*
  activation of, 1048–50
  antigen receptors of, 1045–47
  discovery of, 1042–43
  in gene therapy for severe combined immunodeficiency (SCID), 384
  HIV and helper, 713–14, 717–18
  as lymphocytes, 1042–43
  memory cells and, 1053–55
T-DNA, 385–86
T (transverse) tubules, **977**
Tabin, Clifford, 458–59
Table salt. *See* Sodium chloride (table salt)
Tagging, 324, B:15–B:16, B:20

Tagmata, **670**
Taiga, **1072**
Tails, mRNA, 323–24
Tails, post-anal, 686
Tails, vestigial, 448
Talking trees hypothesis, 817–18
Tannins, 115, 815
Tapeworms, 666
Taproots, **733**, 734–35
Taq polymerase, 374–76, 380*t*, 531
Target crops, agricultural biotechnology and, 385
Tarsiers, 704
Taste (gustation), 646, 647*t*, 964–65
Taste buds, **964**–65
TATA box, **320**, **354**
TATA box-binding protein (TBP), 47, **354**
Tatum, Edward, 305
Taxol drug, 607
Taxon (taxa), **8**, B:10–B:11
Taxon-specific surveys, 1175
Taxonomic bias, fossil record, 512
Taxonomic diversity, 1174–75
Taxonomy, **8**
  fossil record and taxonomic bias, 511
  scientific names for organisms in, 8–9
Taxonomy, Bloom's, B:29–B:30*t*
Taylor, Richard, 986, 987*f*
Tectorial membrane, **956**–57
Teeth
  adaptations of, 884, 885*f*
  in fossil record, 512
  mammalian, 701
  reptilian, 703
Teleosts, 699
Telomerase, **296**–97
Telomeres, 295–97
Telophase, 225*f*, **226**, 245*f*
Telophase I, 243–44, 245*f*
Telophase II, 243*f*, 244–45
Temperate forests, **1071**, 1158–59
Temperate grasslands, **1071**, 1184–87
Temperate zone, **1071**
Temperature, **31**
  as abiotic factor in geographic distribution and abundance of organisms, 1062
  animal regulation of body (*see* Thermoregulation)
  animal sensing of, 646, 647*t*
  in aquatic biomes, 1076
  blood flow and body, 924
  in climates, 1066–67
  in decomposition rates, 1157
  effect of, on hemoglobin, 914
  effects of, on reaction rates of chemical reactions, 139–40*f*
  global climate change and, 1073, 1163, 1164–66, 1182
  in mating, 1087–88
  as measure of thermal energy, 31
  membrane permeability and, 90–91
  metric units and conversions for, B:1*t*
  in net primary production of terrestrial biomes, 1155
  optimal, for enzyme function, 148
  spontaneous chemical reactions and, 138–39
  stratification of, in oceans, 1168
  in terrestrial biomes, 1068–72
  thermoreception of, 967, 968*f*
  tropical species richness and, 1145
  of water for gas solubility, 904
Template strands, DNA, **64**, **318**, 319–20*f*

Temporal avoidance mechanisms, plant self-fertilization, 830
Temporal bias, fossil record, 512–13
Temporal lobe, **944**, 945–46
Temporal prezygotic isolation, 491*t*
Tendon tissue, 846*t*
Tendons, **981**–82, 984, 986
Tendrils, 739
Tension-resisting elements, 201
Tentacles, 649*t*, 654, **669**
Terminal node, phylogenetic tree, 506*t*, B:10–B:11
Termination phase
  enzyme catalysis, 146*f*–47
  transcription, **320**, 321*f*
  translation, 331, 332*f*
Termites, 571
Terrestrial biomes, 1068–74
  animal locomotion in, 984, 985
  animal transition from aquatic environments to, 659–60
  arctic tundra, 1072
  boreal forests, 1072
  characteristics of natural, 1068–69
  effects of global climate change on, 1073–74
  gas exchange in, 907
  human impacts on, and anthropogenic, 1069
  impacts of global climate change on net primary productivity of, 1168
  insect osmoregulation in, 869–71
  internal fertilization in, 1019
  net primary production of, 1154–56
  nutrient cycling in, 1156–59
  osmoregulation in, 864
  plant adaptations to dry and salty, 754–55, 757–58, 765–66
  plant transition from aquatic environments to, 584–95 (*see also* Land plants)
  protostomes in, 659–60
  species richness of, 1176
  subtropical deserts, 1070
  temperate forests, 1071
  temperate grasslands, 1071
  tropical wet forests, 1070
  vertebrate osmoregulation in, 871–79 (*see also* Kidneys)
Territories, **477**, **1083**–84
Tertiary consumers, 556, **1150**
Tertiary structure, **49**
  DNA vs. RNA, 66*t*
  hemoglobin, 51*t*
  protein, 49–50
  RNA, 66
  tRNA, 327
Test preparation, Bloom's taxonomy and, B:29–B:30*t*
Testcrosses, 258*t*, **265**–66
Testes, **999**, **1016**
  cells of, 118
  hormones of, 994*f*, 999
  male reproductive systems and, 1021–23
  spermatogenesis in mammalian, 1016–17
  variable size of, among species, 1020
Testosterone, **999**, **1025**
  in mammalian reproduction, 1025–26
  in mating, 1087–88
  puberty and, 1026–27
  testes and, 118
Tests, statistical, B:5, B:7
Testudinians, 702
Tetrads, 240*t*, **242**

Boldface page numbers indicate a glossary entry; page numbers followed by an *f* indicate a figure; those followed by *t* indicate a table.

Boldface page numbers indicate a glossary entry; page numbers followed by an *f* indicate a figure; those followed by *t* indicate a table.

Variable number tandem repeats (VNTRs), 396–98*f*
Variable (V) regions, **1045**–46
Variables, experimental, 13
Vas deferens, **1022**–23*t*
Vasa recta, **876**, 877*f*
Vascular bundles, **742**
Vascular cambium, **748**
 auxin and, 807
 secondary growth functions of, 748–50
Vascular plants. *See* Seedless vascular plants; Seed plants
Vascular tissue, plant, **435**, **580**, **759**. *See also* Phloem tissue; Xylem tissue
 elaboration of, into tracheids and vessel elements, 585–86
 formation of, 435, 437*f*
 in land plants, 580–81, 585
 root and shoot systems and, 732
 translocation (sugar transport) and, 766–72
 water transport and, 759–65
Vascular tissue systems, plant, **740**. *See also* Phloem tissue; Xylem tissue
 embryogenesis and, 833
 phloem structure in, 747
 primary plant body and tissues of, 740–41, 742, 748*t*
 secondary growth of, 748–51
 xylem structure in, 746–47
Vectors, cloning, **370**. *See also* Plasmids
Vegetation, emergent, 1077
Vegetative development, **433**, 436–38
 flexibility of cell determination in, 438
 genetic control of leaf shape in, 436–38
 meristems in lifelong growth and, 436, 437*f*
 reproductive development vs., 433 (*see also* Reproductive development, plant)
Veins, 871, **917**–18, 921, 924
Veliger, **668**
Velvet worms, 672–73
Venae cavae, **921**
Venter, Oscar, 1178
Ventilation. *See also* Respiratory systems
 of fish gills, 906
 in gas exchange, 903
 homeostatic control of, 911–12
 of insect tracheae, 907
 of vertebrate lungs, 909–11*f*
Ventral body axis, **410**
Ventricles, **919**, 921
Venules, **917**–18
Venus flytraps, 790, 806
Vertebrae, **688**
Vertebral column, 688
Vertebrates, **645**, **682**, 688–703. *See also* Animal(s)
 adaptive immunity in, 1042 (*see also* Adaptive immune response; Immune systems)
 amphibians, 696
 body plan of, 686–87
 closed circulatory systems of, 917 (*see also* Circulatory systems)
 as deuterostome chordates, 682
 endoskeletons of, 981–83
 evolution of, 688–91
 fishes, 696
 in fossil record, 689–91
 hormones in sexual development and activity of, 998–99
 key lineages of, 696–703
 lungs of, 909–12 (*see also* Respiratory systems)

mammals, 696
molecular phylogenies of, 691
morphological innovations of, 692–93
nervous systems of, 942–49 (*see also* Nervous systems)
osmoregulation in terrestrial, 871–79 (*see also* Kidneys)
phylogenetic tree of, 690*f*
primates, hominins, and humans, 704–8 (*see also* Human[s])
relative abundance of species of, 691*f*
reproductive innovations of, 693–94
reptiles, 696–97
segmentation in, 645
sensory systems of (*see* Sensory systems, animal)
vertebrae, cranium, and brain structure of, 688
wings and flight in, 694–95
Vervet monkeys, 1098
Vesicles
 artificial membrane-bound, 88–89
 in chemical evolution, 94
 in cytokinesis, 228
 in electron transport chain, 168–69
 in endomembrane protein transport, 123–27
 microtubules and transport by, 130–31
 osmosis across membrane-bound, 92–93
 prokaryotic, 108
 rough ER in, 114
 secretory, 123
 synaptic, 939
Vessel elements, **586**, 746–47
Vessels, angiosperm, 596
Vestigial traits, **448**
 of cetaceans, 453
 as evidence for evolutionary change, 448–49
 as nonadaptive, 461
Vicariance, **494**, 495
Video microscopy
 of ATP synthase, 170
 in biological imaging, B:20
 of transport vesicles, 130*f*
Villi, 852–53*f*, 892
Virchow, Rudolf, 3, 219
Virions, **714**–15. *See also* Replicative growth, viral
Virulent infections, 532, **712**
Virus vaccines, 1055
Viruses, **711**–30
 abundance and diversity of, 712
 analyzing coexistence of, with host cells, 721–22
 analyzing genetic material of, 715–16
 analyzing morphological traits of, 715
 analyzing phases of replicative growth of lytic cycle in, 716–21
 bacteriophage λ, 50
 Baltimore classification system for, 725–26
 biological methods of studying, 714–22
 biological reasons for studying, 712–14
 characteristics of living organisms vs. characteristics of, 712*t*
 diversification themes of, 722–25
 emerging diseases and emerging, 723–25
 eukaryotic translation and, 359–60
 in gene therapy, 383–85
 Hershey–Chase experiment on DNA in genes of, 285–86

human epidemics and pandemics from, 712–14
identification of emerging, 725
infection of host cells by, 711–12
key lineages of, 725–28
living organisms vs., 711–12
nanobiology in isolation of, 714
origins of, 723
photomicrograph of, 711*f*
proteins as defense against, 54
RNA in splitting eukaryotic genes, 321–22
vaccines against, 1055
Visceral mass, **662**–63
Viscosity, animal locomotion and fluid, 988
Visible light, **179**
Vision, animal. *See* Photoreception, animal; Sight
Visual communication, 1093
Visual cues, mating and, 1088
Visualization, molecular, B:15–B:16
Vitamin A deficiency, 385
Vitamin B$_1$, 884*t*
Vitamin B$_6$, 305
Vitamin B$_{12}$, 884*t*
Vitamin C, 884*t*
Vitamin D, 884*t*
Vitamin D-resistant rickets (hypophasphatemia), 279
Vitamins, **883**
 coenzymes and, 147
 as essential nutrients for animals and humans, 883, 884*t*
 lipids as, 87
 transgenic crops and deficiencies of, 385
Vitelline envelope, **1017**–18*f*
Viviparous species, **651**, **694**, **1020**–21
Vivipary, 836
Volcanic gases, 24, 30, 34–35, 45
Volt, **930**
Voltage, **783**, **930**
Voltage clamping, **935**
Voltage-gated channels, 98–99, **934**
Volume, in chemical reactions, 138
Volume, metric units and conversions for, B:1*t*
Volume/surface area, animal body, 850–53
Volume/surface area relationships, plant, 732–33
Voluntary muscles, **977**, 979
Vomeronasal organ, **967**
Von Békésy, Georg, 956–57
Von Frisch, Karl, 1084, 1092–93
von Linné, Karl, 9

W

Waggle dance, honeybee, 1092–93
Waldeyer, Wilhelm, 220
Walking, 986
Wall cress. *See* Mustard plant
Wall pressure, **756**
Wallace, Alfred Russel, 5, 444, 446, 453, 1063
Wallace line, **1063**
Warblers, 1062
Wasps, 485, 818–19
Wasserman, Paul, 79–80
Wastes, 883*f*, 888, 896. *See also* Nitrogenous wastes, animal
Water
 absorption of, in large intestines, 896
 absorption of, in small intestines, 896

in acid-base reactions, 28–30
angiosperm vessels for conducting, 596
animal heat exchange by evaporation of, 855
animal locomotion in, 984, 985, 988
animal requirements for, 842–43
aquaporins and transport of, across plasma membranes, 98
behavior of oxygen and carbon dioxide in, 904–5
cohesion, adhesion, and surface tension of, 26
covalent bonds of, 21–23
in decomposition rates, 1157
density of, as liquid and as solid, 26–27
depth and flow of in aquatic biomes, 1074–75, 1077–79
efficiency of, as solvent, 25–26
energy absorbing capacity of, 27–28
green plant holding of, and moderation of climate, 578
insolubility of lipids in, 85, 86–87
land plant adaptations to prevent loss of, 584–85
large intestine absorption of, 888
lipid bilayers and, 89
molecular structure of, 24*f*, B:12*f*
in net primary production of terrestrial biomes, 1155
nitrate pollution of, 543–44
noncyclic electron flow between, and NADP$^+$ in Z-scheme model, 188–89
in nutrient cycling, 1158
osmosis as diffusion of, across lipid bilayers, 92–93
oxidation of, in photosystem II, 186
photosynthetic oxygen production from, 177–78
plant nutrients from, 780*t*
plant transport of (*see* Water transport, plant)
polarity of amino acid side chains and solubility in, 44
production of, by cellular respiration, 165
production of, by electron transport chain, 166*f*–67
properties of, 26–28
purification of, as ecosystem regulating service, 1188
in seed germination, 837
specific heat of, 27*t*
Water balance, animal. *See* Osmoregulation
Water bears, 672–73
Water cycle, global, 1159–60, 1166
Water depth, 1074, 1075*f*, 1077–79
Water flea, 1167
Water flow, 1074–75, 1077–79
Water loss, animal
 terrestrial adaptations to prevent, 662–63
 insect minimization of, 869–70
Water loss, plant
 biochemical pathway adaptations to, 766
 regulation of, 754–55, 765–66
 seed maturation and, 834
 stomata and regulation of, 743–44
 turgor pressure, wilting, and plant, 758
Water molds, 574

Boldface page numbers indicate a glossary entry; page numbers followed by an *f* indicate a figure; those followed by *t* indicate a table.